Student Study Guide and Solutions Manual

for

Organic Chemistry

David R. Klein
Johns Hopkins University

JOHN WILEY & SONS, INC.

ASSOCIATE PUBLISHER Petra Recter
DIRECTOR OF DEVELOPMENT Barbara Heaney
SENIOR DEVELOPMENT EDITOR Leslie Kraham
SPONSORING EDITOR Joan Kalkut
MARKETING MANAGER Kristine Ruff

This book is printed on acid free paper. ∞

Founded in 1807, John Wiley & Sons, Inc. has been a valued source of knowledge and understanding for more than 200 years, helping people around the world meet their needs and fulfill their aspirations. Our company is built on a foundation of principles that include responsibility to the communities we serve and where we live and work. In 2008, we launched a Corporate Citizenship Initiative, a global effort to address the environmental, social, economic, and ethical challenges we face in our business. Among the issues we are addressing are carbon impact, paper specifications and procurement, ethical conduct within our business and among our vendors, and community and charitable support. For more information, please visit our website: www.wiley.com/go/citizenship.

ISBN 978-0-471-75739-9
Binder-Ready Version ISBN 978-0-470-92661-1

Printed in the United States of America

10 9 8 7 6 5 4 3

CONTENTS

PREFACE

This book contains more than just solutions to all of the problems in the textbook. Each chapter of this book also contains a series of exercises that will help you review the concepts, skills and reactions presented in the corresponding chapter of the textbook. These exercises are designed to serve as study tools that can help you identify your weak areas. Each chapter of this solutions manual/study guide has the following parts:

- ***Review of Concepts***. These exercises are designed to help you identify which concepts are the least familiar to you. Each section contains sentences with missing words (blanks). Your job is to fill in the blanks, demonstrating mastery of the concepts. To verify that your answers are correct, you can open your textbook to the end of the corresponding chapter, where you will find a section entitled *Review of Concepts and Vocabulary*. In that section, you will find each of the sentences, verbatim.
- ***Review of Skills***. These exercises are designed to help you identify which skills are the least familiar to you. Each section contains exercises in which you must demonstrate mastery of the skills developed in the *SkillBuilders* of the corresponding textbook chapter. To verify that your answers are correct, you can open your textbook to the end of the corresponding chapter, where you will find a section entitled *SkillBuilder Review*. In that section, you will find the answers to each of these exercises.
- ***Review of Reactions***. These exercises are designed to help you identify which reagents are not at your fingertips. Each section contains exercises in which you must demonstrate familiarity with the reactions covered in the textbook. Your job is to fill in the reagents necessary to achieve each reaction. To verify that your answers are correct, you can open your textbook to the end of the corresponding chapter, where you will find a section entitled *Review of Reactions*. In that section, you will find the answers to each of these exercises.
- ***Solutions***. At the end of each chapter, you'll find solutions to all problems in the textbook, including all Skillbuilders, conceptual checkpoints, additional problems, integrated problems, and challenge problems.

The sections described above have been designed to serve as useful tools as you study and learn organic chemistry. Good luck!

David Klein
Department of Chemistry
Johns Hopkins University

Chapter 1
Electrons, Bonds and Molecular Properties

Review of Concepts

Fill in the blanks below. To verify that your answers are correct, look in your textbook at the end of Chapter 1. Each of the sentences below appears verbatim in the section entitled *Review of Concepts and Vocabulary.*

- _____________ **isomers** share the same molecular formula but have different connectivity of atoms and different physical properties.
- Second-row elements generally obey the _______ **rule**, bonding to achieve noble gas electron configuration.
- A pair of unshared electrons is called a ______________.
- A **formal charge** occurs when an atom does not exhibit the appropriate number of ____________________________.
- An **atomic orbital** is a region of space associated with ____________________, while a **molecular orbital** is a region of space associated with _______________.
- Methane's tetrahedral geometry can be explained using four degenerate _____-**hybridized orbitals** to achieve its four single bonds.
- Ethylene's planar geometry can be explained using three degenerate _____-**hybridized orbitals.**
- Acetylene's linear geometry is achieved via _____-**hybridized** carbon atoms.
- The geometry of small compounds can be predicted using valence shell electron pair repulsion (**VSEPR**) theory, which focuses on the number of _______ bonds and _______________ exhibited by each atom.
- The physical properties of compounds are determined by _________________ forces, the attractive forces between molecules.
- **London dispersion forces** result from the interaction between transient __________________ and are stronger for larger alkanes due to their larger surface area and ability to accommodate more interactions.

Review of Skills

Fill in the blanks and empty boxes below. To verify that your answers are correct, look in your textbook at the end of Chapter 1. The answers appear in the section entitled *SkillBuilder Review.*

SkillBuilder 1.1 Determining the Constitution of Small Molecules

STEP 1 *- DETERMINE THE VALENCY (NUMBER OF EXPECTED BONDS) FOR EACH ATOM IN* C_2H_5Cl	***STEP 2*** *- DRAW THE STRUCTURE OF* C_2H_5Cl *BY PLACING ATOMS WITH THE HIGHEST VALENCY AT THE CENTER, AND PLACING MONOVALENT ATOMS AT THE PERIPHERY*
Each carbon atom is expected to form ___ bonds. Each hydrogen atom is expected to form ___ bonds. The chlorine atom is expected to form ___ bonds.	

SkillBuilder 1.2 Drawing the Lewis Dot Structure of an Atom

STEP 1 - DETERMINE THE NUMBER OF VALENCE ELECTRONS	STEP 2 - PLACE ONE ELECTRON BY ITSELF ON EACH SIDE OF THE ATOM	STEP 3 - IF THE ATOM HAS MORE THAN FOUR VALENCE ELECTRONS, PAIR THE REMAINING ELECTRONS WITH THE ELECTRONS ALREADY DRAWN
Nitrogen is in Group ___ of the periodic table, and is expected to have ___ valence electrons.	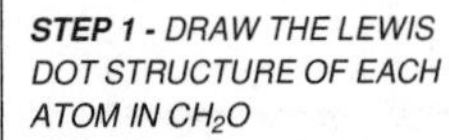	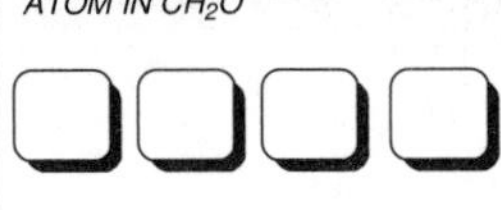

SkillBuilder 1.3 Drawing the Lewis Structure of a Small Molecule

STEP 1 - DRAW THE LEWIS DOT STRUCTURE OF EACH ATOM IN CH_2O	STEP 2 - FIRST CONNECT ATOMS THAT FORM MORE THAN ONE BOND	STEP 3 - CONNECT THE HYDROGEN ATOMS	STEP 4 - PAIR ANY UNPAIRED ELECTRONS, SO THAT EACH ATOM ACHIEVES AN OCTET
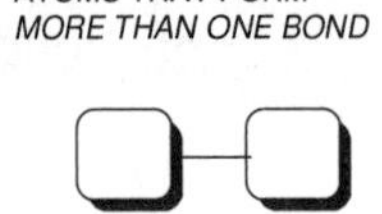		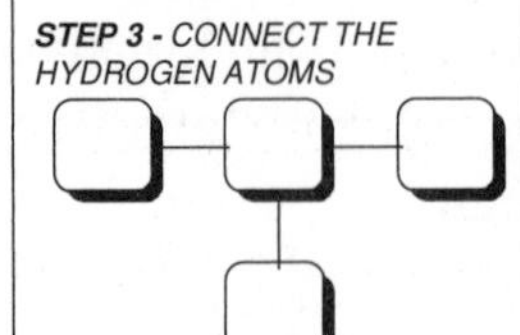	

SkillBuilder 1.4 Calculating Formal Charge

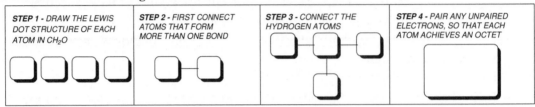

STEP 1 - DETERMINE THE APPROPRIATE NUMBER OF VALENCE ELECTRONS	STEP 2 - DETERMINE THE NUMBER OF VALENCE ELECTRONS IN THIS CASE	STEP 3 - ASSIGN A FORMAL CHARGE TO THE NITROGEN ATOM IN THIS CASE
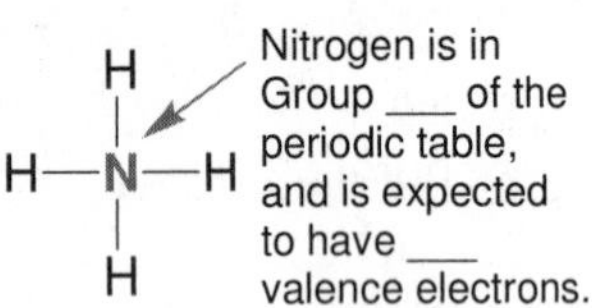 Nitrogen is in Group ___ of the periodic table, and is expected to have ___ valence electrons.	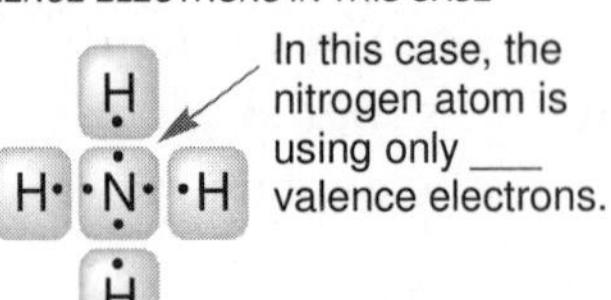 In this case, the nitrogen atom is using only ___ valence electrons.	

SkillBuilder 1.5 Locating Partial Charges

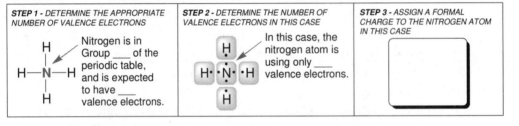

STEP 1 - CIRCLE THE BONDS BELOW THAT ARE POLAR COVALENT	STEP 2 - FOR EACH POLAR COVALENT BOND, DRAW AN ARROW THAT SHOWS THE DIRECTION OF THE DIPOLE MOMENT	STEP 3 - INDICATE THE LOCATION OF ALL PARTIAL CHARGES (δ+ and δ-)
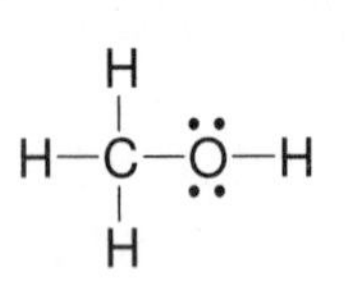	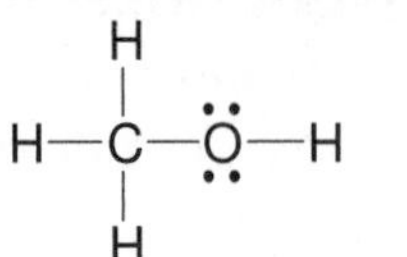	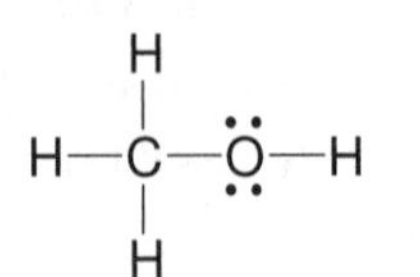

SkillBuilder 1.6 Identifying Electron Configurations

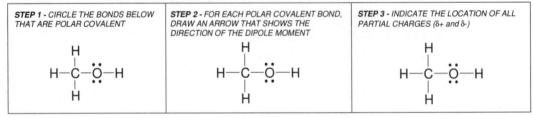

STEP 1 - IN THE ENERGY DIAGRAM SHOWN HERE, DRAW THE ELECTRON CONFIGURATION OF NITROGEN (USING ARROWS TO REPRESENT ELECTRONS).	STEP 2 - FILL IN THE BOXES BELOW WITH THE NUMBERS THAT CORRECTLY DESCRIBE THE ELECTRON CONFIGURATION OF NITROGEN
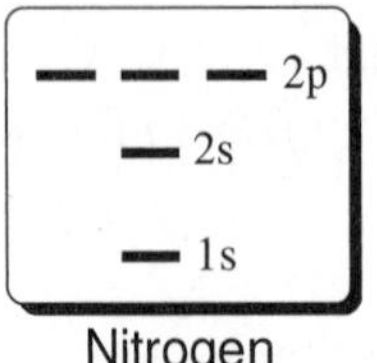	

SkillBuilder 1.7 Identifying Hybridization States

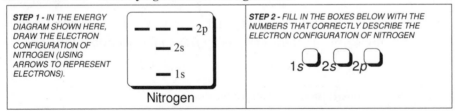

A CARBON ATOM WITH FOUR SINGLE BONDS WILL BE _____ HYBRIDIZED	A CARBON ATOM WITH ONE DOUBLE BOND WILL BE _____ HYBRIDIZED	A CARBON ATOM WITH A TRIPLE BOND WILL BE _____ HYBRIDIZED
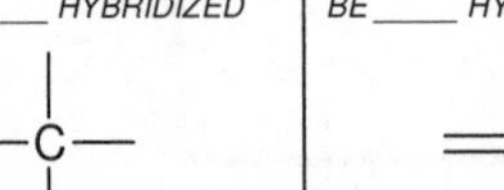	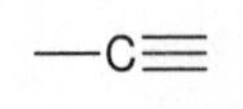	—C≡

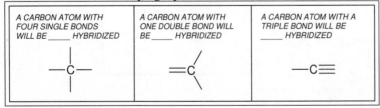

SkillBuilder 1.8 Predicting Geometry

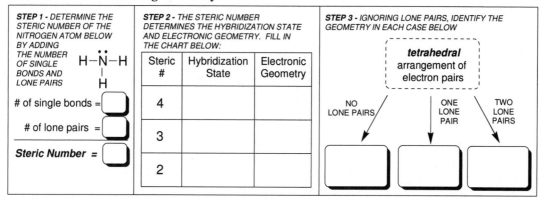

STEP 1 *- DETERMINE THE STERIC NUMBER OF THE NITROGEN ATOM BELOW BY ADDING THE NUMBER OF SINGLE BONDS AND LONE PAIRS*

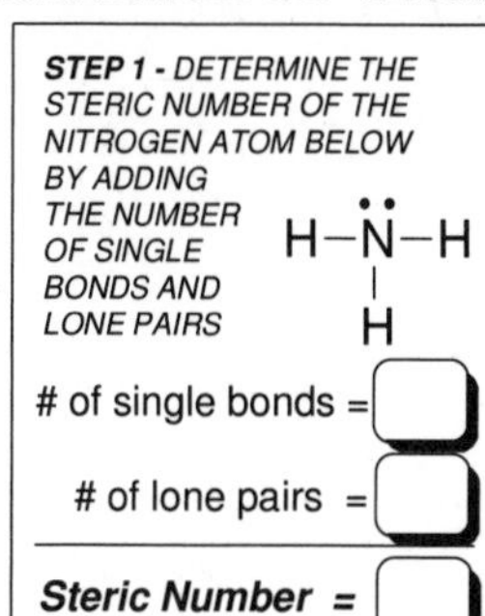

\# of single bonds = ☐

\# of lone pairs = ☐

Steric Number = ☐

STEP 2 *- THE STERIC NUMBER DETERMINES THE HYBRIDIZATION STATE AND ELECTRONIC GEOMETRY. FILL IN THE CHART BELOW:*

Steric #	Hybridization State	Electronic Geometry
4		
3		
2		

STEP 3 *- IGNORING LONE PAIRS, IDENTIFY THE GEOMETRY IN EACH CASE BELOW*

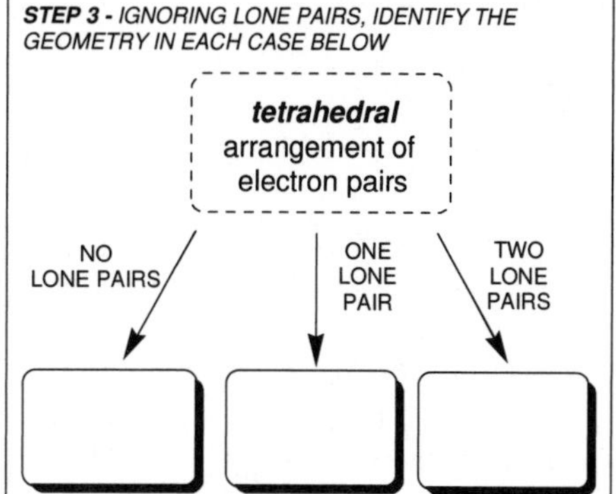

SkillBuilder 1.9 Identifying Molecular Dipole Moments

STEP 1 *- IDENTIFY THE GEOMETRY OF THE OXYGEN ATOM BELOW*

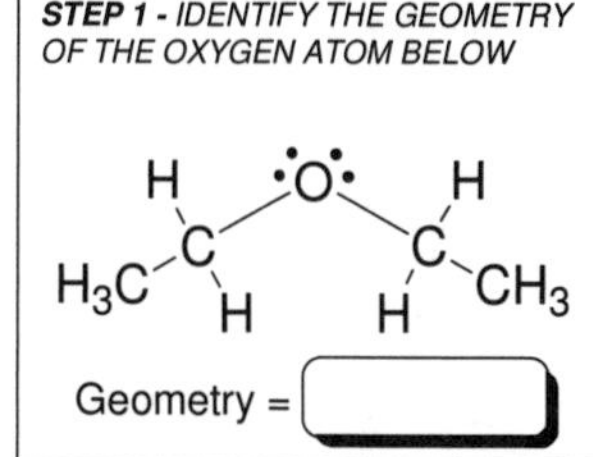

Geometry = ☐

STEP 2 *- REDRAW THE COMPOUND. FOR EACH POLAR COVALENT BOND, DRAW AN ARROW THAT SHOWS THE DIRECTION OF THE DIPOLE MOMENT*

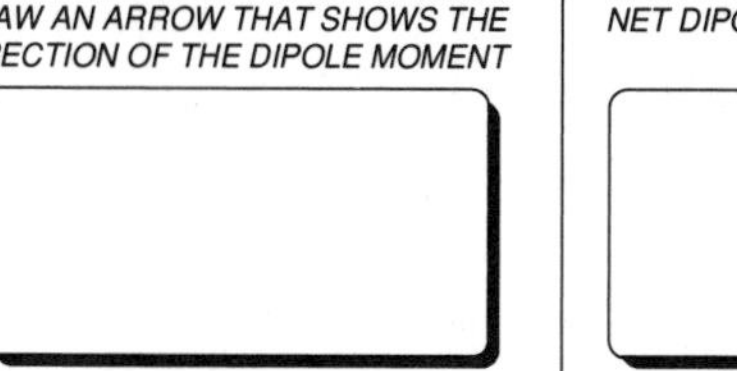

STEP 3 *- REDRAW THE COMPOUND, AND DRAW THE NET DIPOLE MOMENT*

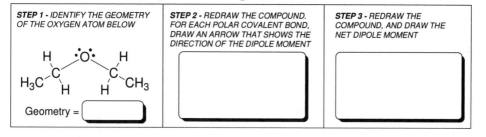

SkillBuilder 1.10 Predicting Physical Properties

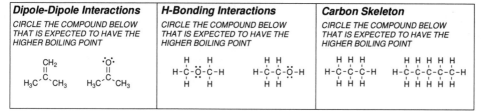

Dipole-Dipole Interactions	***H-Bonding Interactions***	***Carbon Skeleton***
CIRCLE THE COMPOUND BELOW THAT IS EXPECTED TO HAVE THE HIGHER BOILING POINT	*CIRCLE THE COMPOUND BELOW THAT IS EXPECTED TO HAVE THE HIGHER BOILING POINT*	*CIRCLE THE COMPOUND BELOW THAT IS EXPECTED TO HAVE THE HIGHER BOILING POINT*
$H_3C-C(=CH_2)-CH_3$ $H_3C-C(=O)-CH_3$	$H-CH_2-O-CH_2-H$ $H-CH_2-CH_2-O-H$	$H-(CH_2)_3-H$ $H-(CH_2)_5-H$

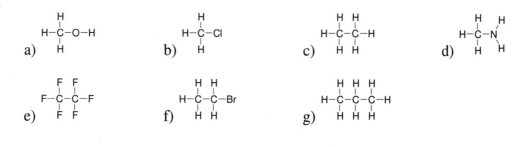

Solutions

1.1.

a) H–C(H)(H)–O–H

b) H–C(H)(H)–Cl

c) H–C(H)(H)–C(H)(H)–H

d) H–C(H)(H)–N(H)–H

e) F–C(F)(F)–C(F)(F)–F

f) H–C(H)(H)–C(H)(H)–Br

g) H–C(H)(H)–C(H)(H)–C(H)(H)–H

1.2.

H–C(H)(H)–C(H)(H)–C(H)(H)–Cl or H–C(H)(H)–C(Cl)(H)–C(H)(H)–H

1.3.

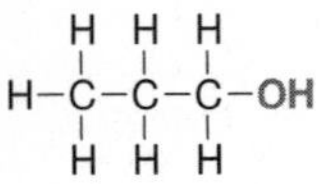 or 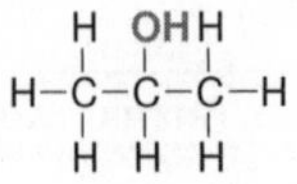 or 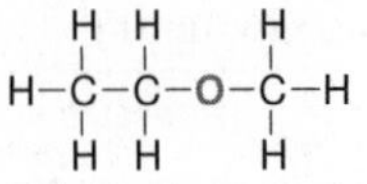

1.4.

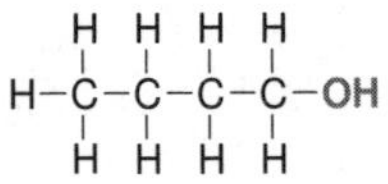 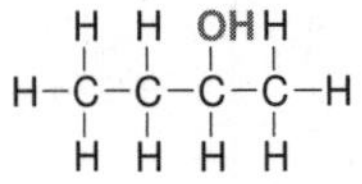 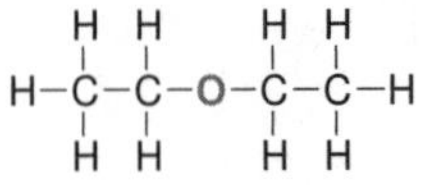

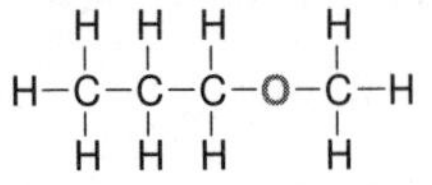 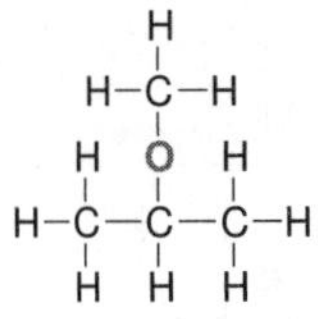 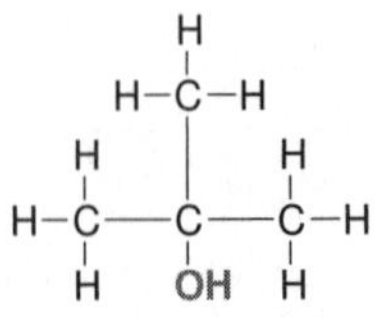 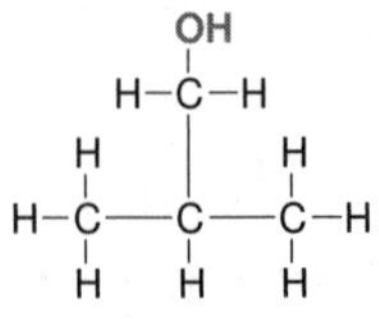

1.5.

a)  b)  c)  d) 

e) 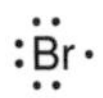 f)  g)  h) 

1.6. Both nitrogen and phosphorous belong to column 5A of the periodic table, and therefore, each of these atoms has five valence electrons. In order to achieve an octet, we expect each of these elements to form three bonds.

1.7. Aluminum is directly beneath boron on the periodic table (Column 3A), and therefore both elements exhibit three valence electrons.

1.8.  resembles boron because it exhibits three valence electrons.

1.9. 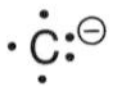 resembles nitrogen because it exhibits five valence electrons.

1.10.

a) 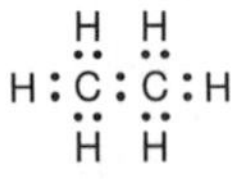 b) 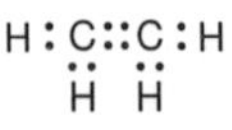 c)  d) 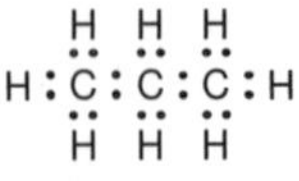

e) 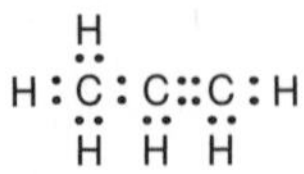 f) 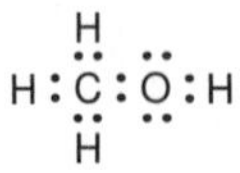

1.11 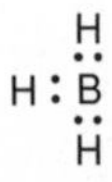 The central boron atom lacks an octet of electrons.

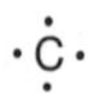

1.12

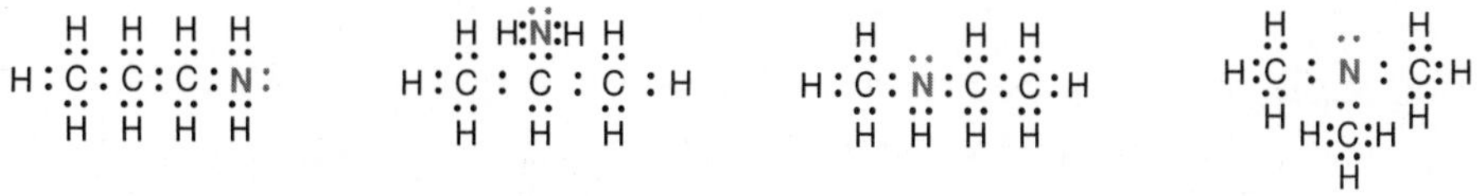

In all of the constitutional isomers above, the nitrogen atom has one lone pair.

1.13.

(a) 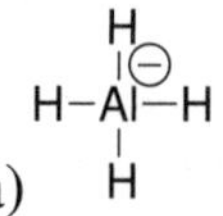 (b) 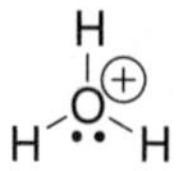 (c) 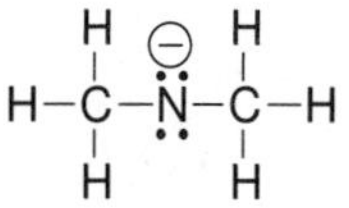 (d) 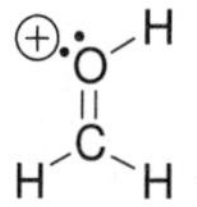 (e)

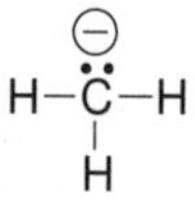

(f) 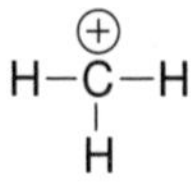 (g) 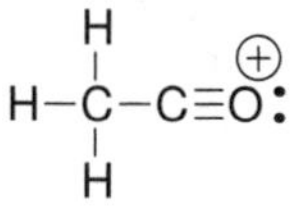 (h) 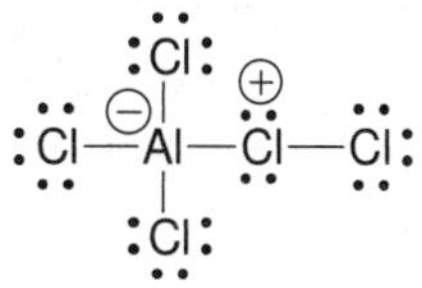 (i)

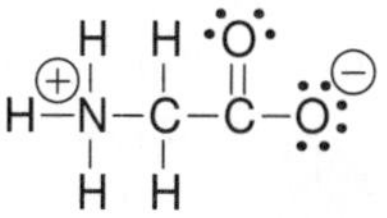

1.14.

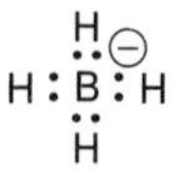

a) *Boron has a formal charge*

b) *Nitrogen has a formal charge*

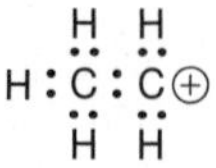

c) *Carbon has a formal charge*

1.15.

(a) 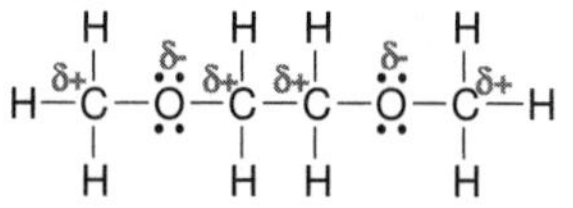 (b) 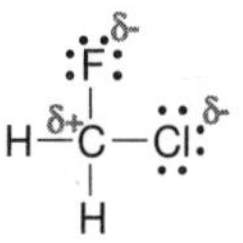 (c)

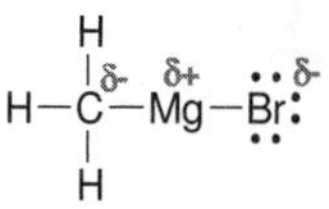

(d) 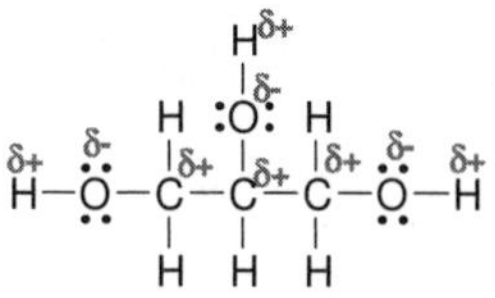 (e) 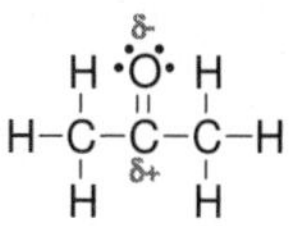 (f)

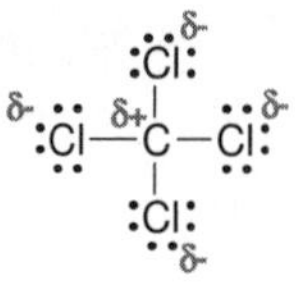

1.16.

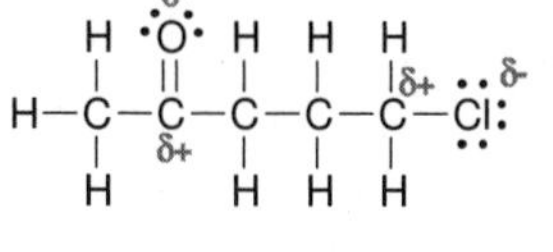

1.17.

a) $\mathbf{1s^22s^22p^2}$ b) $\mathbf{1s^22s^22p^4}$ c) $\mathbf{1s^22s^22p^1}$

d) $\mathbf{1s^22s^22p^5}$ e) $\mathbf{1s^22s^22p^63s^1}$ f) $\mathbf{1s^22s^22p^63s^23p^1}$

1.18.

a) $\mathbf{1s^22s^22p^3}$ b) $\mathbf{1s^22s^22p^1}$ c) $\mathbf{1s^22s^22p^2}$ d) $\mathbf{1s^22s^22p^5}$

1.19. The bond angles of an equilateral triangle are 60°, but each bond angle of cyclopropane is supposed to be 109.5°. Therefore, each bond angle is severely strained, causing an increase in energy. This form of strain, called ring strain, will be discussed in Chapter 4. The ring strain associated with a three-membered ring is greater than the ring strain of larger rings, because larger rings do not require bond angles of 60°.

1.20

a) The C=O bond of formaldehyde is comprised of one sigma bond and one pi bond.
b) Each C-H bond is formed from the interaction between an sp^2 hybridized orbital from carbon and an s orbital from hydrogen.
c) The oxygen atom is sp^2 hybridized, so the lone pairs occupy sp^2 hybridized orbitals.

1.21. Rotation of a single bond does not cause a reduction in the extent of orbital overlap, because the orbital overlap occurs on the bond axis. In contrast, rotation of a pi bond results in a reduction in the extent of orbital overlap, because the orbital overlap is NOT on the bond axis.

1.22.

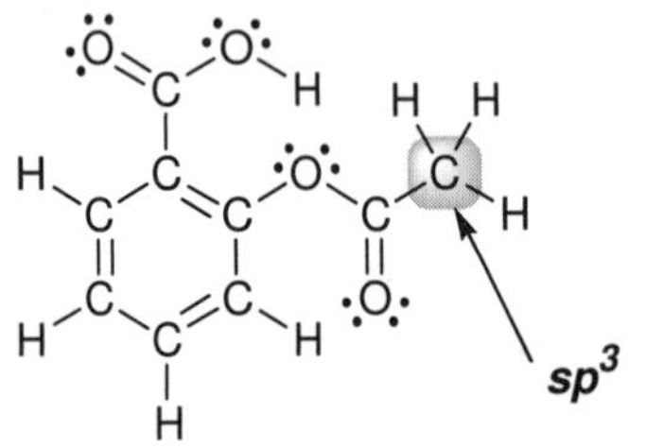

sp^3

a) All carbon atoms in this molecule are sp^2 hybridized, except for the carbon atom highlighted above, which is sp^3 hybridized

b) The carbon atoms highlighted above are sp^3 hybridized. All other carbon atoms in this compound are sp^2 hybridized

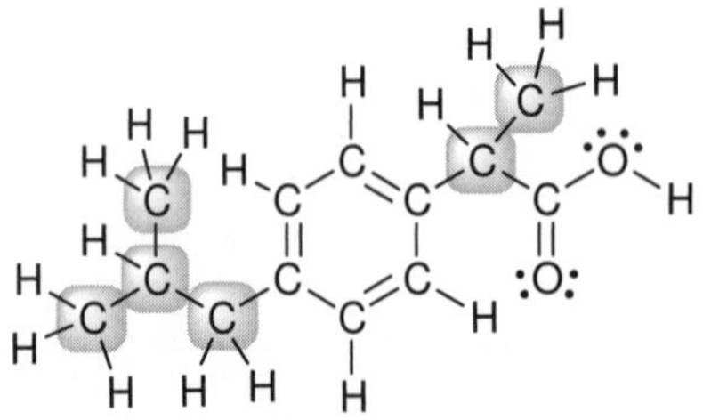

1.23.

a) sp^2 … sp

b) sp … sp^2

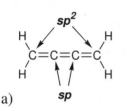

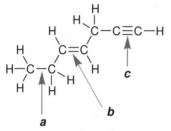

1.24.

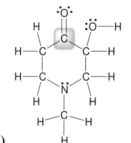

c < b < a

a is the longest bond and c is the shortest bond

1.25.

a) The nitrogen atom has three bonds and one lone pair, and is therefore trigonal pyramidal.
b) The oxygen atom has three bonds and one lone pair, and is therefore trigonal pyramidal.
c) The boron atom has four bonds and no lone pairs, and is therefore tetrahedral.
d) The boron atom has three bonds and no lone pairs, and is therefore trigonal planar.
e) The boron atom has four bonds and no lone pairs, and is therefore tetrahedral.
f) The carbon atom has four bonds and no lone pairs, and is therefore tetrahedral.
g) The carbon atom has four bonds and no lone pairs, and is therefore tetrahedral.
h) The carbon atom has four bonds and no lone pairs, and is therefore tetrahedral.

1.26.

(a)

All carbon atoms in this molecule are tetrahedral except for the highlighted carbon atom, which is trigonal planar.

The oxygen atom (of the OH group) has bent geometry, and the nitrogen atom is trigonal pyramidal.

(b)

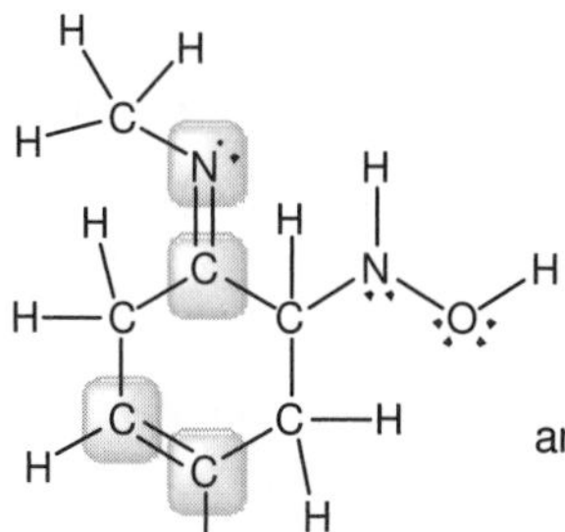

All carbon atoms are tetrahedral except for the carbon atoms highlighted, which are trigonal planar.

The oxygen atom and the highlighted nitrogen atom have bent geometry, and the other nitrogen atom is trigonal pyramidal.

(c)

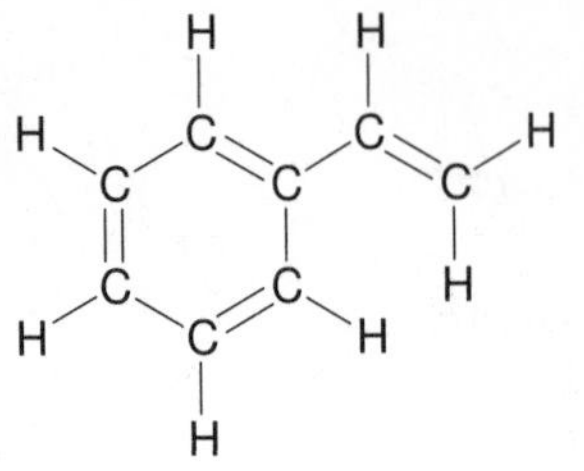

H H H H H H H H — C C C C C C C C

All carbon atoms are trigonal planar.

1.27. The carbon atom of the carbocation has three bonds and no lone pairs, and is therefore trigonal planar. The carbon atom of the carbanion has three bonds and one lone pair, and is therefore trigonal pyramidal.

1.28. Every carbon atom in benzene is sp^2 hybridized and trigonal planar. Therefore, the entire molecule is planar (all of the atoms in this molecule occupy the same plane).

1.29.

a) 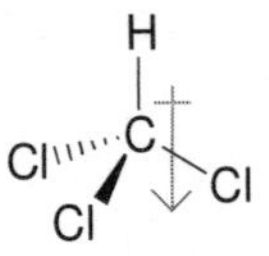 Cl Cl Cl C H

b) H_3C O CH_3

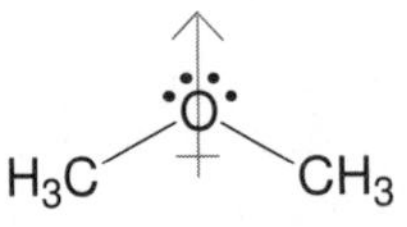

c) H H H N

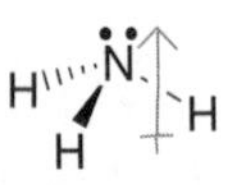

d) 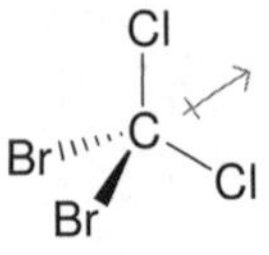 Br Br Cl Cl C

(e) 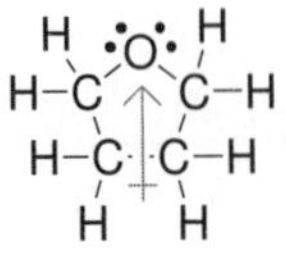 H H H H H H C C C C O

(f) none

(g) 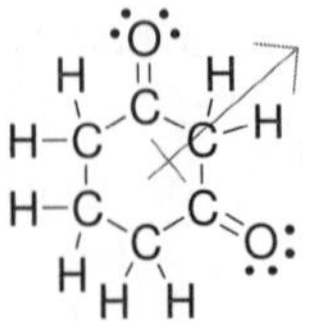 O H H H H H H C C C C C O

(h) none

(i) 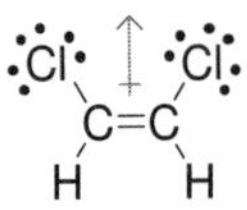 Cl Cl C=C H H

(j) none

(k) 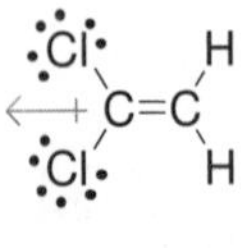 Cl Cl C=C H H

(l) none

1.30. $CHCl_3$ is expected to have a larger dipole moment than $CBrCl_3$, because the bromine atom in the latter compound serves to nearly cancel out the effects of the other three chlorine atoms (as is the case in CCl_4).

1.31. The carbon atom of CO_2 has a steric number of two, and therefore has linear geometry. As a result, the individual dipole moments of each C=O bond cancel each other completely to give no overall molecular dipole moment. In contrast, the sulfur atom in SO_2 has a steric number of three (because it also has a lone pair, in addition to the two S=O bonds), which means that it has bent geometry. As a result, the individual dipole moments of each S=O bond do NOT cancel each other completely, and the molecule does have a molecular dipole moment.

1.32.

a) The latter, because it is less branched.
b) The latter, because it has more carbon atoms.
c) The latter, because it has an OH bond.
d) The former, because it is less branched.

1.33.

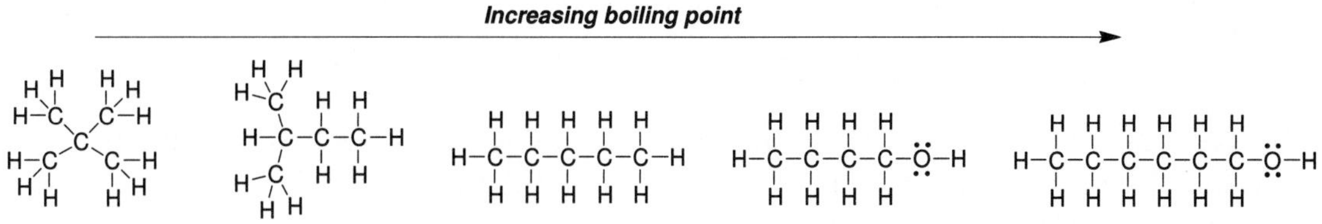

Increasing boiling point →

1.34.

a)

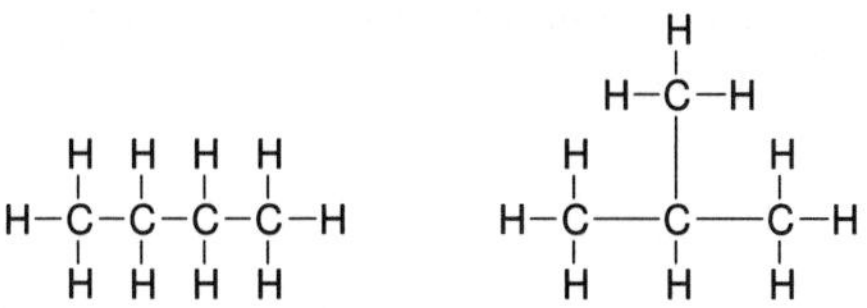

b)

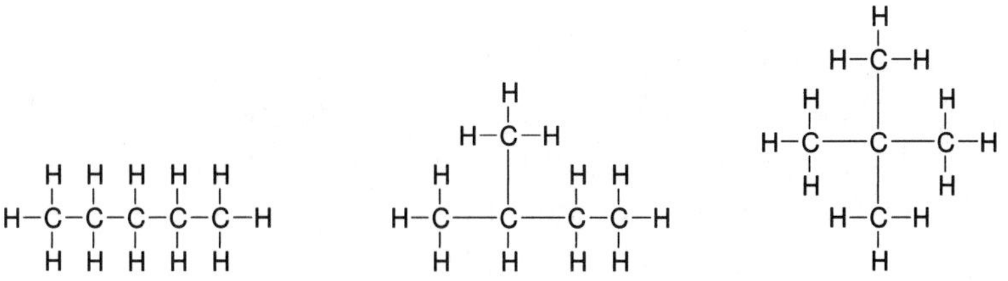

c)

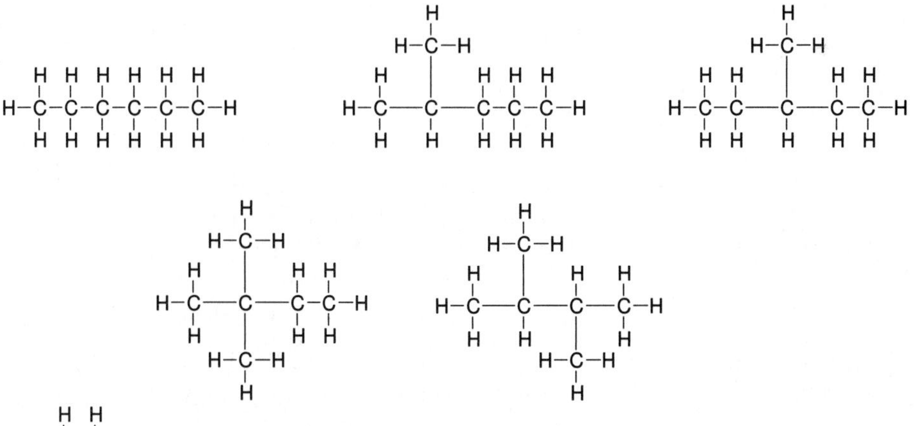

d)

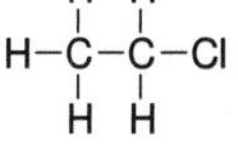

e)

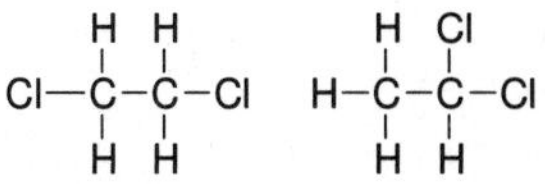

f)

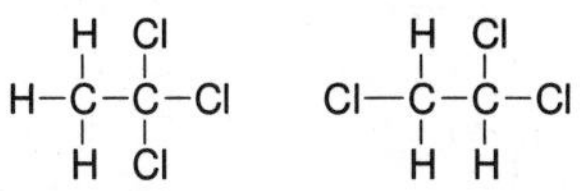

1.35.

a)

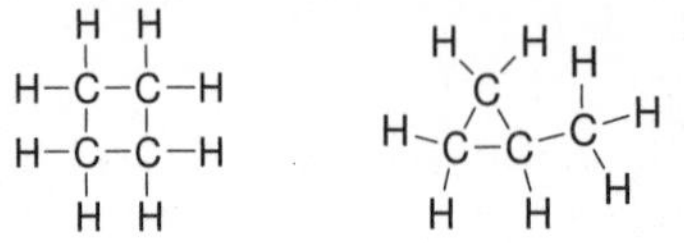

b)

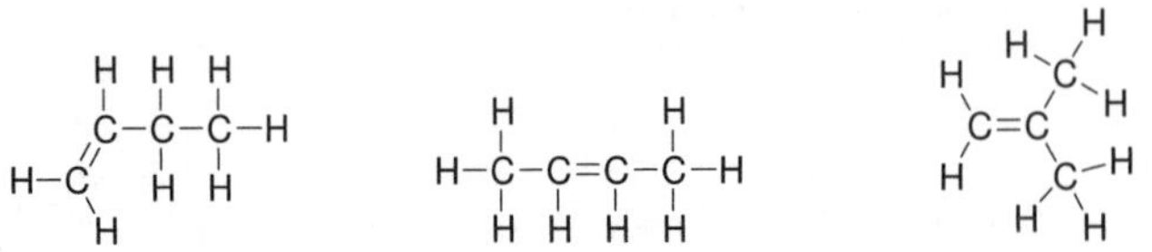

1.36.

a) 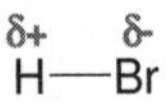 $\delta+$ H—Br $\delta-$ b) 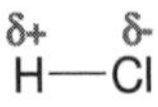 $\delta+$ H—Cl $\delta-$ c) 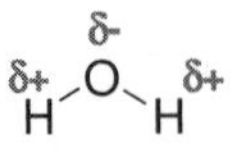 $\delta+$ H–O($\delta-$)–H $\delta+$ d)

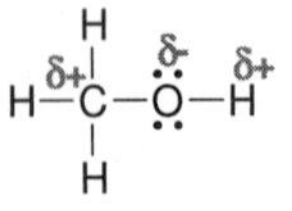

1.37.

a) NaBr, because the difference in electronegativity between Na and Br is greater than the difference in electronegativity between H and Br.
b) FCl, because the disparity in electronegativity between F and Cl is greater than the disparity in electronegativity Br and Cl.

1.38.

a) 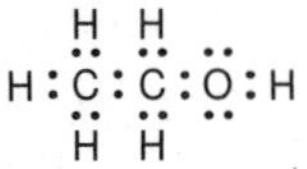 b)

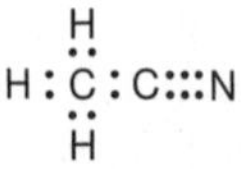

1.39.

a) 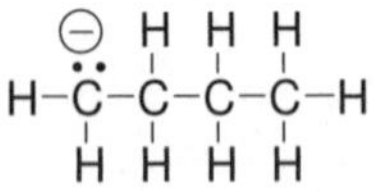 All carbon atoms in this molecule are tetrahedral except for the carbon atom bearing the negative charge, which is trigonal pyramidal.

b) 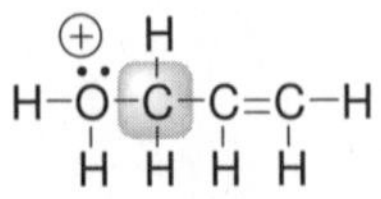 The highlighted carbon atom is tetrahedral, and the other two carbon atoms are trigonal planar.

The oxygen atom is trigonal pyramidal.

c) 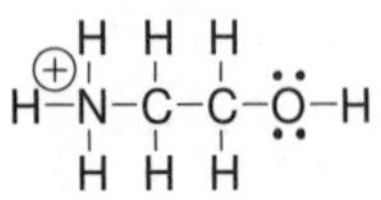 Both carbon atoms and the nitrogen atom are tetrahedral. The oxygen atom is bent.

d) 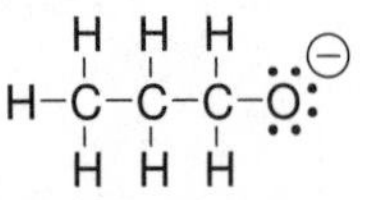 All three carbon atoms in this molecule are tetrahedral. The geometry of the oxygen atom is not relevant because it is only attached to one other group.

1.40.

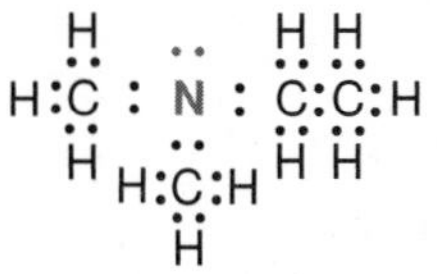

H .. H H
H:C : N : C:C:H
H H:C:H H H
H

The nitrogen atom has trigonal pyramidal geometry. The compound is expected to have the following molecular dipole moment:

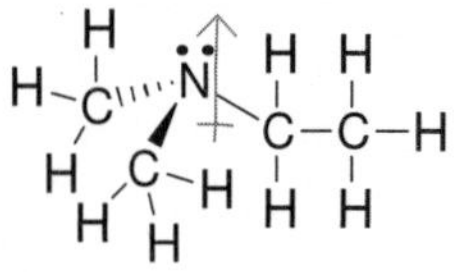

1.41.

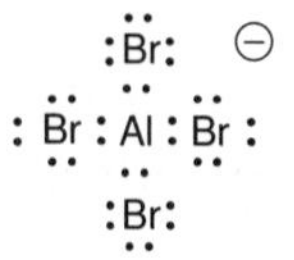

:Br: ⊖
: Br : Al : Br :
:Br:

The central aluminum has tetrahedral geometry.

1.42.

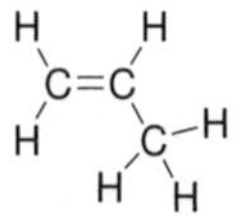

1.43.

a) No b) Yes c) Yes d) No 5) No 6)Yes

1.44.

a) Oxygen b) Fluorine c) Carbon d) Nitrogen e) Chlorine

1.45.

a) ionic
b) Na-O is ionic, and O-H is polar covalent
c) Na-O is ionic, O-C is polar covalent, and each C-H bond is covalent
d) The O-H and C-O bonds are polar covalent, and each C-H bond is covalent
e) The C=O bond is polar covalent, and each C-H bond is covalent

1.46.

a)

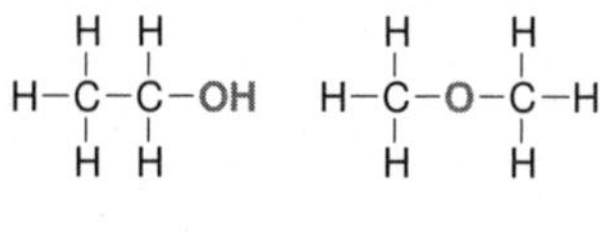

b)

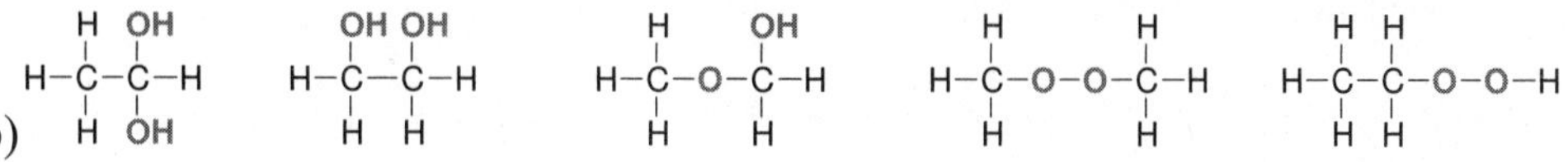

c)

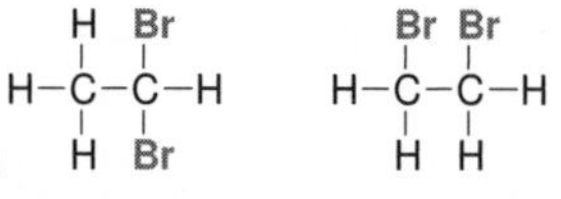

1.47.

```
   H OH           OH OH           OH    OH               H         OH              H     OH
   |  |            |  |            |     |               |          |              |      |
H–C–C–OH        H–C–C–OH        H–C–O–C–H            H–C–O–O–C–H               H–C–O–C–H
   |  |            |  |            |     |               |          |              |      |
   H OH            H  H            H     H               H          H              H     OH
```

1.48.

a) $\overset{+\!\!\longrightarrow}{\text{C—O}}$ b) $\overset{\longleftarrow\!\!+}{\text{C—Mg}}$ c) $\overset{+\!\!\longrightarrow}{\text{C—N}}$ d) $\overset{\longleftarrow\!\!+}{\text{C—Li}}$

e) $\overset{+\!\!\longrightarrow}{\text{C—Cl}}$ f) $\overset{\longleftarrow\!\!+}{\text{C—H}}$ g) $\overset{\longleftarrow\!\!+}{\text{O—H}}$ h) $\overset{\longleftarrow\!\!+}{\text{N—H}}$

1.49.

a) All bond angles are approximately 109.5º, except for the C-O-H bond angle which is expected to be less than 109.5º as a result of the repulsion of the lone pairs on the oxygen.
b) All bond angles are approximately 120º.
c) All bond angles are approximately 120º.
d) All bond angles are 180º.
e) All bond angles are approximately 109.5º, except for the C-O-C bond angle which is expected to be less than 109.5º as a result of the repulsion of the lone pairs on the oxygen.
f) All bond angles are approximately 109.5º.
g) All bond angles are approximately 109.5º.
h) All bond angles are approximately 109.5º except for the C-C≡N bond angle which is 180º.

1.50.

a) sp^3, trigonal pyramidal
b) sp^2, trigonal planar
c) sp^2, trigonal planar
d) sp^3, trigonal pyramidal
e) sp^3, trigonal pyramidal

1.51. Sixteen sigma bonds and three pi bonds.

1.52.

a) the second, because it possesses an O-H bond.
b) the second, because it has more carbon atoms.
c) the first, because it has a polar bond.

1.53.

a) yes
b) no (this compound can serve as a hydrogen bond acceptor, but not a hydrogen bond donor)
c) no

d) no
e) no (this compound can serve as a hydrogen bond acceptor, but not a hydrogen bond donor)
f) yes
g) no
h) yes

1.54.
a) 3 b) 4 c) 3 d) 2

1.55.
a) 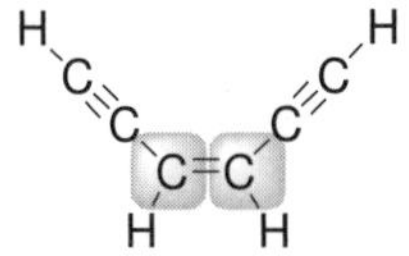 The highlighted carbon atoms are sp^2 hybridized and trigonal planar. The remaining four carbon atoms are *sp* hybridized and linear.

b) 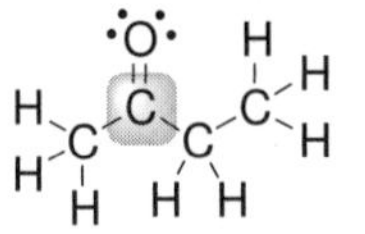 The highlighted carbon atom is sp^2 hybridized and trigonal planar. The remaining three carbon atoms are sp^3 hybridized and tetrahedral.

c) 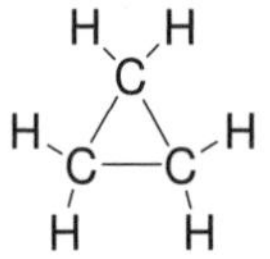 All carbon atoms are sp^3 hybridized and tetrahedral.

1.56. 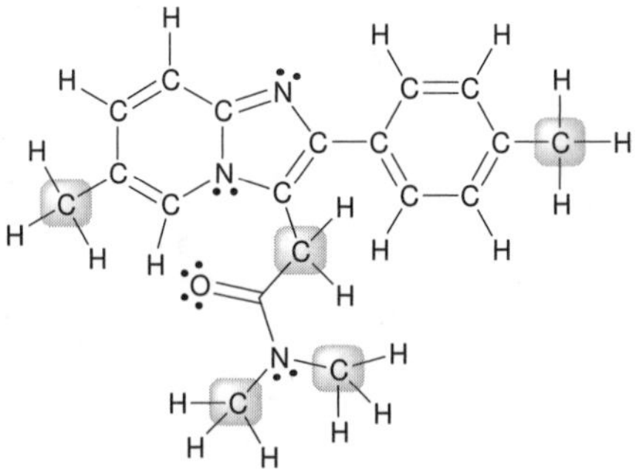
The highlighted carbon atoms are sp^3 hybridized and tetrahedral. The remaining carbon atoms are sp^2 hybridized and trigonal planar.

1.57.
a) oxygen b) fluorine c) carbon

1.58.
sp^2, bent

sp^3, trigonal pyramidal

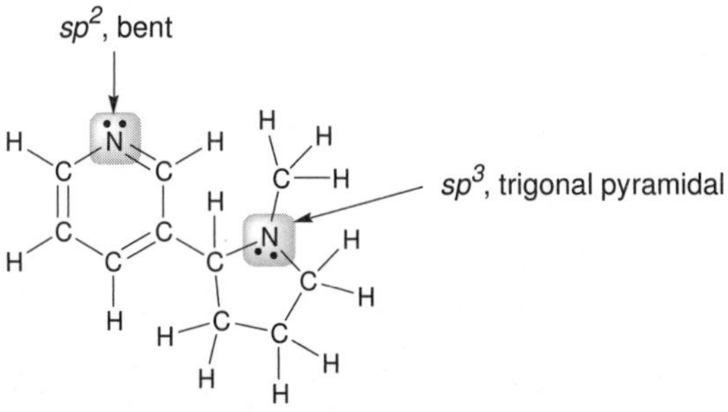

nicotine

1.59.

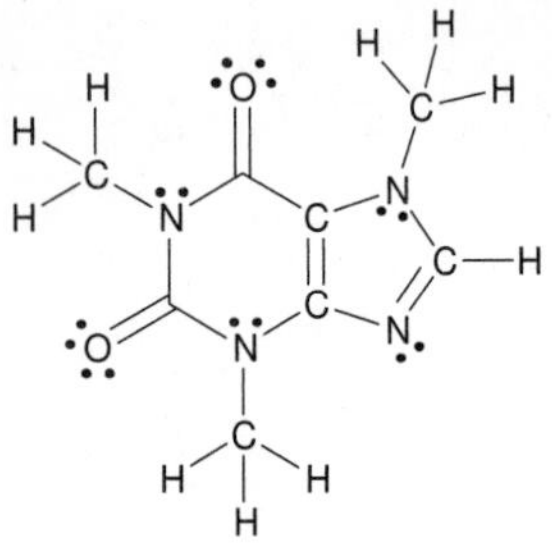

caffeine

1.60. The two isomers are:

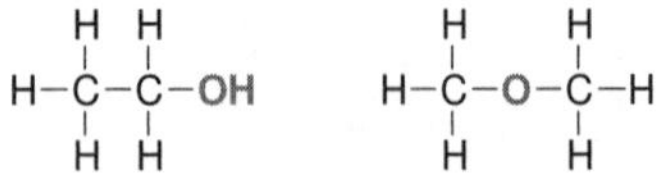

H H
H–C–C–OH
H H

H H
H–C–O–C–H
H H

The first will have a higher boiling point because it possesses an OH group which can form hydrogen bonds.

1.61.

a)

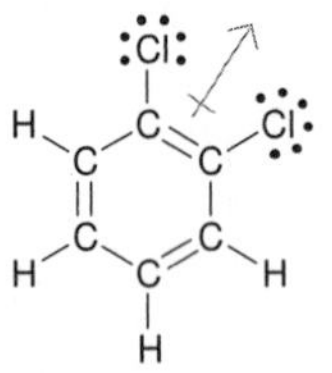

b)

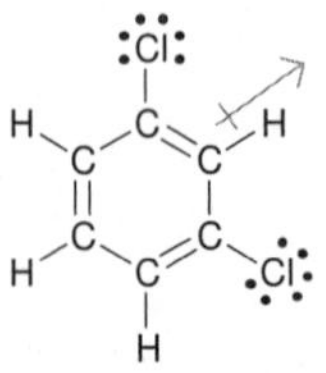

c)

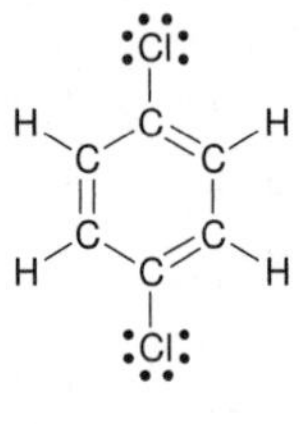

there is no molecular dipole moment

d)

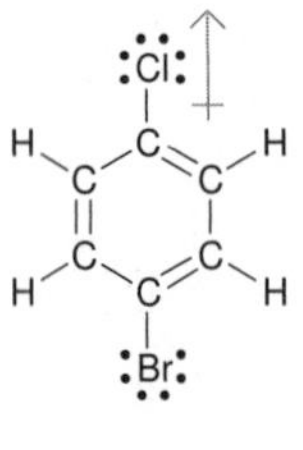

Cl is more electronegative than Br

1.62. The third chlorine atom in chloroform partially cancels the effects of the other two chlorine atoms, thereby reducing the molecular dipole moment relative to methylene chloride.

1.63.

a) Compound A and Compound B
b) Compound B
c) Compound B
d) Compound C
e) Compound C
f) Compound A
g) Compound B
h) Compound A is capable of hydrogen bonding

1.64.

a)

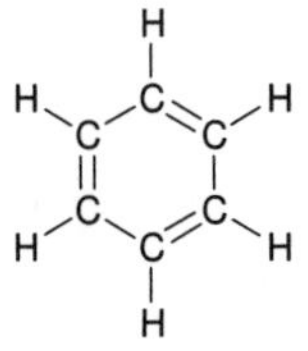

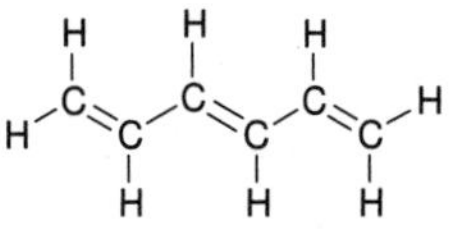

b)

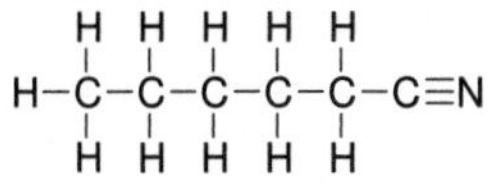

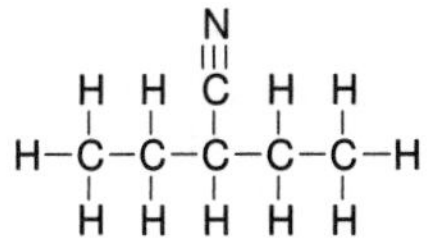

```
  H H H H H
  | | | | |
H-C-C-C-C-C-C≡N
  | | | | |
  H H H H H
```

```
        N
        ≡
  H  H  C  H  H
  |  |  |  |  |
H-C--C--C--C--C-H
  |  |  |  |  |
  H  H  H  H  H
```

c)

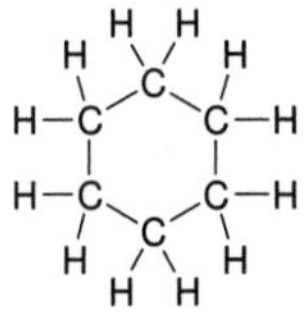

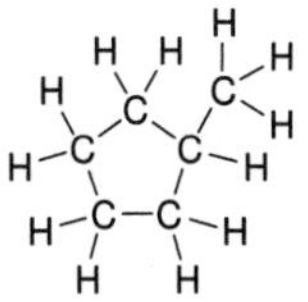

d)

N≡C−C≡C−C≡C−C≡N

F−C≡C−C≡C−C≡C−F

1.65.

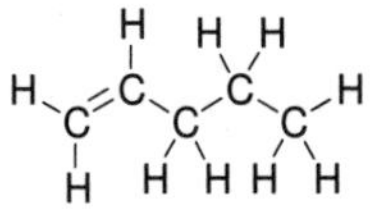

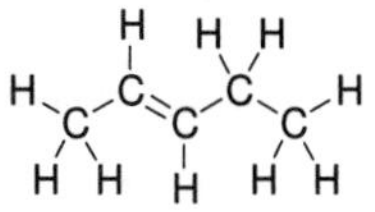

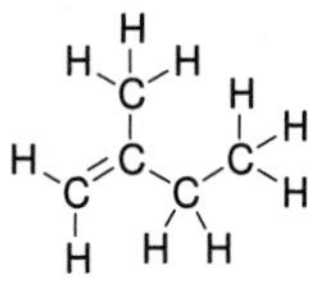

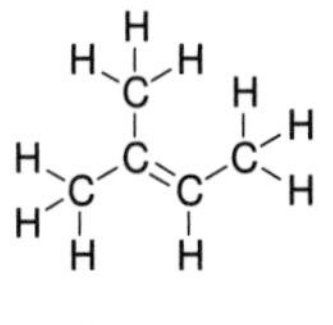

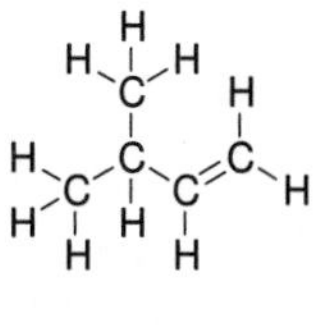

1.66.

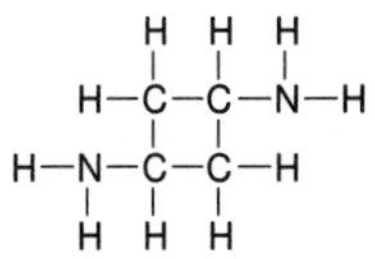

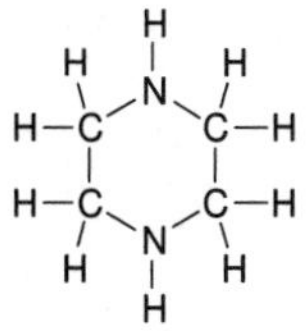

1.67.

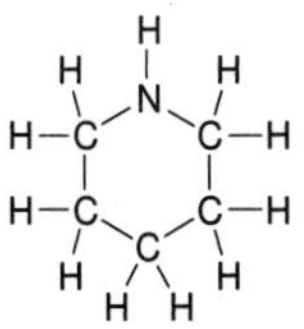

Chapter 2
Molecular Representations

Review of Concepts

Fill in the blanks below. To verify that your answers are correct, look in your textbook at the end of Chapter 2. Each of the sentences below appears verbatim in the section entitled *Review of Concepts and Vocabulary.*

- In **bond-line structures**, _______atoms and most ________ atoms are not drawn.
- A ________________ is a characteristic group of atoms/bonds that show a predictable behavior.
- When a carbon atom bears either a positive charge or a negative charge, it will have ___________, rather than four, bonds.
- In bond-line structures, a **wedge** represents a group coming ______ the page, while a **dash** represents a group _________ the page.
- ___________ **arrows** are tools for drawing resonance structures.
- When drawing curved arrows for resonance structures, avoid breaking a _______ bond and never exceed _____________ for second-row elements.
- There are three rules for identifying significant resonance structures:
    1. Minimize ____________.
    2. Electronegative atoms can bear a positive charge, but only if they possess an ________ of electrons.
    3. Avoid drawing a resonance structure in which two carbon atoms bear _____________ charges.
- A ______________ lone pair participates in resonance and is said to occupy a ____ orbital.
- A _____________ lone pair does not participate in resonance.

Review of Skills

Fill in the blanks and empty boxes below. To verify that your answers are correct, look in your textbook at the end of Chapter 2. The answers appear in the section entitled *SkillBuilder Review.*

SkillBuilder 2.1 Converting Between Different Drawing Styles

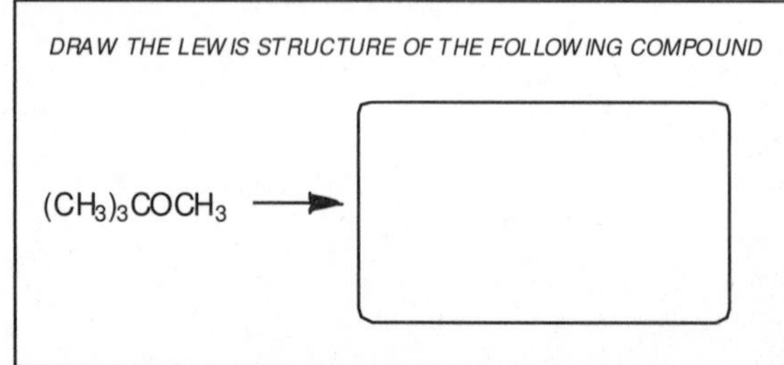

SkillBuilder 2.2 Reading Bond-Line Structures

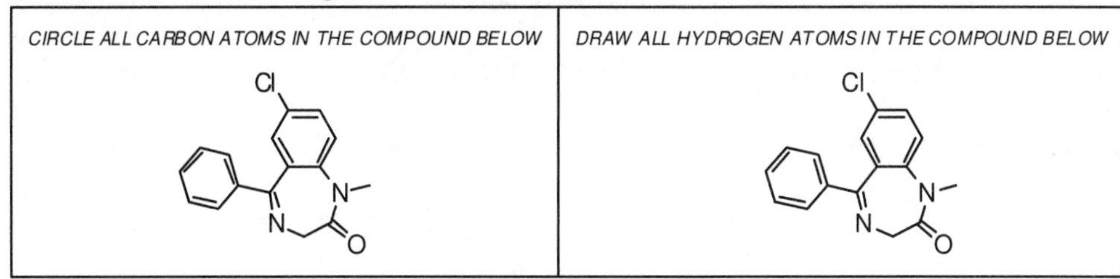

CIRCLE ALL CARBON ATOMS IN THE COMPOUND BELOW
DRAW ALL HYDROGEN ATOMS IN THE COMPOUND BELOW
Cl
N
N
O

SkillBuilder 2.3 Drawing Bond-Line Structures

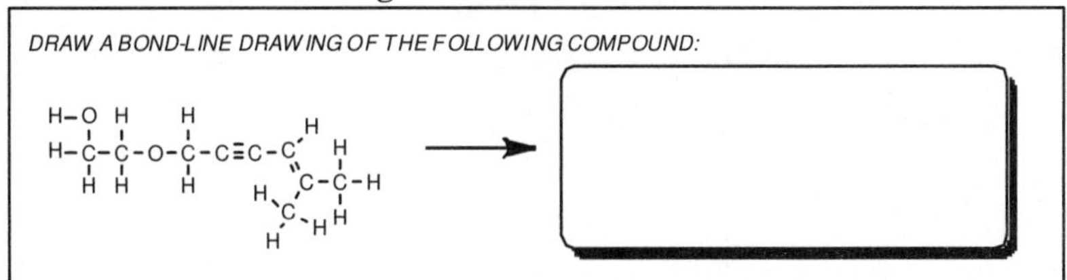

DRAW A BOND-LINE DRAWING OF THE FOLLOWING COMPOUND:

SkillBuilder 2.4 Identifying Lone Pairs on Oxygen Atoms

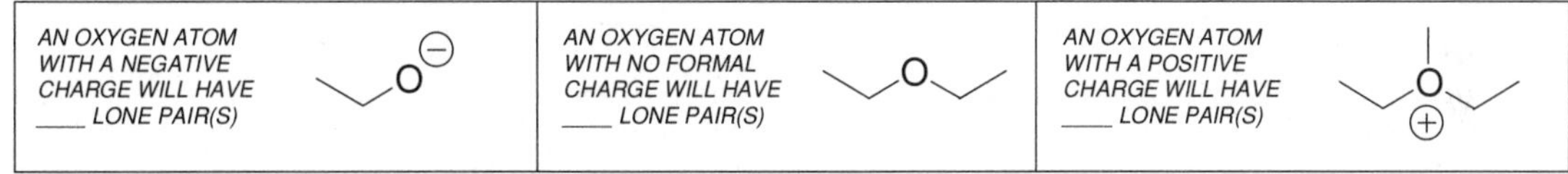

AN OXYGEN ATOM WITH A NEGATIVE CHARGE WILL HAVE ____ LONE PAIR(S)
AN OXYGEN ATOM WITH NO FORMAL CHARGE WILL HAVE ____ LONE PAIR(S)
AN OXYGEN ATOM WITH A POSITIVE CHARGE WILL HAVE ____ LONE PAIR(S)

SkillBuilder 2.5 Identifying Lone Pairs on Nitrogen Atoms

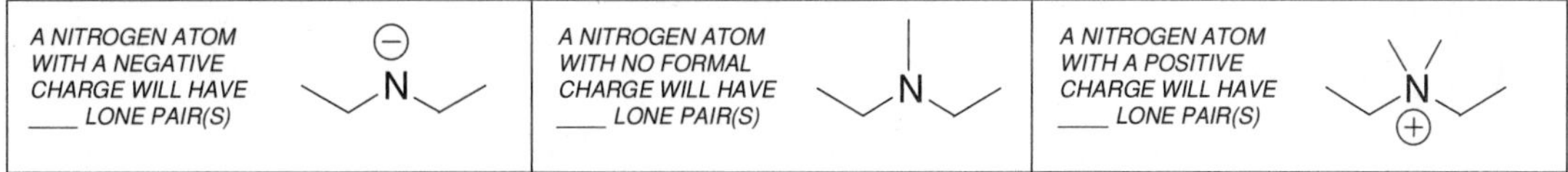

A NITROGEN ATOM WITH A NEGATIVE CHARGE WILL HAVE ____ LONE PAIR(S)
A NITROGEN ATOM WITH NO FORMAL CHARGE WILL HAVE ____ LONE PAIR(S)
A NITROGEN ATOM WITH A POSITIVE CHARGE WILL HAVE ____ LONE PAIR(S)

SkillBuilder 2.6 Identifying Valid Resonance Arrows

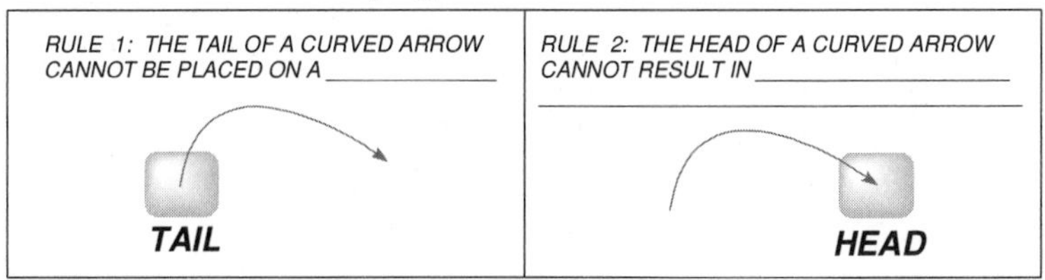

RULE 1: THE TAIL OF A CURVED ARROW CANNOT BE PLACED ON A ____________
RULE 2: THE HEAD OF A CURVED ARROW CANNOT RESULT IN ____________
TAIL
HEAD

SkillBuilder 2.7 Assigning Formal Charges in Resonance Structures

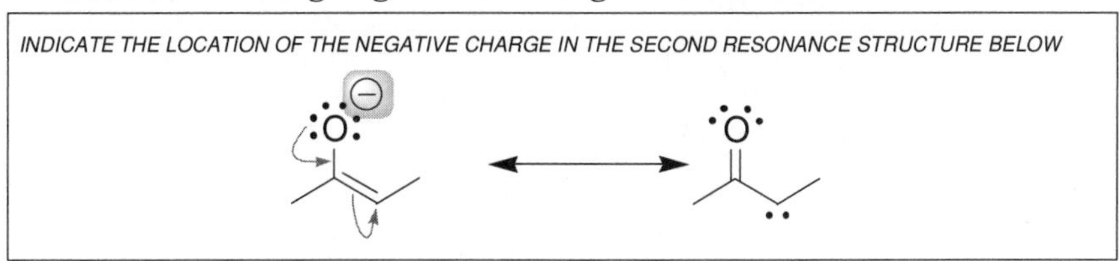

INDICATE THE LOCATION OF THE NEGATIVE CHARGE IN THE SECOND RESONANCE STRUCTURE BELOW
O
O

SkillBuilder 2.8 Drawing Significant Resonance Structures

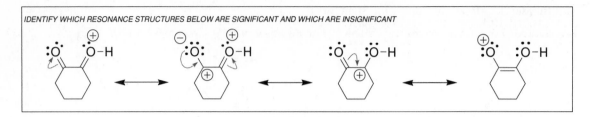

IDENTIFY WHICH RESONANCE STRUCTURES BELOW ARE SIGNIFICANT AND WHICH ARE INSIGNIFICANT

SkillBuilder 2.9 Identifying Localized and Delocalized Lone Pairs

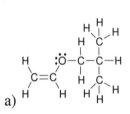

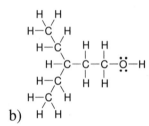

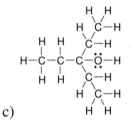

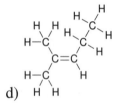

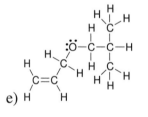

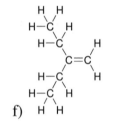

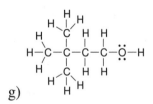

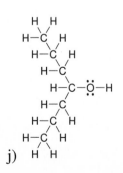

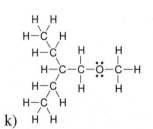

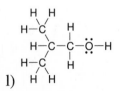

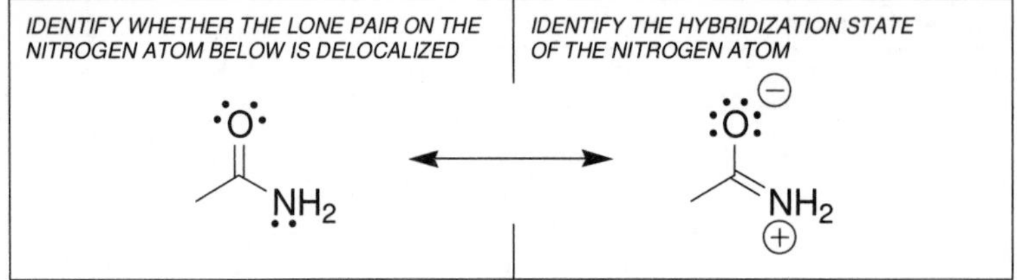
IDENTIFY WHETHER THE LONE PAIR ON THE NITROGEN ATOM BELOW IS DELOCALIZED
IDENTIFY THE HYBRIDIZATION STATE OF THE NITROGEN ATOM

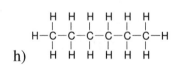

Solutions

2.1.

a)

b)

c)

d)

e)

f)

g)

h)

i)

j)

k)

l)

2.2 $(CH_3)_3C\ddot{\underset{\cdot\cdot}{O}}CH_3$ and $(CH_3)_2CH\ddot{\underset{\cdot\cdot}{O}}CH_2CH_3$

2.3 Six

2.4 $H_2C{=}CHCH_3$

2.5.

a) 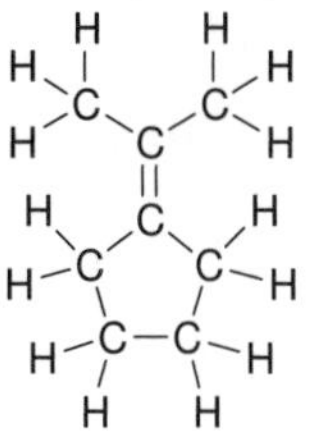 b) 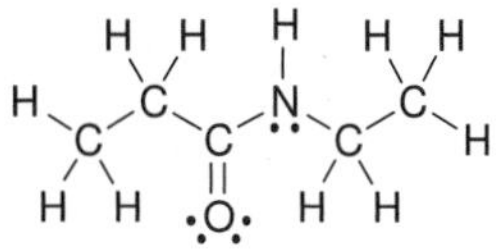 c)

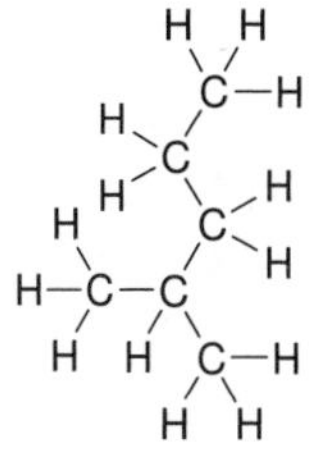

d) 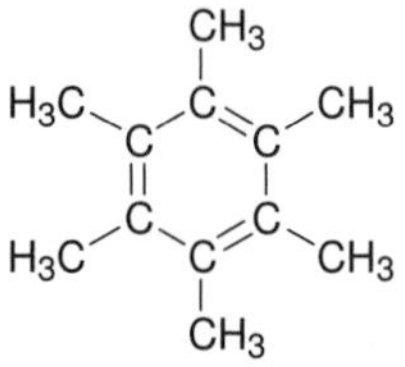 e) 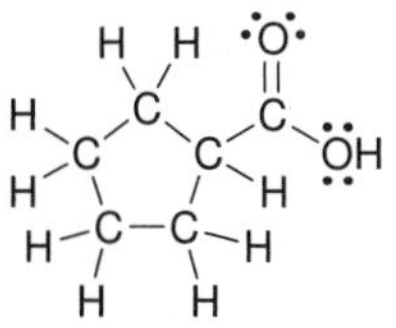 f)

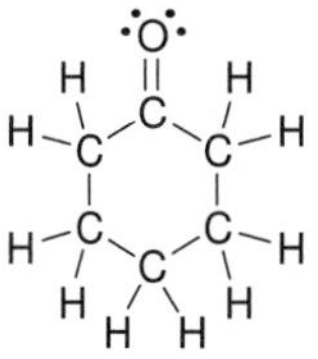

g) 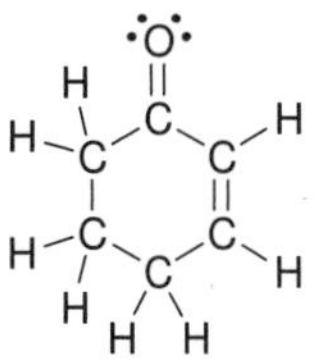 h) 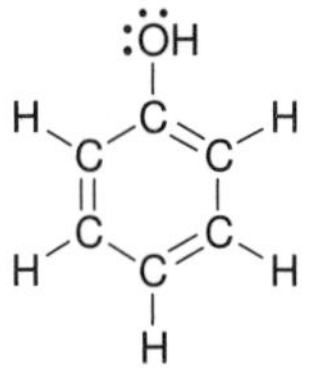 i)

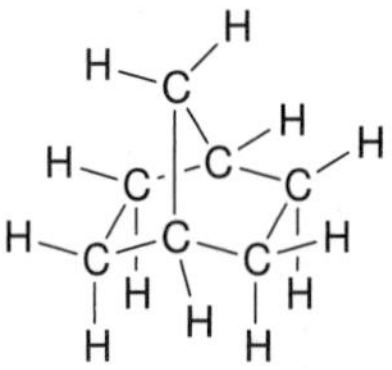

j) 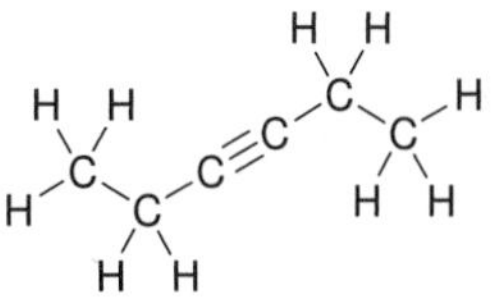 k) 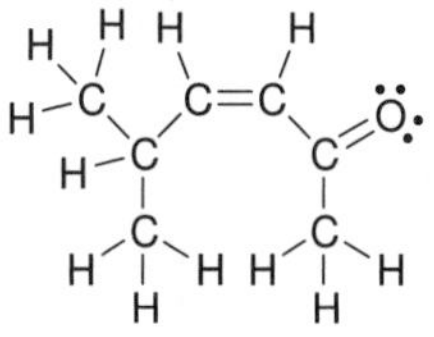 l)

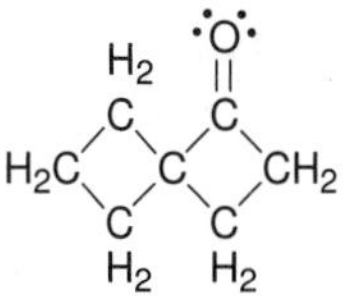

2.6

a) decrease (7→6) b) no change (8→8)
c) no change (8→8) d) increase (5→7)

2.7

a) increase (12→14) b) decrease (8→6)

2.8.

a) 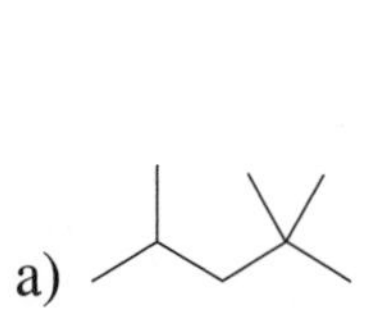 b) 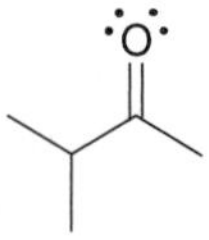 c) 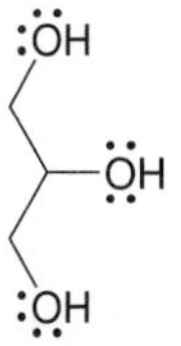 d)

e) 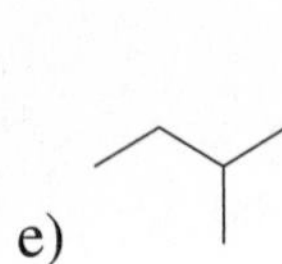

f) g) h)

i) j) k) l)

m) n) 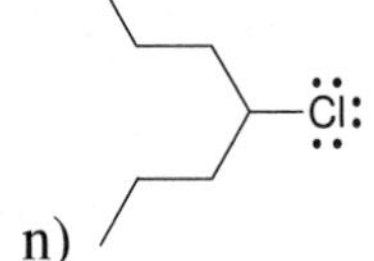

o) 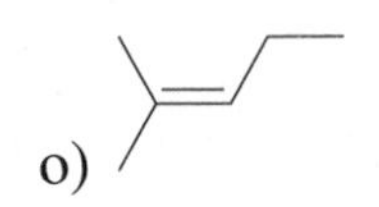

p)

q) r) 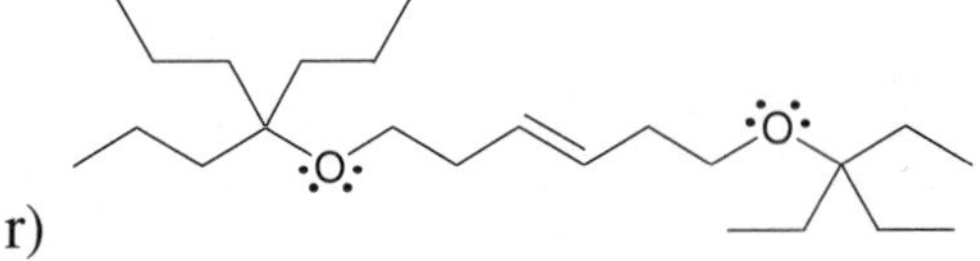

2.9.

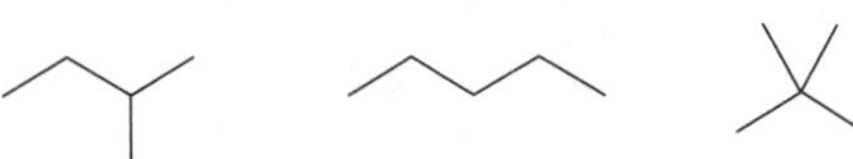

2.10.

a) 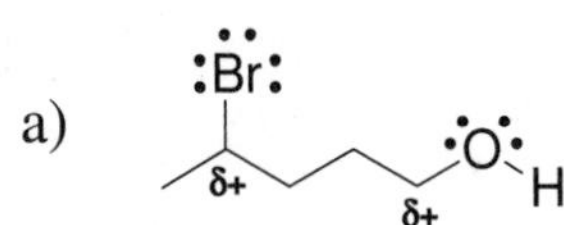

b) 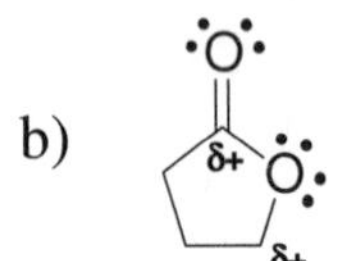

c)

2.11.

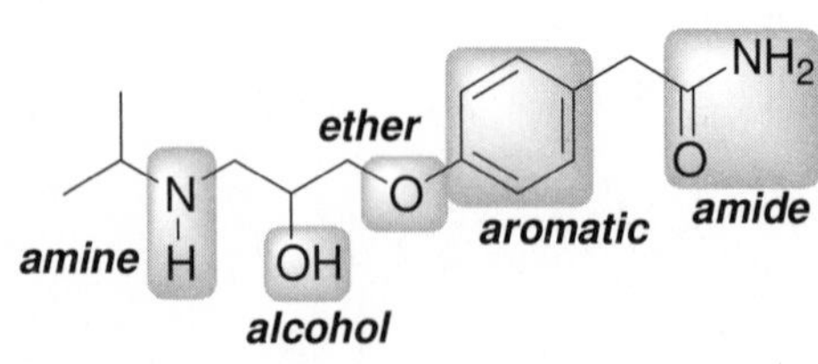

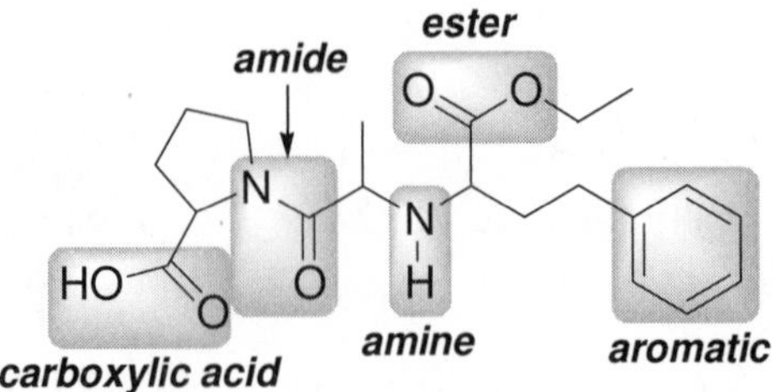

2.12.

a) 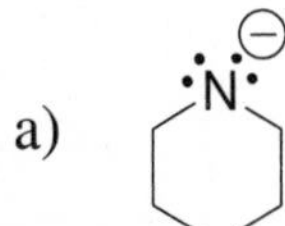

b) No charge

c) 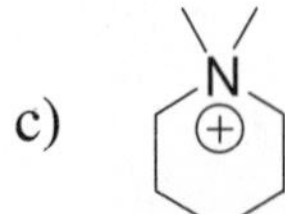

d) No charge

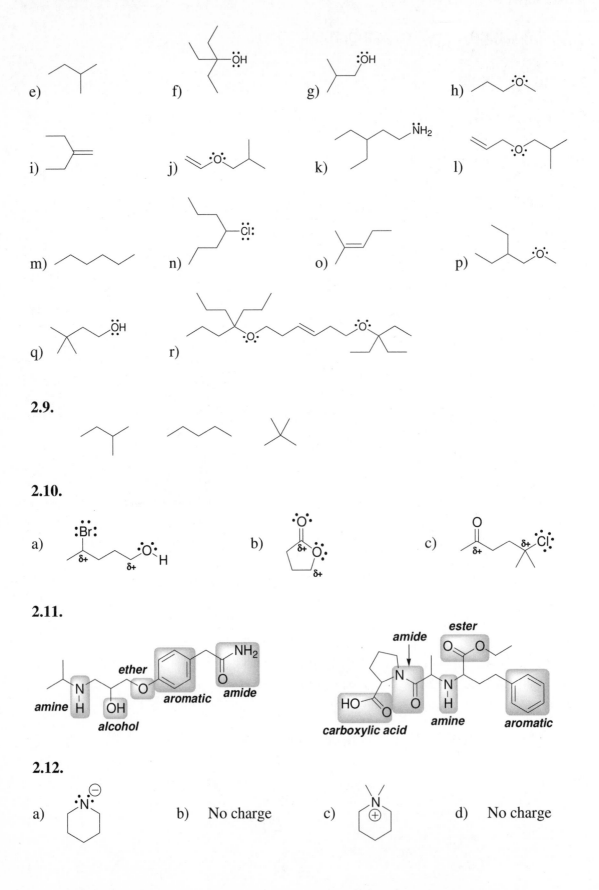

2.13.

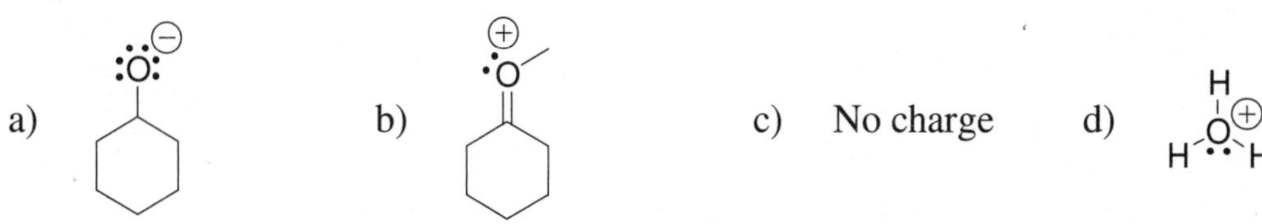

2.14.

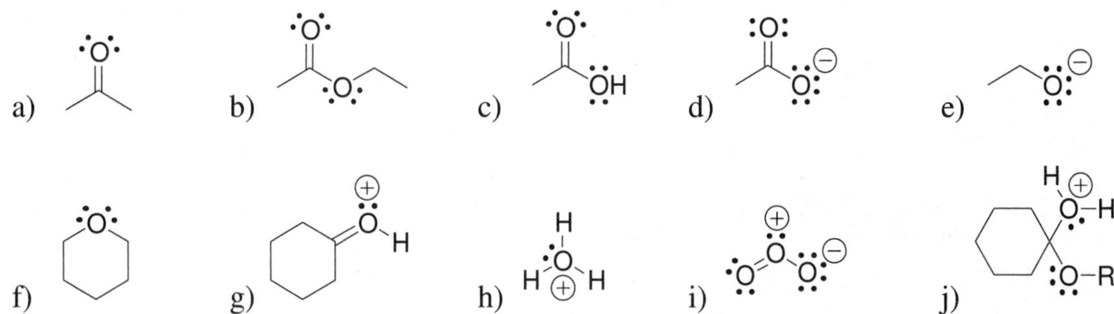

2.15. There are no hydrogen atoms attached to the central carbon atom. The carbon atom has four valence electron. Two valence electrons are being used to form bonds, and the remaining two electrons are a lone pair. This carbon atom is using the appropriate number of valence electrons.

2.16.

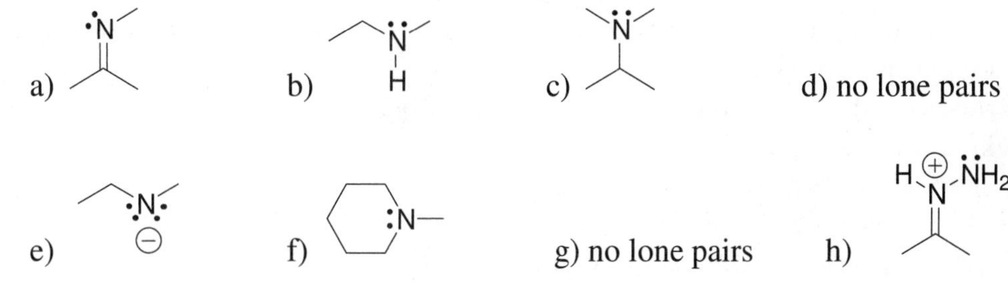

2.17.

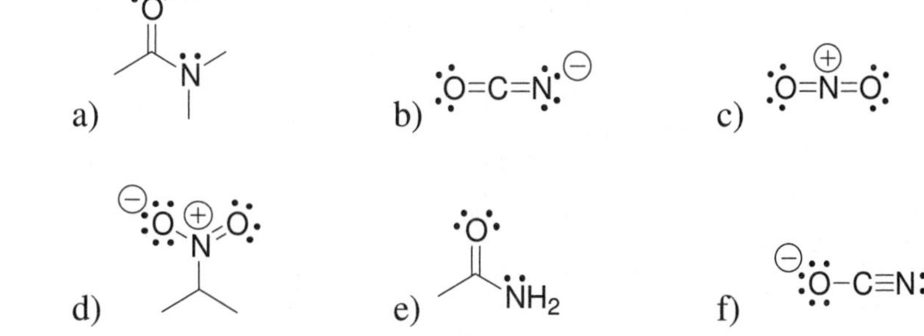

2.18.

a) one b) zero c) one d) five

2.19 Five lone pairs:

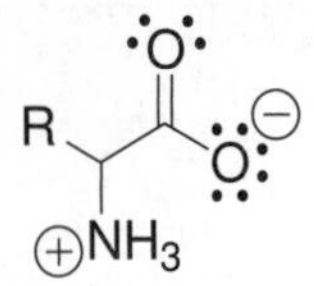

2.20

a)

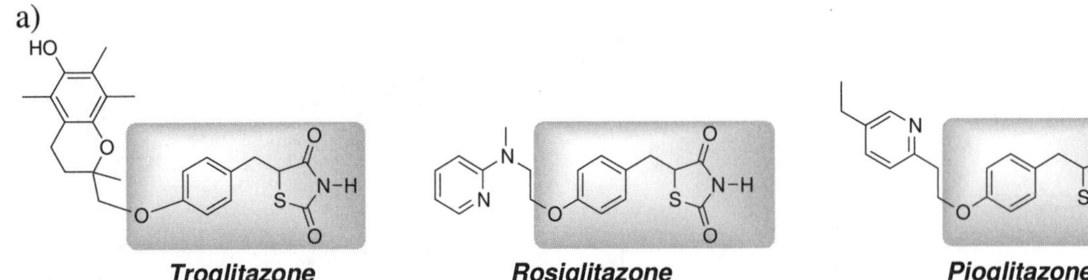

Troglitazone ***Rosiglitazone*** ***Pioglitazone***

b) Yes, it contains the likely pharmacophore highlighted above.

2.21

a) Violates second rule by giving a fifth bond to a nitrogen atom.
b) Does not violate either rule.
c) Violates second rule by giving five bonds to a carbon atom.
d) Violates second rule by giving three bonds and two lone pairs to an oxygen atom.
e) Violates second rule by giving five bonds to a carbon atom.
f) Violates second rule by giving five bonds to a carbon atom.
g) Violates second rule by giving five bonds to a carbon atom, and violates second rule by breaking a single bond.
h) Violates second rule by giving five bonds to a carbon atom, and violates second rule by breaking a single bond.
i) Does not violate either rule.
j) Does not violate either rule.
k) Violates second rule by giving five bonds to a carbon atom.
l) Violates second rule by giving five bonds to a carbon atom.

2.22.

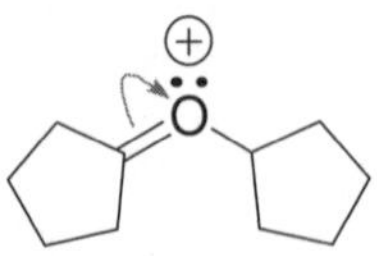

2.23.

a)

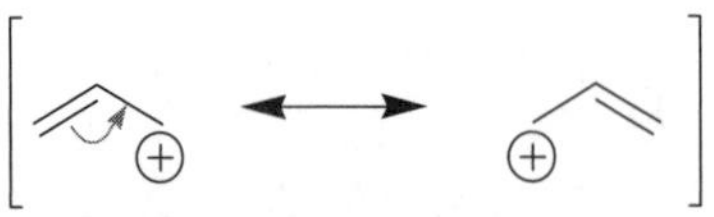

b)

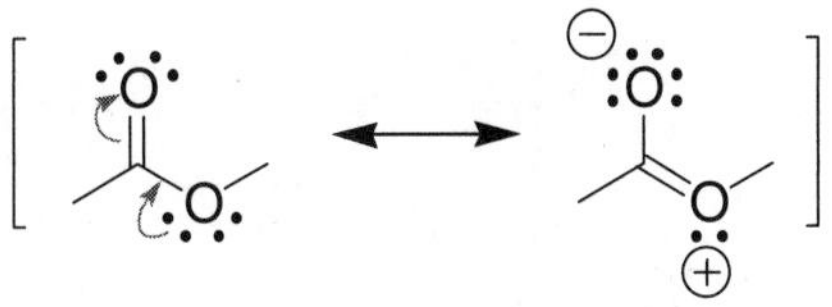

c)

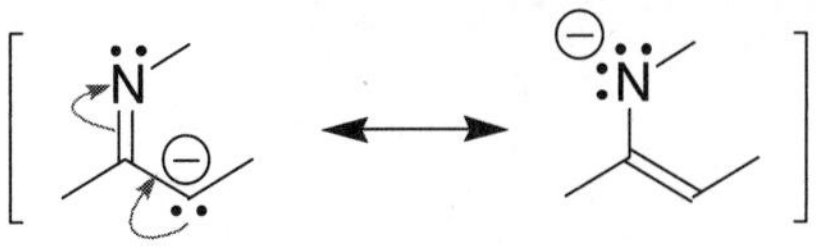

d)

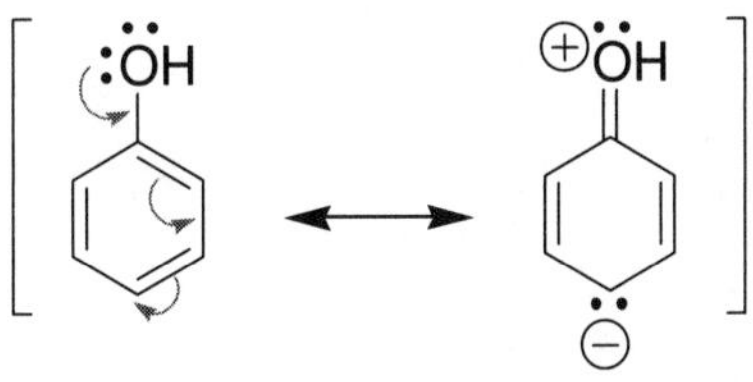

e)

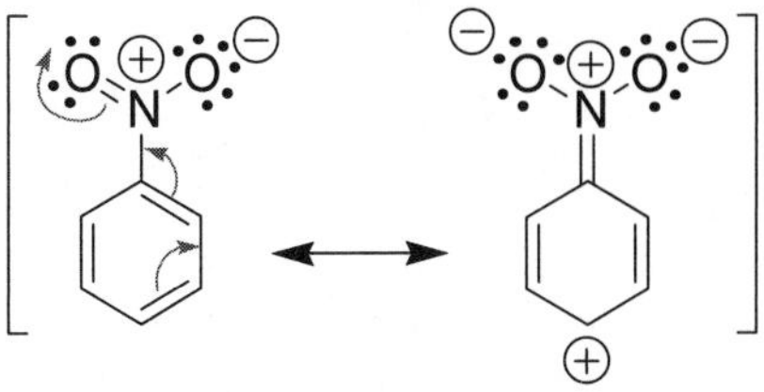

f)

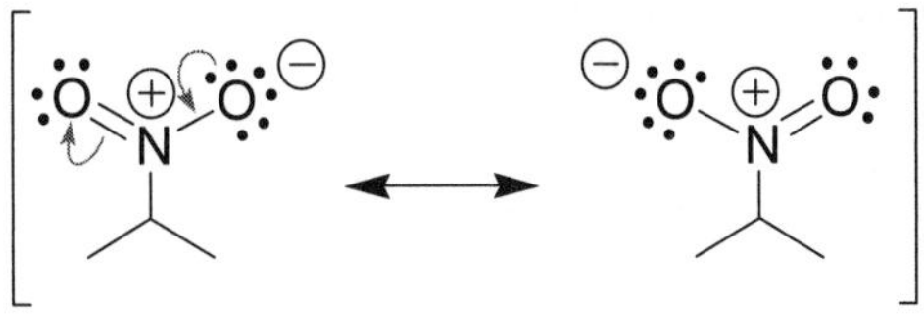

g)

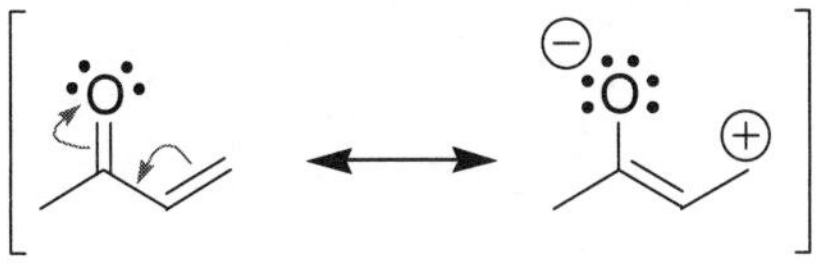

h)

2.24.

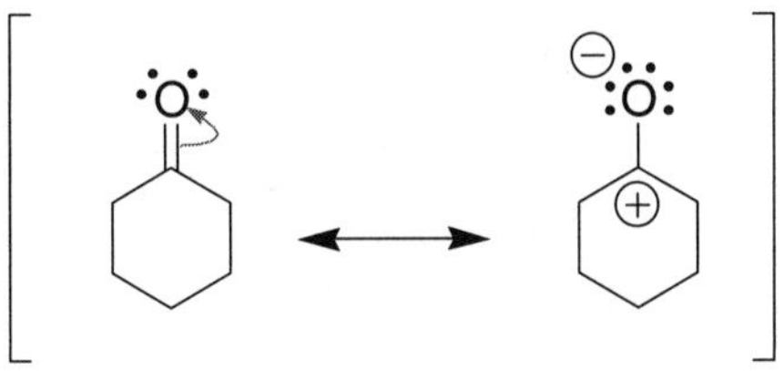

a)

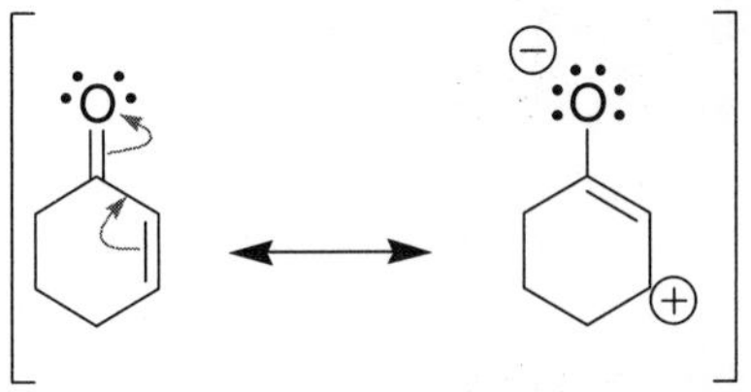

b)

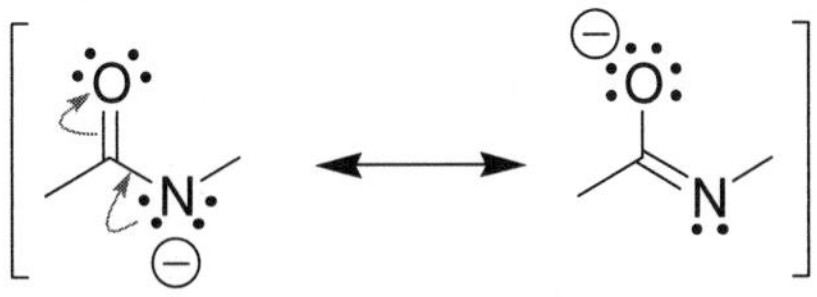

c)

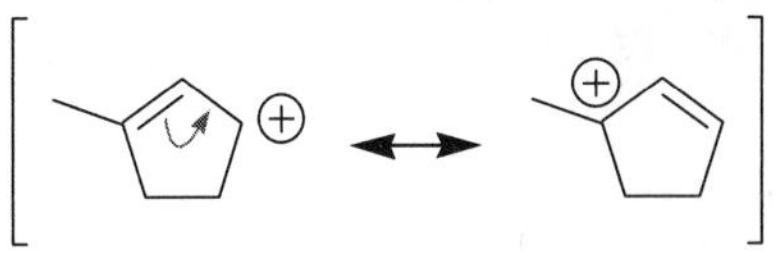

b)

2.25.

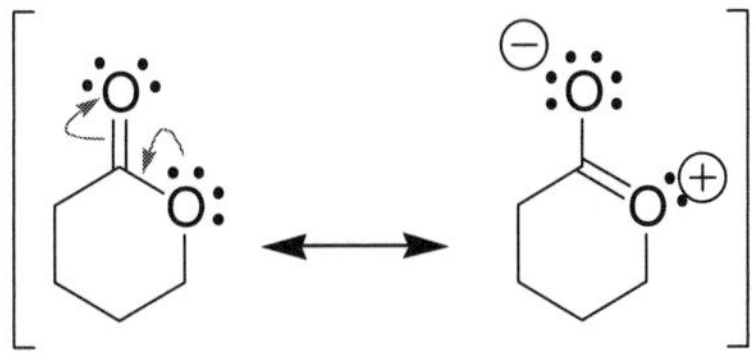

a)

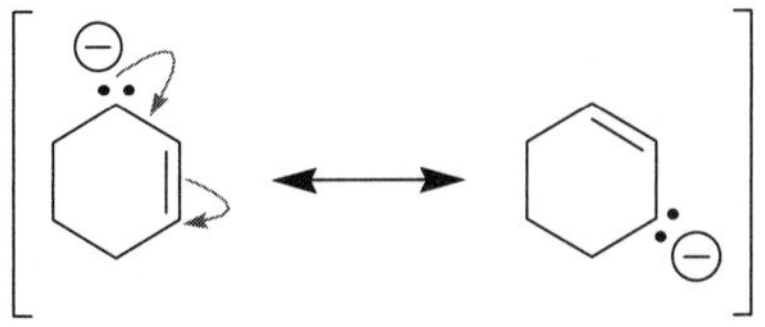

b)

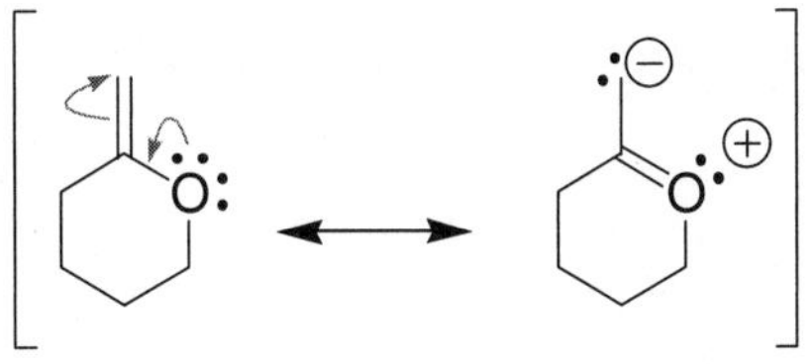

c)

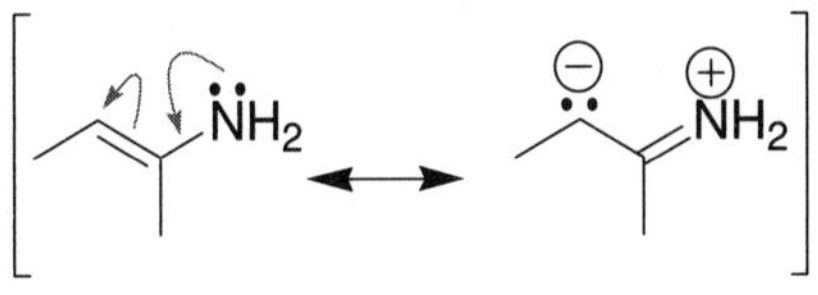

d)

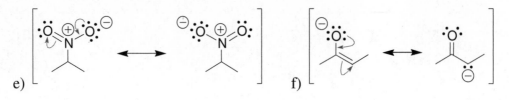

e)

f)

g)

h)

2.26.

a)

b)

c)

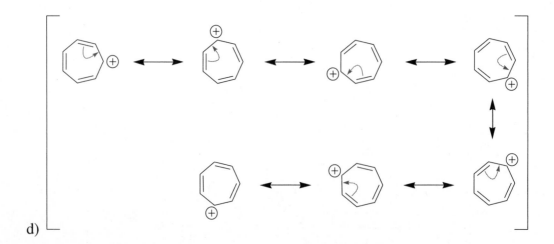

d)

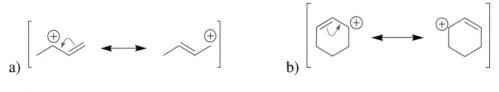

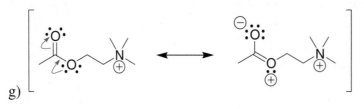

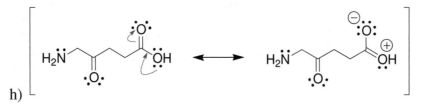

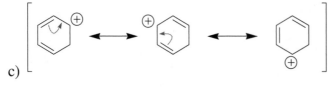

2.27.

a)  :N: O O

b) 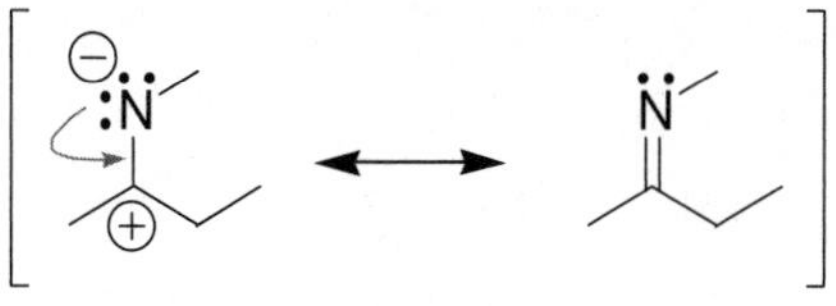 :N: N: O O

c) 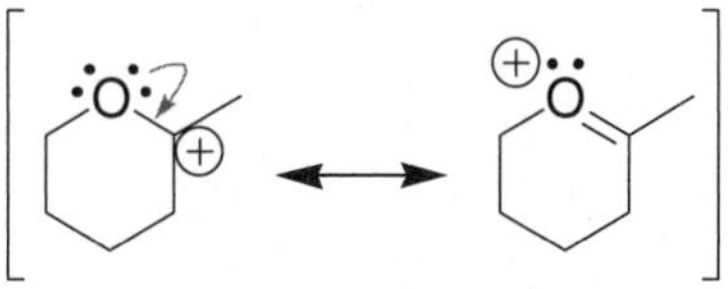 :O: O O:

2.28.

a) 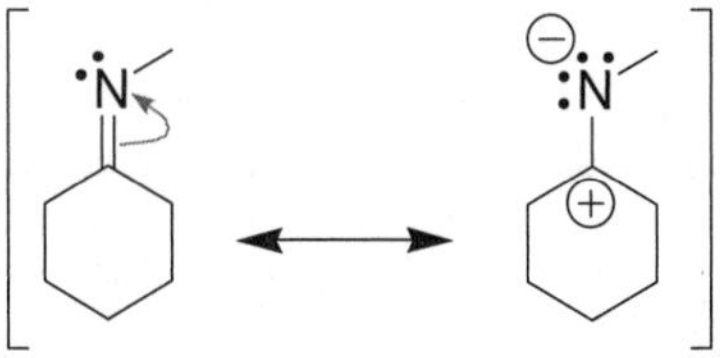 :N: :N: O O

b) 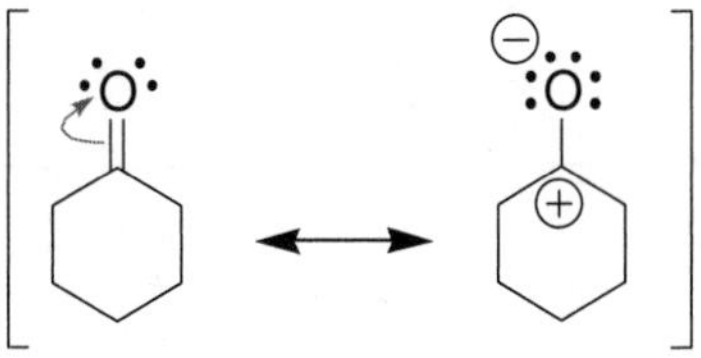 :O: :O: O O

c) 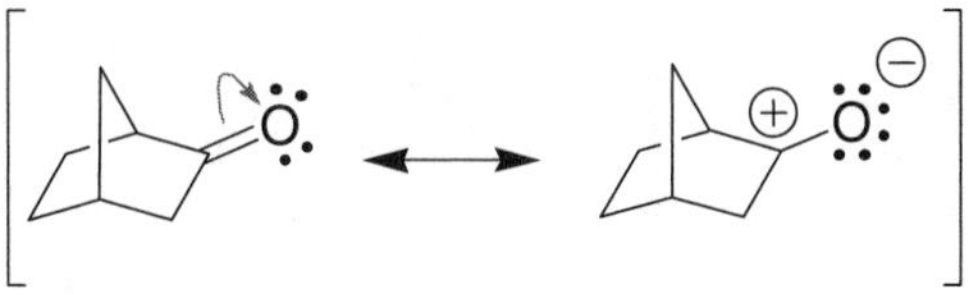 :O: O :O: O

2.29.

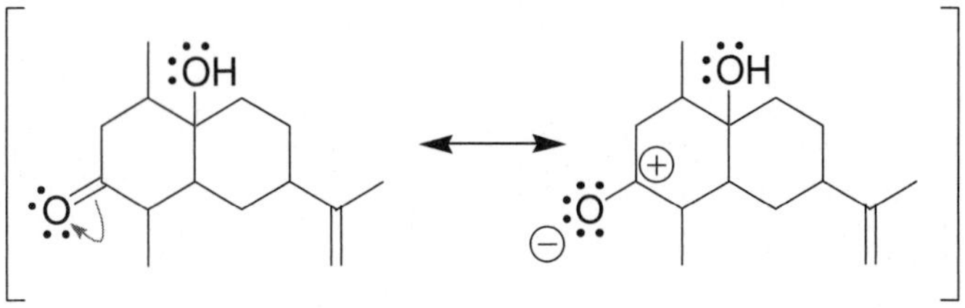

:OH :O: :OH O :O: O

2.30.

:O: :O: O O

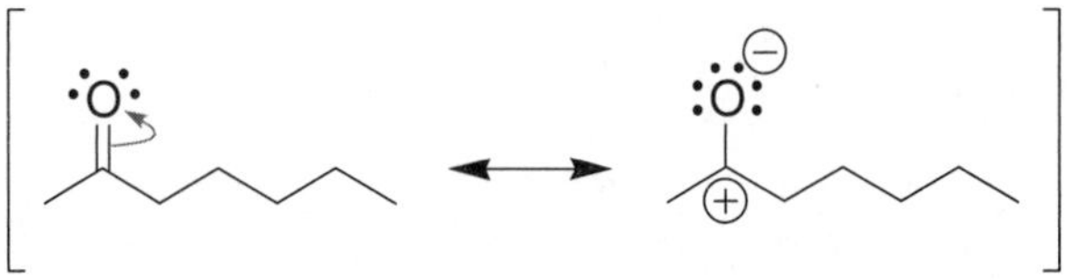

2.31.

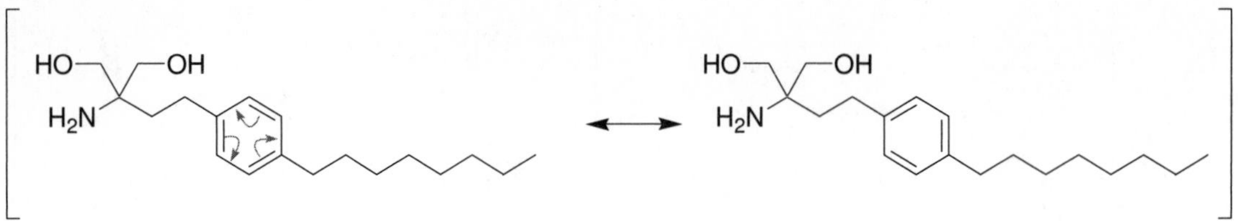

HO OH
H_2N

HO OH
H_2N

2.32.

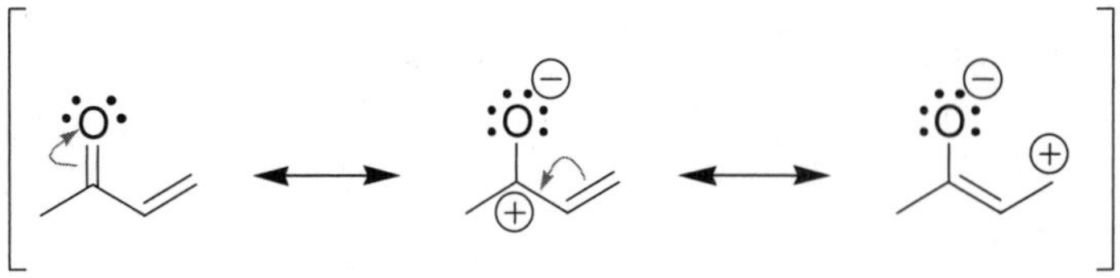

a)

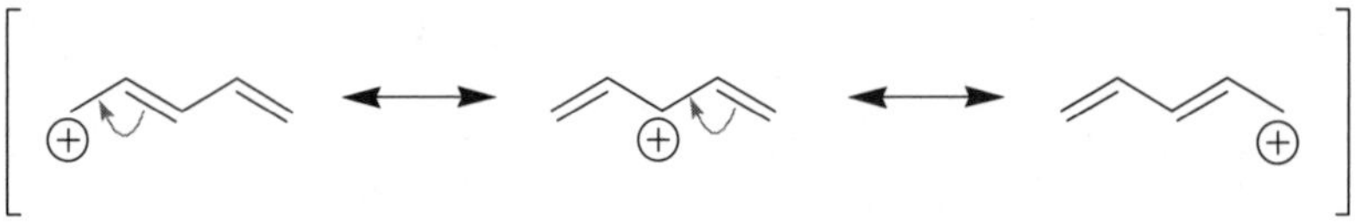

b)

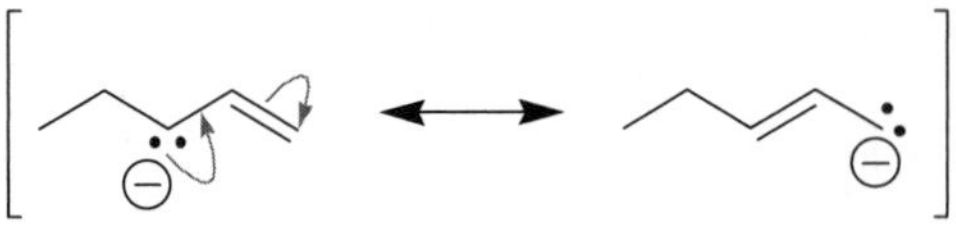

c)

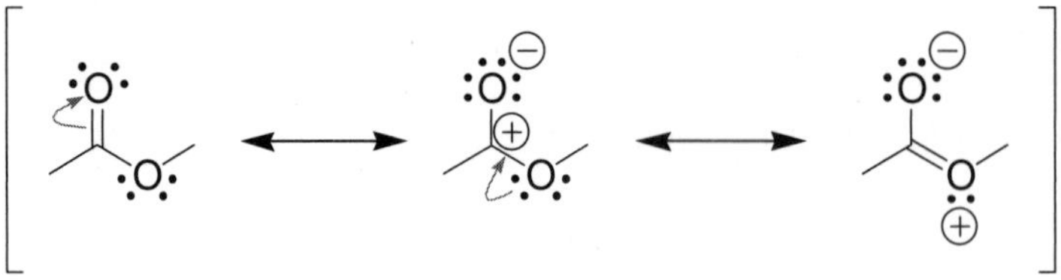

d)

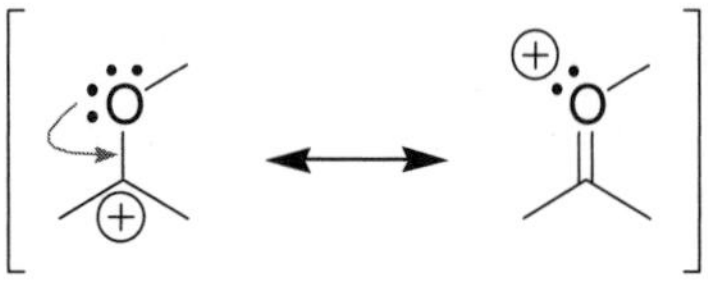

e)

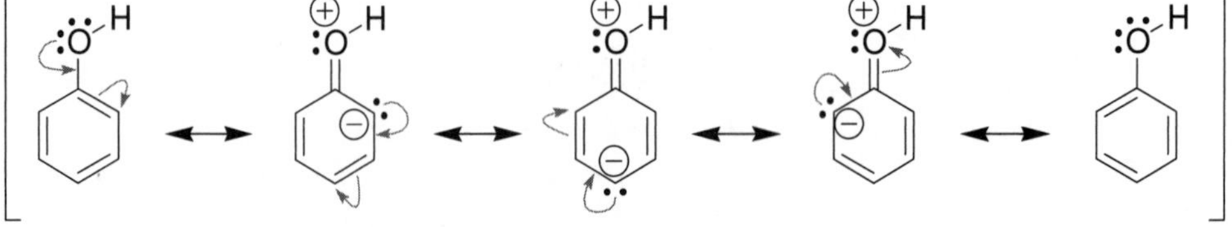

f)

g) 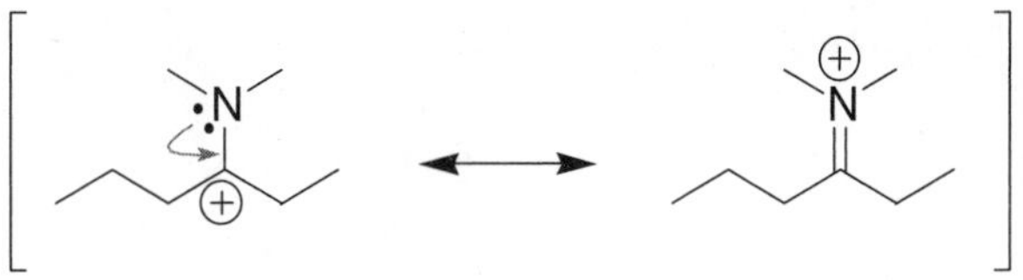 [N–C⁺ resonance structures]

h)

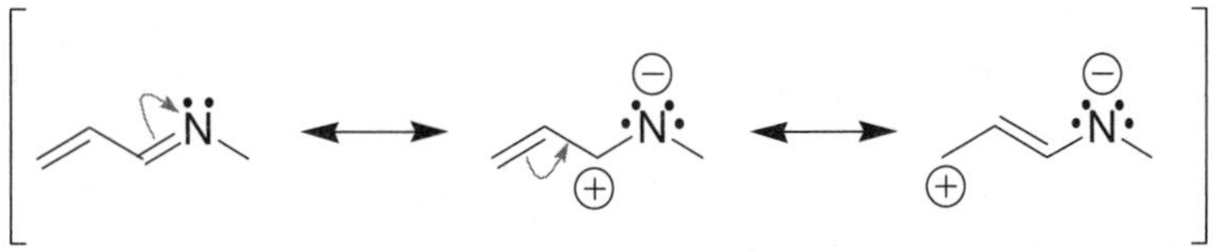

i)

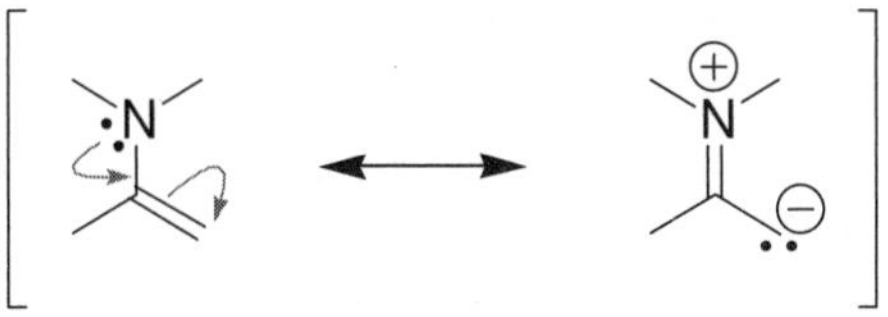

j)

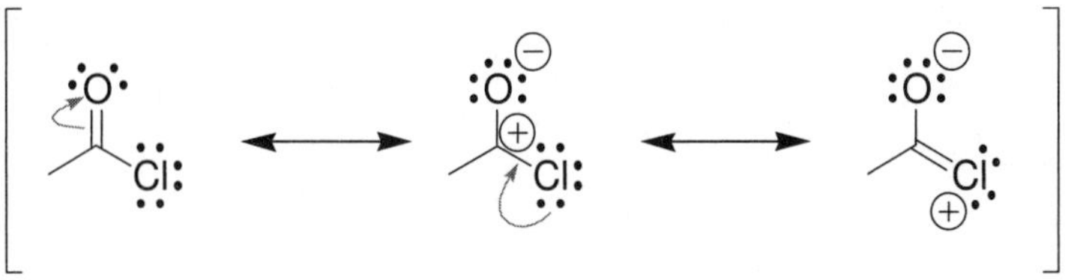

2.33.

a)

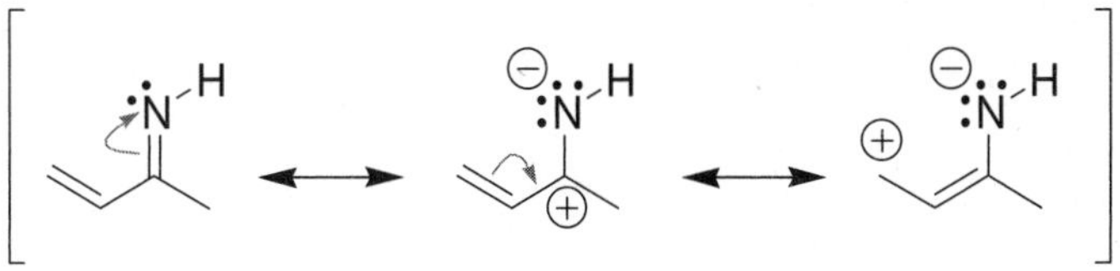

b)

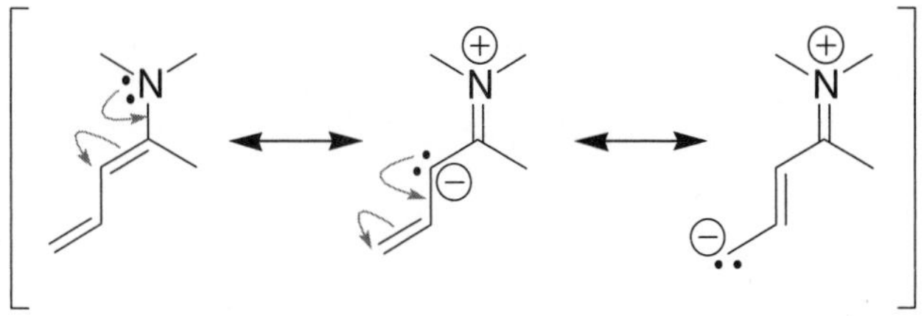

c)

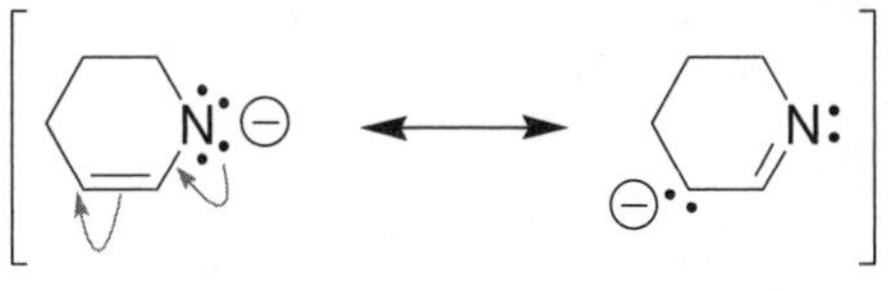

d)

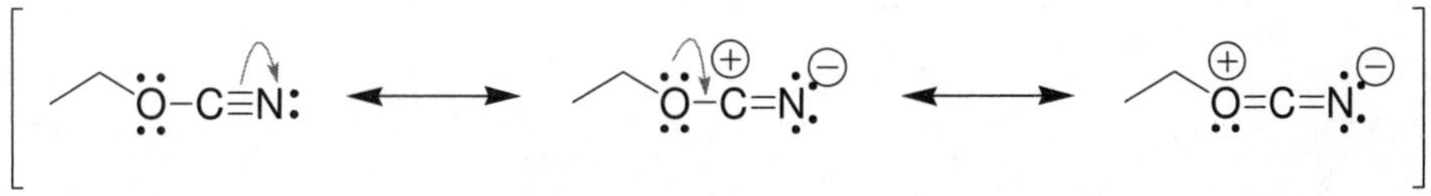

e)

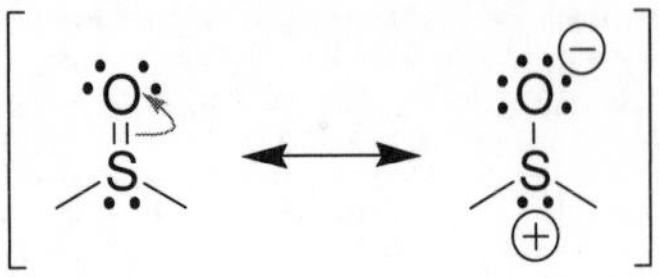

f)

g)

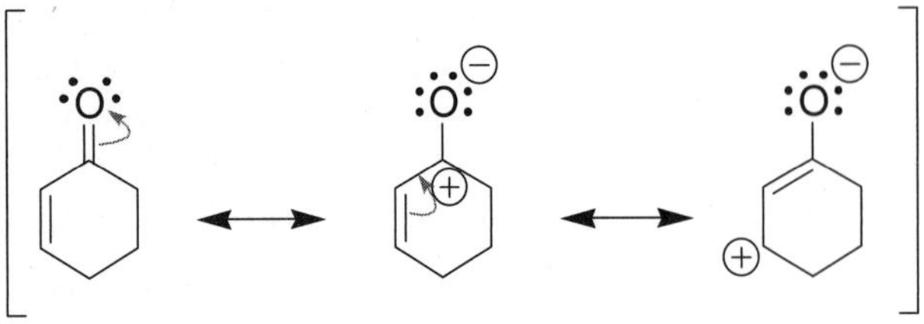

h)

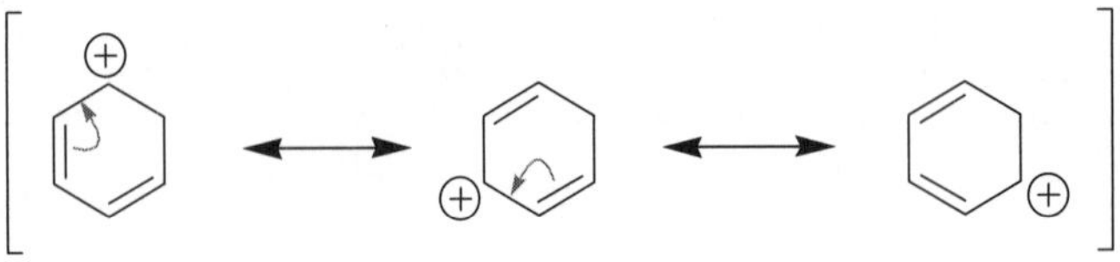

i)

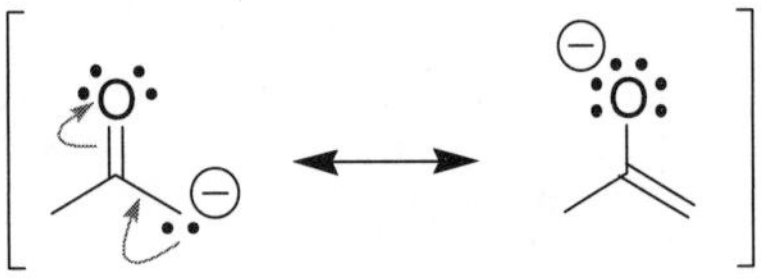

j)

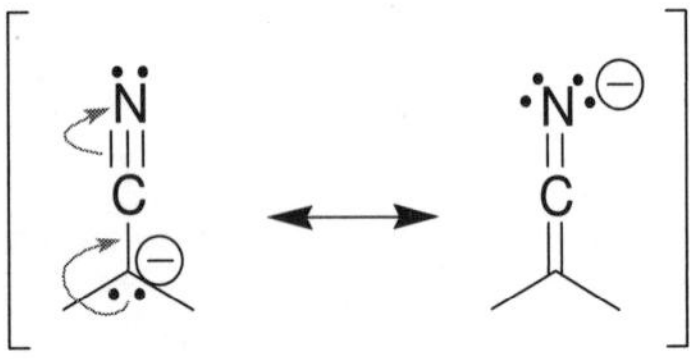

k)

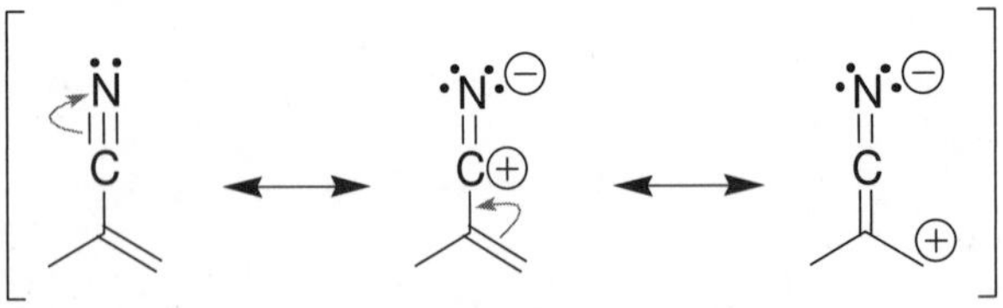

l)

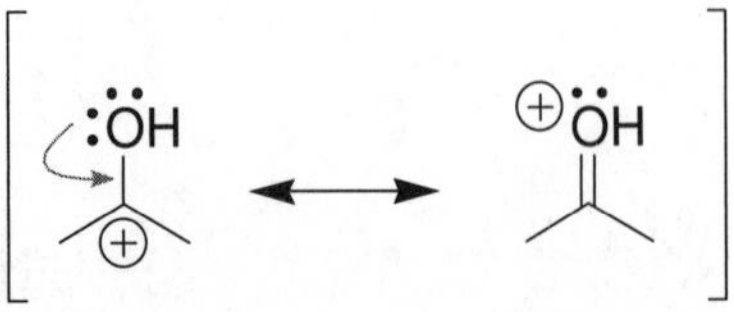

2.34.

2.35.

2.36.

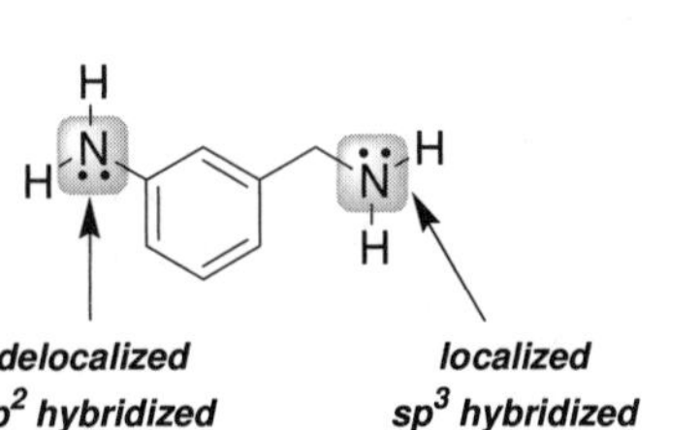

a)

localized
sp^2 hybridized
geometry not relevant
(connected to only one atom)

One of these lone pairs is delocalized. The oxygen atom is therefore sp^2 hybridized and has bent geometry.

delocalized
sp^2 hybridized
trigonal planar

b)

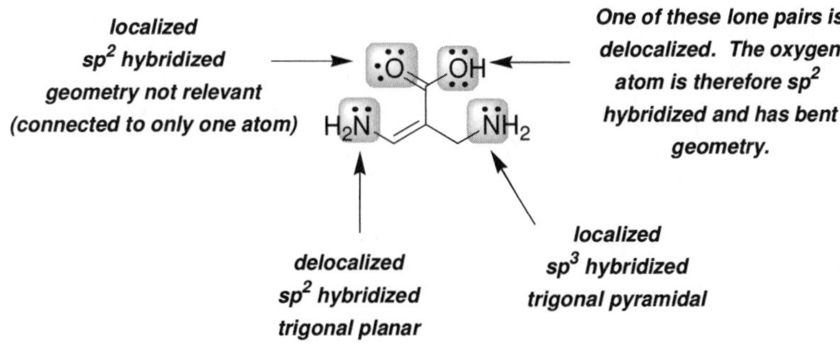

c)

localized
sp^2 hybridized
bent

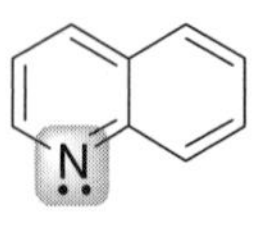

d)

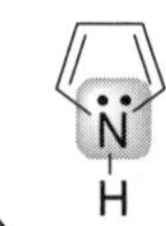

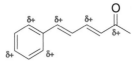

delocalized
sp^2 hybridized
trigonal planar

e)

f)

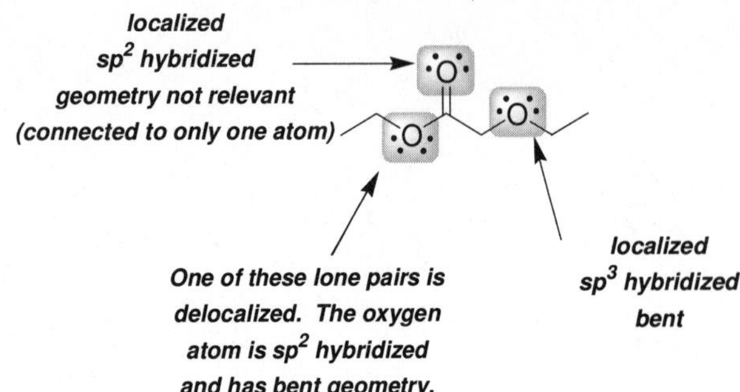

2.37. Both lone pairs are localized and, therefore, both are expected to be reactive.

2.38.

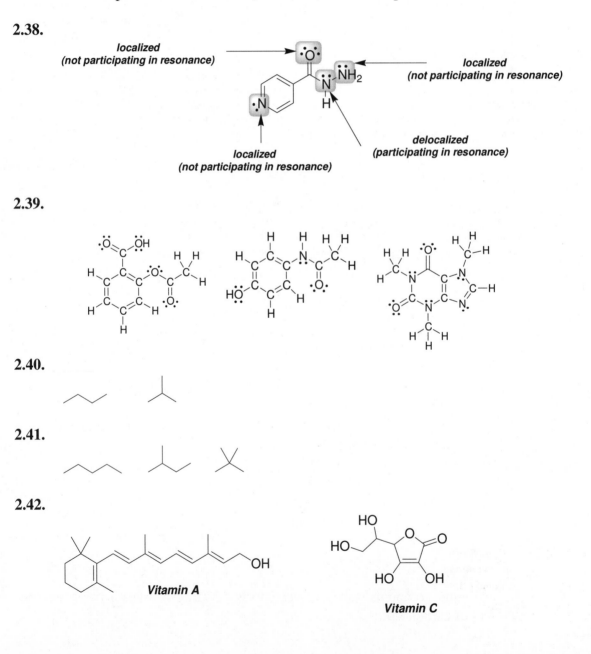

localized (not participating in resonance)

localized (not participating in resonance)

localized (not participating in resonance)

delocalized (participating in resonance)

2.39.

2.40.

2.41.

2.42.

Vitamin A

Vitamin C

2.43. Twelve (each oxygen atom has two lone pairs)

2.44.

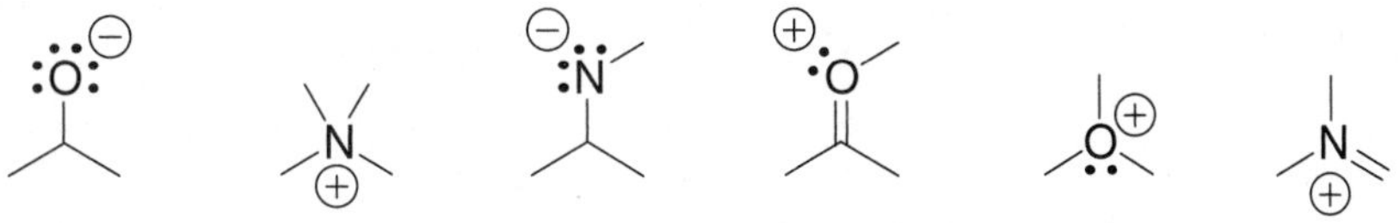

2.45.

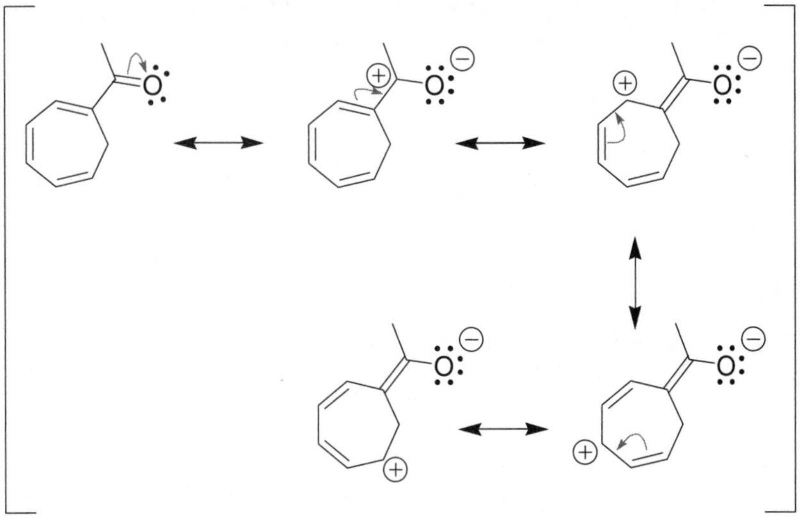

2.46.

a)

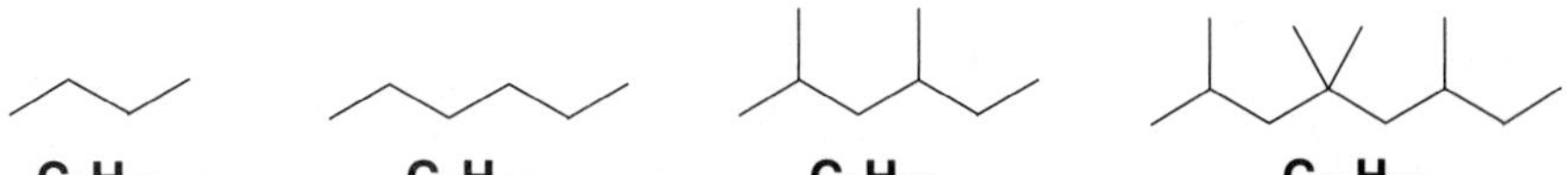

C_4H_{10} C_6H_{14} C_8H_{18} $C_{12}H_{26}$

In each of the compounds above, the number of hydrogen atoms is equal to two times the number of carbon atoms, plus two.

b)

C_4H_8 C_7H_{14} C_7H_{14} $C_{12}H_{24}$

In each of the compounds above, the number of hydrogen atoms is two times the number of carbon atoms.

c)

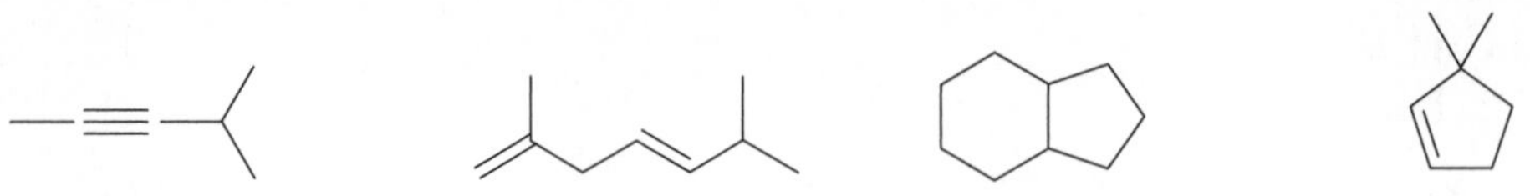

C_6H_{10} C_9H_{16} C_9H_{16} C_7H_{12}

In each of the compounds above, the number of hydrogen atoms is two times the number carbon atoms, minus two.

d) A compound with molecular formula $C_{24}H_{48}$ must have either one double bond or one ring. It cannot have a triple bond, but it may have a double bond.

e)

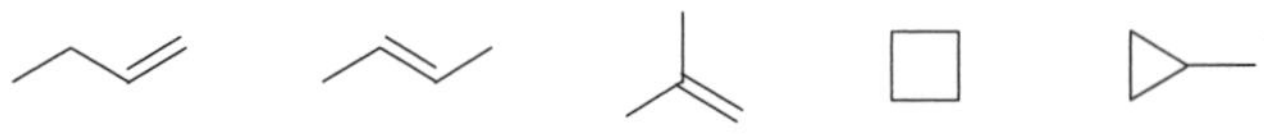

2.47.

a) an sp^2 hybridized atomic orbital
b) a *p* orbital
c) a *p* orbital

2.48.

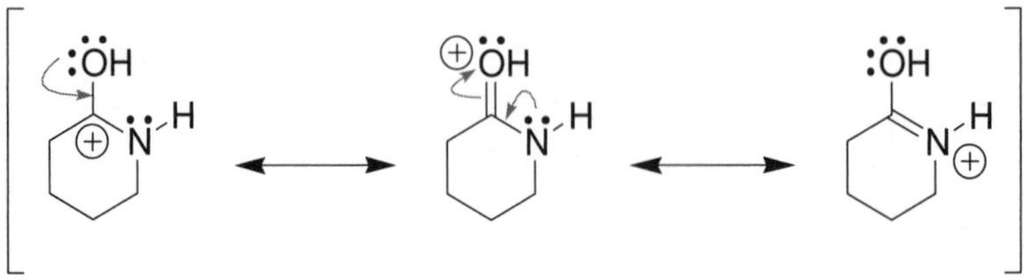

a)

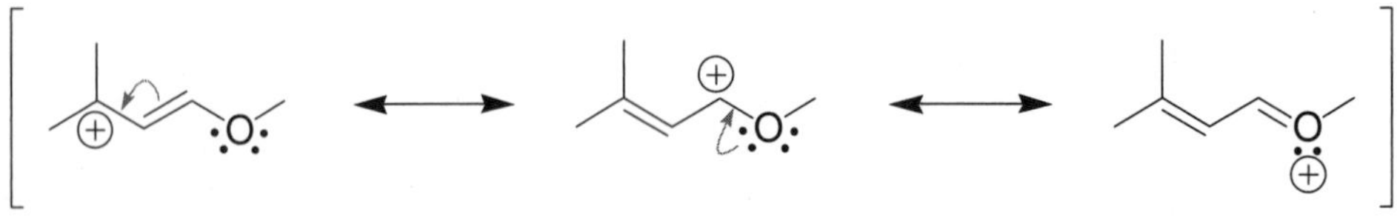

b)

c)

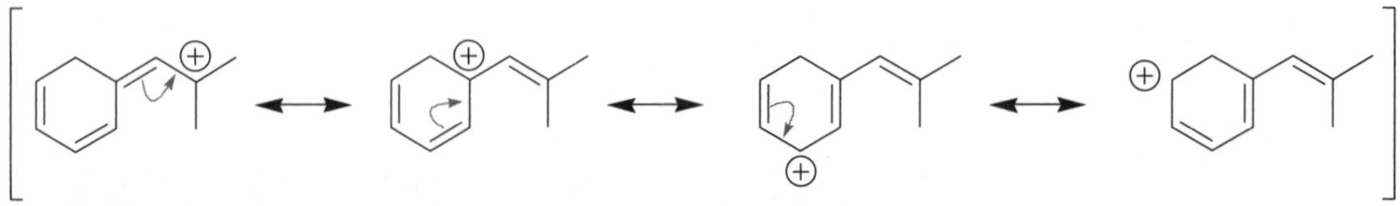

2.49.

a) $(CH_3)_3CCH_2CH_2CH(CH_3)_2$
b) $(CH_3)_2CHCH_2CH_2CH_2OH$
c) $CH_3CH_2CH{=}C(CH_2CH_3)_2$

2.50.

a) C_9H_{20}
b) $C_6H_{14}O$
c) C_8H_{16}

2.51.
(d) is not a valid resonance structure, because it violates the octet rule. The nitrogen atom has five bonds in this drawing, which is not possible, because the nitrogen atom only has four orbitals with which it can form bonds.

2.52. 15 carbon atoms and 18 hydrogen atoms:

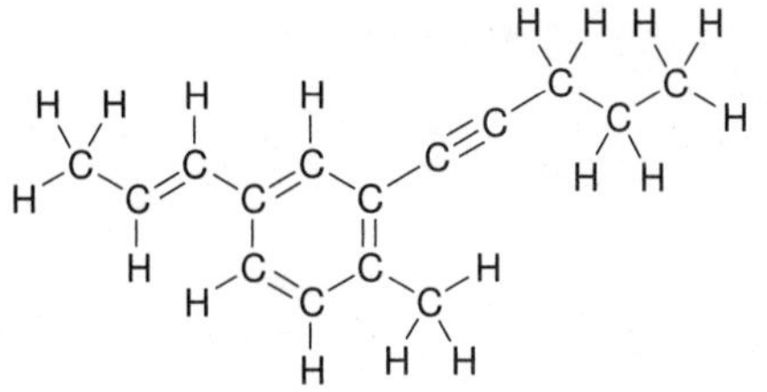

2.53.

a) 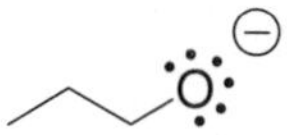 b) 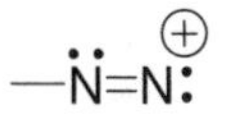 c) 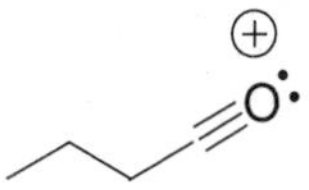 d)

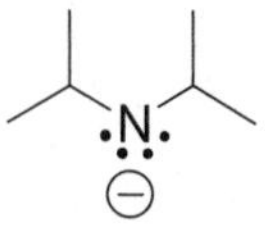

2.54.

Cl Cl Cl Cl

2.55.

a) 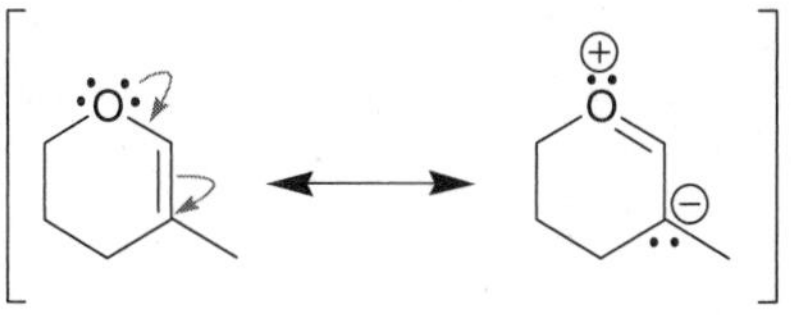 b)

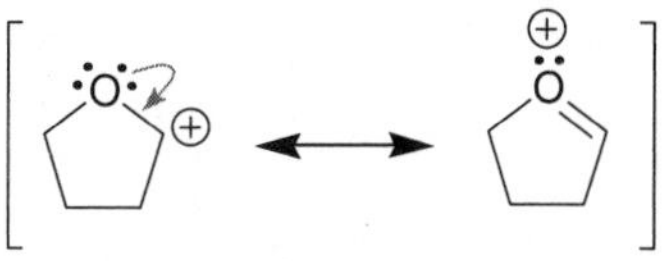

c) 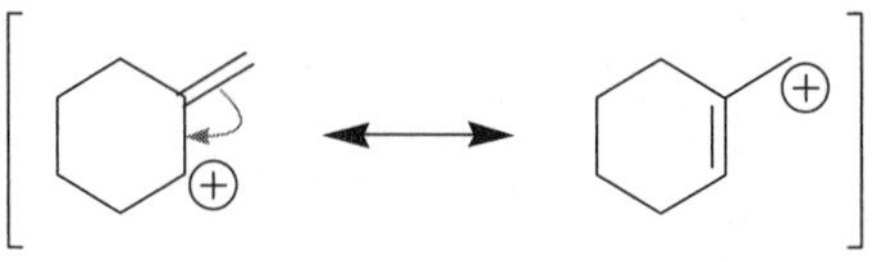 d)

e)

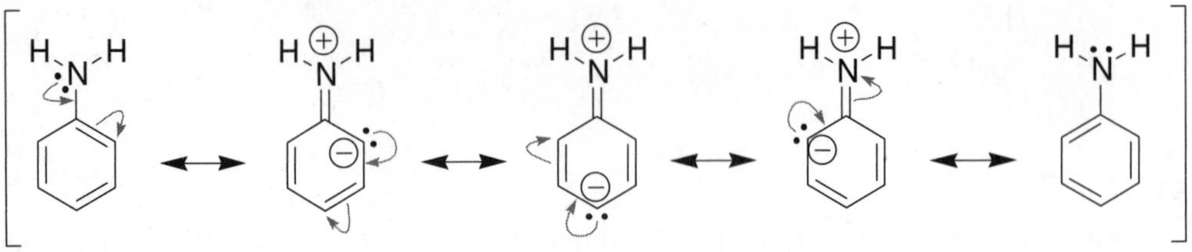

f)

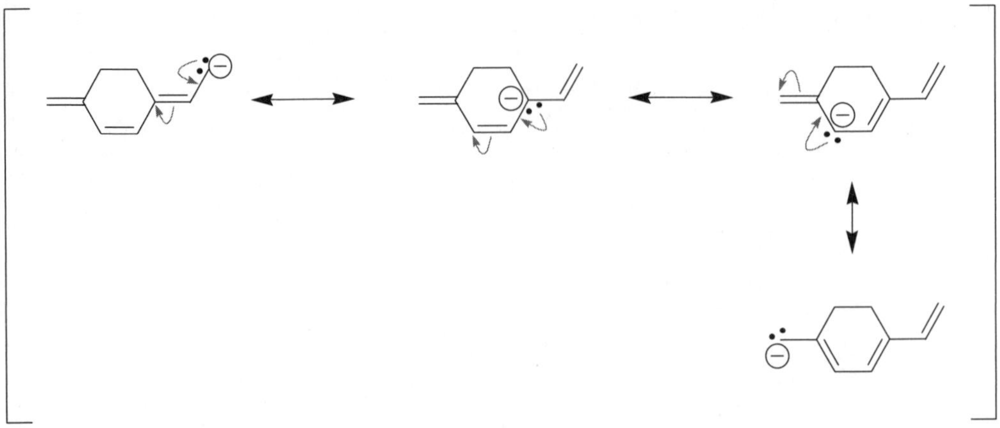

g)

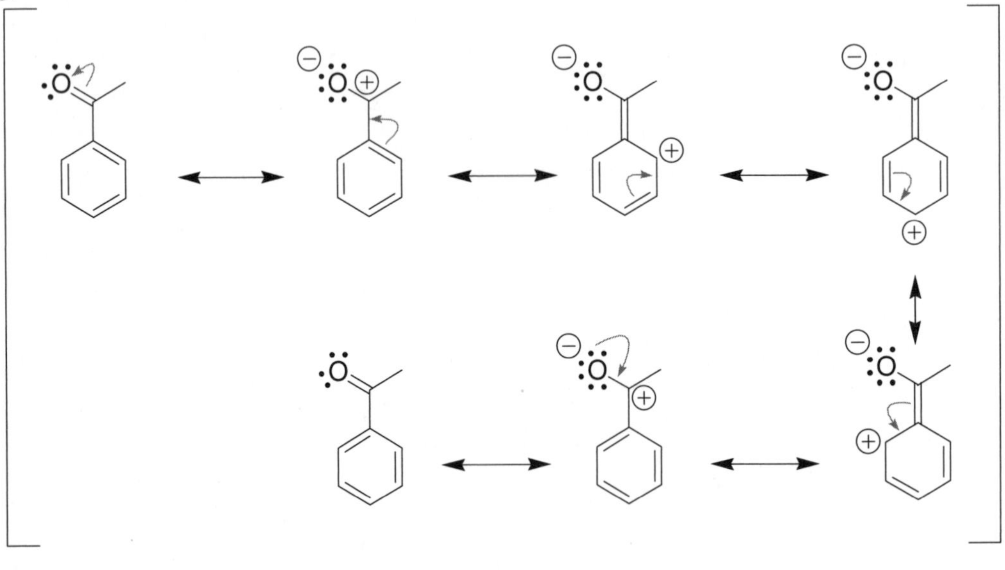

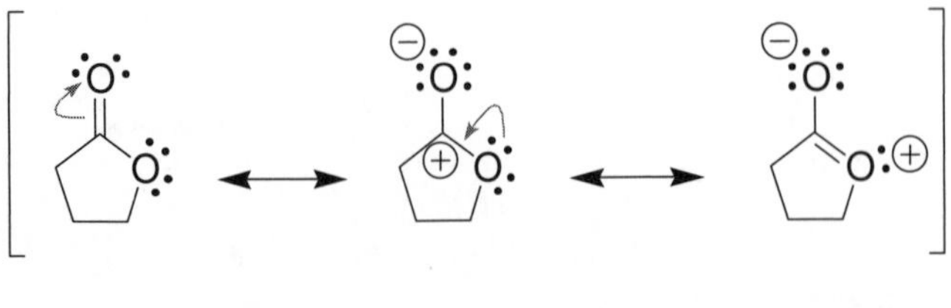

h)

i)

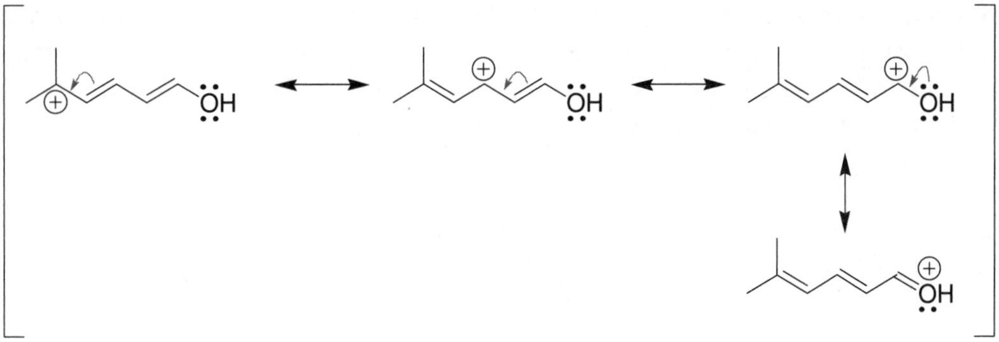

j)

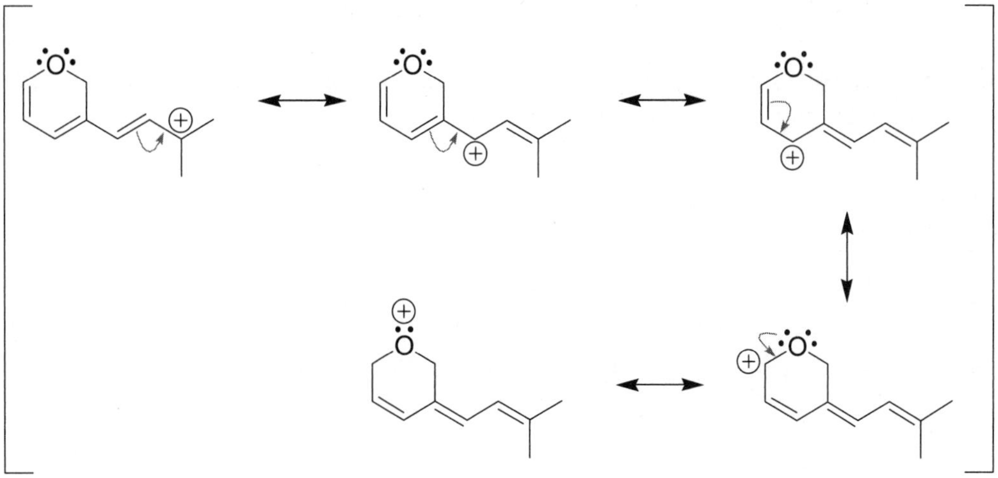

2.56. These structures do not differ in their connectivity of atoms. They differ only in the placement of electrons, and are therefore resonance structures.

2.57.

a) constitutional isomers
b) same compound
c) different compounds that are not isomeric
d) constitutional isomers

2.58.

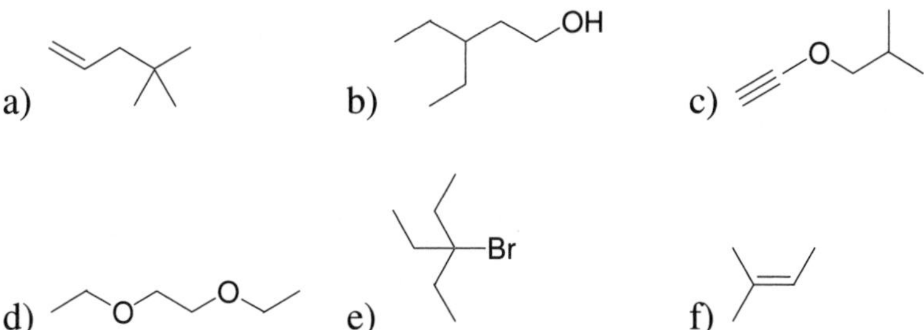

2.59. The nitronium ion does *not* have any significant resonance structures because any attempts to draw a resonance structure will either 1) exceed an octet for the nitrogen atom or 2) generate a nitrogen atom with less than an octet of electrons, or 3) generate a structure with three charges. The first of these would not be a valid resonance structure, and the latter two would not give significant resonance structures.

2.60.

2.61. Both nitrogen atoms are sp^2 hybridized and trigonal planar, because in each case, the lone pair participates in resonance.

2.62.

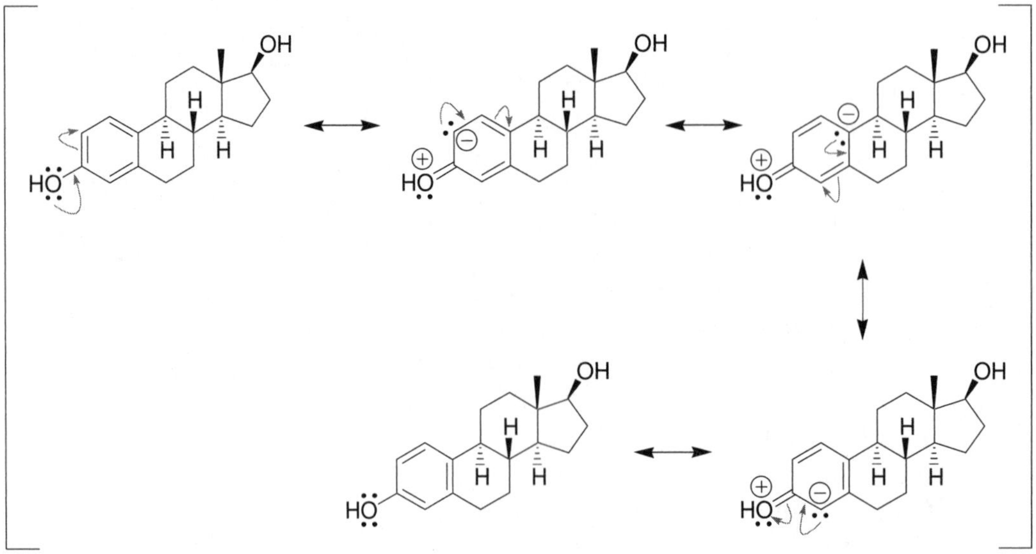

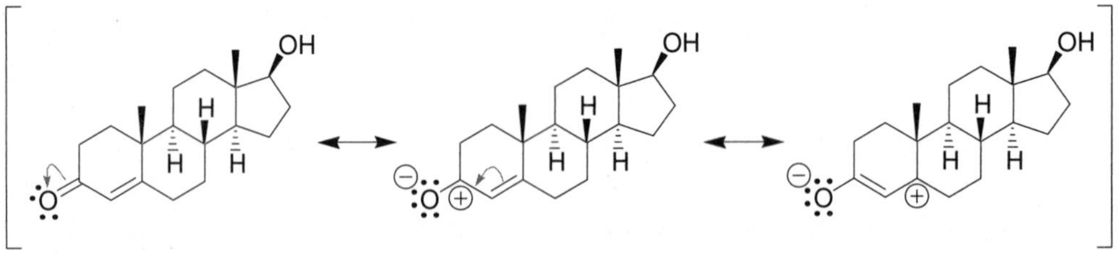

2.63.

a) The molecular formula is $C_3H_6N_2O_2$
b) There are two sp^3 hybridized carbon atoms
c) There is one sp^2 hybridized carbon atom
d) There are no *sp* hybridized carbon atoms
e) There are six lone pairs (each nitrogen atom has one lone pair and each oxygen atom has two lone pairs)
f)

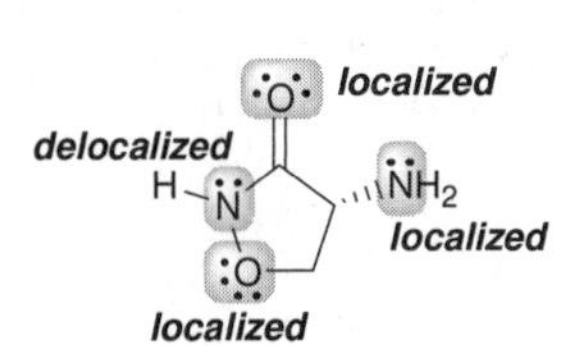

g)

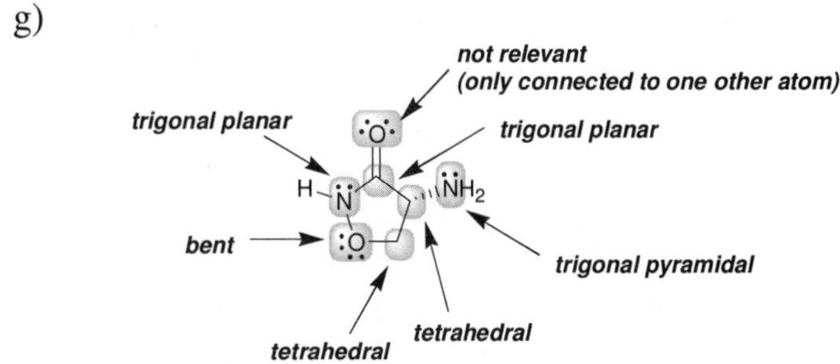

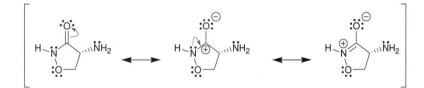

h)

2.64.

a) The molecular formula is $C_{16}H_{21}NO_2$
b) There are nine sp^3 hybridized carbon atoms
c) There is seven sp^2 hybridized carbon atoms
d) There are no *sp* hybridized carbon atoms
e) There are five lone pairs (the nitrogen atom has one lone pair and each oxygen atom has two lone pairs)
f) The lone pairs on the oxygen of the C=O bond are localized. One of the lone pairs on the other oxygen atom is delocalized. The lone pair on the nitrogen atom is delocalized.
g) All sp^2 hybridized carbon atoms are trigonal planar. All sp^3 hybridized carbon atoms are tetrahedral. The nitrogen atom is trigonal planar. The oxygen atom of the C=O bond does not have a geometry because it is connected to only one other atom, and the other oxygen atom has bent geometry.

2.65.

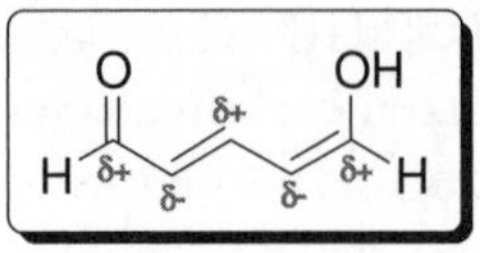

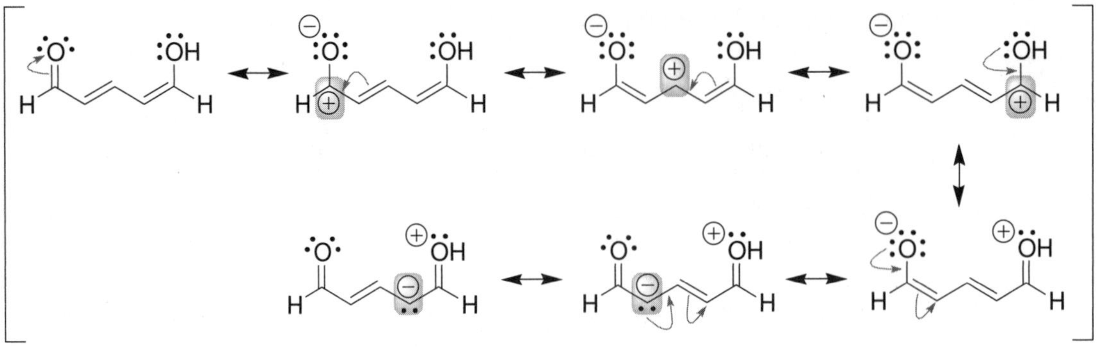

2.66.

a) Compound B has one additional resonance structure that Compound A lacks, because of the relative positions of the two groups on the aromatic ring. Specifically, Compound B has a resonance structure in which one oxygen atom has a negative charge and the other oxygen atom has a positive charge:

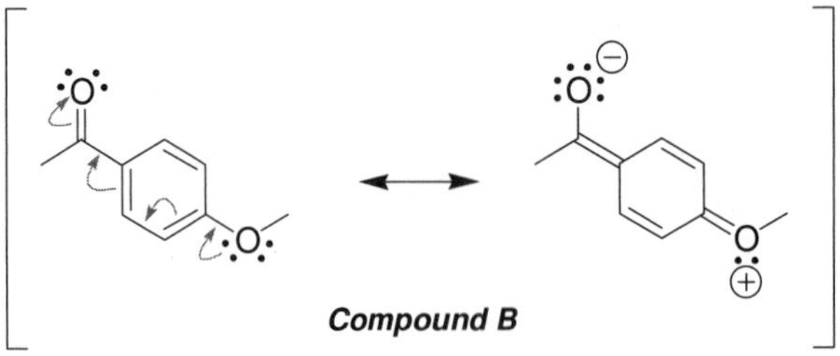

Compound B

Compound A does *not* have a resonance structure in which one oxygen atom has a negative charge and the other oxygen atom has a positive charge. That is, Compound A has fewer resonance structures than Compound B. Accordingly, Compound B has greater resonance stabilization.

b) Compound C is expected to have resonance stabilization similar to that of Compound B, because Compound C also has a resonance structure in which one oxygen atom has a negative charge and the other oxygen atom has a positive charge:

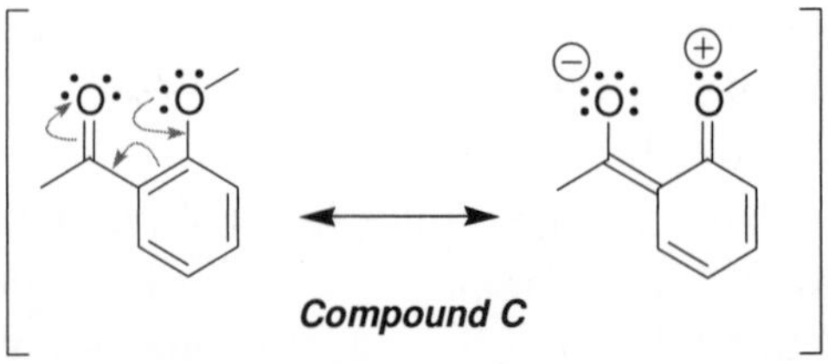

Compound C

2.67.

The single bond mentioned in this problem has some double bond character, as a result of resonance:

Each of the carbon atoms of this single bond uses an atomic p orbital to form a conduit (as described in Section 2.7):

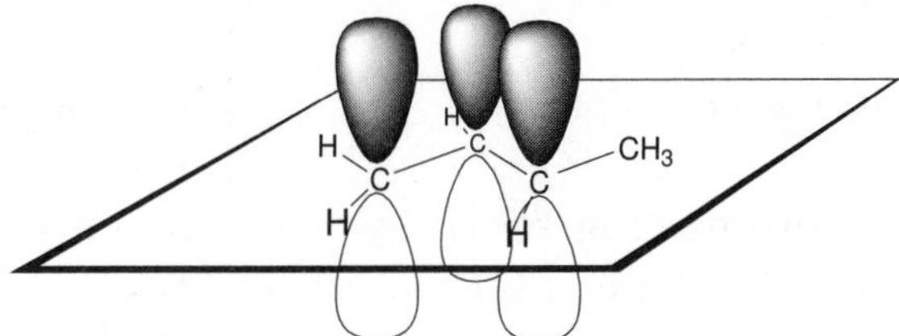

Rotation about this single bond will destroy the overlap of the p orbitals, thereby destroying the resonance stabilization. This single bond therefore exhibits a large barrier to rotation.

Chapter 3
Acids and Bases

Review of Concepts

Fill in the blanks below. To verify that your answers are correct, look in your textbook at the end of Chapter 3. Each of the sentences below appears verbatim in the section entitled *Review of Concepts and Vocabulary.*

- A **Brønsted-Lowry acid** is a proton ________, while a **Brønsted-Lowry base** is a proton ________.
- The mechanism of **proton transfer** always involves at least _____ curved arrows.
- A strong acid has a _____ pK_a, while a weak acid has a _____ pK_a.
- There are four factors to consider when comparing the ___________ of conjugate bases.
- The equilibrium of an acid-base reaction always favors the more ___________ negative charge.
- A **Lewis acid** is an electron ___________, while a **Lewis base** is an electron ___________.

Review of Skills

Fill in the blanks and empty boxes below. To verify that your answers are correct, look in your textbook at the end of Chapter 3. The answers appear in the section entitled *SkillBuilder Review.*

SkillBuilder 3.1 Drawing the Mechanism of a Proton Transfer

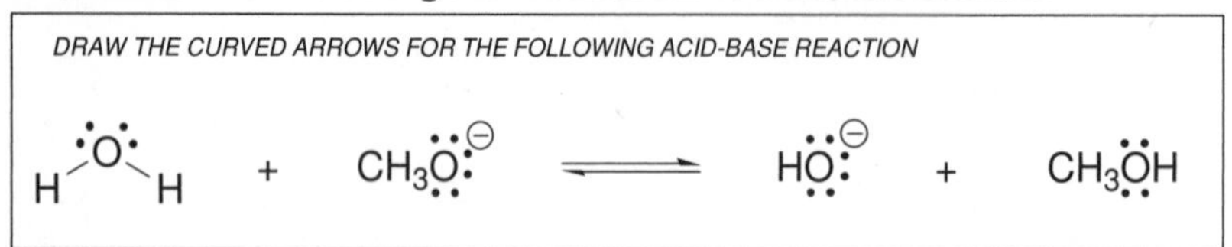

SkillBuilder 3.2 Using pK_a Values to Compare Acids

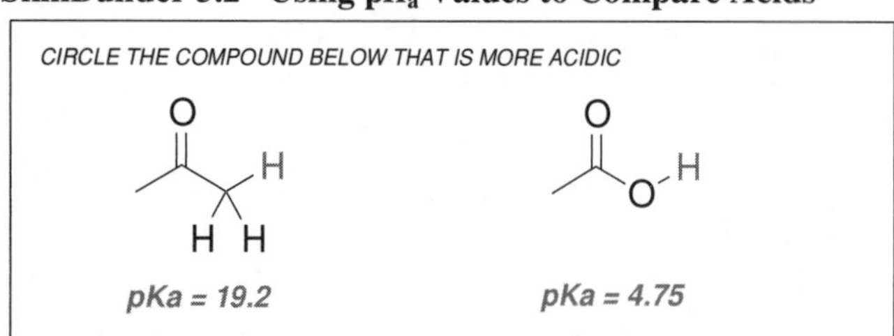

SkillBuilder 3.3 Using pK_a Values to Compare Basicity

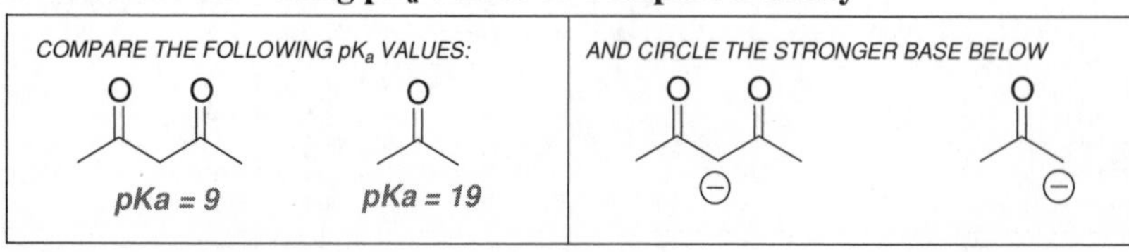

SkillBuilder 3.4 Using pK_a Values to Predict the Position of Equilibrium

CIRCLE THE SIDE OF THE EQUILIBRIUM THAT IS FAVORED:

pKa = 9.0

pKa = 15.7

SkillBuilder 3.5 Assessing Relative Stability. Factor #1: Atom

COMPARE THE TWO PROTONS SHOWN IN THE FOLLOWING COMPOUND, AND CIRCLE THE ONE THAT IS MORE ACIDIC. USE THE EXTRA SPACE TO DRAW THE TWO POSSIBLE CONJUGATE BASES.

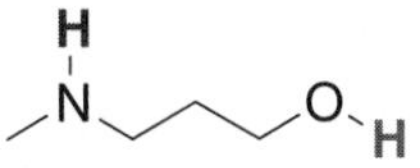

SkillBuilder 3.6 Assessing Relative Stability. Factor #2: Resonance

COMPARE THE TWO PROTONS SHOWN IN THE FOLLOWING COMPOUND, AND CIRCLE THE ONE THAT IS MORE ACIDIC. USE THE EXTRA SPACE TO DRAW THE TWO POSSIBLE CONJUGATE BASES.

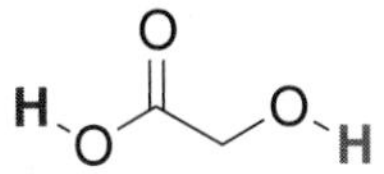

SkillBuilder 3.7 Assessing Relative Stability. Factor #3: Induction

COMPARE THE TWO PROTONS SHOWN IN THE FOLLOWING COMPOUND, AND CIRCLE THE ONE THAT IS MORE ACIDIC. USE THE EXTRA SPACE TO DRAW THE TWO POSSIBLE CONJUGATE BASES.

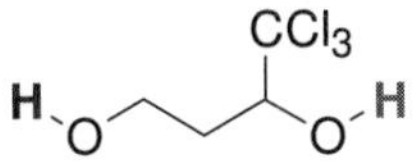

SkillBuilder 3.8 Assessing Relative Stability. Factor #4: Orbital

COMPARE THE TWO PROTONS SHOWN IN THE FOLLOWING COMPOUND, AND CIRCLE THE ONE THAT IS MORE ACIDIC. USE THE EXTRA SPACE TO DRAW THE TWO POSSIBLE CONJUGATE BASES.

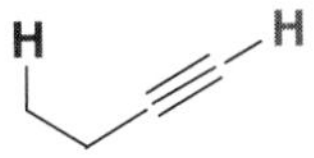

SkillBuilder 3.9 Assessing Relative Stability. Using all Four Factors

COMPARE THE TWO PROTONS SHOWN IN THE FOLLOWING COMPOUND, AND CIRCLE THE ONE THAT IS MORE ACIDIC. USE THE EXTRA SPACE TO DRAW THE TWO POSSIBLE CONJUGATE BASES.

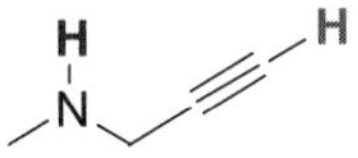

SkillBuilder 3.10 Predicting the Position of Equilibrium Without the Use of pK_a Values

CIRCLE THE SIDE OF THE EQUILIBRIUM THAT IS FAVORED:

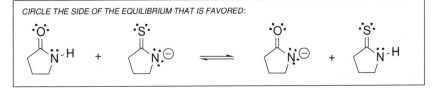

SkillBuilder 3.11 Choosing the appropriate reagent for a proton transfer reaction

DETERMINE WHETHER WATER IS A SUITABLE PROTON SOURCE TO PROTONATE THE ACETATE ION, AS SHOWN BELOW:

SkillBuilder 3.12 Identifying Lewis Acids and Lewis Bases

IDENTIFY THE LEWIS ACID AND THE LEWIS BASE IN THE FOLLOWING REACTION:

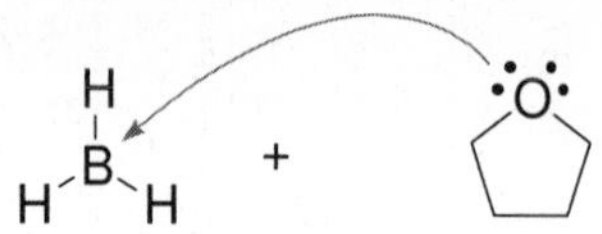

Solutions

3.1.

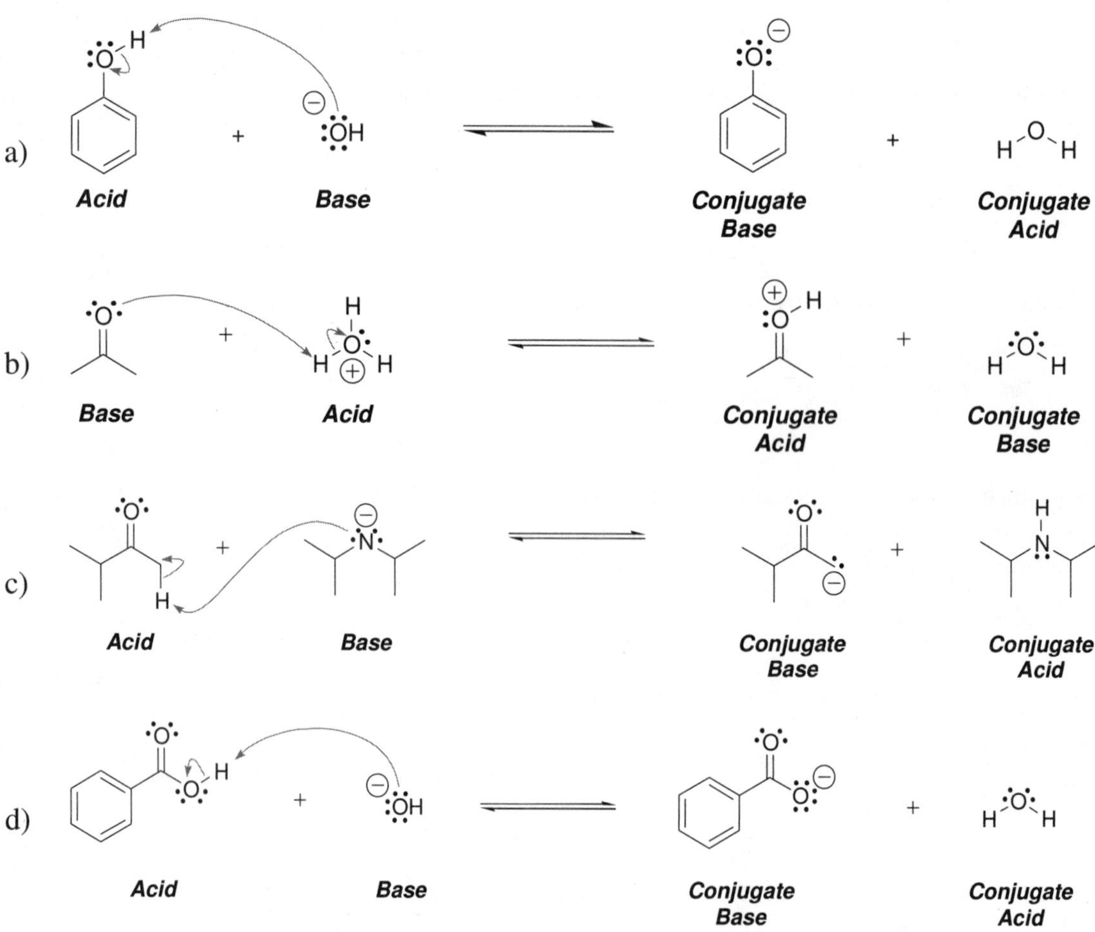

3.2.

a) There is only one arrow, and it is going in the wrong direction. The tail has been placed on the hydrogen atom, and this is incorrect. Curved arrows do not show the motion of atoms, but the motion of electrons. The tail of this curved arrow should be on the lone pair of the nitrogen atom, and the head of the curved arrow should be on the proton. In addition, a second curved arrow is also required. It should look like this:

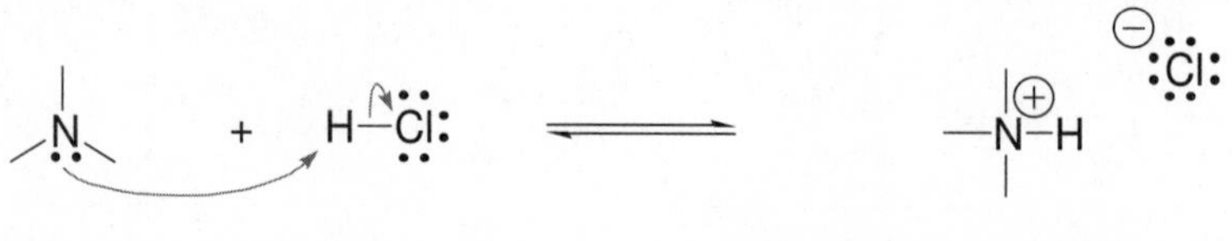

b) The first arrow (from the lone pair of nitrogen to the proton) is correct, but the second curved arrow is not correct. Specifically, the tail is placed on the proton, and instead should be placed on the bond between the proton and the oxygen atom. This bond must be drawn in order to properly place the second curved arrow:

c) The second curved arrow is missing:

3.3.

3.4.

a) 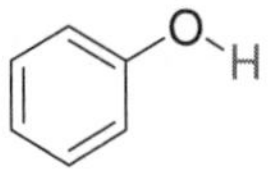 b) 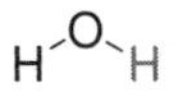 c) H–C≡C–H d) Cl—H

e) H–C≡C–H f)

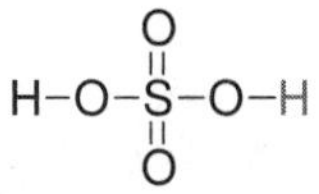

3.5.

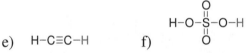

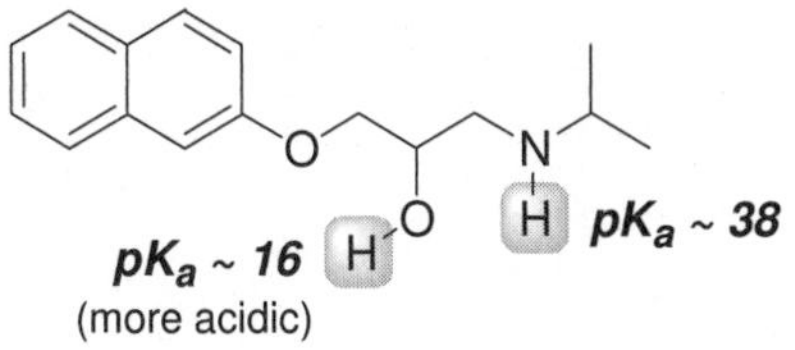

pK_a ~ 16 (more acidic) pK_a ~ 38

3.6.

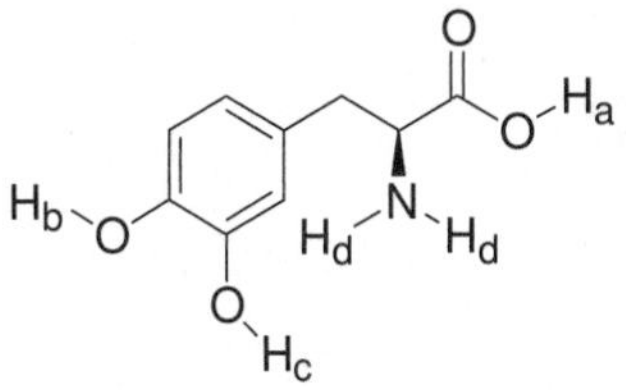

$H_d < H_b \sim H_c < H_a$

3.7.

a) 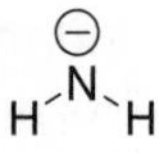 H_2N^- b) 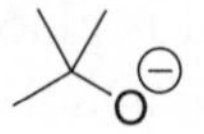 $(CH_3)_3CO^-$ c) 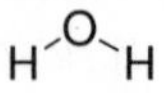 H_2O d) 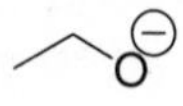 $CH_3CH_2O^-$

e) 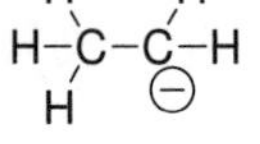 $CH_3CH_2^-$ f) 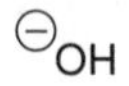 ^-OH

3.8.

strongest base

weakest base

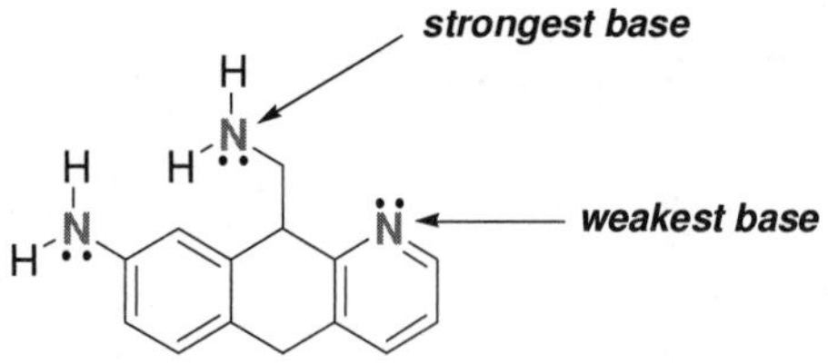

3.9.
a) A lone pair on a nitrogen atom will be more basic than a lone pair on an oxygen atom.
b) The lone pair on the nitrogen atom is thirteen orders of magnitude more basic than the lone pair on the oxygen atom.

3.10.
a) left side
b) right side
c) right side
d) right side

3.11. The equilibrium does not favor deprotonation of acetylene by hydroxide, because water is more acidic than acetylene. The equilibrium will favor the weaker acid (acetylene). A suitable base would be one whose conjugate acid is less acidic than acetylene. For example, H_2N^- would be a suitable base, because ammonia (NH_3) is less acidic than acetylene.

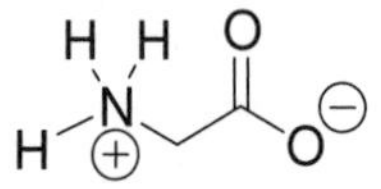

3.12. *glycine*

3.13.

a) 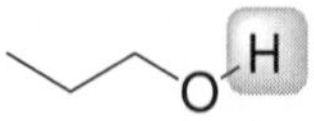 b) 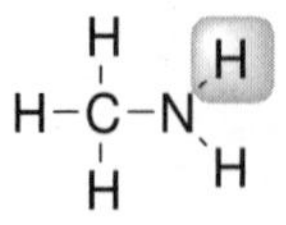 c) 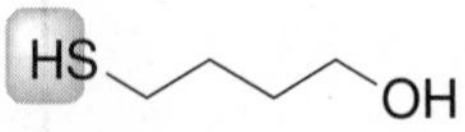 d)

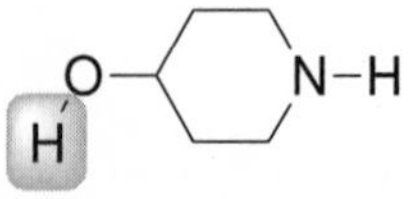

3.14. A proton connected to a sulfur atom will be more acidic than a proton connected to an oxygen atom, which will be more acidic than a proton connected to a nitrogen atom. Therefore, the proton on the sulfur atom will definitely be more acidic than the proton on the oxygen atom.

3.15.

a) 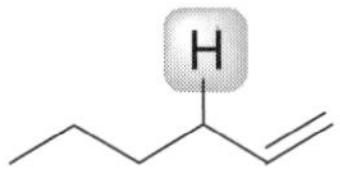 b) 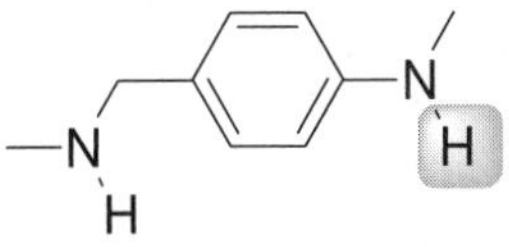 c)

d) 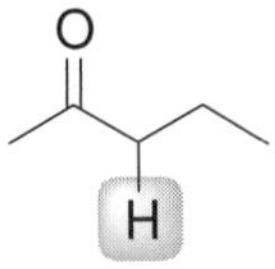 e) 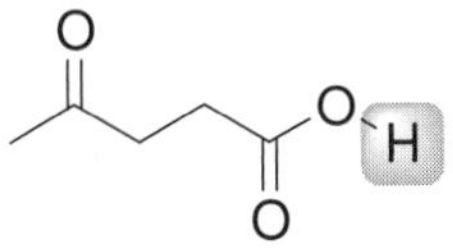 f)

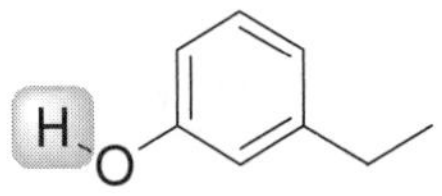

3.16.

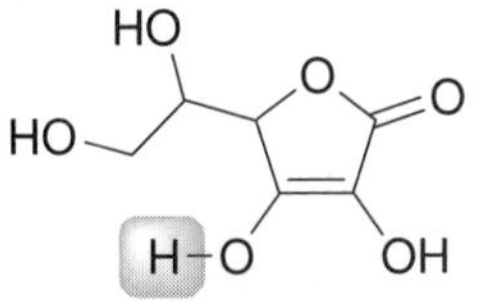

The proton highlighted above is the most acidic proton in the structure, because deprotonation at that location generates a resonance-stabilized anion, in which the negative charge is spread over two oxygen atoms and one carbon atom:

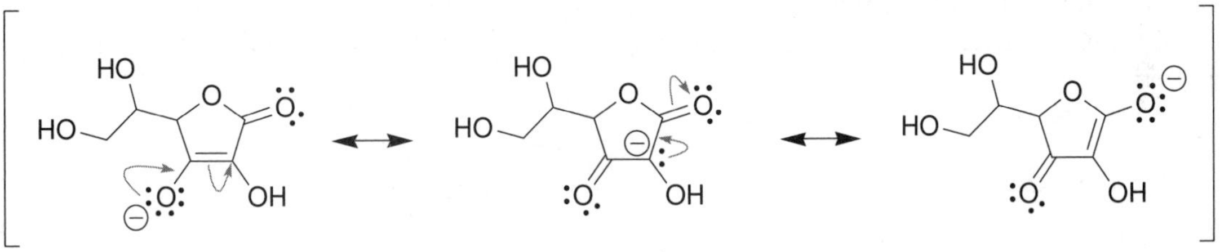

3.17.

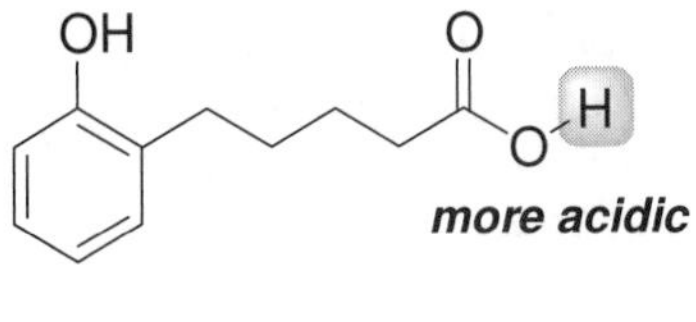

more acidic

3.18.

a) The highlighted proton is more acidic. When this location is deprotonated, the conjugate base that is formed is stabilized by the electron-withdrawing effects of the electronegative fluorine atoms:

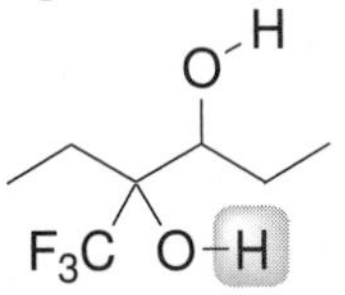

F_3C O–H

b) The highlighted proton is more acidic. When this location is deprotonated, the conjugate base that is formed is stabilized by the electron-withdrawing effects of the electronegative chlorine atoms, which are closer to this proton than the other proton:

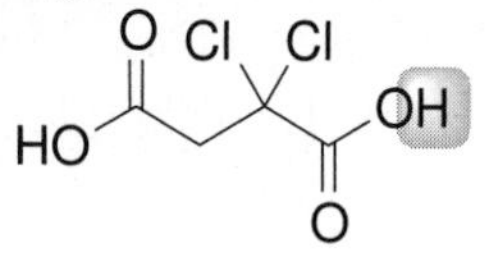

3.19.

a) The compound with two chlorine atoms is more acidic, because of the electron-withdrawing effects of the additional chlorine atom, which help stabilize the conjugate base that is formed when the proton is removed:

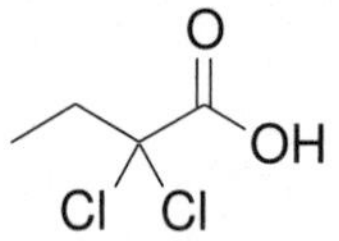

b) The more acidic compound is the one in which the bromine atom is closer to the acidic proton. The electron-withdrawing effects of the bromine atom stabilize the conjugate base that is formed when the proton is removed:

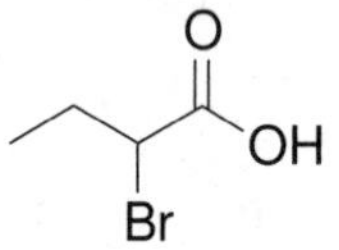

3.20.

a) In the compound below, one of the chlorine atoms has been moved closer to the acidic proton, which further stabilizes the conjugate base that is formed when the proton is removed:

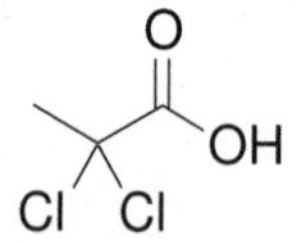

b) In the compound below, one of the chlorine atoms has been moved farther away from the acidic proton, which destabilizes the conjugate base that is formed when the proton is removed:

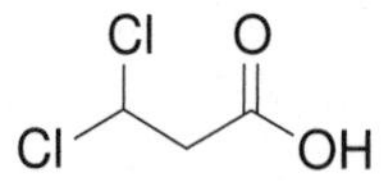

c) The compound below is less acidic than the compounds above, because this compound is not a carboxylic acid. That is, the conjugate base of this compound is NOT resonance stabilized:

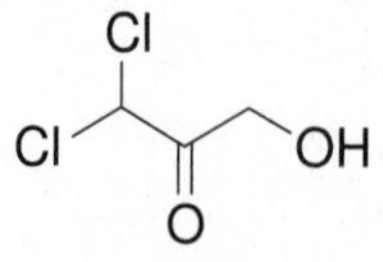

3.21. Both protons are the same distance from the fluorine atom, and both protons are the same distance from the chlorine atom. Accordingly, these protons are expected to be of equivalent acidity.

3.22. The compound below (acetylene) is more acidic. The conjugate base of this compound has a negative charge associated with a lone pair in an *sp* hybridized orbital, which is more stable than a negative charge associated with a lone pair in an sp^2 hybridized orbital.

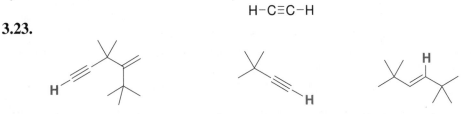

H−C≡C−H

3.23.

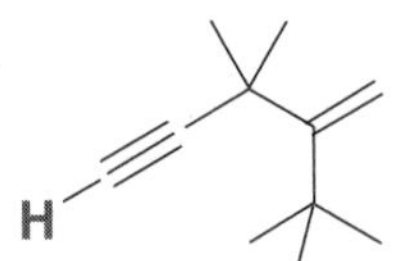

3.24. Most imines will have a pK_a below 35, because imines are expected to be more acidic than amines. This prediction derives from a comparison of the conjugate bases of amines and imines. The former has a negative charge in an sp^3 hybridized orbital, while the latter has a negative charge in an sp^2 hybridized orbital. The latter is expected to be more stable, and therefore, imines are expected to be more acidic.

3.25.

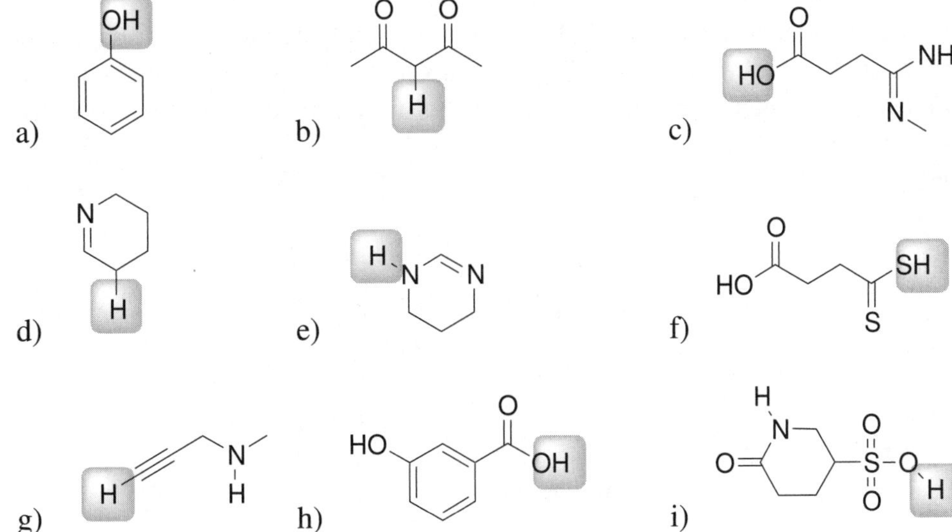

3.26.

a) HBr b) H_2S c) NH_3 d) H−≡−H e) Cl_3C–CH(OH)–CCl_3

3.27.

a) When the proton is removed, the resulting conjugate base is highly resonance stabilized because the negative charge is spread over four nitrogen atoms and seven oxygen atoms. In addition, the inductive effects of the trifluoromethyl groups (-CF_3) further stabilize the negative charge.

b) The OH group can be replaced with an SH group. Sulfur is larger than oxygen and more capable of stabilizing a negative charge:

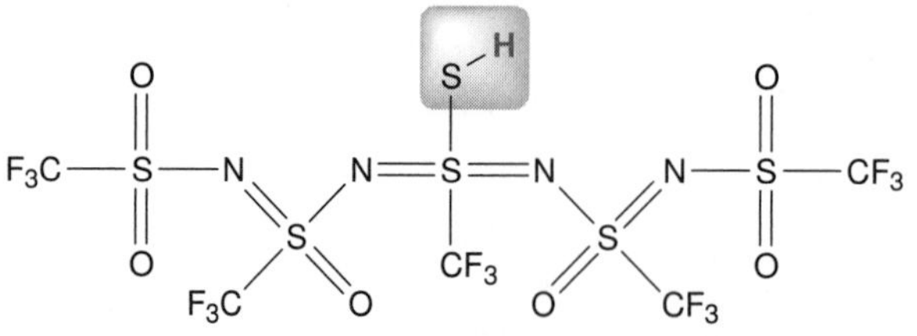

Alternatively, the conjugate base could be further stabilized by spreading the charge over a larger number of nitrogen and oxygen atoms, for example:

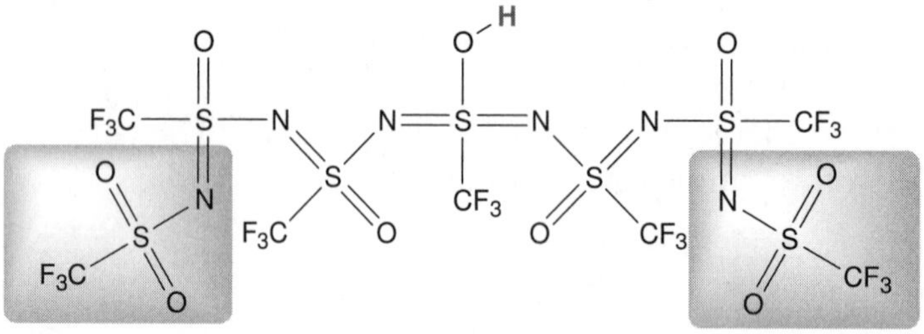

The additional structural units (highlighted above) would enable the conjugate base to spread its negative charge over six nitrogen atoms and nine oxygen atoms, which should be even more stable than being spread over four nitrogen atoms and seven oxygen atoms.

3.28.

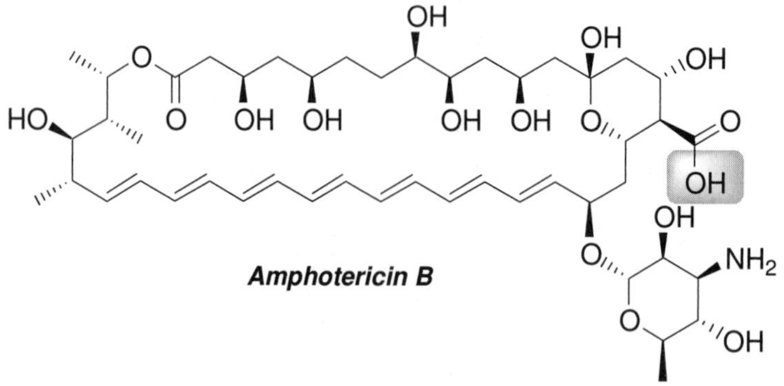

Amphotericin B

3.29.

a) the right side

b) the left side

c) the right side

3.30.

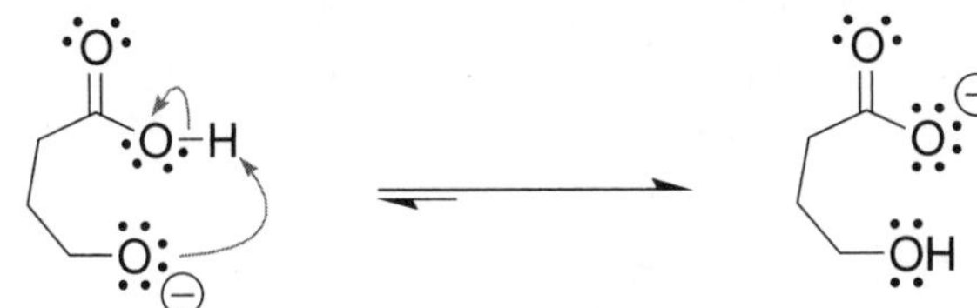

The equilibrium favors the right side because the negative charge is resonance stabilized.

3.31.

a) Yes, because a negative charge on an oxygen atom will be more stable than a negative charge on a nitrogen atom.

b) Yes, because a negative charge on a nitrogen atom will be more stable than a negative charge on an sp^3 hybridized carbon atom.

c) No, because a negative charge on an sp^2 hybridized carbon atom will be less stable than a negative charge on a nitrogen atom.

d) No, because this base is resonance-stabilized, with the negative charge spread over two oxygen atoms and one carbon atom. Protonating this base with water would result in the formation of a hydroxide ion, which is less stable because the negative charge is localized on one oxygen atom.

e) Yes, because a negative charge on an oxygen atom will be more stable than a negative charge on a carbon atom.

f) Yes, because a negative charge on an *sp* hybridized carbon atom will be more stable than a negative charge on a nitrogen atom.

3.32.

a) Yes. This negative charge is less stable than hydroxide.

b) No. This negative charge is resonance stabilized and is more stable than hydroxide.

c) No. This negative charge is resonance stabilized and is more stable than hydroxide.

3.33. Water is more acidic than ethanol. Indeed, the pK_a of water (15.7) is lower than the pK_a of ethanol (16).

3.34.

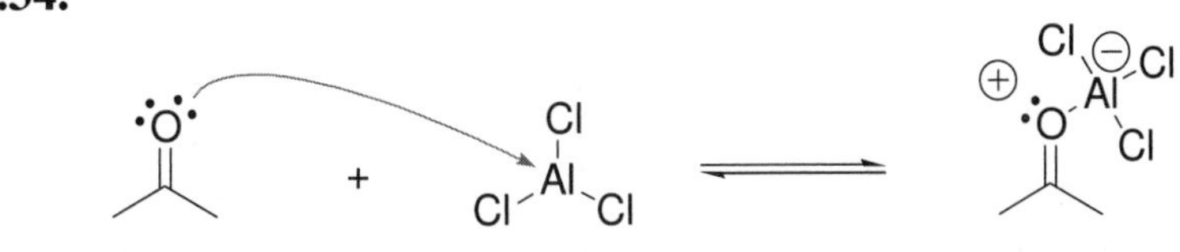

a) *Lewis Base* *Lewis Acid*

b) *Lewis Base* *Lewis Acid*

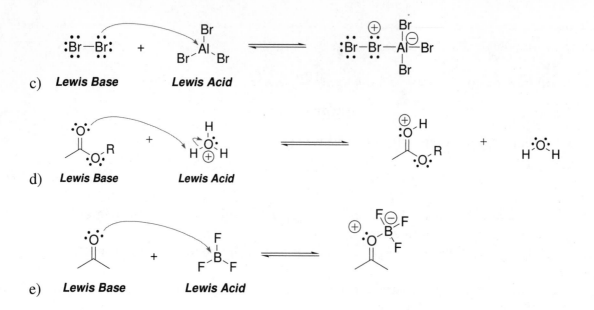

c) ***Lewis Base*** ***Lewis Acid***

d) ***Lewis Base*** ***Lewis Acid***

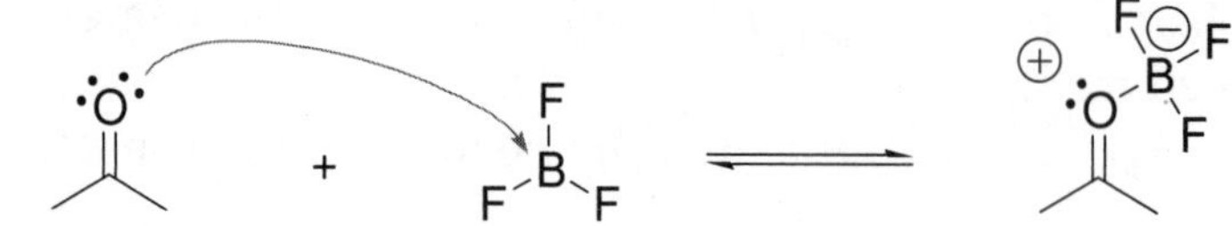

e) ***Lewis Base*** ***Lewis Acid***

3.35.

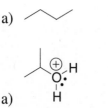

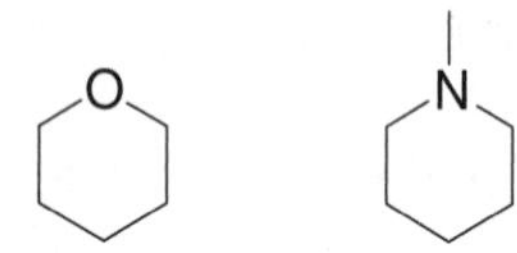

3.36.

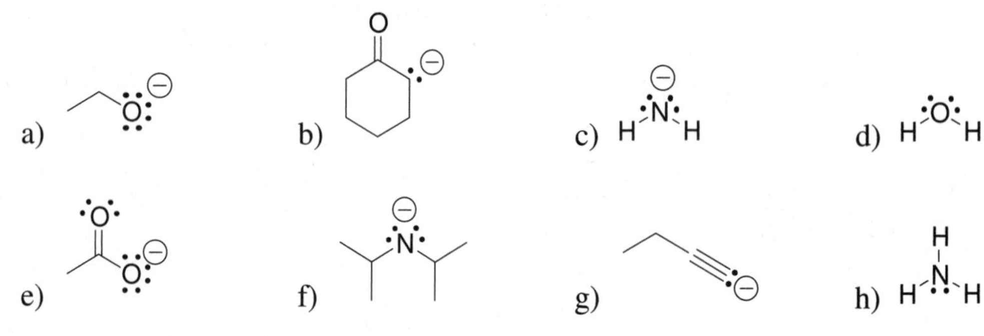

3.37.

a) b) c) d)

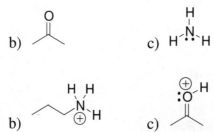

a) b) c) d)

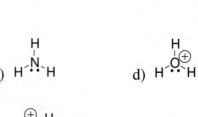

3.38. Compound A is 1000 times more acidic than compound B.

3.39. In each reaction below, identify the Lewis acid and the Lewis base:

a) ***Lewis Base*** ***Lewis Acid***

b) ***Lewis Base*** ***Lewis Acid***

c) ***Lewis Base*** ***Lewis Acid***

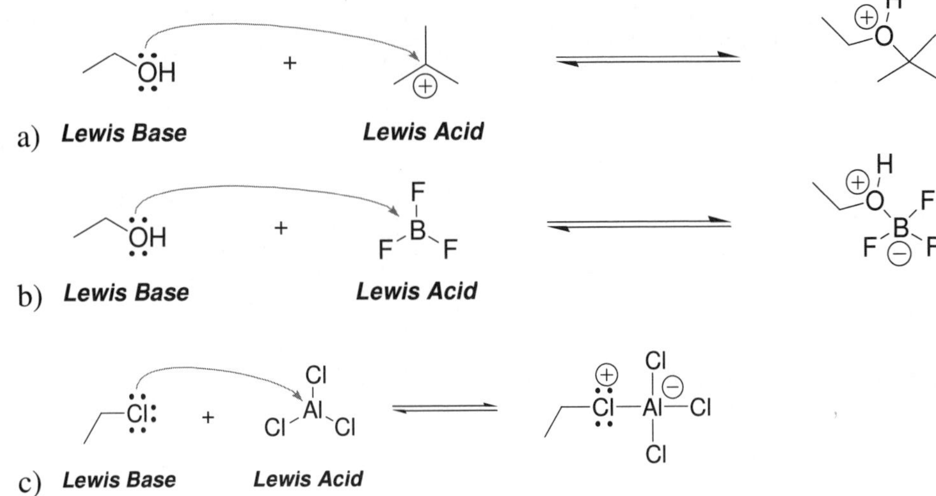

3.40.

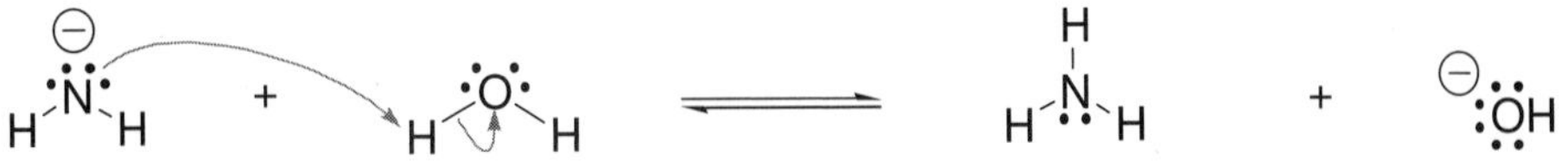

3.41. No, because the leveling effect would cause the deprotonation of ethanol to form ethoxide ions, and the desired anion would not be formed under these conditions.

3.42. No, water would not be a suitable proton source in this case. This anion is the conjugate base of a carboxylic acid. The negative charge is resonance stabilized and is more stable than hydroxide.

3.43.

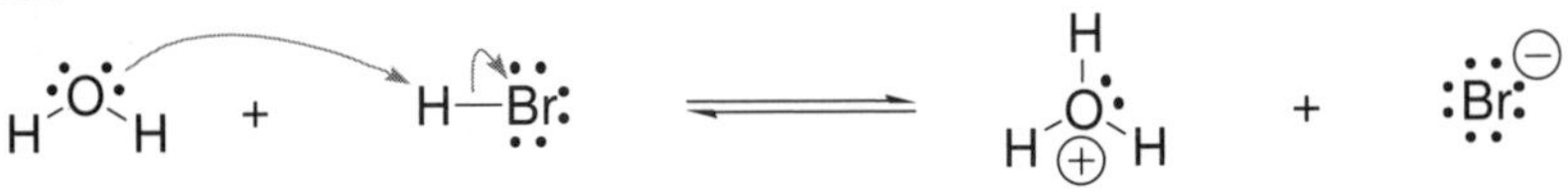

a)

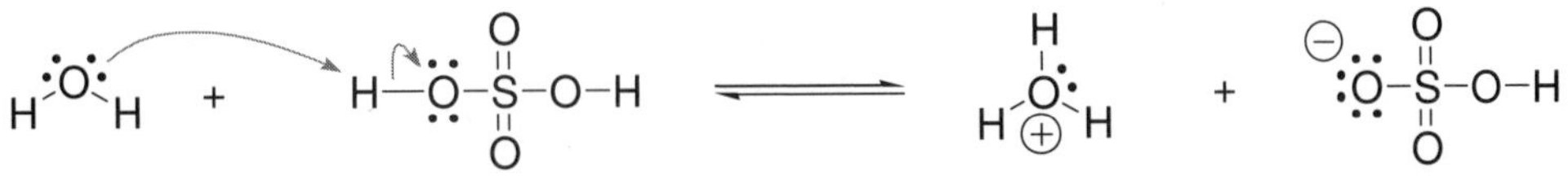

b)

c)

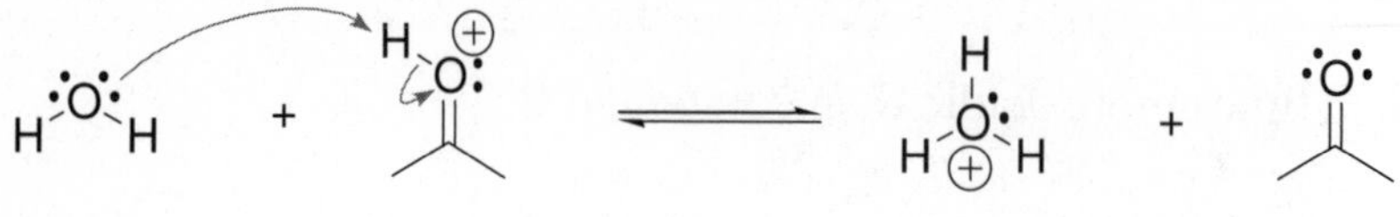

d)

3.44.

a)

b)

c)

d)

3.45.

a) The second anion is more stable because it is resonance stabilized.
b) The second anion is more stable because the negative charge is on a nitrogen atom, rather than an sp^3 hybridized carbon atom.
c) The second anion is more stable because the negative charge is on an *sp* hybridized carbon atom, rather than an sp^3 hybridized carbon atom.

3.46.

a) 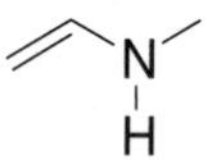 b)

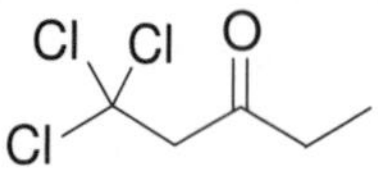

3.47.

a) 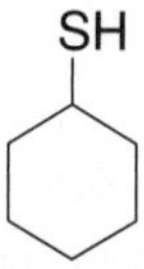 b) 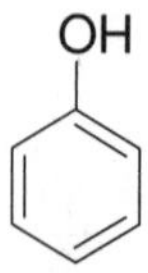 c) 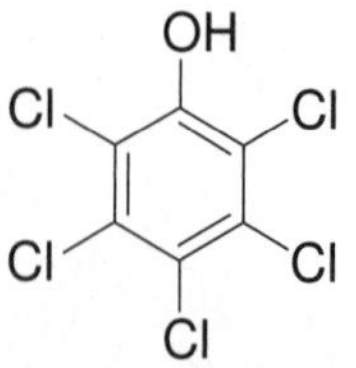 d)

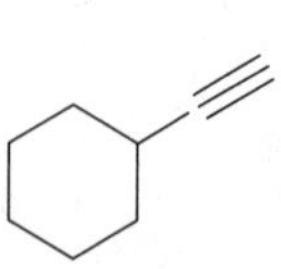

e) 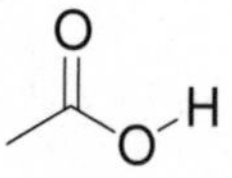 f)  g)  h)

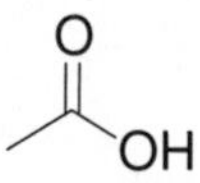

3.48.

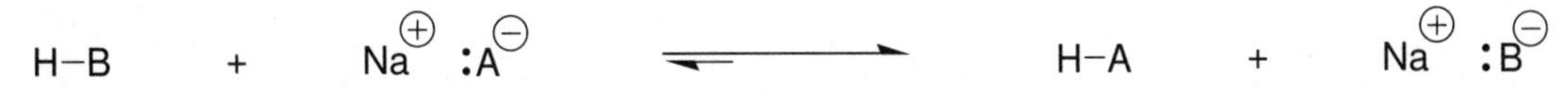

$$H{-}B \quad + \quad Na^{\oplus} \; {:}A^{\ominus} \rightleftharpoons H{-}A \quad + \quad Na^{\oplus} \; {:}B^{\ominus}$$

pK_a = 5 (H–B) pK_a = 15 (H–A)

The equilibrium will favor the weaker acid (the acid with the higher pK_a value). In this case, the equilibrium favors formation of HA.

3.49.

a)

b)

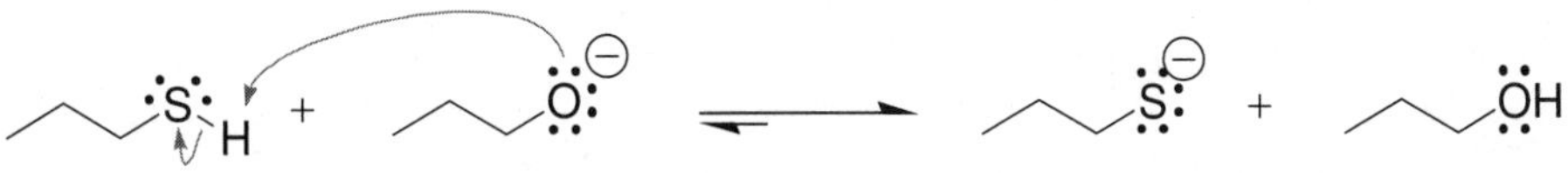

c)

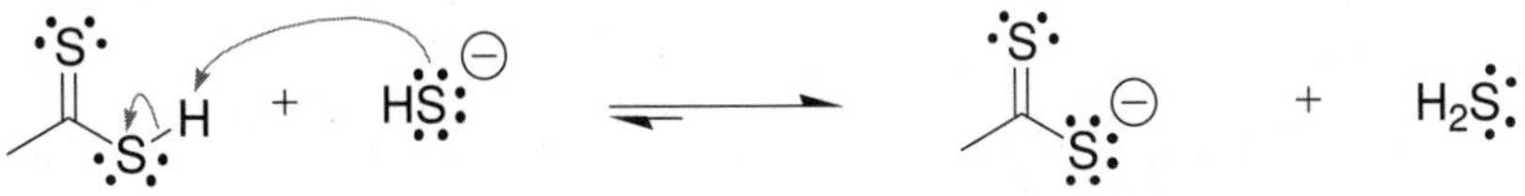

d)

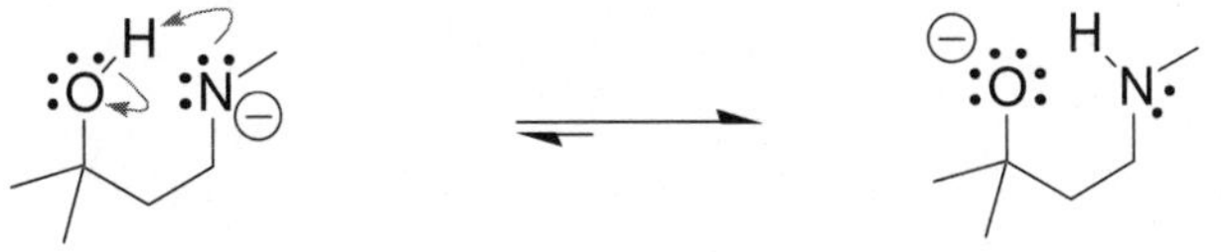

3.50.

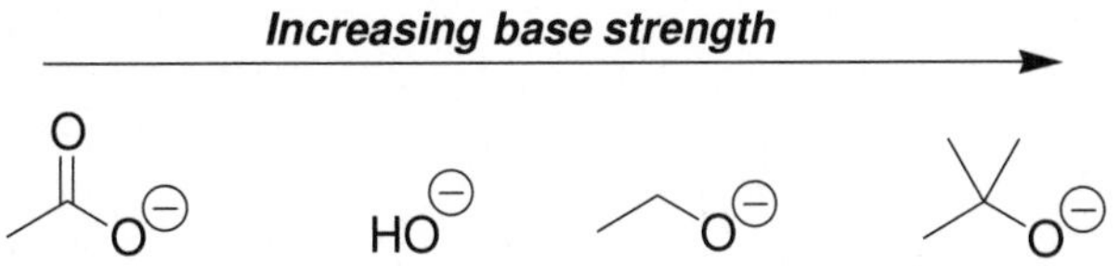

Increasing base strength

3.51.

a)

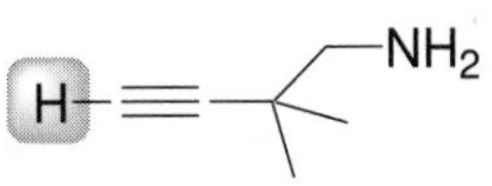

b)

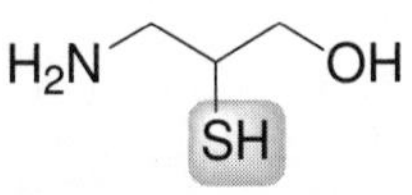

c)

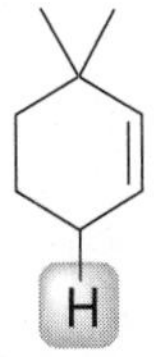

d) 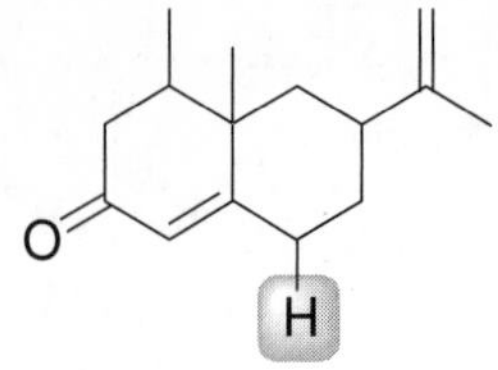 H

e) 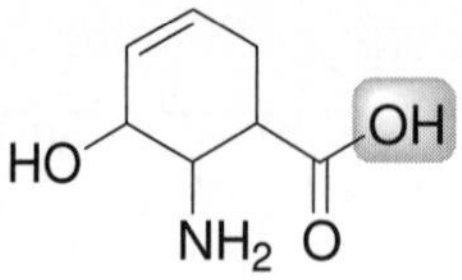 HO, OH, NH_2, O

f) 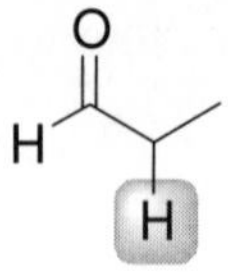 O, H, H

g) 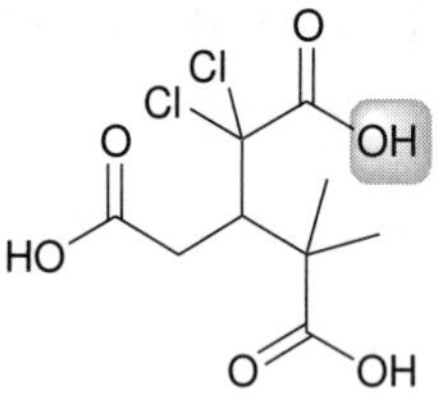 Cl, Cl, O, OH, O, HO, O, OH

h) HS, OH

3.52.

a)

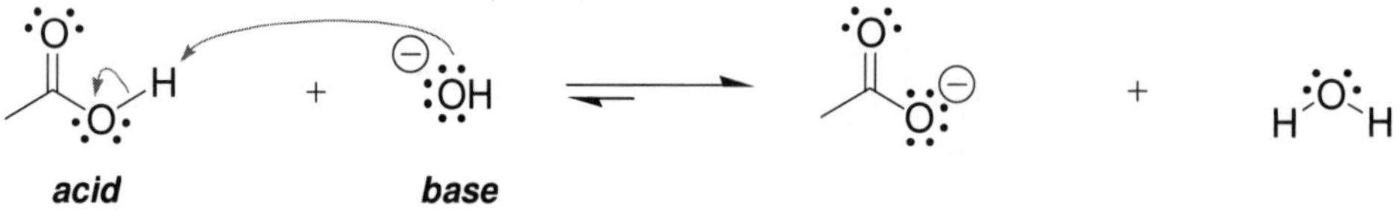

acid + base ⇌ H_2O

b)

base + acid ⇌ + :OH⁻

c)

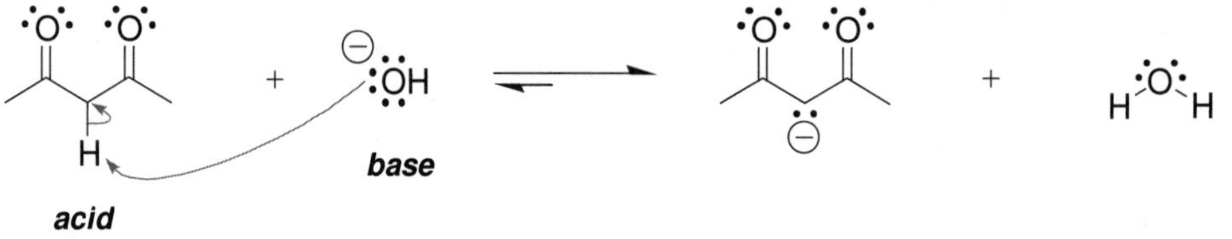

acid + base ⇌ + H_2O

3.53.

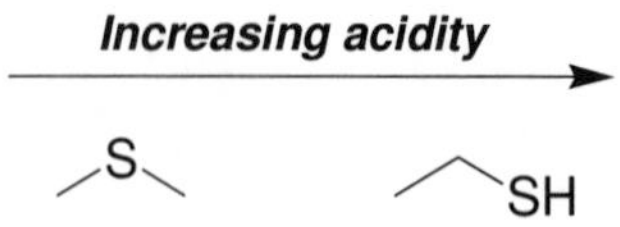

Increasing acidity →

S, SH

3.54.

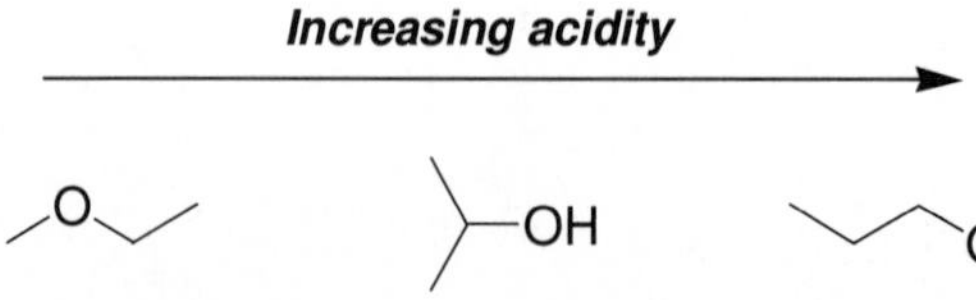

Increasing acidity →

O, OH, OH

3.55.

a) There is only one sp^3 hybridized carbon atom in cyclopentadiene.

b) The most acidic proton in cyclopentadiene is highlighted below:

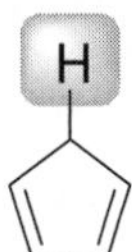

The corresponding conjugate base is highly resonance stabilized. In addition, the conjugate base is further stabilized by yet another factor that we will discuss in Chapter 18.

c)

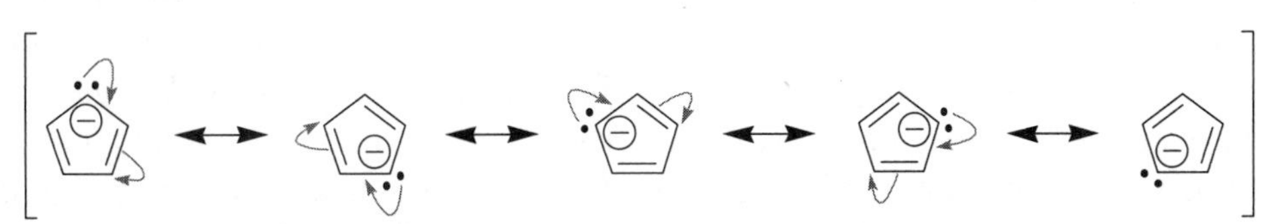

d) There are no sp^3 hybridized carbon atoms in the conjugate base.

e) All carbon atoms are sp^2 hybridized and trigonal planar. Therefore, the entire compound has planar geometry.

f) There are five hydrogen atoms in the conjugate base.

g) There is one lone pair in the conjugate base, and it is highly delocalized.

3.56. When salicylic acid is deprotonated, the resulting conjugate base is further stabilized by intramolecular hydrogen bonding:

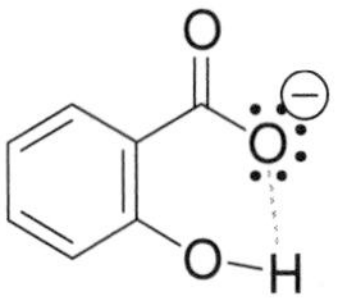

3.57.

a) 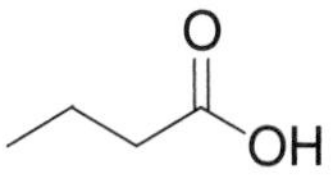 or

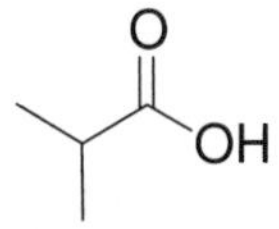

b) 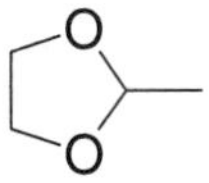 or 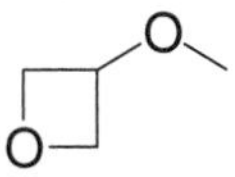 (there are other possibilities as well)

c) 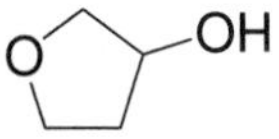 or 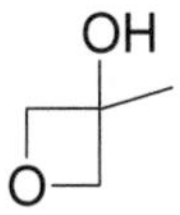 (there are other possibilities as well)

3.58. The four constitutional isomers are shown below.

The last compound is expected to have the highest pK_a because its conjugate base is not resonance stabilized. The other three compounds have resonance-stabilized conjugate bases, for example:

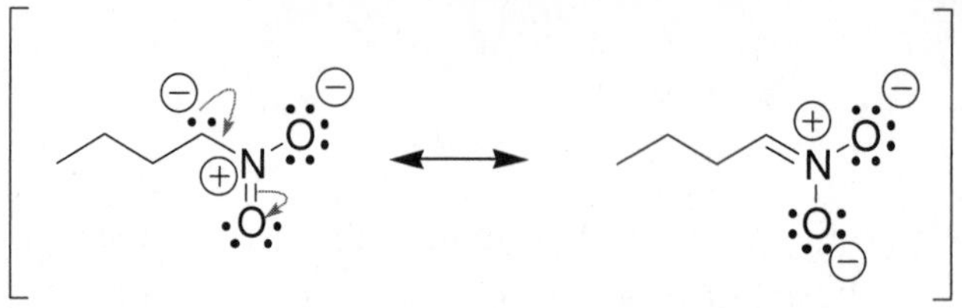

3.59. Compare the conjugate bases. Both are resonance stabilized. But the conjugate base of the first compound has a negative charge spread over two nitrogen atoms and two carbon atoms, while the conjugate base of the second compound has a negative charge spread over one nitrogen atom and three carbon atoms. Since nitrogen is more electronegative than carbon, nitrogen is more capable of stabilizing a negative charge. Therefore, the conjugate base of the first compound is more stable than the conjugate base of the second compound. As a result, the first compound will be more acidic.

3.60.

a) The two most acidic protons are labeled H_a and H_b:

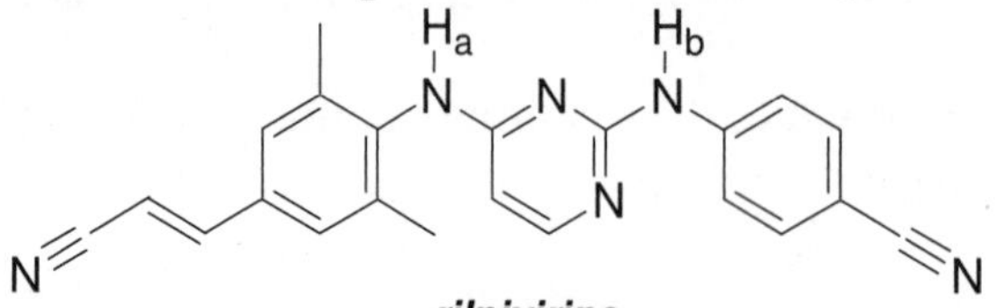

rilpivirine

b) H_a is expected to be slightly more acidic than H_b, because removal of H_a produces a conjugate base that has one more resonance structure than the conjugate base formed from removal of H_b. The former has the negative charge spread over four nitrogen atoms and *five* carbon atoms, while the latter has the negative charge spread over four nitrogen atoms and *four* carbon atoms.

3.61.

a) When R is a cyano group, the conjugate base is resonance stabilized:

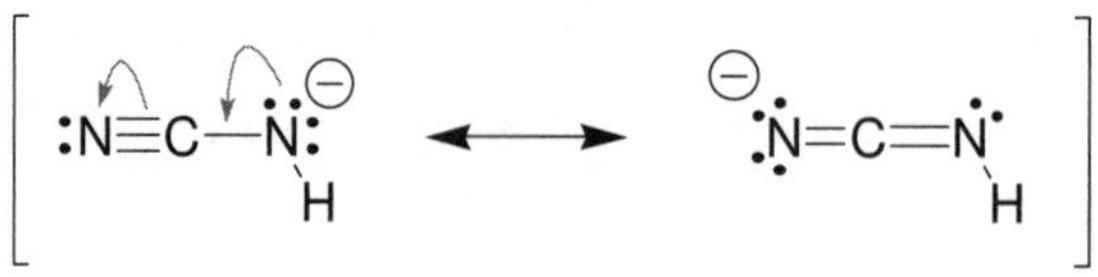

b) There are many possible answers. Here is one example, for which the conjugate base has the negative charge spread over three nitrogen atoms, rather than just two nitrogen atoms:

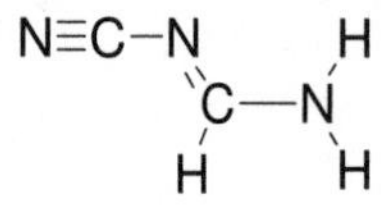

Chapter 4
Alkanes and Cycloalkanes

Review of Concepts

Fill in the blanks below. To verify that your answers are correct, look in your textbook at the end of Chapter 4. Each of the sentences below appears verbatim in the section entitled *Review of Concepts and Vocabulary.*

- Hydrocarbons that lack ____________ are called **saturated hydrocarbons**, or ___________.
- _________________ provide a systematic way for naming compounds.
- Rotation about C-C single bonds allows a compound to adopt a variety of _________________.
- ___________ **projections** are often used to draw the various conformations of a compound.
- _____________ **conformations** are lower in energy, while ____________ **conformations** are higher in energy.
- The difference in energy between staggered and eclipsed conformations of ethane is referred to as _____________ **strain**.
- ________ **strain** occurs in cycloalkanes when bond angles deviate from the preferred _____°.
- The _______ conformation of cyclohexane has no torsional strain and very little angle strain.
- The term **ring flip** is used to describe the conversion of one ____________ conformation into the other. When a ring has one substituent…the equilibrium will favor the chair conformation with the substituent in the _____________ position.

Review of Skills

Fill in the blanks and empty boxes below. To verify that your answers are correct, look in your textbook at the end of Chapter 4. The answers appear in the section entitled *SkillBuilder Review.*

SkillBuilder 4.1 Identifying the Parent

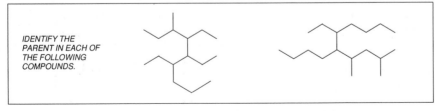

IDENTIFY THE PARENT IN EACH OF THE FOLLOWING COMPOUNDS.

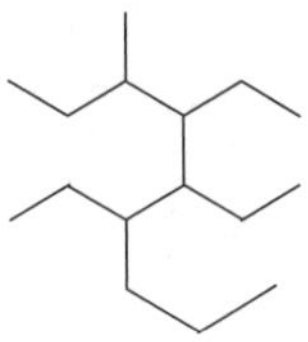

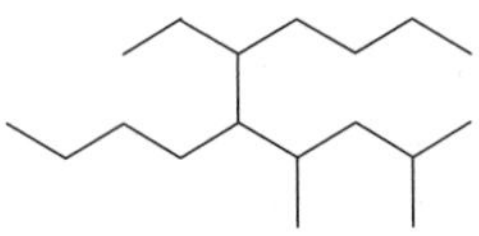

SkillBuilder 4.2 Identifying and Naming Substituents

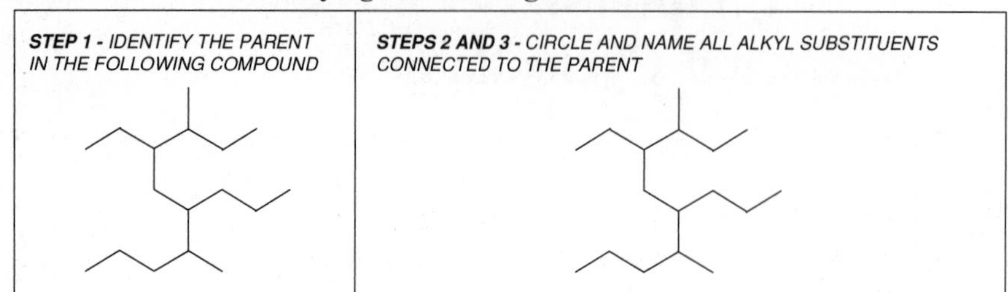

STEP 1 - IDENTIFY THE PARENT IN THE FOLLOWING COMPOUND
STEPS 2 AND 3 - CIRCLE AND NAME ALL ALKYL SUBSTITUENTS CONNECTED TO THE PARENT

SkillBuilder 4.3 Identifying and Naming Complex Substituents

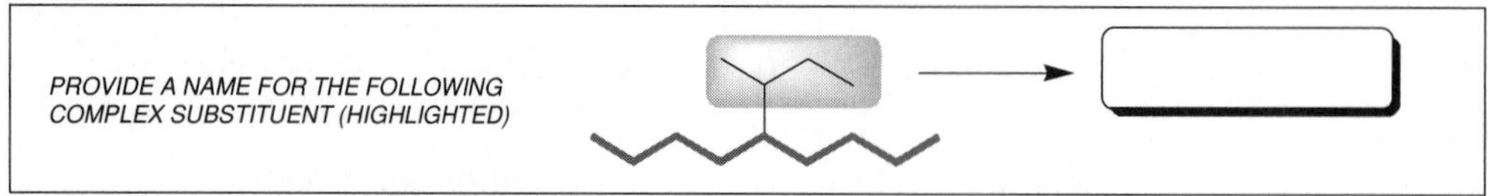

PROVIDE A NAME FOR THE FOLLOWING COMPLEX SUBSTITUENT (HIGHLIGHTED)

SkillBuilder 4.4 Assembling the Systematic Name of an Alkane

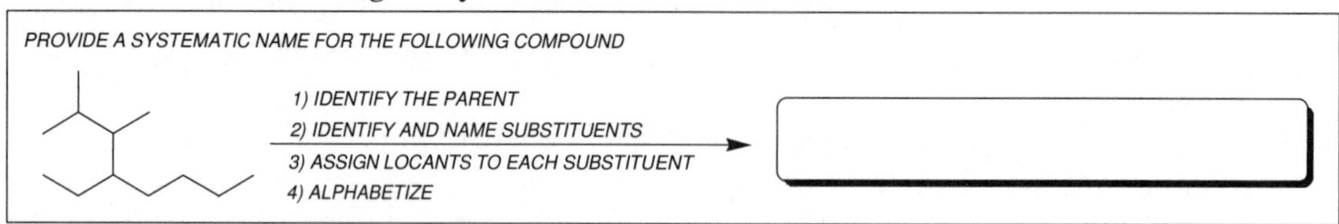

SkillBuilder 4.5 Assembling the Name of a Bicyclic Compound

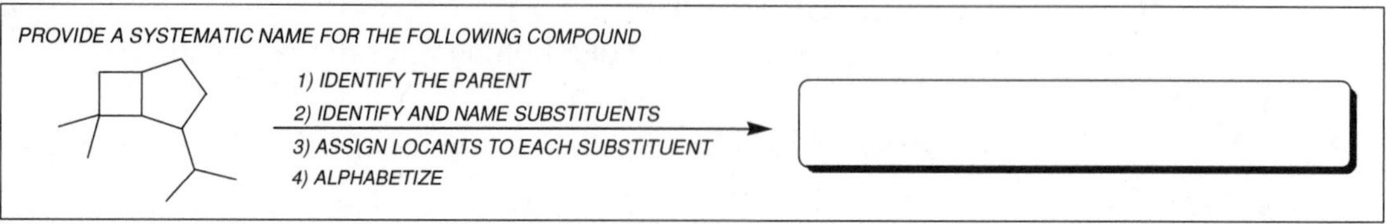

SkillBuilder 4.6 Identifying Constitutional Isomers

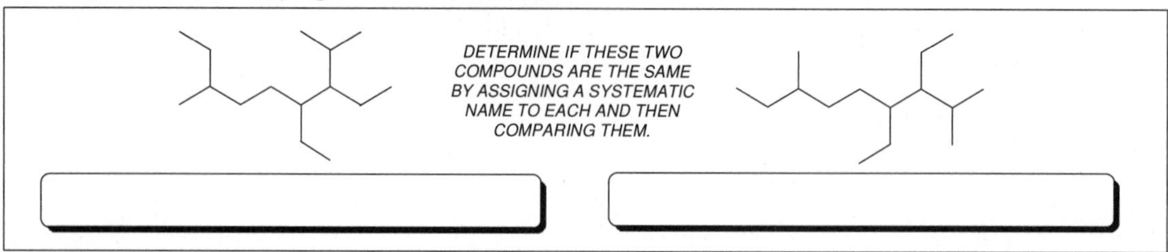

DETERMINE IF THESE TWO COMPOUNDS ARE THE SAME BY ASSIGNING A SYSTEMATIC NAME TO EACH AND THEN COMPARING THEM.

SkillBuilder 4.7 Drawing Newman Projections

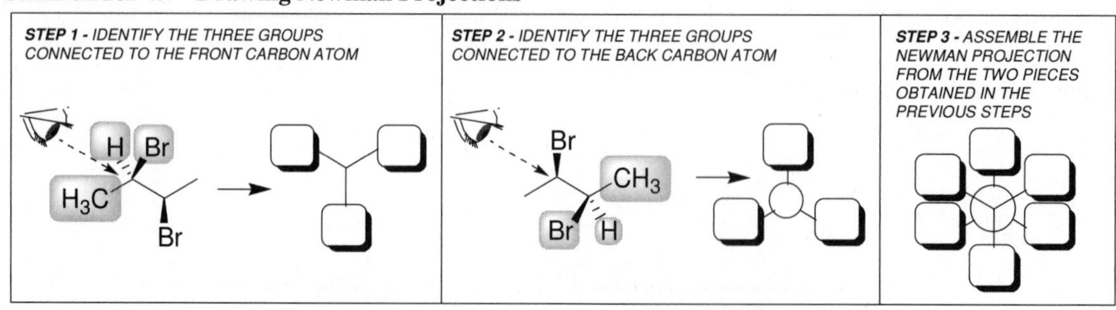

SkillBuilder 4.8 Identifying Relative Energy of Conformations

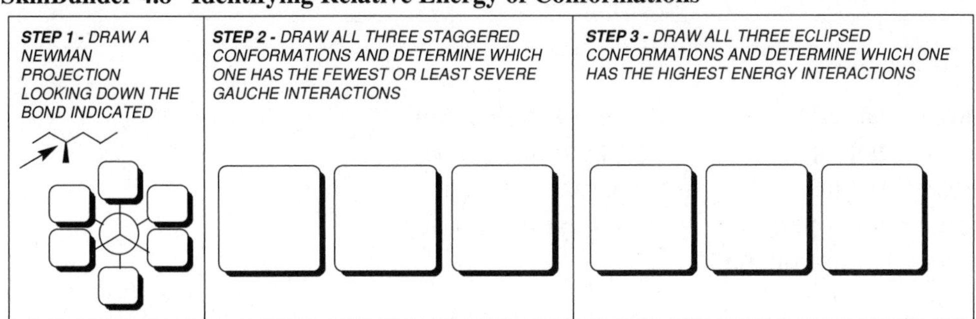

STEP 1 - DRAW A NEWMAN PROJECTION LOOKING DOWN THE BOND INDICATED
STEP 2 - DRAW ALL THREE STAGGERED CONFORMATIONS AND DETERMINE WHICH ONE HAS THE FEWEST OR LEAST SEVERE GAUCHE INTERACTIONS
STEP 3 - DRAW ALL THREE ECLIPSED CONFORMATIONS AND DETERMINE WHICH ONE HAS THE HIGHEST ENERGY INTERACTIONS

SkillBuilder 4.9
Drawing a Chair Conformation

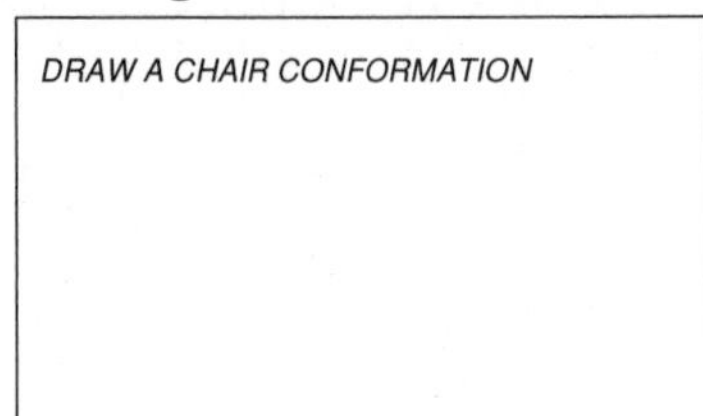

SkillBuilder 4.10
Drawing Axial and Equatorial Positions

DRAW A CHAIR CONFORMATION SHOWING ALL SIX AXIAL POSITIONS AND ALL SIX EQUATORIAL POSITIONS

SkillBuilder 4.11 Drawing Both Chair Conformations of a Monosubstituted Cyclohexane

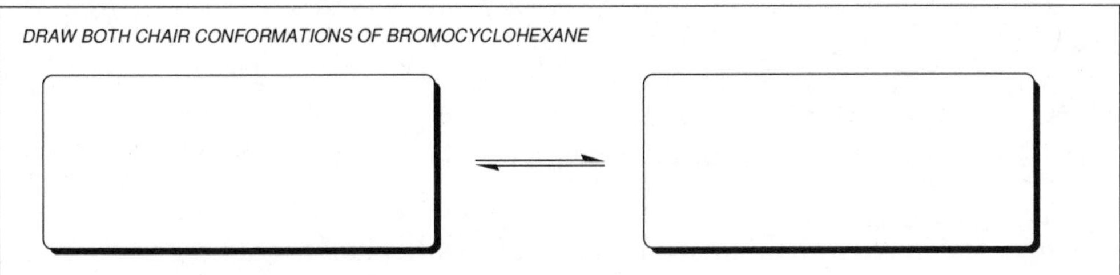

SkillBuilder 4.12 Drawing Both Chair Conformations of Disubstituted Cyclohexanes

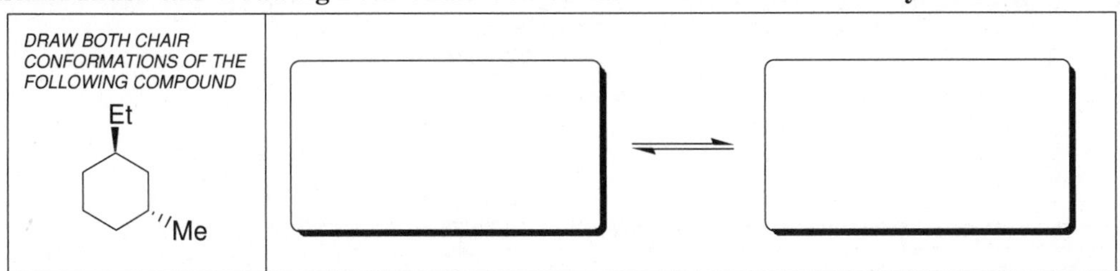

SkillBuilder 4.13 Drawing the More Stable Chair Conformation of Polysubstituted Cyclohexanes

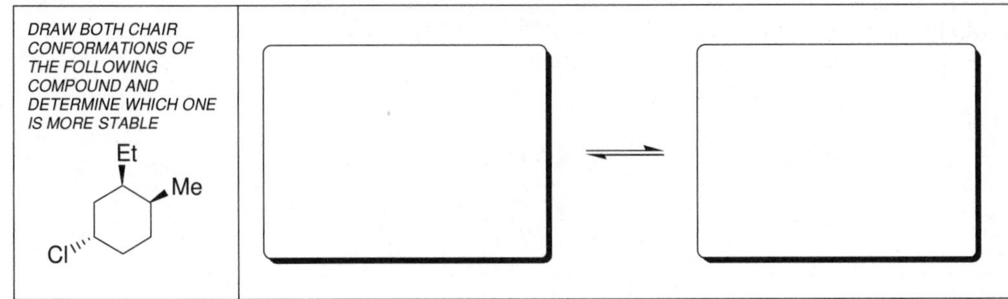

DRAW BOTH CHAIR CONFORMATIONS OF THE FOLLOWING COMPOUND AND DETERMINE WHICH ONE IS MORE STABLE
Et
Me
Cl

Solutions

4.1.

a) parent = hexane
b) parent = heptane
c) parent = heptanes
d) parent = nonane
e) parent = octane
f) parent = heptane
g) parent = cyclopentane
h) parent = cycloheptene
i) parent = cyclopropane

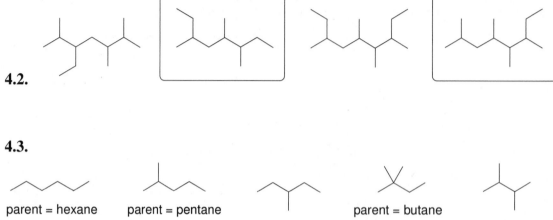

4.2.

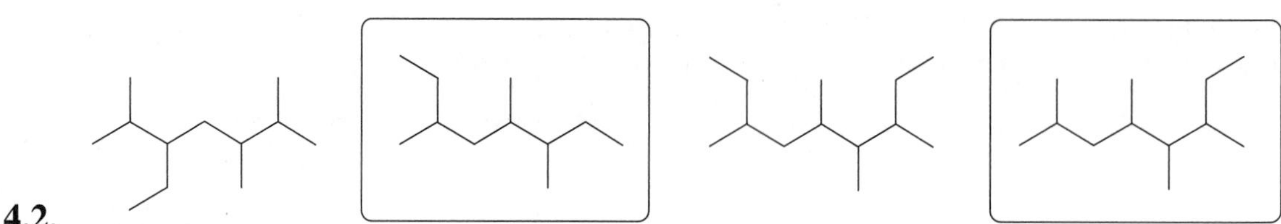

4.3.

parent = hexane

parent = pentane

parent = pentane

parent = butane

parent = butane

4.4. Only three of the isomers will have a parent name of heptane:

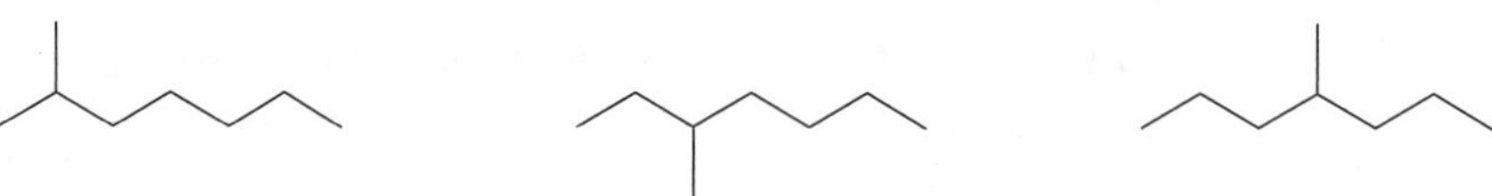

4.5.

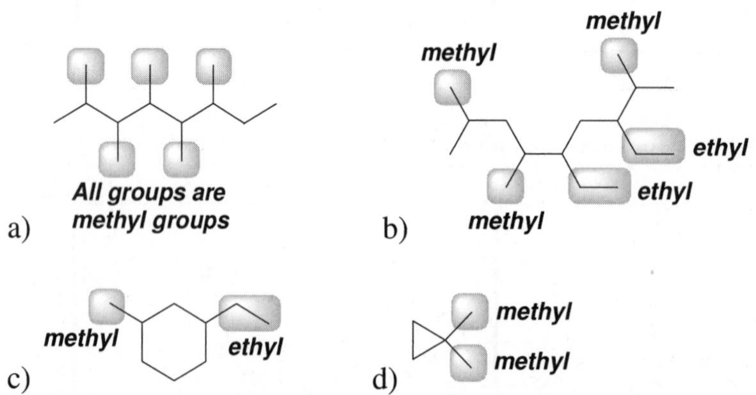

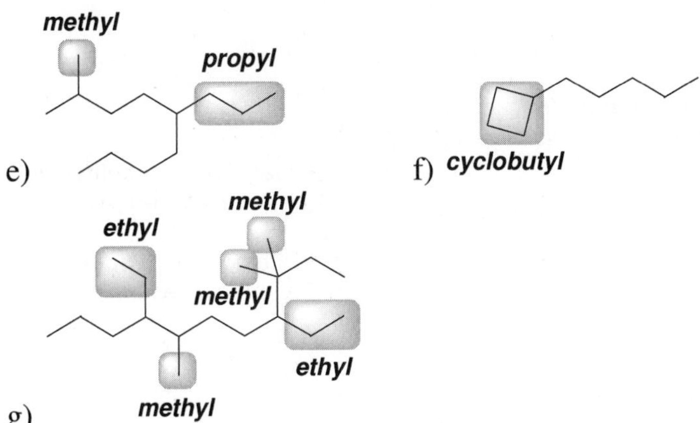
methyl
propyl
e)
f) cyclobutyl
methyl
ethyl
methyl
g)
methyl
ethyl

4.6.

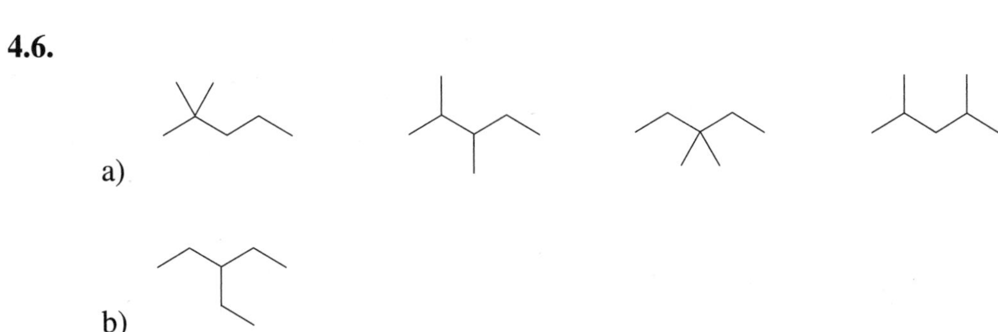
a)
b)

4.7.

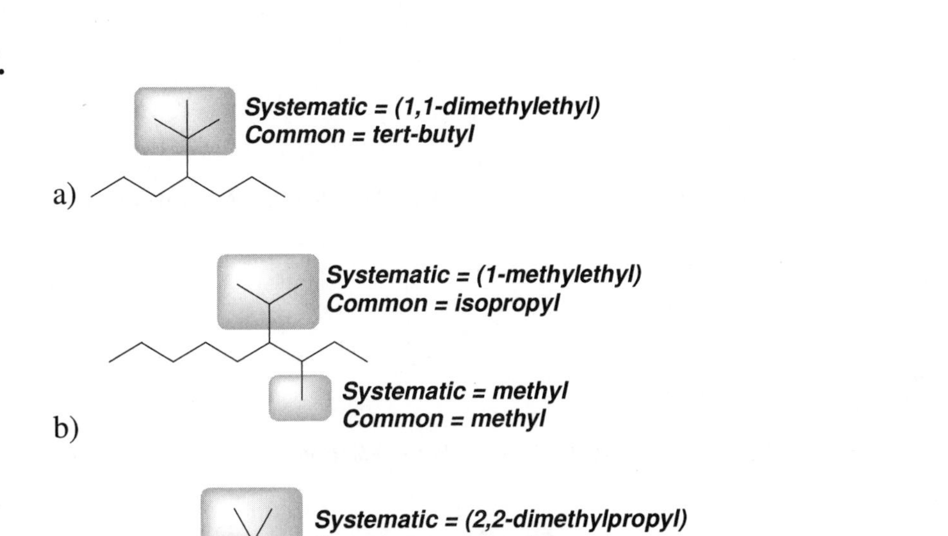
a)
Systematic = (1,1-dimethylethyl)
Common = tert-butyl
b)
Systematic = (1-methylethyl)
Common = isopropyl
Systematic = methyl
Common = methyl
c)
Systematic = (2,2-dimethylpropyl)
Common = neopentyl

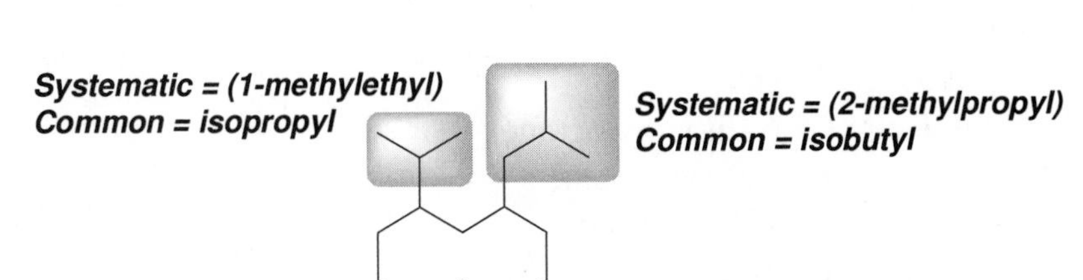
d)
Systematic = (1-methylethyl)
Common = isopropyl
Systematic = (2-methylpropyl)
Common = isobutyl

e)

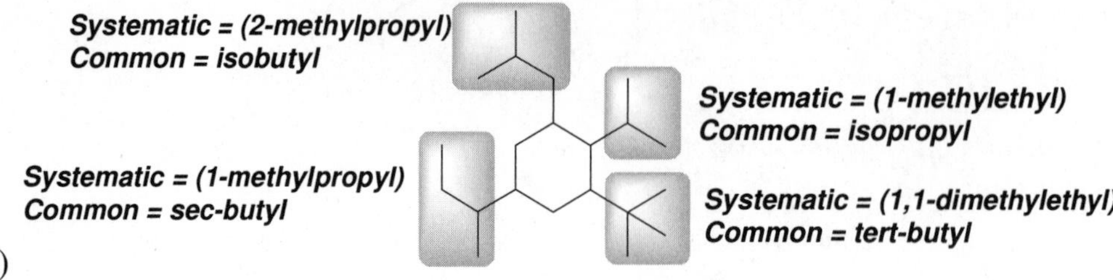

4.8.

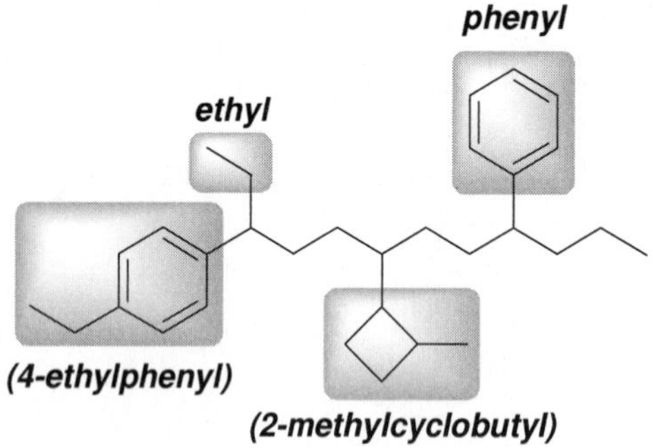

4.9.

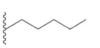

pentyl *(1-methylbutyl)* *(2-methylbutyl)* *(3-methylbutyl)*

(1,1-dimethylpropyl) *(1,2-dimethylpropyl)* *(2,2-dimethylpropyl)* *(1-ethylpropyl)*

4.10.

a) 3,4,6-trimethyloctane
b) sec-butylcyclohexane
c) 3-ethyl-2-methylheptane
d) 3-isopropyl-2,4-dimethylpentane
e) 3-ethyl-2,2-dimethylhexane
f) 2-cyclohexyl-4-ethyl-5,6-dimethyloctane
g) 3-ethyl-2,5-dimethyl-4-propylheptane
h) 5-sec-butyl-4-ethyl-2-methyldecane
i) 2,2,6,6,7,7-hexamethylnonane
j) 4,5-dimethylnonane

k) 2,4,4,6-tetramethylheptane
l) 2,2,5-trimethylpentane
m) 4-tert-butylheptane
n) 3-ethyl-6-isopropyl-2,4-dimethyldecane
o) 3,5-diethyl-2-methyloctane
p) 1,3-diisopropylcyclopentane
q) 3-ethyl-2,5-dimethylheptane

4.11.

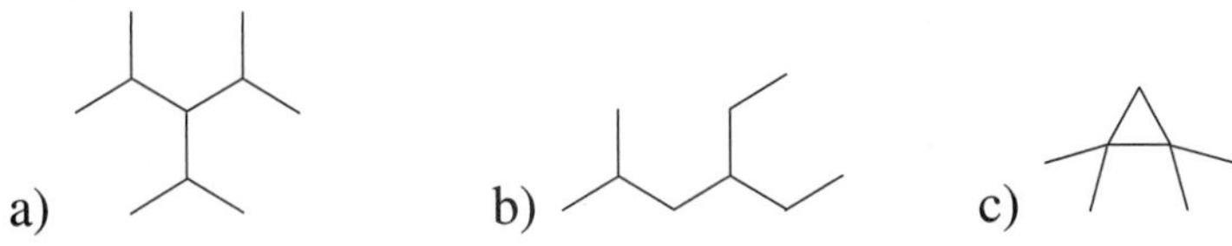

4.12.

a) 4-ethyl-1-methylbicyclo[3.2.1]octane
b) 2,2,5,7-tetramethylbicyclo[4.2.0]octane
c) 2,7,7-trimethylbicyclo[4.2.2]decane
d) 3-*sec*-butyl-2-methylbicyclo[3.1.0]hexane
e) 2,2-dimethylbicyclo[2.2.2]octane
f) 2,7-dimethylbicyclo[3.3.0]octane
g) bicyclo[1.1.0]butane
h) 5,5-dimethylbicyclo[2.1.1]hexane
i) 3-(3-methylbutyl)bicyclo[4.4.0]decane

4.13.

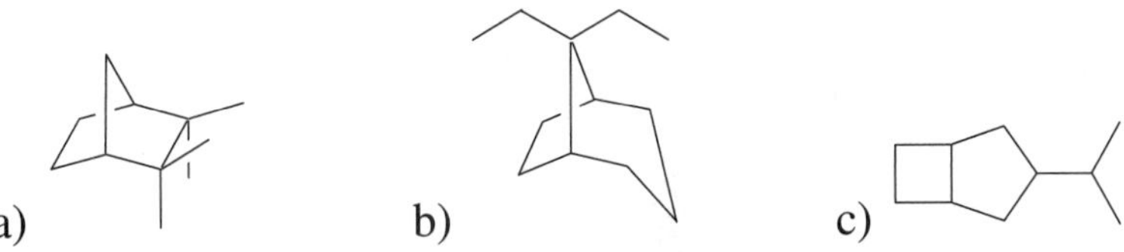

4.14.

a) same compound
b) same compound
c) same compound
d) constitutional isomers

4.15.

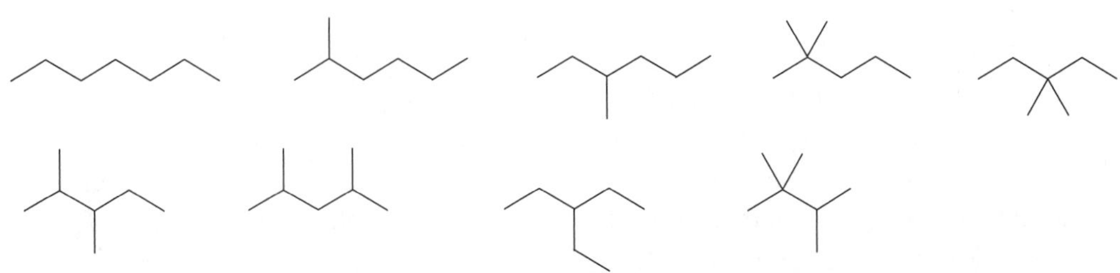

4.16.

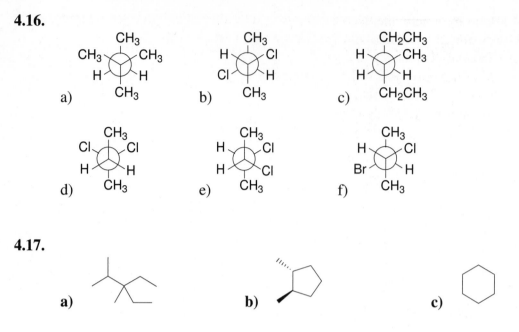

a) b) c)

d) e) f)

4.17.

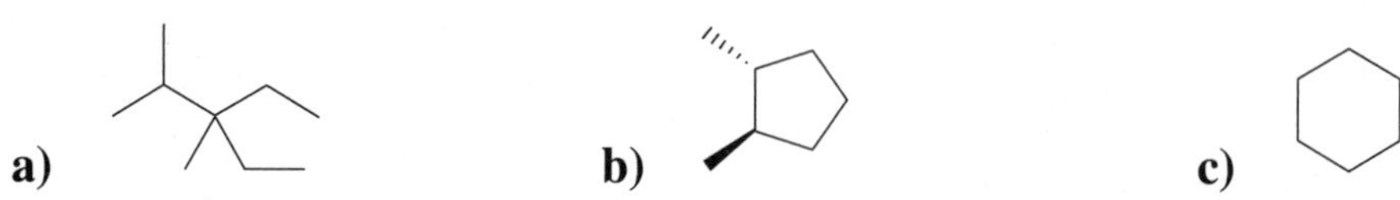

4.18. The compounds are not constitutional isomers. They are just two different representations of the same compound. They are both 2,3-dimethylbutane.

4.19.

a) The energy barrier is expected to be approximately 18 kJ / mol (calculation below):

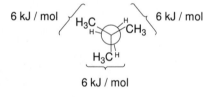

6 kJ / mol 6 kJ / mol 6 kJ / mol

b) The energy barrier is expected to be approximately 16 kJ / mol (calculation below):

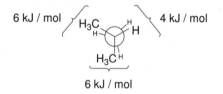

6 kJ / mol 4 kJ / mol 6 kJ / mol

4.20.

a) ***Lowest Energy*** ***Highest Energy***

b) ***Lowest Energy*** ***Highest Energy***

c) ***Lowest Energy*** ***Highest Energy***

d) ***Lowest Energy*** ***Highest Energy***

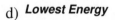

4.21. The gauche conformations are capable of intramolecular hydrogen bonding, as shown below. The anti conformation lacks this stabilizing effect.

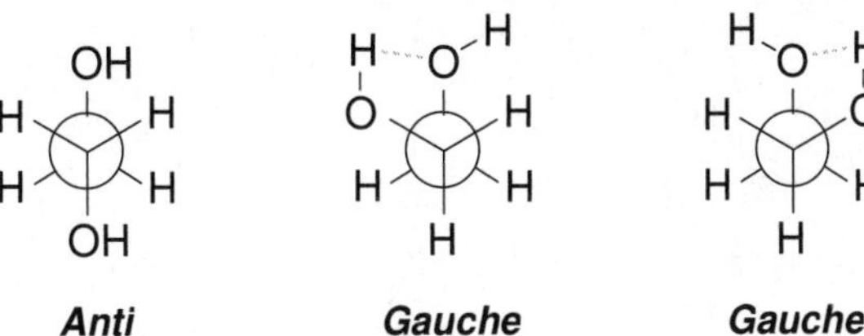

4.22.

4.23.

a)

b)

4.24.

4.25.

4.26.

4.27. There are eight hydrogen atoms in axial positions and seven hydrogen atoms in equatorial positions.

4.28.

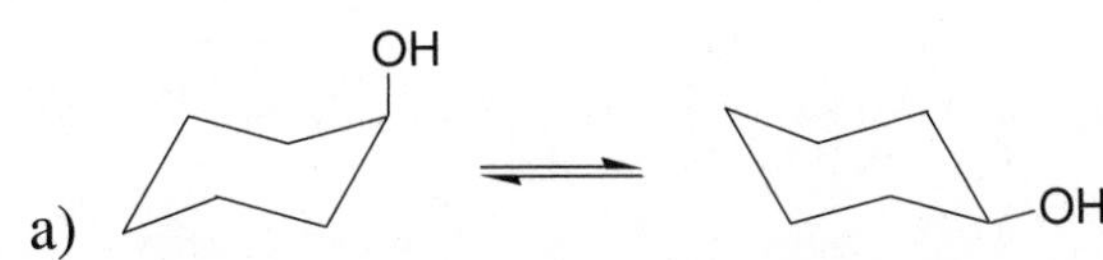

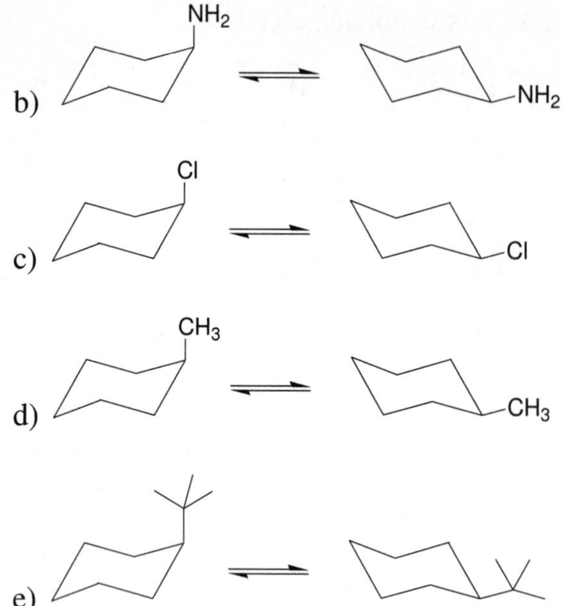

4.29.

a) The bromine atom occupies an equatorial position.

4.30. Although the OH group is in an axial position, nevertheless, this conformation is capable of intramolecular hydrogen bonding, which is a stabilizing effect:

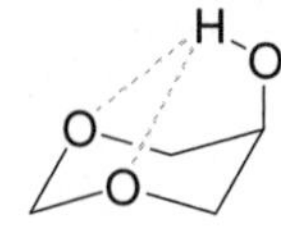

4.31.

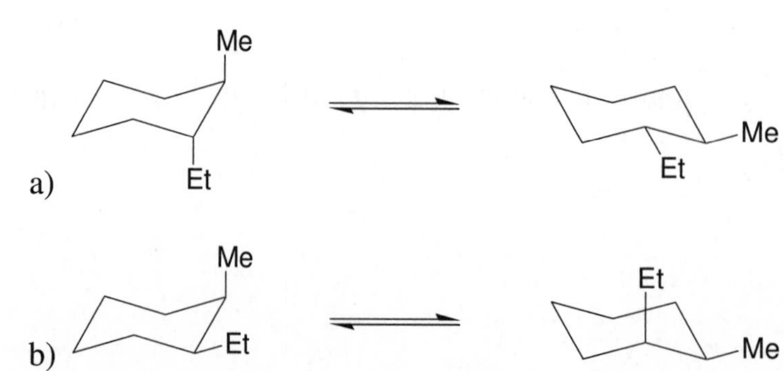

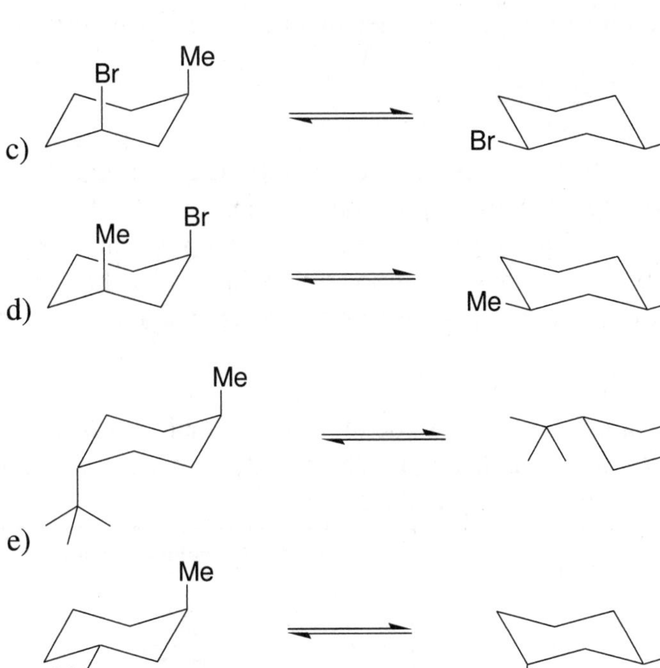
c)
Br
Me
Br
Me
d)
Me
Br
Me
Br
e)
Me
Me
f)
Me
Me
Me
Me

g)

h)
Me
Me
Me
Me

4.32.

4.33.

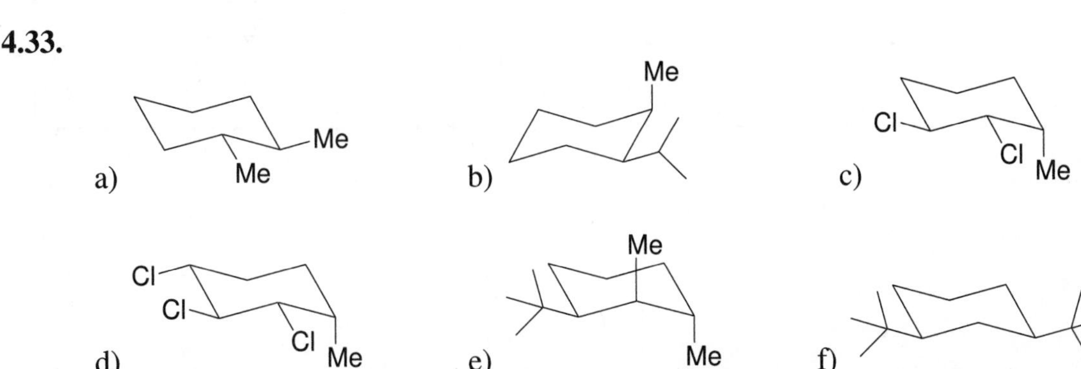
a)
Me
Me
b)
Me
c)
Cl
Cl
Me
d)
Cl
Cl
Cl
Me
e)
Me
Me
f)

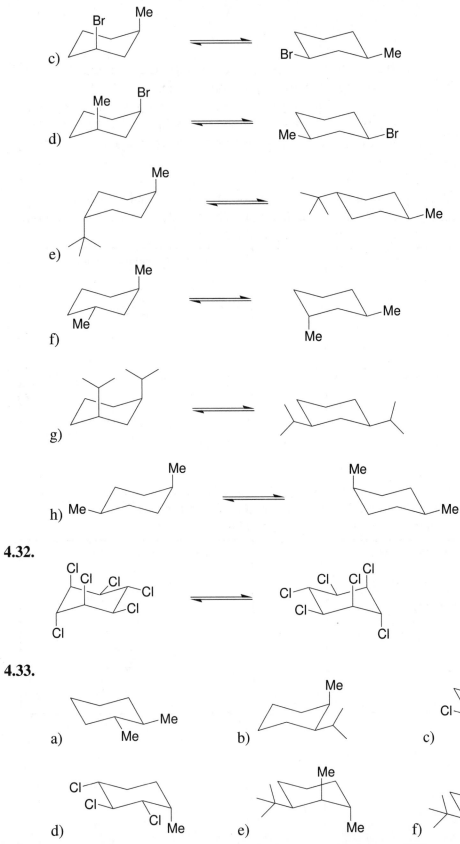

4.34. The two chair conformations of lindane are degenerate. There is no difference in energy between them.

4.35. *trans*-1,4-di-*tert*-butylcyclohexane exists predominantly in a chair conformation, because both substituents can occupy equatorial positions. In contrast, *cis*-1,4-di-*tert*-butylcyclohexane cannot have both of its substituents in equatorial positions. Each chair conformation has one of the substituents in an axial position, which is too high in energy. The compound can achieve a lower energy state by adopting a twist boat conformation.

4.36. *cis*-1,3-dimethylcyclohexane is expected to be more stable than *trans*-1,3-dimethylcyclohexane because the former can adopt a chair conformation in which both substituents are in equatorial positions (highlighted below):

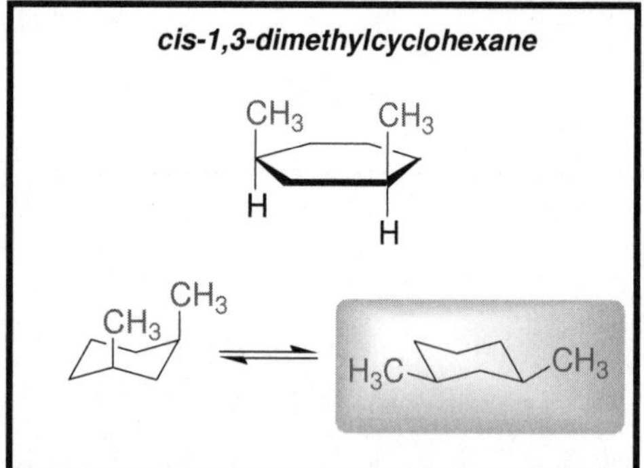

trans-1,3-dimethylcyclohexane

H CH3 CH3 H CH3 H3C CH3 CH3

4.37. *trans*-1,4-dimethylcyclohexane is expected to be more stable than *cis*-1,4-dimethylcyclohexane because the latter can adopt a chair conformation in which both substituents are in equatorial positions (highlighted below):

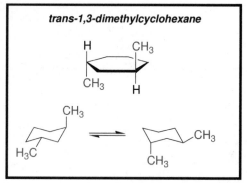

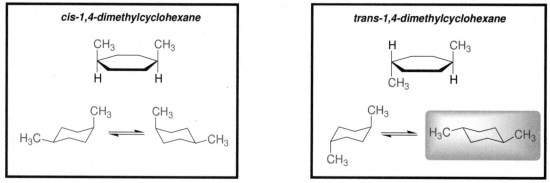

cis-1,4-dimethylcyclohexane

CH3 CH3 H H CH3 H3C CH3 CH3

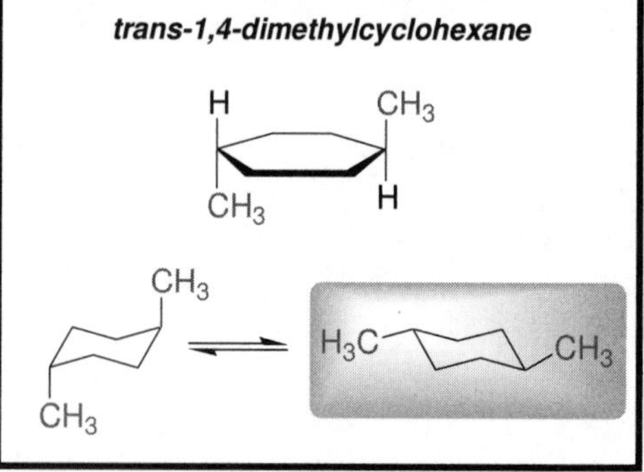

4.38. *cis*-1,3-di-*tert*-butylcyclohexane can adopt a chair conformation in which both tert-butyl groups occupy equatorial positions (highlighted below), and as a result, it is expected to exist primarily in that conformation. In contrast, *trans*-1,3-di-*tert*-butylcyclohexane cannot adopt a chair conformation in which both tert-butyl groups occupy equatorial positions. In either chair conformation, one of the tert-butyl groups occupies an axial position. This compound can achieve a lower energy state by adopting a twist-boat conformation.

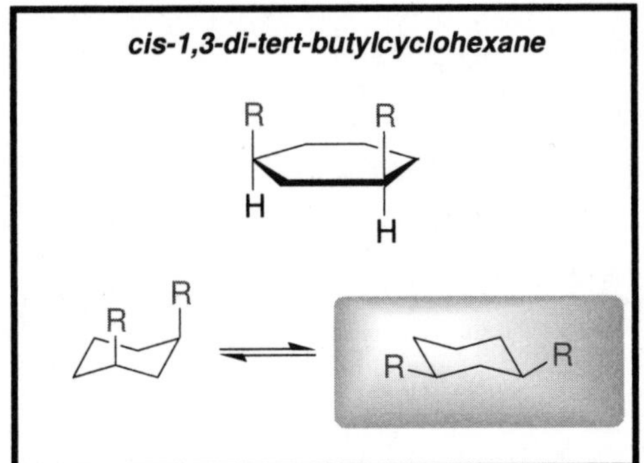

trans-1,3-di-tert-butylcyclohexane

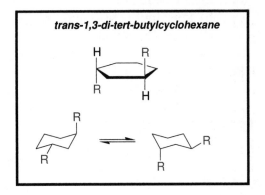

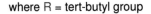

H R R H R R R R

where R = tert-butyl group

4.39.

a) parent = octane
b) parent = nonane
c) parent = octane
d) parent = heptane

4.40.

a)

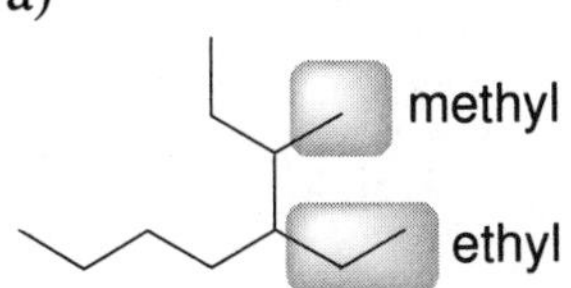

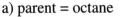

b) isopropyl or (1-methylethyl)

c)

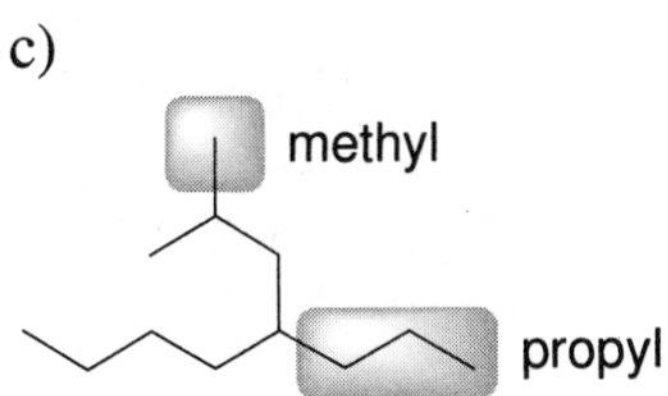

d) *tert*-butyl or (1,1-dimethylethyl)

4.41.

a) 2,3,5-trimethyl-4-propylheptane
b) 1,2,4,5-tetramethyl-3-propylcyclohexane
c) 2,3,5,9-tetramethylbicyclo[4.4.0]decane
d) 1,4-dimethylbicyclo[2.2.2]octane

4.42.

a) same compound
b) constitutional isomers
c) same compound

4.43.

Me
Me H
H Et
H

4.44.

4.45.

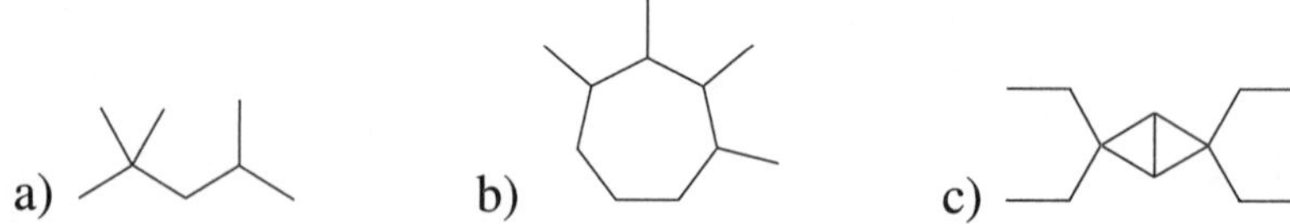

4.46. The energy diagram more closely resembles the shape of the energy diagram for the conformational analysis of ethane.

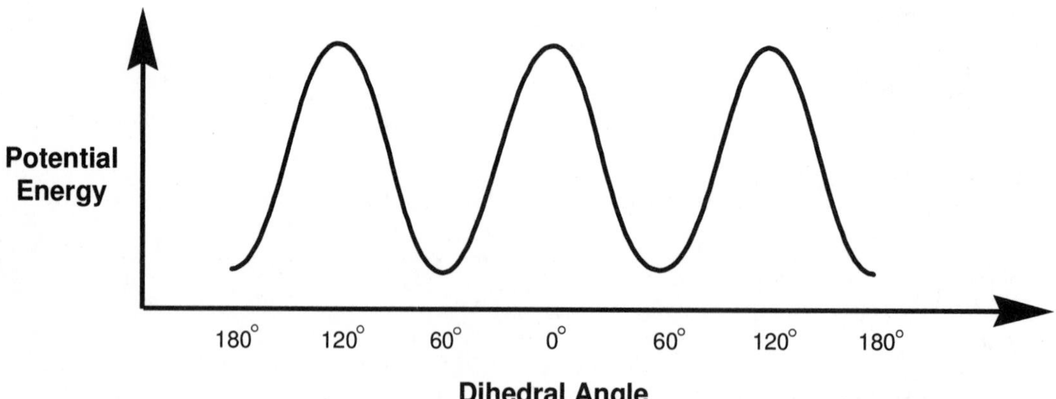

4.47. Two of the staggered conformations are degenerate. The remaining staggered conformation is lower in energy than the other two, as shown below:

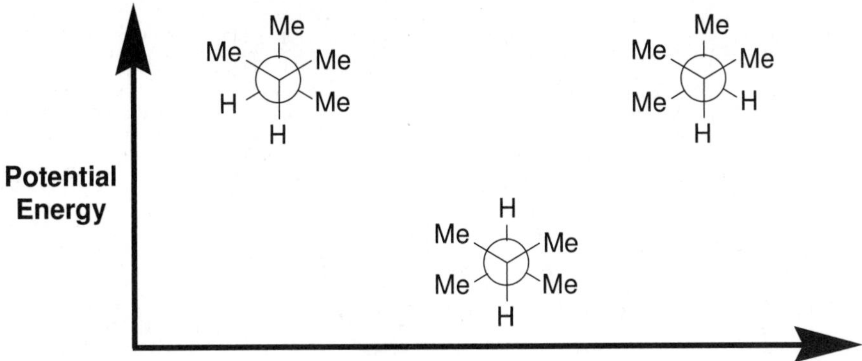

4.48.

a) Cl OH ⇌ Cl OH

b) Cl OH ⇌ OH Cl

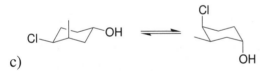

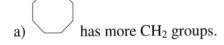

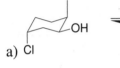

c) Cl OH ⇌ Cl OH

4.49.

a) has more CH_2 groups.

b) cannot adopt a chair conformation in which both groups occupy equatorial positions.

c) cannot adopt a chair conformation in which both groups occupy equatorial positions.

d) cannot adopt a chair conformation in which both groups occupy equatorial positions.

4.50.

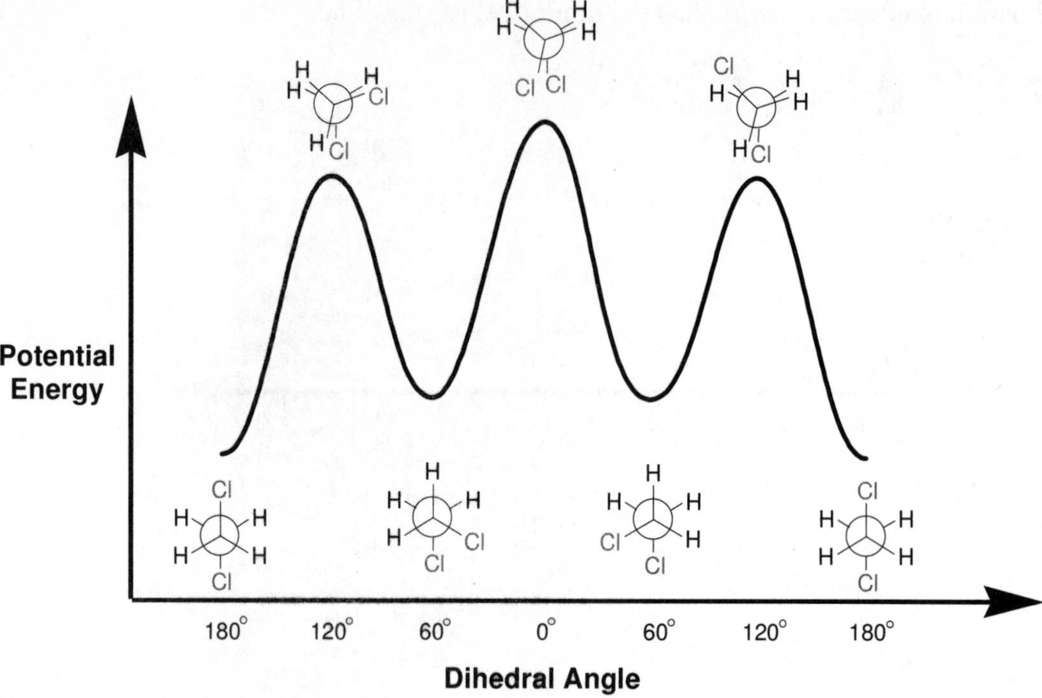

4.51.

a) hexane
b) methylcyclohexane
c) methylcyclopentane
d) *trans*-1,2-dimethylcyclopentane

4.52. Each H-H eclipsing interaction is 4 kJ / mol, and there are two of them (for a total of 8 kJ / mol). The remaining energy cost is associated with the Br-H eclipsing interaction: 15 – 8 = 7 kJ / mol.

4.53.

OH

HO

more stable
(all groups are equatorial)

4.54.

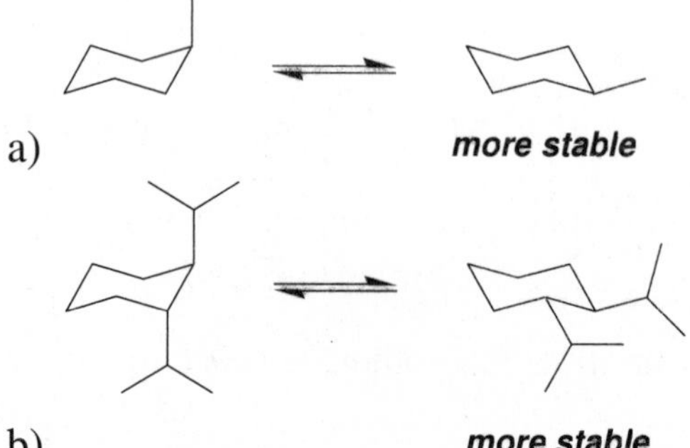

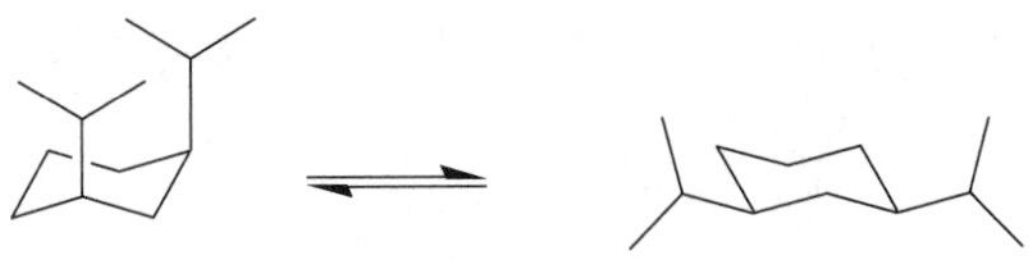

c) *more stable*

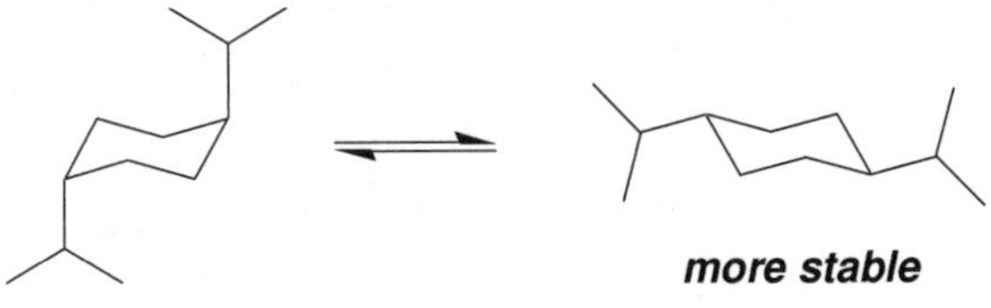

d) *more stable*

4.55.

a) The second compound can adopt a chair conformation in which all three substituents occupy equatorial positions. Therefore, the second compound is expected to be more stable.
b) The first compound can adopt a chair conformation in which all three substituents occupy equatorial positions. Therefore, the first compound is expected to be more stable.
c) The first compound can adopt a chair conformation in which both substituents occupy equatorial positions. Therefore, the first compound is expected to be more stable.
d) The first compound can adopt a chair conformation in which all four substituents occupy equatorial positions. Therefore, the first compound is expected to be more stable.

4.56.

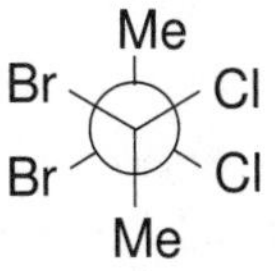

Me
Br Cl
Br Cl
Me

4.57. All groups are in equatorial positions.

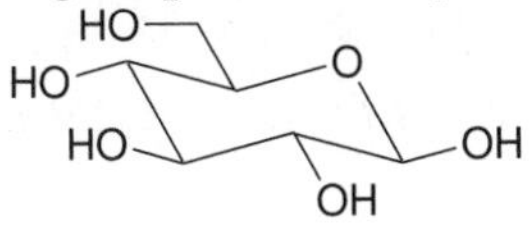

HO
HO O
HO OH
OH

4.58.

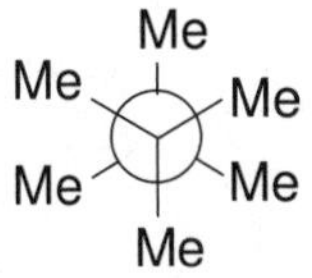

Me
Me Me
Me Me
Me

2,2,4,4-tetramethylbutane

All staggered conformations are degenerate, and the same is true for all eclipsed conformations. The energy diagram has a shape that is similar to the energy diagram for the conformational analysis of ethane:

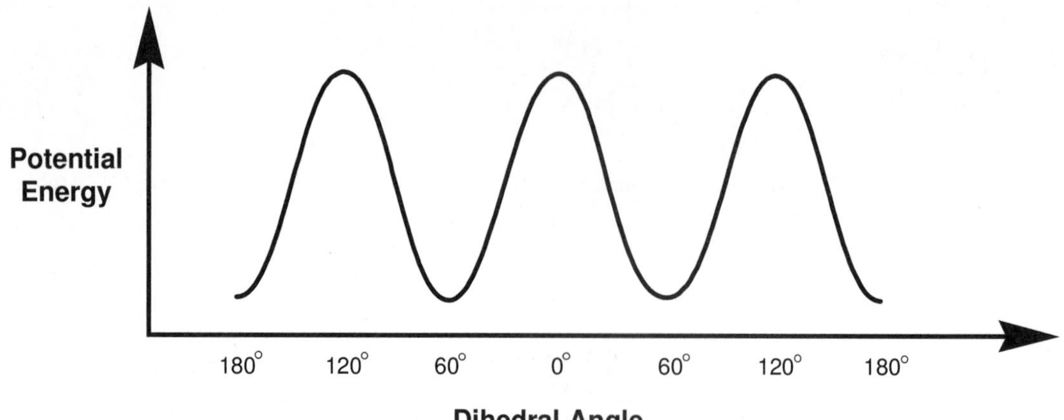

The staggered conformations have six gauche interactions, each of which has an energy cost of 3.8 kJ / mol. Therefore, each staggered conformation has an energy cost of 22.8 kJ / mol. The eclipsed conformations have three methyl-methyl eclipsing interactions, each of which has an energy cost of 11 kJ / mol. Therefore, each eclipsed conformation has an energy cost of 33 kJ / mol. The difference in energy between staggered and eclipsed conformations is therefore expected to be approximately 10.2 kJ / mol.

4.59.

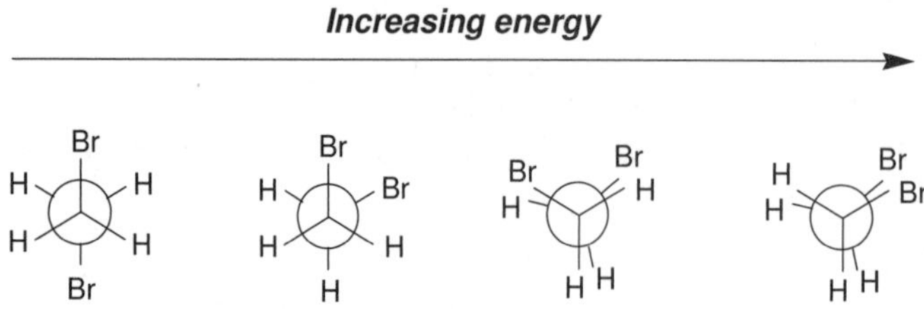

4.60.

a) This conformation has three gauche interactions, each of which has an energy cost of 3.8 kJ / mol. Therefore, this conformation has a total energy cost of 11.4 kJ / mol associated with torsional strain and steric strain.

b) This conformation has two methyl-H eclipsing interactions, each of which has an energy cost of 6 kJ / mol. In addition, it also has one methyl-methyl eclipsing interaction, which has an energy cost of 11 kJ / mol. Therefore, this conformation has a total energy cost of 23 kJ / mol associated with torsional strain and steric strain.

4.61.

OH
HO
HO
OH
OH
OH

4.62.

a) equatorial
b) equatorial
c) axial
d) equatorial
e) equatorial
f) axial

4.63.

cyclopropane

4.64. As mentioned in Section 4.9, cyclobutene adopts a slightly puckered conformation in order to alleviate some of the torsional strain associated with the eclipsing hydrogen atoms:

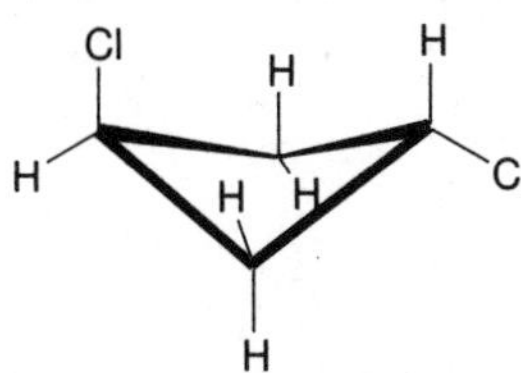

In this non-planar conformation, the individual dipole moments of the C-Cl bonds in *trans*-1,3-dichlorocyclobutane do not fully cancel each other, giving rise to a small molecular dipole moment.

4.65. Cyclohexene cannot adopt a chair conformation because two of the carbon atoms are sp^2 hybridized and trigonal planar. A chair conformation can only be achieved when all six carbon atoms are sp^3 hybridized and tetrahedral (with bond angles of 109.5°).

4.66.

a) identical compounds
b) constitutional isomers
c) identical compounds
d) constitutional isomers
e) identical compounds
f) stereoisomers
g) stereoisomers
h) stereoisomers
i) constitutional isomers
j) different conformations of the same compound
k) stereoisomers
l) constitutional isomers

4.67.

a) the trans isomer s expected to be more stable, because the *cis* isomer has a very high energy methyl-methyl eclipsing interaction (11 kJ / mol). See calculation below.

b) We calculate the energy cost associated with all eclipsing interactions in both compounds. Let's begin with the *trans* isomer. It has the following eclipsing interactions, below the ring and above the ring, giving a total of 32 kJ / mol:

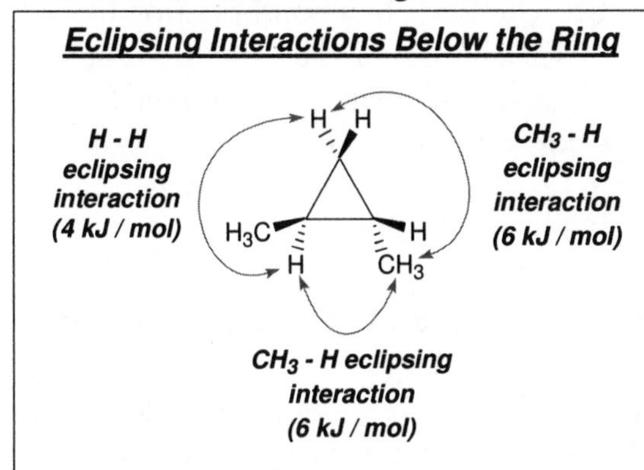

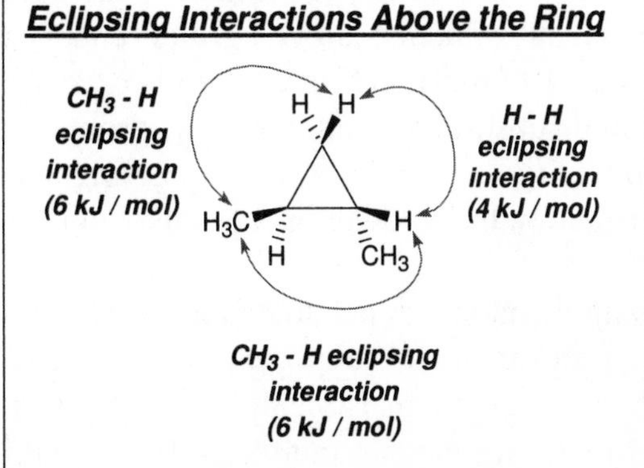

Now let's focus on the *cis* isomer. It has the following eclipsing interactions, below the ring and above the ring, giving a total of 35 kJ / mol:

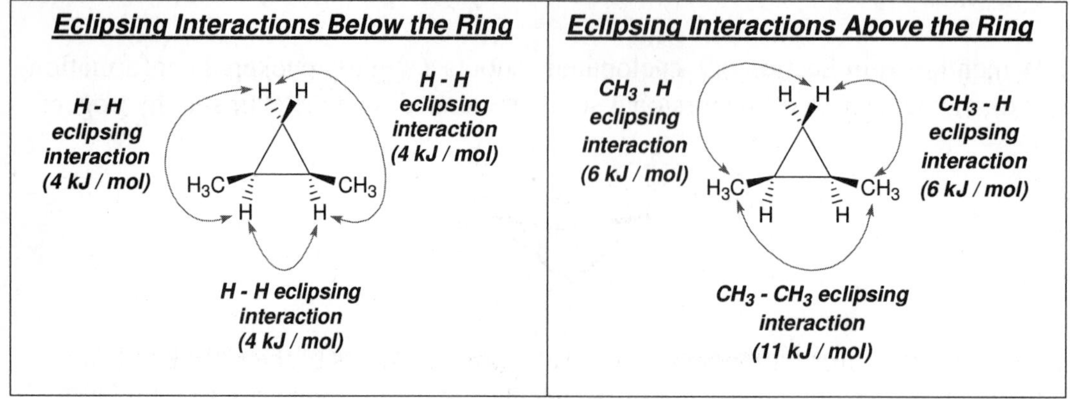

The difference between these two isomers is therefore predicted to be (35 kJ / mol) – (32 kJ / mol) = 3 kJ / mol.

4.68. With increasing halogen size, the bond length also increases. That is, the C-I bond is longer than the C-Br bond, which is longer than the C-Cl bond. So, although iodine is much larger than the other halogens, the longer bond length helps to accommodate the additional steric bulk. These two factors (increased steric bulk and increased bond length) mostly offset each other.

4.69.

a)

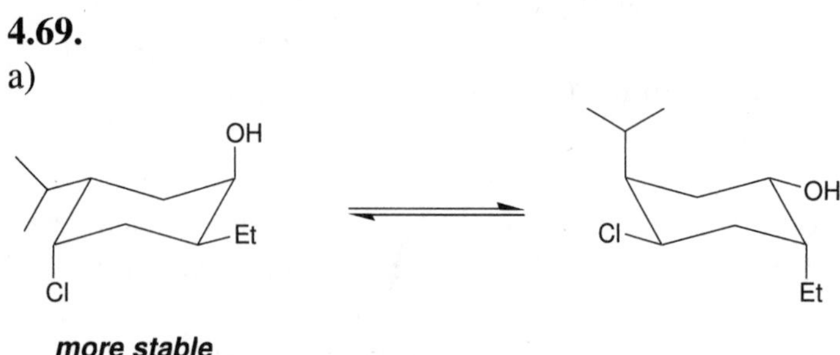

b) Comparison of these chair conformations requires a comparison of the energy costs associated with all axial substituents (see Table 4.8). The first chair conformation has two axial substituents: an OH group (energy cost = 4.2 kJ / mol) and a Cl group (energy cost = 2.0 kJ / mol), giving a total of 6.2 kJ / mol. The second chair conformation has two axial substituents: an isopropyl group (energy cost = 9.2 kJ / mol) and an ethyl group (energy cost = 8.0 kJ / mol), giving a total of 17.2 kJ / mol. The first chair conformation has a lower energy cost, and is therefore more stable.

c) Using the numbers calculated in part b, the difference in energy between the these two chair conformations is expected to be (17.2 kJ / mol) – (6.2 kJ / mol) = 11 kJ / mol. Using the numbers in Table 4.8, we see that a difference of 9 kJ / mol corresponds with a ratio of 97:3 for the two conformations. In this case, the difference in energy is more than 9 kJ / mol, so the ratio should be even higher (more than 97%). Therefore, we do expect the compound to spend more than 95% of its time in the more stable chair conformation.

4.70.

a) *cis*-Decalin has three gauche interactions, while *trans*-decalin has only two gauche interactions.

***cis*-decalin** ***trans*-decalin**

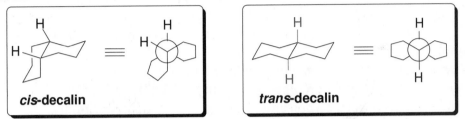

b) *trans*-Decalin is incapable of ring flipping, because a ring flip of one ring would cause its two alkyl substituents (which comprise the second ring) to be too far apart to accommodate the second ring.

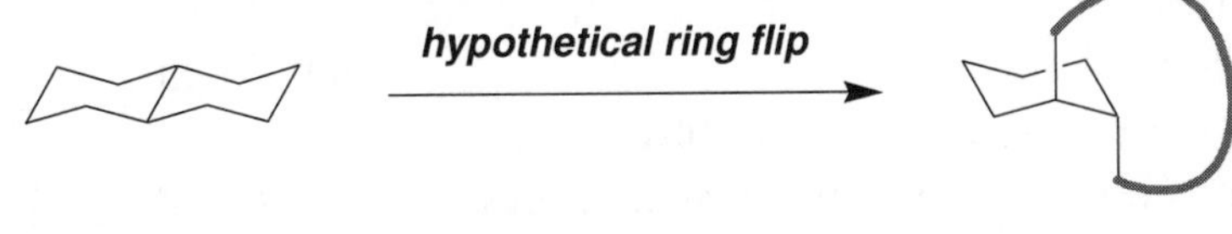

Chapter 5
Stereoisomerism

Review of Concepts

Fill in the blanks below. To verify that your answers are correct, look in your textbook at the end of Chapter 5. Each of the sentences below appears verbatim in the section entitled *Review of Concepts and Vocabulary.*

- ______**isomers** have the same connectivity of atoms but differ in their spatial arrangement.
- **Chiral** objects are not **superimposable** on their ____________________. The most common source of molecular chirality is the presence of a ________________, a carbon atom bearing ______ different groups.
- A compound with one chirality center will have one non-superimposable mirror image, called its ______________.
- The Cahn-Ingold-Prelog system is used to assign the ______________ of a chirality center.
- A **polarimeter** is a device used to measure the ability of chiral organic compounds to rotate the plane of ____________________ light. Such compounds are said to be ____________ **active**.
- A solution containing equal amounts of both enantiomers is called a __________ **mixture**. A solution containing a pair of enantiomers in unequal amounts is described in terms of **enantiomeric** _________ (*ee*).
- For a compound with multiple chirality centers, a family of stereoisomers exists. Each stereoisomer will have at most one enantiomer, with the remaining members of the family being _____________.
- A ______ **compound** contains multiple chirality centers but is nevertheless achiral because it possesses reflectional symmetry.
- __________ **projections** are drawings that convey the configuration of chirality centers, without the use of wedges and dashes.

Review of Skills

Fill in the blanks and empty boxes below. To verify that your answers are correct, look in your textbook at the end of Chapter 5. The answers appear in the section entitled *SkillBuilder Review.*

SkillBuilder 5.1 Identifying *cis-trans* Stereoisomerism

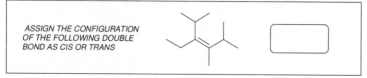

ASSIGN THE CONFIGURATION OF THE FOLLOWING DOUBLE BOND AS CIS OR TRANS

SkillBuilder 5.2 Locating Chirality Centers

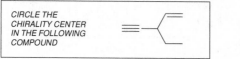

CIRCLE THE CHIRALITY CENTER IN THE FOLLOWING COMPOUND

SkillBuilder 5.3 Drawing an Enantiomer

SHOW THREE WAYS TO DRAW THE ENANTIOMER OF THE FOLLOWING COMPOUND. PLACE YOUR ANSWERS IN THE BOXES SHOWN.

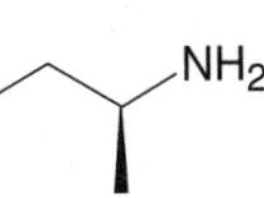

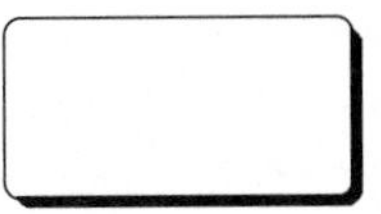

SkillBuilder 5.4 Assigning Configuration

ASSIGN THE CONFIGURATION OF THE CHIRALITY CENTER IN THE FOLLOWING COMPOUND

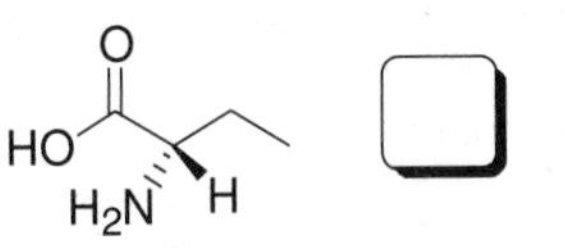

SkillBuilder 5.5 Calculating specific rotation

CALCULATE THE SPECIFIC ROTATION GIVEN THE FOLLOWING INFORMATION:

0.300 grams sucrose dissolved in 10.0 mL of water
sample cell = 10.0 cm
observed rotation = +1.99°

$$\text{specific rotation} = [\alpha] = \frac{\alpha}{c \times l} = \frac{\square}{\square \times \square} = \square$$

SkillBuilder 5.6 Calculating *% ee*

CALCULATE THE ENANTIOMERIC EXCESS GIVEN THE FOLLOWING INFORMATION:

The specific rotation of optically pure adrenaline is -53 . A mixture of (*R*)- and (*S*)- adrenaline was found to have a specific rotation of - 45 . Calculate the % *ee* of the mixture

$$\%\,ee = \frac{\text{observed [ ]}}{\text{[ ] of pure enantiomer}} \times 100\,\%$$

$$= \frac{\square}{\square} \times 100\,\% = \square$$

SkillBuilder 5.7 Determining Stereoisomeric Relationship

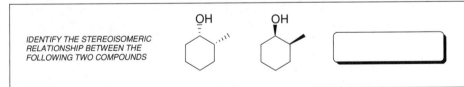

IDENTIFY THE STEREOISOMERIC RELATIONSHIP BETWEEN THE FOLLOWING TWO COMPOUNDS

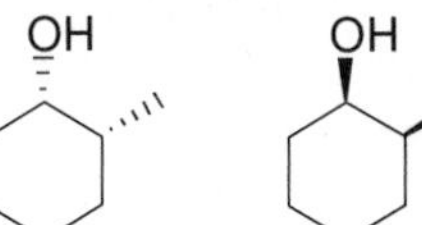

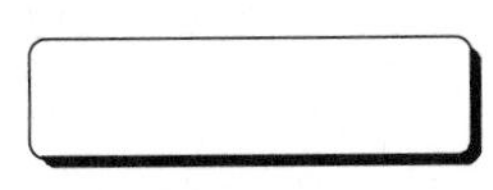

SkillBuilder 5.8 Identifying Meso Compounds

OH
OH

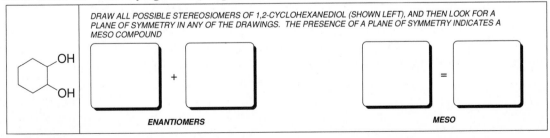

DRAW ALL POSSIBLE STEREOSIOMERS OF 1,2-CYCLOHEXANEDIOL (SHOWN LEFT), AND THEN LOOK FOR A PLANE OF SYMMETRY IN ANY OF THE DRAWINGS. THE PRESENCE OF A PLANE OF SYMMETRY INDICATES A MESO COMPOUND

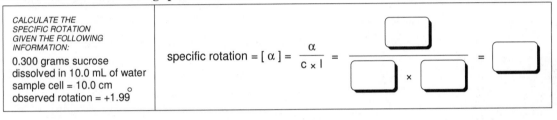

ENANTIOMERS

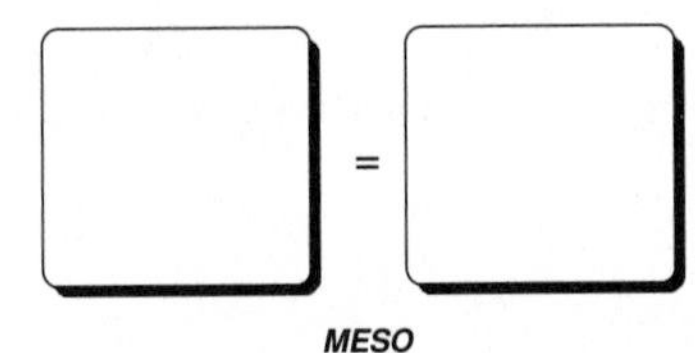

MESO

SkillBuilder 5.9 Assigning configuration from a Fischer projection

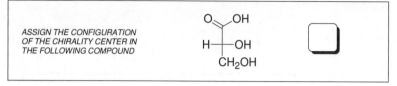

ASSIGN THE CONFIGURATION OF THE CHIRALITY CENTER IN THE FOLLOWING COMPOUND

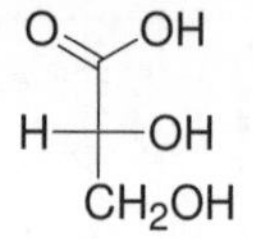

O OH
H—+—OH
CH_2OH

Solutions

5.1.

a) *trans* b) not stereoisomeric
c) *trans* d) *trans*
e) *trans* f) not stereoisomeric
g) *cis*

5.2. $H_2CCHCH_2CH_2CH_2CHCH_2$ =

Neither double bond exhibits stereoisomerism, so this compound does not have any stereoisomers.

5.3.

a)

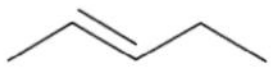

b)

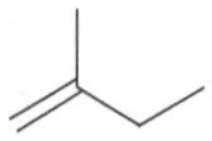

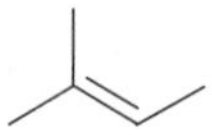

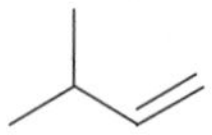

5.4. All chirality centers are highlighted below:

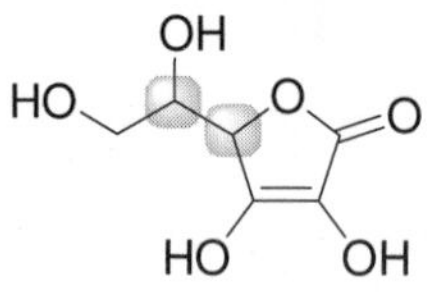

a) OH HO O O HO OH

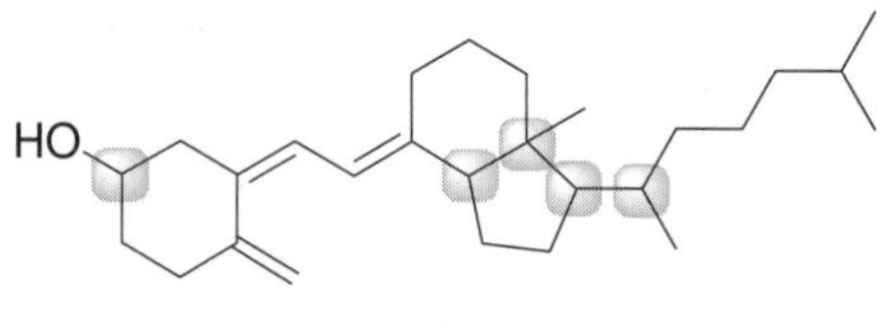

b) HO

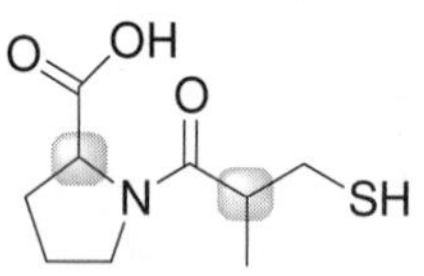

c) O OH O N SH

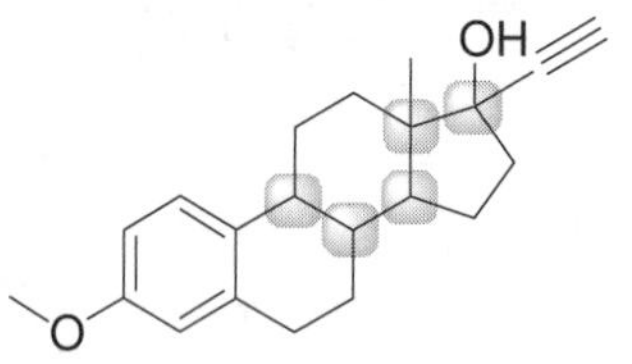

d) OH O

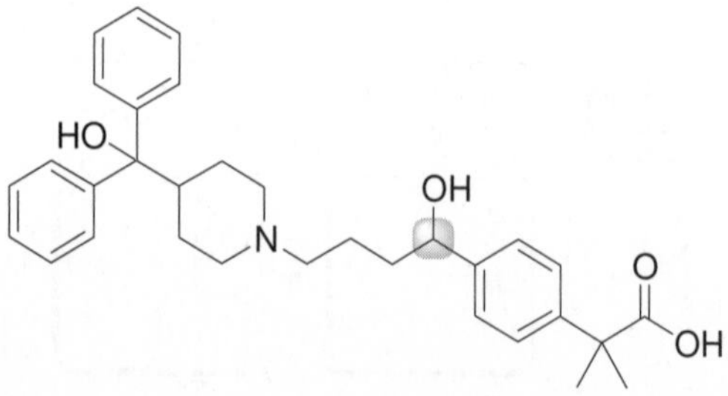

e) HO N OH O OH

f)

5.5.

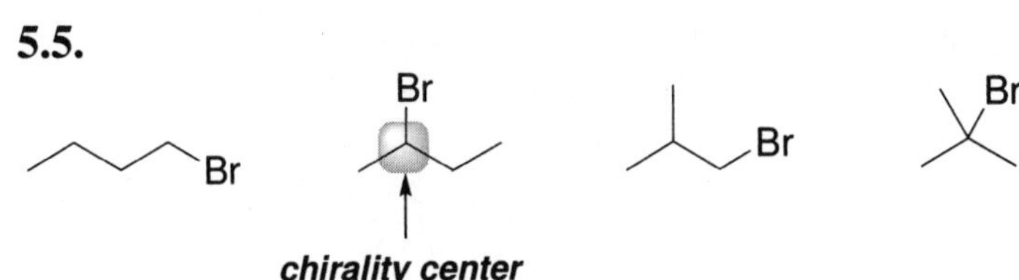

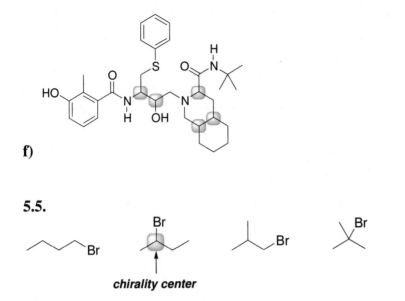

5.6. The phosphorus atom has four different groups attached to it (a methyl group, an ethyl group, a phenyl group, and a lone pair). This phosphorous atom therefore represents a chirality center. This compound is not superimposable on its mirror image, as can be seen clearly by building and comparing molecular models.

5.7.

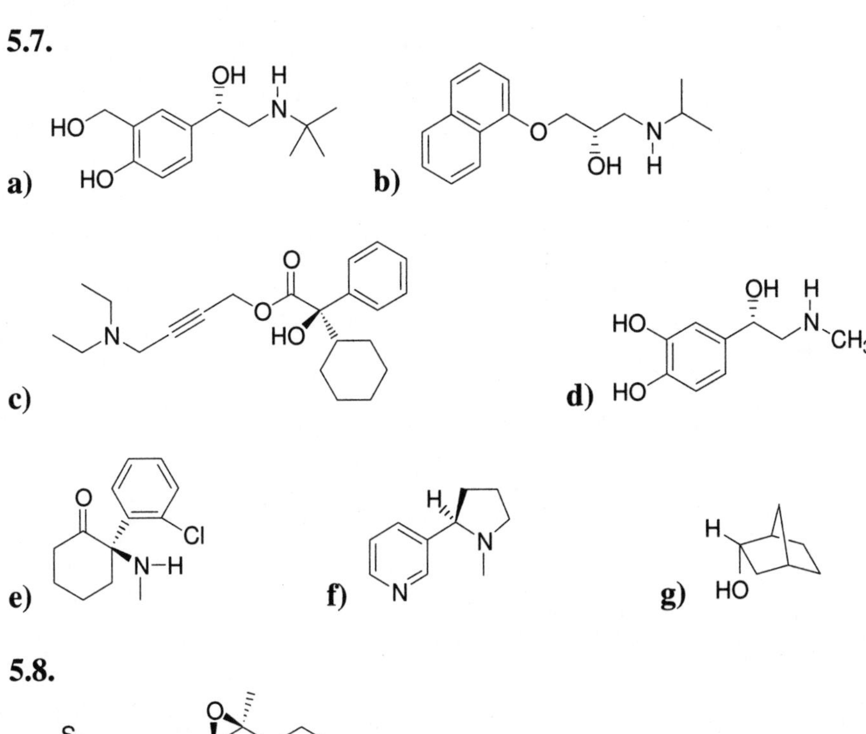

5.8.

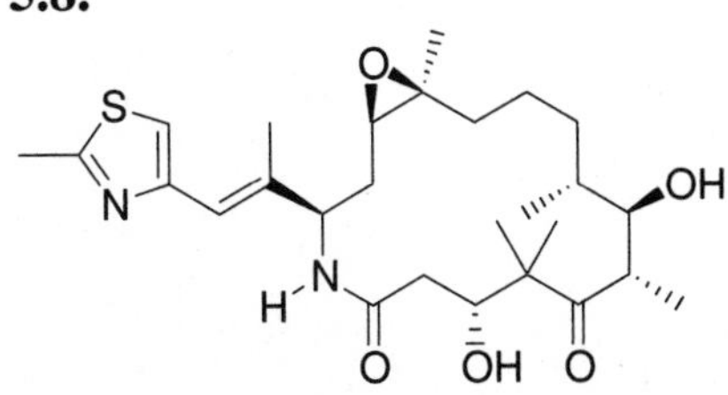

5.9.

a)

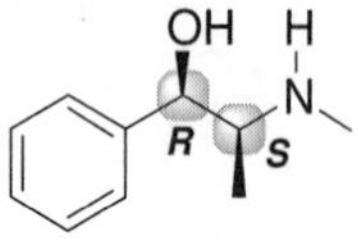

b)

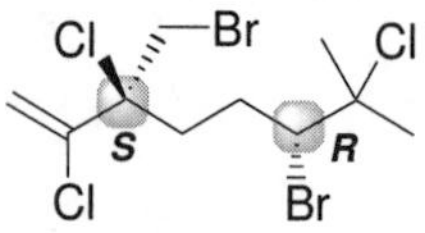

c)

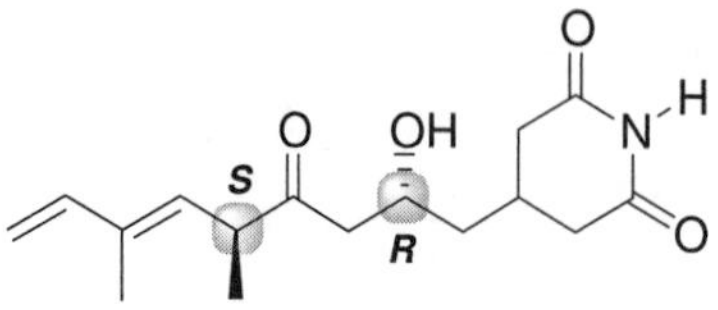

d)

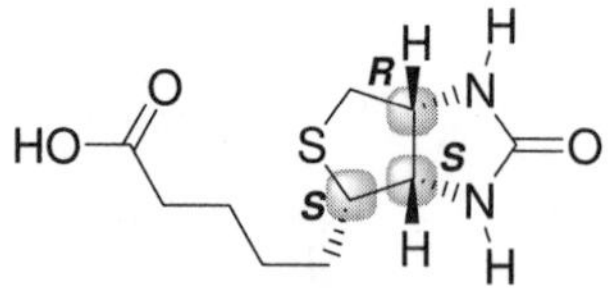

e)

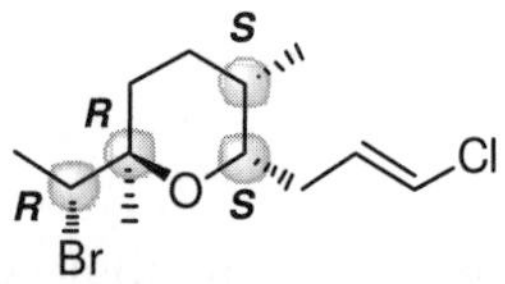

f)

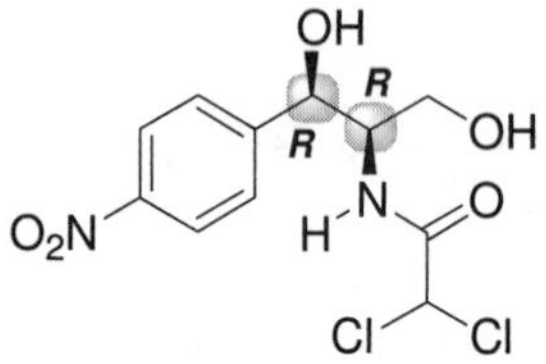

5.10.

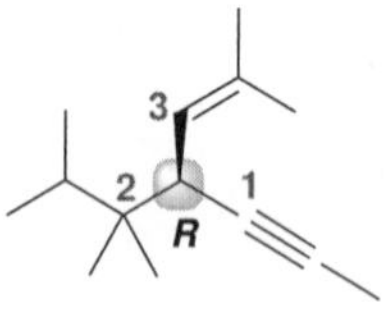

5.11.

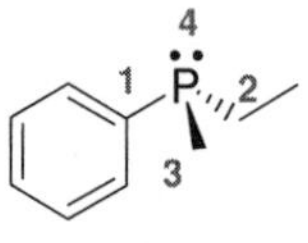

S

5.12. specific rotation $= [\alpha] = \dfrac{\alpha}{c \times l} = \dfrac{(+1.47^\circ)}{(0.0575\ \text{g/mL}) \times (1.00\ \text{dm})} = \mathbf{+25.6}$

5.13. specific rotation $= [\alpha] = \dfrac{\alpha}{c \times l} = \dfrac{(-2.99^\circ)}{(0.095\ \text{g/mL}) \times (1.00\ \text{dm})} = \mathbf{-31.5}$

5.14. specific rotation $= [\alpha] = \dfrac{\alpha}{c \times l} = \dfrac{(+0.57^\circ)}{(0.260\ \text{g/mL}) \times (1.00\ \text{dm})} = \mathbf{+2.2}$

5.15. This compound does not have a chirality center, because two of the groups are identical:

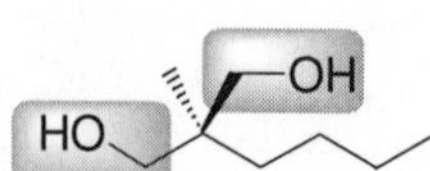

Accordingly, the compound is achiral and is not optically active.

5.16.

$$[\alpha] = \frac{\alpha}{c \times l}$$

$$\alpha = [\alpha] \times c \times l = (+13.5)(0.100 \text{ g / mL})(1.00 \text{ dm}) = +1.35\,°$$

5.17.

$$\%\,ee = \frac{\text{observed } [\alpha]}{[\alpha] \text{ of pure enantiomer}} \times 100\,\%$$

$$= \frac{(-37^{\circ})}{(-39.5^{\circ})} \times 100\,\%$$

$$= 94\,\%$$

5.18.

$$\%\,ee = \frac{\text{observed } [\alpha]}{[\alpha] \text{ of pure enantiomer}} \times 100\,\%$$

$$= \frac{(-6.0^{\circ})}{(-6.3^{\circ})} \times 100\,\%$$

$$= 95\,\%$$

5.19.

$$\%\,ee = \frac{\text{observed } [\alpha]}{[\alpha] \text{ of pure enantiomer}} \times 100\,\%$$

$$= \frac{(85^{\circ})}{(92^{\circ})} \times 100\,\%$$

$$= 92\,\%$$

5.20. Observed $[\alpha] = \dfrac{\alpha}{c \times 1} = \dfrac{(+0.78°)}{(0.350 \text{ g / mL}) \times (1.00 \text{ dm})} = \mathbf{+2.2}$

$$\% \, ee = \frac{\text{observed } [\alpha]}{[\alpha] \text{ of pure enantiomer}} \times 100 \%$$

$$= \frac{(2.2°)}{(2.8°)} \times 100 \%$$

$$= 79 \%$$

5.21.

a) enantiomers
b) diastereomers
c) diastereomers
d) diastereomers
e) diastereomers
f) enantiomers

5.22. There are three chirality centers, and only one of these chirality centers has a different configuration in these two compounds. The other two chirality centers have the same configuration in both compounds. Therefore, these compounds are diastereomers.

5.23.
a) yes b) yes c) no d) yes e) yes f) no

5.24. 5.23f has three planes of symmetry.

5.25.

a) 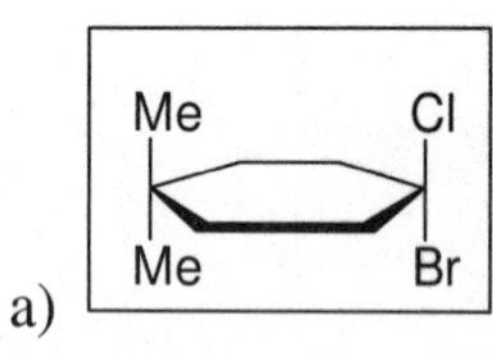

b) 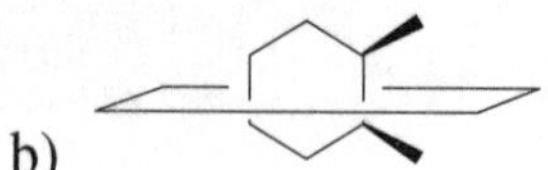

c) 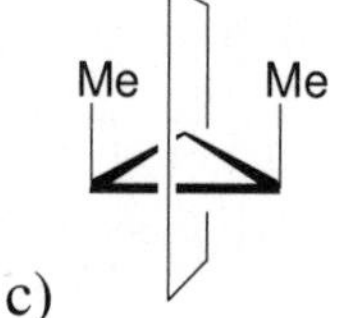

d) 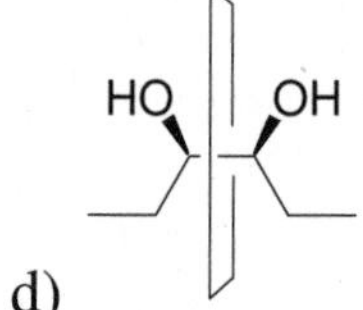

e) 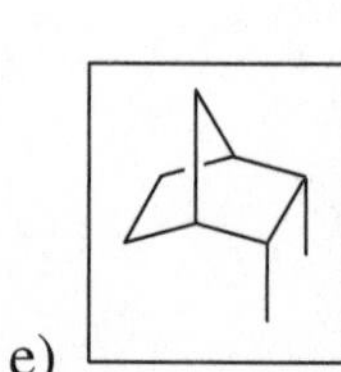

f) 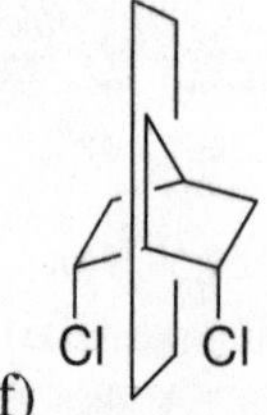

5.26.

a)

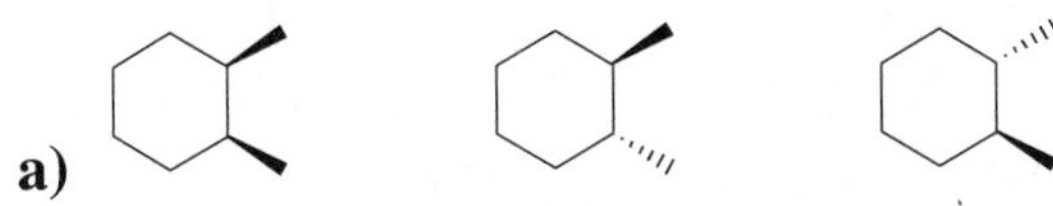

b)

c)

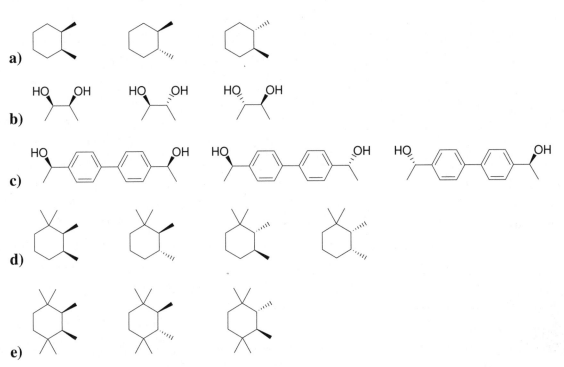

d)

e)

5.27. Each of these compounds is a meso compound and does not have an enantiomer.

5.28 There are only four stereoisomers:

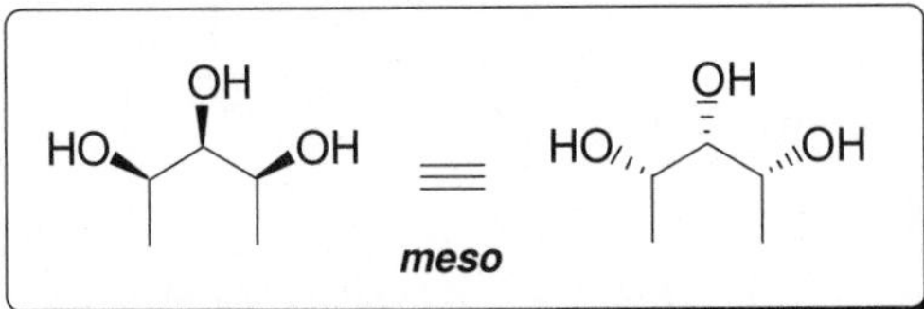

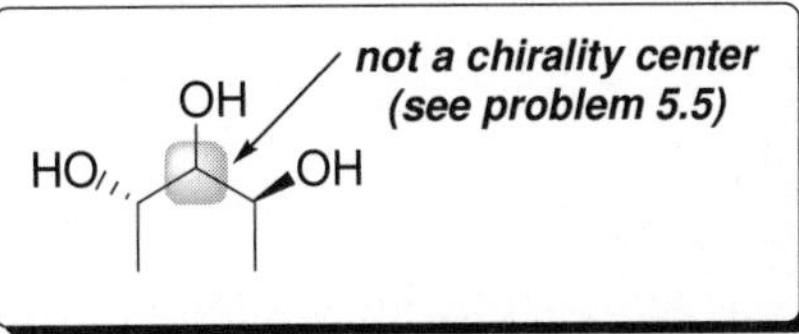

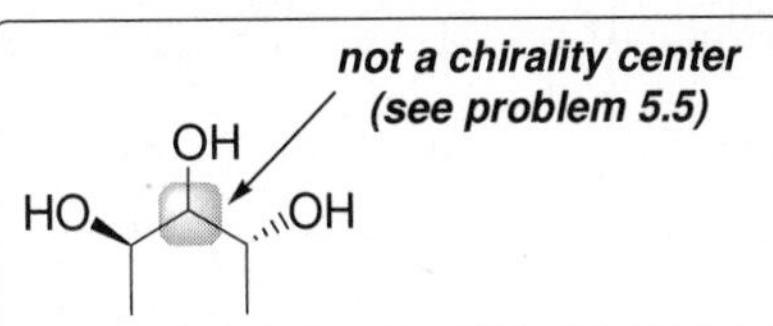

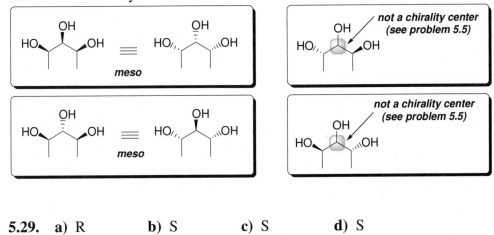

5.29. **a)** R **b)** S **c)** S **d)** S

5.30.

a) 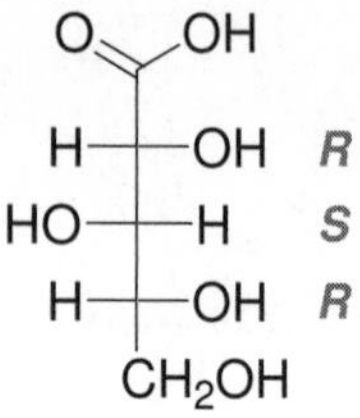 b) 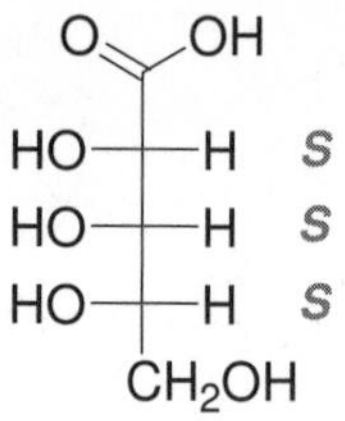 c)

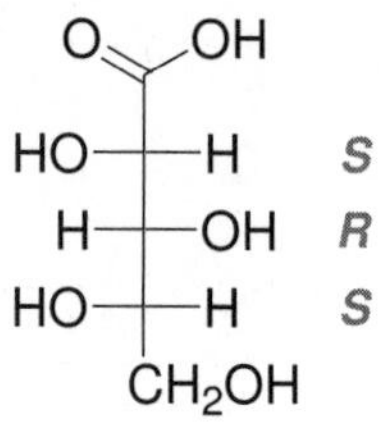

5.31.

a) 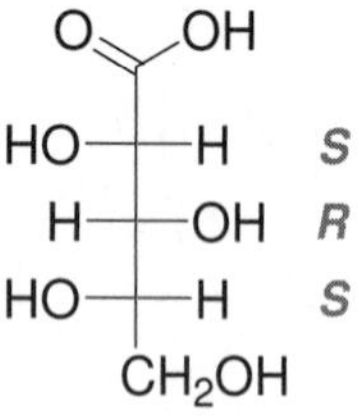 b) 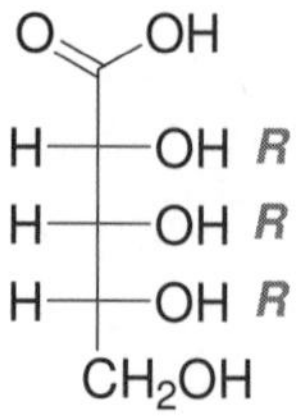 c)

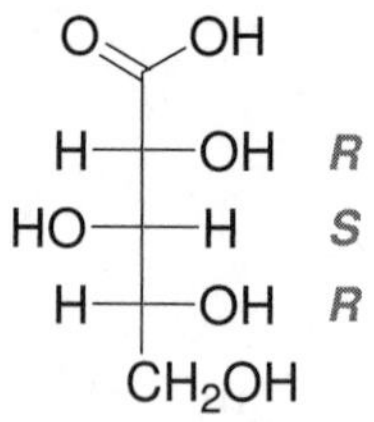

5.32.

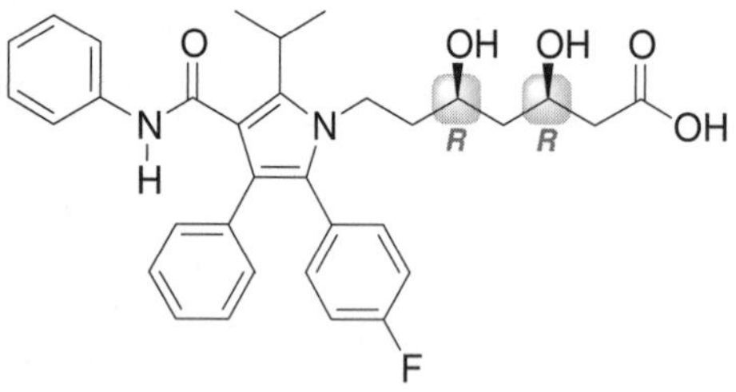

5.33.

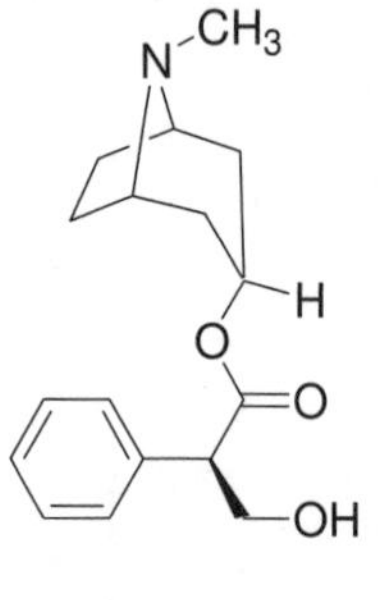

5.34.

a) Paclitaxel has eleven chirality centers.

b) The enantiomer of paclitaxel is shown below:

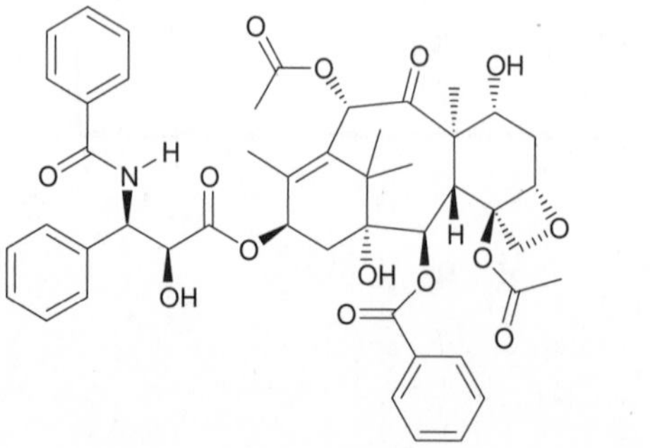

5.35.

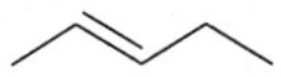

trans 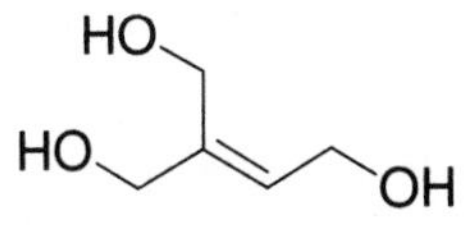 | *not stereoisomeric* 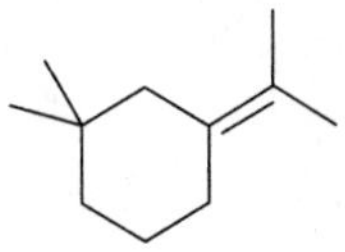 | *not stereoisomeric*

5.36.

a) enantiomers
b) same compound
c) constitutional isomers
d) constitutional isomers
e) diastereomers
f) same compound
g) enantiomers
h) diastereomers
i) same compound
j) same compound
k) same compound
l) same compound

5.37. a) 8 b) 3 c) 16 d) 3 e) 3 f) 32

5.38.

a) 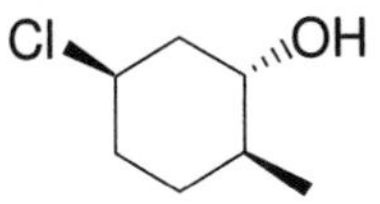 b) 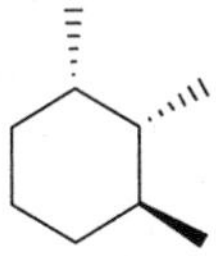 c)

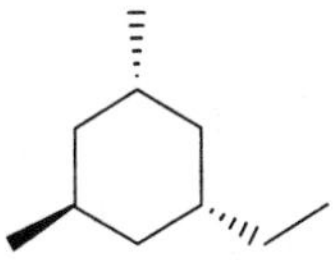

d) 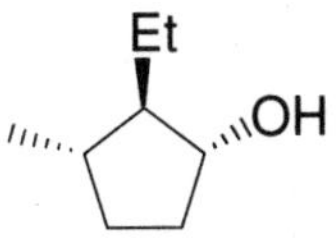 e) 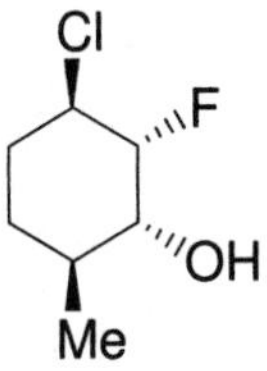 f)

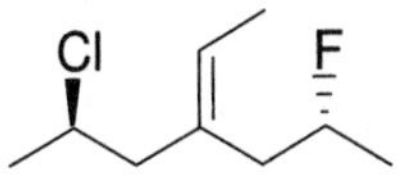

g) 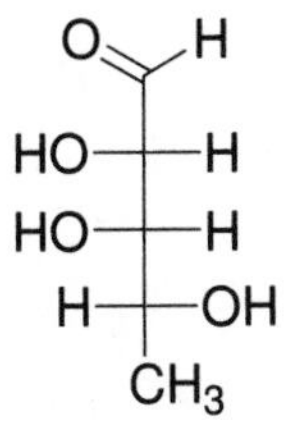 h) 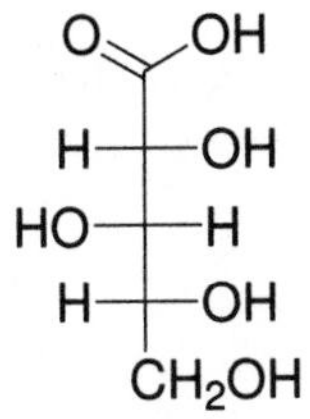 i)

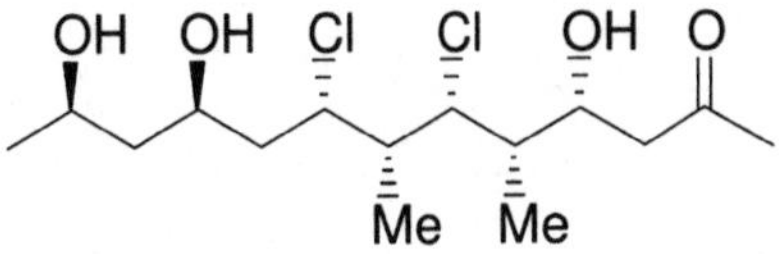

j) 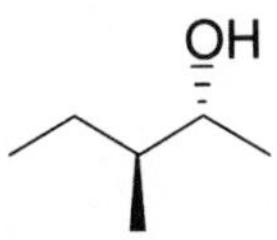 k) 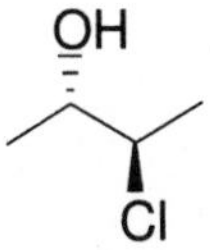 l)

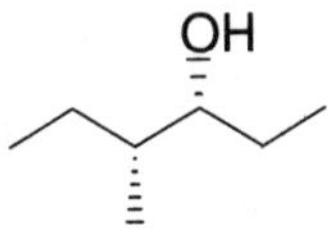

5.39.

a) 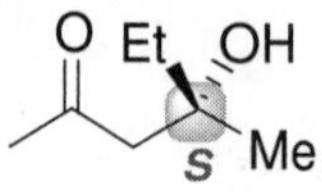 O Et OH, S, Me

b) 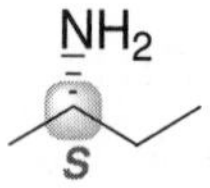 NH_2, S

c) 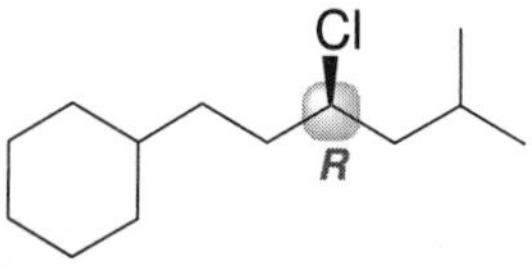 Cl, R

d) 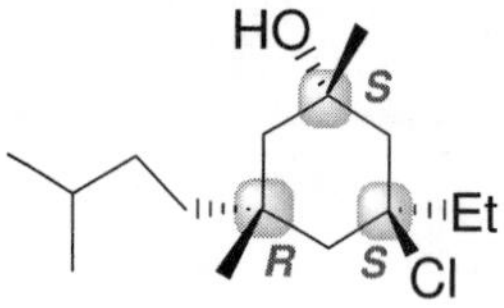 HO, S, R, S, Et, Cl

e) 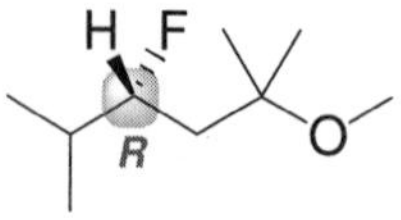 H F, R

f) 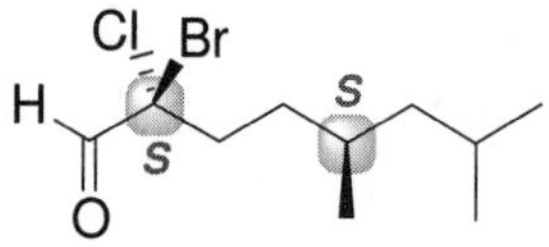 Cl Br, H, S, S, O

g) 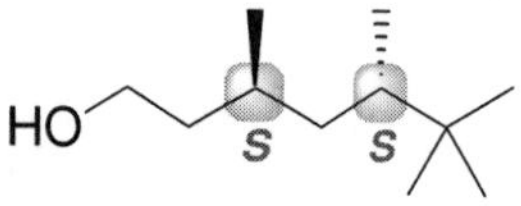 HO, S, S

h)

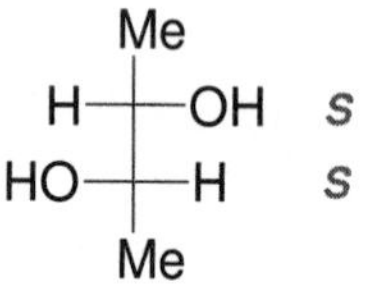

	Me		
H	—	OH	S
HO	—	H	S
	Me		

i) 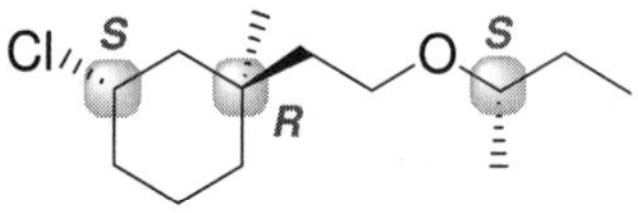 Cl, S, R, O, S

5.40. 96% *ee*

5.41.
a) diastereomers
b) diastereomers
c) enantiomers
d) same compound
e) enantiomers
f) diastereomers
g) enantiomers
h) diastereomers
i) enantiomers
j) same compound
k) enantiomer
l) diastereomers

5.42.

$$\%\, ee = \frac{\text{observed } [\alpha]}{[\alpha] \text{ of pure enantiomer}} \times 100\ \%$$

$$= \frac{(-55^\circ)}{(-61^\circ)} \times 100\ \%$$

$$= 90\ \%$$

5.43.

a) True.
b) False.
c) True.

5.44. specific rotation = $[\alpha] = \dfrac{\alpha}{c \times l} = \dfrac{(-0.47°)}{(0.0075\ g\ /\ mL) \times (1.00\ dm)}$ = **-63**

5.45.

a) (*S*)-limonene b) (*R*)-limonene c) (*S*)-limonene d) (*R*)-limonene

5.46.

a) 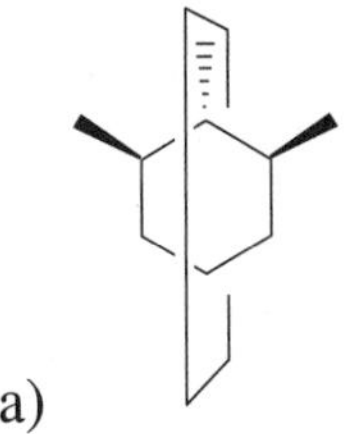

b) HO OH

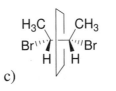

c) H_3C CH_3 Br Br H H

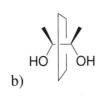

d) Cl Cl

5.47.

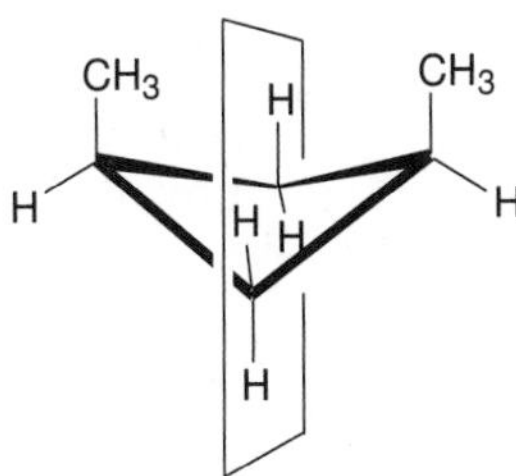

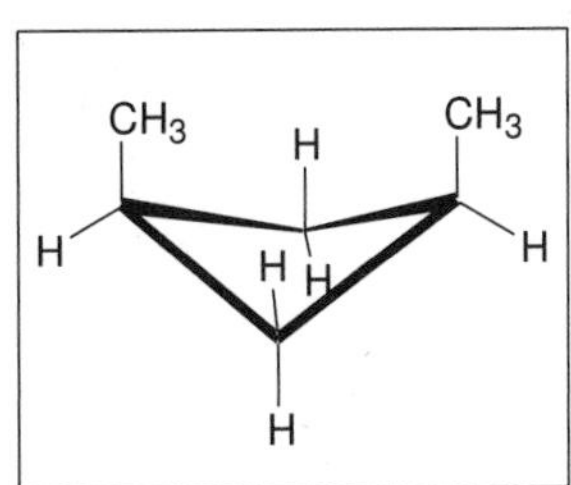

5.48. The first compound has three chirality centers:

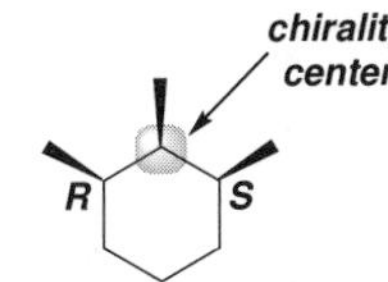

three chirality centers

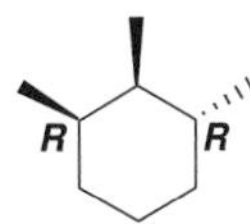

two chirality centers

This is apparent if we assign the configuration at C1 and C3 of the cyclohexane ring. In the first compound, the configuration at C1 is different than the configuration at C3. As a result, there are four different groups attached to the C2 position. That is, C1 and C3 represent two different groups: one with the R configuration and the other with the S configuration. In contrast, consider the configuration at C1 and C3 in the second compound. Both of these positions have the same configuration, and therefore, the C2 position in that compound does not have four different groups. Two of the groups are identical, so C2 is not a chirality center.

5.49.

a) enantiomers
b) diastereomers
c) enantiomers
d) same compound
e) enantiomers
f) diastereomers
g) same compound
h) constitutional isomers
i) diastereomers
j) diastereomers
k) same compound
l) enantiomers

5.50.

a) -61
b) 90 % *ee*
c) 95 % of the mixture is (*S*)-carvone

5.51.

a) chiral	b) chiral	c) achiral	d) achiral
e) chiral	f) achiral	g) achiral	h) chiral
i) chiral	j) achiral	k) chiral	l) chiral
l) achiral	m) chiral	n) achiral	o) achiral

5.52.

$$[\alpha] = \frac{\alpha}{c \times l}$$

$$\alpha = [\alpha] \times c \times l = (+24)(0.0100 \text{ g / mL})(1.00 \text{ dm}) = +0.24\,°$$

5.53.

a) optically inactive (meso)
b) optically active
c) optically active
d) optically inactive
e) optically active
f) optically inactive (3-methylpentane has no chirality centers)
g) optically inactive (meso)
h) optically inactive

5.54.

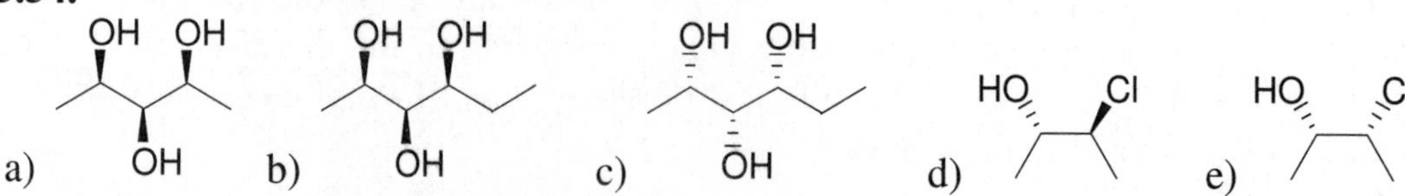

5.55.

a)

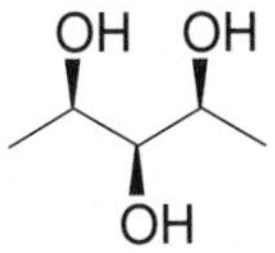

b) No. A racemic mixture is not optically active.

c) Yes, because d and e are not enantiomers. They are diastereomers, which are not expected to exhibit equal and opposite rotations.

5.56.

a) 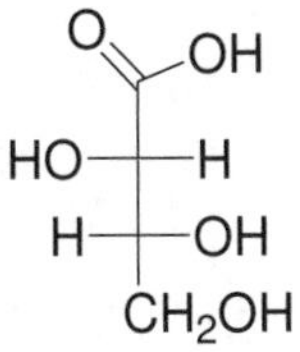 b) 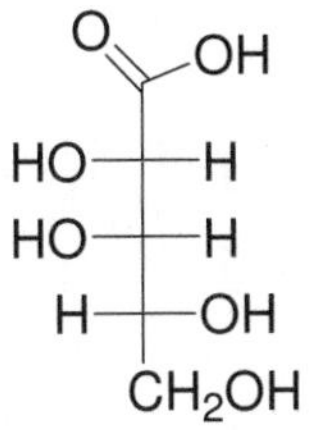 c)

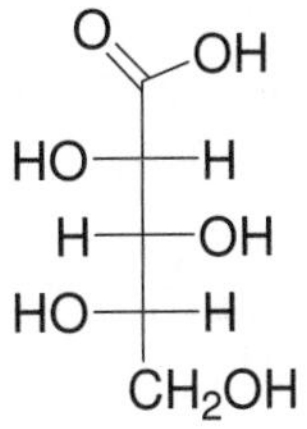

5.57.

a) 3-methylpentane and 2-methylpentane are constitutional isomers.

b) *trans*-1,2-dimethylcyclohexane and *cis*-1,2-dimethylcyclohexane are diastereomers.

5.58. The following two compounds are enantiomers because they are nonsuperimposable mirror images. You may find it helpful to construct molecular models to help visualize the mirror image relationship between these two compounds.

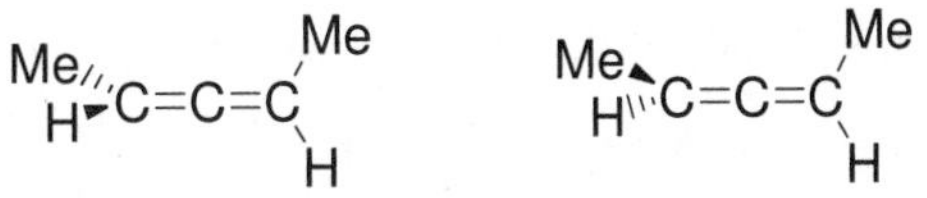

5.59. This compound will be achiral.

5.60.

a) This compound cannot be completely planar because steric hindrance prevents the two ring systems from rotating with respect to each other. The compound is locked in a particular conformation that is chiral.

b) This ring system cannot be planar because of steric hindrance, and must therefore adopt a spiral shape (like a spiral staircase). The spiral can be right handed or left handed, and the relationship between these two forms is enantiomeric.

5.61. The compound is chiral because it is not superimposable on its mirror image.

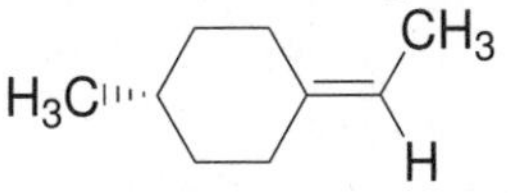

5.62. This compound has a center of inversion, which is a form of reflection symmetry. As a result, this compound is superimposable on its mirror image and is therefore optically inactive.

Chapter 6
Chemical Reactivity and Mechanisms

Review of Concepts

Fill in the blanks below. To verify that your answers are correct, look in your textbook at the end of Chapter 6. Each of the sentences below appears verbatim in the section entitled *Review of Concepts and Vocabulary*.

- _______________ reactions involve a transfer of energy from the system to the surroundings, while ______________ reactions involve a transfer of energy from the surroundings to the system.
- Each type of bond has a unique ______________________ **energy**, which is the amount of energy necessary to accomplish **homolytic bond cleavage**.
- **Entropy** is loosely defined as the ___________ of a system.
- In order for a process to be spontaneous, the change in ____________________ must be negative.
- The study of relative energy levels and equilibrium concentrations is called ___________________.
- ___________ is the study of reaction rates.
- _____________ speed up the rate of a reaction by providing an alternate pathway with a lower **energy of activation**.
- On an energy diagram, each peak represents a _________________, while each valley represents _______________________.
- A ________________ has an electron-rich atom that is capable of donating a pair of electrons.
- An _________________ has an electron-deficient atom that is capable of accepting a pair of electrons.
- For **ionic reactions**, there are four characteristic arrow-pushing patterns: 1) __________________, 2) __________________, 3) __________________, and 4) __________________.
- As a result of **hyperconjugation**, ___________ carbocations are more stable than secondary carbocations, which are more stable than ___________ carbocations.

Review of Skills

Fill in the empty boxes below. To verify that your answers are correct, look in your textbook at the end of Chapter 6. The answers appear in the section entitled *SkillBuilder Review.*

SkillBuilder 6.1 Predicting $\Delta H°$ of a Reaction

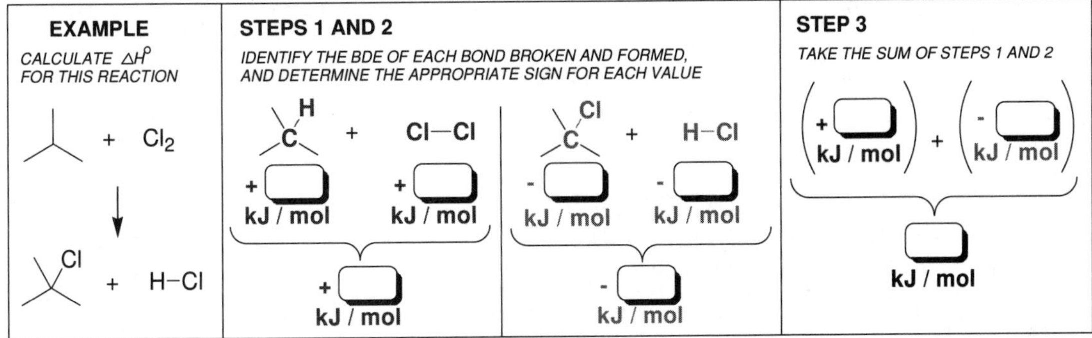

SkillBuilder 6.2 Identifying Nucleophilic and Electrophilic Centers

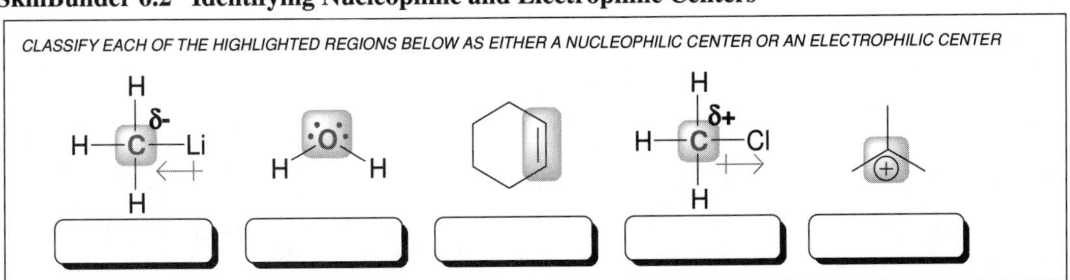

SkillBuilder 6.3 Identifying an Arrow Pushing Pattern

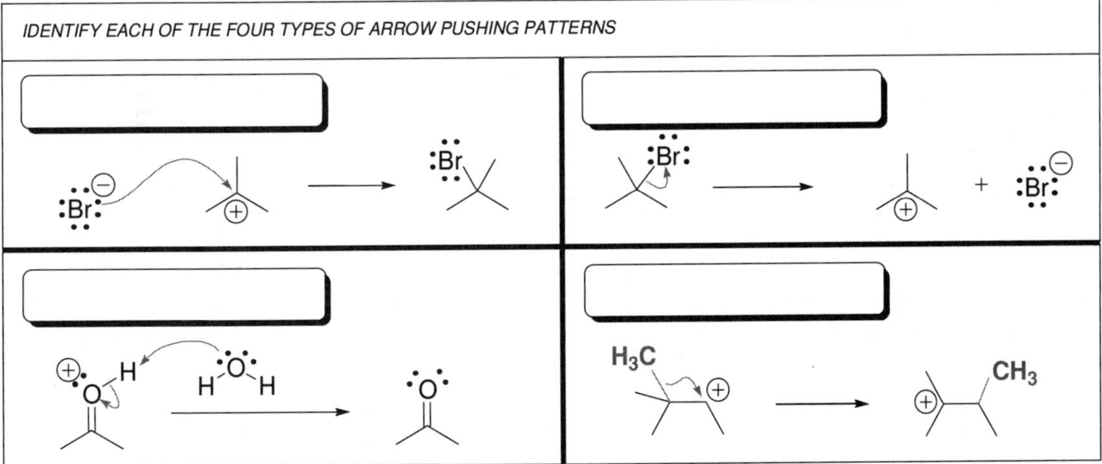

SkillBuilder 6.4 Identifying a Sequence of Arrow Pushing Patterns

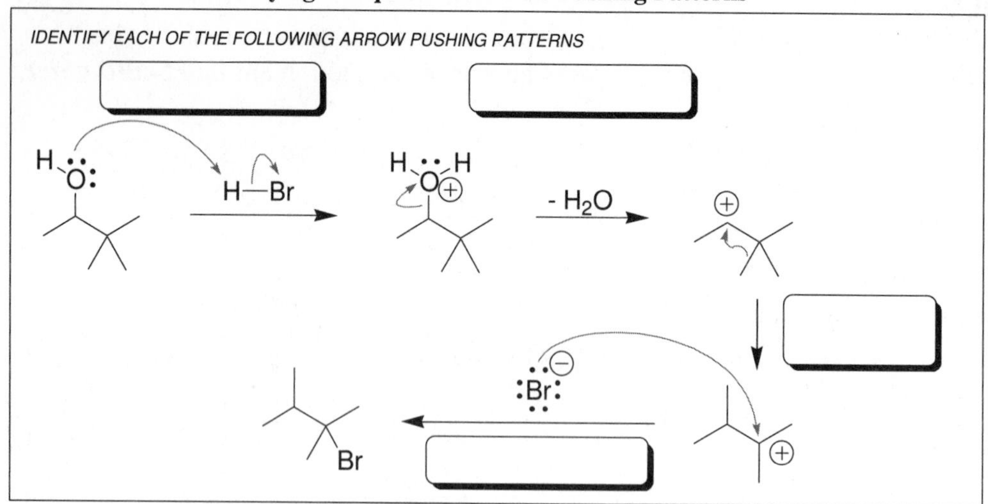

SkillBuilder 6.5 Drawing Curved Arrows

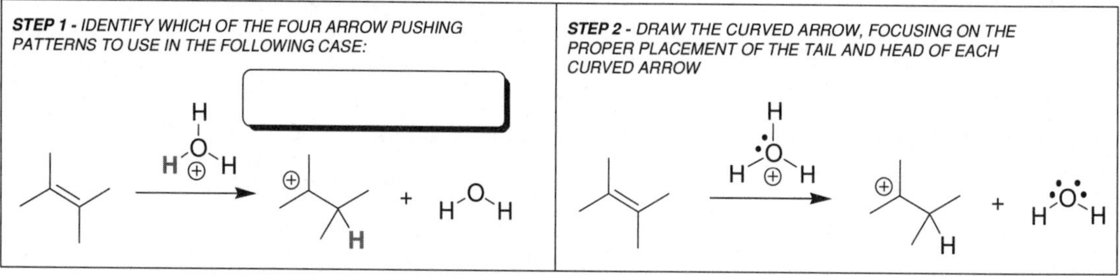

SkillBuilder 6.6 Predicting Carbocation Rearrangements

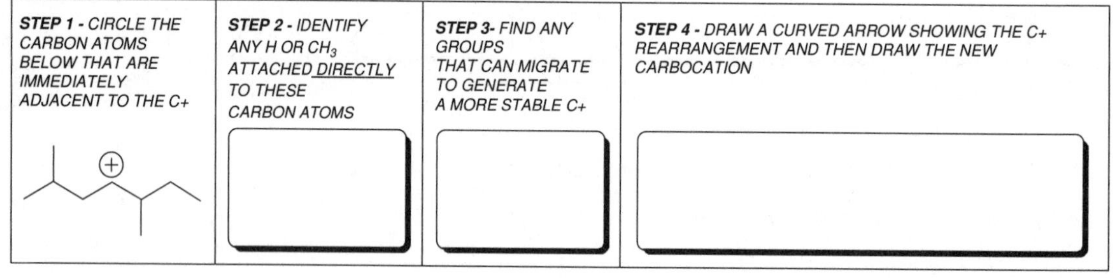

Solutions

6.1.

a)

Bonds Broken	**kJ/mol**	**Bonds Formed**	**kJ/mol**
H—CH(CH_3)$_2$	**+ 397**	(CH_3)$_2$CH—Br	**– 285**
Br—Br	**+ 192**	H—Br	**– 368**

Sum = – 64 kJ/mol.
ΔH° for this reaction is negative, which means that the system is losing energy. It is giving off energy to the environment, so the reaction is exothermic.

b)

Bonds Broken	**kJ/mol**	**Bonds Formed**	**kJ/mol**
(CH_3)$_3$C—Cl	**+ 331**	(CH_3)$_3$C—OH	**– 381**
H—OH	**+ 498**	H—Cl	**– 431**

Sum = + 17 kJ/mol.
ΔH° for this reaction is positive, which means that the system is gaining energy. It is receiving energy from the environment, so the reaction is endothermic.

c)

Bonds Broken	**kJ/mol**	**Bonds Formed**	**kJ/mol**
(CH_3)$_3$C—Br	**+ 272**	(CH_3)$_3$C—OH	**– 381**
H—OH	**+ 498**	H—Br	**– 368**

Sum = + 21 kJ/mol.
ΔH° for this reaction is positive, which means that the system is gaining energy. It is receiving energy from the environment, so the reaction is endothermic.

d)

Bonds Broken	**kJ/mol**	**Bonds Formed**	**kJ/mol**
(CH_3)$_3$C—I	**+ 209**	(CH_3)$_3$C—OH	**– 381**
H—OH	**+ 498**	H—I	**– 297**

Sum = + 29 kJ/mol.
ΔH° for this reaction is positive, which means that the system is gaining energy. It is receiving energy from the environment, so the reaction is endothermic.

6.2.

The C-C bond of CH_3—CH_3 has a bond dissociation energy of = +368 kJ/mol. If a C=C bond has a total bond dissociation energy of +632 kJ/mol, then the pi component of the double bond can be estimated to be (632 kJ/mol) – (368 kJ/mol) = 264 kJ/mol. In other words, the pi component of the C=C bond is not as strong as the sigma component of the C=C bond. In the reaction shown in this problem, the pi component of the C=C bond is broken but the sigma component remains intact. Accordingly, the calculation is as follows:

Bonds Broken	kJ/mol	Bonds Formed	kJ/mol
C=C (just the pi component)	**+ 264**	CH_3CH_2—OH	**– 381**
H—OH	**+ 498**	H—CH_2R	~ **– 410**

Sum = - 29 kJ/mol.
ΔH° for this reaction is predicted to be negative, which means that the system is losing energy. It is giving off energy to the environment, so the reaction is exothermic.

6.3.

a) ΔS_{sys} is expected to be negative (a decrease in entropy) because two molecules are converted into one molecule.
b) ΔS_{sys} is expected to be negative (a decrease in entropy) because an acylic compound is converted into a cyclic compound.
c) ΔS_{sys} is expected to be positive (an increase in entropy) because one molecule is converted into two molecules.
d) ΔS_{sys} is expected to be positive (an increase in entropy) because one molecule is converted into two ions.
e) ΔS_{sys} is expected to be negative (a decrease in entropy) because two chemical entities are converted into one.
f) ΔS_{sys} is expected to be positive (an increase in entropy) because a cyclic compound is converted into an acyclic compound.

6.4.

a) There is a competition between the two terms contributing to ΔG. In this case, the reaction is endothermic, which contributes to a positive value for ΔG, but the second term contributes to a negative value for ΔG:

$$\Delta G = \underset{\oplus}{\Delta H} + \underset{\ominus}{(-T\Delta S)}$$

The sign of ΔG will therefore depend on the competition between these two terms, which is affected by temperature. A high temperature will cause the second term to dominate, giving rise to a positive value of ΔG. A low temperature will render the second term insignificant, and the first term will dominate, giving rise to a negative value of ΔG.

b) In this case, both terms contribute to a negative value for ΔG, so ΔG will definitely be negative (the process will be spontaneous).

c) In this case, both terms contribute to a positive value for ΔG, so ΔG will definitely be positive (the process will not be spontaneous).

d) There is a competition between the two terms contributing to ΔG. In this case, the reaction is exothermic, which contributes to a negative value for ΔG, but the second term contributes to a positive value for ΔG:

$$\Delta G = \underset{\ominus}{\Delta H} + \underset{\oplus}{(-T\Delta S)}$$

The sign of ΔG will therefore depend on the competition between these two terms, which is affected by temperature. A high temperature will cause the second term to dominate, giving rise to a negative value of ΔG. A low temperature will render the second term insignificant, and the first term will dominate, giving rise to a positive value of ΔG.

6.5. A system can only achieve a lower energy state by transferring energy to its surroundings (conservation of energy). This increases the entropy of the surroundings, which more than offsets the decrease in entropy of the system. As a result, ΔS_{tot} increases.

6.6.

a) A positive value of ΔG favors reactants.
b) A reaction for which $K_{eq} < 1$ will favor reactants.
c) $\Delta G = \Delta H - T\Delta S = (33 \text{ kJ/mol}) - (298 \text{ K})(0.150 \text{ kJ/mol} \bullet \text{K}) = -11.7 \text{ kJ/mol}$
A negative value of ΔG favors products.
d) Both terms contribute to a negative value of ΔG, which favors products.
e) Both terms contribute to a positive value of ΔG, which favors reactants.

6.7.

a) Process D will occur more rapidly because it has a lower energy of activation than process A.
b) Process A will more greatly favor products at equilibrium than process B, because the former is exergonic (the products are lower in energy than the reactants) while the latter is not exergonic.
c) None of these processes exhibits an intermediate, because none of the energy diagrams has a local minimum (a valley). But all of the processes proceed via a transition state, because all of the energy diagrams have a local maximum (a peak).

d) In process A, the transition state resembles the reactants more than products because the transition state is closer in energy to the reactant than the products (the Hammond postulate).

e) Process A will occur more rapidly because it has a lower energy of activation than process B.

f) Process D will more greatly favor products at equilibrium than process B, because the former is exergonic (the products are lower in energy than the reactants) while the latter is not exergonic.

g) In process C, the transition state resembles the products more than reactants because the transition state is closer in energy to the products than the reactants (the Hammond postulate).

6.8.

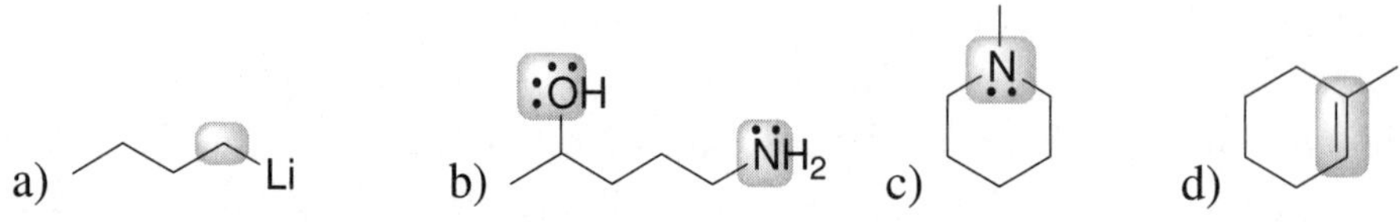

6.9.

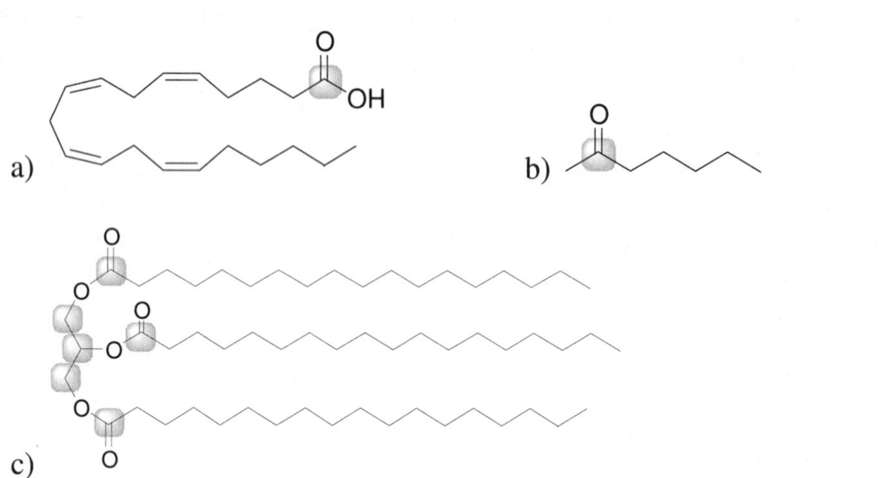

6.10.

6.11.

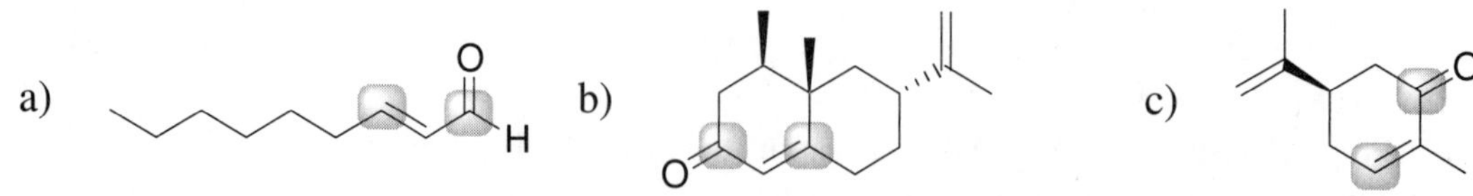

6.12.

a) loss of a leaving group
b) proton transfer
c) rearrangement

d) nucleophilic attack
e) proton transfer
f) nucleophilic attack
g) rearrangement
h) loss of a leaving group
i) nucleophilic attack

6.13. The pi bond functions as a nucleophile and attacks the electrophilic carbocation. This step is therefore a nucleophilic attack.

6.14.

a) proton transfer; nucleophilic attack; proton transfer
b) nucleophilic attack; proton transfer; proton transfer
c) proton transfer; nucleophilic attack; loss of a leaving group
d) proton transfer; loss of a leaving group; nucleophilic attack; proton transfer
e) proton transfer; nucleophilic attack; proton transfer

6.15. Both reactions have the same sequence: 1) nucleophilic attack, followed by 2) loss of a leaving group. In both cases, a hydroxide ion functions as a nucleophile and attacks a compound that can accept the negative charge and store it temporarily. The charge is then expelled as a chloride ion in both cases.

6.16.

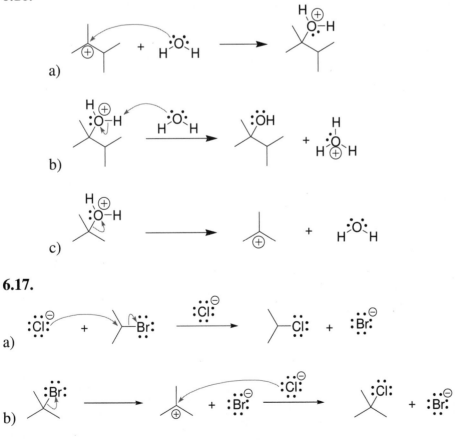

a) ⊕ + H_2O → $\overset{\oplus}{O}H_2$

b) $\overset{\oplus}{O}H_2$ + H_2O → OH + H_3O^+

c) $\overset{\oplus}{O}H_2$ → ⊕ + H_2O

6.17.

a) Cl^- + Br —(Cl^-)→ Cl + Br^-

b) Br → ⊕ + Br^- —(Cl^-)→ Cl + Br^-

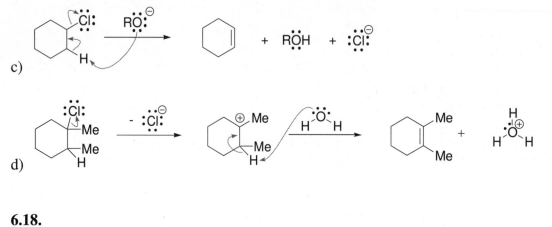

c)

d)

6.18.

a)

b) This carbocation is tertiary and will not rearrange

c) *tertiary* *tertiary allylic*

d) This carbocation is secondary, but it cannot rearrange to form a tertiary carbocation.

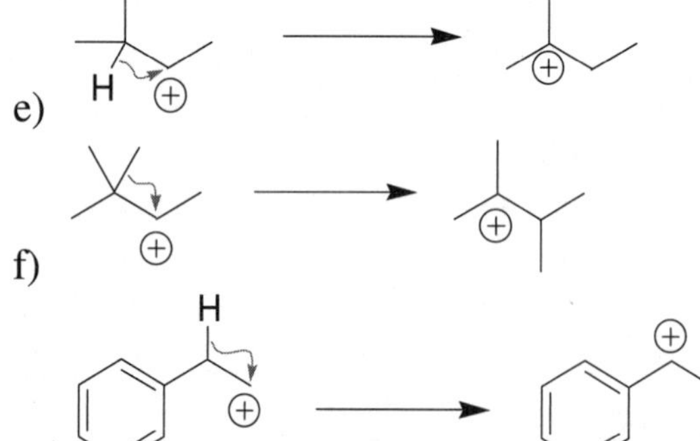

e)

f)

g)

h) This carbocation is tertiary and it is resonance stabilized (we will see in Chapter 7 that this carbocation is called a benzylic carbocation). It will not rearrange.

6.19.

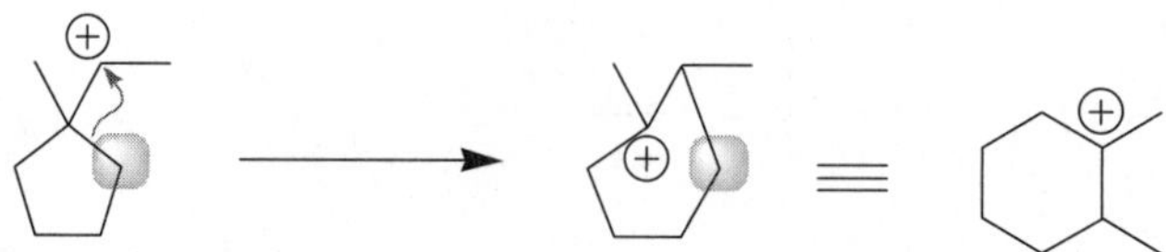

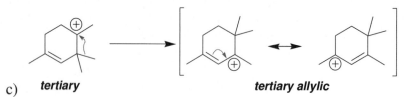

6.20.

a) a carbon-carbon triple bond is comprised of one sigma bond and two pi bonds, and is therefore stronger than a carbon-carbon double bond (one sigma and one pi bond) or a carbon-carbon single bond (only one sigma bond).

b) Using the data in Table 6.1, the C-F will have the largest bond dissociation energy.

6.21.

a)

Bonds Broken	kJ/mol	Bonds Formed	kJ/mol
RCH_2—Br	**+ 285**	RCH_2—OR	**- 381**
RCH_2O—H	**+ 435**	H—Br	**- 368**

Sum = – 29 kJ/mol.
ΔH° for this reaction is negative, which means that the system is losing energy. It is giving off energy to the environment, so the reaction is exothermic.

b) ΔS of this reaction is positive because one mole of reactant is converted into two moles of product.
c) Both terms (ΔH) and (-TΔS) contribute to a negative value of ΔG.
d) No.
e) Yes.

6.22.

a) A reaction for which $K_{eq} > 1$ will favor products.
b) A reaction for which $K_{eq} < 1$ will favor reactants.
c) A positive value of ΔG favors reactants.
d) Both terms contribute to a negative value of ΔG, which favors products.
e) Both terms contribute to a positive value of ΔG, which favors reactants.

6.23. $K_{eq} = 1$ when $\Delta G = 0$ kJ/mol (See Table 6.2).

6.24. $K_{eq} < 1$ when ΔG has a positive value. The answer is therefore "a" (+1 kJ/mol)

6.25.

a) ΔS_{sys} is expected to be negative (a decrease in entropy) because two moles of reactant are converted into one mole of product.
b) ΔS_{sys} is expected to be positive (an increase in entropy) because one mole of reactant is converted into two moles of product.

c) ΔS_{sys} is expected to be approximately zero, because two moles of reactant are converted into two moles of product.
d) ΔS_{sys} is expected to be negative (a decrease in entropy) because an acylic compound is converted into a cyclic compound.
e) ΔS_{sys} is expected to be approximately zero, because one mole of reactant is converted into one mole of product, and both the reactant and the product are acyclic.

6.26.

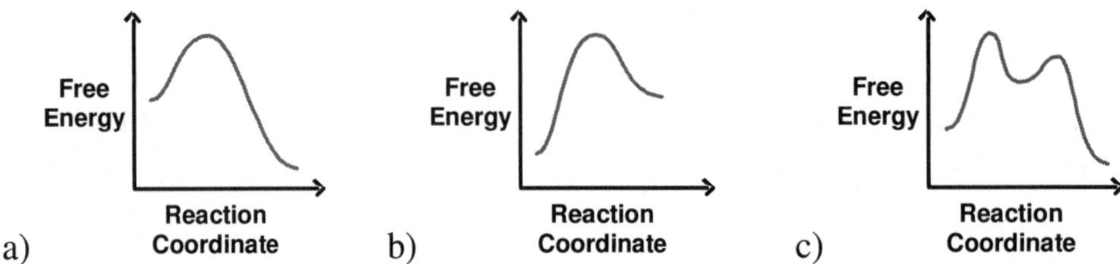

6.27.
a) B and D
b) A and C
c) C
d) A
e) D
f) D
g) A and B
h) C

6.28. All local minima (valleys) represent intermediates, while all local maxima (peaks) represent transition states:

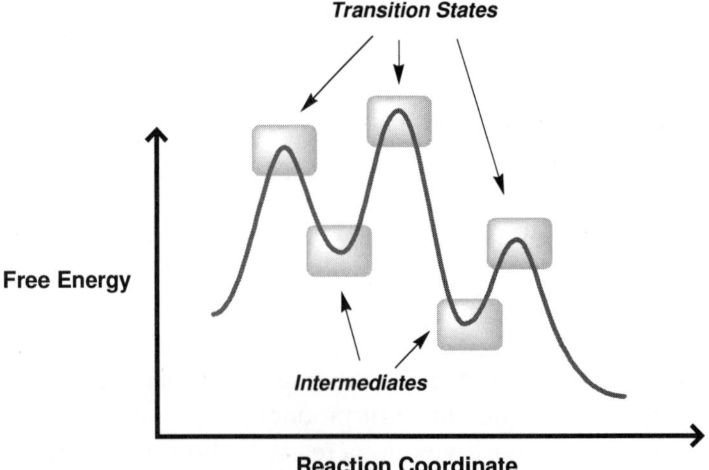

6.29.

a) Rate = k[nucleophile][substrate].

b) The rate will be tripled, because the rate is linearly dependent on the concentration of the nucleophile.

c) The rate will be tripled, because the rate is linearly dependent on the concentration of the substrate.

d) As a rule of thumb, the rate doubles for every increase of 10° C. Therefore an increase of 40° C will correspond to increase in rate of approximately 16-fold (2 x 2 x 2 x 2)

6.30.

a) loss of a leaving group

b) carbocation rearrangement

c) nucleophilic attack

d) proton transfer

6.31.

a)

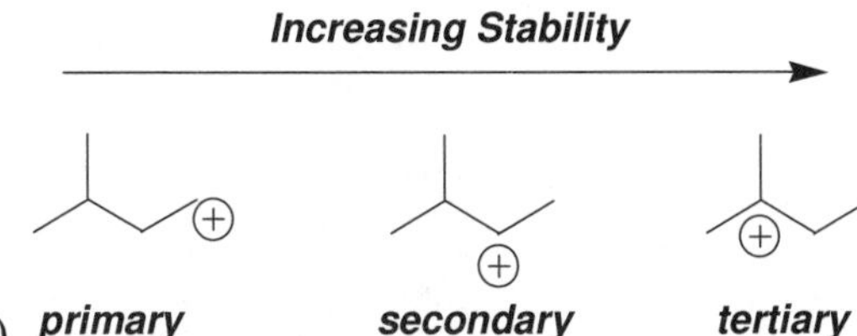

b)

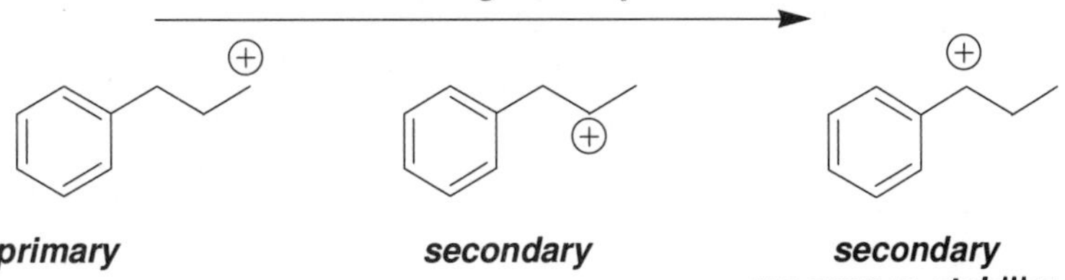

6.32.

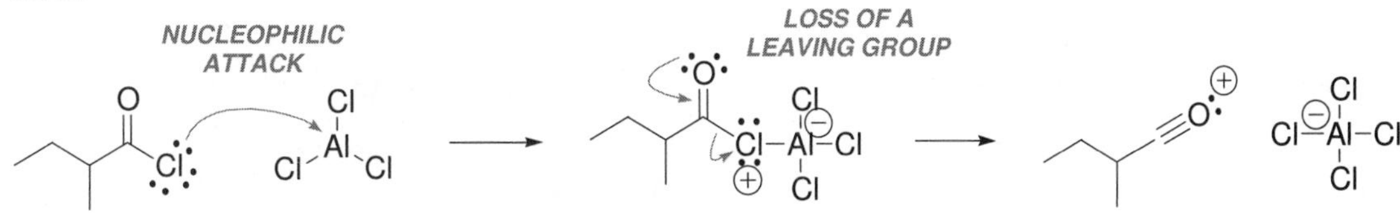

6.33.

NUCLEOPHILIC ATTACK

PROTON TRANSFER

6.34.

PROTON TRANSFER

PROTON TRANSFER

+ SO_3

6.35.

NUCLEOPHILIC ATTACK

LOSS OF A LEAVING GROUP

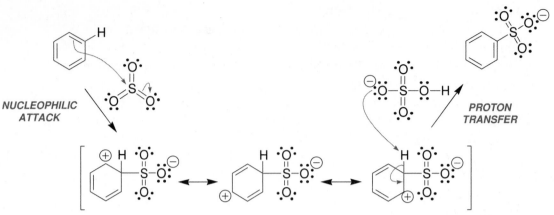

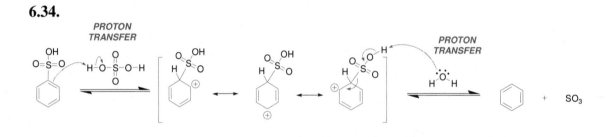

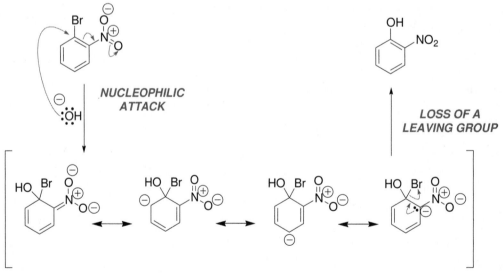

6.36.

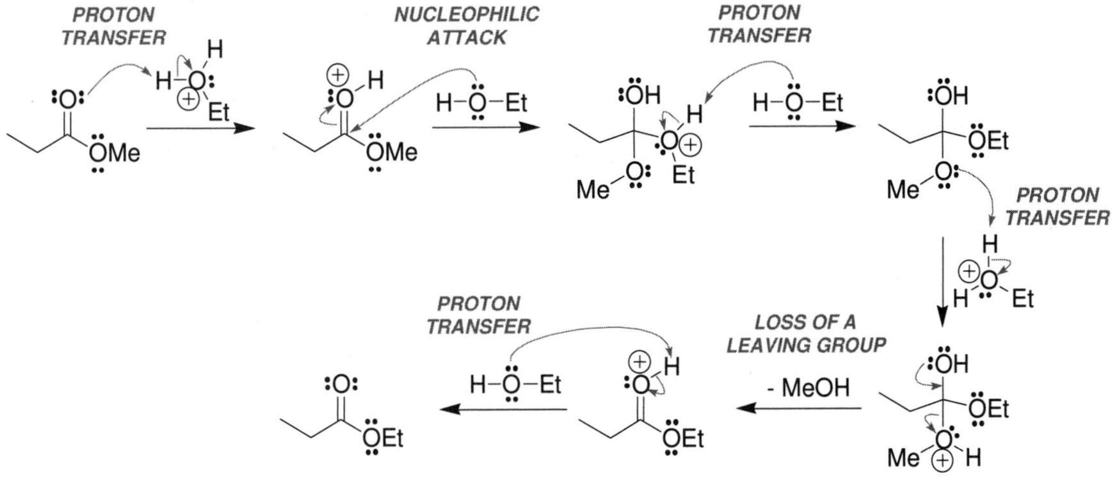

PROTON TRANSFER

NUCLEOPHILIC ATTACK

PROTON TRANSFER

PROTON TRANSFER

LOSS OF A LEAVING GROUP

- MeOH

PROTON TRANSFER

6.37.

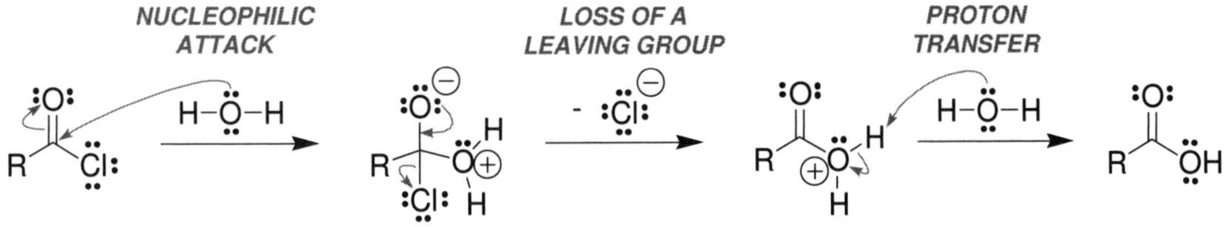

NUCLEOPHILIC ATTACK

LOSS OF A LEAVING GROUP

PROTON TRANSFER

6.38.

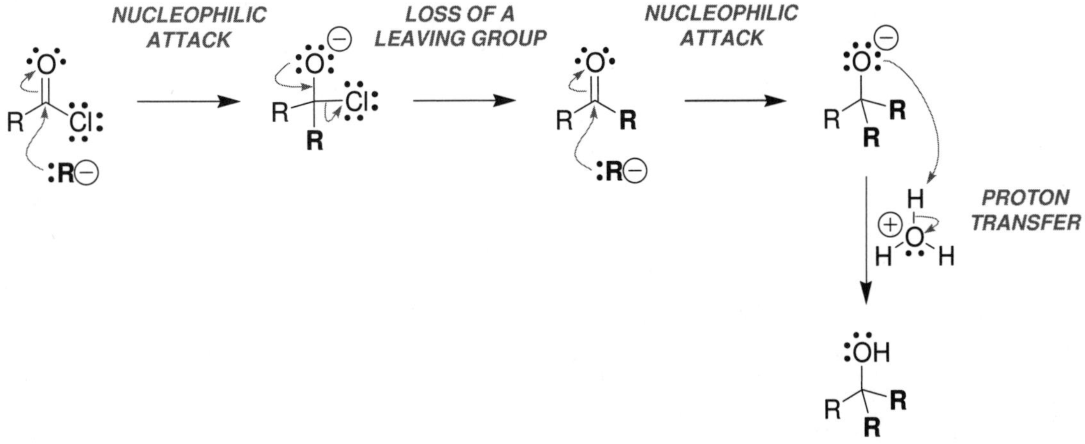

NUCLEOPHILIC ATTACK

LOSS OF A LEAVING GROUP

NUCLEOPHILIC ATTACK

PROTON TRANSFER

6.39.

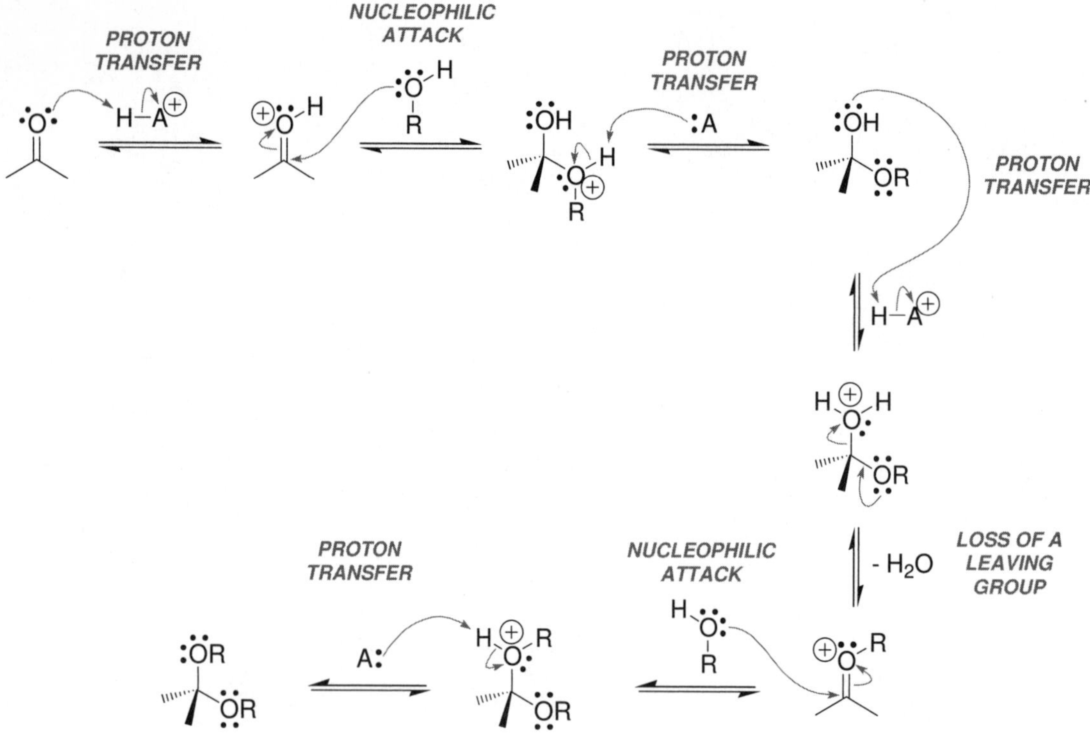

PROTON TRANSFER
NUCLEOPHILIC ATTACK
PROTON TRANSFER
PROTON TRANSFER
LOSS OF A LEAVING GROUP
- H_2O
NUCLEOPHILIC ATTACK
PROTON TRANSFER

6.40.

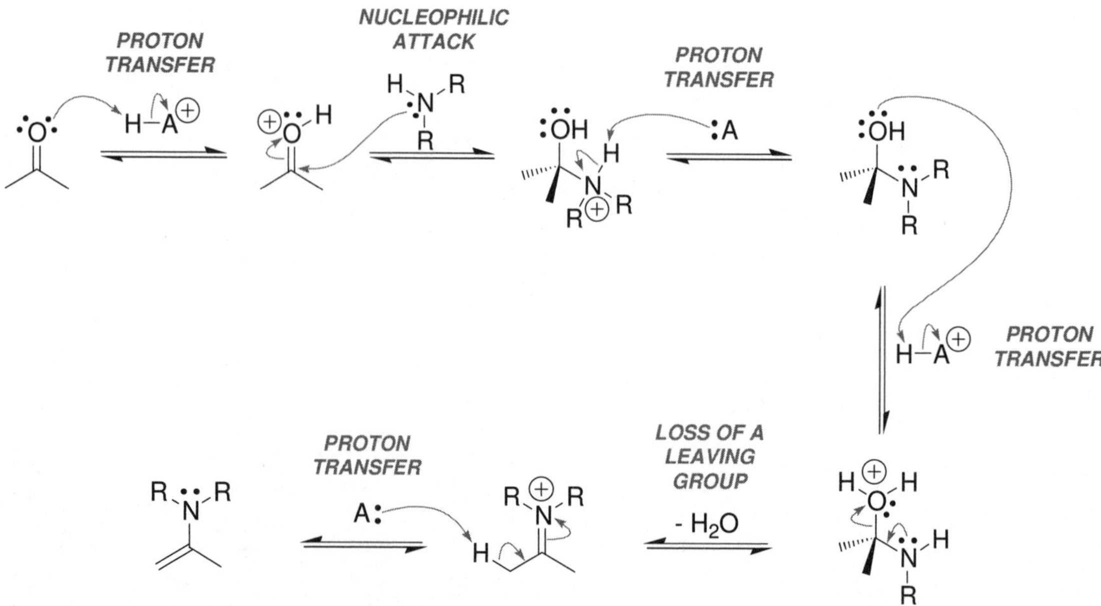

PROTON TRANSFER
NUCLEOPHILIC ATTACK
PROTON TRANSFER
PROTON TRANSFER
LOSS OF A LEAVING GROUP
- H_2O
PROTON TRANSFER

6.41.

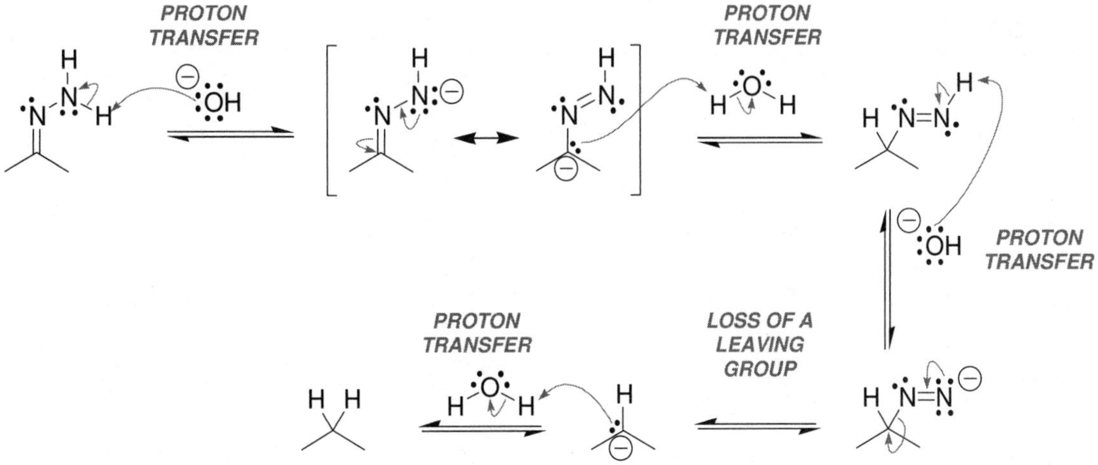

PROTON TRANSFER

PROTON TRANSFER

PROTON TRANSFER

LOSS OF A LEAVING GROUP

PROTON TRANSFER

6.42.

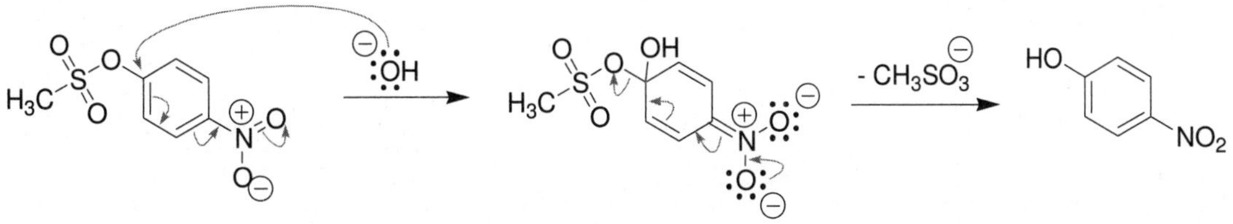

$-CH_3SO_3^{-}$

HO

NO_2

6.43.

H–OSO_3H

+

$\ominus$ OSO_3H

H–OSO_3H

- H_2O

$\ominus$ OSO_3H

H–OSO_3H

$\ominus$ OSO_3H

H–OSO_3H

- H_2O

$\ominus$ OSO_3H

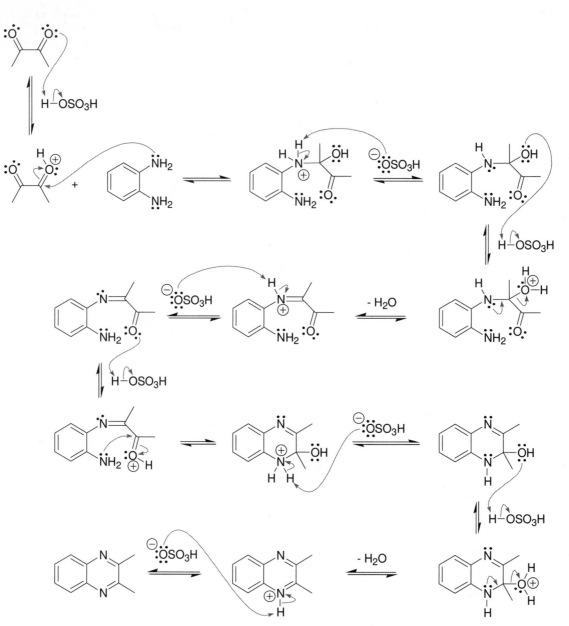

6.44.

H H O H H H O H N N H N H H O H ⊕ O H H N H OH H O H H

H N H H O H O H H N H O H H ⊕ N H H OH

6.45.

O O H H H O H H H O ⊕ O H H O H H H O H ⊕ O H O H H O H H

H HO ⊕ O OH H H HO OH H ⊕ O H H H O H H HO OH O H H

H O H

HO O OH

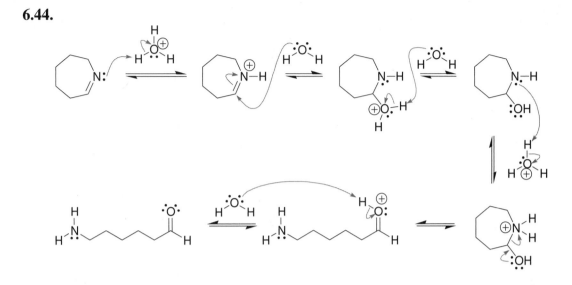

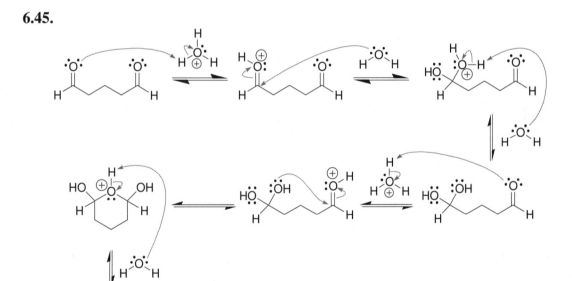

6.46.

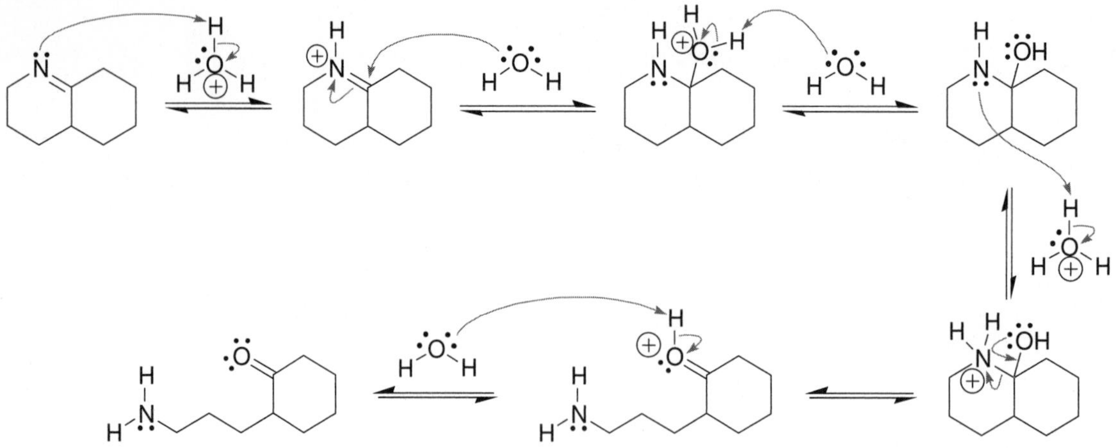

6.47.

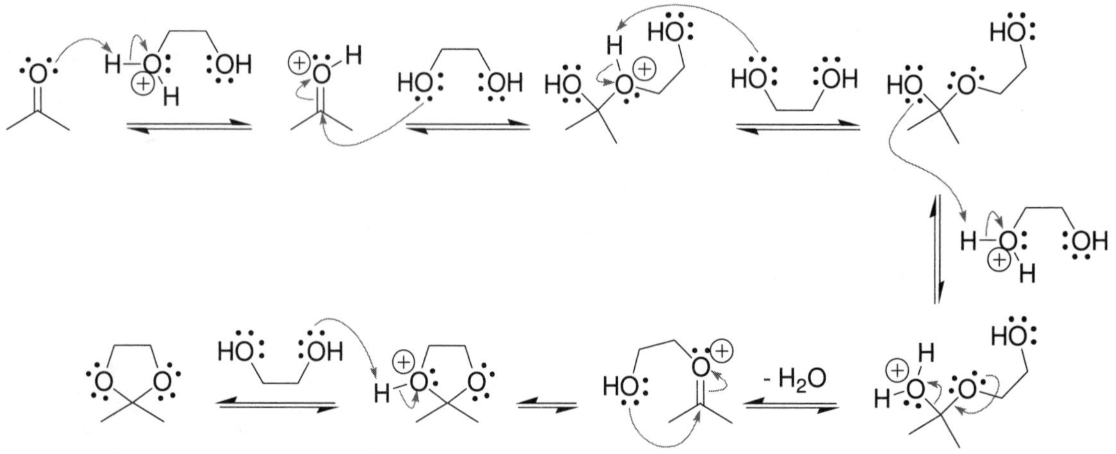

6.48.

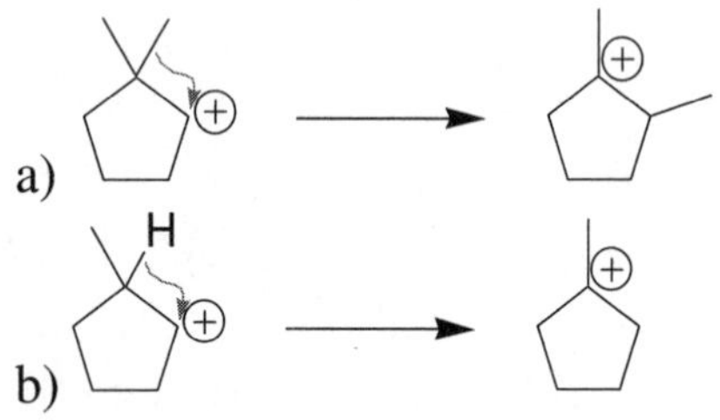

a)

b)

c) This carbocation is secondary, but it cannot rearrange to generate a tertiary carbocation.

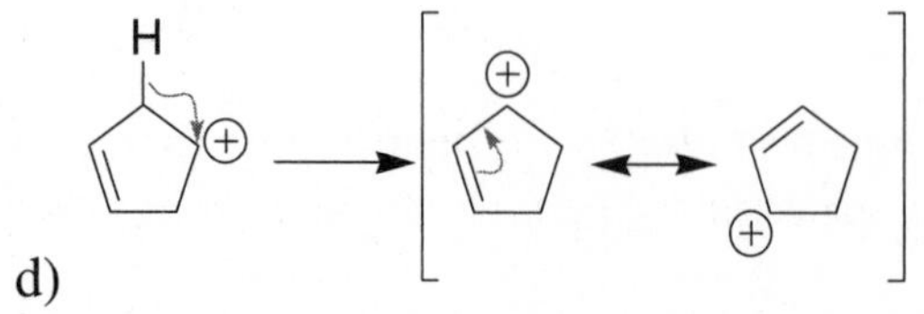

d)

e)

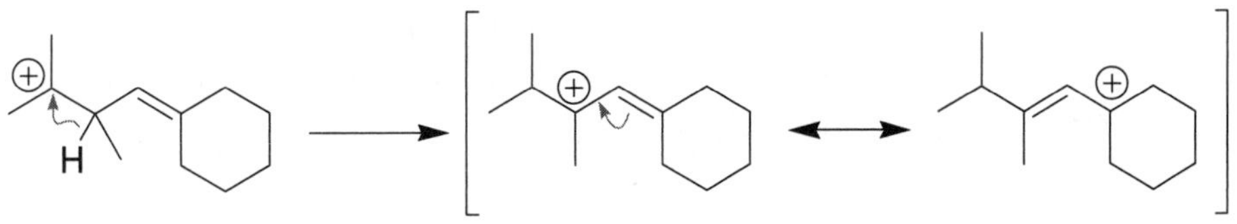

f)  H

g) This carbocation is tertiary and will not rearrange.

6.49.

a)

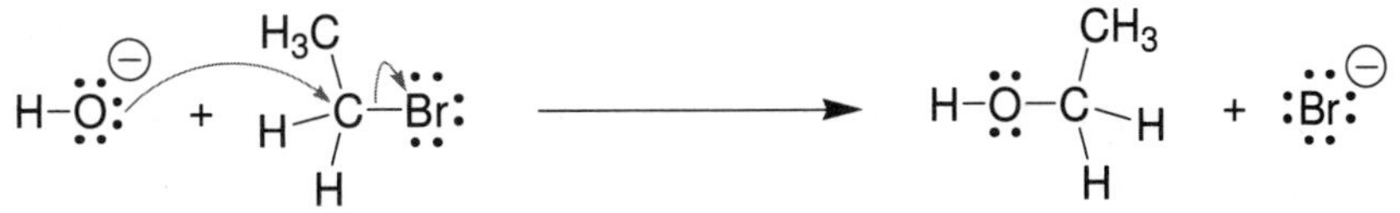

b) Nucleophilic attack and loss of a leaving group.

c) A CH_3CH_2—Br is broken, and a CH_3CH_2—OH is formed. Using the data in Table 6.1, ΔH for this reaction is expected to be approximately (285 kJ/mol) – (381 kJ/mol). The sign of ΔH is therefore predicted to be negative, which means that the reaction should be exothermic.

d) Two chemical entities are converted into two chemical entities. Both the reactants and products are acyclic. Therefore, ΔS for this process is expected to be approximately zero.

e) ΔG has two components: (ΔH) and (-TΔS). Based on the answers to the previous questions, the first term has a negative value and the second term is insignificant. Therefore, ΔG is expected to have a negative value. This is confirmed by the energy diagram, which shows the products having lower free energy than the reactants.

f) The position of equilibrium is dependent on the sign and value of ΔG. As mentioned in part e, ΔG is comprised of two terms. The effect of temperature appears in the second term (-TΔS), which is insignificant because ΔS is approximately zero. Therefore, an increase (or decrease) in temperature is not expected to have a significant impact on the position of equilibrium.

g) This transition state corresponds with the peak of the curve, and has the following structure:

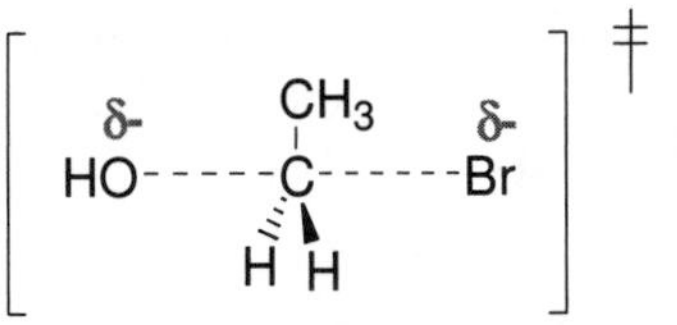

h) The transition state in this case is closer in energy to the reactants than the products, and therefore, it is closer in structure to the reactants than the products (the Hammond postulate).

i) The reaction is second order.

j) According to the rate equation, the rate is linearly dependent on the concentration of hydroxide. Therefore, the rate will be doubled if the concentration of hydroxide is doubled.
k) Yes, the rate will increase with increasing temperature.

6.50.

a) K_{eq} does not affect the rate of the reaction. It only affects the equilibrium concentrations.
b) ΔG does not affect the rate of the reaction. It only affects the equilibrium concentrations.
c) Temperature does affect the rate of the reaction, by increasing the number of collisions that result in a reaction.
d) ΔH does not affect the rate of the reaction. It only affects the equilibrium concentrations.
e) E_a greatly affects the rate of the reaction. Lowering the E_a will increase the rate of reaction.
f) ΔS does not affect the rate of the reaction. It only affects the equilibrium concentrations.

6.51. In order to determine if reactants or products are favored at high temperature, we must consider the effect of temperature on the sign of ΔG. Recall that ΔG has two components: (ΔH) and (-TΔS). The reaction is exothermic, so the first term (ΔH) has a negative value, which contributes to a negative value of ΔG. This favors products. At low temperature, the second term will be insignificant and the first term will dominate. Therefore, the process will be thermodynamically favorable, and the reaction will favor the formation of products. However, at high temperature, the second term becomes more significant. In this case, two moles of reactants are converted into one mole of product. Therefore, ΔS for this process is negative, which means that (-TΔS) is positive. At high enough temperature, the second term (-TΔS) should dominate over the first term (ΔH), generating a positive value for ΔG. Therefore, the reaction will favor reactants at high temperature.

6.52. Recall that ΔG has two components: (ΔH) and (-TΔS). We must analyze each term separately. The first term is expected to have a negative value, because three pi bonds are being converted into one pi bond and two sigma bonds. A sigma bond is stronger (lower in energy) than the pi component of a double bond (see problems 6.2 and 6.20). Therefore, reaction is expected to release energy to the environment, which means the reaction should be exothermic. In other words, the first term (ΔH) has a negative value, which contributes to a negative value of ΔG. This favors products. Now let's consider the second term (-TΔS) contributing to ΔG. In this case, two moles of reactants are converted into one mole of product. Therefore, ΔS for this process is negative, which means that (-TΔS) is positive. At low temperature, the second term will be insignificant and the first term will dominate. Therefore, the process will be thermodynamically favorable, and the reaction will favor the formation of products. However, at high temperature, the second term becomes more significant. At high enough temperature, the second term ($-T\Delta S$) should dominate over the first term (ΔH), generating a positive value for ΔG. Therefore, the reaction will favor reactants at high temperature.

6.53. The nitrogen atom of an ammonium ion is positively charged, but that does not render it electrophilic. In order to be electrophilic, it must have an empty orbital that can be attacked by a nucleophile. The nitrogen atom in this case does not have an empty orbital, because nitrogen is a second row element and therefore only has four orbitals with which to form bonds. All four orbitals are being used for bonding, leaving none of the orbitals vacant. As a result, the nitrogen atom is not electrophilic, despite the fact that is positively charged.
In contrast, an iminium ion is resonance stabilized:

An iminium ion

The second resonance structure exhibits a positive charge on a carbon atom, which serves as an electrophilic center (a carbocation is an empty p orbital). Therefore, an iminium ion is an electrophile and is subject to attack by a nucleophile:

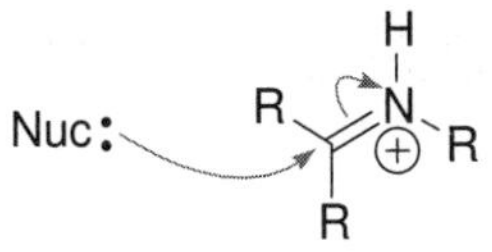

Nuc:

6.54.

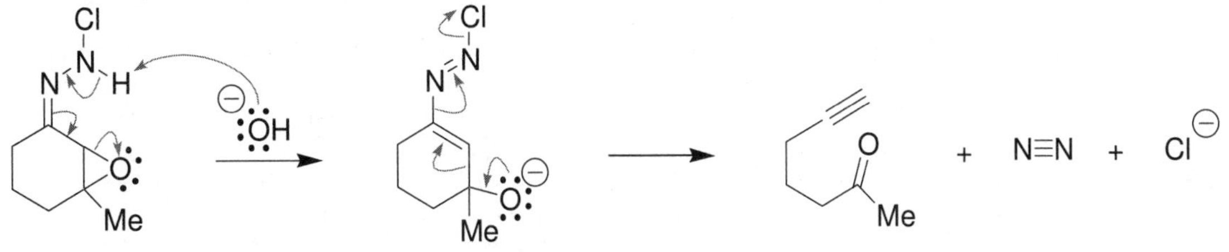

$N \equiv N$ + $Cl^{\ominus}$

Chapter 7
Substitution Reactions

Review of Concepts

Fill in the blanks below. To verify that your answers are correct, look in your textbook at the end of Chapter 7. Each of the sentences below appears verbatim in the section entitled *Review of Concepts and Vocabulary.*

- Substitution reactions exchange one ____________ for another.
- Evidence for the concerted mechanism, called **S_N2**, includes the observation of a __________**-order** rate equation. The reaction proceeds with ____________ **of configuration**.
- S_N2 reactions are said to be _________________ because the configuration of the product is determined by the configuration of the substrate.
- Evidence for the stepwise mechanism, called **S_N1**, includes the observation of a __________**-order** rate equation.
- The __________ step of an S_N1 process is the **rate-determining step.**
- An **S_N1 reaction** is a stepwise process with a **first-order** rate equation.
- There are four factors that impact the competition between the S_N2 mechanism and S_N1: 1) the _____________, 2) the ________________, 3) the ___________ _______________, and 4) the _______________.
- _______________________ solvents favor S_N2.

Review of Skills

Follow the instructions below. To verify that your answers are correct, look in your textbook at the end of Chapter 7. The answers appear in the section entitled *SkillBuilder Review.*

SkillBuilder 7.1 Drawing the Curved Arrows of a Substitution Reaction

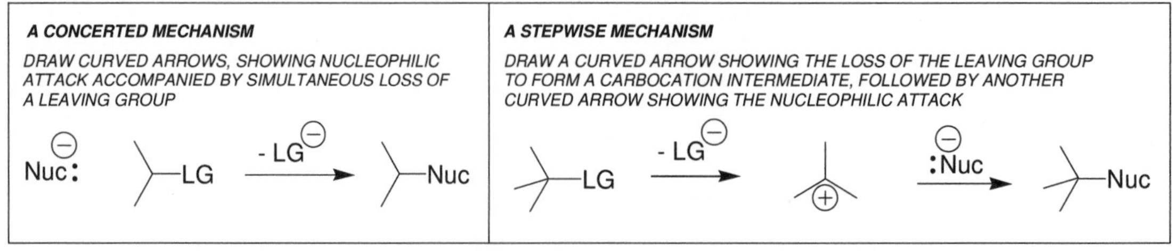

A CONCERTED MECHANISM
DRAW CURVED ARROWS, SHOWING NUCLEOPHILIC ATTACK ACCOMPANIED BY SIMULTANEOUS LOSS OF A LEAVING GROUP
Nuc:
LG
- LG
Nuc
A STEPWISE MECHANISM
DRAW A CURVED ARROW SHOWING THE LOSS OF THE LEAVING GROUP TO FORM A CARBOCATION INTERMEDIATE, FOLLOWED BY ANOTHER CURVED ARROW SHOWING THE NUCLEOPHILIC ATTACK
LG
- LG
:Nuc
Nuc

SkillBuilder 7.2 Drawing the Product of an S_N2 Process

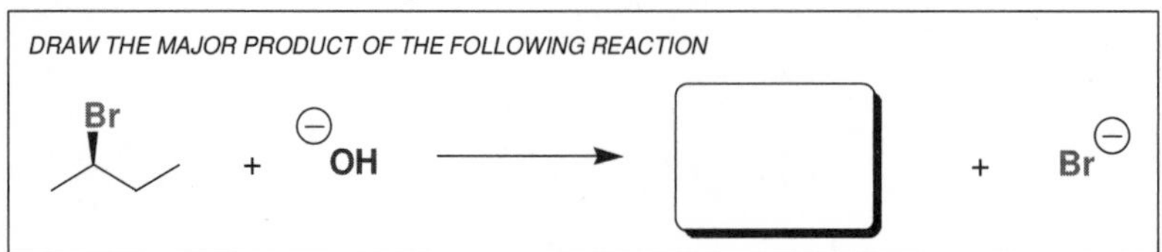

SkillBuilder 7.3 Drawing the Transition State of an S_N2 Process

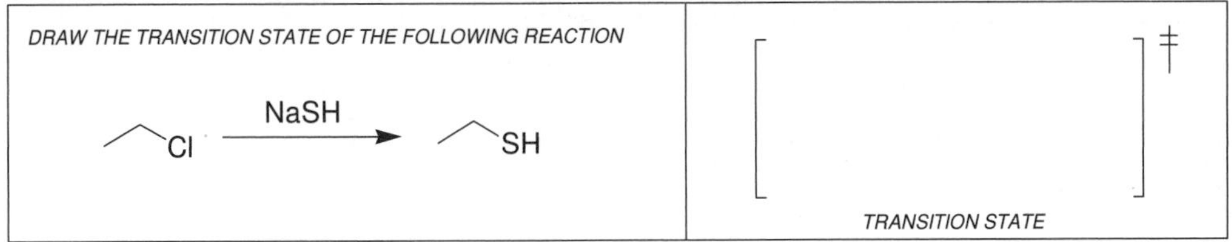

SkillBuilder 7.4 Drawing the Carbocation Intermediate of an S_N1 Process

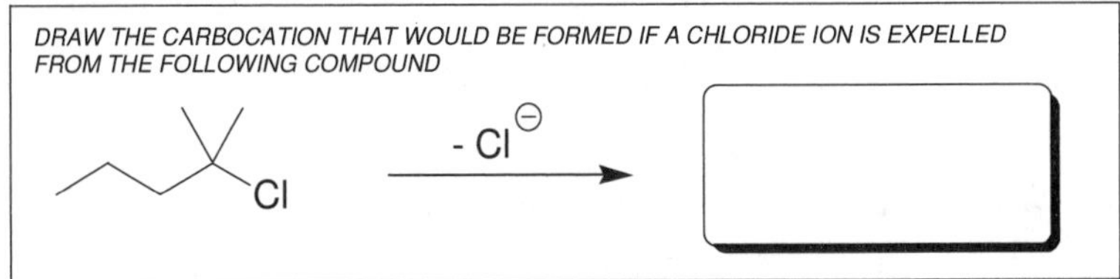

SkillBuilder 7.5 Drawing the Products of an S_N1 Process

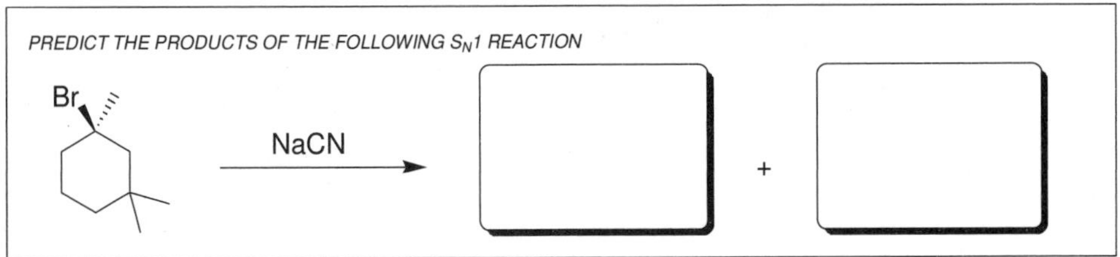

SkillBuilder 7.6 Drawing the Complete Mechanism of an S_N1 Process

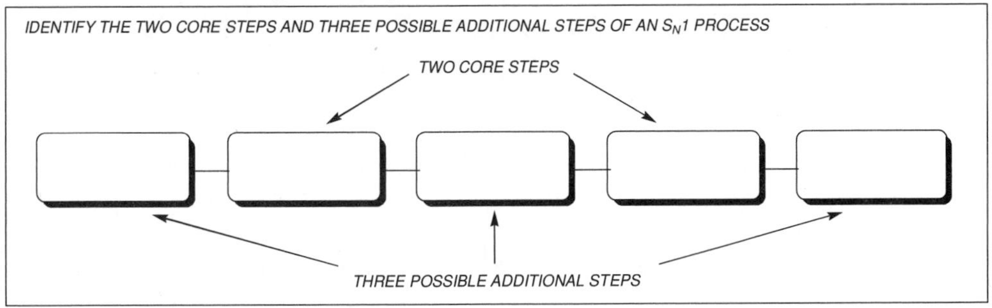

SkillBuilder 7.7 Drawing the Complete Mechanism of an S_N2 Process

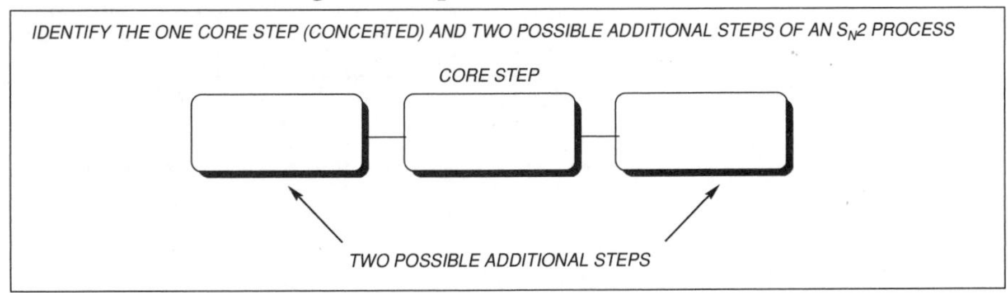

SkillBuilder 7.8 Determining whether a Reaction Proceeds via an S_N1 Mechanism or an S_N2 Mechanism

FILL IN THE TABLE BELOW, SHOWING THE FEATURES THAT FAVOR S_N2 OR S_N1 REACTIONS

	S_N2	S_N1
SUBSTRATE		
NUC		
LG		
SOLVENT		

SkillBuilder 7.9 Identifying the Reagents Necessary for a Substitution Reaction

IDENTIFY THE REAGENTS NECESSARY TO ACHIEVE THE FOLLOWING TRANSFORMATION

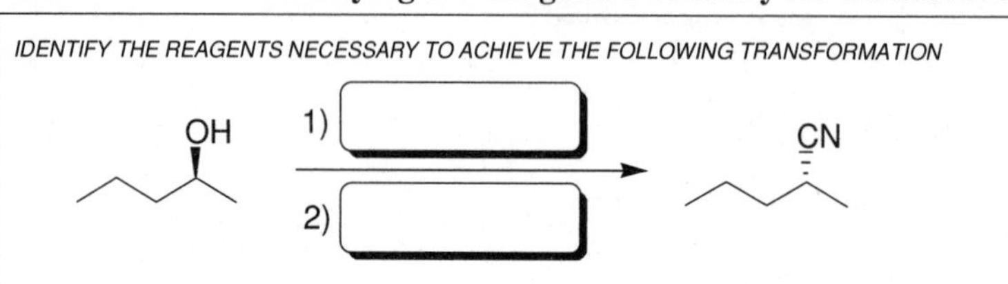

Review of Reactions

Follow the instructions below. To verify that your answers are correct, look in your textbook at the end of Chapter 7. The answers appear in the section entitled *Review of Reactions*.

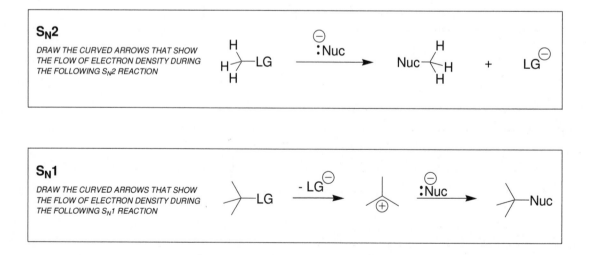

S_N2

DRAW THE CURVED ARROWS THAT SHOW THE FLOW OF ELECTRON DENSITY DURING THE FOLLOWING S_N2 REACTION

H_3C–LG + :Nuc$^{\ominus}$ ⟶ Nuc–CH_3 + LG$^{\ominus}$

S_N1

DRAW THE CURVED ARROWS THAT SHOW THE FLOW OF ELECTRON DENSITY DURING THE FOLLOWING S_N1 REACTION

$(CH_3)_3C$–LG $\xrightarrow{-\,LG^{\ominus}}$ $(CH_3)_3C^{\oplus}$ $\xrightarrow{:Nuc^{\ominus}}$ $(CH_3)_3C$–Nuc

Solutions

7.1.

a) 4-chloro-4-ethylheptane
b) 1-bromo-1-methylcyclohexane
c) 4,4-dibromo-1-chloropentane
d) (*S*)-5-fluoro-2,2-dimethylhexane

7.2.

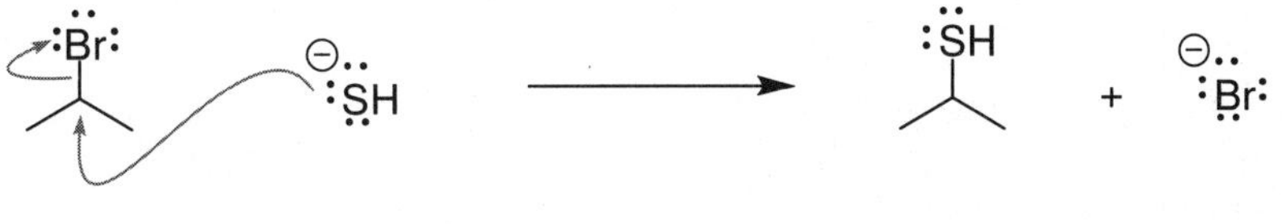

a)

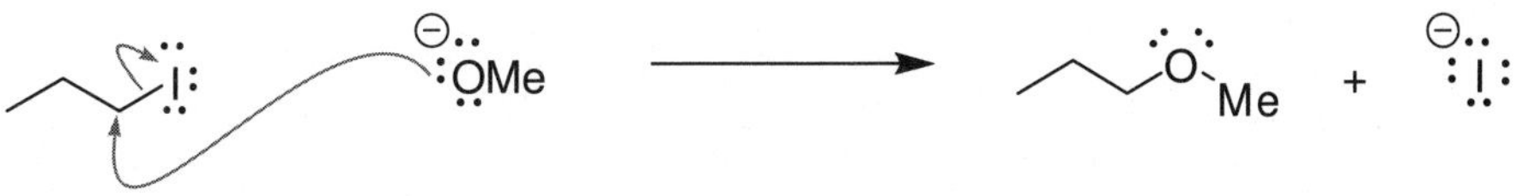

:Br: :SH → :SH + :Br:

b)

:I: :OMe → O Me + :I:

7.3.

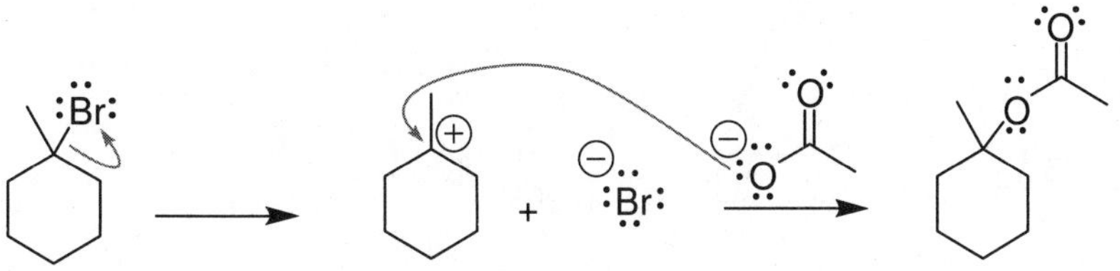

a)

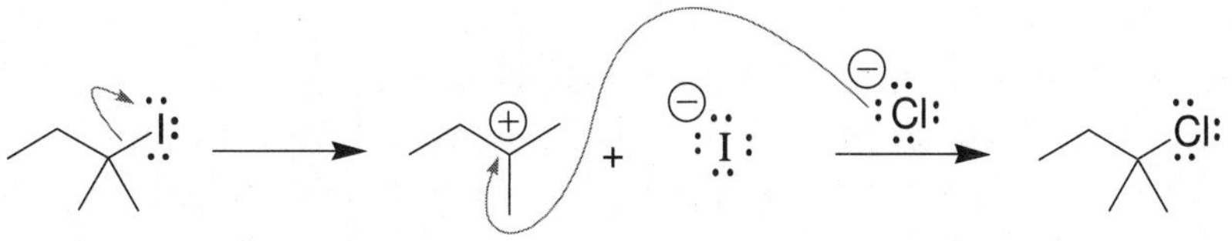

:Br: → + :Br: :O: O → :O: O

b)

:I: → + :I: :Cl: → :Cl:

7.4.

:O: Br: → :O: + :Br:

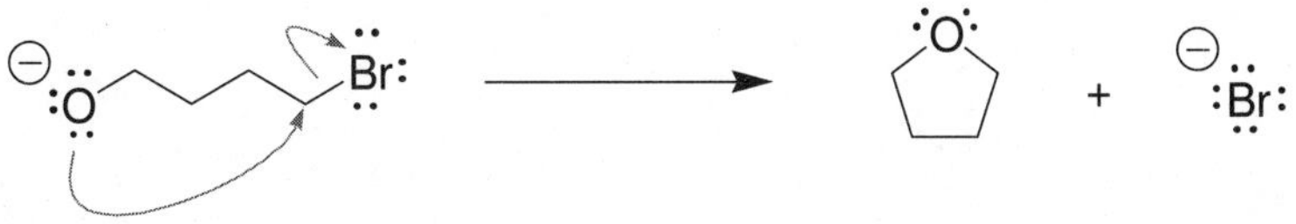

7.5.

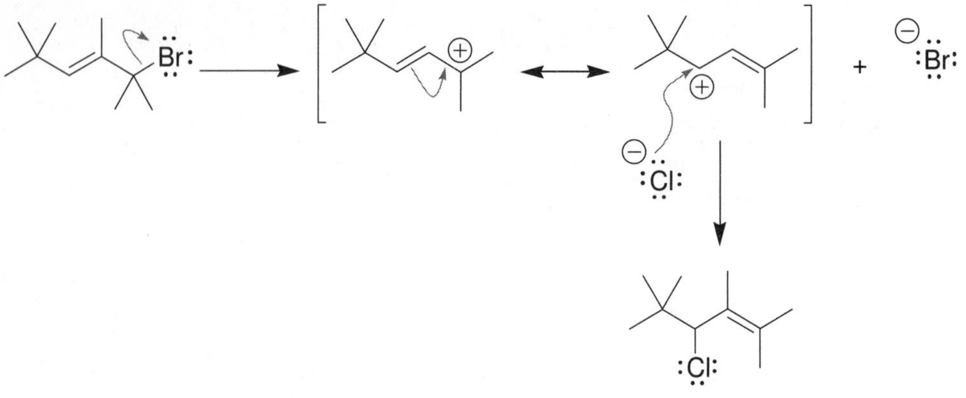

:Br: + :Br:

:Cl:

:Cl:

7.6.

a) the rate of the reaction is tripled.
b) the rate of the reaction is doubled.
c) the rate of the reaction will be six times faster.

7.7.

a) 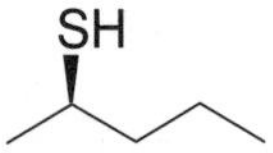 SH

b) 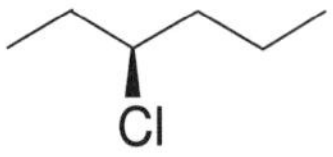 Cl

c) 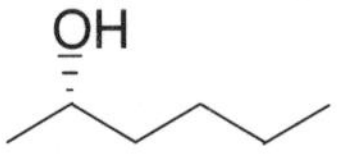 OH

7.8.

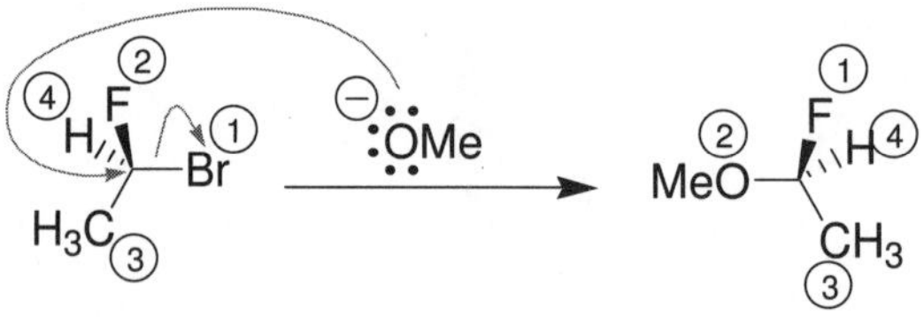

H F Br H_3C :OMe MeO F H CH_3

S ***S***

The reaction does proceed with inversion of configuration. However, the Cahn-Ingold-Prelog system for assigning a stereodescriptor (*R* or *S*) is based on a prioritization scheme. Specifically, the four groups connected to a chirality center are ranked (one through four). In the reactant (above left), the highest priority group is the leaving group (bromide) which is then replaced by a group that does not receive the highest priority. In the product, the fluorine atom has been promoted to the highest priority as a result of the reaction, and as such, the prioritization scheme has changed. In this way, the stereodescriptor (*S*) remains unchanged, despite the fact that chirality center undergoes inversion.

7.9.

a) 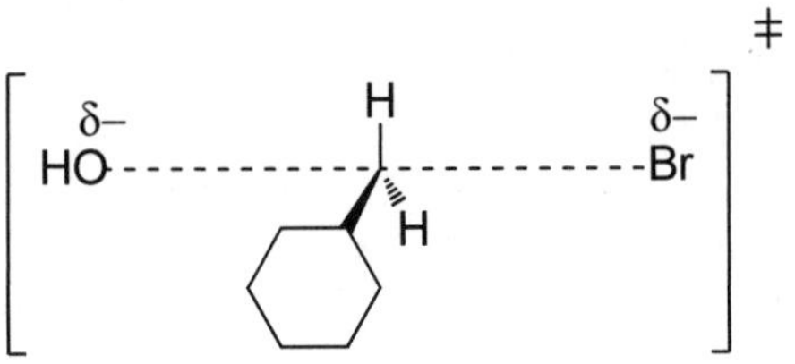 [$\overset{\delta-}{HO}$ - - - C(H)(H)(cyclohexyl) - - - $\overset{\delta-}{Br}$]‡

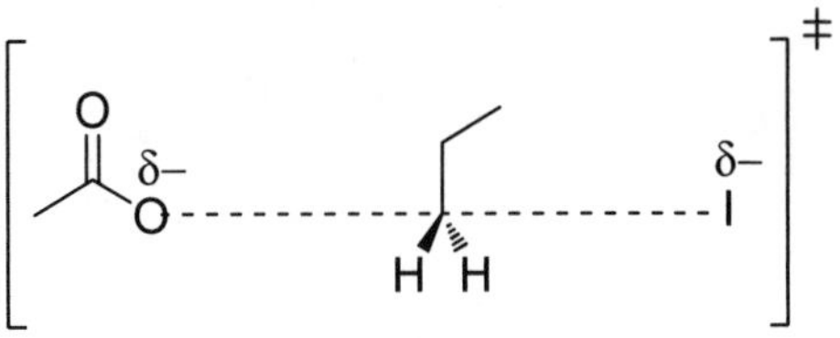

b) [$CH_3C(=O)\overset{\delta-}{O}$ - - - C(H)(H)(CH_2CH_3) - - - $\overset{\delta-}{I}$]‡

c) [$\overset{\delta-}{HO}$ - - - C(H)(H)($CH_2CH_2CH_3$) - - - $\overset{\delta-}{Cl}$]‡

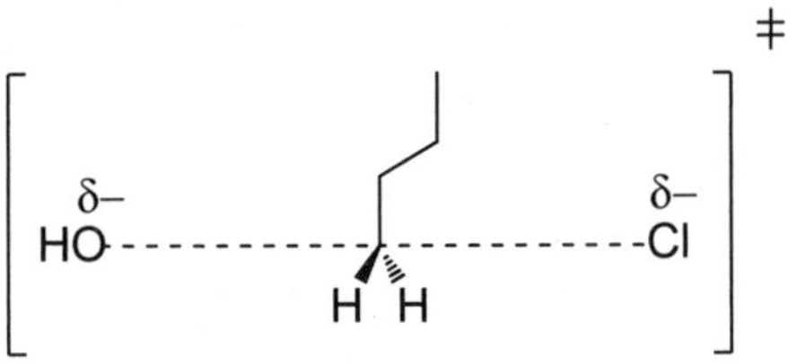

d) 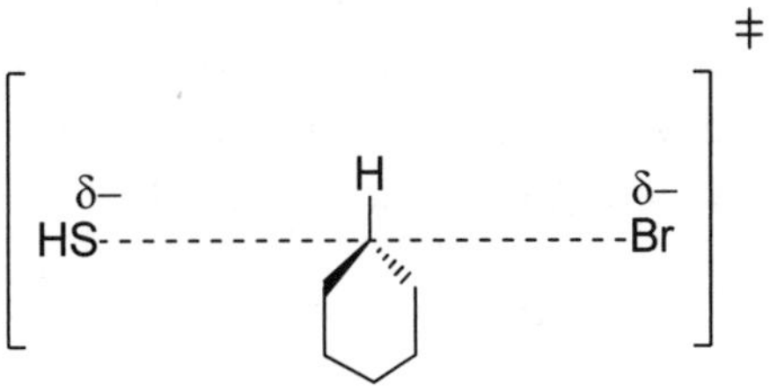 [$\overset{\delta-}{HS}$ - - - C(H)(cyclohexyl) - - - $\overset{\delta-}{Br}$]‡

7.10.

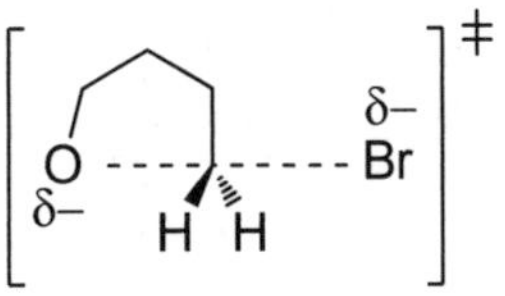

[$\overset{\delta-}{O}$ - - - C(H)(H) - - - $\overset{\delta-}{Br}$]‡

7.11.

[H–O $\delta+$ - - - C(H) - - - Br $\delta+$]‡

Being formed

Being broken

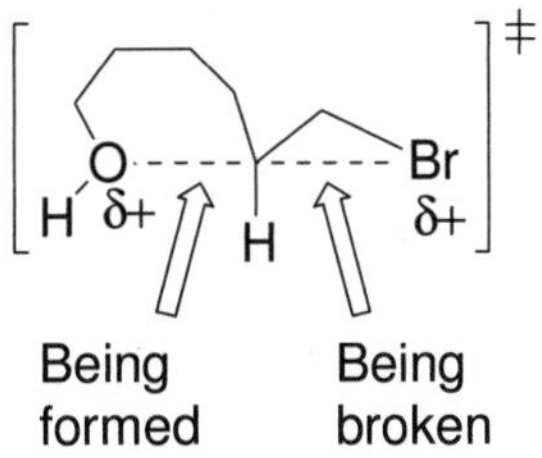

This step is favorable (downhill in energy) because ring strain is alleviated when the three-membered ring is opened.

7.12.

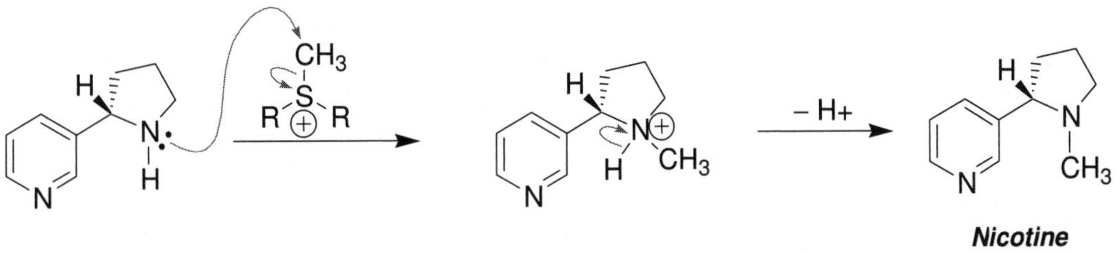

a)

$R-\overset{+}{S}(CH_3)-R$

– H+

Nicotine

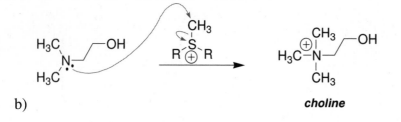

b)

choline

7.13. a) The rate of the reaction will be doubled, because the change in concentration of sodium chloride will not affect the rate.
b) The rate of the reaction will remain the same, because the change in concentration of sodium chloride will not affect the rate.

7.14. Draw the carbocation intermediate generated by each of the following substrates in an S_N1 reaction:

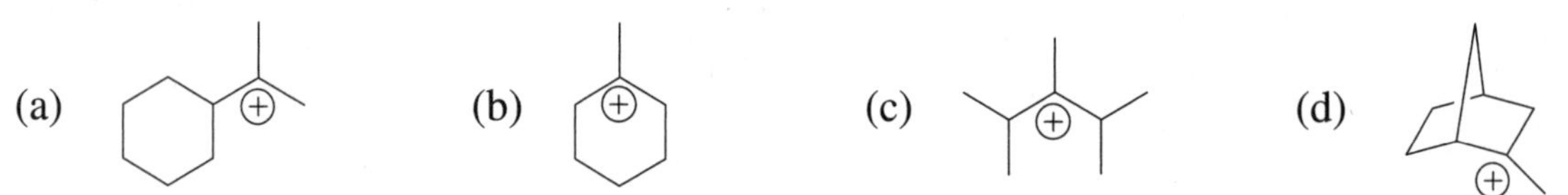

7.15.

Br

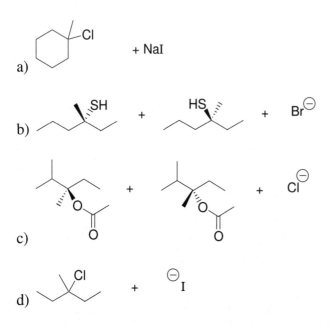

The first compound will generate a tertiary carbocation, while the second compound will generate a tertiary benzylic carbocation that is resonance stabilized. The second compound leads to a more stable carbocation, so that compound will lose its leaving group more rapidly than the first compound.

7.16.

a) Cl + NaI

b) SH + HS + $Br^{\ominus}$

c) + + $Cl^{\ominus}$

d) Cl + $^{\ominus}I$

7.17.

SH HS

Diastereomers

7.18.

a) No b) Yes c) No d) Yes e) Yes f) No

7.19.

a) No b) Yes c) Yes d) Yes e) No f) No
g) No h) Yes i) No j) No k) Yes l) No

7.20.

a) No b) Yes c) Yes d) No e) No f) No

7.21.

a)

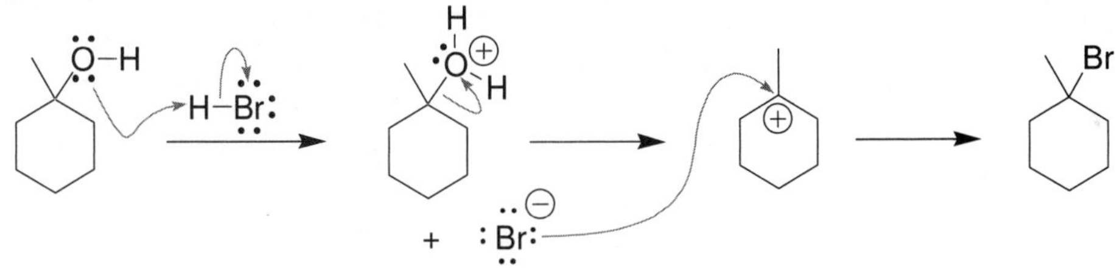

b)

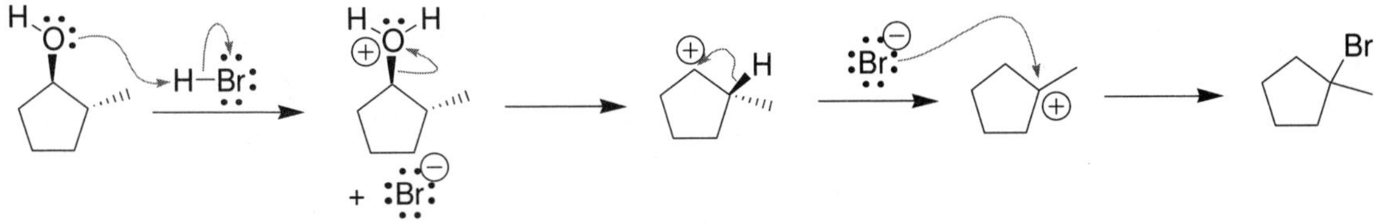

c)

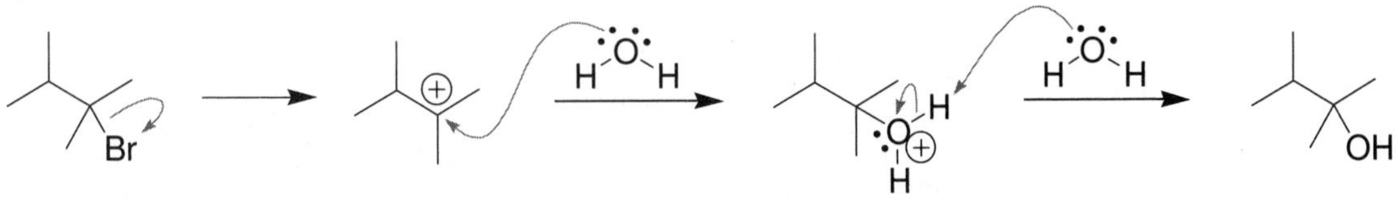

d)

EtÖH

EtÖH

OEt

e)

MeÖH

MeÖH

OMe

f)

MeÖH

MeÖH

MeO

g)

SH

h)

EtÖH

EtÖH

OEt

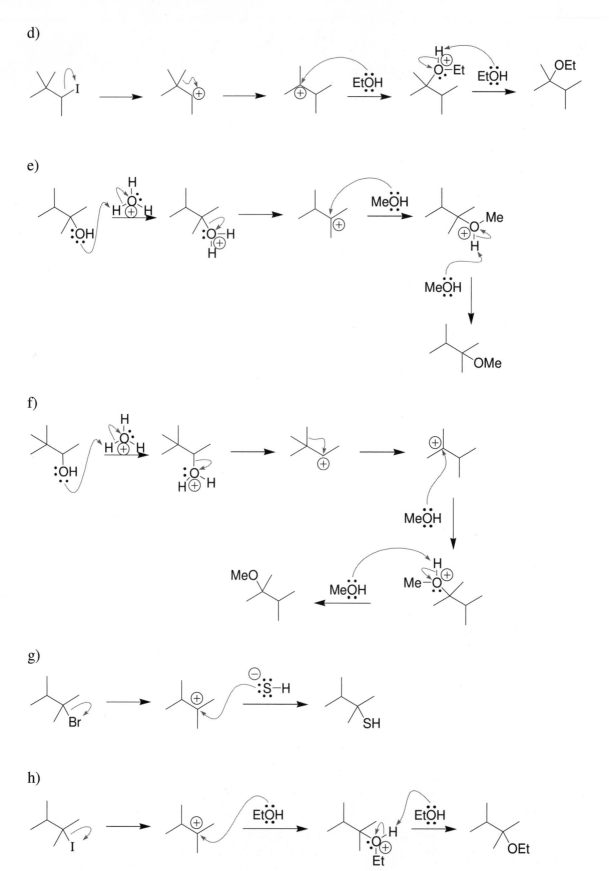

7.22.

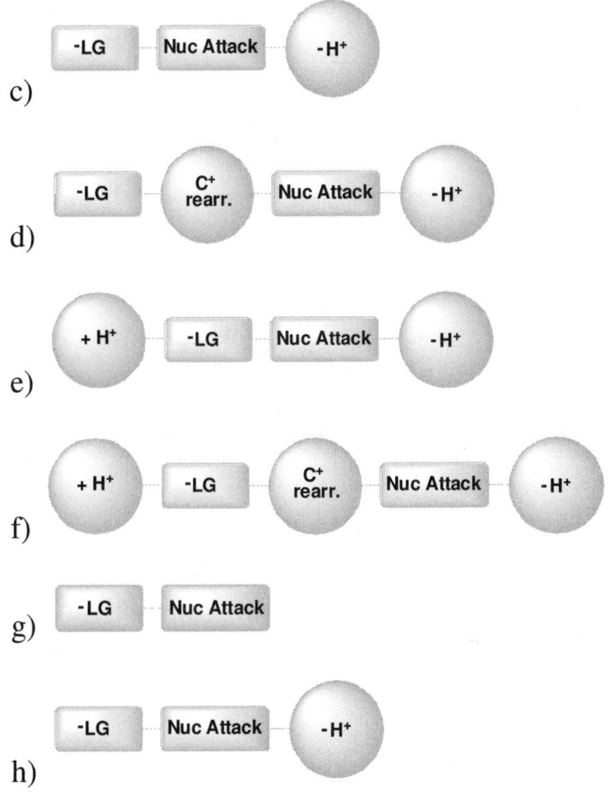

Problem 7.20c and 7.20h exhibit the same pattern. Both problems are characterized by three mechanistic steps: 1) loss of a leaving group, 2) nucleophilic attack, and 3) proton transfer.

7.23.

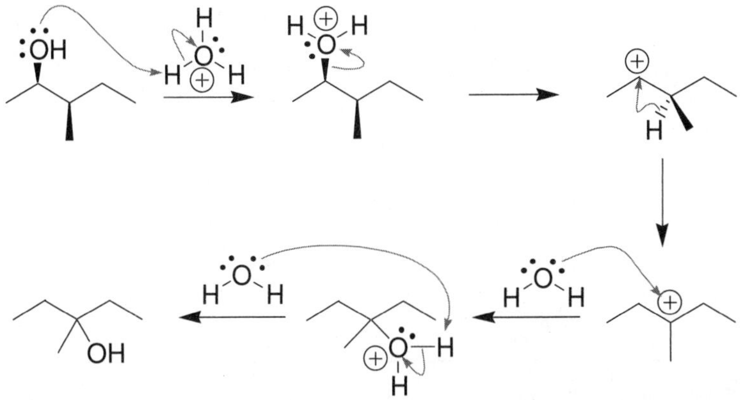

The chirality center at C2 is lost when the leaving group leaves to form a carbocation with trigonal planar geometry. The chirality center at C3 is lost during the hydride shift in the following step. Once again, the chirality center is converted into a trigonal planar sp^2 hybridized center (which is no longer a chirality center).

7.24.

a)

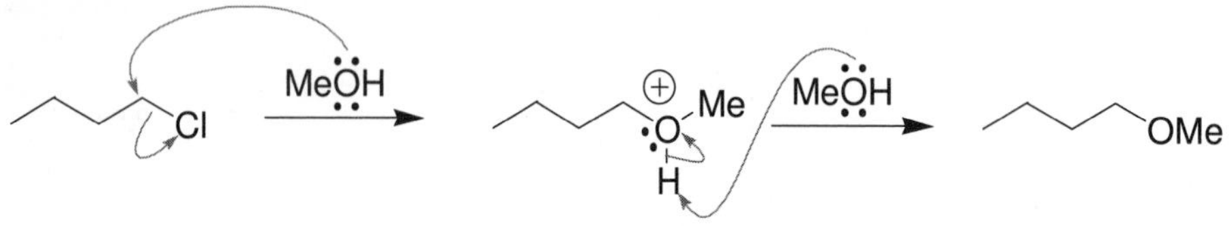

MeÖH → MeÖH → OMe

b)

EtÖH → EtÖH →

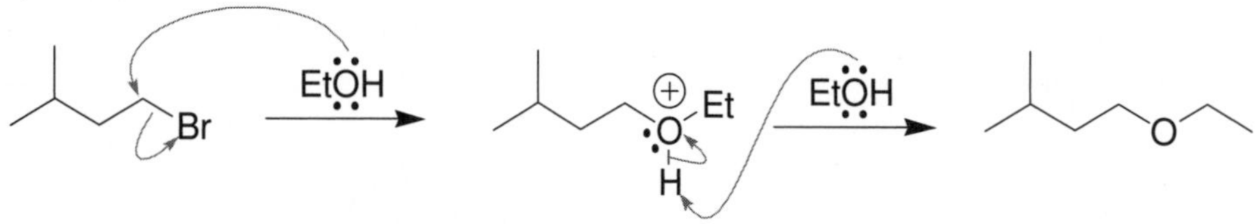

c)

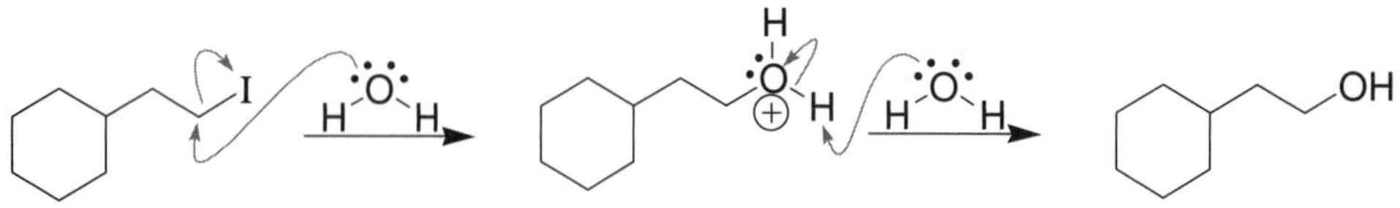

d)

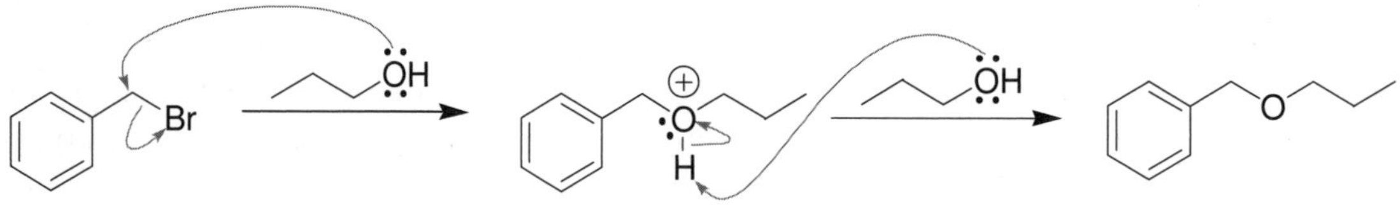

7.25.

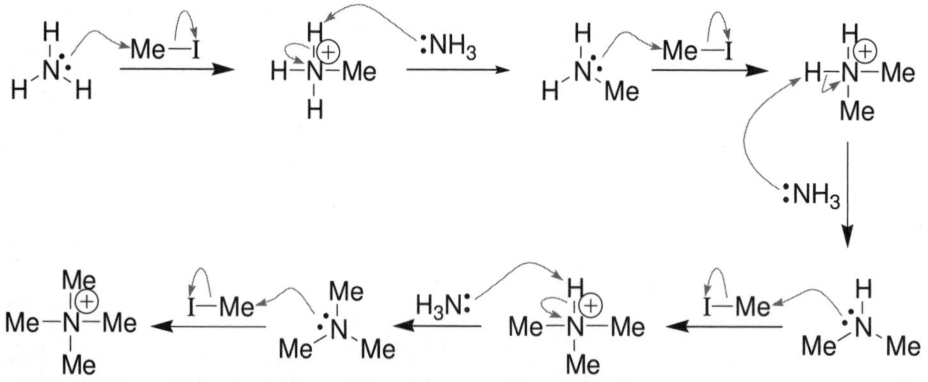

7.26.

a) S_N1 b) S_N2 c) Neither d) S_N1

e) Both f) Neither g) Both

7.27.

a) S_N1 b) S_N2 c) S_N2 d) S_N2 e) S_N2

7.28.

a)

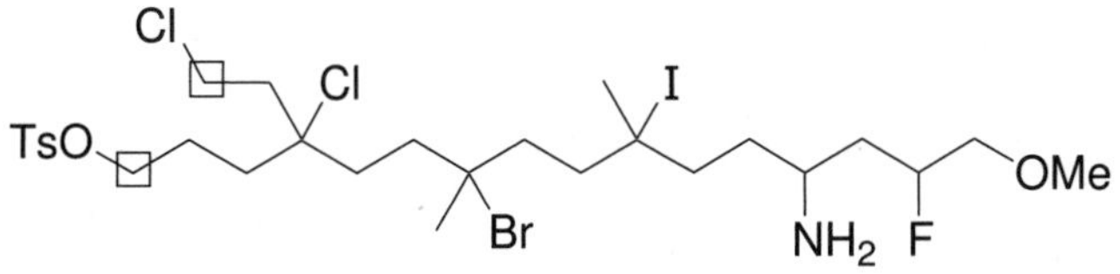

Cl Cl I TsO OMe Br NH_2 F

b)

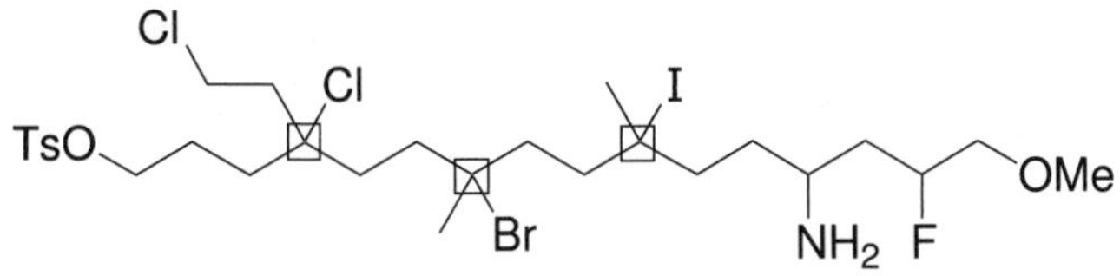

Cl Cl I TsO OMe Br NH_2 F

7.29.

a) S_N1 b) S_N2 c) S_N1 d) S_N2
e) S_N1 f) S_N2 g) S_N2 h) S_N1

7.30.

Acetone is a polar aprotic solvent and will favor S_N2 by raising the energy of the nucleophile, giving a smaller E_a.

7.31.

a) I → (MeOH) → OMe S_N1

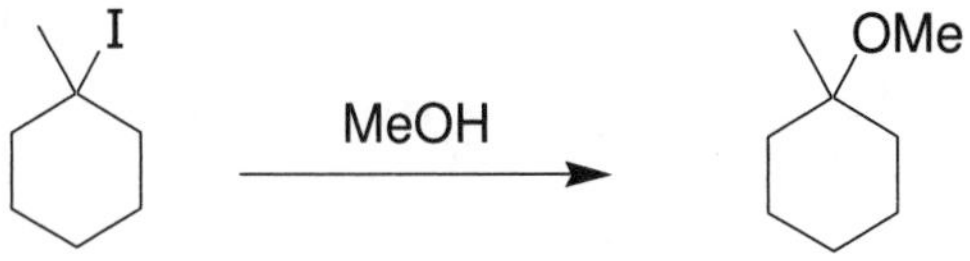

b) Br → ($Cl^{\ominus}$, HMPA) → Cl S_N2

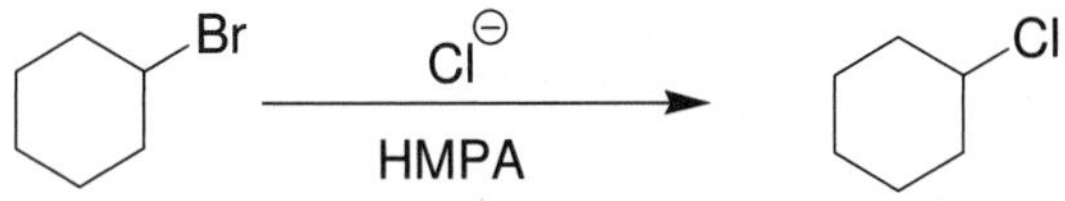

c) O–H → (H–Br) → Br S_N1

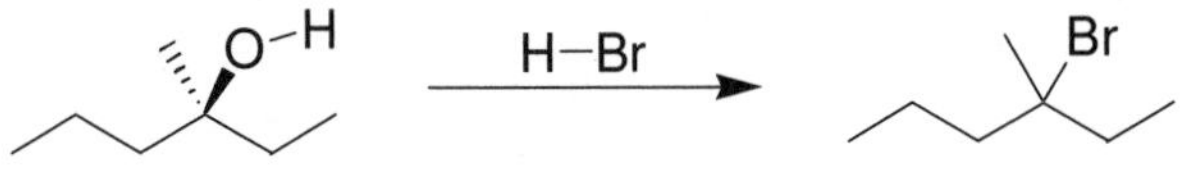

Racemic

d) OTs → (NaCN, DMF) → CN S_N2

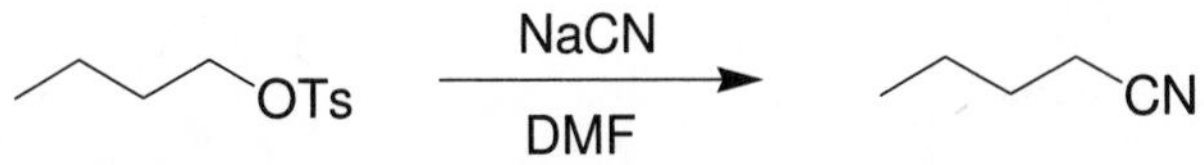

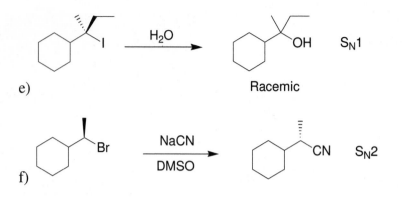

e) H_2O → Racemic S_N1

f) NaCN, DMSO → S_N2

7.32.

No. Preparation of this amine via the Gabriel synthesis would require the use of a tertiary alkyl halide, which will not undergo an S_N2 process.

7.33.

(a) NaOH → OH

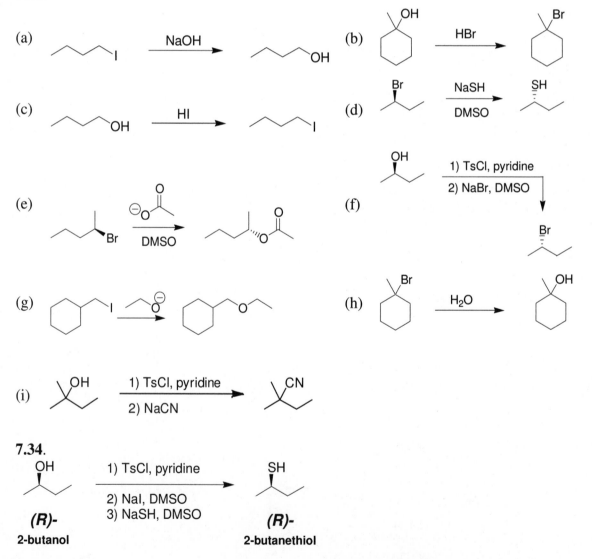

(b) OH, HBr → Br

(c) OH, HI → I

(d) Br, NaSH, DMSO → SH

(e) Br, DMSO → O

(f) OH, 1) TsCl, pyridine 2) NaBr, DMSO → Br

(g) I → O

(h) Br, H_2O → OH

(i) OH, 1) TsCl, pyridine 2) NaCN → CN

7.34.

OH, 1) TsCl, pyridine 2) NaI, DMSO 3) NaSH, DMSO → SH

***(R)*-** **2-butanol** → ***(R)*-** **2-butanethiol**

7.35.

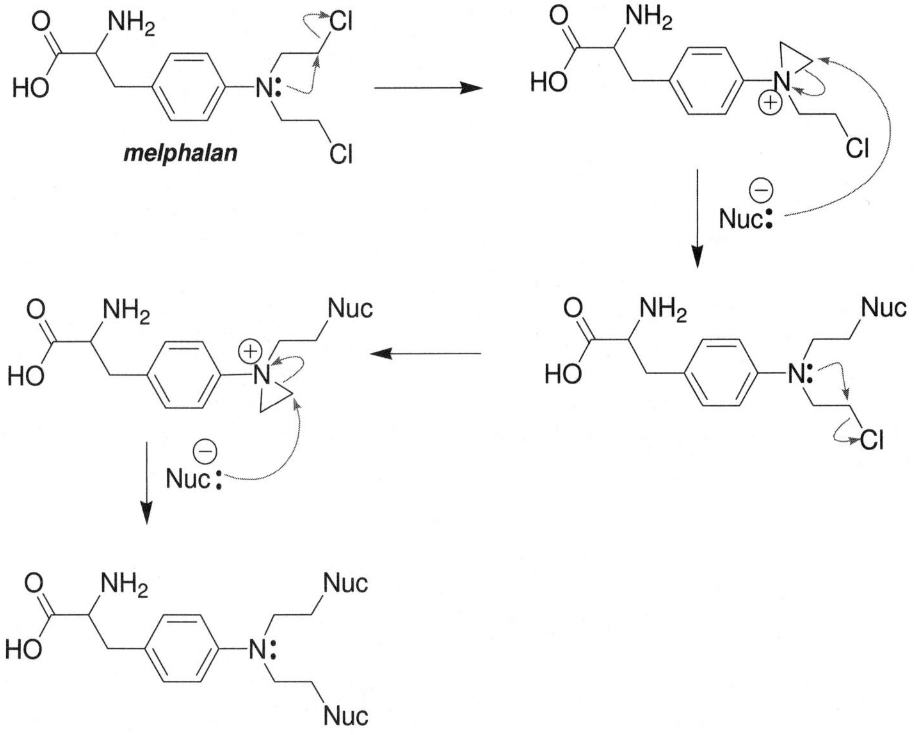

melphalan

7.36.

a) Systematic Name = 2-chloropropane
   Common Name = isopropyl chloride

b) Systematic Name = 2-bromo-2-methylpropane
   Common Name = *tert*-butyl bromide

c) Systematic Name = 1-iodopropane
   Common Name = propyl iodide

d) Systematic Name = 2-chlorobutane
   Common Name = propyl iodide

d) Systematic Name = *(R)*-2-bromobutane
   Common Name = *(R)*-*sec*-butyl bromide

e) Systematic Name = 1-chloro-2,2-dimethylpropane
   Common Name = neopentyl chloride

f) Systematic Name = chlorocyclohexane
   Common Name = cyclohexyl chloride

7.37.

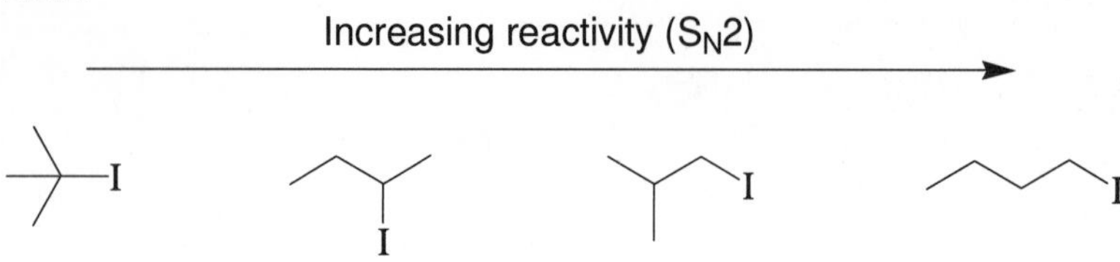

7.38.

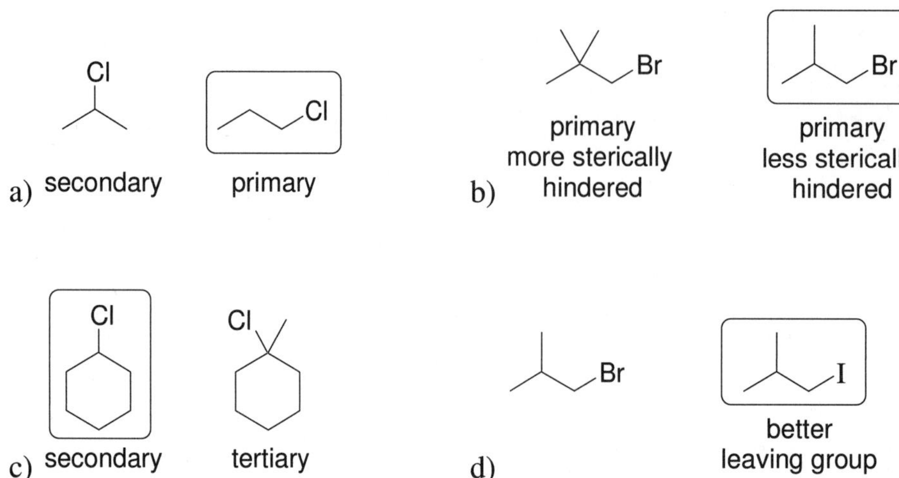

7.39.
No. Preparation of this compound via the process above would require the use of a tertiary alkyl halide, which will not undergo an S_N2 process.

7.40.
a) NaSH
b) sodium hydroxide
c) methoxide dissolved in DMSO

7.41.

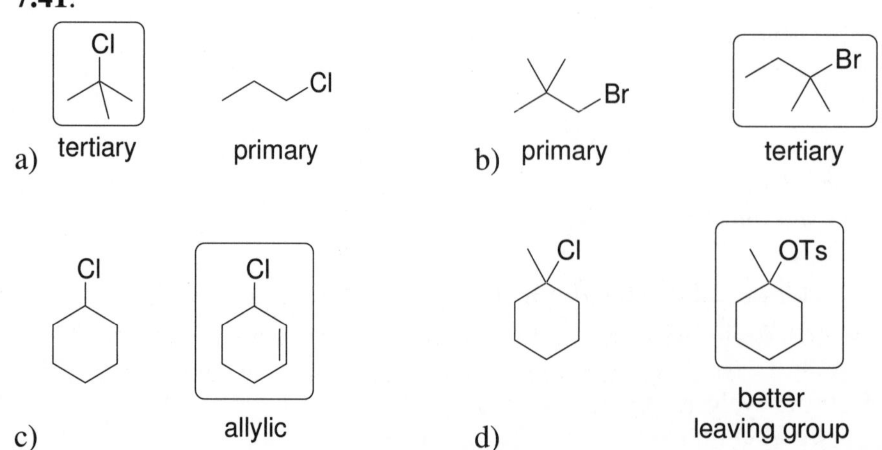

7.42. a) The rate of the reaction is doubled.
b) The rate of the reaction is doubled.

7.43. a) The rate of the reaction is doubled
b) The rate of the reaction will remain the same.

7.44. a) aprotic
b) protic
c) aprotic
d) protic
e) protic

7.45.

a) 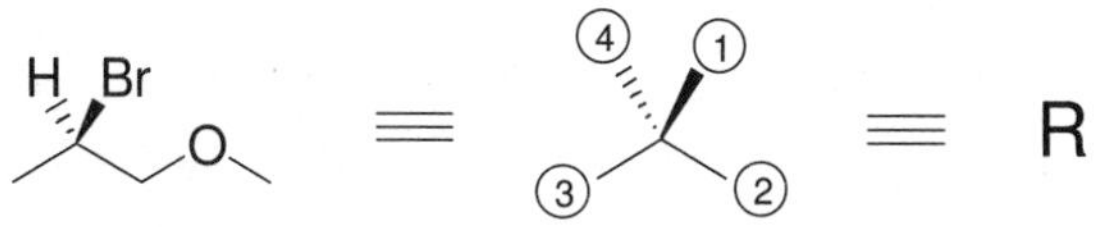 H Br ≡ ④ ① ③ ② ≡ R

b) 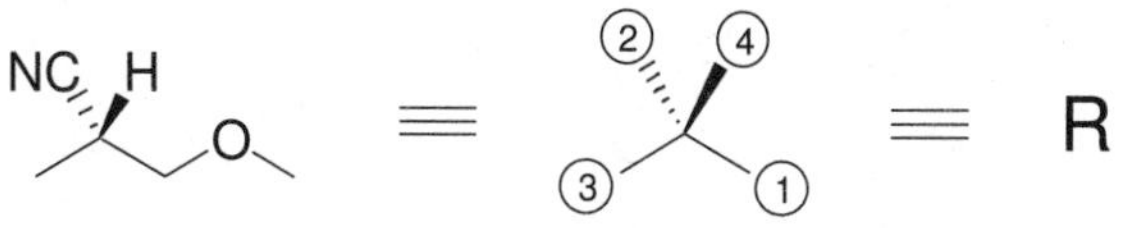 NC H ≡ ② ④ ③ ① ≡ R

c) The reaction is an S_N2 process, and it does proceed with inversion of configuration. However, the prioritization scheme changes when bromide (#1) is replaced with a cyano group (#2). As a result, the Cahn-Ingold-Prelog system assigns the same configuration to the reactant and the product.

7.46.

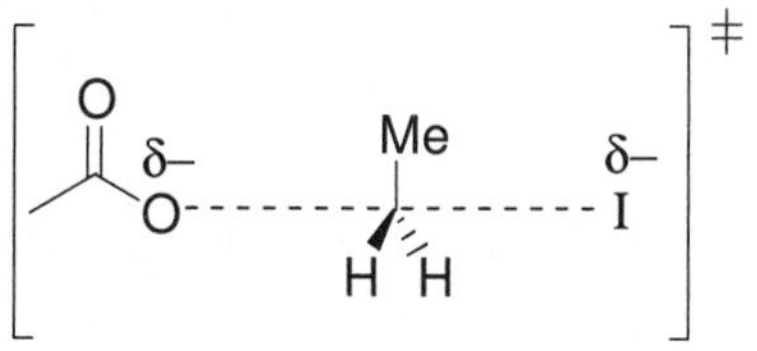

$$\left[\; CH_3C(=O)\overset{\delta-}{O} \cdots\cdots C(Me)(H)(H) \cdots\cdots \overset{\delta-}{I} \; \right]^{\ddagger}$$

7.47.
Iodide functions as a nucleophile and attacks (*S*)-2-iodopentane, displacing iodide as a leaving group. The reaction is an S_N2 process, and therefore proceeds via inversion of configuration. The product is (*R*)-2-iodopentane. The reaction continues until a racemic mixture is obtained.

7.48.

H–Br:
– H_2O
:Br:
Br
Racemic

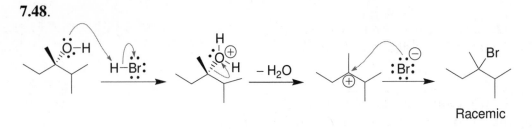

The chirality center is lost when the leaving group leaves to form a carbocation with trigonal planar geometry. The nucleophile can then attack either face of the planar carbocation, leading to a racemic mixture.

7.49.

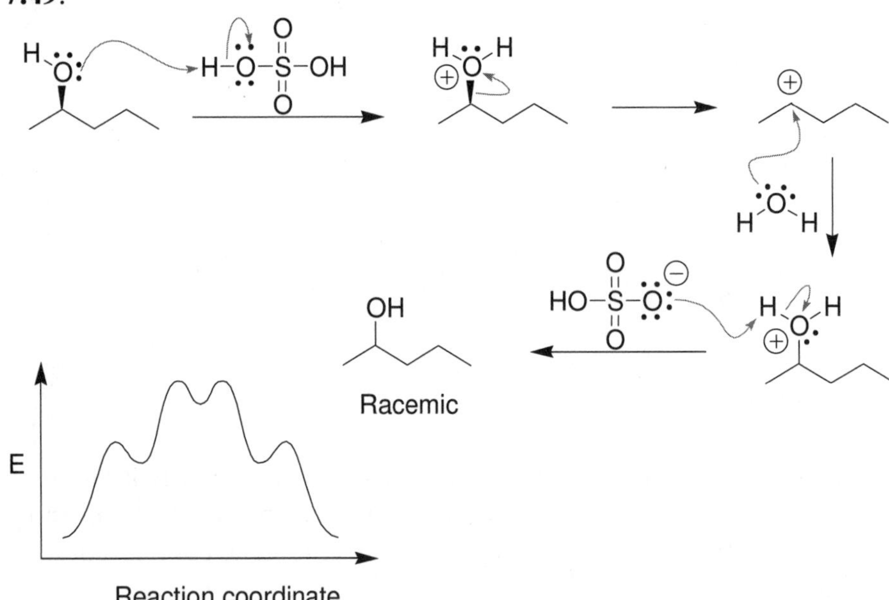

7.50.

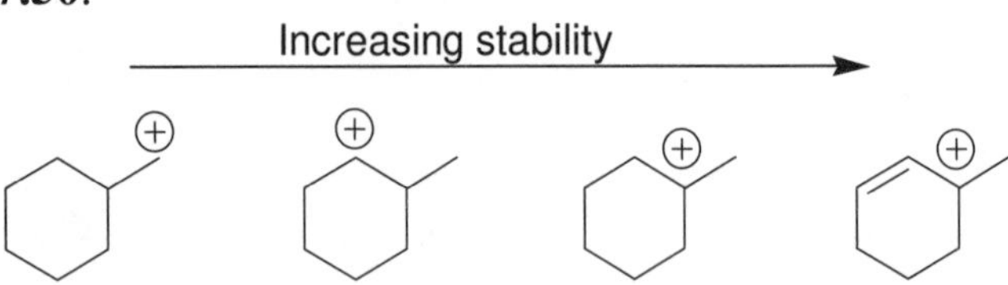

7.51.

7.52.

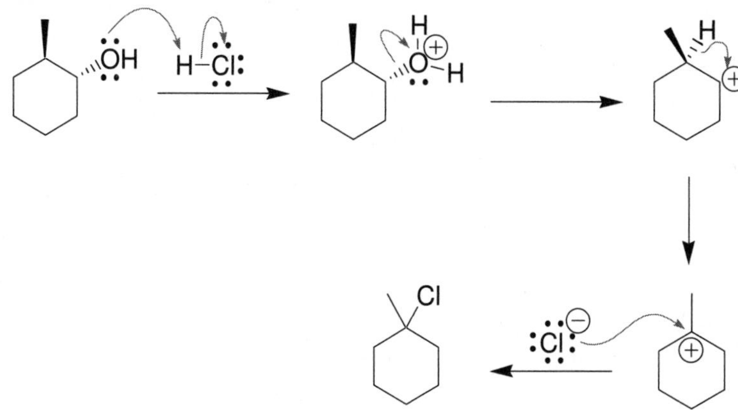

7.53.

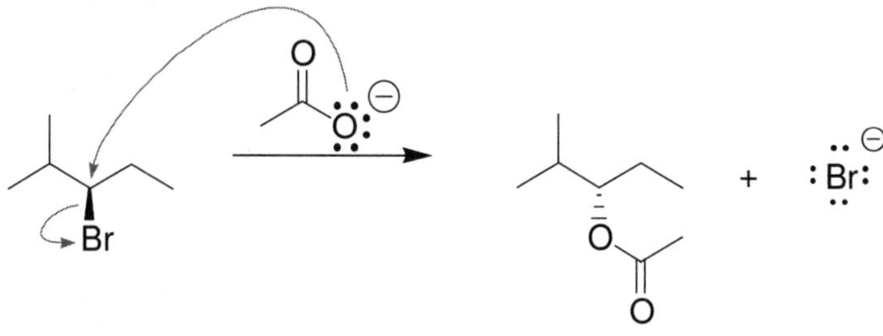

7.54.

a)

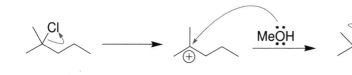

Cl

MeOH

H

O

Me

MeOH

OMe

3 Steps

b)

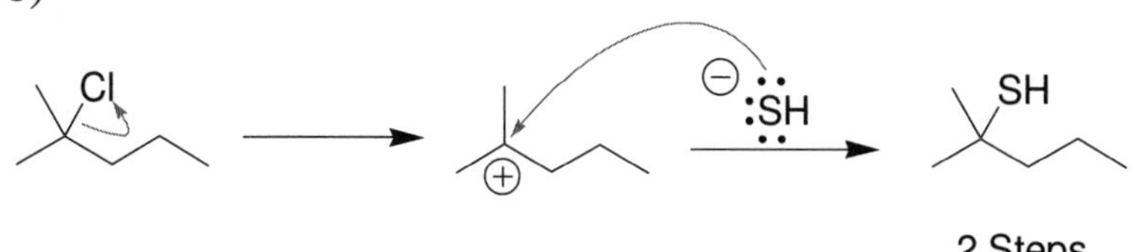

c)

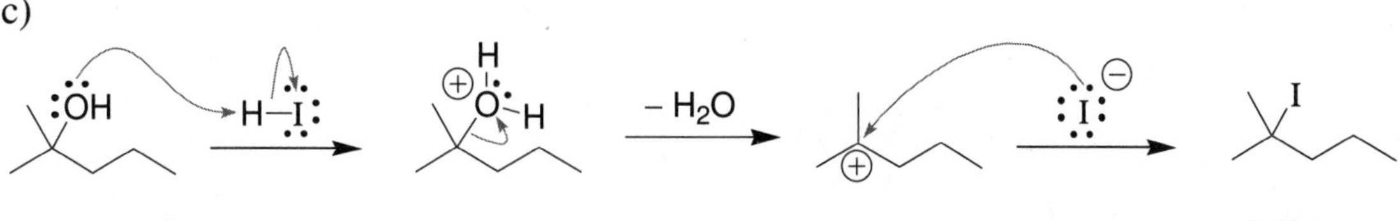

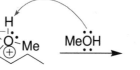

d)

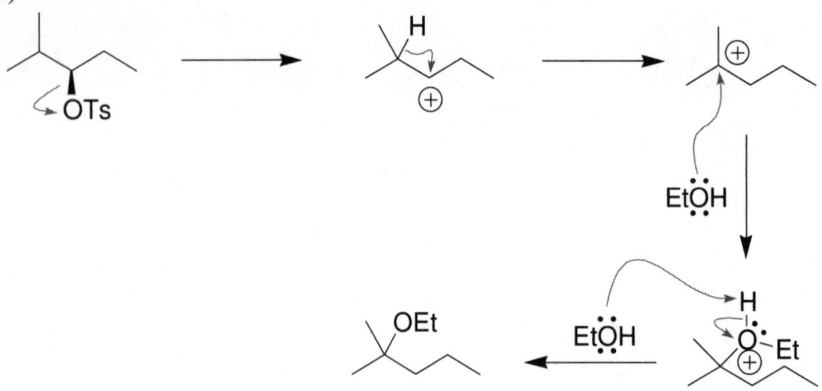

OTs

H

EtOH

OEt

EtOH

Et

4 Steps

7.55.

a) Br EtOH OEt

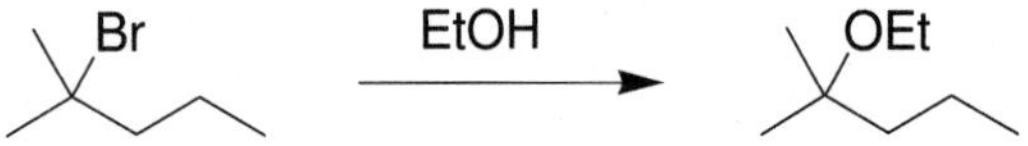

b) OTs NaBr Br

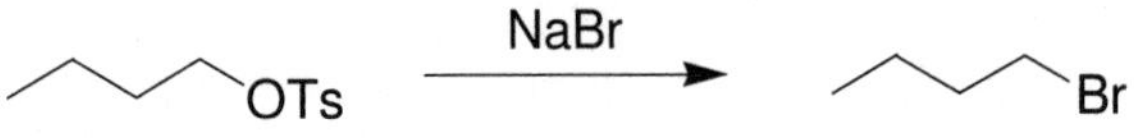

c) OH HCl Cl

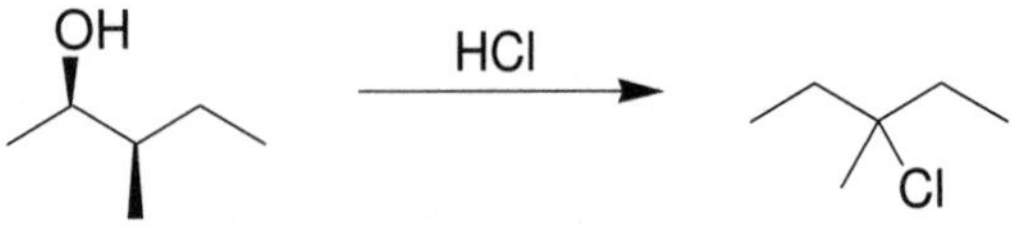

d) I NaCN DMSO CN

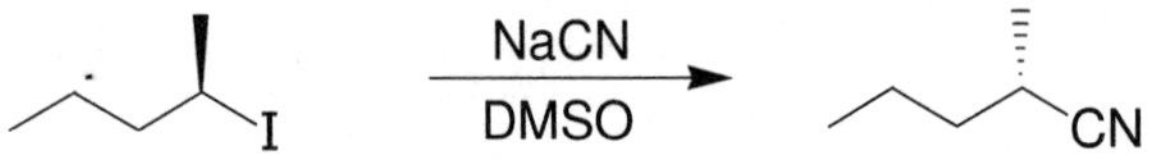

7.56.

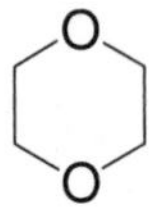

O O

7.57.

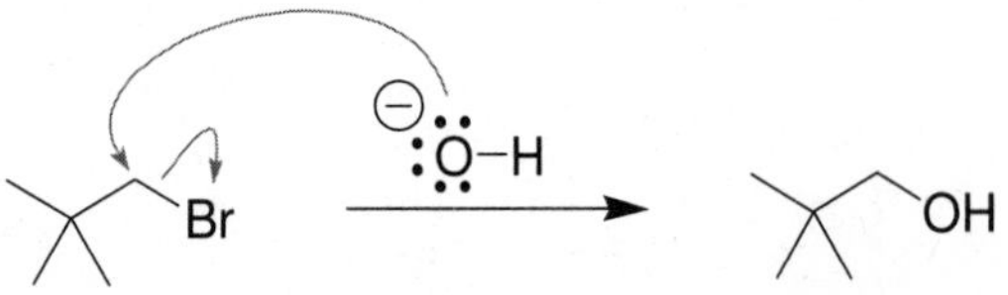

Br :O–H OH

Although the substrate is primary, it is still sterically hindered. As a result, S_N2 reactions at neopentyl halides do not occur at an appreciable rate.

7.58.

a)

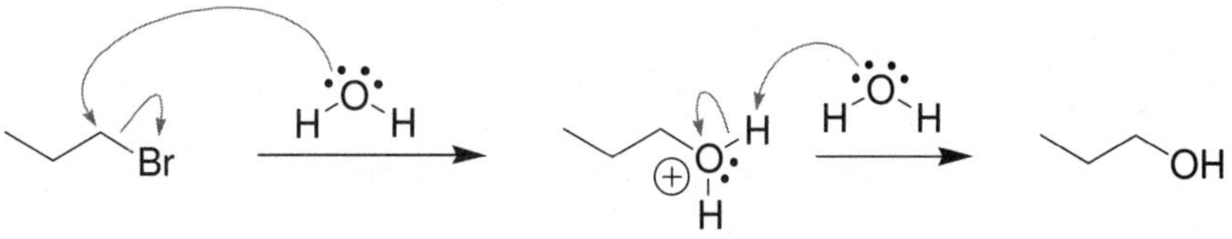

H Br H–O–H ⊕O–H H–O–H OH

b) The substrate is primary, and therefore, the reaction must proceed via an S_N2 process. S_N2 reactions are highly sensitive to the strength of the nucleophile, and the nucleophile (water) is a weak nucleophile. As a result, the reaction occurs slowly.

Br ⊖:Ö–H OH + :Br:⊖

c)

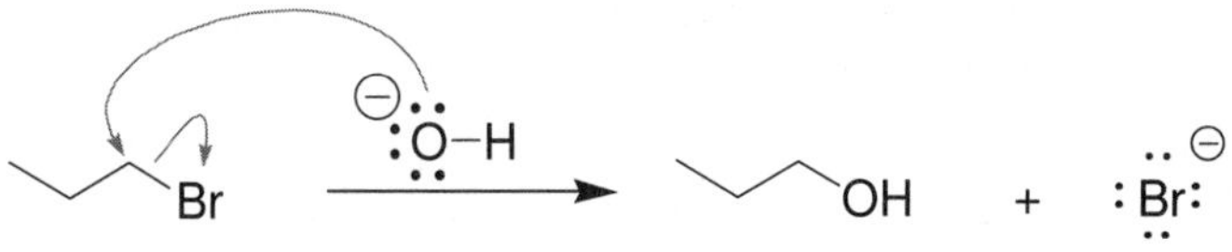

Hydroxide is a strong nucleophile, which favors the S_N2 process.

7.59.

a) OTs —NaOH→ OH

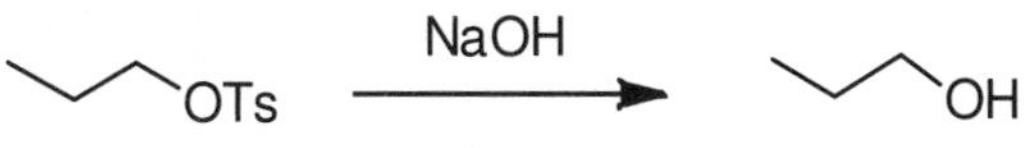

b) OH —1) TsCl, py; 2) NaCN→ CN

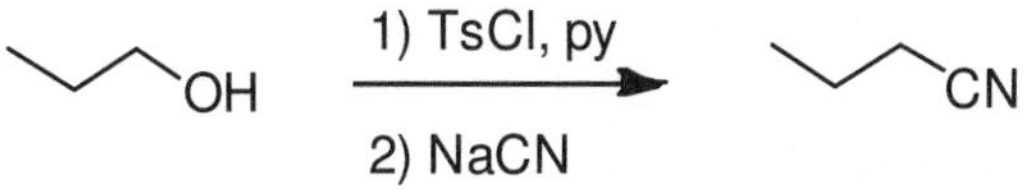

c) OH —HBr→ Br

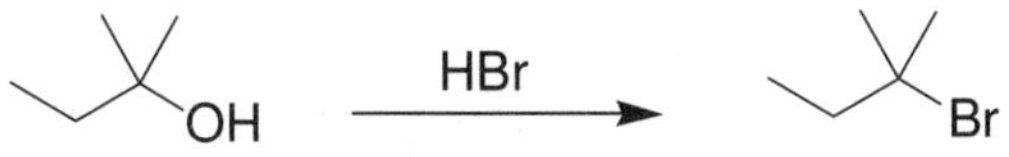

d) Cl —NaSH→ SH

e) Br —NaO(C=O)CH₃→ O(C=O)

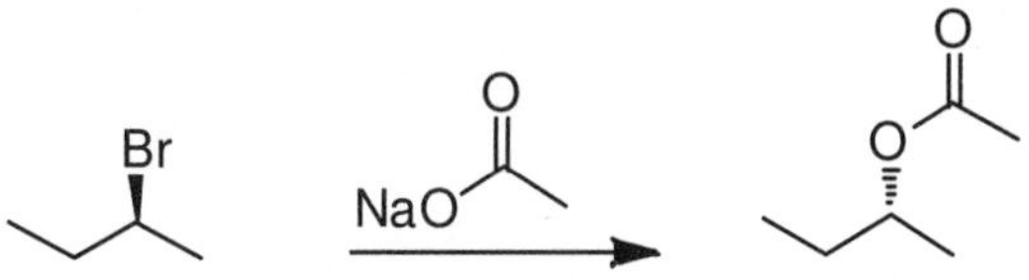

7.60.

a) 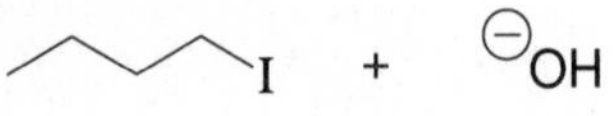 I + $^{\ominus}$OH

b) 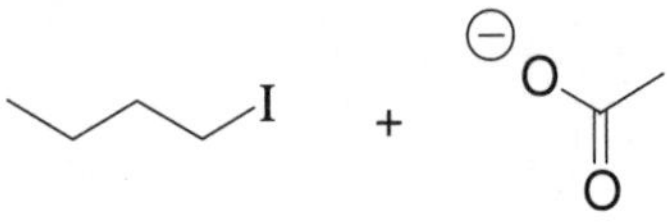 I + $^{\ominus}$O (acetate)

c) 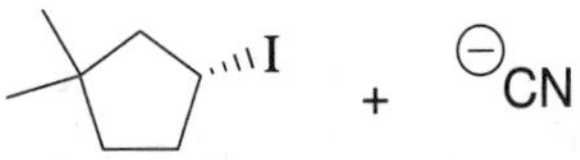 I + $^{\ominus}$CN

d) 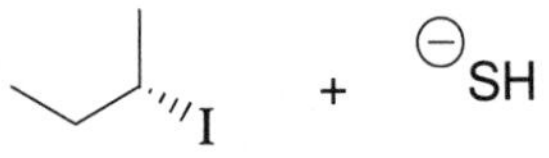 I + $^{\ominus}$SH

e) 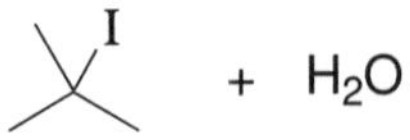 I + H_2O

f) 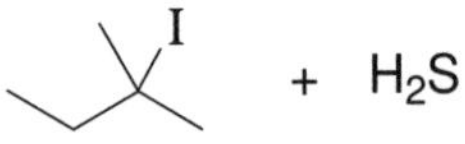 I + H_2S

7.61.

a) 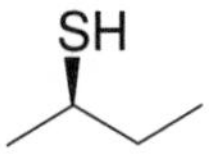 SH b) 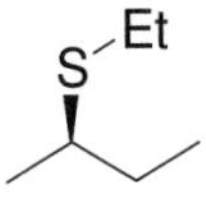 S-Et c) 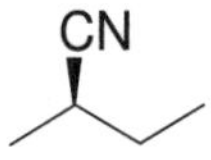 CN

7.62.

The second method is more efficient because the alkyl halide (methyl iodide) is not sterically hindered. The first method is not efficient because it employs a tertiary alkyl halide, and S_N2 reactions do not occur at tertiary substrates.

7.63.

a) 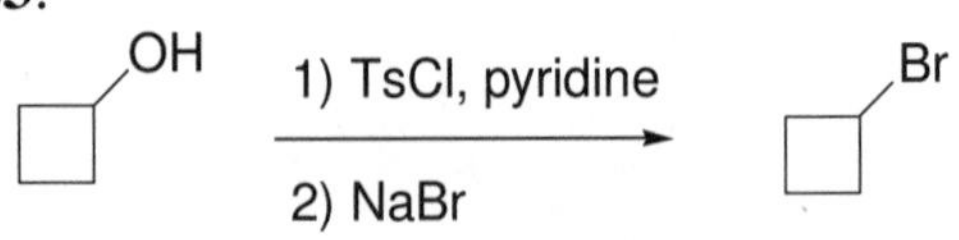 OH → 1) TsCl, pyridine 2) NaBr → Br

b) 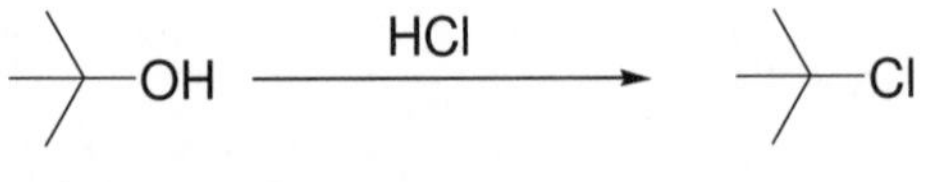 OH → HCl → Cl

c) 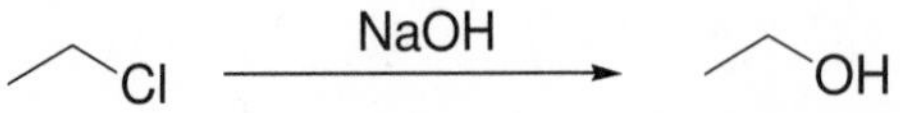 Cl → NaOH → OH

7.64.

a) Br —($^{\ominus}$:S̈–H)→ SH + :B̈r:$^{\ominus}$

b) Rate = k [Br] [NaSH]

c) The rate would be slower.

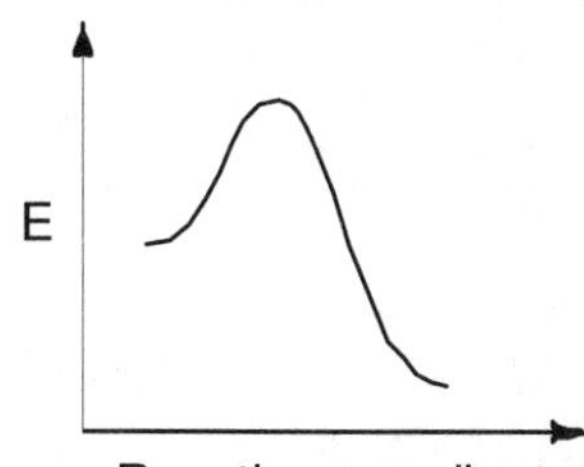

d)

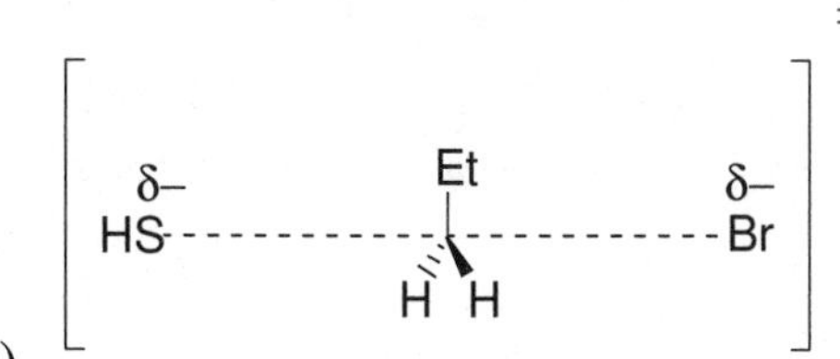

e)

7.65.

a) S_N1 (tertiary substrate)

b) Ö–H —(H–B̈r:)→ ⊕Ö–H (H) ⟶ ⊕ —(:B̈r:$^{\ominus}$)→ Br

c) Rate = k [OH]

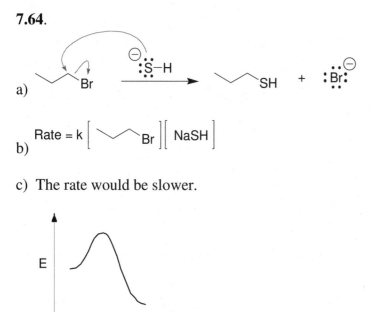

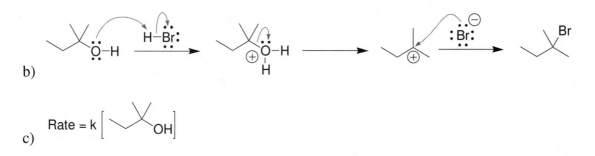

d) No. The rate is not dependent on the concentration or strength of the nucleophile.

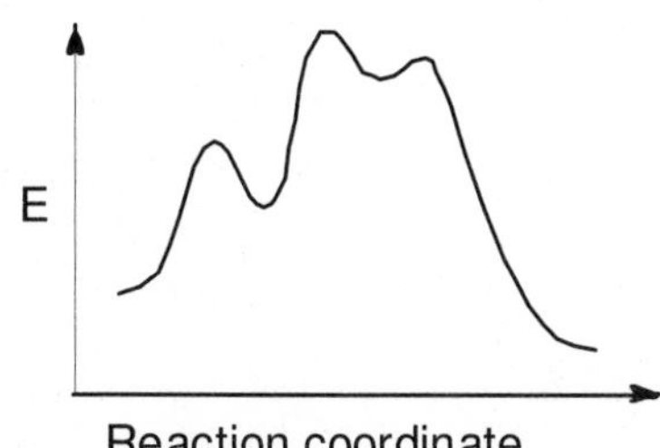

e)

7.66.

a) S_N2

b)

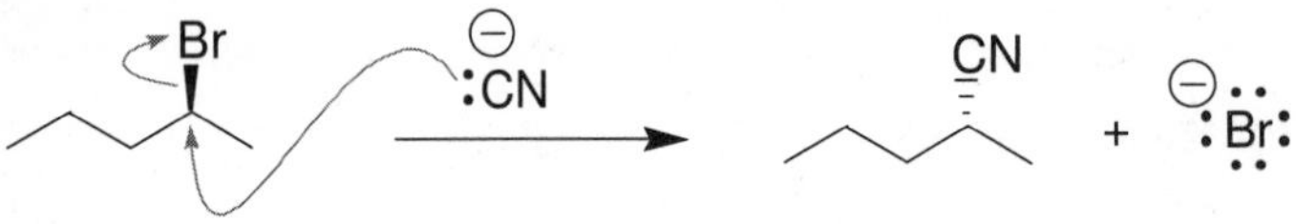

c)

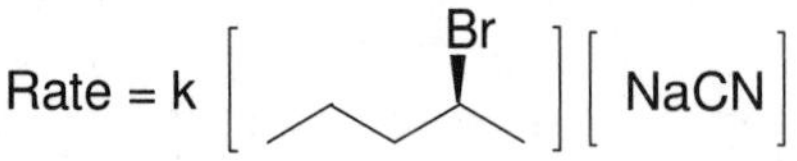

d) Yes. The reaction rate would double.

e)

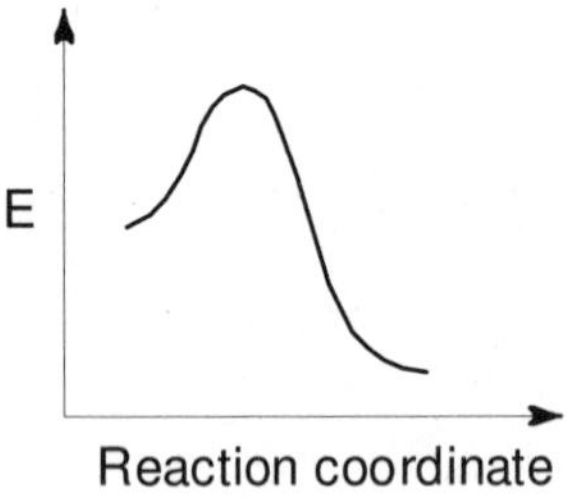

7.67.

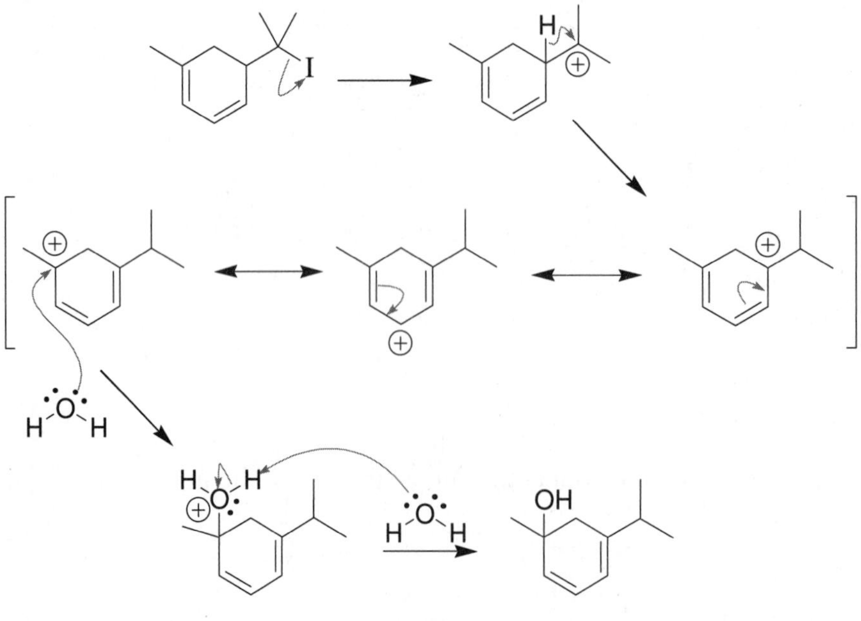

7.68.

a)

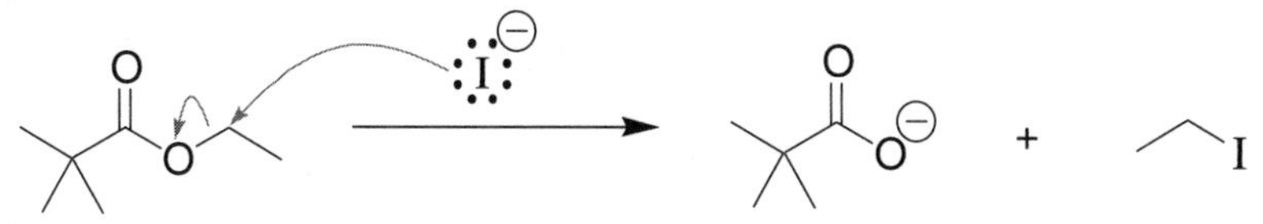

b) This reaction occurs via an S_N2 process. As such, the rate of the reaction is highly sensitive to the nature of the substrate. The reaction will be faster in this case, because the methyl ester is less sterically hindered than the ethyl ester.

7.69.

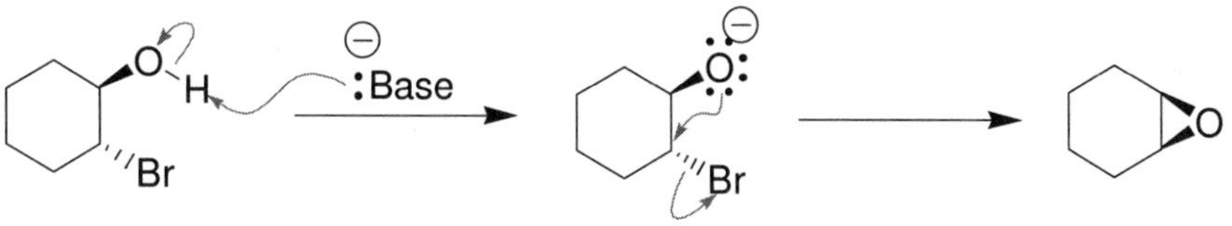

7.70.

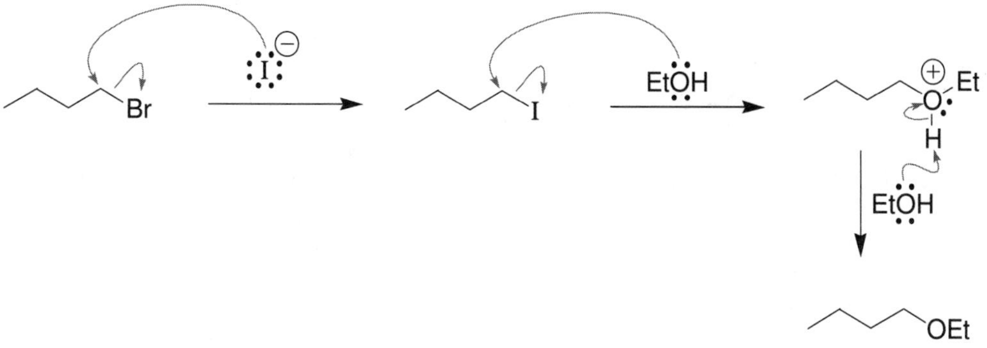

7.71.
When the leaving group leaves, the carbocation formed is resonance stabilized:

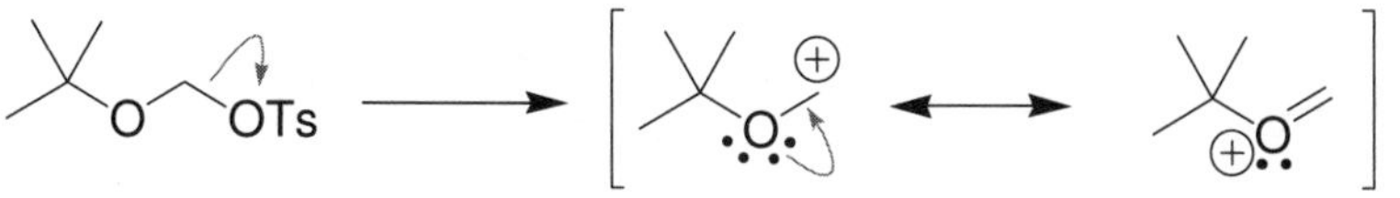

Resonance stabilized

7.72.
Iodide is a very good nucleophile (because it is polarizable), and it is also a very good leaving group (because it can stabilize the negative charge by spreading the charge over a large volume of space). As such, iodide will function as a nucleophile to displace the chloride ion. Once installed, the iodide group is a better leaving group than chloride, thereby increasing the rate of the reaction.

7.73.

OH

H
O
H H

H H
O

H
O
H H

H H
O

OH

OH

≡

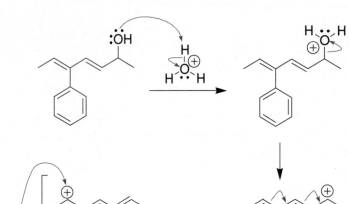

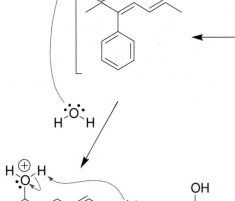

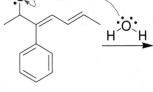

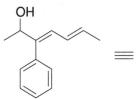

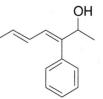

Chapter 8
Alkenes: Structure and Preparation via Elimination Reactions

Review of Concepts

Fill in the blanks below. To verify that your answers are correct, look in your textbook at the end of Chapter 8. Each of the sentences below appears in the section entitled *Review of Concepts and Vocabulary.*

- Alkene stability increases with increasing degree of _______________.
- E2 reactions are said to be **regioselective**, because the more substituted alkene, called the ___________ **product**, is generally the major product.
- When both the substrate and the base are sterically hindered, the less substituted alkene, called the __________ **product,** is the major product.
- E2 reactions are **stereospecific** because they generally occur via the __________________ conformation.
- Substituted cyclohexanes only undergo E2 reactions from the chair conformation in which the leaving group and the proton both occupy _________ positions.
- E1 reactions exhibit a regiochemical preference for the ___________ product.
- E1 reactions are not stereospecific, but they are stereo___________.
- Strong nucleophiles are compounds that contain a _____________ and/or are _______________.
- Strong bases are compounds whose conjugate acids are _______________.

Review of Skills

Fill in the blanks and empty boxes below. To verify that your answers are correct, look in your textbook at the end of Chapter 8. The answers appear in the section entitled *SkillBuilder Review.*

8.1 Assembling the Systematic Name of an Alkene

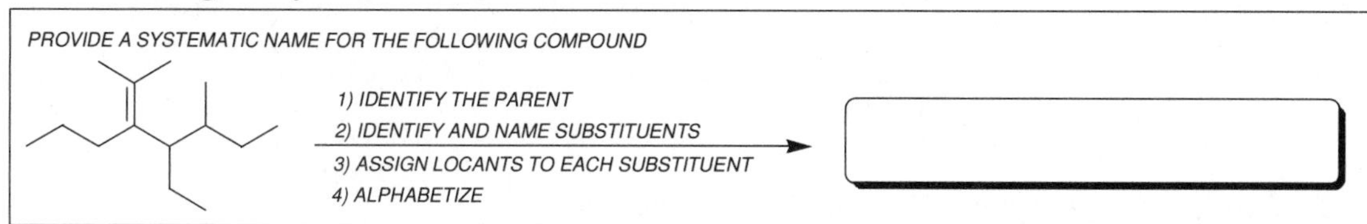

8.2 Assigning the Configuration of a double bond

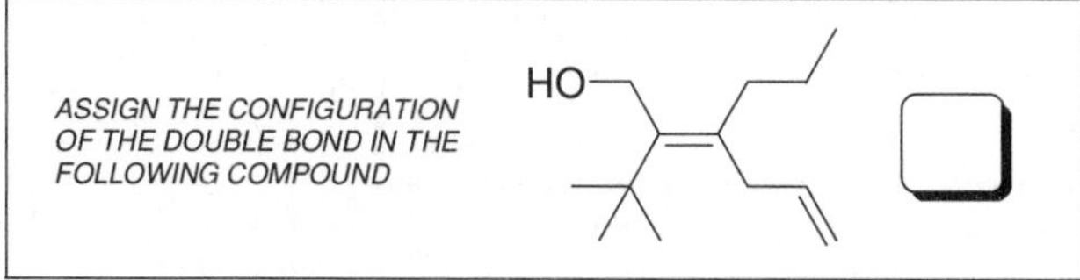

ASSIGN THE CONFIGURATION OF THE DOUBLE BOND IN THE FOLLOWING COMPOUND
HO

8.3 Comparing the Stability of Isomeric Alkenes

CIRCLE THE MOST STABLE ALKENE BELOW

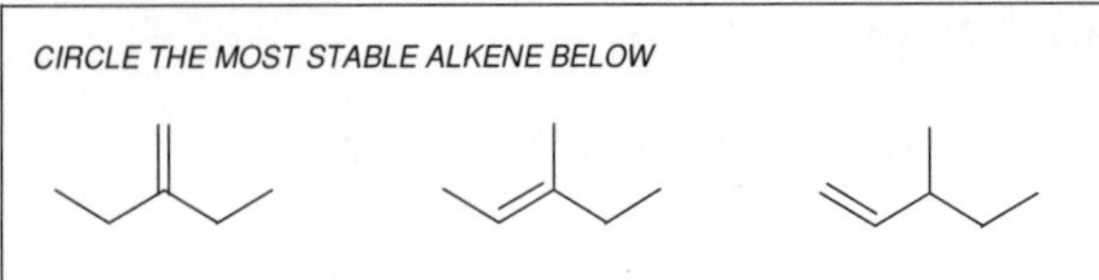

8.4 Drawing the Curved Arrows of an Elimination Reaction

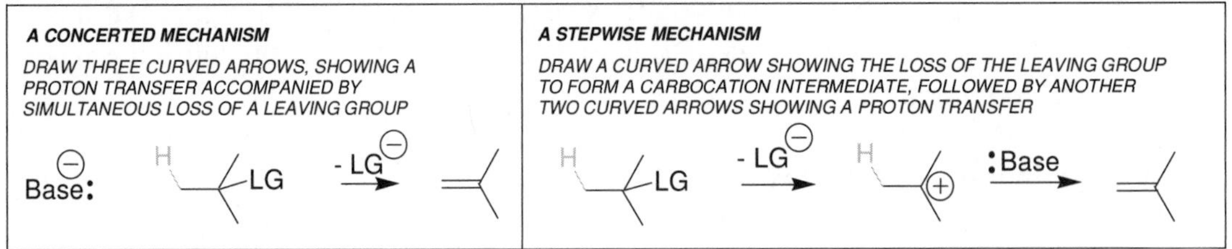
A CONCERTED MECHANISM
DRAW THREE CURVED ARROWS, SHOWING A PROTON TRANSFER ACCOMPANIED BY SIMULTANEOUS LOSS OF A LEAVING GROUP
Base:
H
LG
- LG
A STEPWISE MECHANISM
DRAW A CURVED ARROW SHOWING THE LOSS OF THE LEAVING GROUP TO FORM A CARBOCATION INTERMEDIATE, FOLLOWED BY ANOTHER TWO CURVED ARROWS SHOWING A PROTON TRANSFER
H
LG
- LG
H
:Base

8.5 Predicting the Regiochemical Outcome of an E2 Reaction

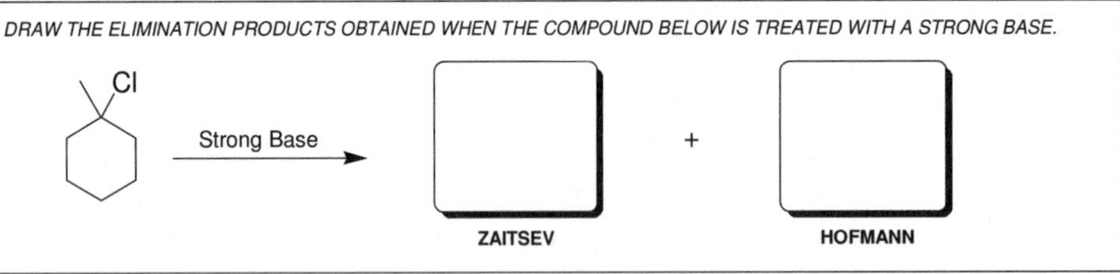
DRAW THE ELIMINATION PRODUCTS OBTAINED WHEN THE COMPOUND BELOW IS TREATED WITH A STRONG BASE.
Cl
Strong Base
+
ZAITSEV
HOFMANN

8.6 Predicting the Stereochemical Outcome of an E2 Reaction

PREDICT THE STEREOCHEMICAL OUTCOME OF THE FOLLOWING REACTION, AND DRAW THE PRODUCT.

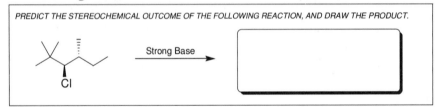

Cl

Strong Base

8.7 Drawing the Products of an E2 Reaction

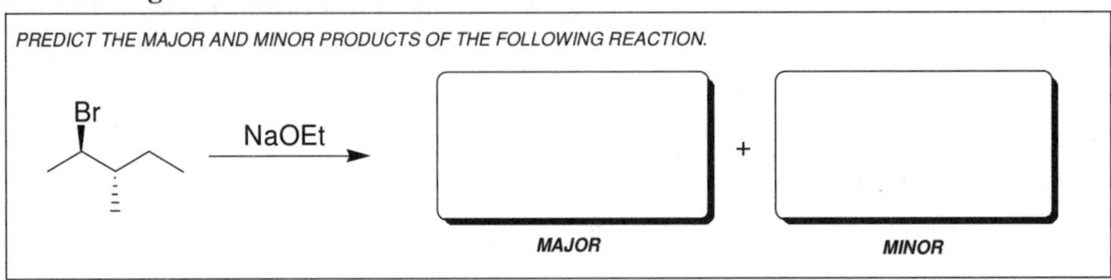
PREDICT THE MAJOR AND MINOR PRODUCTS OF THE FOLLOWING REACTION.
Br
NaOEt
+
MAJOR
MINOR

8.8 Predicting the Regiochemical Outcome of an E1 Reaction

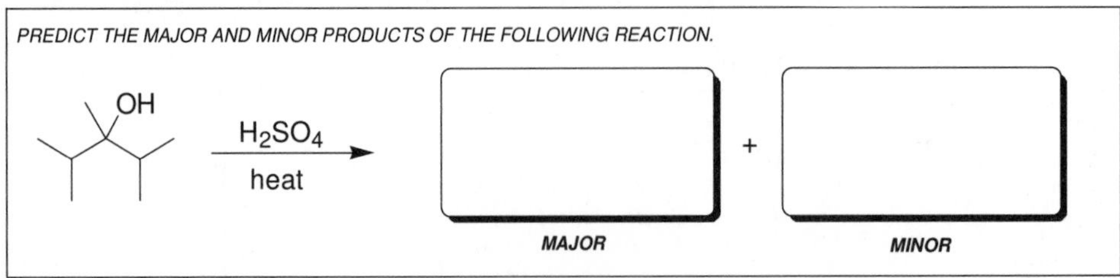
PREDICT THE MAJOR AND MINOR PRODUCTS OF THE FOLLOWING REACTION.
OH
H2SO4
heat
+
MAJOR
MINOR

8.9 Drawing the Complete Mechanism of an E1 Reaction

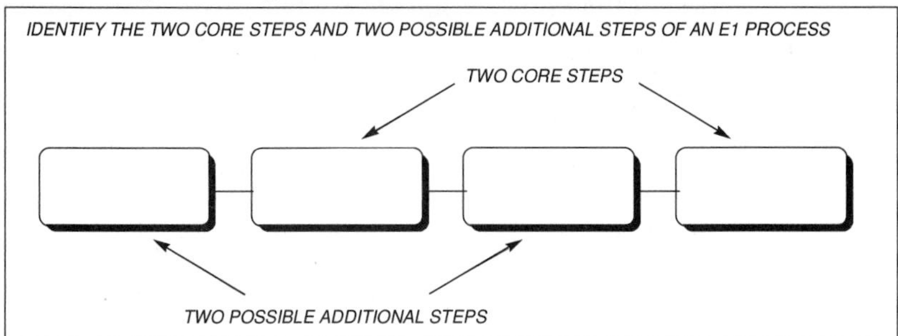

IDENTIFY THE TWO CORE STEPS AND TWO POSSIBLE ADDITIONAL STEPS OF AN E1 PROCESS
TWO CORE STEPS
TWO POSSIBLE ADDITIONAL STEPS

8.10 Determining the Function of a Reagent

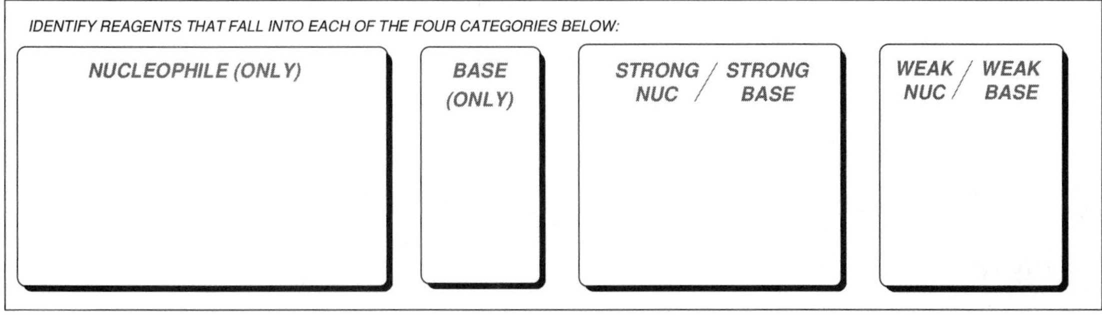

IDENTIFY REAGENTS THAT FALL INTO EACH OF THE FOUR CATEGORIES BELOW:
NUCLEOPHILE (ONLY)
BASE (ONLY)
IDENTIFY THE MECHANISM(S) THAT OPERATE IN EACH OF THE CASES BELOW:
STRONG NUC / STRONG BASE
WEAK NUC / WEAK BASE

8.11 Identifying the Expected Mechanism(s)

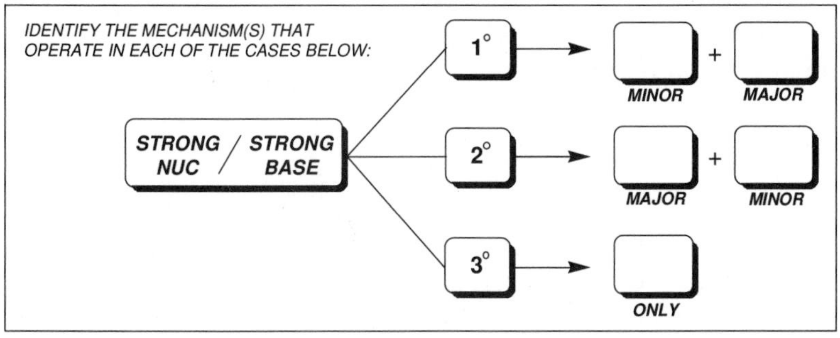

8.12 Predicting the Products of Substitution and Elimination Reactions

FILL IN THE BLANKS BELOW:		
STEP 1 DETERMINE THE FUNCTION OF THE ________________	**STEP 2** ANALYZE THE ________________ AND DETERMINE THE EXPECTED MECHANISM(S).	**STEP 3** CONSIDER ANY RELEVANT REGIOCHEMICAL AND ________________ REQUIREMENTS

Review of Synthetically Useful Elimination Reactions

Identify reagents that will achieve each of the transformations below. To verify that your answers are correct, look in your textbook at the end of Chapter 8. The answers appear in the section entitled *Review of Synthetically Useful Elimination Reactions*.

OH

Br

Br

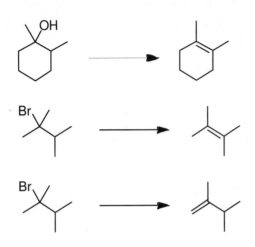

Solutions

8.1.

a) 2,3,5-trimethyl-2-heptene
b) 3-ethyl-2-methyl-2-heptene
c) 3-isopropyl-2,4-dimethyl-1-pentene
d) 4-*tert*-butyl-1-heptene

8.2.

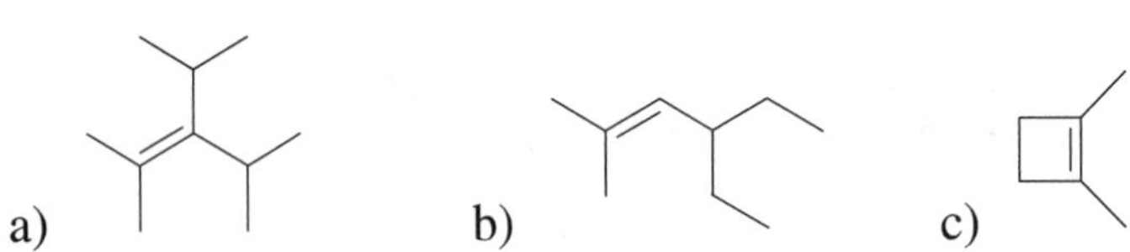

8.3. 2,3-dimethylbicyclo[2.2.1]hept-2-ene

8.4.

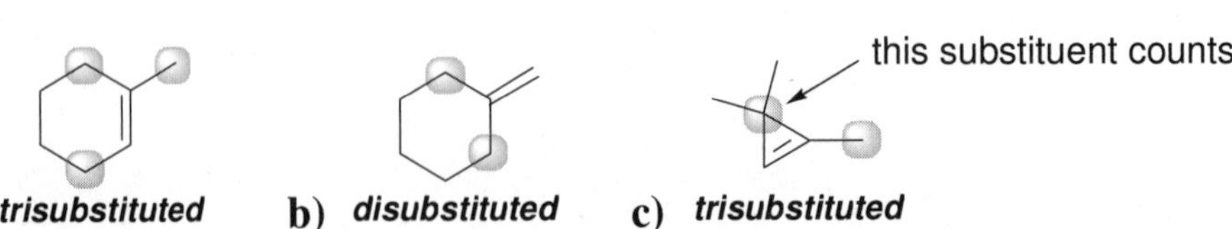

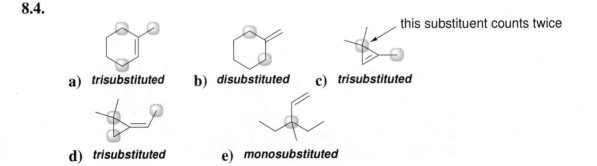

a) ***trisubstituted*** **b)** ***disubstituted*** **c)** ***trisubstituted***

d) ***trisubstituted*** **e)** ***monosubstituted***

8.5.

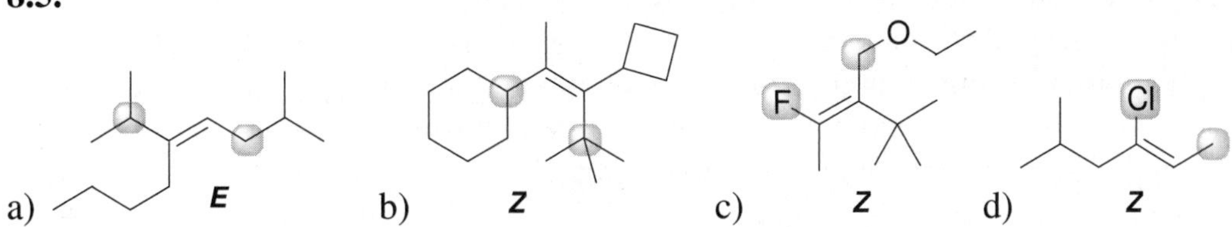

8.6. When using *cis-trans* terminology, we look for two identical groups. In this case, there are two ethyl groups that are in the *trans* configuration:

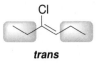

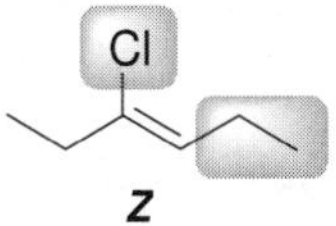

However, when using E-Z terminology, we look for the highest priority at each vinylic position. Chlorine receives a higher priority than ethyl, so in this case, the highest priority groups are on the same side of the pi bond:

Cl

Z

Below are two other examples of alkenes that have the trans configuration, but nevertheless have the Z configuration:

F Br

8.7.

a)

increasing stability

di-substituted	*tri-substituted*	*tetra-substituted*

b)

increasing stability

mono-substituted	*di-substituted*	*tri-substituted*

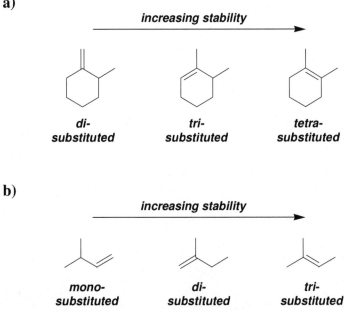

8.8. In the first compound, all of the carbon atoms of the ring are sp^3 hybridized and tetrahedral. As a result, they are supposed to have bond angles of approximately 109.5°, but their bond angles are compressed due to the ring (and are almost 90°). In other words, the compound exhibits angle strain characteristic of small rings. In the second compound, two of the carbon atoms are sp^2 hybridized and trigonal planar. As a result, they are supposed to have bond angles of approximately 120°, but their bond angles are compressed due to the ring (and are almost 90°). The resulting angle strain (120° → 90°) is greater than the angle strain in the first compound (109.5° → 90°). Therefore, the second compound is higher in energy, despite the fact that it has a more highly substituted double bond.

8.9.

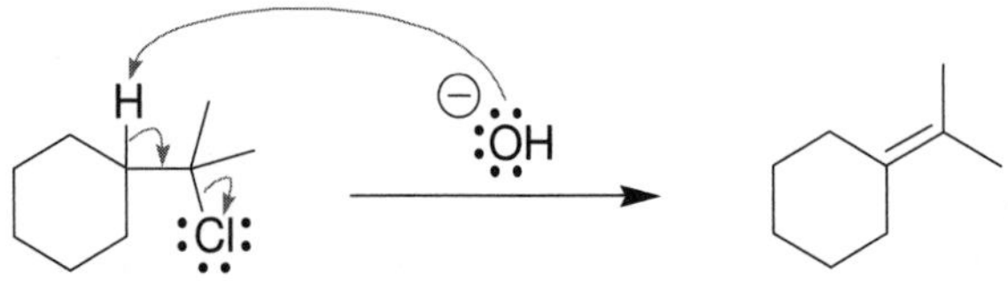

a) H, :Cl:, :OH ⊖

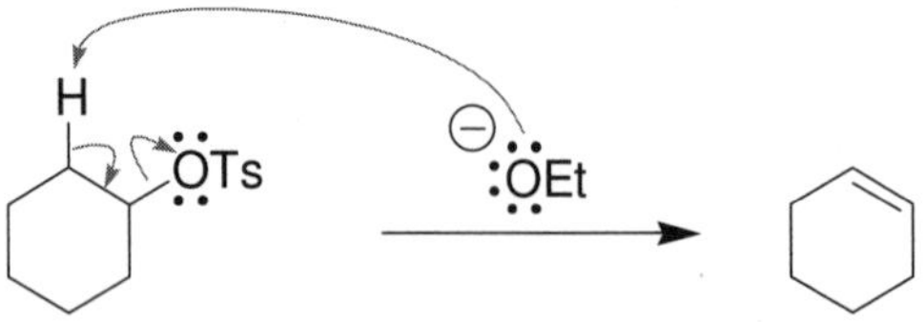

b) H, :OTs, :OEt ⊖

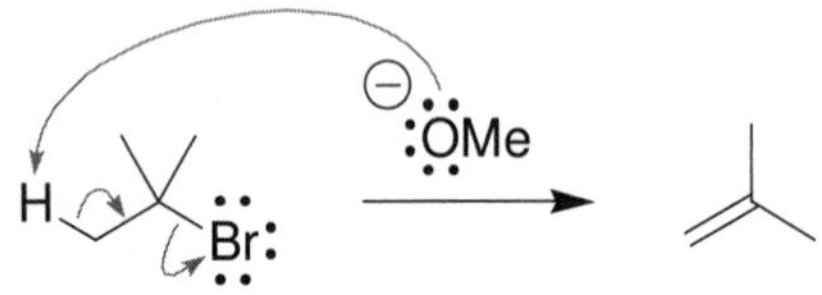

c) H, :Br:, :OMe ⊖

8.10.

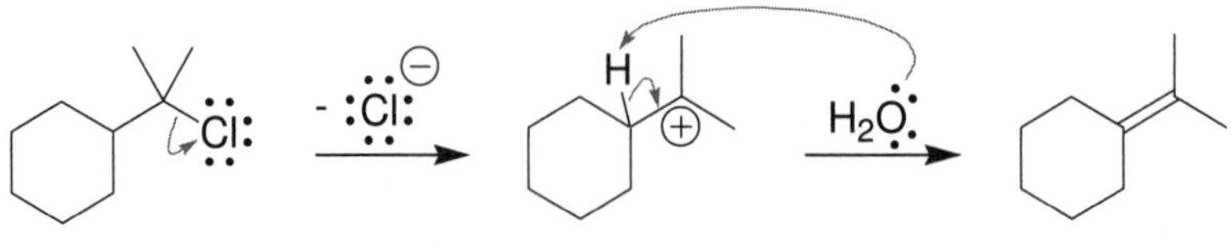

a) :Cl:, - :Cl: ⊖, H, ⊕, H_2O:

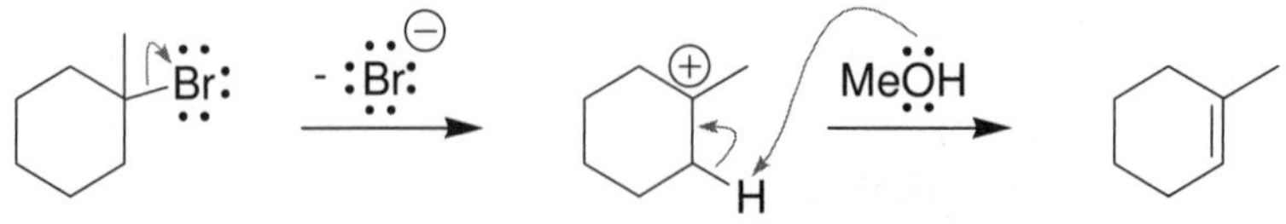

b) :Br:, - :Br: ⊖, ⊕, H, MeOH

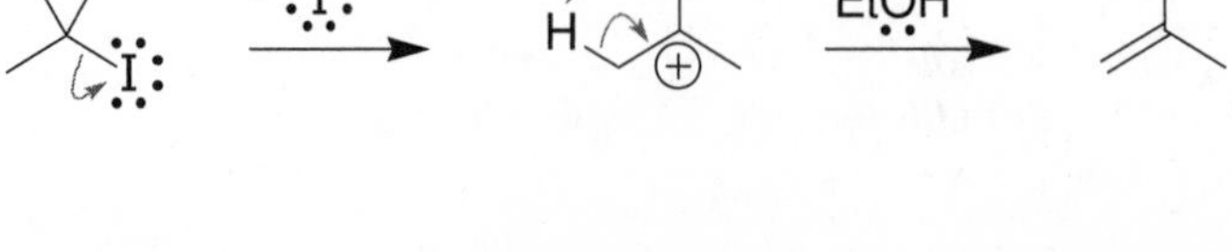

c) :I:, - :I: ⊖, H, ⊕, EtOH

8.11. This mechanism is concerted:

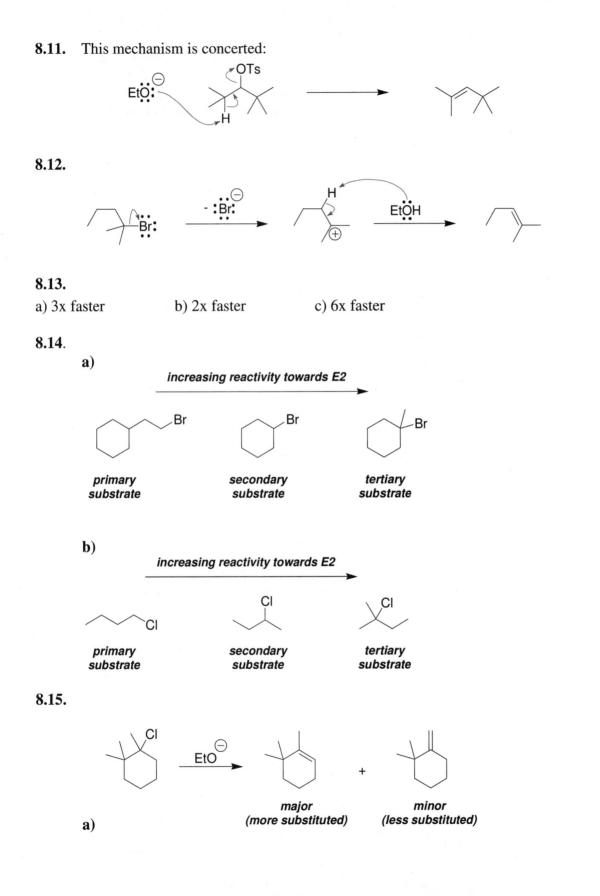

8.12.

8.13.

a) 3x faster b) 2x faster c) 6x faster

8.14.

a)

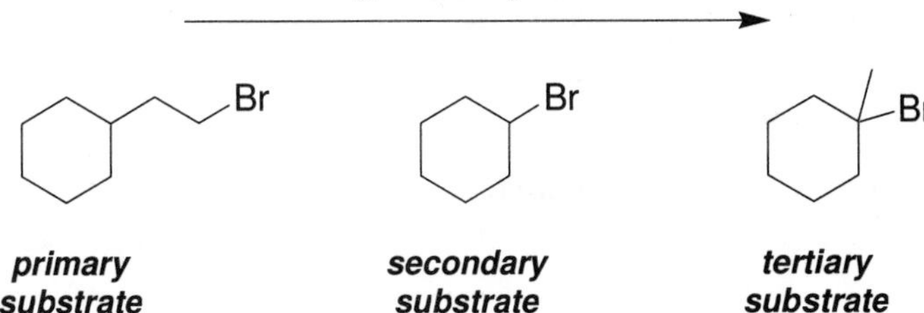

b)

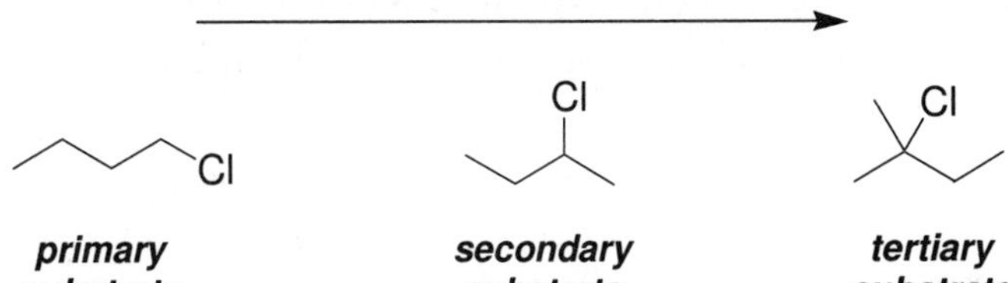

8.15.

a)

major (more substituted) *minor (less substituted)*

b)

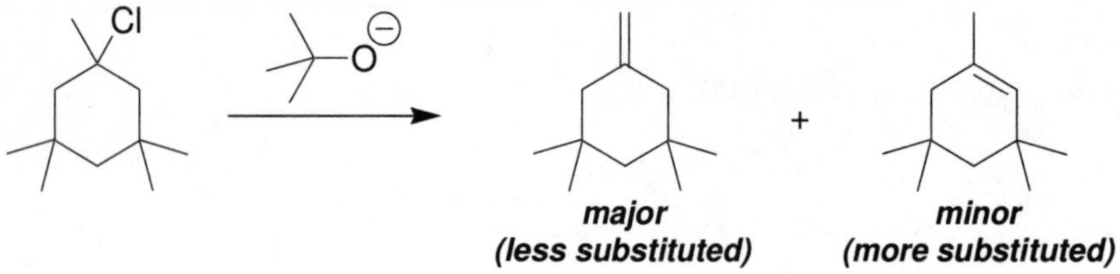

major (less substituted) + minor (more substituted)

c)

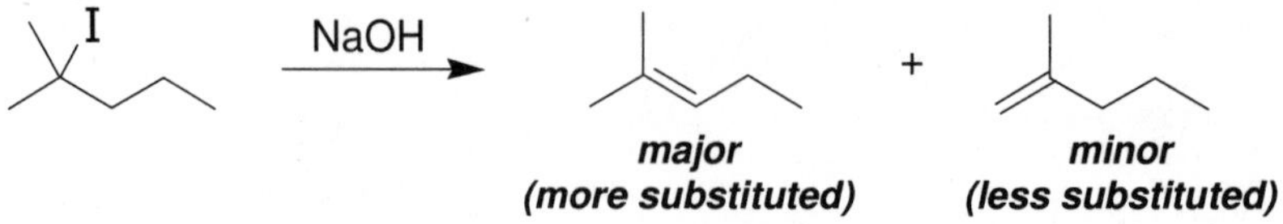

NaOH

major (more substituted) + minor (less substituted)

d)

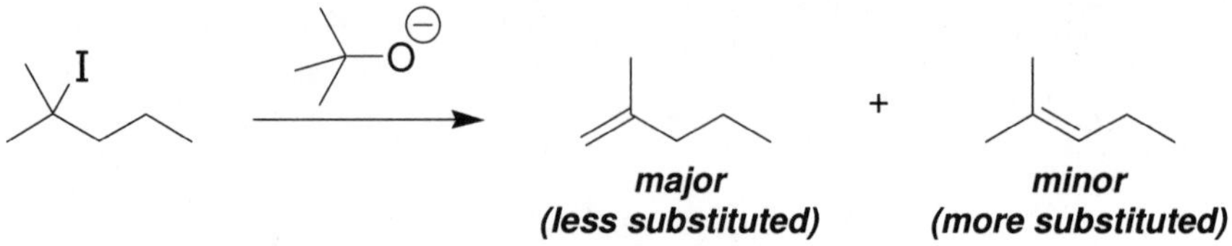

major (less substituted) + minor (more substituted)

e)

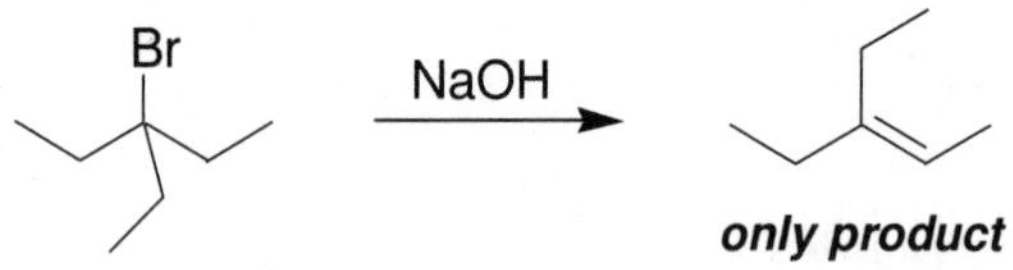

NaOH

only product

f)

only product

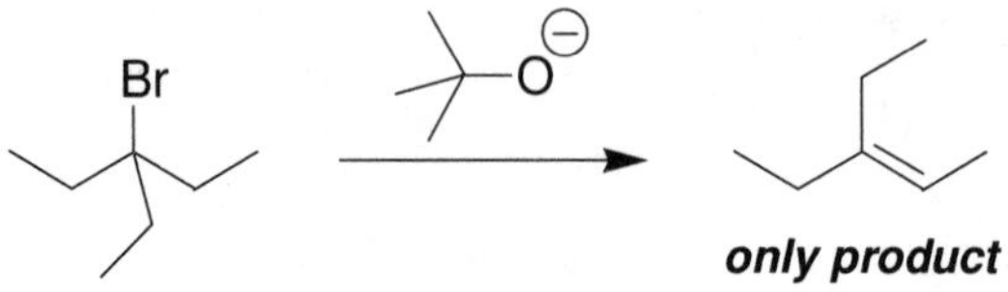

8.16.

a) The more substituted alkene is desired, so hydroxide should be used.
b) The less substituted alkene is desired, so *tert*-butoxide should be used.

8.17.

a)

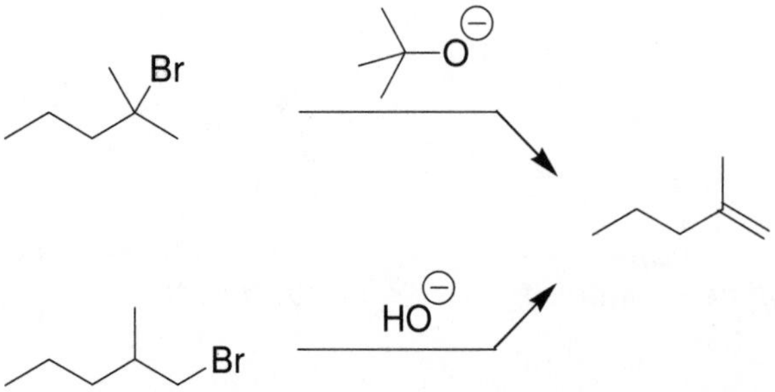

b)

Br

HO

Br

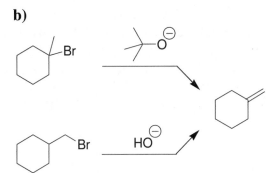

8.18.

+

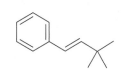

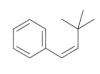

a) *major* *minor*

b) *the only E2 product*

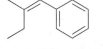

c) *the only E2 product*

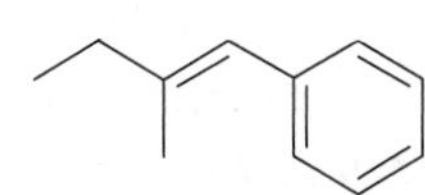

d) *the only E2 product*

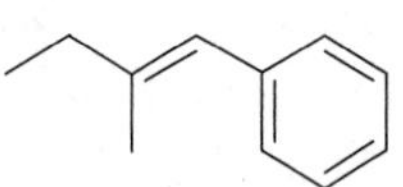

e) *the only E2 product*

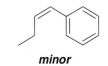

f) *the only E2 product*

g) *major* + *minor*

h) *major* + *minor*

8.19.

Br

Strong Base

8.20. The leaving group in menthyl chloride can only achieve antiperiplanarity with one beta proton, so only one elimination product is observed. In contrast, the leaving group in neomenthyl chloride can achieve antiperiplanarity with two beta protons, giving rise to two possible products:

Cl
H
menthyl chloride

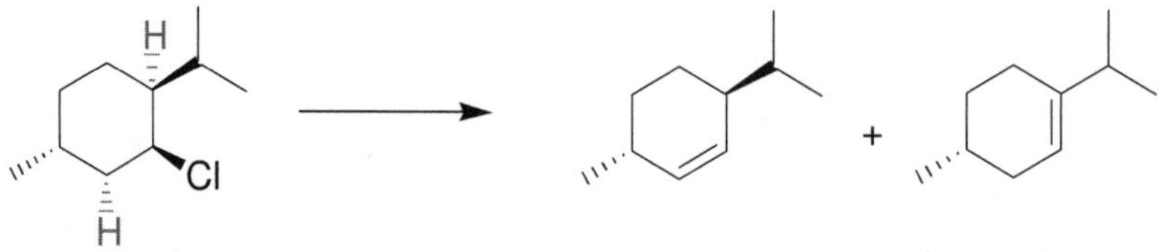

H
Cl
H
neomenthyl chloride

8.21. Because of the bulky *tert*-butyl group, the first compound is essentially locked in a chair conformation in which the chlorine occupies an equatorial position. This conformation cannot undergo an E2 reaction because the leaving group is not antiperiplanar to a proton. However, the second compound is locked in a chair conformation in which the chlorine occupies an axial position. This conformation rapidly undergoes an E2 reaction. Therefore, the second compound is expected to be more reactive towards an E2 process than the first compound.

8.22.

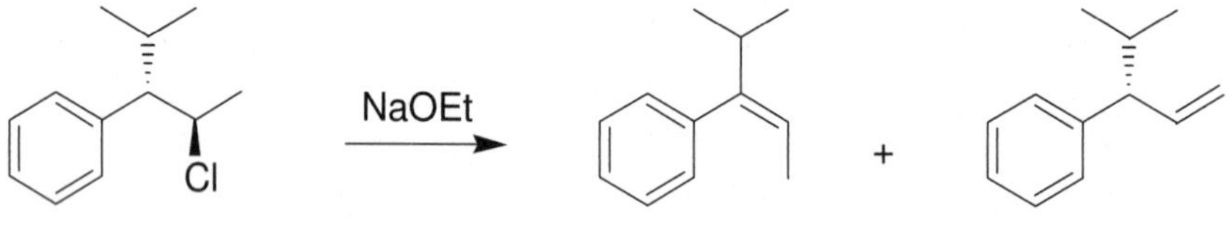

a) Cl — NaOEt → ***major*** + ***minor***

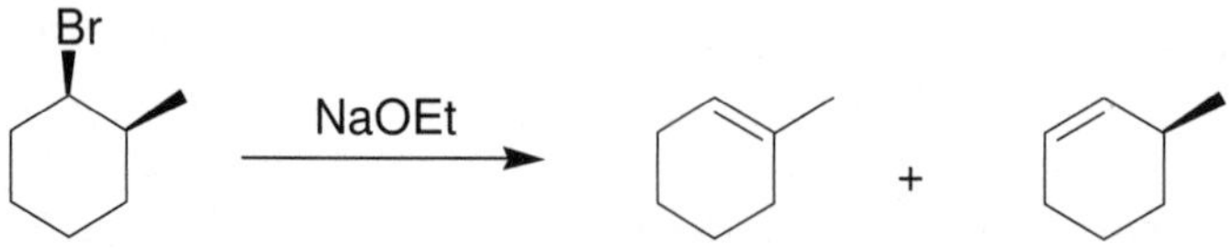

b) Br — NaOEt → ***major*** + ***minor***

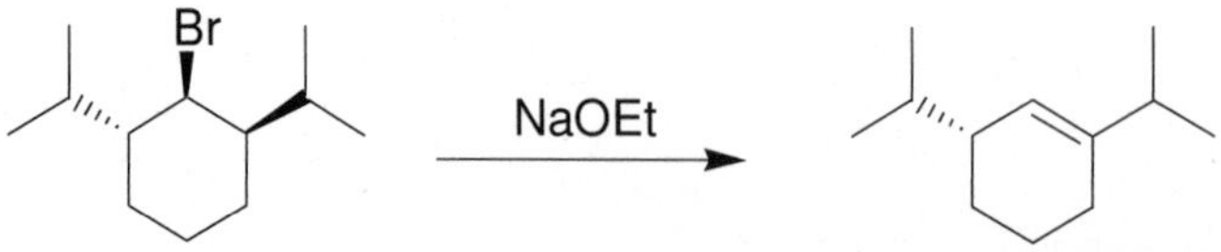

c) Br — NaOEt → ***only E2 product***

I NaOEt + major minor

d)

Br + major minor

e)

Br only E2 product

f)

8.23.

X R R strong base R R R R R R

all R groups are identical

8.24.

X strong base +

8.25.

X strong base +

8.26.

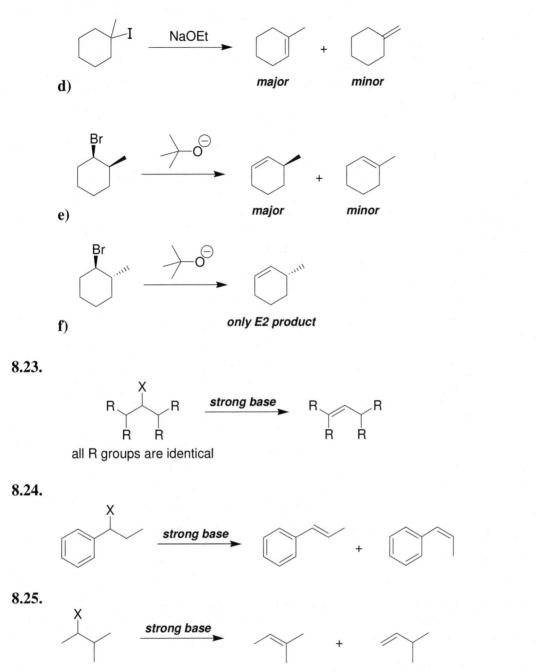

a) Only the concentration of *tert*-butyl iodide affects the rate, so the rate will double.
b) Only the concentration of *tert*-butyl iodide affects the rate, so the rate will remain the same.

8.27.

8.28.

a) 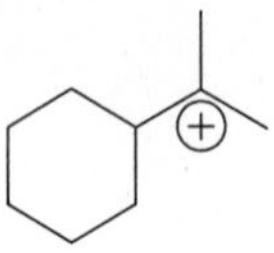 b) 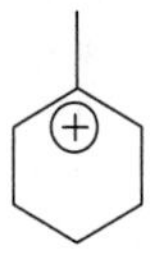 c) 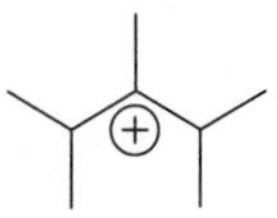 d)

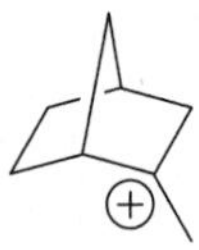

8.29.

a) 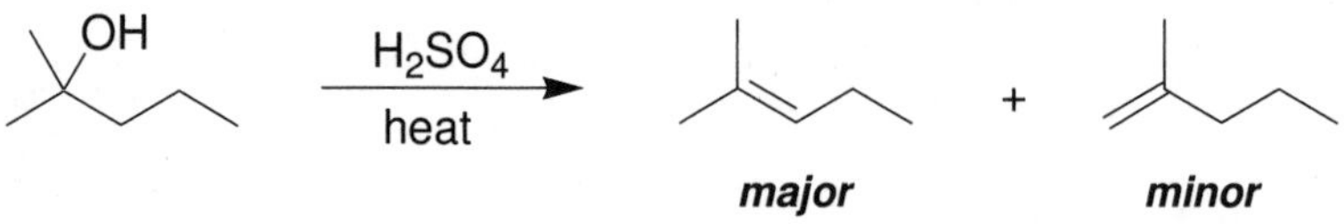 OH; H_2SO_4, heat → *major* + *minor*

b) 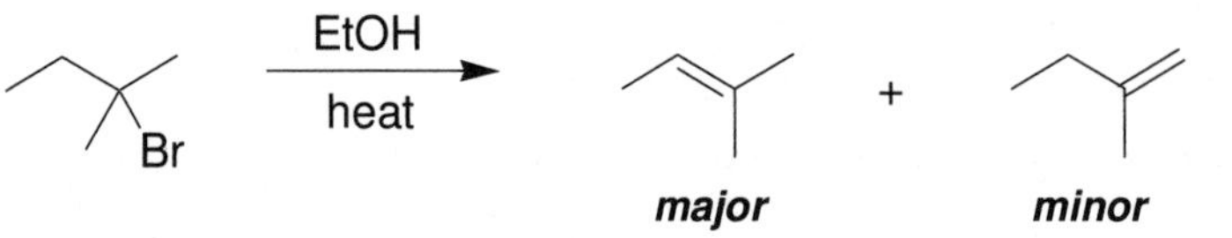 Br; EtOH, heat → *major* + *minor*

c) 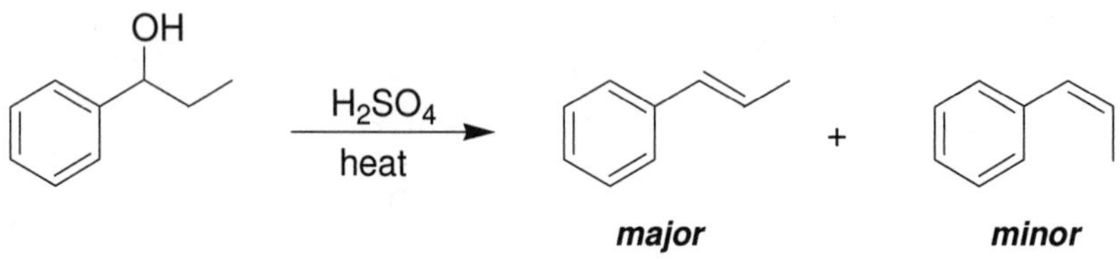 OH; H_2SO_4, heat → *major* + *minor*

d) 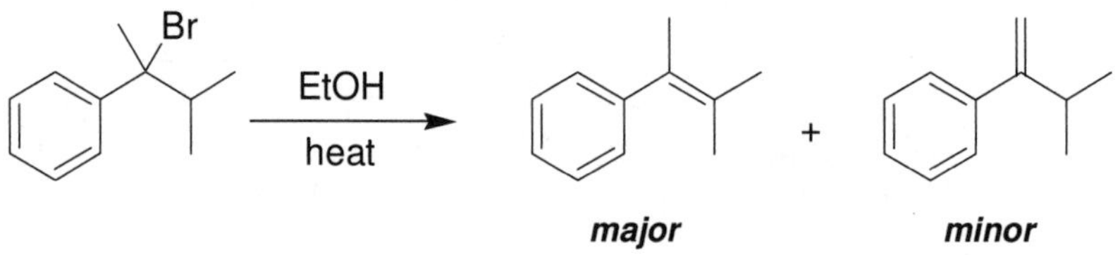 Br; EtOH, heat → *major* + *minor*

8.30. Both alcohols below can be used to form the product. The tertiary alcohol below will react more rapidly because the rate determining step involves formation of a tertiary carbocation rather than a secondary carbocation.

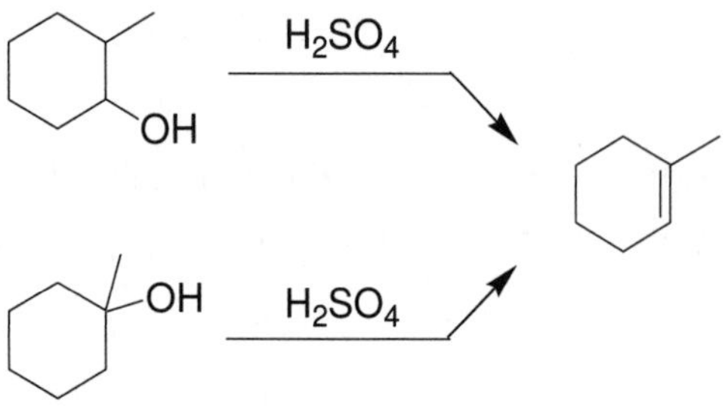

OH H_2SO_4

OH H_2SO_4

8.31.

a) OH; H_2SO_4, heat

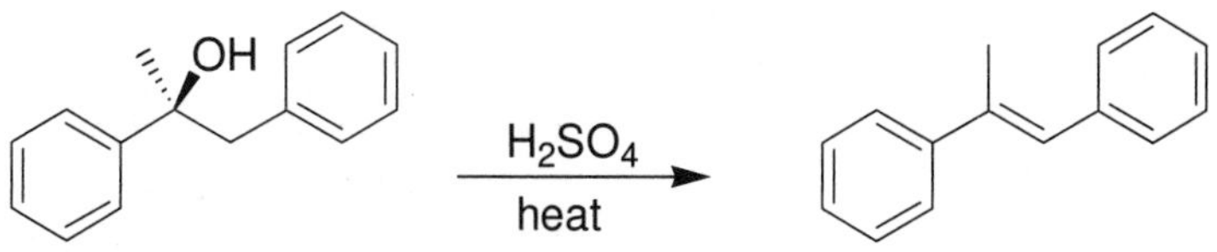

b) HO; H_2SO_4, heat

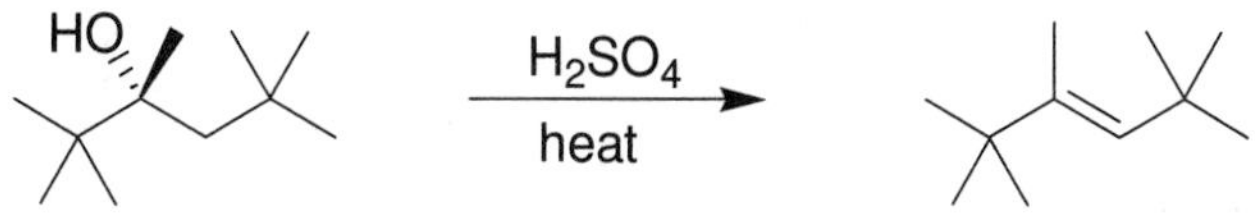

8.32.

a) No, the leaving group is not OH
b) Yes, the leaving group is OH
c) No, the leaving group is not OH
d) Yes, the leaving group is OH
e) No, the leaving group is not OH
f) No, the leaving group is not OH

8.33.

a) No. Loss of the leaving group forms a tertiary carbocation, which will not rearrange.
b) Yes. Loss of the leaving group forms a secondary carbocation, which can undergo a methyl shift to form a more stable tertiary carbocation.
c) Yes. Loss of the leaving group forms a secondary carbocation, which can undergo a hydride shift to form a more stable tertiary carbocation.
d) No. Loss of the leaving group forms a secondary carbocation, which cannot rearrange in this case to form a tertiary carbocation.
e) No. Loss of the leaving group forms a tertiary carbocation, which will not rearrange.
f) No. Loss of the leaving group forms a secondary carbocation, which cannot rearrange in this case to form a tertiary carbocation.

8.34.

a)

$H-O-SO_2-O-H$

$-H_2O$

H_2O

b)

$-Br^{\ominus}$

EtOH

c)

$-Cl^{\ominus}$

EtOH

d)

$H-O-SO_2-O-H$

$-H_2O$

H_2O

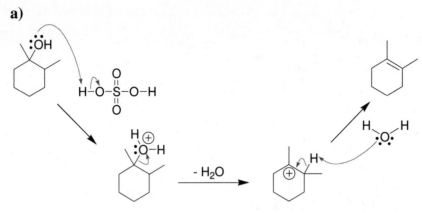

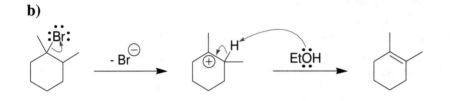

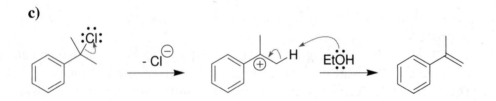

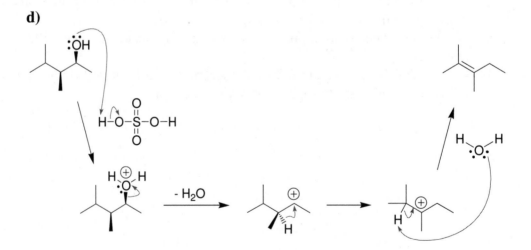

8.35. Problem 8.34b and 8.34c exhibit the same pattern because both have leaving groups that can leave without being protonated, and both do not exhibit a carbocation rearrangement. As a result both mechanisms involve only two steps: 1) loss of a leaving group and 2) proton transfer.

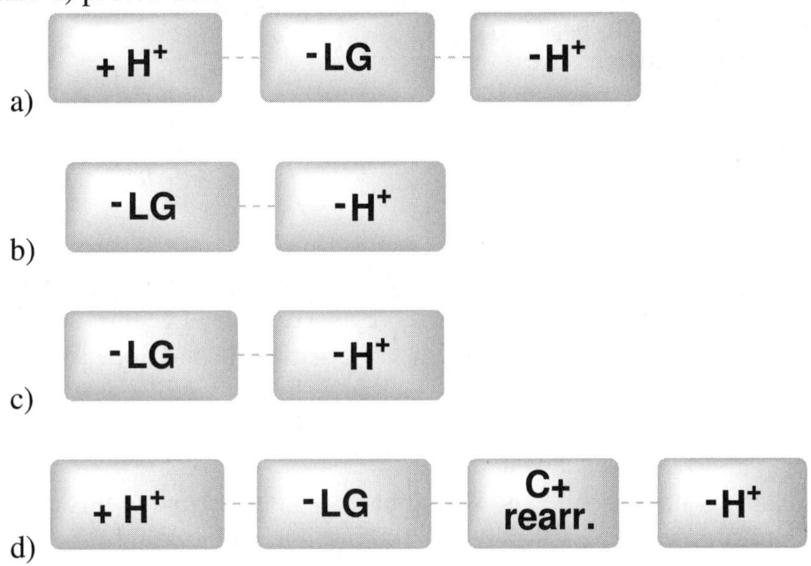

8.36. The first method is more efficient because it employs a strong base to promote an E2 process for a secondary substrate bearing a good leaving group. The second method relies on an E1 process occurring at a secondary substrate, which will be slow and will involve a carbocation rearrangement to produce a different product.

8.37.

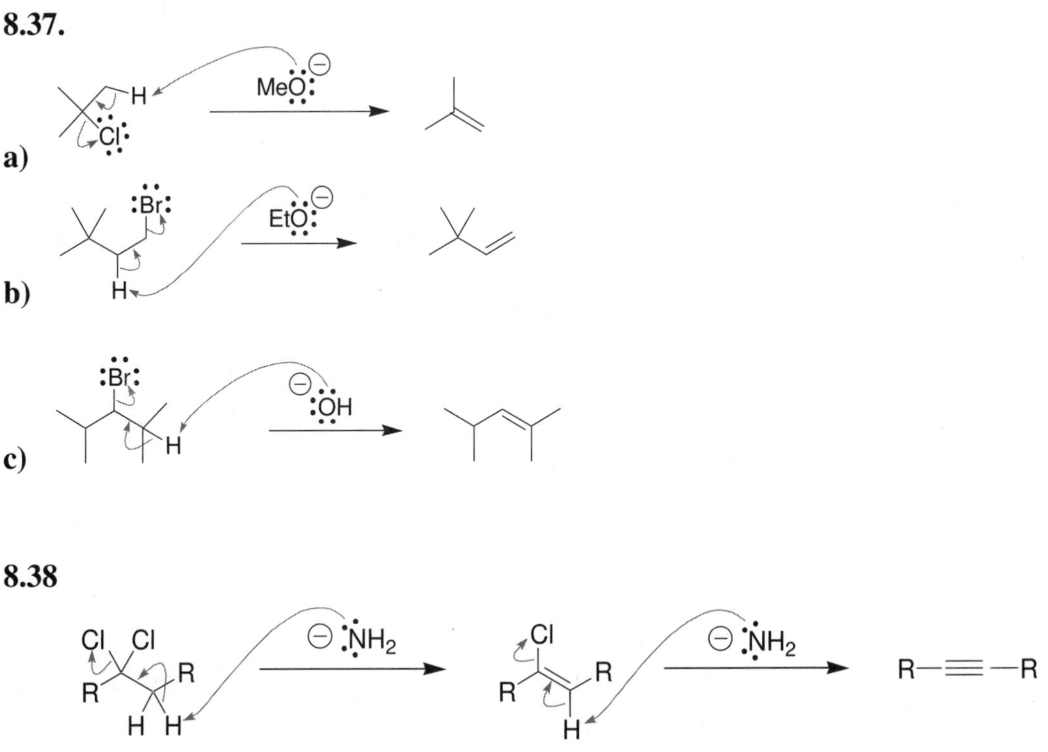

8.38

8.39.

a) weak nucleophile, weak base
b) strong nucleophile, weak base
c) strong nucleophile, strong base
d) strong nucleophile, weak base
e) strong nucleophile, strong base
f) weak nucleophile, weak base
g) strong nucleophile, strong base
h) weak nucleophile, strong base

8.40. Aluminum is a larger atom and is polarizable. Therefore, the entire complex can function as a strong nucleophile, and can serve as a delivery agent of a hydride ion. In contrast, the hydride ion by itself is not polarizable and does not function as a nucleophile.

8.41.

a) NaOH is a strong nucleophile and strong base. The substrate in this case is primary. Therefore, we expect S_N2 (giving the major product) and E2 (giving the minor product).
b) NaSH is a strong nucleophile and weak base. The substrate in this case is primary. Therefore, we expect only S_N2.
c) When a primary alkyl halide is treated with *t*-BuOK, the predominant pathway is expected to be E2.
d) DBN is a weak nucleophile and a strong base. Therefore, we expect only E2.
e) NaOMe is a strong nucleophile and strong base. The substrate in this case is primary. Therefore, we expect S_N2 (giving the major product) and E2 (giving the minor product).

8.42.

a) NaOEt is a strong nucleophile and strong base. The substrate in this case is secondary. Therefore, we expect E2 (giving the major product) and S_N2 (giving the minor product).
b) NaI is a strong nucleophile and weak base. DMSO is a polar aprotic solvent. The substrate is secondary. Under these conditions, only S_N2 can occur.
c) DBU is a weak nucleophile and a strong base. Therefore, we expect only E2.
d) NaOH is a strong nucleophile and strong base. The substrate in this case is secondary. Therefore, we expect E2 (giving the major product) and S_N2 (giving the minor product).
e) *t*-BuOK is a strong, sterically hindered base. Therefore, we expect only E2.

8.43.

a) EtOH is a weak nucleophile and weak base. The substrate in this case is tertiary. Therefore, we expect both S_N1 and E1.

b) *t*-BuOK is a strong, sterically hindered base. Therefore, we expect only E2.
c) NaI is a strong nucleophile and weak base. The substrate in this case is tertiary. Therefore, we expect only S_N1.
d) NaOEt is a strong nucleophile and strong base. The substrate in this case is tertiary. Therefore, we expect only E2.
e) NaOH is a strong nucleophile and strong base. The substrate in this case is tertiary. Therefore, we expect only E2.

8.44.

a) An E2 reaction does not readily occur because the base is weak.
b) An E1 reaction does not readily occur because the substrate is primary.
c) Replacing the weak base (EtOH) with a strong base (such as NaOEt) would greatly enhance the rate of an E2 process.
d) Replacing the primary substrate with a tertiary substrate (such as 1-chloro-1,1-dimethylbutane) would greatly enhance the rate of an E1 process.

8.45. The substrate is tertiary, so S_N2 cannot occur at a reasonable rate. There are no beta protons, so E2 also cannot occur.

8.46.

a)

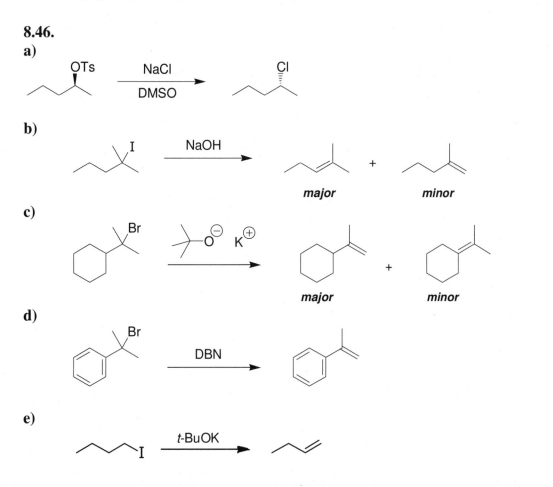

OTs NaCl DMSO Cl

b)

I NaOH

major *minor*

c)

Br K

major *minor*

d)

Br DBN

e)

I *t*-BuOK

f)

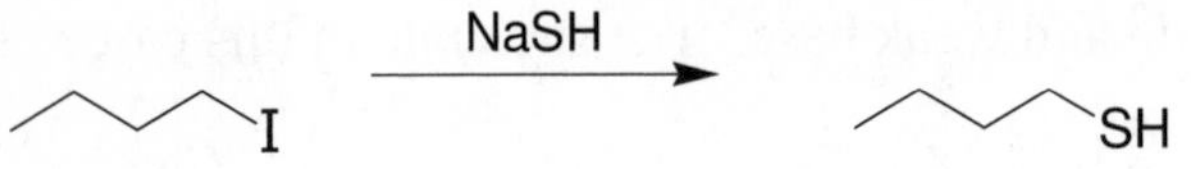

NaSH

I

SH

g)

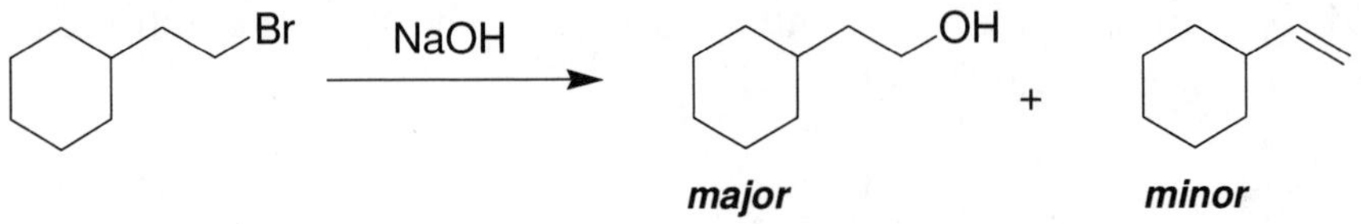

Br

NaOH

OH

+

major

minor

h)

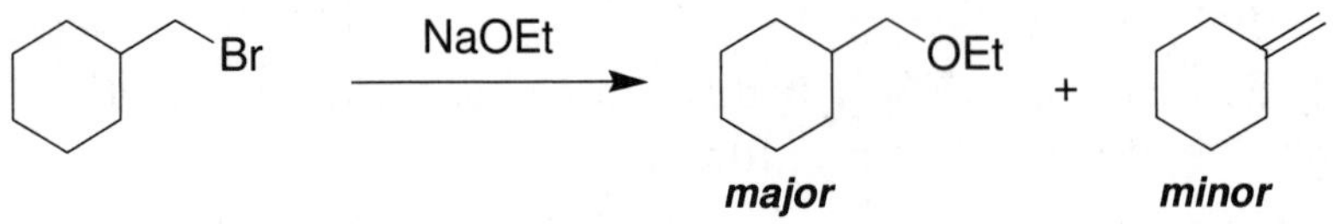

Br

NaOEt

OEt

+

major

minor

i)

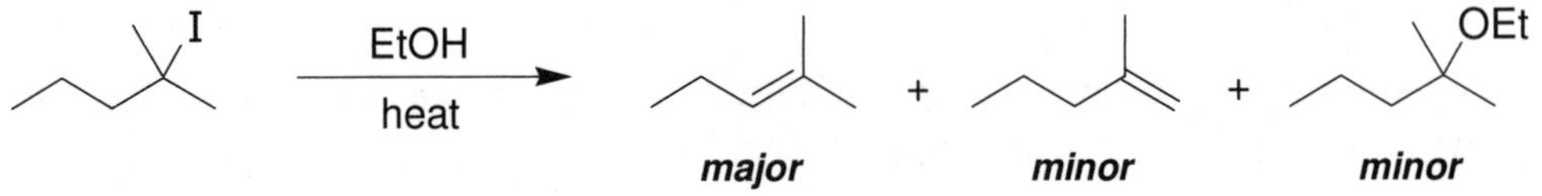

I

EtOH

heat

+

+

OEt

major

minor

minor

j)

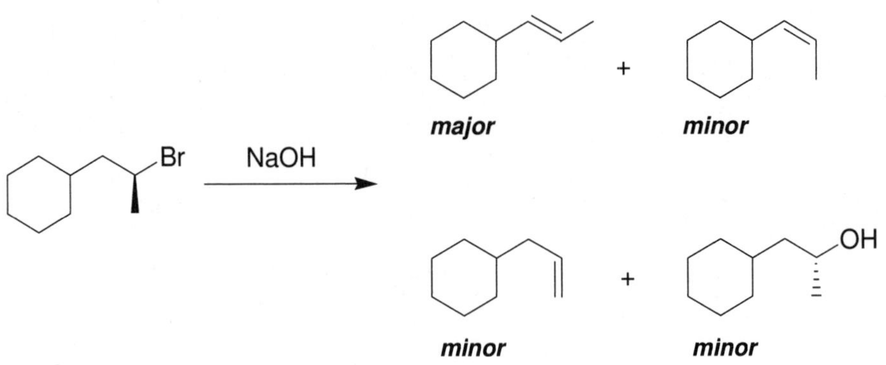

+

major

minor

Br

NaOH

+

OH

minor

minor

k)

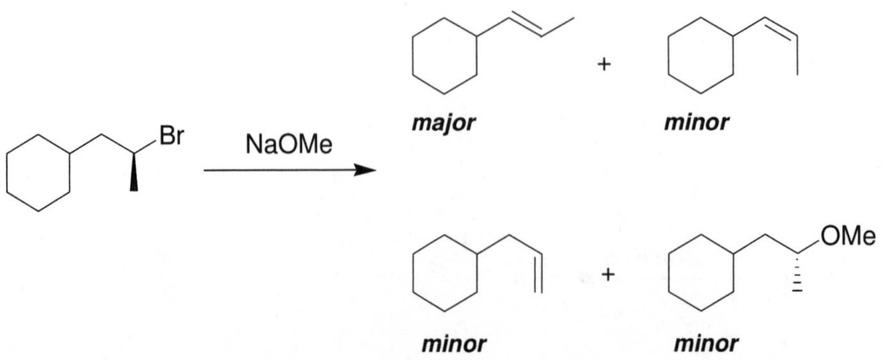

+

major

minor

Br

NaOMe

+

OMe

minor

minor

l)

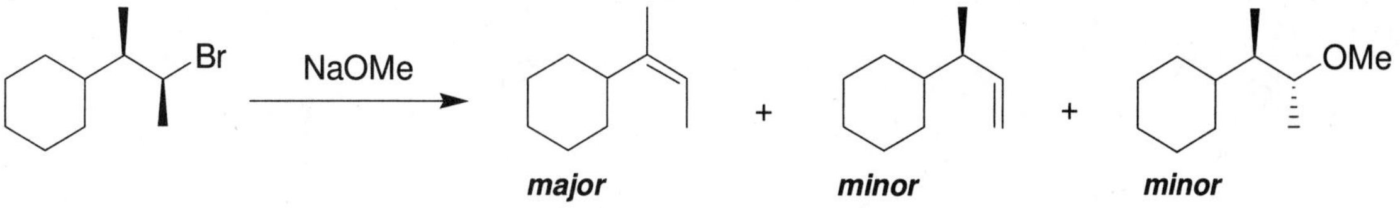
Br
NaOMe
OMe
major
minor
minor

m)

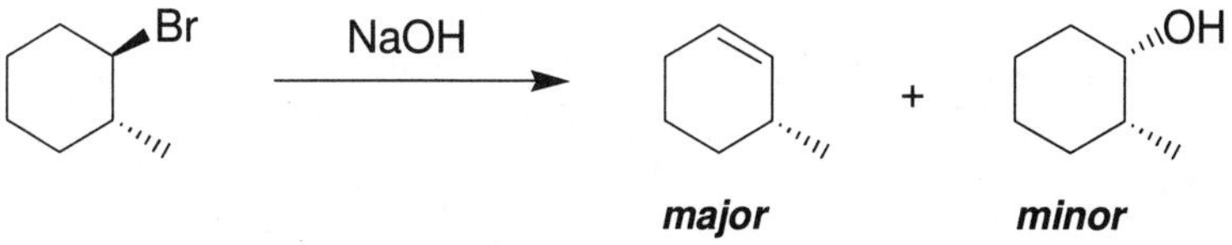
Br
NaOH
OH
major
minor

n)

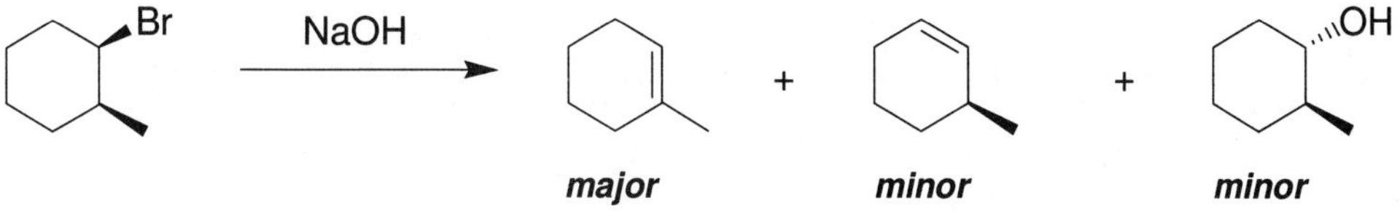
Br
NaOH
OH
major
minor
minor

8.47. There are only two constitutional isomers with molecular formula C_3H_7Cl:

Cl
Cl

a primary alkyl halide *a secondary alkyl halide*

Sodium methoxide is both a strong nucleophile and a strong base. When compound A is treated with sodium methoxide, a substitution reaction predominates. Therefore, compound A must be the primary alkyl chloride above. When compound B is treated with sodium methoxide, an elimination reaction predominates. Therefore, compound B must be the secondary alkyl chloride:

Cl
Cl

Compound A *Compound B*

8.48.

a)

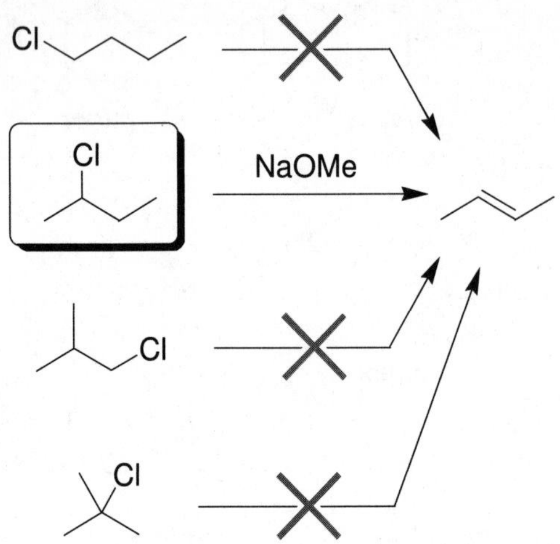

b)

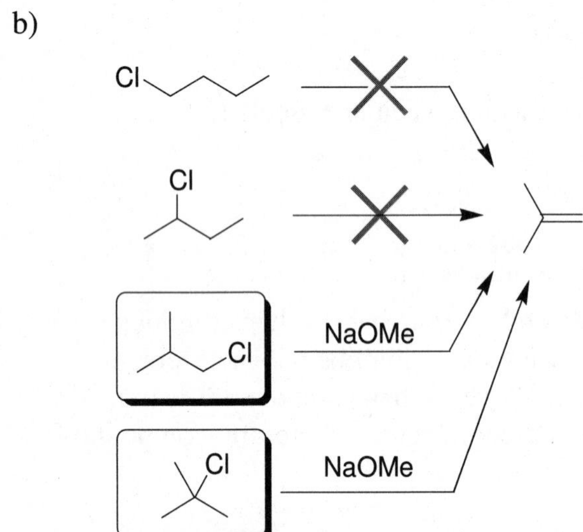

8.49.

Cl NaOEt

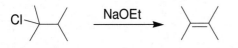

8.50.

a) *trans*-3,4,5,5-tetramethyl-3-heptene
b) 1-ethylcyclohexene
c) 2-methylbicyclo[2.2.2]oct-2-ene

8.51.

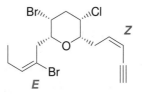

8.52. Because of the bulky *tert*-butyl group, the *trans* isomer is essentially locked in a chair conformation in which the chlorine occupies an equatorial position. This conformation cannot readily undergo an E2 reaction because the leaving group is not antiperiplanar to a proton. However, the *cis* isomer is locked in a chair conformation in which the chlorine occupies an axial position. This conformation rapidly undergoes an E2 reaction.

8.53.

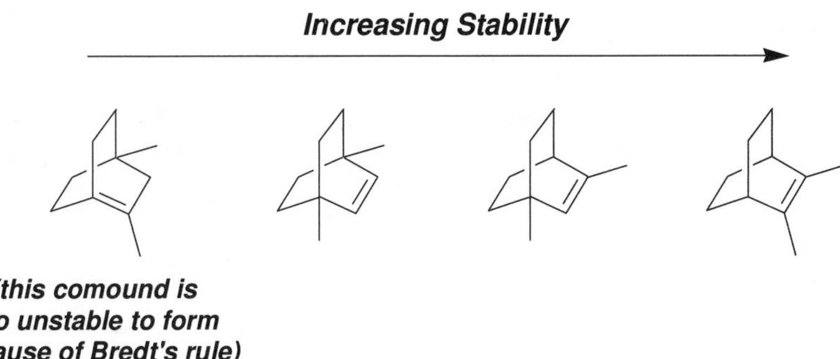

8.54.

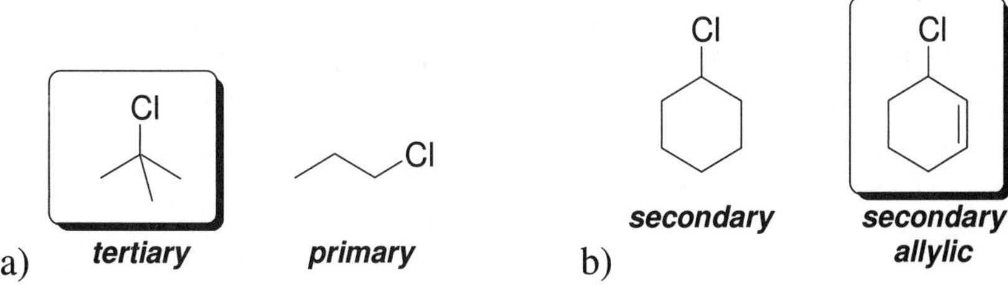

8.55.

a) NaOH, because hydroxide bears a negative charge.
b) sodium ethoxide, because ethoxide bears a negative charge.
c) trimethylamine, because of the electron donating effects of the alkyl groups.

8.56.

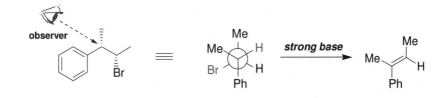

8.57.

a) The rate of an E2 process is dependent on the concentrations of the substrate and the base. Therefore, the rate will be doubled if the concentration of *tert*-butyl bromide is doubled.

b) The rate of an E2 process is dependent on the concentrations of the substrate and the base. Therefore, the rate will be doubled if the concentration of sodium ethoxide is doubled.

8.58.

a) The rate of an E1 process is dependent only on the concentration of the substrate (not the base). Therefore, the rate will be doubled if the concentration of *tert*-butyl bromide is doubled.

b) The rate of an E1 process is dependent only on the concentration of the substrate (not the base). Therefore, the rate will remain the same if the concentration of ethanol is doubled.

8.59.

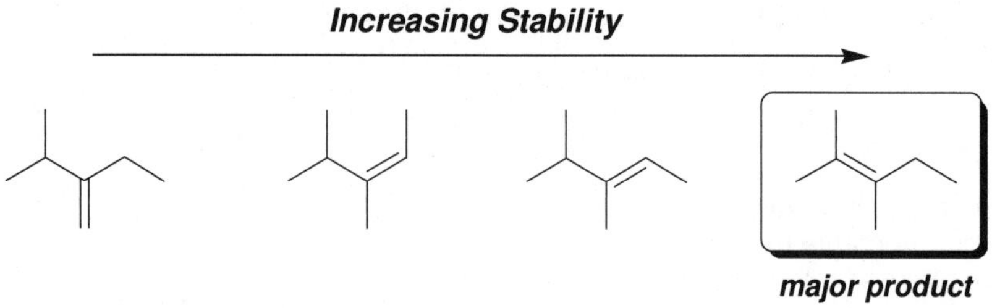

Increasing Stability

major product

8.60. There are only two beta protons to abstract: one at C2 and the other at C4. Abstraction of either proton leads to the same product.

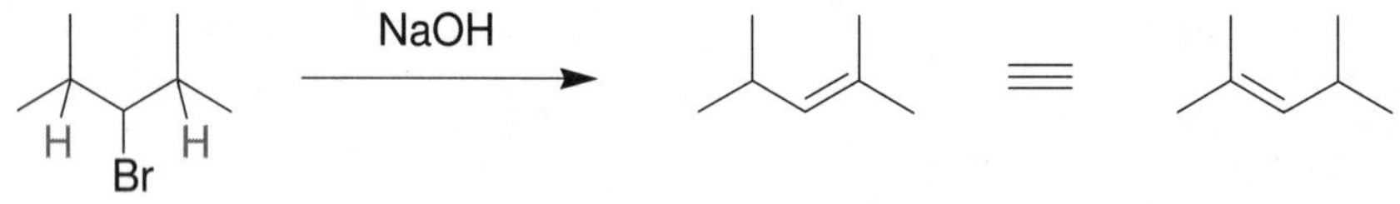

H H Br NaOH

8.61.

a) 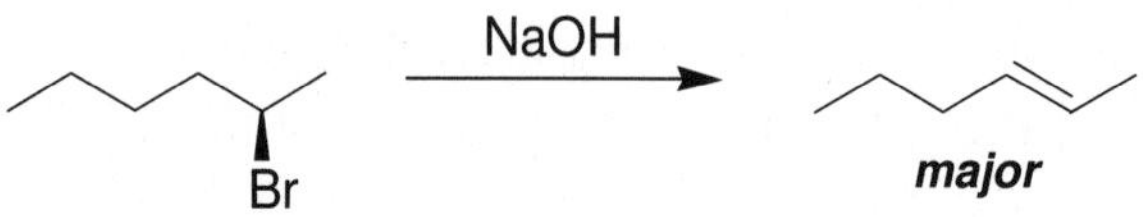 Br NaOH ***major***

b) 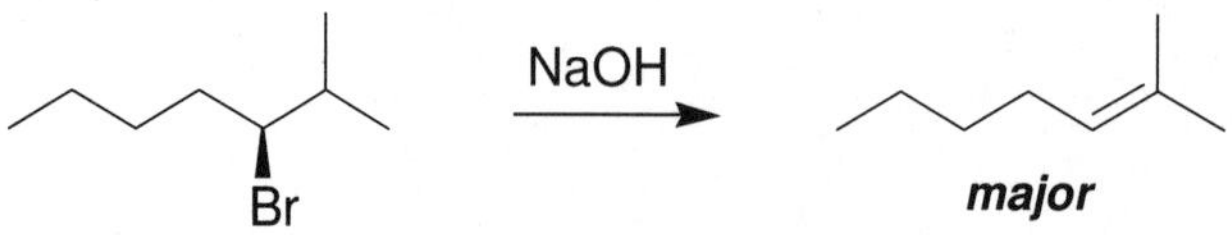 Br NaOH ***major***

c) 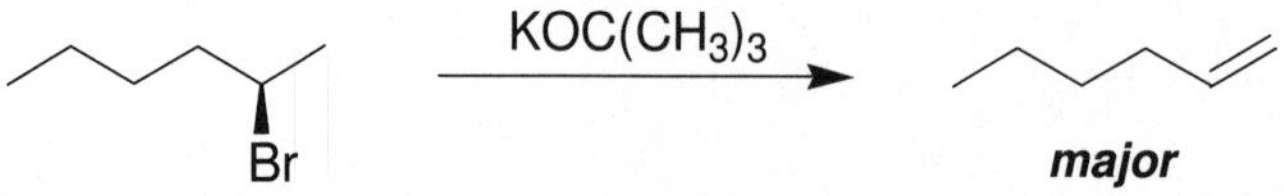 Br $KOC(CH_3)_3$ ***major***

d) 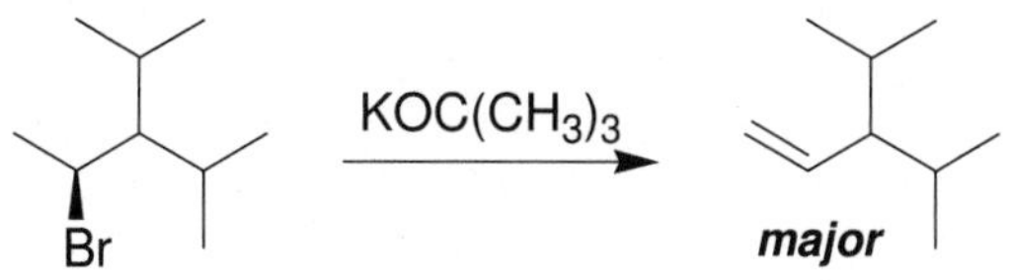 Br $\xrightarrow{KOC(CH_3)_3}$ *major*

8.62.

a) 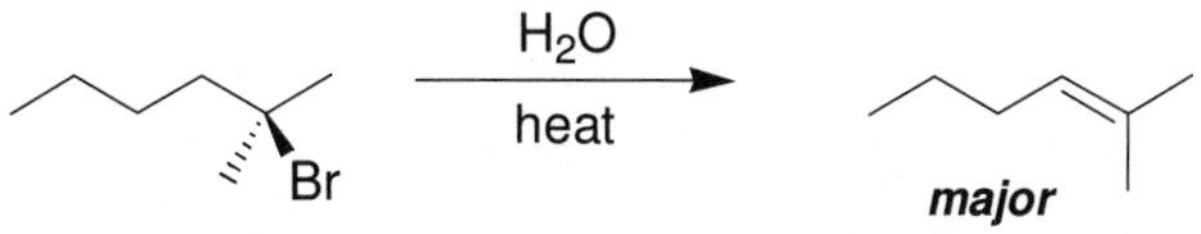 Br $\xrightarrow[\text{heat}]{H_2O}$ *major*

b) 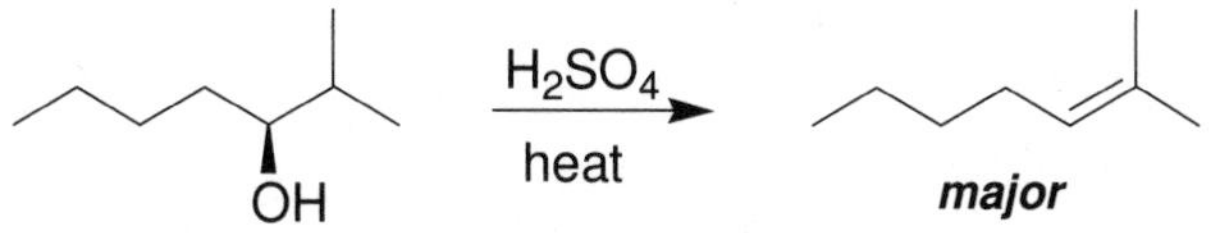 OH $\xrightarrow[\text{heat}]{H_2SO_4}$ *major*

8.63.

a) 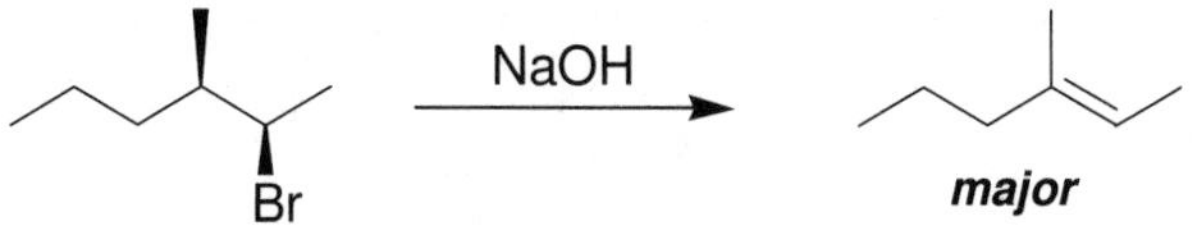 Br $\xrightarrow{NaOH}$ *major*

b) 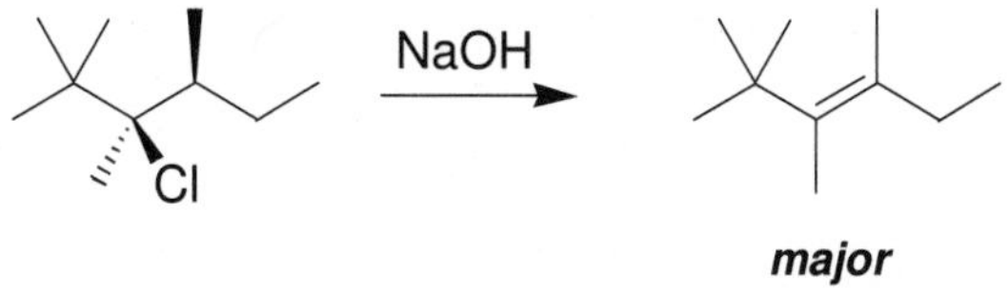 Cl $\xrightarrow{NaOH}$ *major*

8.64. The reagent is a strong nucleophile and a strong base, so we expect a bimolecular reaction. The substrate is tertiary so only E2 can operate (S_N2 is too sterically hindered to occur). There is only one possible regiochemical outcome for the E2 process, because the other beta positions lack protons.

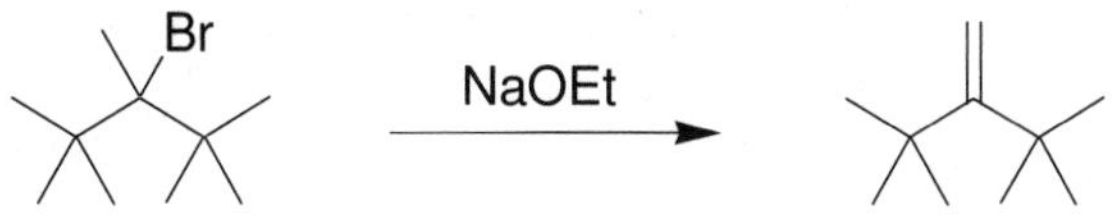

Br $\xrightarrow{NaOEt}$

8.65.

a) 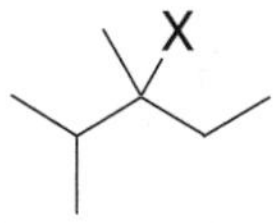 X b) 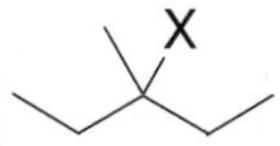 X c)  X d) 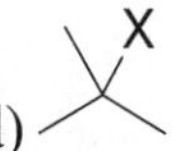 X

8.66.

a) one
b) three
c) two
d) two
e) five

8.67.

a)

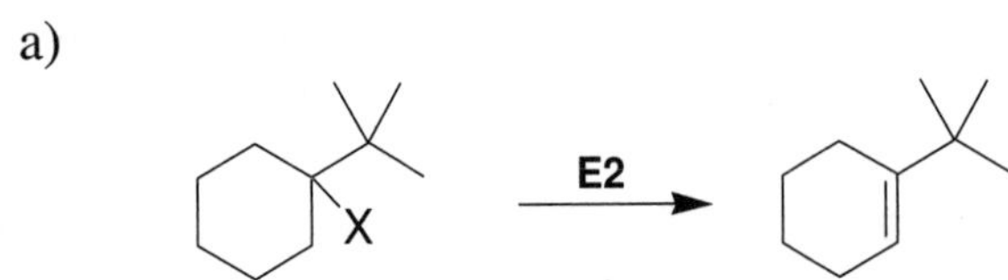

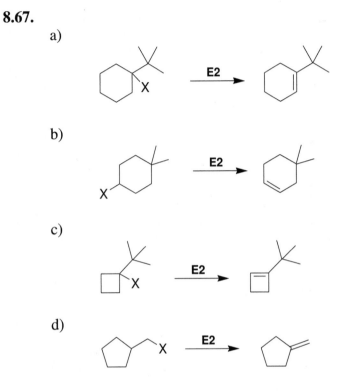

b)

c)

d)

8.68.

a)

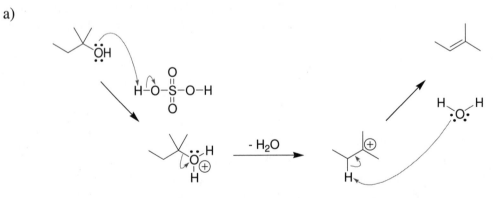

b) This is an E1 process, so the rate is dependent only on the substrate:
Rate = k[substrate]

c)

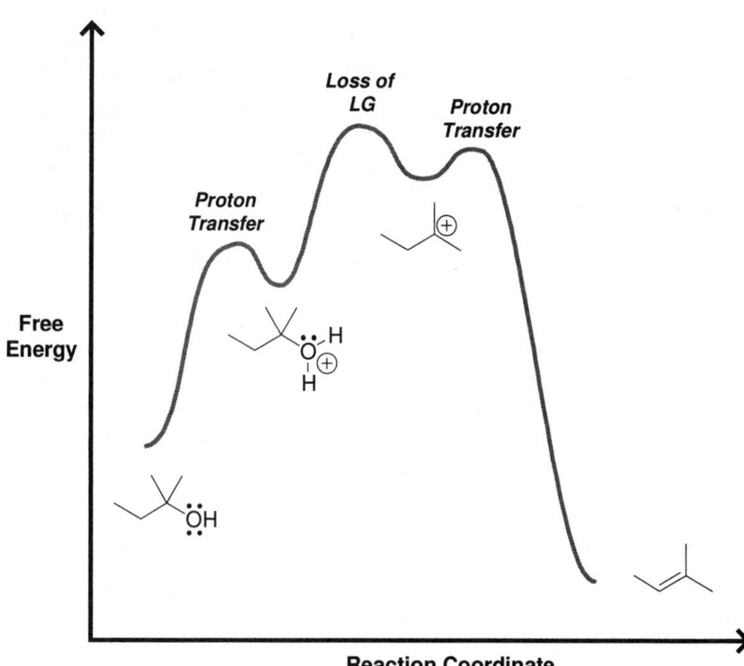

8.69. The primary substrate will not undergo an E1 reaction because primary carbocations are too high in energy to form readily.

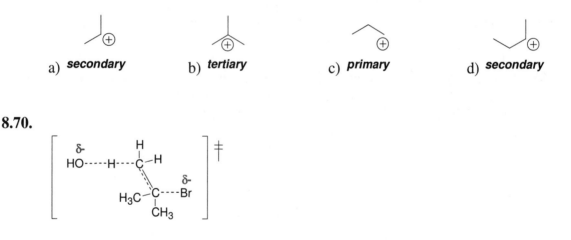

a) ***secondary***

b) ***tertiary***

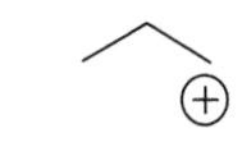

c) ***primary***

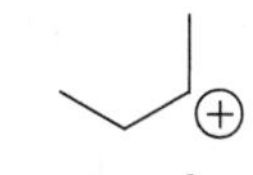

d) ***secondary***

8.70.

δ- HO-----H----C(H)(H)----C(H$_3$C)(CH$_3$)----Br δ- ‡

8.71. The first compound produces a trisubstituted alkene, while the second compound produces a monosubstituted alkene. As such, the transition state for the reaction of the first compound will be lower in energy than the transition state for the reaction of the second compound.

8.72. The first reaction is very slow, because the *tert*-butyl group effectively locks the ring in a chair conformation in which the leaving group (Br) occupies an equatorial position. In this conformation, the leaving group cannot be antiperiplanar to a beta proton. So the reaction can only occur from the other chair conformation, which the compound does not readily adopt. The second reaction is very rapid, because the *tert*-butyl group effectively locks the ring in a chair conformation in which the leaving group (Br) occupies an axial position. In this conformation, the leaving group is antiperiplanar to a beta proton. The third reaction does not occur at all because there are no beta protons that are antiperiplanar to the leaving group (on a cyclohexane ring, there must be at least one beta proton that is *trans* to the leaving group in order to be able to adopt an antiperiplanar conformation).

8.73. Pi bonds cannot be formed at the bridgehead of a bicyclic compound, unless one of the rings is large (at least eight carbon atoms). This rule is known as Bredt's rule.

8.74.

a) The first compound will react more rapidly because it is tertiary
b) The second compound will react more rapidly in an E2 reaction because the first compound does not have any beta protons (and therefore cannot undergo E2 at all).

8.75.

a) The Zaitsev product is desired, so sodium hydroxide should be used.
b) The Hofmann product is desired, so potassium *tert*-butoxide should be used.
c) The Zaitsev product is desired, so sodium hydroxide should be used.
d) The Hofmann product is desired, so potassium *tert*-butoxide should be used.

8.76. There is only one beta proton that can be abstracted so as to form the Zaitsev product. This proton is *cis* to the leaving group, and therefore, it cannot be antiperiplanar to the leaving group (not in either chair conformation). As a result, only the Hofmann product can be formed.

8.77.

a)

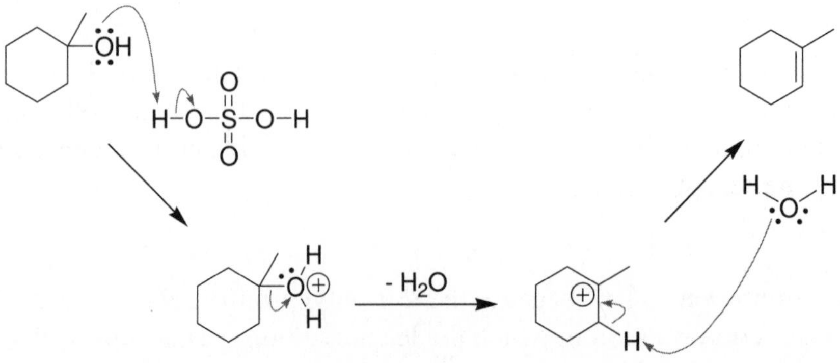

H-O-S-O-H

- H_2O

b)

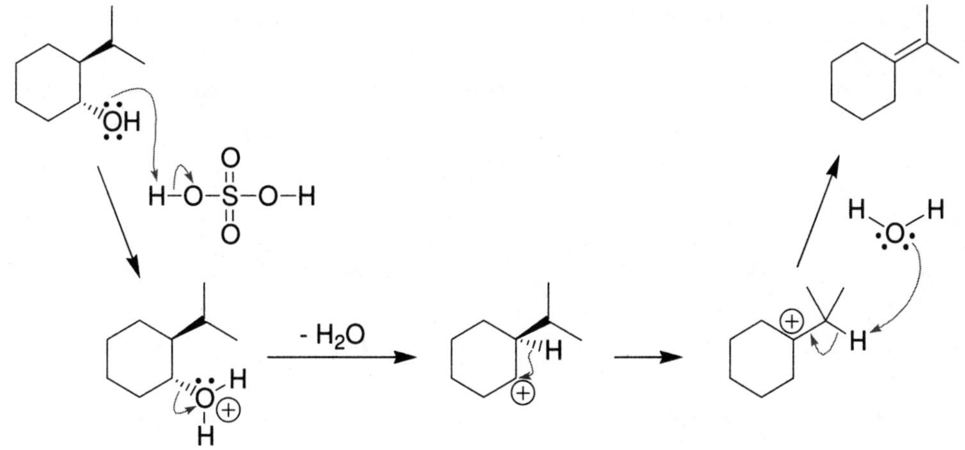

- H_2O

c)

- H_2O

d)

- $I^{\ominus}$

EtOH

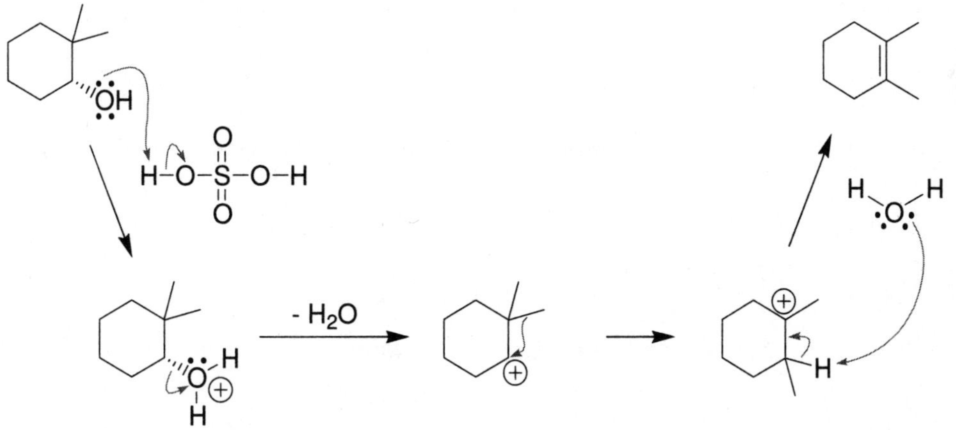

e)

OEt

8.78.

Cl

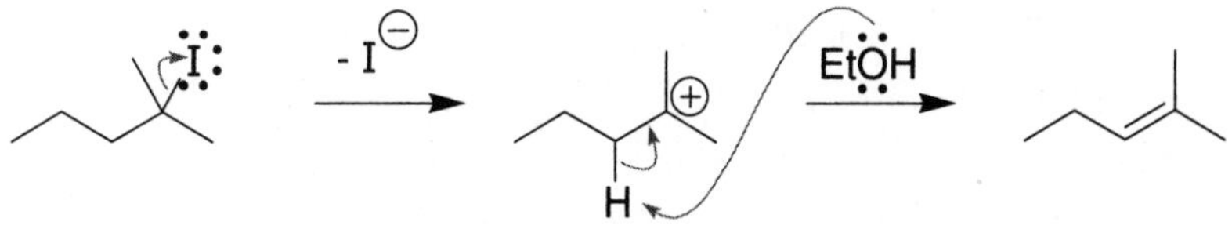

t-BuOK

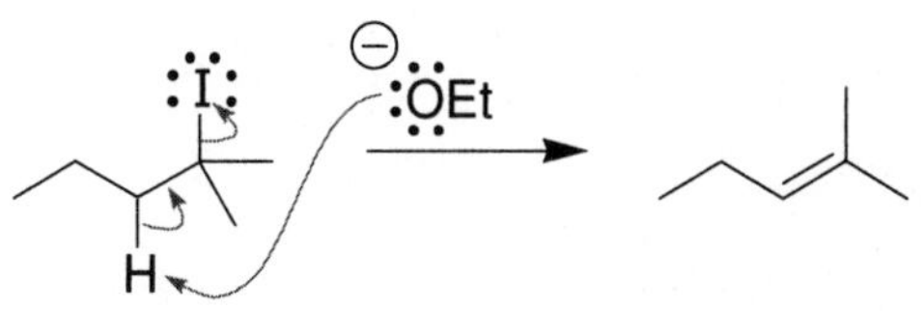

a)

major *minor* *minor*

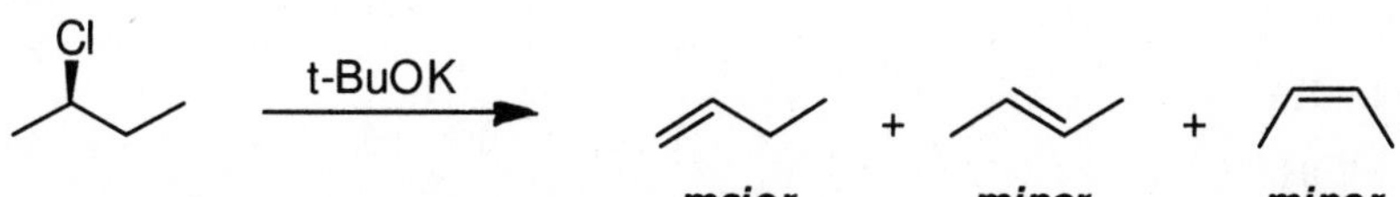

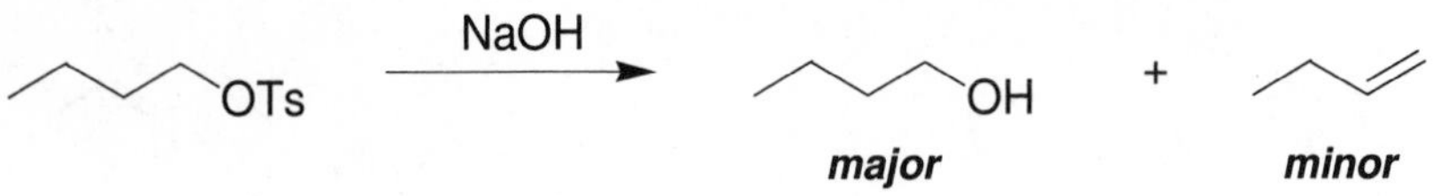

b) NaOH — *major*, *minor*

c) H_2SO_4, heat — *major*, *minor*, *minor*

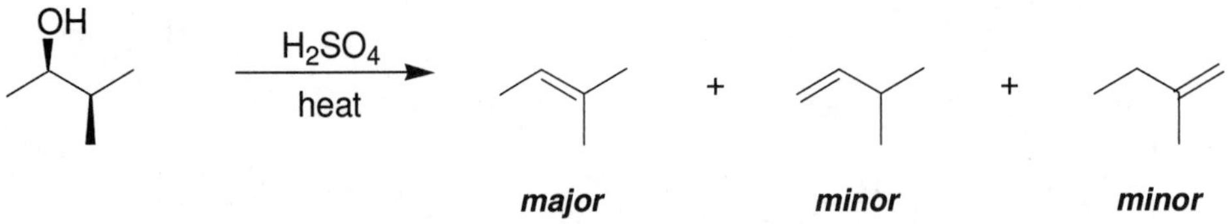

d) OTs — NaCl, DMSO — Cl

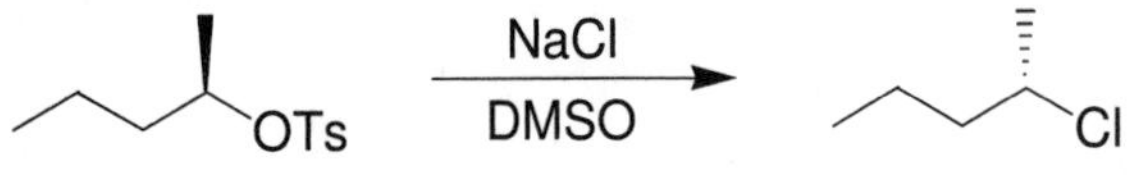

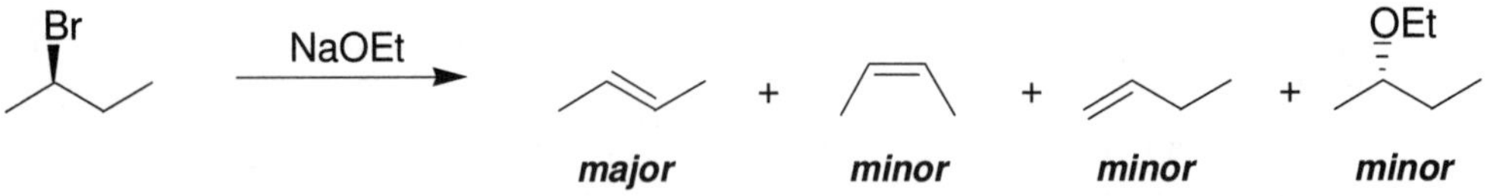

e) Br — NaOEt — *major*, *minor*, *minor*, OEt *minor*

f) Br — NaOEt — *major*, *minor*

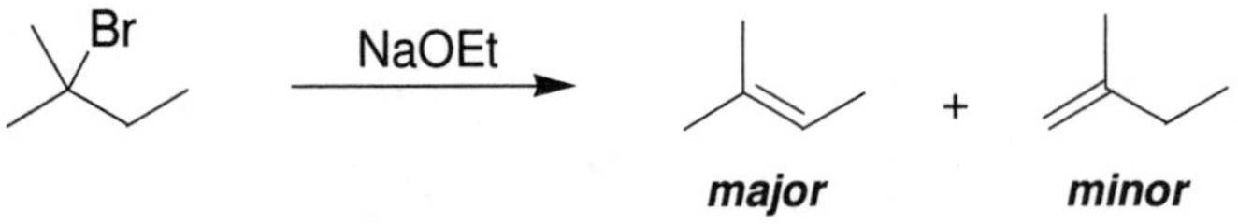

g) Br — NaOEt, EtOH — *major*, *minor*, OEt *minor*

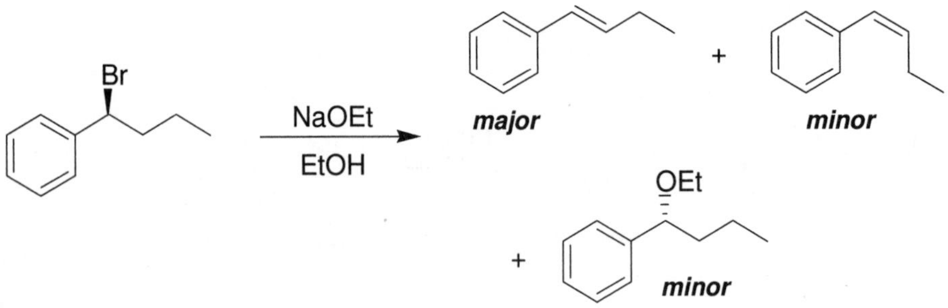

h) Br — NaOEt, EtOH — *major*, OEt *minor*

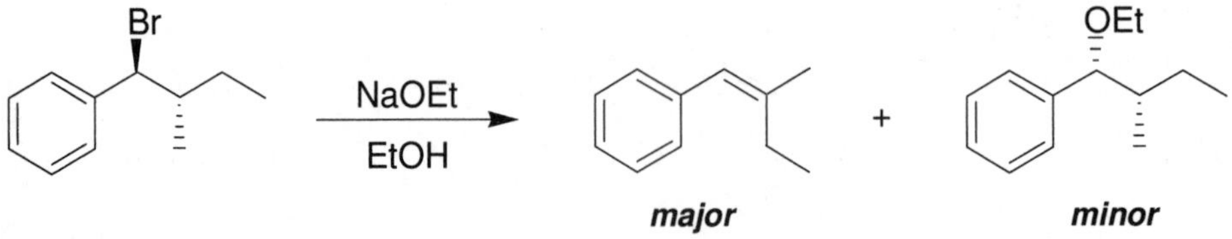

i) Br — NaOEt, EtOH — *major*, OEt *minor*

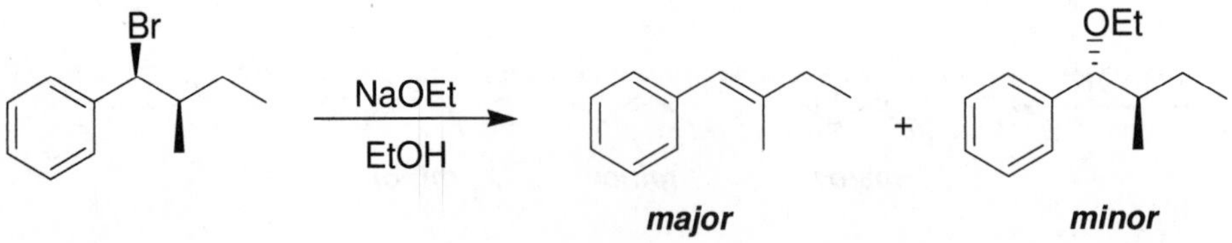

j) $KOC(CH_3)_3$ — *major* + *minor*

g) NaOMe — *major* + *minor*

h) Cl — NaOH — *major* + *minor* + *minor* + OH *minor*

8.79.

a) :Br: — $EtO^{\ominus}$ — H

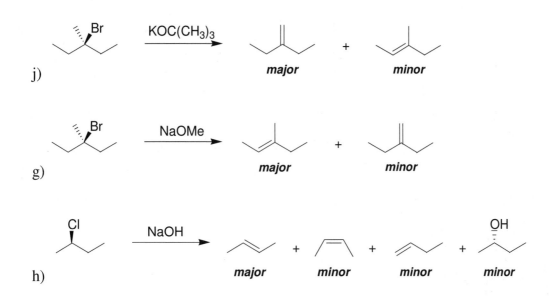

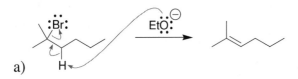

b) For an E2 process, the rate is dependent on the concentrations of the substrate and the nucleophile: Rate = k[substrate][base]

c) If the concentration of base is doubled, the rate will be doubled.

d)

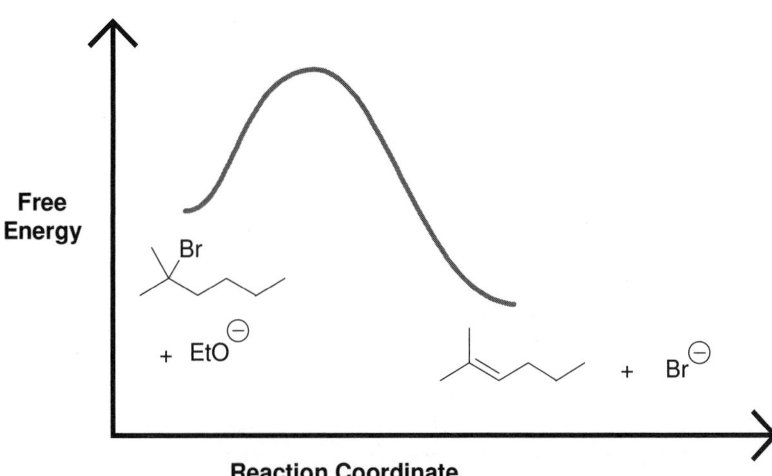

e)

Br δ-

H

EtO δ-

‡

8.80.

a) NaSH

Cl → SH

b) OTs → DBN

c) I → NaOH / H_2O

d) Br → H_2O / heat

e) OTs → $O^{\ominus}$ $K^{\oplus}$

f) I → $O^{\ominus}$ $K^{\oplus}$

g) OTs → NaOH / H_2O

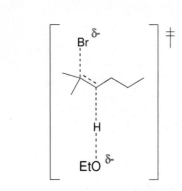

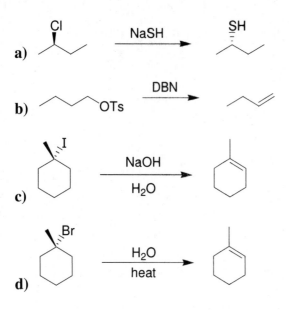

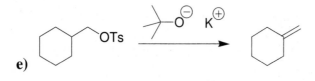

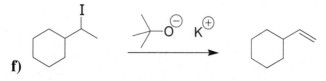

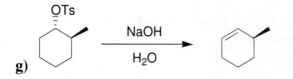

h) 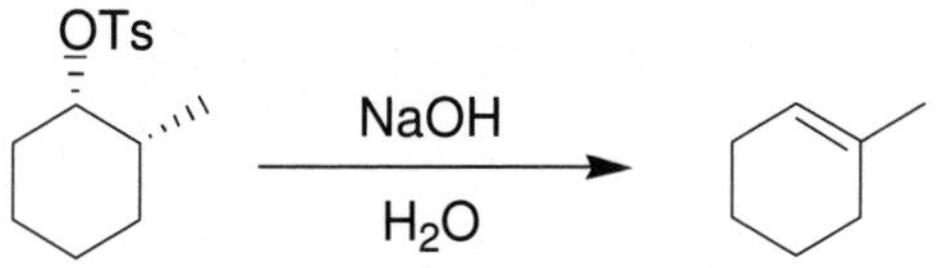 OTs / NaOH, H_2O

i) 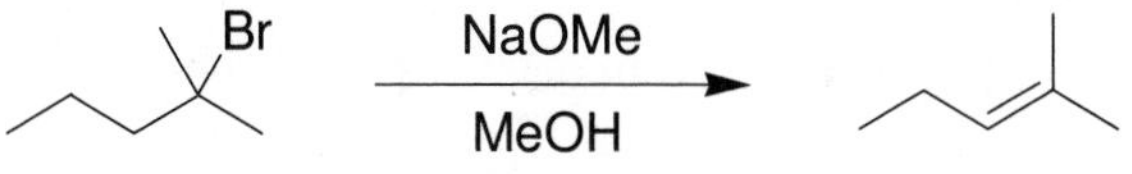 Br / NaOMe, MeOH

j) 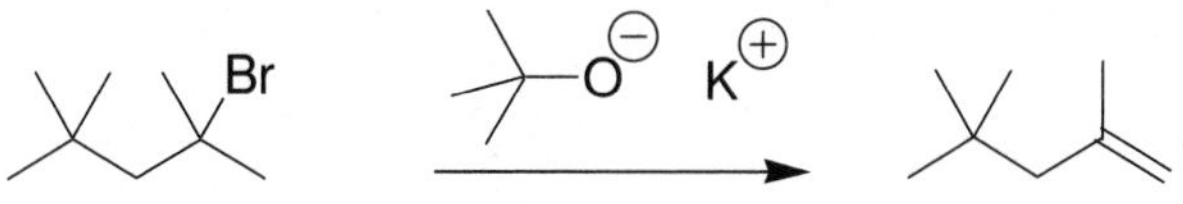 Br / $^{-}$O-t-Bu K^{+}

8.81.

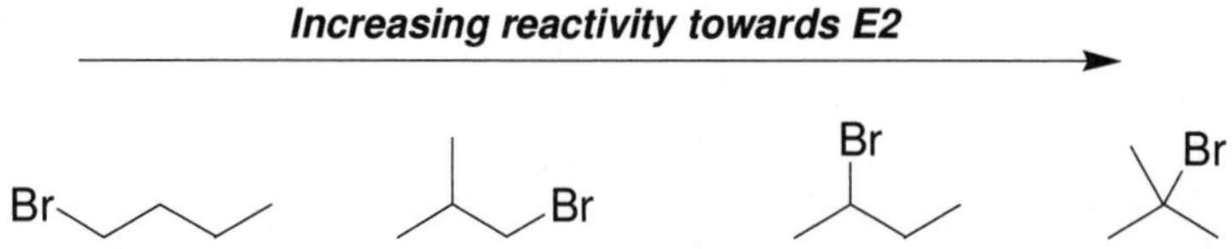

Increasing reactivity towards E2

Br Br Br Br

8.82.

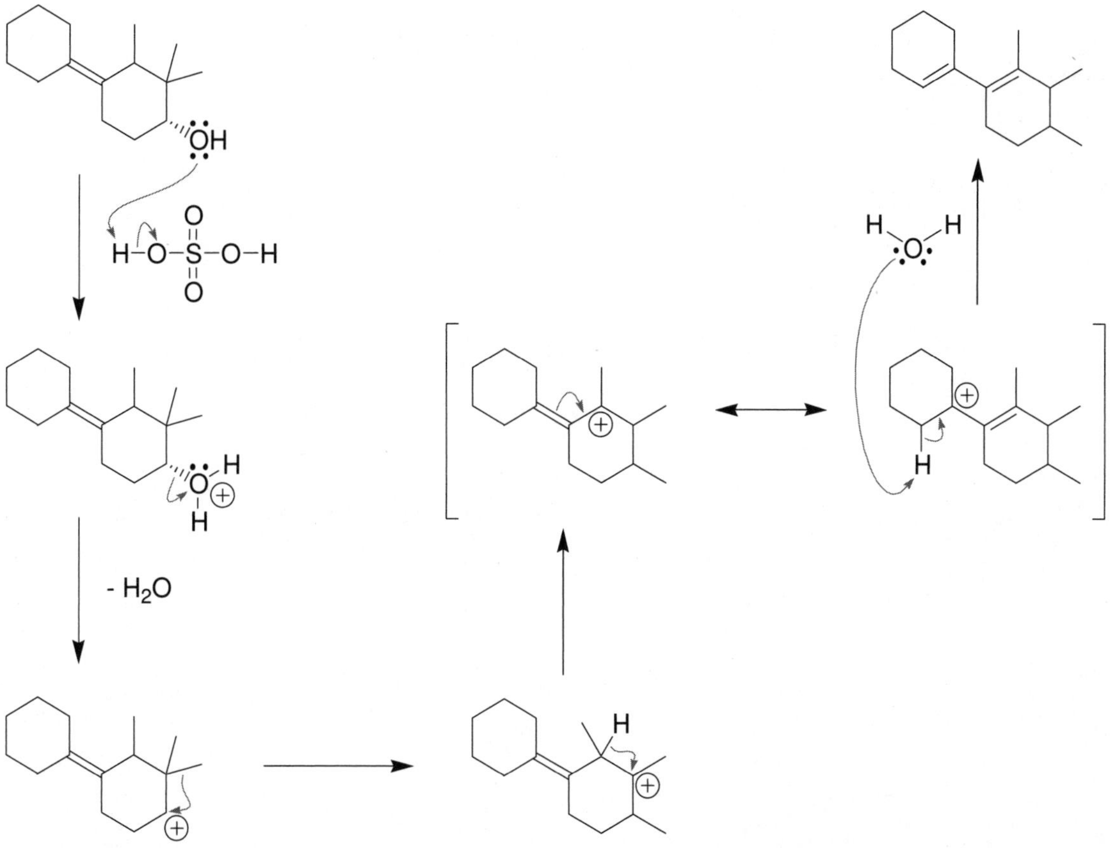

OH

H–O–S(=O)(=O)–O–H

H O H H

- H_2O

H

H H–O–H

8.83.

a)

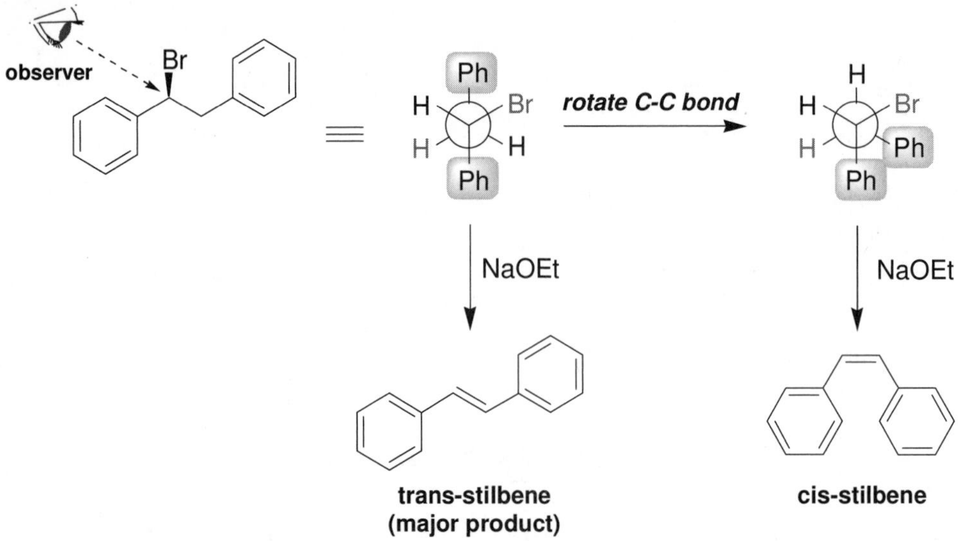

observer

Br

Ph H Br H H Ph

rotate C-C bond

H H Br H Ph Ph

NaOEt

NaOEt

trans-stilbene (major product)

cis-stilbene

b) There are still two beta protons that can be abstracted in a beta elimination, and both products are still possible. The reaction will still proceed via the conformation with the least steric hinderance. That conformation will lead to the formation of *trans*-stilbene.

8.84.

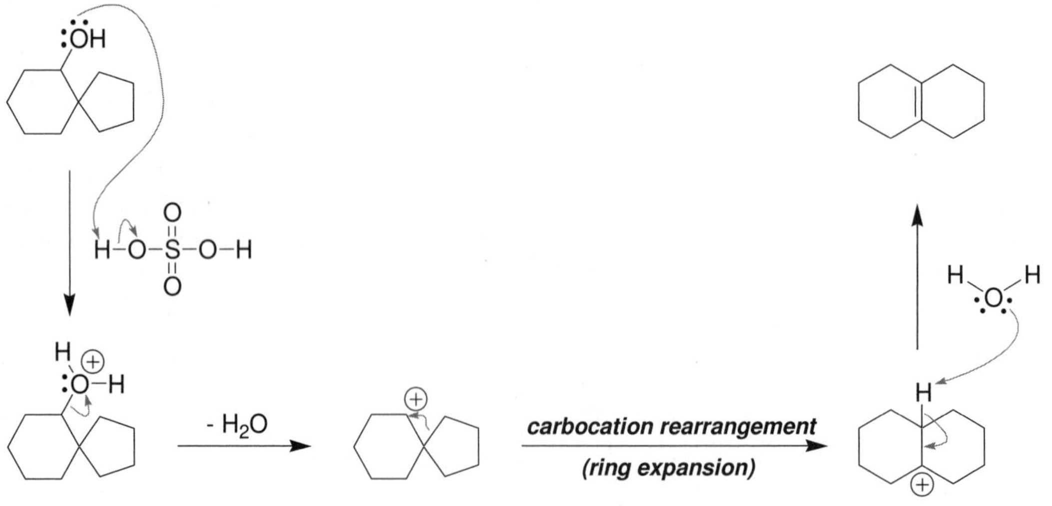

- H_2O

carbocation rearrangement

(ring expansion)

8.85.

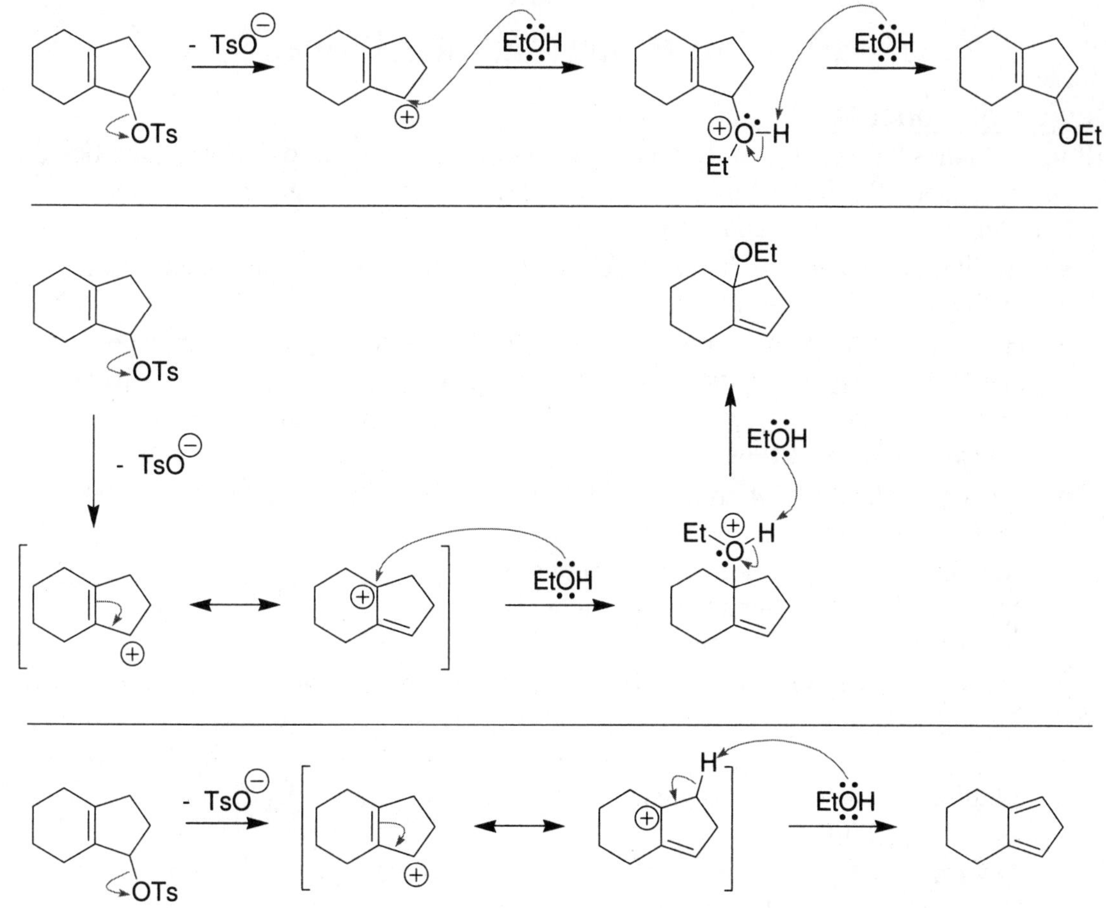

- TsO⁻ EtÖH EtÖH OTs OEt Et H

OEt OTs - TsO EtÖH Et H EtÖH

- TsO EtÖH OTs H

8.86. The stereoisomer shown below does not readily undergo E2 elimination because none of the chlorine atoms can be antiperiplanar to a beta proton in a chair conformation. Recall that for substituted cyclohexanes, the leaving group must be *trans* to a beta proton in order to achieve antiperiplanarity. In the isomer below, none of the chlorine atoms are *trans* to a beta proton.

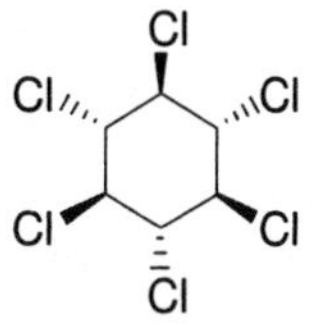

Cl Cl Cl Cl Cl Cl

8.87. The first compound is a tertiary substrate. The second compound is a tertiary allylic substrate. The latter will undergo E1 more rapidly because a tertiary allylic carbocation is more highly stabilized than a tertiary carbocation. The rate-determining step (loss of the leaving group) will therefore occur more rapidly for the second compound.

Chapter 9
Addition Reactions of Alkenes

Review of Concepts

Fill in the blanks below. To verify that your answers are correct, look in your textbook at the end of Chapter 9. Each of the sentences below appears verbatim in the section entitled *Review of Concepts and Vocabulary.*

- Addition reactions are thermodynamically favorable at ____ temperature and disfavored at _____ temperature.
- Hydrohalogenation reactions are **regioselective**, because the halogen is generally placed at the ______ substituted position, called _______________ **addition.**
- In the presence of _____________, addition of HBr proceeds via an ***anti*-Markovnikov addition.**
- The regioselectivity of an ionic addition reaction is determined by the preference for the reaction to proceed through _____________________________________.
- Acid-catalyzed hydration is inefficient when _____________________________ are possible. Dilute acid favors formation of the ___________ and while concentrated acid favors the ___________.
- **Oxymercuration-demercuration** achieves hydration of an alkene without ____.
- _____________-_______________ can be used to achieve an *anti*-Markovnikov addition of water across an alkene. The reaction is stereospecific and proceeds via a _____ **addition.**
- **Asymmetric hydrogenation** can be achieved with a ________ catalyst.
- Bromination proceeds through a bridged intermediate, called a ________________ ______, which is opened by an S_N2 process that produces an _____ **addition.**
- A two-step procedure for *anti* dihydroxylation involves conversion of an alkene to an _________, followed by acid-catalyzed ring opening.
- Ozonolysis can be used to cleave a double bond and produce two ______ groups.
- The position of a leaving group can be changed via ______________ followed by ________________.
- The position of a π bond can be changed via ______________ followed by ________________.

Review of Skills

Fill in the blanks and empty boxes below. To verify that your answers are correct, look in your textbook at the end of Chapter 9. The answers appear in the section entitled *SkillBuilder Review.*

9.1 Drawing a Mechanism for Hydrohalogenation

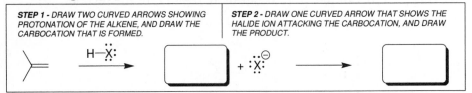

STEP 1 - *DRAW TWO CURVED ARROWS SHOWING PROTONATION OF THE ALKENE, AND DRAW THE CARBOCATION THAT IS FORMED.*	***STEP 2*** - *DRAW ONE CURVED ARROW THAT SHOWS THE HALIDE ION ATTACKING THE CARBOCATION, AND DRAW THE PRODUCT.*
H—X: → [ ]	+ :X:⊖ → [ ]

9.2 Drawing a Mechanism for Hydrohalogenation with a Carbocation Rearrangement

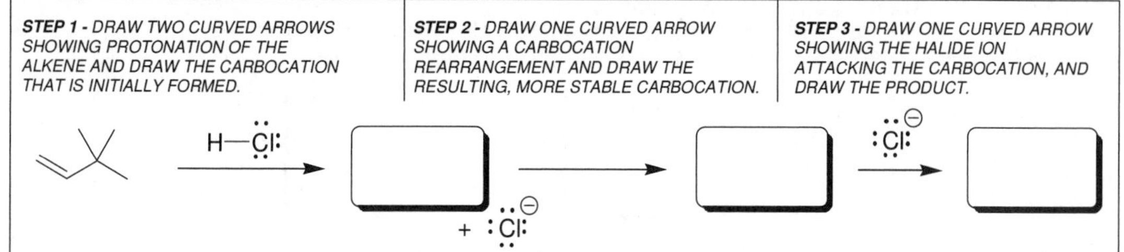

STEP 1 - DRAW TWO CURVED ARROWS SHOWING PROTONATION OF THE ALKENE AND DRAW THE CARBOCATION THAT IS INITIALLY FORMED.
STEP 2 - DRAW ONE CURVED ARROW SHOWING A CARBOCATION REARRANGEMENT AND DRAW THE RESULTING, MORE STABLE CARBOCATION.
STEP 3 - DRAW ONE CURVED ARROW SHOWING THE HALIDE ION ATTACKING THE CARBOCATION, AND DRAW THE PRODUCT.
H—Cl:
+ :Cl:
:Cl:

9.3 Drawing a Mechanism for an Acid-Catalyzed Hydration

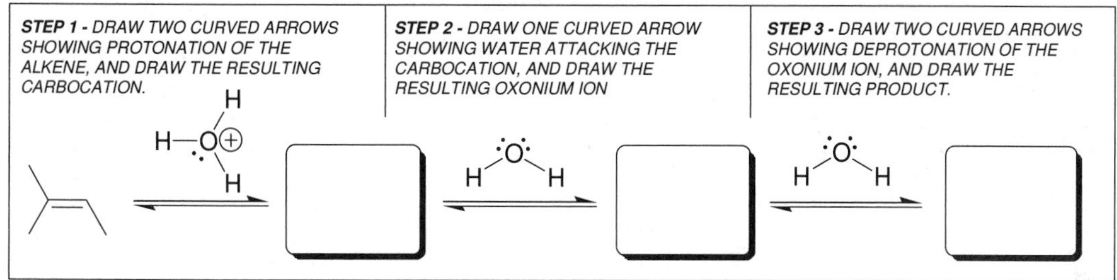

STEP 1 - DRAW TWO CURVED ARROWS SHOWING PROTONATION OF THE ALKENE, AND DRAW THE RESULTING CARBOCATION.
STEP 2 - DRAW ONE CURVED ARROW SHOWING WATER ATTACKING THE CARBOCATION, AND DRAW THE RESULTING OXONIUM ION
STEP 3 - DRAW TWO CURVED ARROWS SHOWING DEPROTONATION OF THE OXONIUM ION, AND DRAW THE RESULTING PRODUCT.
H
H—O⊕
H
H O H
H O H

9.4 Predicting the Products of Hydroboration-Oxidation

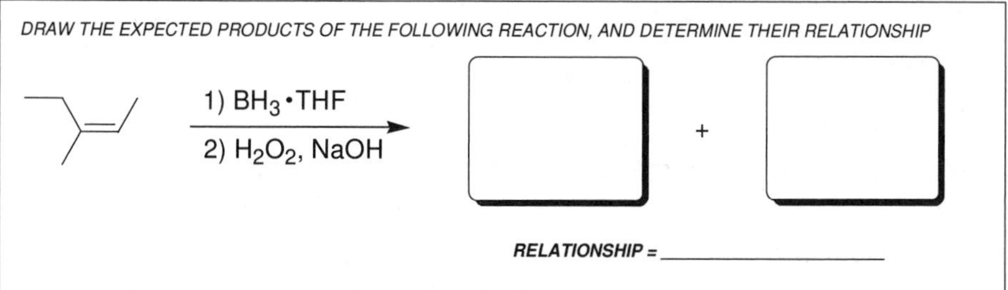

DRAW THE EXPECTED PRODUCTS OF THE FOLLOWING REACTION, AND DETERMINE THEIR RELATIONSHIP
1) BH3•THF
2) H2O2, NaOH
+
RELATIONSHIP = ____________

9.5 Predicting the Products of Catalytic Hydrogenation

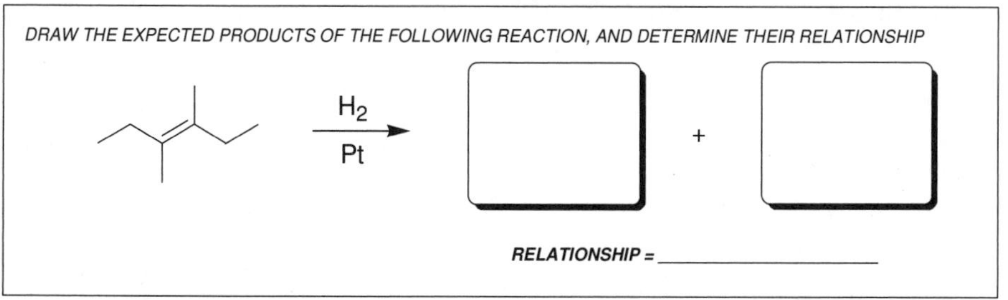

DRAW THE EXPECTED PRODUCTS OF THE FOLLOWING REACTION, AND DETERMINE THEIR RELATIONSHIP
H2
Pt
+
RELATIONSHIP = ____________

9.6 Predicting the Products of Halohydrin Formation

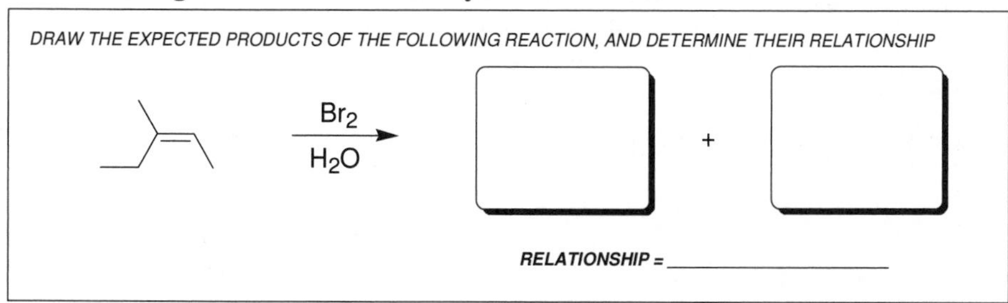

DRAW THE EXPECTED PRODUCTS OF THE FOLLOWING REACTION, AND DETERMINE THEIR RELATIONSHIP
Br2
H2O
+
RELATIONSHIP = ____________

9.7 Drawing the Products of *Anti* Dihydroxylation

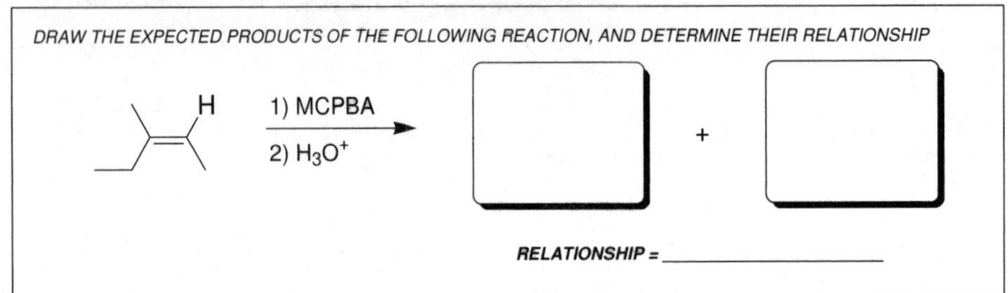

9.8 Predicting the Products of Ozonolysis

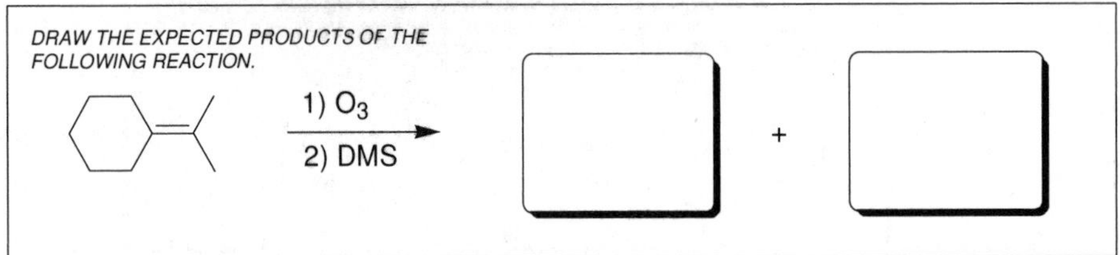

9.9 Predicting the Products of an Addition Reaction

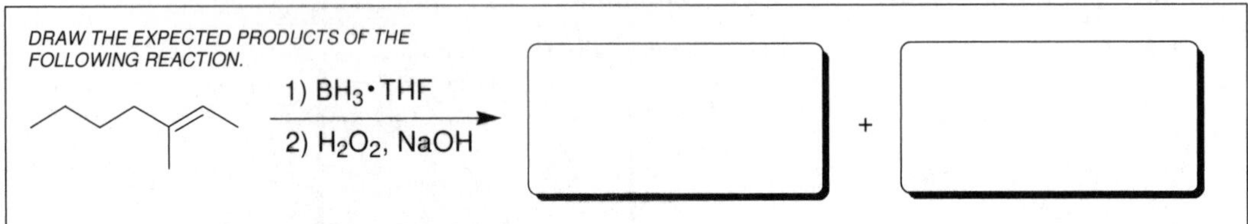

9.10 Proposing a One-Step Synthesis

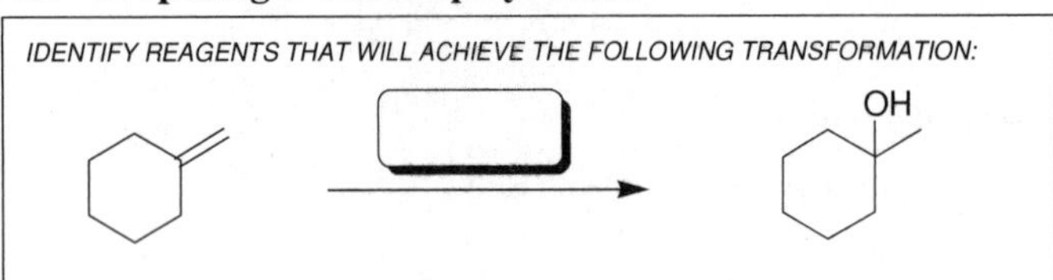

IDENTIFY REAGENTS THAT WILL ACHIEVE THE FOLLOWING TRANSFORMATION:
OH

9.11 Changing the Position of a Leaving Group

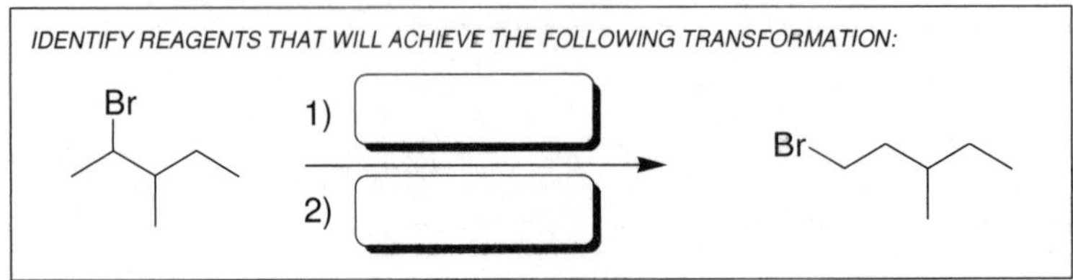

9.12 Changing the Position of a π Bond

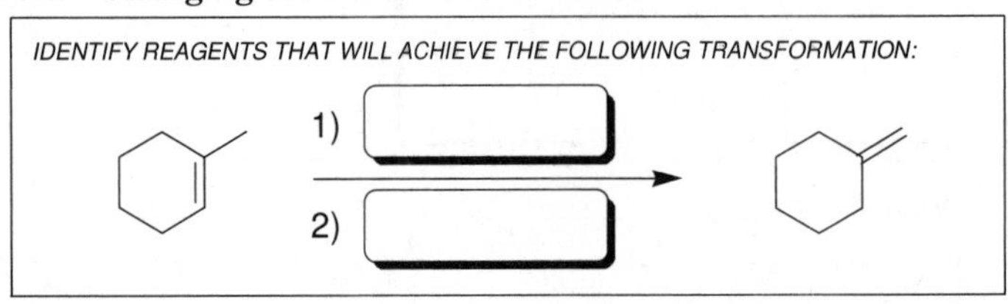

IDENTIFY REAGENTS THAT WILL ACHIEVE THE FOLLOWING TRANSFORMATION:
1)
2)

Review of Reactions

Identify the reagents necessary to achieve each of the following transformations. To verify that your answers are correct, look in your textbook at the end of Chapter 9. The answers appear in the section entitled *Review of Reactions*.

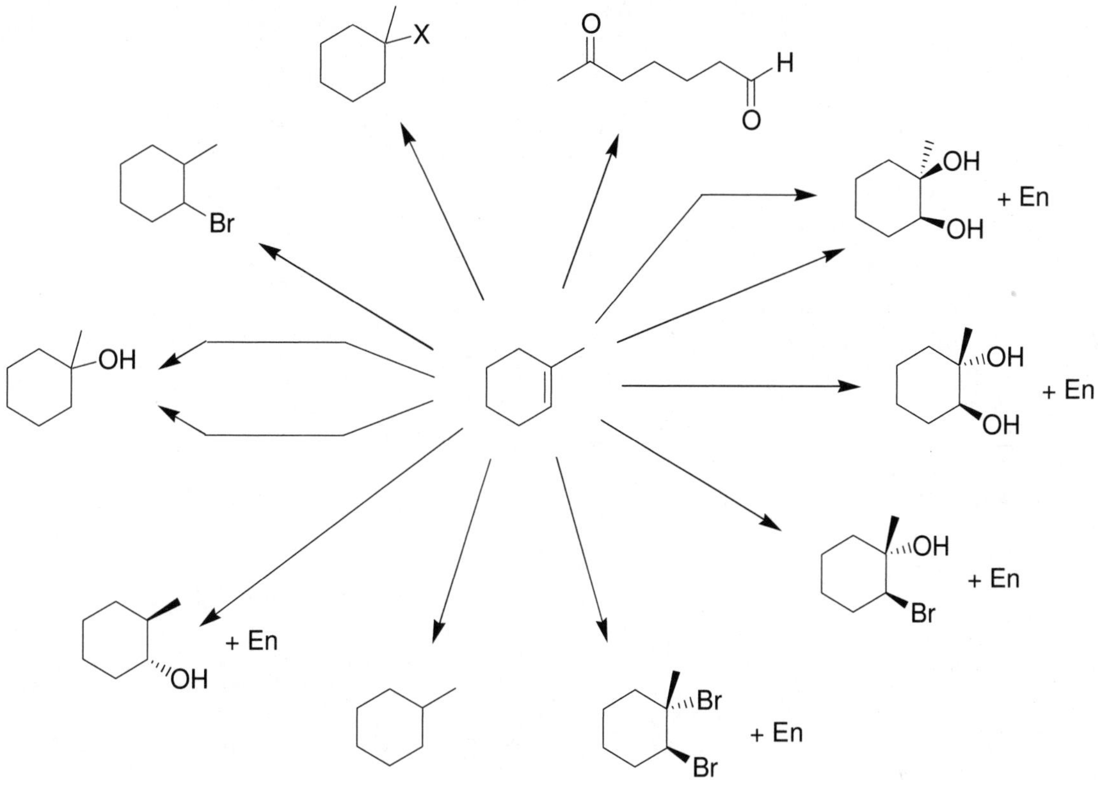

X

OH

H

O

Br

OH + En

OH

OH + En

OH

OH + En

OH

OH + En

Br

Br + En

Br

Solutions

9.1.

a) HBr Br

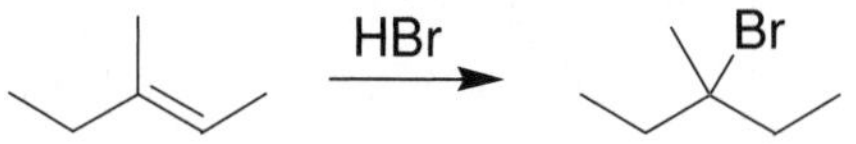

b) HBr ROOR Br

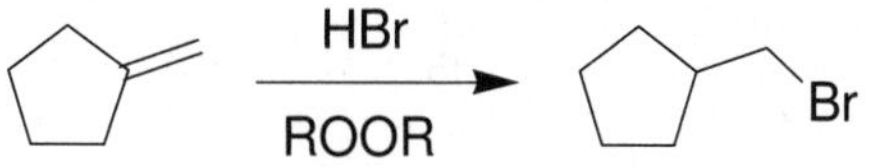

c) HBr Br

d) HCl Cl

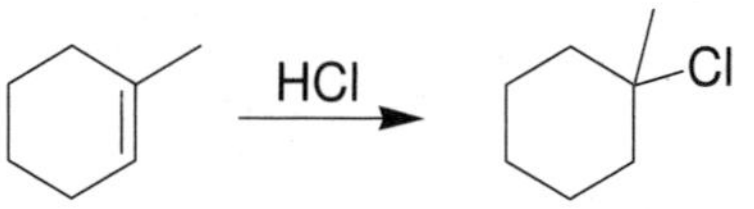

e) HI I

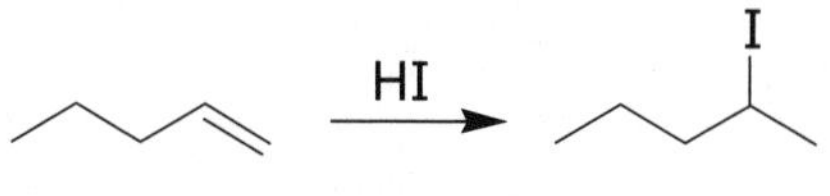

f) HBr ROOR Br

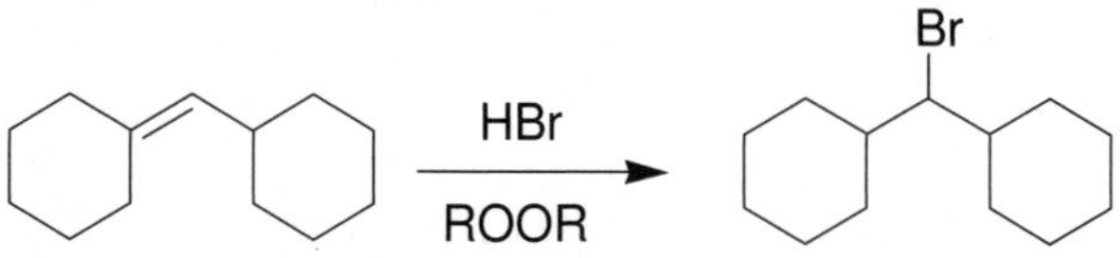

9.2.

a) 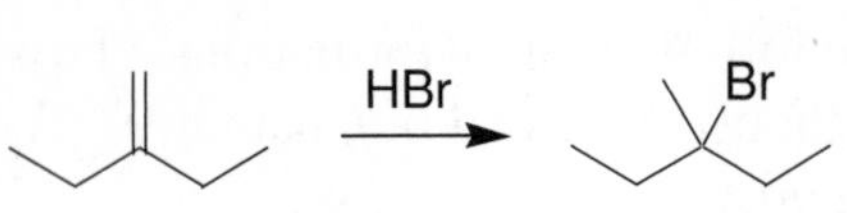 HBr → Br

b) 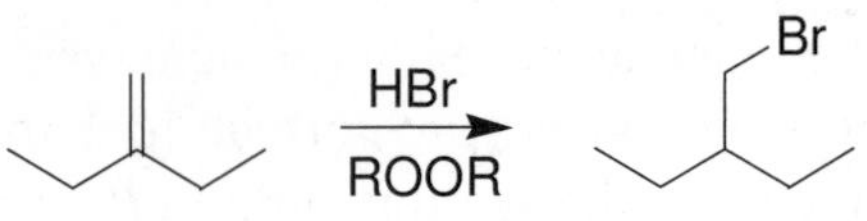 HBr / ROOR → Br

9.3.

a) 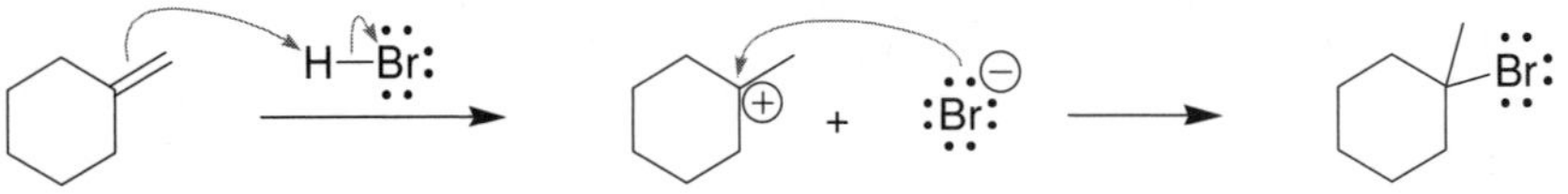 H–Br: → + :Br: → Br:

b) 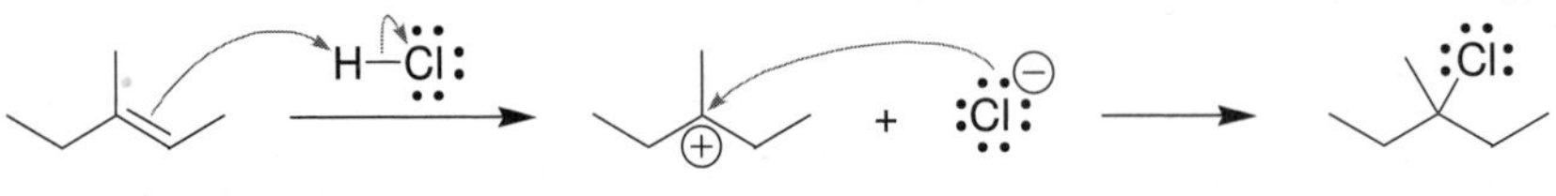 H–Cl: → + :Cl: → :Cl:

c) 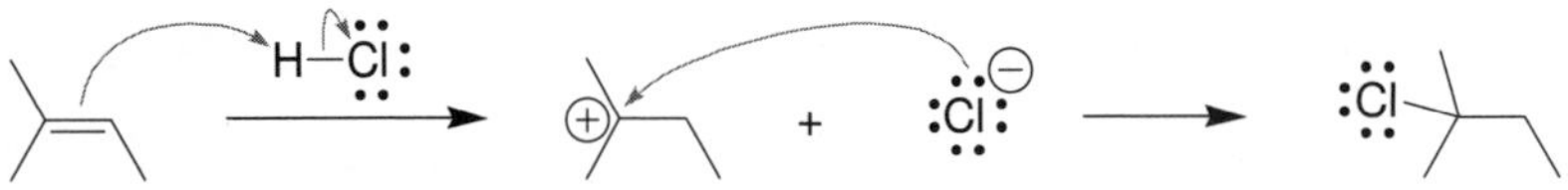 H–Cl: → + :Cl: → :Cl

9.4.

a) 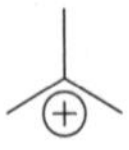 b) 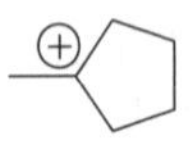 c) 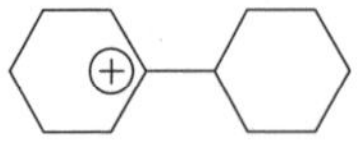 d)

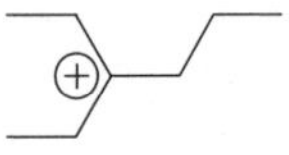

9.5. In this case, the less-substituted carbocation is more stable because it is resonance-stabilized:

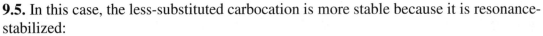

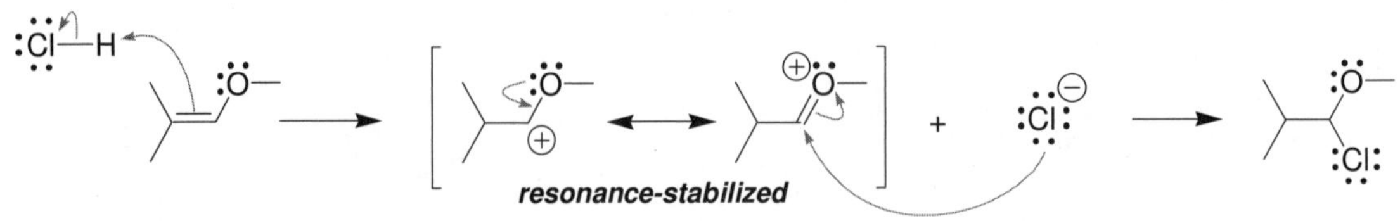

:Cl–H :O– → :O– ⟷ O– + :Cl: → :O– :Cl:

resonance-stabilized

9.6.

a) 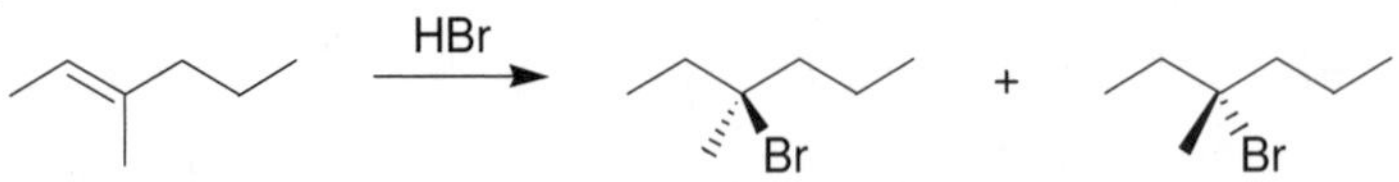 HBr → Br + Br

b) 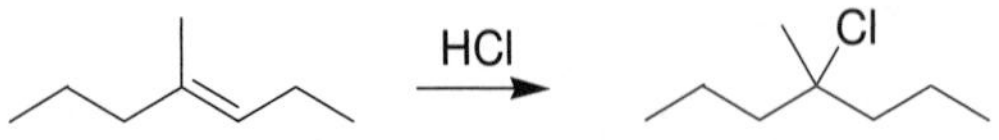 HCl → Cl

c) 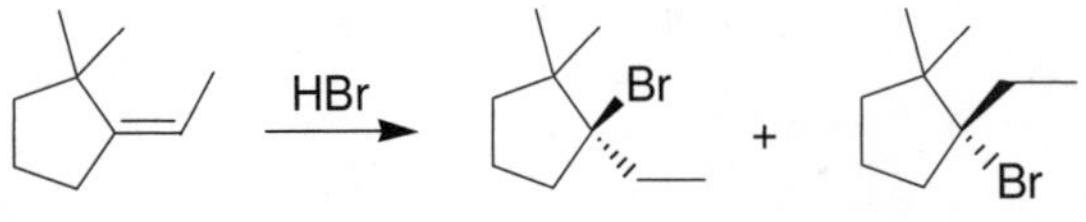 HBr → Br + Br

d) HI

e) HCl

f) HCl

9.7.

a) Hydride Shift

b) Hydride Shift

c) Methyl Shift

9.8.

Ring Expansion

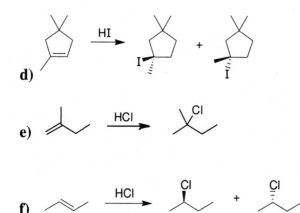

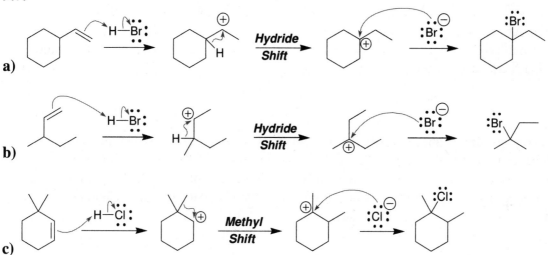

9.9.

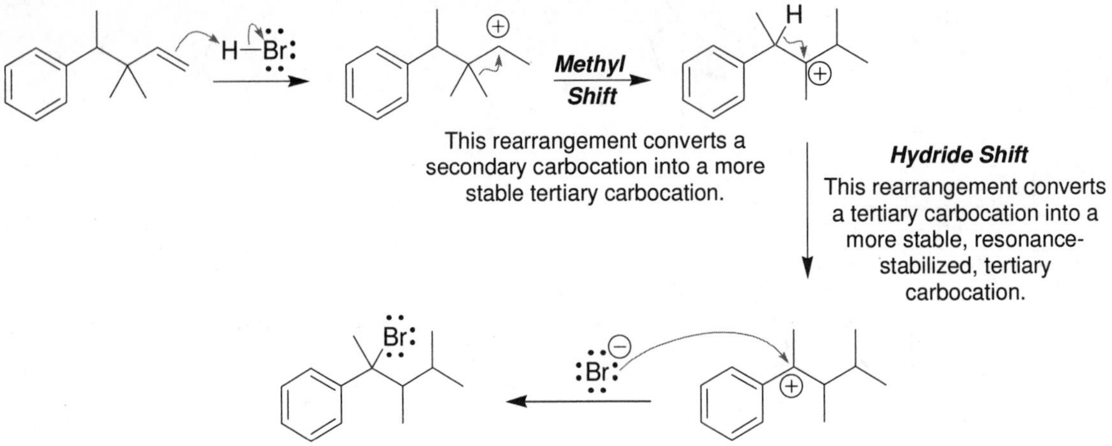

Methyl Shift

This rearrangement converts a secondary carbocation into a more stable tertiary carbocation.

Hydride Shift

This rearrangement converts a tertiary carbocation into a more stable, resonance-stabilized, tertiary carbocation.

9.10.

a) 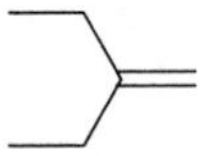 , because the reaction proceeds via a tertiary carbocation, rather than a secondary carbocation.
b) 2-methyl-2-butene, because the reaction proceeds via a tertiary carbocation, rather than a secondary carbocation.

9.11.

a) To favor the alcohol, dilute sulfuric acid (mostly water) is used. Having a high concentration of water favors the alcohol according to Le Chatelier's principle.
b) To favor the alkene, concentrated sulfuric acid (which has very little water) is used. Having a low concentration of water favors the alkene according to Le Chatelier's principle.

9.12.

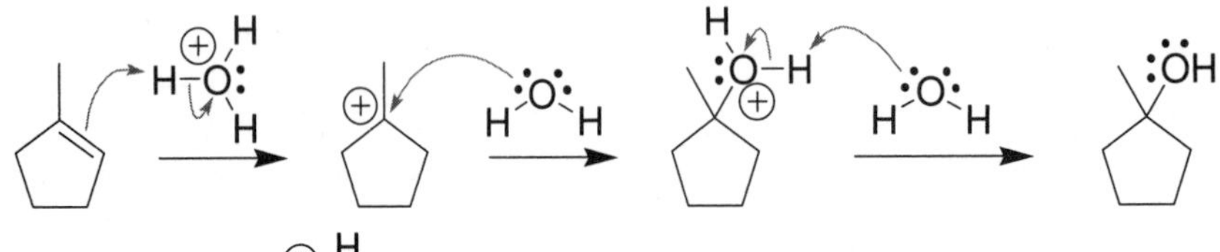

a)

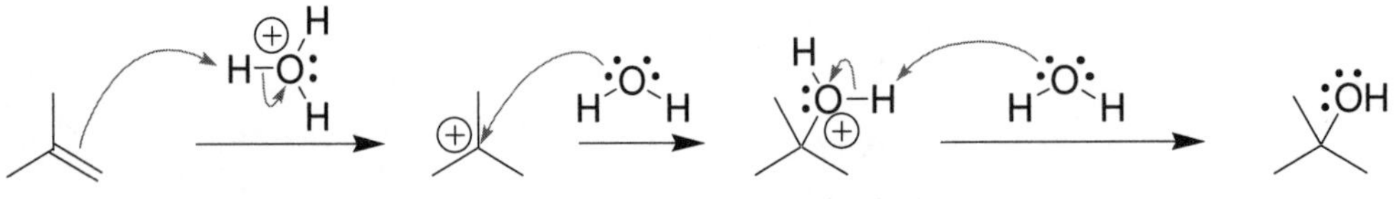

b)

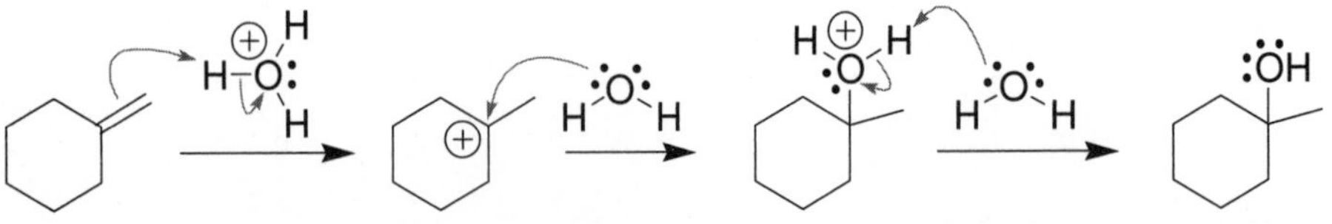

c)

9.13.

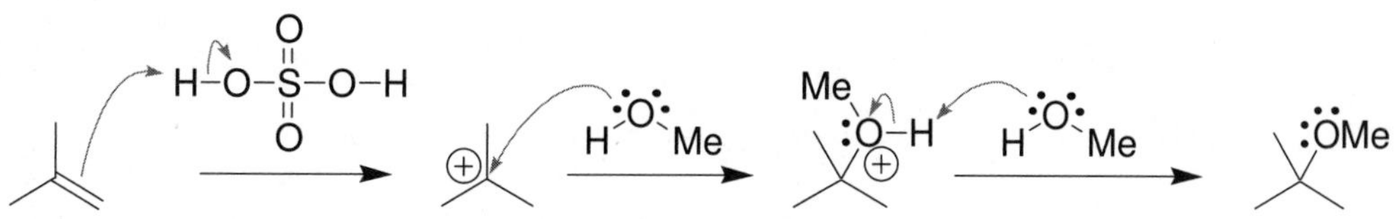

H–O–S(=O)(=O)–O–H

H–O–Me

Me

O–H

H–O–Me

OMe

9.14.

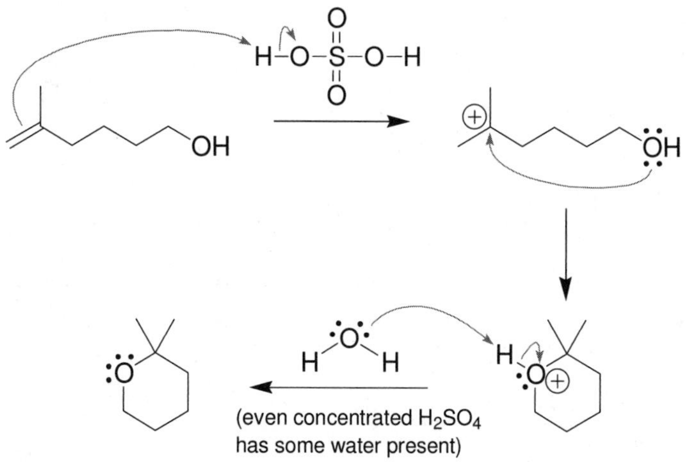

H–O–S(=O)(=O)–O–H

OH

OH

H–O–H

H

O

(even concentrated H_2SO_4 has some water present)

O

9.15.

a)

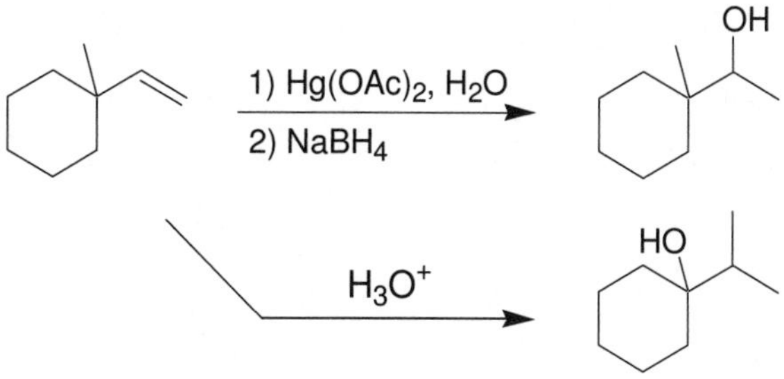

1) $Hg(OAc)_2$, H_2O

2) $NaBH_4$

OH

H_3O^+

HO

b)

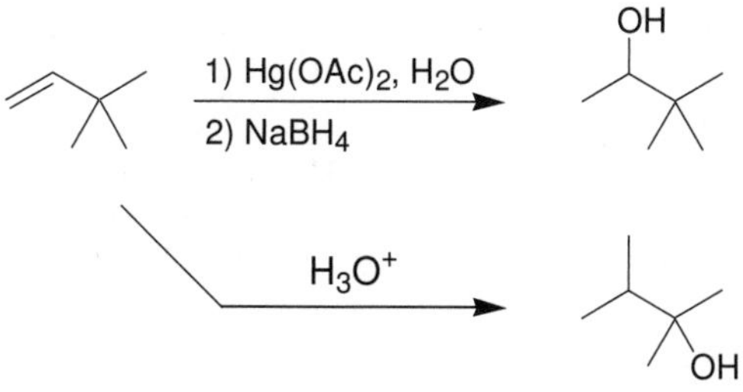

1) $Hg(OAc)_2$, H_2O

2) $NaBH_4$

OH

H_3O^+

OH

c)

1) $Hg(OAc)_2$, H_2O
2) $NaBH_4$

OH

H_3O^+

9.16.

1) $Hg(OAc)_2$, EtOH
2) $NaBH_4$

a)

OEt

1) $Hg(OAc)_2$, $EtNH_2$
2) $NaBH_4$

H N Et

b)

9.17.

1) $BH_3 \cdot THF$
2) H_2O_2, NaOH

OH

a)

1) $BH_3 \cdot THF$
2) H_2O_2, NaOH

OH

b)

1) $BH_3 \cdot THF$
2) H_2O_2, NaOH

OH

c)

9.18.

1) $BH_3 \cdot THF$
2) H_2O_2, NaOH

OH

9.19.

1) $BH_3 \cdot THF$
2) H_2O_2, NaOH

+ En

OH

a)

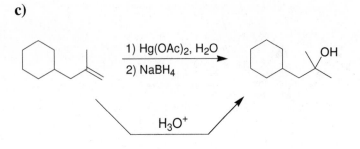

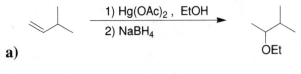

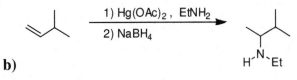

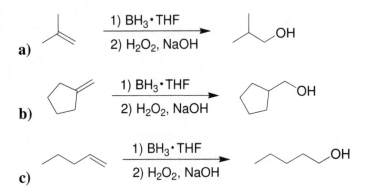

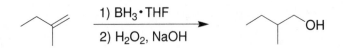

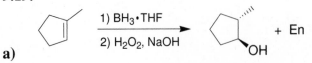

b) 1) $BH_3 \cdot THF$ 2) H_2O_2, NaOH → OH + En

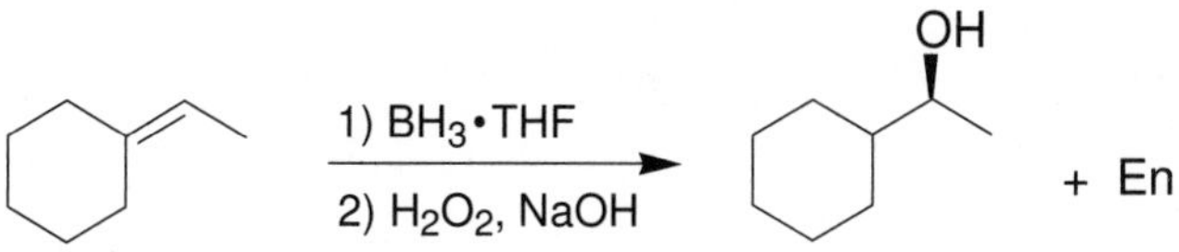

c) 1) $BH_3 \cdot THF$ 2) H_2O_2, NaOH → OH

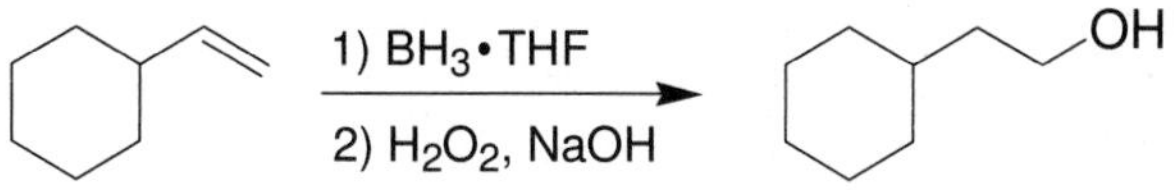

d) 1) $BH_3 \cdot THF$ 2) H_2O_2, NaOH → HO + En

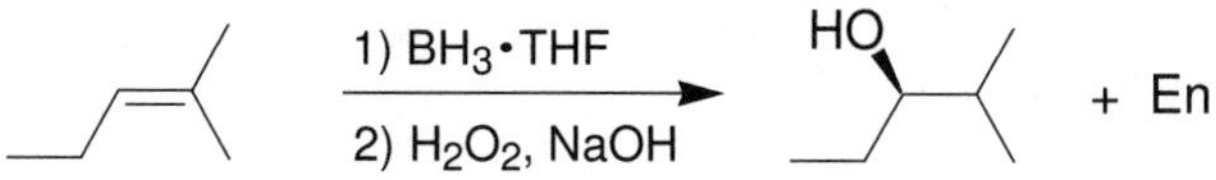

e) 1) $BH_3 \cdot THF$ 2) H_2O_2, NaOH → HO

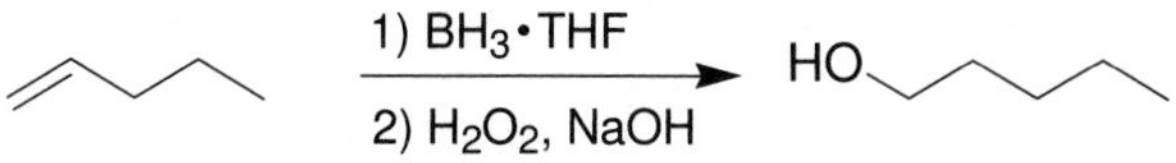

f) 1) $BH_3 \cdot THF$ 2) H_2O_2, NaOH → H, OH + En

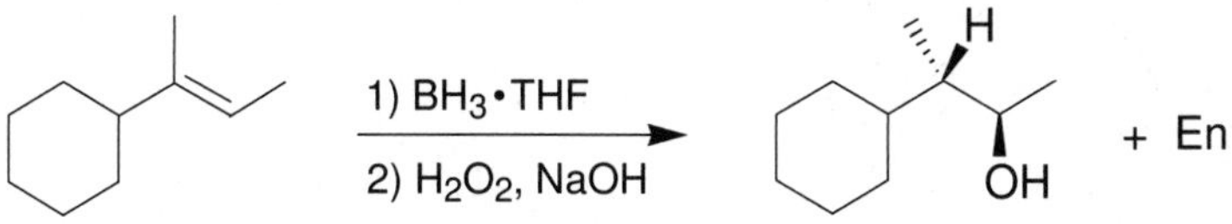

9.20.

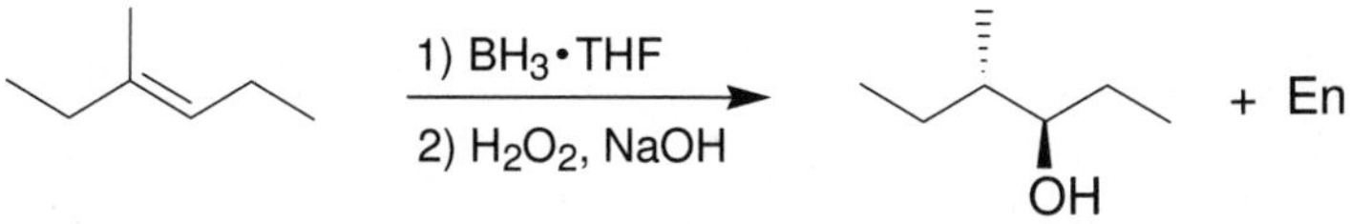

1) $BH_3 \cdot THF$ 2) H_2O_2, NaOH → OH + En

9.21. Only one chirality center is formed, so both possible stereoisomers (enantiomers) are obtained, regardless of the configuration of the starting alkene:

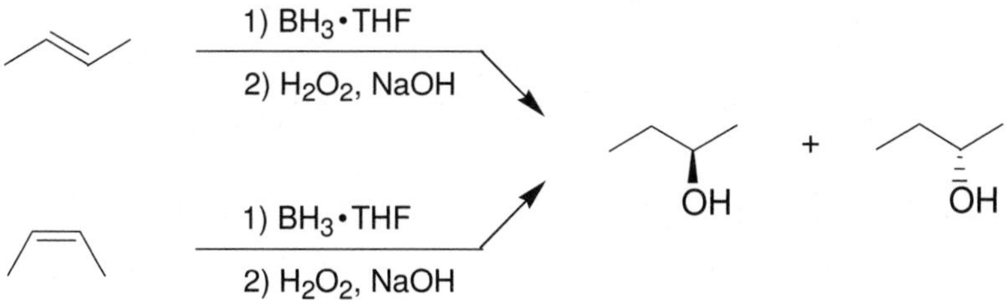

1) $BH_3 \cdot THF$ 2) H_2O_2, NaOH

1) $BH_3 \cdot THF$ 2) H_2O_2, NaOH

OH + OH

9.22.

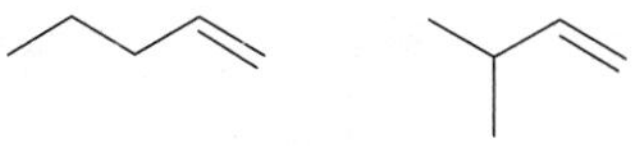

9.23.

a) 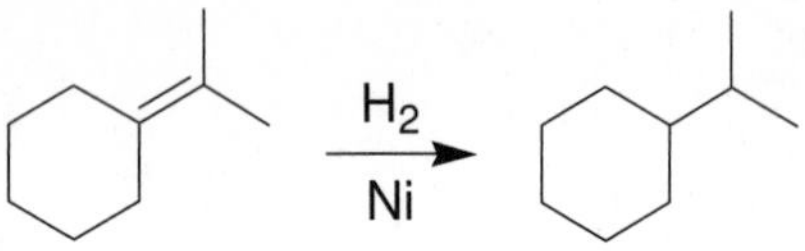 $\xrightarrow[\text{Ni}]{H_2}$

b) 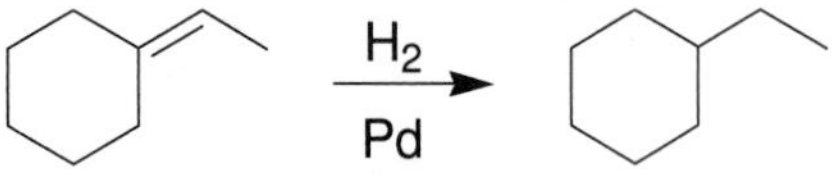 $\xrightarrow[\text{Pd}]{H_2}$

c) 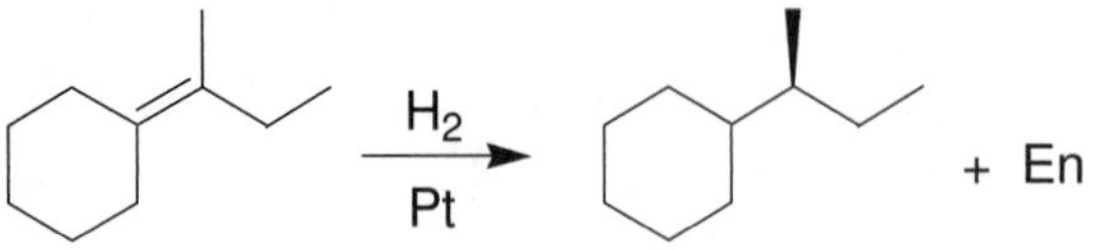 $\xrightarrow[\text{Pt}]{H_2}$ + En

d)  $\xrightarrow[\text{Ni}]{H_2}$ + En

e) 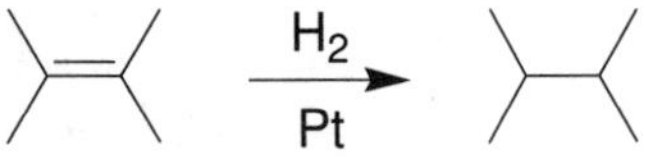 $\xrightarrow[\text{Pt}]{H_2}$

f) 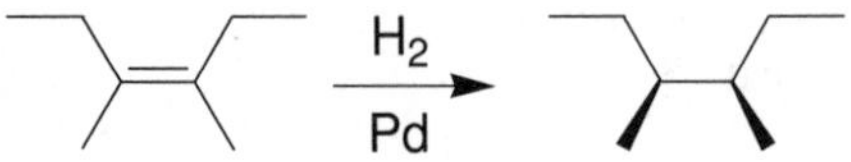 $\xrightarrow[\text{Pd}]{H_2}$ (meso)

9.24.

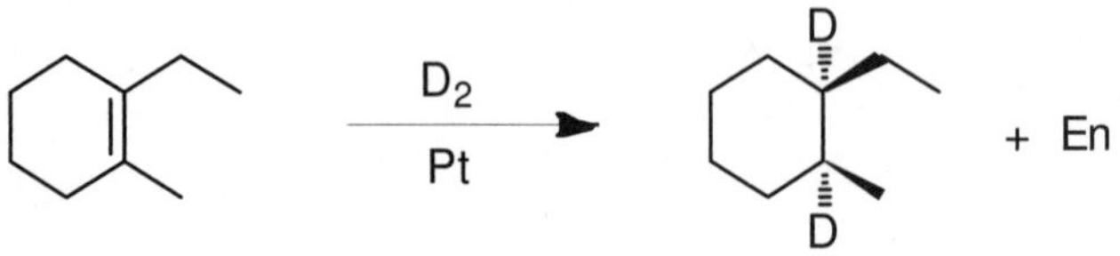

$\xrightarrow[\text{Pt}]{D_2}$ D D + En

9.25.

a)

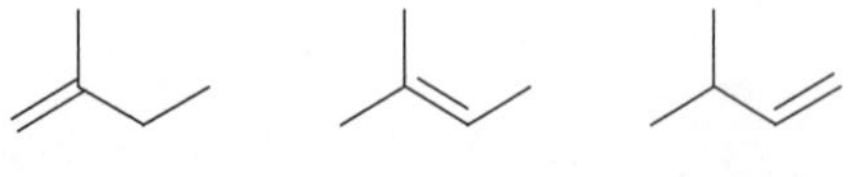

b) 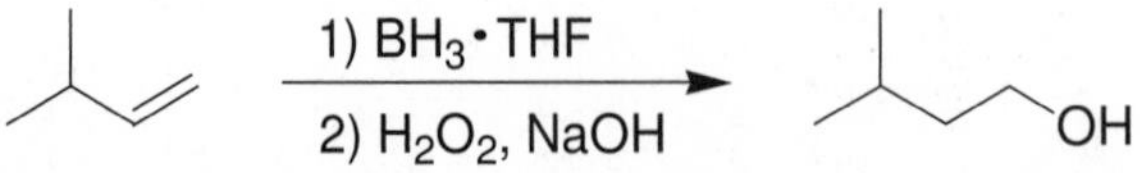 $\xrightarrow[\text{2) } H_2O_2\text{, NaOH}]{\text{1) } BH_3\cdot\text{THF}}$ OH

9.26.

a) 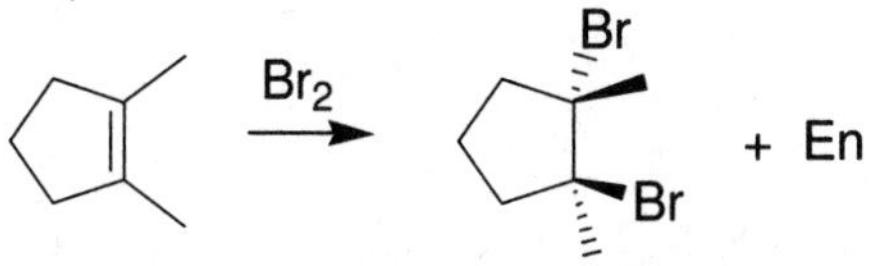 Br_2 → Br, Br + En

b) 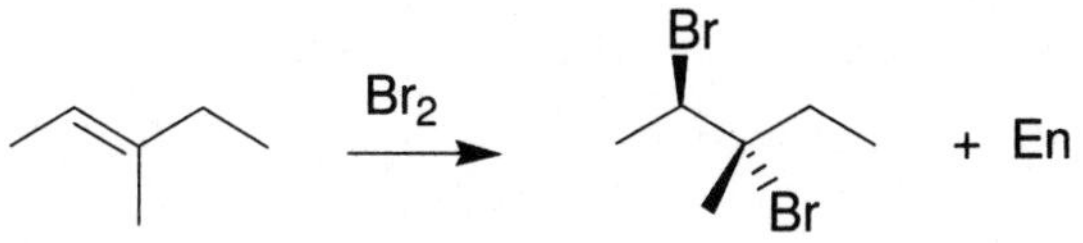 Br_2 → Br, Br + En

c) 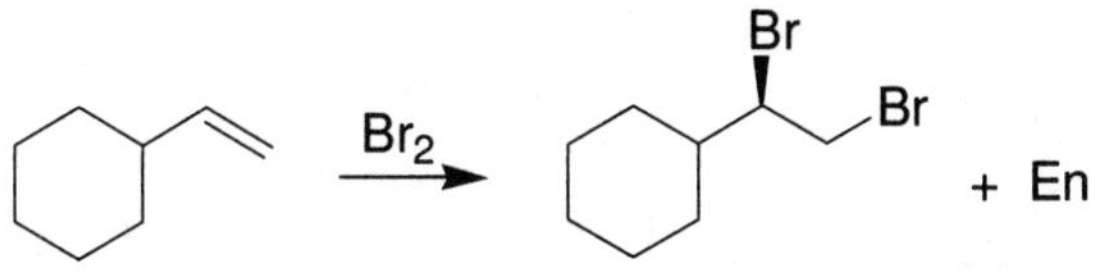 Br_2 → Br, Br + En

d) 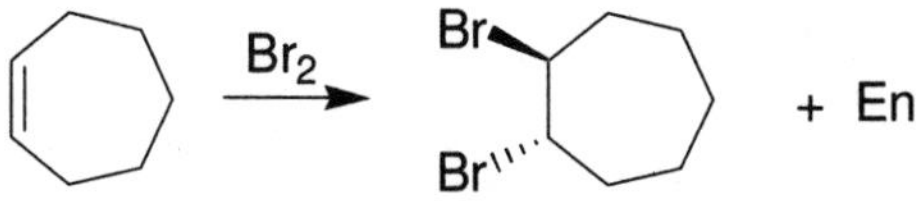 Br_2 → Br, Br + En

9.27.

a) 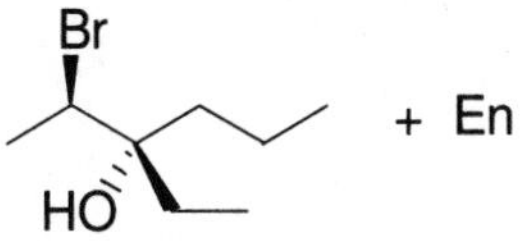 Br, HO + En

b) 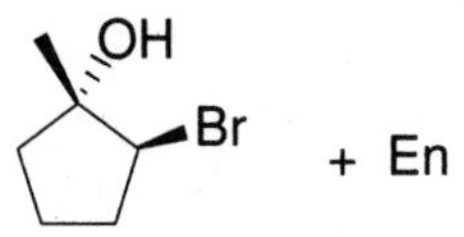 OH, Br + En

c) 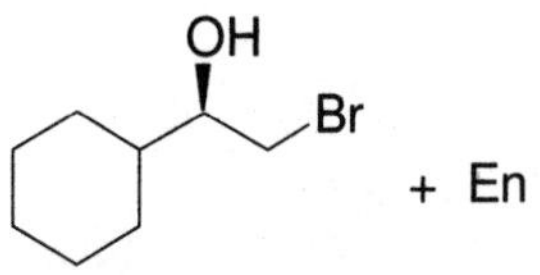 OH, Br + En

d) 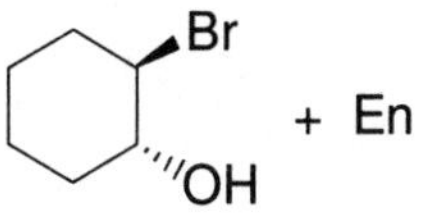 Br, OH + En

9.28.

a) 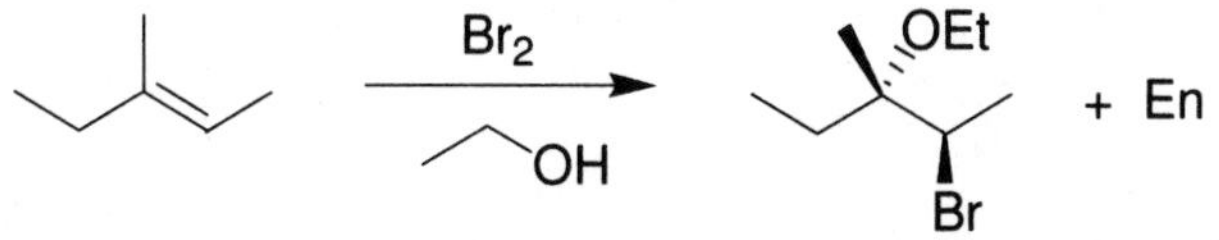 Br_2, OH → OEt, Br + En

b) 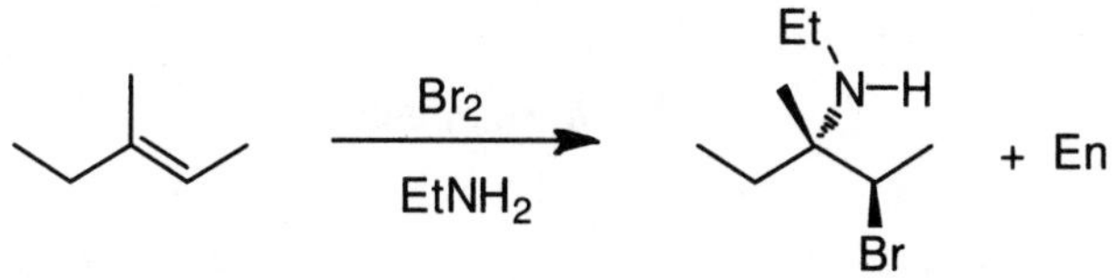 Br_2, $EtNH_2$ → Et, N–H, Br + En

9.29. The bromonium ion can open (before a bromide ion attacks), forming a resonance stabilized carbocation. This carbocation is trigonal planar and can be attacked from either side:

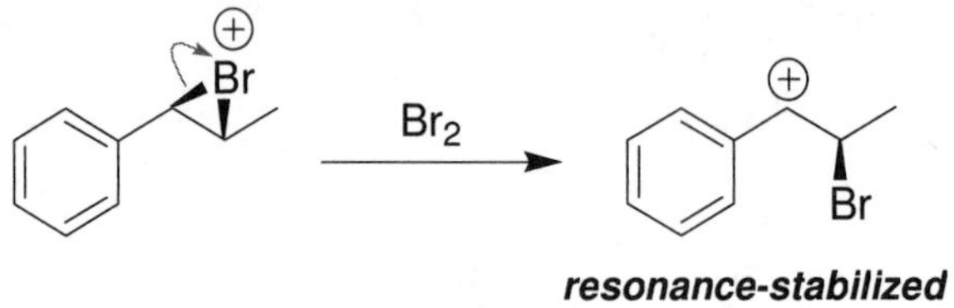

Br Br_2 Br

resonance-stabilized

9.30.

a) 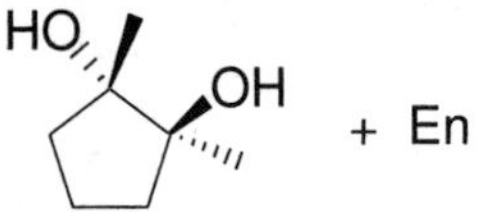 HO OH + En **b)** 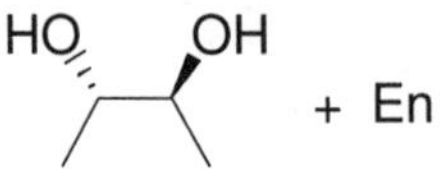 HO OH + En **c)** 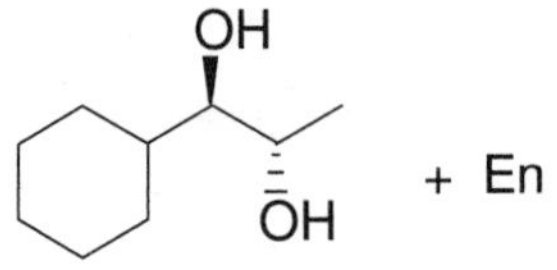 OH OH + En

d) 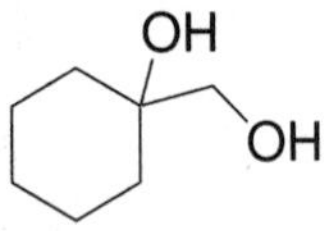 OH OH **e)** 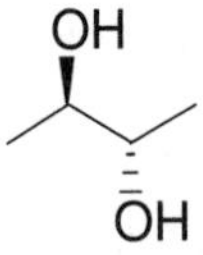 OH OH (meso) **f)** 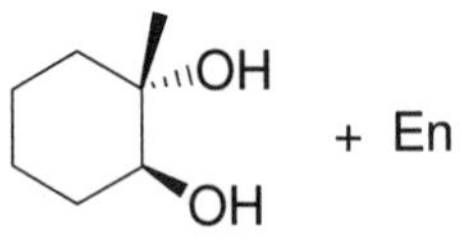 OH OH + En

9.31.

a) 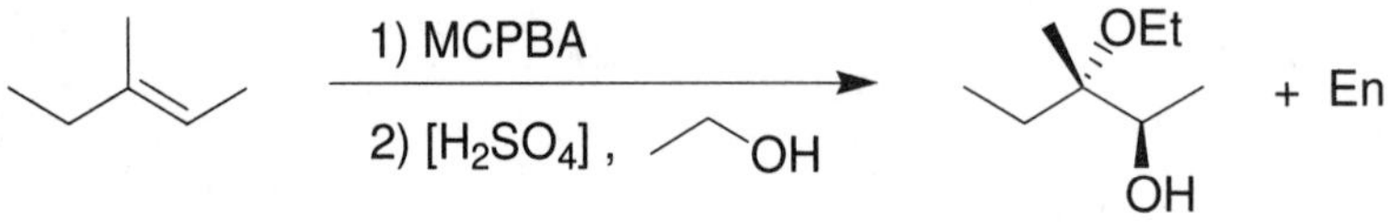 1) MCPBA 2) $[H_2SO_4]$, OH → OEt OH + En

b) 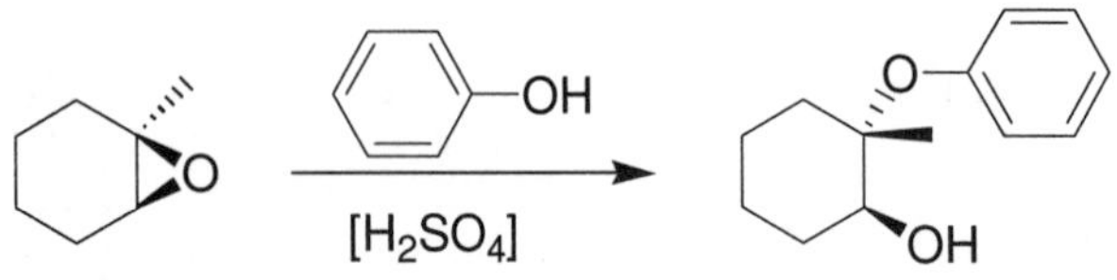 O OH $[H_2SO_4]$ → O OH

9.32.
a)

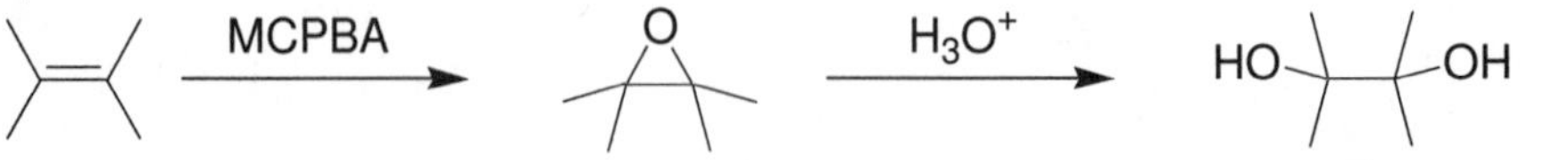

MCPBA → O H_3O^+ → HO OH

no chirality centers

MCPBA → O H_3O^+ → OH OH

no chirality centers

b)

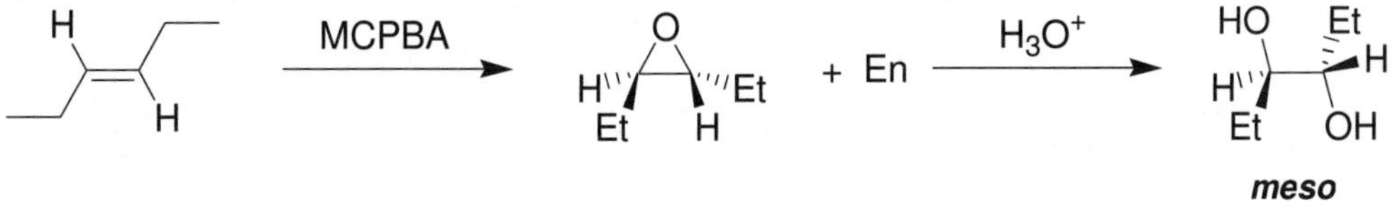
H
H
MCPBA
O
H
Et
Et
H
+ En
H_3O^+
HO
Et
H
H
Et
OH
meso

9.33.

a)

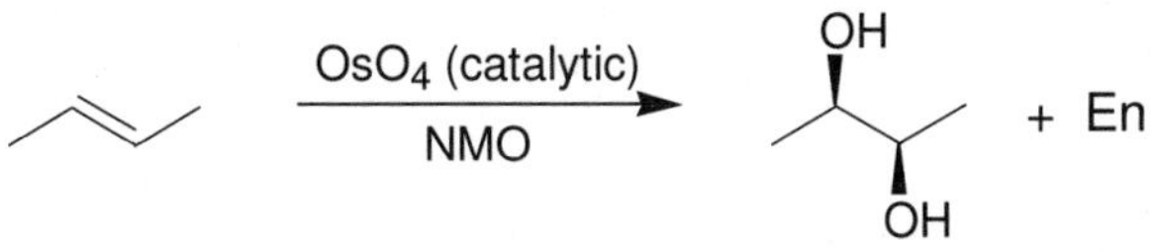
OsO_4 (catalytic)
NMO
OH
OH
+ En

b)

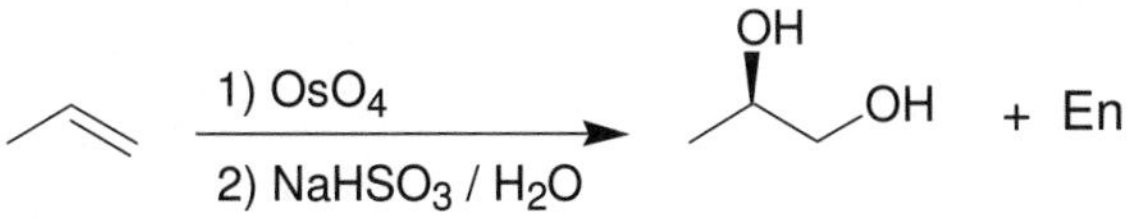
1) OsO_4
2) $NaHSO_3$ / H_2O
OH
OH
+ En

c)

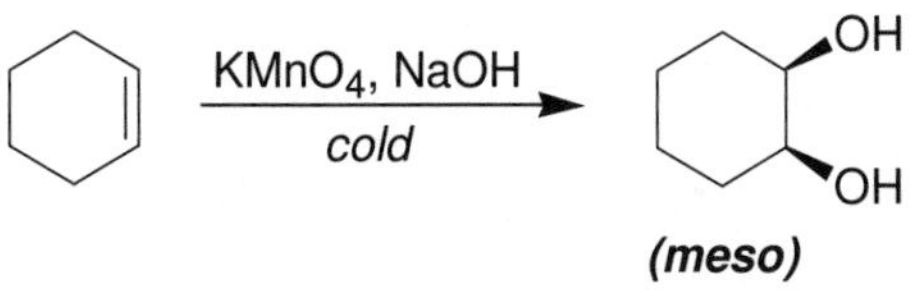
$KMnO_4$, NaOH
cold
OH
OH
(meso)

d)

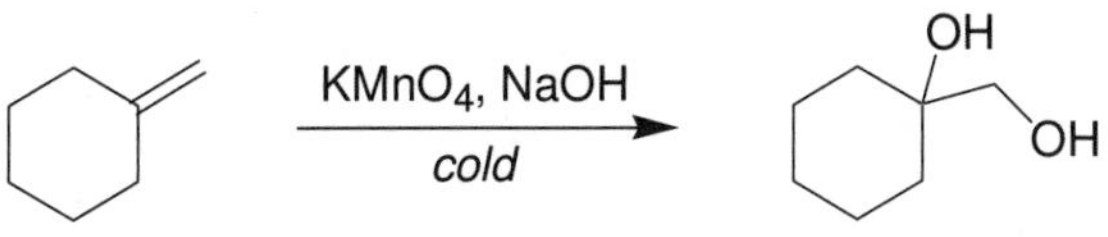
$KMnO_4$, NaOH
cold
OH
OH

e)

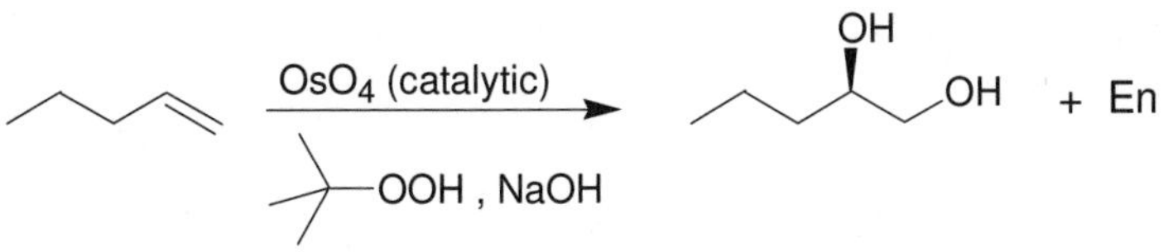
OsO_4 (catalytic)
OOH , NaOH
OH
OH
+ En

f)

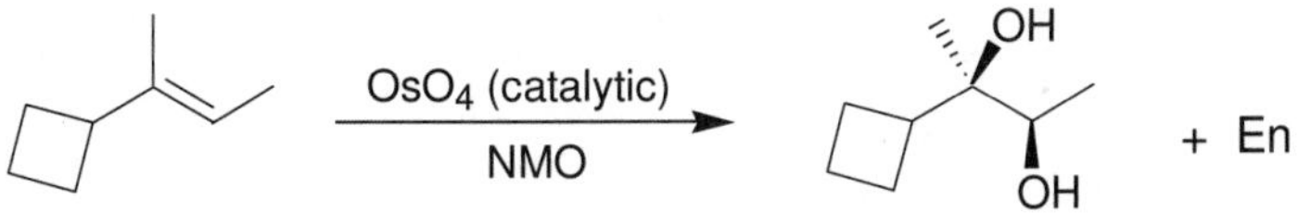
OsO_4 (catalytic)
NMO
OH
OH
+ En

9.34.

a)

1) O_3
2) DMS

H O O O O

b)

1) O_3
2) DMS

O O O O

c)

1) O_3
2) DMS

H O O H

d)

1) O_3
2) DMS

H O O H

e)

1) O_3
2) DMS

H O O H ≡ H O *meso* O H

f)

1) O_3
2) DMS

O O

9.35.

a)

1) O_3
2) DMS

O O

b)

1) O_3
2) DMS

O

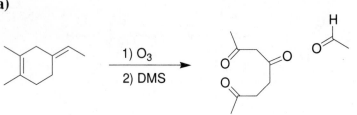

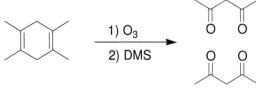

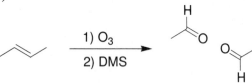

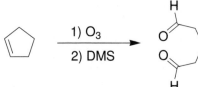

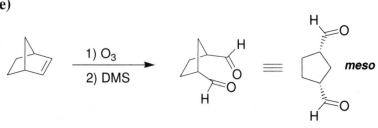

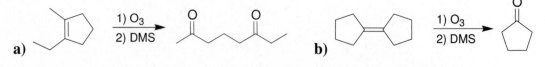

c) 1) O_3 2) DMS

9.36.

a) 1) $BH_3 \cdot THF$ 2) H_2O_2, NaOH → OH

b) H_2 / Pt → + En

c) 1) CH_3CO_3H 2) H_3O^+ → HO, OH + En

d) 1) OsO_4 2) $NaHSO_3$ / H_2O → OH, OH + En

e) H_3O^+ → OH + En

f) HBr → Br

g) 1) MCPBA 2) H_3O^+ → OH, OH + En

h) 1) $BH_3 \cdot THF$ 2) H_2O_2, NaOH → OH + En

i) OsO_4 (catalytic) / NMO → OH, HO + En

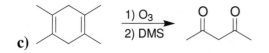

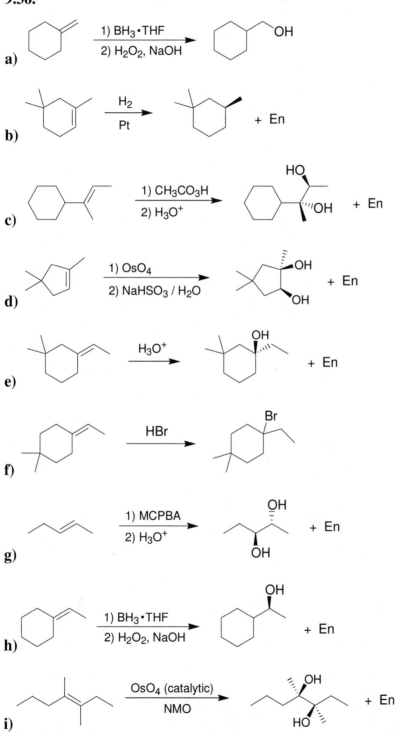

9.37.

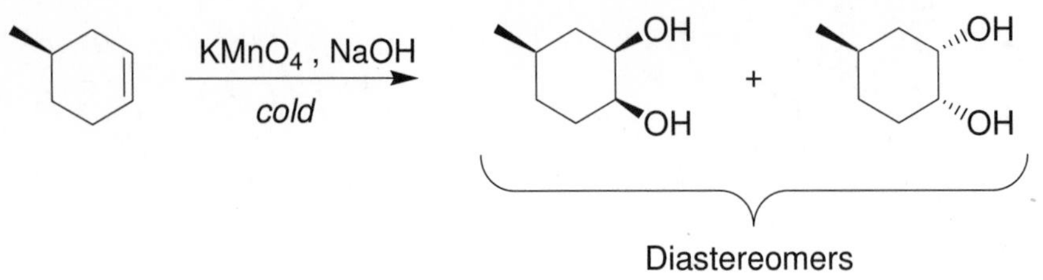

$KMnO_4$, NaOH / *cold*

OH OH + OH OH

Diastereomers

9.38. The products are the same:

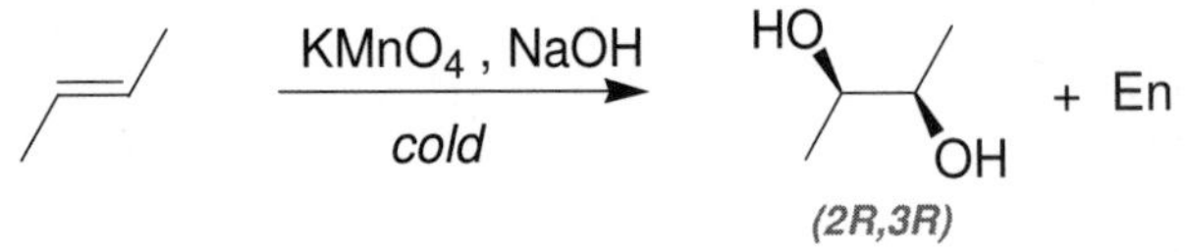

$KMnO_4$, NaOH / *cold*

HO OH + En

(2R,3R)

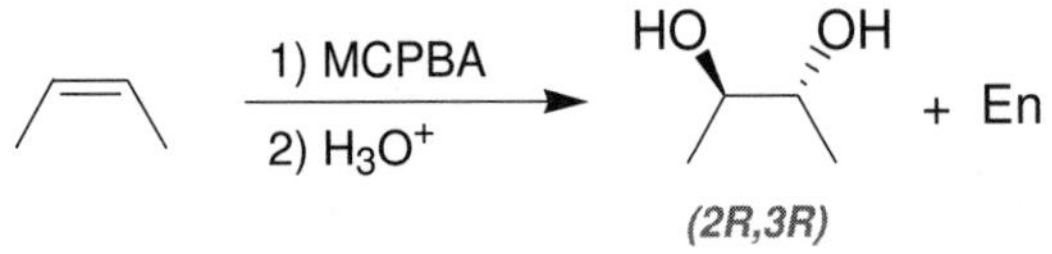

1) MCPBA 2) H_3O^+

HO OH + En

(2R,3R)

9.39.

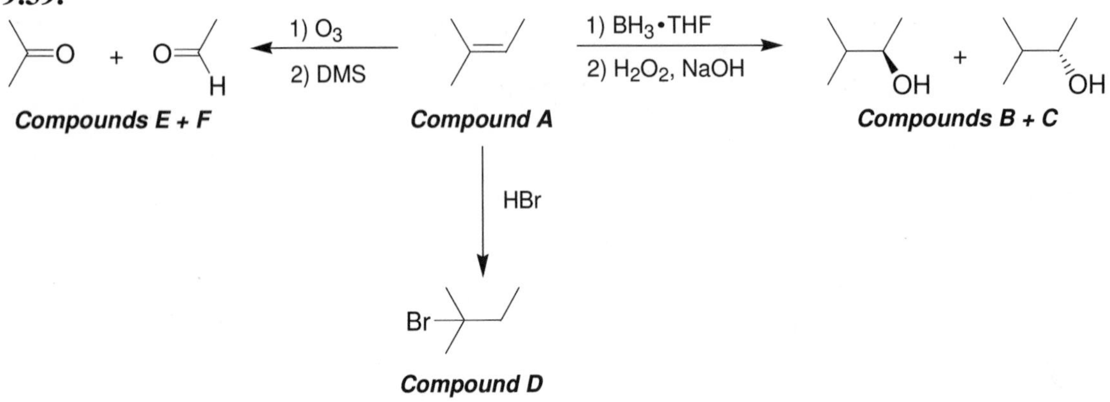

O + O H

Compounds E + F

1) O_3 2) DMS

Compound A

1) BH_3•THF 2) H_2O_2, NaOH

OH + OH

Compounds B + C

HBr

Br

Compound D

9.40.

a) 1) BH_3•THF 2) H_2O_2, NaOH

OH + En

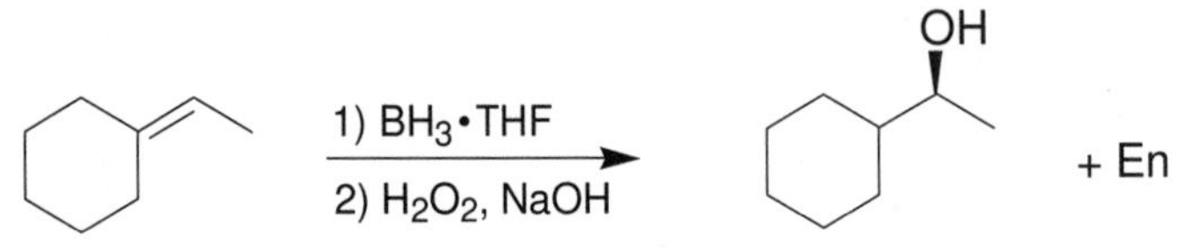

b) Br *t*-BuOK

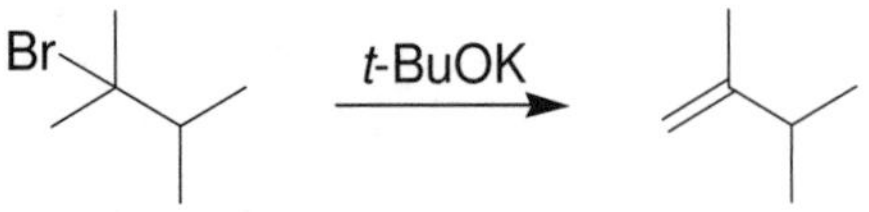

c) HBr ROOR Br

d) $\xrightarrow[Pt]{H_2}$

e) $\xrightarrow{HCl}$ Cl

f) $\xrightarrow[2)\ H_2O_2,\ NaOH]{1)\ BH_3\cdot THF}$ OH + En

g) Br $\xrightarrow{NaOEt}$

h) $\xrightarrow{HBr}$ Br

9.41.

a) $\xrightarrow[ROOR]{HBr}$ Br

b) $\xrightarrow{HBr}$ Br

c) $\xrightarrow[cold]{KMnO_4,\ NaOH}$ HO OH

d) $\xrightarrow[2)\ H_3O^+]{1)\ MCPBA}$ HO OH + En

9.42.

a) Cl $\xrightarrow[2)\ HCl]{1)\ NaOMe}$ Cl

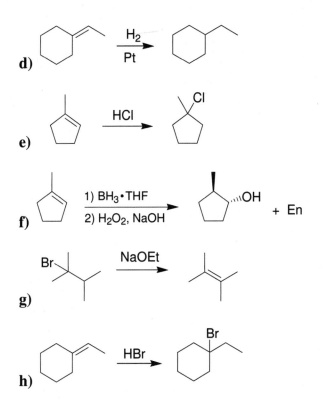

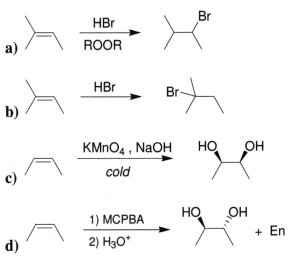

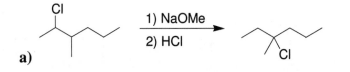

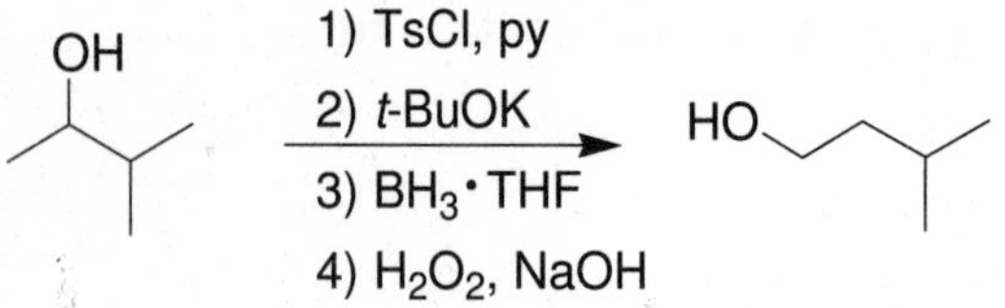

b) 1) TsCl, py; 2) *t*-BuOK; 3) $BH_3 \cdot THF$; 4) H_2O_2, NaOH

OH

HO

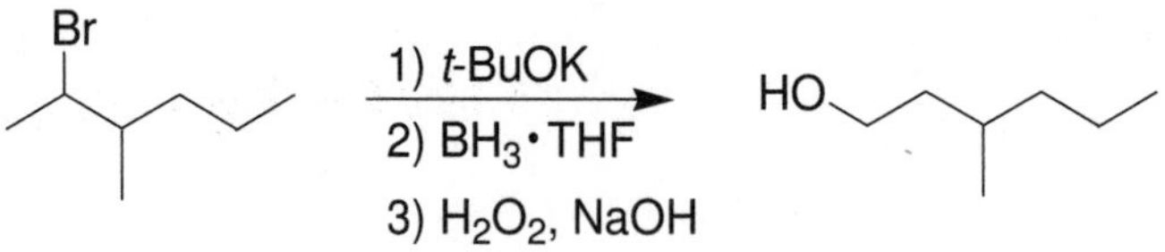

c) 1) *t*-BuOK; 2) $BH_3 \cdot THF$; 3) H_2O_2, NaOH

Br

HO

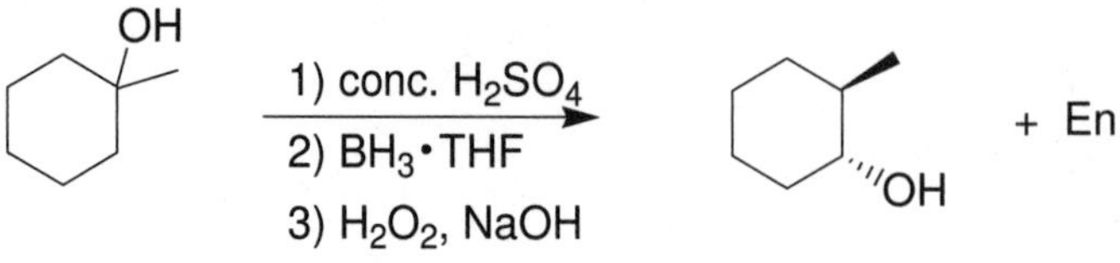

d) 1) conc. H_2SO_4; 2) $BH_3 \cdot THF$; 3) H_2O_2, NaOH

OH

OH + En

9.43.

a) 1) NaOMe; 2) HBr, ROOR

Br

Br

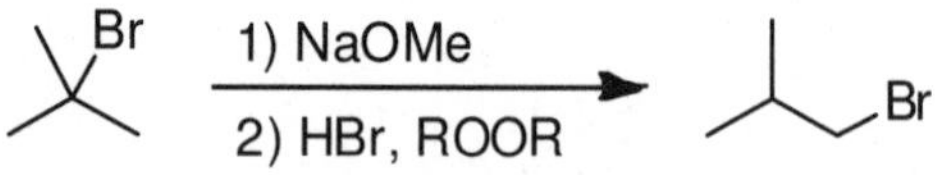

b) 1) NaOMe; 2) HBr, ROOR

Br

Br

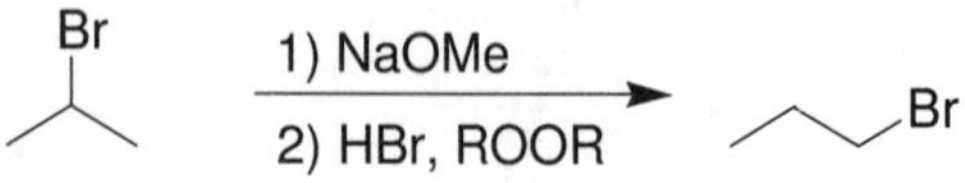

9.44.

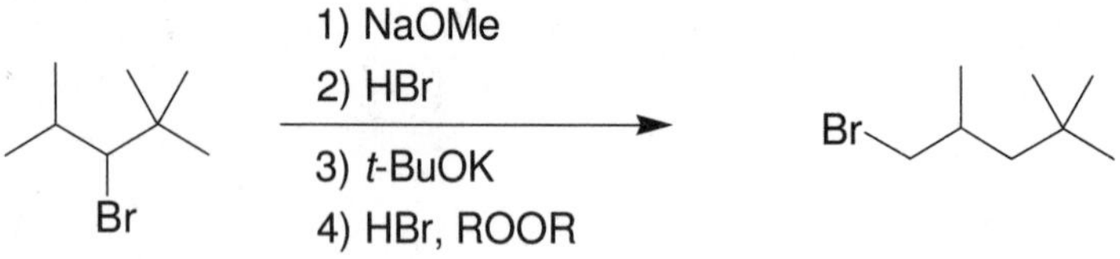

1) NaOMe
2) HBr
3) *t*-BuOK
4) HBr, ROOR

Br

Br

9.45.

a)

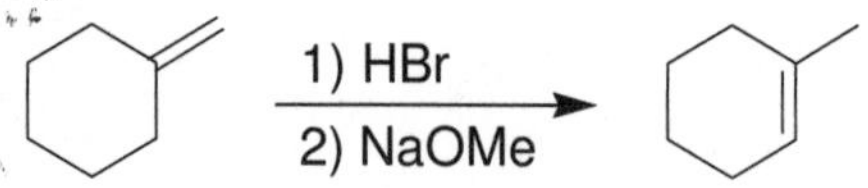

1) HBr
2) NaOMe

b)

1) HBr
2) *t*-BuOK

1) HBr
2) NaOMe

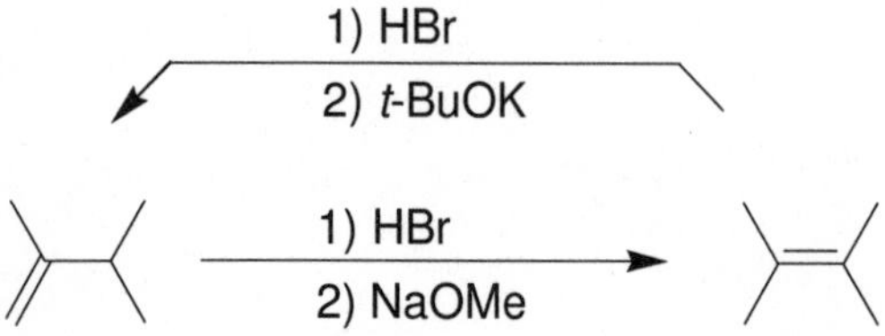

9.46.

a) 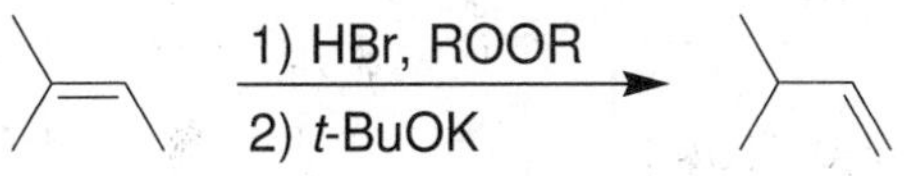 1) HBr, ROOR; 2) *t*-BuOK

b) 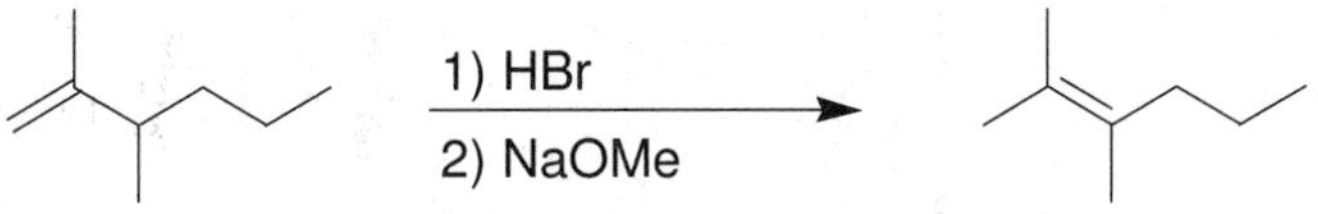 1) HBr; 2) NaOMe

9.47.

a) 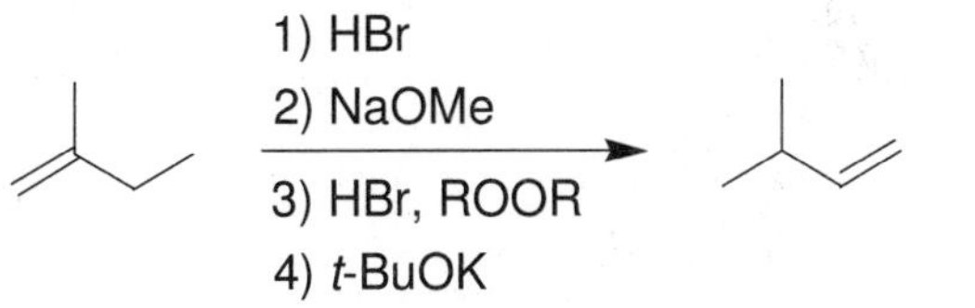 1) HBr; 2) NaOMe; 3) HBr, ROOR; 4) *t*-BuOK

b) 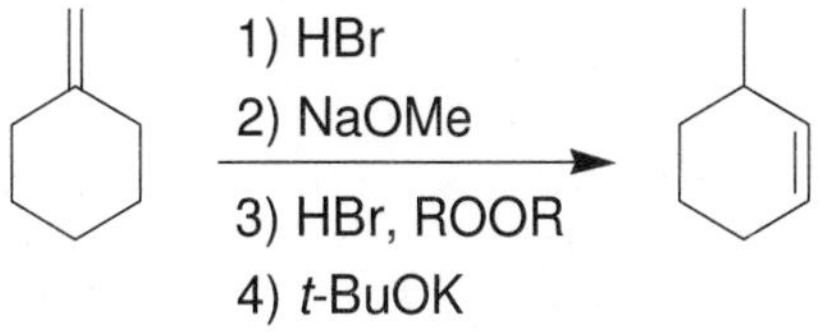 1) HBr; 2) NaOMe; 3) HBr, ROOR; 4) *t*-BuOK

9.48. A reaction is only favorable if ΔG is negative. Recall that ΔG has two components: (ΔH) and (-TΔS). The first term (ΔH) is positive for this reaction (two sigma bonds are converted into one sigma bond and one pi bond). The second term (-TΔS) is negative because ΔS is positive (one molecule is converted into two molecules). Therefore, the reaction is only favorable if the second term is greater in magnitude than the first term. This only occurs at high temperature.

9.49.

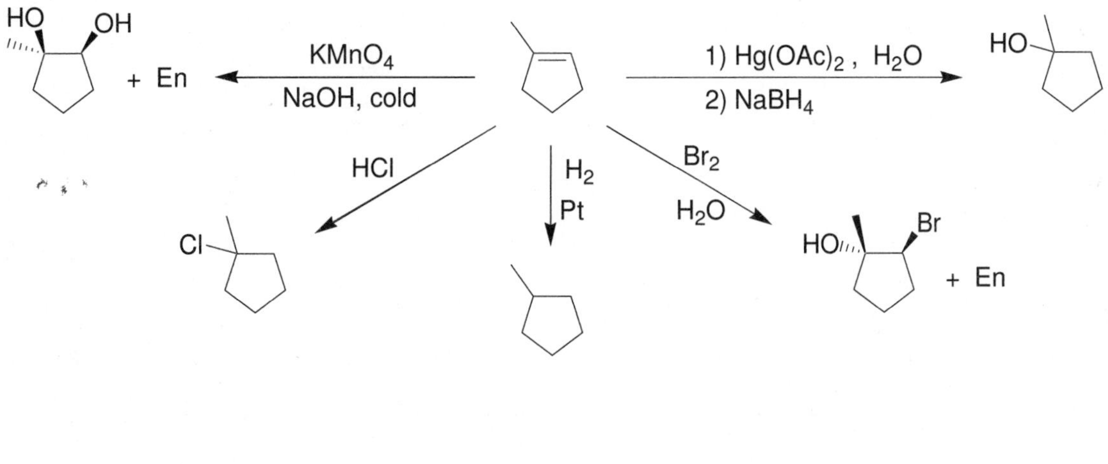

HO OH + En ← $KMnO_4$, NaOH, cold

1) $Hg(OAc)_2$, H_2O; 2) $NaBH_4$ → HO

HCl → Cl

H_2, Pt

Br_2, H_2O → HO Br + En

9.50.

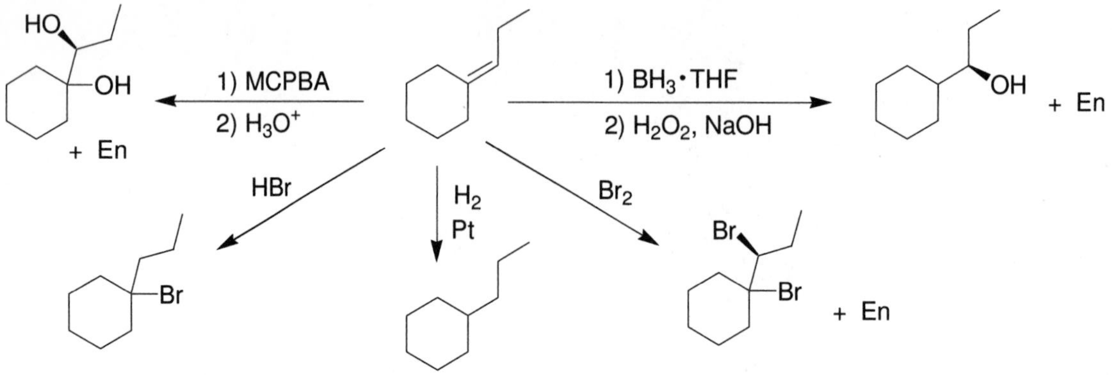

HO
OH
+ En
1) MCPBA
2) H_3O^+
1) $BH_3 \cdot THF$
2) H_2O_2, NaOH
OH
+ En
HBr
H_2
Pt
Br_2
Br
Br
Br
+ En

9.51.

a)

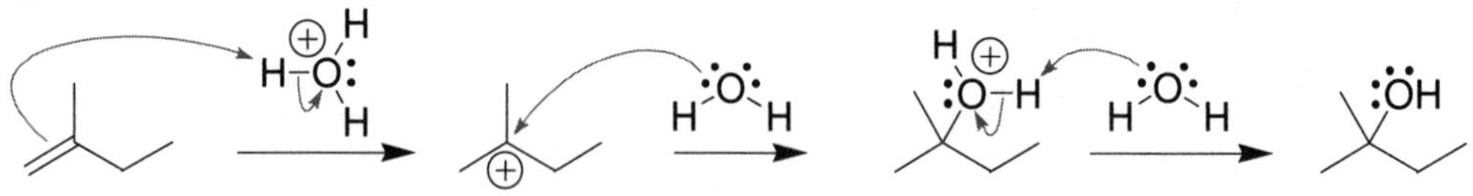

H
H–O:
H
H
:O:
H
H
:O–H
H
:O:
H
:ÖH

b)

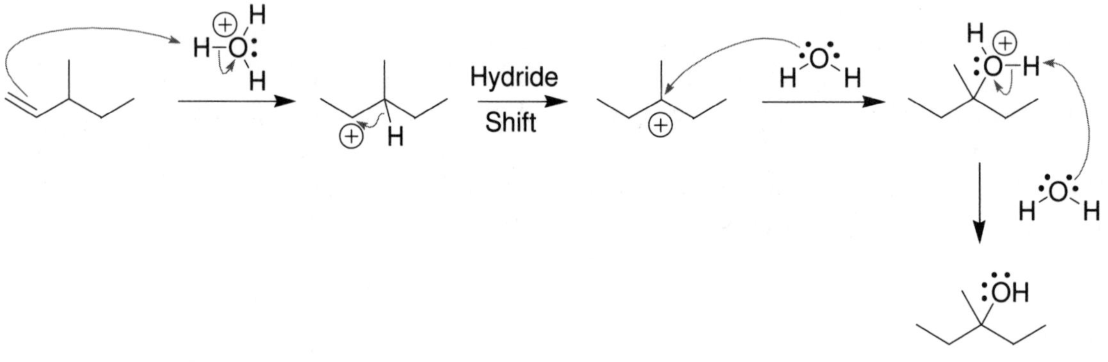

H
H–O:
H
H
Hydride
Shift
:O:
H H
H
:O–H
:O:
H H
:ÖH

c)

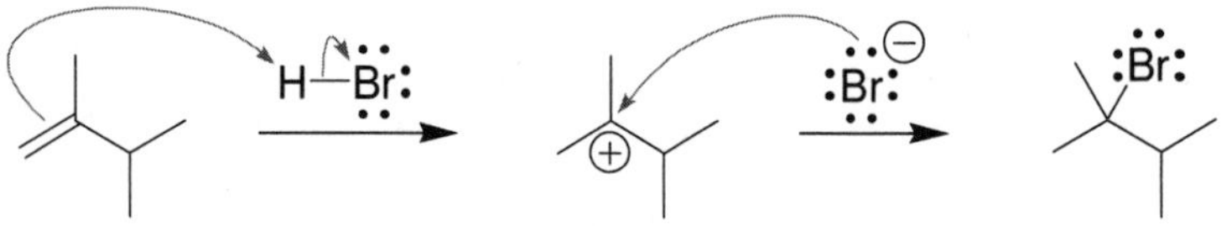

H–Br:
:Br:
:Br:

d)

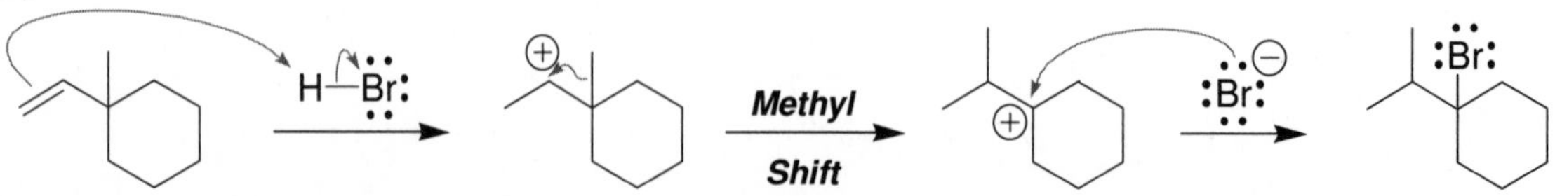

H–Br:
Methyl
Shift
:Br:
:Br:

9.52.

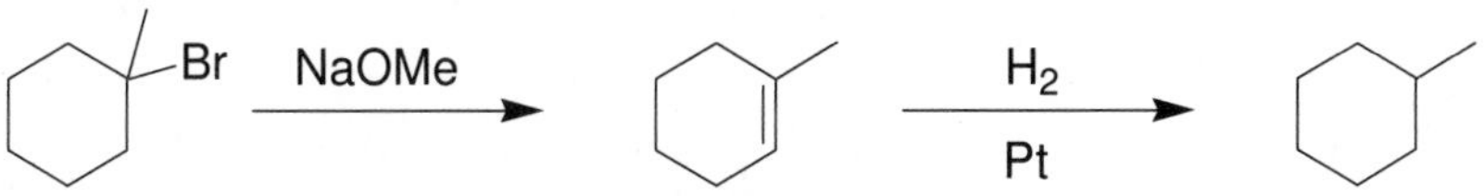

Br NaOMe

Compound A

H_2

Pt

9.53.

a)

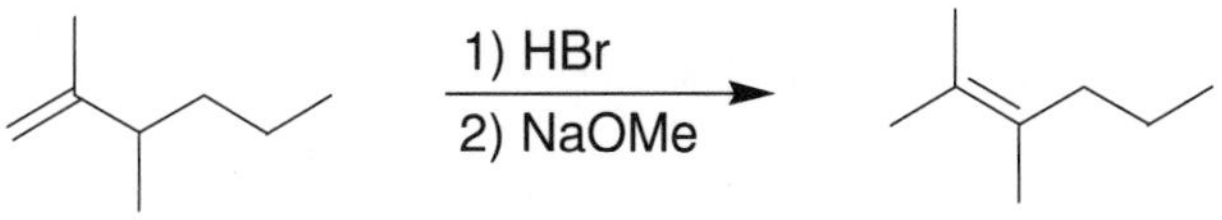

1) HBr

2) NaOMe

b)

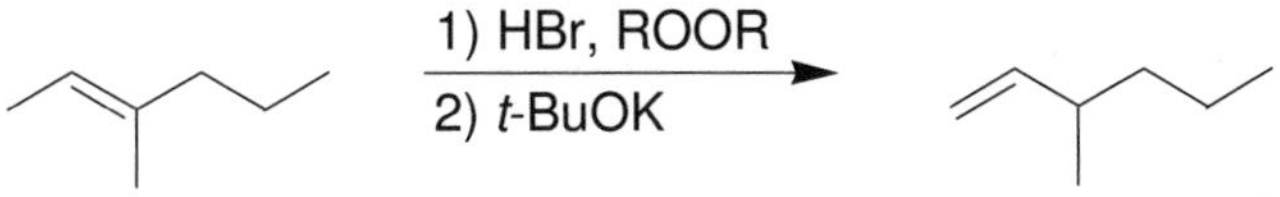

1) HBr, ROOR

2) *t*-BuOK

9.54.

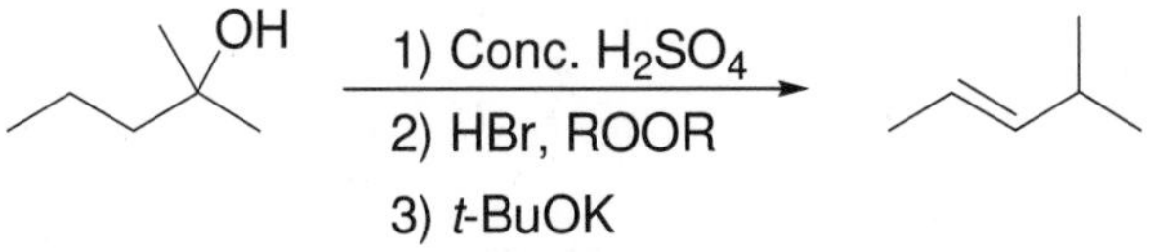

OH

1) Conc. H_2SO_4

2) HBr, ROOR

3) *t*-BuOK

9.55. Two different alkenes will produce 2,4-dimethylpentane upon hydrogenation:

9.56.

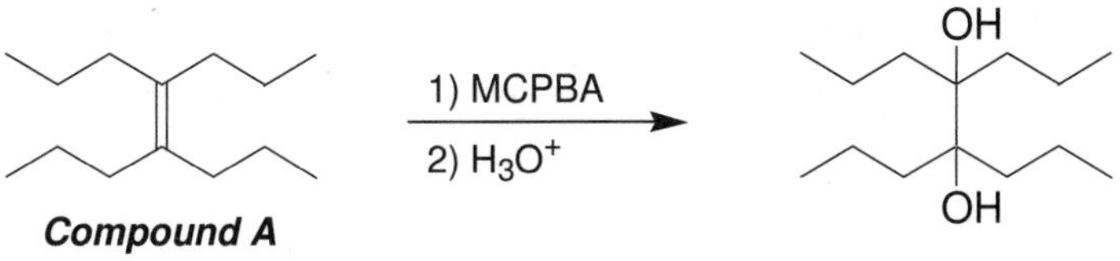

Compound A

1) MCPBA

2) H_3O^+

OH

OH

9.57.

a)

1) Conc. H_2SO_4

2) dilute H_2SO_4

OH

OH

1) Conc. H_2SO_4

2) $BH_3 \cdot THF$

3) H_2O_2, NaOH

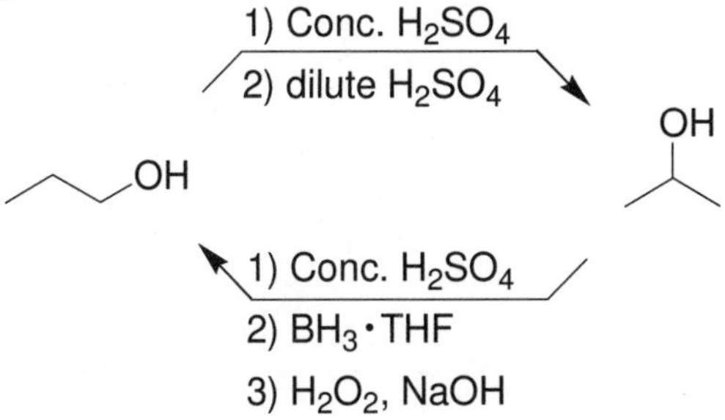

b)

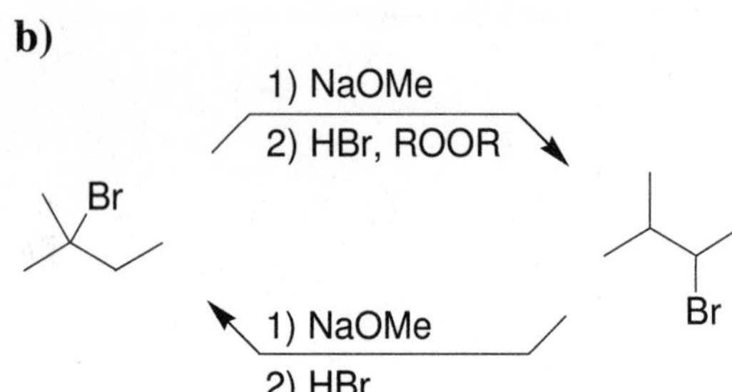

1) NaOMe
2) HBr, ROOR
Br
Br
1) NaOMe
2) HBr

c)

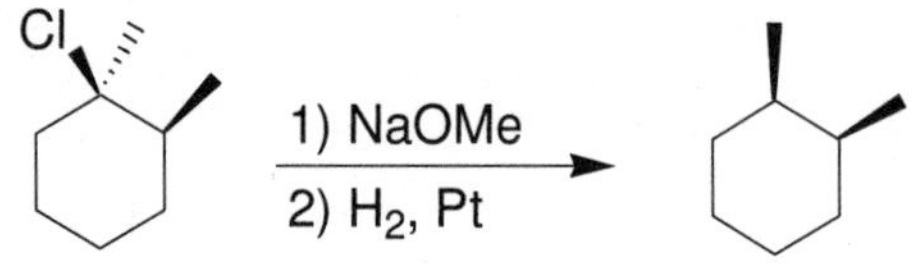

Cl
1) NaOMe
2) H2, Pt

d)

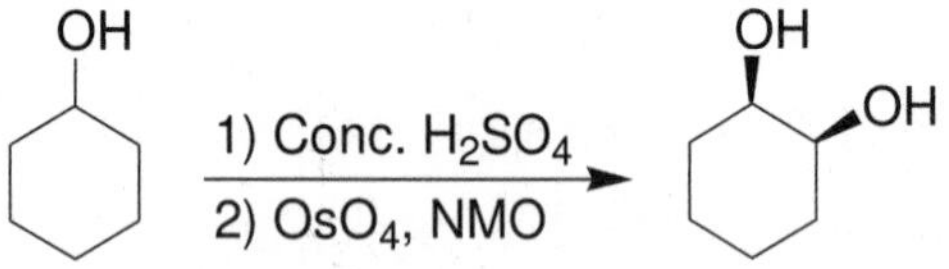

OH
1) Conc. H2SO4
2) OsO4, NMO
OH
OH

9.58.

a)

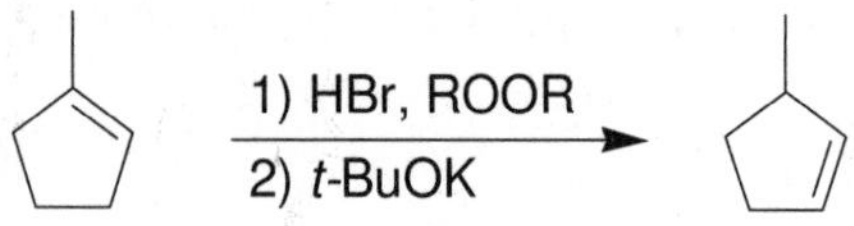

1) HBr, ROOR
2) t-BuOK

b)

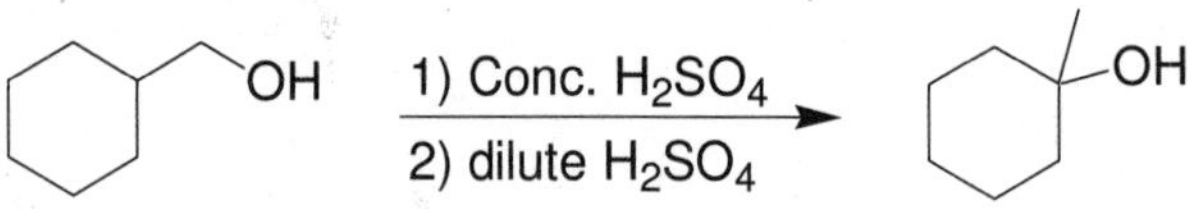

OH
1) Conc. H2SO4
2) dilute H2SO4
OH

9.59.

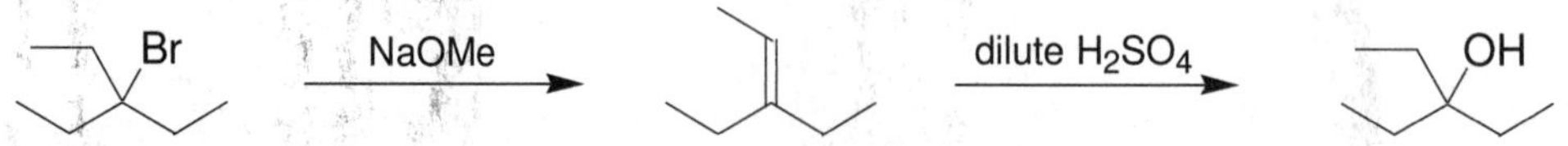

Br
NaOMe
dilute H2SO4
OH

Compound A *Compound B* *Compound C*

9.60.

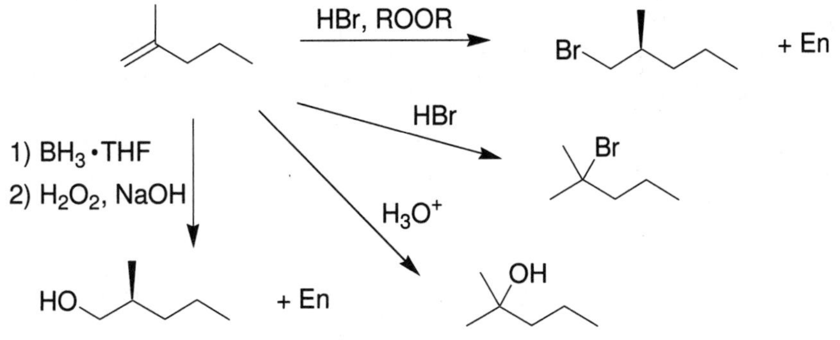

HBr, ROOR

Br + En

HBr

Br

1) $BH_3 \cdot THF$
2) H_2O_2, NaOH

H_3O^+

HO + En

OH

9.61.

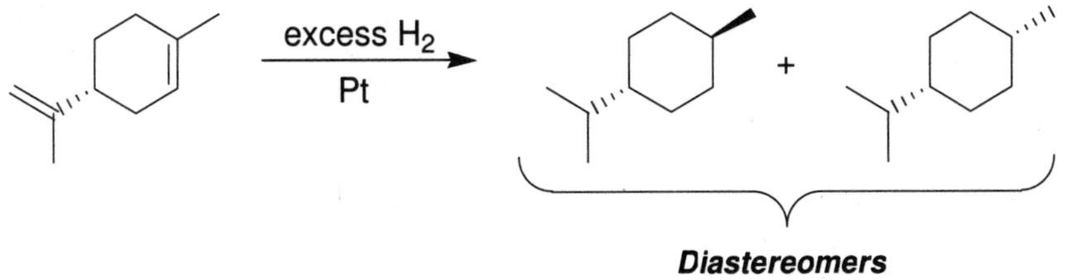

excess H_2
Pt

+

Diastereomers

9.62. Markovnikov addition of water without carbocation rearrangements can be achieved via oxymercuration-demercuration:

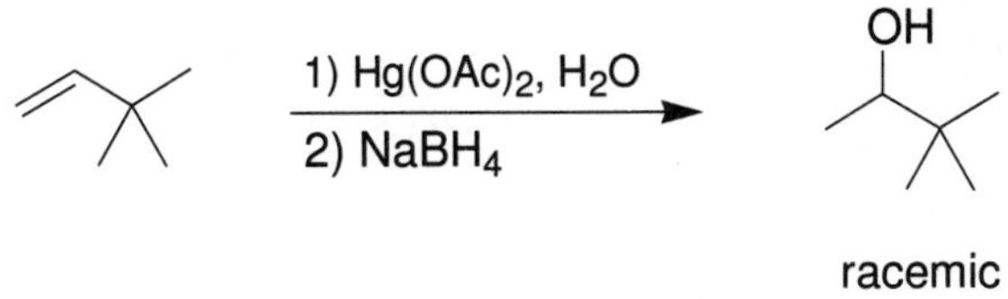

1) $Hg(OAc)_2$, H_2O
2) $NaBH_4$

OH

racemic

9.63.

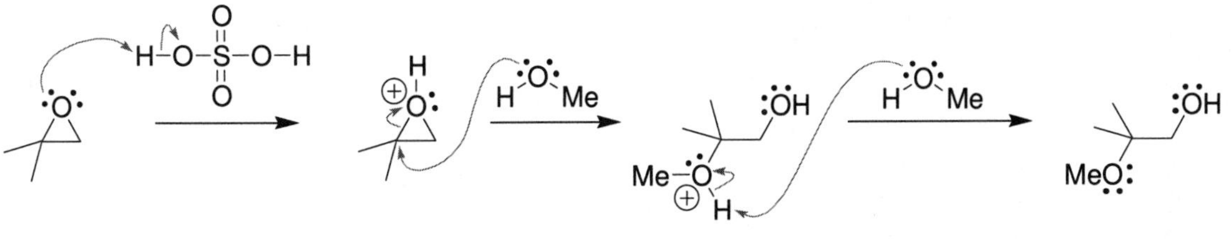

H–O–S–O–H

H O Me

OH

H O Me

Me–O H

MeO OH

9.64.

a)

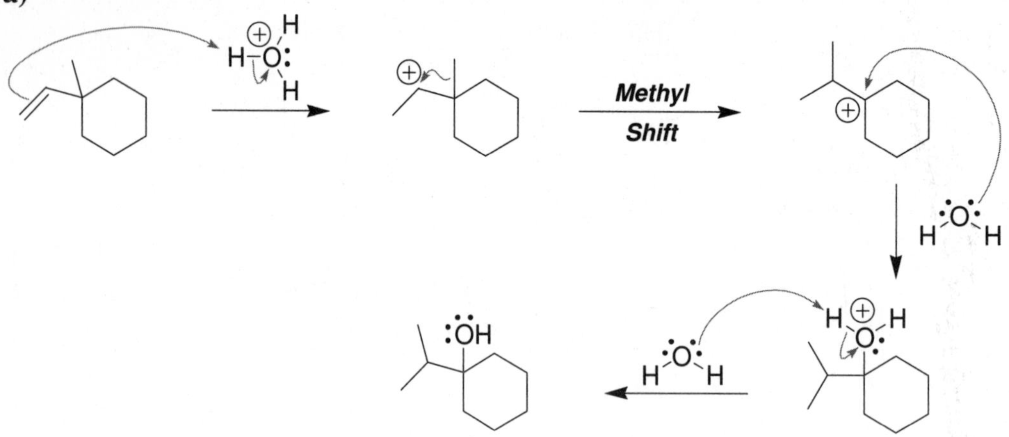

b)

H—Br:

\+ :Br:

Br

9.65.

H_2

$(PPh_3)_3RhCl$

a)

(meso)

b)

H_3O^+

OH

c)

1) $BH_3 \cdot THF$

2) H_2O_2, NaOH

OH

d)

1) MCPBA

2) H_3O^+

HO OH

\+ En

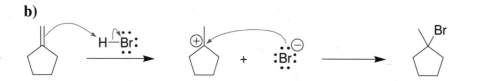

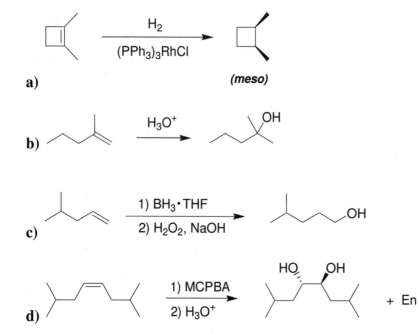

9.66.

a) Hydroboration-oxidation gives an anti-Markovnikov addition. If 1-propene is the starting material, the OH group will not be installed in the correct location. Acid-catalyzed hydration of 1-propene would give the desired product.

b) Hydroboration-oxidation gives a syn addition of H and OH across a double bond. This compound does not have a proton that is *cis* to the OH group, and therefore, hydroboration-oxidation cannot be used to make this compound.

c) Hydroboration-oxidation gives an anti-Markovnikov addition. There is no starting alkene that would yield the desired product via an anti-Markovnikov addition.

9.67.

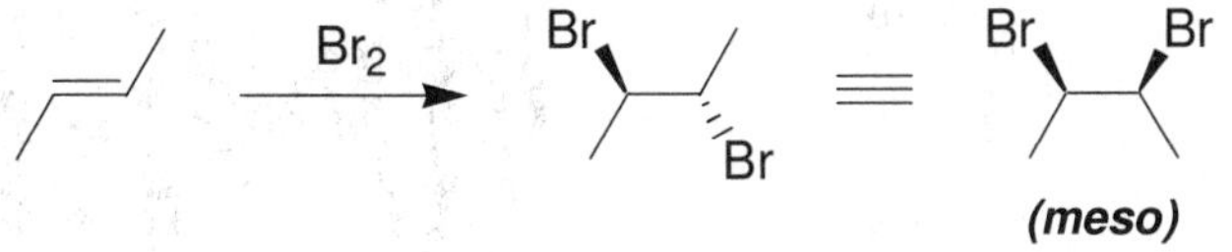

Br_2

Br Br ≡ Br Br

(meso)

9.68.

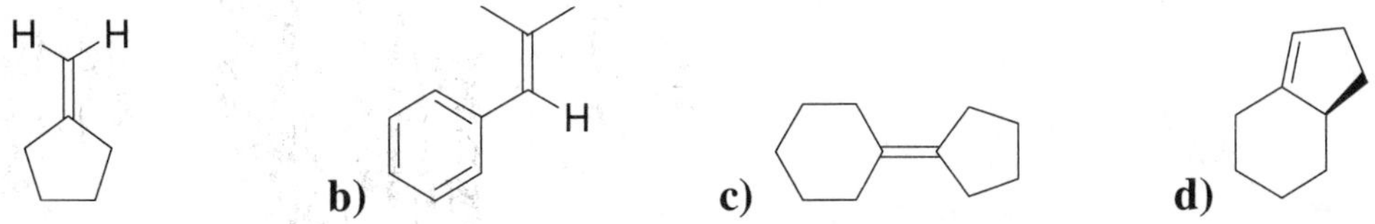

a) H H **b)** H **c)** **d)**

9.69. The reaction proceeds via a resonance-stabilized carbocation, which is even lower in energy than a tertiary carbocation:

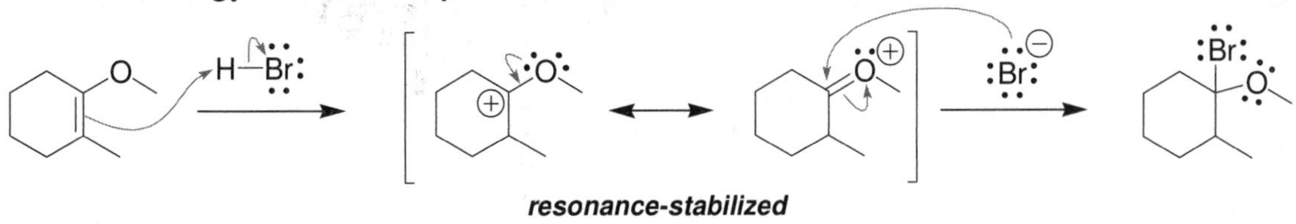

H—Br: :Br:

resonance-stabilized

9.70.

1) $BH_3 \cdot THF$
2) H_2O_2, NaOH

HBr
ROOR

OsO_4, NMO

HBr

Br_2
H_2O

Cl_2

OH

Br

OH OH

OH Br

Cl Cl

Br

NaOMe

H_2
Pt

Br_2
H_2O

OH
Br
+ En

1) O_3
2) DMS

O
H
O

1) $BH_3 \cdot THF$
2) H_2O_2, NaOH

OH
+ En

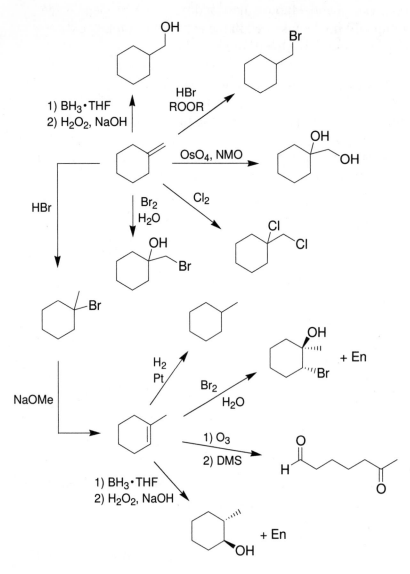

9.71.

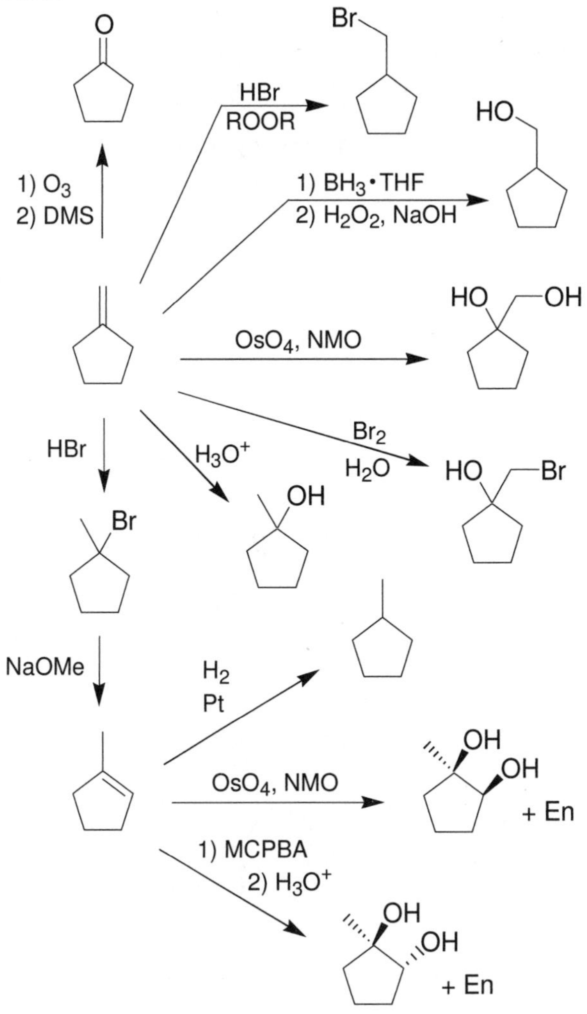
O
1) O_3
2) DMS
HBr
ROOR
Br
1) BH_3•THF
2) H_2O_2, NaOH
HO
OsO_4, NMO
HO
OH
HBr
H_3O^+
Br_2
H_2O
HO
Br
OH
Br
NaOMe
H_2
Pt
OH
OH
OsO_4, NMO
+ En
1) MCPBA
2) H_3O^+
OH
OH
+ En

9.72. Addition of HBr to 2-methyl-2-pentene should be more rapid because the reaction can proceed via a tertiary carbocation. In contrast, addition of HBr to 4-methyl-1-pentene proceeds via a less stable, secondary carbocation.

9.73.

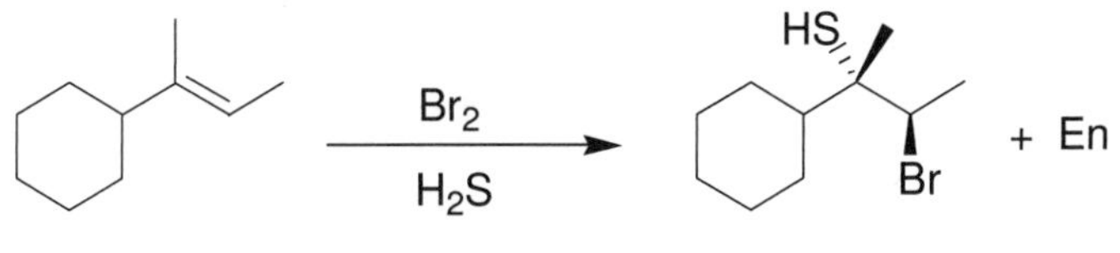
Br_2
H_2S
HS
Br
+ En

9.74.

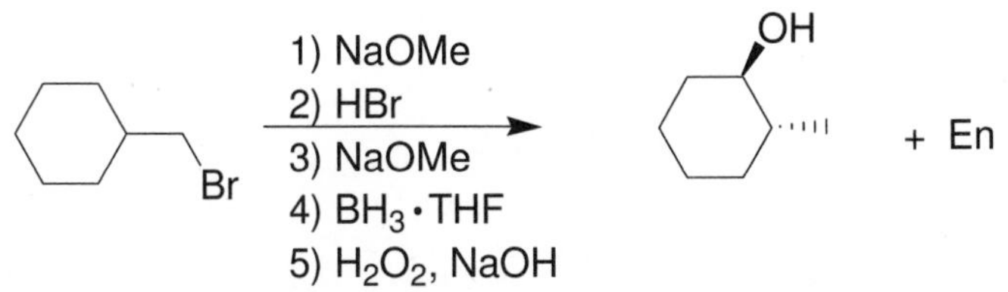
Br
1) NaOMe
2) HBr
3) NaOMe
4) BH_3•THF
5) H_2O_2, NaOH
OH
+ En

9.75.

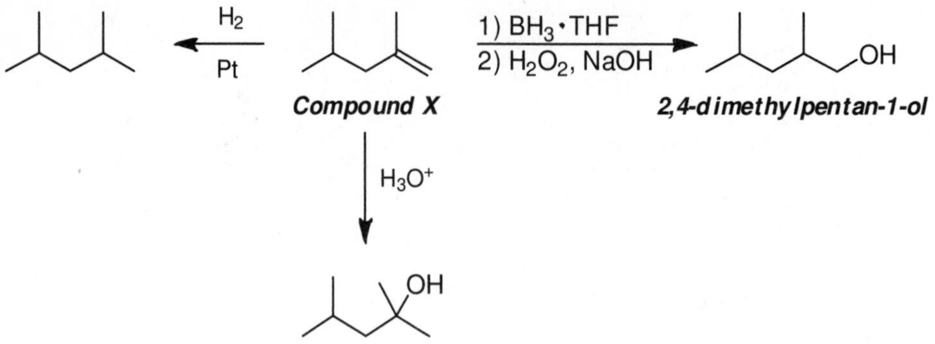

H_2 / Pt

1) $BH_3 \cdot THF$
2) H_2O_2, NaOH

OH

Compound X

2,4-dimethylpentan-1-ol

H_3O^+

OH

9.76.

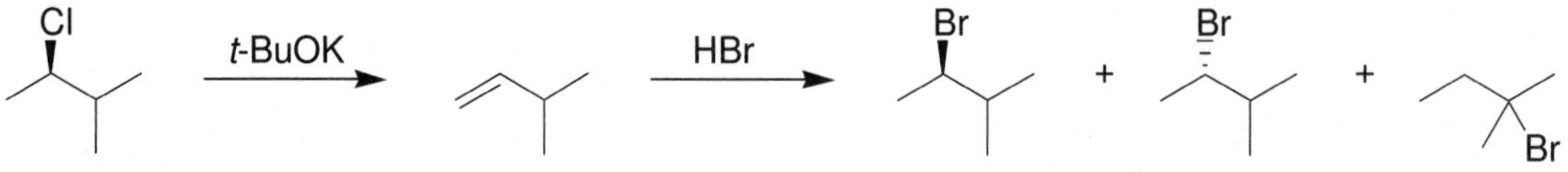

Cl

t-BuOK

HBr

Br + Br + Br

9.77.

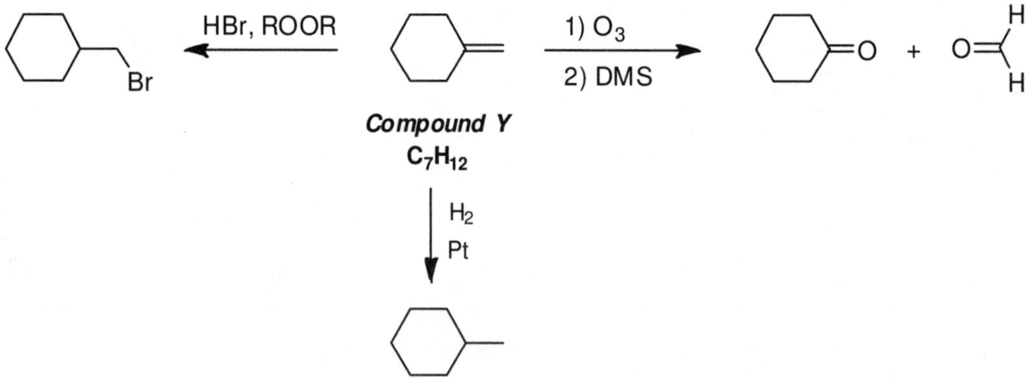

HBr, ROOR

1) O_3
2) DMS

Br

O + O H H

Compound Y
C_7H_{12}

H_2
Pt

9.78.

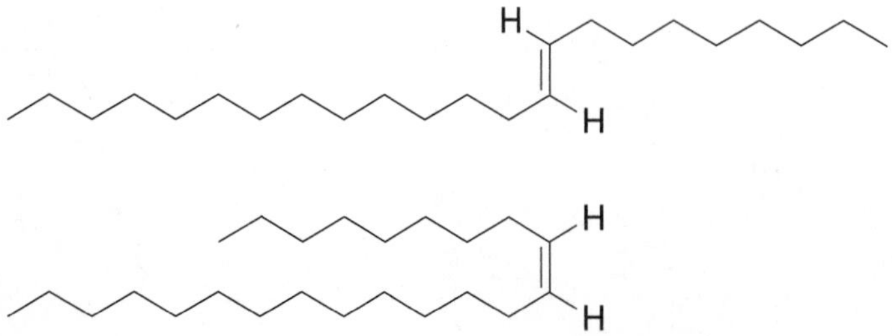

H
H

H
H

9.79.

a)

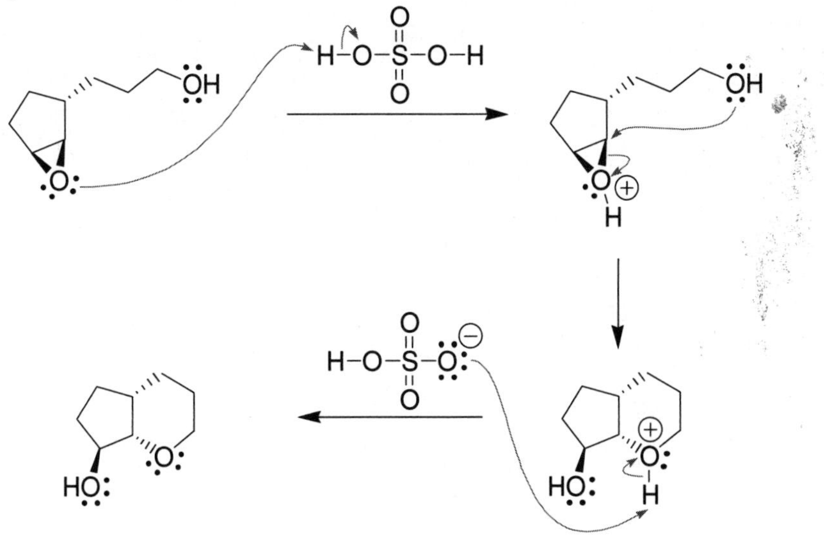

H–O–S(=O)(=O)–O–H

HO:

H–O–S(=O)(=O)–O:⊖

b)

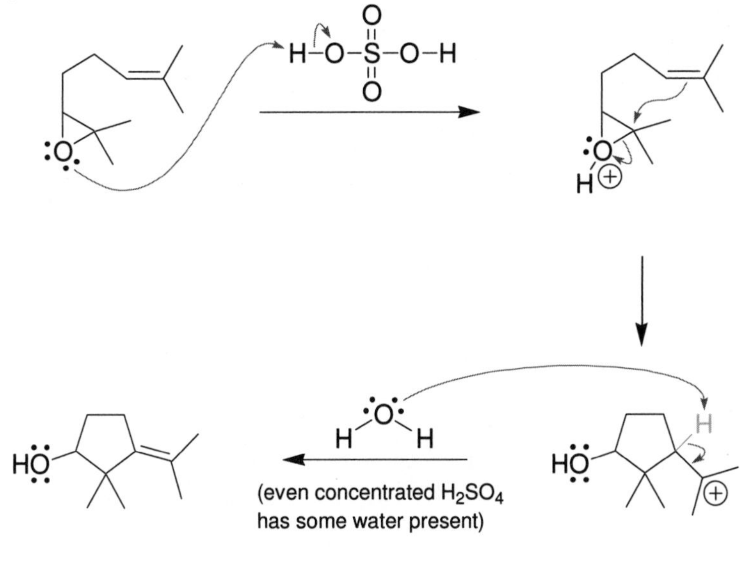

H–O–S(=O)(=O)–O–H

H–O–H

(even concentrated H_2SO_4 has some water present)

9.80.

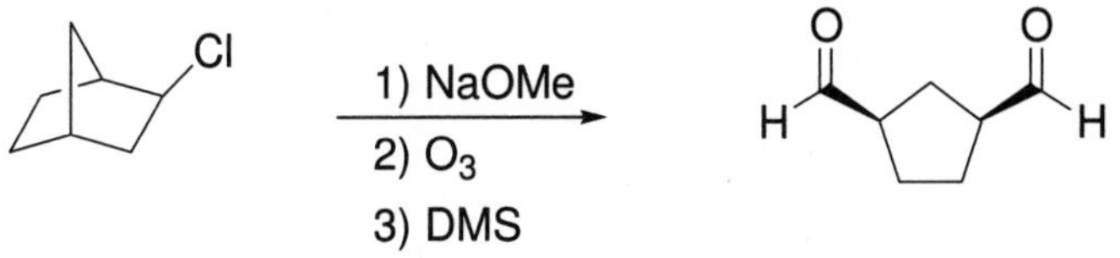

Cl

1) NaOMe

2) O_3

3) DMS

H H

9.81.

a)

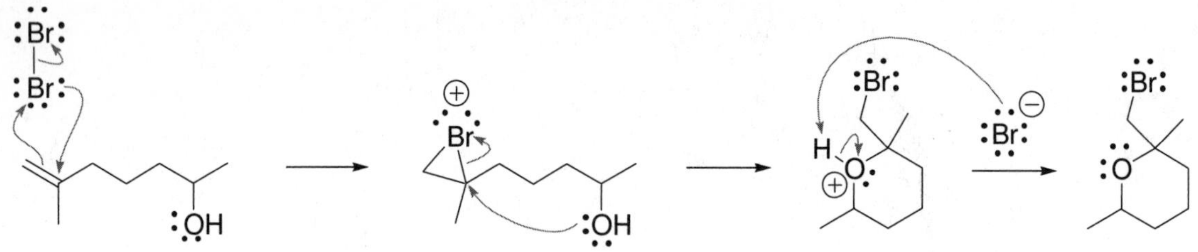

:Br: Br: OH Br Br: Br: O Br: Br: O H

b)

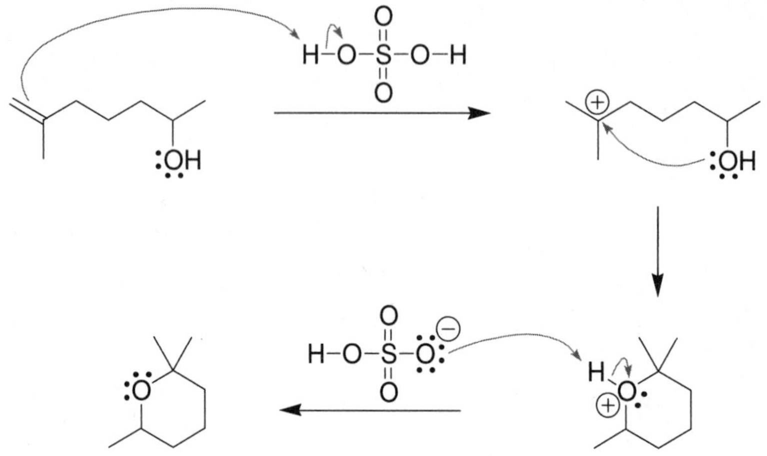

OH H–O–S–O–H O O OH H–O–S–O O O O H O

9.82.

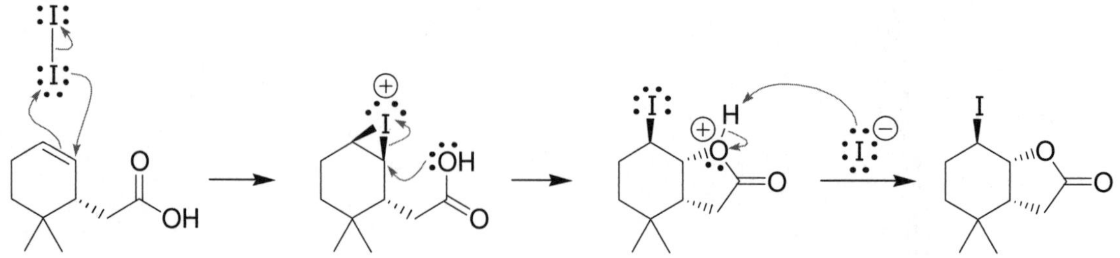

I I O OH I OH O I H O O I I O O

9.83.

Br: + En H–Br: Br: + En Br: + En Br: Br Br + En

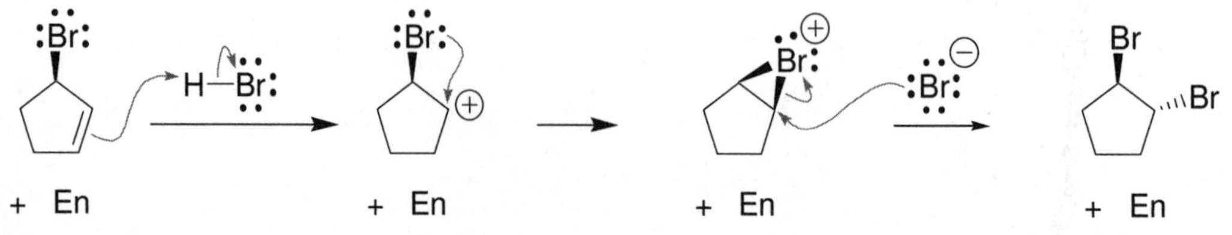

Chapter 10
Alkynes

Review of Concepts

Fill in the blanks below. To verify that your answers are correct, look in your textbook at the end of Chapter 10. Each of the sentences below appears verbatim in the section entitled *Review of Concepts and Vocabulary.*

- A triple bond is comprised of three separate bonds: one ____ bond and two ____ bonds.
- Alkynes exhibit __________ geometry and can function either as bases or as ____________________.
- Monosubstituted alkynes are **terminal alkynes**, while disubstituted alkynes are _________ **alkynes.**
- Catalytic hydrogenation of an alkyne yields an __________.
- A **dissolving metal reduction** will convert an alkyne into a _______ alkene.
- Acid-catalyzed hydration of alkynes is catalyzed by mercuric sulfate to produce an ________ that cannot be isolated because it is rapidly converted into a ketone.
- Enols and ketones are ____________, which are constitutional isomers that rapidly interconvert via the migration of a proton.
- When treated with ozone, followed by water, internal alkynes undergo oxidative cleavage to produce ______________________.
- Alkynide ions undergo ______________ when treated with an alkyl halide (methyl or primary).

Review of Skills

Fill in the blanks and empty boxes below. To verify that your answers are correct, look in your textbook at the end of Chapter 10. The answers appear in the section entitled *SkillBuilder Review.*

10.1 Assembling the Systematic Name of an Alkyne

PROVIDE A SYSTEMATIC NAME FOR THE FOLLOWING COMPOUND

1) IDENTIFY THE PARENT
2) IDENTIFY AND NAME SUBSTITUENTS
3) ASSIGN LOCANTS TO EACH SUBSTITUENT
4) ALPHABETIZE

10.2 Predicting the Position of Equilibrium for the Deprotonation of a Terminal Alkyne

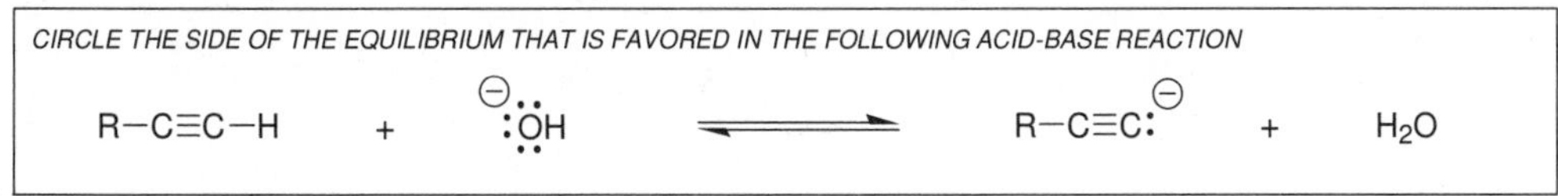

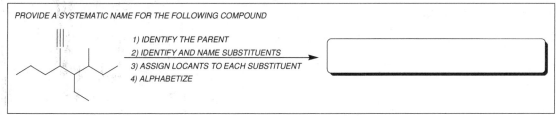

10.3 Drawing the Mechanism of Acid-Catalyzed Keto-Enol Tautomerization

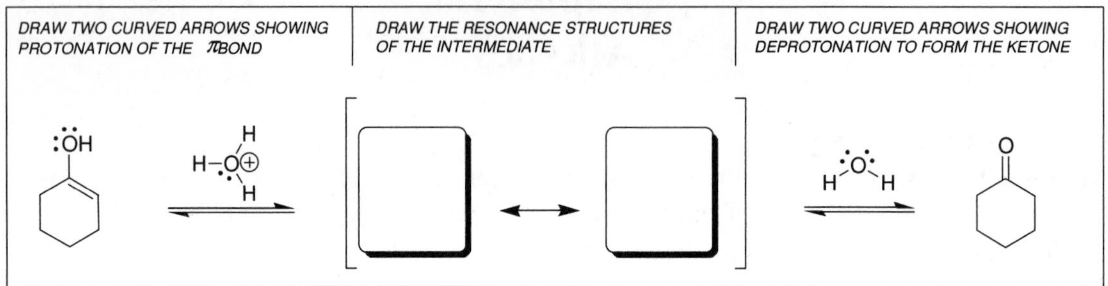
DRAW TWO CURVED ARROWS SHOWING PROTONATION OF THE π BOND
DRAW THE RESONANCE STRUCTURES OF THE INTERMEDIATE
DRAW TWO CURVED ARROWS SHOWING DEPROTONATION TO FORM THE KETONE
OH
H
H–O⊕
H
O
H H
O

10.4 Choosing the Appropriate Reagents for the Hydration of an Alkyne

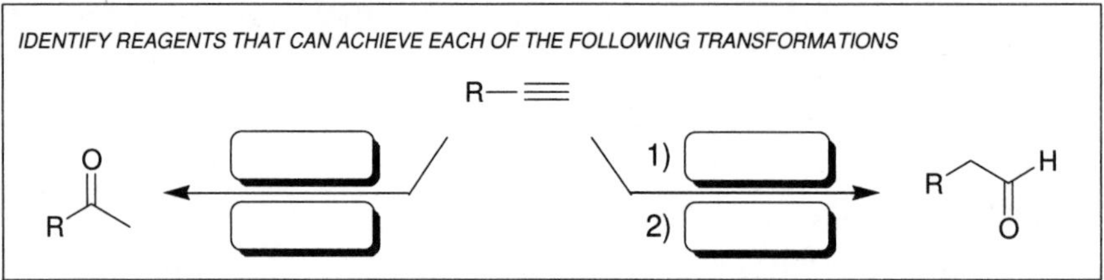
IDENTIFY REAGENTS THAT CAN ACHIEVE EACH OF THE FOLLOWING TRANSFORMATIONS
R—≡
O
R
1)
2)
R
H
O

10.5 Alkylating Terminal Alkynes

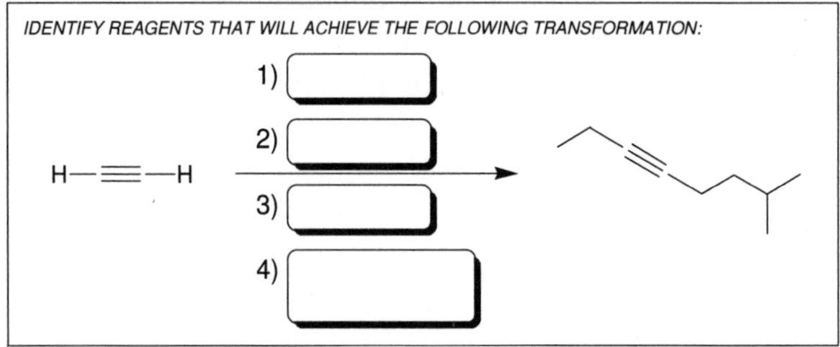
IDENTIFY REAGENTS THAT WILL ACHIEVE THE FOLLOWING TRANSFORMATION:
1)
2)
3)
4)
H—≡—H

10.6 Interconverting Alkanes, Alkenes, and Alkyne

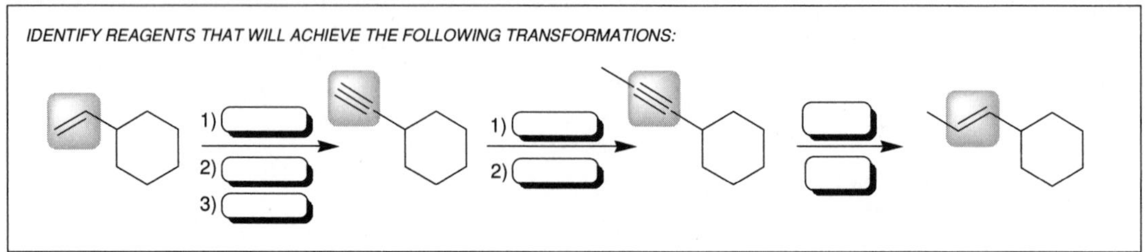
IDENTIFY REAGENTS THAT WILL ACHIEVE THE FOLLOWING TRANSFORMATIONS:
1)
2)
3)
1)
2)

Review of Reactions

Identify the reagents necessary to achieve each of the following transformations. To verify that your answers are correct, look in your textbook at the end of Chapter 10. The answers appear in the section entitled *Review of Reactions*.

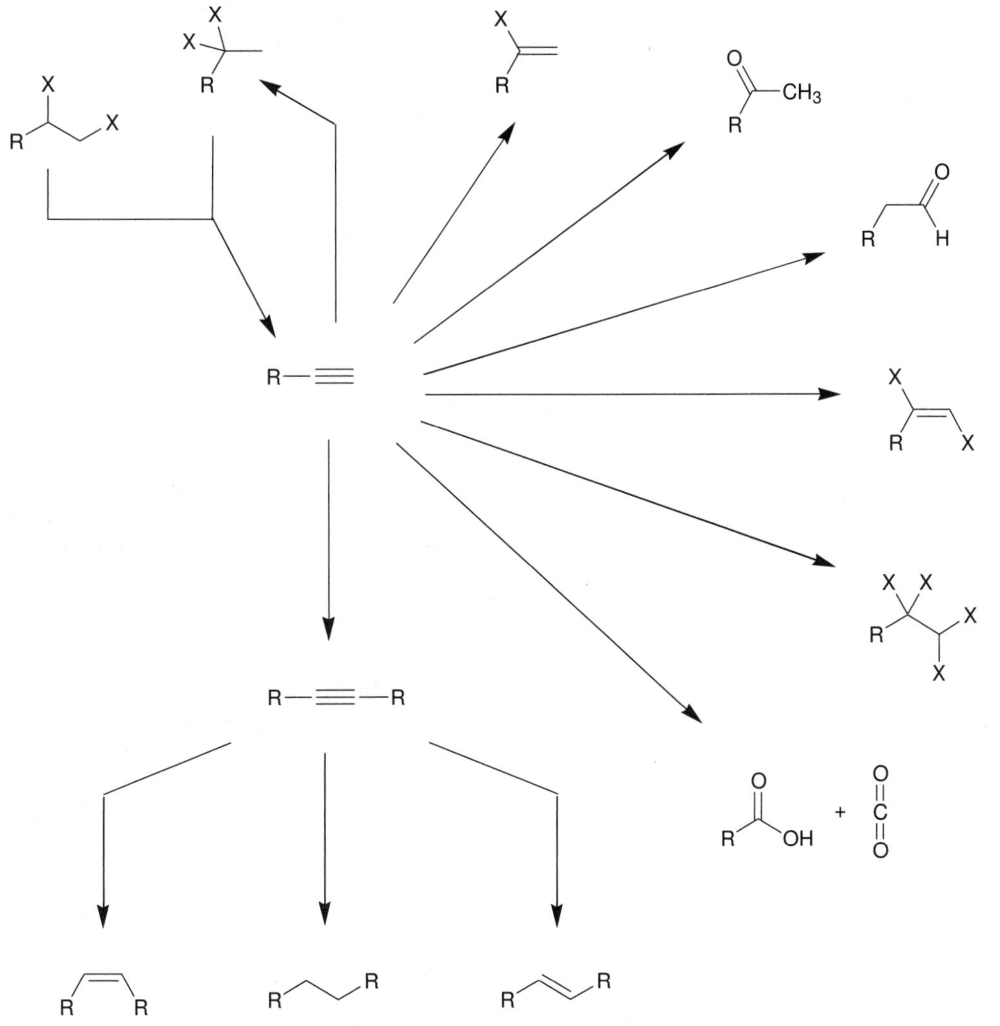

Solutions

10.1.

a) 3-hexyne
b) 2-methyl-3-hexyne
c) 3-octyne
d) 3,3-dimethyl-1-butyne

10.2.

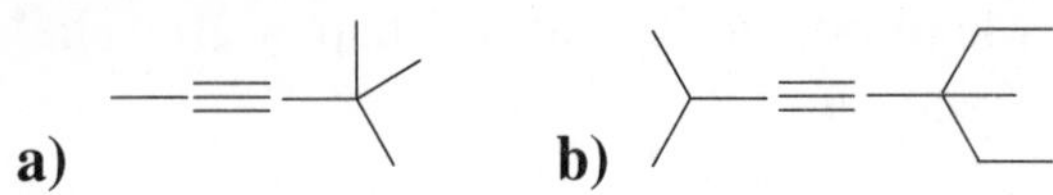

10.3.

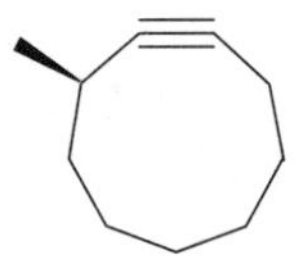

10.4.

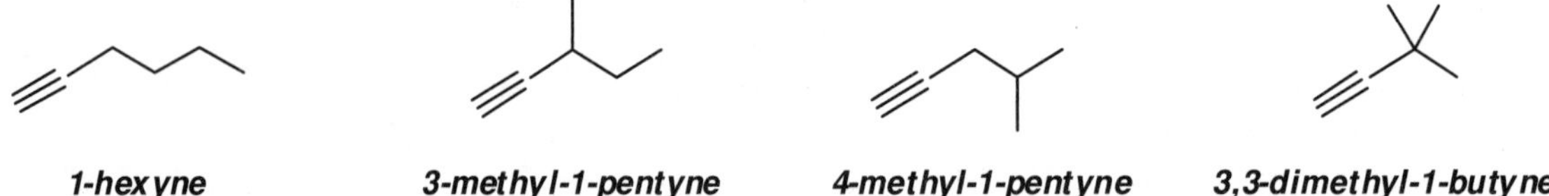

10.5.

a) Yes, $NaNH_2$ is strong enough of a base to deprotonate a terminal alkyne.
b) No, NaOEt is not strong enough of a base to deprotonate a terminal alkyne.
c) No, NaOH is not strong enough of a base to deprotonate a terminal alkyne.
d) Yes, BuLi is strong enough of a base to deprotonate a terminal alkyne.
e) Yes, NaH is strong enough of a base to deprotonate a terminal alkyne.
f) No, *t*-BuOK is not strong enough of a base to deprotonate a terminal alkyne.

10.6.

a) In the conjugate base of methyl amine (CH_3NH_2), the negative charge is associated with an *sp*^{3} hybridized nitrogen atom. In the conjugate base of HCN, the negative charge is associated with an *sp* hybridized carbon atom. The latter is more stable, because the charge is closer to the positively charged nucleus. As a result, HCN is a stronger acid than methyl amine.
b) The pK_a of HCN is lower than the pK_a of a terminal alkyne. Therefore, cyanide cannot be used as a base to deprotonate a terminal alkyne, as it would involve the formation of a stronger acid.

$$\underset{\textit{weaker acid}}{R{-}C{\equiv}C{-}H} \;+\; NaCN \;\rightleftharpoons\; R{-}C{\equiv}C{:}^{\ominus}\;Na^{\oplus} \;+\; \underset{\textit{stronger acid}}{HCN}$$

10.7.

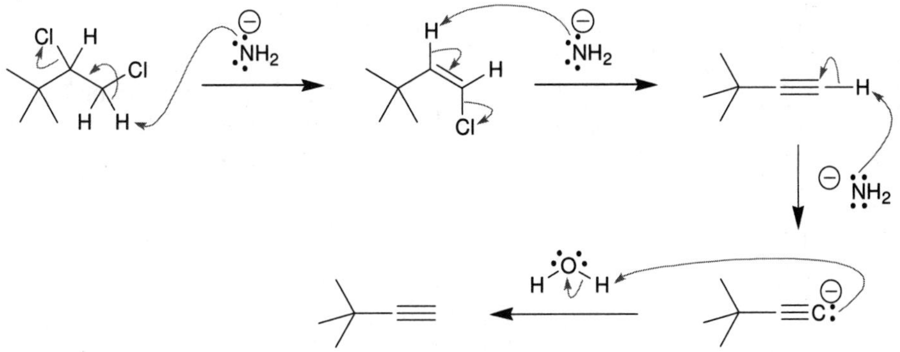

a)

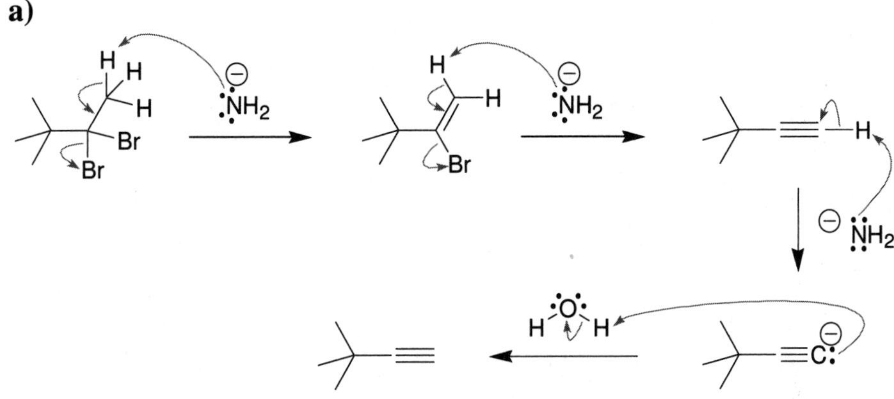

b)

10.8.

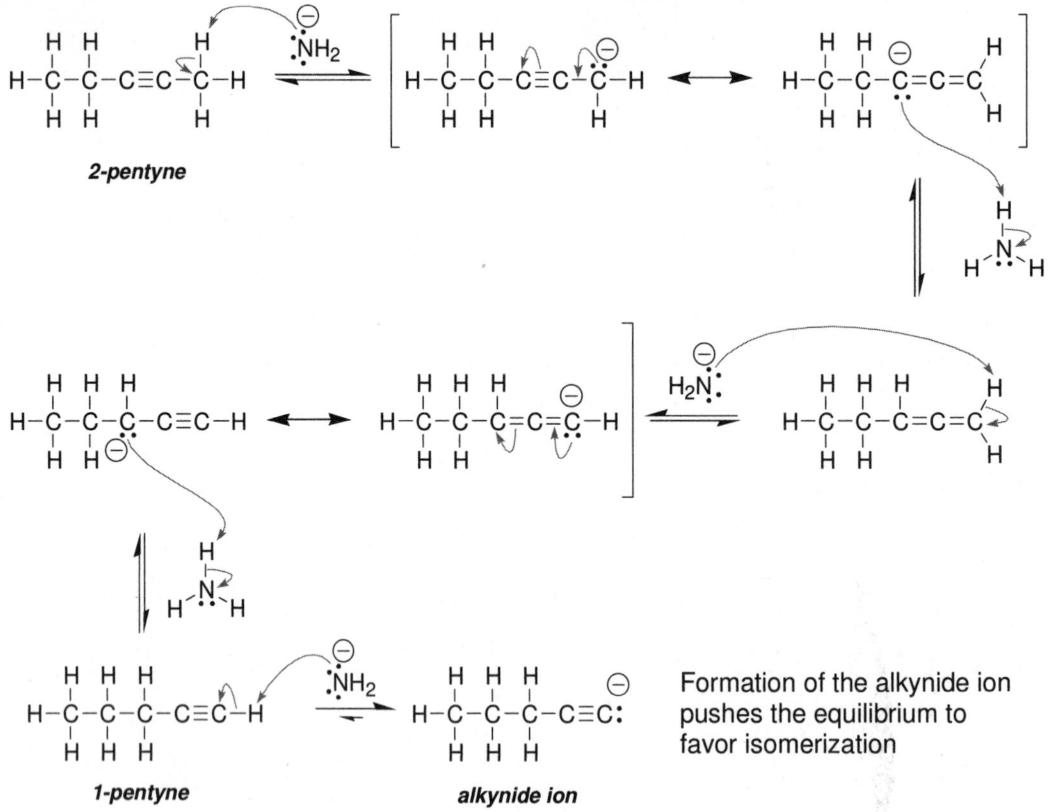

2-pentyne

1-pentyne

alkynide ion

Formation of the alkynide ion pushes the equilibrium to favor isomerization

10.9.

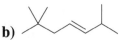

a) 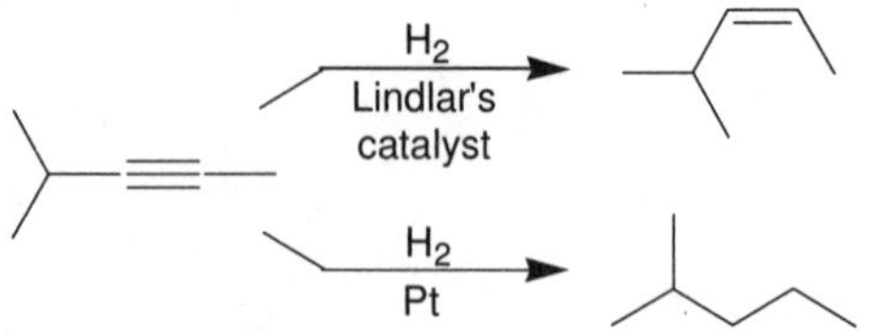 H_2, Lindlar's catalyst; H_2, Pt

b) 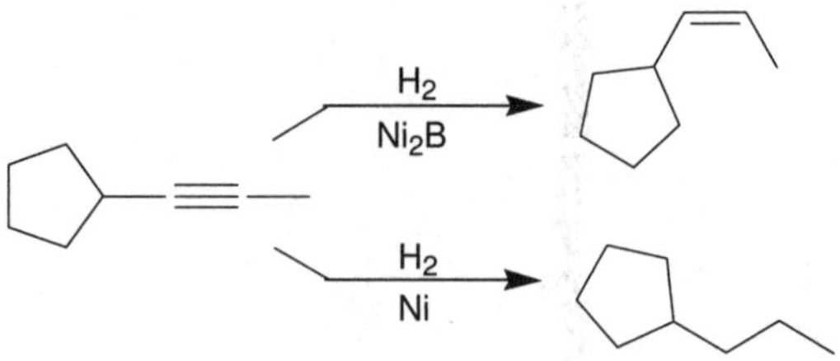 H_2, Ni_2B; H_2, Ni

10.10.

a) 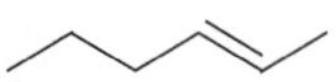 b) 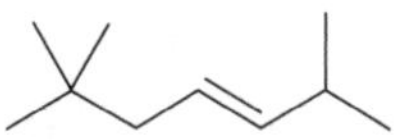 c) 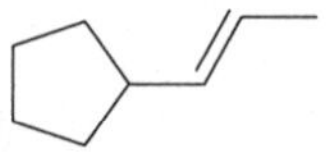 d)

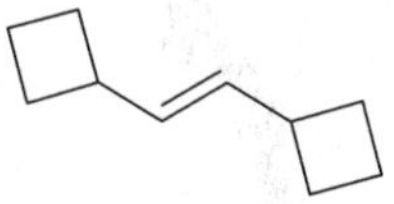

10.11.

a)

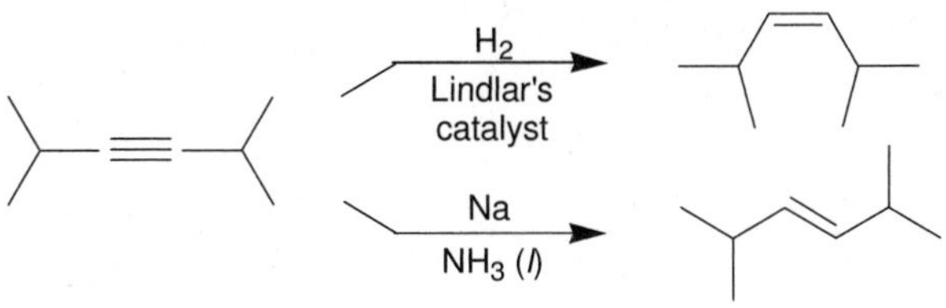

H_2 / Lindlar's catalyst

Na / NH_3 (*l*)

b)

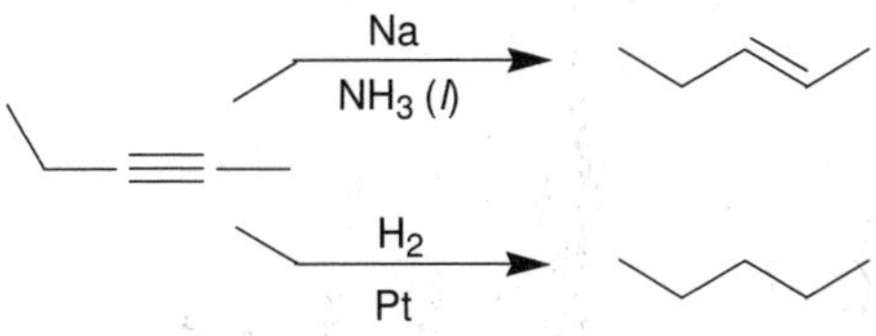

Na / NH_3 (*l*)

H_2 / Pt

10.12.

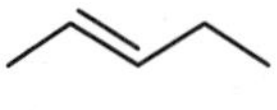

10.13.

a)

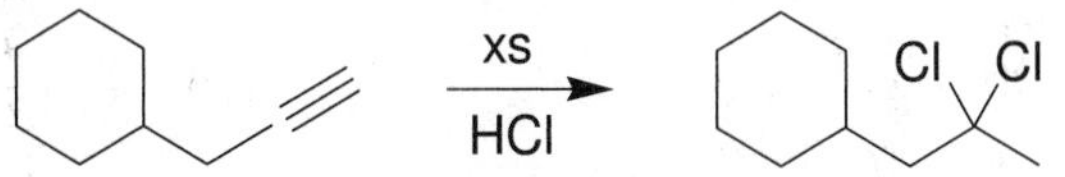

xs HCl

Cl Cl

b)

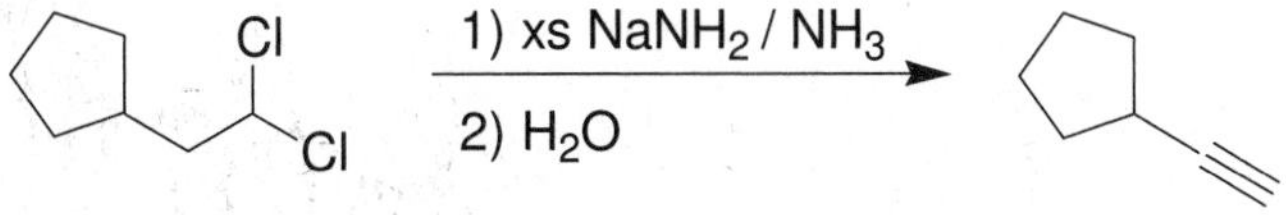

Cl Cl

1) xs $NaNH_2$ / NH_3

2) H_2O

c)

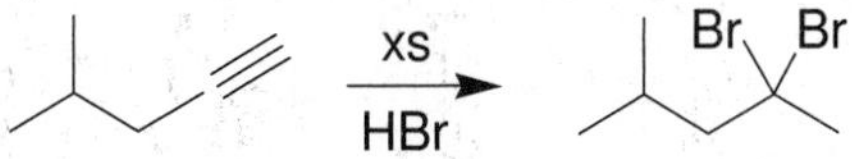

xs HBr

Br Br

d)

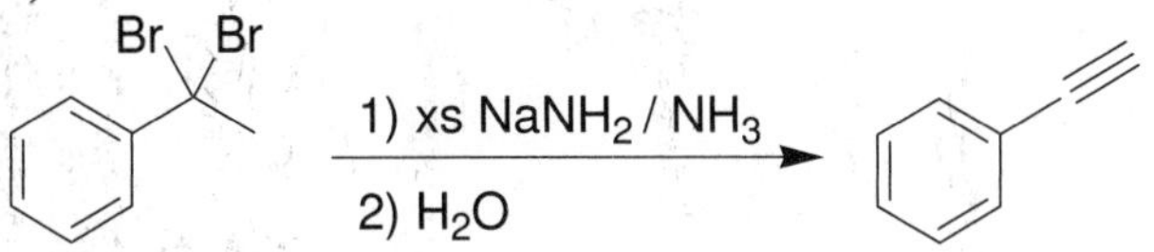

Br Br

1) xs $NaNH_2$ / NH_3

2) H_2O

e)

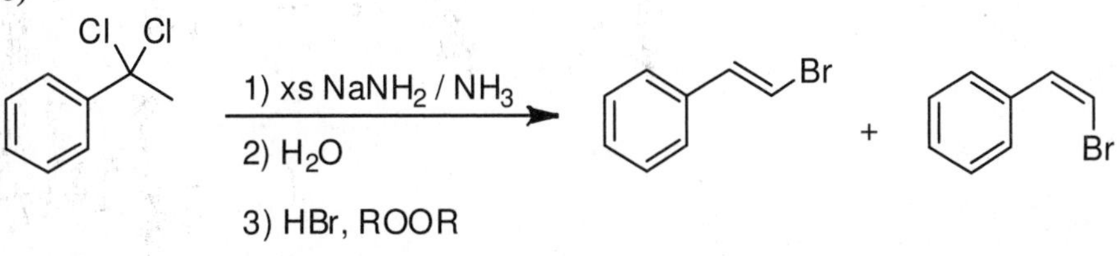

Cl Cl

1) xs $NaNH_2$ / NH_3

2) H_2O

3) HBr, ROOR

Br + Br

f)

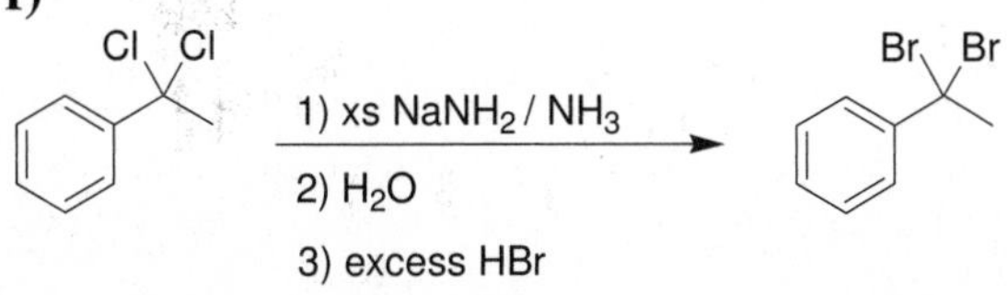

Cl Cl

1) xs $NaNH_2$ / NH_3

2) H_2O

3) excess HBr

Br Br

10.14.

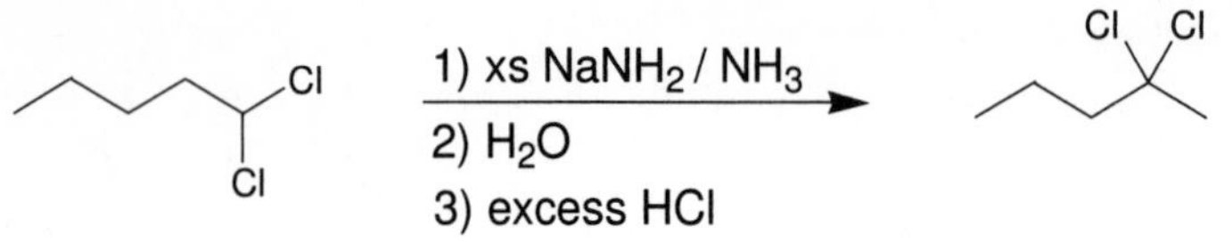

1) xs $NaNH_2$ / NH_3
2) H_2O
3) excess HCl

10.15. If two products are obtained, then the alkyne must be internal and unsymmetrical. There is only one such alkyne with molecular formula C_5H_8:

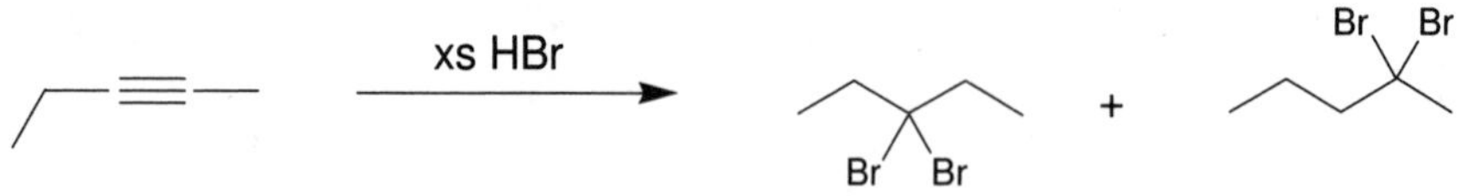

xs HBr

10.16.

a)

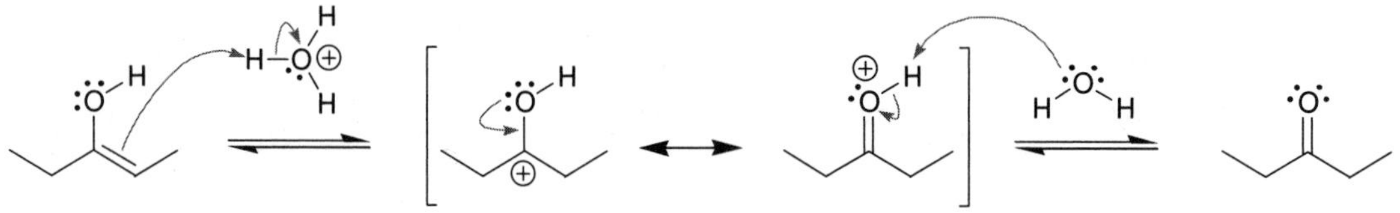

b)

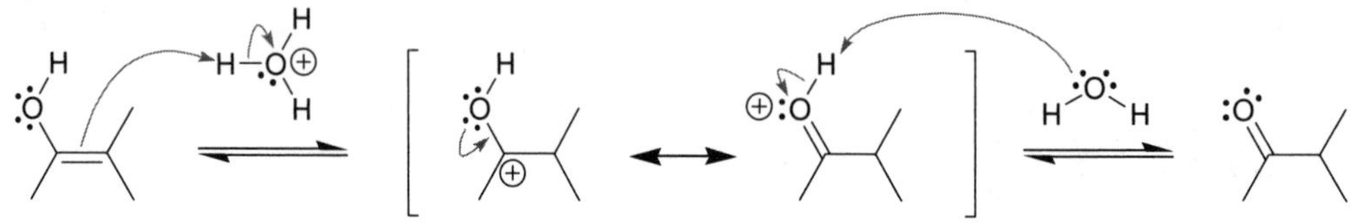

c)

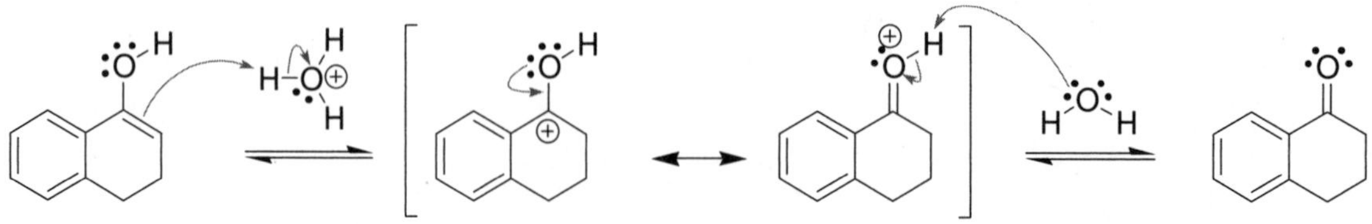

d)

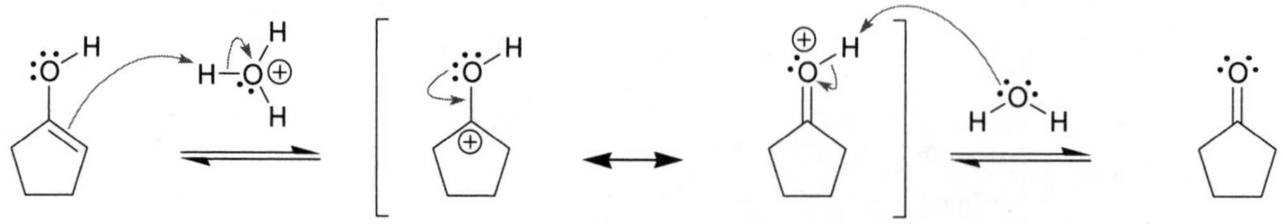

10.17.

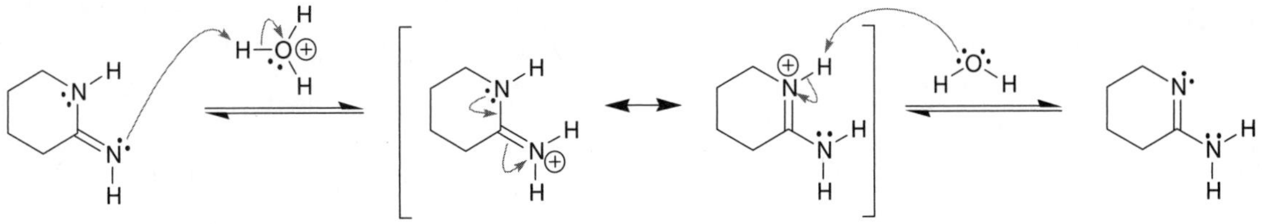

H
H–O⊕
H
N–H
N
H
N–H
N⊕
H
H
⊕
N–H
N–H
H
O
H H
N
N–H
H

10.18.

a) 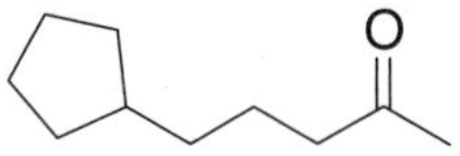 O b) 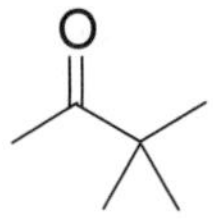 O

c) 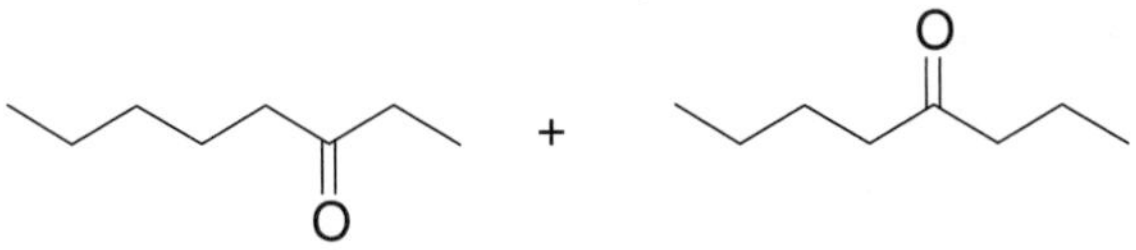 O + O

d) 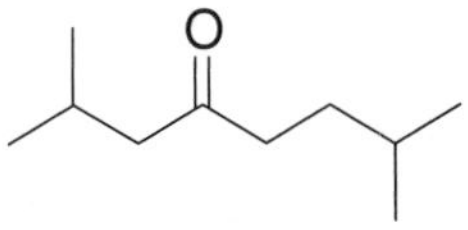 O e) 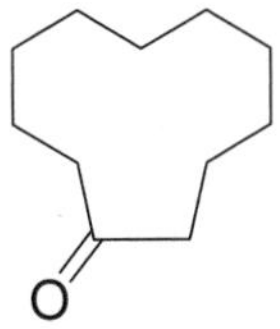 O

10.19.

a) 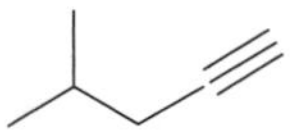 b)  c)

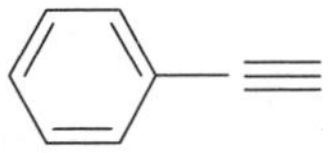

10.20.

a)

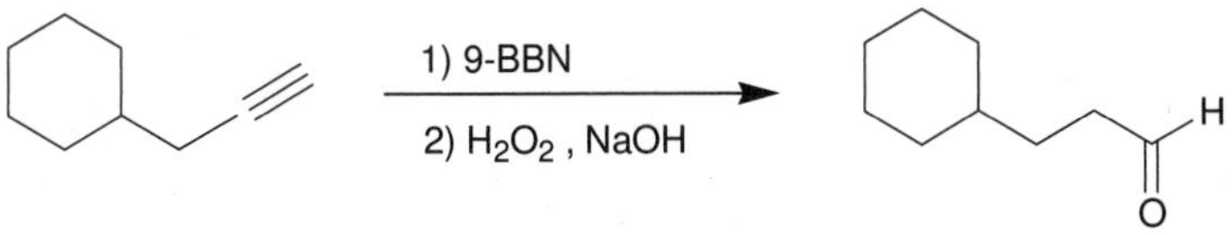

1) 9-BBN
2) H_2O_2 , NaOH

H
O

b)

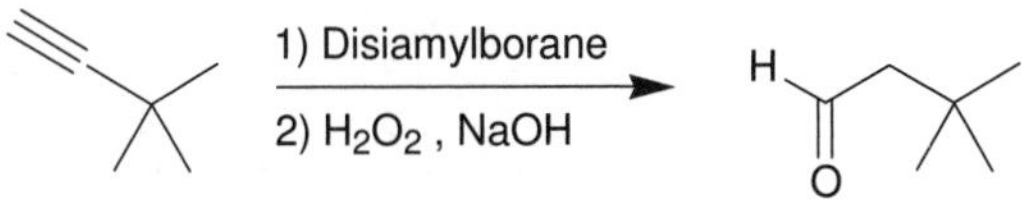

1) Disiamylborane
2) H_2O_2 , NaOH

H
O

c)

1) 9-BBN
2) H_2O_2 , NaOH

O

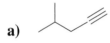

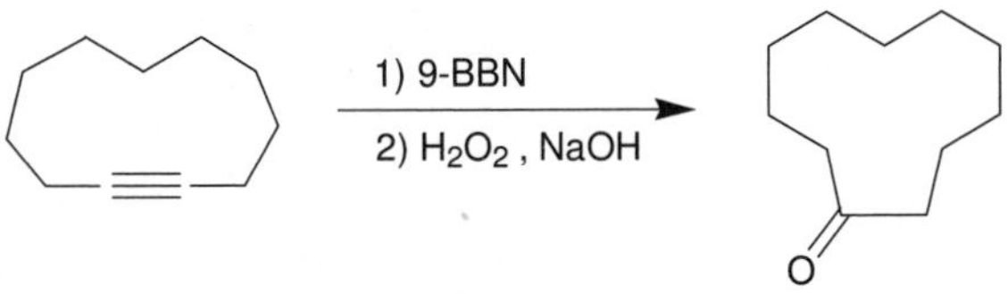

10.21.

a) 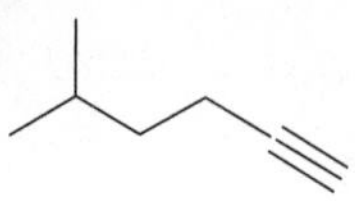 b)  c)

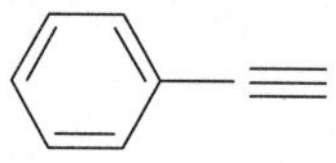

10.22.

a) 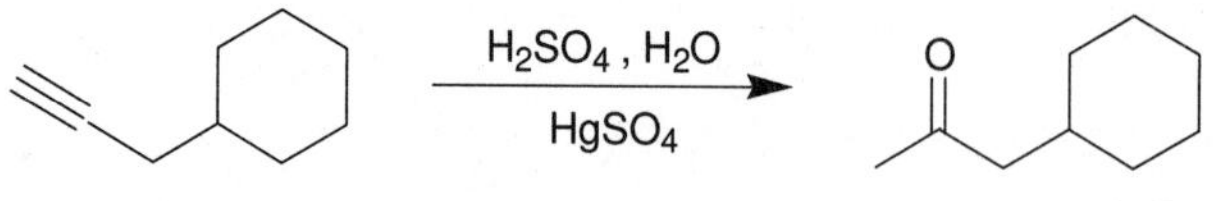 H_2SO_4 , H_2O / $HgSO_4$

O

b) 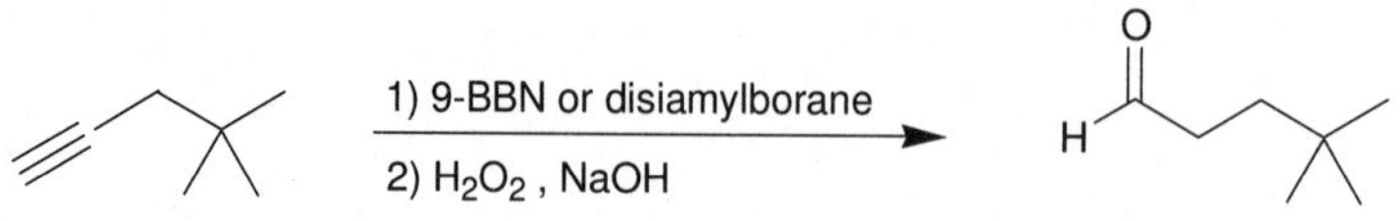 1) 9-BBN or disiamylborane

2) H_2O_2 , NaOH

O

H

10.23.

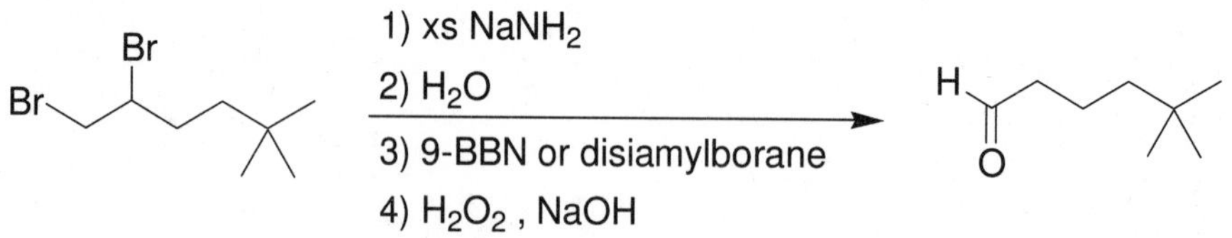

Br

Br

1) xs $NaNH_2$

2) H_2O

3) 9-BBN or disiamylborane

4) H_2O_2 , NaOH

H

O

a)

1) xs $NaNH_2$

2) H_2O

3) H_2SO_4, H_2O, $HgSO_4$

Cl

Cl

O

b)

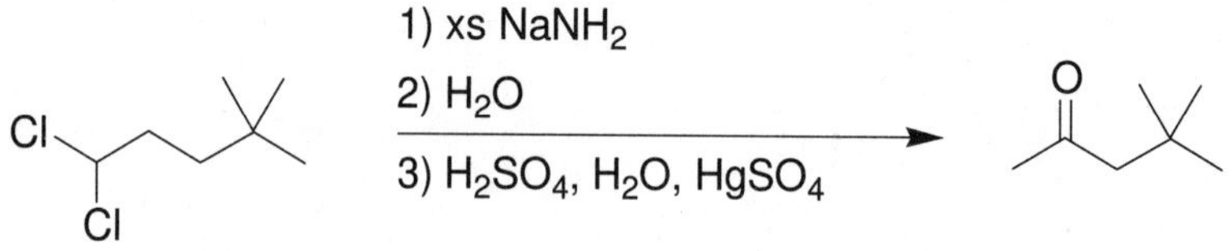

10.24.

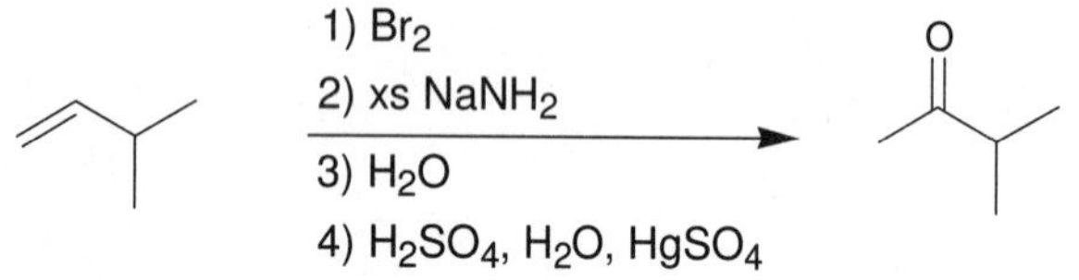

1) Br_2

2) xs $NaNH_2$

3) H_2O

4) H_2SO_4, H_2O, $HgSO_4$

O

10.25.

a)

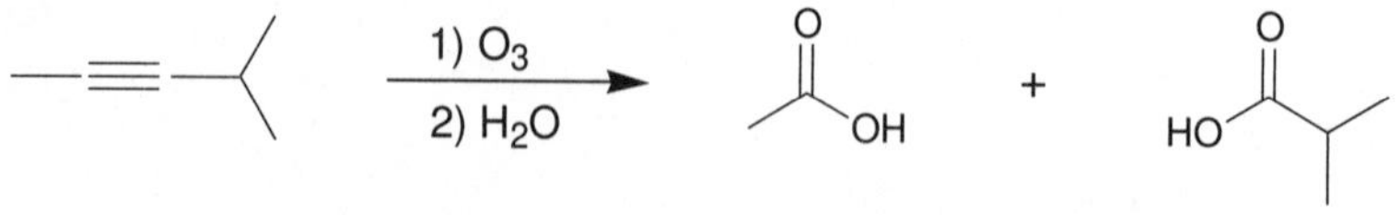

1) O_3

2) H_2O

O

OH

+

O

HO

b)

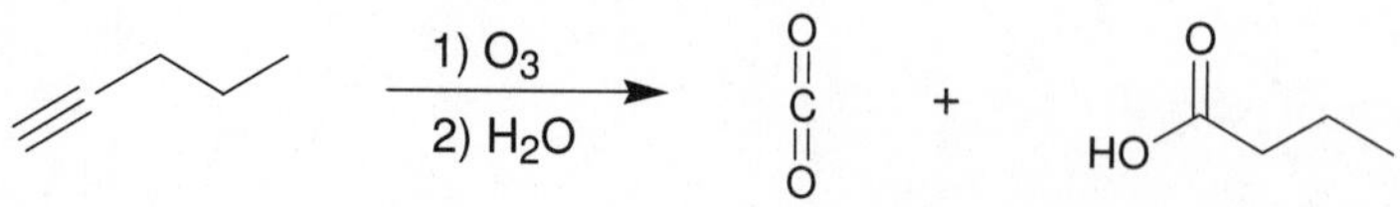

1) O_3

2) H_2O

O=C=O

+

O

HO

c)

1) O_3
2) H_2O

OH + HO

d)

1) O_3
2) H_2O

OH HO

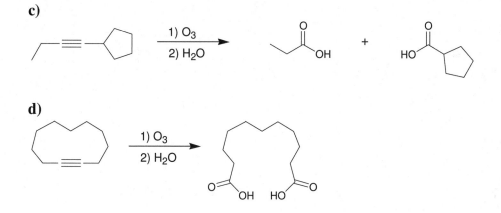

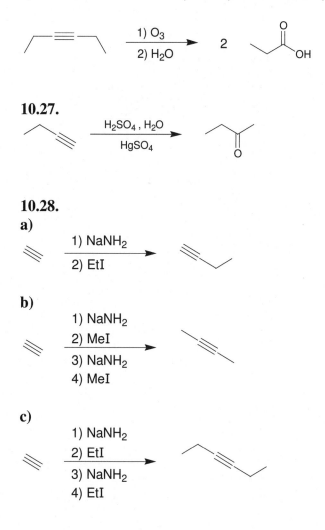

10.26. If ozonolysis produces only one product, then the starting alkyne must be symmetrical. There is only one symmetrical alkyne with molecular formula C_6H_{10}:

1) O_3
2) H_2O
2 OH

10.27.

H_2SO_4, H_2O
$HgSO_4$

10.28.

a)

1) $NaNH_2$
2) EtI

b)

1) $NaNH_2$
2) MeI
3) $NaNH_2$
4) MeI

c)

1) $NaNH_2$
2) EtI
3) $NaNH_2$
4) EtI

d)

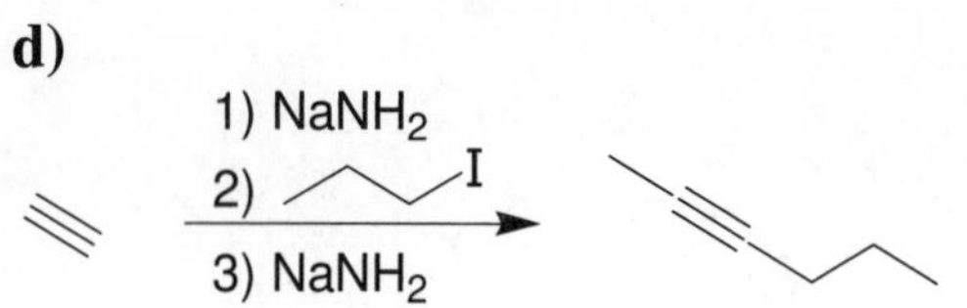

e)

1) $NaNH_2$

2) I

f)

1) $NaNH_2$

2) I

3) $NaNH_2$

4) MeI

g)

1) $NaNH_2$

2) I

3) $NaNH_2$

4) EtI

h)

1) $NaNH_2$

2) I

3) $NaNH_2$

4) MeI

i)

1) $NaNH_2$

2) EtI

3) $NaNH_2$

4) MeI

j)

1) $NaNH_2$

2) EtI

3) $NaNH_2$

4) I

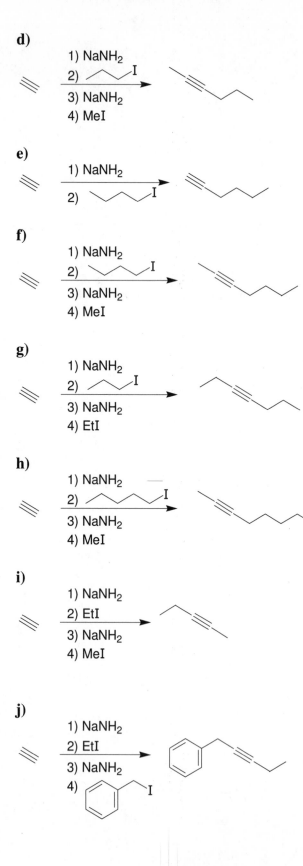

k)

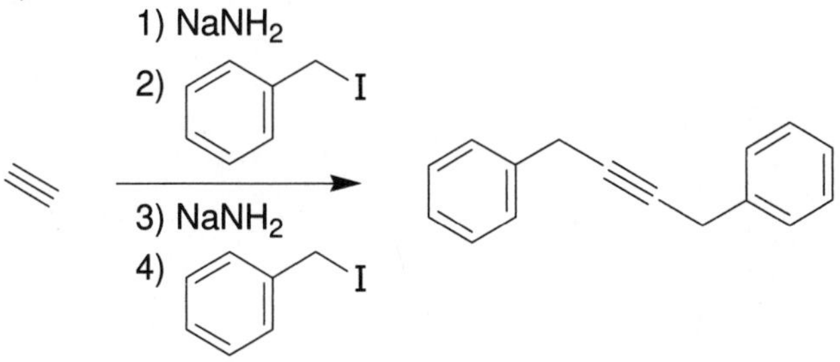

1) $NaNH_2$
2) (benzyl iodide)
3) $NaNH_2$
4) (benzyl iodide)

10.29. This process would require the used of a tertiary substrate, which is not reactive toward S_N2.

10.30. 4-octyne

10.31.

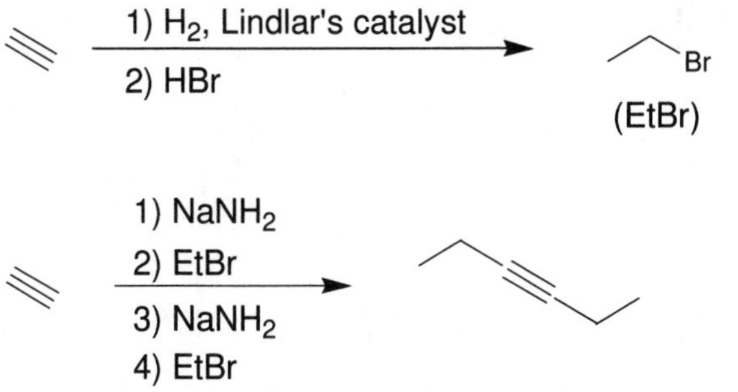

1) H_2, Lindlar's catalyst
2) HBr

Br

(EtBr)

1) $NaNH_2$
2) EtBr
3) $NaNH_2$
4) EtBr

10.32.

a)

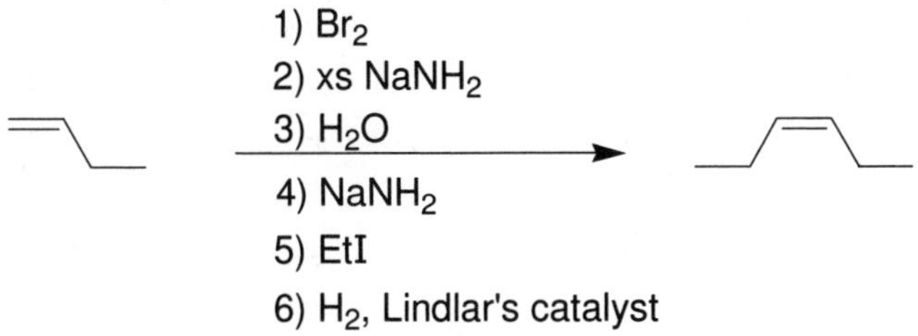

1) Br_2
2) xs $NaNH_2$
3) H_2O
4) $NaNH_2$
5) EtI
6) H_2, Lindlar's catalyst

Note: The alkyne produced after step 3 does not need to be isolated and purified, and therefore, steps 3 and 4 can be omitted.

b)

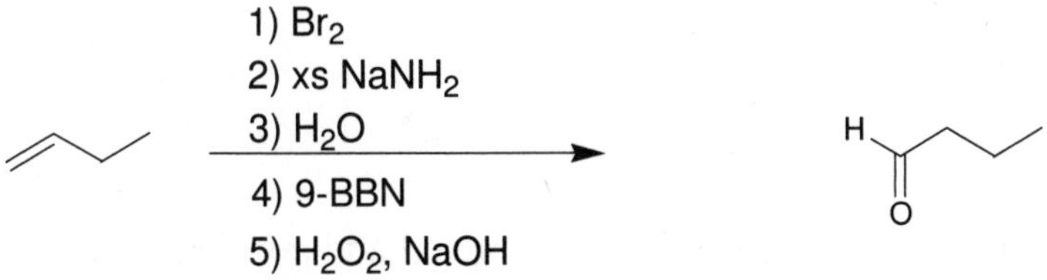

1) Br_2
2) xs $NaNH_2$
3) H_2O
4) 9-BBN
5) H_2O_2, NaOH

H

O

c) 1) H_2, Lindlar's catalyst; 2) dilute H_2SO_4 → OH

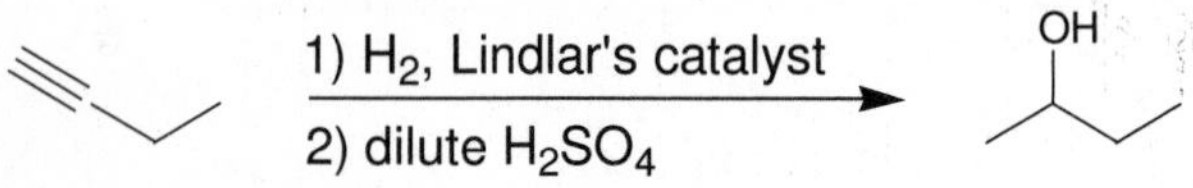

d) 1) H_2, Lindlar's catalyst; 2) $BH_3 \cdot THF$; 3) H_2O_2, NaOH → HO

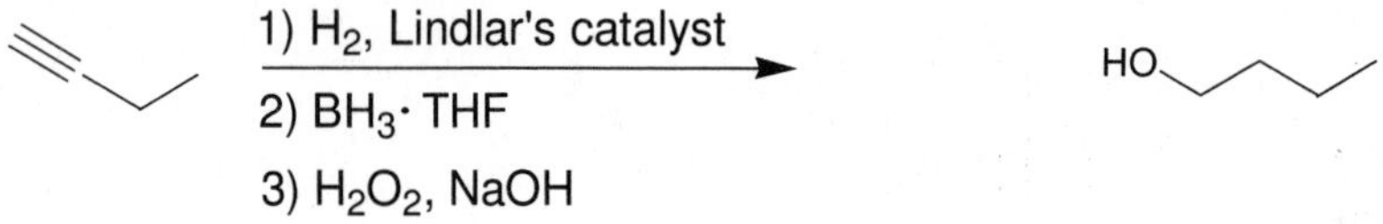

e) 1) $NaNH_2$; 2) EtI; 3) Na, NH_3 (*l*); 4) Br_2 → Br, Br

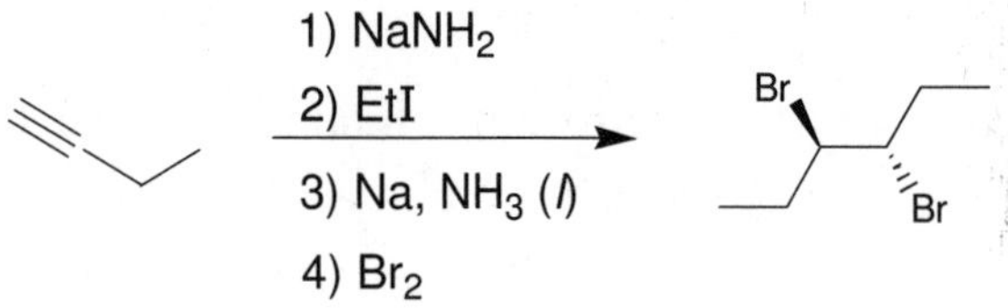

f) 1) $NaNH_2$; 2) EtI; 3) H_2, Lindlar's catalyst; 4) Br_2 → Br, Br + En

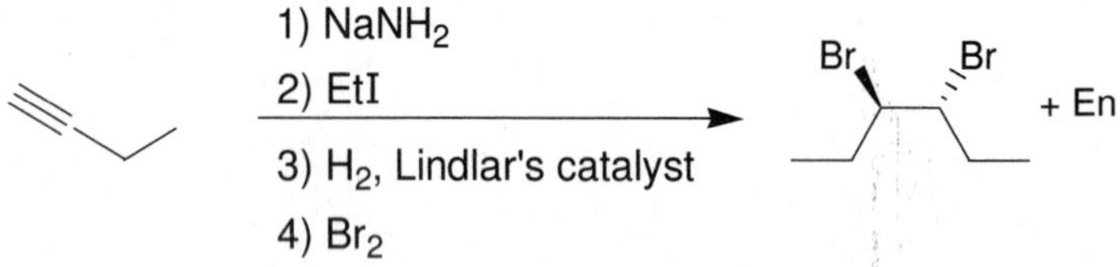

10.33.

a)

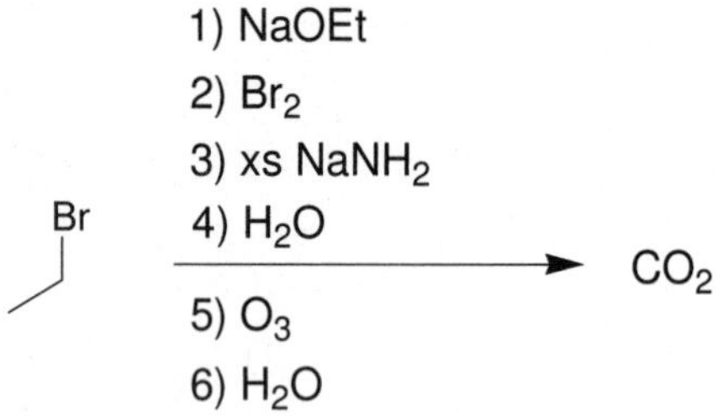

Br

1) NaOEt
2) Br_2
3) xs $NaNH_2$
4) H_2O
5) O_3
6) H_2O

→ CO_2

b)

Br

1) NaOEt
2) Br_2
3) xs $NaNH_2$
4) H_2O
5) $NaNH_2$
6) MeI
7) O_3
8) H_2O

→ O, OH

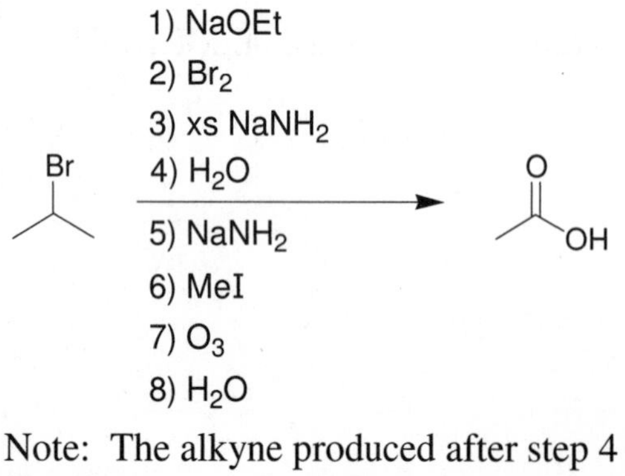

Note: The alkyne produced after step 4 does not need to be isolated and purified, and therefore, steps 4 and 5 can be omitted.

10.34.

1) H_2, Lindlar's catalyst
2) HBr

Br
(EtBr)

1) Br_2
2) xs $NaNH_2$
3) H_2O

1) $NaNH_2$
2) EtBr
3) $NaNH_2$
4) EtBr

H_2SO_4, H_2O
$HgSO_4$

O

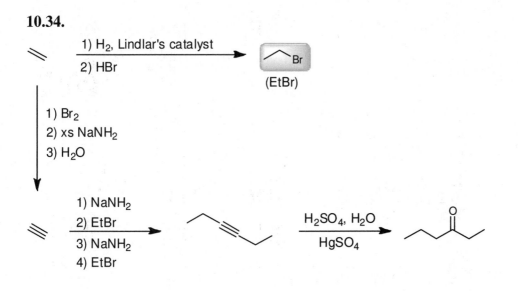

10.35.
a) 2,2,5-trimethyl-3-hexyne
b) 4,4-dichloro-2-hexyne
c) 1-hexyne
d) 3-bromo-3-methyl-1-butyne

10.36.
a)

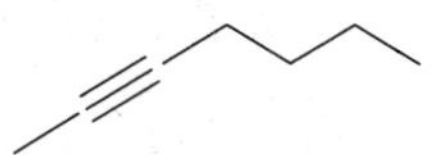

b)

c)

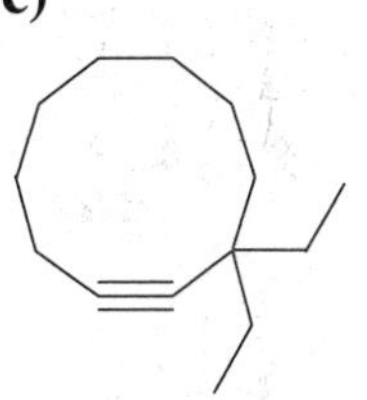

10.37.

a)

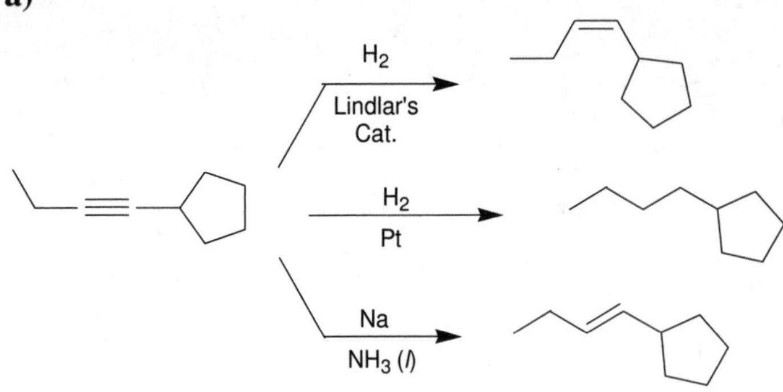

b)

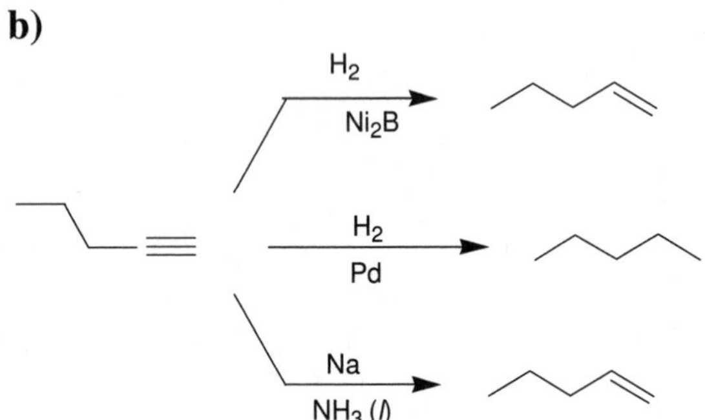

10.38.

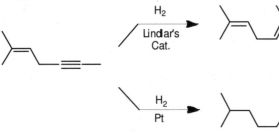

H_2
Lindlar's
Cat.

H_2
Pt

10.39.

a)

b)

H + Na :H

Na + H_2

10.40.

a) H_2SO_4, H_2O / $HgSO_4$

b) 1) 9-BBN / 2) H_2O_2, NaOH

c) HBr (2 eq)

d) HCl (1 eq)

e) Br_2 (2 eq) / CCl_4

f) 1) $NaNH_2$, NH_3 / 2) MeI

g) H_2 / Pt

10.41.

H_2 / Pt

HCl (1 eq)

H_2 / Lindlar's catalyst

HBr (xs)

1) O_3 / 2) H_2O

1) 9-BBN / 2) H_2O_2, NaOH

H_2SO_4 / H_2O / $HgSO_4$

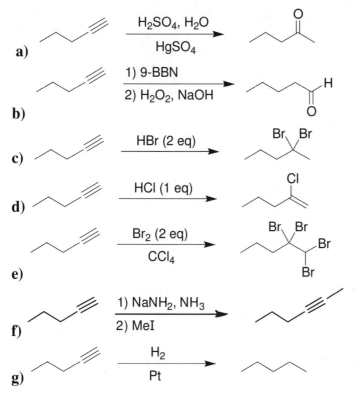

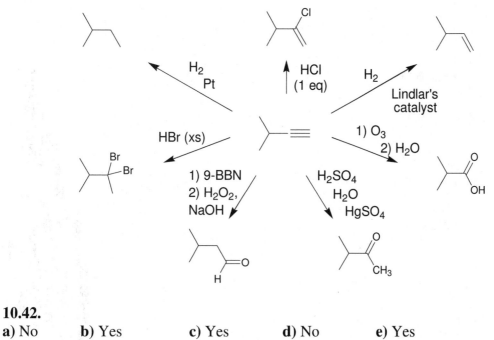

10.42.

a) No **b)** Yes **c)** Yes **d)** No **e)** Yes

10.43.

a) No. These compounds are constitutional isomers, but they are not keto-enol tautomers because the pi bond is not adjacent to the OH group.

b) Yes

c) Yes

d) Yes

10.44.

a) 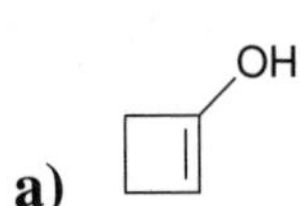

b) OH

c) HO

10.45.

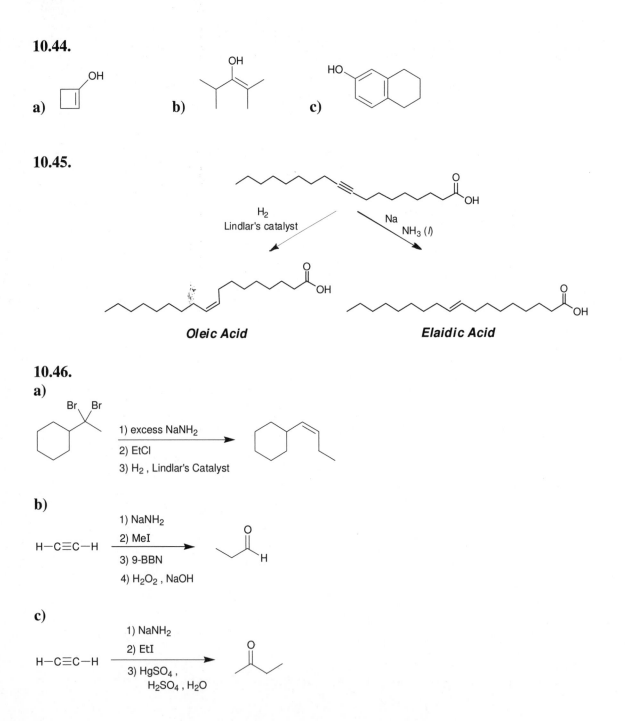

H_2
Lindlar's catalyst

Na
NH_3 (*l*)

Oleic Acid

Elaidic Acid

10.46.

a)

Br Br

1) excess $NaNH_2$
2) EtCl
3) H_2, Lindlar's Catalyst

b)

H−C≡C−H

1) $NaNH_2$
2) MeI
3) 9-BBN
4) H_2O_2, NaOH

c)

H−C≡C−H

1) $NaNH_2$
2) EtI
3) $HgSO_4$, H_2SO_4, H_2O

d)

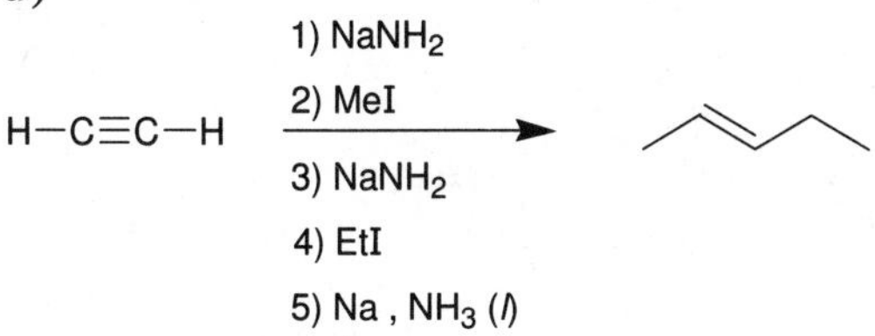

H−C≡C−H

1) $NaNH_2$
2) MeI
3) $NaNH_2$
4) EtI
5) Na , NH_3 (*l*)

10.47. When *(R)*-4-bromohept-2-yne is treated with H_2 in the presence of Pt, the asymmetry is destroyed and C4 is no longer a chirality center:

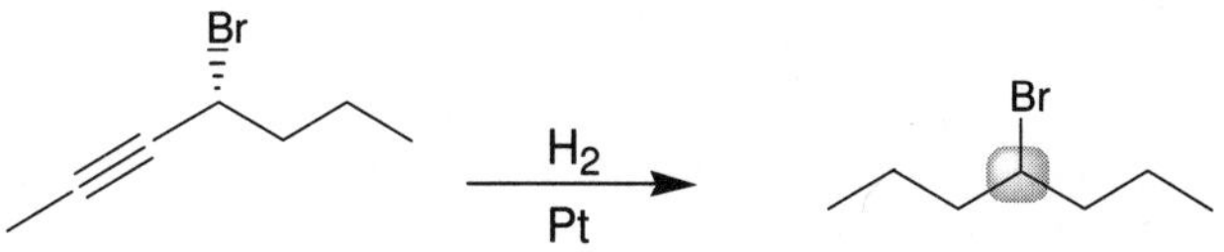

Br

H_2
Pt

Br

not a chirality center

This is not the case for *(R)*-4-bromohex-2-yne.

10.48.

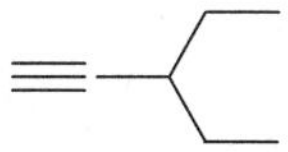

3-ethyl-1-pentyne

10.49.

a)

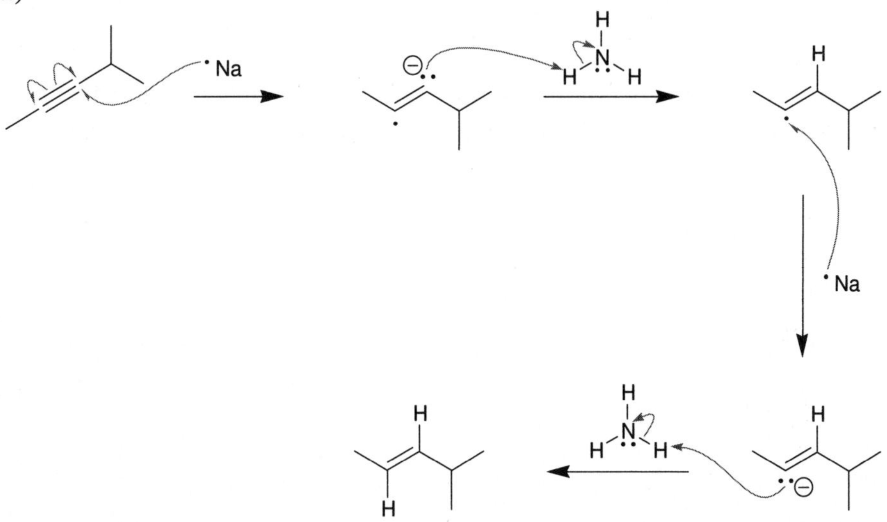

Na

H
N
H H

H

Na

H
N
H H

H

H

H

b)

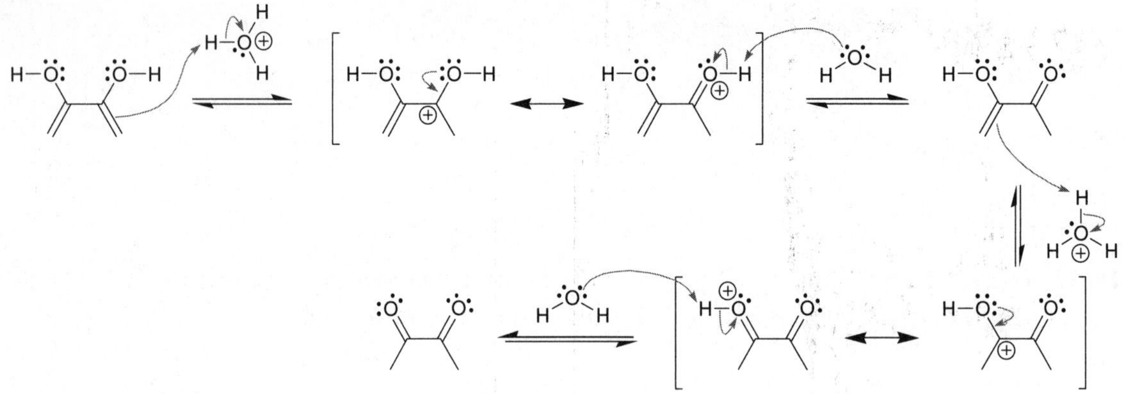

H–Ö: :Ö–H H–O⊕ H H

H–Ö: :Ö–H ⊕ ⟷ H–Ö: O–H ⊕

H H :O: H–Ö: :Ö:

H H :O: H ⊕ H

:Ö: :Ö: H H :O: H–O⊕ :Ö: ⟷ H–Ö: :Ö: ⊕

10.50.

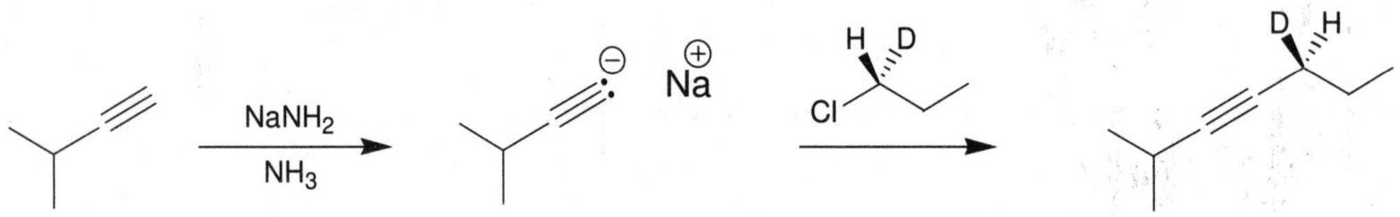

NaNH$_2$ NH$_3$ ⊖ ⊕ Na H D Cl D H

10.51.

a)

H$_2$ Pd

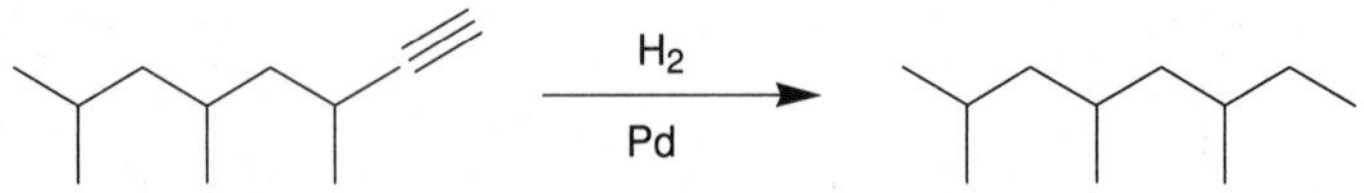

Compound A ***2,4,6-trimethyloctane***

b) Compound A has two chirality centers:

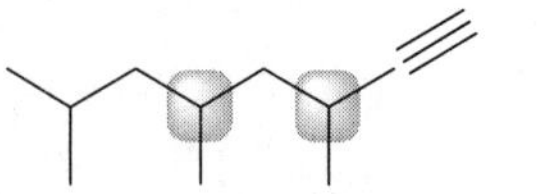

c) The locants for the methyl groups in Compound A are 3, 5, and 7, because locants are assigned in a way that gives the triple bond the lower possible number (1 rather than 7).

10.52.

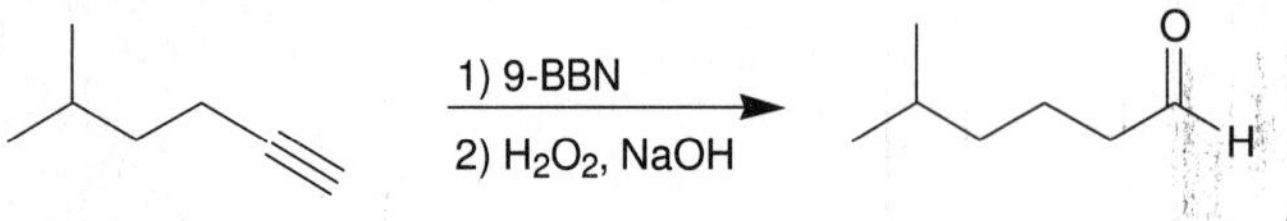

1) 9-BBN 2) H_2O_2, NaOH O H

Compound A

10.53.

a)

1) $NaNH_2$
2) EtI
3) H_2SO_4, H_2O, $HgSO_4$

O

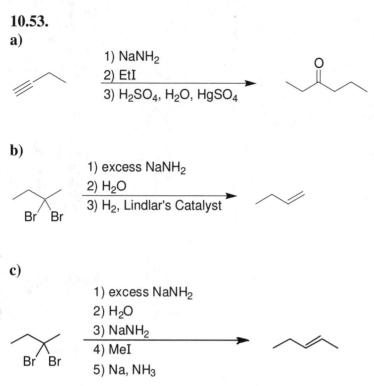

b)

1) excess $NaNH_2$
2) H_2O
3) H_2, Lindlar's Catalyst

Br Br

c)

1) excess $NaNH_2$
2) H_2O
3) $NaNH_2$
4) MeI
5) Na, NH_3

Br Br

Note: The alkyne produced after step 2 does not need to be isolated and purified, and therefore, steps 2 and 3 can be omitted.

d)

1) excess $NaNH_2$
2) H_2O
3) H_2SO_4, H_2O, $HgSO_4$

Cl Cl

O

e)

1) excess $NaNH_2$
2) H_2O
3) Br_2 (1 eq)

Br Br

Br

Br

f)

1) excess $NaNH_2$
2) H_2O
3) H_2, Lindlar's Catalyst
4) dilute H_2SO_4

Cl Cl

OH

10.54.

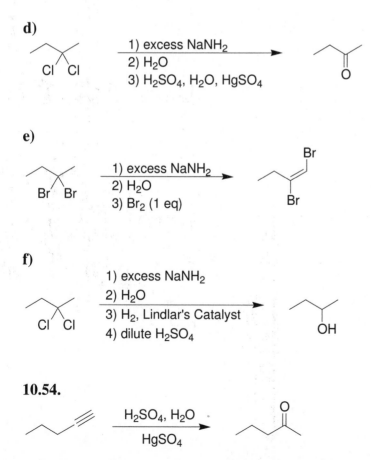

H_2SO_4, H_2O

$HgSO_4$

O

10.55. $H_3C{-}C{\equiv}C{-}H$

10.56. If two products are obtained, then the alkyne must be internal and unsymmetrical. There is only one such alkyne with molecular formula C_5H_8:

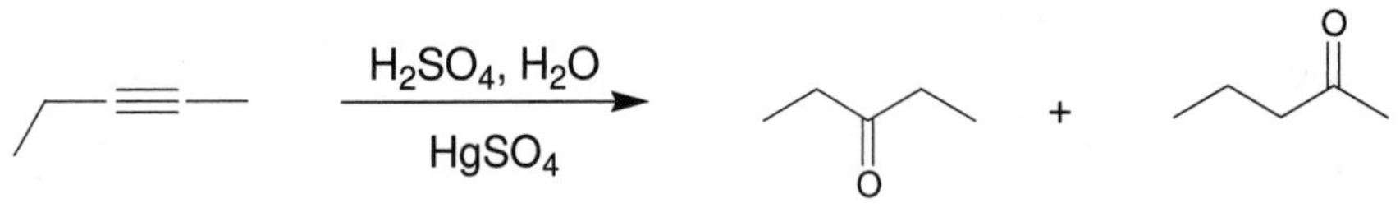

H_2SO_4, H_2O

$HgSO_4$

+

O

O

10.57.

a)

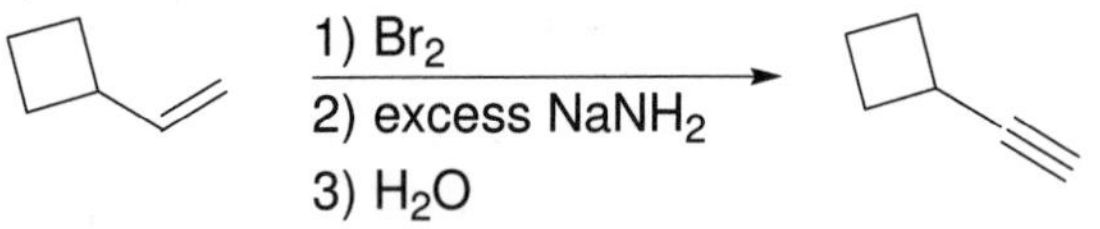

1) Br_2
2) excess $NaNH_2$
3) H_2O

b)

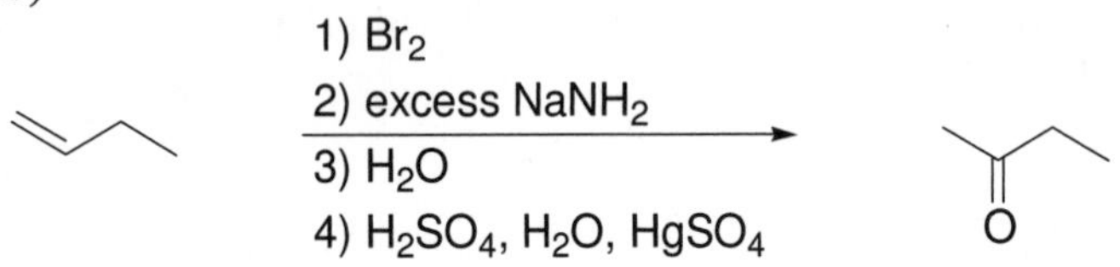

1) Br_2
2) excess $NaNH_2$
3) H_2O
4) H_2SO_4, H_2O, $HgSO_4$

O

c)

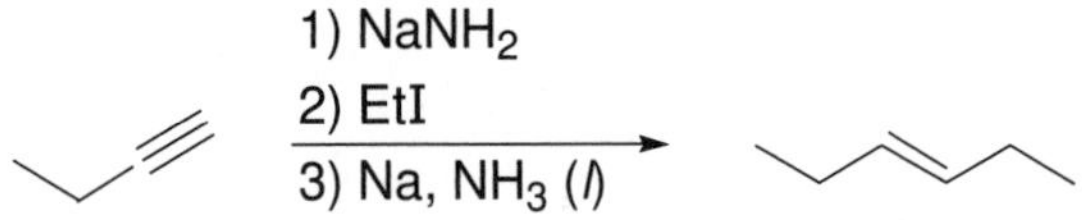

1) $NaNH_2$
2) EtI
3) Na, NH_3 (*l*)

d)

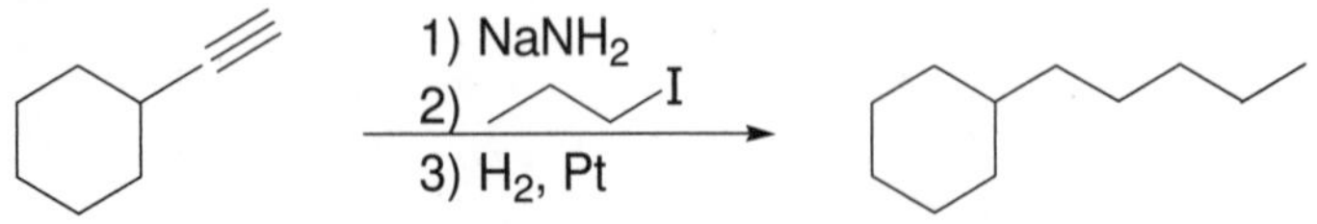

1) $NaNH_2$
2) I
3) H_2, Pt

10.58.

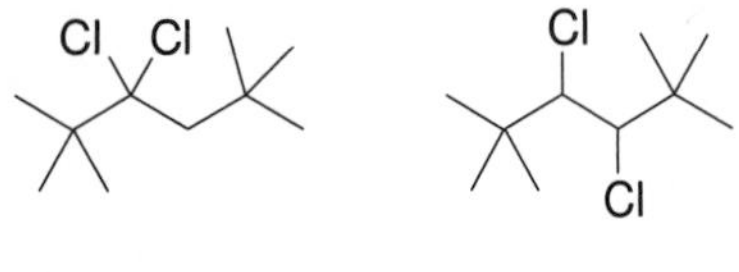

Cl Cl

Cl

Cl

10.59.

$\xrightarrow{Br_2}$ (dibromide) $\xrightarrow[\text{2) } H_2O]{\text{1) excess } NaNH_2}$ (terminal alkyne) $\xrightarrow{NaNH_2}$ (acetylide, Na⁺) $\xrightarrow{\text{ethyl iodide}}$ (internal alkyne)

10.60.

a)

$\xrightarrow[\text{2) } Br_2]{\text{1) Na, } NH_3\ (l)}$ (Br, Br product)

b)

$\xrightarrow[\text{2) } Br_2]{\text{1) } H_2\text{, Lindlar's Catalyst}}$ (Br, Br product) + En

c)

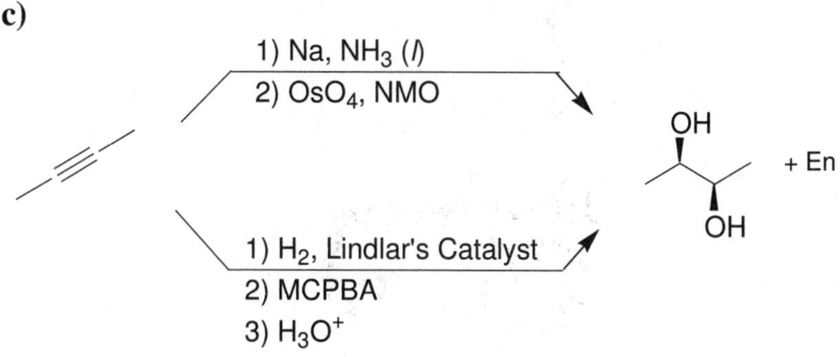

d)

1) Na, NH_3 (l)
2) MCPBA
3) H_3O^+

OH, OH product

1) H_2, Lindlar's Catalyst
2) OsO_4, NMO

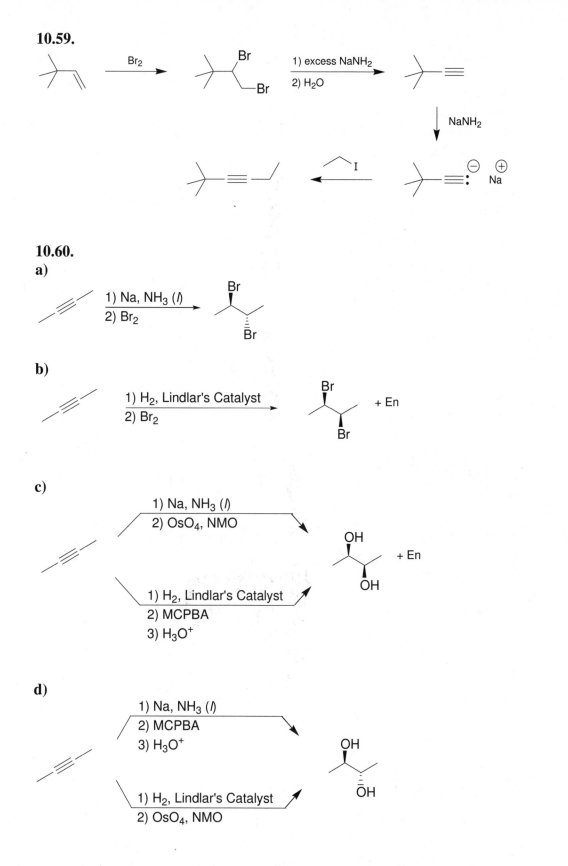

e)

1) $NaNH_2$
2) MeI
3) $NaNH_2$
4) MeI
5) Na, NH_3 (*l*)
6) OsO_4, NMO

1) $NaNH_2$
2) MeI
3) $NaNH_2$
4) MeI
5) H_2, Lindlar's Catalyst
6) MCPBA
7) H_3O^+

OH
OH
+ En

f)

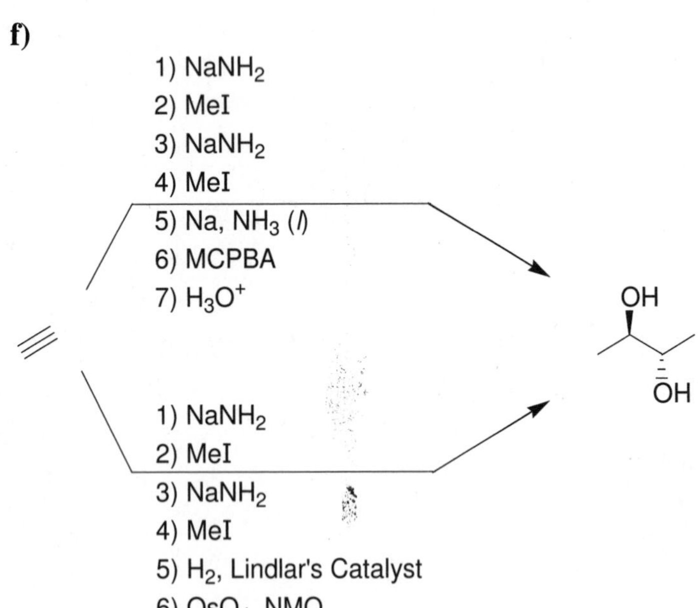

10.61.

a) D_2 / Lindlar's Catalyst

D D

b) Na / ND_3 (*l*)

D D

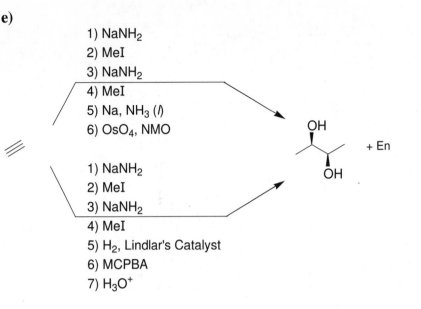

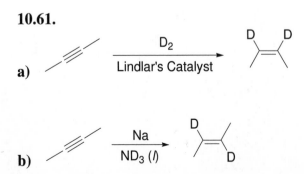

c) 1) $NaNH_2$ 2) D_2O

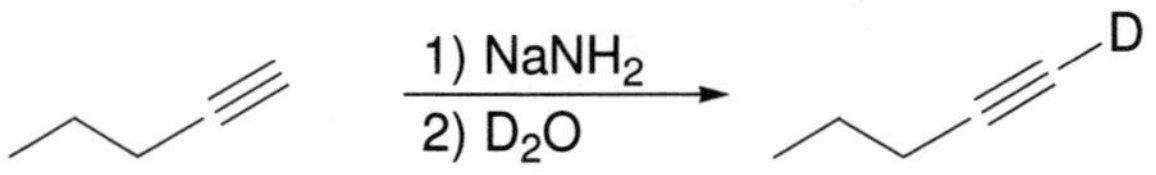

10.62.

$^{\ominus}NH_2$

$H_3C—I$

H_3C

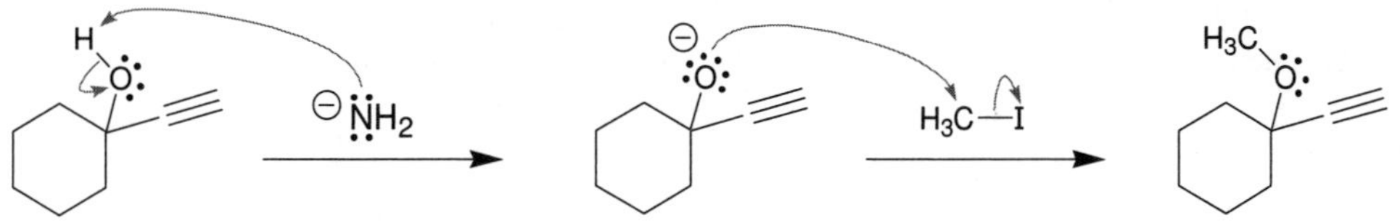

10.63.

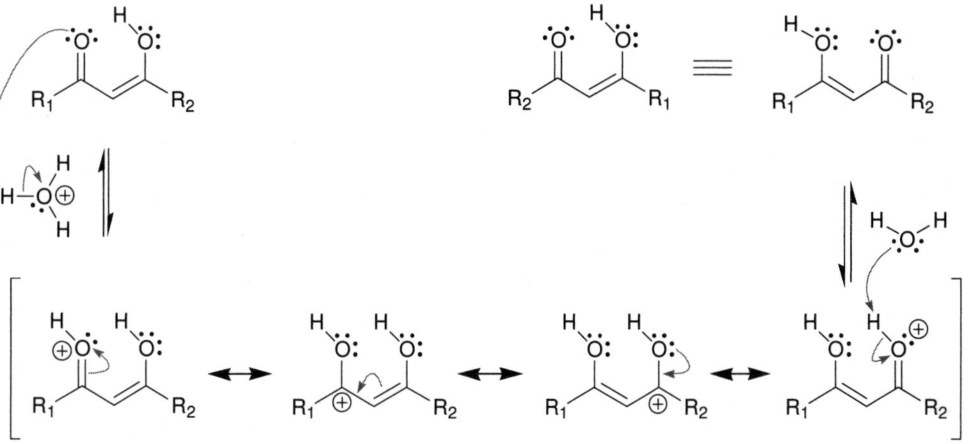

10.64.

$H-N-CH_3$

CH_3

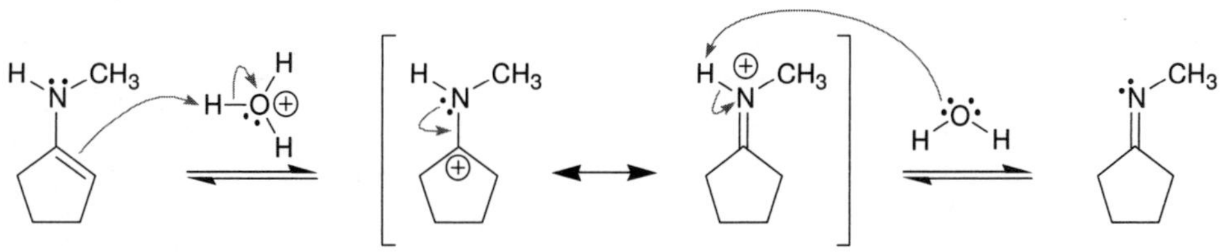

10.65.

a)

1) H_2, Lindlar's catalyst
2) HBr

(EtBr)

1) $NaNH_2$
2) EtBr
3) $NaNH_2$
4) MeBr

1) Na, NH_3 (*l*)
2) MCPBA

Et Me + En

b)

1) H_2, Lindlar's catalyst
2) HBr

(EtBr)

1) $NaNH_2$
2) EtBr
3) $NaNH_2$
4) MeBr

1) H_2, Lindlar's Catalyst
2) MCPBA

Et Me + En

10.66.

Br_2
H_3O^+

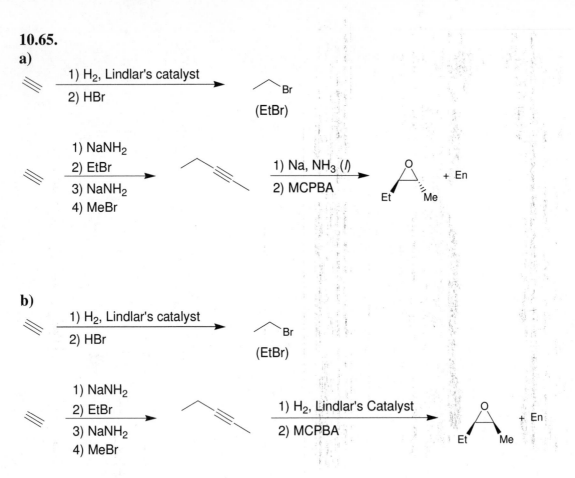

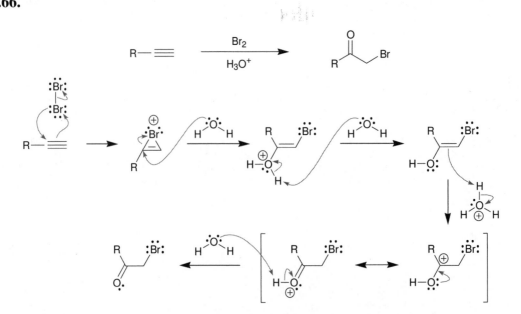

10.67.

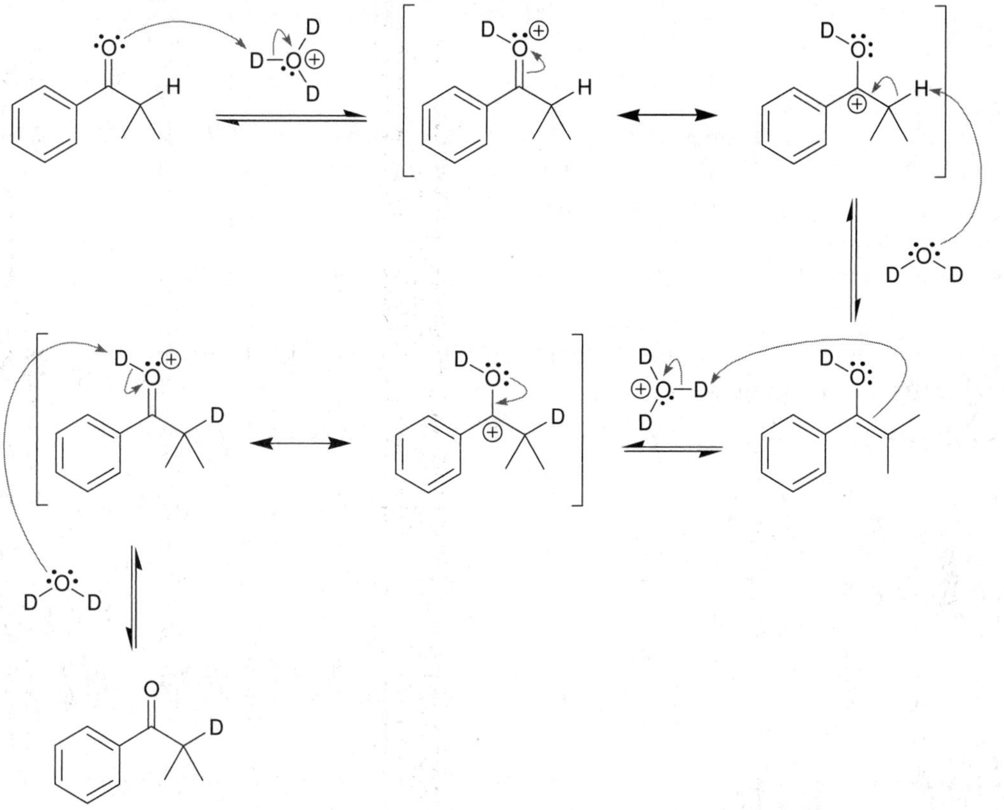

Chapter 11
Radical Reactions

Review of Concepts

Fill in the blanks below. To verify that your answers are correct, look in your textbook at the end of Chapter 11. Each of the sentences below appears verbatim in the section entitled *Review of Concepts and Vocabulary*.

- Radical mechanisms utilize **fishhook arrows**, each of which represents the flow of ____________________.
- Every step in a radical mechanism can be classified as **initiation**, ___________, or **termination**.
- A **radical initiator** is a compound with a weak bond that readily undergoes ______________________.
- A __________________, also called a radical scavenger, is a compound that prevents a chain process from either getting started or continuing.
- _________________ is more selective than chlorination.
- When a new chirality center is created during a radical halogenation process, a ____________ mixture is obtained.
- ________________ can undergo **allylic bromination**, in which bromination occurs at the allylic position.
- Organic compounds undergo oxidation in the presence of atmospheric oxygen to produce **hydroperoxides**. This process, called __________________, is believed to proceed via a ____________ mechanism.
- **Antioxidants**, such as BHT and BHA, are used as food preservatives to prevent autooxidation of ________________ oils.
- When vinyl chloride is polymerized, _______________________ is obtained.
- Radical halogenation provides a method for introducing _______________ into an alkane.

Review of Skills

Fill in the blanks and empty boxes below. To verify that your answers are correct, look in your textbook at the end of Chapter 11. The answers appear in the section entitled *SkillBuilder Review*.

11.1 Drawing Resonance Structures of Radicals

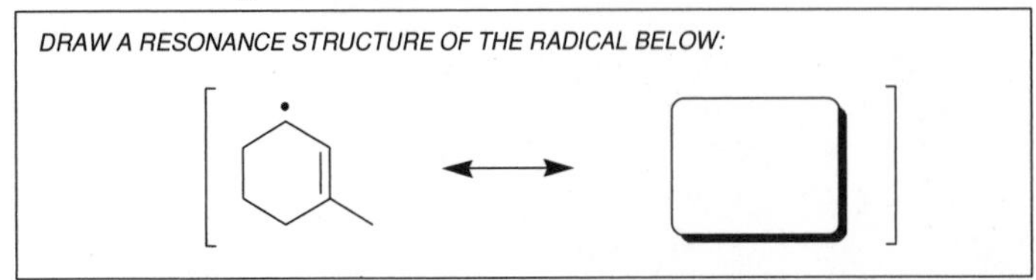

11.2 Identifying the Weakest C-H Bond in a Compound

IDENTIFY THE WEAKEST C-H BOND IN THE FOLLOWING COMPOUND:

11.3 Identifying a Radical Pattern and Drawing Fishhook Arrows

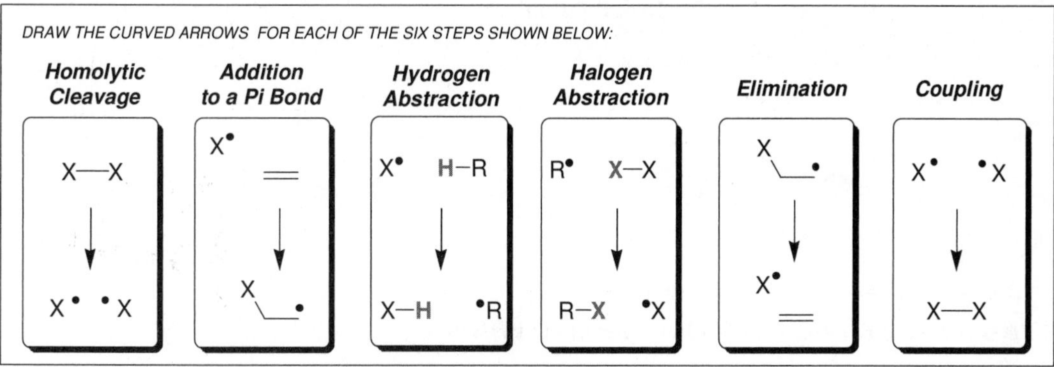

11.4 Drawing a Mechanism for Radical Halogenation

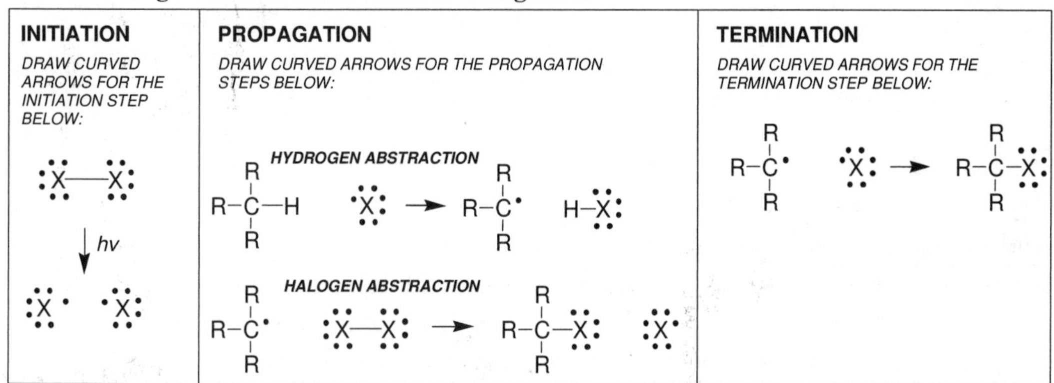

11.5 Predicting the Regiochemistry of Radical Bromination

DRAW THE EXPECTED PRODUCTS OF THE FOLLOWING MONOBROMINATION REACTION:

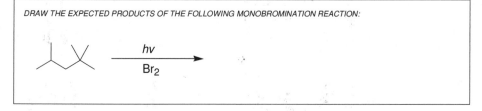

hν

Br_2

11.6 Predicting the Stereochemical Outcome of Radical Bromination

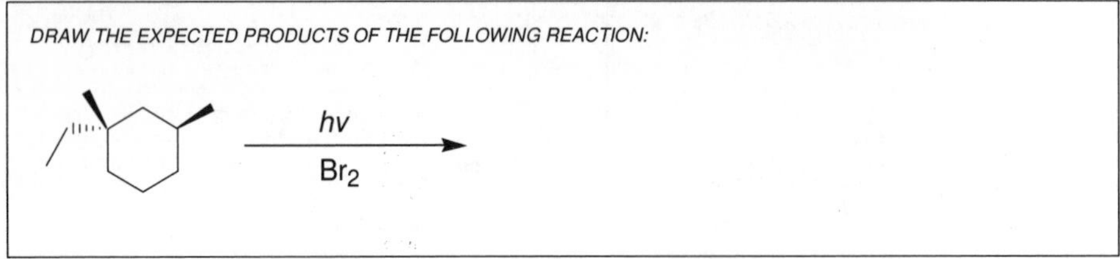
DRAW THE EXPECTED PRODUCTS OF THE FOLLOWING REACTION:
hv
Br_2

11.7 Predicting the Products of Allylic Bromination

DRAW THE EXPECTED PRODUCTS OF THE FOLLOWING REACTION:

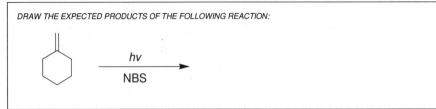

hv

NBS

11.8 Predicting the Products for Radical Addition of HBr

DRAW THE EXPECTED PRODUCTS OF THE FOLLOWING REACTION:

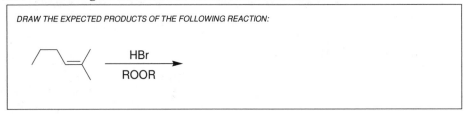

HBr

ROOR

Review of Reactions

Identify the reagents necessary to achieve each of the following transformations. To verify that your answers are correct, look in your textbook at the end of Chapter 11. The answers appear in the section entitled *Review of Synthetically Useful Radical Reactions*.

Br

Br

Br

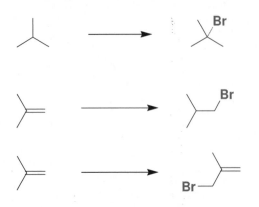

Solutions

11.1.

a) The tertiary radical is the most stable and the primary radical is the least stable.

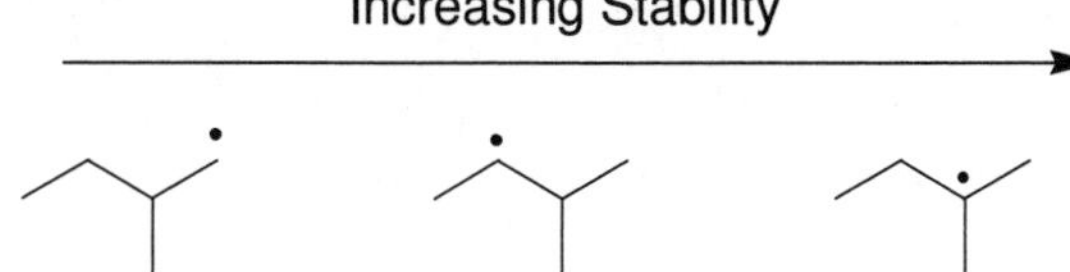

b)

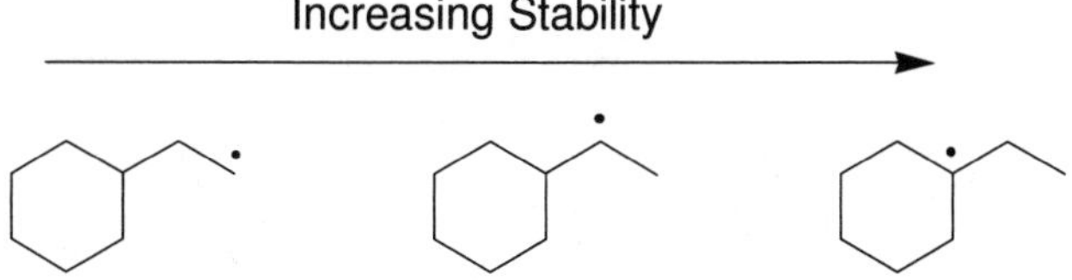

11.2.

a)

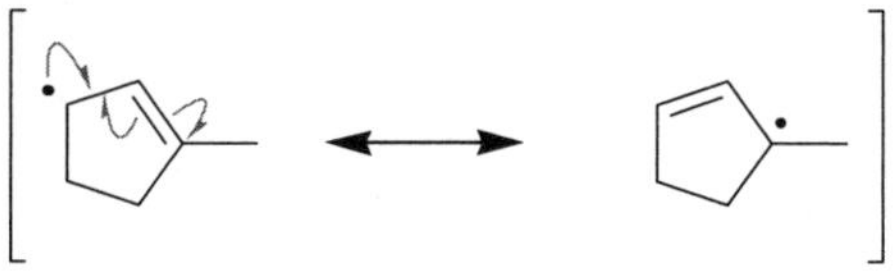

b)

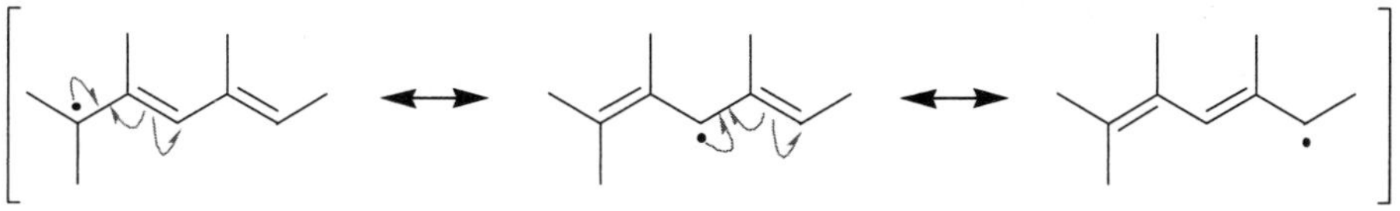

c)

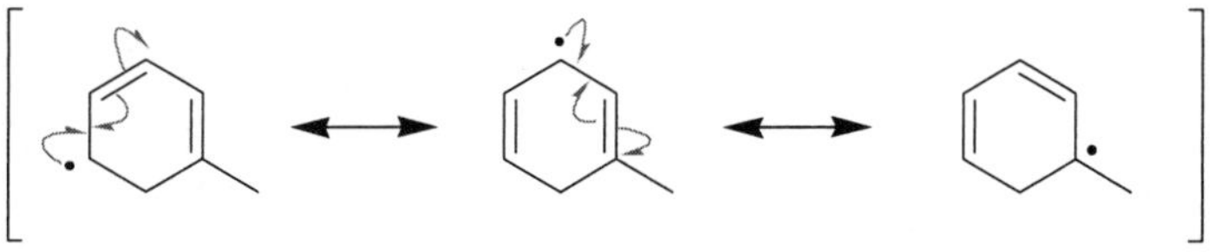

d)

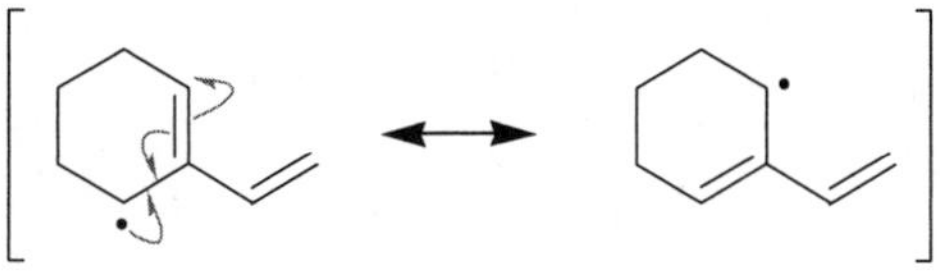

11.3. This radical is highly stabilized by resonance:

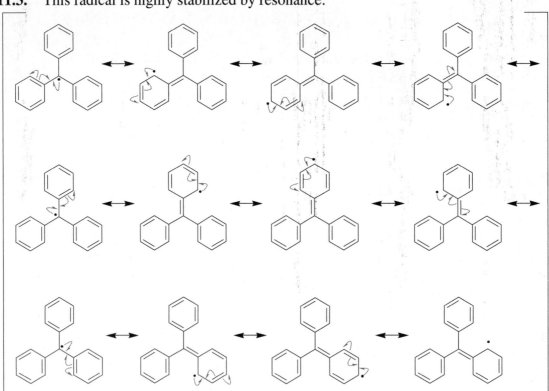

11.4.

H This hydrogen atom is removed

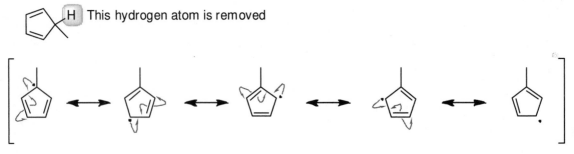

11.5.

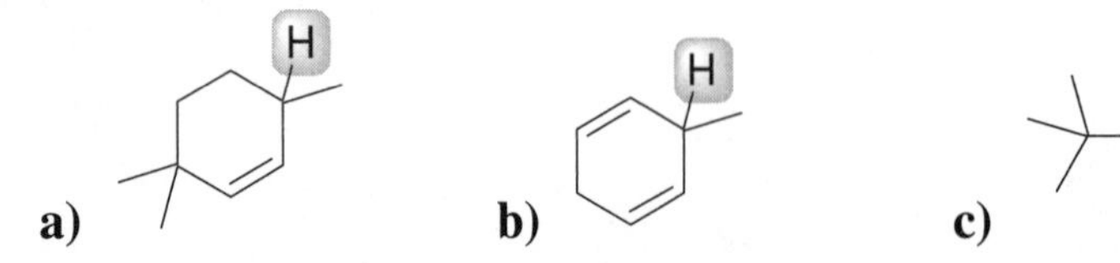

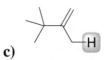

11.6. Draw the resonance structures of the radical that is formed when H_a is abstracted, and then draw the resonance structures of the radical that is formed when H_b is abstracted:

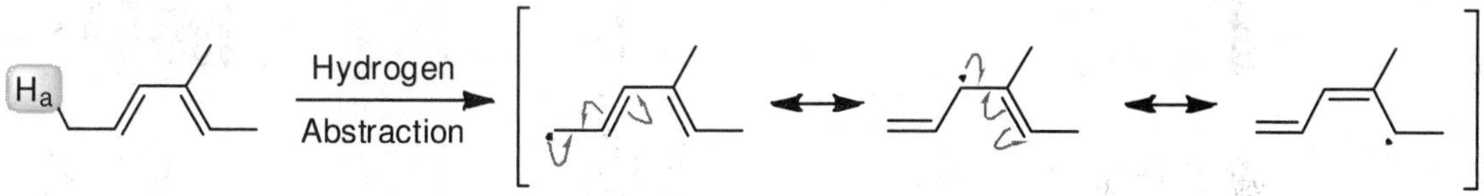

Ha Hydrogen Abstraction

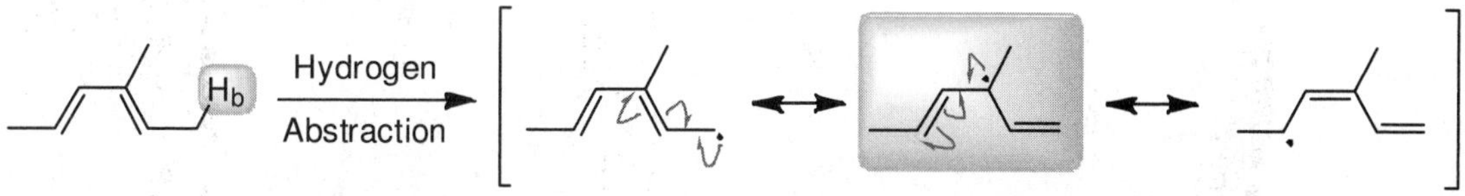

Hb Hydrogen Abstraction

Compare the resonance structures in each case. Specifically, look at the middle resonance structure in each case. When H_b is abstracted, the middle resonance structure is tertiary, and the methyl group stabilizes the radical via electron donation. This stabilizing factor is not present when H_a is abstracted. Therefore, we expect the C-H_b bond to be slightly weaker than the C-H_a bond.

11.7.

a)

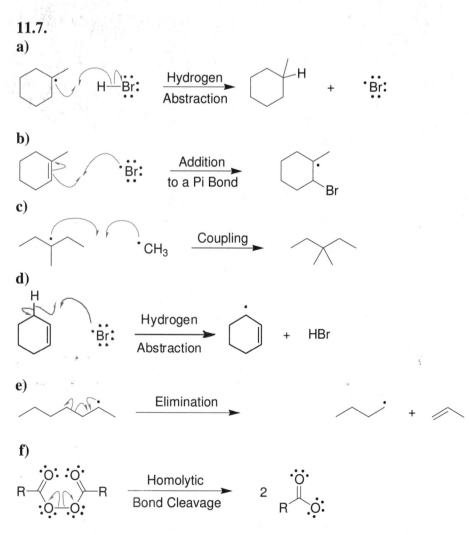

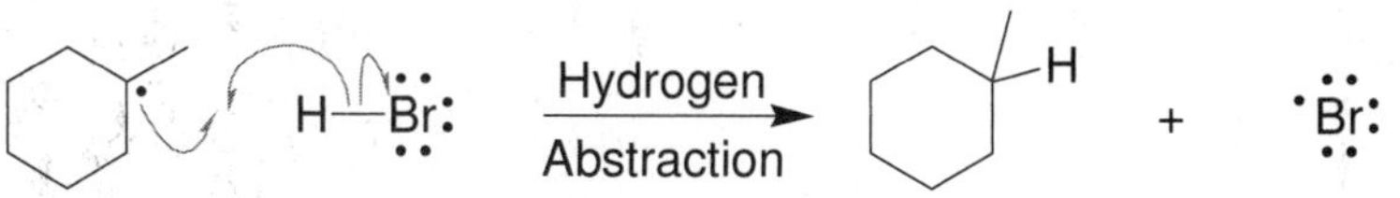

H—Br: Hydrogen Abstraction H + •Br:

b)

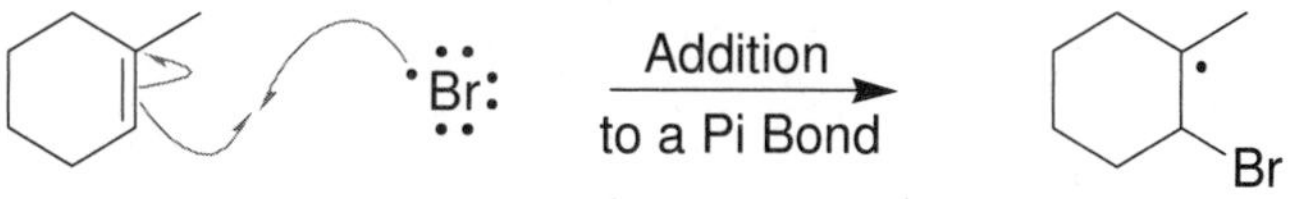

•Br: Addition to a Pi Bond Br

c)

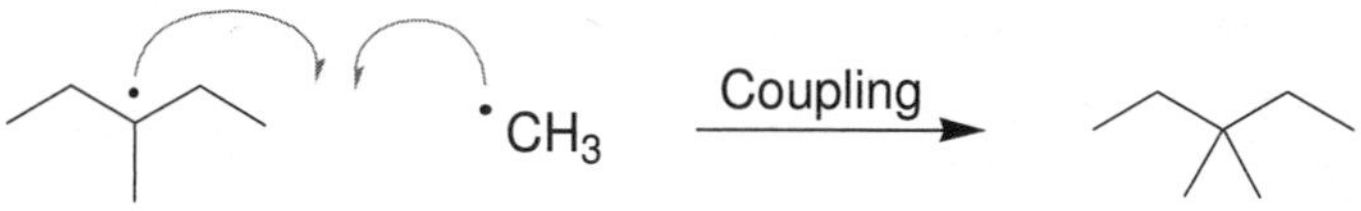

•CH3 Coupling

d)

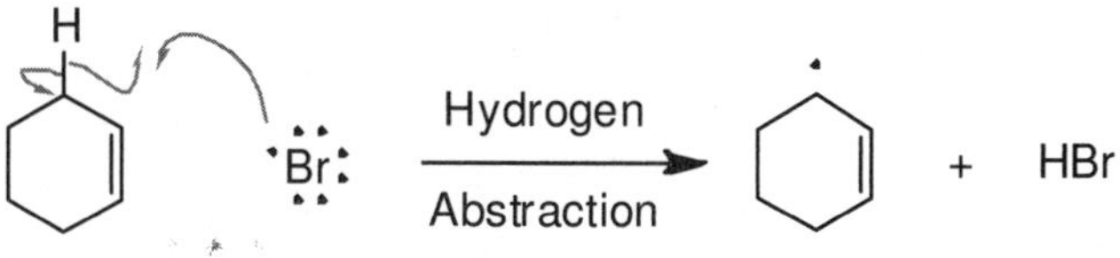

H •Br: Hydrogen Abstraction + HBr

e)

Elimination +

f)

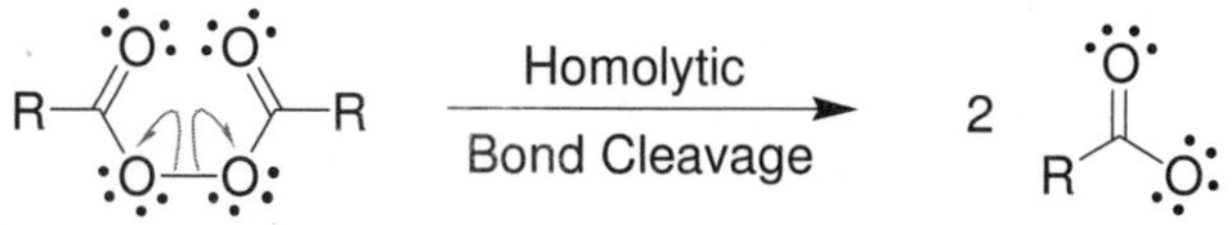

R R Homolytic Bond Cleavage 2 R

11.8.

11.9.

Coupling

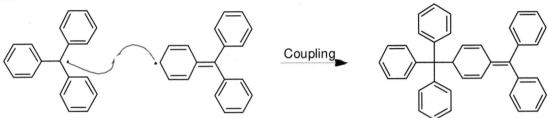

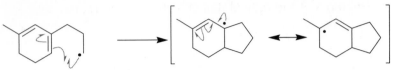

11.10.

a) Chlorination of methylene chloride to produce chloroform:

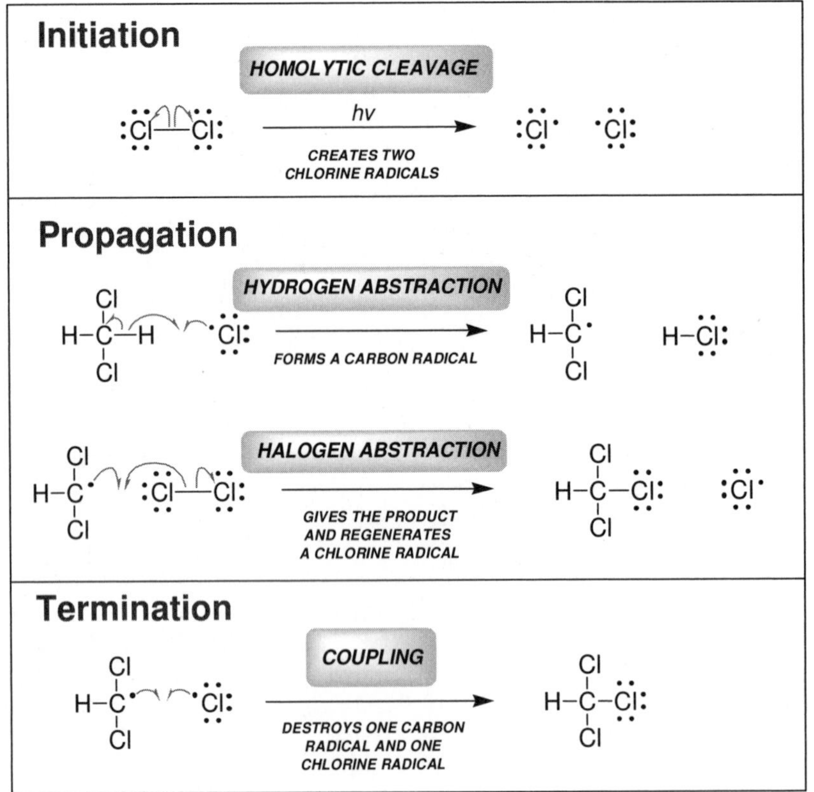

b) Chlorination of chloroform to produce carbon tetrachloride:

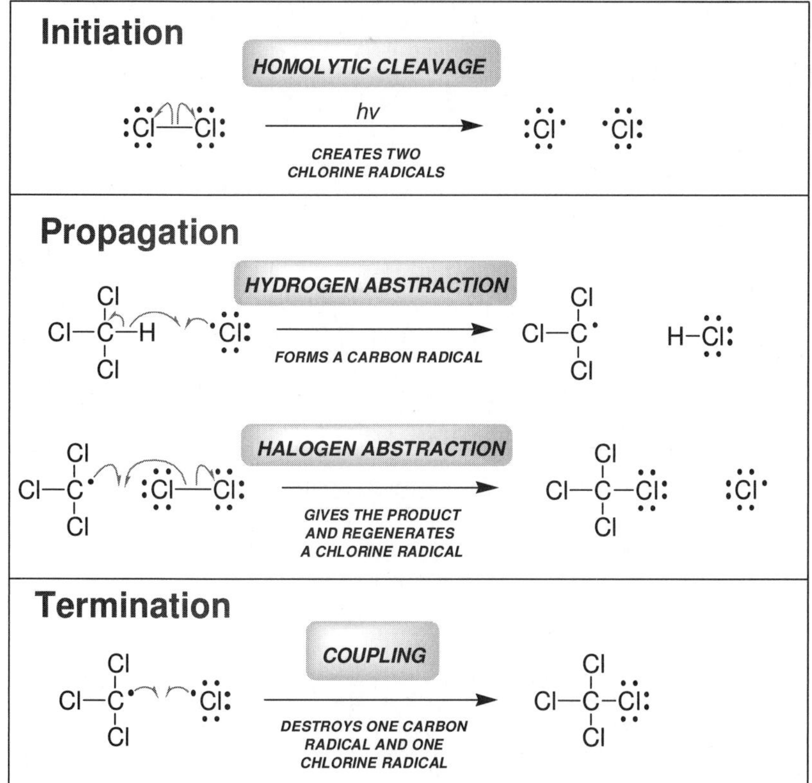

c) Chlorination of ethane to produce ethyl chloride

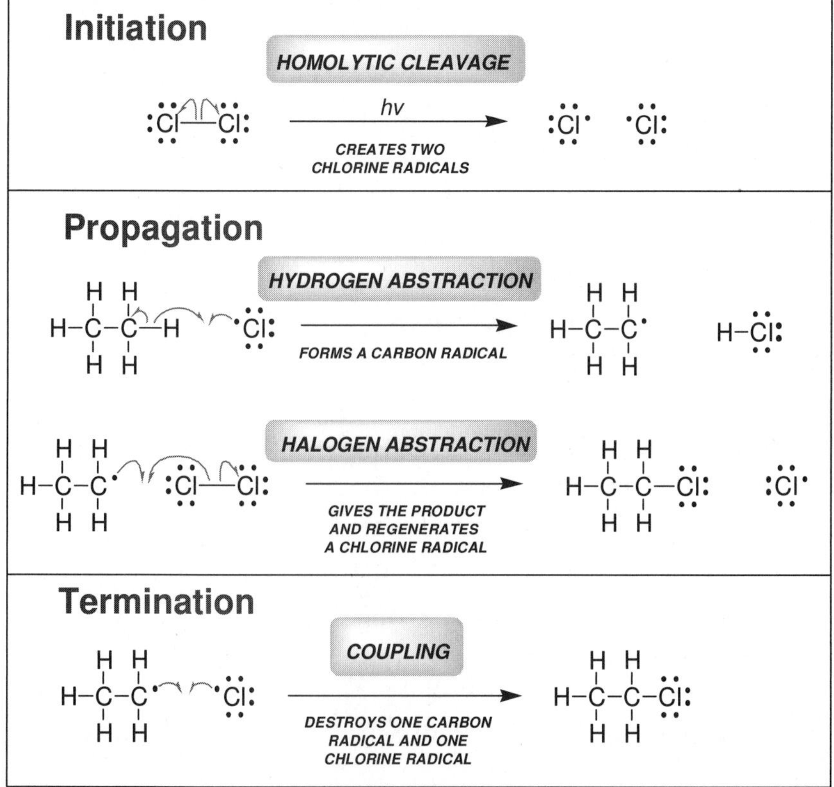

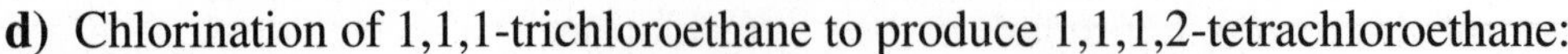

d) Chlorination of 1,1,1-trichloroethane to produce 1,1,1,2-tetrachloroethane:

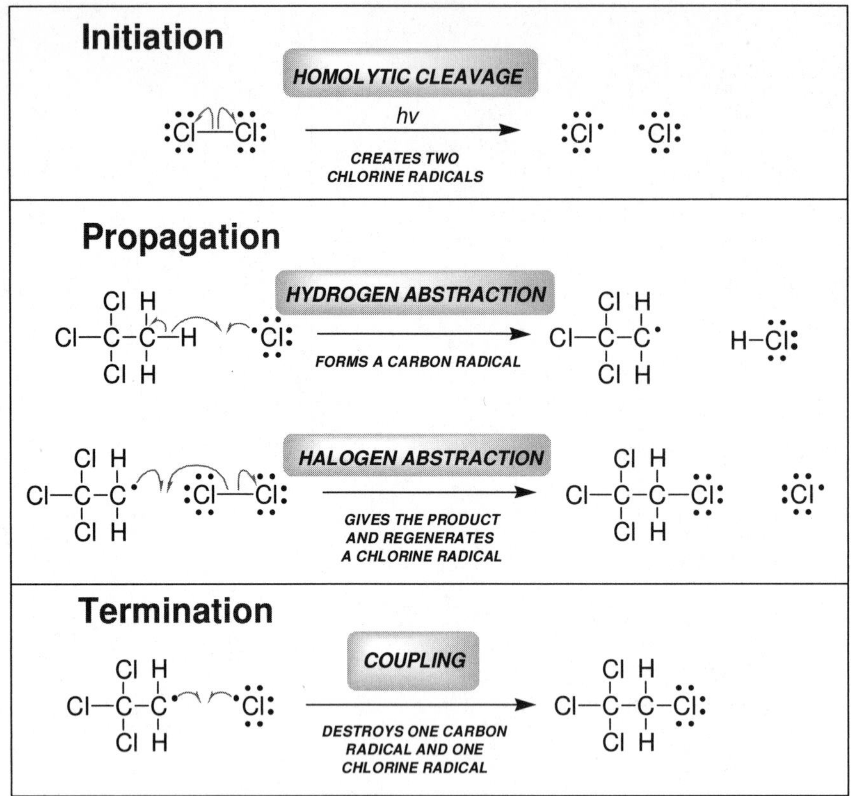

e) Chlorination of 2,2-dichloropropane to produce 1,2,2-trichloropropane:

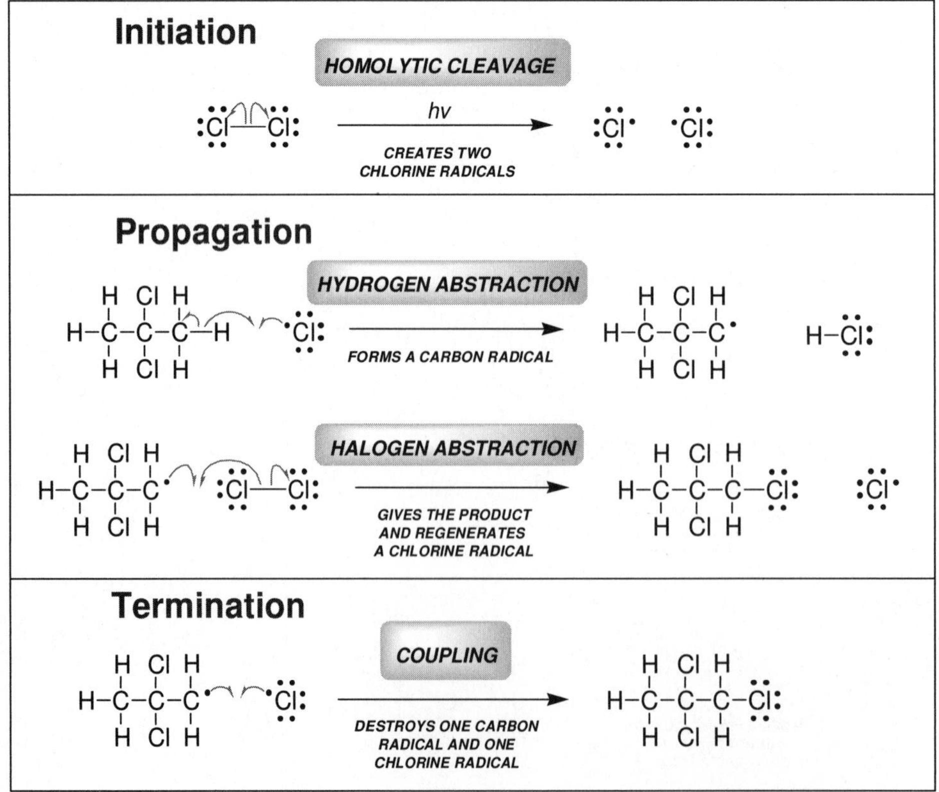

11.11. During the chlorination of methane, methyl radicals are generated. Two of these methyl radicals can couple together to form ethane:

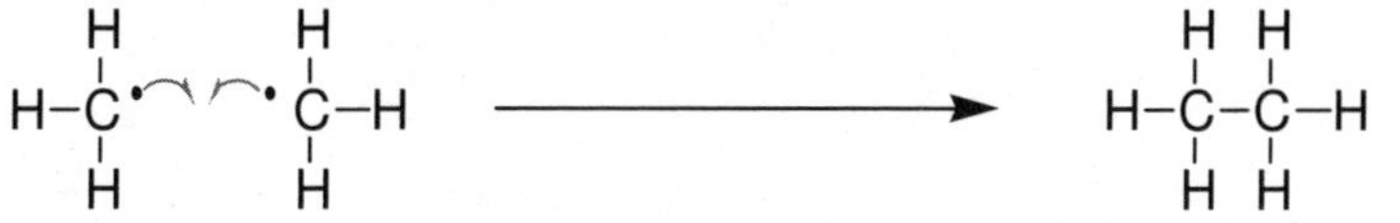

Ethane can then undergo hydrogen abstraction, followed by halogen abstraction to generate ethyl chloride:

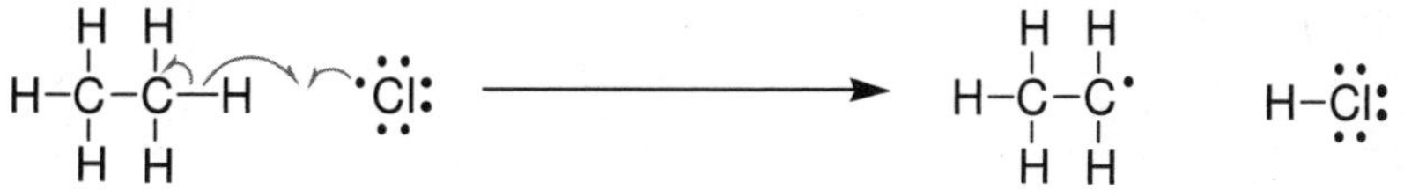

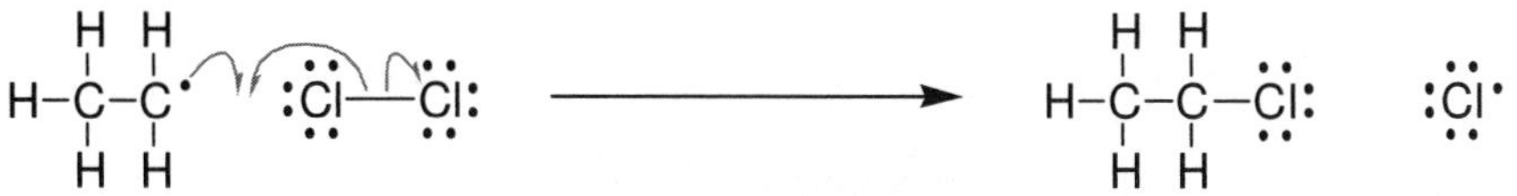

11.12.

a)

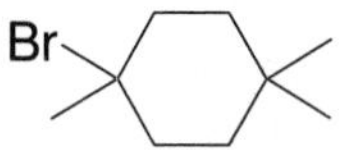

b)

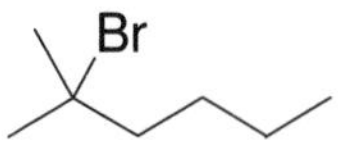

c)

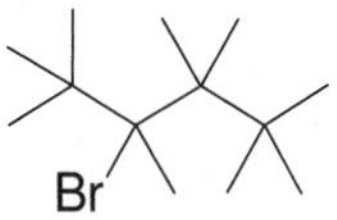

11.13.

a)

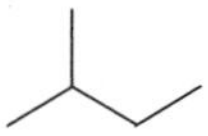

b)

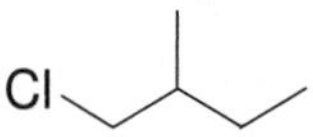

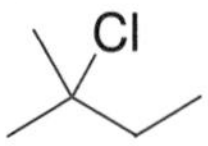

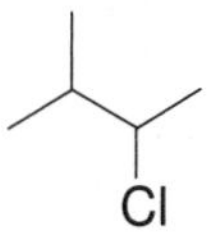

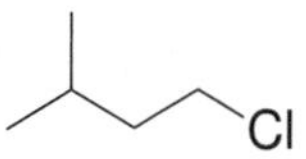

c)

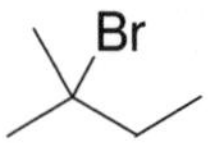

11.14.

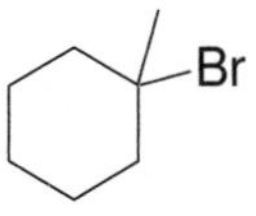

a) no chirality center

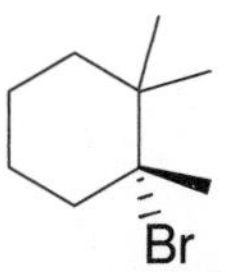

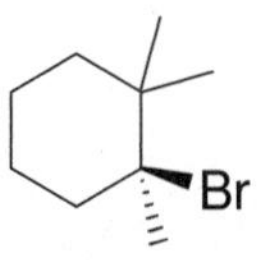

b)

c)

Br + Br

d)

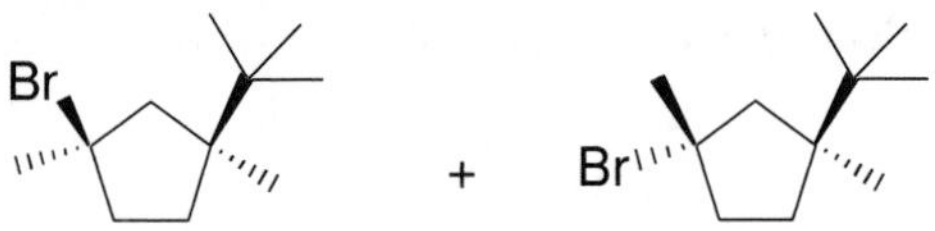

Br + Br

11.15.

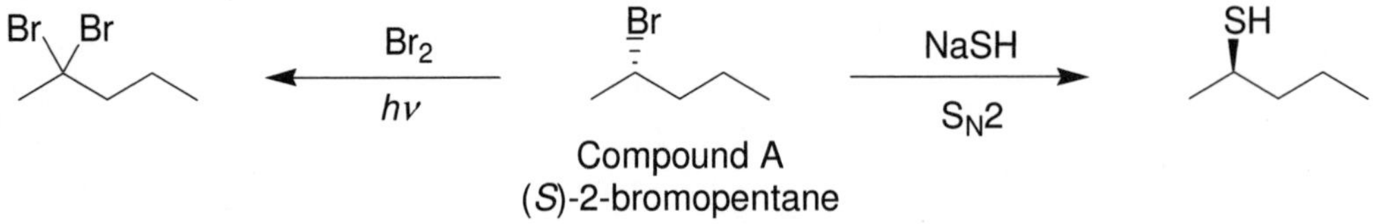

Br Br ← Br_2 / $h\nu$ — Br — NaSH / S_N2 → SH

Compound A
(*S*)-2-bromopentane

11.16.

a)

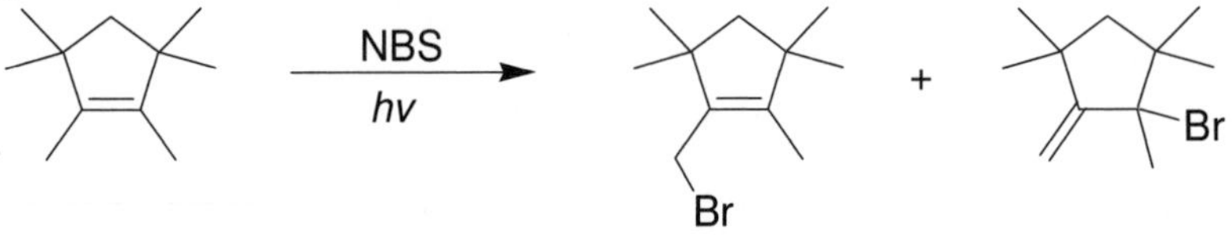

NBS / $h\nu$ → Br + Br

b)

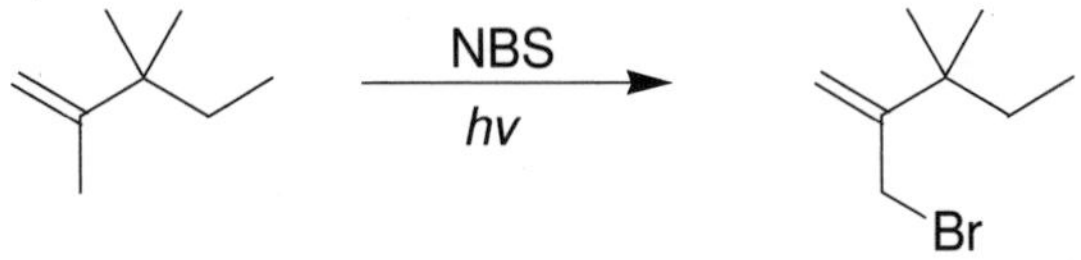

NBS / $h\nu$ → Br

c)

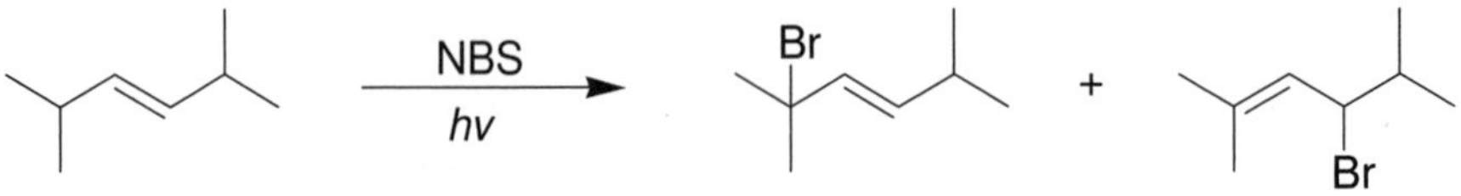

NBS / $h\nu$ → Br + Br

d)

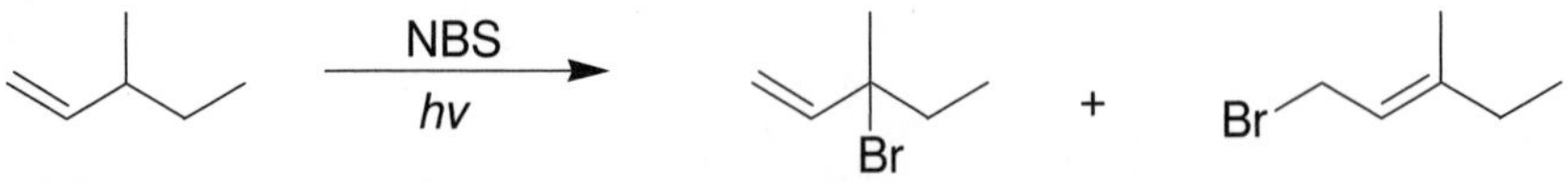

NBS / $h\nu$ → Br + Br

11.17.

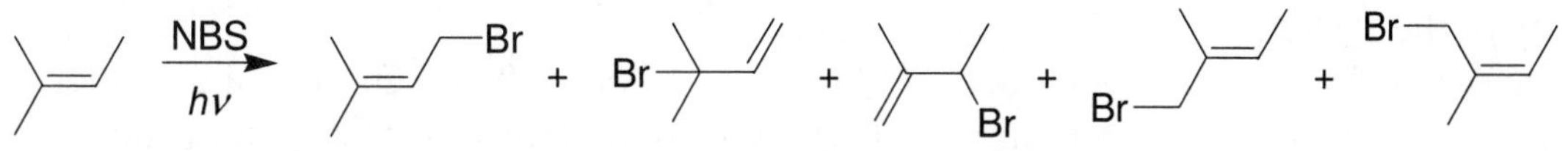

NBS, $h\nu$ + + + +

(racemic mixture)

11.18.

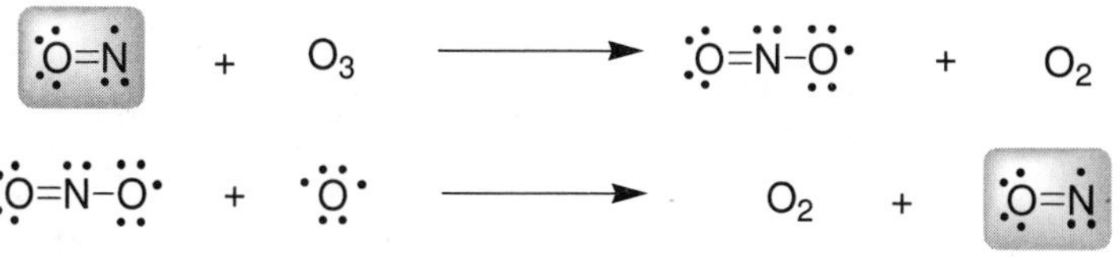

O_3 O_2

O_2

11.19.

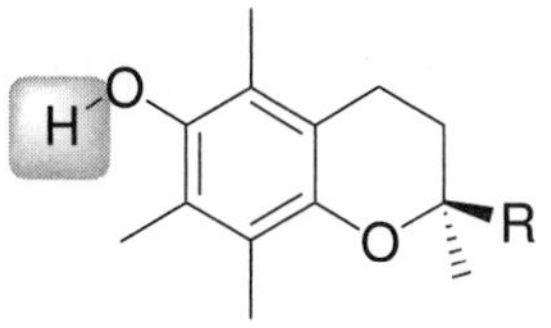

Vitamin E

11.20.

a) HBr, ROOR +

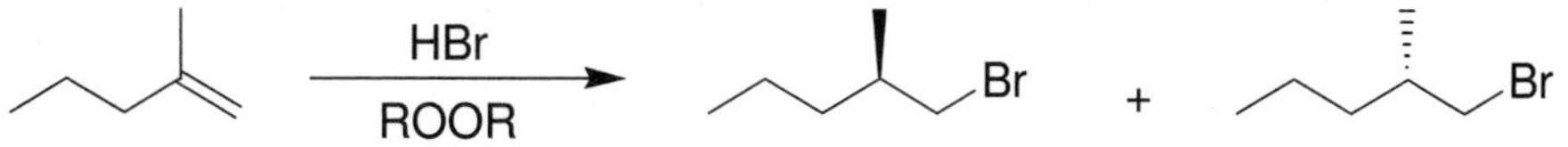

b) HBr, ROOR

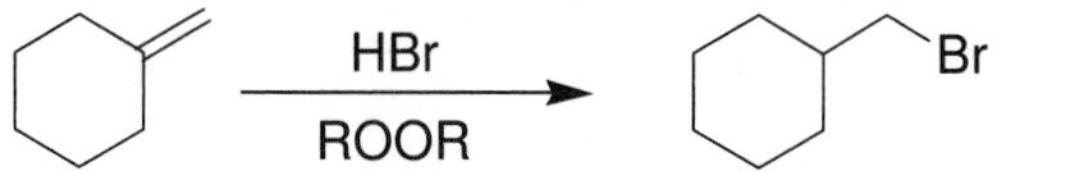

(no chirality centers)

c) HBr, ROOR +

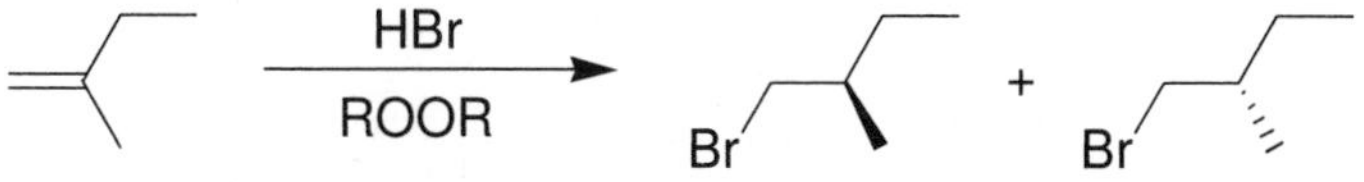

d) HBr, ROOR +

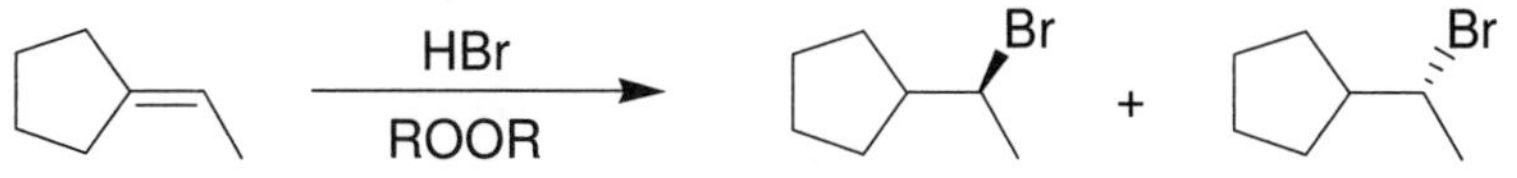

e) HBr, ROOR

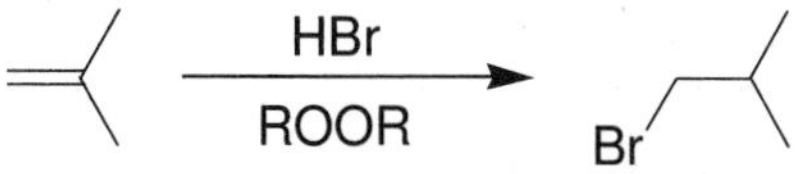

(no chirality centers)

f) HBr, ROOR

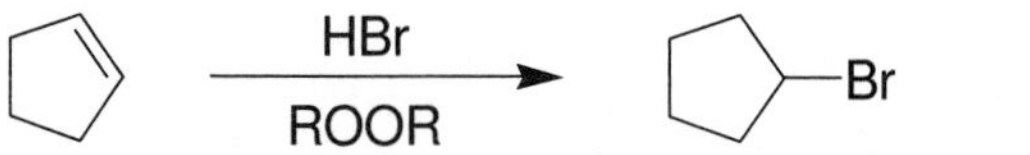

(no chirality centers)

11.21.

a) One chemical entity is being converted into two chemical entities, which increases the entropy of the system.

b) Recall that ΔG has two components: (ΔH) and ($-T\Delta S$). The magnitude of the latter term is dependent on the temperature. At high temperature, the latter term dominates over the former, and the reaction is thermodynamically favorable. However, at low temperature, the first term (enthalpy) dominates, and the reaction is no longer thermodynamically favored.

11.22.

a)

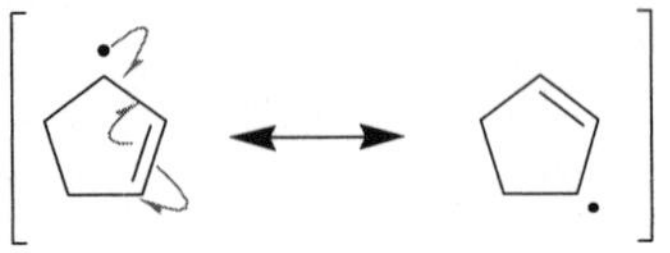

b)

c)

d)

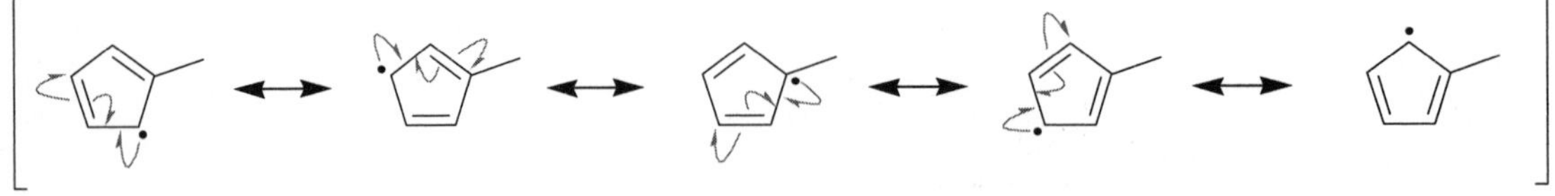

e)

11.23.

H_b H_c H_a

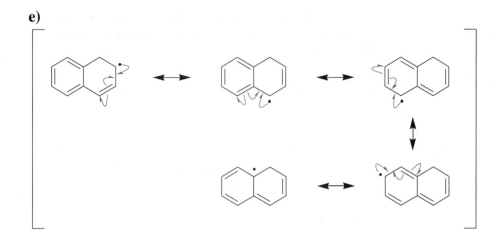

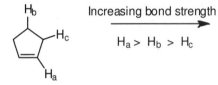

Increasing bond strength

$H_a > H_b > H_c$

abstraction of H_a generates an unstable vinyl radical

abstraction of H_c generates a resonance-stabilized radical

11.24.

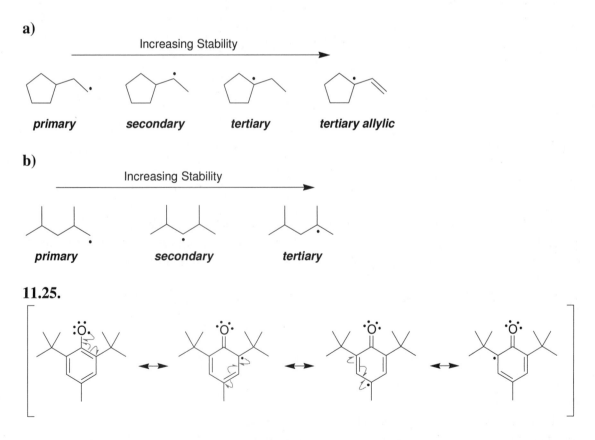

a)

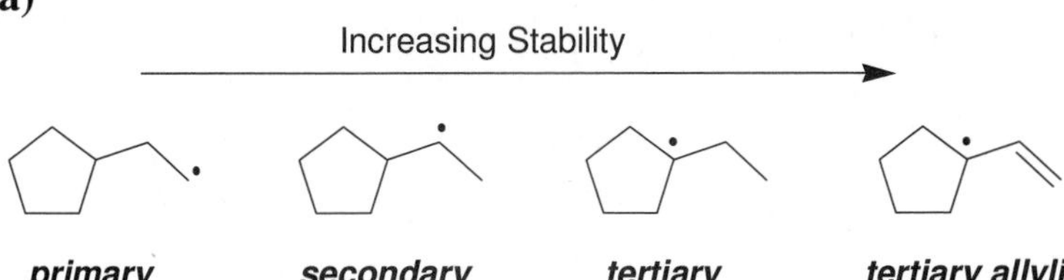

b)

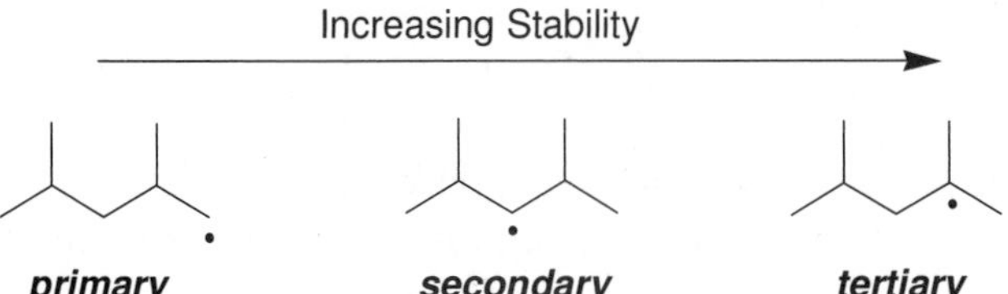

11.25.

11.26. The benzylic position is selectively brominated because the benzylic C-H bond is the weakest bond. The benzylic hydrogen atom is the only hydrogen atom that can be abstracted to generate a resonance-stabilized radical.

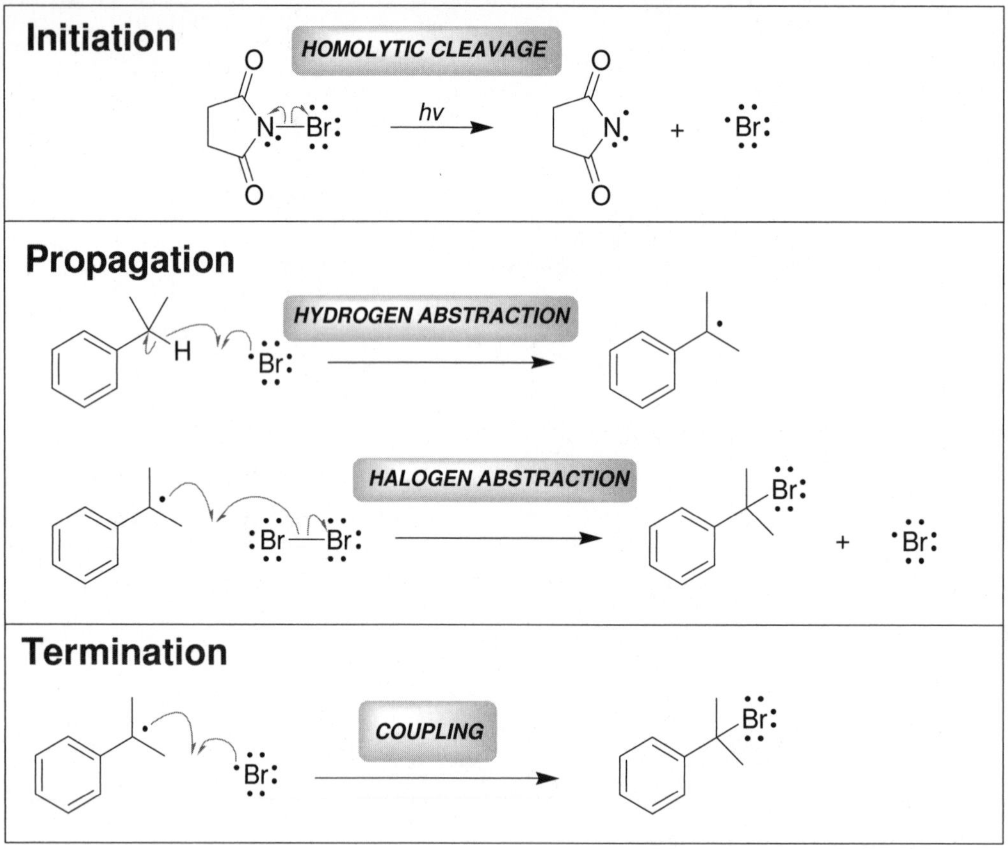

Initiation

HOMOLYTIC CLEAVAGE

hv

Propagation

HYDROGEN ABSTRACTION

HALOGEN ABSTRACTION

Termination

COUPLING

11.27.

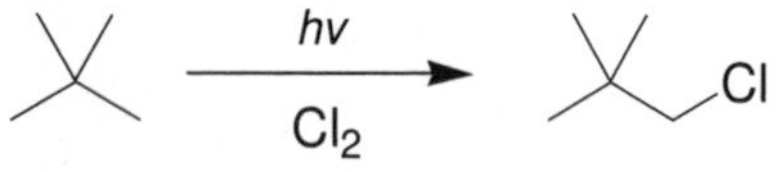

hv

Cl_2

11.28. Selective bromination at the benzylic position generates a new chirality center. The intermediate benzylic radical is expected to be attacked from either face of the planar radical with equal likehood, giving rise to a racemic mixture of enantiomers:

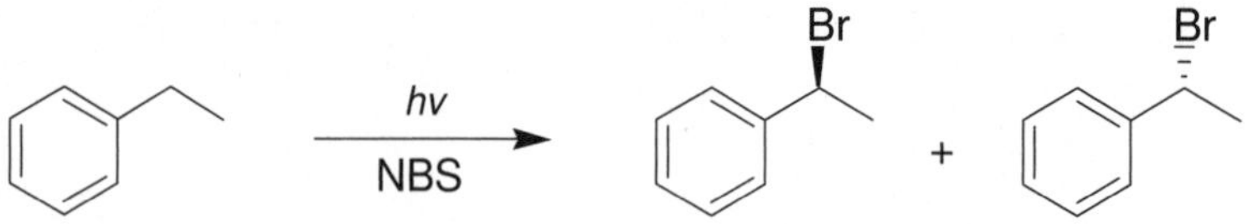

hv

NBS

11.29.

a) These radicals are tertiary, and they are stabilized by resonance.

b) Loss of nitrogen gas would result in the formation of vinyl radicals, which are too unstable to form under normal conditions:

11.30.

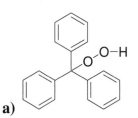

a)

b) Hydrogen abstraction leads to an exceptionally stable radical, with many, many resonance structures (see problem 11.3).

c) Phenol acts as a radical scavenger, thereby preventing the chain process from taking off.

11.31.

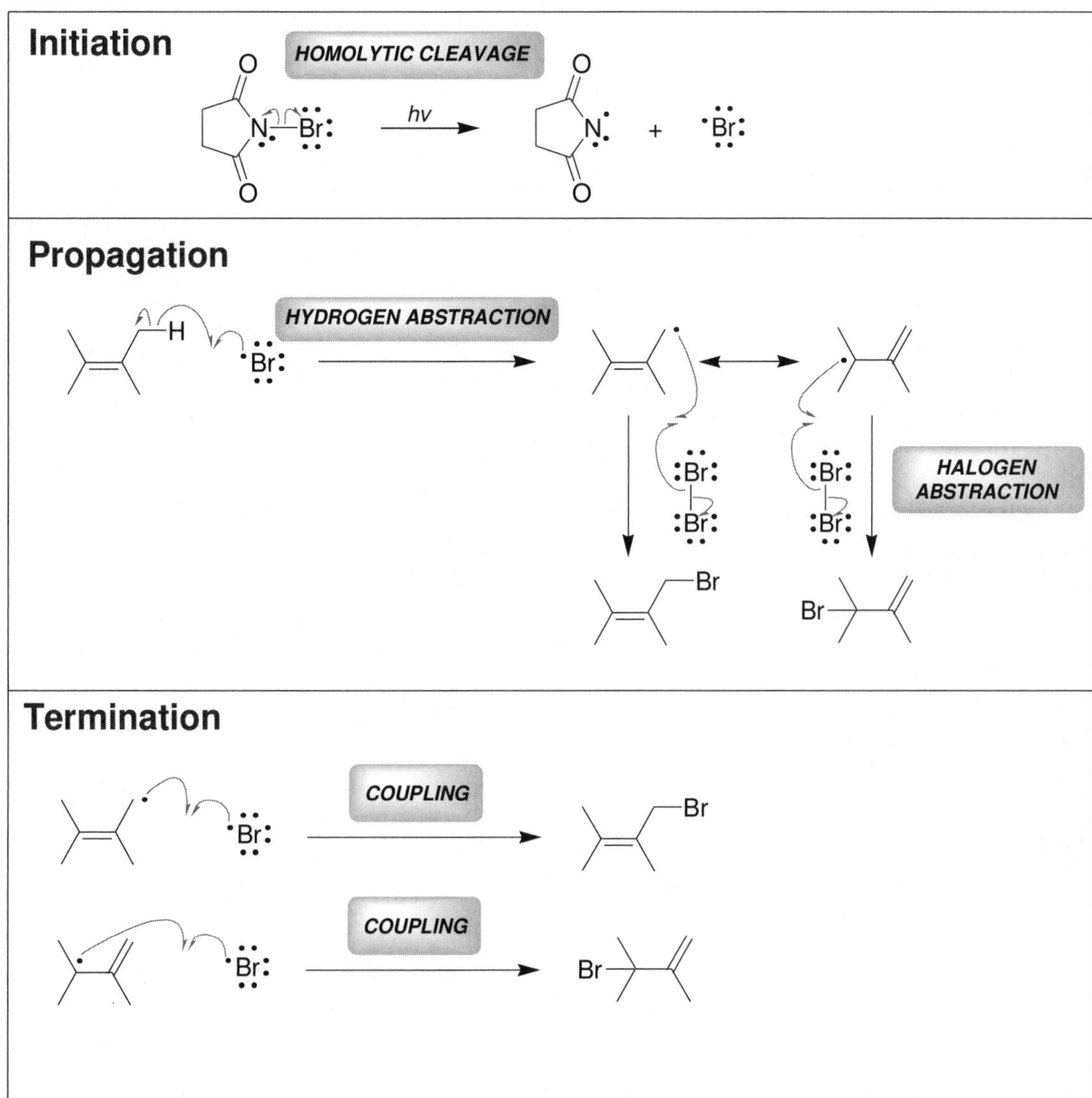

11.32.

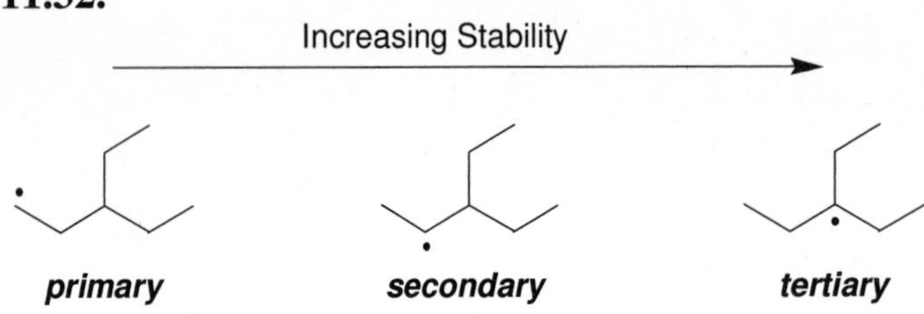

11.33.

a)

Br_2, *hv* → Br

b)

I_2, *hv* → no reaction

c)

Cl_2, *hv* → Cl + Cl (racemic mixture) + Cl

d)

NBS, *hv* → Br

e)

NBS, *hv* → Br + Br

f)

Br_2, *hv* → Br

11.34.

Cl Cl_2, *hv* → Cl Cl + Cl Cl + Cl Cl + Cl Cl

Cl Cl + Cl Cl + Cl Cl

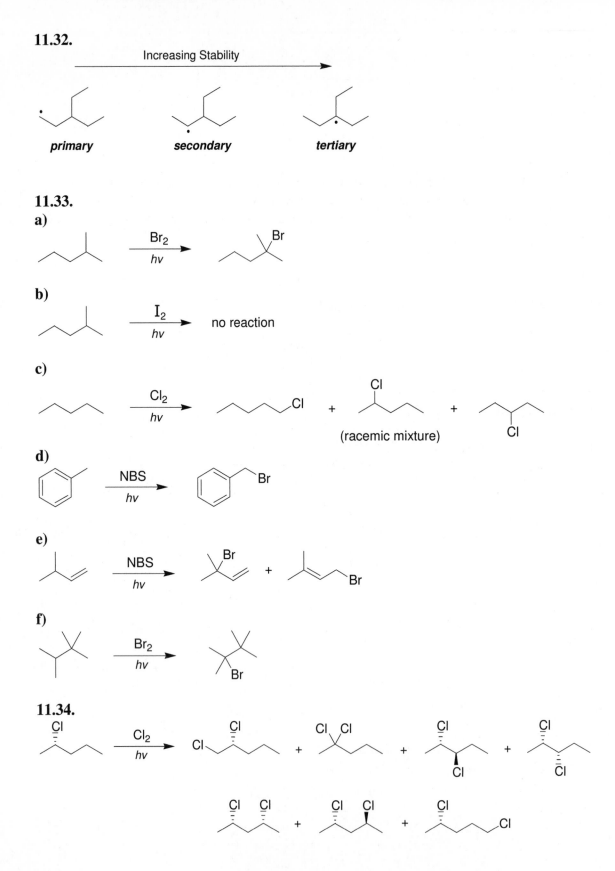

11.35.

Br_2 / *hv*

Br + Br

11.36.
a)

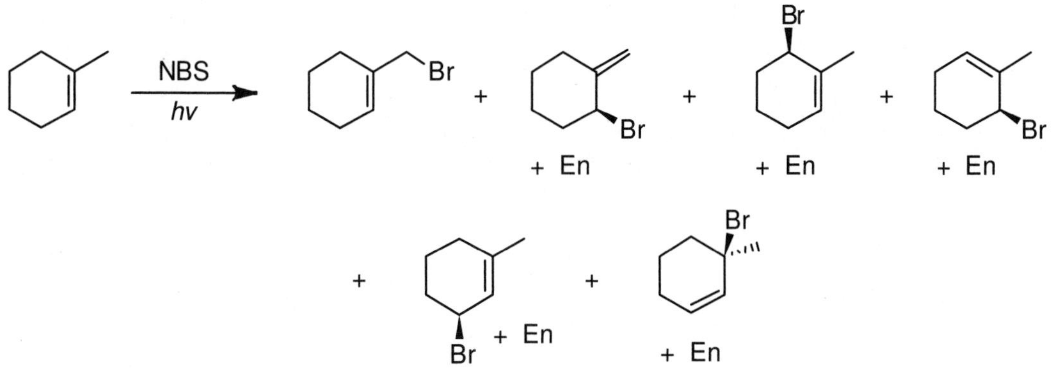

NBS / *hv*

Br + Br + Br + Br
+ En + En + En

+ Br + En + Br + En

b)

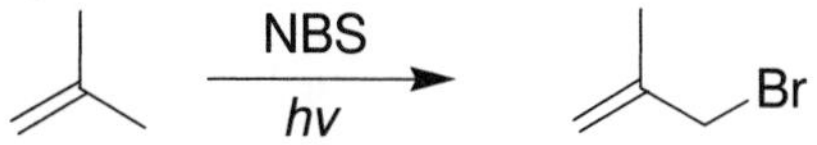

NBS / *hv*

Br

c)

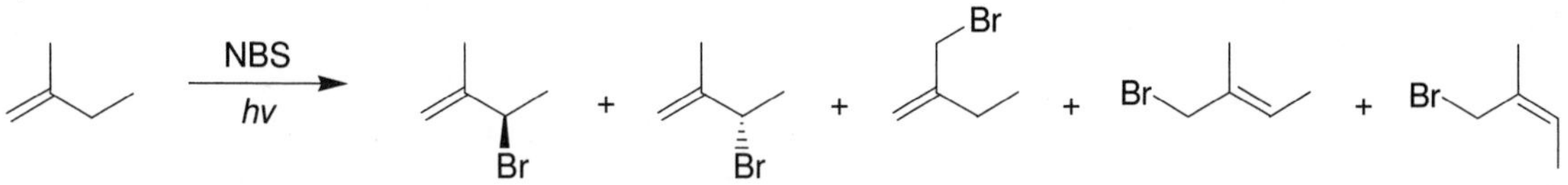

NBS / *hv*

Br + Br + Br + Br + Br

d)

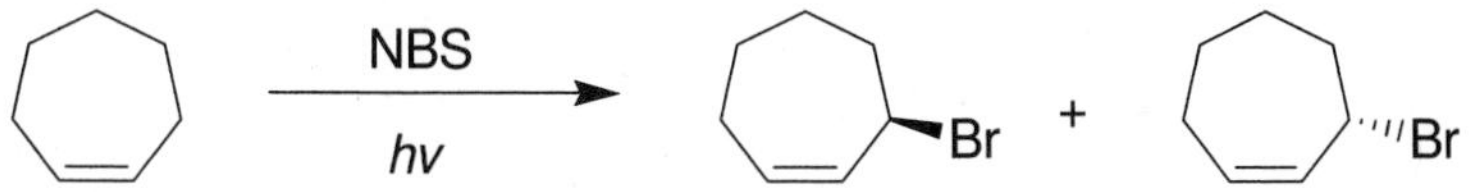

NBS / *hv*

Br + Br

11.37.

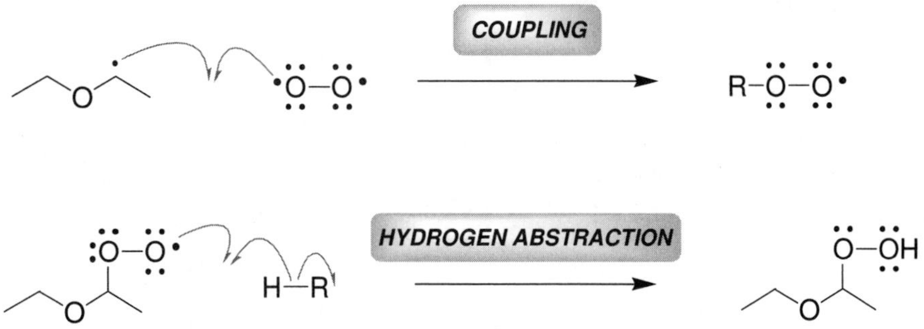

COUPLING

R–O–O•

HYDROGEN ABSTRACTION

H–R

O–OH R•

11.38. When Compound B is treated with a sterically hindered base, the Hofmann product (Compound D) is favored. When treated with sodium ethoxide, the Zaitsev product (Compound B) is favored:

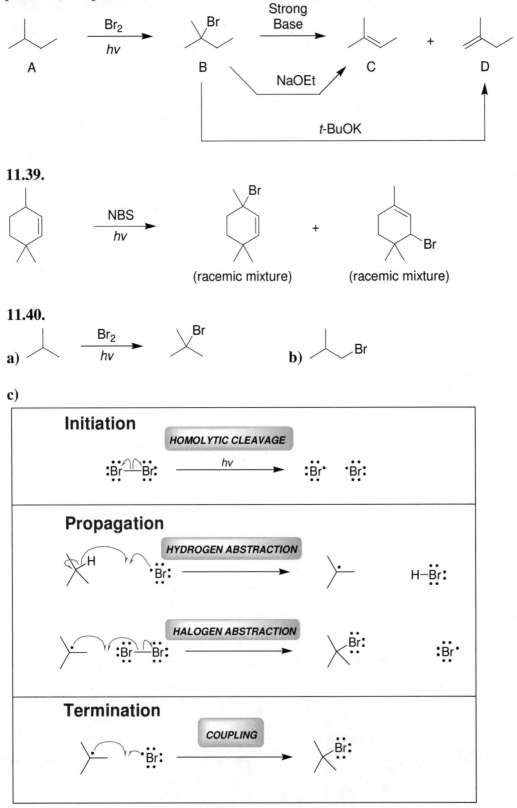

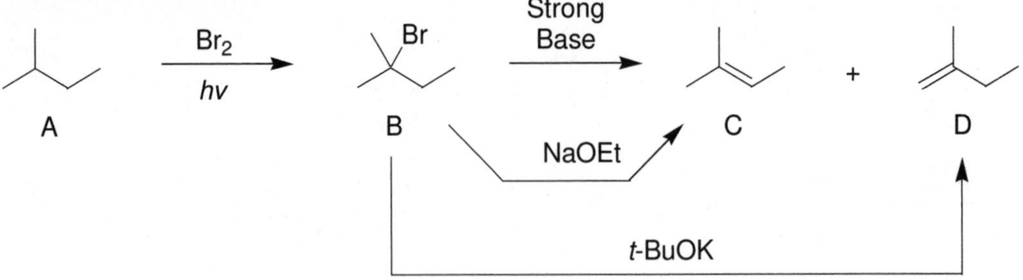

11.39.

NBS
hv
Br
+
Br
(racemic mixture)
(racemic mixture)

11.40.

a) Br_2 hv Br

b) Br

c)

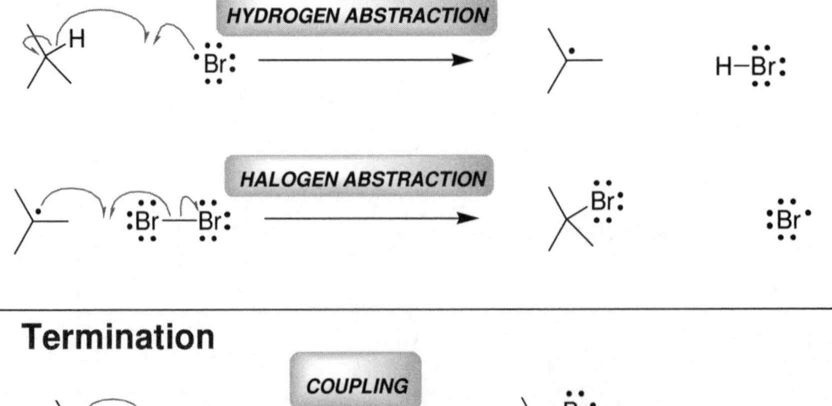

d)

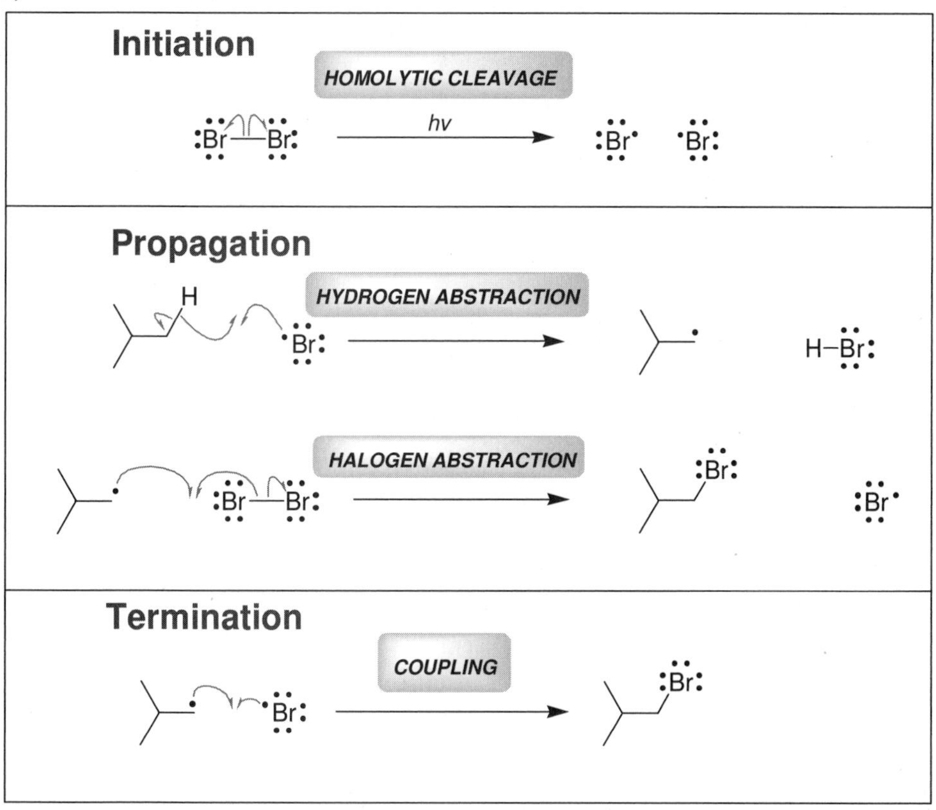

e) The minor product is only formed via a primary radical, which does not readily form under bromination conditions. The tertiary radical is selectively formed, which leads to the tertiary alkyl bromide as the major product.

11.41.

a) The two products are diastereomers

Br_2

$h\nu$

Br +

b)

enantiomers

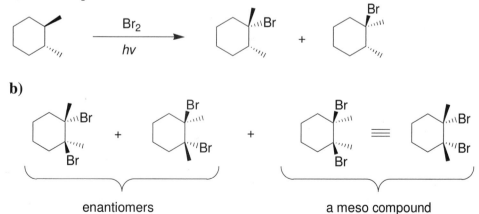

a meso compound

There are two chirality centers, so we might expect four products. However, one product is a meso compound, so there are only three products, rather than four. For a review of meso compounds, see Section 5.6.

11.42. Methyl radicals are less stable and less hindered than *tert*-butyl radicals.

11.43.

a) 1 **b)** 4 **c)** 3 **d)** 4 **e)** 2 **f)** 1 **g)** 5 **h)** 1 **i)** 3 **j)** 3

11.44.

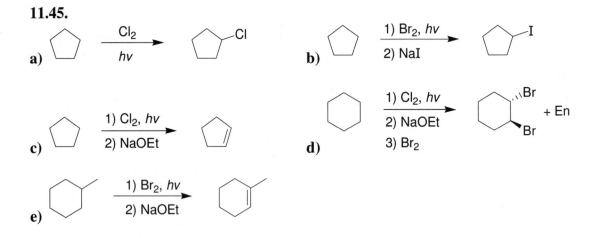

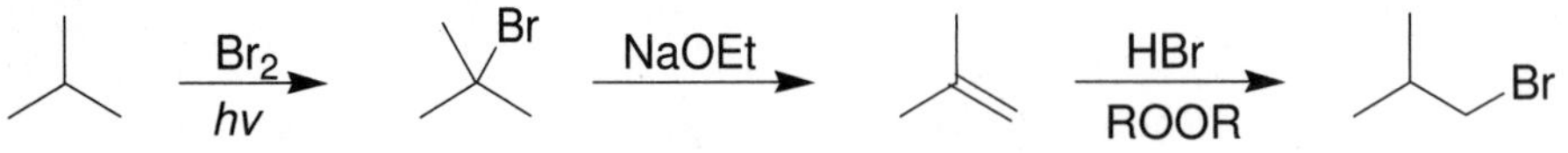

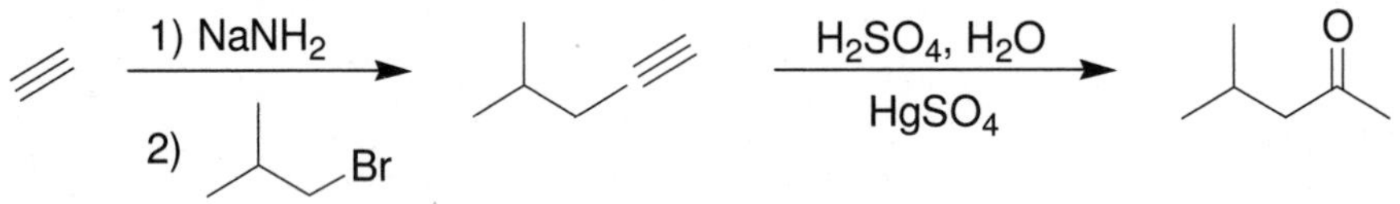

Br_2, $h\nu$ → Br; NaOEt →; HBr, ROOR → Br

1) $NaNH_2$ 2) Br → ; H_2SO_4, H_2O, $HgSO_4$ → O

11.45.

a) Cl_2, $h\nu$ → Cl

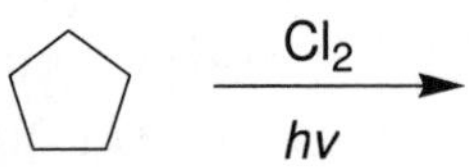

b) 1) Br_2, $h\nu$ 2) NaI → I

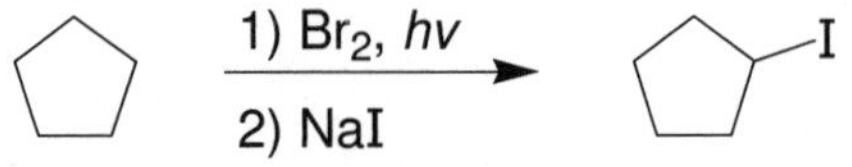

c) 1) Cl_2, $h\nu$ 2) NaOEt →

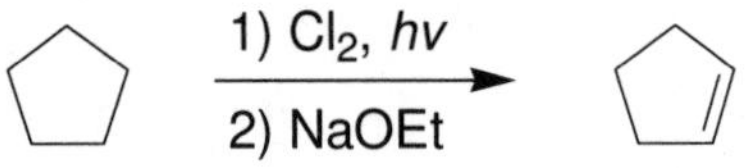

d) 1) Cl_2, $h\nu$ 2) NaOEt 3) Br_2 → Br, Br + En

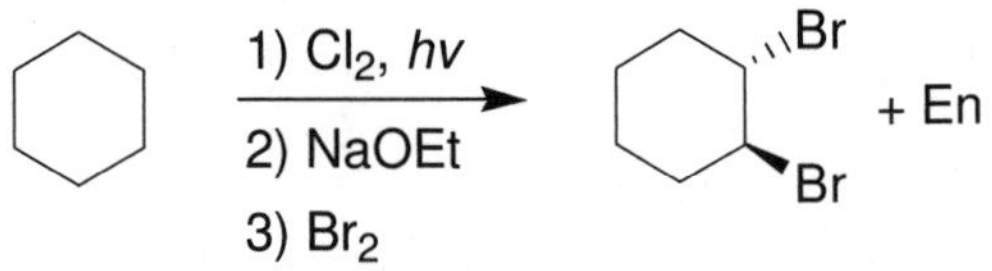

e) 1) Br_2, $h\nu$ 2) NaOEt →

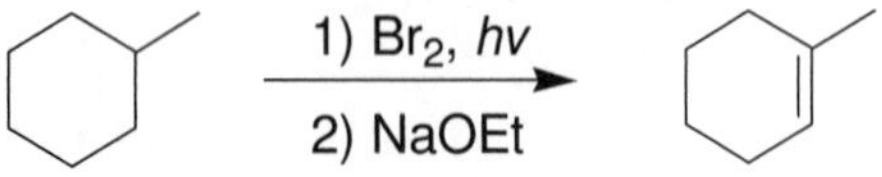

11.46. *cis*-1,2-Dimethylcyclopentane produces six pairs of compounds, where each pair have a diastereomeric relationship. In contrast, *trans*-dimethylcyclopentane produces only six different compounds, as shown below:

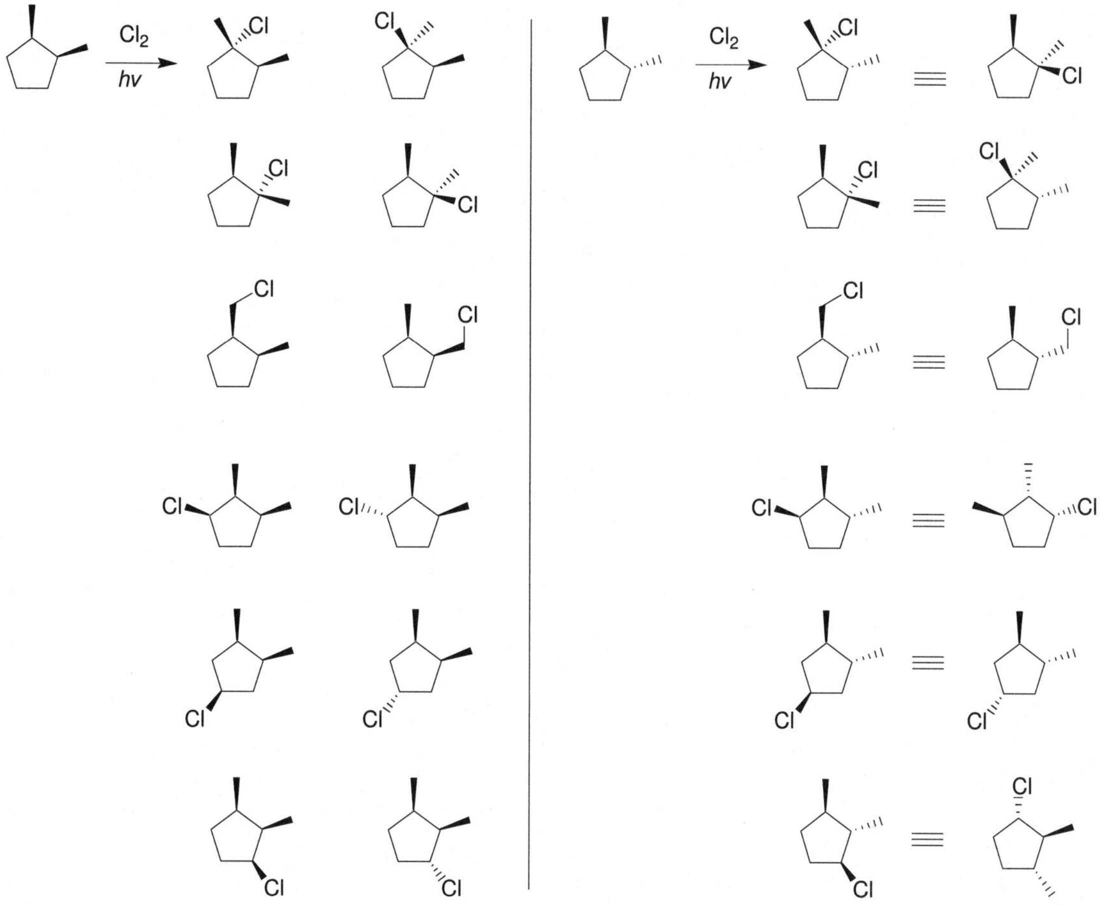

Cl_2
hv

Cl Cl Cl Cl Cl Cl Cl Cl Cl Cl Cl Cl

Cl_2
hv

Cl Cl Cl Cl Cl Cl Cl Cl Cl Cl Cl Cl

11.47. The first propagation step in a bromination process is generally slow and selective. In fact, this is the source of the regioselectivity for this reaction. A pathway via a tertiary radical will be significantly lower in energy than a pathway via a secondary or primary radical. As a result, bromination occurs predominantly at the more substituted position. However, when chlorine is present, chlorine radicals can perform the first propagation step (hydrogen abstraction) very rapidly, and with little selectivity. Under these conditions, secondary and primary radicals are formed almost as easily as tertiary radicals. The resulting radicals then react with bromine in the second propagation step to yield monobrominated products. Therefore, in the presence of chlorine, the selectivity normally observed for bromination is lost.

11.48.

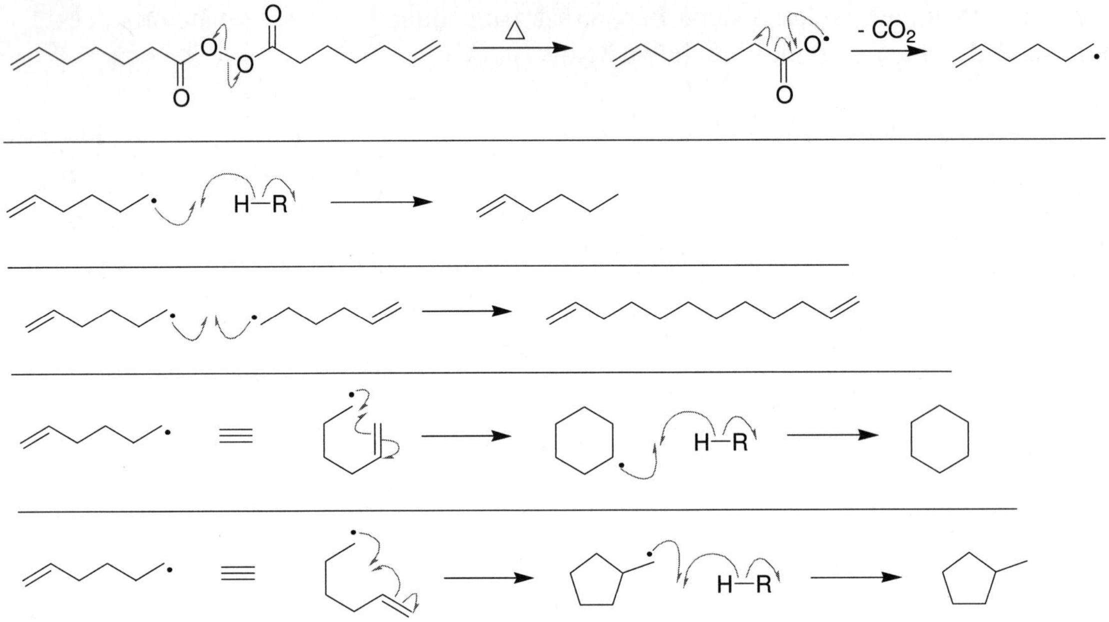

Δ

$- CO_2$

H—R

H—R

H—R

Chapter 12
Synthesis

Review of Concepts

Fill in the blanks below. To verify that your answers are correct, look in your textbook at the end of Chapter 12. Each of the sentences below appears verbatim in the section entitled *Review of Concepts and Vocabulary.*

- The position of a halogen can be moved by performing __________ followed by ______________.
- The position of a π bond can be moved by performing ___________ followed by ______________.
- An alkane can be functionalized via radical ________________.
- Every synthesis problem should be approached by asking the following two questions:
    1. Is there any change in the ____________________?
    2. Is there any change in the identity or location of the _________________?
- In a _________________ **analysis**, the last step of the synthetic route is first established, and the remaining steps are determined, working backwards from the product.

Review of Skills

Fill in the blanks and empty boxes below. To verify that your answers are correct, look in your textbook at the end of Chapter 12. The answers appear in the section entitled *SkillBuilder Review.*

12.1 Changing the Identity or Position of a Functional Group

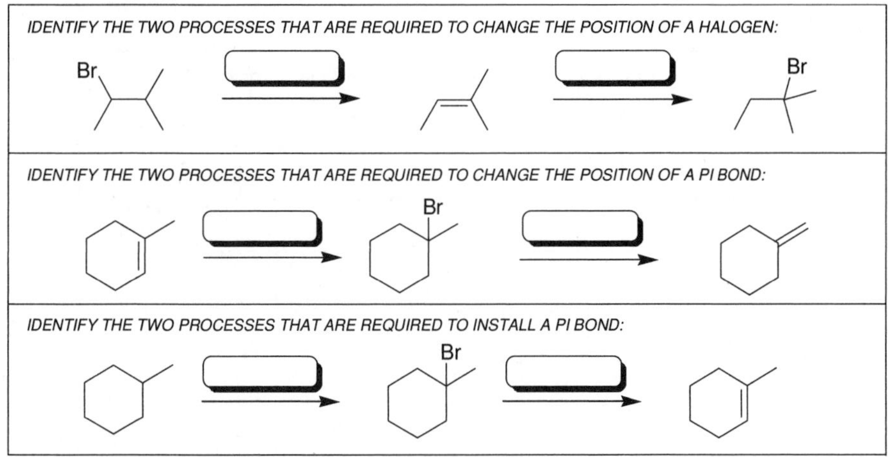

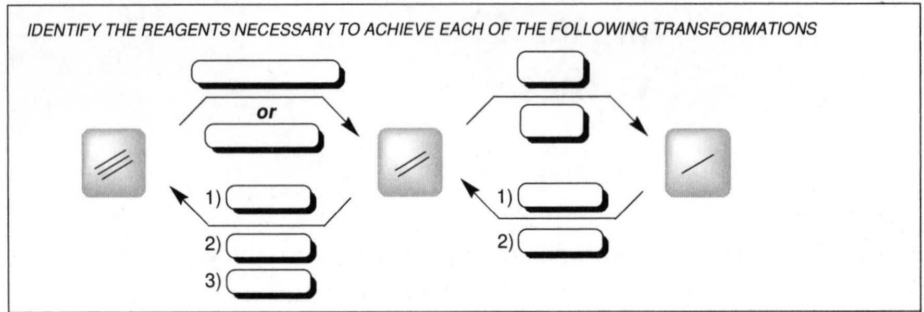

12.2 Changing the Carbon Skeleton

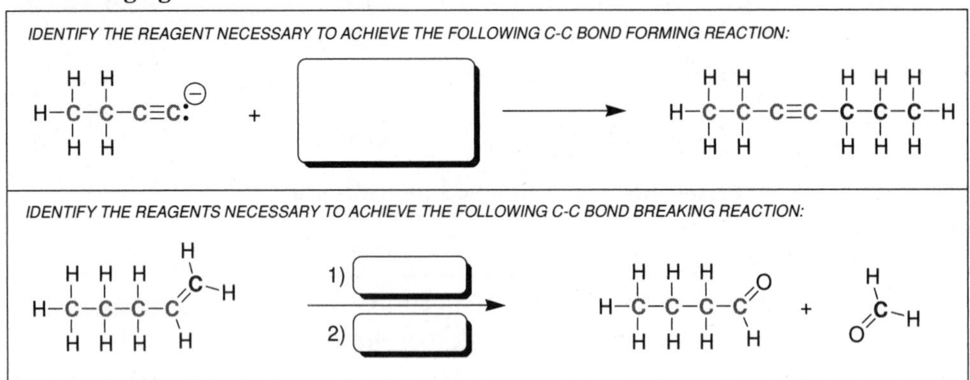

12.3 Approaching a Synthesis Problem by Asking Two Questions

IDENTIFY THE TWO QUESTIONS TO ASK WHEN APPROACHING A SYNTHESIS PROBLEM:

1) ____?

2) ____?

12.4 Retrosynthetic Analysis

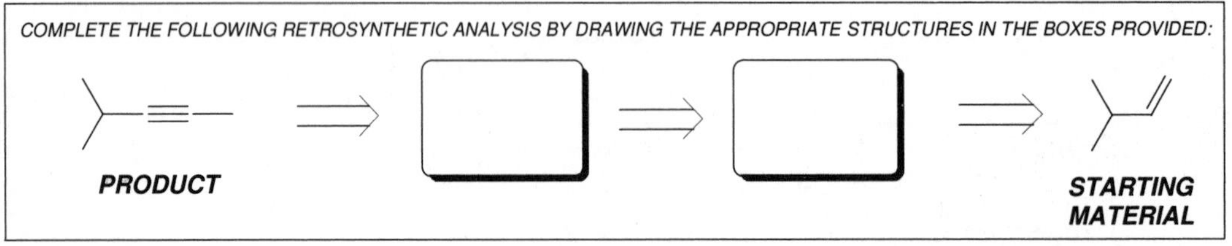

Solutions

12.1.

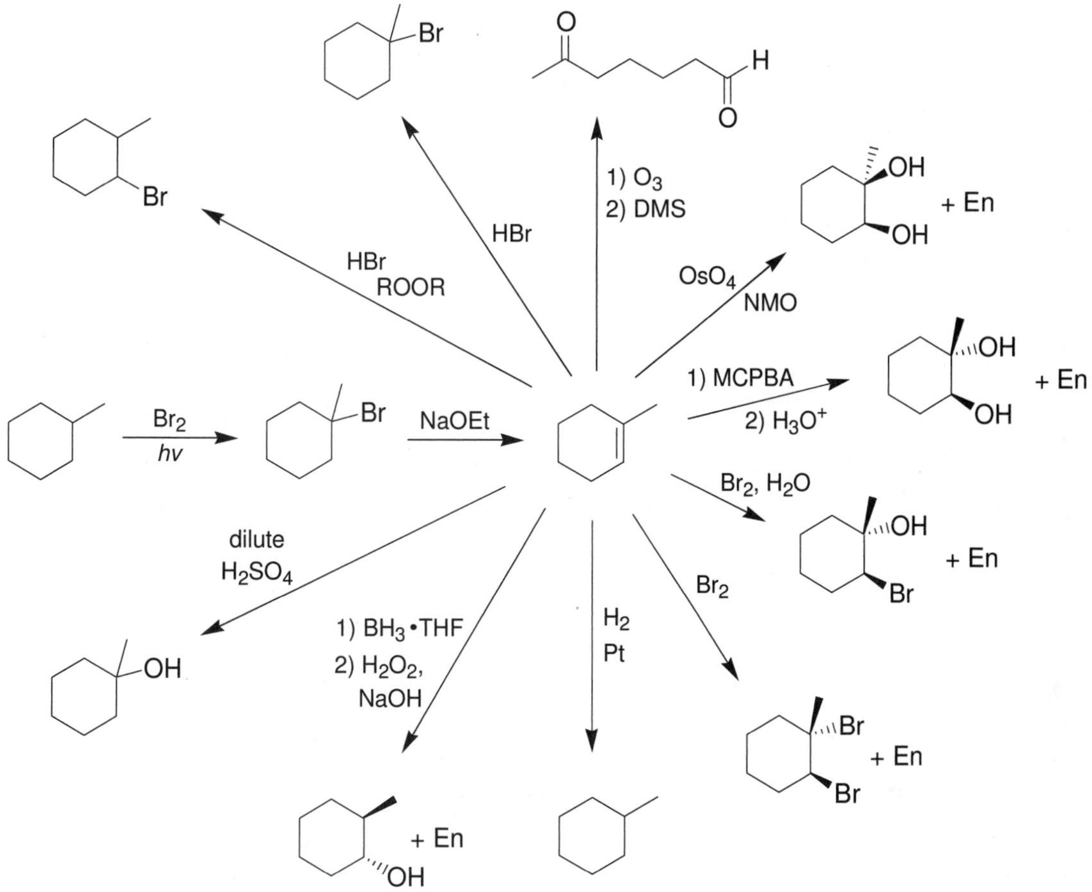

Br
O
H
O
1) O_3
2) DMS
OH
+ En
OH
Br
HBr
HBr
ROOR
OsO_4
NMO
OH
+ En
OH
1) MCPBA
2) H_3O^+
Br
Br_2
hv
Br
NaOEt
Br_2, H_2O
OH
+ En
Br
dilute
H_2SO_4
Br_2
1) BH_3•THF
2) H_2O_2,
NaOH
H_2
Pt
OH
Br
+ En
Br
+ En
OH

12.2.

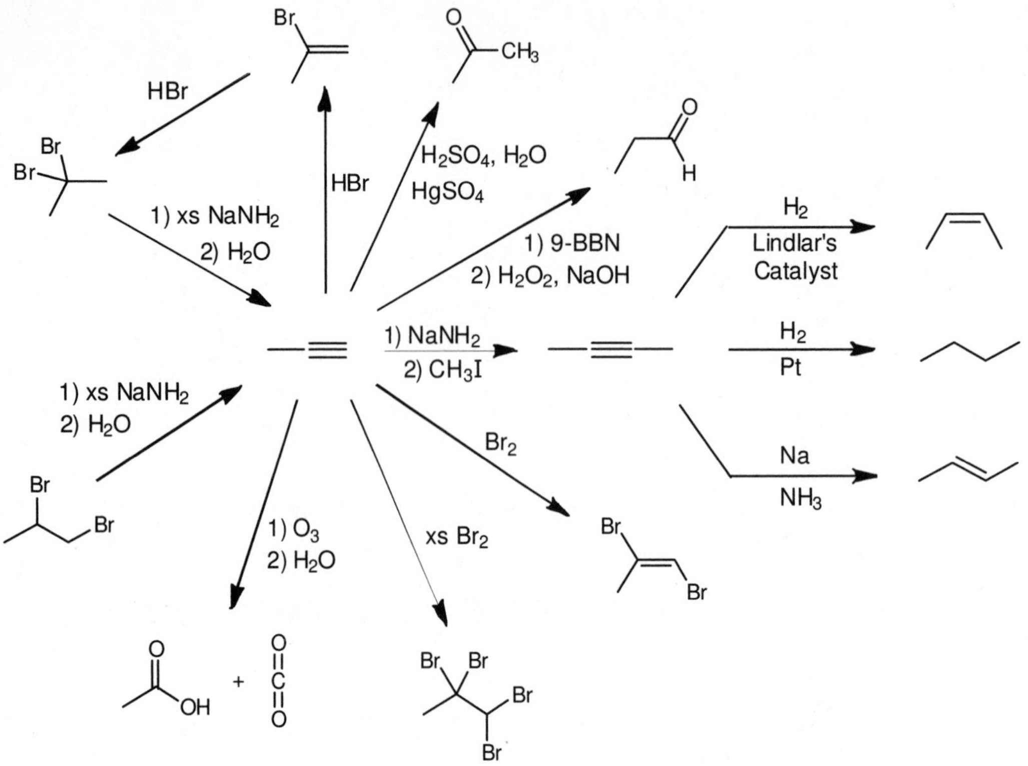

HBr

Br

H_2SO_4, H_2O

$HgSO_4$

HBr

1) xs $NaNH_2$

2) H_2O

1) 9-BBN

2) H_2O_2, NaOH

H_2

Lindlar's Catalyst

1) $NaNH_2$

2) CH_3I

H_2

Pt

1) xs $NaNH_2$

2) H_2O

Br_2

Na

NH_3

1) O_3

2) H_2O

xs Br_2

OH + O=C=O

Br Br Br Br

12.3.

a)

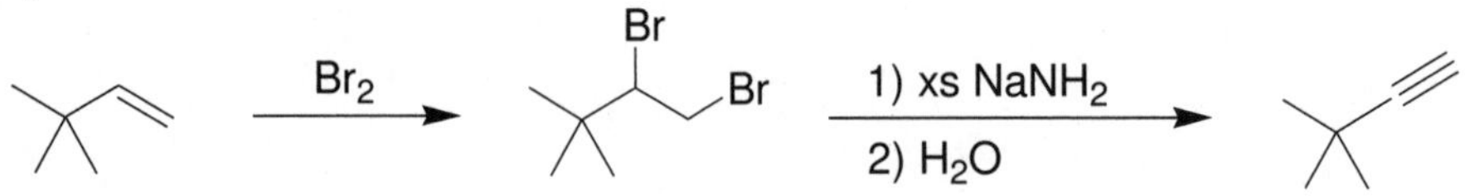

Br_2 → (Br, Br) → 1) xs $NaNH_2$ 2) H_2O →

b)

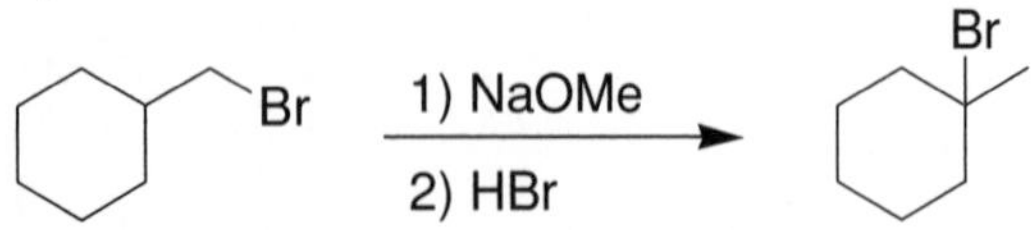

Br → 1) NaOMe 2) HBr → Br

c)

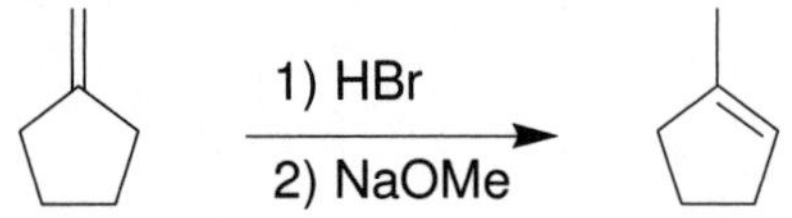

1) HBr 2) NaOMe →

d)

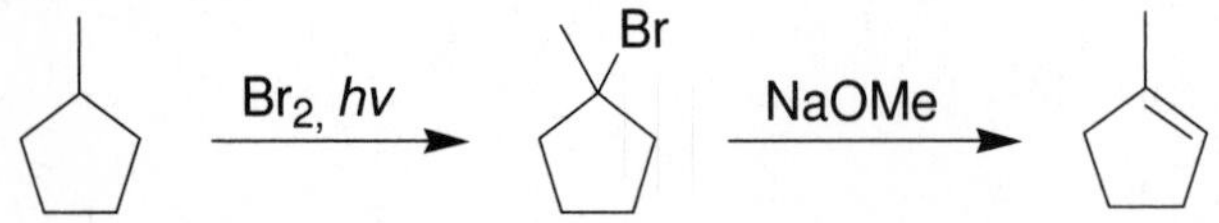

Br_2, *hv* → Br → NaOMe →

e)

conc. H_2SO_4, heat

1) TsCl, py
2) NaOEt

dilute H_2SO_4

OH

f)

Br

1) NaOMe
2) dilute H_2SO_4

OH

g)

OH

1) TsCl, py
2) NaOEt

HBr, ROOR

Br

h)

OH

1) TsCl, py
2) NaOEt

1) $BH_3 \cdot THF$
2) H_2O_2, NaOH

1) HBr, ROOR
2) NaOH (S_N2)

OH

12.4.

Br

1) NaOMe
2) HBr, ROOR

Br

1) *t*-BuOK
2) Br_2

Br Br

1) xs $NaNH_2$
2) H_2O

12.5.

Br_2

Br Br

1) xs $NaNH_2$
2) H_2O

excess HBr

Br Br

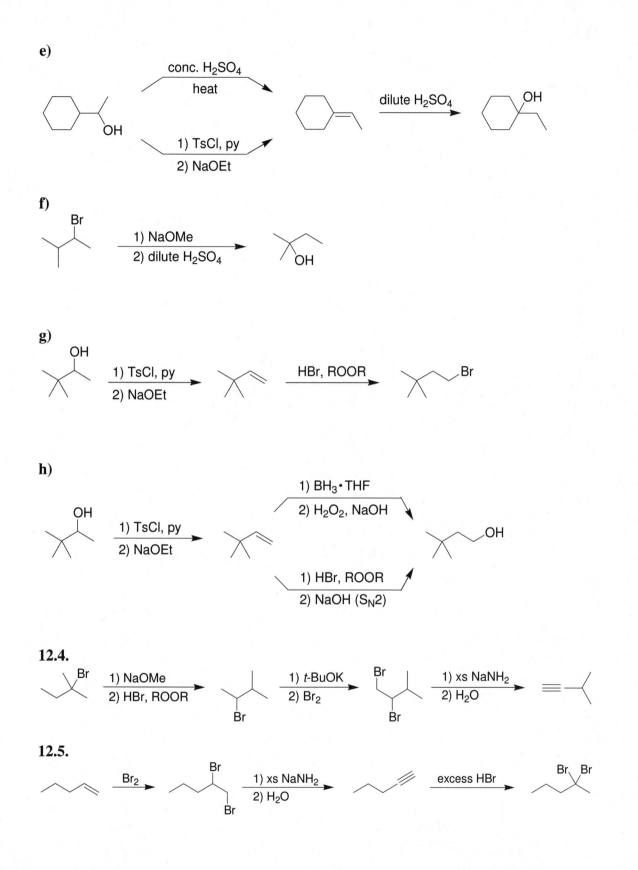

12.6.

a)

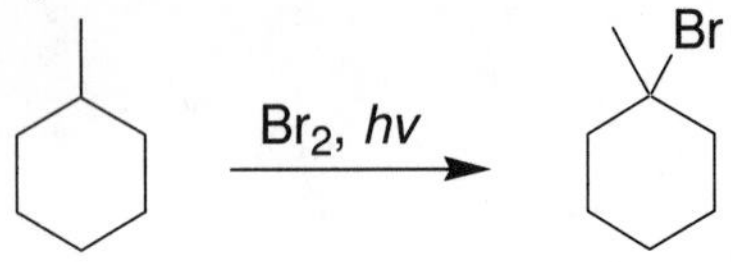

$\xrightarrow{Br_2,\ h\nu}$

Br

b)

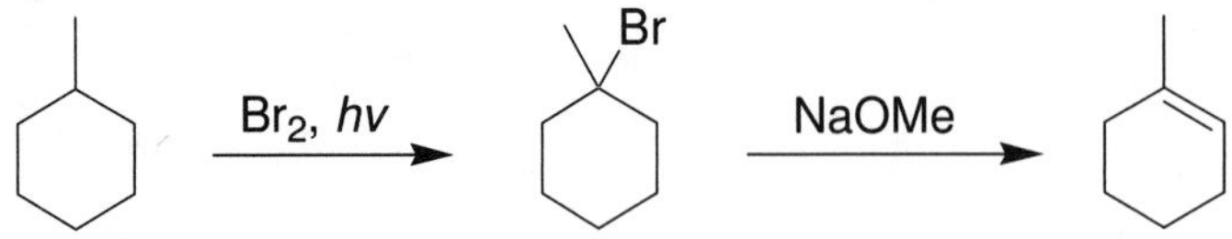

Br

$\xrightarrow{Br_2,\ h\nu}$ $\xrightarrow{NaOMe}$

c)

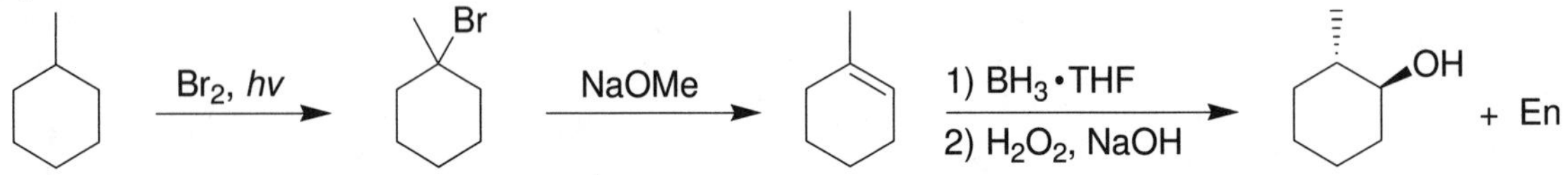

Br

$\xrightarrow{Br_2,\ h\nu}$ $\xrightarrow{NaOMe}$

1) $BH_3 \cdot THF$

2) H_2O_2, NaOH

OH

\+ En

d)

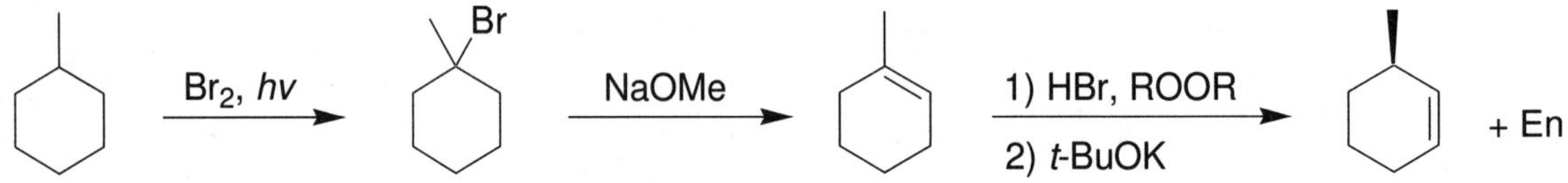

Br

$\xrightarrow{Br_2,\ h\nu}$ $\xrightarrow{NaOMe}$

1) HBr, ROOR

2) *t*-BuOK

\+ En

12.7.

a)

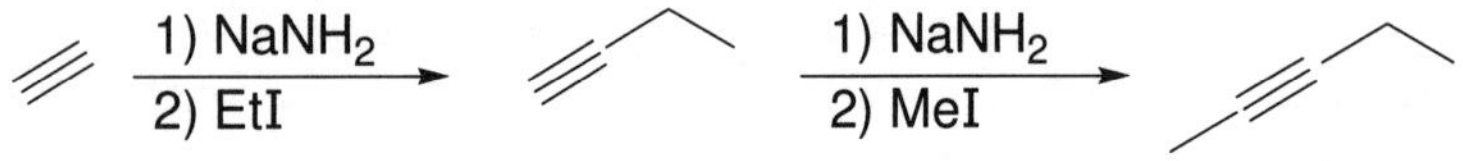

1) $NaNH_2$

2) EtI

1) $NaNH_2$

2) MeI

b)

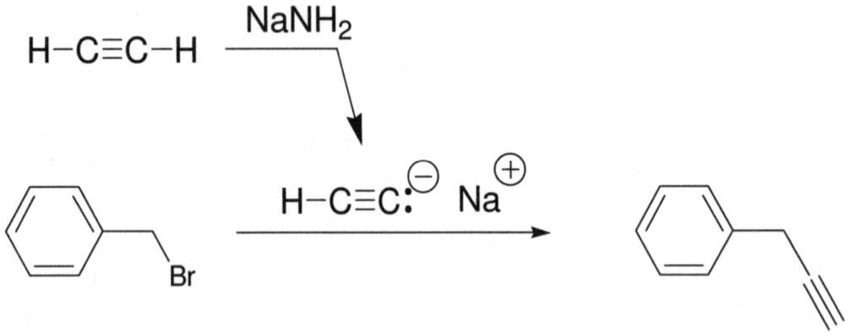

H−C≡C−H

$NaNH_2$

Br

$H-C\equiv C:^{\ominus}\ Na^{\oplus}$

c)

O

1) O_3

2) DMS

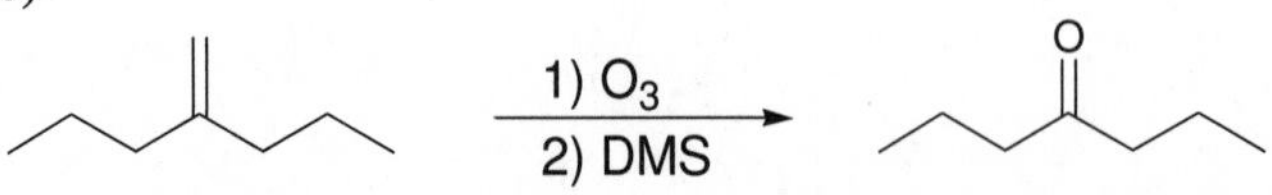

12.8.

a)

H−C≡C−H, $NaNH_2$ → H−C≡C:$^{\ominus}$ $Na^{\oplus}$

Br; H−C≡C:$^{\ominus}$ $Na^{\oplus}$ → ; H_2, Lindlar's catalyst →

b)

Br; *t*-BuOK → ; 1) O_3 2) DMS → O, H

c)

H−C≡C−H, $NaNH_2$ → H−C≡C:$^{\ominus}$ $Na^{\oplus}$

Br; H−C≡C:$^{\ominus}$ $Na^{\oplus}$ → ; xs HBr → Br Br

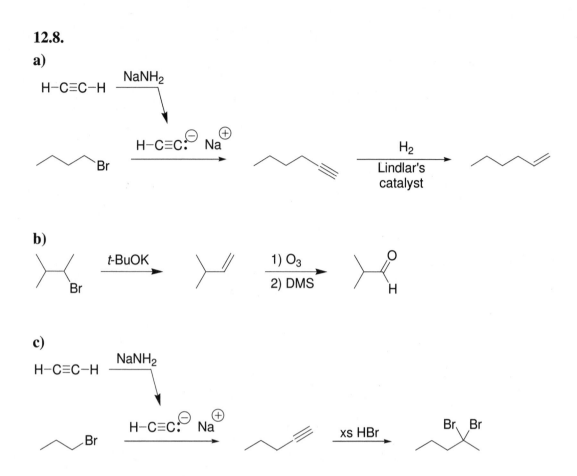

12.9. The alkyl halide is a tertiary substrate and does not readily undergo S_N2. Under these conditions, the acetylide ion functions as a base, rather than a nucleophile, giving an E2 reaction, instead of S_N2:

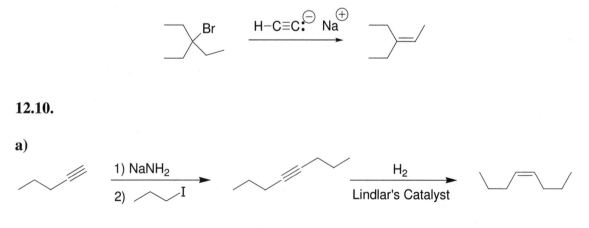

Br; H−C≡C:$^{\ominus}$ $Na^{\oplus}$ →

12.10.

a)

1) $NaNH_2$ 2) I → ; H_2, Lindlar's Catalyst →

b)

H_2 / Lindlar's Catalyst

Na / NH_3 (*l*)

HBr, ROOR

Br

$H{-}C{\equiv}C{-}H$ $\xrightarrow{NaNH_2}$ $H{-}C{\equiv}C:^{\ominus}\ Na^{\oplus}$

c)

Br

Br

1) xs $NaNH_2$
2) H_2O

1) O_3
2) H_2O

O

OH

d)

HBr / ROOR

Br

$H_3C{-}C{\equiv}C{-}H$ $\xrightarrow{NaNH_2}$ $H_3C{-}C{\equiv}C:^{\ominus}\ Na^{\oplus}$

Na / NH_3 (*l*)

e)

1) HBr, ROOR
2) *t*-BuOK

HBr / ROOR

Br

$H{-}C{\equiv}C{-}H$ $\xrightarrow{NaNH_2}$ $H{-}C{\equiv}C:^{\ominus}\ Na^{\oplus}$

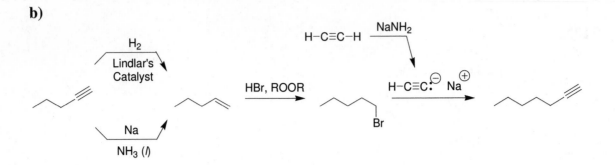

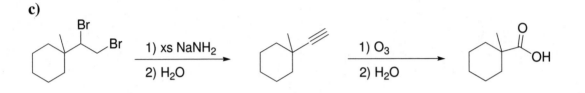

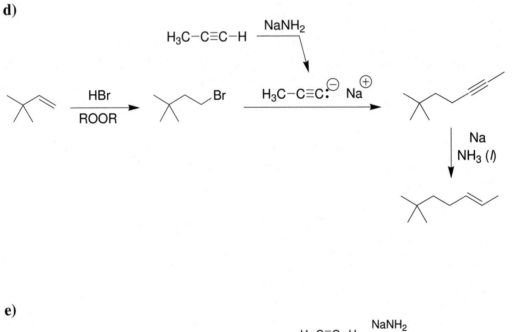

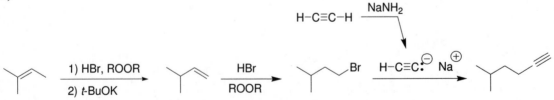

f)

OH

1) TsCl, py
2) *t*-BuOK

1) O_3
2) DMS

H
O

12.11.

H–C≡C–H

$NaNH_2$

HO

Br

H–C≡C:$^{\ominus}$ Na$^{\oplus}$

1) TsCl, py
2) *t*-BuOK

HBr
ROOR

H_2
Lindlar's
Catalyst

OH

1) BH_3• THF
2) H_2O_2, NaOH

12.12.

H–C≡C–H

$NaNH_2$

H–C≡C:$^{\ominus}$ Na$^{\oplus}$

HBr, ROOR

Br

H_2
Lindlar's
Catalyst

O
H

1) O_3
2) DMS

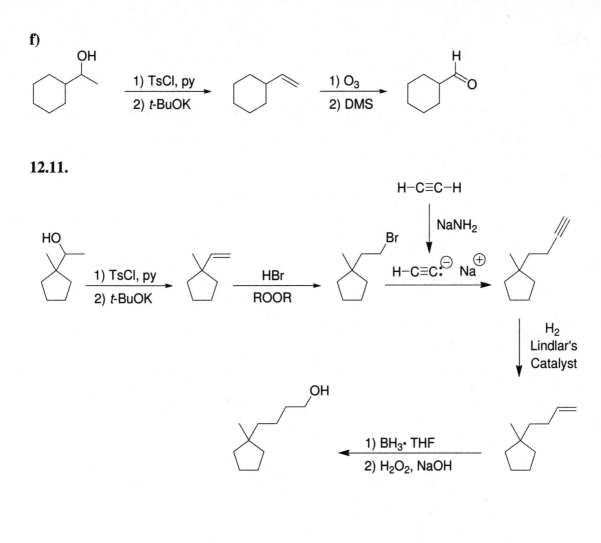

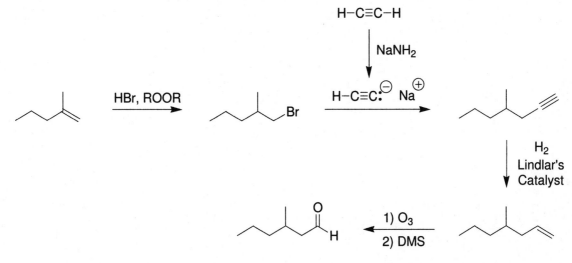

12.13.

a)

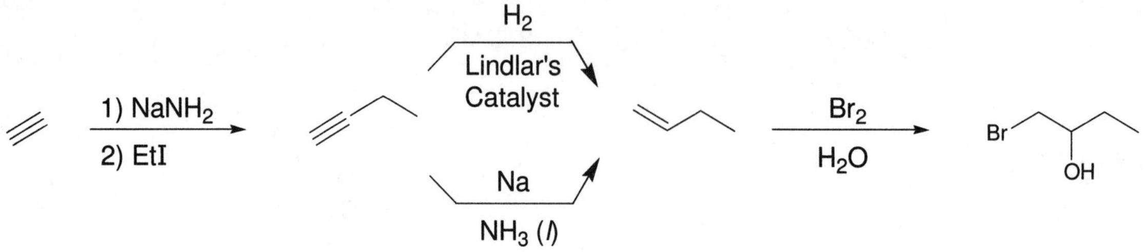

1) $NaNH_2$
2) EtI

H_2
Lindlar's Catalyst

Na
NH_3 (*l*)

Br_2
H_2O

Br
OH

b)

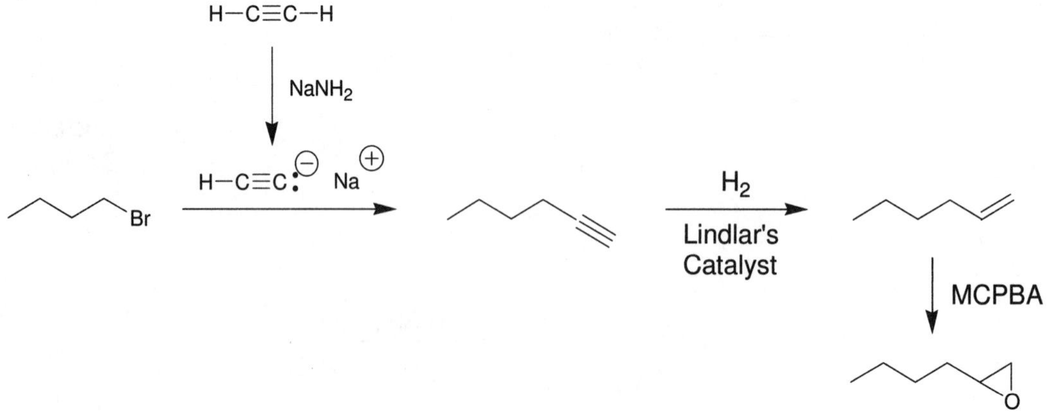

H—C≡C—H

$NaNH_2$

H—C≡C:⁻ Na⁺

Br

H_2
Lindlar's Catalyst

MCPBA

O

c)

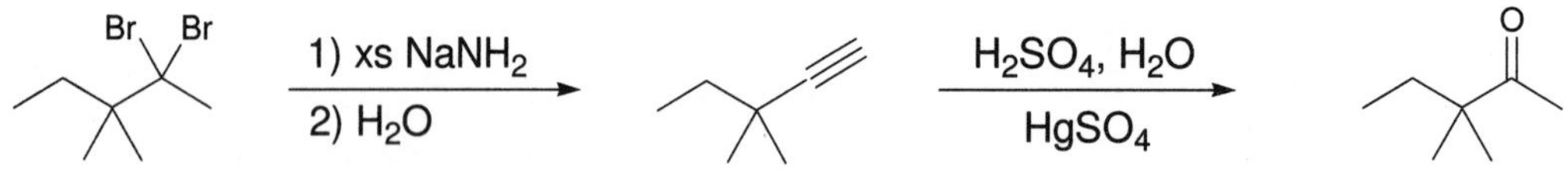

Br Br

1) xs $NaNH_2$
2) H_2O

H_2SO_4, H_2O
$HgSO_4$

O

d)

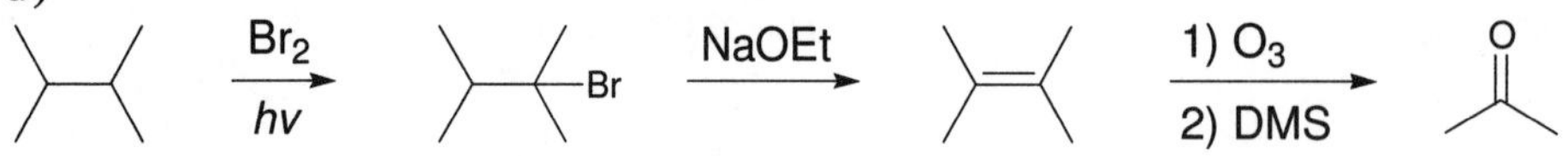

Br_2
hv

Br

NaOEt

1) O_3
2) DMS

O

e)

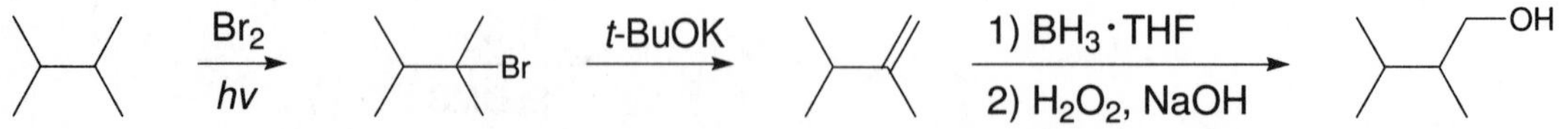

Br_2
hv

Br

t-BuOK

1) $BH_3 \cdot THF$
2) H_2O_2, NaOH

OH

f)

Br Br

1) xs $NaNH_2$
2) H_2O

1) O_3
2) H_2O

O
OH

g)

1) $NaNH_2$
2) EtI

1) $NaNH_2$
2) EtI

Na
NH_3 (*l*)

Br_2
H_2O

Br
OH
+ En

h)

Br_2
hv

Br

NaOEt

1) HBr, ROOR
2) *t*-BuOK

1) Br_2
2) xs $NaNH_2$
3) H_2O

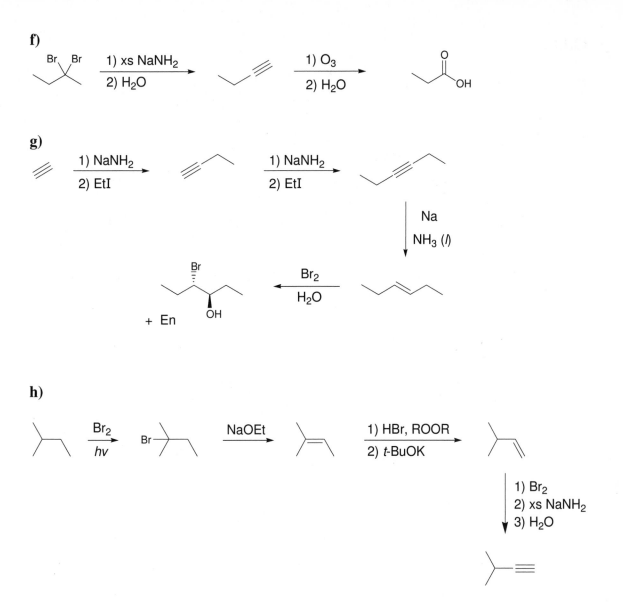

12.14.

H−C≡C−H

$NaNH_2$

H−C≡C:⊖ Na⊕

1) H_2, Lindlar's Catalyst

2) HBr

Br

H−C≡C:⊖ Na⊕

H_2
Lindlar's
Catalyst

HBr, ROOR

Br

H−C≡C:⊖ Na⊕

1) $NaNH_2$

2) Br

Na

NH_3 (*l*)

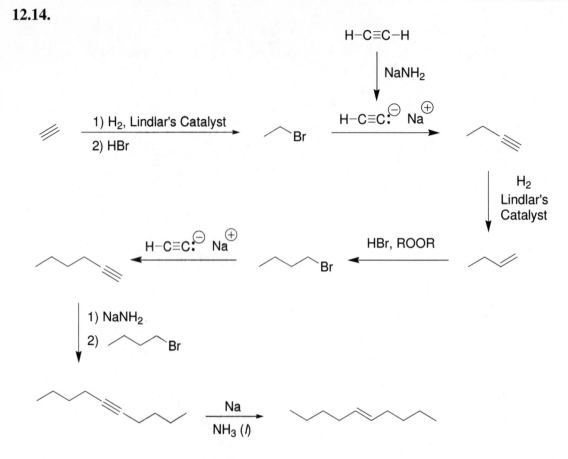

12.15.

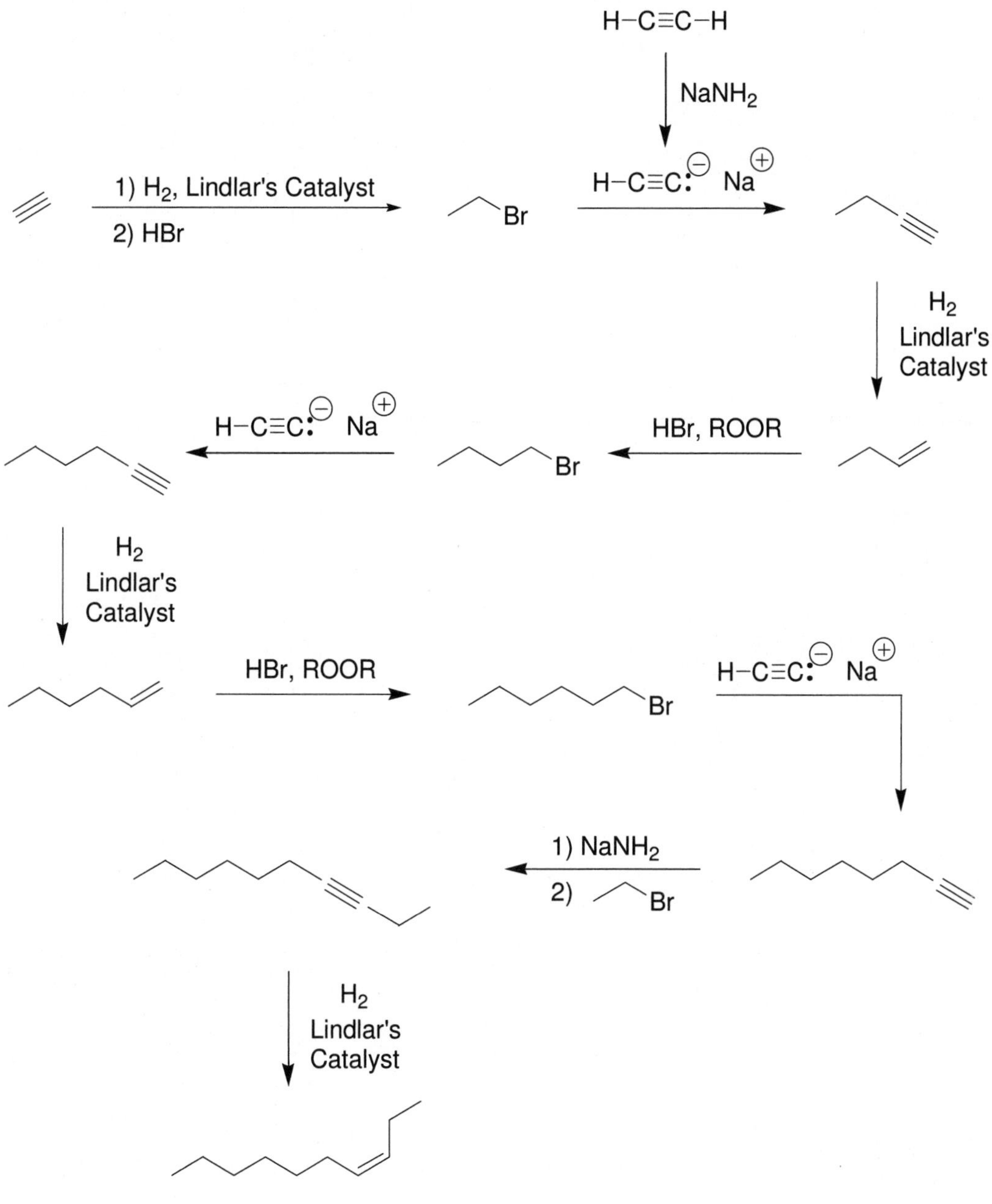

$H-C\equiv C-H$

$NaNH_2$

$H-C\equiv C:^{\ominus}\ Na^{\oplus}$

1) H_2, Lindlar's Catalyst

2) HBr

Br

$H-C\equiv C:^{\ominus}\ Na^{\oplus}$

H_2 Lindlar's Catalyst

HBr, ROOR

Br

$H-C\equiv C:^{\ominus}\ Na^{\oplus}$

H_2 Lindlar's Catalyst

HBr, ROOR

Br

$H-C\equiv C:^{\ominus}\ Na^{\oplus}$

1) $NaNH_2$

2) Br

H_2 Lindlar's Catalyst

12.16.

H−C≡C−H

$NaNH_2$

$H-C\equiv C:^{\ominus}$ $Na^{\oplus}$

1) H_2, Lindlar's Catalyst
2) HBr

Br

$H-C\equiv C:^{\ominus}$ $Na^{\oplus}$

H_2
Lindlar's Catalyst

HBr, ROOR

Br

$H-C\equiv C:^{\ominus}$ $Na^{\oplus}$

H_2
Lindlar's Catalyst

1) O_3
2) DMS

O
H

12.17.

Br_2
$h\nu$

Br

t-BuOK

H_2
Pt

Br

HBr
ROOR

H_3O^+

OH

OsO_4
NMO

OH
OH

1) $BH_3 \cdot THF$
2) H_2O_2, NaOH

OH

Br_2
H_2O

OH
Br

Br_2

Br
Br

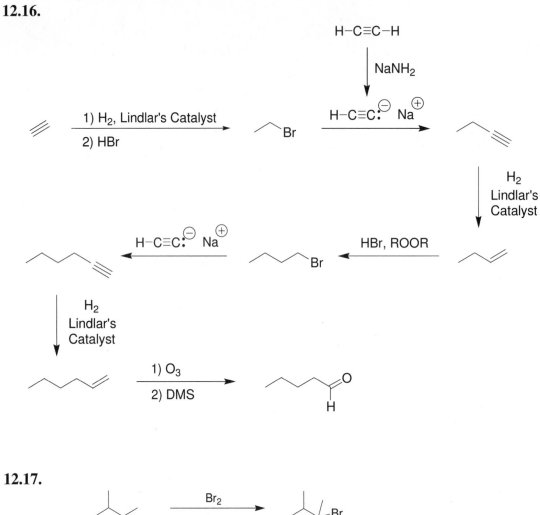

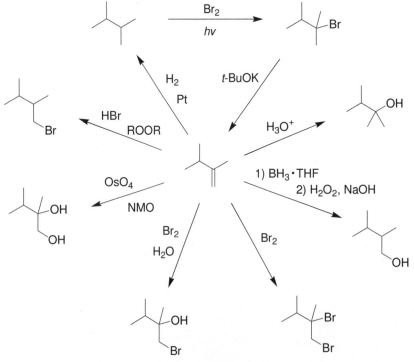

12.18.

H_2SO_4, H_2O
$HgSO_4$
HBr
HBr
1) 9-BBN
2) H_2O_2, NaOH
1) xs $NaNH_2$
2) H_2O
Br_2
xs HBr
xs Br_2
1) xs $NaNH_2$
2) H_2O
1) O_3
2) H_2O
1) $NaNH_2$
2) MeI
H_2
Lindlar's
Catalyst
H_2, Pt
Na, NH_3 (*l*)

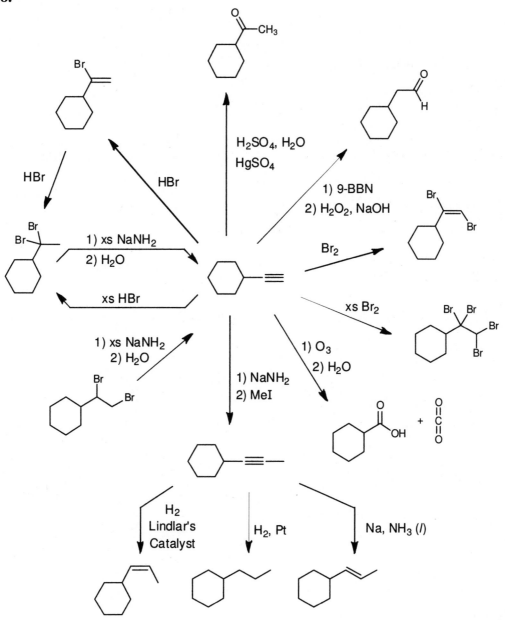

12.19.

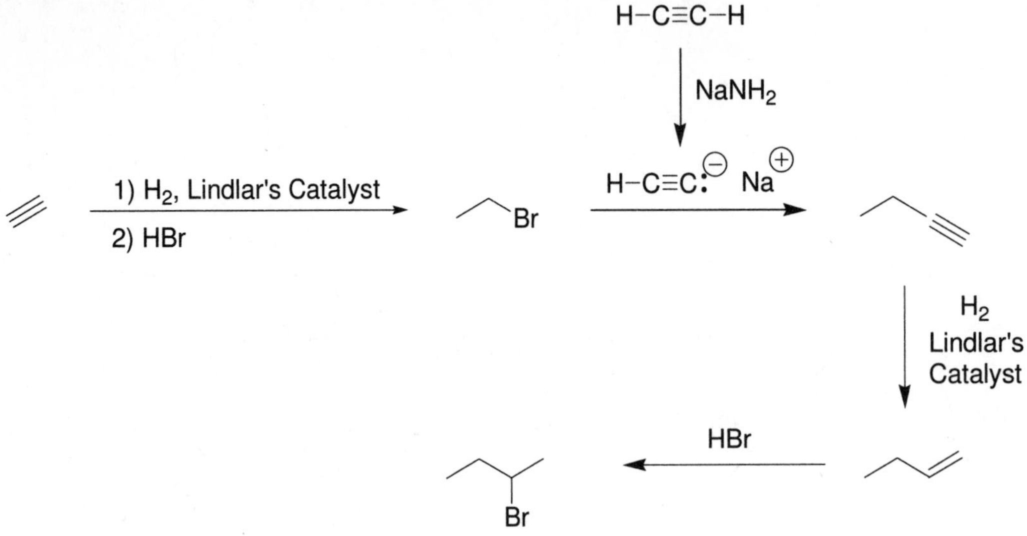

12.20.

$H-C\equiv C-H$

$NaNH_2$

$H-C\equiv C:^{\ominus}\ Na^{\oplus}$

1) H_2, Lindlar's Catalyst

2) HBr

Br

H_2 Lindlar's Catalyst

HBr, ROOR

Br

12.21.

a) OH — 1) TsCl, py 2) *t*-BuOK → 1) $BH_3 \cdot THF$ 2) H_2O_2, NaOH → OH

b) 1) Br_2 2) xs $NaNH_2$ 3) H_2O → H_2SO_4, H_2O $HgSO_4$ → O

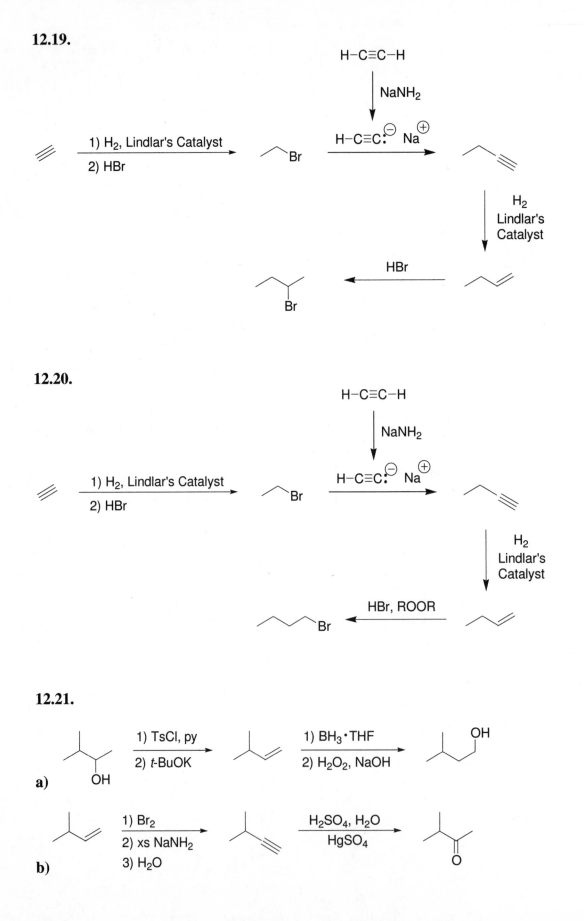

12.22.

1) HBr, ROOR
2) *t*-BuOK

12.23.

a)

1) Br_2
2) xs $NaNH_2$
3) H_2O

1) O_3
2) H_2O

OH
O

b)

Br
HC≡CNa

1) O_3
2) H_2O

O
OH

c)

Br
HC≡CNa

H_2
Lindlar's Catalyst

1) O_3
2) DMS

O
H

d)

Br_2
hv

Br

1) NaOMe
2) HBr, ROOR

Br

HC≡CNa

H_2
Lindlar's Catalyst

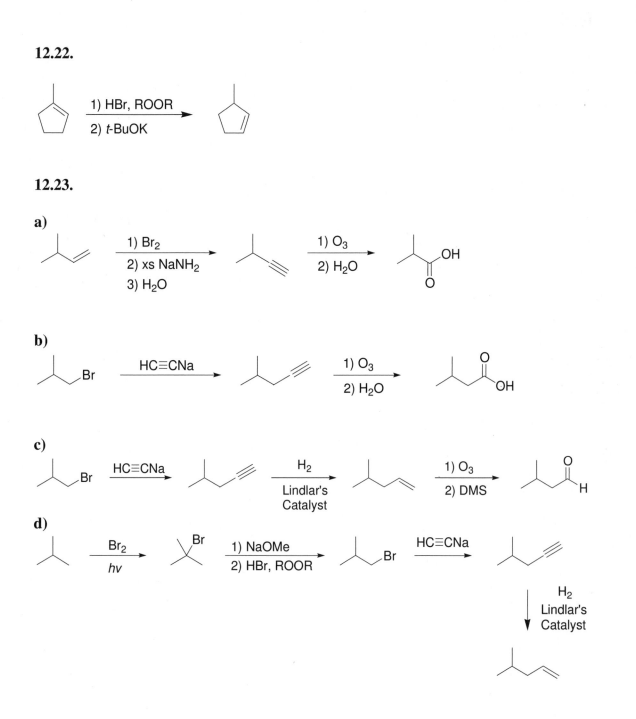

12.24.

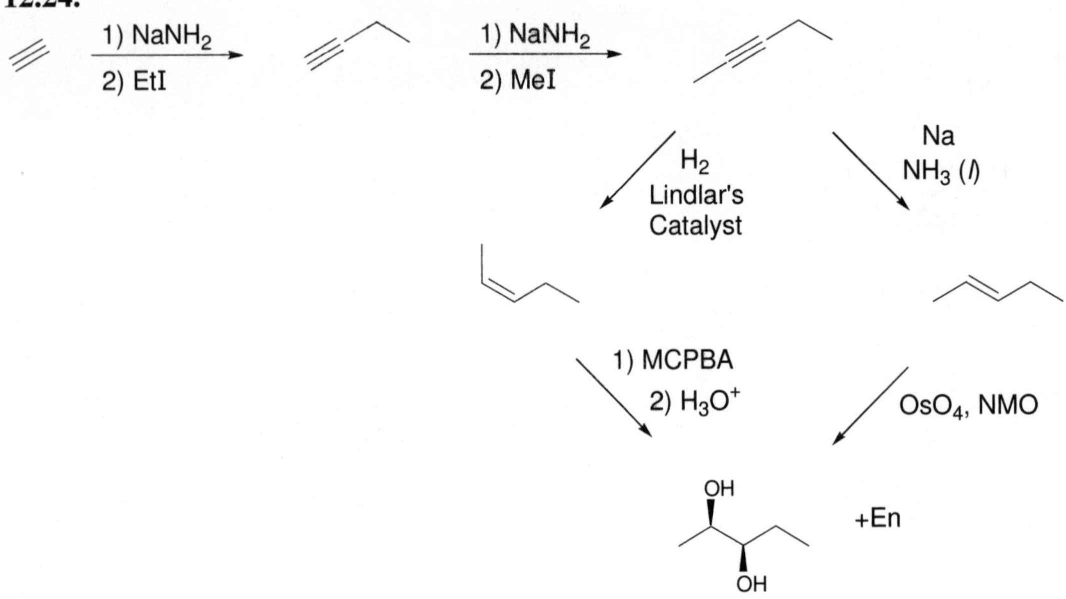

12.25.

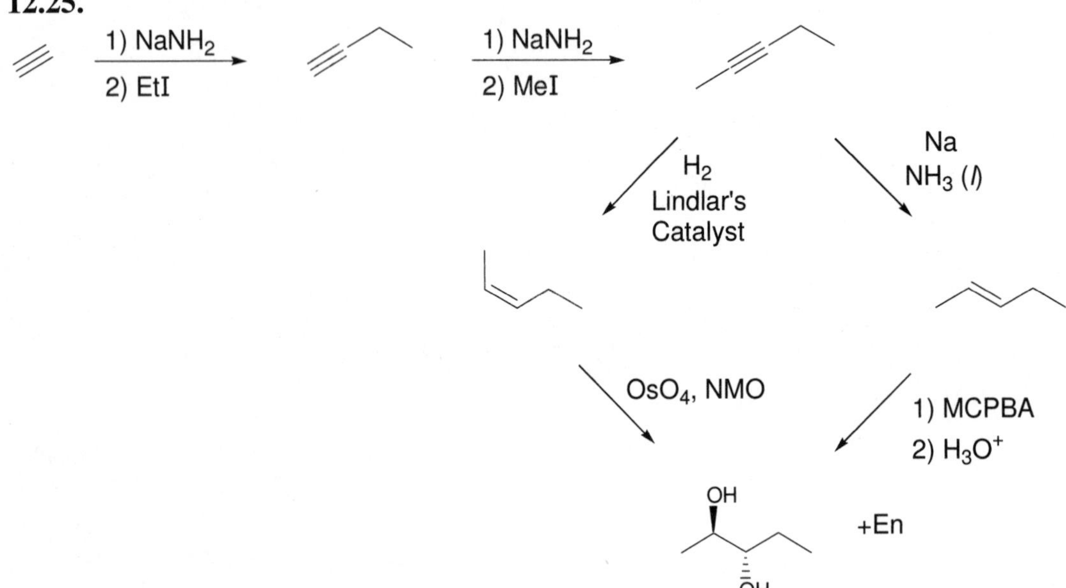

12.26.

a)

HBr / ROOR → Br → HC≡CNa → H_2SO_4, H_2O / $HgSO_4$ → O

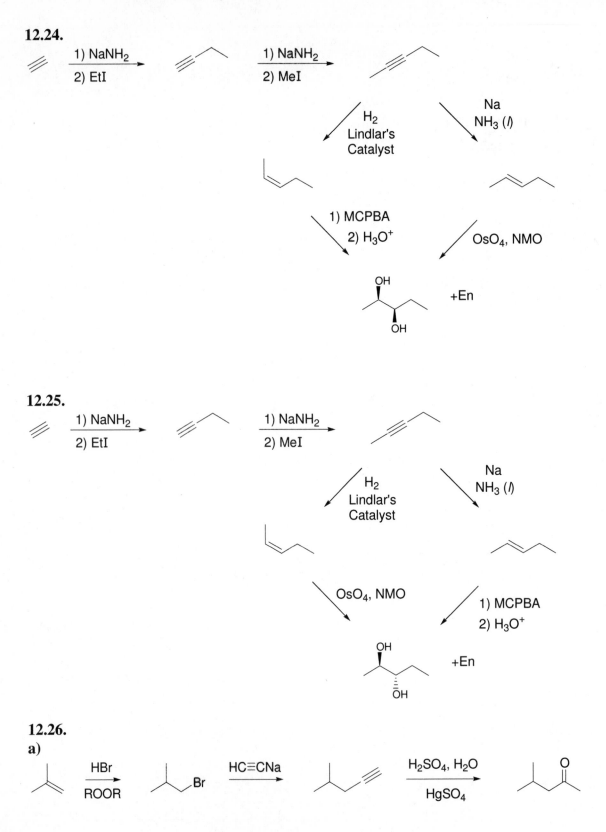

b)

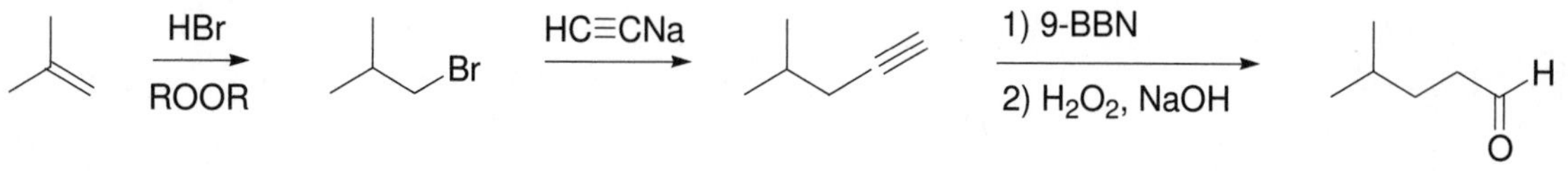

HBr / ROOR → Br; HC≡CNa → ; 1) 9-BBN 2) H_2O_2, NaOH → H, O

c)

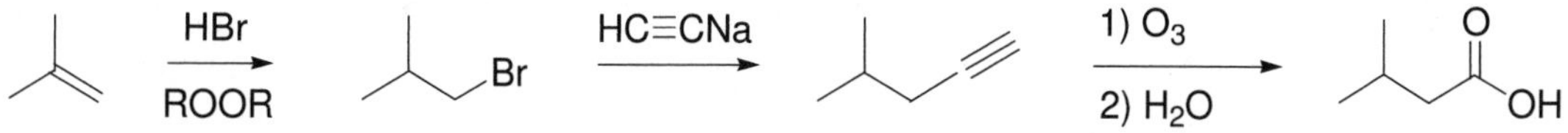

HBr / ROOR → Br; HC≡CNa → ; 1) O_3 2) H_2O → O, OH

d)

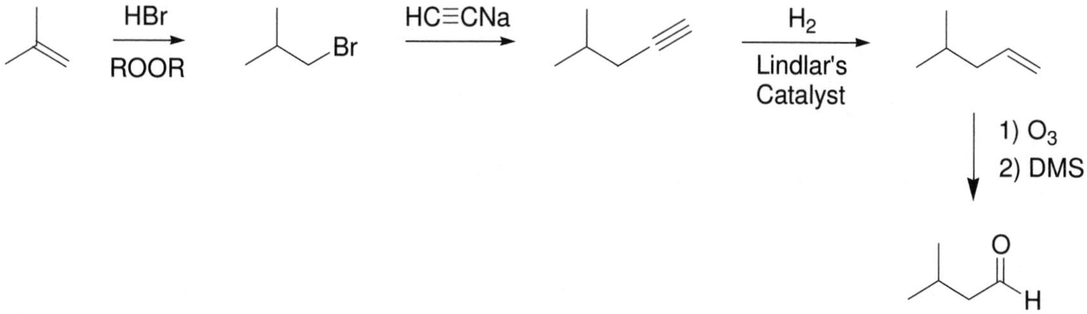

HBr / ROOR → Br; HC≡CNa → ; H_2 / Lindlar's Catalyst → ; 1) O_3 2) DMS → O, H

e)

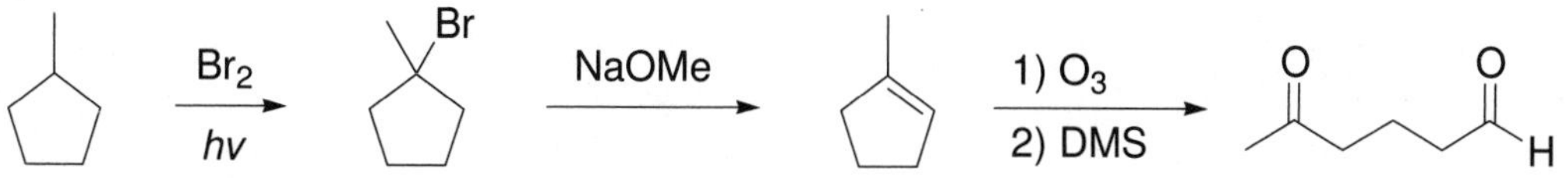

Br_2 / *hv* → Br; NaOMe → ; 1) O_3 2) DMS → O, O, H

12.27.

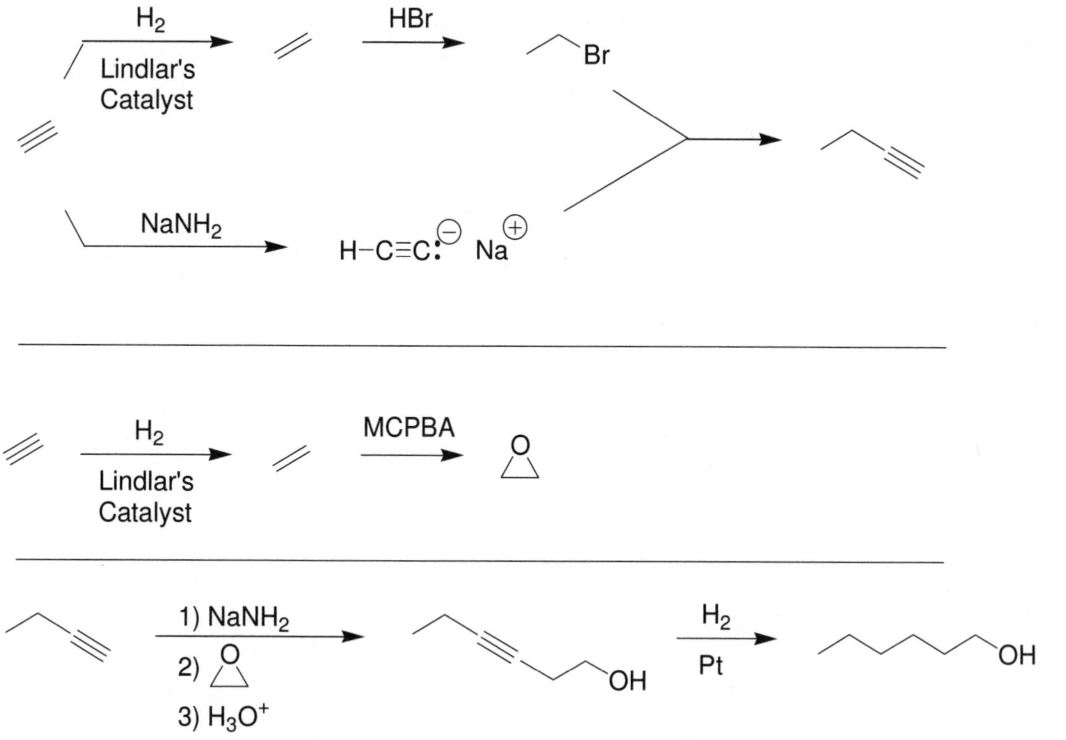

H_2 / Lindlar's Catalyst → ; HBr → Br

$NaNH_2$ → $H-C\equiv C:^{\ominus}$ $Na^{\oplus}$

H_2 / Lindlar's Catalyst → ; MCPBA → O

---

1) $NaNH_2$ 2) O 3) H_3O^+ → OH; H_2 / Pt → OH

12.28.

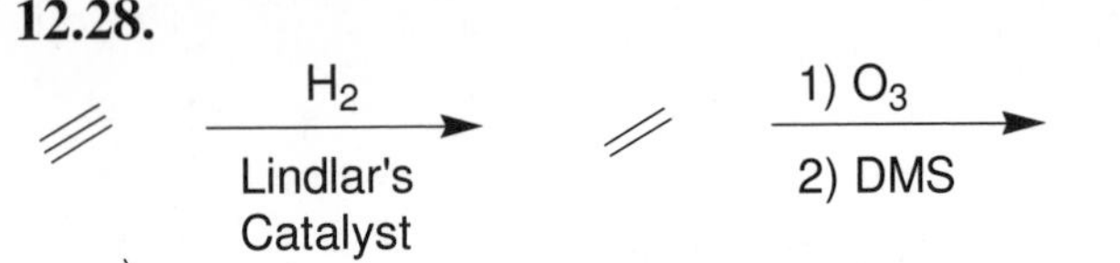

NaNH$_2$ → $H{-}C{\equiv}C{:}^{\ominus}$ $Na^{\oplus}$

O, H H → 1) $HC{\equiv}CNa$, 2) H_3O^+ → OH → H_2, Lindlar's Catalyst → OH

12.29.

HO → H_2, Lindlar's Catalyst → HO ≡ OH

↓ Br_2

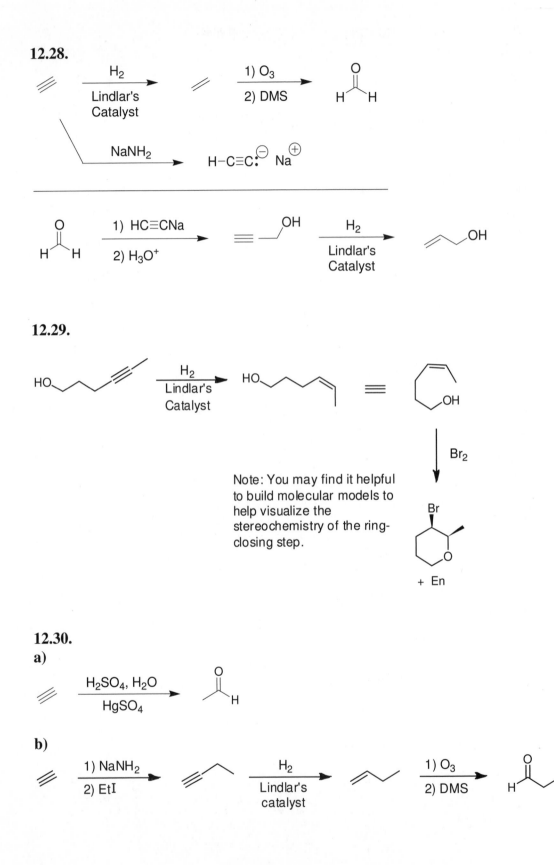

Note: You may find it helpful to build molecular models to help visualize the stereochemistry of the ring-closing step.

Br, O

+ En

12.30.

a)

→ H_2SO_4, H_2O, $HgSO_4$ → O, H

b)

→ 1) $NaNH_2$, 2) EtI → → H_2, Lindlar's catalyst → → 1) O_3, 2) DMS → O, H

c)

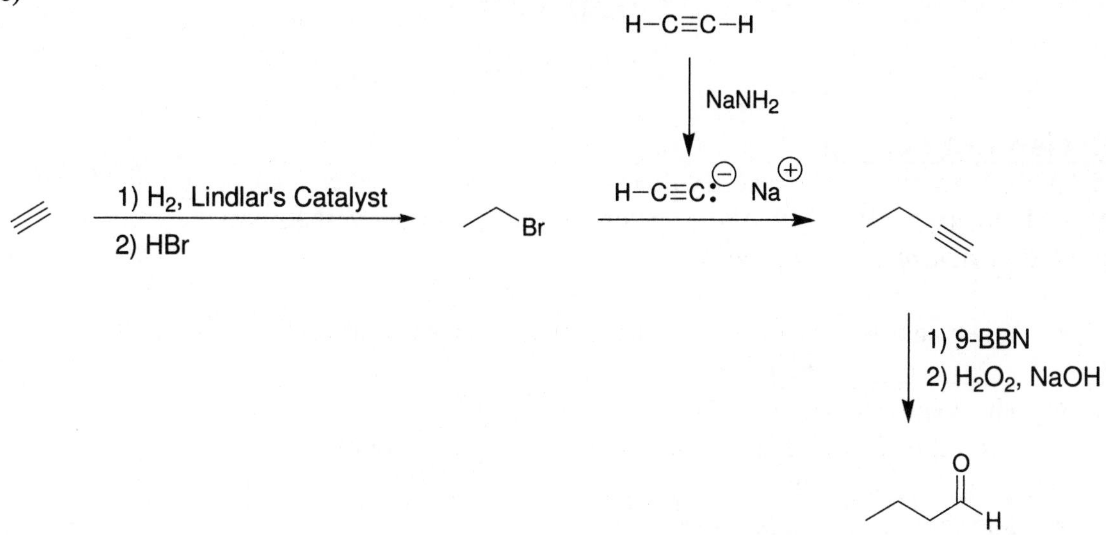

$H-C\equiv C-H$

$NaNH_2$

$H-C\equiv C:^{\ominus}\ Na^{\oplus}$

1) H_2, Lindlar's Catalyst

2) HBr

Br

1) 9-BBN

2) H_2O_2, NaOH

O

H

d)

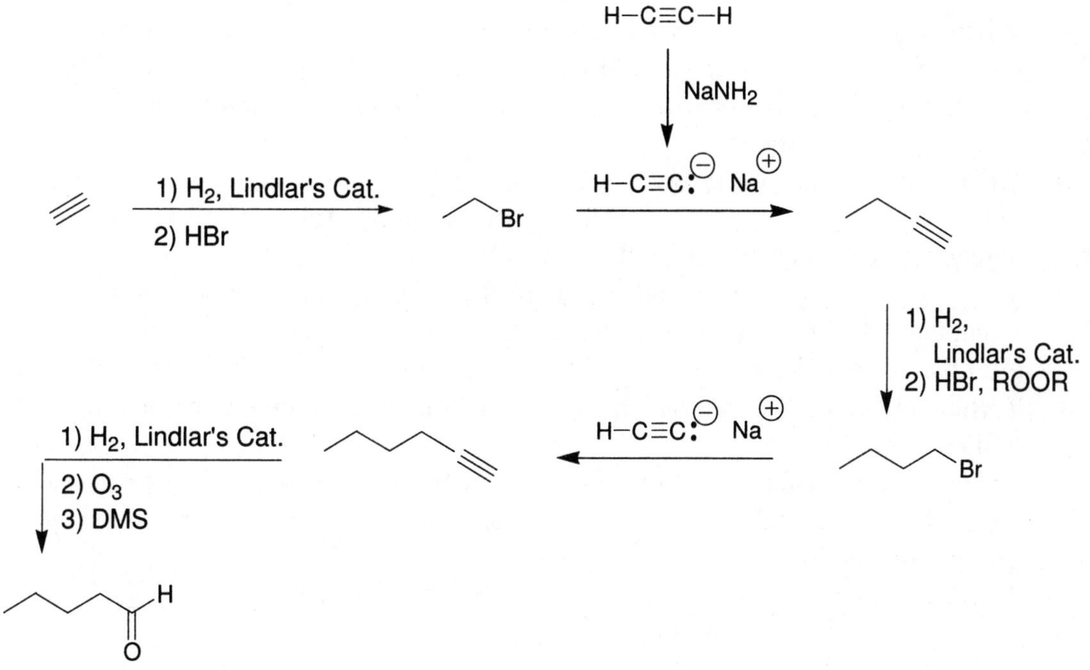

$H-C\equiv C-H$

$NaNH_2$

$H-C\equiv C:^{\ominus}\ Na^{\oplus}$

1) H_2, Lindlar's Cat.

2) HBr

Br

1) H_2, Lindlar's Cat.

2) HBr, ROOR

$H-C\equiv C:^{\ominus}\ Na^{\oplus}$

Br

1) H_2, Lindlar's Cat.

2) O_3

3) DMS

H

O

12.31.

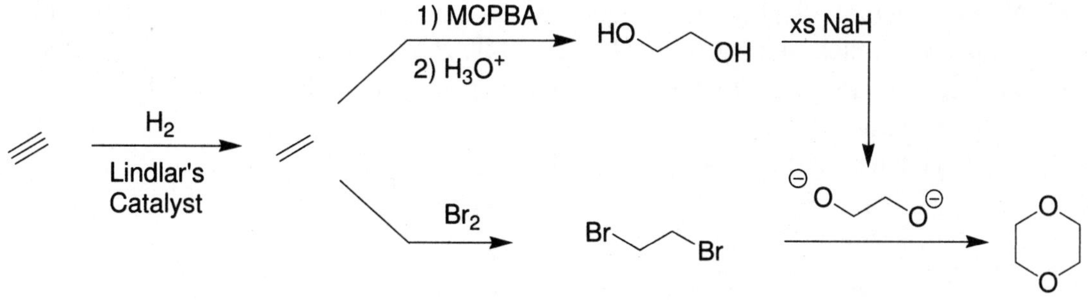

H_2

Lindlar's Catalyst

1) MCPBA

2) H_3O^+

HO OH

xs NaH

Br_2

Br Br

$^{\ominus}O$ $O^{\ominus}$

O

O

Chapter 13
Alcohols

Review of Concepts

Fill in the blanks below. To verify that your answers are correct, look in your textbook at the end of Chapter 13. Each of the sentences below appears verbatim in the section entitled *Review of Concepts and Vocabulary.*

- When naming an alcohol, the parent is the longest chain containing the __________ group.
- The conjugate base of an alcohol is called an ____________ ion.
- Several factors determine the relative acidity of alcohols, including ___________, ____________, and ______________________.
- The conjugate base of phenol is called a ____________, or _____________ ion.

- When preparing an alcohol via a substitution reaction, primary substrates will require S_N___ conditions, while tertiary substrates will require S_N___ conditions.
- Alcohols can be formed by treating a **carbonyl group** (C=O bond) with a ______________ **agent**.
- **Grignard reagents** are carbon nucleophiles that are capable of attacking a wide range of ________________, including the carbonyl group of ketones or aldehydes, to produce an alcohol.
- ______________ **groups**, such as the trimethylsilyl group, can be used to circumvent the problem of Grignard incompatibility and can be easily removed after the desired Grignard reaction has been performed.
- Tertiary alcohols will undergo an S_N___ reaction when treated with a hydrogen halide.
- Primary and secondary alcohols will undergo an S_N___ process when treated with either HX, $SOCl_2$, PBr_3, or when the hydroxyl group is converted into a tosylate group followed by nucleophilic attack.
- Tertiary alcohols undergo E1 elimination when treated with __________.
- Primary alcohols undergo **oxidation** twice to give a ____________________.
- Secondary alcohols are oxidized only once to give a ___________
- PCC is used to convert a primary alcohol into an _____________.
- NADH is a biological reducing agent that functions as a ____________ delivery agent (very much like $NaBH_4$ or LAH), while NAD^+ is an _____________ agent.
- The are two key issues to consider when proposing a synthesis is whether there is:
  1. a change in the __________________.
  2. a change in the ___________________.

Review of Skills

Fill in the blanks and empty boxes below. To verify that your answers are correct, look in your textbook at the end of Chapter 13. The answers appear in the section entitled *SkillBuilder Review*.

13.1 Naming an Alcohol

PROVIDE A SYSTEMATIC NAME FOR THE FOLLOWING COMPOUND

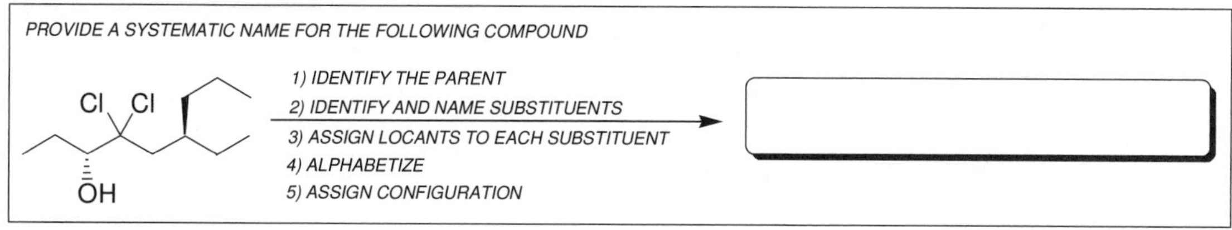

Cl
Cl
OH
1) IDENTIFY THE PARENT
2) IDENTIFY AND NAME SUBSTITUENTS
3) ASSIGN LOCANTS TO EACH SUBSTITUENT
4) ALPHABETIZE
5) ASSIGN CONFIGURATION

13.2 Comparing the Acidity of Alcohols

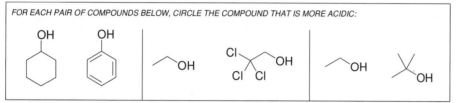

FOR EACH PAIR OF COMPOUNDS BELOW, CIRCLE THE COMPOUND THAT IS MORE ACIDIC:

13.3 Identifying Oxidation and Reduction Reactions

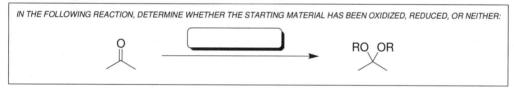

IN THE FOLLOWING REACTION, DETERMINE WHETHER THE STARTING MATERIAL HAS BEEN OXIDIZED, REDUCED, OR NEITHER:

13.4 Drawing a Mechanism, and Predicting the Products of Hydride Reductions

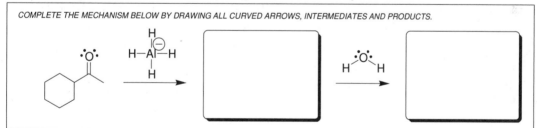

COMPLETE THE MECHANISM BELOW BY DRAWING ALL CURVED ARROWS, INTERMEDIATES AND PRODUCTS.

13.5 Preparing an Alcohol via a Grignard Reaction

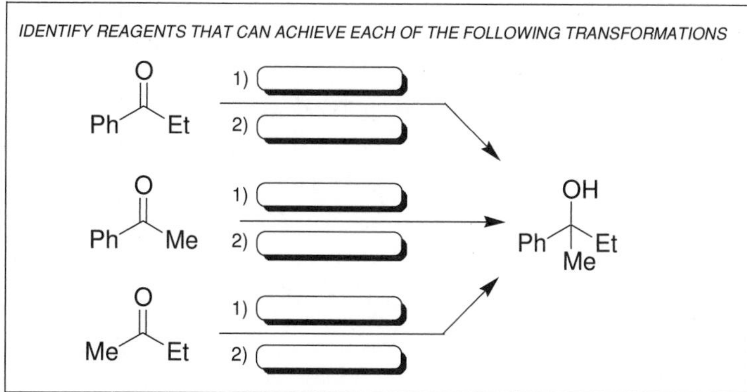

13.6 Proposing Reagents for the Conversion of an Alcohol into an Alkyl Halide

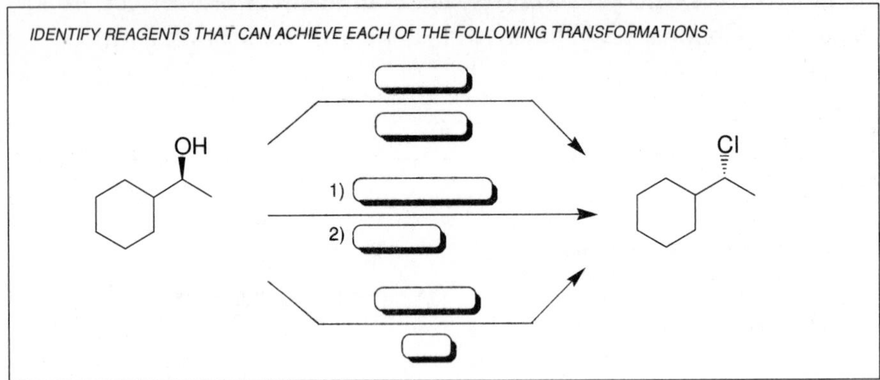
IDENTIFY REAGENTS THAT CAN ACHIEVE EACH OF THE FOLLOWING TRANSFORMATIONS
OH
1)
2)
Cl

13.7 Predicting the Products of an Oxidation Reaction

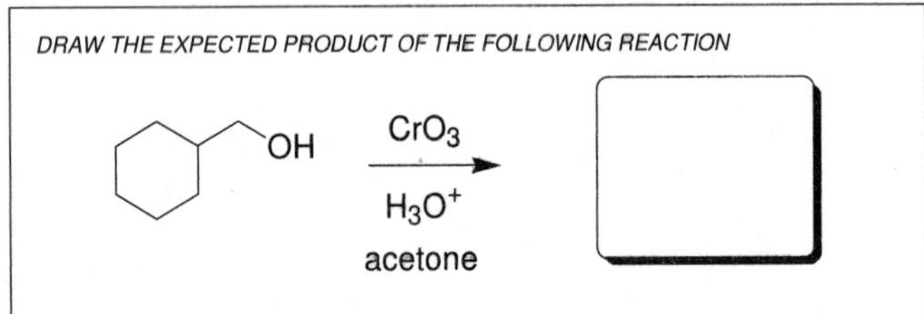
DRAW THE EXPECTED PRODUCT OF THE FOLLOWING REACTION
OH
CrO3
H3O+
acetone

13.8 Converting Functional Groups

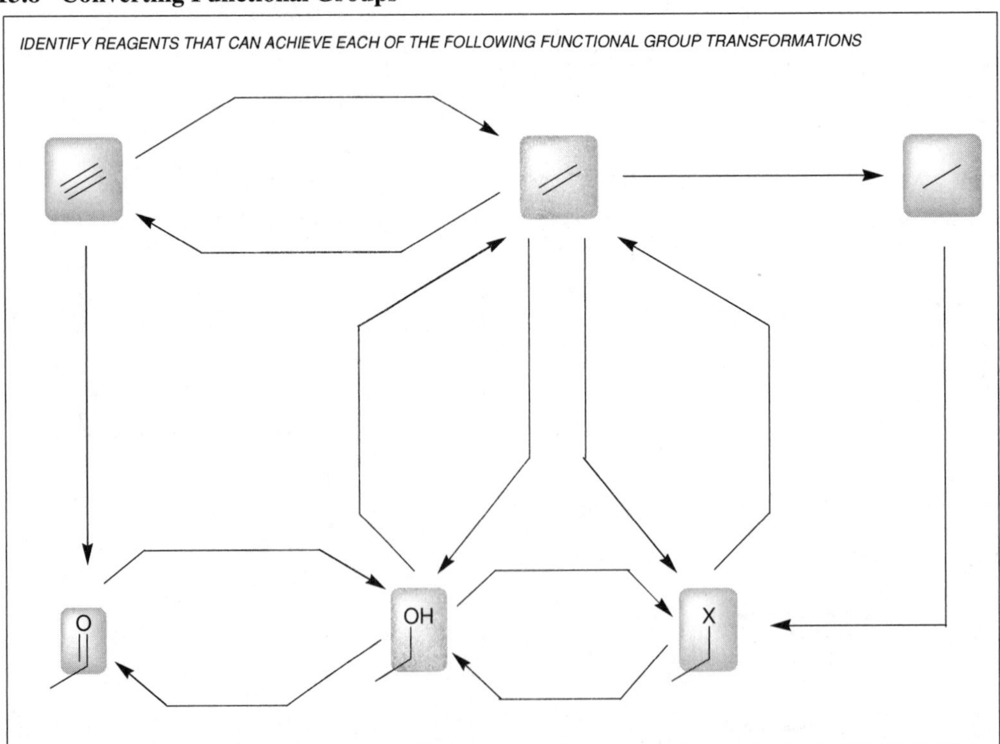
IDENTIFY REAGENTS THAT CAN ACHIEVE EACH OF THE FOLLOWING FUNCTIONAL GROUP TRANSFORMATIONS
O
OH
X

13.9 Proposing a Synthesis

AS A GUIDE FOR PROPOSING A SYNTHESIS, ASK THE FOLLOWING TWO QUESTIONS:
1) IS THERE A CHANGE IN THE ____________ SKELETON?
2) IS THERE A CHANGE IN THE LOCATION OR IDENTITY OF THE ____________________?
AFTER PROPOSING A SYNTHESIS, USE THE FOLLOWING TWO QUESTIONS TO ANALYZE YOUR ANSWER:
1) IS THE ____________________ OUTCOME OF EACH STEP CORRECT?
2) IS THE ____________________ OUTCOME OF EACH STEP CORRECT?

Review of Reactions

Identify the reagents necessary to achieve each of the following transformations. To verify that your answers are correct, look in your textbook at the end of Chapter 13. The answers appear in the section entitled *Review of Reactions*.

Preparation of Alkoxides

RÖH ⟶ RÖ:⊖ Na⊕

Preparation of Alcohols via Reduction

R(C=O)H ⟶ RCH₂OH

R(C=O)R ⟶ R–CH(OH)–R (two routes)

R(C=O)OH ⟶ RCH₂OH

R(C=O)OMe ⟶ RCH₂OH + MeOH

Preparation of Alcohols via Grignard Reagents

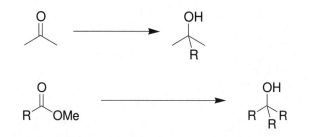

(CH₃)₂C=O ⟶ (CH₃)₂C(OH)R

R(C=O)OMe ⟶ R₃C–OH

Protection and Deprotection of Alcohols

R–OH ⇄ R–O–TMS

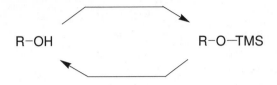

S_N1 Reactions with Alcohols

R_3C–OH → R_3C–X + H_2O

S_N2 Reactions with Alcohols

OH → Br

OH → Cl

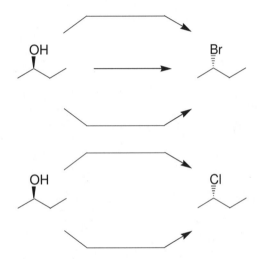

E1 and E2 Reactions with Alcohols

–OH → + H_2O

–OH → –OTs →

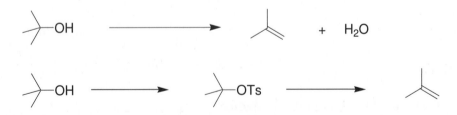

Oxidation of Alcohols and Phenols

OH, R, R → O, R, R

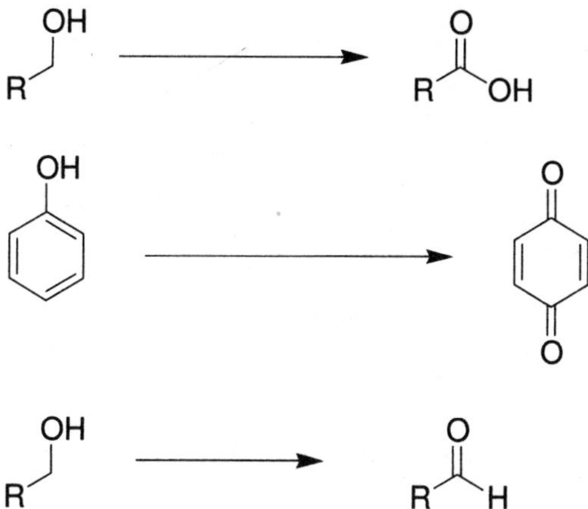

Solutions

13.1.

a) 5,5-dibromo-2-methylhexan-2-ol
b) (2*S*,3*R*)-2,3,4-trimethylpentan-1-ol
c) 2,2,5,5-tetramethylcyclopentanol
d) 2,6-diethylphenol
e) (*S*)-2,2,4,4-tetramethylcyclohexanol

13.2.

OH

Br
a) Br

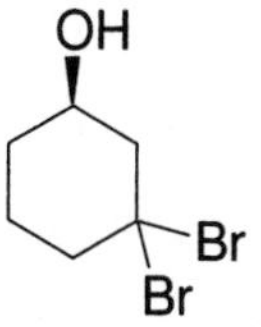

b) OH

OH
c)

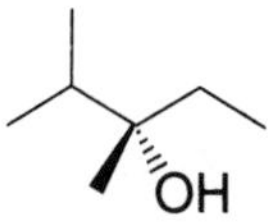

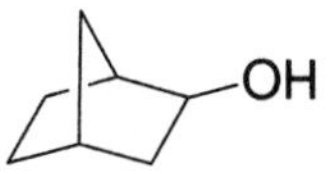

13.3. Nonyl mandelate has a longer alkyl chain than octyl mandelate and is therefore more effective at penetrating cell membranes, rendering it a more potent agent. Nonyl mandelate has a shorter alkyl chain than decyl mandelate and is therefore more water-soluble, enabling it to be transported through aqueous media and to reach its target destination more effectively.

13.4.

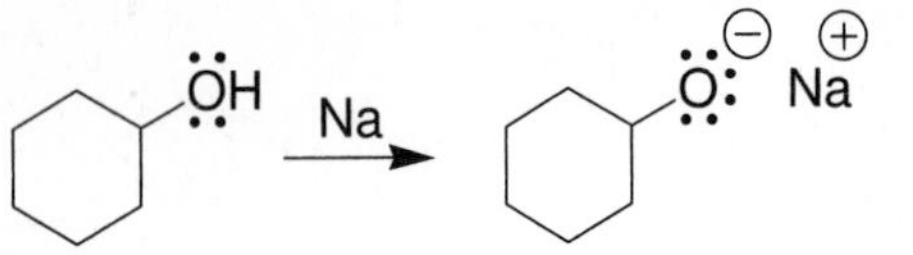

a) Cyclohexanol + Na → sodium cyclohexoxide (O^{-} Na^{+})

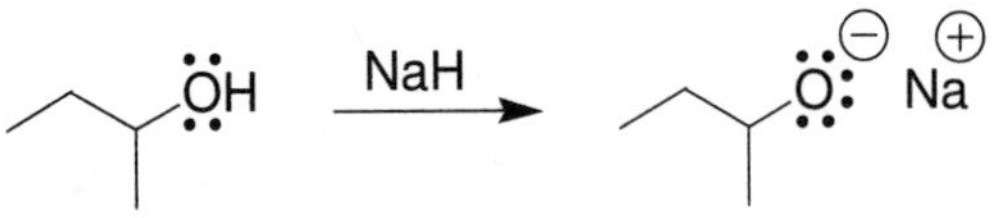

b) Secondary alcohol + NaH → alkoxide (O^{-} Na^{+})

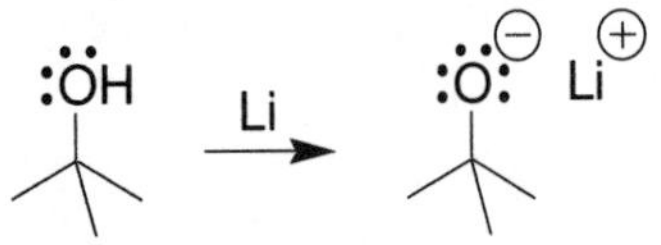

c) tert-Butanol + Li → lithium tert-butoxide (O^{-} Li^{+})

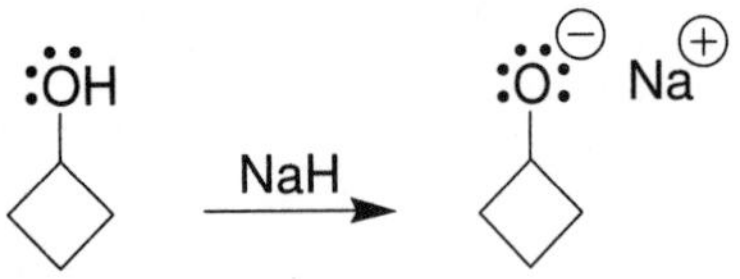

d) Cyclobutanol + NaH → sodium cyclobutoxide (O^{-} Na^{+})

13.5.

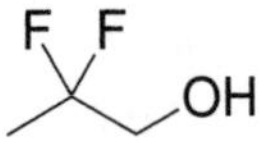

a) (2,2-difluoro alcohol: F, F, OH)

The electron-withdrawing effects of the fluorine atoms stabilize the conjugate base.

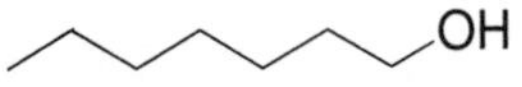

b) (primary alcohol, OH)

The conjugate base of a primary alcohol will be more easily solvated than the conjugate base of a tertiary alcohol.

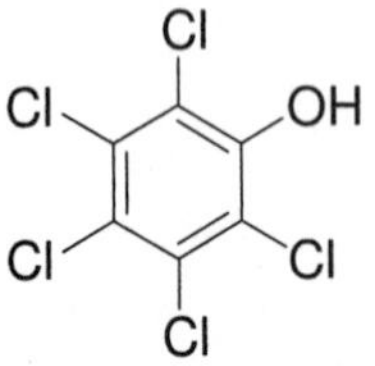

c) (pentachlorophenol: Cl, Cl, Cl, Cl, Cl, OH)

The electron-withdrawing effects of the chlorine atoms stabilize the conjugate base.

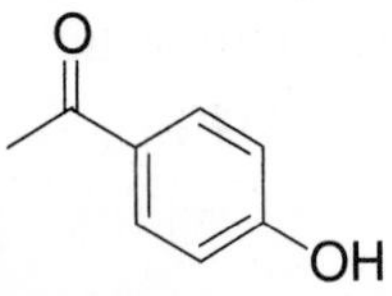

d) (4-hydroxyacetophenone: O, OH)

The conjugate base is more highly stabilized by resonance, with the negative charge spread over two oxygen atoms, rather than just one oxygen atom.

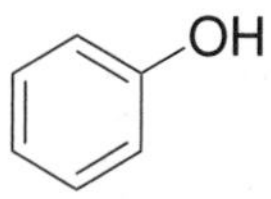

OH

e)

The conjugate base is stabilized by resonance.

13.6. 2-nitrophenol is expected to be more acidic (lower pK_a) because the conjugate base has a resonance structure in which the negative charge is spread onto an oxygen atom of the nitro group, shown below. In contrast, 3-nitrophenol does not have such a resonance structure:

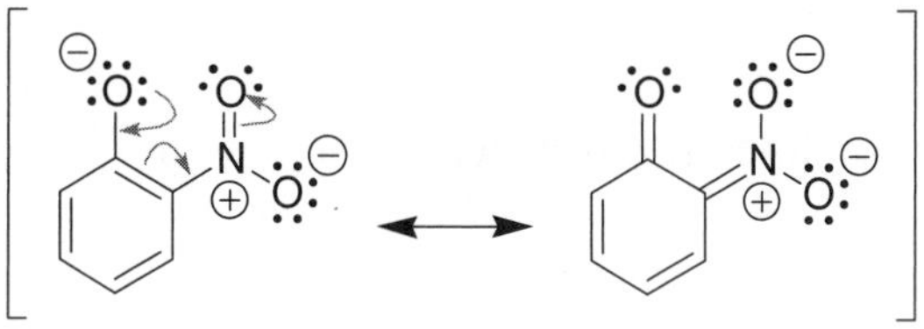

13.7.

a) Br $\xrightarrow{H_2O}$ OH

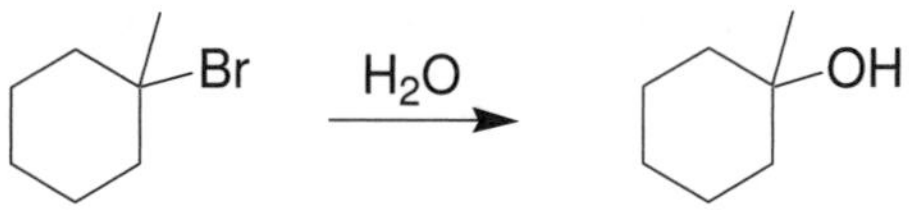

b) Br $\xrightarrow{NaOH}$ OH

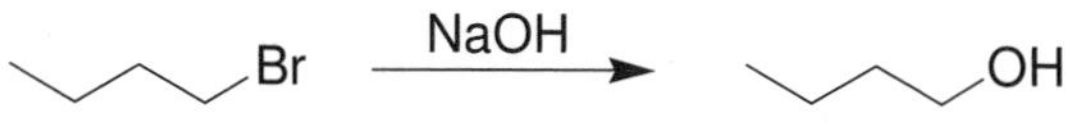

c) $\xrightarrow{\text{dilute } H_2SO_4}$ OH

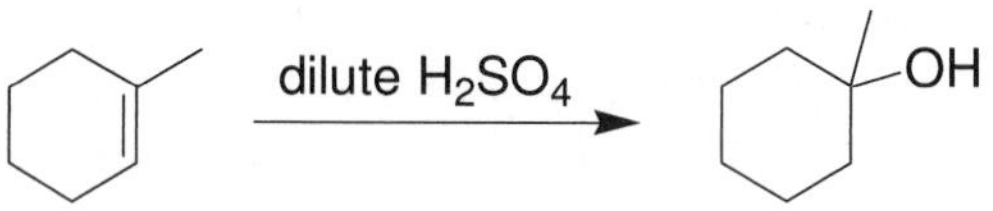

d) 1) $BH_3 \cdot THF$ 2) H_2O_2, NaOH → OH

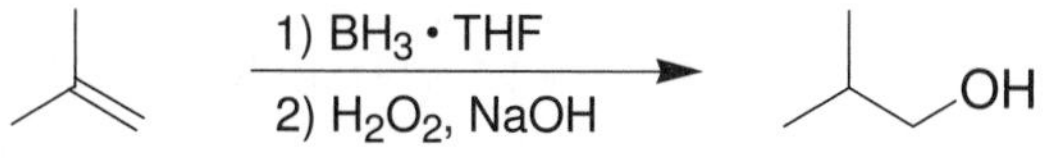

e) 1) $Hg(OAc)_2$, H_2O 2) $NaBH_4$ → OH

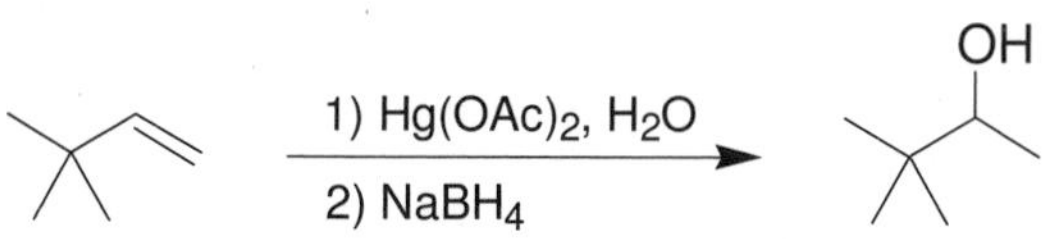

f) 1) $BH_3 \cdot THF$ 2) H_2O_2, NaOH → HO

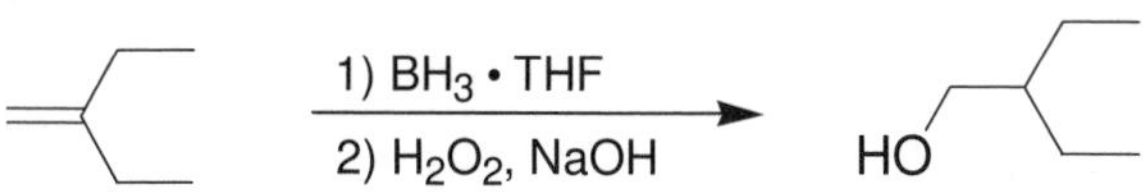

13.8

a) 1) $BH_3 \cdot THF$ 2) H_2O_2, NaOH → OH

b) 1) $Hg(OAc)_2$, H_2O 2) $NaBH_4$ → OH

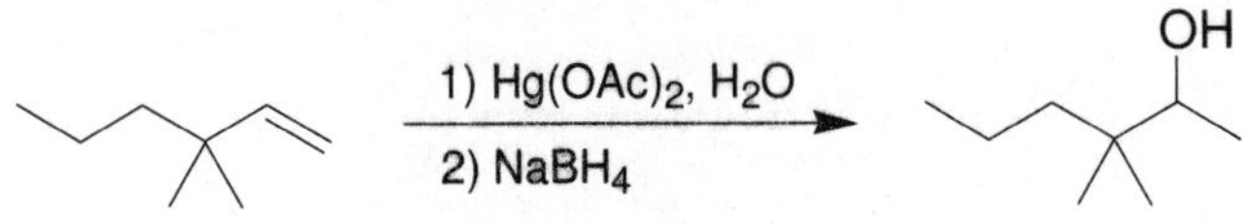

c) dilute H_2SO_4 → OH

13.9.

a) (+2) → (+2). The starting material is neither oxidized nor reduced.

b) (+1) → (+3). The starting material is oxidized.

c) (+3) → (-1). The starting material is reduced.

d) (+3) → (+3). The starting material is neither oxidized nor reduced.

e) (0) → (+2). The starting material is oxidized.

f) (+2) → (+3). The starting material is oxidized.

13.10. One carbon atom is reduced from an oxidation state of 0 to an oxidation state of -1, while the other carbon atom is oxidized from an oxidation state of 0 to an oxidation state of +1. Overall, the starting material does not undergo a net change in oxidation state and is, therefore, neither reduced nor oxidized.

13.11. One carbon atom is reduced from an oxidation state of 0 to an oxidation state of -2, while the other carbon atom is oxidized from an oxidation state of 0 to an oxidation state of +2. Overall, the starting material does not undergo a net change in oxidation state and is, therefore, neither reduced nor oxidized.

13.12.

a)

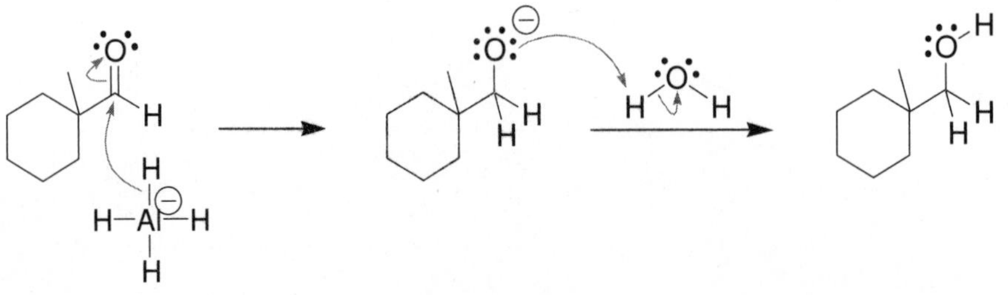

:O: H H H–Al–H H → :O:⊖ H H H–O–H → :O–H H H

b)

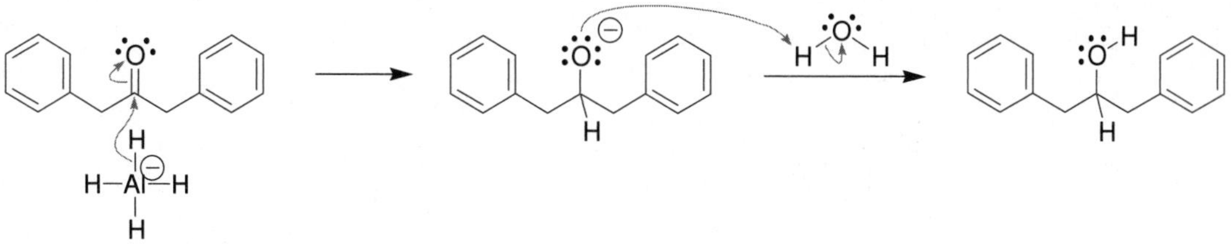

c)

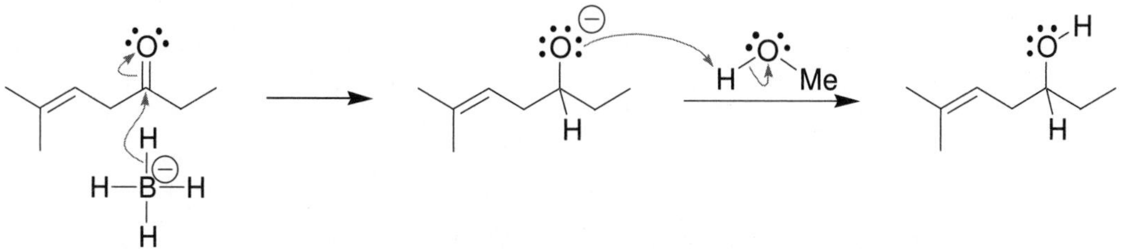

d)

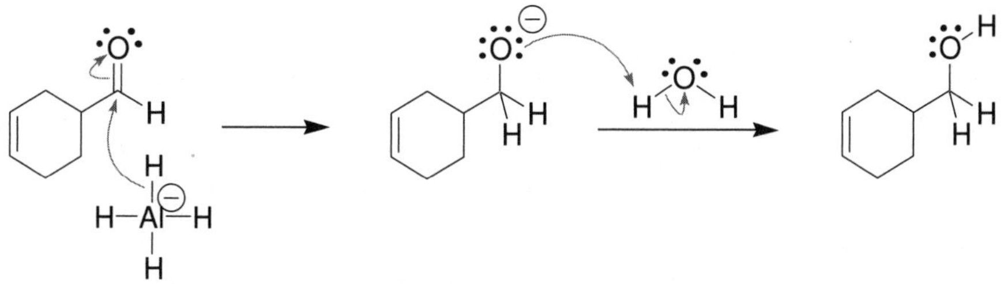

e)

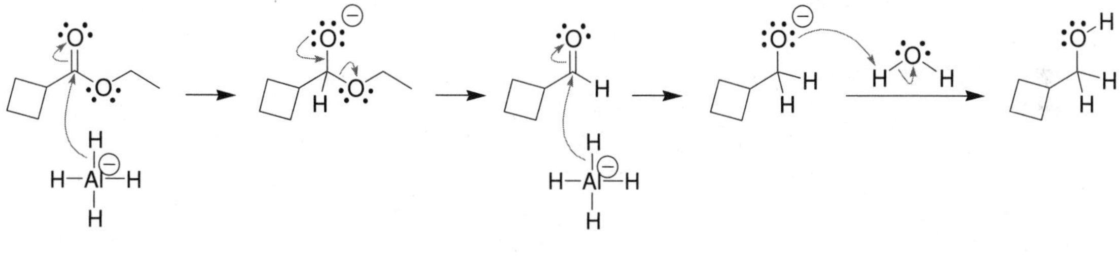

f)

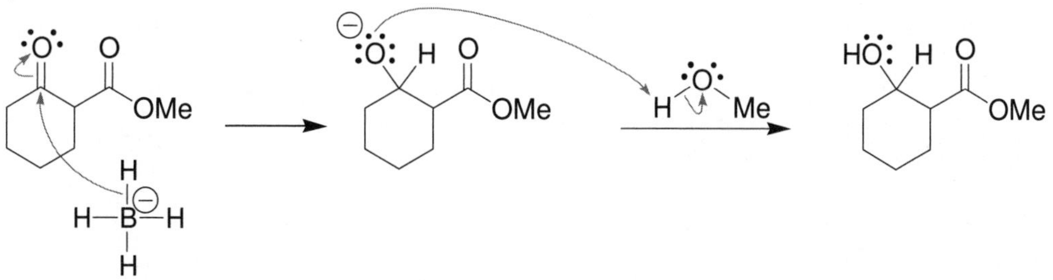

13.13.

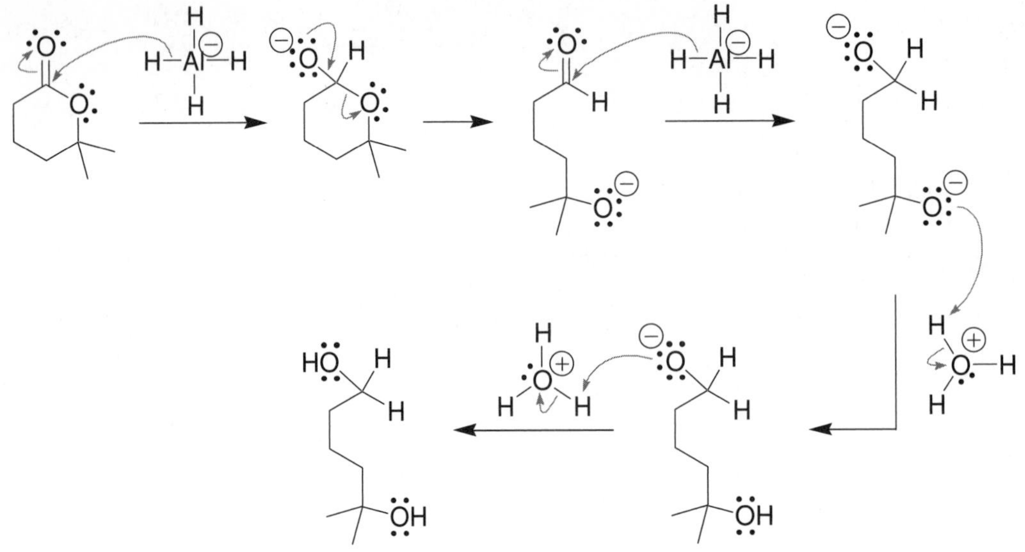

H–Al–H (H, H) ⊖

13.14.

a)

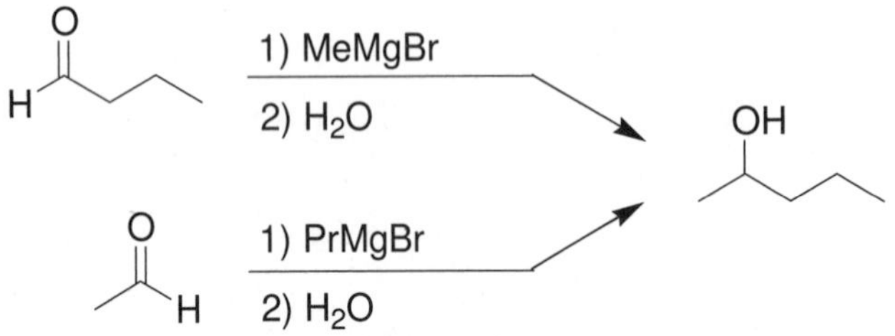

1) MeMgBr
2) H_2O

OH

1) PrMgBr
2) H_2O

b)

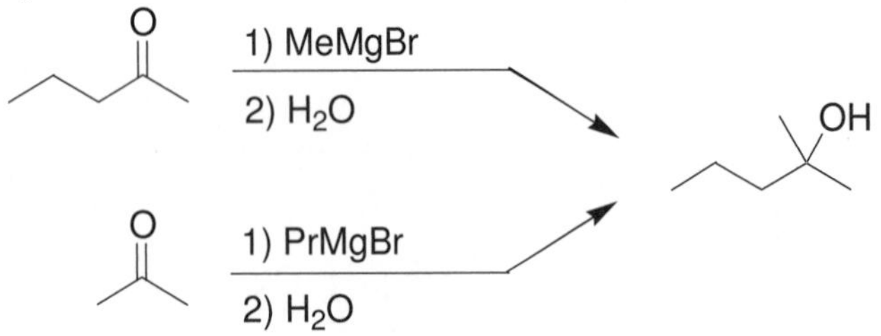

1) MeMgBr
2) H_2O

OH

1) PrMgBr
2) H_2O

c)

1) PrMgBr
2) H_2O

OH

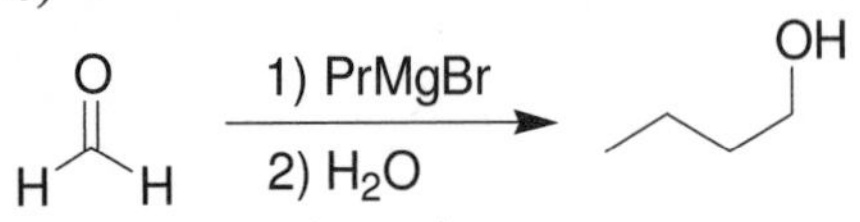

d)

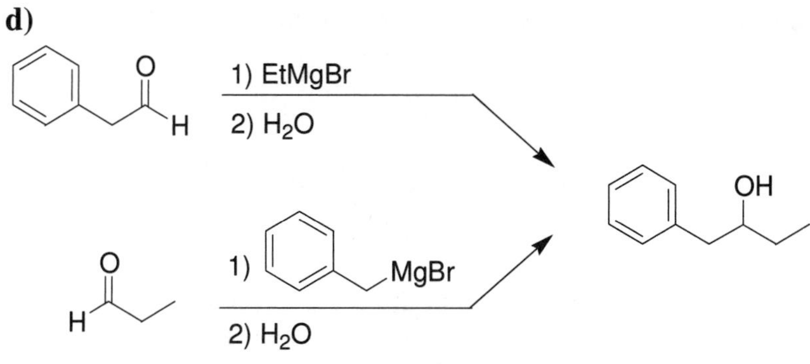

e)

O
1) MeMgBr
2) H_2O

O
1) EtMgBr
2) H_2O

OH

O
1) BuMgBr
2) H_2O

f)

O
1) MeMgBr
2) H_2O

OH

O
1)
MgBr
2) H_2O

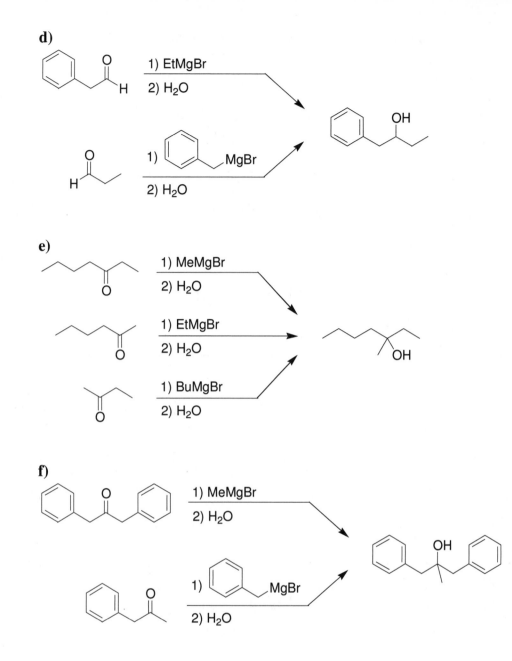

13.15 Each of the following two compounds can be prepared from the reaction between a Grignard reagent and an ester, because each of these compounds has two identical groups connected to the α position:

Me OH
Me

OH

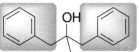

The other four compounds from Problem 13.14 do not contain two identical groups connected to the α position, and cannot be prepared from the reaction between an ester and a Grignard reagent.

13.16 Each of the following three compounds can be prepared from the reaction between a hydride reducing agent ($NaBH_4$ or LAH) and a ketone or aldehyde, because each of these compounds has a hydrogen atom connected to the α position:

OH OH OH

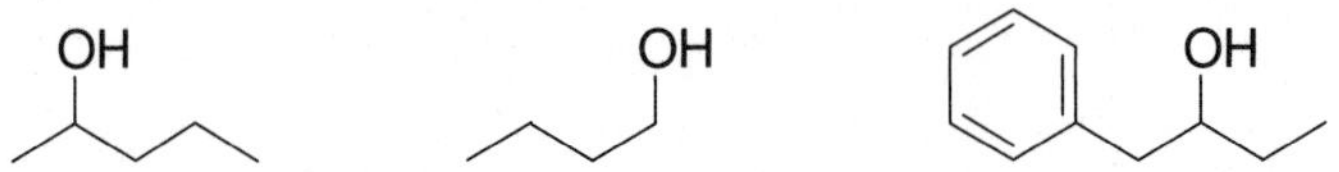

The other three compounds from Problem 13.14 do not contain a hydrogen atom connected to the α position and, therefore, cannot be prepared from the reaction between a hydride reducing agent ($NaBH_4$ or LAH) and a ketone or aldehyde.

13.17.

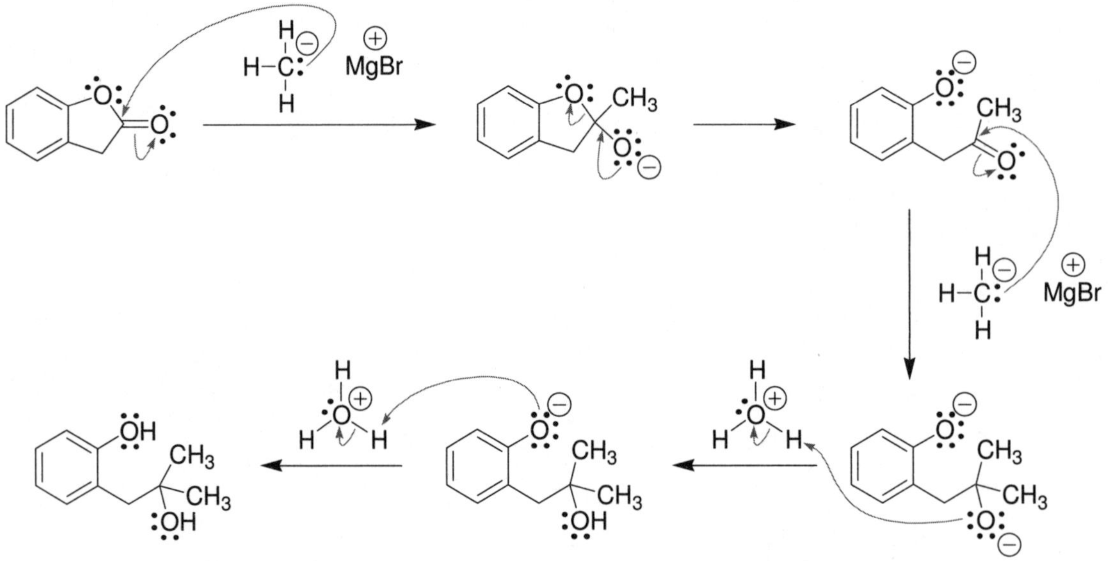

H–C–H (H) MgBr ⊕ CH₃ CH₃ CH₃ CH₃ OH

In this case, H_3O^+ must be used as a proton source because water is not sufficiently acidic to protonate a phenolate ion (see Section 13.2, Acidity of Alcohols and Phenols).

13.18.
a)

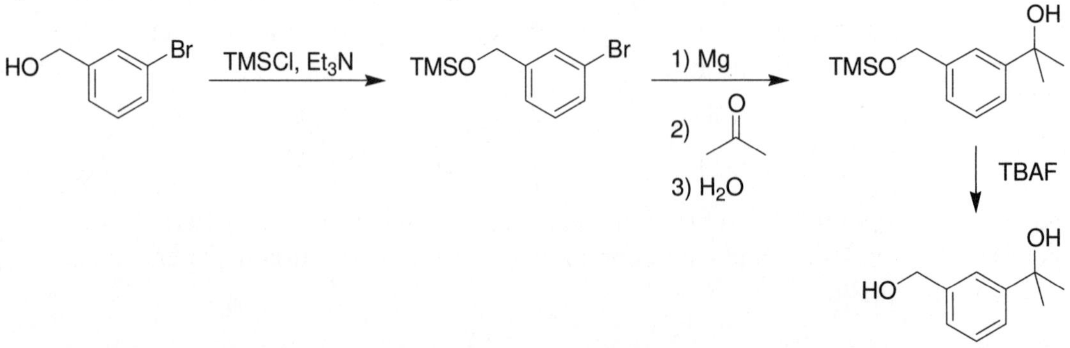

HO Br → TMSCl, Et_3N → TMSO Br → 1) Mg 2) acetone 3) H_2O → TMSO OH → TBAF → HO OH

b)

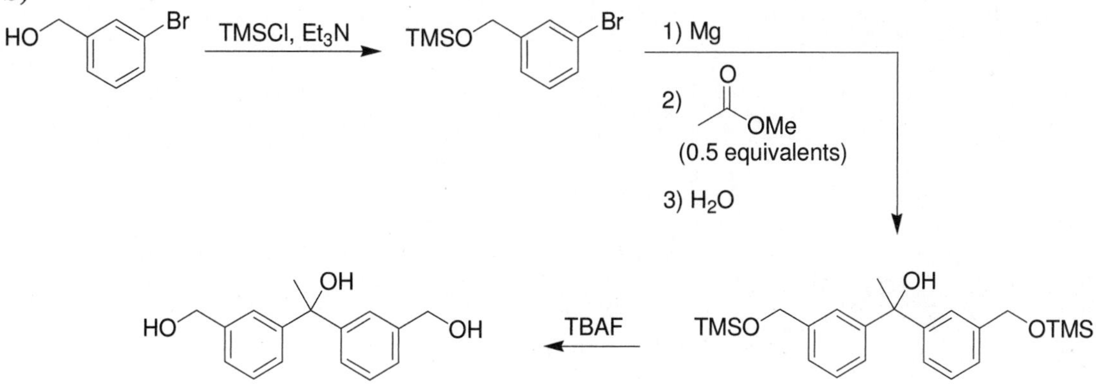

HO ... Br —TMSCl, Et_3N→ TMSO ... Br —1) Mg; 2) methyl acetate (0.5 equivalents); 3) H_2O→ TMSO ... OH ... OTMS —TBAF→ HO ... OH ... OH

13.19.

a)

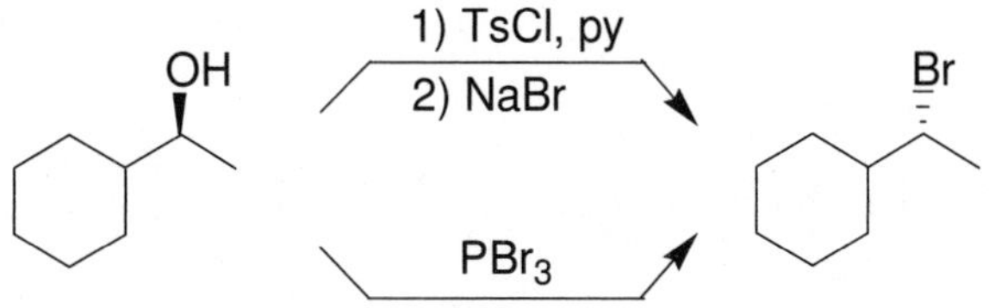

OH → Br

1) TsCl, py
2) NaBr

PBr_3

b)

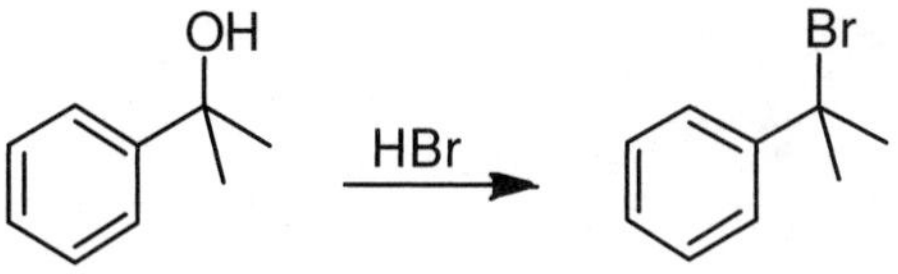

OH —HBr→ Br

c)

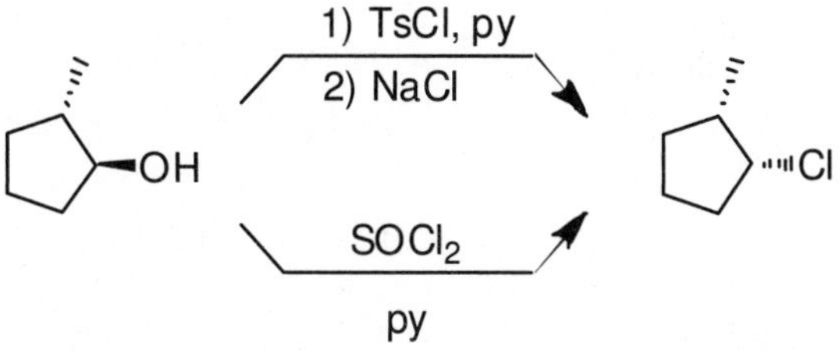

OH → Cl

1) TsCl, py
2) NaCl

$SOCl_2$
py

d)

1) TsCl, py
2) NaBr

PBr_3

e)

1) TsCl, py
2) NaCl

$SOCl_2$
py

f)

1) TsCl, py
2) NaBr

HBr

PBr_3

13.20.

1) TsCl, py
2) NaI

NaCl

13.21.

a)

H_2SO_4
heat

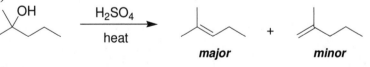

major + *minor*

b)

1) TsCl, py
2) NaOEt

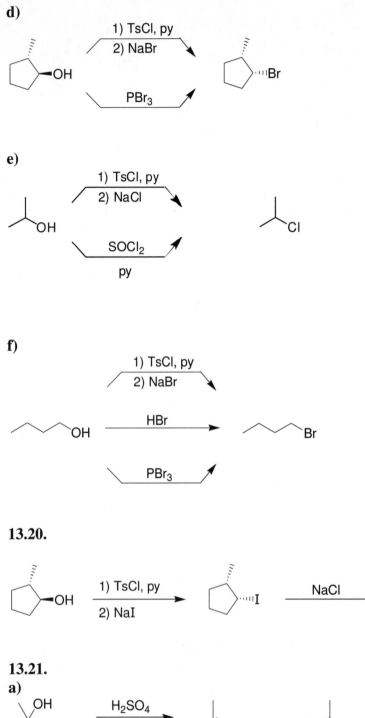

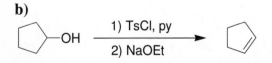

13.22.

a)

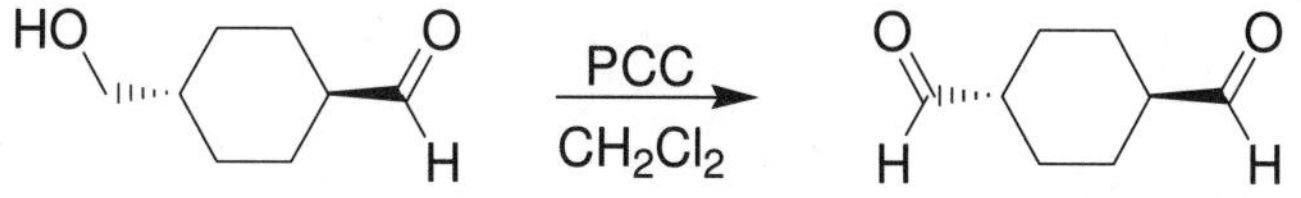

PCC, CH_2Cl_2

b)

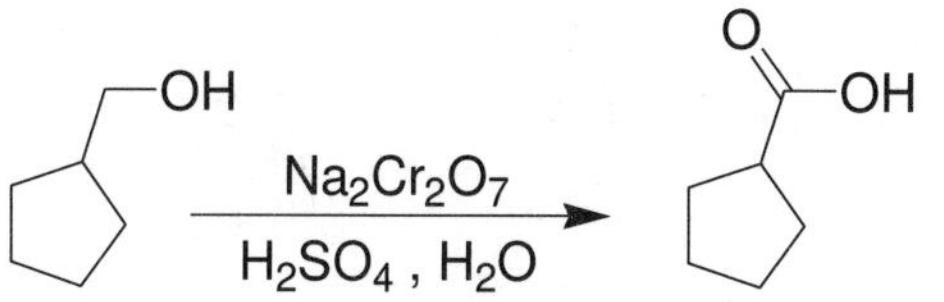

$Na_2Cr_2O_7$, H_2SO_4, H_2O

c)

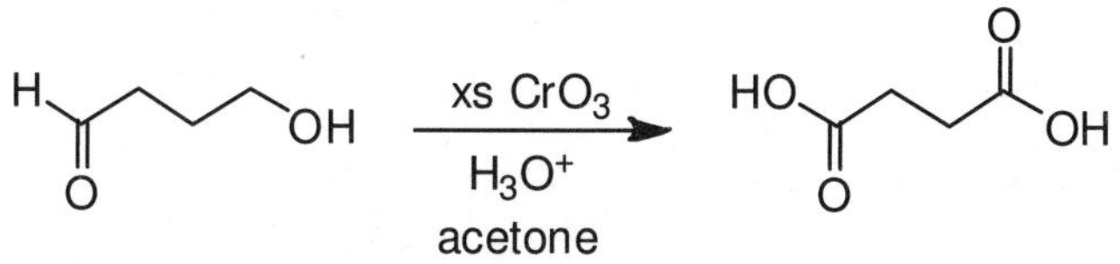

xs CrO_3, H_3O^+, acetone

c)

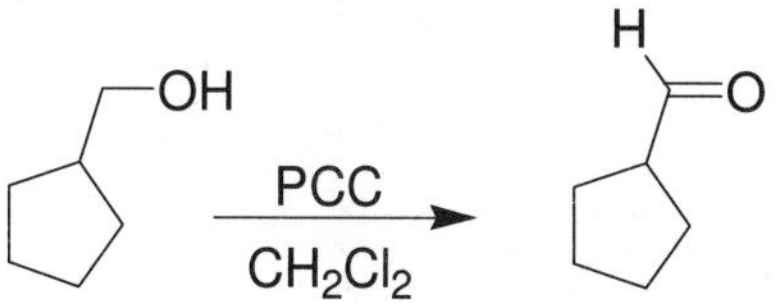

PCC, CH_2Cl_2

e)

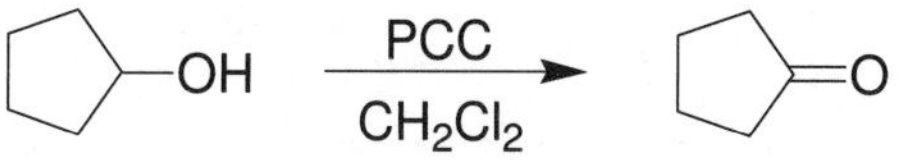

PCC, CH_2Cl_2

f)

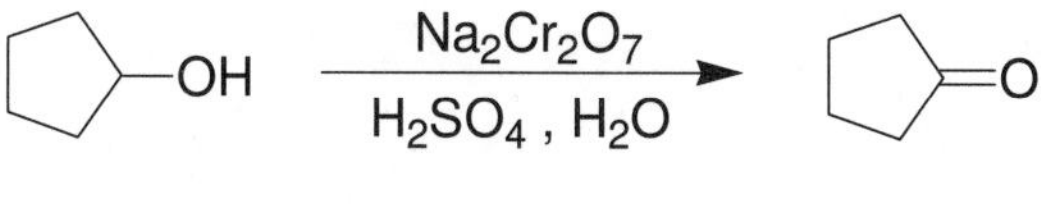

$Na_2Cr_2O_7$, H_2SO_4, H_2O

13.23.

a)

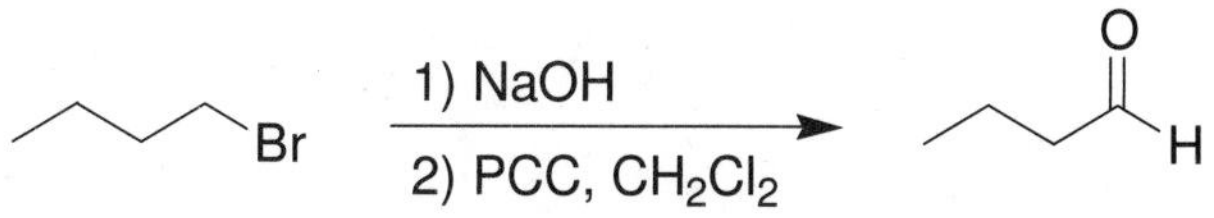

1) NaOH
2) PCC, CH_2Cl_2

b)

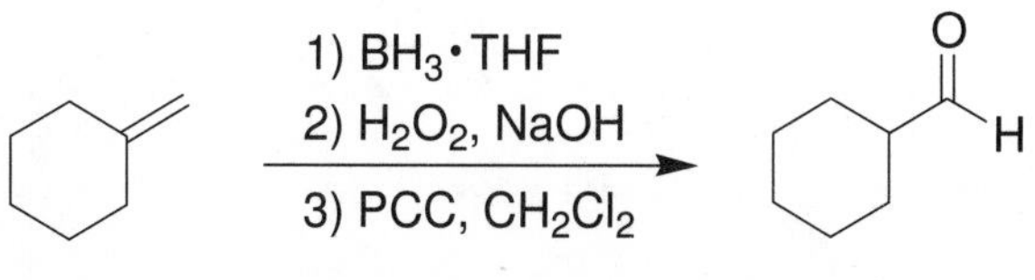

1) $BH_3 \cdot THF$
2) H_2O_2, NaOH
3) PCC, CH_2Cl_2

c)

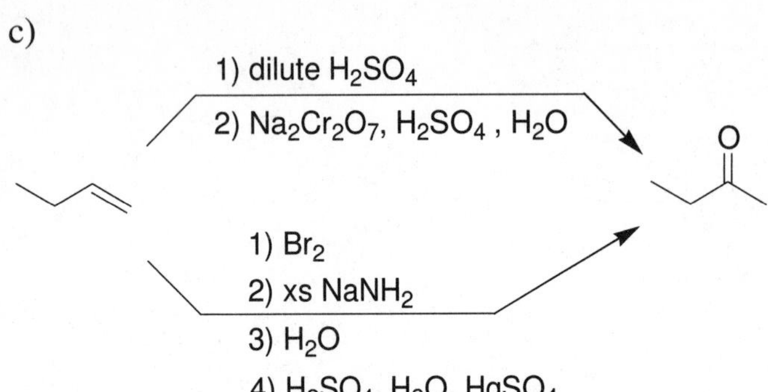

d)

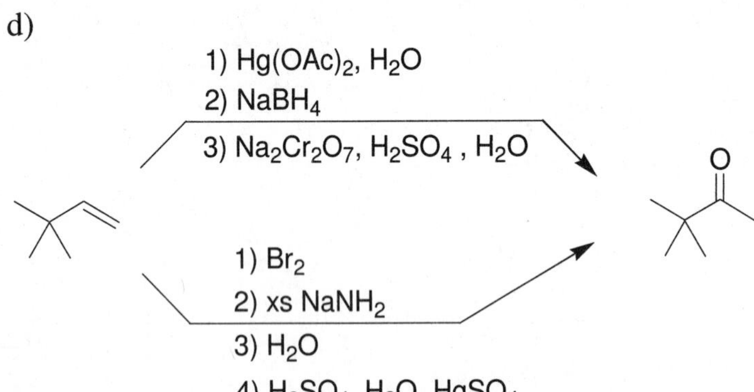

13.24.

a)

1) H_2, Lindlar's Catalyst
2) $BH_3 \cdot THF$
3) H_2O_2, NaOH

OH

1) 9-BBN
2) H_2O_2, NaOH
3) LAH
4) H_2O

b)

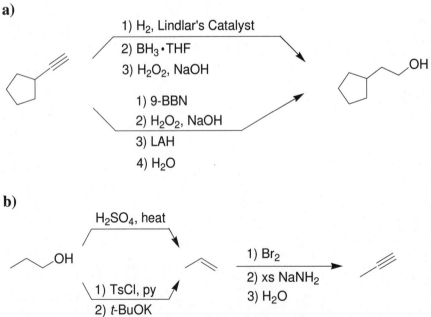

OH

H_2SO_4, heat

1) TsCl, py
2) *t*-BuOK

1) Br_2
2) xs $NaNH_2$
3) H_2O

d)

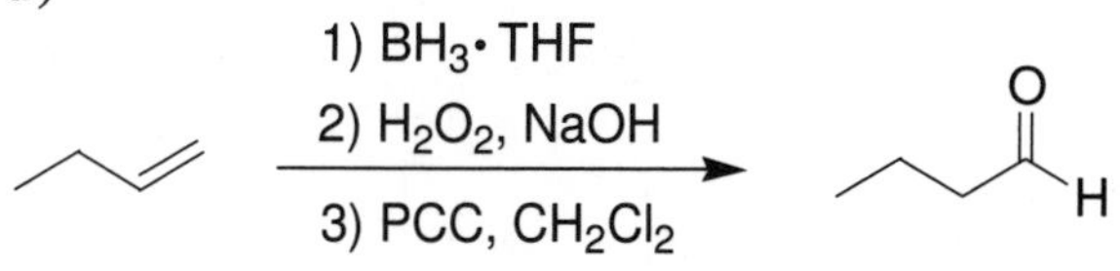

1) $BH_3 \cdot THF$
2) H_2O_2, NaOH
3) PCC, CH_2Cl_2

O
H

e)

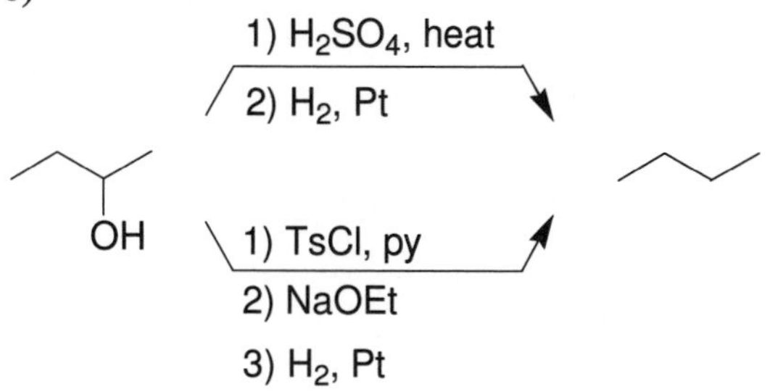

1) H_2SO_4, heat
2) H_2, Pt

OH

1) TsCl, py
2) NaOEt
3) H_2, Pt

f)

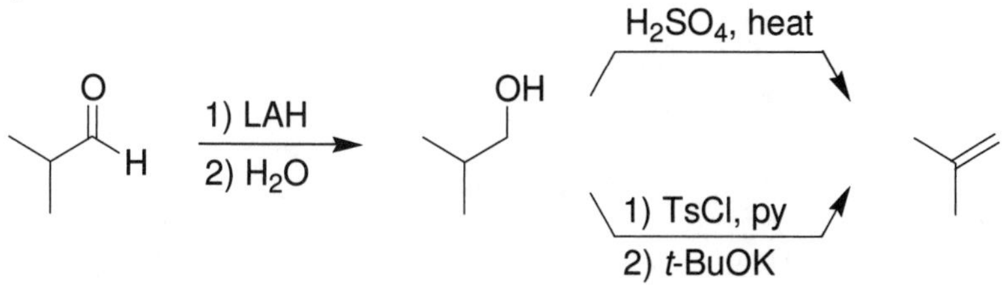

O
H

1) LAH
2) H_2O

OH

H_2SO_4, heat

1) TsCl, py
2) *t*-BuOK

g)

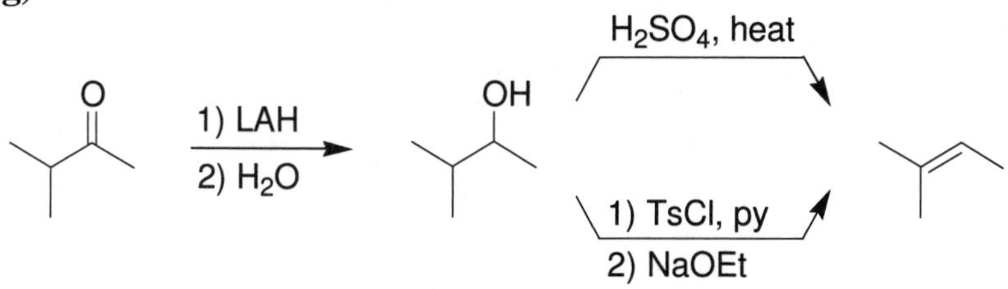

O

1) LAH
2) H_2O

OH

H_2SO_4, heat

1) TsCl, py
2) NaOEt

13.25.

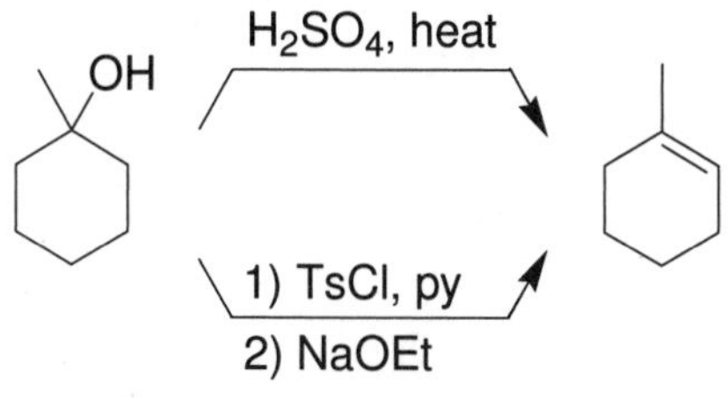

OH

H_2SO_4, heat

1) TsCl, py
2) NaOEt

13.26.

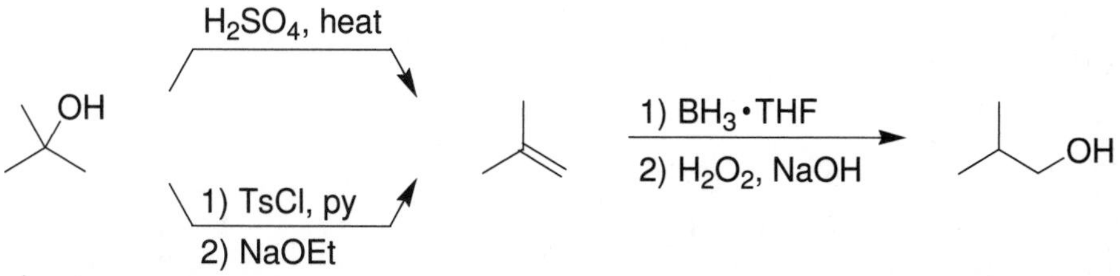

OH

H_2SO_4, heat

1) TsCl, py
2) NaOEt

1) $BH_3 \cdot THF$
2) H_2O_2, NaOH

OH

13.27.

a) Reagents: 1) EtMgBr 2) H_2O 3) $Na_2Cr_2O_7$, H_2SO_4, H_2O

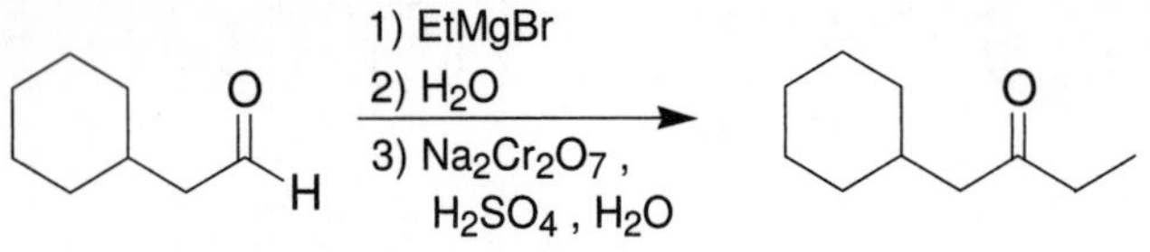

b) Reagents: 1) MeMgBr 2) H_2O 3) $Na_2Cr_2O_7$, H_2SO_4, H_2O

13.28.

a)

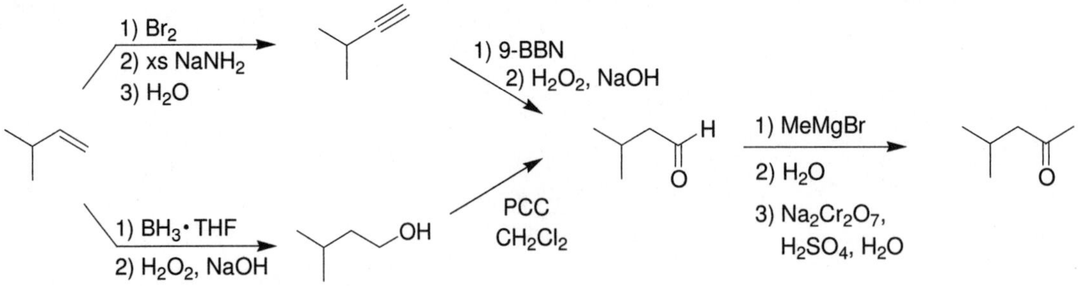

1) Br_2 2) xs $NaNH_2$ 3) H_2O

1) 9-BBN 2) H_2O_2, NaOH

H

1) MeMgBr 2) H_2O 3) $Na_2Cr_2O_7$, H_2SO_4, H_2O

1) $BH_3 \cdot THF$ 2) H_2O_2, NaOH

OH

PCC CH_2Cl_2

b)

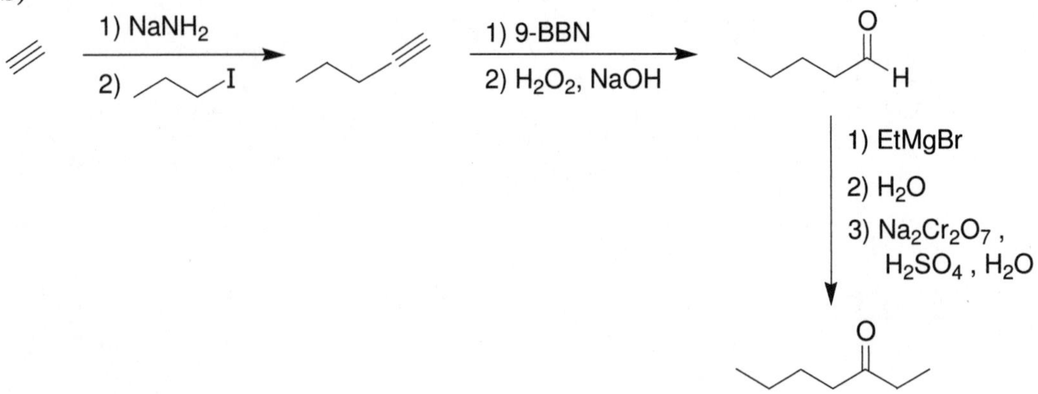

1) $NaNH_2$ 2) I

1) 9-BBN 2) H_2O_2, NaOH

H

1) EtMgBr 2) H_2O 3) $Na_2Cr_2O_7$, H_2SO_4, H_2O

c)

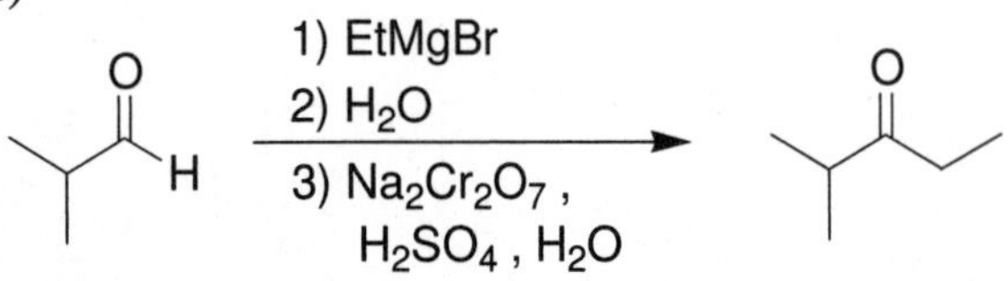

H

1) EtMgBr 2) H_2O 3) $Na_2Cr_2O_7$, H_2SO_4, H_2O

d)

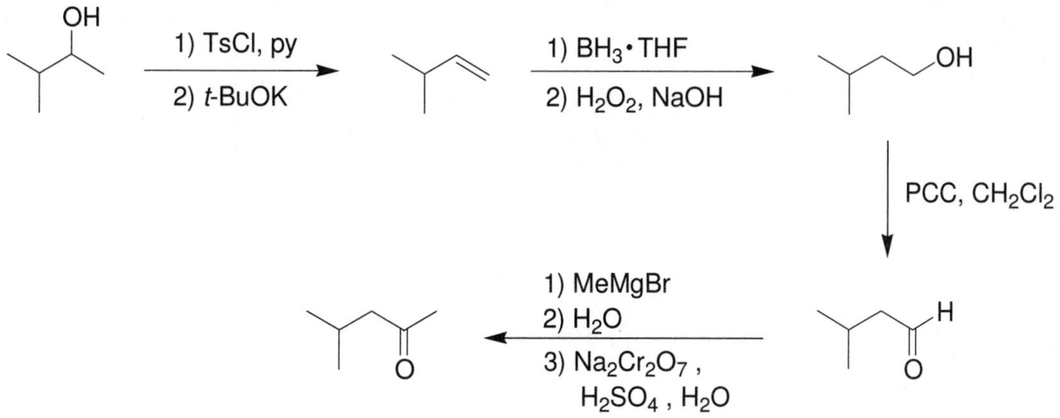

1) TsCl, py; 2) *t*-BuOK

1) $BH_3 \cdot THF$; 2) H_2O_2, NaOH

PCC, CH_2Cl_2

1) MeMgBr; 2) H_2O; 3) $Na_2Cr_2O_7$, H_2SO_4, H_2O

e)

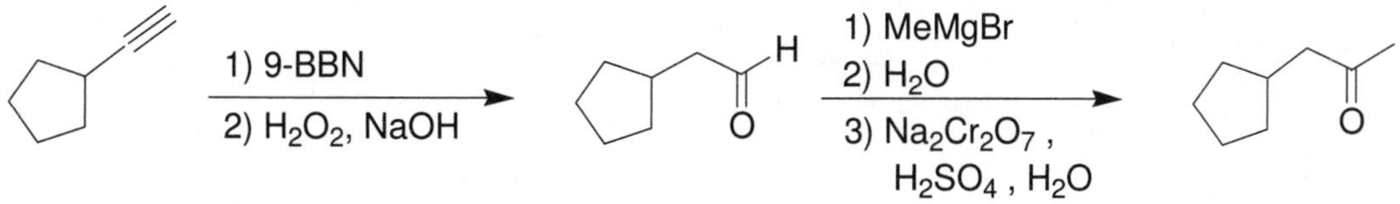

1) 9-BBN; 2) H_2O_2, NaOH

1) MeMgBr; 2) H_2O; 3) $Na_2Cr_2O_7$, H_2SO_4, H_2O

f)

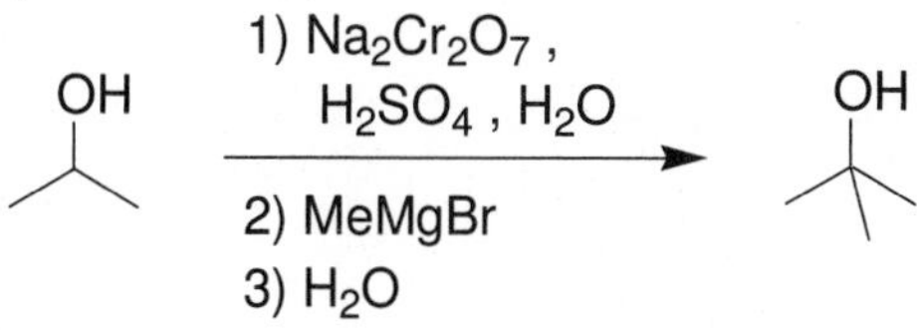

1) $Na_2Cr_2O_7$, H_2SO_4, H_2O; 2) MeMgBr; 3) H_2O

13.29.

a)

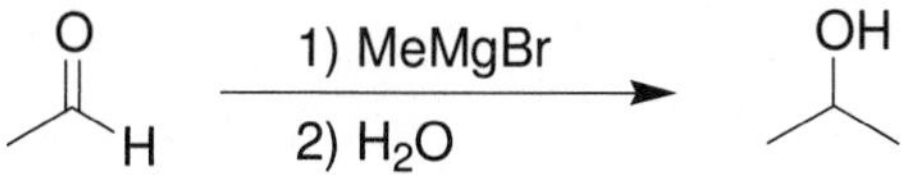

1) MeMgBr; 2) H_2O

b)

$H{-}C{\equiv}C{:}^{\ominus}\ Na^{\oplus}$

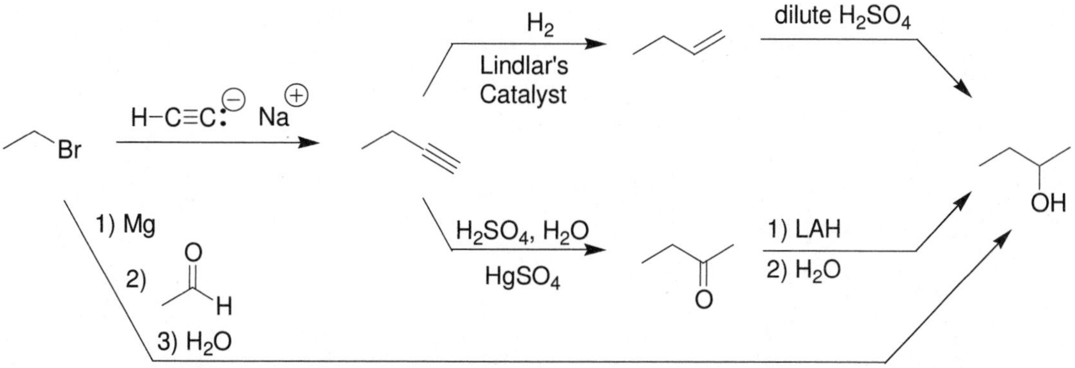

H_2, Lindlar's Catalyst

dilute H_2SO_4

H_2SO_4, H_2O, $HgSO_4$

1) LAH; 2) H_2O

1) Mg; 2) CH3CHO; 3) H_2O

c)

Br → ($H-C{\equiv}C:^{\ominus}\ Na^{\oplus}$) → 1-butyne

1-butyne → (H_2, Lindlar's Catalyst) → 1-butene → (1) HBr, ROOR; 2) NaOH) → 1-butanol (OH)

1-butyne → (1) 9-BBN; 2) H_2O_2, NaOH) → butanal (O, H) → (1) LAH; 2) H_2O) → 1-butanol (OH)

d)

Br → ($H-C{\equiv}C:^{\ominus}\ Na^{\oplus}$) → 1-butyne → (1) 9-BBN; 2) H_2O_2, NaOH) → butanal (O, H) → (1) EtMgBr; 2) H_2O) → 3-hexanol (OH)

1-butyne → (1) $NaNH_2$; 2) EtBr) → 3-hexyne → (H_2SO_4, H_2O, $HgSO_4$) → 3-hexanone (O) → (1) LAH; 2) H_2O) → 3-hexanol

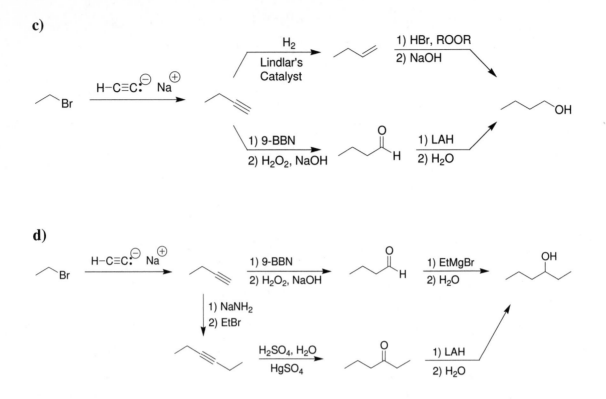

13.30.

a) 2-propyl-1-pentanol
b) (*R*)-4-methyl-2-pentanol
c) 2-bromo-4-methylphenol
d) (*1R,2R*)-2-methylcyclohexanol

13.31.

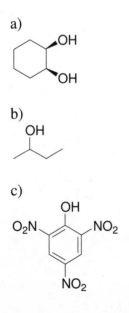

a)

cyclohexane-1,2-diol (OH, OH)

b)

OH (2-butanol)

c)

OH, O_2N, NO_2, NO_2 (2,4,6-trinitrophenol)

d)

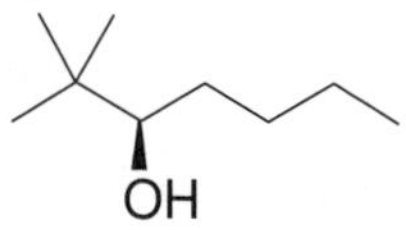

OH

e) HO OH

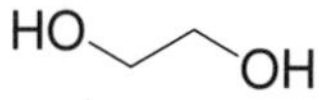

f) HO

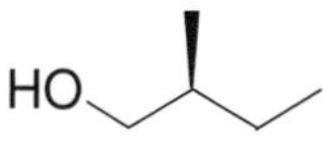

13.32.

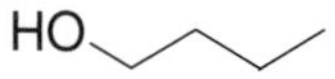

HO OH OH OH

1-butanol 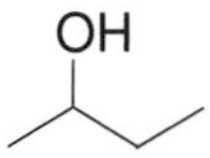 2-butanol 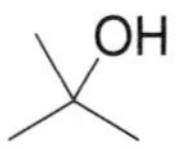 2-methyl-2-propanol 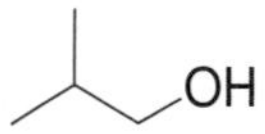 2-methyl-1-propanol

13.33.

a)

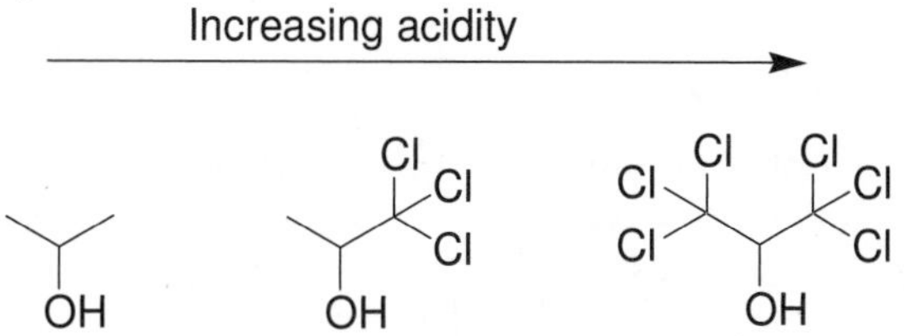

Increasing acidity

OH | OH Cl Cl Cl | OH Cl Cl Cl Cl Cl Cl

b)

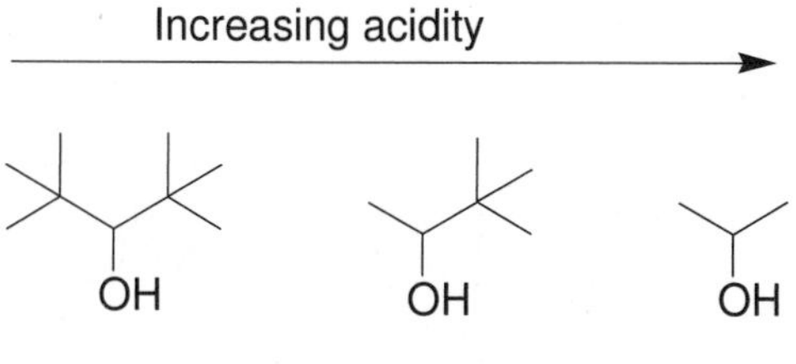

Increasing acidity

OH OH OH

c)

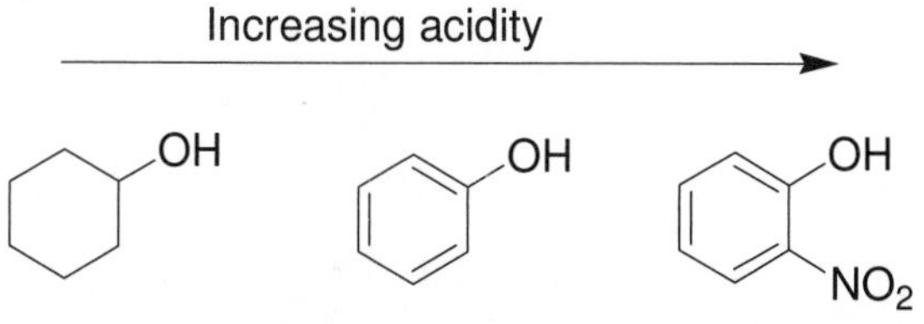

Increasing acidity

OH OH OH NO_2

13.34.

a)

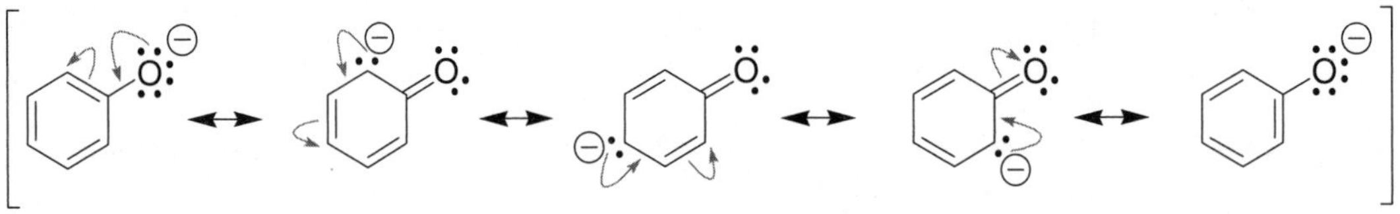

O⁻ … ↔ … ↔ … ↔ … ↔ … O⁻

b)

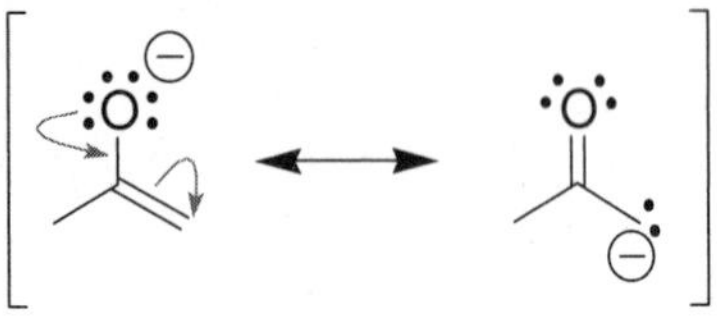

O⁻ ↔ O

c)

O⁻ ↔ O ↔ O

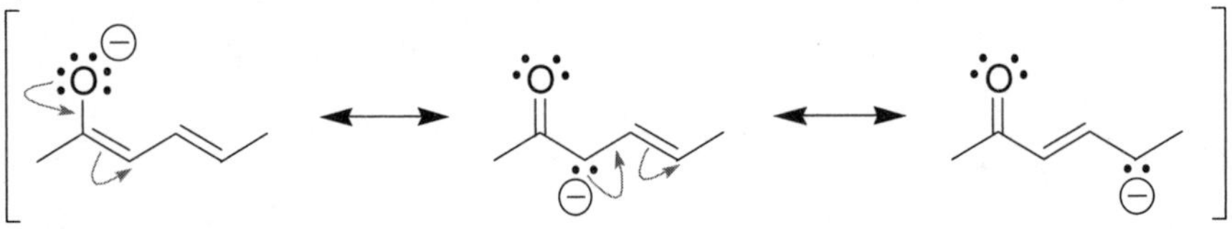

13.35.

a) 1-bromobutane

b) 1-chlorobutane

c) 1-chlorobutane

d) *trans*-2-butene

e) 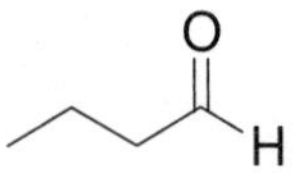 H

f) 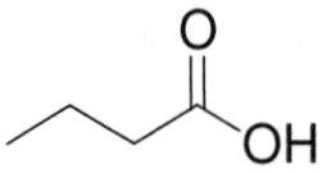 OH

g) 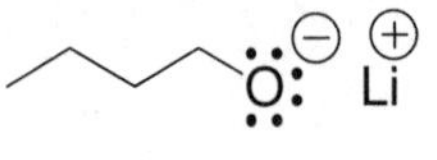 O⁻ Li⁺

h) O⁻ Na⁺

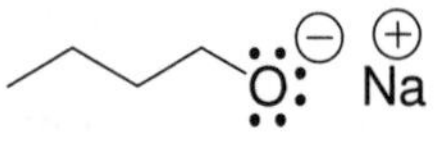

i) 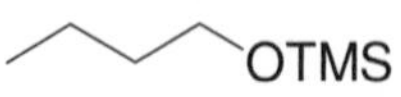 OTMS

j) 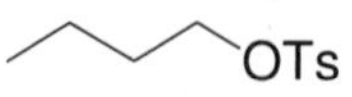 OTs

k) 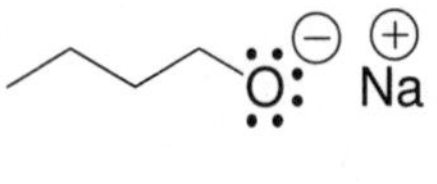 O⁻ Na⁺

l) 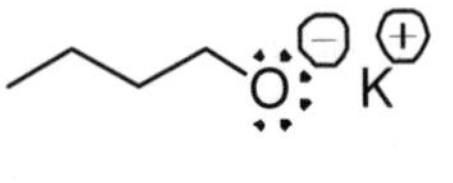 O⁻ K⁺

13.36.

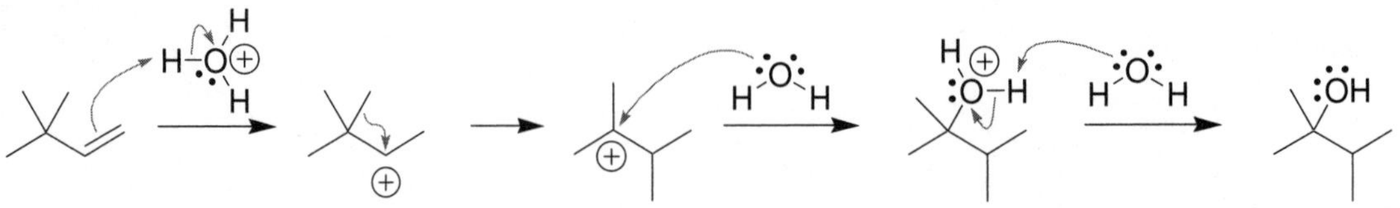

H–O⁺(H)H → ⁺ → ⁺ —H_2O→ H–O⁺–H —H_2O→ OH

13.37.

a)

OH —PCC / CH_2Cl_2→ O, H

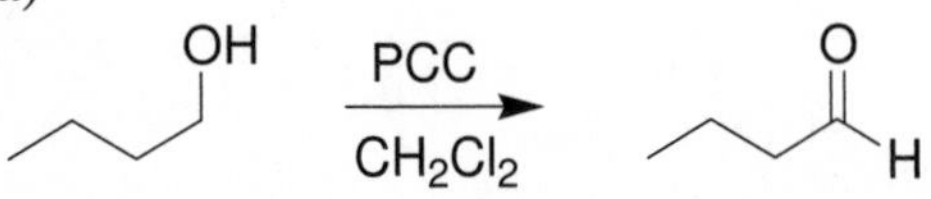

b)

OH — $Na_2Cr_2O_7$ / H_2SO_4, H_2O → OH (carboxylic acid, =O)

c)

OH — 1) PCC, CH_2Cl_2 2) EtMgBr 3) H_2O → OH

d)

OH — 1) PCC, CH_2Cl_2 2) EtMgBr 3) H_2O 4) $Na_2Cr_2O_7$, H_2SO_4, H_2O 5) MeMgBr 6) H_2O → OH

e)

OH — 1) PCC, CH_2Cl_2 2) PrMgBr 3) H_2O 4) $Na_2Cr_2O_7$, H_2SO_4, H_2O → O

13.38.

a)

O, H, H — 1) (cyclohexyl)–MgBr 2) H_2O → OH

b)

O — 1) MeMgBr 2) H_2O

O — 1) PrMgBr 2) H_2O → OH

O — 1) (phenyl)–MgBr 2) H_2O

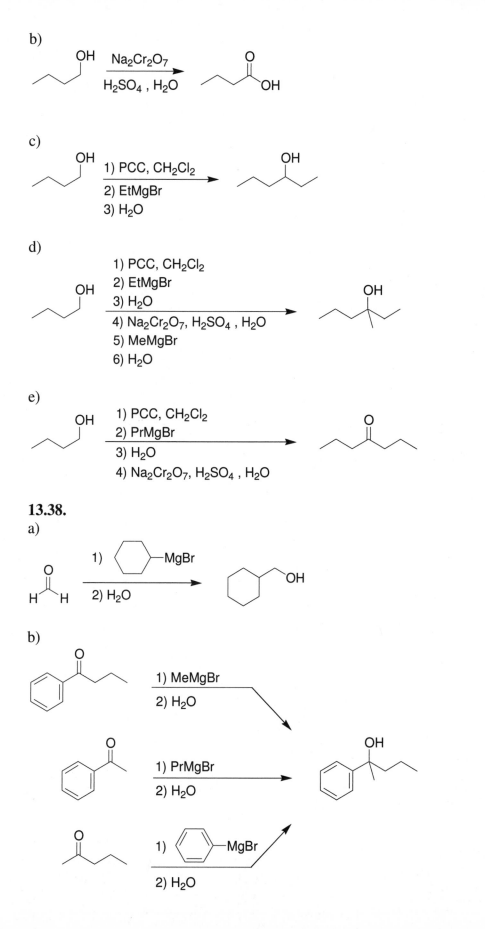

c)

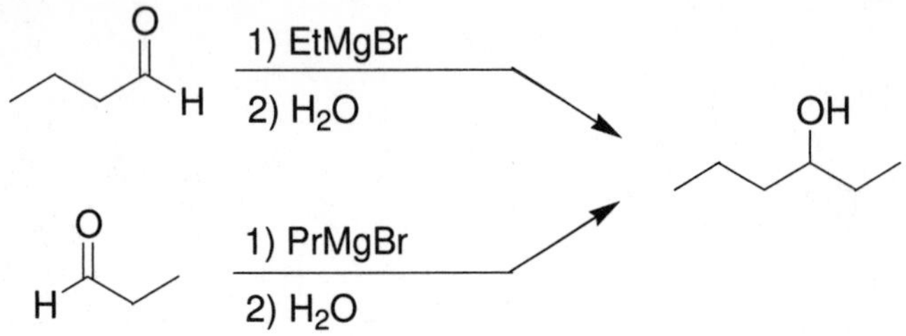

1) EtMgBr
2) H_2O

OH

1) PrMgBr
2) H_2O

d)

1) MeMgBr
2) H_2O

1) EtMgBr
2) H_2O

OH

1) PrMgBr
2) H_2O

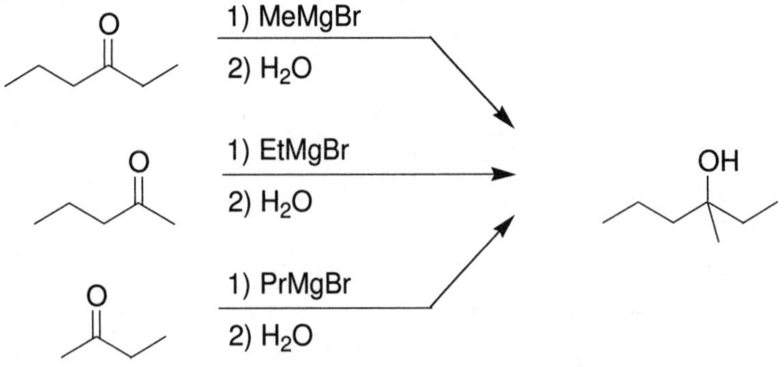

13.39.

a) 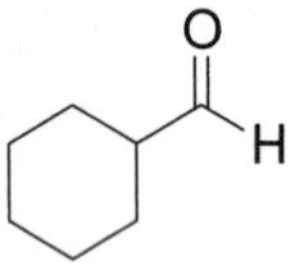 b) 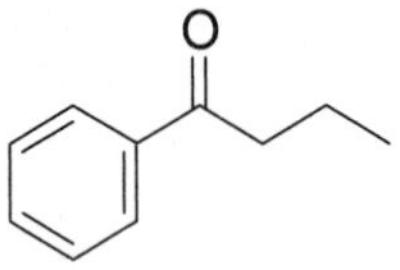 c) 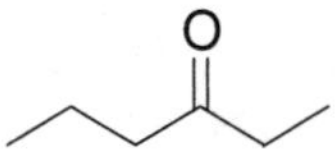 d)

13.40.

a)

1) EtMgBr
2) H_2O
3) $Na_2Cr_2O_7$, H_2SO_4, H_2O

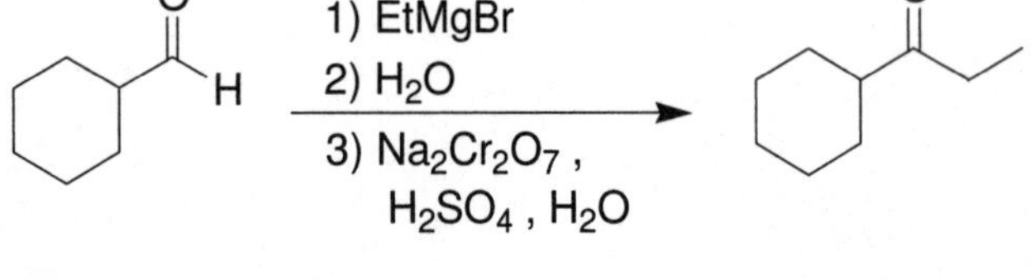

b)

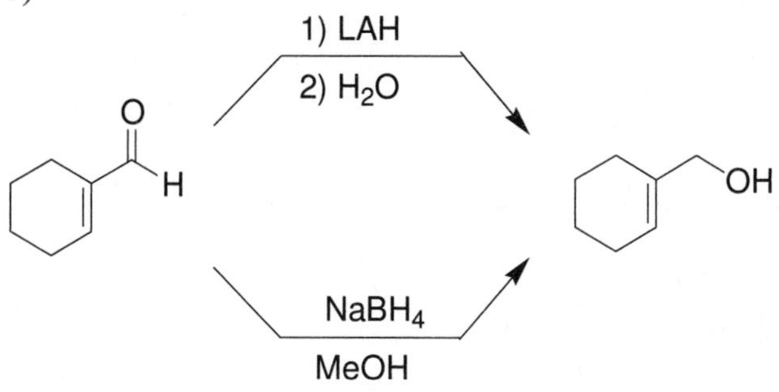
1) LAH
2) H_2O
O
H
OH
$NaBH_4$
MeOH

13.41.

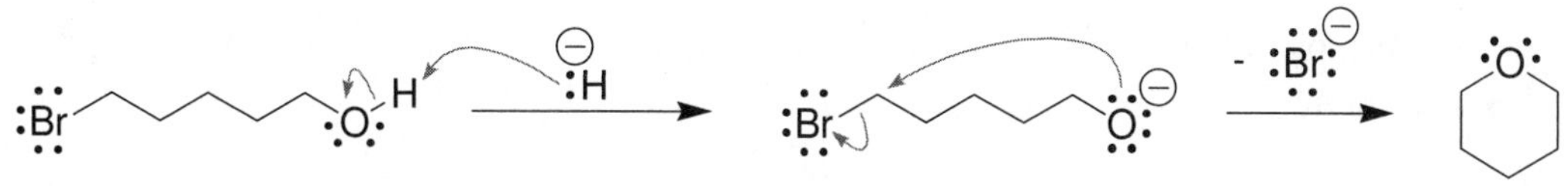
Br
O
H
H
Br
O
- Br
O

13.42. The major product is 1-methylcyclohexanol (resulting from Markvonikov addition), which is a tertiary alcohol. Tertiary alcohols do not generally undergo oxidation. The minor product (2-methylcyclohexanol) is a secondary alcohol and can undergo oxidation to yield a ketone.

13.43.

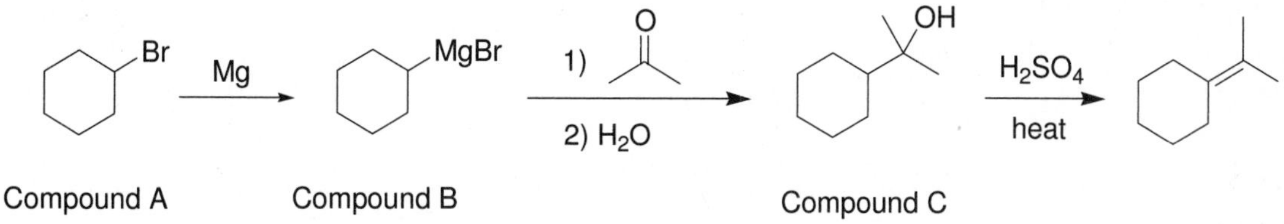
Br
Mg
MgBr
1)
O
2) H_2O
OH
H_2SO_4
heat
Compound A
Compound B
Compound C

13.44.

1) MeMgBr
2) H_2O

$Na_2Cr_2O_7$
H_2SO_4, H_2O

1) PhMgBr
2) H_2O

1) TsCl, py
2) *t*-BuOK

1) $BH_3 \cdot THF$
2) H_2O_2, NaOH

PBr_3

1) Mg
2) HCHO
3) H_2O

PCC
CH_2Cl_2

13.45.

1) TsCl
2) NaOEt

1) HBr, ROOR
2) Mg

$Na_2Cr_2O_7$
H_2SO_4, H_2O

1) PrMgBr
2) H_2O

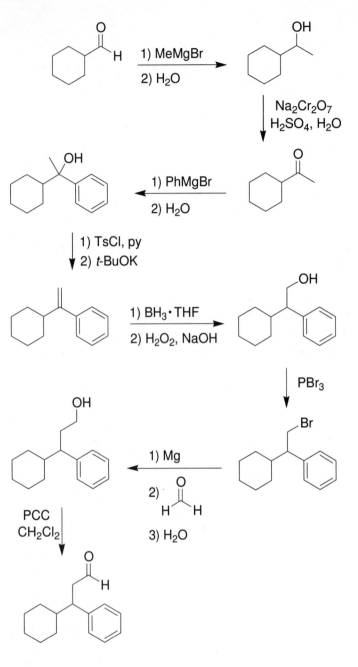

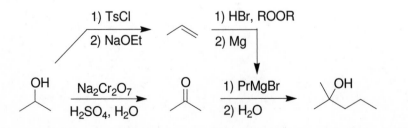

13.46.

a)

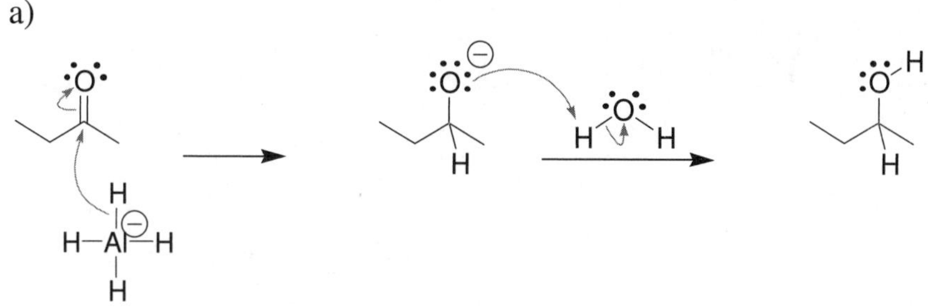

b)

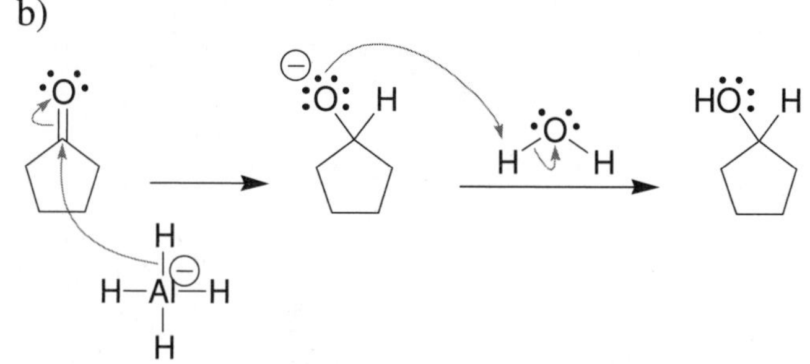

c)

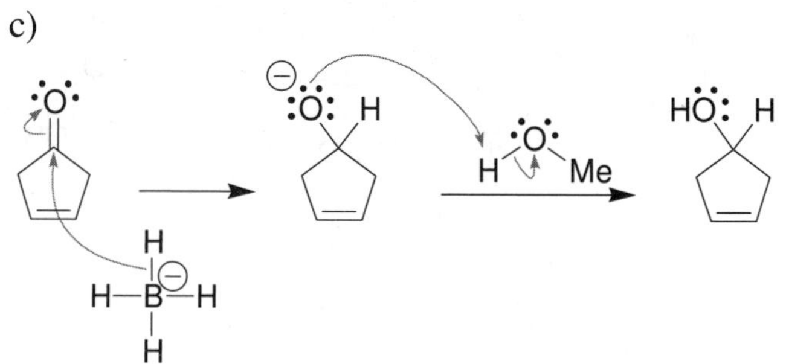

13.47.

a)

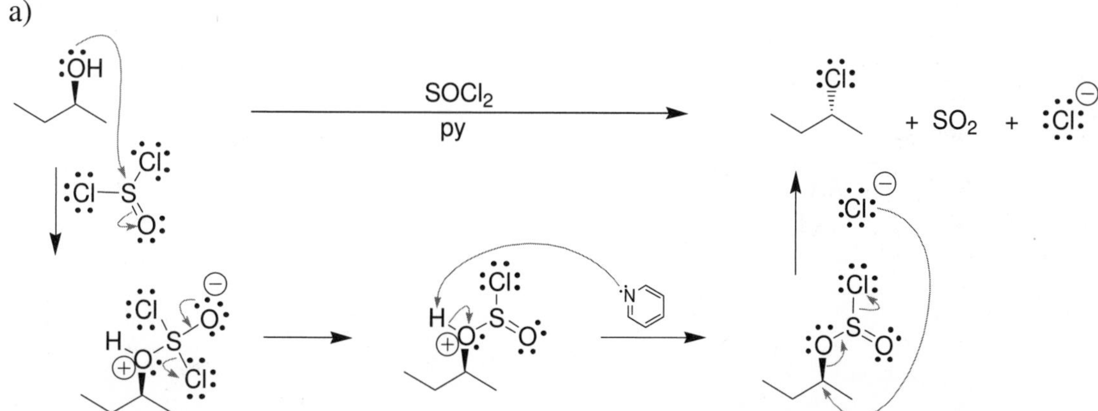

b)

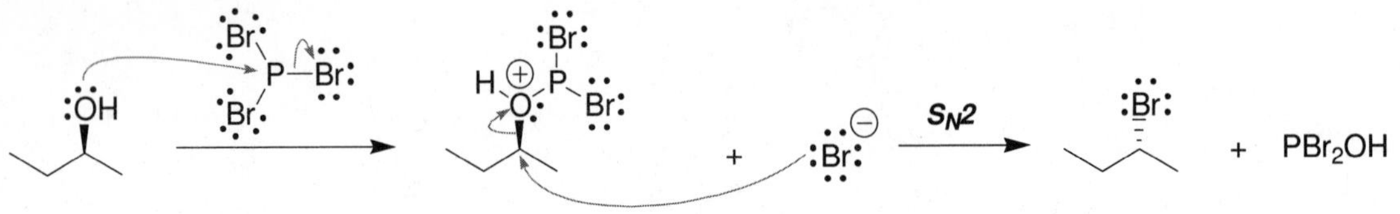

:ÖH, :Br–PBr–Br: → S_N2 → :Br: + PBr_2OH

c)

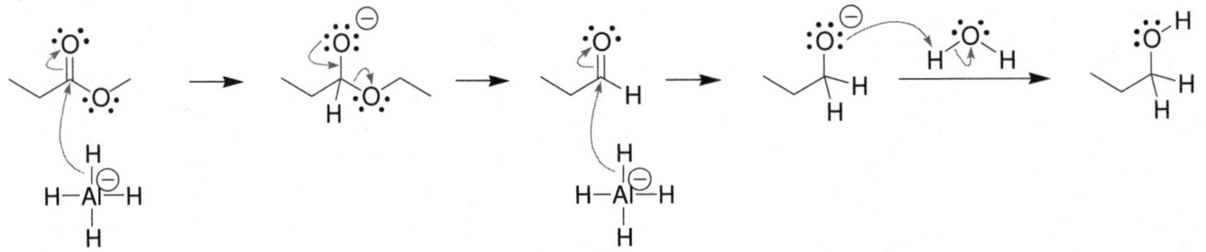

H–Al–H (H, H)

13.48.

a) OH → $Na_2Cr_2O_7$ / H_2SO_4, H_2O → O

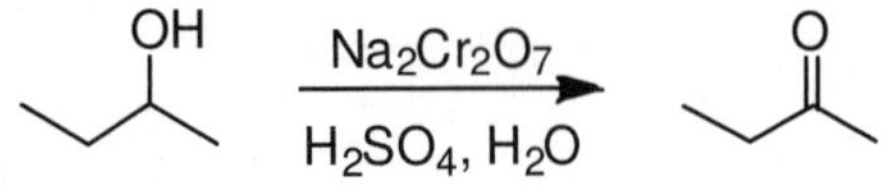

b) OH → PCC / CH_2Cl_2 → O

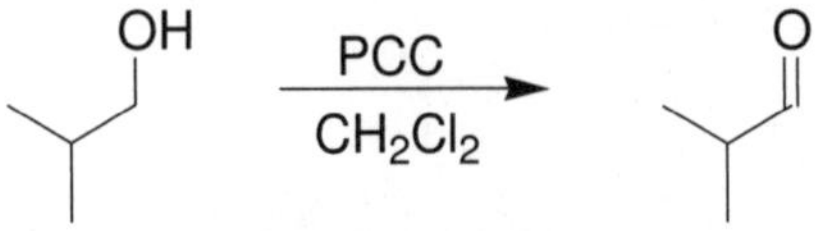

c) OH → PCC / CH_2Cl_2 → O, H

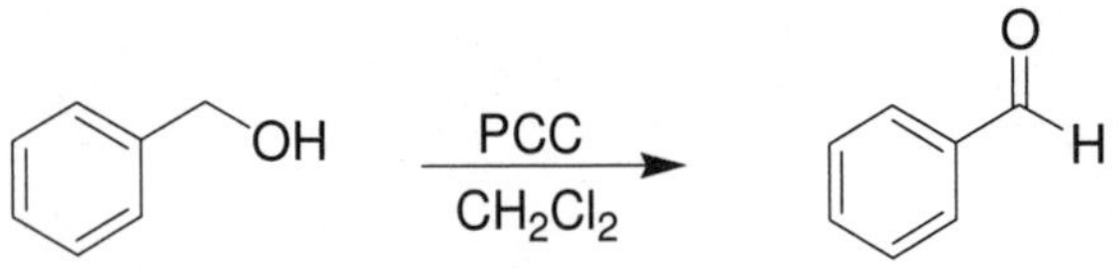

d) OH → $Na_2Cr_2O_7$ / H_2SO_4, H_2O → O, OH

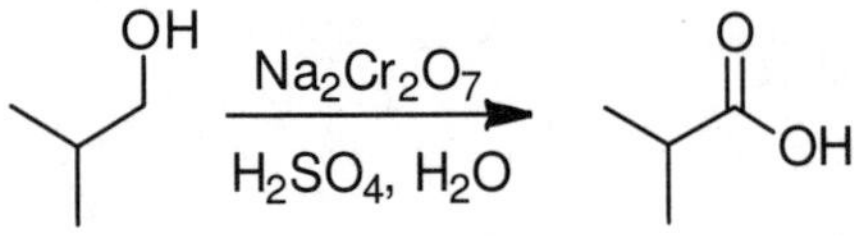

e) O, H → 1) LAH / 2) H_2O → OH

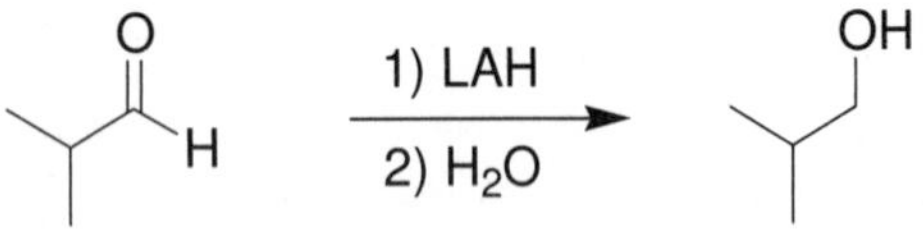

f) O → 1) LAH / 2) H_2O → OH

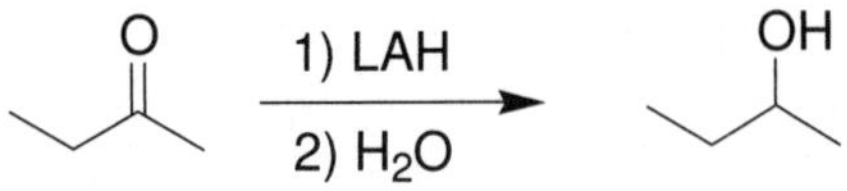

13.49

a)

1) O_3
2) DMS
3) Excess LAH
4) H_2O

→ HO OH

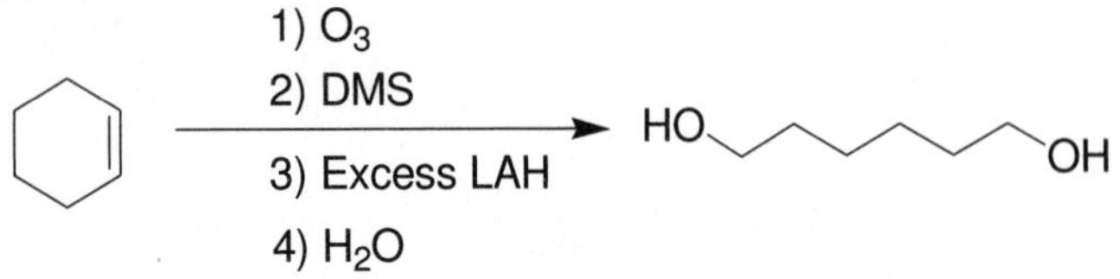

b)

1) O_3
2) DMS
3) Excess LAH
4) H_2O

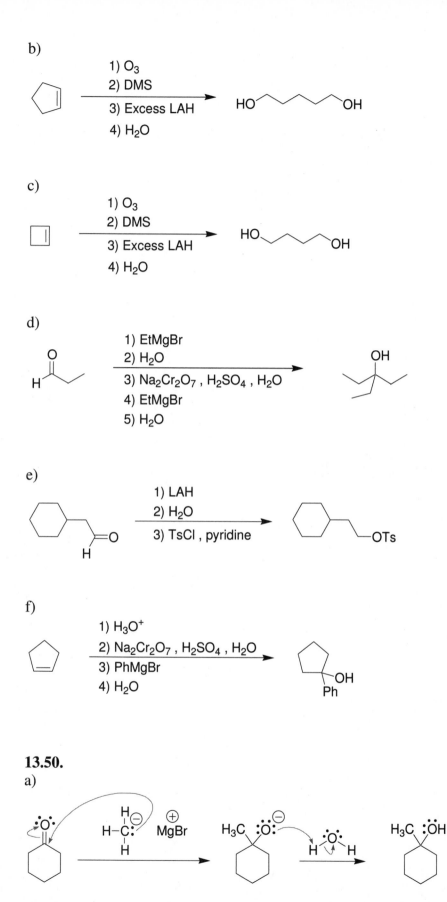

HO OH

c)

1) O_3
2) DMS
3) Excess LAH
4) H_2O

HO OH

d)

O
H

1) EtMgBr
2) H_2O
3) $Na_2Cr_2O_7$, H_2SO_4 , H_2O
4) EtMgBr
5) H_2O

OH

e)

H O

1) LAH
2) H_2O
3) TsCl , pyridine

OTs

f)

1) H_3O^+
2) $Na_2Cr_2O_7$, H_2SO_4 , H_2O
3) PhMgBr
4) H_2O

OH
Ph

13.50.

a)

:O:

H
H–C:
H
⊖ ⊕ MgBr

H_3C :O: ⊖

H–O–H

H_3C :OH

b)

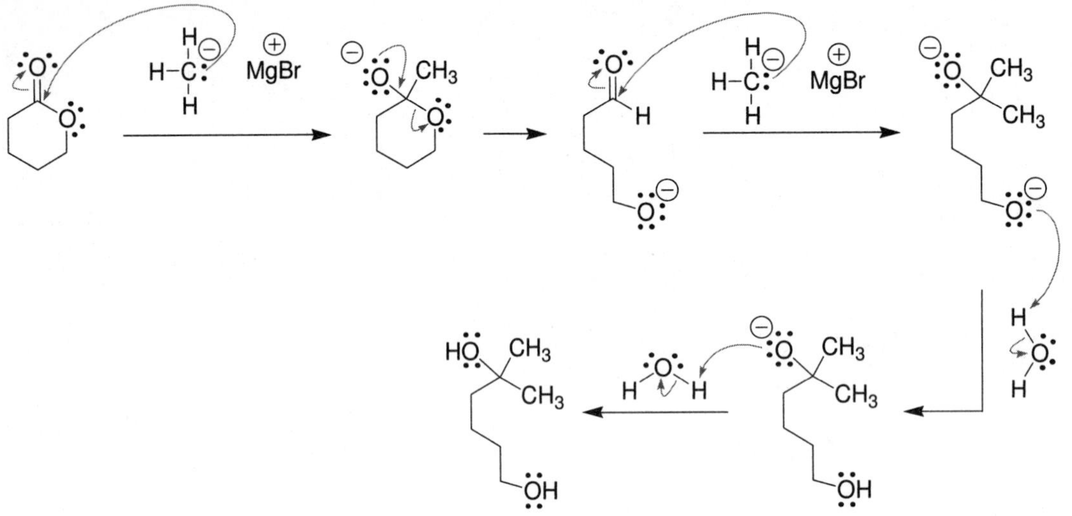

H–C: MgBr
CH₃
H
CH₃
CH₃
H–O–H
HO CH₃
CH₃
OH

13.51.

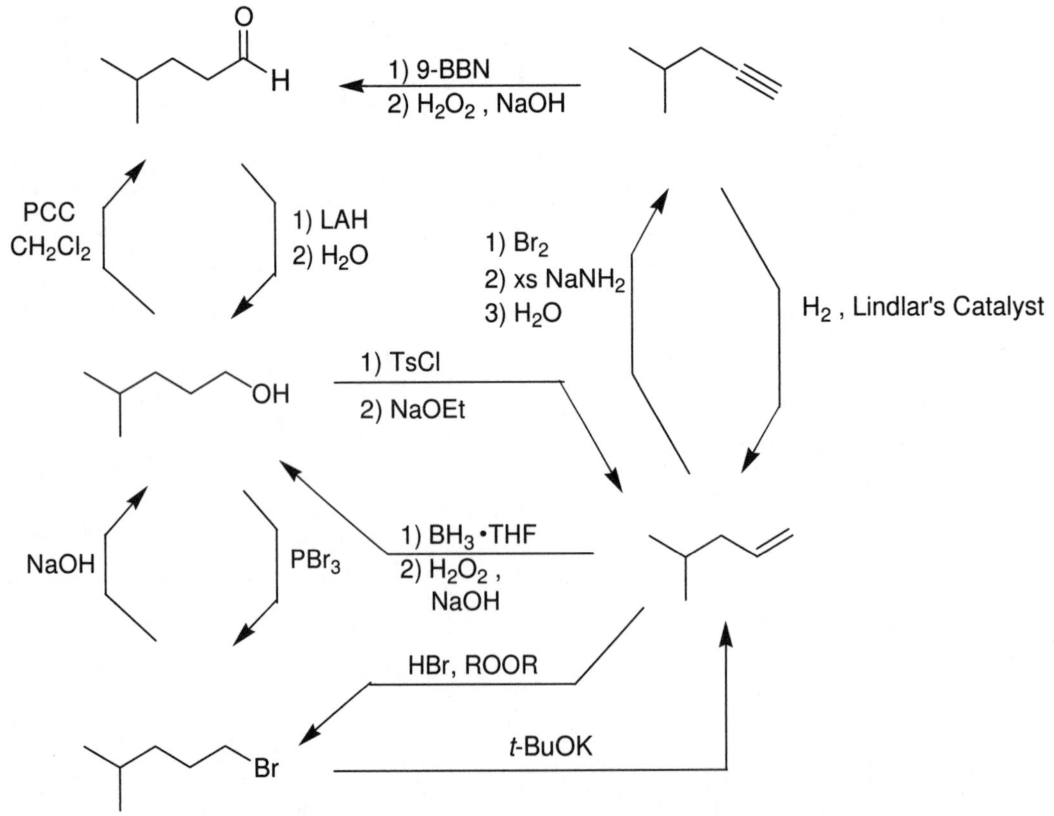

1) 9-BBN
2) H_2O_2 , NaOH

PCC
CH_2Cl_2

1) LAH
2) H_2O

1) Br_2
2) xs $NaNH_2$
3) H_2O

H_2 , Lindlar's Catalyst

OH

1) TsCl
2) NaOEt

NaOH

PBr_3

1) BH_3•THF
2) H_2O_2 ,
NaOH

HBr, ROOR

Br

t-BuOK

13.52.

a)

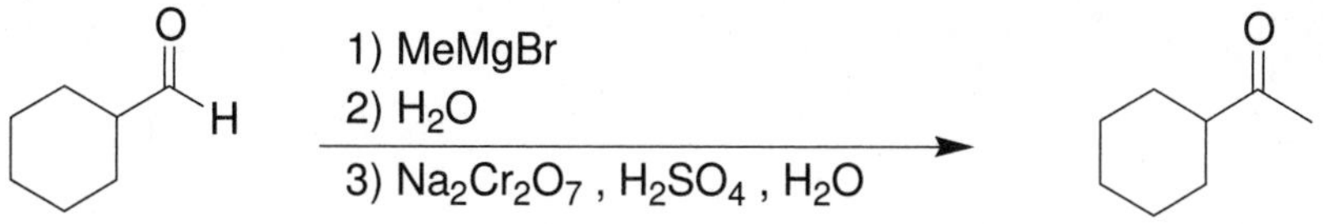

1) MeMgBr
2) H_2O
3) $Na_2Cr_2O_7$, H_2SO_4 , H_2O

b)

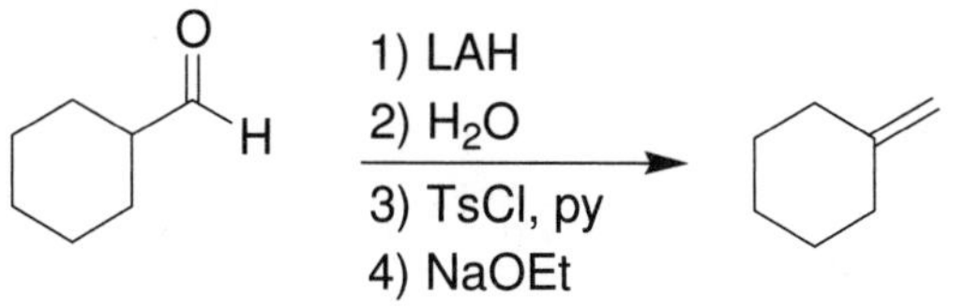

1) LAH
2) H_2O
3) TsCl, py
4) NaOEt

c)

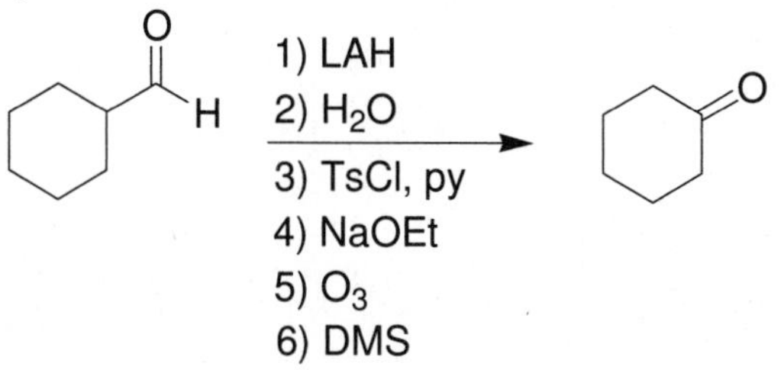

1) LAH
2) H_2O
3) TsCl, py
4) NaOEt
5) O_3
6) DMS

d)

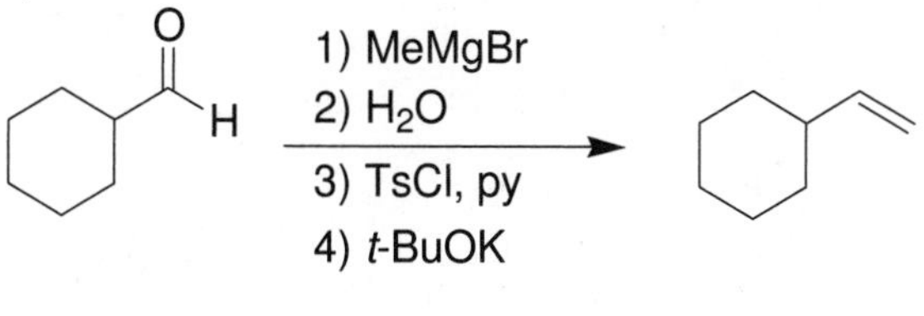

1) MeMgBr
2) H_2O
3) TsCl, py
4) *t*-BuOK

e)

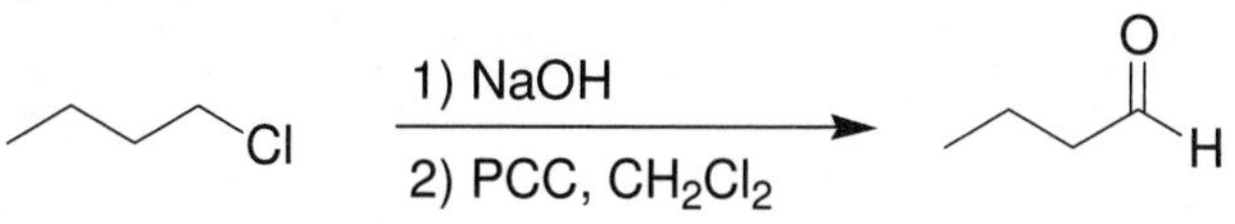

1) NaOH
2) PCC, CH_2Cl_2

f)

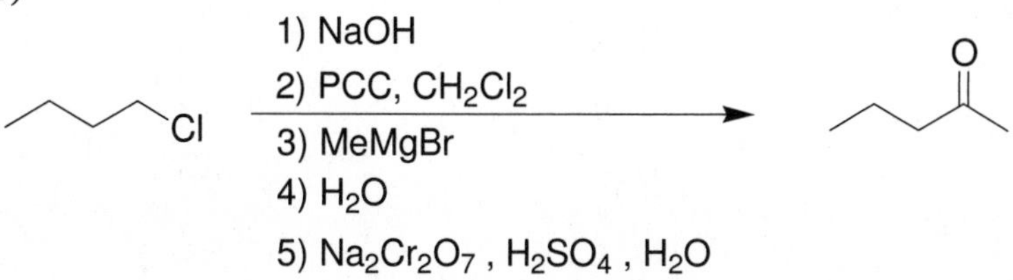

1) NaOH
2) PCC, CH_2Cl_2
3) MeMgBr
4) H_2O
5) $Na_2Cr_2O_7$, H_2SO_4 , H_2O

g)

1) dilute H_2SO_4
2) $Na_2Cr_2O_7$, H_2SO_4 , H_2O

O

h)

1) dilute H_2SO_4
2) $Na_2Cr_2O_7$, H_2SO_4 , H_2O
3) MeMgBr
4) H_2O

OH

i)

1) dilute H_2SO_4
2) $Na_2Cr_2O_7$, H_2SO_4 , H_2O
3) MeMgBr
4) conc. H_2SO_4, heat

j)

1) $HgSO_4$, H_2SO_4, H_2O
2) MeMgBr
3) H_2O

OH

k)

1) dilute H_2SO_4
2) $Na_2Cr_2O_7$, H_2SO_4, H_2O
3) MeMgBr
4) H_2O

OH

l)

O
H

1) EtMgBr
2) H_2O

OH

m)

O

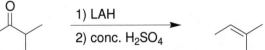

1) LAH
2) conc. H_2SO_4

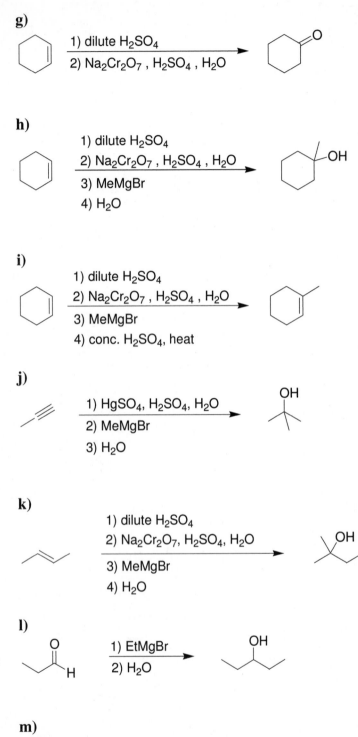

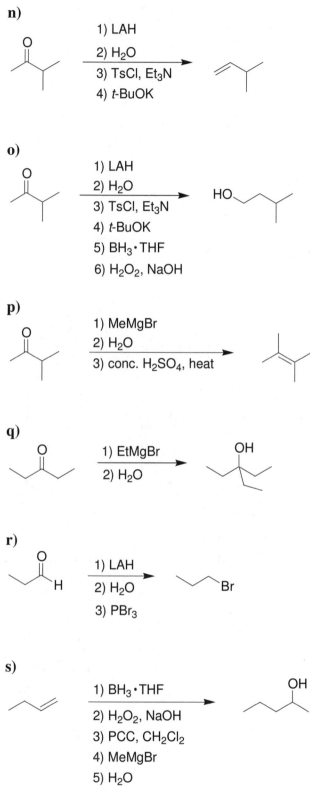

n)

1) LAH
2) H_2O
3) TsCl, Et_3N
4) *t*-BuOK

o)

1) LAH
2) H_2O
3) TsCl, Et_3N
4) *t*-BuOK
5) $BH_3 \cdot THF$
6) H_2O_2, NaOH

HO

p)

1) MeMgBr
2) H_2O
3) conc. H_2SO_4, heat

q)

O

1) EtMgBr
2) H_2O

OH

r)

O

H

1) LAH
2) H_2O
3) PBr_3

Br

s)

1) $BH_3 \cdot THF$
2) H_2O_2, NaOH
3) PCC, CH_2Cl_2
4) MeMgBr
5) H_2O

OH

13.53.

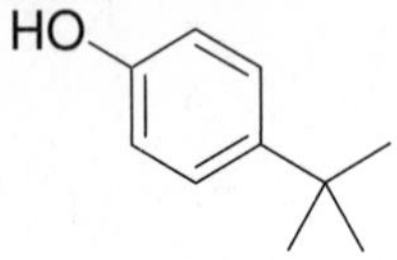

HO

13.54.

OH

13.55.

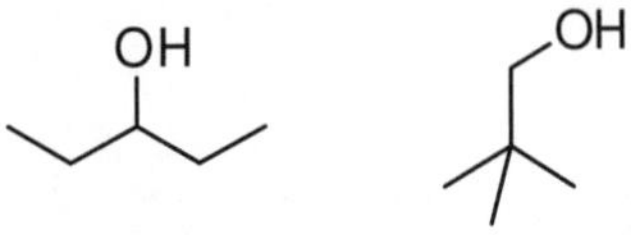

OH

OH

13.56.

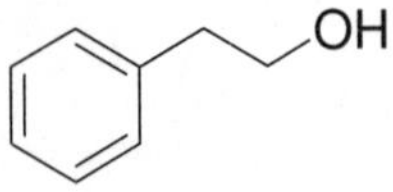

OH

13.57

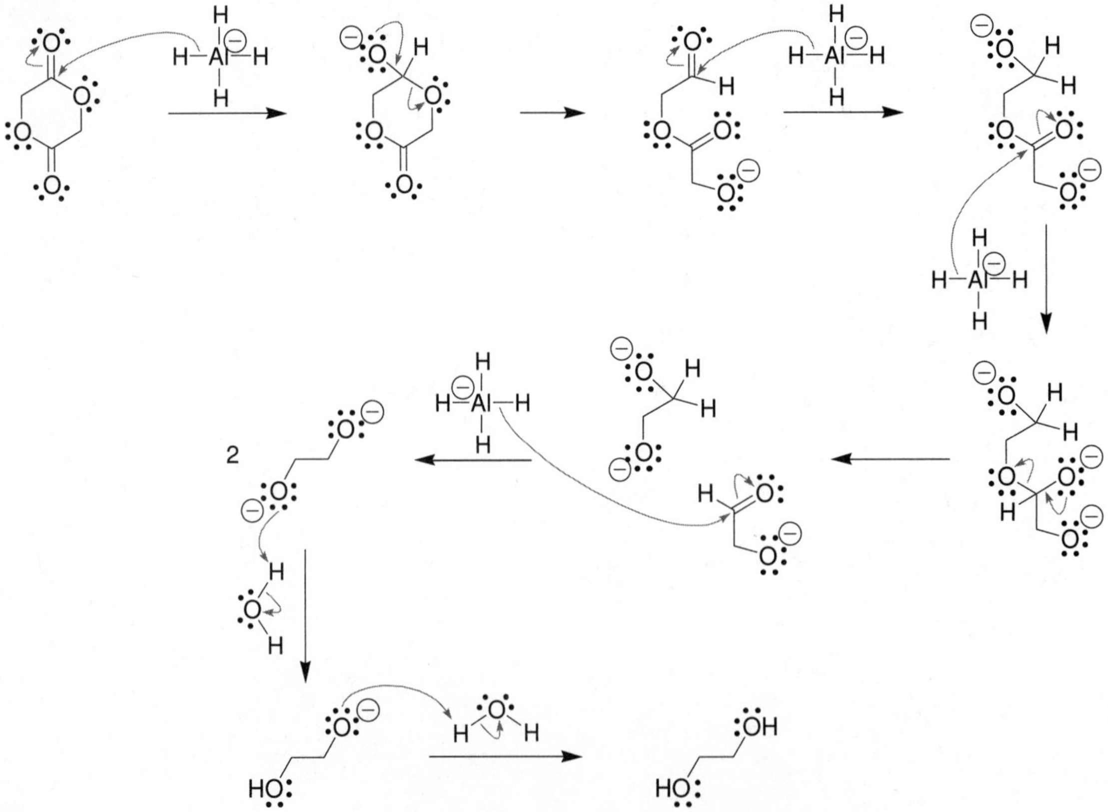

H

H–Al–H

H

2

HO:

HO:

:OH

13.58

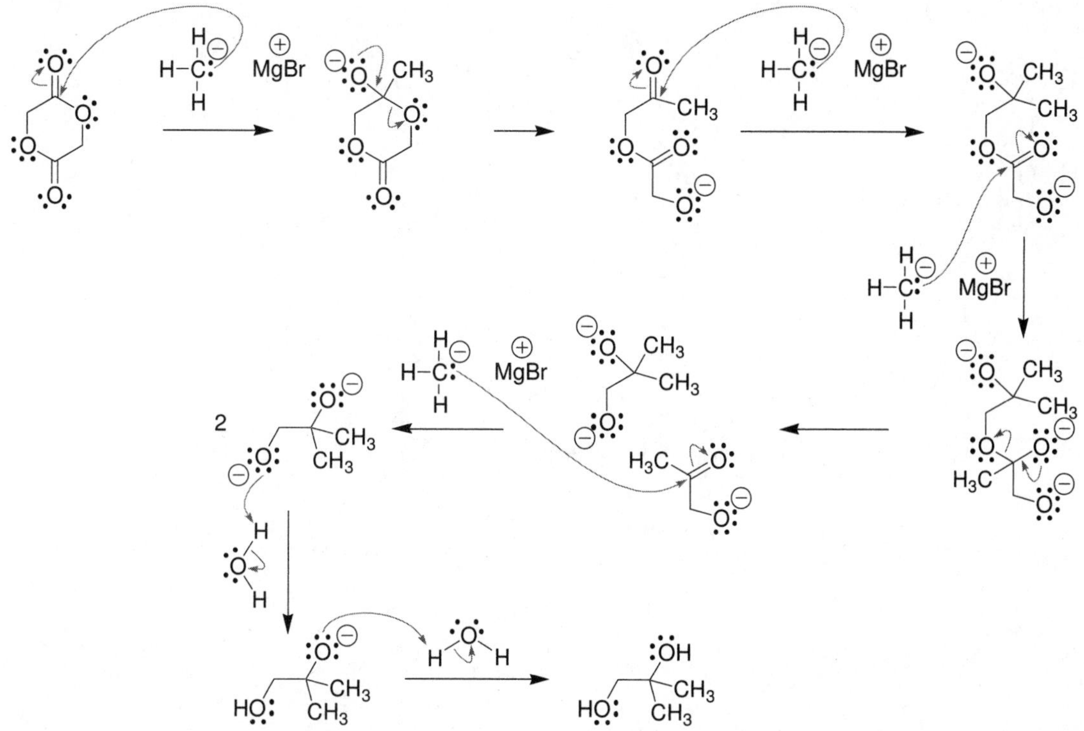

H–C:⁻ ⁺MgBr

2

13.59

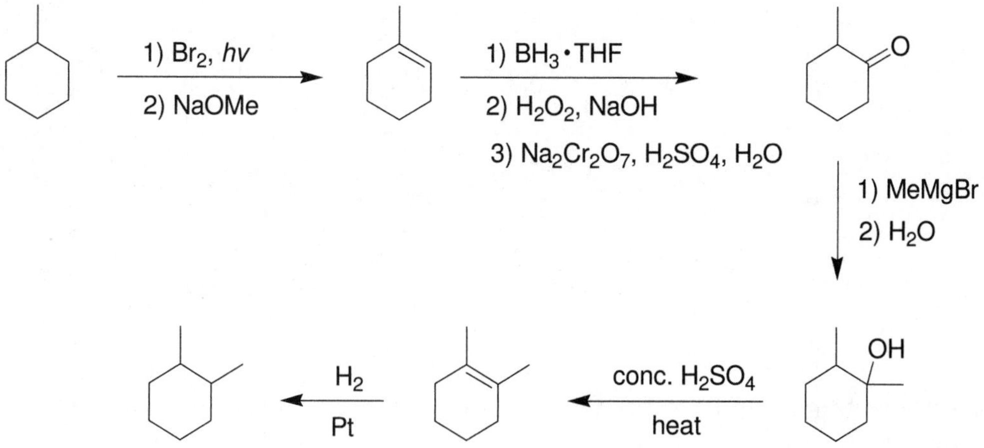

1) Br_2, *hv*
2) NaOMe

1) $BH_3 \cdot THF$
2) H_2O_2, NaOH
3) $Na_2Cr_2O_7$, H_2SO_4, H_2O

1) MeMgBr
2) H_2O

conc. H_2SO_4
heat

H_2
Pt

13.60.

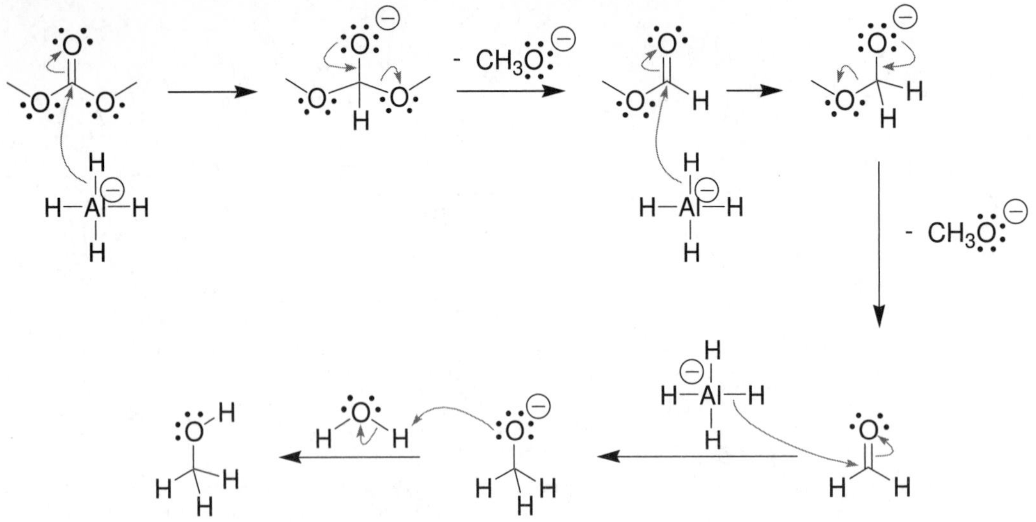

- CH_3O^-

- CH_3O^-

13.61.

- H_2O

methyl shift

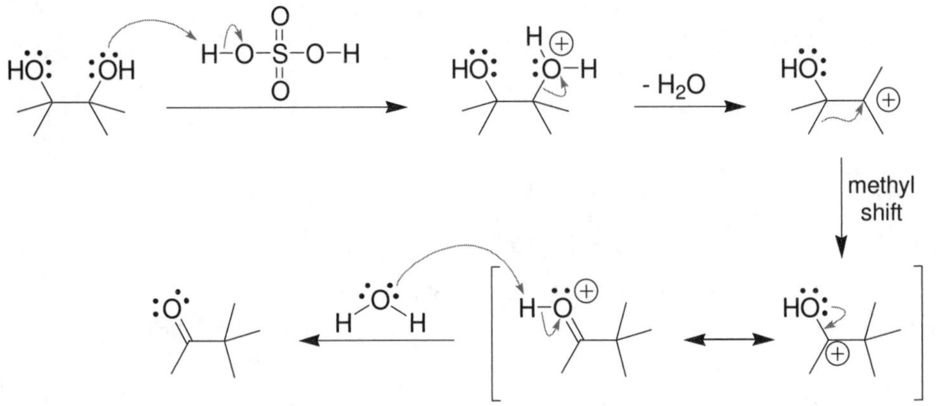

13.62. One carbon atom is oxidized from an oxidation state of +1 to an oxidation state of +2, while the other carbon atom is reduced from an oxidation state of +1 to an oxidation state of 0. Overall, the starting material does not undergo a net change in oxidation state and is, therefore, neither reduced nor oxidized.

Chapter 14
Ethers and Epoxides; Thiols and Sulfides

Review of Concepts

Fill in the blanks below. To verify that your answers are correct, look in your textbook at the end of Chapter 14. Each of the sentences below appears verbatim in the section entitled *Review of Concepts and Vocabulary*.

- Ethers are often used as ____________ for organic reactions.
- Cyclic polyethers, or __________ **ethers**, are capable of solvating metal ions in organic (nonpolar) solvents.
- Ethers can be readily prepared from the reaction between an alkoxide ion and an ______________________, a process called a **Williamson ether synthesis**. This process works best for _________ or _____________ alkyl halides. ____________ alkyl halides are significantly less efficient, and _____________ alkyl halides cannot be used.
- When treated with a strong acid, an ether will undergo **acidic** _____________ in which it is converted into two alkyl halides.
- When a phenyl ether is cleaved under acidic conditions, the products are ____________ and an alkyl halide.
- Ethers undergo autooxidation in the presence of atmospheric oxygen to form _______________________.
- Substituted oxiranes are also called _______________.
- _____________ can be converted into epoxides by treatment with peroxy acids or via halohydrin formation and subsequent epoxidation.
- ____________ catalysts can be used to achieve the enantioselective epoxidation of allylic alcohols.
- Epoxides will undergo **ring-opening reactions** in: 1) conditions involving a strong nucleophile, or under 2) _____-catalyzed conditions. When a strong nucleophile is used, the nucleophile attacks at the ______-substituted position.
- Sulfur analogs of alcohols contain an SH group rather than an OH group, and are called ____________.
- Thiols can be prepared via an S_N2 reaction between sodium hydrosulfide (NaSH) and a suitable _______________________.
- The sulfur analogs of ethers (thioethers) are called ____________**.**
- Sulfides can be prepared from thiols in a process that is essentially the sulfur analog of the Williamson ether synthesis, involving a ___________ ion, rather than an alkoxide.

Review of Skills

Fill in the blanks and empty boxes below. To verify that your answers are correct, look in your textbook at the end of Chapter 14. The answers appear in the section entitled *SkillBuilder Review*.

14.1 Naming an Ether

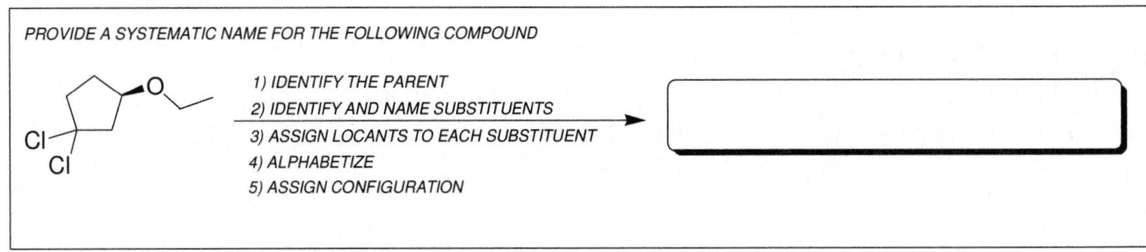

PROVIDE A SYSTEMATIC NAME FOR THE FOLLOWING COMPOUND
Cl
Cl
O
1) IDENTIFY THE PARENT
2) IDENTIFY AND NAME SUBSTITUENTS
3) ASSIGN LOCANTS TO EACH SUBSTITUENT
4) ALPHABETIZE
5) ASSIGN CONFIGURATION

14.2 Preparing an Ether via a Williamson Ether Synthesis

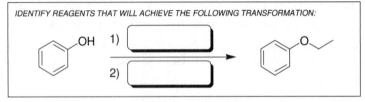

IDENTIFY REAGENTS THAT WILL ACHIEVE THE FOLLOWING TRANSFORMATION:

OH 1) 2) O

14.3 Preparing Epoxides

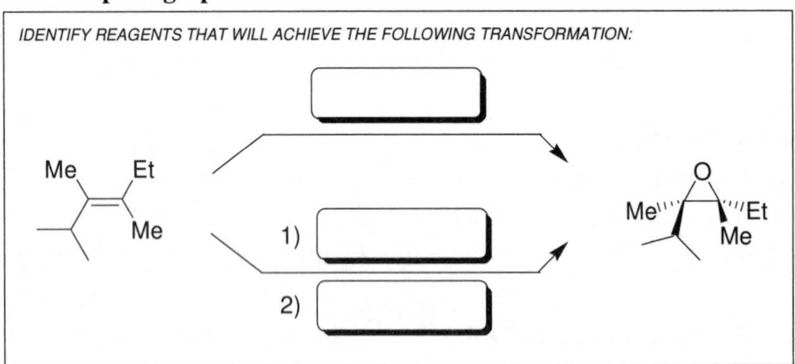

14.4 Drawing the Mechanism and Predicting the Product of the Reaction between a Strong Nucleophile and an Epoxide

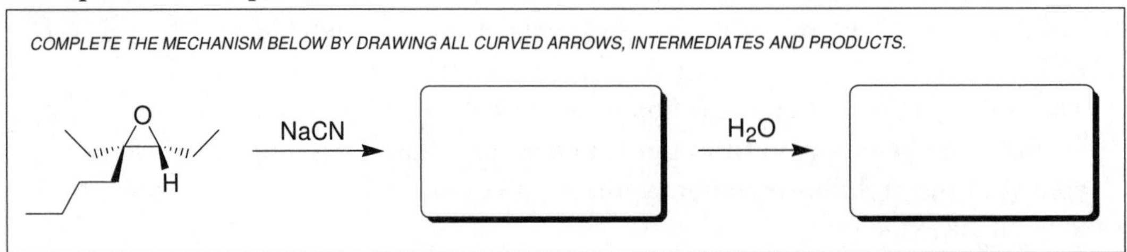

14.5 Drawing the Mechanism and Predicting the Product of Acid-Catalyzed Ring-Opening

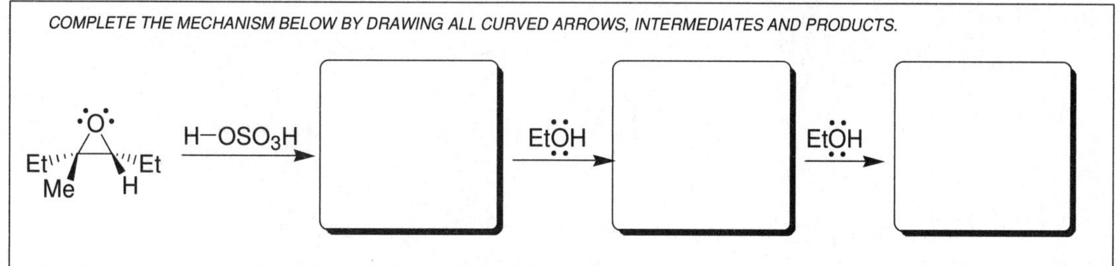
COMPLETE THE MECHANISM BELOW BY DRAWING ALL CURVED ARROWS, INTERMEDIATES AND PRODUCTS.
O
Et
Et
Me
H
H–OSO3H
EtOH
EtOH

14.6 Installing Two Adjacent Functional Groups

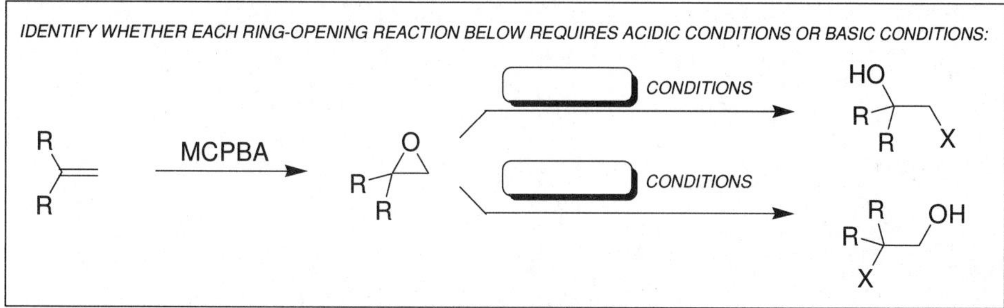
IDENTIFY WHETHER EACH RING-OPENING REACTION BELOW REQUIRES ACIDIC CONDITIONS OR BASIC CONDITIONS:
MCPBA
CONDITIONS
CONDITIONS
HO
X
OH
X

14.7 Choosing the Appropriate Grignard Reaction

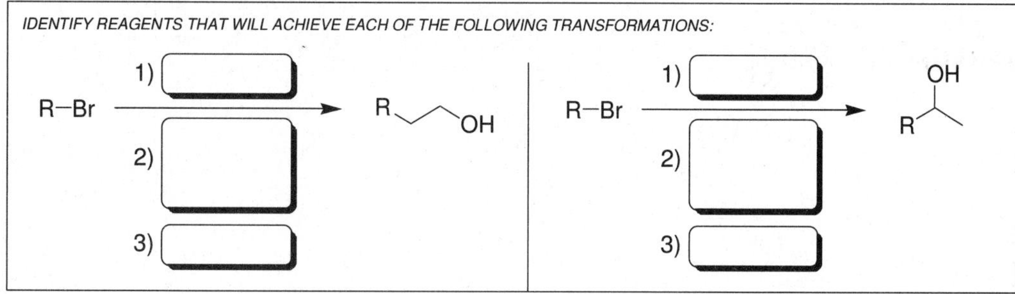
IDENTIFY REAGENTS THAT WILL ACHIEVE EACH OF THE FOLLOWING TRANSFORMATIONS:
R–Br
1)
2)
3)
OH
R–Br
1)
2)
3)
OH

Review of Reactions

Identify the reagents necessary to achieve each of the following transformations. To verify that your answers are correct, look in your textbook at the end of Chapter 14. The answers appear in the section entitled *Review of Reactions.*

Preparation of Ethers

Williamson ether synthesis

R–OH ⟶ R–O–R

Alkoxymercuration-demercuration

$R_2C{=}CHR$ ⟶ $R_2C(OR)–CH_2R$

Reactions of Ethers

Acidic cleavage

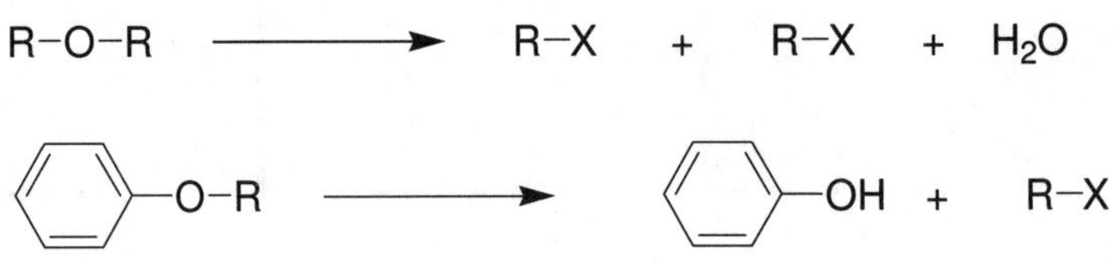

R–O–R ⟶ R–X + R–X + H_2O

Ph–O–R ⟶ Ph–OH + R–X

Autooxidation

OOH

O ⟶ O

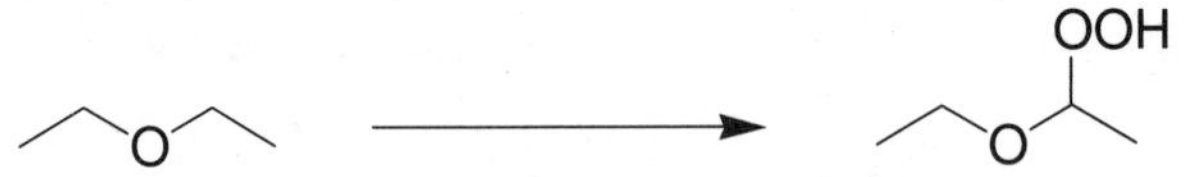

Preparation of Epoxides

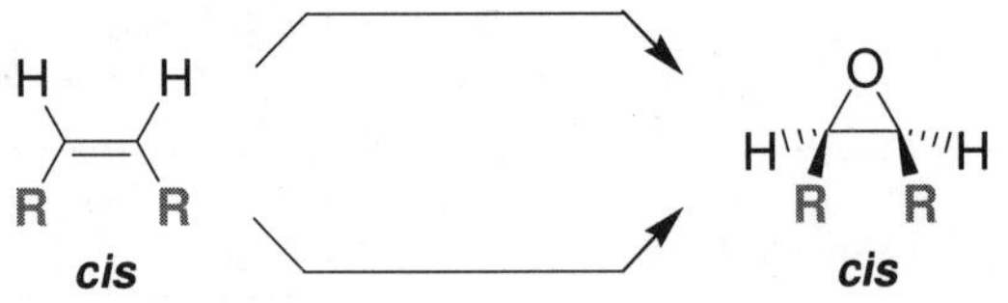

H H R R *cis* ⟶ O H H R R *cis*

Enantioselective Epoxidation

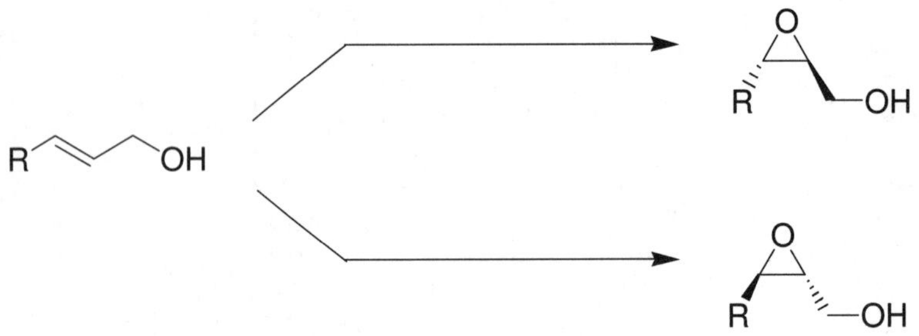

R–CH=CH–CH2–OH ⟶ O R OH

⟶ O R OH

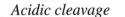

Ring-Opening Reactions of Epoxides

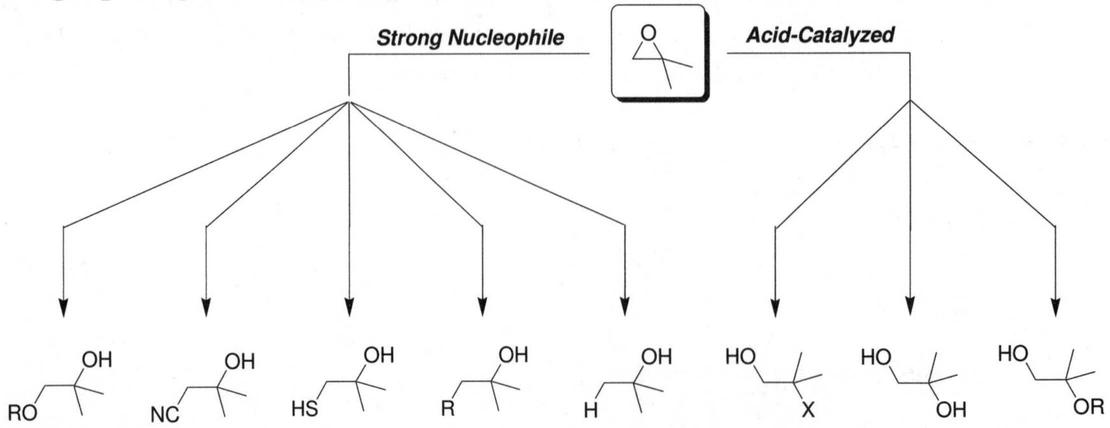

Strong Nucleophile | *Acid-Catalyzed*

OH RO | OH NC | OH HS | OH R | OH H | HO X | HO OH | HO OR

Thiols and Sulfides

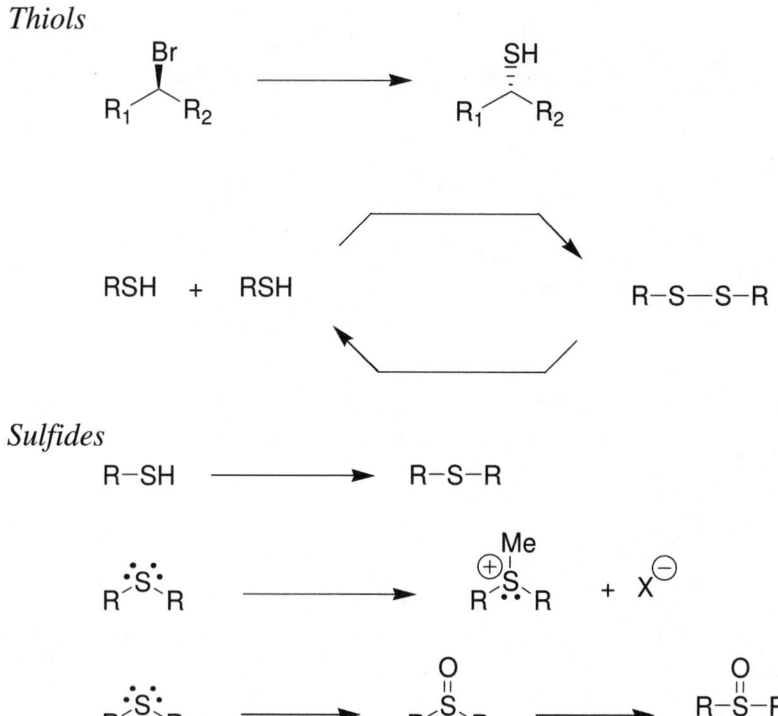

Solutions

14.1.

a) 2-ethoxypropane
b) (*S*)-2-chloro-1-ethoxypropane
c) 2,4-dichloro-1-ethoxybenzene
d) (*1R,2R*)-2-ethoxycyclohexanol
e) 1-ethoxycyclohexene

14.2.

a)

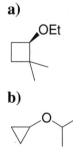

OEt

b)

O

14.3.

Common names in parentheses:

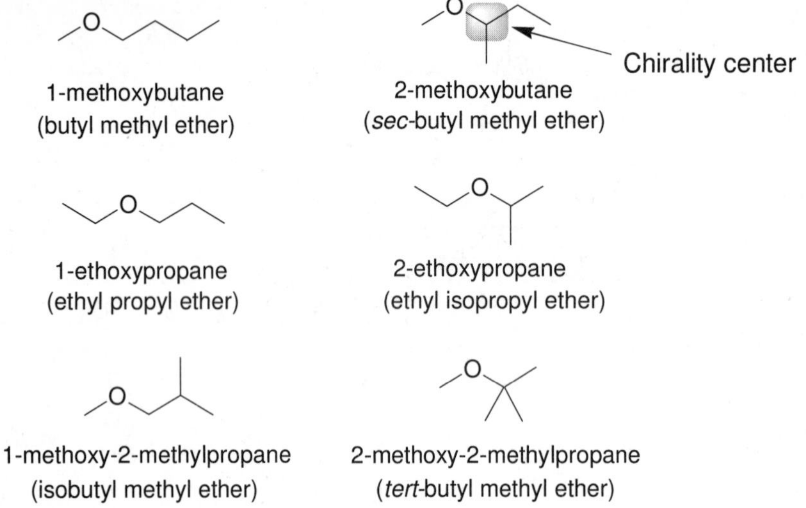

14.4.

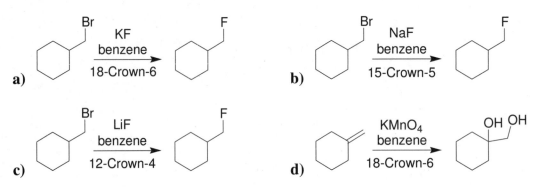

a) Br → KF, benzene, 18-Crown-6 → F

b) Br → NaF, benzene, 15-Crown-5 → F

c) Br → LiF, benzene, 12-Crown-4 → F

d) → $KMnO_4$, benzene, 18-Crown-6 → OH OH

14.5.

a) A Williamson ether synthesis will be more efficient with a less sterically hindered substrate, since the process involves an S_N2 reaction. Therefore, in this case, it is better to start with a secondary alcohol and a primary alkyl halide, rather than a primary alcohol and a secondary alkyl halide:

OH → 1) Na; 2) I → O

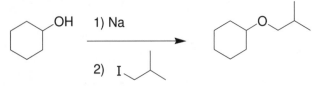

b) In this case, it is better to start with a secondary alcohol and a primary alkyl halide, rather than a primary alcohol and a secondary alkyl halide:

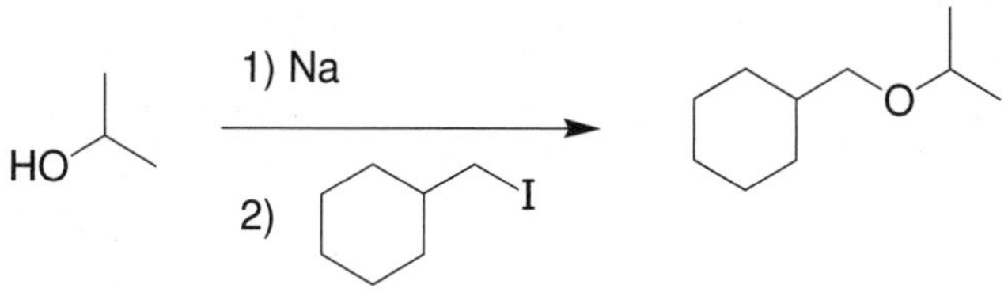

HO — 1) Na, 2) (iodomethyl)cyclohexane (I) → ether (O)

c) In this case, it is better to start with a tertiary alcohol and a methyl halide, rather than methanol and a tertiary alkyl halide:

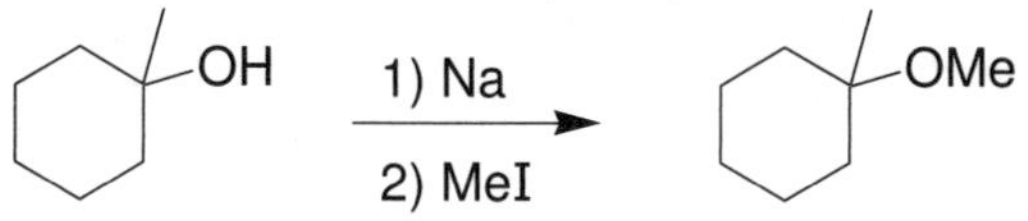

OH — 1) Na, 2) MeI → OMe

14.6.

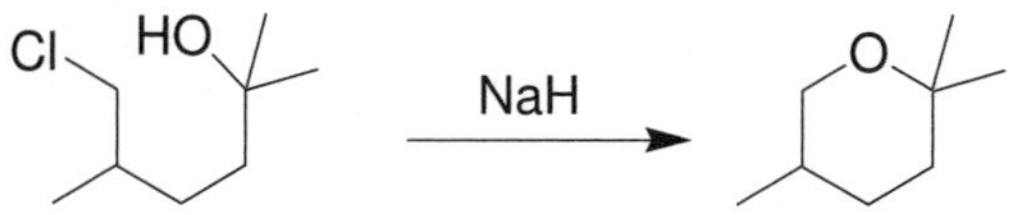

Cl, HO — NaH → O

14.7. No. The Williamson ether synthesis employs an S_N2 process, which cannot occur readily at tertiary or vinylic positions. Making this ether would require at least one of these two processes, neither of which can be used.

14.8.

a)

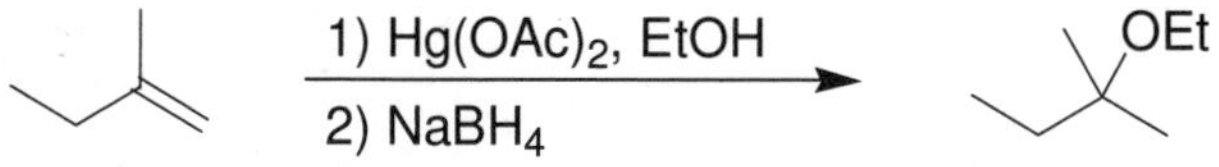

1) $Hg(OAc)_2$, EtOH
2) $NaBH_4$
→ OEt

b)

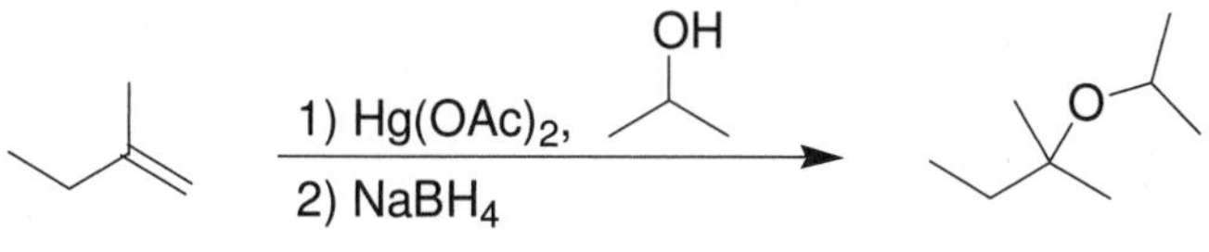

1) $Hg(OAc)_2$, (OH)
2) $NaBH_4$
→ O

c)

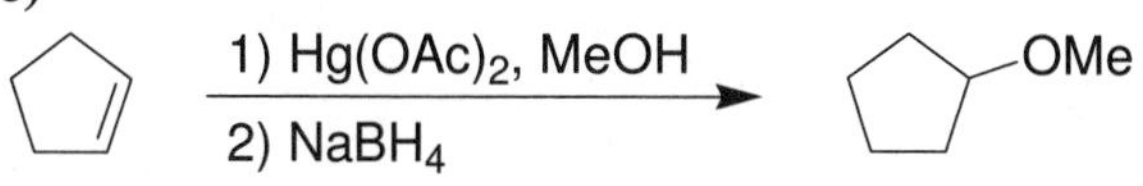

1) $Hg(OAc)_2$, MeOH
2) $NaBH_4$
→ OMe

d)

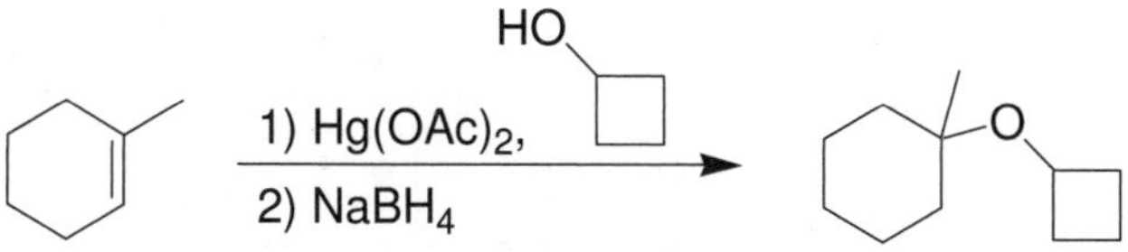

1) $Hg(OAc)_2$, (HO)
2) $NaBH_4$
→ O

14.9.

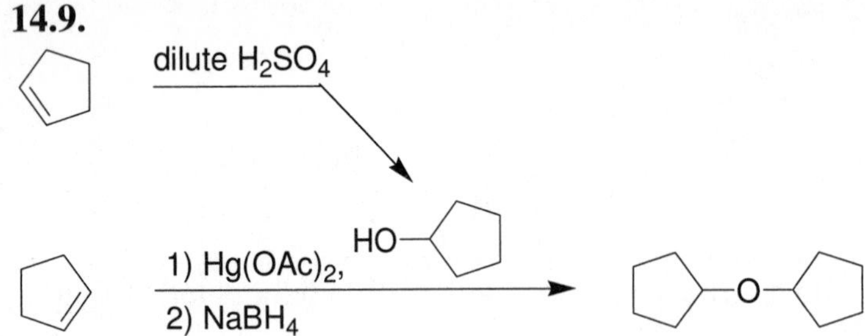

14.10.

1) $BH_3 \cdot THF$

2) H_2O_2, NaOH

1) $Hg(OAc)_2$, OH

2) $NaBH_4$

O

14.11.

a)

O HBr → 2 Br + H_2O

b)

O HI → I I + H_2O

c)

O HBr → OH + EtBr

d)

O HI → OH I

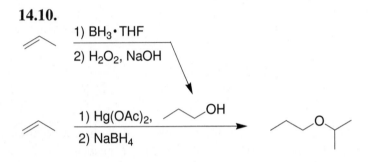

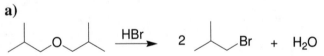

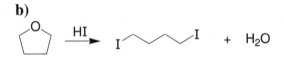

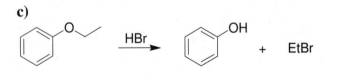

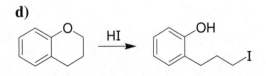

e)

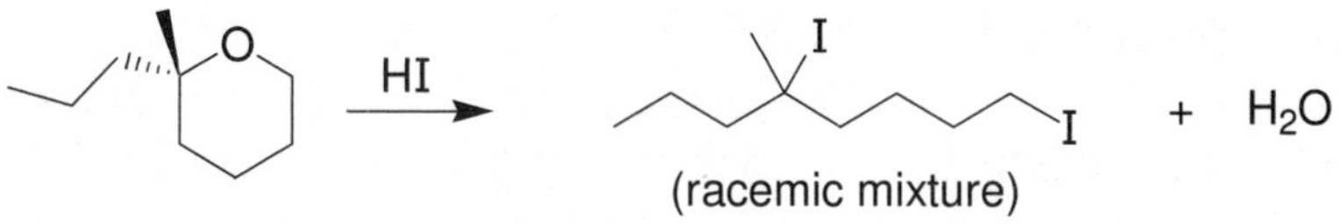

HI → (racemic mixture) + H_2O

f)

HBr → + EtBr + H_2O

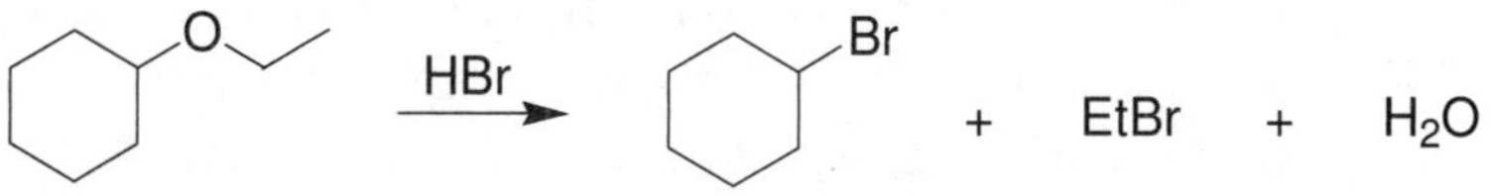

14.12.

a) 2-methyl-1,2-epoxypropane or 1,1-dimethyloxirane
b) 1,1-diphenyl-1,2-epoxyethane or 1,1-diphenyloxirane
c) 1,2-epoxycyclohexane

14.13.

a) (*S*)-2-phenyl-1,2-epoxypropane or (*S*)-1-methyl-1-phenyloxirane
b) (*3R,4R*)-3,4-epoxyheptane or (*1R,2R*)-1-ethyl-2-propyloxirane
c) (*2R,3S*)-4-methyl-2,3-epoxypentane or (*1S,2R*)-1-isopropyl-2-methyloxirane

14.14.

a)

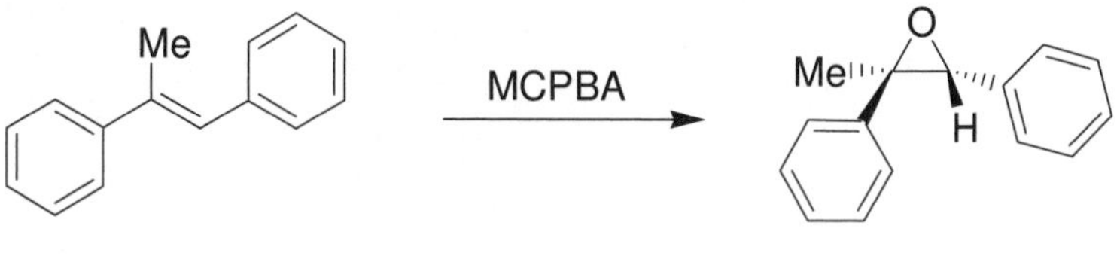

Me MCPBA → Me O H

b)

MCPBA → O H H Me

c)

O MCPBA → O O H Me

d)

MCPBA → O H H

14.15. This process for epoxide formation involves deprotonation of the hydroxyl group, followed by an intramolecular S_N2 attack. The S_N2 step requires back-side attack, which can only be achieved when both the hydroxyl group and the bromine occupy axial positions. Due to the steric bulk of a tert-butyl group, Compound A spends most of time in a chair conformation that has the tert-butyl group in an equatorial position. In this conformation, the OH and Br are indeed in axial positions, so the reaction can occur quite rapidly. In contrast, Compound B spends most of its time in a chair conformation in which the OH and Br occupy equatorial positions. The S_N2 process cannot occur from this conformation.

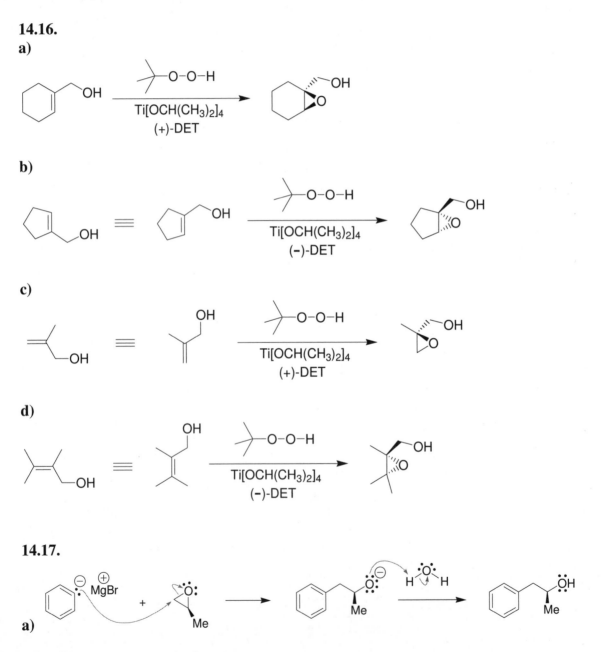

14.16.

a)

O-O-H
OH
Ti[OCH(CH3)2]4
(+)-DET
OH
O

b)

OH ≡ OH
O-O-H
Ti[OCH(CH3)2]4
(-)-DET
OH
O

c)

OH ≡ OH
O-O-H
Ti[OCH(CH3)2]4
(+)-DET
OH
O

d)

OH ≡ OH
O-O-H
Ti[OCH(CH3)2]4
(-)-DET
OH
O

14.17.

a)

MgBr + O: Me → O: Me → H O H → OH Me

b)

N≡C: + (epoxide, Me) → NC–CH₂–CH(Me)–O⁻ + H_2O → NC–CH₂–CH(Me)–OH

c)

HS:⁻ + (epoxide, Me) → HS–CH₂–CH(Me)–O⁻ + H_2O → HS–CH₂–CH(Me)–OH

d)

AlH_4^- + (epoxide, Me) → (CH₃)₂CH–O⁻ ≡ → + H_2O → (CH₃)₂CH–OH

e)

(epoxide: Ph, Et, H, Me) + ⁻:SH → (alkoxide, SH, Me, H, Et) + H_2O → (alcohol, SH, Me, H, Et)

f)

(epoxide: Ph, Et, H, Me) + AlH_4^- → (alkoxide, Et, Me, H, H) + H_2O → (alcohol) ≡ (3-phenyl-3-pentanol)

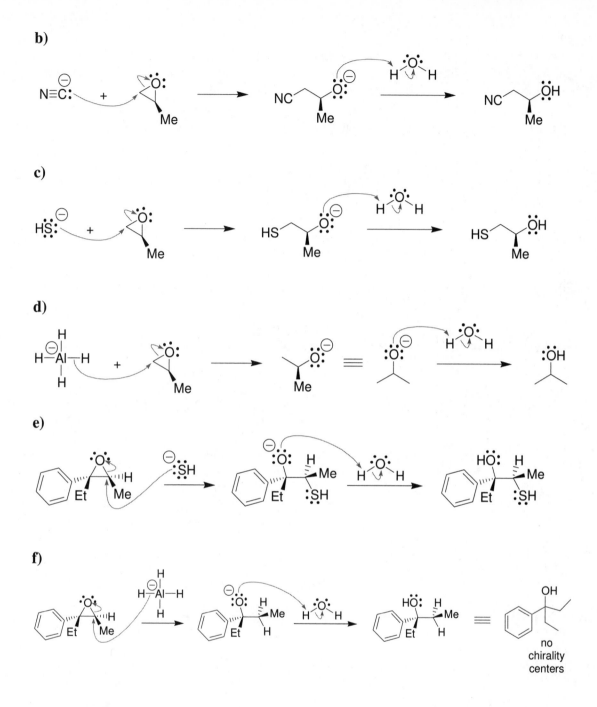

no chirality centers

14.18. The reaction yields a meso compound, regardless of which electrophilic position is attacked by hydroxide.

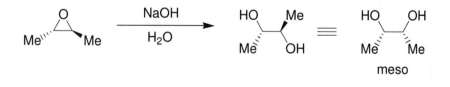

(cis-2,3-dimethyloxirane) $\xrightarrow[H_2O]{NaOH}$ HO, Me / Me, OH ≡ HO, OH / Me, Me

meso

14.19.

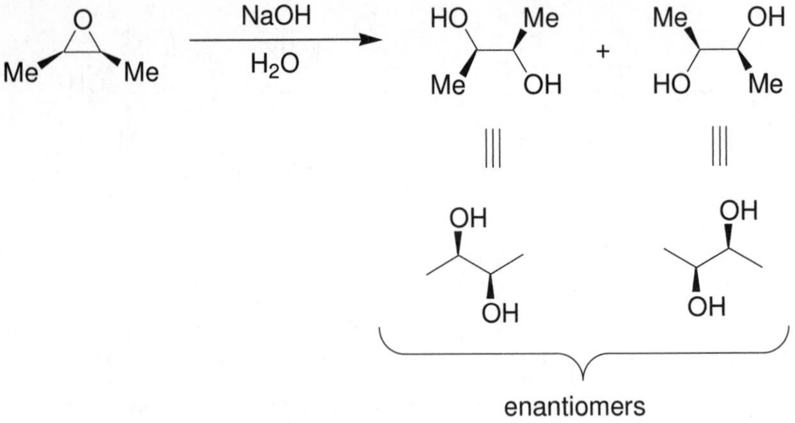

NaOH
H_2O

HO Me Me OH
Me OH + HO Me

OH OH
OH OH

enantiomers

14.20.

a)

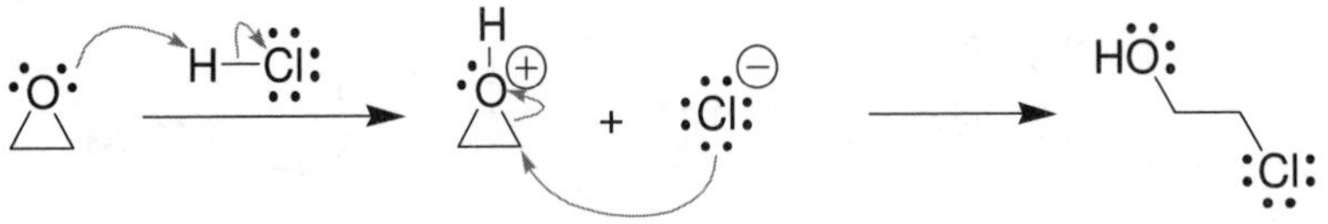

H—Cl H HO
O O + Cl Cl

b)

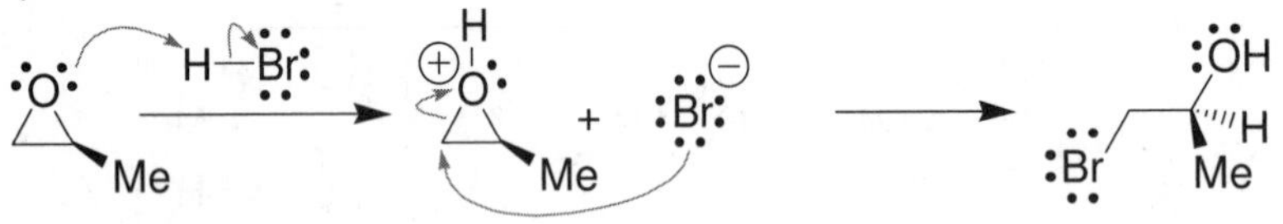

H—Br H OH
O O + Br H
Me Me Br Me

c)

H—OSO_3H H EtOH Et OH
O O H
H H Et—O Me
Et Me Et Me H
EtOH

Et OH
H
EtO Me

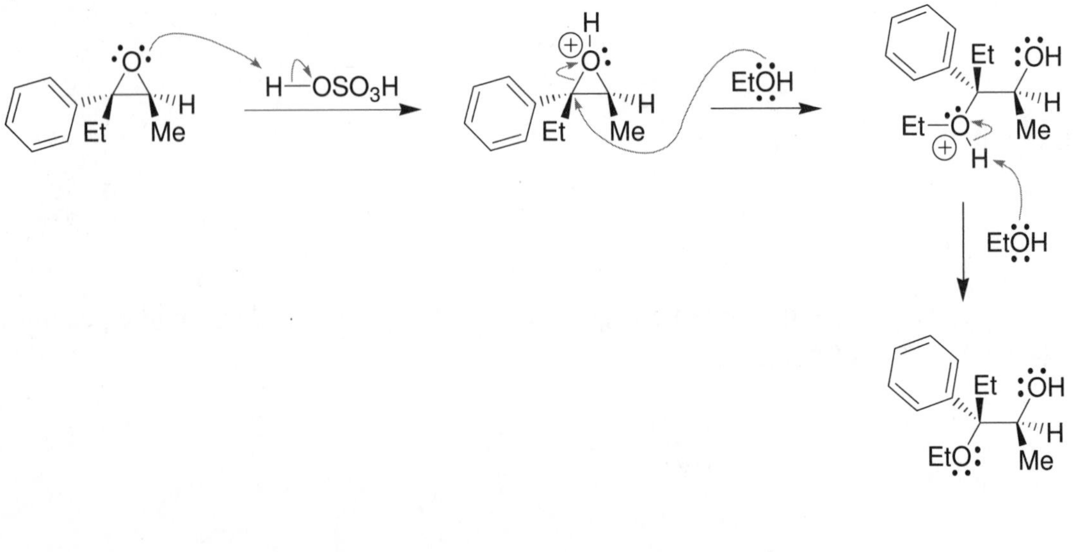

d)

Me, Et, H, Me; $H-OSO_3H$; H, Me, Et, H, Me; H_2O; Me, Et, :OH, H, H–O, Me, H; H_2O; Me, Et, :OH, H, HO:, Me

e)

H, Me; $H-OSO_3H$; H, H, Me; MeOH; :OH, H, Me–O, Me, H; MeOH; :OH, H, MeO:, Me

f)

Et, H, Me; H–Br:; H, Et, H, Me, + :Br:; Et, :OH, H, :Br:, Me

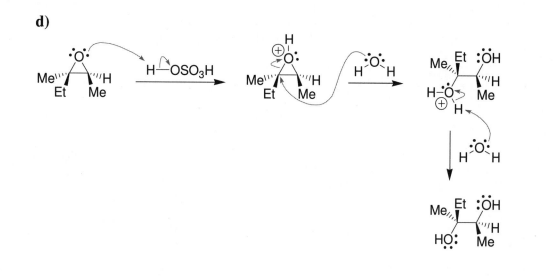

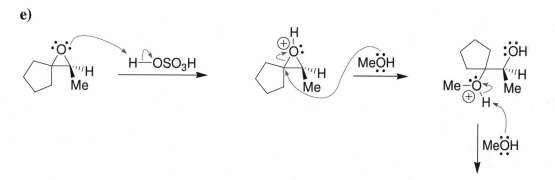

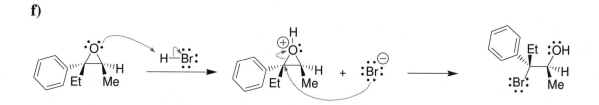

14.21.

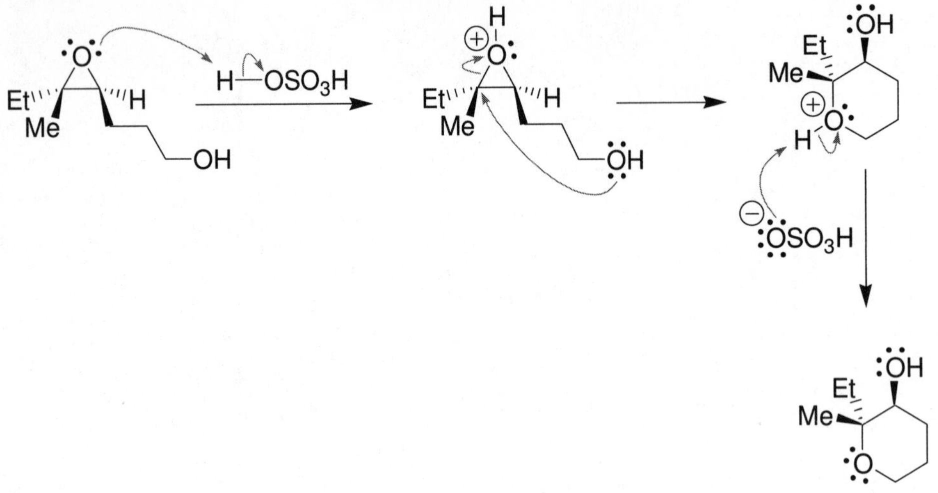

Et, Me, H, OH, H—OSO$_3$H

Et, Me, H, H, OH

Et, Me, OH, H

$^{\ominus}$OSO$_3$H

Et, Me, OH, O

14.22.

a)

Br —NaSH→ SH

b)

Br —NaSH→ SH

c)

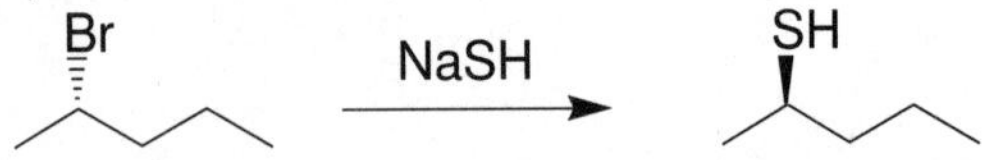

Br —NaSH→ SH

14.23.

a)

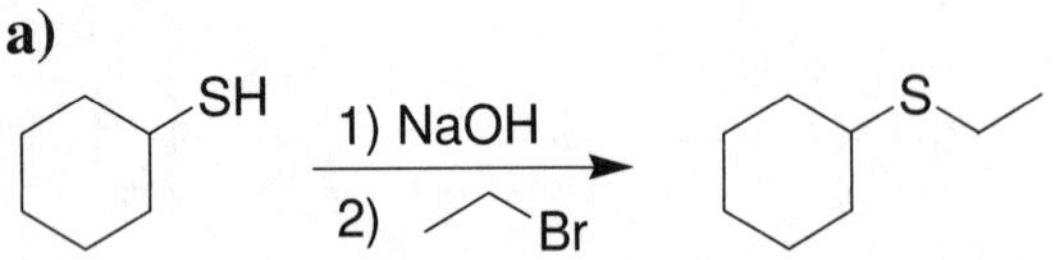

SH —1) NaOH; 2) Br→ S

b)

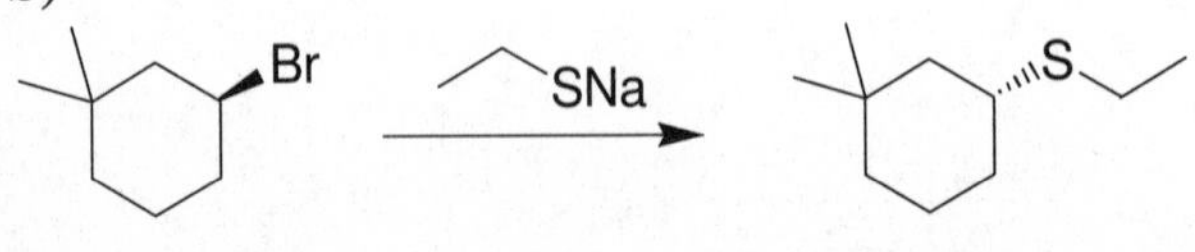

Br —SNa→ S

c)

$NaIO_4$

d)

2 H_2O_2

14.24.

a)

1) MCPBA
2) NaCN
3) H_2O

HO, CN

b)

1) Br_2, *hv*
2) NaOEt

1) MCPBA
2) NaOH
3) H_2O

1) MCPBA
2) H_3O^+

OH, OH + En

c)

1) MCPBA
2) NaSH
3) H_2O

OH, SH

d)

1) MCPBA
2) [H_2SO_4], H_2S

SH, OH

e)

O

1) LAH
2) H_2O

OH

$Na_2Cr_2O_7$
H_2SO_4, H_2O

O

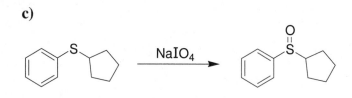

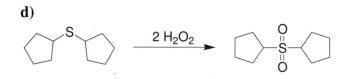

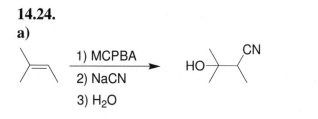

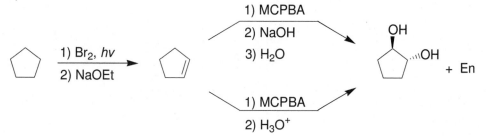

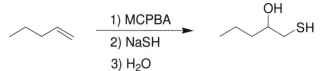

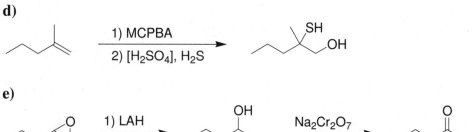

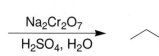

14.25.

(ethyl iodide, I) —t-BuOK→ (ethylene) —1) MCPBA; 2) H_3O^+→ HO–CH$_2$CH$_2$–OH

14.26.

a)

(bromocyclohexane, Br) —1) Mg; 2) (ethylene oxide, O); 3) H_2O→ (2-cyclohexylethanol, OH)

b)

(bromocyclohexane, Br) —1) Mg; 2) (acetaldehyde, O, H); 3) H_2O→ (1-cyclohexylethanol, OH)

c)

(bromocyclohexane, Br) —1) Mg; 2) (2,2-dimethyloxirane, O); 3) H_2O→ (1-cyclohexyl-2-methyl-2-propanol, OH)

d)

(bromocyclohexane, Br) —1) Mg; 2) (2-methylpropanal, O, H); 3) H_2O→ (1-cyclohexyl-2-methyl-1-propanol, OH) —$Na_2Cr_2O_7$; H_2SO_4, H_2O→ (cyclohexyl isopropyl ketone, O)

e)

(chlorobenzene, Cl) —1) Mg; 2) (ethylene oxide, O); 3) H_2O→ (2-phenylethanol, OH) —$SOCl_2$; py→ (2-phenylethyl chloride, Cl)

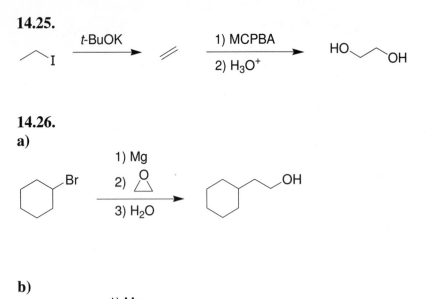

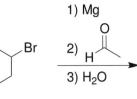

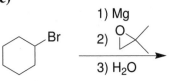

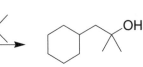

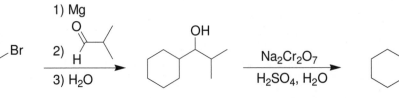

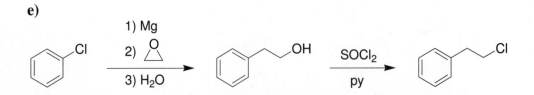

f)

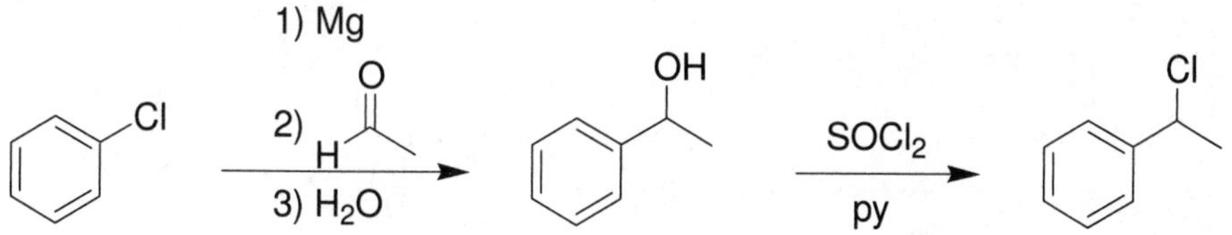

1) Mg

2) CH_3CHO

3) H_2O

$SOCl_2$

py

OH

Cl

g)

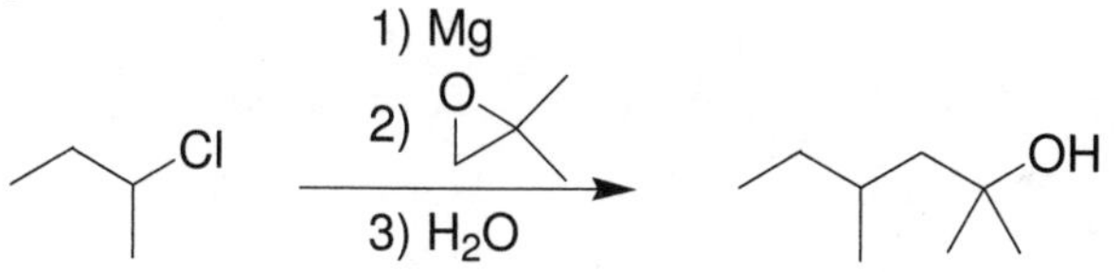

1) Mg

2)

3) H_2O

Cl

OH

h)

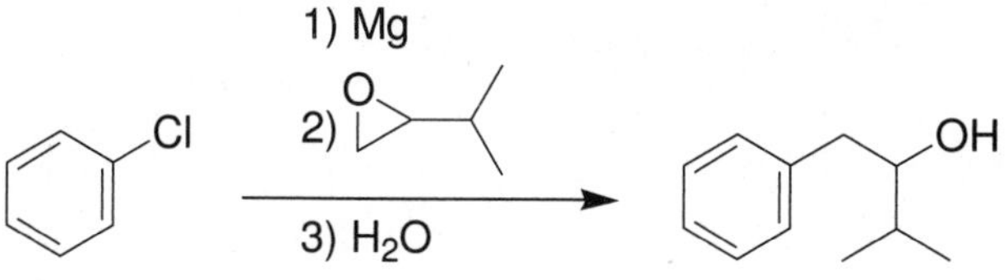

1) Mg

2)

3) H_2O

Cl

OH

i)

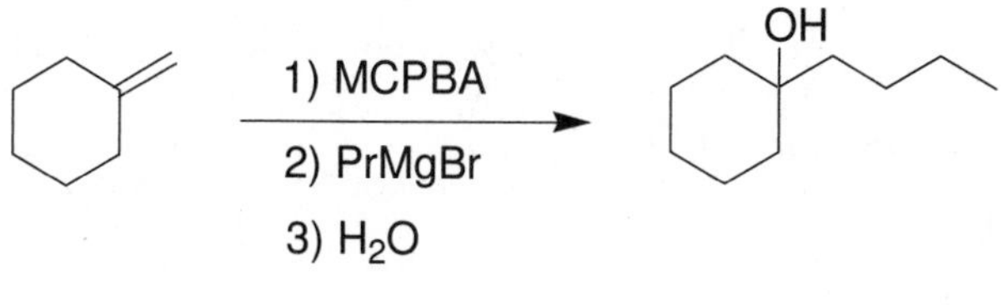

1) MCPBA

2) PrMgBr

3) H_2O

OH

14.27.

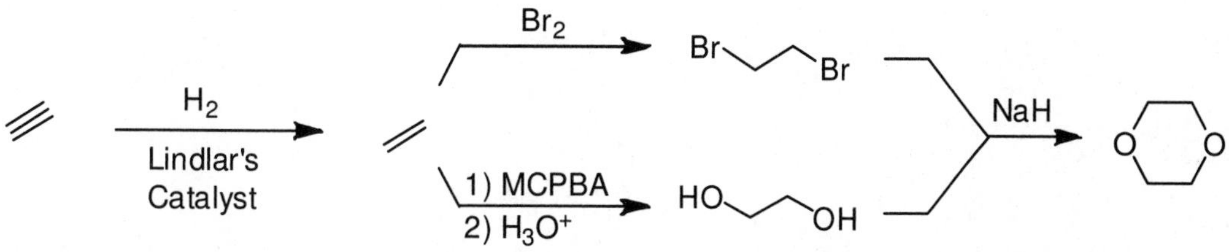

H_2

Lindlar's Catalyst

Br_2

Br Br

1) MCPBA

2) H_3O^+

HO OH

NaH

O O

14.28.

Acetylene $\xrightarrow[\text{Lindlar's Catalyst}]{H_2}$ ethylene $\xrightarrow[\text{NMO}]{OsO_4}$ HO–CH₂CH₂–OH $\xrightarrow[\text{2) } CH_3I\text{ (2 eq)}]{\text{1) Na (2 eq)}}$ H_3CO–CH₂CH₂–OCH_3

14.29.

Ethylene oxide $\xrightarrow[\text{2) } H_2O]{\text{1) MeMgBr}}$ 1-propanol (OH) $\xrightarrow[CH_2Cl_2]{PCC}$ propanal (O, H) $\xrightarrow[\text{2) } H_2O \text{ 3) } Na_2Cr_2O_7, H_2SO_4, H_2O]{\text{1) EtMgBr}}$ 3-pentanone (O) $\xrightarrow[\text{2) } H_2O]{\text{1) EtMgBr}}$ 3-ethyl-3-pentanol (OH) $\xrightarrow[\text{2) EtI}]{\text{1) NaH}}$ ether (O)

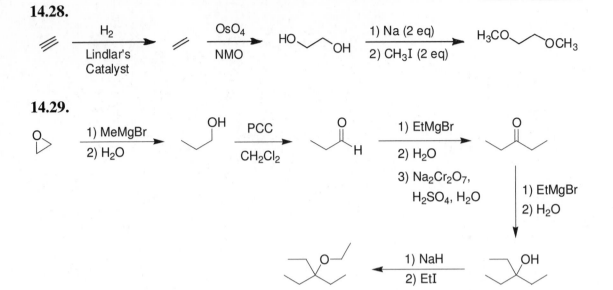

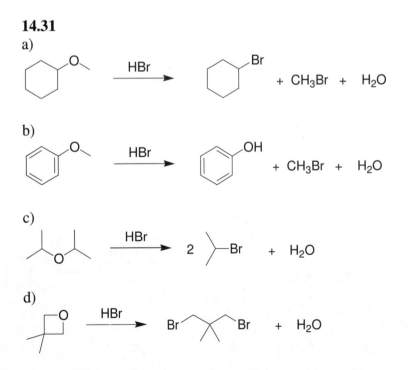

14.30

a) (*1S, 2S*)-1-ethoxy-2-methylcyclohexane
b) (*R*)-2-ethoxybutane
c) (S)-3-hexanethiol
d) ethyl propyl sulfoxide
e) (*E*)-2-ethoxy-3-methyl-2-pentene
f) 1,2-dimethoxybenzene
g) ethyl propyl sulfide

14.31

a)

Methoxycyclohexane $\xrightarrow{HBr}$ bromocyclohexane (Br) + CH_3Br + H_2O

b)

Anisole $\xrightarrow{HBr}$ phenol (OH) + CH_3Br + H_2O

c)

Diisopropyl ether (O) $\xrightarrow{HBr}$ 2 isopropyl bromide (Br) + H_2O

d)

3,3-dimethyloxetane (O) $\xrightarrow{HBr}$ Br–CH₂C(CH₃)₂CH₂–Br + H_2O

14.32

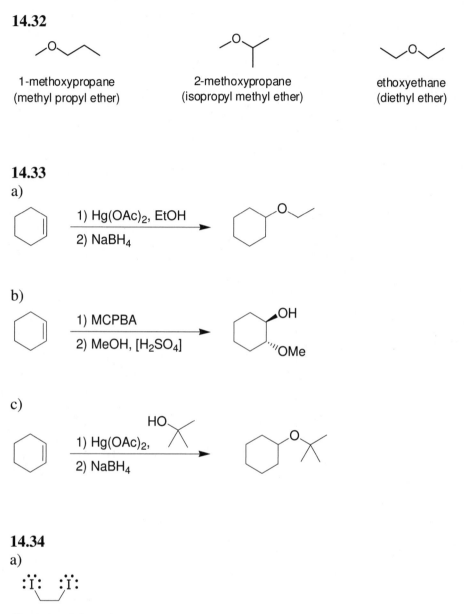

1-methoxypropane
(methyl propyl ether)

2-methoxypropane
(isopropyl methyl ether)

ethoxyethane
(diethyl ether)

14.33

a)

1) $Hg(OAc)_2$, EtOH
2) $NaBH_4$

b)

1) MCPBA
2) MeOH, $[H_2SO_4]$

OH
OMe

c)

HO
1) $Hg(OAc)_2$,
2) $NaBH_4$

O

14.34

a)

:I: :I:

Compound A

b) Two moles of Compound A are produced for every one mole of 1,4-dioxane.

c)

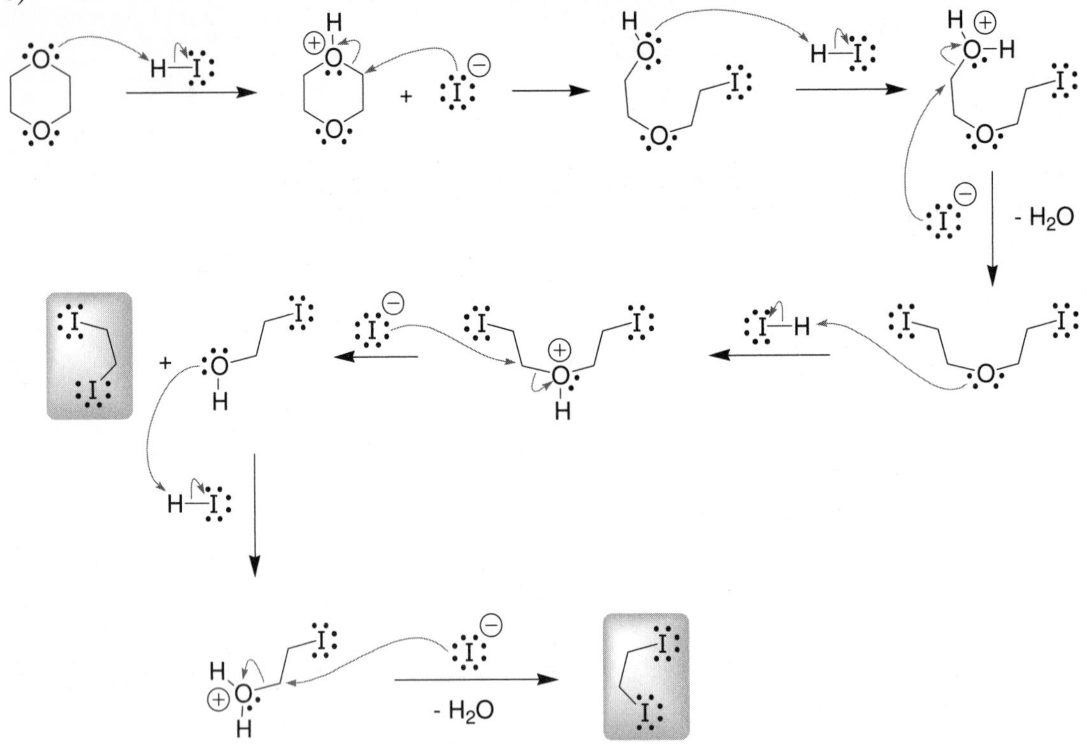

H–I, I⊖, - H_2O, - H_2O

14.35

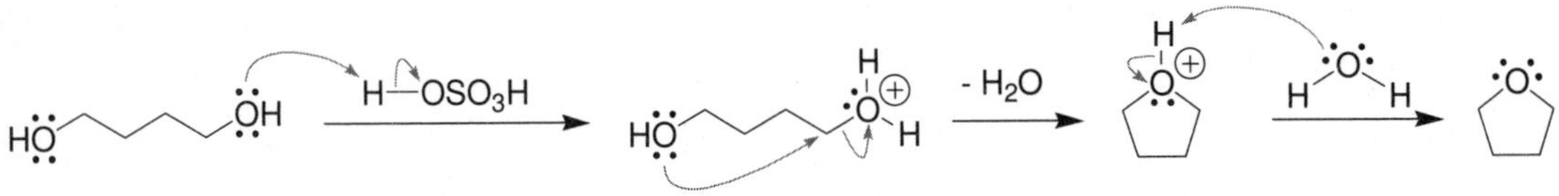

H–OSO_3H, - H_2O

14.36

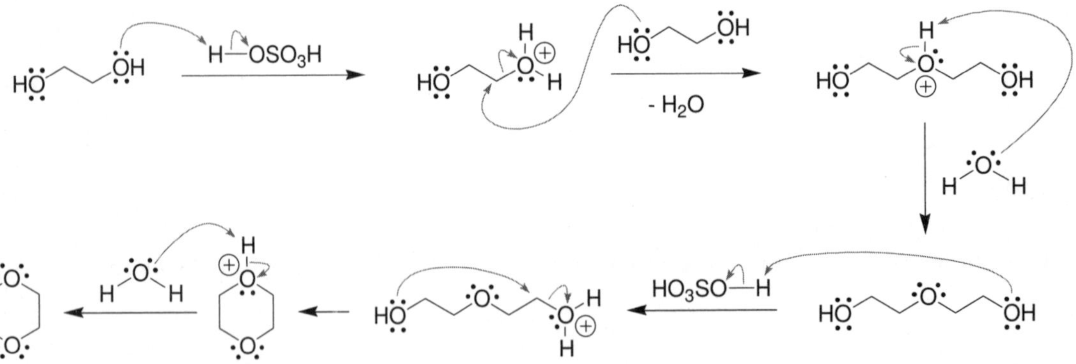

H–OSO_3H, - H_2O, HO_3SO–H

14.37

a) Neither alkyl group (on either side of the oxygen atom) can be installed via a Williamson ether synthesis. Installation of the *tert*-butyl group would require a tertiary alkyl halide, which is too sterically hindered to serve as an electrophile for an S_N2 process. Installation of the phenyl group would require an S_N2 reaction taking place at an sp^2 hybridized center, which does not readily occur.

b) Oxymercuration-demercuration can be used to prepare *tert*-butyl phenyl ether:

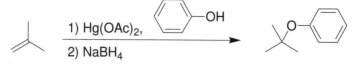

1) $Hg(OAc)_2$, [phenol: OH]
2) $NaBH_4$

14.38 Ethylene oxide has a high degree of ring strain, and readily functions as an electrophile in an S_N2 reaction. The reaction opens the ring and alleviates the ring strain. Oxetane has less ring strain and is, therefore, less reactive as an electrophile towards S_N2. The reaction can still occur, albeit at a slower rate, to alleviate the ring strain associated with the four membered ring. THF has almost no ring strain (very little) and does not function as an electrophile in an S_N2 reaction.

14.39

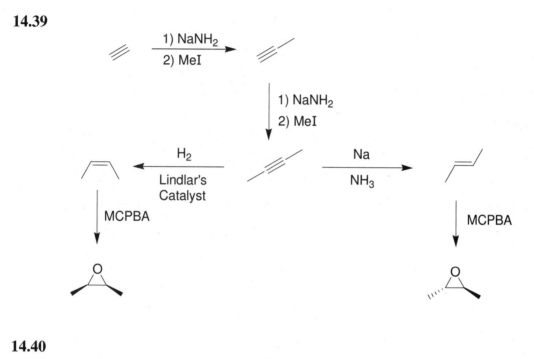

1) $NaNH_2$
2) MeI

1) $NaNH_2$
2) MeI

H_2
Lindlar's Catalyst

Na
NH_3

MCPBA

MCPBA

14.40

HO Br

NaH

14.41

a)

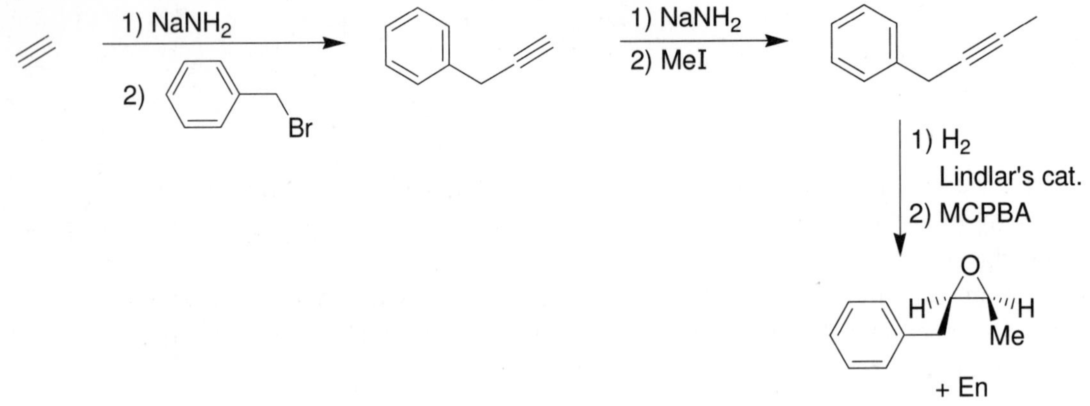

(ethyne) → 1) $NaNH_2$ 2) benzyl bromide (Br) → (3-phenylpropyne) → 1) $NaNH_2$ 2) MeI → (1-phenyl-2-butyne)

↓ 1) H_2 Lindlar's cat. 2) MCPBA

(cis-epoxide: benzyl, H, O, H, Me) + En

b)

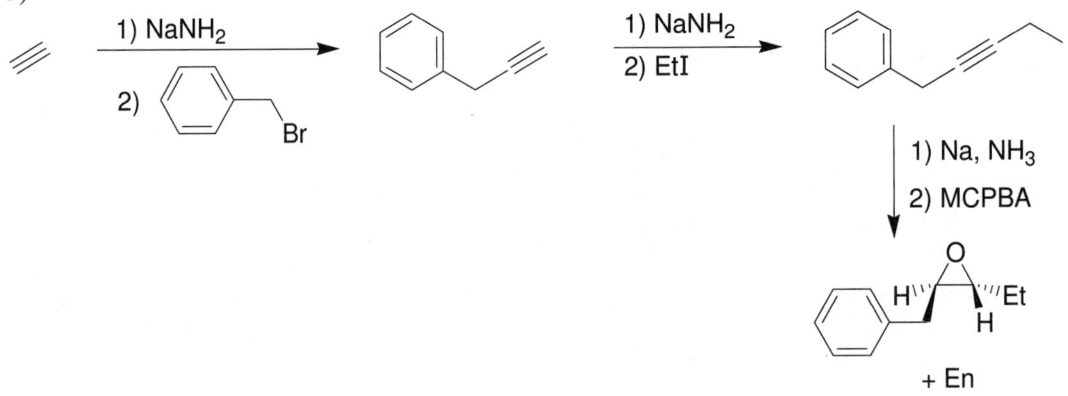

(ethyne) → 1) $NaNH_2$ 2) benzyl bromide (Br) → (3-phenylpropyne) → 1) $NaNH_2$ 2) EtI → (1-phenyl-2-pentyne)

↓ 1) Na, NH_3 2) MCPBA

(trans-epoxide: benzyl, H, O, Et, H) + En

c)

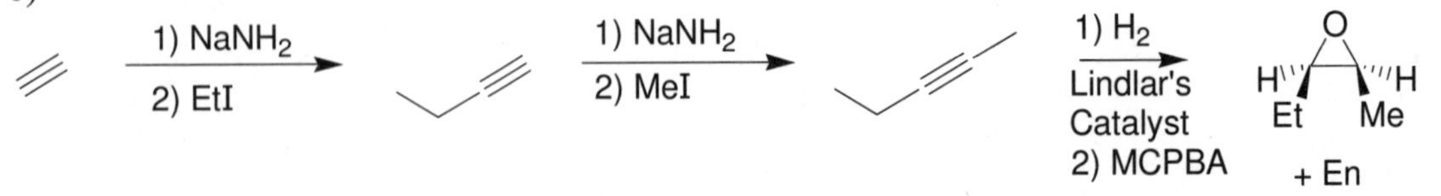

(ethyne) → 1) $NaNH_2$ 2) EtI → (1-butyne) → 1) $NaNH_2$ 2) MeI → (2-pentyne) → 1) H_2 Lindlar's Catalyst 2) MCPBA → (epoxide: H, O, H, Et, Me) + En

d)

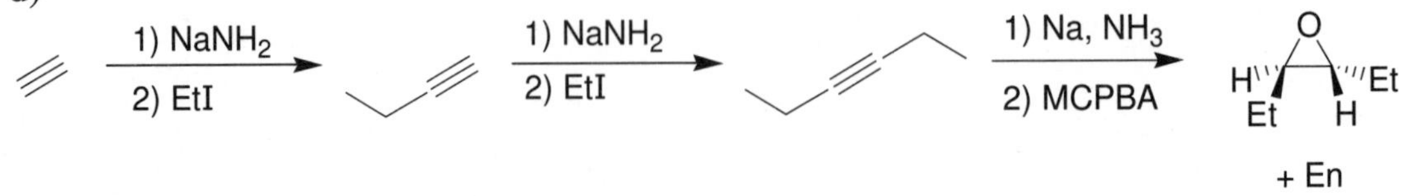

(ethyne) → 1) $NaNH_2$ 2) EtI → (1-butyne) → 1) $NaNH_2$ 2) EtI → (3-hexyne) → 1) Na, NH_3 2) MCPBA → (epoxide: H, O, Et, Et, H) + En

14.42

a)

1) RCO_3H
2) MeMgBr
3) H_2O

OH

b)

1) $Hg(OAc)_2$, MeOH
2) $NaBH_4$

OMe
(racemic)

c)

1) MCPBA
2) NaSH

SH
OH
(racemic)

d)

OH
1) Na
2) EtCl
OEt

e)

OH
1) Na
2) O
3) H_2O
O
OH

f)

Cl
1) Mg, diethyl ether
2) O
3) H_2O
OH

g)

OH
1) Na
2) O
3) H_2O
O
OH

h)

Cl
1) Mg, diethyl ether
2) O
3) H_2O
OH

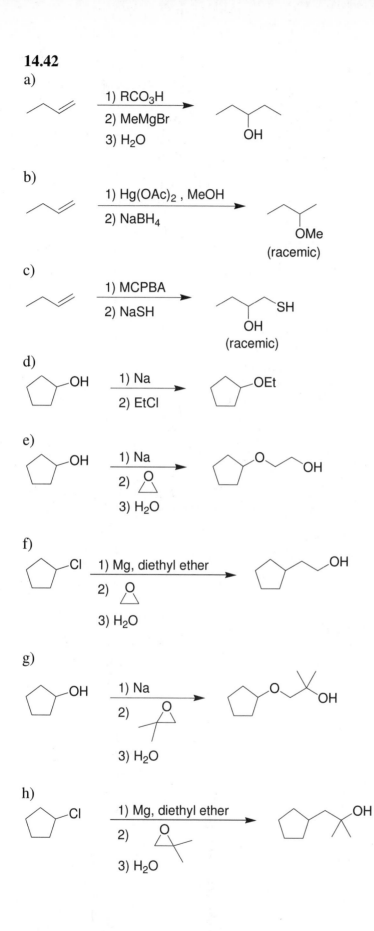

14.43

a)

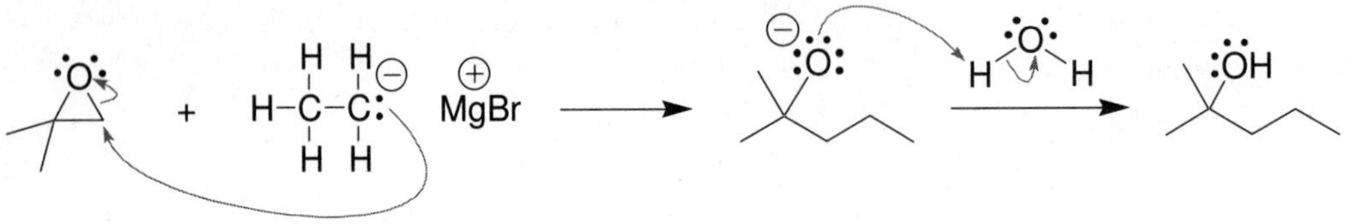

H–C–C: MgBr ⊕ ⊖

:OH

b)

:H ⊖

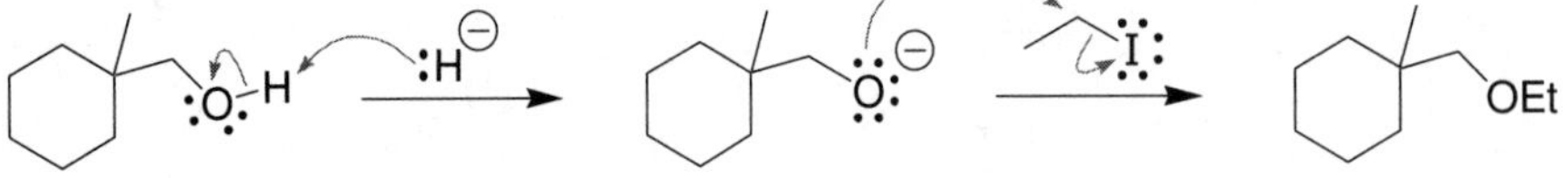

OEt

c)

⊖ :C≡C–H

HO

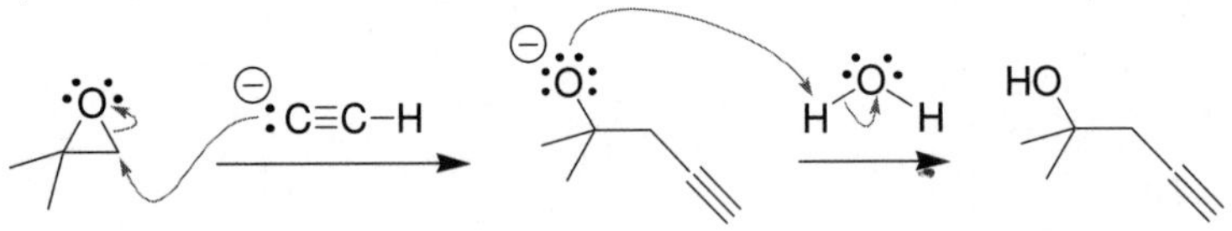

d)

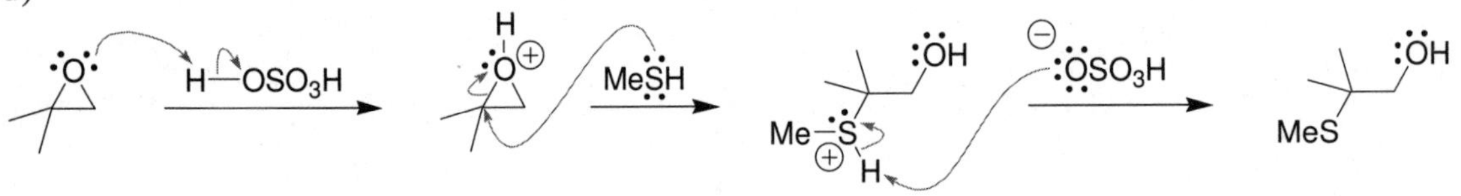

H–OSO$_3$H

MeSH

:OH

Me–S ⊕ H

⊖ :OSO$_3$H

MeS

e)

Cl

:H ⊖

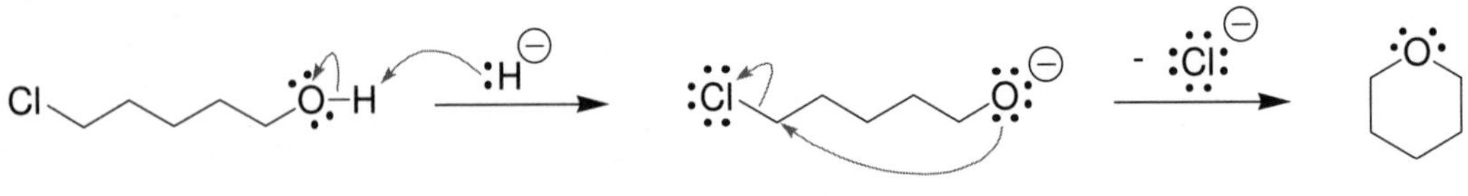

:Cl

- :Cl: ⊖

f)

:Cl

⊖ :OH

:Cl:

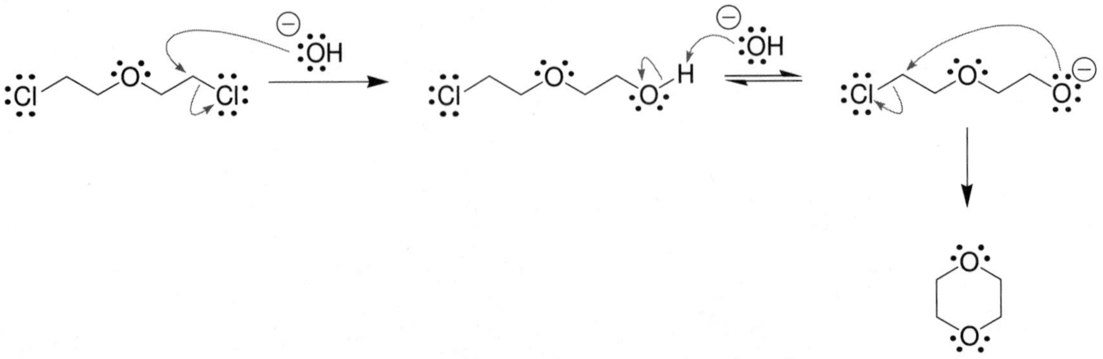

14.44

HBr → Br(CH$_2$)$_4$Br + H_2O

14.45

HBr

Br Br

14.46

H H H−C−C: H H MgBr

Et Me O Cl

Et Me O

H H H−C−C: H H MgBr

Et Me OH Et

H O H

Et Me O Et

14.47

a)

1) $Hg(OAc)_2$, MeOH

2) $NaBH_4$

OMe

b)

1) $Hg(OAc)_2$, MeOH

2) $NaBH_4$

OMe

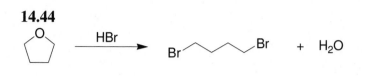

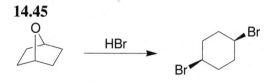

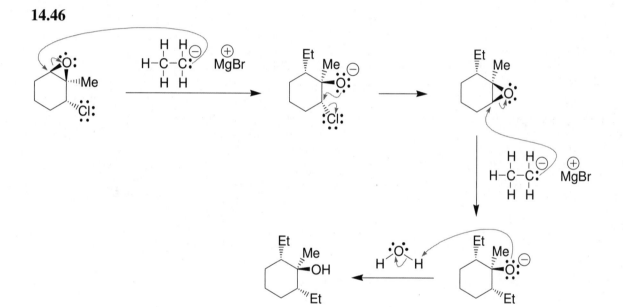

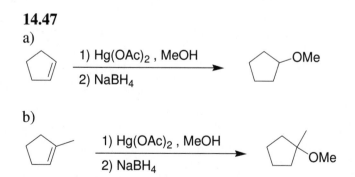

c)

1) $Hg(OAc)_2$, cyclopentanol (OH)
2) $NaBH_4$

cyclopentene → dicyclopentyl ether (–O–)

d)

1) $Hg(OAc)_2$, cyclopentanol (OH)
2) $NaBH_4$

1-methylcyclopentene → 1-methylcyclopentyl cyclopentyl ether (–O–)

14.48

a)

1) $NaNH_2$
2) ethylene oxide (O)
3) H_2O

→ OH

1) TMSCl, Et_3N
2) $NaNH_2$
3) ethylene oxide (O)
4) H_2O
5) TBAF

→ HO … OH

H_2
Lindlar's
Catalyst

→ HO … OH

b)

1) $NaNH_2$
2) ethylene oxide (O)
3) H_2O

→ OH

1) TMSCl, Et_3N
2) $NaNH_2$
3) ethylene oxide (O)
4) H_2O
5) TBAF

→ HO … OH

H_2
Pt

→ HO … OH

PCC
CH_2Cl_2

→ H(C=O) … (C=O)H

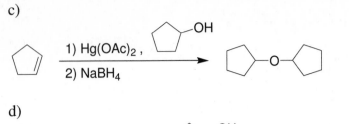

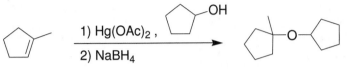

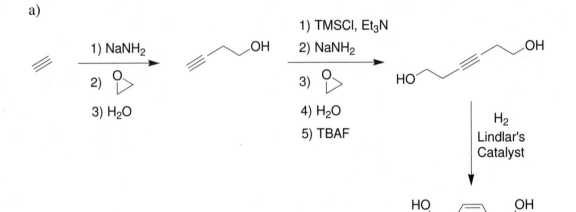

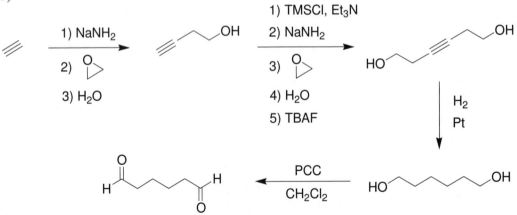

14.49

MCPBA

1) EtMgBr
2) H_2O

OH

1) NaOEt
2) H_2O

OEt

O

$[H_2SO_4]$
MeOH

O–Me
OH

1) NaH
2) EtI

HBr

1) NaSH
2) H_2O

1) NaH
2) CH_3I

Br
OH

1) NaCN
2) H_2O

1) LAH
2) H_2O

SH

OEt
OEt

CN

OMe

14.50

1) $Hg(OAc)_2$, EtOH
2) $NaBH_4$

OEt

HI
heat

I

+ EtI
+ H_2O

MCPBA

1) $H-C{\equiv}C:^{\ominus}$ $\overset{\oplus}{Na}$
2) H_2O

OH

+ En

$C{\equiv}C-H$

O

1) NaSH
2) H_2O

OH
SH

+ En

HBr

OH
Br

+ En

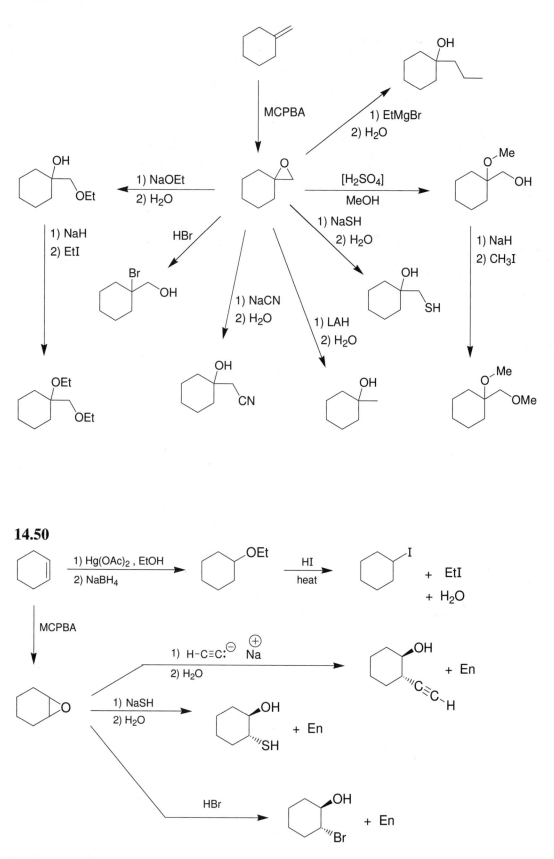

14.51

a)

MCPBA

1) MeMgBr

2) H_2O

OH

b)

MCPBA

1) MeMgBr

2) H_2O

OH

$Na_2Cr_2O_7$

H_2SO_4, H_2O

O

c)

Cl

1) Mg

2) O

3) H_2O

OH

$SOCl_2$

py

Cl

d)

Cl

1) Mg

2) O

3) H_2O

OH

$SOCl_2$

py

Cl

e)

OH

1) NaH

2) EtI

O

f)

OH

1) NaH

2) O

3) H_2O

O

OH

g)

1) $BH_3 \cdot THF$

2) H_2O_2, NaOH

OH

1) NaH

2) EtI

O

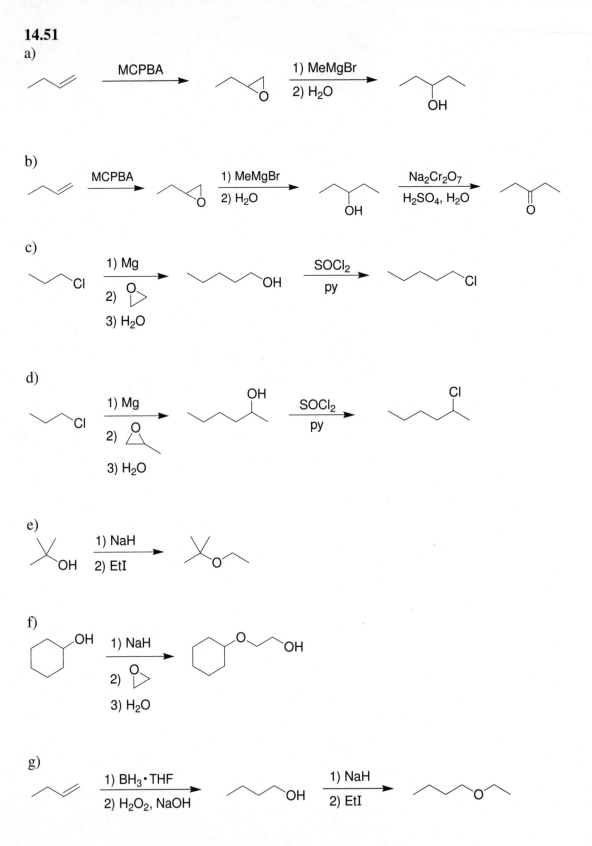

h)

dilute H_2SO_4 → OH → 1) NaH 2) EtI → O + En

i)

H_2 Lindlar's Catalyst → MCPBA → O + En

j)

H_2 Lindlar's Catalyst → MCPBA → O

k)

Na NH_3 → MCPBA → O + En

l)

H_2 Lindlar's Catalyst → MCPBA → O → 1) EtMgBr 2) H_2O → OH

m)

O → HI → OH → 1) NaOH 2) EtI → O

n)

O → HBr → Br → 1) Mg 2) O 3) H_2O → OH

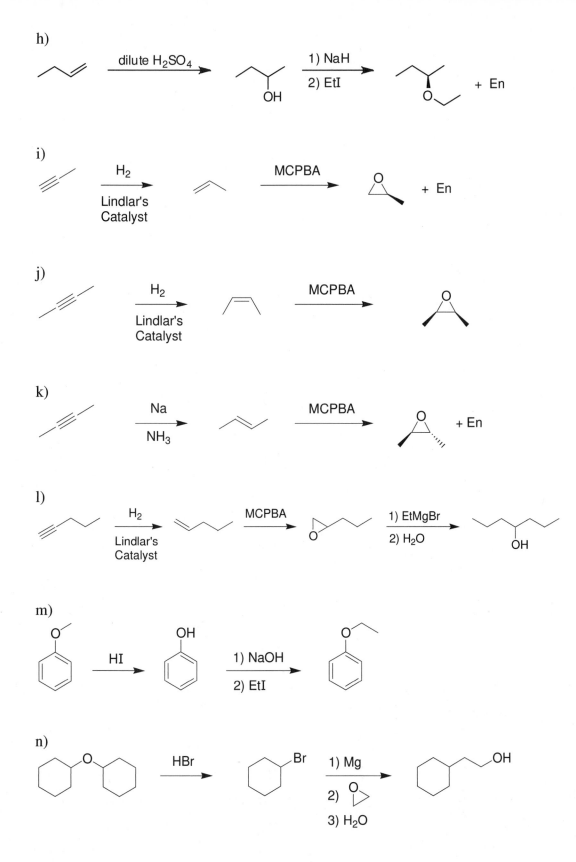

o)

1) NaH
2) (epoxide)
3) H_2O

OH → O OH

p)

Cl

1) Mg
2) (epoxide)
3) H_2O

OH

q)

Cl

1) Mg
2) (epoxide)
3) H_2O

OH

PCC
CH_2Cl_2

O
H

r)

OH

$[H_2SO_4]$

O

O OH

s)

OH

1) NaH
2) O
3) H_2O

O OH

t)

1) Br_2, *hv*
2) NaOEt

MCPBA

O + En

u)

1) Br_2, *hv*
2) NaOEt
3) MCPBA

O + En

1) NaOMe
2) H_2O

OH
OMe
+ En

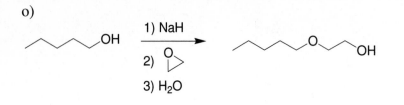

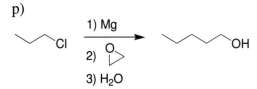

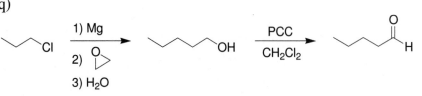

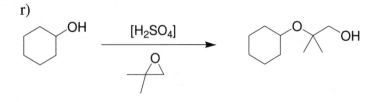

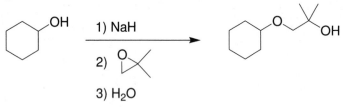

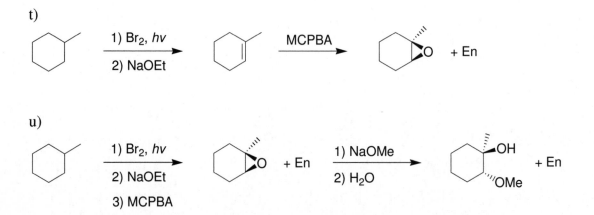

14.52.

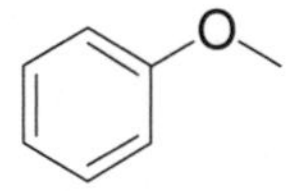

14.53.

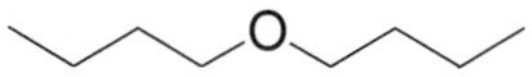

14.54.

14.55.

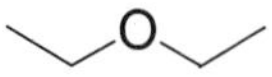

14.56

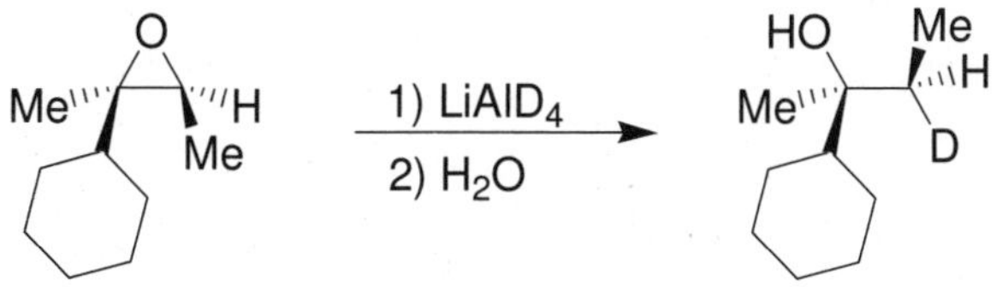

1) $LiAlD_4$
2) H_2O

14.57

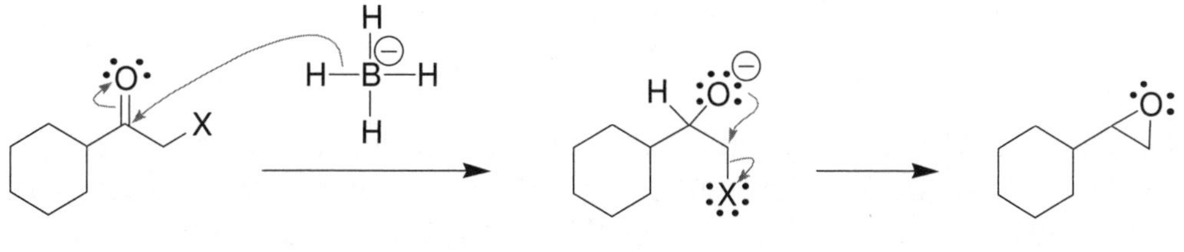

14.58 When methyloxirane is treated with HBr, the regiochemical outcome is determined by a competition between steric and electronic factors, with steric factors prevailing – the Br is positioned at the less substituted position. However, when phenyloxirane is treated with HBr, electronic factors prevail in controlling the regiochemical outcome. Specifically, the position next to the phenyl group is a benzylic position and can stabilize a large partial positive charge. In such a case, electronic factors are more powerful than steric factors, and the Br is positioned at the more substituted position.

14.59.

1) Br_2, *hv*
2) NaOEt

1) MCPBA
2) PrMgBr
3) H_2O

OH

conc. H_2SO_4
heat

1) MCPBA
2) H_3O^+

OH
OH

+ Enantiomer

14.60

Br

1) Mg
2) O
3) H_2O

OH

conc. H_2SO_4
heat

1) MCPBA
2) PhMgBr
3) H_2O

OH

conc. H_2SO_4
heat

MCPBA

O

+ Enantiomer

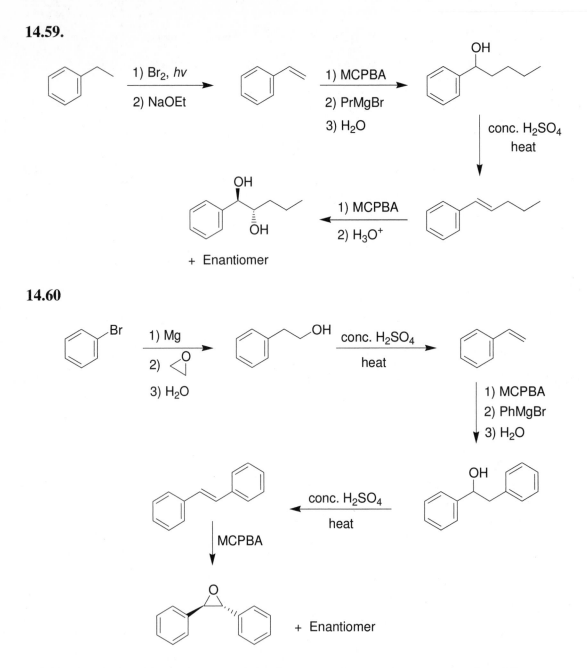

14.61 Since the Grignard reagent is both a strong base and a strong nucleophile, substitution and elimination can both occur. Indeed, they compete with each other. As we discussed in Chapter 8, elimination will be favored when the substrate is secondary. Both electrophilic positions in this epoxide are secondary, and so, elimination predominates:

H H
H–C–C:
H H
MgBr
E2

OH

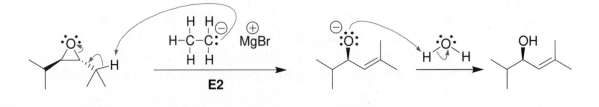

Chapter 15
Infrared Spectroscopy and Mass Spectrometry

Review of Concepts

Fill in the blanks below. To verify that your answers are correct, look in your textbook at the end of Chapter 15. Each of the sentences below appears verbatim in the section entitled *Review of Concepts and Vocabulary.*

- **Spectroscopy** is the study of the interaction between _______ and ________.
- The difference in energy (ΔE) between vibrational energy levels is determined by the nature of the bond. If a photon of light possesses exactly this amount of energy, the bond can absorb the photon to promote a _________________ **excitation.**
- IR spectroscopy can be used to identify which _____________________ are present in a compound.
- The location of each signal in an IR spectrum is reported in terms of a frequency-related unit called _________________.
- The wavenumber of each signal is determined primarily by bond __________ and the __________ of the atoms sharing the bond.
- The intensity of a signal is dependent on the ______________ of the bond giving rise to the signal.
- _________________ C=C bonds do not produce signals.
- Primary amines exhibit two signals resulting from ___________ **stretching** and _____________ **stretching.**
- **Mass spectrometry** is used to determine the ___________________ and _________________________ of a compound.
- **Electron impact ionization** (**EI**) involves bombarding the compound with high energy ________________, generating a radical cation that is symbolized by $(M)^{+\bullet}$ and is called the **molecular ion**, or the __________ **ion.**
- Only the molecular ion and the cationic fragments are deflected, and they are then separated by their ____________________ (*m/z*).
- The tallest peak in a mass spectrum is assigned a relative value of 100% and is called the __________ **peak.**
- The relative heights of the $(M)^{+\bullet}$ peak and the $(M+1)^{+\bullet}$ peak indicates the number of ____________________.
- A signal at M-15 indicates the loss of a _________ group; a signal at M-29 indicates the loss of an _________ group.
- ______________ alkanes have a molecular formula of the form C_nH_{2n+2}.
- Each double bond and each ring represents one **degree of** _______________.

Review of Skills

Fill in the blanks and empty boxes below. To verify that your answers are correct, look in your textbook at the end of Chapter 15. The answers appear in the section entitled *SkillBuilder Review.*

15.1 Analyzing an IR Spectrum

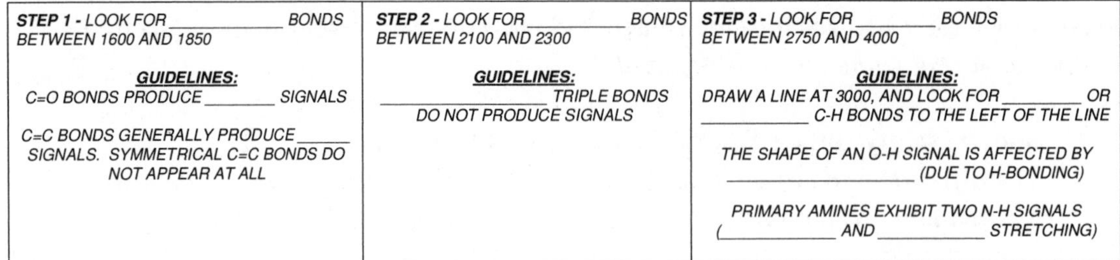

STEP 1 - LOOK FOR __________ BONDS BETWEEN 1600 AND 1850	STEP 2 - LOOK FOR __________ BONDS BETWEEN 2100 AND 2300	STEP 3 - LOOK FOR __________ BONDS BETWEEN 2750 AND 4000
GUIDELINES: C=O BONDS PRODUCE ________ SIGNALS C=C BONDS GENERALLY PRODUCE ______ SIGNALS. SYMMETRICAL C=C BONDS DO NOT APPEAR AT ALL	GUIDELINES: ______________ TRIPLE BONDS DO NOT PRODUCE SIGNALS	GUIDELINES: DRAW A LINE AT 3000, AND LOOK FOR ________ OR __________ C-H BONDS TO THE LEFT OF THE LINE THE SHAPE OF AN O-H SIGNAL IS AFFECTED BY ______________ (DUE TO H-BONDING) PRIMARY AMINES EXHIBIT TWO N-H SIGNALS (__________ AND __________ STRETCHING)

15.2 Distinguishing Two Compounds Using IR Spectroscopy

STEP 1 - WORK METHODICALLY THROUGH THE EXPECTED ______________ OF EACH COMPOUND	STEP 2 - DETERMINE IF ANY __________ WILL BE PRESENT FOR ONE COMPOUND BUT ABSENT FOR THE OTHER	STEP 3 - FOR EACH EXPECTED SIGNAL, COMPARE FOR ANY POSSIBLE DIFFERENCES IN __________, __________, OR __________.

15.3 Using the Relative Abundance of the $(M+1)^{+\bullet}$ Peak to Propose a Molecular Formula

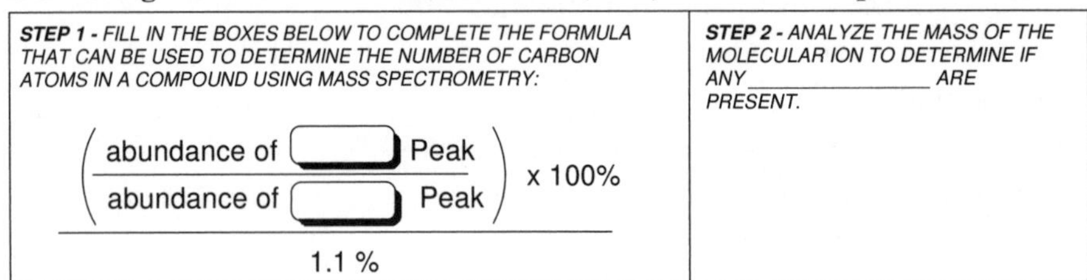

STEP 1 - FILL IN THE BOXES BELOW TO COMPLETE THE FORMULA THAT CAN BE USED TO DETERMINE THE NUMBER OF CARBON ATOMS IN A COMPOUND USING MASS SPECTROMETRY:	STEP 2 - ANALYZE THE MASS OF THE MOLECULAR ION TO DETERMINE IF ANY __________ ARE PRESENT.
$$\frac{\left(\frac{\text{abundance of } \square \text{ Peak}}{\text{abundance of } \square \text{ Peak}}\right) \times 100\%}{1.1\ \%}$$	

15.4 Calculating HDI

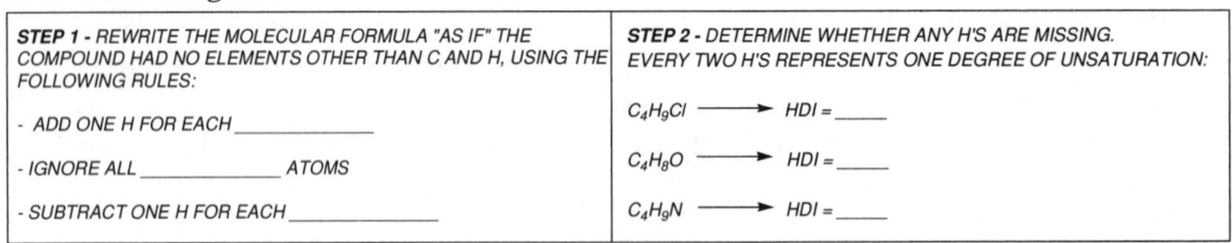

STEP 1 - REWRITE THE MOLECULAR FORMULA "AS IF" THE COMPOUND HAD NO ELEMENTS OTHER THAN C AND H, USING THE FOLLOWING RULES:	STEP 2 - DETERMINE WHETHER ANY H'S ARE MISSING. EVERY TWO H'S REPRESENTS ONE DEGREE OF UNSATURATION:
- ADD ONE H FOR EACH __________	C_4H_9Cl ⟶ HDI = _____
- IGNORE ALL __________ ATOMS	C_4H_8O ⟶ HDI = _____
- SUBTRACT ONE H FOR EACH __________	C_4H_9N ⟶ HDI = _____

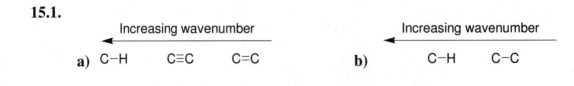

Solutions

15.1.

15.2.

a)

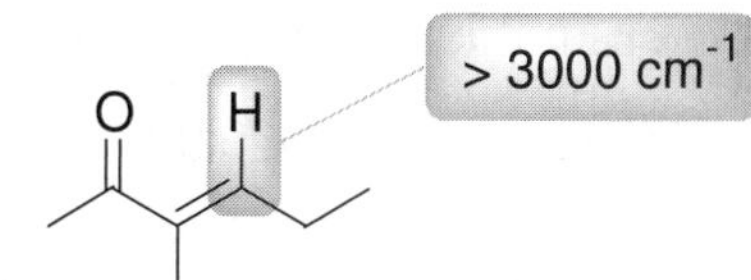

b) No **c)** No

d)

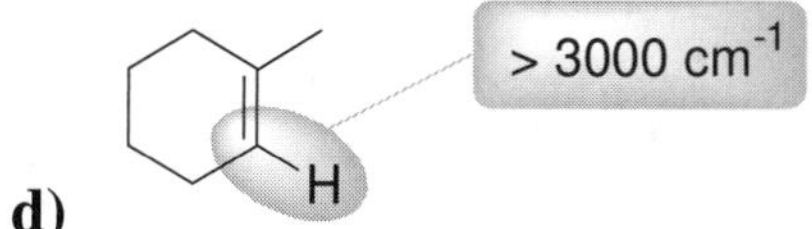

e) No **f)**

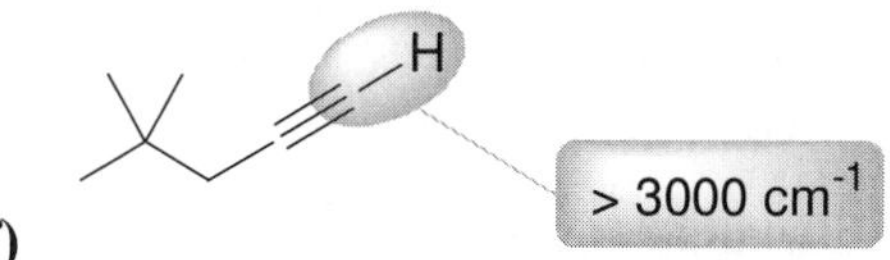

15.3.

a)

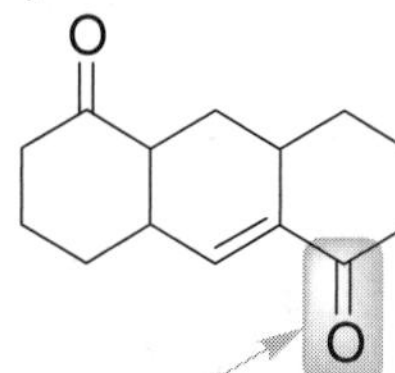

lower wavenumber
(carbonyl group is conjugated)

b)

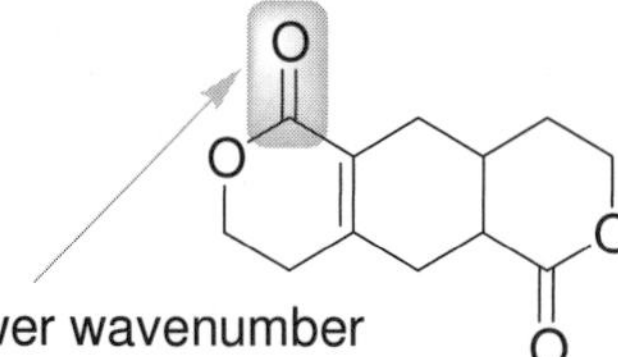

lower wavenumber
(carbonyl group is conjugated)

c)

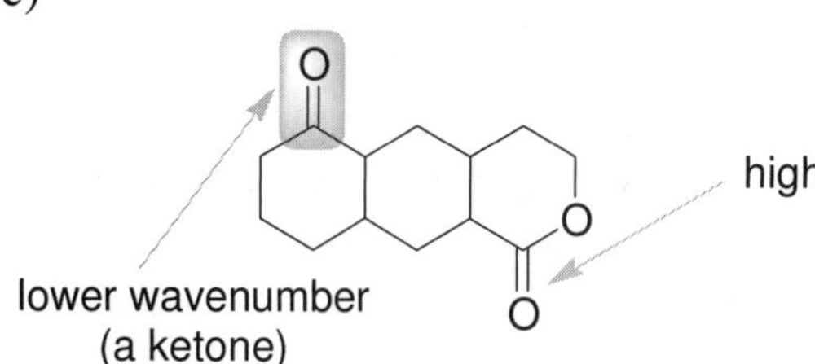

lower wavenumber
(a ketone)

higher wavenumber
(an ester)

15.4. The C=C bond in the conjugated compound produces a signal at lower wavenumber because it has some single bond character, as seen in the third resonance structure below:

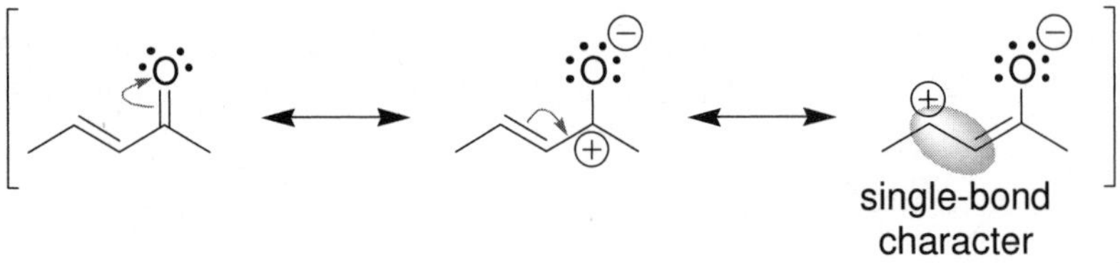

single-bond character

15.5.

a) 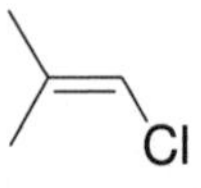 Cl The presence of the Cl will cause the C=C bond in this compound to have a larger dipole moment than the C=C bond in the other compound.

b) 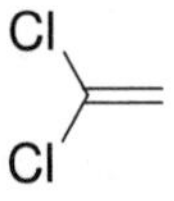 Cl Cl The C=C bond in this compound will have a larger dipole moment than the C=C bond in the other compound.

15.6. The C=C bond in 2-cyclohexenone has a large dipole moment, as can be rationalized with the third resonance structure below:

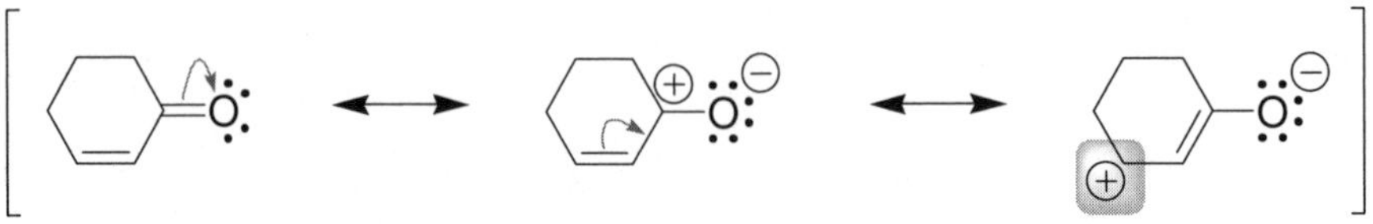

15.7. The vinylic C-H bond should produce a signal above 3000 cm^{-1}.

15.8. The narrow signal is produced by the O-H stretching in the absence of hydrogen bonding effect. The broad signal is produced by O-H stretching when hydrogen bonding is present. Hydrogen bonding effectively lowers the bond strength of the O-H bonds, because each hydrogen atom is slightly pulled away from the oxygen to which it is connected. A longer bond length (albeit temporary) corresponds with a weaker bond, which corresponds with a lower wavenumber.

15.9.

a) ROH b) neither c) RCO_2H d) neither e) ROH f) RCO_2H

15.10.

a) ketone b) RCO_2H c) R_2NH d) RNH_2 e) ROH f) ketone

15.11. The C_{sp^3}—H bonds can stretch symmetrically, asymmetrically, or in a variety of ways with respect to each other. Each one of these possible stretching modes is associated with a different wavenumber of absorption.

15.12.

a)  b) 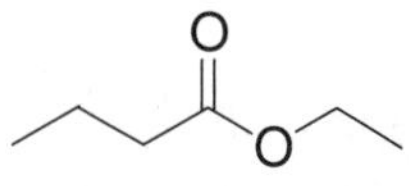 c)

d) 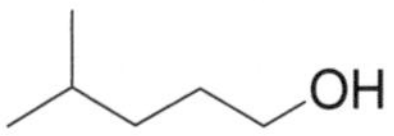 e) 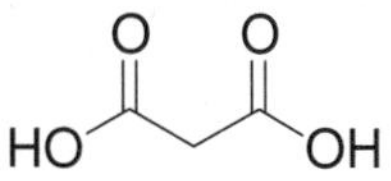 f)

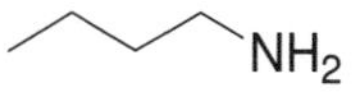

15.13.

1) The O-H bond of the carboxylic acid moiety (expected to be 2200 - 3600 cm^{-1})
2) The vinylic C-H bond (expected to be ~ 3100 cm^{-1})
3) All other C-H bonds (expected to be <3000 cm^{-1})
4) The C=O bond of the carboxylic acid moiety (expected to be ~ 1720 cm^{-1})
5) The C=C bond (expected to be ~ 1650 cm^{-1})

15.14.

a) The starting material is an alcohol and is expected to produce a typical signal for an O-H stretch –a broad signal between 3200 - 3600 cm^{-1}. In contrast, the product is a carboxylic acid and is expected to produce an even broader O-H signal (2200 - 3600 cm^{-1}) as a result of more extensive hydrogen bonding. Alternatively, the product can be differentiated from the starting material by looking for a signal at around 1720 cm^{-1}. The product has a C=O bond and should exhibit this signal. The starting material lacks a C=O bond and will not show a signal at 1720 cm^{-1}.

b) The starting material is secondary amine and is expected to produce a typical signal for an N-H stretch at around 3400 cm^{-1}. In contrast, the product is a tertiary amine and is not expected to produce a signal above 3000 cm^{-1}.

c) The starting material is an unsymmetrical alkyne and is expected to produce a signal at around 2200 cm^{-1}. In contrast, the product is an unsymmetrical alkene and is expected to produce a signal at around 1600 cm^{-1}. Also, the product has vinylic C-H bonds that are absent in the starting material. The product is expected to have a signal at around 3100 cm^{-1}, and the starting material will have no signal in that region.

d) The C≡C bond in the starting material and the C=C bond in the product are both symmetrical and will not produce signals. However the product has vinylic C-H bonds that are absent in the starting material. The product is expected to have a signal at around 3100 cm^{-1}, and the starting material will have no signal in that region.

e) The starting material will have two signals in the double-bond region: one for the C=O bond and one for the C=C bond. The product only has one signal in the double-bond region. It only has the signal for the C=O bond, which is now at higher wavenumber because it is no longer in conjugation.

15.15. The starting material has a cyano group (C≡N) and is expected to produce a signal at around 2200 cm^{-1}. In contrast, the product is a carboxylic acid and is expected to produce a broad signal from 2200 – 3600 cm^{-1}, as well as a signal at 1720 cm^{-1} for the C=O bond.

15.16. The C≡C bond in the starting material (1-butyne) is unsymmetrical and produces a signal at 2200 cm^{-1}, corresponding with the C≡C stretch. In contrast, the C≡C bond in the product (3-hexyne) is symmetrical and does not produce a signal at 2200 cm^{-1}.

15.17. The starting material should have a C=O signal at 1720 cm^{-1}, while the product should have an O-H signal at 3200 – 3600 cm^{-1}.

15.18. 1-chlorobutane is primary substrate. When treated with sodium hydroxide, substitution is expected to dominate over elimination (see Chapter 8), but both products are expected to be obtained:

NaOH

Cl → OH +

major minor

The substitution product is an alcohol and should have a broad signal from 3200 – 3600 cm^{-1}. The elimination product is an unsymmetrical alkene and is expected to give a C=C signal at approximately 1650 cm^{-1}, as well as a vinylic C-H signal at 3100 cm^{-1}.

15.19.

a) m/z = 68 m/z = 66 **b)** m/z = 78 m/z = 79

15.20.

a) This compound does not have any nitrogen atoms. According to the nitrogen rule, this compound should have an even molecular weight (*m/z* = 86).
b) This compound does not have any nitrogen atoms. According to the nitrogen rule, this compound should have an even molecular weight (*m/z* = 100).
c) This compound has one nitrogen atom. According to the nitrogen rule, this compound should have an odd molecular weight (*m/z* = 101).
d) This compound has two nitrogen atoms. According to the nitrogen rule, this compound should have an even molecular weight (*m/z* = 102).

15.21.

a) There must be four carbon atoms, and the molecular weight must be 72. The molecular formula could be C_4H_8O.

b) There must be four carbon atoms, and the molecular weight must be 68. The molecular formula could be C_4H_4O.

c) There must be four carbon atoms, and the molecular weight must be 54. The molecular formula could be C_4H_6.

d) There must be seven carbon atoms, and the molecular weight must be 96. The molecular formula could be C_7H_{12}.

15.22. Each nitrogen atom in the molecular formula of a compound should contribute 0.37% to the $(M+1)^{+\bullet}$ peak. Three nitrogen atoms therefore contribute the same amount (1.1%) as one carbon atom. A compound with molecular formula $C_8H_{11}N_3$ should have an $(M+1)^{+\bullet}$ peak that is 9.9% as tall as the molecular ion peak. If the molecular ion peak is 24% of the base peak, then the $(M+1)^{+\bullet}$ peak must be 2.4% of the base peak.

15.23.

a) This fragment is M – 79, which is formed by loss of a Br. So the fragment does not contain Br.

b) 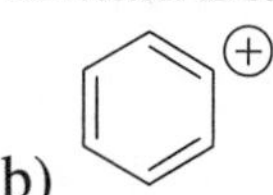

15.24.

a) There is not a significant $(M+2)^{+\bullet}$ peak, so neither bromine nor chlorine are present.
b) There is not a significant $(M+2)^{+\bullet}$ peak, so neither bromine nor chlorine are present.
c) The $(M+2)^{+\bullet}$ peak is approximately equivalent in height to the molecular ion peak, indicating the presence of a bromine atom.
d) The $(M+2)^{+\bullet}$ peak is approximately one-third as tall as the molecular ion peak, indicating the presence of a chlorine atom.

15.25.

a)

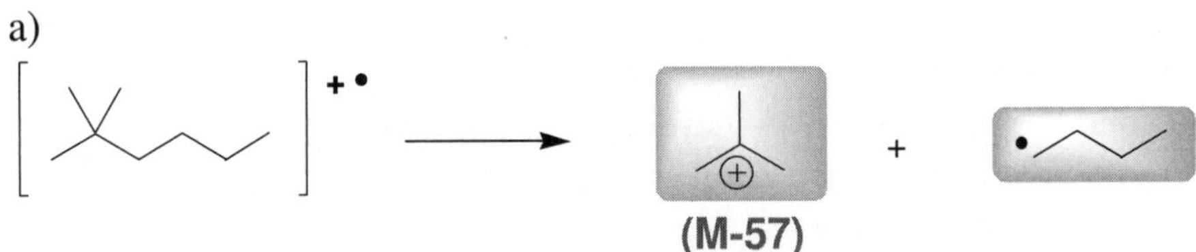

b) This carbocation is tertiary, and its formation is favored over the other possible secondary and primary carbocations.
c) They readily fragment to produce tertiary carbocations.
d) M-15 corresponds with loss of a methyl group. Indeed, loss of methyl group would also produce a tertiary carbocation, but at the expense of forming a methyl radical. That pathway is less favorable.

15.26. A fragment at M-29 should result from α cleavage:

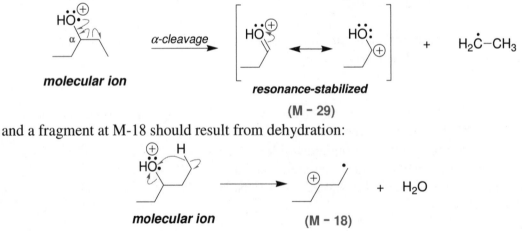

HÖ⊕ α — *α-cleavage* → [HÖ⊕ ⟷ HÖ: ⊕] + $H_2\dot{C}-CH_3$

molecular ion — ***resonance-stabilized*** **(M - 29)**

and a fragment at M-18 should result from dehydration:

⊕ HÖ: H → ⊕ • + H_2O

molecular ion — **(M - 18)**

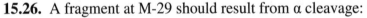

15.27. The fragment at M-43 is expected to be the base peak because it corresponds with formation of a tertiary carbocation:

(M-43)

15.28. In the first spectrum, the base peak appears at M-29, signifying the loss of an ethyl group. This spectrum is likely the mass spectrum of ethylcyclohexane. The second spectrum has a base peak at M-15, signifying the loss of a methyl group. The second spectrum is likely the mass spectrum of 1,1-dimethylcyclohexane.

15.29.

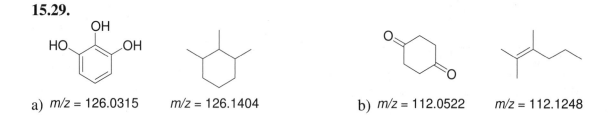

a) m/z = 126.0315 m/z = 126.1404

b) m/z = 112.0522 m/z = 112.1248

15.30.

a) The first compound should have a very broad signal between 3200 and 3600 cm^{-1}, corresponding with O-H stretching. The second compound will not have such a signal.

b) The first compound should have a pair of strong signals around 1720 cm^{-1}, corresponding with symmetric and asymmetric stretching of the C=O bonds. In contrast, the second compound will have a weak signal at around 1650cm^{-1}, corresponding to the C=C bond.

15.31.

a) 2	b) 1	c) 2	d) 1	e) 1
f) 1	g) 4	h) 0	i) 1	j) 0

15.32. Both $C_3H_5ClO_2$ and C_3H_6 have one degree of unsaturation.

15.33. The compound has one degree and unsaturation, which is a C=O bond:. The following are all of the possible structures that have molecular formula C_4H_8O and contain a C=O bond:

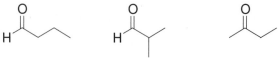

15.34. The compound must have one degree of unsaturation. The broad signal between 3200-3600 cm^{-1} indicates an OH group, and the absence of signals between 1600 and 1850 cm^{-1} indicates the absence of a double bond (either a C=O bond or an unsymmetrical C=C bond). This implies the presence of a ring (to achieve one degree of unsaturation). Below are possible structures:

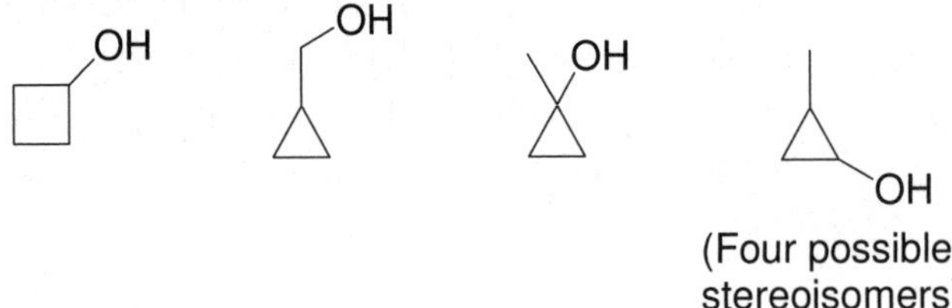

15.35. A signal at 2200cm^{-1} signifies the presence of a C≡C bond. There are only two possible constitutional isomers: 1-butyne or 2-butyne. The latter is symmetrical and would not produce a signal at 2200cm^{-1}. The compound must be 1-butyne.

15.36. The compound has exactly one degree of unsaturation, which means that it must contain either one ring or one double bond.

15.37.

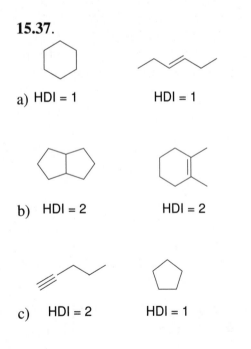

a) HDI = 1 HDI = 1

b) HDI = 2 HDI = 2

c) HDI = 2 HDI = 1

15.38.

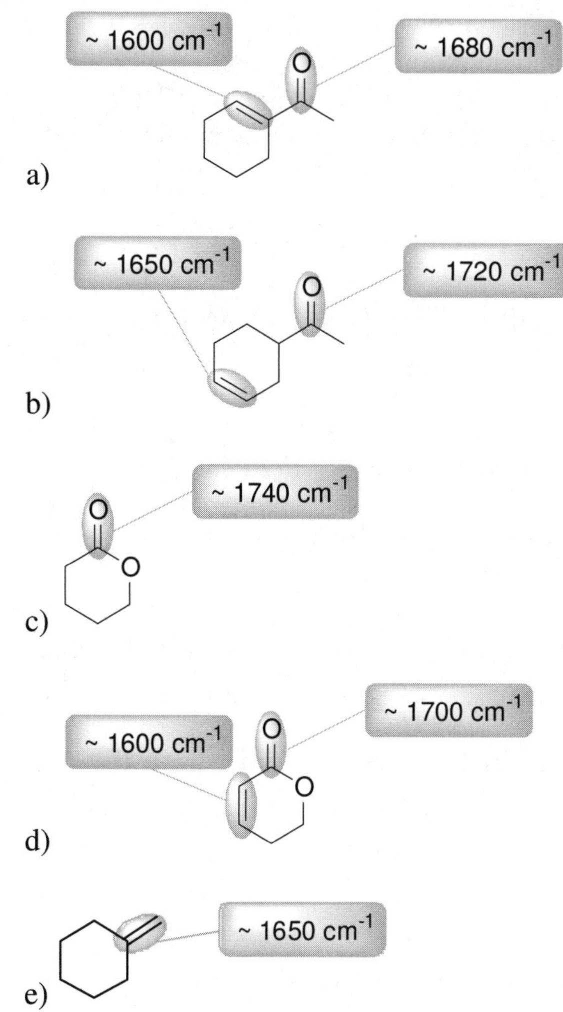

15.39.

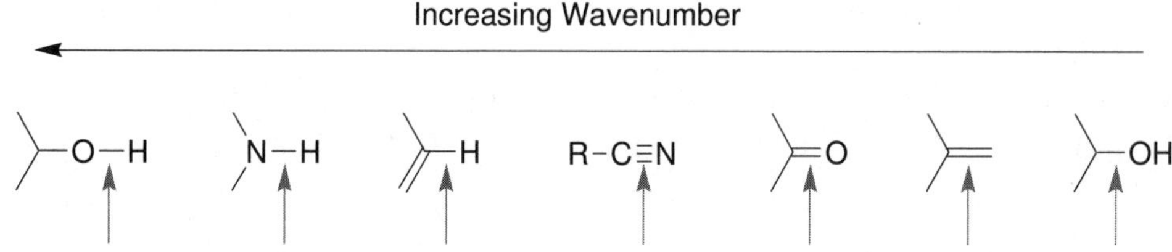

15.40.

a) The C=N bond and the C=O bond should each produce a signal in the double bond region, 1600 - 1850 cm^{-1}

b) The C=C bond should produce a signal in the double bond region, 1600 - 1850 cm^{-1}

c) The C=C bond and the C=O bond should each produce a signal in the double bond region, 1600 - 1850 cm^{-1}, In addition, the two C≡C bonds should produce two signals around 2200 cm^{-1}, and the C_{sp}-H bond should produce a signal around 3300 cm^{-1}.
d) The C=O bond should produce a signal in the double bond region, 1720 cm^{-1}, and the O-H of the carboxylic acid moiety should produce a very broad signal from 2200 – 3600 cm^{-1}.

15.41.
a) The reactant should have signals at 1650 cm^{-1} and 3100 cm^{-1}, while the product should not have either signals.

b) The reactant should have a broad signal from 3200 - 3600 cm^{-1}, while the product should lack this signal and instead should have a signal at ~ 1720 cm^{-1}.

c) The C=O bond of an ester should appear at higher wavenumber (~1740 cm^{-1}) than the C=O bond of a ketone (~1720 cm^{-1}).

d) The reactant should have signals at 1650 cm^{-1} and 3100 cm^{-1}, while the products should both have a signal at ~ 1720 cm^{-1}.

e) The reactant should not have signals at 1650 cm^{-1} or 3100 cm^{-1}, while the product should have such signals.

15.42.
a) a C=O signal at ~ 1720 cm^{-1}
b) a C=O signal at ~ 1680 cm^{-1} (conjugated) and a C=C signal at ~ 1600 cm^{-1} (conjugated) and a C_{sp^2}-H signal at 3100 cm^{-1}.
c) a C=O signal at ~ 1720 cm^{-1} and a C=C signal at ~ 1650 cm^{-1} and a C_{sp^2}-H signal at 3100 cm^{-1}.
d) a C=C signal at ~ 1650 cm^{-1} and an O-H signal at ~ 3200 – 3600 cm^{-1} and a C_{sp^2}-H signal at 3100 cm^{-1}.
e) a C=O signal at ~ 1720 cm^{-1} and an O-H signal at ~ 2200 – 3600 cm^{-1}
f) a C=O signal at ~ 1720 cm^{-1} and an O-H signal at ~ 3200 – 3600 cm^{-1}

15.43.
a) C_7H_8 (m/z = 92)
b) C_6H_6O (m/z = 94)
c) $C_6H_{10}O$ (m/z = 98)
d) C_6H_8O (m/z = 96)
e) $C_6H_{15}N$ (m/z = 101)

15.44. $C_5H_{10}O$

15.45. There are nine carbon atoms in the compound (10 / 1.1).

15.46.

a) 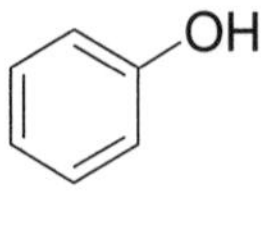 b) 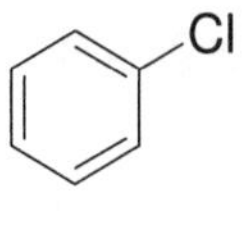 c) 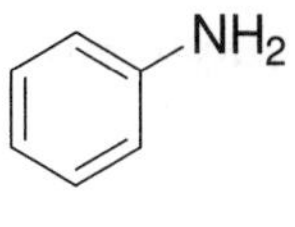 d)

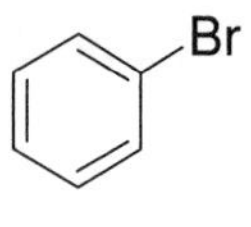

15.47.

a) an OH group and double bonds.

b) $\frac{3.9\ \%}{27.2\ \%} \times 100\% = 14.3\%$

$$\#\text{ of C} = \frac{14.3\ \%}{1.1\ \%} = 13$$

c) $C_{13}H_{24}O$ has thirteen carbon atoms and two degrees of unsaturation, consistent with the information that there are two double bonds.

15.48.
a) Both compounds are C_6H_{12}
b) Both compounds have an HDI of 1.
c) No, both compounds have exactly six carbon atoms and twelve hydrogen atoms, so they should have the same *m/z* even with high resolution mass spectrometry.
d) The IR spectrum of the alkene would have a signal at ~ 1650 cm^{-1} for the C=C bond and another signal at ~ 3100 cm^{-1} for the vinylic C-H bond. The IR spectrum of cyclohexane lacks both of these signals.

15.49. The signal at *m/z* = 111 is (M-15) which corresponds with loss of a methyl group. The signal at *m/z* = 97 is (M-29) which corresponds with loss of an ethyl group. Both fragmentations lead to a tertiary carbocation:

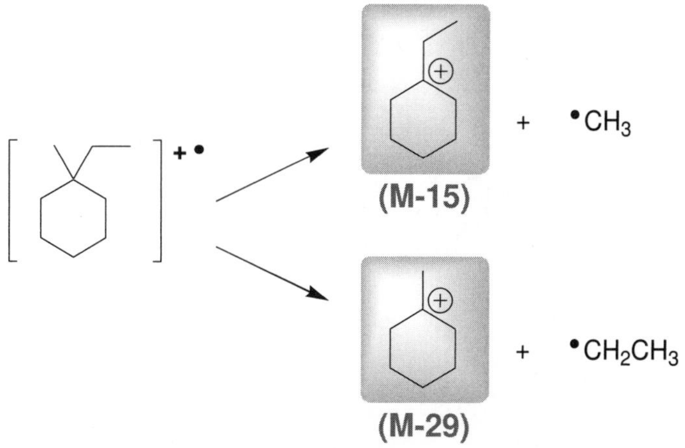

+ $\bullet CH_3$

(M-15)

+ $\bullet CH_2CH_3$

(M-29)

15.50.

a) No, because this fragment does not contain the bromine atom.
b) Yes, because this fragment still contains the bromine atom.
c) Yes, because this fragment still contains the bromine atom.

15.51. The more substituted alkene will not produce a signal at 1650 cm^{-1} nor will it produce a signal at 3100 cm^{-1}. The less substituted alkene will display both signals in its IR spectrum:

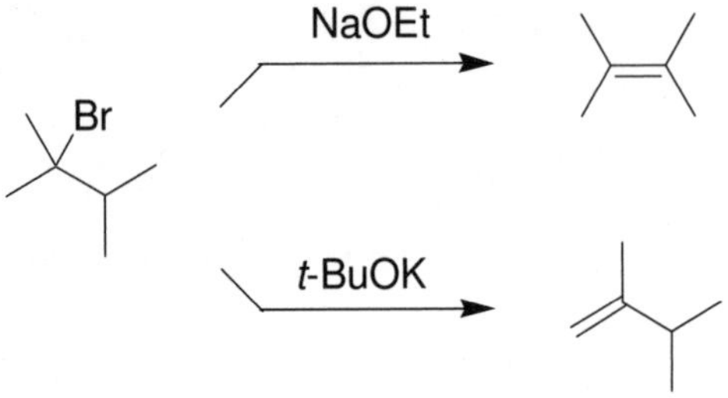

Br

NaOEt

t-BuOK

15.52.

a) C_5H_6
b) C_4H_6O

15.53.

a) the molecular ion peak appears at *m/z*= 114
b) the base peak appears at *m/z*= 43
c)

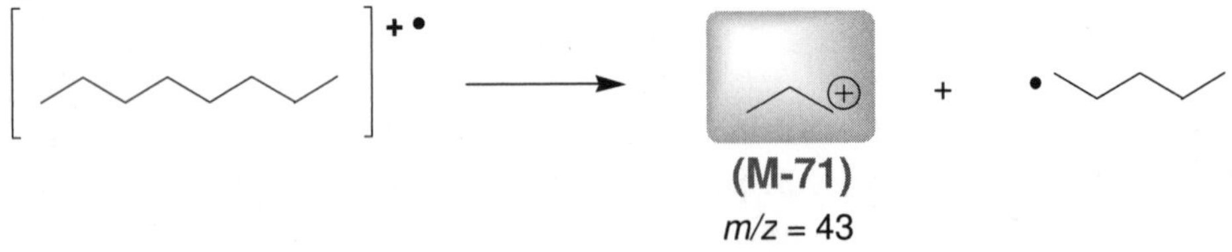

+•

+

(M-71)
m/z = 43

15.54.

a) 2	b) 2	c) 2	d) 36	e) 4
f) 4	g) 5	h) 1	i) 4	j) 1

15.55

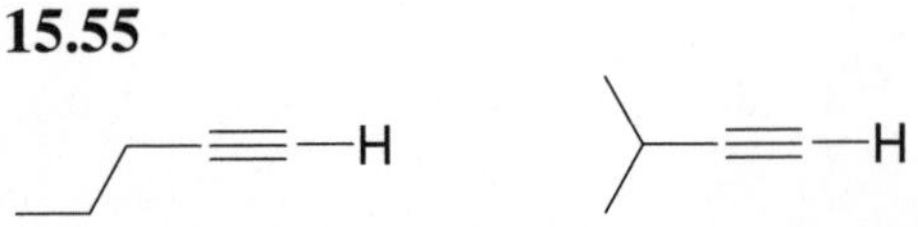

—H —H

15.56. Limonene has three degrees of unsaturation and is comprised of only carbon and hydrogen atoms. The molecular weight is 136, so the molecular formula of limonene must be $C_{10}H_{16}$.

15.57. The IR spectrum of *trans*-3-hexene is expected to have a signal at 3100 cm^{-1} as a result of the vinylic C-H stretch. In contrast, 2,3-dimethyl-2-butene lacks a vinylic C-H group and does not exhibit that signal.

15.58. The compound exhibits intramolecular hydrogen bonding even in dilute solutions.

15.59. The IR spetrum indicates that the compound is a ketone. The mass spectrum indicates a molecular weight of 86. The base peak is at M-43, indicating the loss of a propyl group. The compound likely has a three carbon chain (either a propyl group or isopropyl group) on one side of the ketone:

O O

15.60. The IR spectrum indicates the presence of a triple bond as well as a C-H bond indicating that the triple bond is terminal. The mass spectrum indicates a molecular weight of 68. The following two structures are consistent with these data:

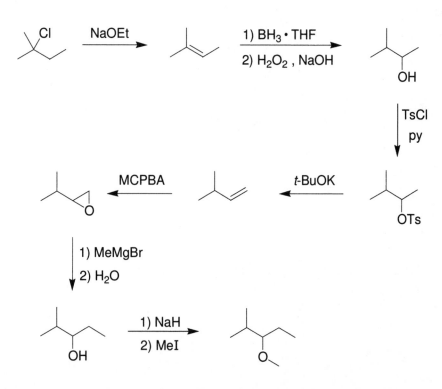

—≡—H —≡—H

15.61.

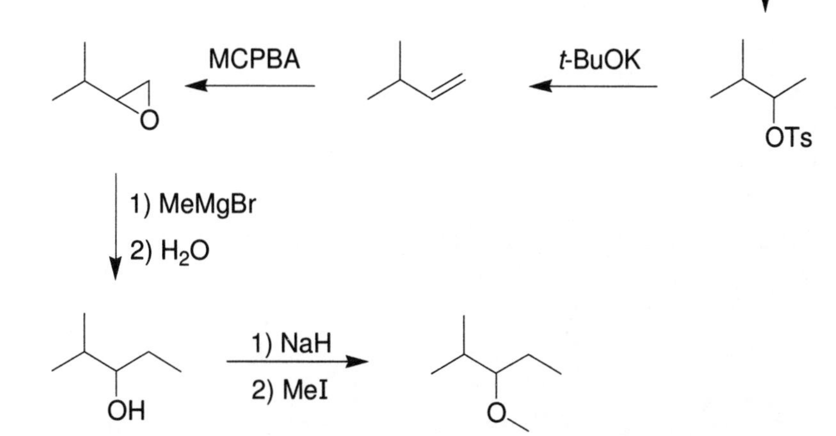

Cl NaOEt → 1) $BH_3 \cdot THF$ 2) H_2O_2 , NaOH → OH

TsCl py ↓

OTs t-BuOK → MCPBA → O

↓ 1) MeMgBr 2) H_2O

OH 1) NaH 2) MeI → O

a) Compound F is an alcohol and its IR spectrum will exhibit a broad signal between 3200 and 3600 cm^{-1}. Compound G is an ether and its IR spectrum will not exhibit the same signal.
b) Compound D is an alkene and its IR spectrum will exhibit a signal at approximately 1650 cm^{-1} (for the C=C bond), as well as a signal at 3100 cm^{-1} (for vinylic C-H stretching). Compound E is an epoxide and its IR spectrum will not have these two signals.
c) IR spectroscopy would not be helpful to distinguish these two compounds because they are both alcohols. Mass spectrometry could be used to differentiate these two compounds because they have different molecular weights.
d) No, they both have the same molecular formula, although a trained expert might be able to distinguish these compounds based on their fragmentation patterns.

15.62. 1-butene can lose a methyl group to form a resonance stabilized carbocation

+• → + •CH_3

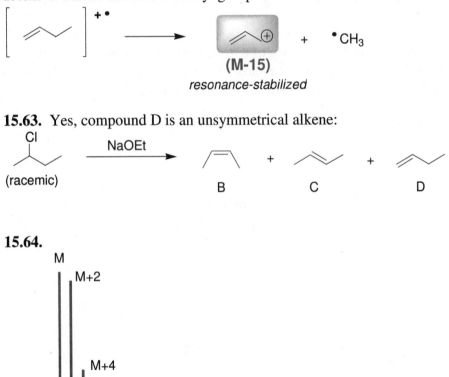

(M-15)

resonance-stabilized

15.63. Yes, compound D is an unsymmetrical alkene:

Cl

NaOEt

(racemic)

B C D

15.64.

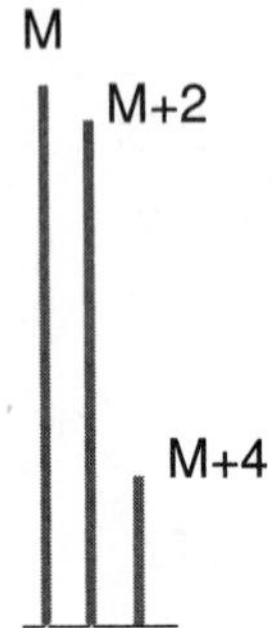

15.65. The OH group in ephedrine can engage in intramolecular hydrogen bonding, even in dilute solutions.

OH H N ≡ O H H N

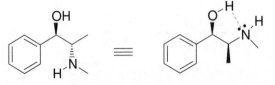

15.66.

15.67.

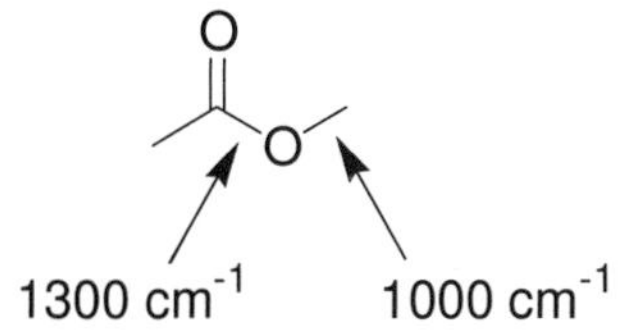

1300 cm^{-1} 1000 cm^{-1}

Explanation #1) One of the C-O bonds of the ester has some double bond character, as can be seen in the third resonance structure below:

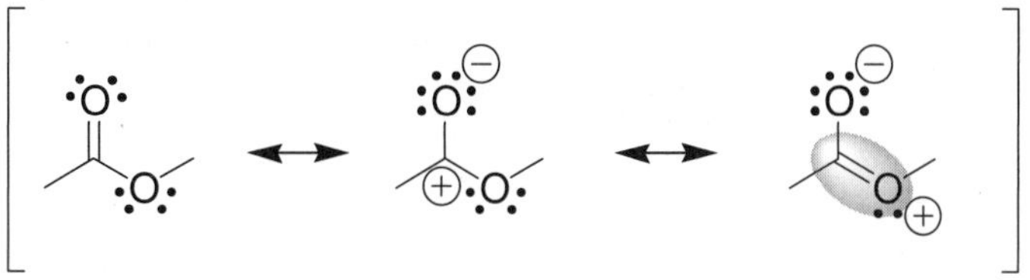

This C-O bond is a stronger bond than the other C-O single bond, which does not have any double bond character. As a result, the stronger C-O bond (highlighted above) appears at higher wavenumber.

Explanation #2) The C-O bond at 1300 cm^{-1} involves an sp^2 hybridized carbon atom, rather than an sp^3 hybridized carbon atom. The former has more *s*-character and holds its electrons closer to the positively charged nucleus. A $C_{sp2}-O$ bond is therefore stronger than a $C_{sp3}-O$ bond.

15.68.

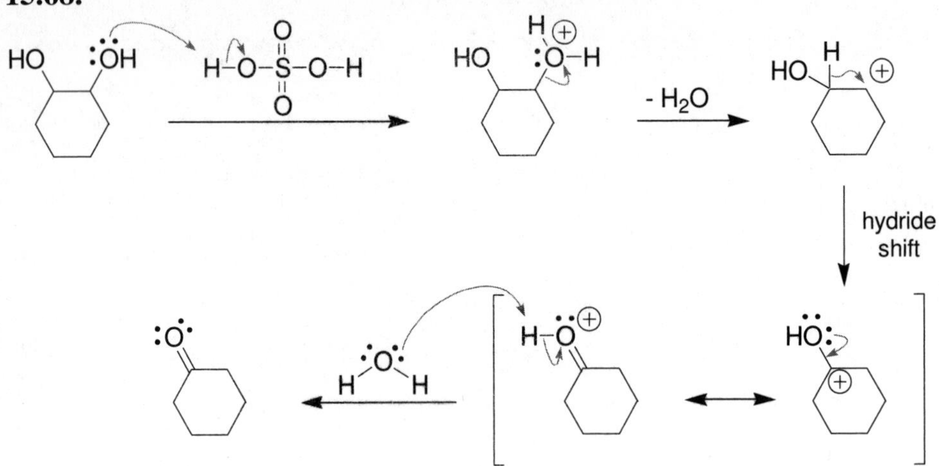

15.69.

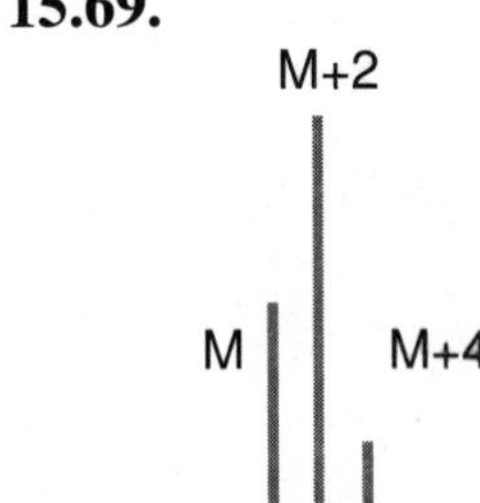

Chapter 16
Nuclear Magnetic Resonance Spectroscopy

Review of Concepts

Fill in the blanks below. To verify that your answers are correct, look in your textbook at the end of Chapter 16. Each of the sentences below appears verbatim in the section entitled *Review of Concepts and Vocabulary.*

- A spinning proton generates a **magnetic** _____________, which must align either with or against an imposed external magnetic field.
- All protons do not absorb the same frequency because of ___________________, a weak magnetic effect due to the motion of surrounding electrons that either **shield** or **deshield** the proton.
- ________________ solvents are generally used for acquiring NMR spectra.
- In a ^{1}H NMR spectrum, each signal has three important characteristics: location, area and shape.
- When two protons are interchangeable by rotational symmetry, the protons are said to be ___________.
- When two protons are interchangeable by rotational symmetry, the protons are said to be ___________.
- The left side of an NMR spectrum is described as _____**field**, and the right side is described as _____**field**.
- In the absence of inductive effects, a methyl group (CH_3) will produce a signal near _____ppm, a **methylene group** (CH_2) will produce a signal near _____, and a __________ **group** (CH) will produce a signal near _____. The presence of nearby groups increases these values somewhat predictably.
- The ________________, or area under each signal, indicates the number of protons giving rise to the signal.
- ________________represents the number of peaks in a signal. A __________has one peak, a __________has two, a __________has three, a __________has four, and a __________has five.
- Multiplicity is the result of **spin-spin splitting**, also called __________, which follows **the n+1 rule**.
- When signal splitting occurs, the distance between the individual peaks of a signal is called the **coupling constant**, or _____ **value**, and is measured in Hz.
- Complex splitting occurs when a proton has two different kinds of neighbors, often producing a __________.
- ^{13}C is an __________of carbon, representing _____% of all carbon atoms.
- All ^{13}C-^{1}H splitting is suppressed with a technique called **broadband** ______________, causing all of the ^{13}C signals to collapse to ___________.

Review of Skills

Fill in the blanks and empty boxes below. To verify that your answers are correct, look in your textbook at the end of Chapter 16. The answers appear in the section entitled *SkillBuilder Review*.

16.1 Determining the Relationship between Two Protons in a Compound

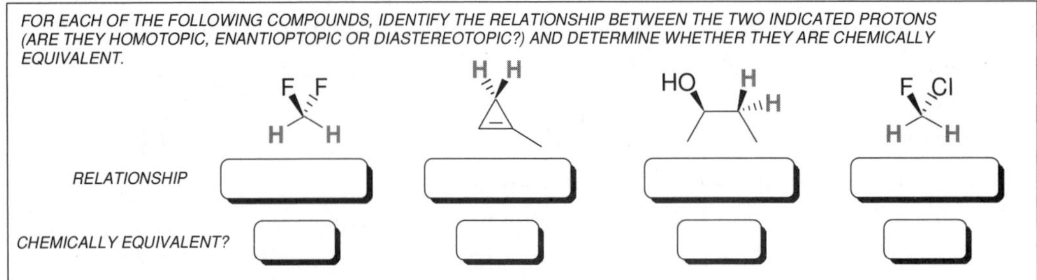

FOR EACH OF THE FOLLOWING COMPOUNDS, IDENTIFY THE RELATIONSHIP BETWEEN THE TWO INDICATED PROTONS (ARE THEY HOMOTOPIC, ENANTIOPTOPIC OR DIASTEREOTOPIC?) AND DETERMINE WHETHER THEY ARE CHEMICALLY EQUIVALENT.
RELATIONSHIP
CHEMICALLY EQUIVALENT?

16.2 Identifying the Number of Expected Signals in a ^{1}H NMR Spectrum

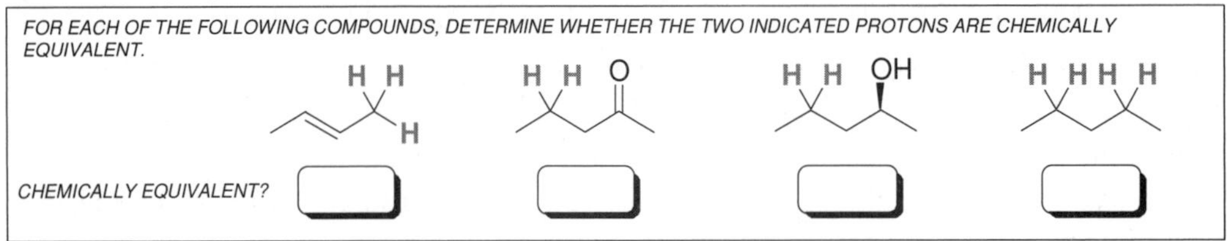

FOR EACH OF THE FOLLOWING COMPOUNDS, DETERMINE WHETHER THE TWO INDICATED PROTONS ARE CHEMICALLY EQUIVALENT.
CHEMICALLY EQUIVALENT?

16.3 Predicting Chemical Shifts

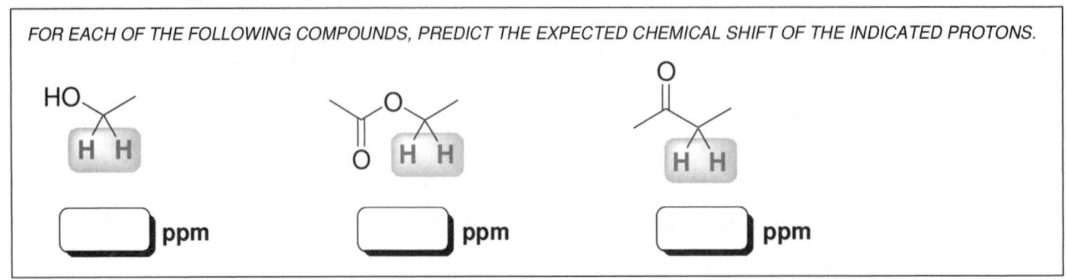

FOR EACH OF THE FOLLOWING COMPOUNDS, PREDICT THE EXPECTED CHEMICAL SHIFT OF THE INDICATED PROTONS.
ppm
ppm
ppm

16.4 Determining the Number of Protons Giving Rise to a Signal

STEP 1 - COMPARE THE RELATIVE ______ VALUES, AND CHOOSE THE LOWEST NUMBER	**STEP 2** - DIVIDE ALL INTEGRATION VALUES BY THE NUMBER FROM STEP #1, WHICH GIVES THE RATIO OF ______	**STEP 3** - IDENTIFY THE NUMBER OF PROTONS IN THE COMPOUND (FROM THE MOLECULAR FORMULA) AND THEN ADJUST THE RELATIVE INTEGRATION VALUES SO THAT THE SUM TOTAL EQUALS THE NUMBER OF ______.

16.5 Predicting the Multiplicity of a Signal

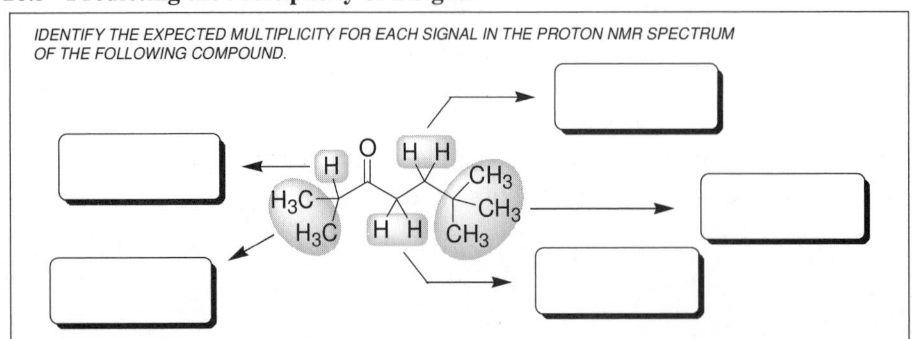

IDENTIFY THE EXPECTED MULTIPLICITY FOR EACH SIGNAL IN THE PROTON NMR SPECTRUM OF THE FOLLOWING COMPOUND.

16.6 Drawing the Expected ^{1}H NMR Spectrum of a Compound

STEP 1 - IDENTIFY THE NUMBER OF ________	**STEP 2**- PREDICT THE ________ ________ OF EACH SIGNAL	**STEP 3** - DETERMINE THE ________ OF EACH SIGNAL BY COUNTING THE NUMBER OF ________ GIVING RISE TO EACH SIGNAL	**STEP 4**- PREDICT THE ________ OF EACH SIGNAL	**STEP 5** - DRAW EACH SIGNAL

16.7 Using ^{1}H NMR Spectroscopy to Distinguish Between Compounds

STEP 1 - IDENTIFY THE NUMBER OF ________ THAT EACH COMPOUND WILL PRODUCE.	**STEP 2** - IF EACH COMPOUND IS EXPECTED TO PRODUCE THE SAME NUMBER OF SIGNALS, THEN DETERMINE THE ________, ________, AND ________ OF EACH SIGNAL IN BOTH COMPOUNDS	**STEP 3** - LOOK FOR DIFFERENCES IN THE CHEMICAL SHIFTS, MULTIPLICITIES OR INTEGRATION VALUES OF THE EXPECTED SIGNALS

16.8 Analyzing a ^{1}H NMR Spectrum and Proposing the Structure of a Compound

STEP 1 - USE THE ________ ________ TO DETERMINE THE HDI. AN HDI OF ____ INDICATES THE POSSIBILITY OF AN AROMATIC RING	**STEP 2** - CONSIDER THE NUMBER OF SIGNALS AND INTEGRATION OF EACH SIGNAL (GIVES CLUES ABOUT THE ________ OF THE COMPOUND)	**STEP 3** - ANALYZE EACH SIGNAL (____, ____, AND ____), AND THEN DRAW FRAGMENTS CONSISTENT WITH EACH SIGNAL. THESE FRAGMENTS BECOME OUR PUZZLE PIECES THAT MUST BE ASSEMBLED TO PRODUCE A MOLECULAR STRUCTURE	**STEP 4** - ASSEMBLE THE FRAGMENTS

16.9 Predicting the Number of Signals and Approximate Location of Each Signal in a ^{13}C NMR Spectrum

BELOW ARE SEVEN DIFFERENT TYPES OF CARBON ATOMS. EACH OF THEM IS EXPECTED TO PRODUCE A SIGNAL IN ONE OF FOUR POSSIBLE REGIONS IN A CARBON NMR SPECTRUM. IDENITFY THE EXPECTED REGION FOR EACH TYPE OF CARBON ATOM.

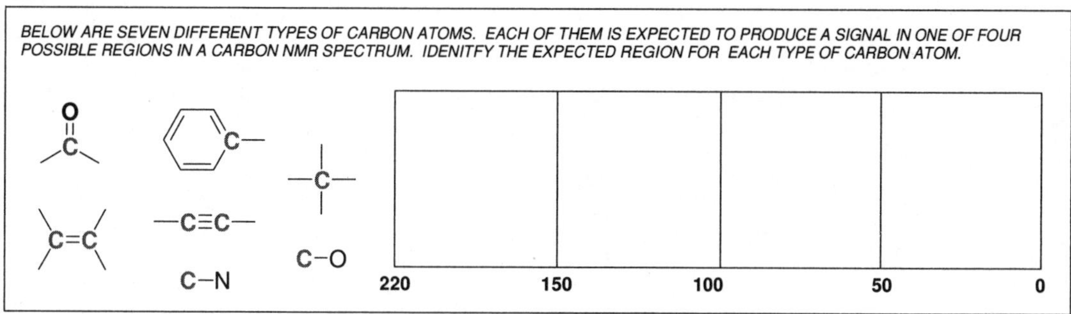

16.10 Determining Molecular Structure using DEPT ^{13}C NMR Spectroscopy

COMPLETE THE FOLLOWING CHART BY DRAWING THE EXPECTED SHAPE OF EACH SIGNAL:

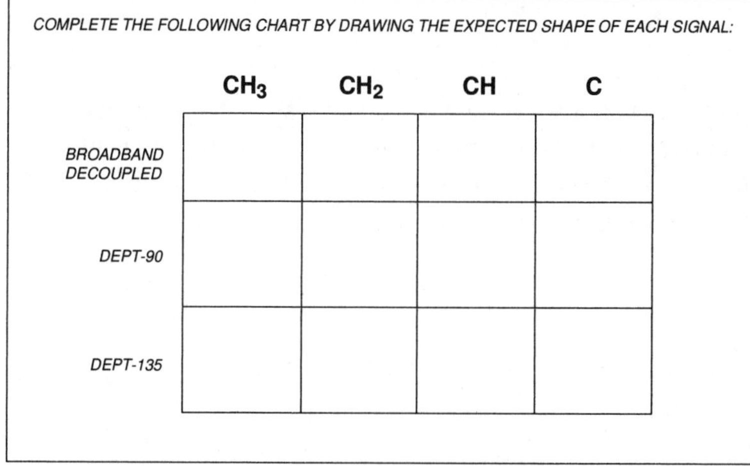

	CH_3	CH_2	CH	C
BROADBAND DECOUPLED				
DEPT-90				
DEPT-135				

Solutions

16.1.

a) homotopic
b) enantiotopic
c) diastereotopic
d) enantiotopic
e) homotopic

16.2.

a) All four protons can be interchanged either via rotation or reflection.
b) The three protons of a methyl group are always equivalent, and in this case, the two methyl groups are equivalent to each other because they can be interchanged by rotation. Therefore, all six protons are equivalent.
c) Three
d) Three
e) Six

16.3.

16.4.

a) 8	b) 4	c) 2	d) 3	e) 5	f) 3
g) 4	h) 2	i) 4	j) 7	k) 4	l) 7

16.5. The presence of the bromine atom does not render C3 a chirality center because there are two ethyl groups connected to C3. Nevertheless, the presence of the bromine atom does prevent the two protons at C2 from being interchangeable by reflection. The replacement test gives a pair of diastereomers, so the protons are diastereotopic.

16.6. This compound will exhibit two signals in its ^{1}H NMR spectrum:

16.7.

a)

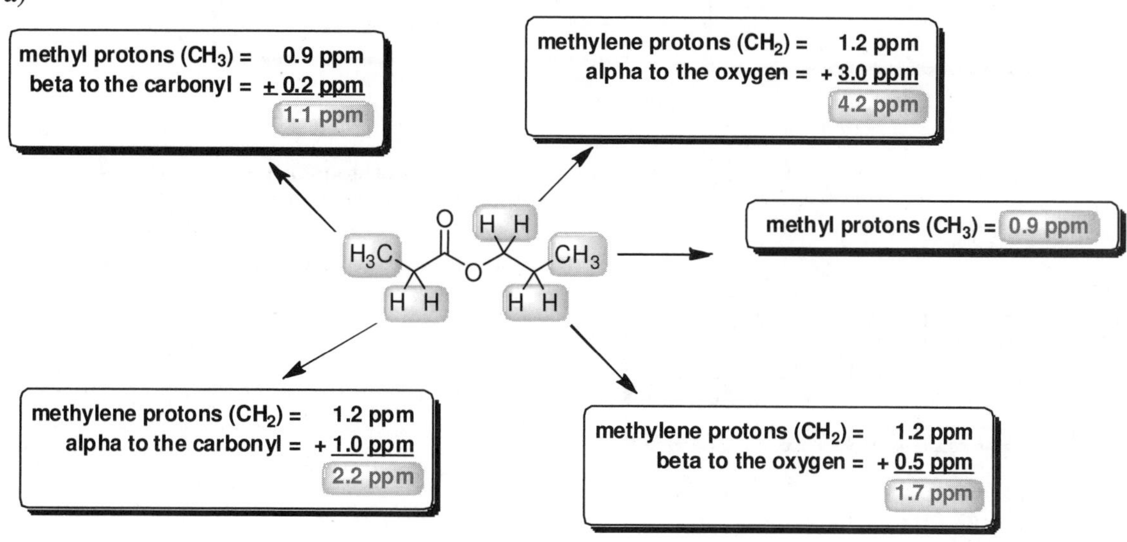
methyl protons (CH_3) = 0.9 ppm
beta to the carbonyl = + 0.2 ppm
1.1 ppm
methylene protons (CH_2) = 1.2 ppm
alpha to the oxygen = + 3.0 ppm
4.2 ppm
methyl protons (CH_3) = 0.9 ppm
methylene protons (CH_2) = 1.2 ppm
alpha to the carbonyl = + 1.0 ppm
2.2 ppm
methylene protons (CH_2) = 1.2 ppm
beta to the oxygen = + 0.5 ppm
1.7 ppm

b)

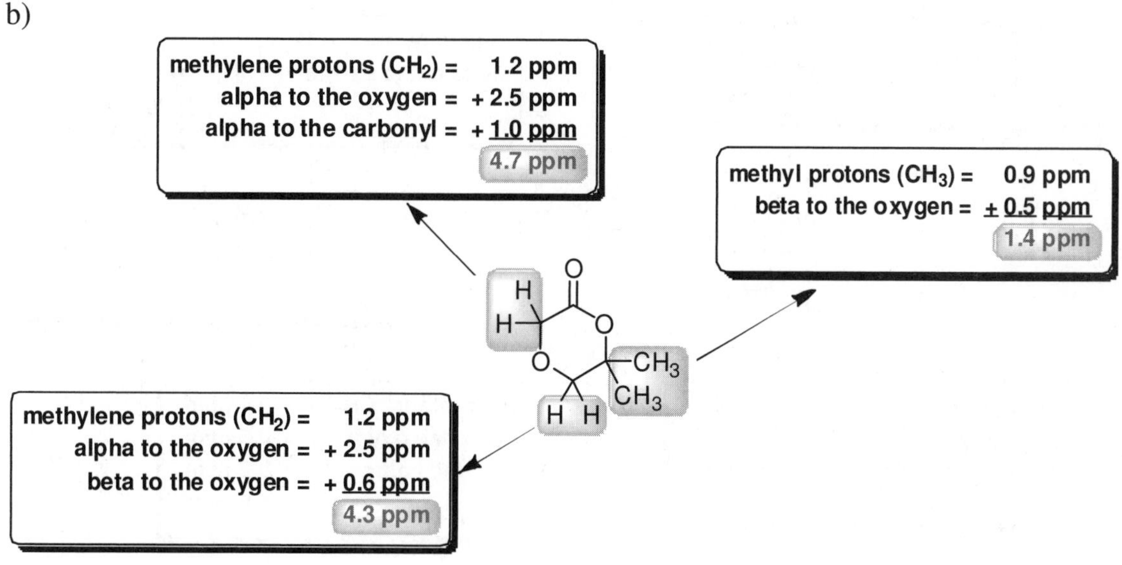
methylene protons (CH_2) = 1.2 ppm
alpha to the oxygen = + 2.5 ppm
alpha to the carbonyl = + 1.0 ppm
4.7 ppm
methyl protons (CH_3) = 0.9 ppm
beta to the oxygen = + 0.5 ppm
1.4 ppm
methylene protons (CH_2) = 1.2 ppm
alpha to the oxygen = + 2.5 ppm
beta to the oxygen = + 0.6 ppm
4.3 ppm

c)

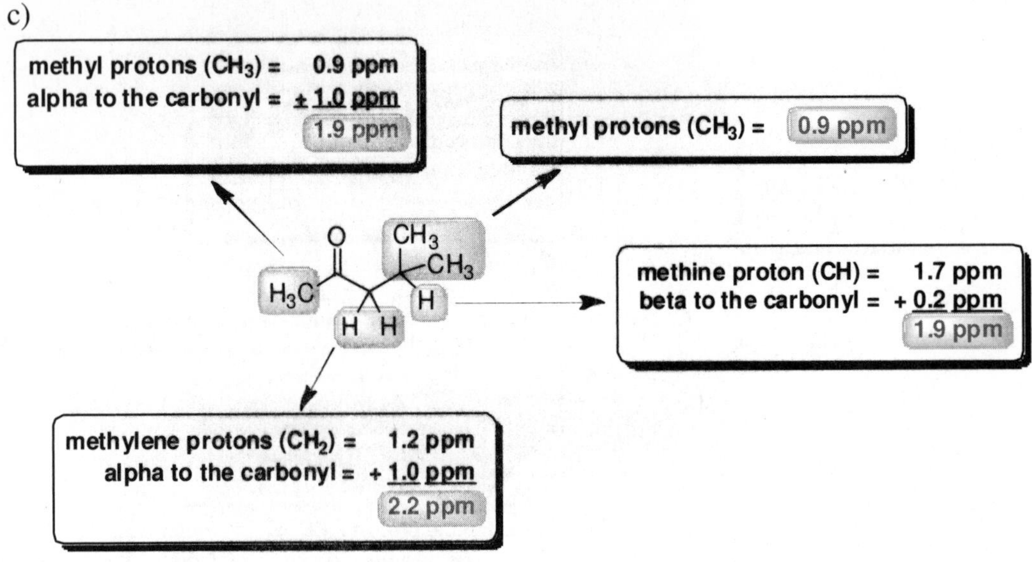

d)

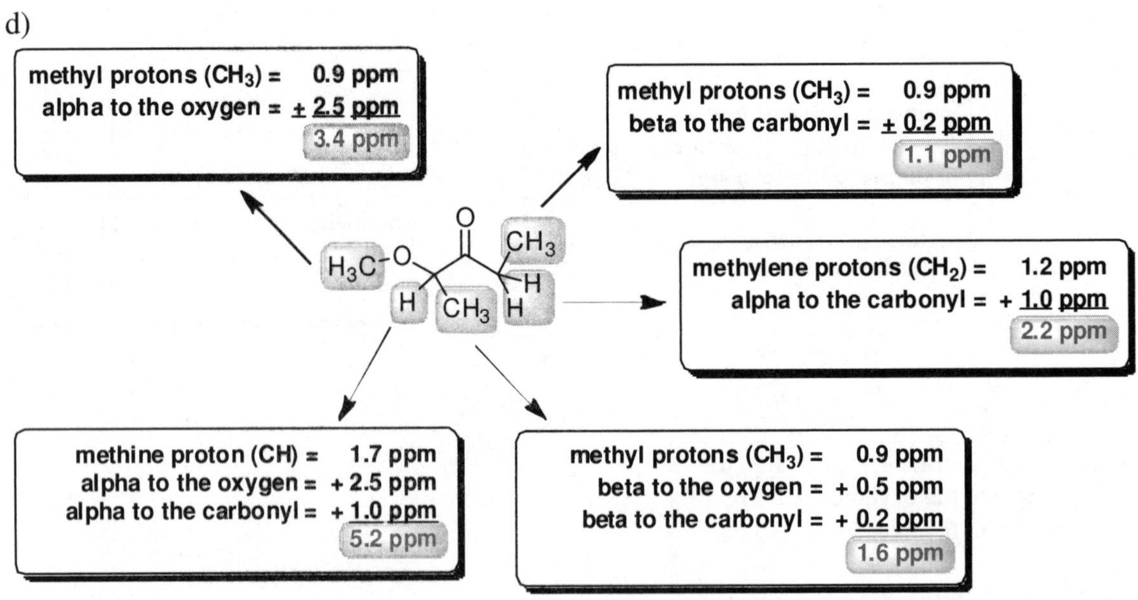

e) All four methylene groups are equivalent, so the compound will have only one signal in its ^{1}H NMR spectrum. That signal is expected to appear at approximately (1.2 + 2.5 + 0.5) = 4.2 ppm.

16.8.

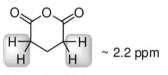

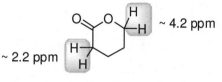

methylene protons (CH_2) = 1.2 ppm
alpha to the oxygen = + 3.0 ppm
alpha to the oxygen = + 2.5 ppm
6.7 ppm

16.9.

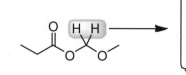

~ 2.2 ppm

Only one signal downfield of 2.0 ppm
(the four highlighted protons are equivalent)

~ 4.2 ppm
~ 2.2 ppm

Two signals downfield of 2.0 ppm

16.10.

a)

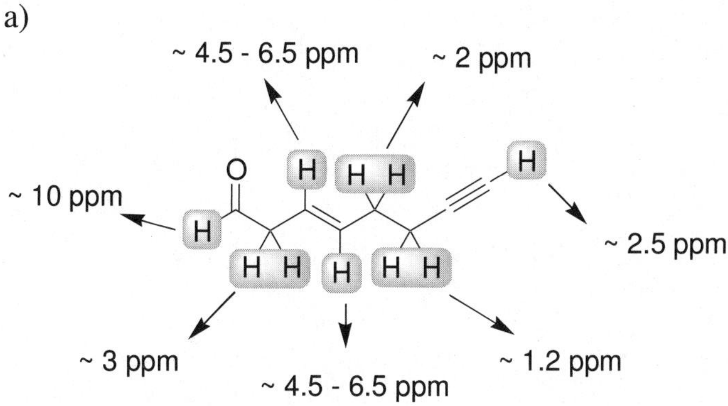

b)

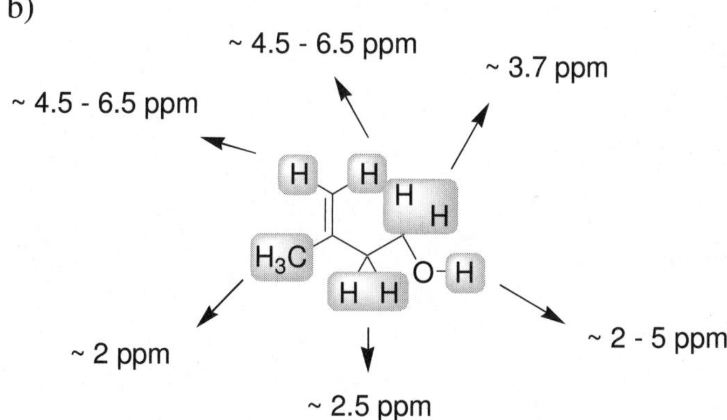

c)

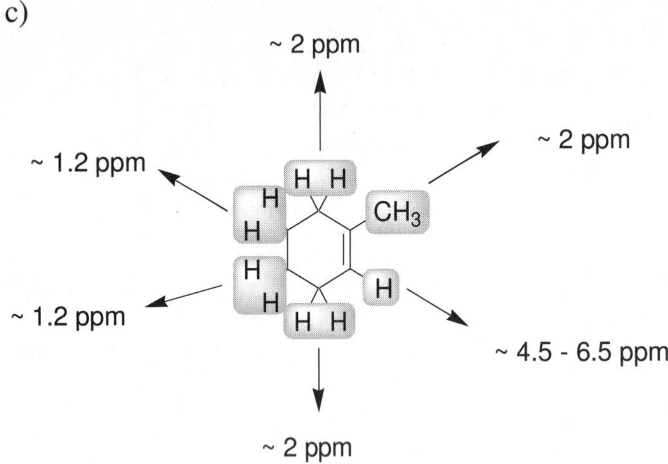

d)

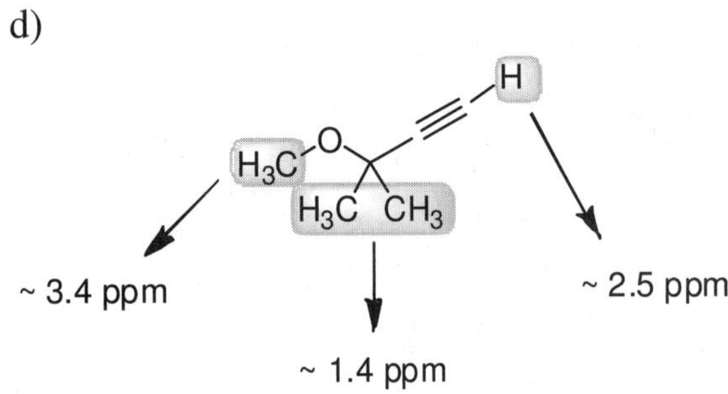

16.11.
The signal at 4.0 ppm represents two protons.
The signal at 2.0 ppm represents three protons.
The signal at 1.6 ppm represents two protons.
The signal at 0.9 ppm represents three protons.

16.12.
The signal at 9.6 ppm represents one proton.
The signal at 7.5 ppm represents five protons.
The signal at 7.3 ppm represents one proton.
The signal at 2.1 ppm represents three protons.

16.13. Each signal represents two protons.

16.14.
Cl

16.15.

a)

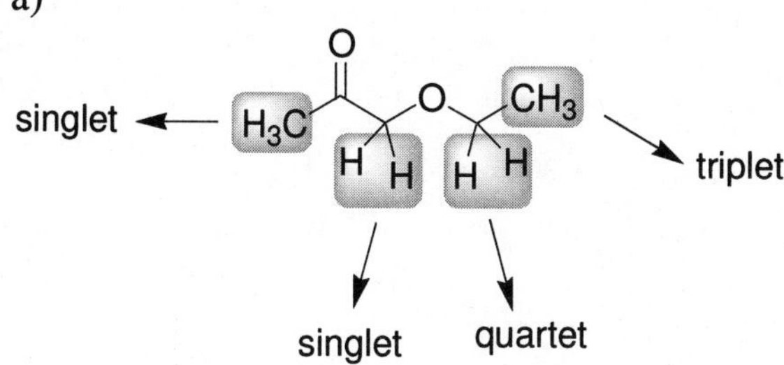

O
singlet
H3C
O
CH3
H
H
H
H
triplet
singlet
quartet

b)

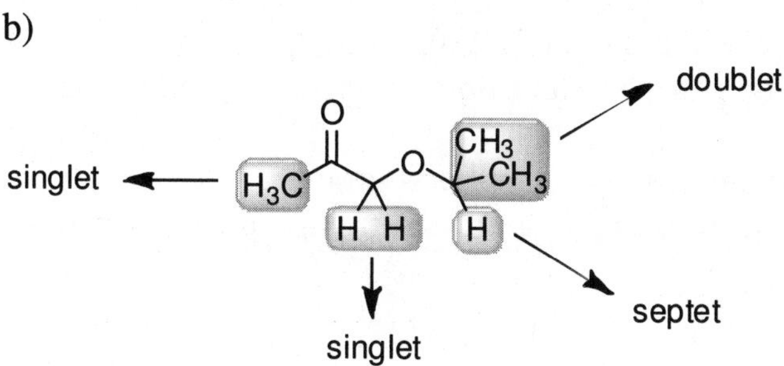

doublet
O
CH3
singlet
H3C
O
CH3
H H
H
septet
singlet

c)

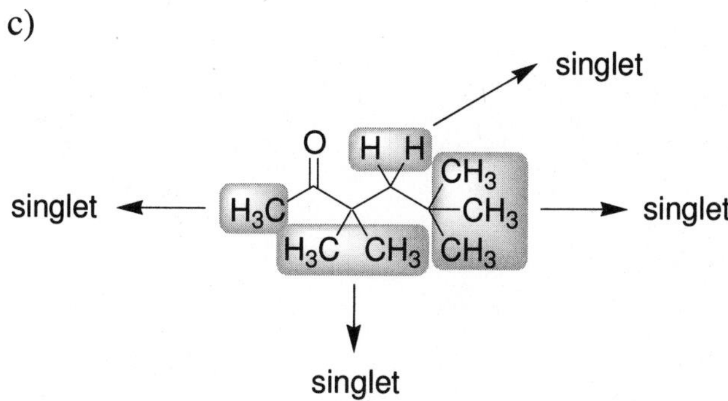

singlet
O
H H
CH3
singlet
H3C
CH3
singlet
H3C
CH3
CH3
singlet

d)

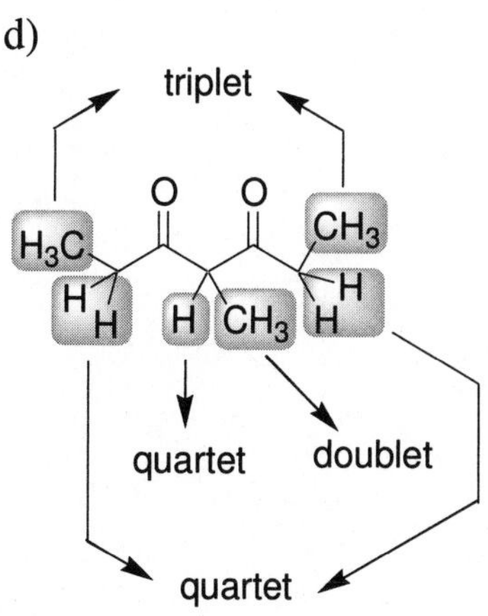

triplet
O
O
H3C
CH3
H
H
H
H
CH3
H
quartet
doublet
quartet

16.16.

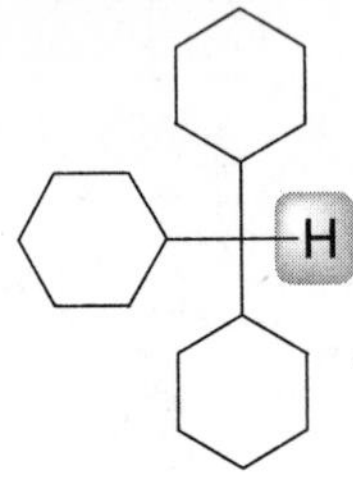

16.17.

a) The spectrum exhibits the characteristic pattern of an isopropyl group.
b) The spectrum exhibits the characteristic pattern of an isopropyl group as well as the characteristic pattern of an ethyl group.
c) The spectrum exhibits the characteristic pattern of a *tert*-butyl group.
d) The spectrum does not exhibit the characteristic pattern of an ethyl group, an isopropyl group, or a *tert*-butyl group.

16.18.

a)

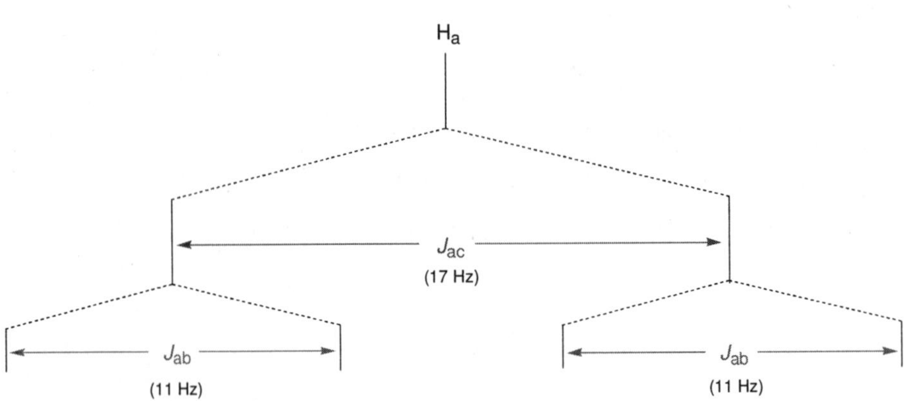

b)

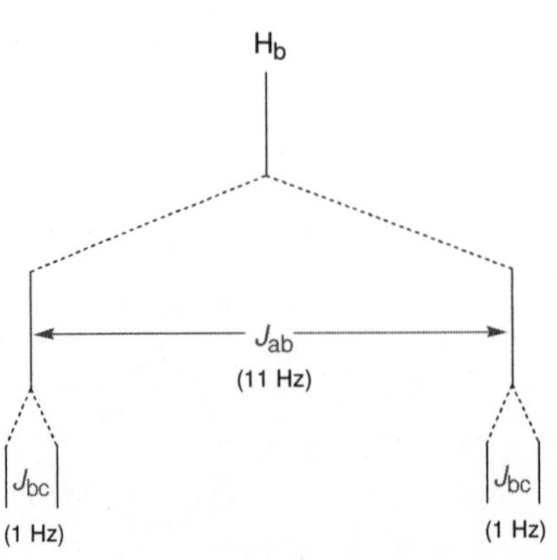

c)

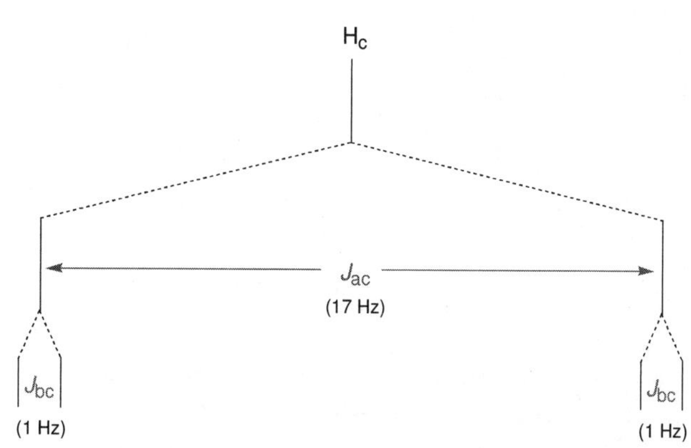

16.19. Draw the expected ^{1}H NMR spectrum for each of the following compounds

a)

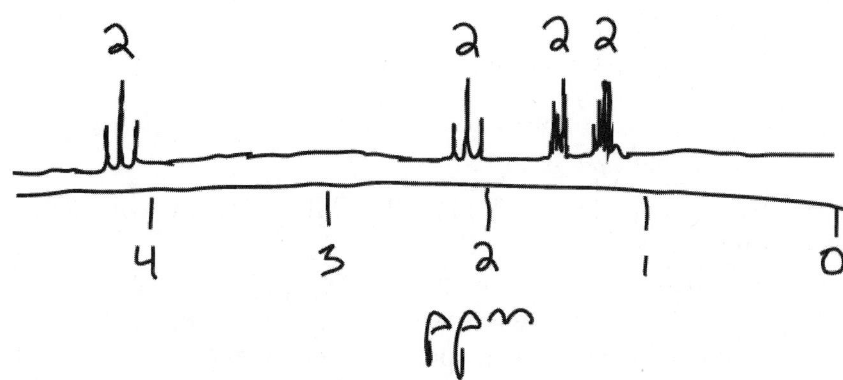

b)

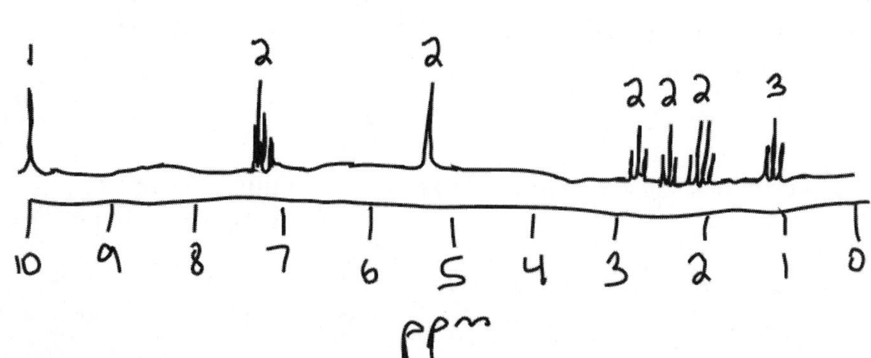

c)

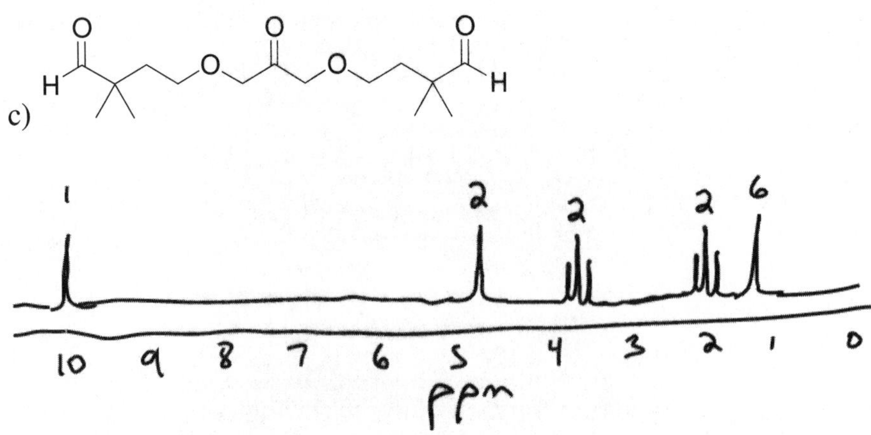

16.20.

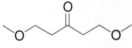

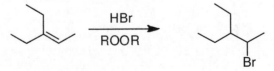

16.21.
a) The first compound will have only three signals in its ^{1}H NMR spectrum, while the second compound will have six signals.
b) Both compounds will exhibit ^{1}H NMR spectra with only two singlets. In each spectrum, the relative integration of the two singlets is 1:3. In the first compound, the singlet with the smaller integration value will be at approximately 2 ppm. In the second compound, the singlet with the smaller integration value will be at approximately 4 ppm.
d) The first compound will have only two signals in its ^{1}H NMR spectrum, while the second compound will have three signals.
e) The first compound will have five signals in its ^{1}H NMR spectrum, while the second compound will have only three signals.
f) The first compound will have only four signals in its ^{1}H NMR spectrum, while the second compound will have five signals.
g) The first compound will have only one signal in its ^{1}H NMR spectrum, while the second compound will have two signals.

16.22. The presence of peroxides caused an *anti*-Markovnikov addition of HBr:

HBr
ROOR
Br

16.23.

a) 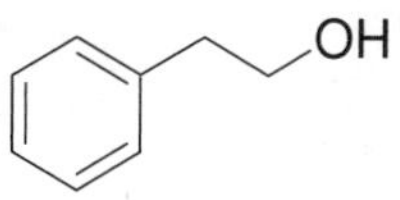 b) 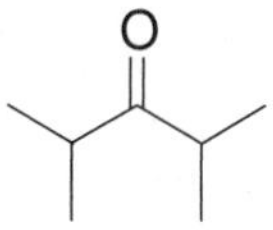 c) 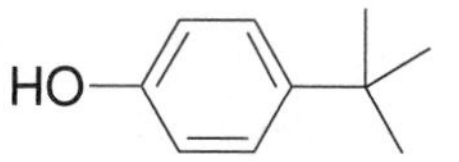 d)

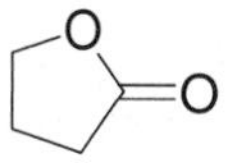

e) 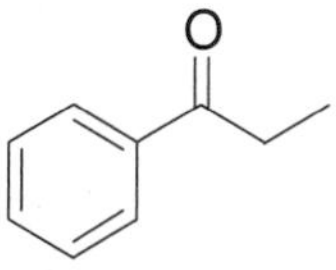 f)

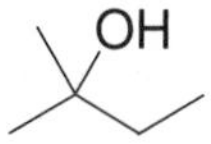

16.24.

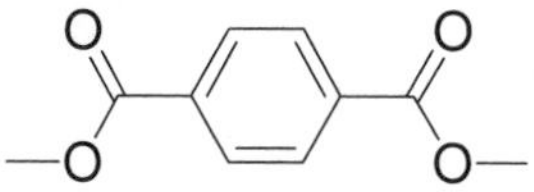

16.25.

a) Four signals. Three appear in the region 0 – 50 ppm, and the fourth signal (the C=O) appears in the region 150 – 220 ppm.

b) Five signals. All five appear in the region 0 – 50 ppm.

c) Six signals. Two appear in the region 0 – 50 ppm, and four signals appear in the region 100-150 ppm.

d) Nine signals. Two appear in the region 0 – 50 ppm, one appears in the region 50 – 100 ppm and six signals appear in the region 100 – 150 ppm.

e) Seven signals. Two appear in the region 0 – 50 ppm, one appears in the region 50 – 100 ppm and four signals appear in the region 100 – 150 ppm.

f) Five signals. Three appear in the region 0 – 50 ppm and two signals appear in the region 100 – 150 ppm.

g) Seven signals. Five appear in the region 0 – 50 ppm and two signals appear in the region 100 – 150 ppm.

h) Two signals. One appears in the region 0 – 50 ppm and the other appears in the region 100 – 150 ppm.

i) One signal appears in the region 50 – 100.
j) Five signals. One appears in the region 0 – 50 ppm, one appears in the region 50 – 100 ppm, two appear in the region 100-150, and one signal appears in the region 150 – 200 ppm.

16.26. The first compound lacks a chirality center. The two methyl groups are enantiotopic and are therefore chemically equivalent. The second compound has a chirality center (the position bearing the OH group). As such, the two methyl groups are diastereotopic and are therefore not chemically equivalent. For this reason, the ^{13}C NMR spectrum of the second compound exhibits six signals, rather than five.

16.27.

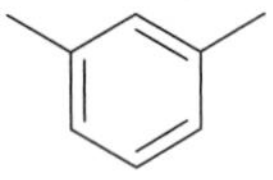

16.28.

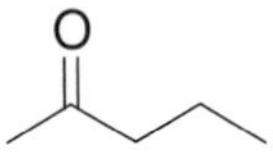

16.29.

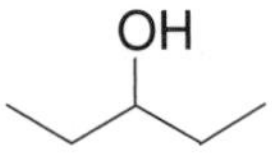

16.30.

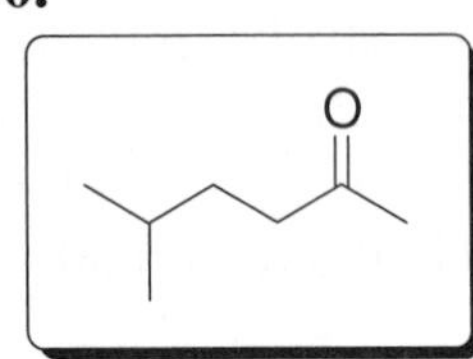

Consistent with ^{1}H NMR and ^{13}C NMR spectra

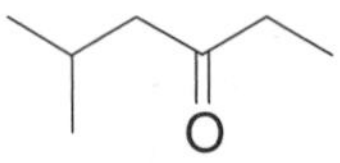

Consistent only with ^{13}C NMR spectrum

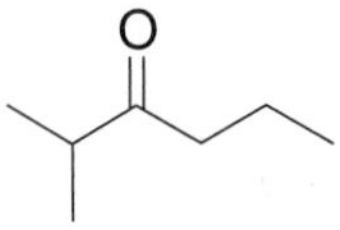

Consistent only with ^{13}C NMR spectrum

16.31.

a) 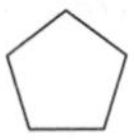 b) 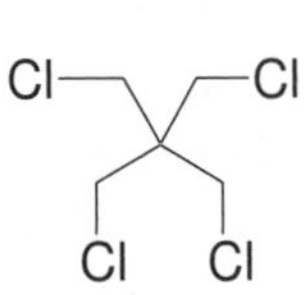 c)

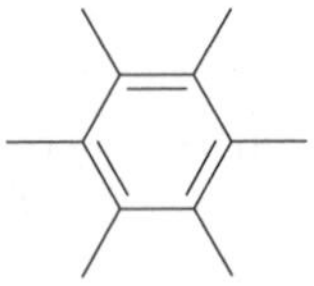

16.32.

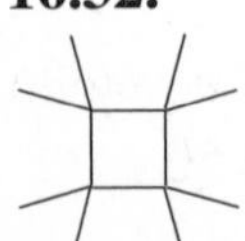

16.33. This compound will exhibit three signals in its ^{13}C NMR spectrum:

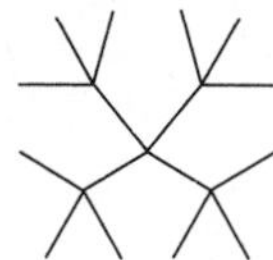

16.34.

a) 2 b) 4 c) 4 d) 2 e) 2 f) 5

16.35.

a) 4 b) 6 c) 6 d) 4 e) 2 f) 4

16.36. The first compound will have five signals in its ^{13}C NMR spectrum, while the second compound will have seven signals.

16.37.

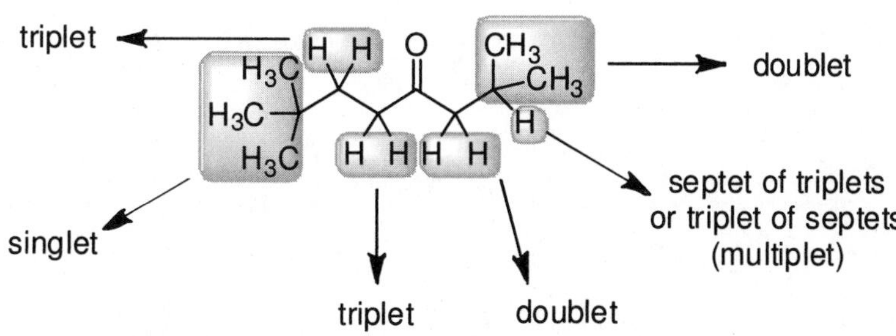

16.38.

a) The first compound will have four signals in its ^{13}C NMR spectrum, while the second compound will have twelve signals.
The first compound will have two signals in its ^{1}H NMR spectrum, while the second compound will have eight signals.

b) The first compound is a meso compound. Two of the protons are enantiotopic (the protons that are alpha to the chlorine atoms) and are therefore chemically equivalent. The first compound will only have two signals in its ^{1}H NMR spectrum, while the second compound will have three signals. For a similar reason, first compound will only have two signals in its ^{13}C NMR spectrum, while the second compound will have three signals.

c) The ^{13}C NMR spectrum of the second compound will have one more signal than the ^{13}C NMR spectrum of the other first compound. The ^{1}H NMR spectra will differ in the following way: the first compound will have a singlet somewhere between 2 and 5 ppm with an integration of 1, while the second compound will have a singlet at approximately 3.4 ppm with an integration of 3.

d) The first compound will have three signals in its ^{13}C NMR spectrum, while the second compound will have five signals.
The first compound will have two signals in its ^{1}H NMR spectrum, while the second compound will have four signals.

16.39. This compound will exhibit two signals in its ^{13}C NMR spectrum:

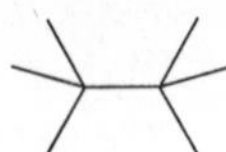

16.40.

a) homotopic b) enantiotopic c) enantiotopic d) homotopic

e) diastereotopic f) homotopic g) diastereotopic h) diastereotopic

i) homotopic j) homotopic k) homotopic l) diastereotopic

m) enantiotopic n) diastereotopic o) homotopic

16.41.

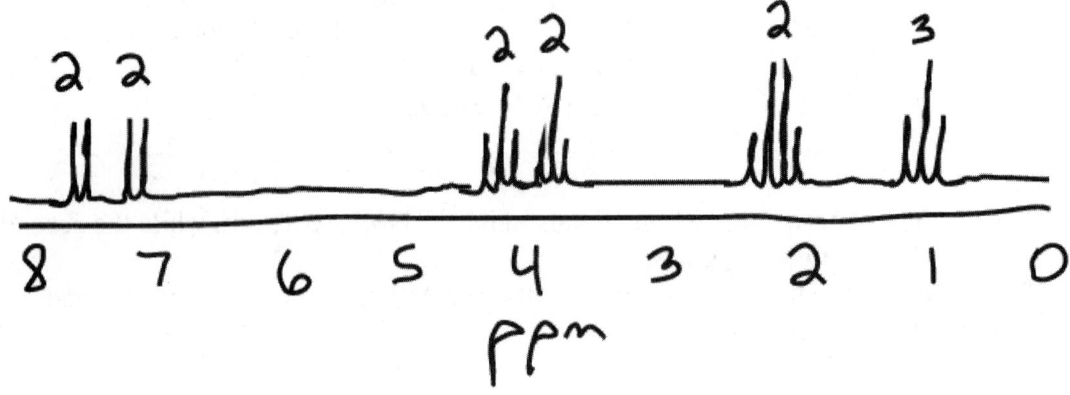

16.42.

a) Four signals are expected in the ^{1}H NMR spectrum of this compound.

F Br
H_a H_c
H_d H_b
I Cl

Increasing chemical shift

b) $H_a > H_b > H_c > H_d$

c) Four signals are expected in the ^{13}C NMR spectrum of this compound.

d) The carbon atoms follow the same trend exhibited by the protons.

16.43.

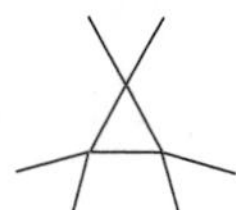

16.44.

a) Nine b) Eight c) Six

16.45.

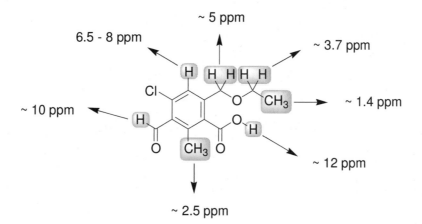

16.46.

a) Six signals, all of which appear in the region 100 – 150 ppm.
b) Seven signals. One appears in the region 150 – 220 ppm, and the remaining six signals appear in the region 0 – 50 ppm.
c) Four signals. One appears in the region 0 – 50 ppm, two appear in the region 50 – 100 ppm, and one signal appears in the region 150 – 200 ppm.

16.47. The ^{1}H NMR spectrum of the Markovnikov product should have only four signals, while the anti-Markovnikov product should have many more signals in its ^{1}H NMR spectrum.

16.48.

a) 2 b) 8 c) 4 d) 2 e) 3 f) 6 g) 2 h) 3

16.49.

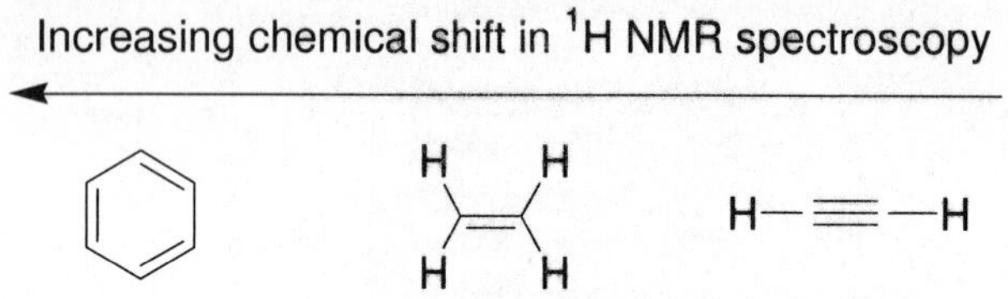

Increasing chemical shift in ¹H NMR spectroscopy ←

benzene, ethylene ($H_2C=CH_2$), acetylene (H—≡—H)

16.50.

$$\delta = \frac{\text{(observed shift from TMS in hertz)} \times 10^6}{\text{(operating frequency of the instrument in hertz)}}$$

(Observed shift from TMS in hertz) = (δ)(operating frequency) / 10^6

16.51.

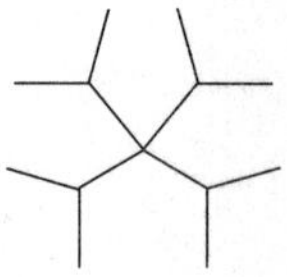

16.52.

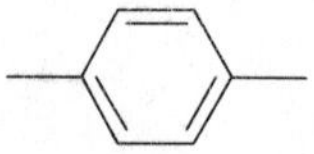

16.53. 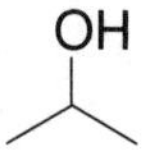 OH

16.54. HO, O, OH, O

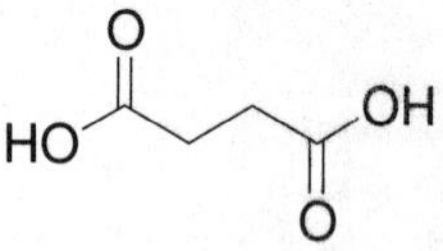

16.55.

a) 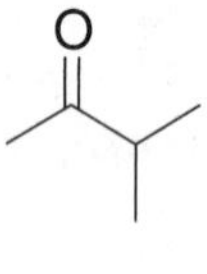 O

b) 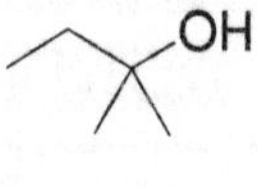 OH

c) 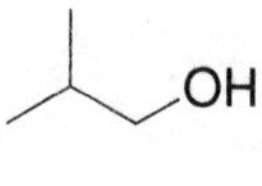 OH

d) 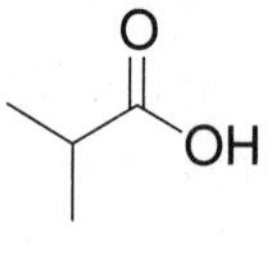 O, OH

16.56. HO

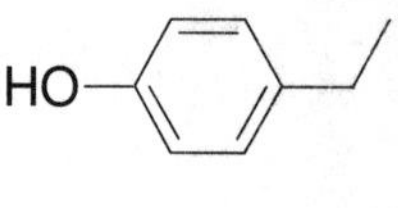

16.57.

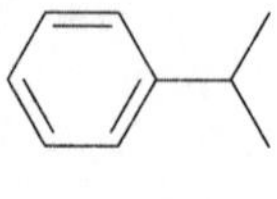

16.58. O, OH

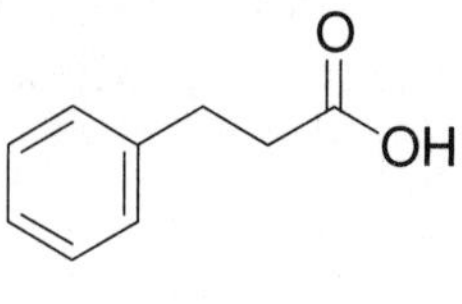

16.59.

a) 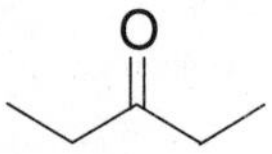 O

b) 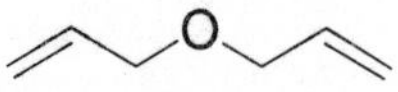 O

16.60. OH

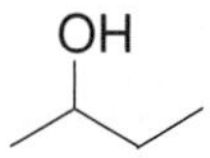

16.61. OH

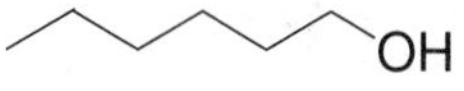

16.62. O OH

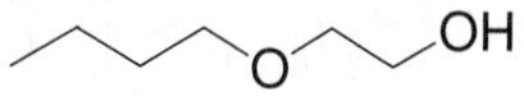

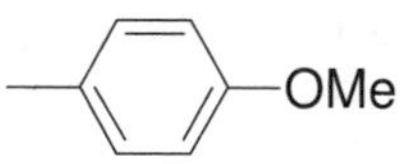

16.63. OMe

16.64. O

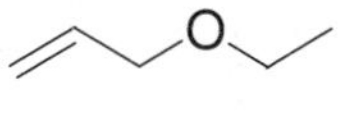

16.65. N,N-dimethylformamide (DMF) has several resonance structures:

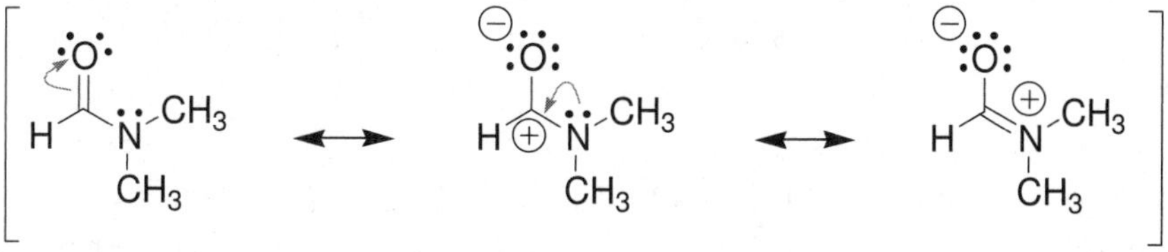

H, :O:, N, CH_3, CH_3 ⟷ ⊖ :O:, H ⊕, N, CH_3, CH_3 ⟷ ⊖ :O:, H, ⊕ N, CH_3, CH_3

Consider the third resonance structure shown above, in which the C-N bond is a double bond. This indicates that this bond is expected to have some double bond character. As such, there is an energy barrier associated with rotation about this bond, such that rotation of this bond occurs at a rate that it slower than the timescale of the NMR spectrometer. At high temperature, more molecules will have the requisite energy to undergo free rotation about the C-N bond, so the process can occur on a time scale that is faster than the timescale of the NMR spectrometer. For this reason, the signals are expected to collapse into one signal at high temperature.

16.66. In a concentrated solution of phenol, the OH groups are engaged in extensive, intermolecular hydrogen-bonding interactions. These interactions cause the average distance to increase between the O and H of each OH group. This effectively deshields the protons of the hydroxyl groups. These protons therefore show up downfield. In a dilute solution, there are fewer hydrogen bonding interactions, and the effect described above is not observed.

16.67. The methyl group on the right side is located in the shielding region of the π bond, so the signal for this proton is moved upfield to 0.8 ppm.

16.68. Bromine is significantly larger than chlorine, and the electron density of a bromine atom partially surrounds any carbon atom attached directly to the bromine, thereby shielding it. In CBr_4, the carbon atom in the center of the compound is significantly shielded because it is positioned within the electron clouds of the four bromine atoms. In fact, it is so strongly shielded that it produces a signal even higher upfield than TMS.

Chapter 17
Conjugated Pi Systems and Pericyclic Reactions

Review of Concepts

Fill in the blanks below. To verify that your answers are correct, look in your textbook at the end of Chapter 17. Each of the sentences below appears verbatim in the section entitled *Review of Concepts and Vocabulary*.

- Conjugated dienes experience free-rotation about the C2-C3 bond, giving rise to two important conformations: *s*-_____ and *s*-_____. The *s-trans* conformation is lower in energy.
- The HOMO and LUMO are referred to as ___________ **orbitals**, and the reactivity of conjugated polyenes can be explained with **frontier orbital theory**.
- An _________ **state** is produced when a π electron in the HOMO absorbs a photon of light bearing the appropriate energy necessary to promote the electron to a higher energy orbital.
- Reactions induced by light are called _______________ **reactions**.
- When butadiene is treated with HBr, two major products are observed, resulting from ______**-addition** and ______**-addition**.
- Conjugated dienes that undergo addition at low temperature are said to be under ____________ **control.** Conjugated dienes that undergo addition at elevated temperature are said to be under _________________ **control**.
- _______________ **reactions** proceed via a concerted process with a cyclic transition state, and they are classified as **cycloaddition reactions**, ______________ **reactions**, and **sigmatropic rearrangements.**
- The Diels–Alder reaction is a [_______] **cycloaddition** in which two C-C bonds are formed simultaneously.
- High temperatures can often be used to achieve the reverse of a Diels–Alder reaction, called a _________ **Diels–Alder**.
- The starting materials for a Diels–Alder reaction are a diene, and a ___________.
- The Diels–Alder reaction only occurs when the diene is in an ____ conformation.
- When cyclopentadiene is used as the starting diene, a bridged bicyclic compound is obtained, and the _____ cycloadduct is favored over the _____ cycloadduct.
- Conservation of orbital symmetry determines whether an electrocyclic reaction occurs in a ___________ fashion or a ____________ fashion.
- A [_________] sigmatropic rearrangement is called a **Cope rearrangement** when all six atoms of the cyclic transition state are carbon atoms.
- Compounds that possess a conjugated π system will absorb UV or visible light to promote an electronic excitation called a __________ transition.
- The most important feature of the absorption spectrum is the _________, which indicates the wavelength of maximum absorption.
- When a compound exhibits a λ_{max} above 400 nm, the compound will absorb __________ light, rather than UV light.

Review of Skills

Fill in the blanks and empty boxes below. To verify that your answers are correct, look in your textbook at the end of Chapter 17. The answers appear in the section entitled *SkillBuilder Review*.

17.1 Proposing the Mechanism and Predicting the Products of Electrophilic Addition to Conjugated Dienes

IN THE SPACE PROVIDED, DRAW THE MECHANISM OF THE REACTION THAT IS EXPECTED TO OCCUR WHEN THE FOLLOWING COMPOUND IS TREATED WITH HBr. MAKE SURE TO DRAW ALL POSSIBLE PRODUCTS.

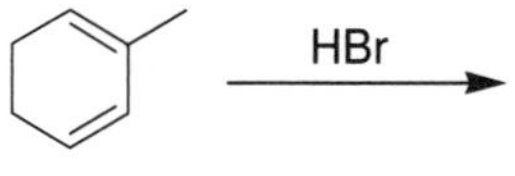

HBr

17.2 Predicting the Major Product of an Electrophilic Addition to Conjugated Dienes

DRAW THE MAJOR PRODUCTS OF THE FOLLOWING REACTION:

HBr

0°C

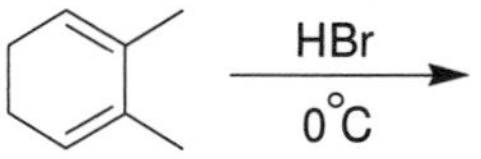

17.3 Predicting the Product of a Diels–Alder Reaction

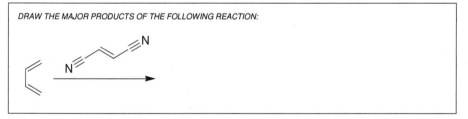

DRAW THE MAJOR PRODUCTS OF THE FOLLOWING REACTION:

N N

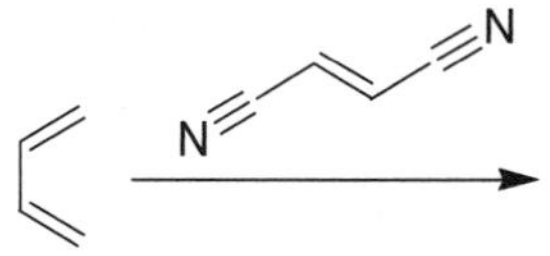

17.4 Predicting the Product of an Electrocyclic Reaction

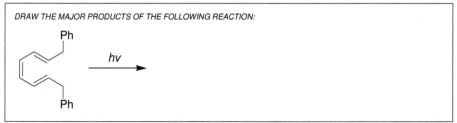

DRAW THE MAJOR PRODUCTS OF THE FOLLOWING REACTION:

Ph

hv

Ph

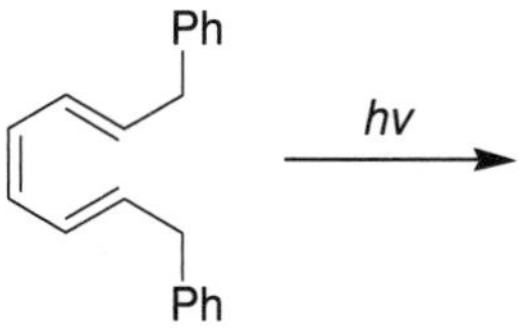

17.5 Using Woodward–Fieser Rules to Estimate λ_{max}

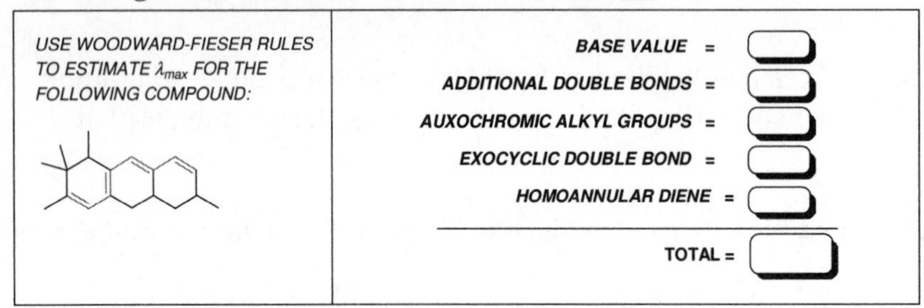

USE WOODWARD-FIESER RULES TO ESTIMATE λ_{max} FOR THE FOLLOWING COMPOUND:
BASE VALUE =
ADDITIONAL DOUBLE BONDS =
AUXOCHROMIC ALKYL GROUPS =
EXOCYCLIC DOUBLE BOND =
HOMOANNULAR DIENE =
TOTAL =

Review of Reactions

Predict the Products for each of the following transformations. To verify that your answers are correct, look in your textbook at the end of Chapter 17. The answers appear in the section entitled *Review of Reactions*.

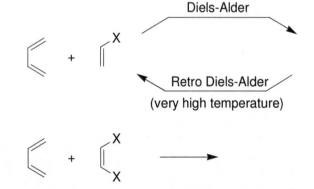

Preparation of Dienes

Br

t-BuOK

Br

t-BuOK

Br

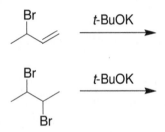

Electrophilic Addition

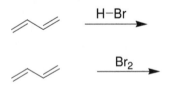

H–Br

Br_2

Diels–Alder Reaction

Diels-Alder

X

+

Retro Diels-Alder

(very high temperature)

X

+

X

X
+
X

O OR
+
O OR

CHO
+
CHO

Electrocyclic Reactions

heat

heat

CH_3 H H CH_3
heat light
CH_3 CH_3
light heat
CH_3 H CH_3 H

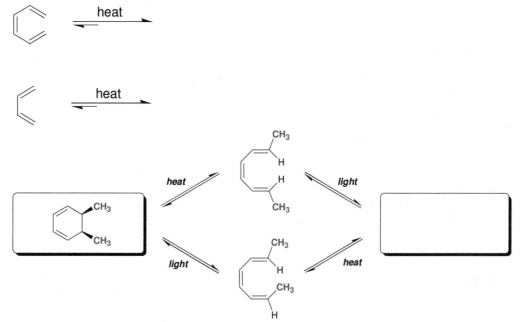

Sigmatropic Rearrangements

Cope Rearrangement

heat

Claisen Rearrangement

O heat

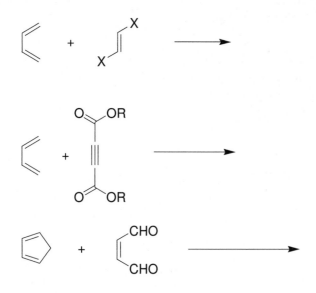

Solutions

17.1.

a)

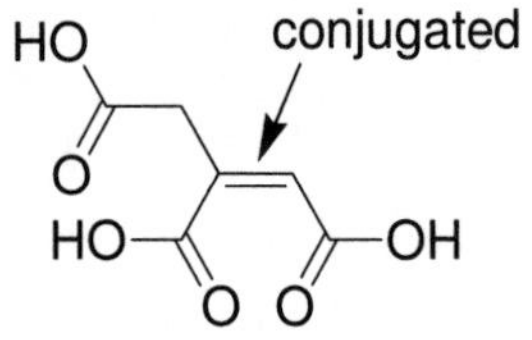

HO conjugated
O
HO— —OH
O O

b)

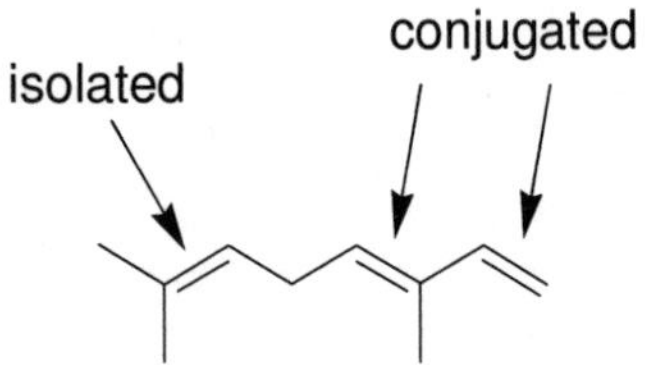

isolated conjugated

c)

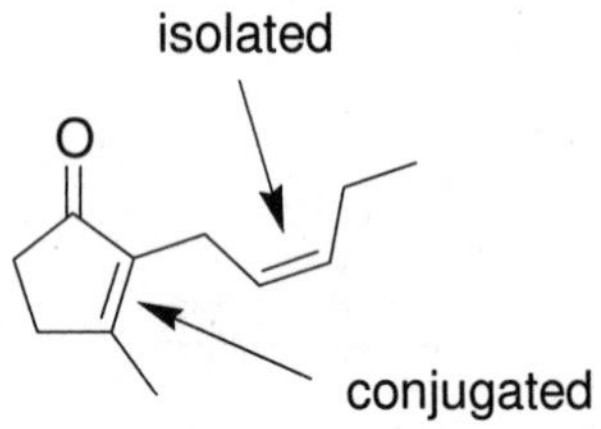

isolated
O
conjugated

d)

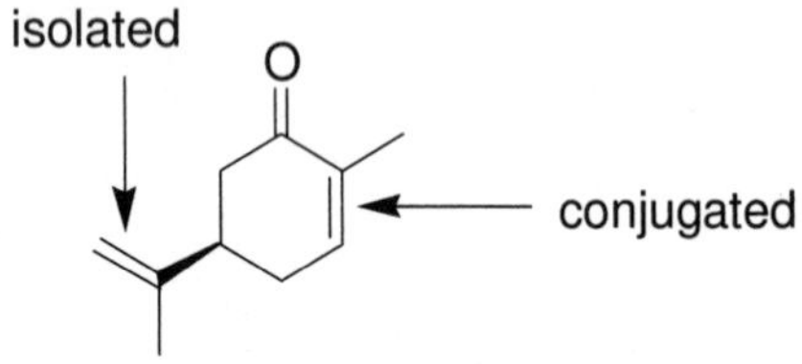

isolated
O
conjugated

17.2.

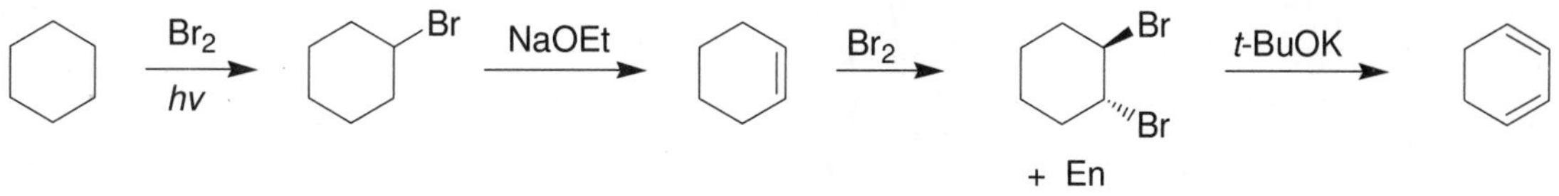

17.3.

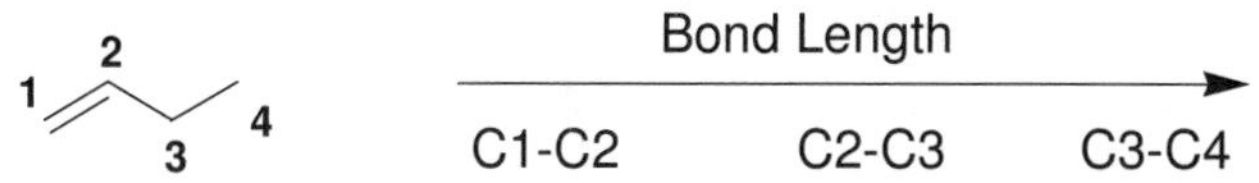

17.4.

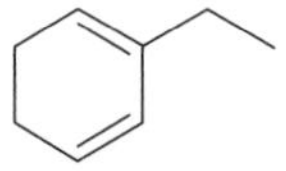

a)
The conjugated diene will liberate the least heat because it is the most stable of the three compounds.

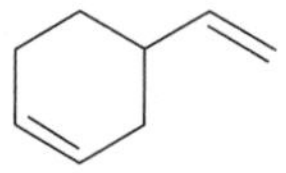

b)
This isolated diene will liberate more heat than the other isolated diene, because the π bonds in this compound are not highly substituted (one π bond is monosubstituted and the other is disubstituted). In the other isolated diene, the π bonds are disubstituted and trisubstituted (and therefore more stable).

17.5. In the compound below, all three π bonds are conjugated:

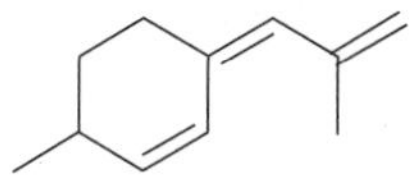

17.6.

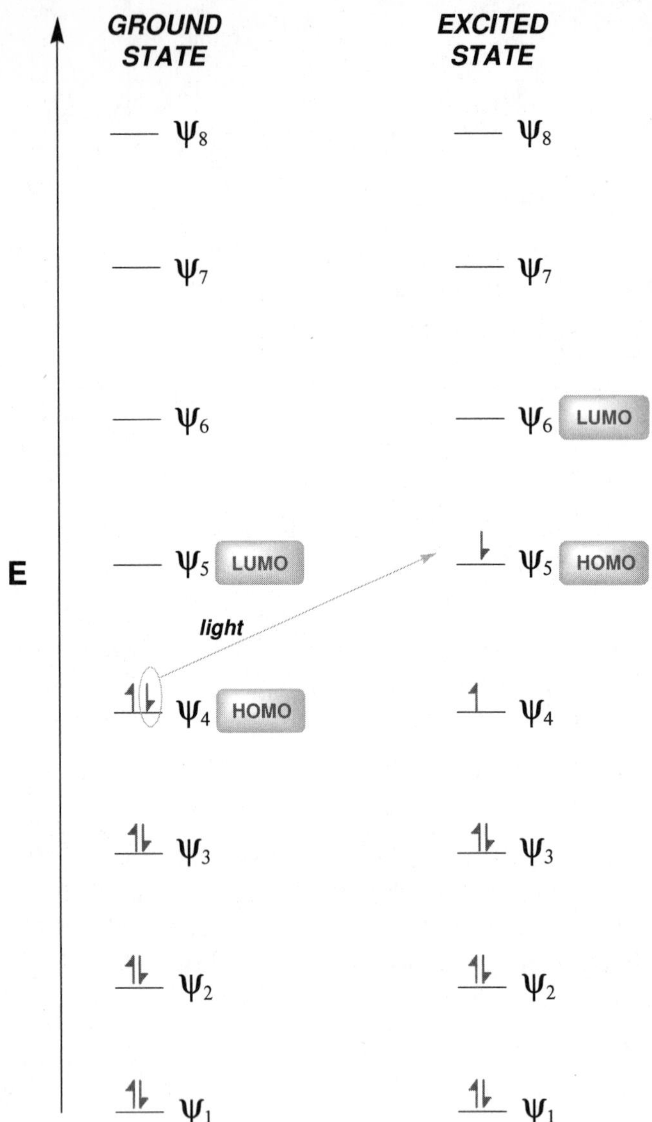
GROUND STATE
EXCITED STATE
E
ψ8
ψ7
ψ6
ψ5
LUMO
light
ψ4
HOMO
ψ3
ψ2
ψ1
ψ8
ψ7
ψ6
LUMO
ψ5
HOMO
ψ4
ψ3
ψ2
ψ1

17.7.

a)

H
:Cl:
:Cl:
:Cl:
(racemic)
(racemic)
+
:Cl:
(racemic)
+
:Cl

b)

H
:Cl:
:Cl:
(racemic)
+
:Cl:
+
:Cl:
(racemic)
+
:Cl

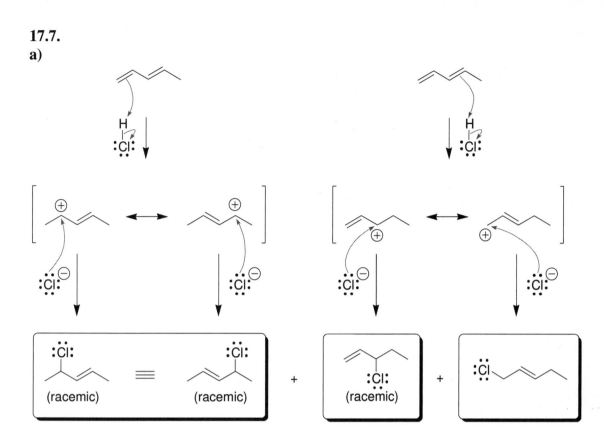

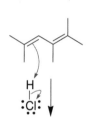

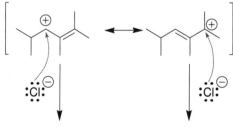

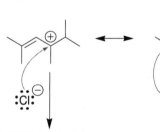

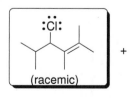

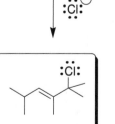

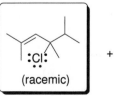

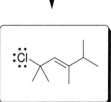

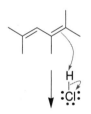

c)

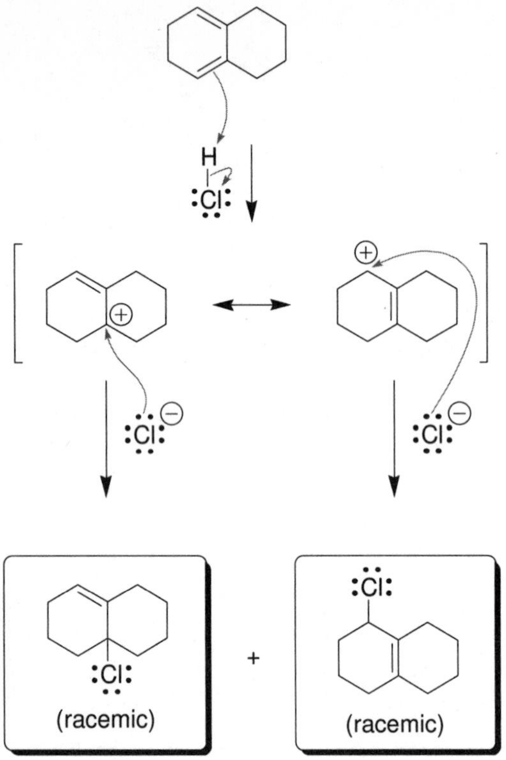

:Cl: :Cl: :Cl: :Cl: :Cl:

(racemic) + (racemic)

d)

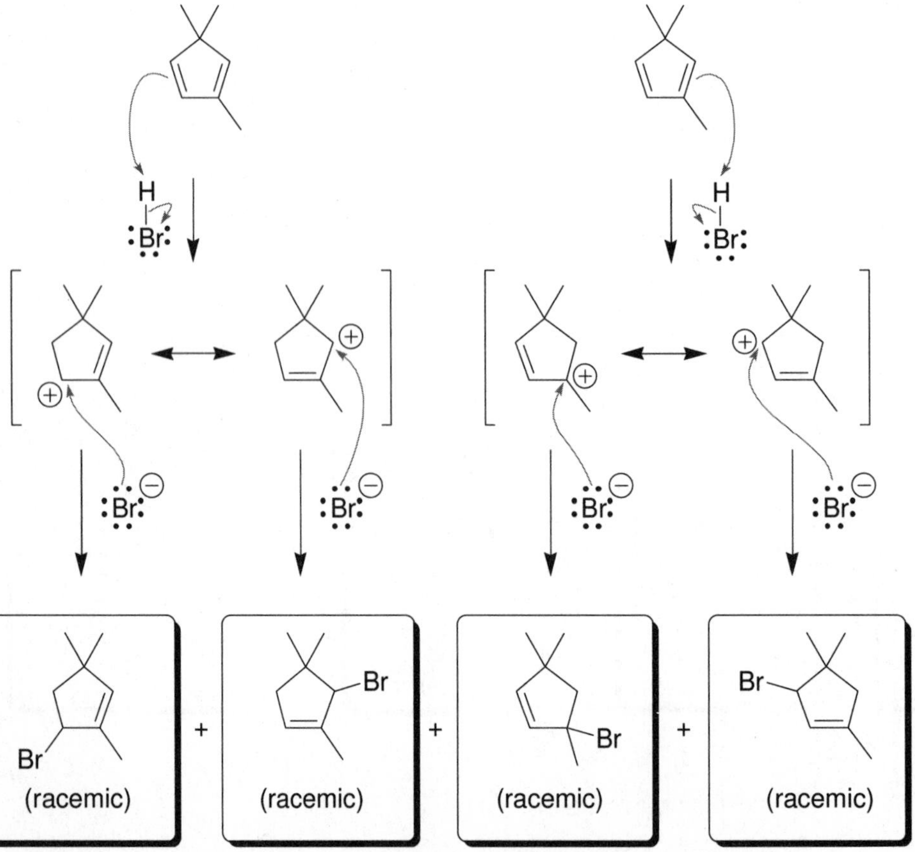

:Br: :Br: :Br: :Br: :Br: :Br:

Br (racemic) + Br (racemic) + Br (racemic) + Br (racemic)

e)

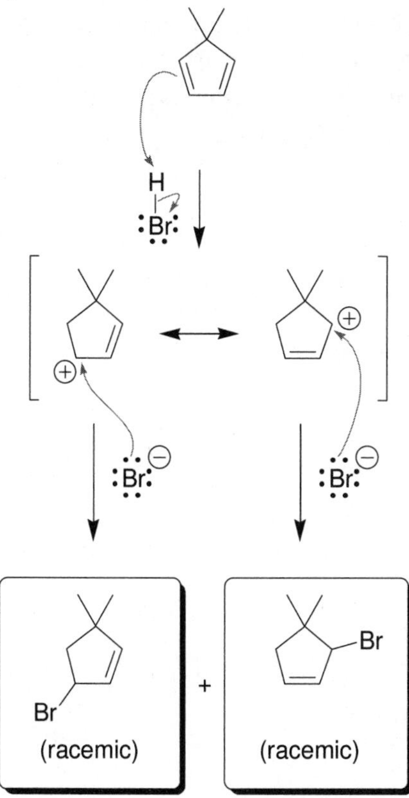

f)

H
Br
Br
Br
Br
(racemic)
+
Br

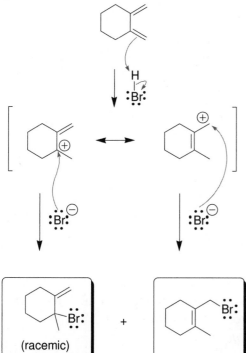

17.8. The first diene can be protonated either at C1 or at C4. Each of these pathways produces a resonance stabilized carbocation. And each of these carbocations can be attacked in two positions, giving rise to four possible products. In contrast, the second diene yields the same carbocation regardless of whether protonation occurs at C1 or at C4. This resonance-stabilized carbocation can be attacked in two positions, giving rise to two products.

17.9.

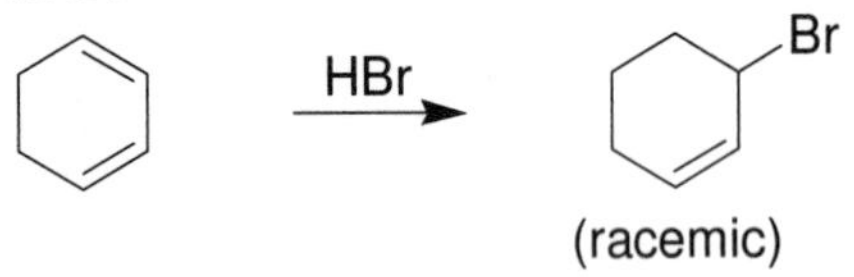

HBr

Br

(racemic)

17.10.

a)

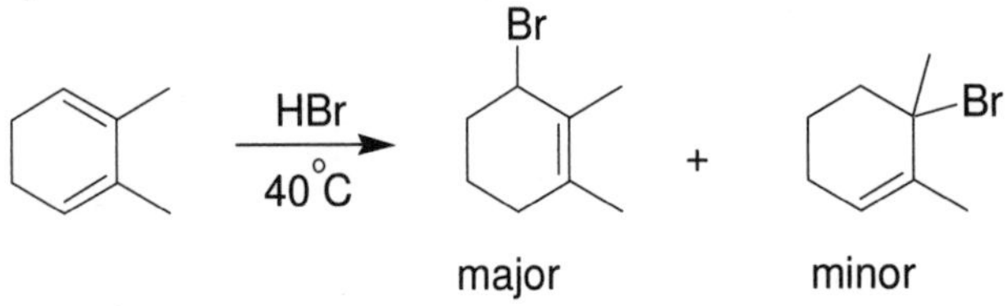

HBr

40°C

Br

+

Br

major

minor

b)

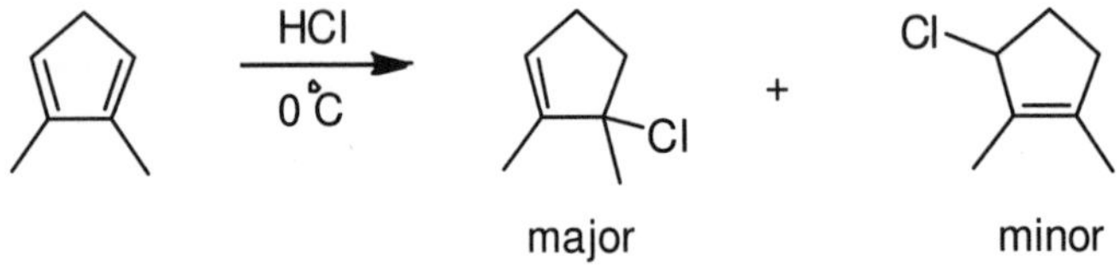

HCl

0°C

Cl

+

Cl

major

minor

c)

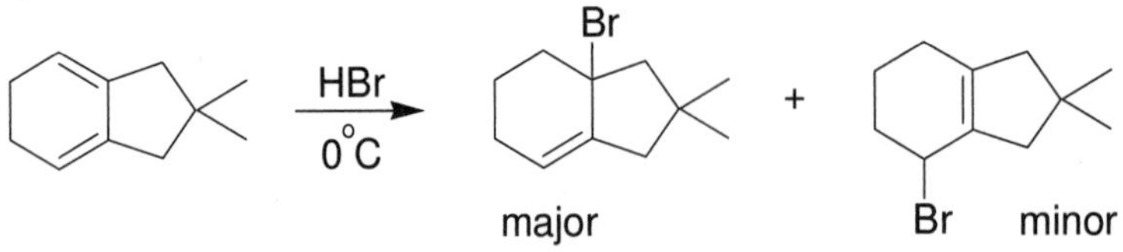

HBr

0°C

Br

+

Br

major

minor

17.11. In this case, the π bond in the 1,2-adduct is more substituted than the π bond in the 1,4-adduct (trisubstituted rather than disubstituted). As a result, the 1,2-adduct predominates at either low temperature or high temperature.

17.12. In this case, 1,2-addition and 1,4-addition yield the same product.

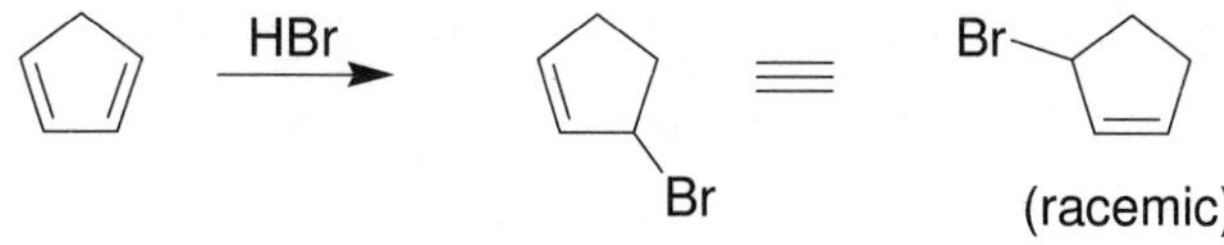

HBr

Br

(racemic)

Br

(racemic)

17.13.

a) 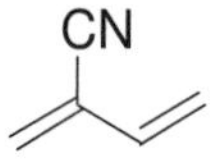 CN

b) 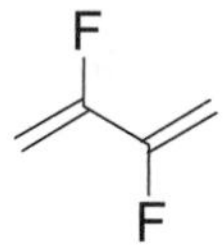 F F

17.14.

a)

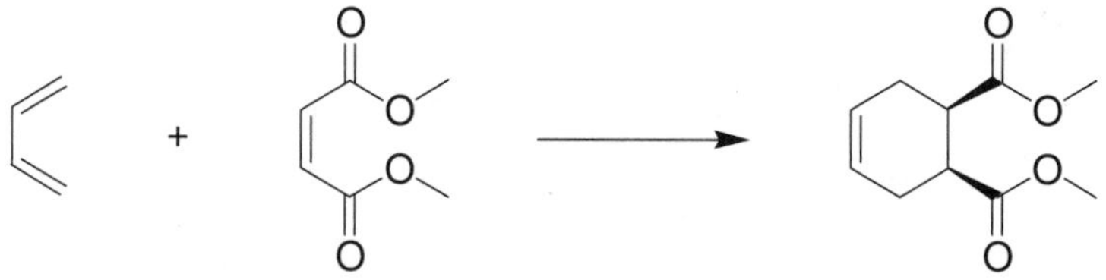

+ O O O O

(meso)

b)

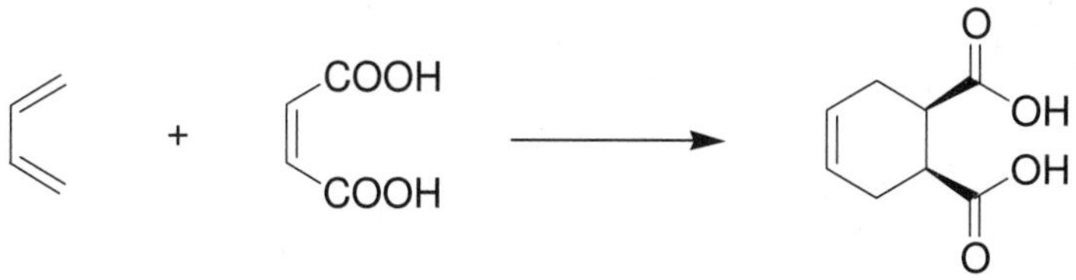

+ COOH COOH

O OH OH O

(meso)

c)

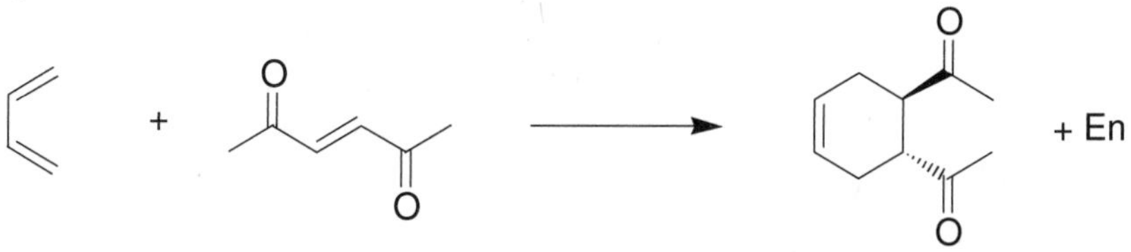

+ O O

O O

+ En

d)

+ O O

O O

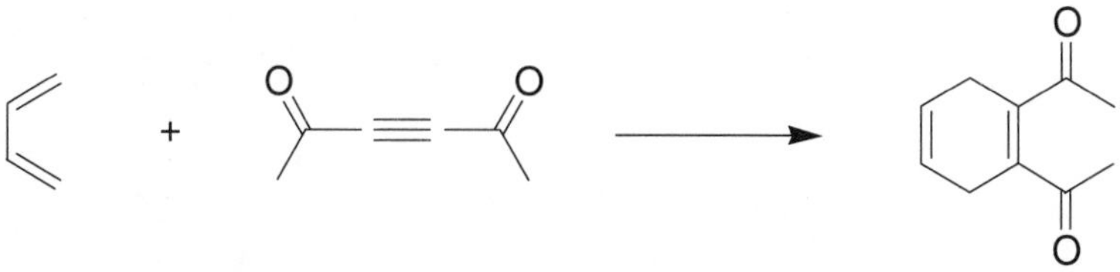

e)

+

(meso)

f)

+ NC—≡—CN

CN

CN

g)

+

O

+ En

h)

+

O

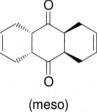

OH

i)

+

O

+ En

17.15.

O

O

+

(excess)

(meso)

(meso)

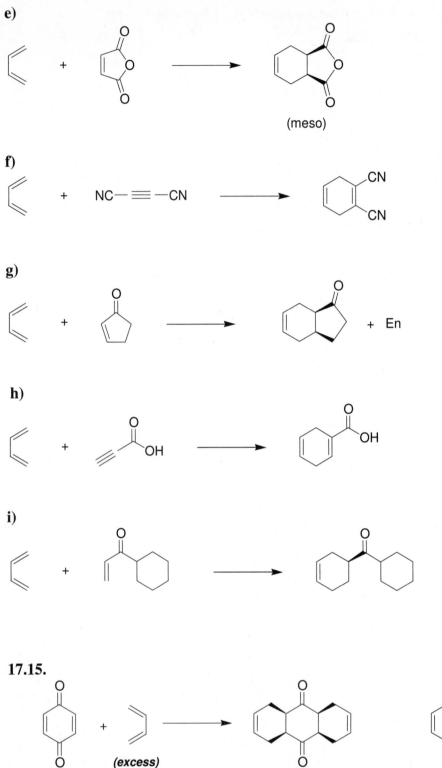

17.16.

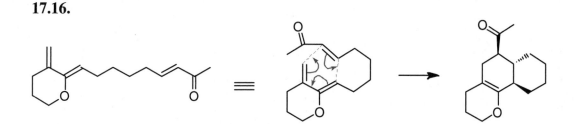

17.17. The *2E,4E* isomer is expected to react more rapidly as a diene in a Diels–Alder reaction, because it can readily adopt an *s-trans* conformation.

(2E,4E)-hexadiene

In contrast, the *2Z,4Z* isomer is expected to react more slowly as a diene in a Diels–Alder reaction, because it cannot readily adopt an *s-trans* conformation, as a result of steric hinderance.

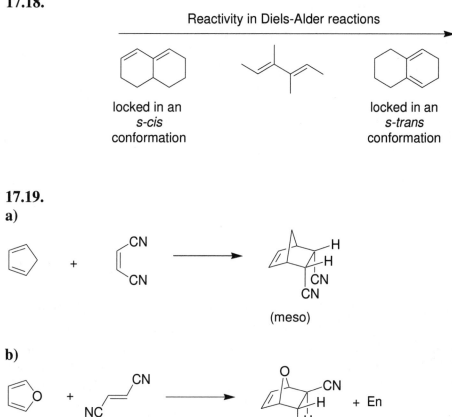

(2Z,4Z)-hexadiene

Me)(Me

17.18.

Reactivity in Diels-Alder reactions

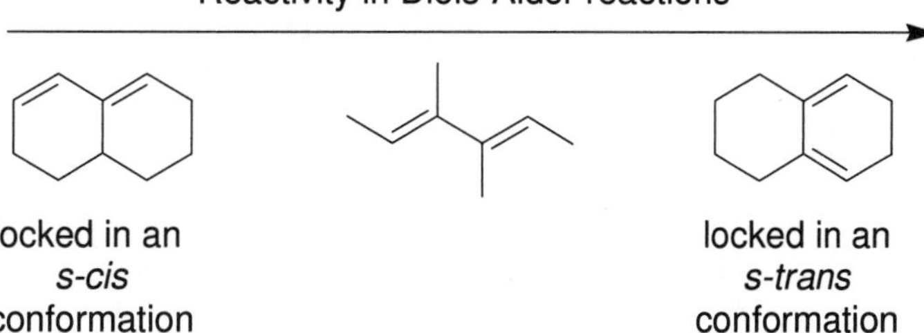

17.19.

a)

+ CN CN → H H CN CN

(meso)

b)

O + NC CN → O CN H H CN + En

c)

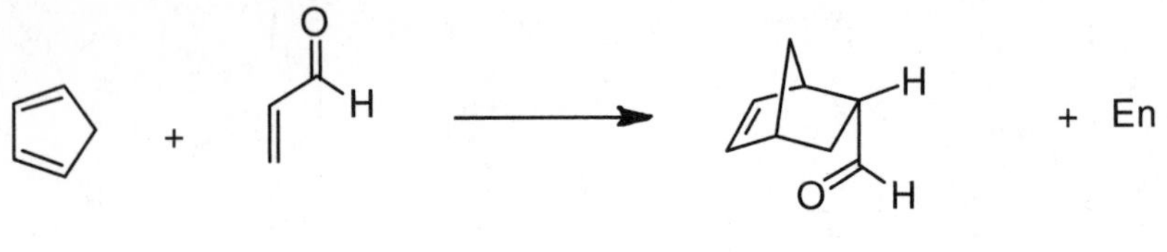

+ H O → H H O + En

d)

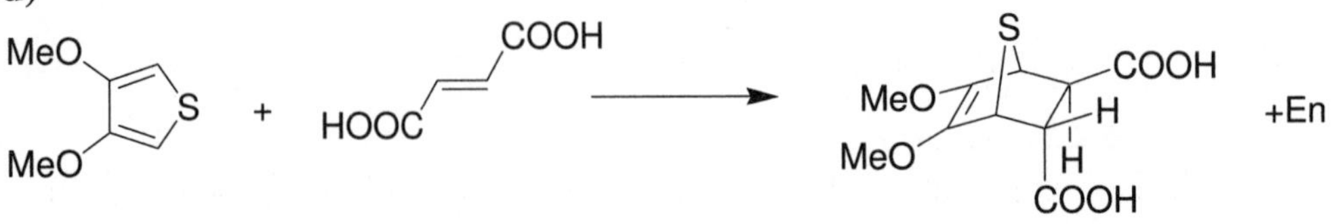

MeO MeO S + HOOC COOH → S COOH H H COOH MeO MeO +En

e)

+ CN → H CN + En

f)

O + O O O → O O O O

(meso)

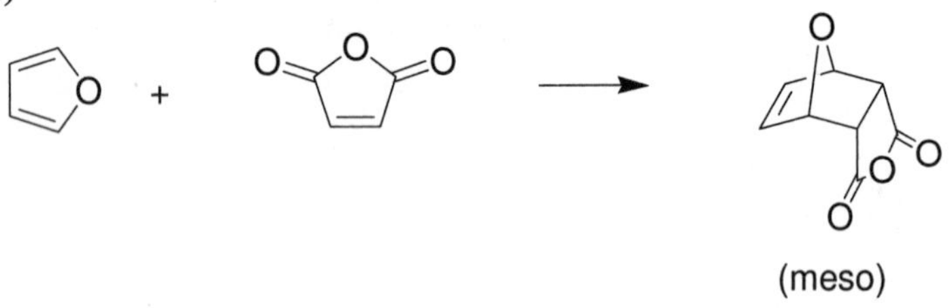

17.20. We first consider the HOMO of one molecule of butadiene and the LUMO of another molecule of butadiene. The phases of these MOs do not align, so a thermal reaction is symmetry-forbidden. However, if one molecule is photochemically excited, the HOMO and LUMO of that molecule are redefined. The phases of the frontier orbitals will align under these conditions, so the reaction is expected to occur photochemically.

17.21.

a)

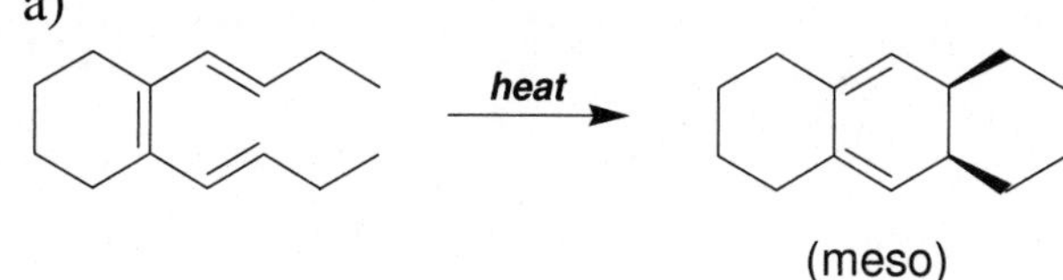

(meso)

b)

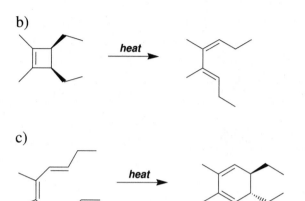

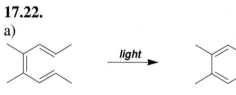

heat

c)

heat

17.22.

a)

light

\+ En

b)

light

c)

heat

\+ En

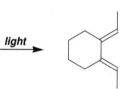

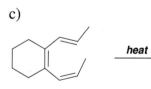

17.23.

a)

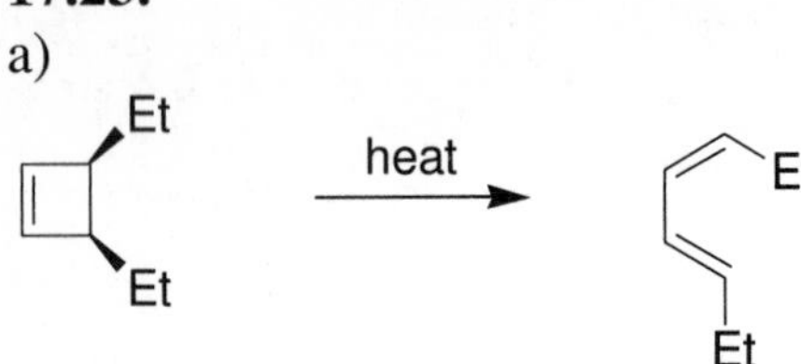

b)

Et

Et

light

Et

Et

+

Et

Et

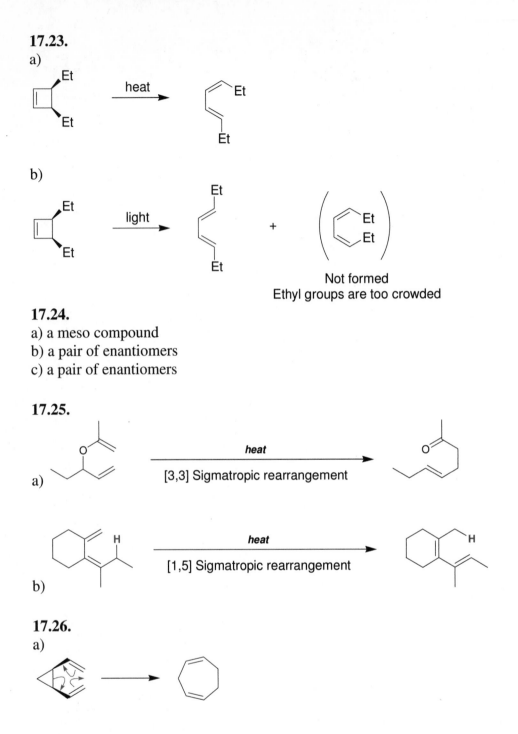

Not formed
Ethyl groups are too crowded

17.24.

a) a meso compound
b) a pair of enantiomers
c) a pair of enantiomers

17.25.

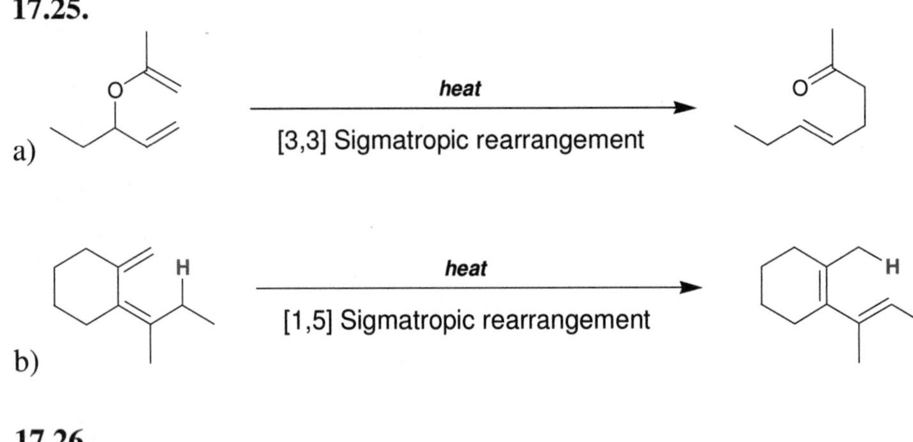

17.26.

a)

b) [3,3] Sigmatropic rearrangement

c) The ring strain associated with the three-membered ring is alleviated. The reverse process would involve forming a high-energy, three-membered ring. The equilibrium disfavors the reverse process.

17.27.

a)

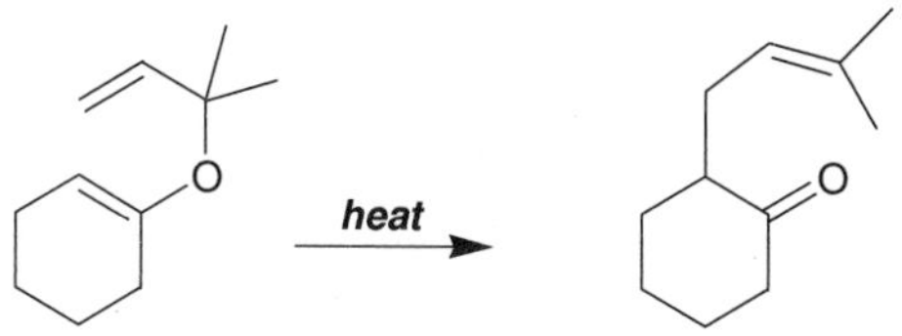

heat

b)

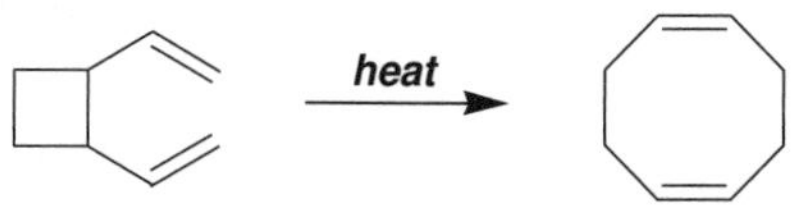

heat

c)

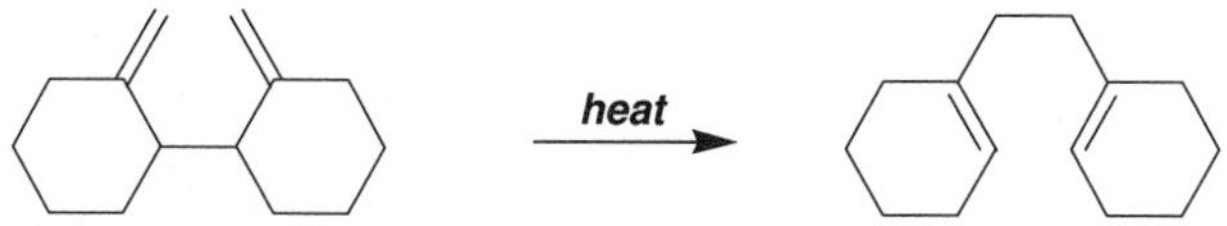

heat

d)

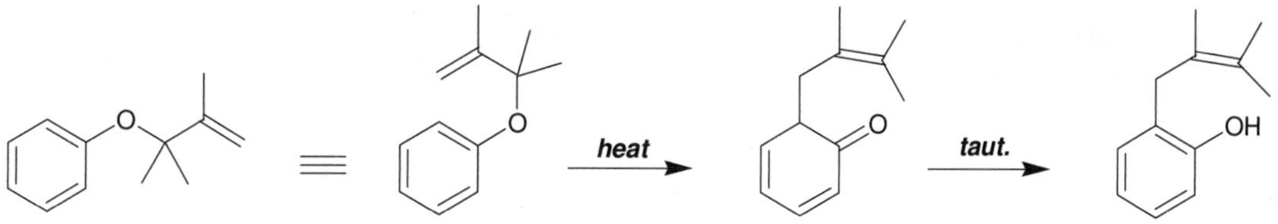

heat

taut.

17.28.

heat

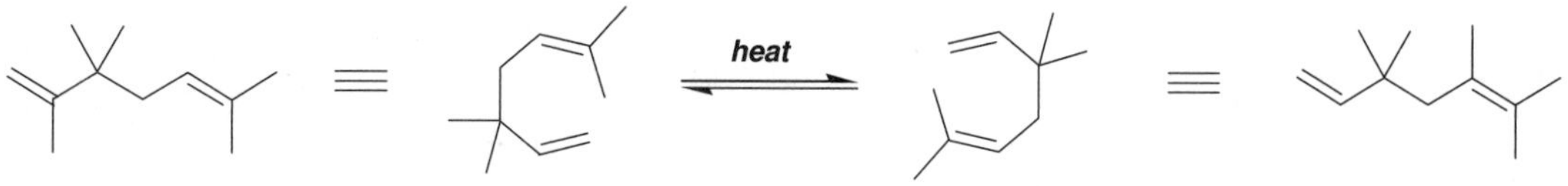

17.29.

a)

Base	=	217
Additional double bonds	=	0
Auxochromic alkyl groups	=	+25
Exocyclic double bond	=	+5
Homoannular diene	=	0
Total =		247 nm

b)

Base	=	217
Additional double bonds	=	+30
Auxochromic alkyl groups	=	+25
Exocyclic double bond	=	+5
Homoannular diene	=	0
Total	=	277 nm

c)

Base	=	217
Additional double bonds	=	+30
Auxochromic alkyl groups	=	+30
Exocyclic double bonds	=	+15
Homoannular diene	=	0
Total	=	292 nm

d)

Base	=	217
Additional double bonds	=	+30
Auxochromic alkyl groups	=	+35
Exocyclic double bonds	=	+5
Homoannular diene	=	+39
Total	=	326 nm

17.30

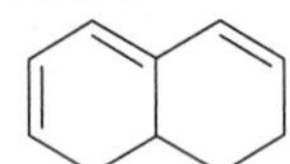

17.31.

a) Blue.
b) Red-Orange.
c) Blue-violet.

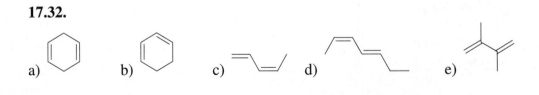

17.32.

a) 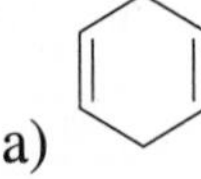 b) c) d) 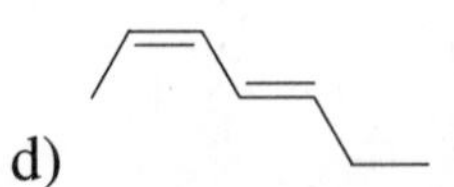 e) 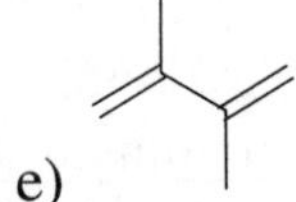

17.33.

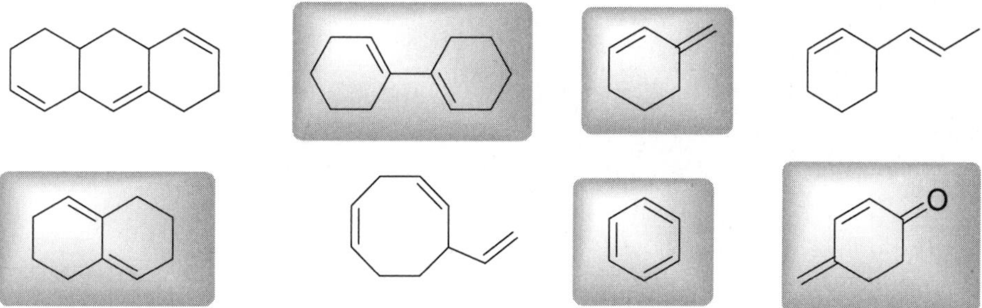

17.34.

a) These drawings represent two different conformations of the same compound: the *s-cis* conformation and the *s-trans* conformation. These two conformations are in equilibrium at room temperature.

b) These drawings represent two different compounds: (*Z*)-1,3,5-hexatriene and (*E*)-1,3,5-hexatriene. These compounds are diastereomers and can be isolated from one another.

c) These drawings represent two different conformations of the same compound: the *s-cis* conformation and the *s-trans* conformation. These two conformations are in equilibrium at room temperature.

17.35.

HBr

Br

(racemic)

17.36.

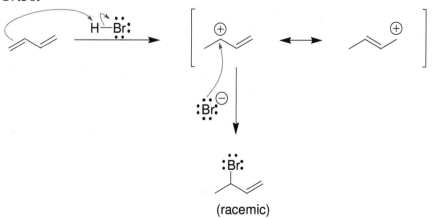

H—Br:

:Br:

:Br:

(racemic)

17.37.

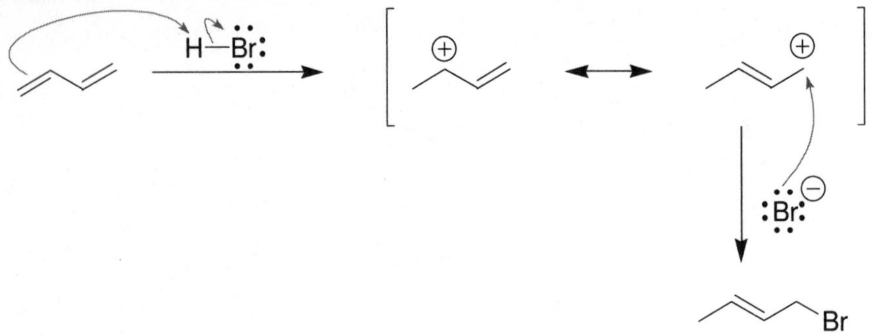

H—Br:

Br

17.38.

H
Br:

(racemic)
+
(racemic)
(racemic)
+
(racemic)

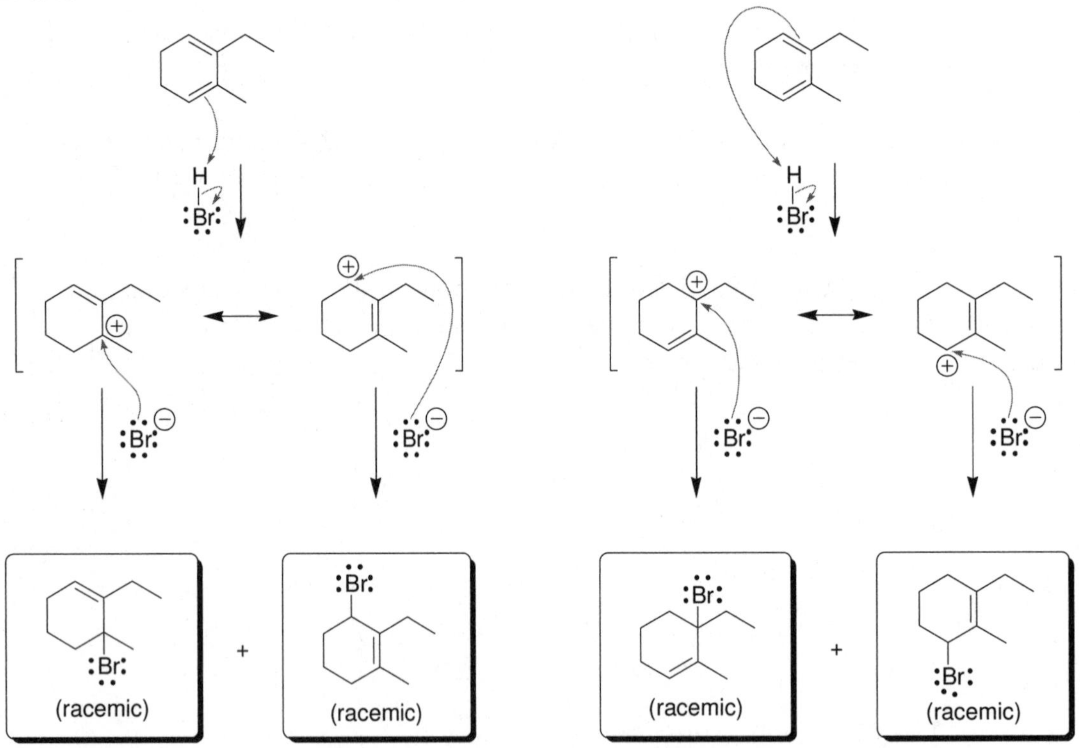

17.39. An increase in temperature allowed the system to reach equilibrium concentrations, which are determined by the relative stability of each product. Under these conditions, the 1,4-adducts predominate. Once at equilibrium, lowering the temperature will not cause a decrease in the concentration of the 1,4-adducts.

17.40.

a) The *tert*-butyl groups provide significant steric hinderance that prevents the compound from adopting an *s-cis* conformation.

b) This diene is not conjugated.

c) The methyl groups provide significant steric hinderance that prevents the compound from adopting an *s-cis* conformation.

d) This diene cannot adopt an *s-cis* conformation

17.41.

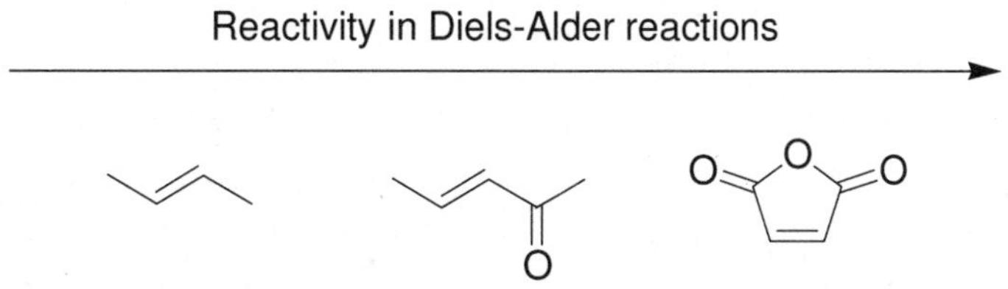

Reactivity in Diels-Alder reactions

17.42. The π bonds in 1,2-butadiene are not conjugated, and λ_{max} is therefore lower than 217 nm. In fact, it is below 200 nm, which is beyond the range used by most UV-VIS spectrometers.

17.43.

a)

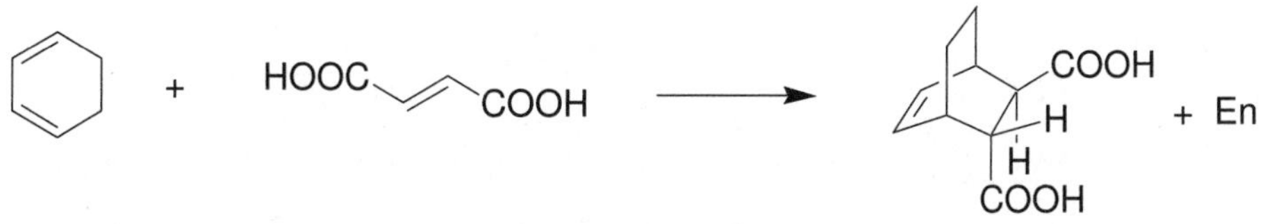

HOOC COOH → COOH H H COOH + En

b)

HOOC CN → H H CN COOH + En

c)

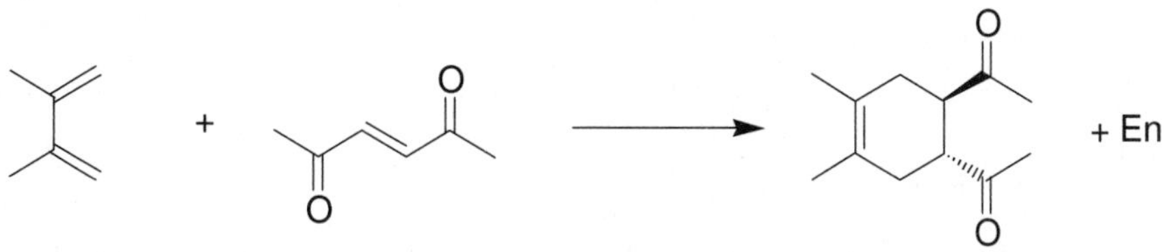

O O → O O + En

d)

S + O O O → S O O O

(meso)

e)

+ NC CN → CN CN

f)

MeO MeO S + CO_2Me CO_2Me → S MeO MeO CO_2Me CO_2Me

17.44.

a)

+ COOH COOH → COOH COOH

b)

+ NC CN NC CN → CN CN CN CN

c)

+ CHO CHO → CHO CHO

d)

O + O O O → O O O O

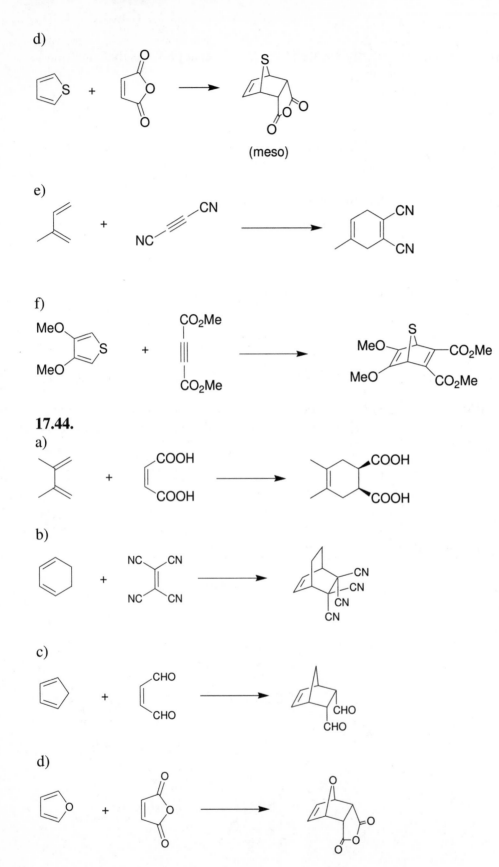

e)

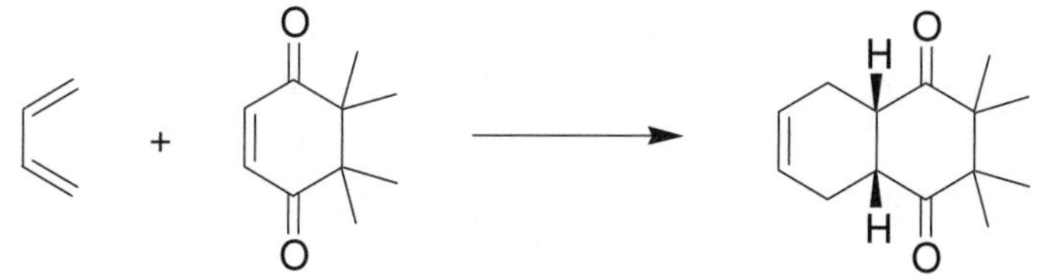

+ →

f)

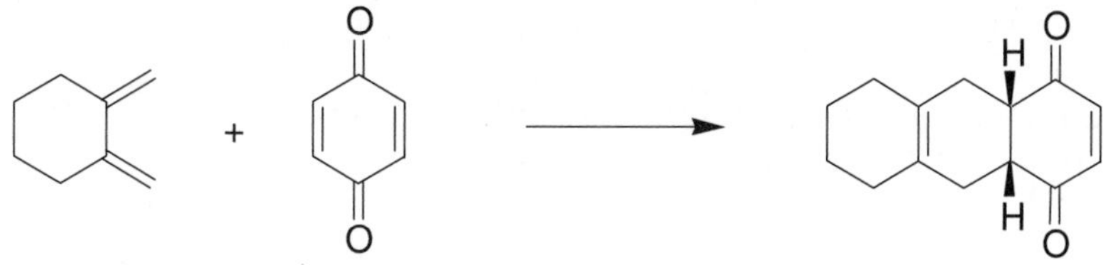

+ →

g)

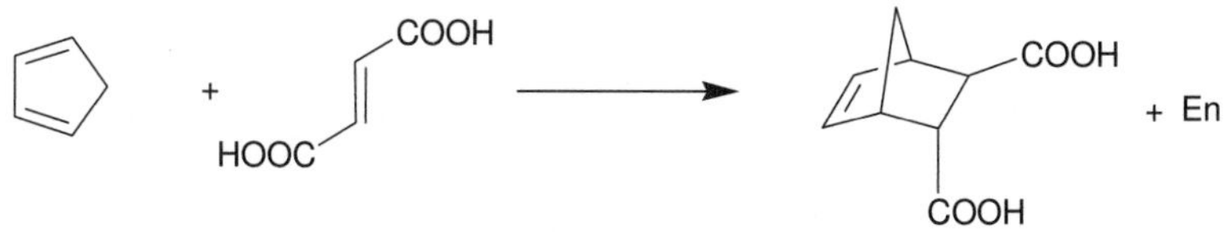

COOH
+
HOOC
→
COOH
COOH
+ En

h)

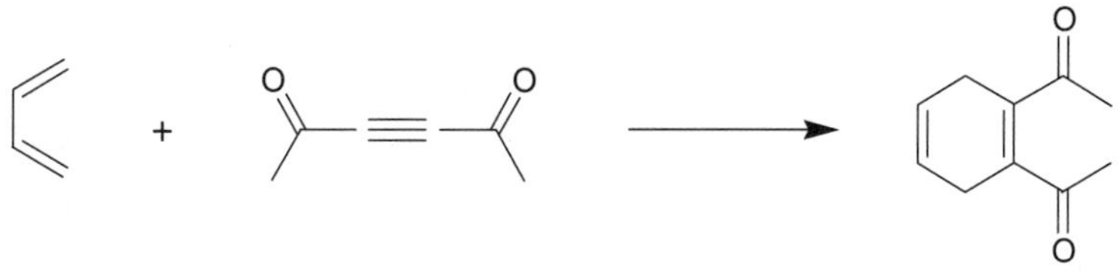

+ →

17.45.

HBr / 40 °C → Br; 1) NaOH / 2) PCC → O H → H / CH_3 + En

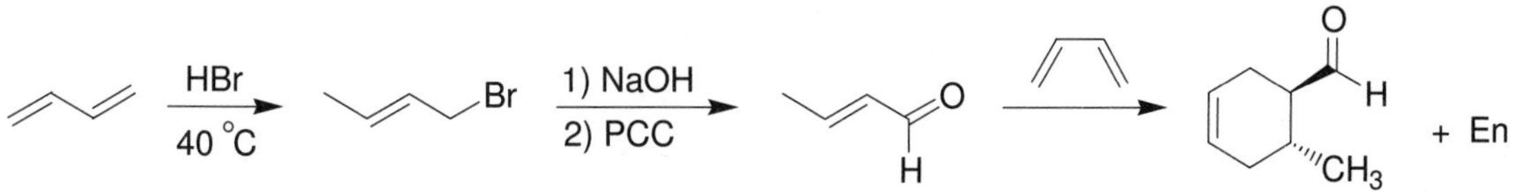

17.46.

+ →

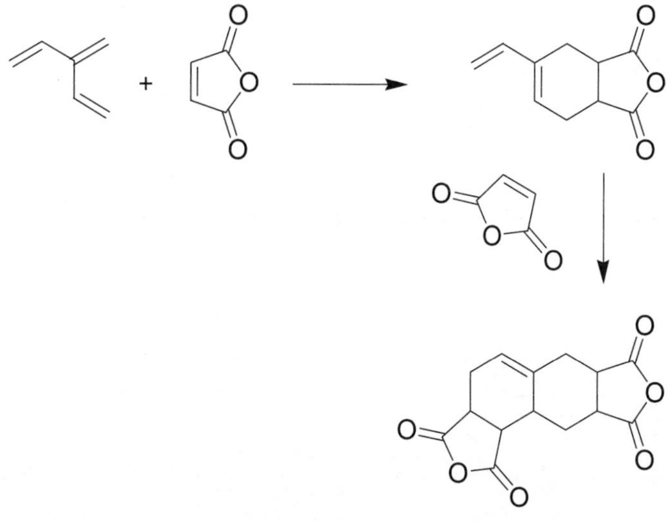

17.47.

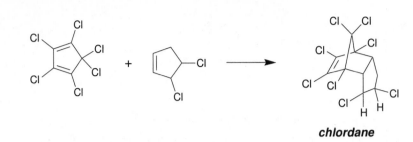

chlordane

17.48. The two ends of the conjugated system are much farther apart in a seven-membered ring than they are in a five-membered ring.

17.49.

17.50.

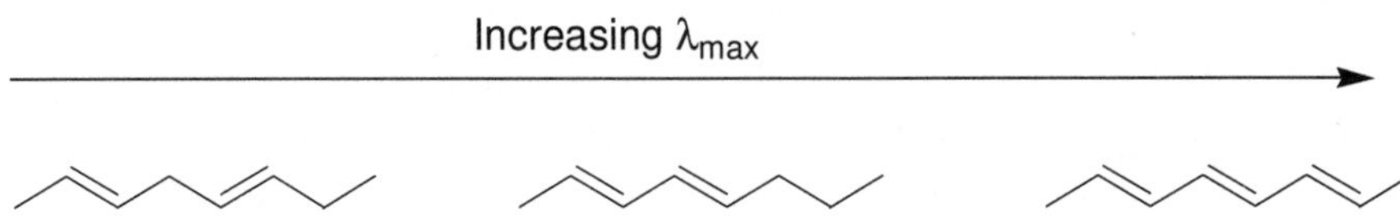

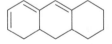

17.51. Two of the π bonds are homoannular in this compound, which adds +39 nm according to Woodward-Fieser rules.

17.52.

Base	=	217
Additional double bonds	=	+60
Auxochromic alkyl groups	=	+35
Exocyclic double bonds	=	+5
Homoannular diene	=	+39
Total	=	356 nm

17.53. Each of these transformations can be explained with a [1,5] sigmatropic rearrangement:

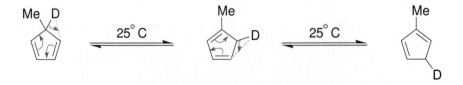

17.54. This transformation can be explained with a [1,5] sigmatropic rearrangement:

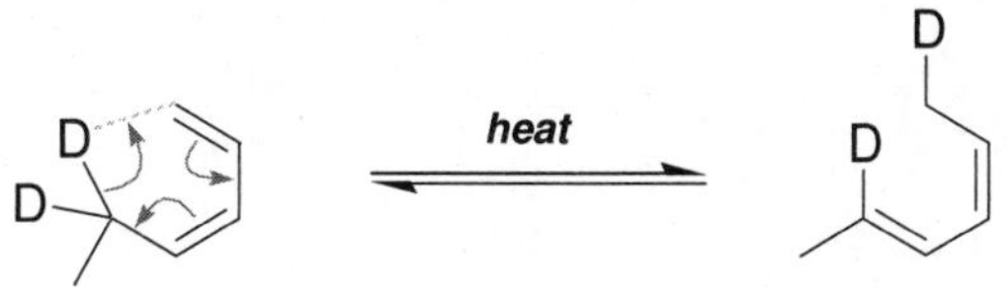

heat

17.55.

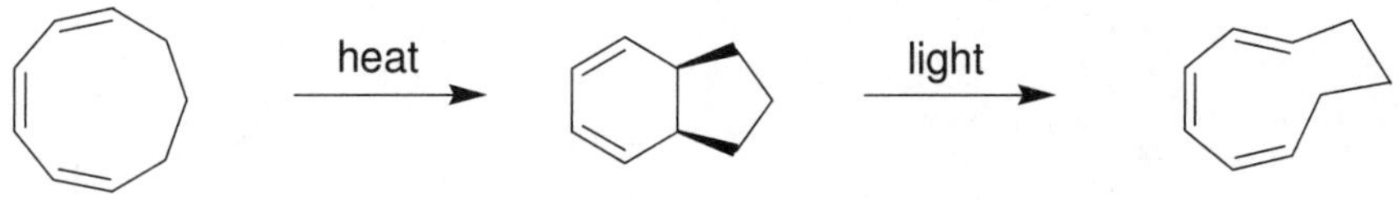

heat

light

17.56.

a)

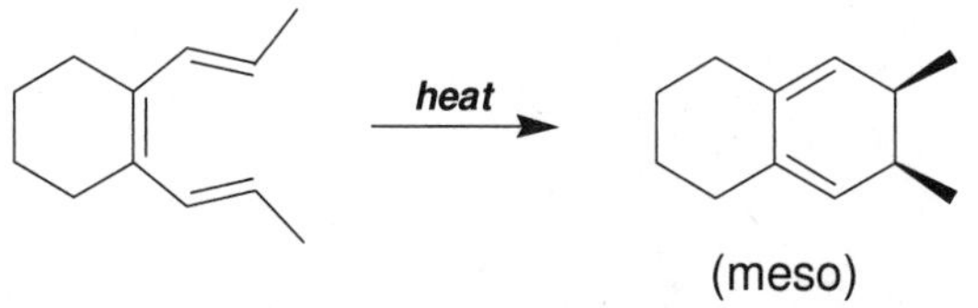

heat

(meso)

b)

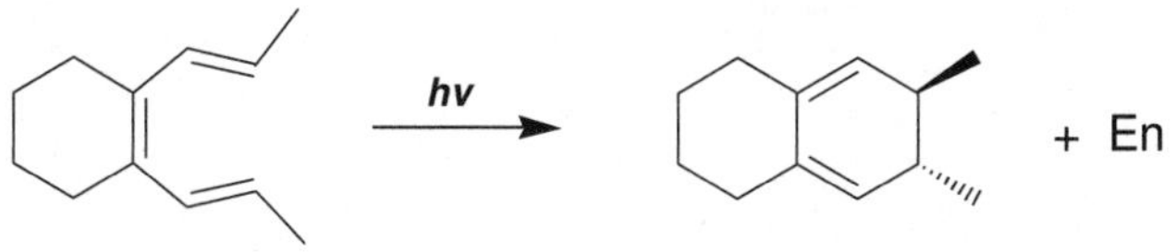

hv

+ En

c)

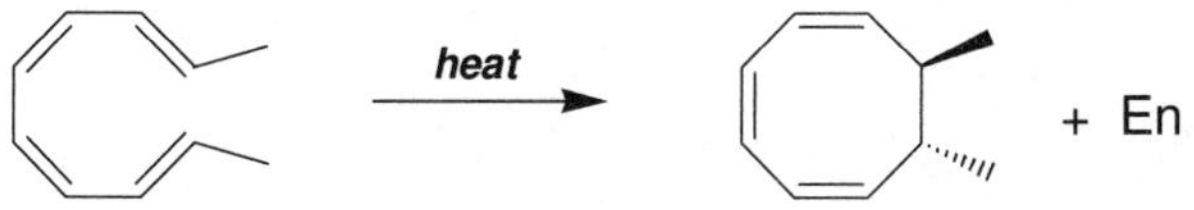

heat

+ En

d)

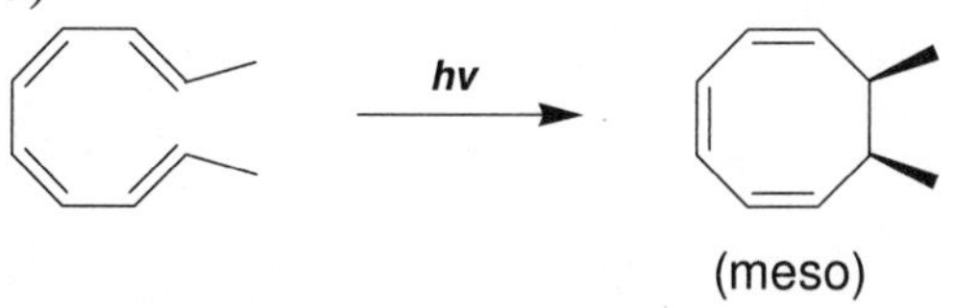

hv

(meso)

17.57. The compound on the right has a π bond in conjugation with the aromatic ring, while the compound on the left does not. Therefore, the compound on the right side of the equilibrium is expected to be more stable, and the equilibrium will favor the compound that is lower in energy.

17.58.

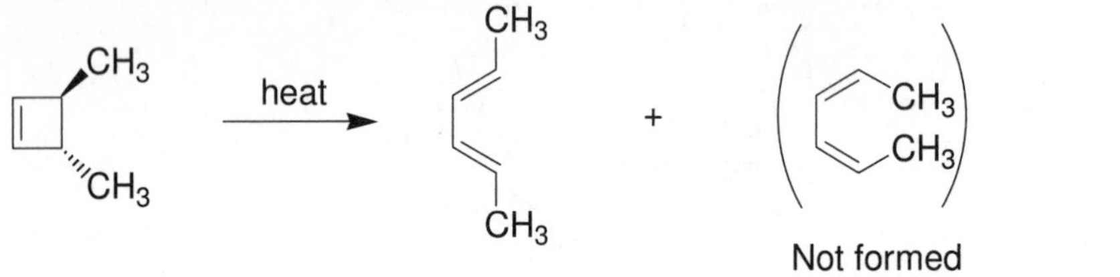

CH_3 heat CH_3 + CH_3 CH_3 CH_3 CH_3

Not formed
Methyl groups are too crowded

17.59.

a) O ≡ O heat O taut OH

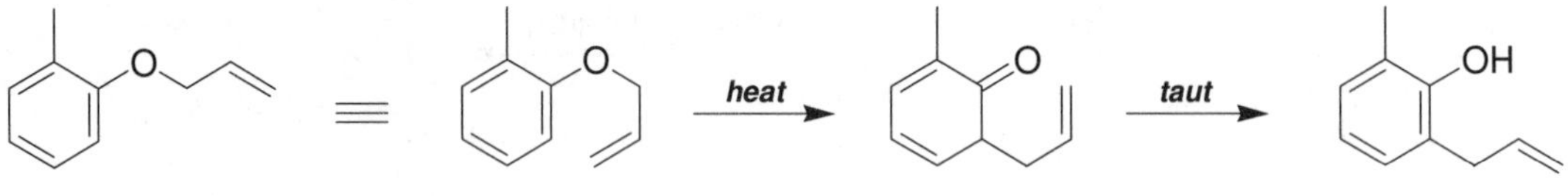

b) O ≡ O heat O

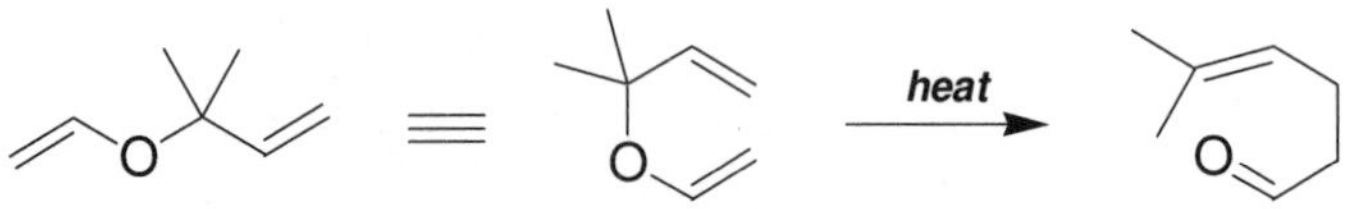

c) ≡ heat

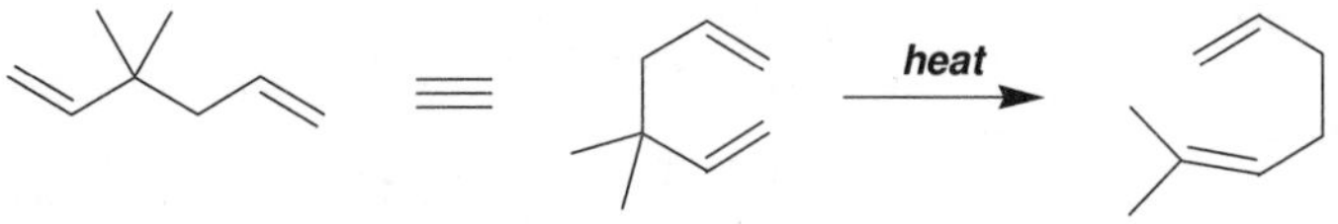

17.60.

a) α-Terpinene has two double bonds.

b)

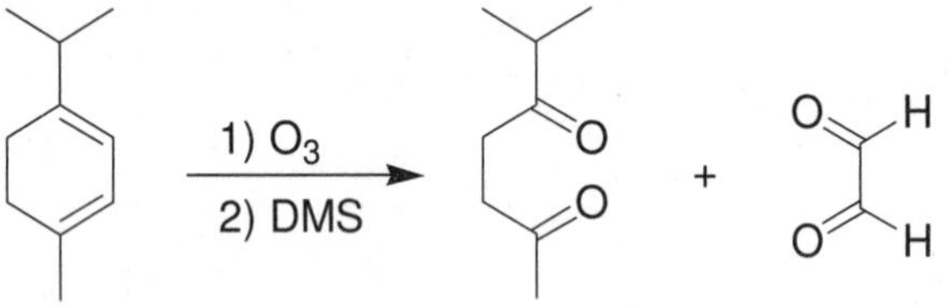

1) O_3
2) DMS

O O + O H O H

α-terpinene

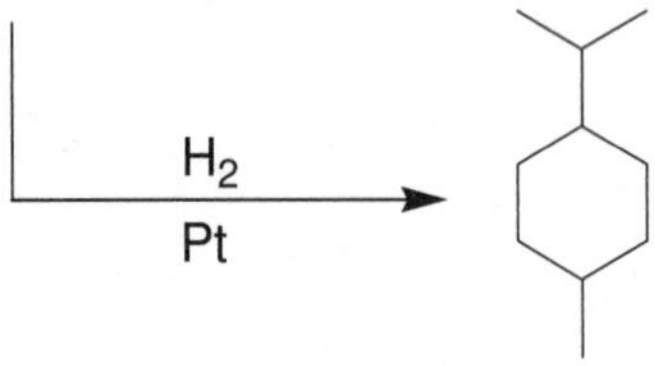

H_2
Pt

c)

Base	=	217
Additional double bonds	=	0
Auxochromic alkyl groups	=	+20
Exocyclic double bonds	=	0
Homoannular diene	=	+39
Total =		276 nm

17.61.

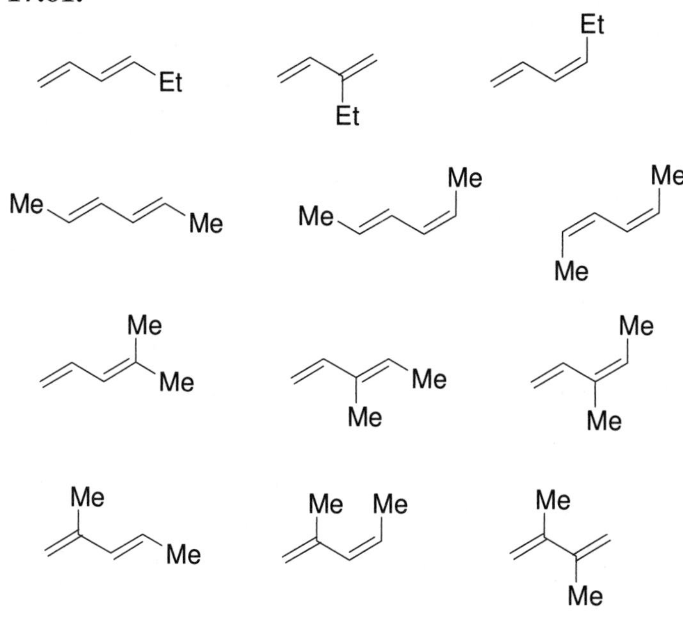

17.62.

Br

Br

t-BuOK

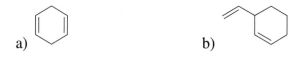

17.63. In each case, the non-conjugated isomer will be higher in energy:

a) 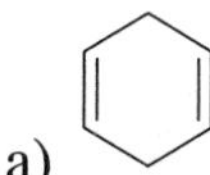

b)

17.64. Nitroethylene should be more reactive than ethylene in a Diels–Alder reaction, because the nitro group is electron-withdrawing, via resonance:

17.65.

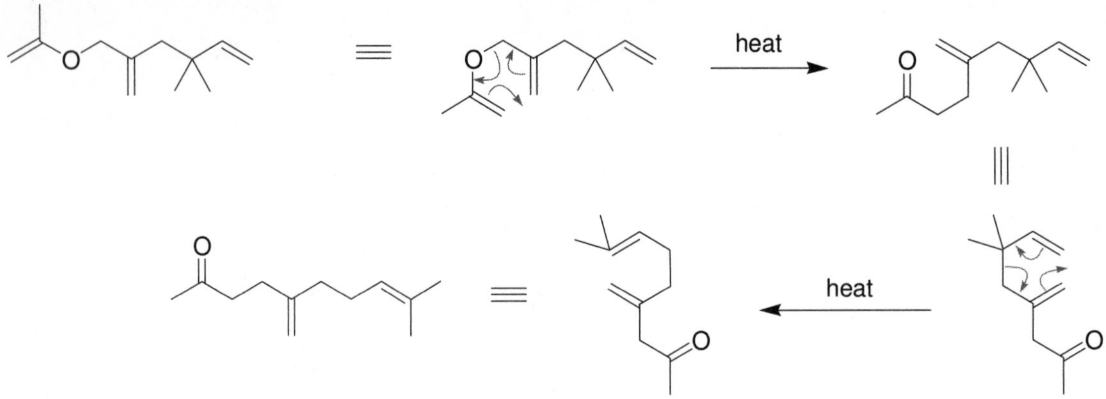

heat

heat

17.66. The diene is electron-rich in one specific location, as seen in the second resonance structure below:

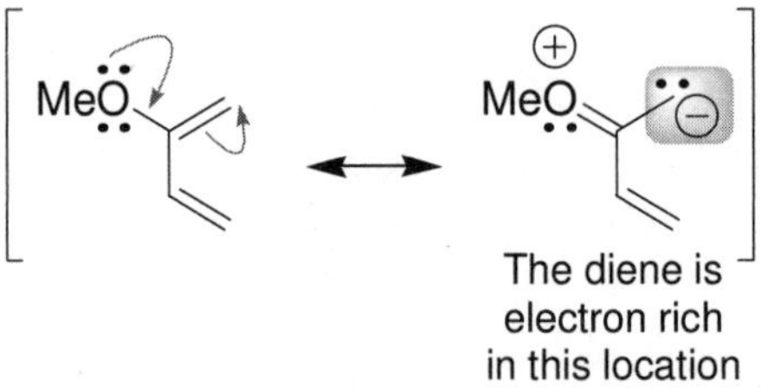

MeÖ

MeO

The diene is electron rich in this location

The dienophile is electron-poor in one specific location, as seen in the third resonance structure below:

The dienophile is electron poor in this location

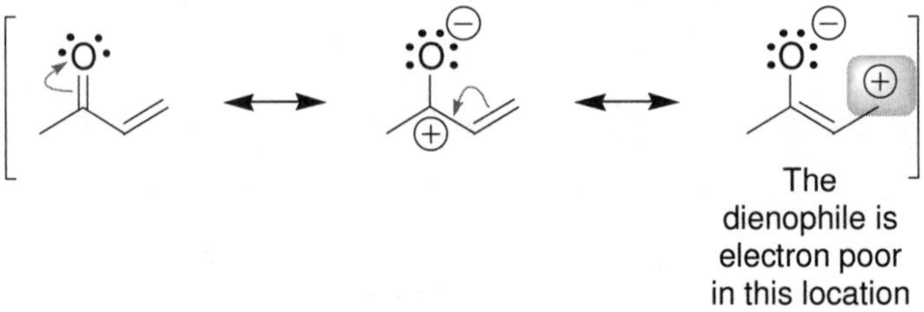

These two compounds will join in such a way that the electron-poor center lines up with the electron-rich center:

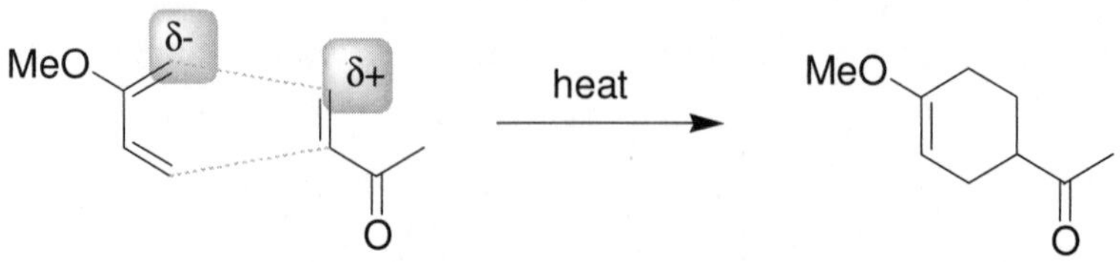

MeO δ- δ+ heat MeO

17.67.

heat +

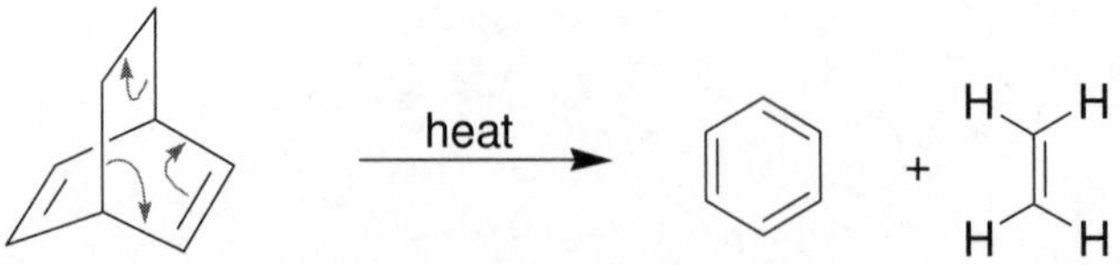

17.68.

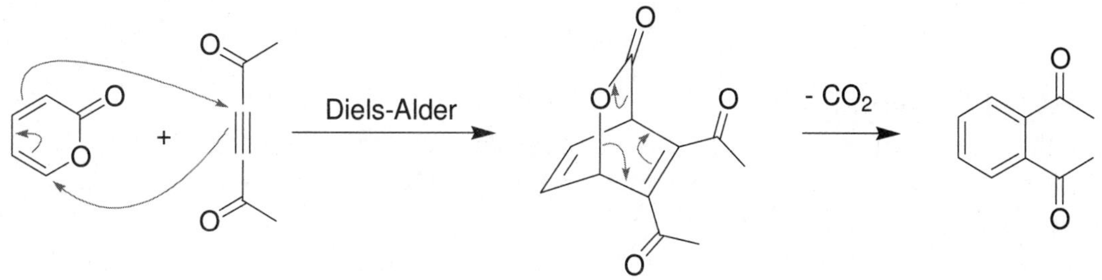

Diels-Alder

$- CO_2$

17.69. The nitrogen atom in divinyl amine is sp^2 hybridized. The lone pair is delocalized, and joins the two neighboring π bonds into one conjugated system. As such, the compound absorbs light above 200 nm (UV light). In contrast, 1,4-pentadiene has two isolated double bonds and therefore does not absorb UV light in the region between 200 and 400 nm.

Chapter 18
Aromatic Compounds

Review of Concepts

Fill in the blanks below. To verify that your answers are correct, look in your textbook at the end of Chapter 18. Each of the sentences below appears verbatim in the section entitled *Review of Concepts and Vocabulary.*

- When a benzene ring is a substituent, it is called a _________ **group**.
- Dimethyl derivatives of benzene can be differentiated by the use of the descriptors _______, ***meta*** and _______, or by the use of locants
- Benzene is comprised of a ring of six identical C-C bonds, each of which has a bond order of _____.
- The **stabilization energy** of benzene can be measured by comparing ________ of hydrogenation.
- The stability of benzene can be explained with MO theory. The six π electrons all occupy ___________ MOs.
- The presence of a fully conjugated ring of π electrons is not the sole requirement for aromaticity. The requirement for an odd number of electron pairs is called ___________ **rule**.
- **Frost circles** accurately predict the relative energy levels of the ________ in a conjugated ring system.
- A compound is aromatic if it contains a ring comprised of __________________ ________________________ and if it has a ________ number π electrons in the ring.
- Compounds that fail the first criterion are called _______________.
- Compounds that satisfy the first criterion, but have 4n electrons (rather than 4n+2) are ______________.
- Cyclic compounds containing hetereoatoms, such as S, N, O are called ____________________.
- Any carbon atom attached directly to a benzene ring is called a _____________ **position**.
- Alkyl benzenes are oxidized at the benzylic position by _____________ or ______________________.
- In a **Birch reduction**, the aromatic moiety is reduced to give a nonconjugated diene. The carbon atom connected to ___________ is not reduced, while the carbon atom connected to _________________________ is reduced.

Review of Skills

Fill in the blanks and empty boxes below. To verify that your answers are correct, look in your textbook at the end of Chapter 18. The answers appear in the section entitled *SkillBuilder Review*.

18.1 Naming a Polysubstituted Benzene

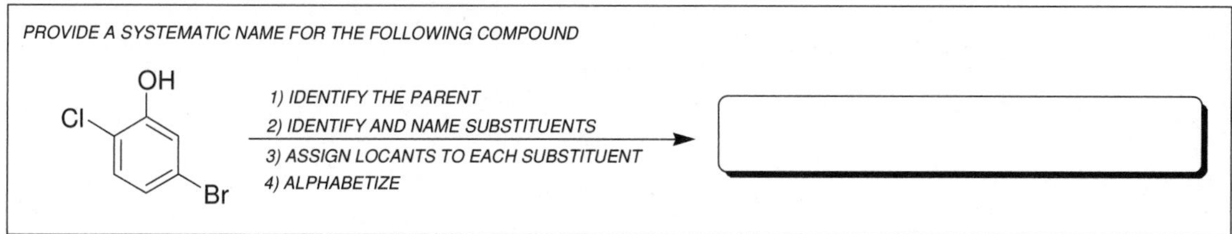

PROVIDE A SYSTEMATIC NAME FOR THE FOLLOWING COMPOUND
OH
Cl
Br
1) IDENTIFY THE PARENT
2) IDENTIFY AND NAME SUBSTITUENTS
3) ASSIGN LOCANTS TO EACH SUBSTITUENT
4) ALPHABETIZE

18.2 Determining Whether a Compound is Aromatic, Nonaromatic, or Antiaromatic

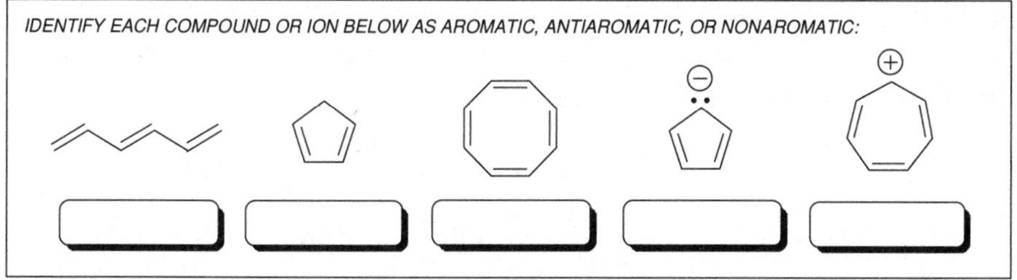

IDENTIFY EACH COMPOUND OR ION BELOW AS AROMATIC, ANTIAROMATIC, OR NONAROMATIC:

18.3 Determining Whether a Lone Pair Participates in Aromaticity

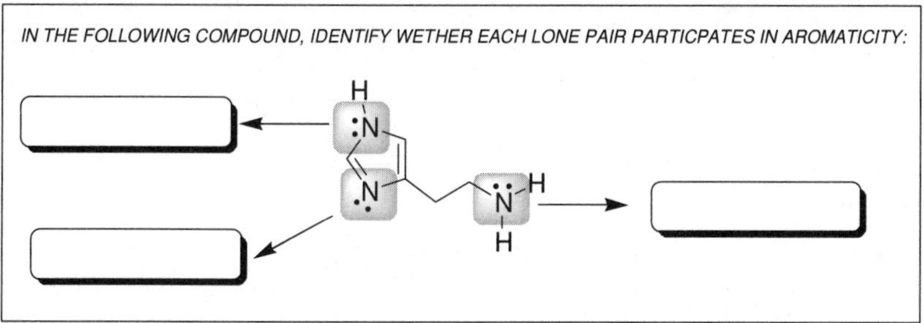

18.4 Manipulating the Side Chain of an Aromatic Compound

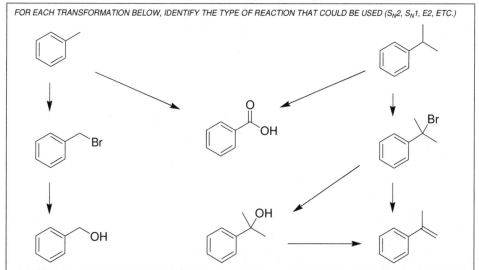

FOR EACH TRANSFORMATION BELOW, IDENTIFY THE TYPE OF REACTION THAT COULD BE USED (S_N2, S_N1, E2, ETC.)

18.5 Predicting the Product of a Birch Reduction

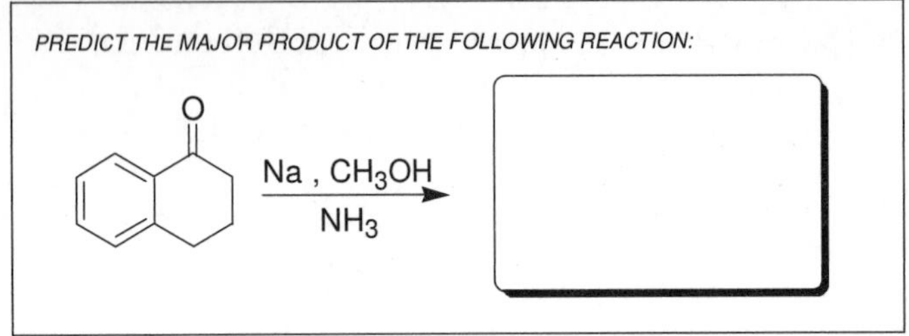

PREDICT THE MAJOR PRODUCT OF THE FOLLOWING REACTION:
O
Na , CH3OH
NH3

Review of Reactions

Identify the reagents necessary to achieve each of the following transformations. To verify that your answers are correct, look in your textbook at the end of Chapter 18. The answers appear in the section entitled *Review of Reactions*.

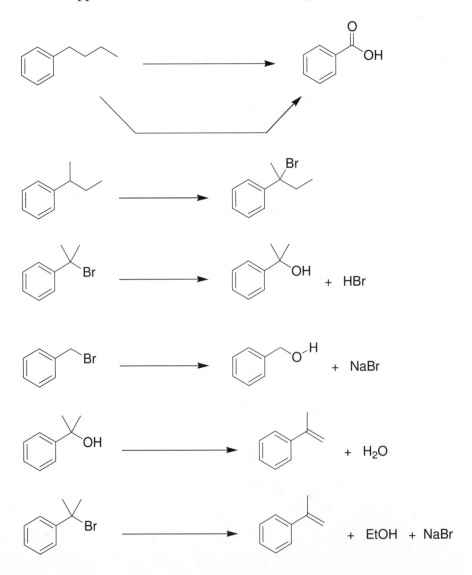

O
OH

Br

Br
OH
+ HBr

Br
O H
+ NaBr

OH
+ H_2O

Br
+ EtOH + NaBr

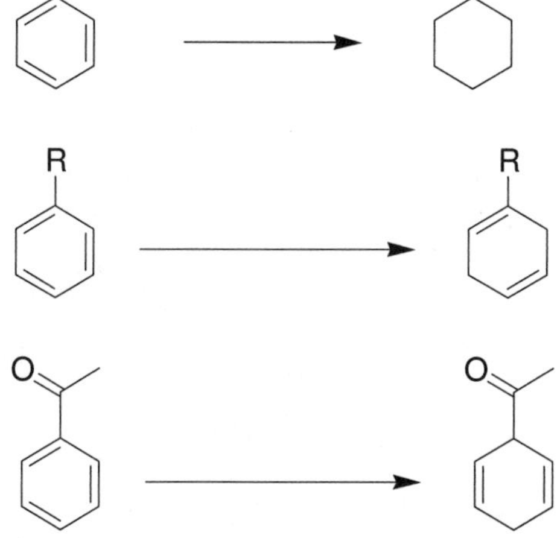

Solutions

18.1.

a) 3-isopropylbenzaldehyde or *meta*-isopropylbenzaldehyde
b) 2-bromotoluene or *ortho*-bromotoluene
c) 2,4-dinitrophenol
d) 2-ethyl-1,4-diisopropylbenzene
f) 2,6-dibromo-4-chloro-3-ethyl-5-isopropylphenol

18.2.
a) 4-bromo-2-methylphenol
b) 2-hydroxy-5-bromotoluene
c) 4-bromo-1-hydroxy-2-methylbenzene

18.3.

a) OMe, Br, Br, Cl

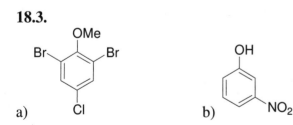

b) OH, NO_2

18.4.
a) meta-xylene
b) 1,3-dimethylbenzene
c) meta-dimethylbenzene
d) meta-methyltoluene
e) 3-methyltoluene

18.5.

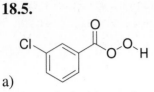

a)

b) 3-methylperbenzoic acid or *meta*-methylperbenzoic acid.

18.6.

Compound A
(C_8H_8)

Compound B
($C_8H_8Br_2$)

18.7.

a) ΔH has a positive value
b) ΔH has a positive value
c) ΔH has a negative value

18.8.

a) No, 12 is not a Hückel number.
b) Yes, 14 is a Hückel number.
c) No, 16 is not a Hückel number.

18.9. The cyclopropenyl cation is expected to exhibit aromatic stabilization.

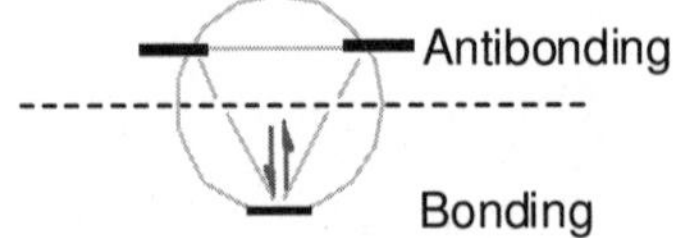

18.10. The compound will be aromatic because there are 22 π electron, and 22 is a Hückel number.

18.11.

a) antiaromatic b) aromatic c) antiaromatic d) aromatic

18.12. Cyclopentadiene is more acidic because its conjugate base is highly stabilized. Deprotonation of cyclopentadiene generates an anion that is aromatic, because it is a continuous system of overlapping p orbitals containing 6 π electrons. In contrast, deprotonation of cycloheptatriene gives an anion with 8 π electrons.

more stable

18.13. The first step of an S_N1 process is loss of a leaving group, forming a carbocation, so we compare the carbocations that would be formed.

The second carbocation is more stable, because it is aromatic, and is therefore lower in energy than the first carbocation. The transition state leading to the second carbocation will be lower in energy than the transition state leading to the first carbocation, and therefore, the second carbocation will be formed more rapidly than the first.

18.14. The first compound is more acidic because deprotonation of the first compound generates a new (second) aromatic ring. Deprotonation of the second compound does not introduce a new aromatic ring:

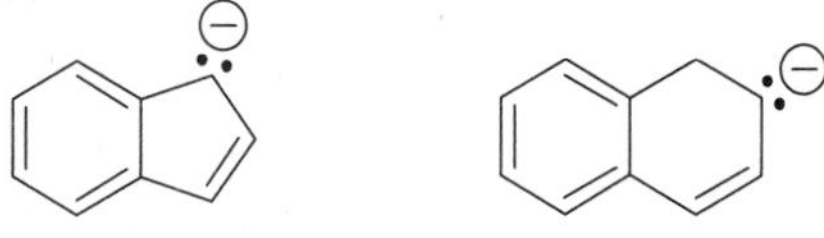

18.15.
a) One of the lone pairs on oxygen
b) One of the lone pairs on sulfur
c) The lone pair on nitrogen is NOT participating in aromaticity (8 pi electrons).
d) One of the lone pairs on sulfur
e) There is only one pair (on oxygen) and it is not participating in aromaticity.
f) Each nitrogen has one lone pair, and neither is participating in aromaticity.
g) The compound is not aromatic. In order to achieve a continuous system of overlapping *p* orbitals, each oxygen atom would need to contribute a lone pair in a *p* orbital, and that would give 8 π electrons (not a Hückel number).
h) One of the lone pairs on oxygen (not the lone pair on the nitrogen)

18.16.

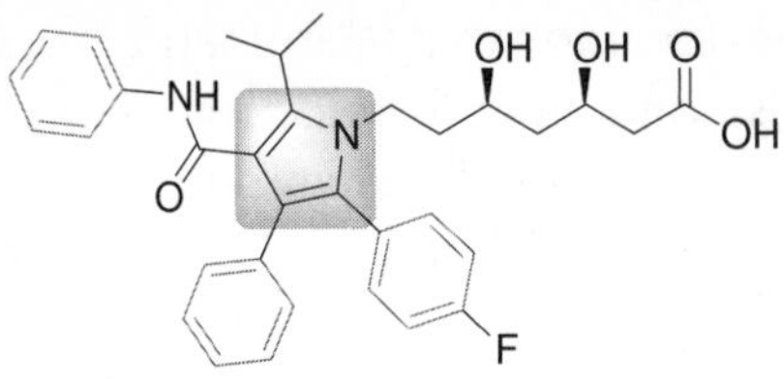

Lipitor™ 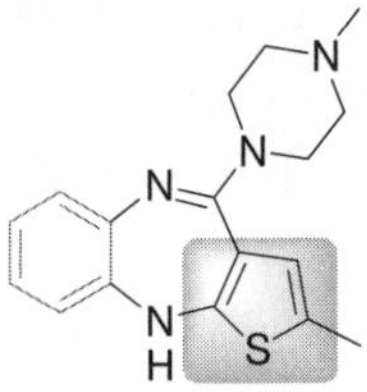 Zyprexa™

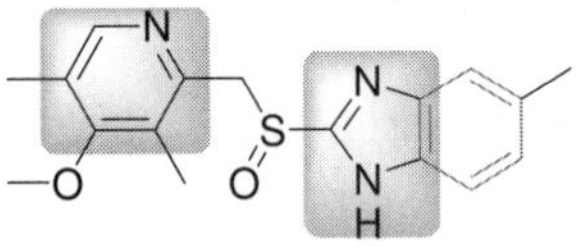

Nexium™ 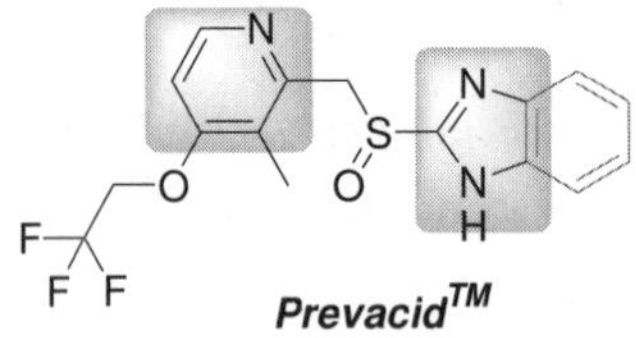 Prevacid™ 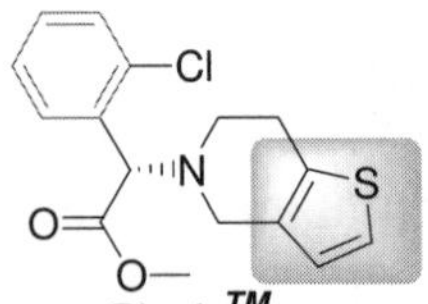 Plavix™

18.17. The first compound is expected to be more acidic (has a lower pK_a), because deprotonation restores aromaticity to the ring. The second compound is already aromatic, even before deprotonation.

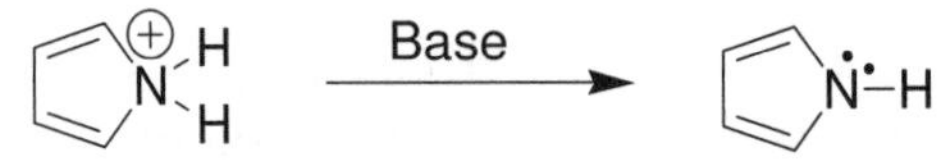

Base

not aromatic **aromatic**

18.18.

a) Yes, it has the required pharmacophore (two aromatic rings separated by one carbon atom, and a tertiary amine.
b) Meclizine crosses the blood-brain barrier and binds with receptors in the central nervous system, causing sedation.
c) Introduce polar functional groups that reduce the ability of the compound to cross the blood-brain barrier.

18.19.

a) 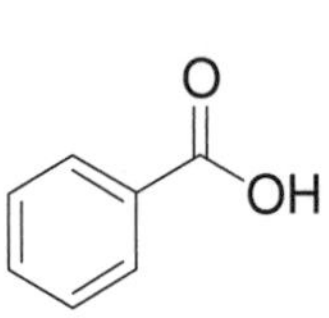 b) 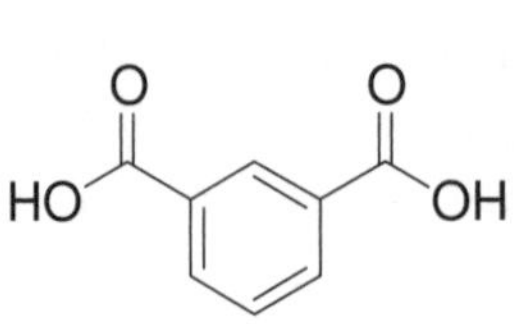 c)

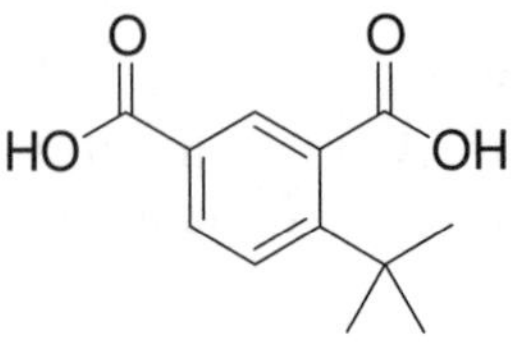

18.20.

a) 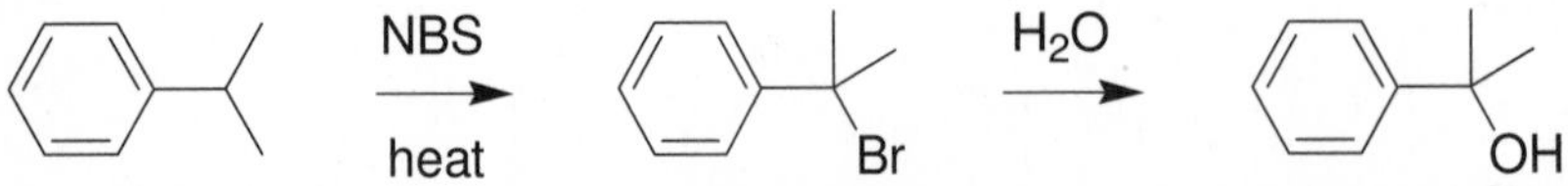 NBS heat H_2O

b) NBS, heat → NaOEt →

c) NBS, heat → Br; NaCN → N

d) NBS, heat → Br; MeOH → O

18.21.

NBS, heat → Br; Mg → MgBr

18.22.

NBS, heat → Br; NaOEt → ; 1) O_3 2) DMS → O

18.23.

1) NBS, heat
2) NaOEt
3) O_3
4) DMS

→ O, H; 1) PrMgBr 2) H_2O → OH

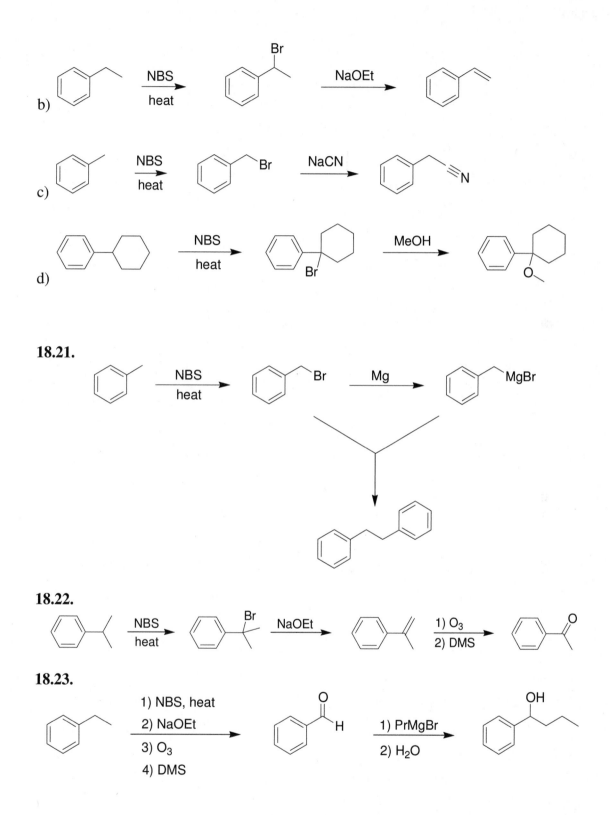

18.24.

a) 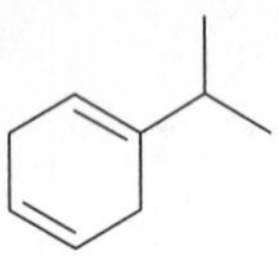 b) 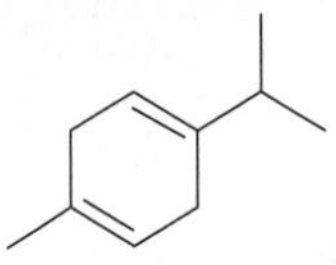 c) 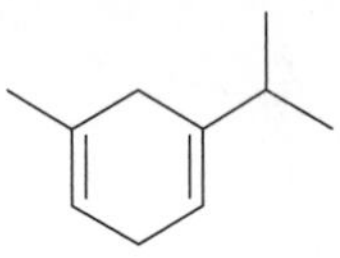 d)

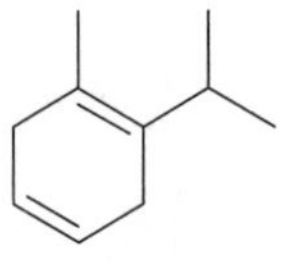

e) 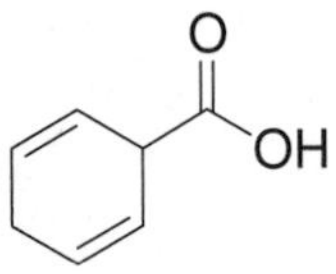 f)

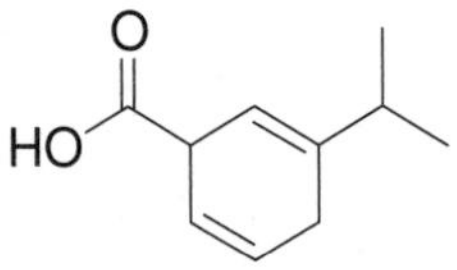

18.25.

a)

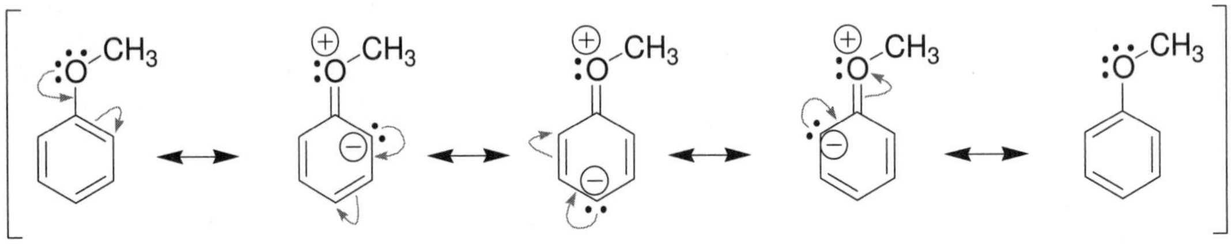

b)

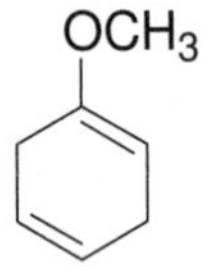

18.26.

a) 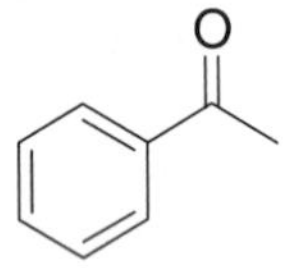 b) acetophenone c)

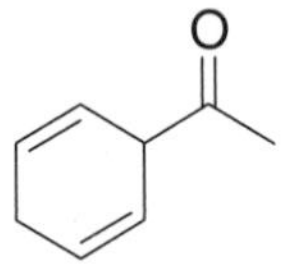

18.27.

a) 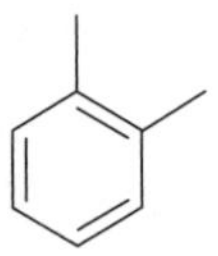 b) *ortho*-xylene c)

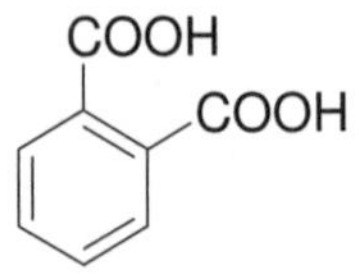

18.28.

a) 4-ethylbenzoic acid or *para*-ethylbenzoic acid
b) 2-bromophenol or *ortho*-bromophenol
c) 2-chloro-4-nitrophenol
d) 2-bromo-5-nitrobenzaldehyde
e) 1,4-diisopropyl benzene or *para*-diisopropyl benzene

18.29.

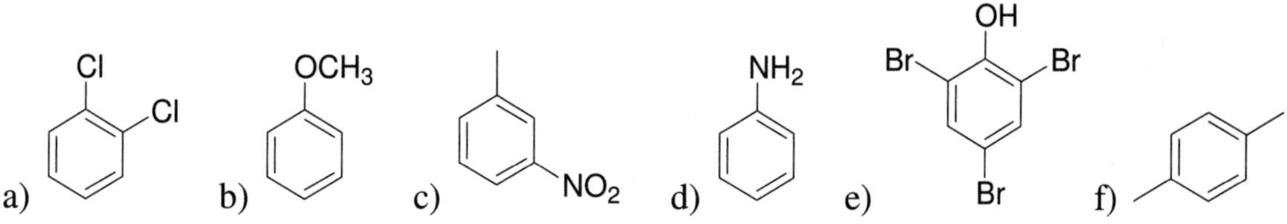

18.30.

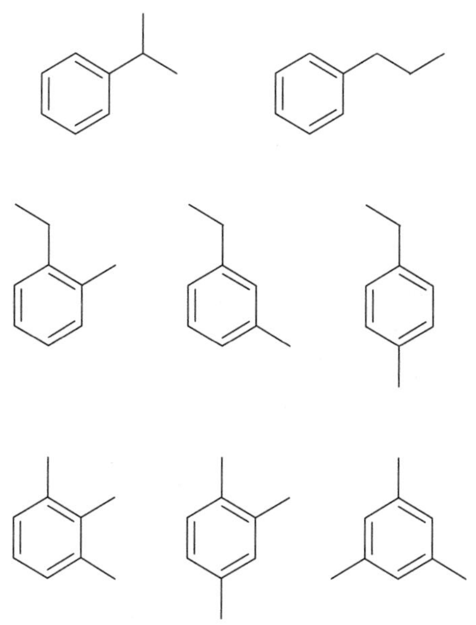

18.31.

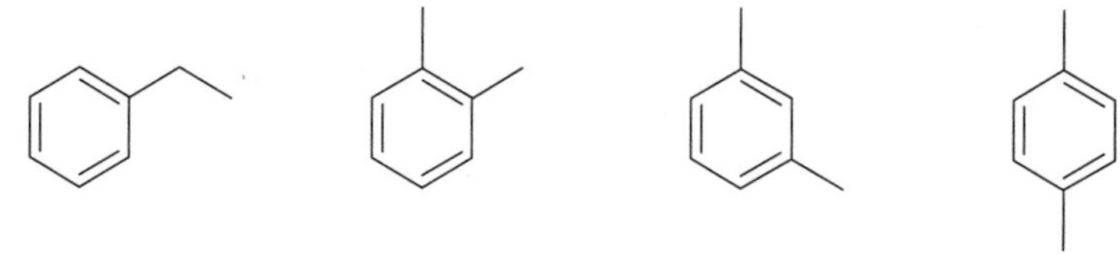

18.32.

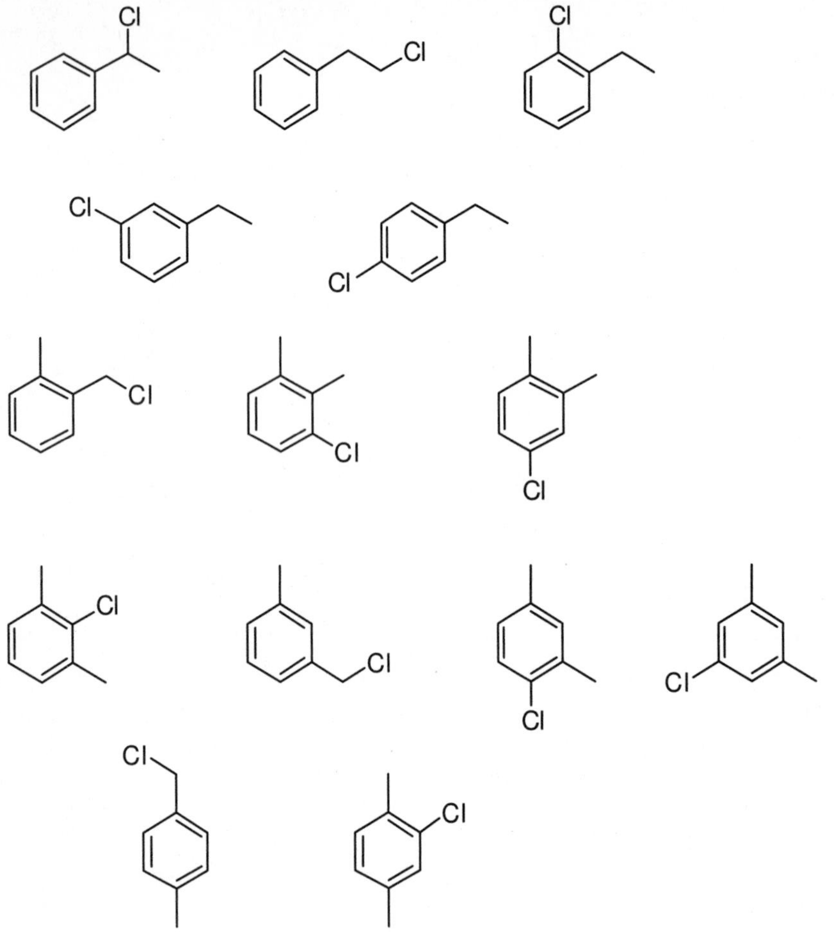

18.33.

2,3,4-trinitrotoluene | 2,3,5-trinitrotoluene | 2,3,6-trinitrotoluene | 2,4,5-trinitrotoluene | 3,4,5-trinitrotoluene

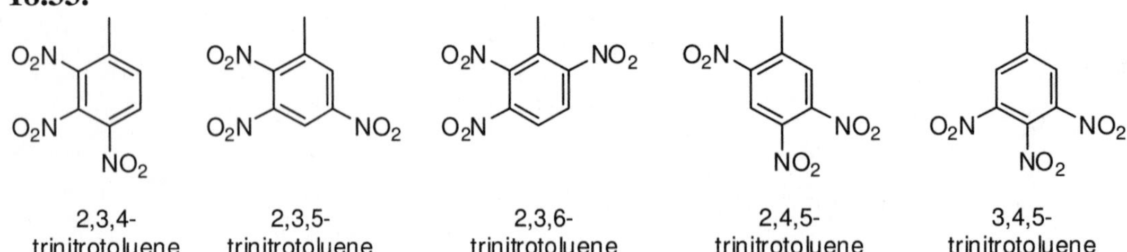

18.34.

a) 10 b) 6 c) 10 d) 4 e) 6

18.35.

a) benzene b) benzene c) benzene
d) cyclohexane e) benzene f) cyclohexane
g) benzene h) benzene i) benzene
j) cyclohexane k) cyclohexane

18.36. a) Yes b) No c) No d) Yes e) No

18.37.

aromatic

HO S N O OH N S

a)

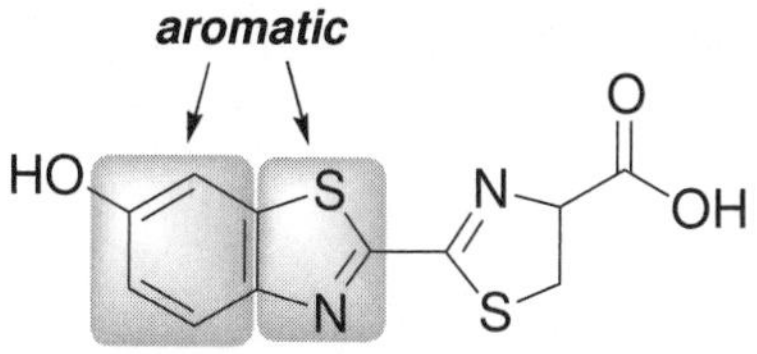

b) One of the lone pairs on the sulfur atom in the five-membered aromatic ring

18.38.

a) Nonaromatic. The lone pairs on the oxygen atom will remain in sp^3 hybridized orbitals in order to avoid anti-aromaticity.

b) Nonaromatic. The lone pair on the nitrogen atom will remain in an sp^3 hybridized orbital in order to avoid anti-aromaticity.

c) Aromatic. One of the lone pairs of the sulfur atom occupies a *p* orbital, thereby establishing a continuous system of overlapping *p* orbitals, containing six π electrons.

d) Aromatic. Both lone pairs occupy sp^2 hybridized orbitals and do not participate in establishing aromaticity.

e) Aromatic. A continuous system of overlapping *p* orbitals, containing six π electrons.

f) Non aromatic. The nitrogen atom does not have a *p* orbital, so there is not a continuous system of overlapping *p* orbitals.

g) Aromatic. The lone pair of the oxygen atom occupies a *p* orbital, thereby establishing a continuous system of overlapping *p* orbitals, containing six π electrons.

h) Aromatic. Both lone pairs occupy *p* orbitals, thereby establishing a continuous system of overlapping *p* orbitals, containing six π electrons.

18.39.

a) 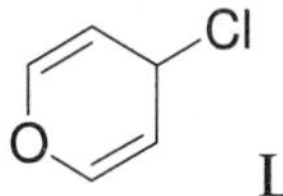 Cl O — Loss of the leaving group generates an aromatic cation.

b) 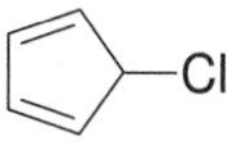 Cl — Loss of the leaving group generates an antiaromatic cation.

18.40.

Deprotonation of cyclopentadiene generates an aromatic anion.

18.41. The second compound is a stronger base, because the lone pair on the nitrogen atom is localized and available to function as a base. However, the nitrogen atom in the first compound is delocalized and is participating in aromaticity. This lone pair is unavailable to function as a base, because that would cause a loss of aromaticity.

18.42. Six π electrons are required in order to achieve aromaticity. This cation only has four electrons.

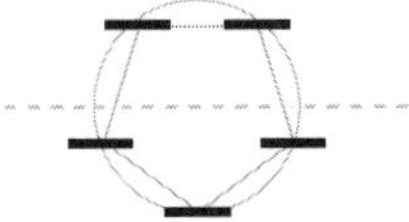

18.43. If both lone pairs occupy *p* orbitals, then there is a continuous system of overlapping *p* orbitals. There are 10 π electrons, so the dianion is aromatic.

18.44. Yes. The lone pairs on the nitrogen atoms do not contribute to aromaticity. They occupy sp^2 hybridized orbitals. One of the lone pairs on the oxygen atom (in the ring) occupies a *p* orbital, giving a continuous system of overlapping *p* orbitals containing six π electrons.

18.45. Steric hindrance forces the rings out of coplanarity.

18.46. Benzene does not have three C-C single bond and three C-C double bonds. In fact, all six C-C bonds of the ring have the same bond order are the same length. However, cyclooctatetraene has four isolated π bonds. The molecule adopts a tub shape to avoid antiaromaticity. Some of the C-C bonds are double bonds (shorter in length), and some of the C-C bonds are single bonds (longer in length). Therefore, the two methyl groups can be separated by a C-C single bond or a C=C double bond. And those two possibilities represent different compounds.

18.47.

a) 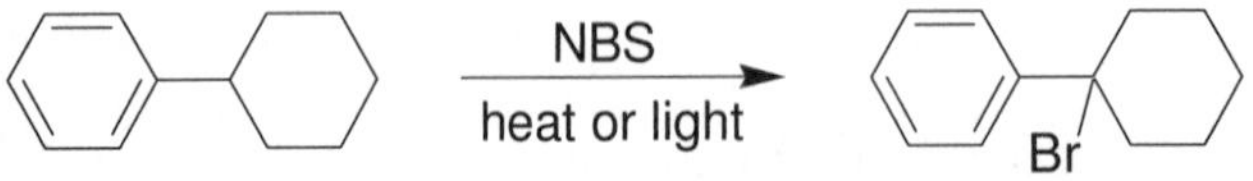 NBS, heat or light → Br

b) 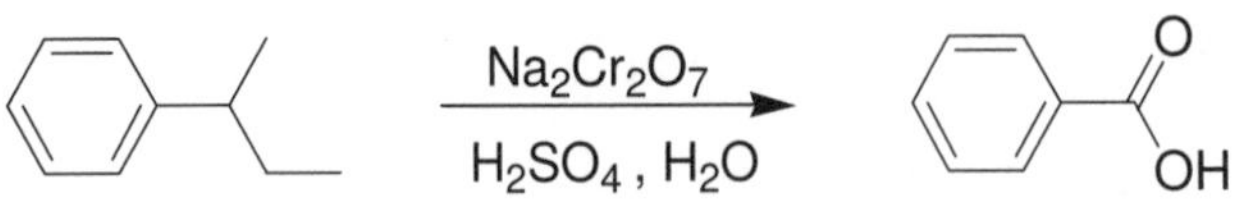 $Na_2Cr_2O_7$, H_2SO_4, H_2O → O, OH

c) 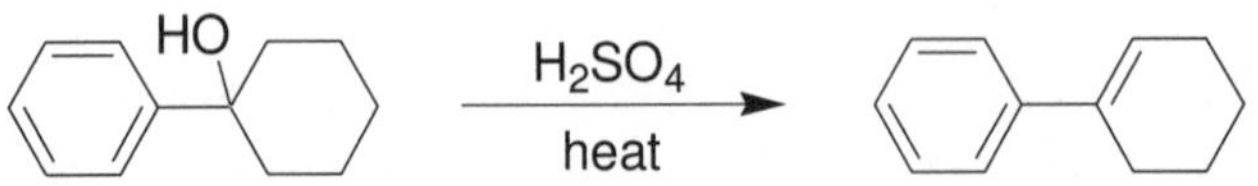 HO; H_2SO_4, heat →

d) 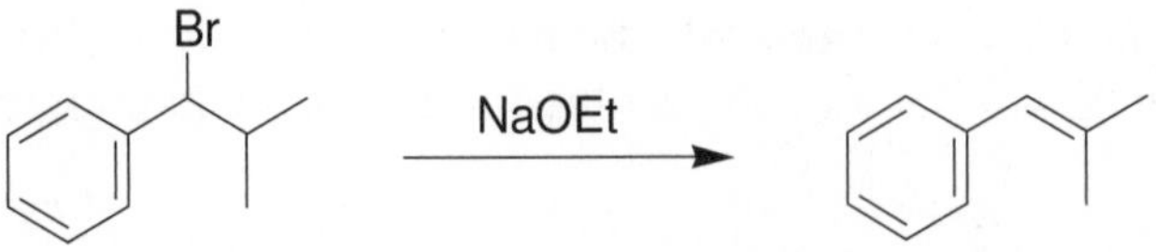 Br; NaOEt →

18.48.

a) 6 b) 5 c) 3 d) 9

18.49.

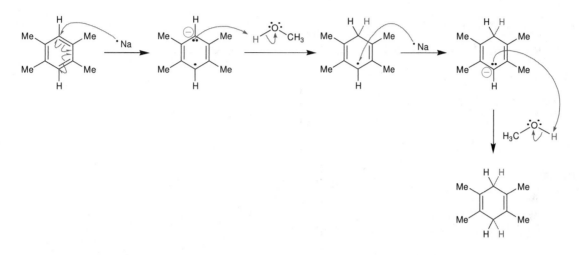

Me Me H Me Me H •Na Me Me H Me Me H H–O–CH3 Me Me H H Me Me H •Na Me Me H H Me Me H H3C–O–H Me Me H H Me Me H H

18.50. *meta*-Xylene.

18.51.

a) The first compound would lack C-H stretching signals just above 3000 cm^{-1}, while the second compound will have C-H stretching signals just above 3000 cm^{-1}.
b) The ^{1}H NMR spectrum of the first compound will have only one signal, while the ^{1}H NMR spectrum of the second compound will have two signals.
c) The ^{13}C NMR spectrum of the first compound will have only two signals, while the ^{13}C NMR spectrum of the second compound will have three signals.

18.52. When either compound is deprotonated, an aromatic anion is generated, which can be drawn with five resonance structures. The resulting anion is the same in either case.

18.53. In cycloheptatrienone, the resonance structures with C+ and O- contribute significant character to the overall resonance hybrid, because these forms are aromatic. Therefore, the oxygen atom of this C=O bond is particularly electron rich. A similar analysis of cyclopentadienone reveals resonance structures with antiaromatic character. These resonance structures contribute very little character to the overall resonance hybrid, and as a result, the oxygen atom of this C=O bond is not as electron rich when compared with most C=O bonds.

18.54.

a) Each of the rings in the following resonance structure is aromatic.

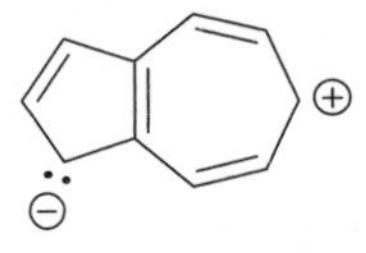

Therefore, this resonance structure contributes significant character to the overall resonance hybrid, which gives the azulene a considerable dipole moment.

b)

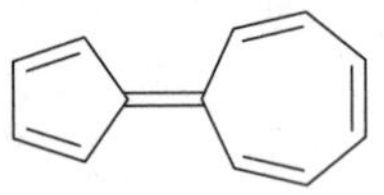

18.55.

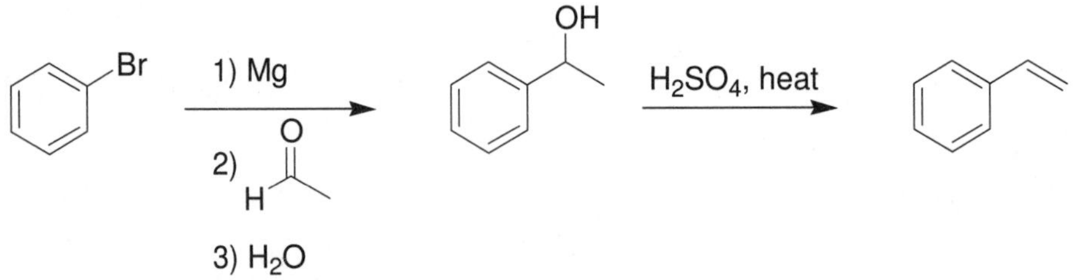

1) Mg
2) H, O
3) H_2O

OH

H_2SO_4, heat

18.56.

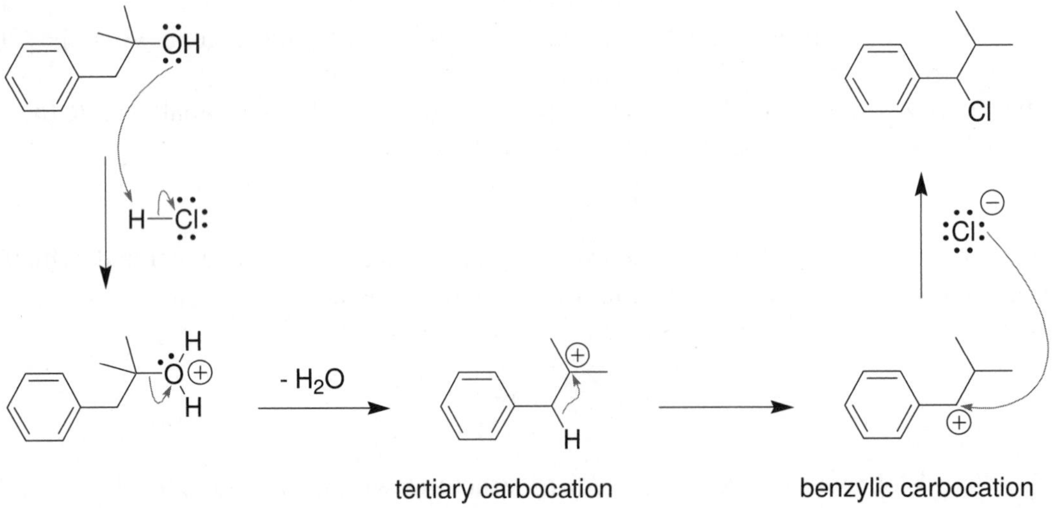

OH

H—Cl

H, O, H

- H_2O

H

tertiary carbocation

Cl

Cl

benzylic carbocation

18.57.

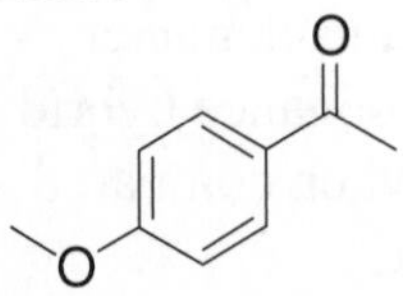

O

O

18.58.

a) 1) NBS, heat
2) NaOH
3) PCC, CH_2Cl_2

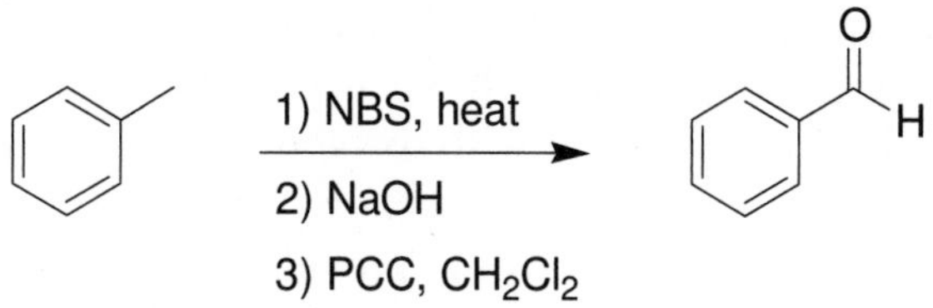

b) 1) NBS, heat
2) NaOEt

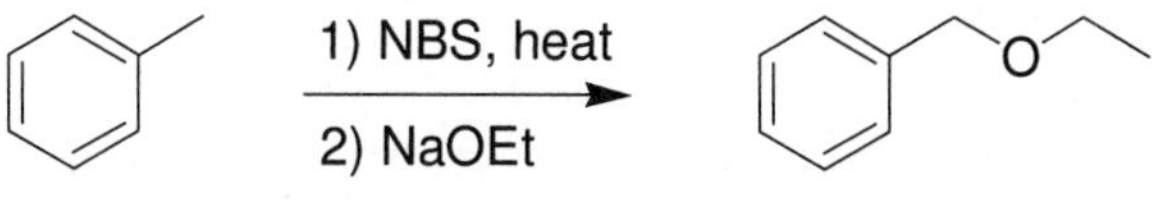

c) OH
1) conc. H_2SO_4, heat
2) BH_3•THF
3) H_2O_2, NaOH
OH

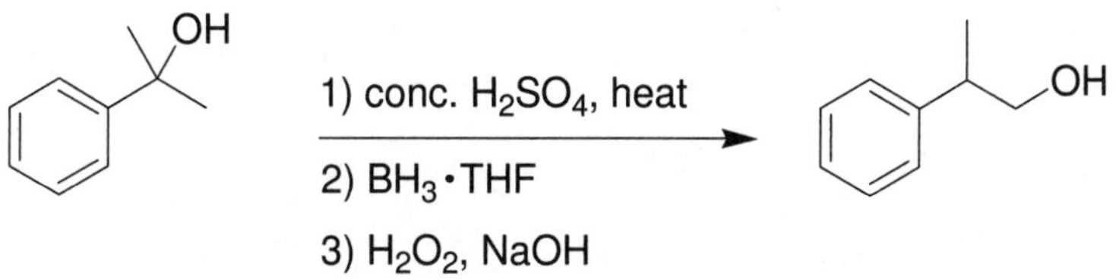

d) 1) NBS, heat
2) HC≡CNa

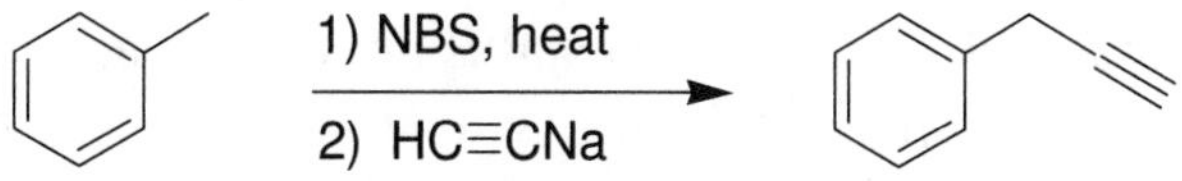

18.59.

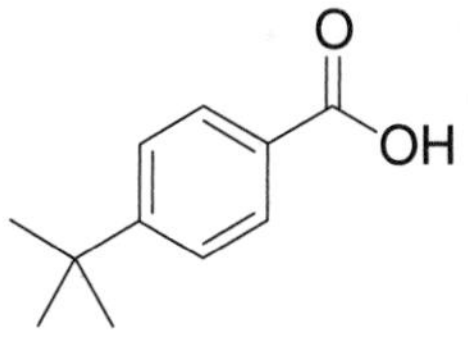

O
OH

18.60.

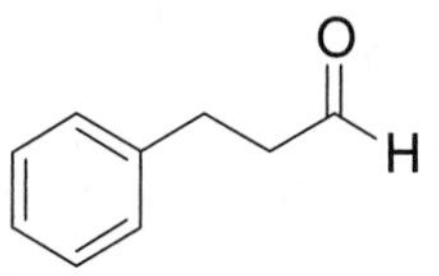

O
H

18.61.

a) The second compound holds greater promise as a potential antihistamine, because it possesses two planar aromatic rings separated from each other by one carbon atom. The first compound has only one aromatic ring. The ring with oxygen is not aromatic and not planar.

b) Yes, because it lacks polar functional groups that would prevent it from crossing the blood-brain barrier.

18.62. No, this compound possesses an allene moiety (C=C=C). The *p* orbitals of one C=C bond of the allene moiety do not overlap with the *p* orbitals of the other C=C bond. This prevents the compound from having one continuous system of overlapping *p* orbitals.

18.63.

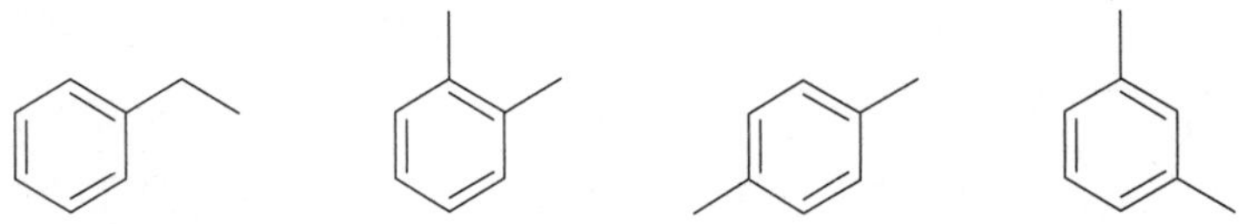

Compound A **Compound B** **Compound C** **Compound D**

18.64. The nitrogen atom in compound A is localized and is not participating in resonance. The nitrogen atom in compound B is delocalized, and some of the resonance structures are aromatic. These resonance structures contribute significant character to the overall resonance hybrid. The nitrogen atom in compound B is not available to function as a base.

18.65.

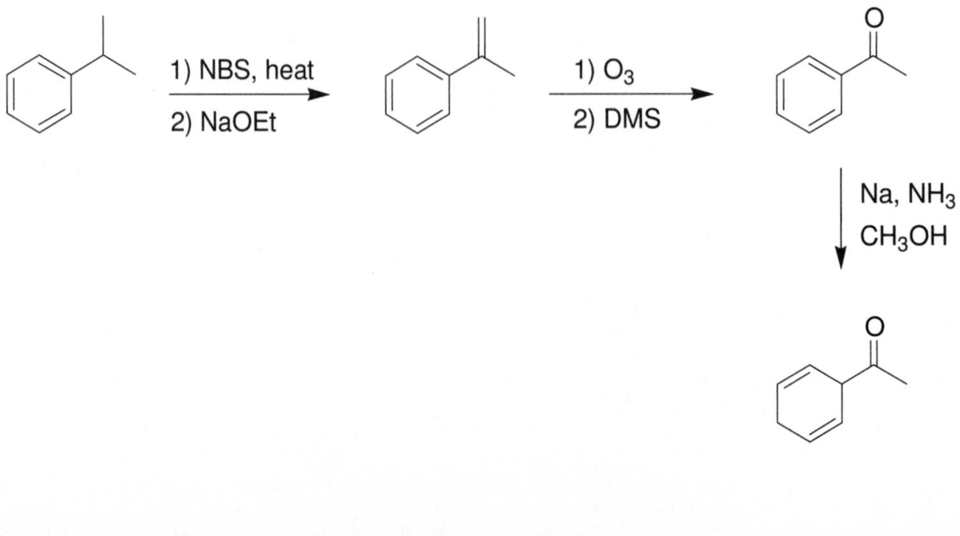

1) NBS, heat
2) NaOEt

1) O_3
2) DMS

O

Na, NH_3
CH_3OH

O

18.66.

NBS / heat

HC≡CNa

1) $NaNH_2$

2)

Br

H_2

Lindlar's catalyst

MCPBA

O

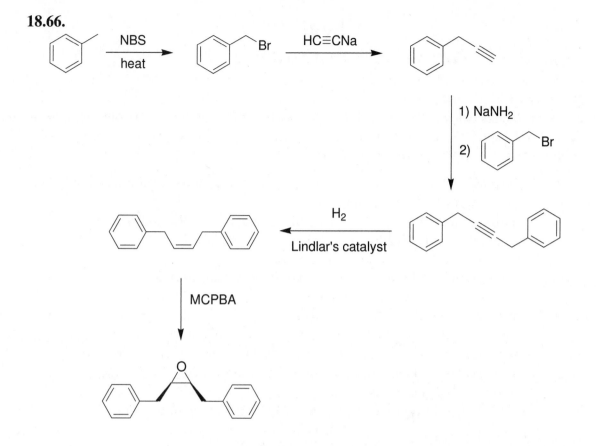

Chapter 19
Aromatic Substitution Reactions

Review of Concepts

Fill in the blanks below. To verify that your answers are correct, look in your textbook at the end of Chapter 19. Each of the sentences below appears verbatim in the section entitled *Review of Concepts and Vocabulary*.

- In the presence of iron, an _____________ **aromatic substitution** reaction is observed between benzene and bromine.
- Iron tribromide is a __________ acid that interacts with Br_2 and generates Br^+, which is sufficiently electrophilic to be attacked by benzene.
- Electrophilic aromatic substitution involves two steps:
  - Formation of the __________ **complex**, or **arenium ion.**
  - Deprotonation, which restores _________________.
- Sulfur trioxide (SO_3) is a very powerful ____________ that is present in fuming sulfuric acid. Benzene reacts with SO_3 in a reversible process called __________.
- A mixture of sulfuric acid and nitric acid produces the **nitronium ion** (NO_2^+). Benzene reacts with the nitronium ion in a process called ____________.
- A nitro group can be reduced to an ______group.
- **Friedel–Crafts alkylation** enables the installation of an alkyl group on ______________________. When choosing an alkyl halide, the carbon atom connected to the halogen must be ______hybridized.
- When treated with a Lewis acid, an acyl chloride will generate an _________ **ion,** which is resonance stabilized and not susceptible to __________ rearrangements.
- When a Friedel–Crafts acylation is followed by a **Clemmensen reduction**, the net result is the installation of an ______ group.
- An aromatic ring is **activated** by a methyl group, which is an ______-______ **director**.
- All activators are ______-______directors
- A nitro group **deactivates** an aromatic ring and is a ______director.
- Most deactivators are ______directors.
- **Strong activators** are characterized by the presence of a ________________ immediately adjacent to the aromatic ring.
- **Strong deactivators** are powerfully electron-withdrawing, either by ____________or ____________.
- When multiple substituents are present, the more powerful _________________ dominates the directing effects.
- In a **nucleophilic aromatic substitution** reaction, the aromatic ring is attacked by a ____________. This reaction has three requirements: 1) the ring must contain a powerful electron-withdrawing group (typically a ______group) 2) the ring must contain a ____________, and 3) the leaving group must be either ______or ______to the electron-withdrawing group.
- An **elimination-addition** reaction occurs via a ____________intermediate.

Review of Skills

Fill in the blanks and empty boxes below. To verify that your answers are correct, look in your textbook at the end of Chapter 19. The answers appear in the section entitled *SkillBuilder Review*.

19.1 Identifying the Effects of a Substituent

PLACE EACH OF THE FOLLOWING GROUPS IN THE CORRECT CATEGORY BELOW.

$-\overset{\oplus}{N}R_3$ $-C(=O)R$ $-\ddot{N}H_2$ $-C(=O)OR$ $-C\equiv N$ $-Br$ $-CX_3$ $-NO_2$ $-\ddot{O}R$ $-F$ $-\ddot{O}C(=O)R$ $-Cl$ $-\ddot{O}H$ $-I$ $-R$

ACTIVATORS			DEACTIVATORS		
STRONG	MODERATE	WEAK	WEAK	MODERATE	STRONG

19.2 Identifying Directing Effects for Disubstituted and Polysubstituted Benzene Rings

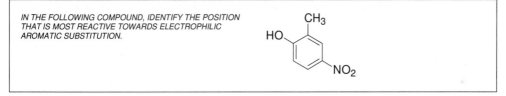

IN THE FOLLOWING COMPOUND, IDENTIFY THE POSITION THAT IS MOST REACTIVE TOWARDS ELECTROPHILIC AROMATIC SUBSTITUTION.

CH_3 HO NO_2

19.3 Identifying Steric Effects for Disubstituted and Polysubstituted Aromatic Benzene Rings

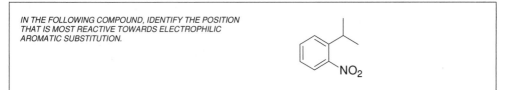

IN THE FOLLOWING COMPOUND, IDENTIFY THE POSITION THAT IS MOST REACTIVE TOWARDS ELECTROPHILIC AROMATIC SUBSTITUTION.

NO_2

19.4 Using Blocking Groups to Control the Regiochemical Outcome of an Electrophilic Aromatic Substitution Reaction

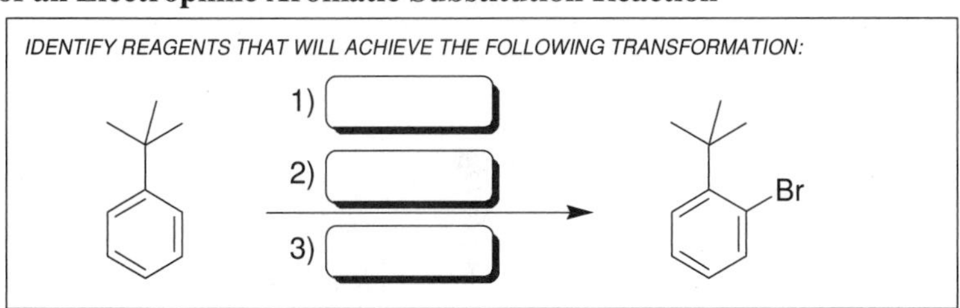

19.5 Proposing a Synthesis for a Disubstituted Benzene Ring

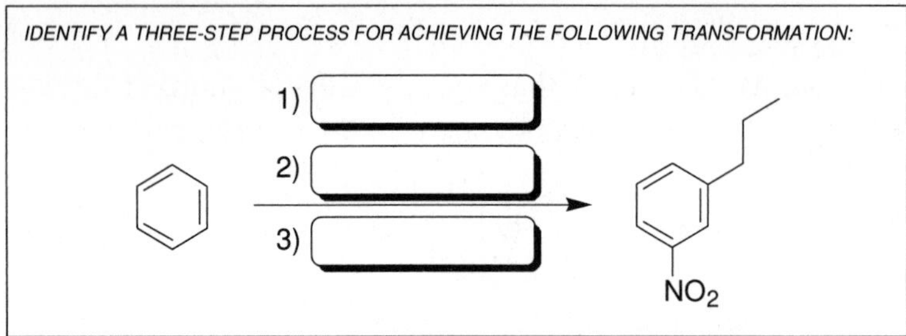

IDENTIFY A THREE-STEP PROCESS FOR ACHIEVING THE FOLLOWING TRANSFORMATION:
1)
2)
3)
NO2

19.6 Proposing a Synthesis for a Polysubstituted Benzene Ring

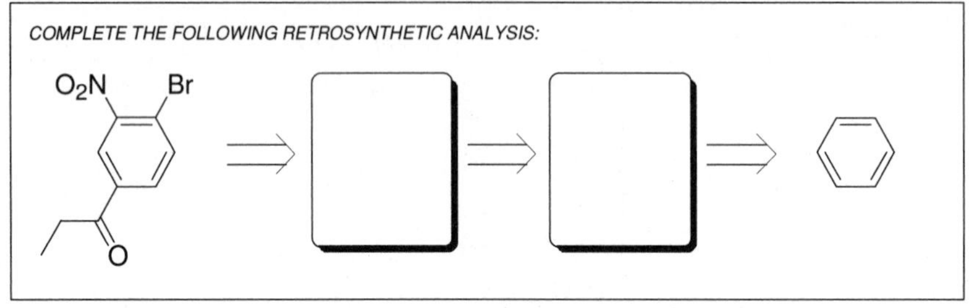

19.7 Determining the Mechanism of an Aromatic Substitution Reaction

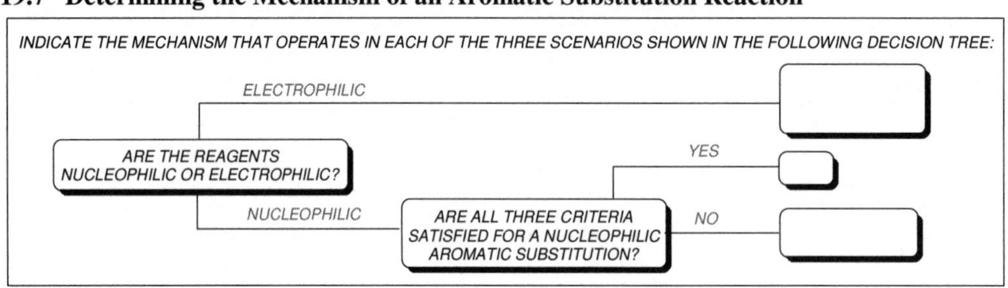

Review of Reactions

Identify the reagents necessary to achieve each of the following transformations. To verify that your answers are correct, look in your textbook at the end of Chapter 19. The answers appear in the section entitled *Review of Reactions*.

Electrophilic Aromatic Substitution

Br

Cl

NO_2

SO_3H

CH_3

O

NH_2

CBr_3

O

OH

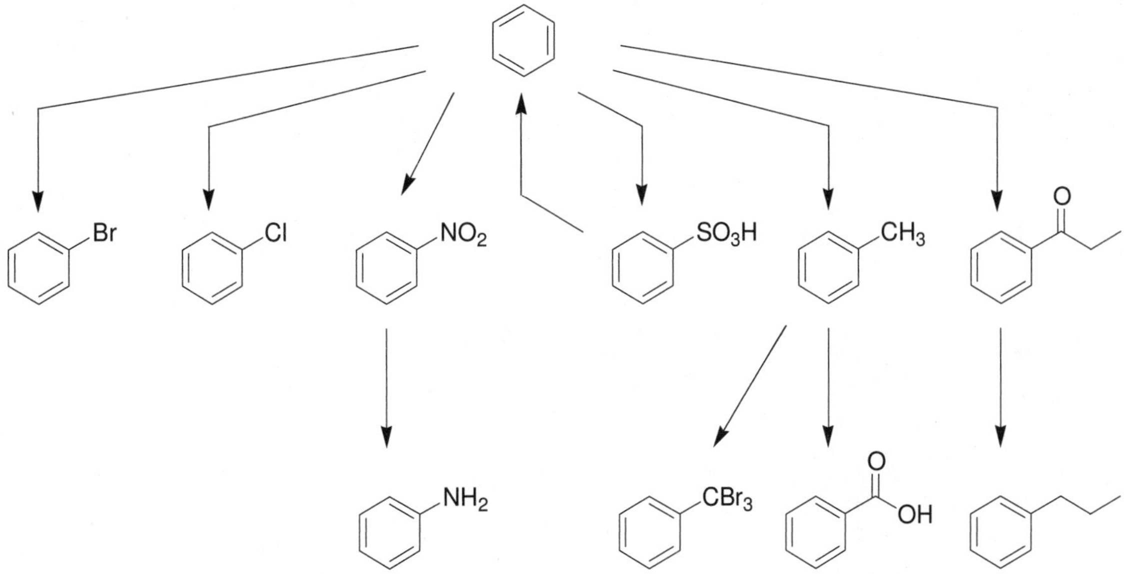

Nucleophilic Aromatic Substitution

Br

NO_2

OH

NO_2

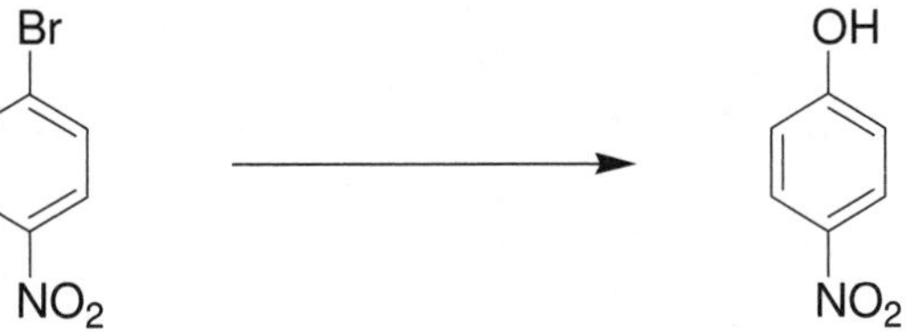

Elimination-Addition

Cl

OH

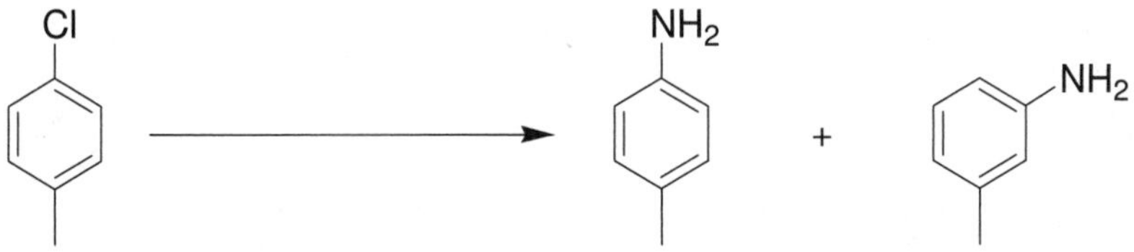

Cl

NH_2

+

NH_2

Solutions

19.1.

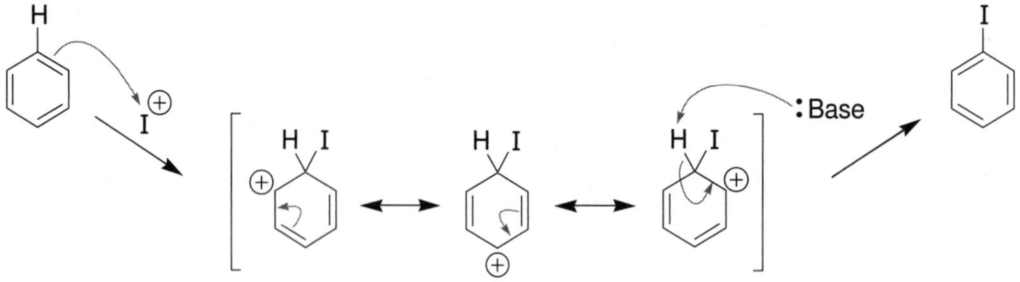

H

I⊕

I

:Base

H I

H I

H I

19.2.

H

- SO_3

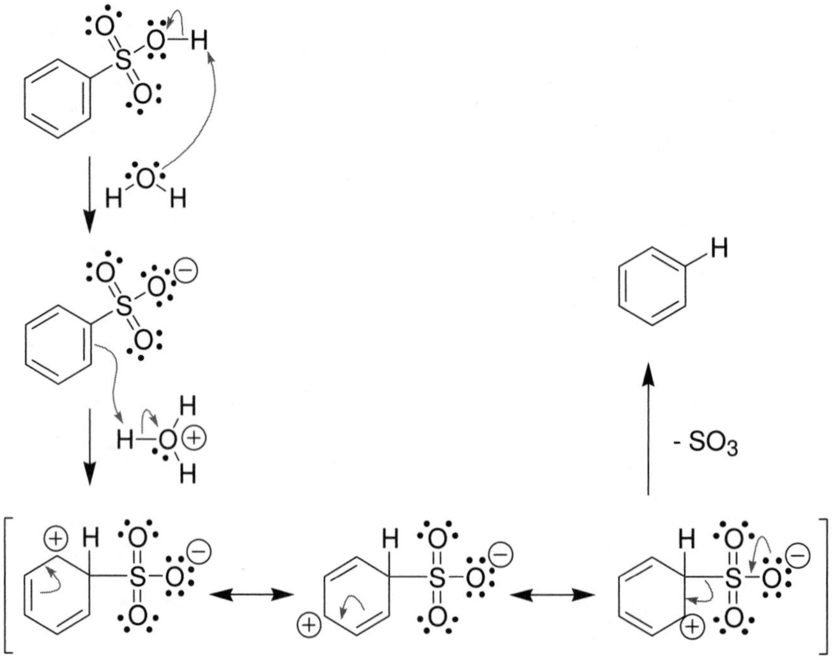

19.3.

19.4.

$- H_2O$

nitronium ion

19.5.

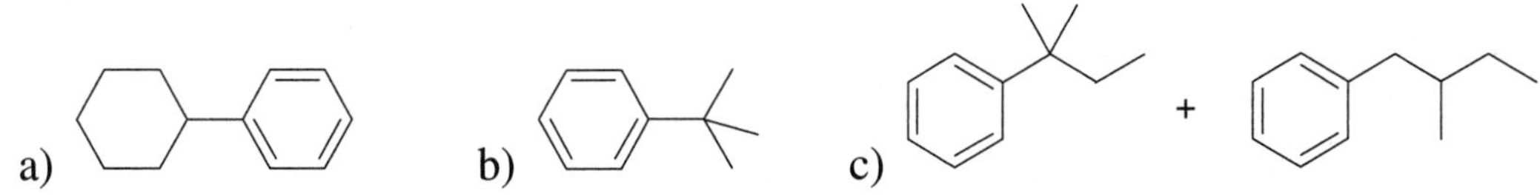

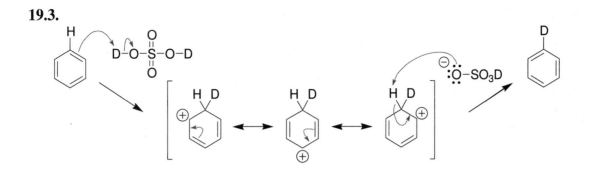

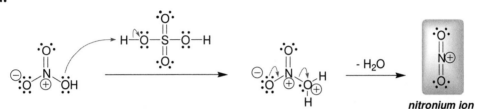

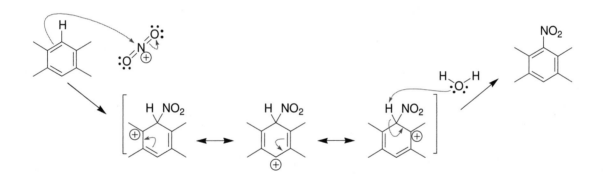

19.6.

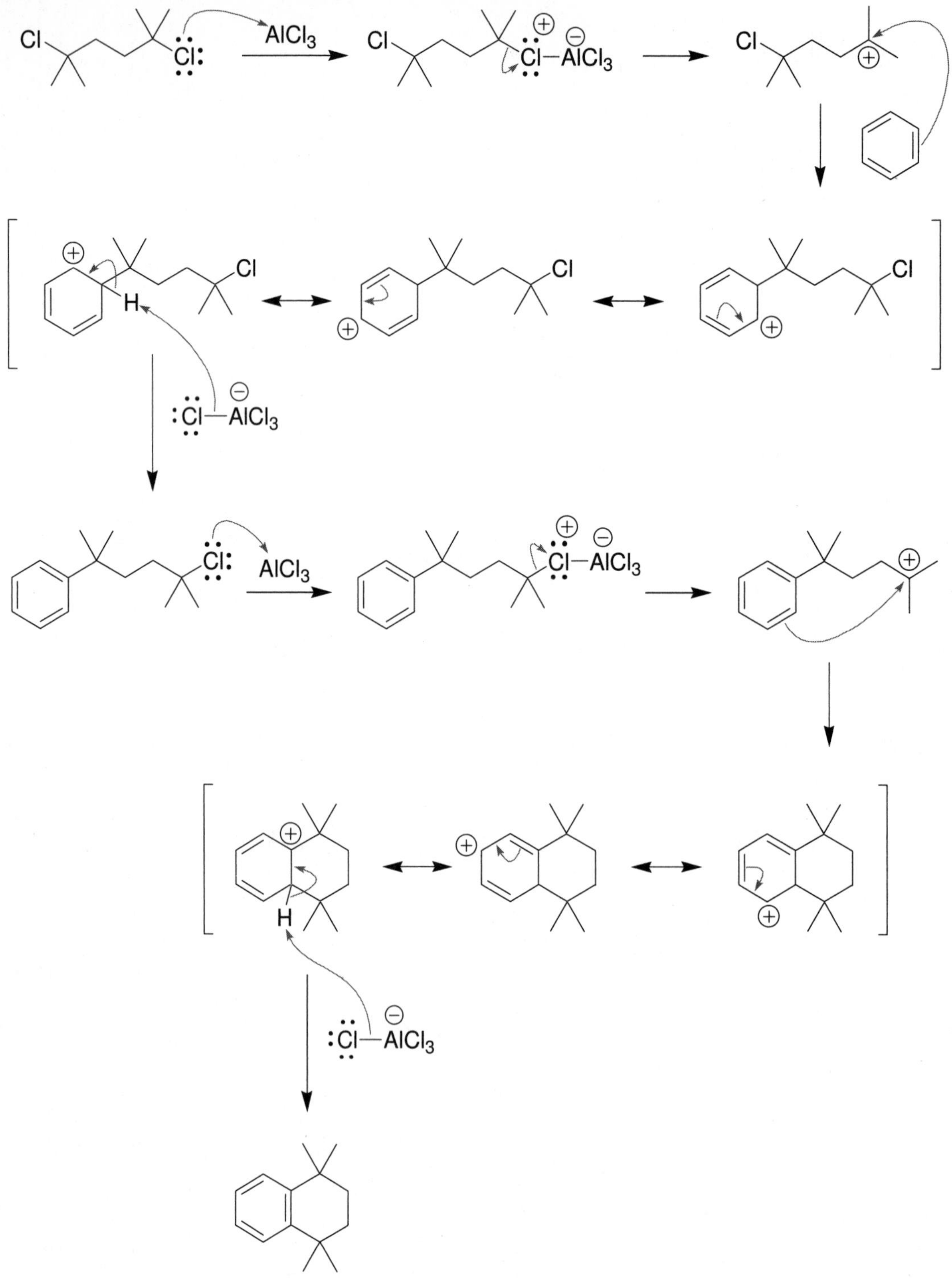

Cl Cl: $AlCl_3$ → Cl Cl–$AlCl_3$ → Cl

Cl Cl Cl H

:Cl–$AlCl_3$

Cl: $AlCl_3$ → Cl–$AlCl_3$ →

H

:Cl–$AlCl_3$

19.7.

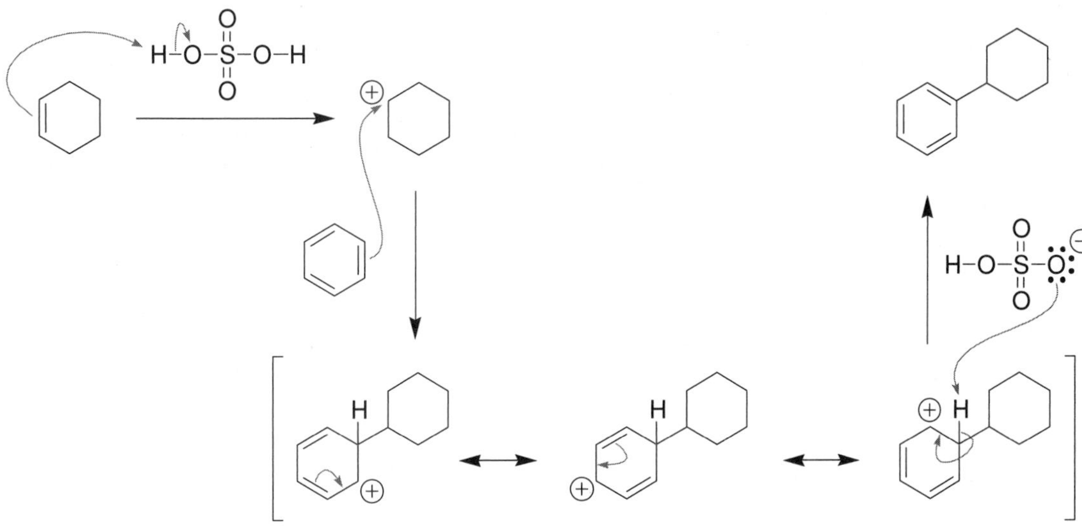

19.8.
a) It is necessary to perform an acylation followed by a Clemmensen reduction to avoid carbocation rearrangements.
b) It is necessary to perform an acylation followed by a Clemmensen reduction to avoid carbocation rearrangements.
c) It is necessary to perform an acylation followed by a Clemmensen reduction to avoid carbocation rearrangements.
d) The compounds can be made using a direct Friedel–Crafts alkylation.

19.9. It cannot be made via alkylation because the carbocation required would undergo a methyl shift to give a tertiary carbocation. It cannot be made via acylation followed by a Clemmensen reduction, because the product of a Clemmensen reduction has two benzylic protons. This compound has only one benzylic proton, which means that it cannot be made via a Clemmensen reduction.

19.10.

AlCl$_3$

AlCl$_3$

:Cl—AlCl$_3$

R

19.11.

Br + Br

19.12.

OEt NO$_2$ + OEt NO$_2$

a)

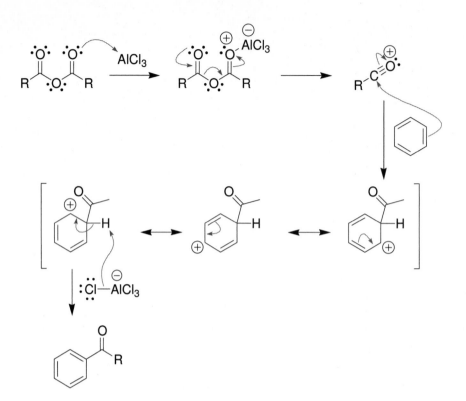

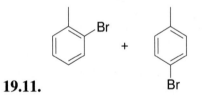

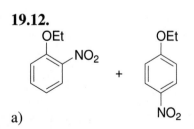

b)

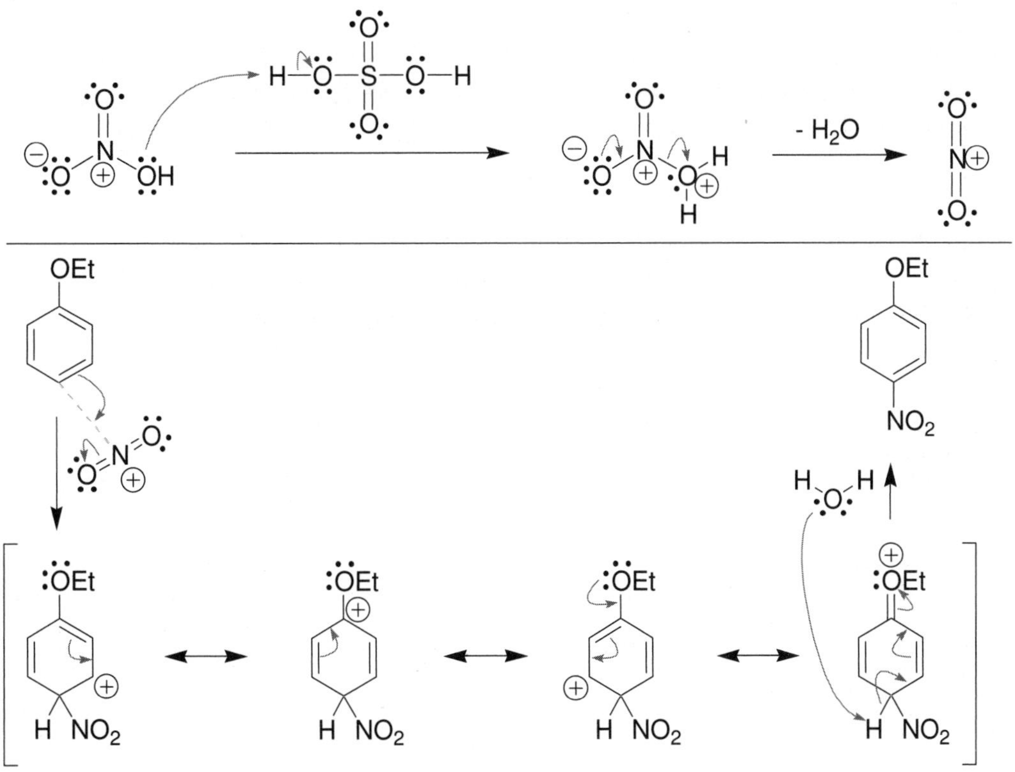

19.13. As show below, attack at C4 or C6 produces a sigma complex in which two of the resonance structures have a positive charge next to an electron-withdrawing group (NO_2). These resonance structures are less contributing to the resonance hybrid, thereby destabilizing the sigma complex. In contrast, attack at C5 produces a sigma complex for which none of the resonance structures have a positive charge next to a nitro group.

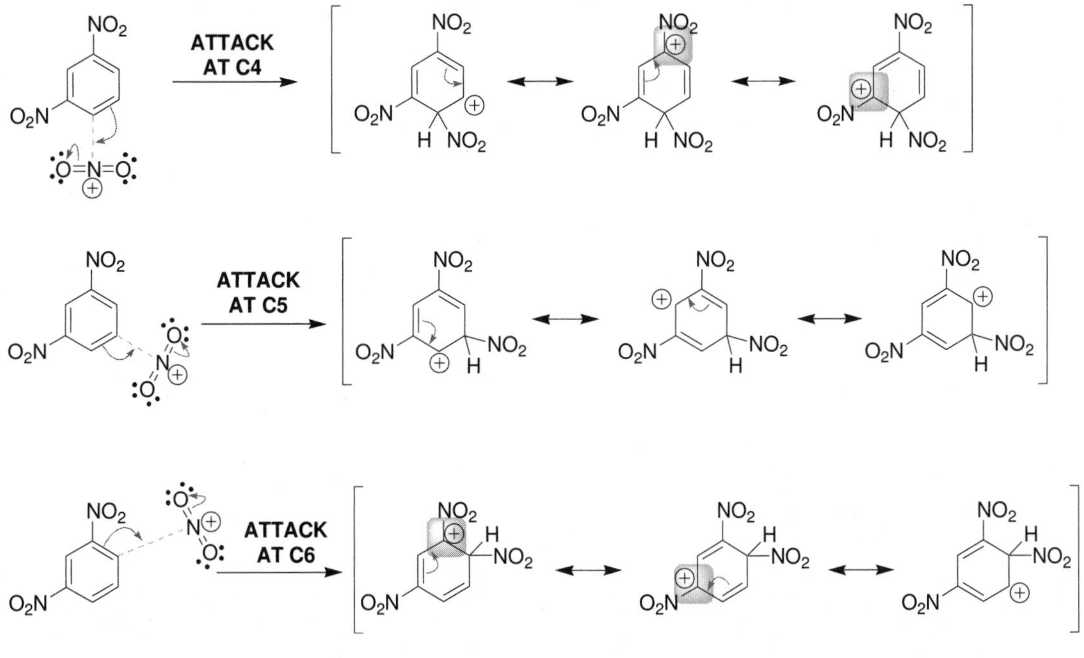

19.14. The chlorine atom in chlorobenzene deactivates the ring relative to benzene. If benzene requires a Lewis acid for chlorination, than chlorobenzene should certainly require a Lewis acid for chlorination.

19.15. *Ortho* attack and *para* attack are preferred because each of these pathways involves a sigma complex with four resonance structures (shown below). Attack at the *meta* position involves formation of a sigma complex with only three resonance structures, which is not as stable as a sigma complex with four resonance structures. The reaction will proceed more rapidly via the lower energy sigma complex, so attack takes place at the *ortho* and *para* positions in preference to the *meta* position.

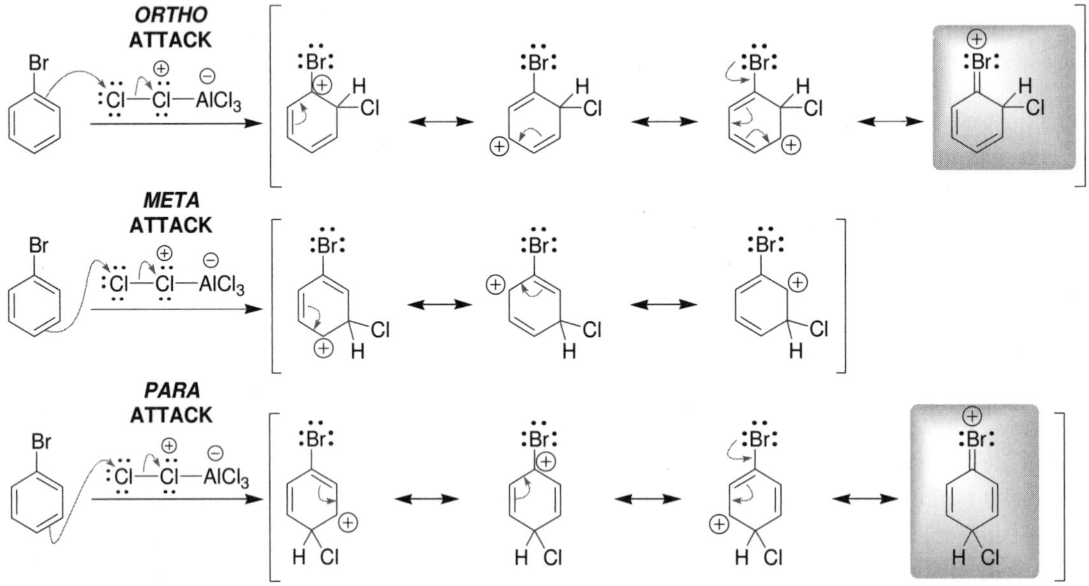

ORTHO
ATTACK

META
ATTACK

PARA
ATTACK

19.16.
a) The nitro is strongly deactivating and *meta*-directing.
b) An acyl group is moderately deactivating and *meta*-directing.
c) A bromine atom weakly deactivating and *ortho, para*-directing.
d) This group is moderately deactivating and *meta*-directing.
e) This group is moderately deactivating and *meta*-directing.
f) This group is moderately activating and *ortho, para* -directing.

19.17.

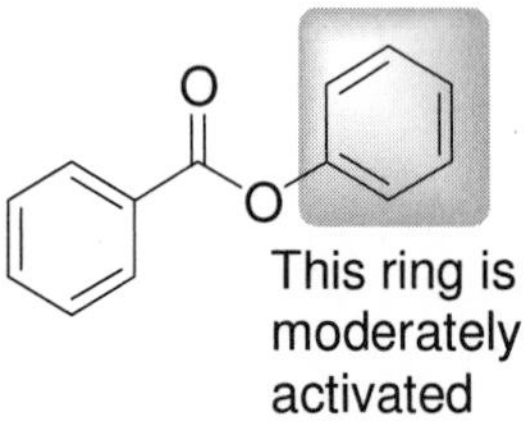

This ring is moderately activated

19.18.

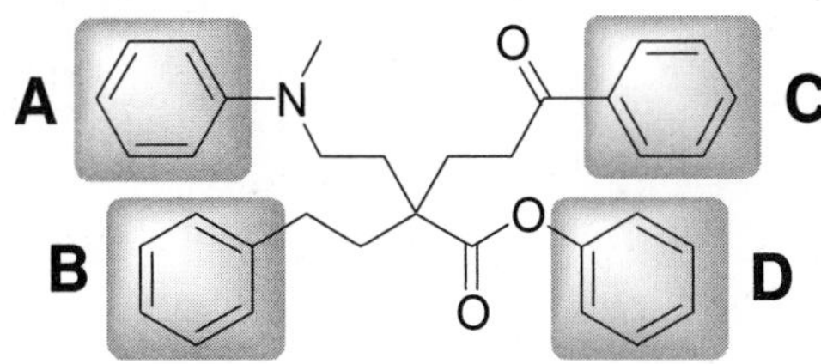

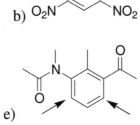

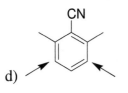

Increasing reactivity toward electrophilic aromatic substitution

C B D A

19.19.

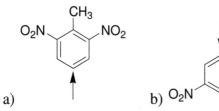

a) b) c)

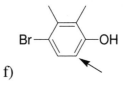

d) e) f)

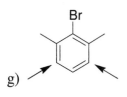

g) h) i)

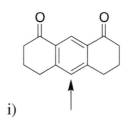

19.20.

a)

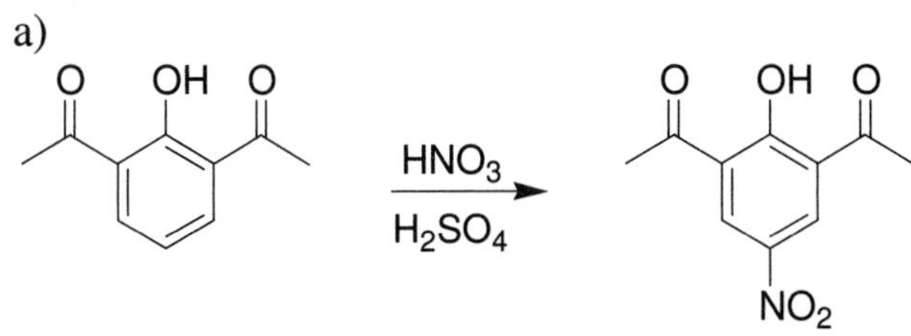

b)

Br_2 / $FeBr_3$

c)

Fuming H_2SO_4

19.21.

Br_2 / Fe

19.22.

a) b) c)

d) e)

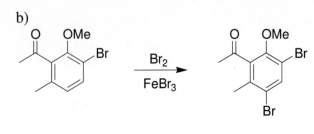

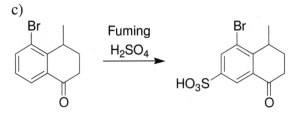

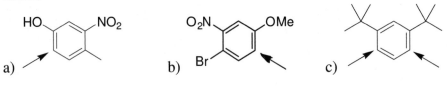

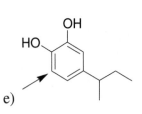

19.23. All three available positions are sterically hindered.

19.24.

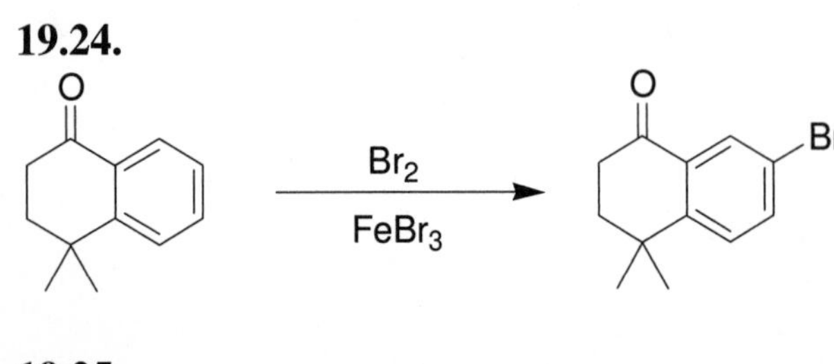

19.25.

a) Yes b) No c) Yes d) No

19.26.

HO_3S, SO_3H, SO_3H (trisulfonated toluene) → dilute H_2SO_4 → toluene

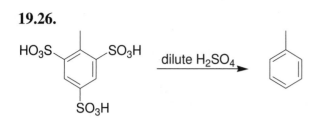

19.27.

a) The nitro group must be installed in a position that is *meta* to each of the OH groups. Even with a blocking group, *meta* attack cannot be achieved on a highly activated ring.
b) The position that must undergo bromination is too sterically hindered because of the presence of the *tert*-butyl groups.

19.28.

a) Cl_2, $AlCl_3$
b) HNO_3, H_2SO_4
c) Br_2, $FeBr_3$
d) CH_3CH_2Cl, $AlCl_3$
e) CH_3CH_2COCl, followed by HCl, Zn[Hg], heat
f) $(CH_3)_2CHCl$, $AlCl_3$
g) HNO_3, H_2SO_4, followed by followed by HCl, Zn
h) CH_3Cl, $AlCl_3$, followed by $KMnO_4$, NaOH, heat, followed by H_3O^+
i) CH_3Cl, $AlCl_3$

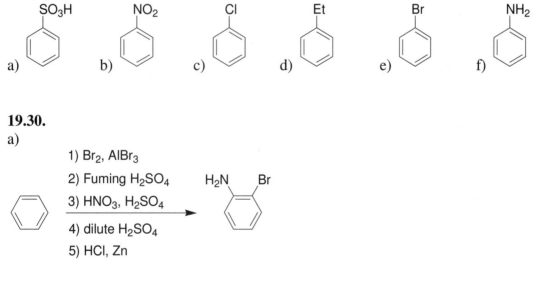

19.29.

a) SO_3H b) NO_2 c) Cl d) Et e) Br f) NH_2

19.30.

a)

1) Br_2, $AlBr_3$
2) Fuming H_2SO_4
3) HNO_3, H_2SO_4
4) dilute H_2SO_4
5) HCl, Zn

H_2N Br

b)

1) HNO_3, H_2SO_4
2) Cl_2, $AlCl_3$

O_2N Cl

c)

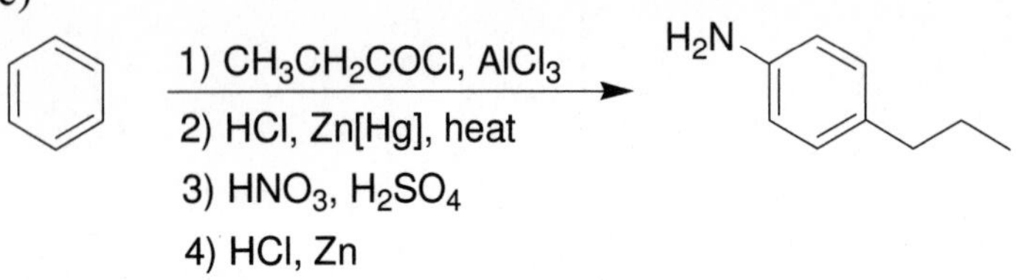
1) CH_3CH_2COCl, $AlCl_3$
2) HCl, Zn[Hg], heat
3) HNO_3, H_2SO_4
4) HCl, Zn
H_2N

d)

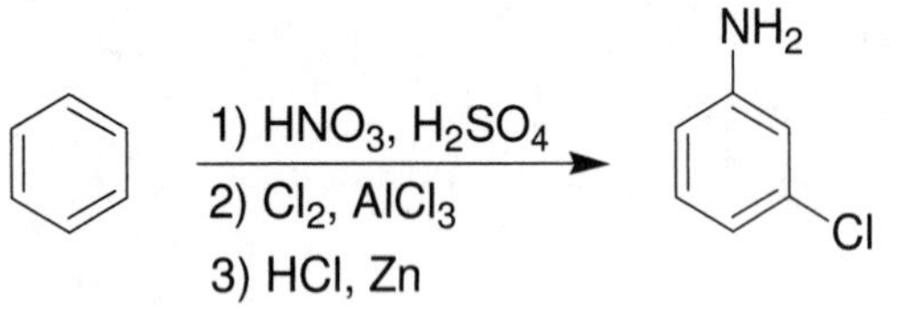
1) HNO_3, H_2SO_4
2) Cl_2, $AlCl_3$
3) HCl, Zn
NH_2
Cl

e)

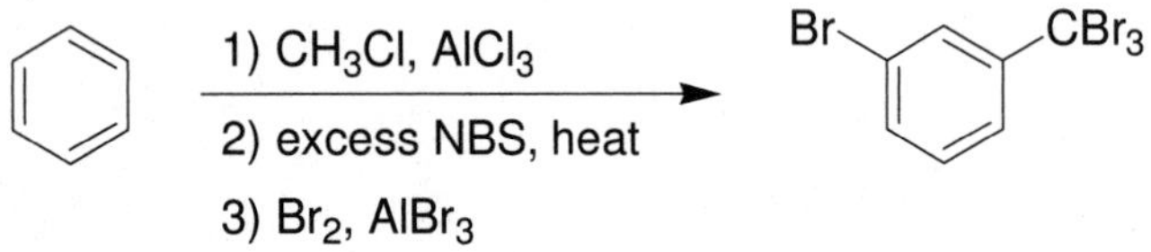
1) CH_3Cl, $AlCl_3$
2) excess NBS, heat
3) Br_2, $AlBr_3$
Br
CBr_3

f)

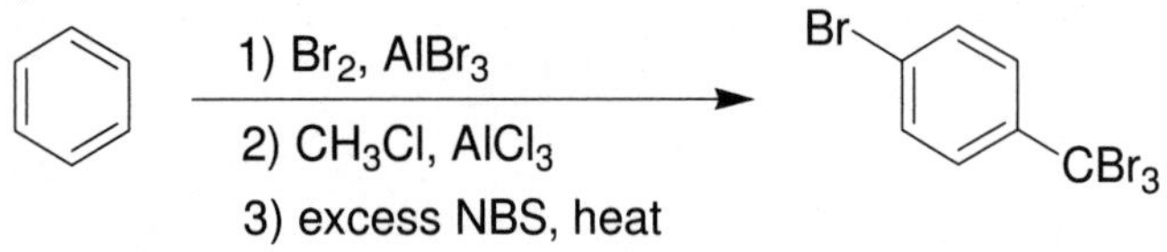
1) Br_2, $AlBr_3$
2) CH_3Cl, $AlCl_3$
3) excess NBS, heat
Br
CBr_3

g)

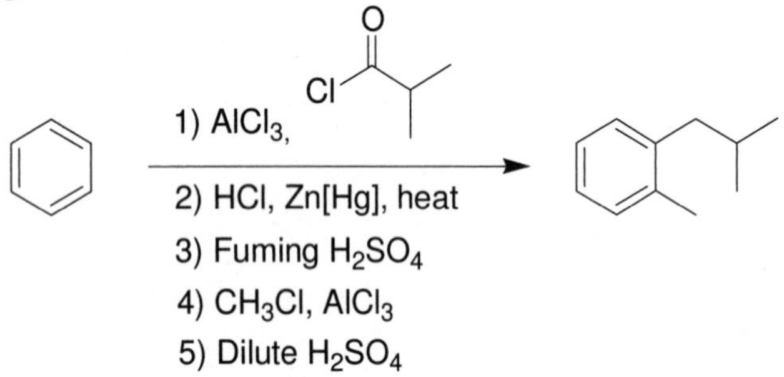
O
Cl
1) $AlCl_3$,
2) HCl, Zn[Hg], heat
3) Fuming H_2SO_4
4) CH_3Cl, $AlCl_3$
5) Dilute H_2SO_4

h)

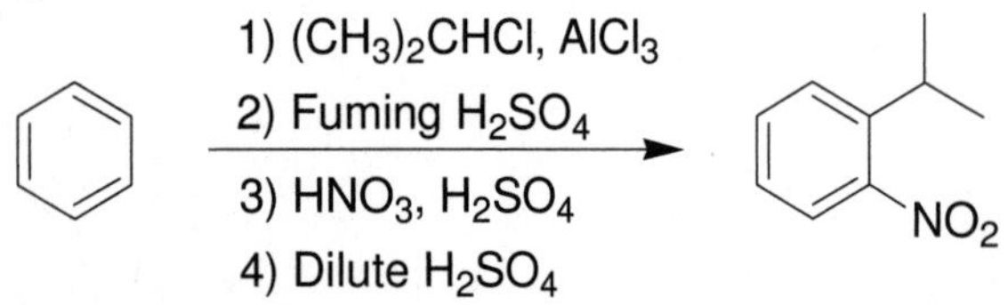
1) $(CH_3)_2CHCl$, $AlCl_3$
2) Fuming H_2SO_4
3) HNO_3, H_2SO_4
4) Dilute H_2SO_4
NO_2

i)

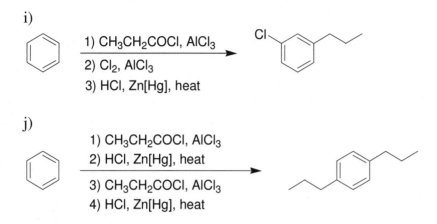

1) CH_3CH_2COCl, $AlCl_3$
2) Cl_2, $AlCl_3$
3) HCl, Zn[Hg], heat

Cl

j)

1) CH_3CH_2COCl, $AlCl_3$
2) HCl, Zn[Hg], heat
3) CH_3CH_2COCl, $AlCl_3$
4) HCl, Zn[Hg], heat

19.31. The *para* product will be more strongly favored over the *ortho* product if the *tert*-butyl group is installed first. The steric hindrance provided by a *tert*-butyl group is greater than the steric hindrance provided by an isopropyl group. Of the following two possible pathways, the first should provide a greater yield of the desired product.

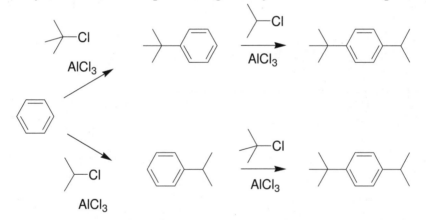

Cl
$AlCl_3$
Cl
$AlCl_3$
Cl
$AlCl_3$
Cl
$AlCl_3$

19.32.

a) Nitration cannot be achieved effectively in the presence of an amino group.
b) Each of the two alkyl groups is *ortho-para* directing, but the two groups are *meta* to each other. A Friedel-Crafts acylation will not work in this case (see solution to problem 19.9)

19.33.

a)

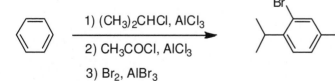

1) $(CH_3)_2CHCl$, $AlCl_3$
2) CH_3COCl, $AlCl_3$
3) Br_2, $AlBr_3$

Br
O

b)

1) CH_3CH_2COCl, $AlCl_3$
2) HCl, Zn[Hg], heat
3) HNO_3, H_2SO_4
4) Br_2, $AlBr_3$
5) HCl, Zn

H_2N Br

c)

1) $(CH_3)_3CCl$, $AlCl_3$
2) HNO_3, H_2SO_4
3) HCl, Zn
4) excess Cl_2

Cl NH_2 Cl

d)

1) Br_2, $AlBr_3$
2) Fuming H_2SO_4
3) Cl_2, $AlCl_3$

Cl Br SO_3H

19.34.

1) $(CH_3)_2CHCl$, $AlCl_3$
2) Fuming H_2SO_4
3) excess Br_2, $AlBr_3$
4) Dilute H_2SO_4
5) $KMnO_4$, NaOH, heat
6) H_3O^+
7) Cl_2, $AlCl_3$

O OH Br Br Cl Cl

a)
b) The sixth position is sterically hindered by the presence of the Cl atoms.
c) The ring is deactivated because all five groups are deactivators.

19.35.

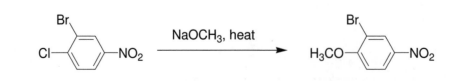

Br Cl NO_2 — NaOCH$_3$, heat → Br H_3CO NO_2

19.36.

1) Cl_2, $AlCl_3$
2) HNO_3, H_2SO_4
3) NaOH, heat
4) H_3O^+
5) HCl, Zn

HO–C₆H₄–NH₂

19.37.

a) Each additional nitro group serves as a reservoir of electron density and provides for an additional resonance structure in the sigma complex, thereby stabilizing the sigma complex and lowering the energy of activation for the reaction.
b) No, a fourth nitro group would not be *ortho* or *para* to the leaving group, and therefore cannot function as a reservoir.

19.38.

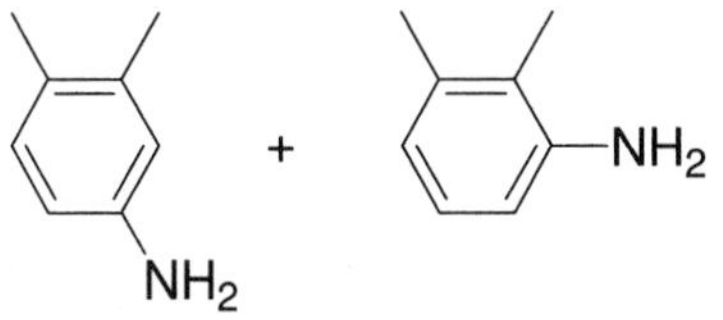

NH_2 + NH_2

19.39.

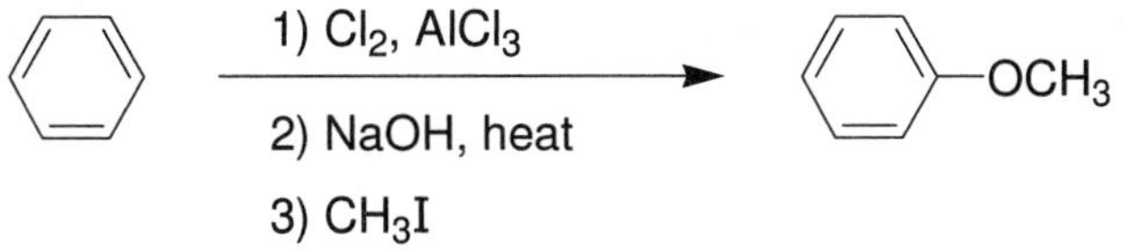

1) Cl_2, $AlCl_3$
2) NaOH, heat
3) CH_3I

OCH_3

19.40.

a)

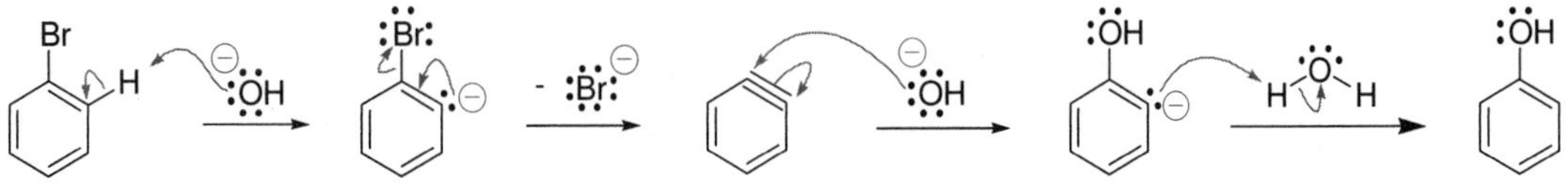

Br, H, :ÖH, :Br:, - :Br:, :ÖH, :ÖH, H–Ö–H, :ÖH

b)

$:Br-Br: \xrightarrow{FeBr_3} :Br-Br^{\oplus}-FeBr_3^{\ominus}$

H

$:Br-Br^{\oplus}-FeBr_3^{\ominus}$

H Br:

H Br:

H Br:

$Br_3Fe^{\ominus}-Br:$

H Br:

H Br:

Br:

+ $FeBr_3$ + HBr

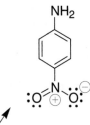

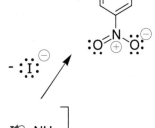

c)

O_2N — I $\xrightarrow[\text{2) } H_3O^+]{\text{1) } NaNH_2}$ O_2N — NH_2

I

$:O=N^{\oplus}-O:^{\ominus}$

$^{\ominus}:NH_2$

I NH2

I NH2

I NH2

I NH2

$- :I:^{\ominus}$

NH_2

$:O=N^{\oplus}-O:^{\ominus}$

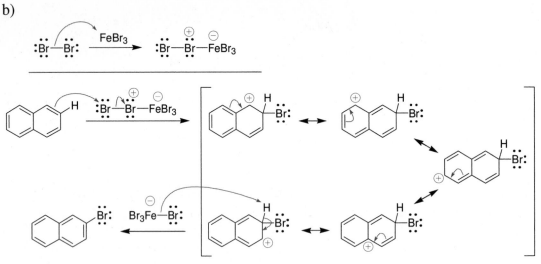

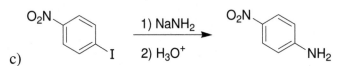

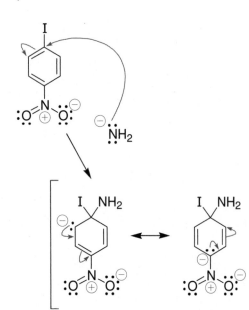

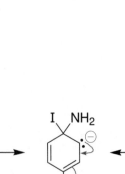

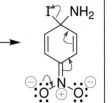

19.41.

NaOH, heat

HO + + HO OH

19.42.

a) NO_2 ONa

b) OH Br ↔ OH Br ↔ OH Br ↔ OH Br

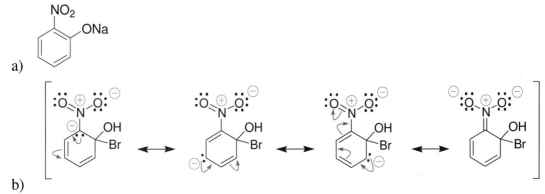

c) No
d) Yes

19.43.

$AlCl_3$

HCl, Zn[Hg], heat

EtCl $AlCl_3$

KMnO$_4$, NaOH heat

OH O

excess NBS, heat

Br_3C

Fuming H_2SO_4

HO_3S

HNO_3, H_2SO_4

O_2N

HCl, Zn

H_2N

Cl_2 $AlCl_3$

Cl

Br_2 $AlBr_3$

Br

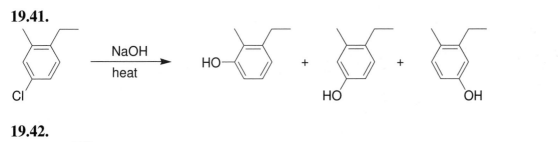

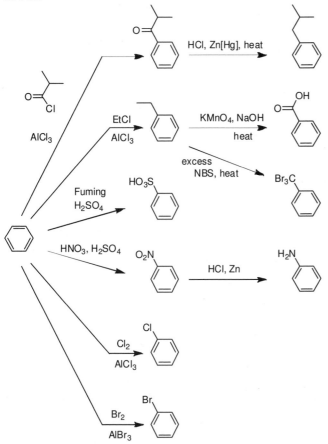

19.44.

Increasing Reactivity toward Electrophilic Aromatic Substitution

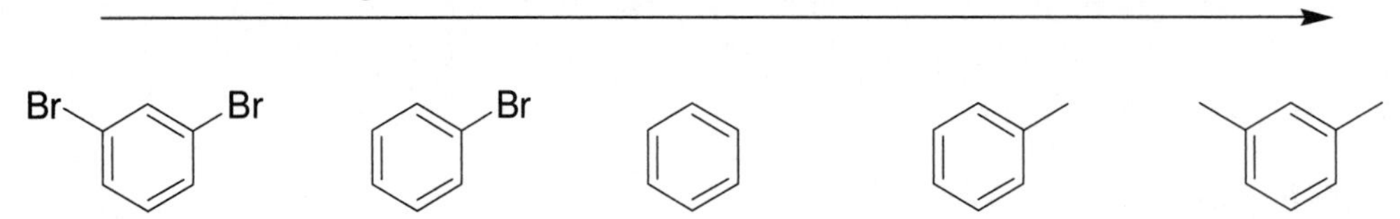

Br Br Br

19.45.

NO_2 HO OMe

least activated most activated

19.46.

a) 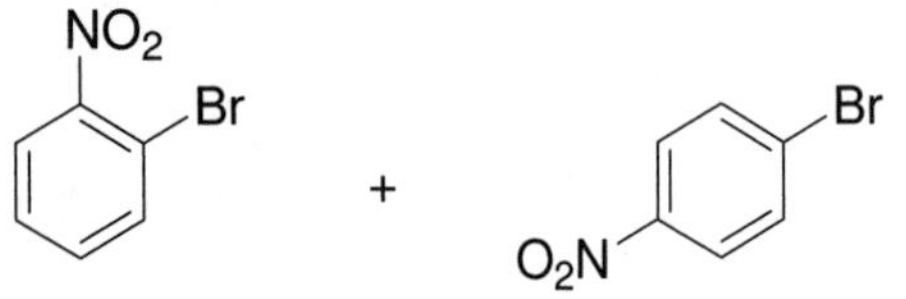 NO_2 Br + Br O_2N

b) 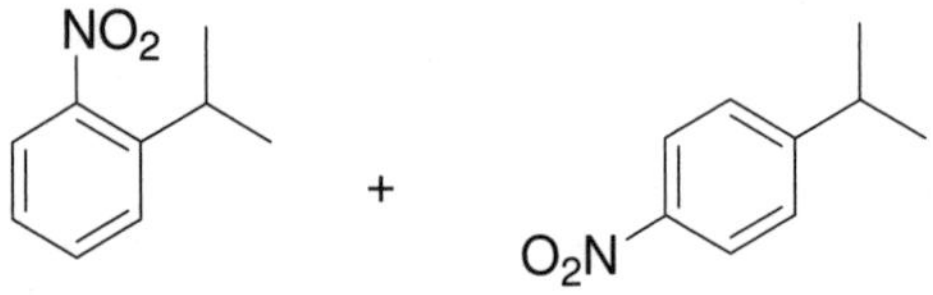 NO_2 + O_2N

c) 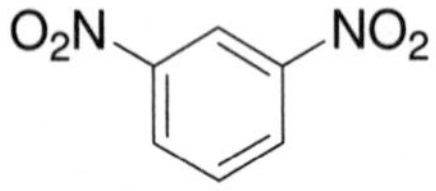 O_2N NO_2

d) 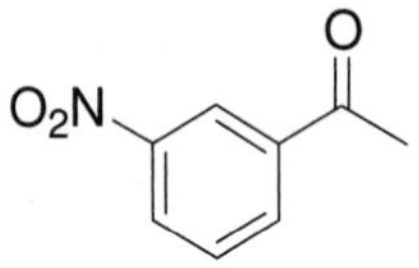 O_2N O

e) 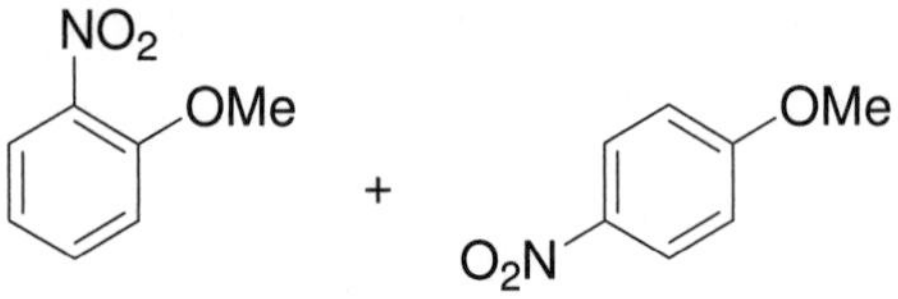 NO_2 OMe + OMe O_2N

19.47.

a) 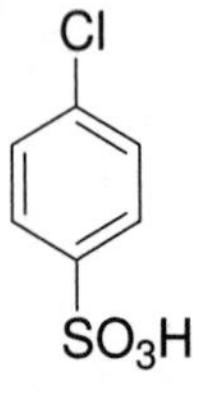 Cl SO_3H

b) 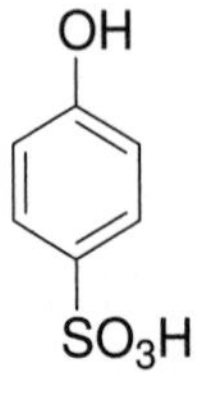 OH SO_3H

c) 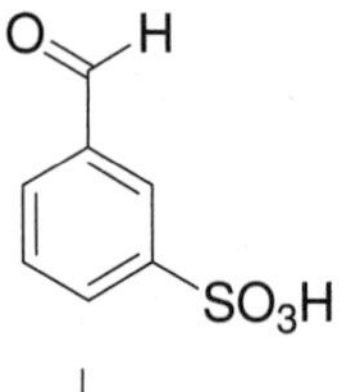 O H SO_3H

d) OH NO_2 SO_3H

e) SO_3H Br

f) O OH SO_3H

g) SO_3H

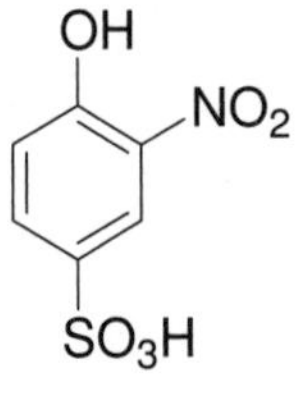

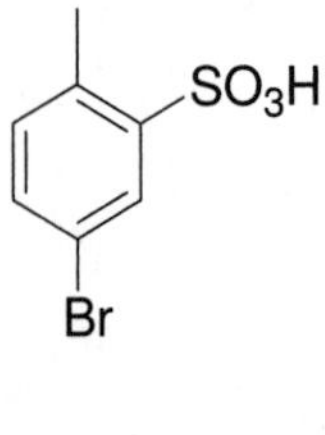

h) SO_3H

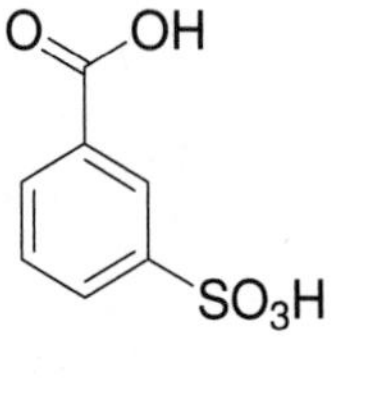

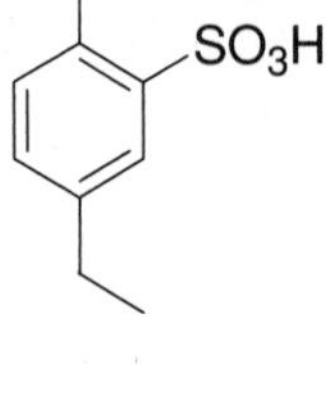

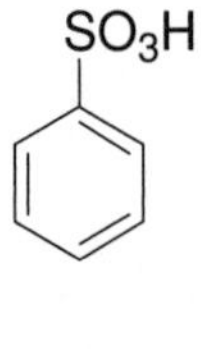

19.48.

a) This group is an activator and an *ortho,para*-director.
b) This group is an activator and an *ortho,para*-director.
c) This group is an activator and an *ortho,para*-director.
d) This group is a deactivator and an *ortho,para*-director.
e) This group is a deactivator and a *meta*-director.
f) This group is a deactivator and a *meta*-director.
g) This group is a deactivator and a *meta*-director.
h) This group is a deactivator and a *meta*-director.
i) This group is a deactivator and an *ortho,para*-director.
j) This group is a deactivator and a *meta*-director.

19.49.

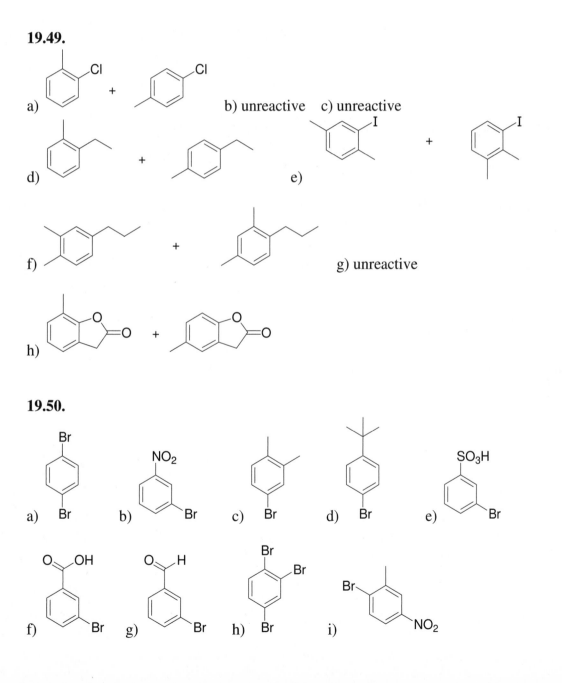

a) 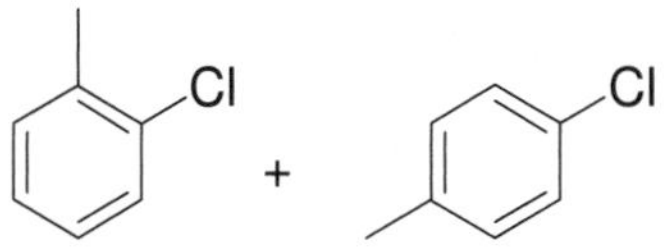 Cl + Cl b) unreactive c) unreactive

d) 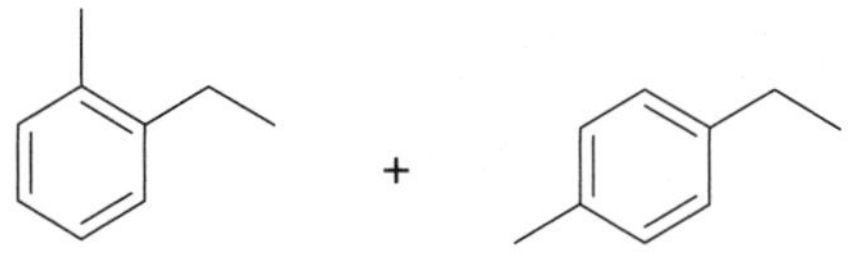 + e)  I + I

f) + g) unreactive

h) 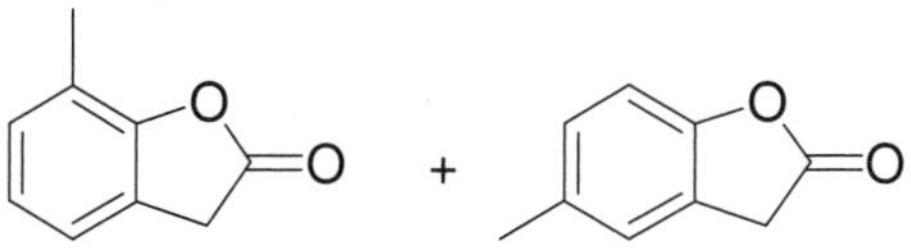 O =O + O =O

19.50.

a)  Br Br b) 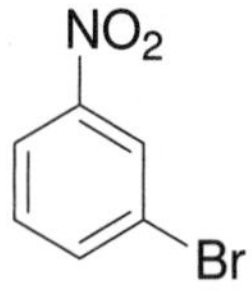 NO_2 Br c) 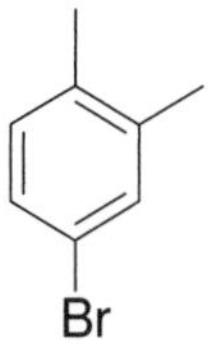 Br d) 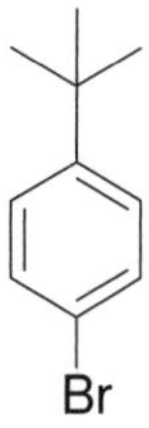 Br e) 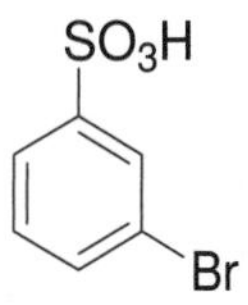 SO_3H Br

f) 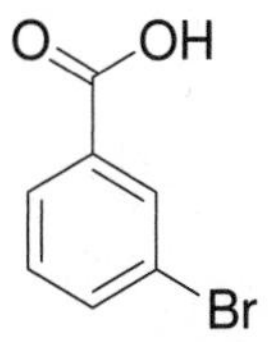 O OH Br g) 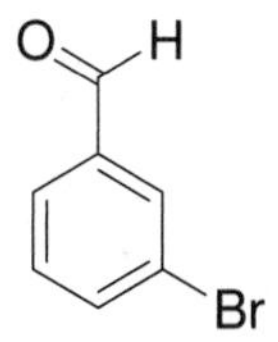 O H Br h) 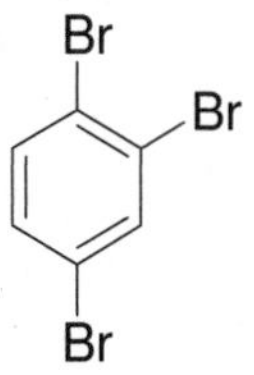 Br Br Br i) 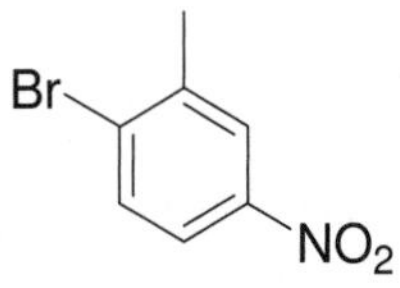 Br NO_2

j) 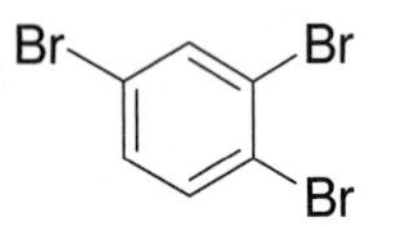 Br, Br, Br

k) 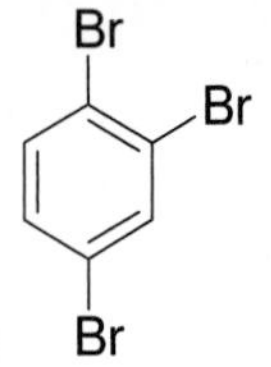 Br, Br, Br

19.51.

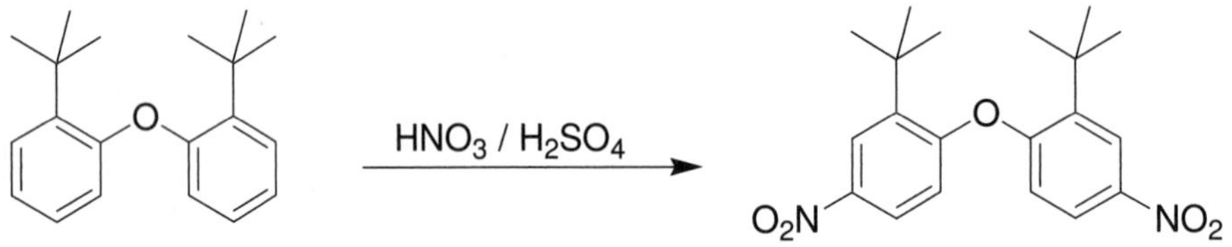

HNO_3 / H_2SO_4

O_2N NO_2

19.52.

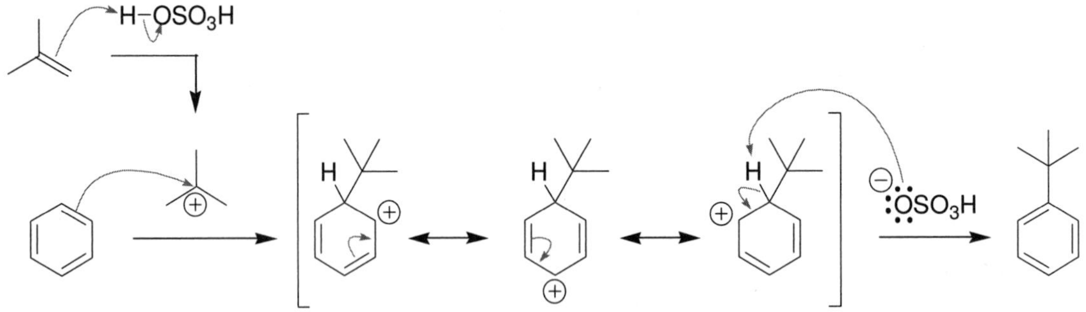

H–OSO_3H

H H H

OSO_3H

19.53.

a)

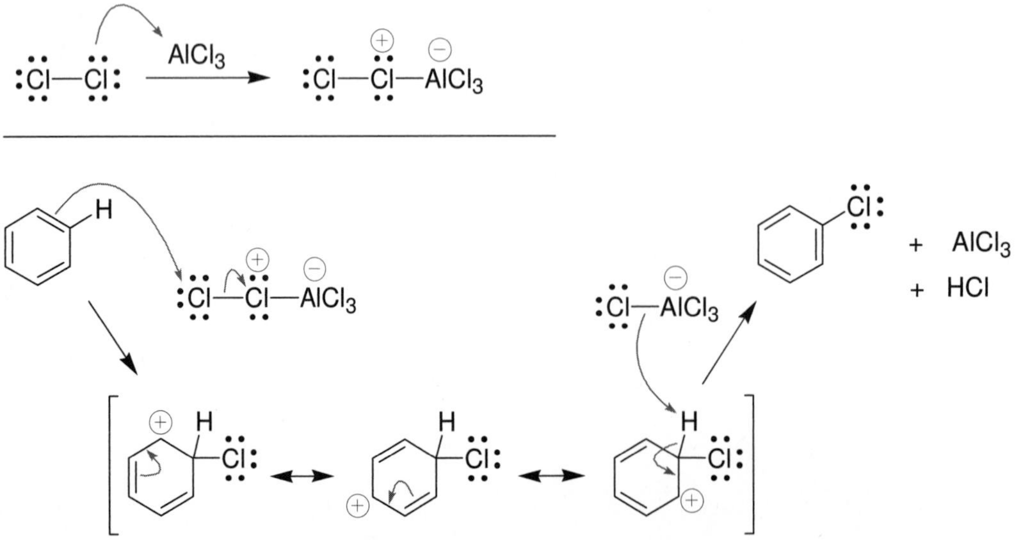

:Cl–Cl: $AlCl_3$:Cl–Cl–$AlCl_3$

H

:Cl–Cl–$AlCl_3$

:Cl–$AlCl_3$

Cl: + $AlCl_3$ + HCl

H Cl: H Cl: H Cl:

b)

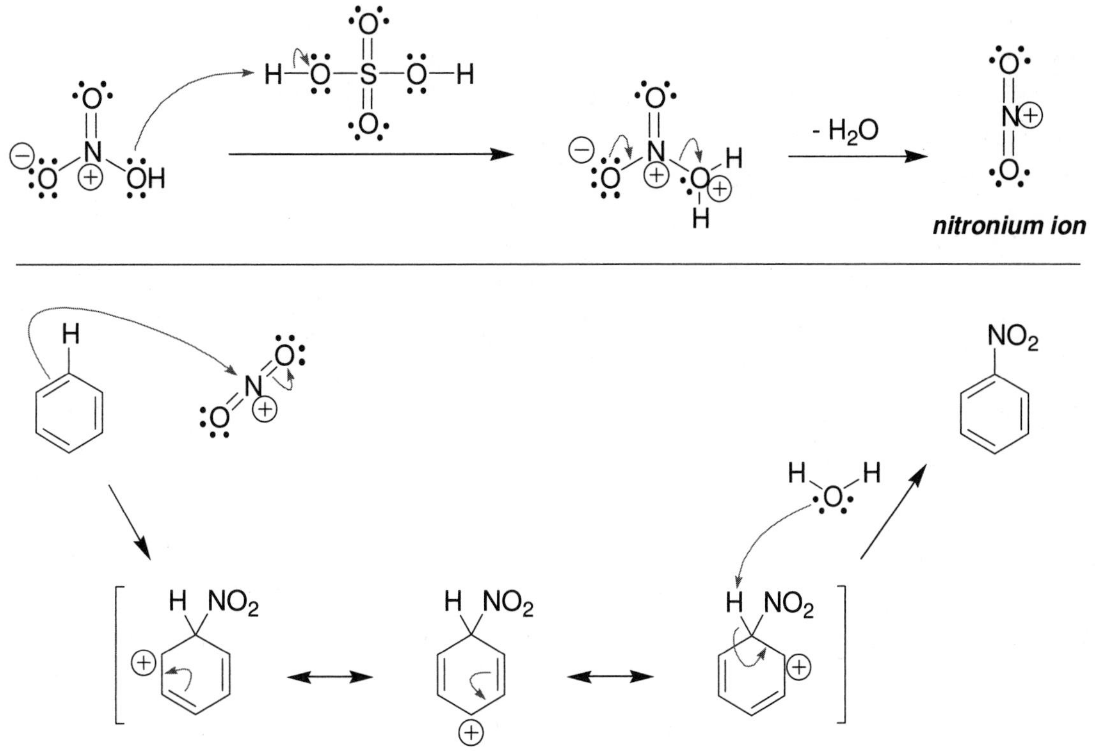

- H_2O

nitronium ion

NO_2

H NO_2

H NO_2

H NO_2

c)

H–O–SO_3H

O–SO_3H

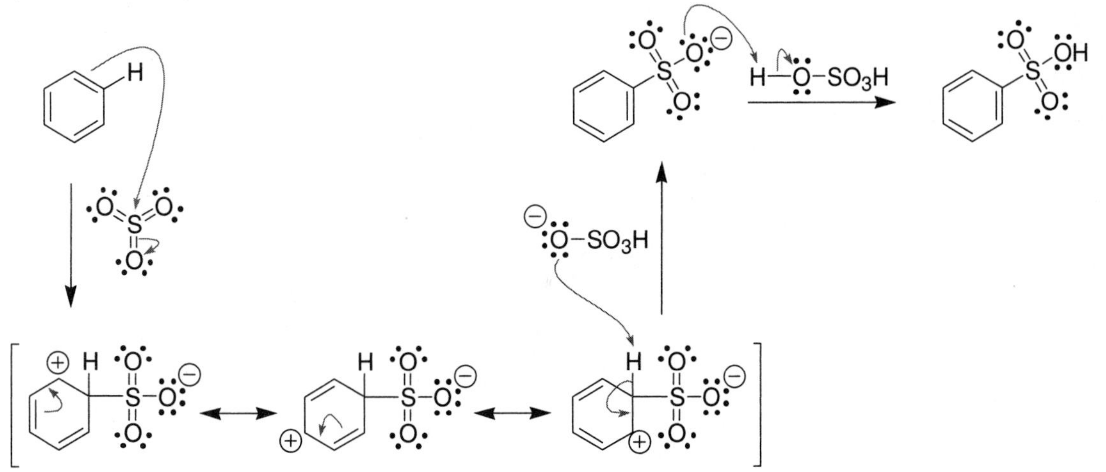

d)

$H_3C{-}Cl$ + $AlCl_3$ → $H_3C{-}Cl^{+}{-}Al^{-}Cl_3$

benzene + $H_3C{-}Cl^{+}{-}Al^{-}Cl_3$ → [arenium ion resonance structures]

$Cl{-}AlCl_3^{-}$

toluene + HCl + $AlCl_3$

e)

Br_2 → Fe → $FeBr_3$

$Br{-}Br$ + $FeBr_3$ → $Br{-}Br^{+}{-}FeBr_3^{-}$

benzene + $Br{-}Br^{+}{-}FeBr_3^{-}$ → [arenium ion resonance structures, H, Br]

$Br{-}FeBr_3^{-}$

bromobenzene + $FeBr_3$ + HBr

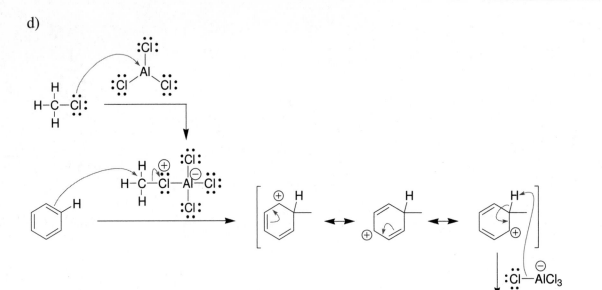

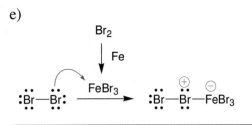

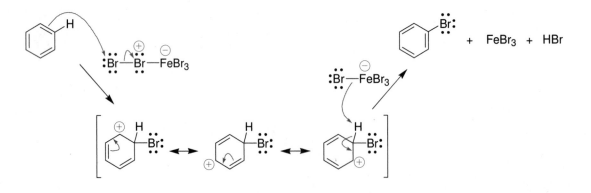

19.54.

a)

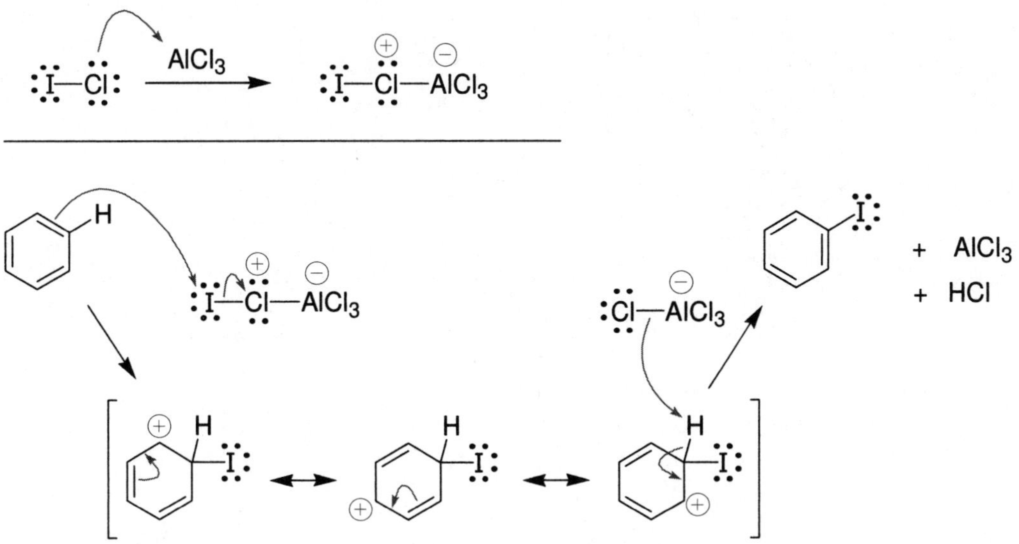

$$:\ddot{\underset{..}{I}}-\ddot{\underset{..}{Cl}}: \xrightarrow{AlCl_3} :\ddot{\underset{..}{I}}-\overset{\oplus}{\ddot{\underset{..}{Cl}}}-\overset{\ominus}{AlCl_3}$$

H

$:\ddot{\underset{..}{I}}-\overset{\oplus}{\ddot{\underset{..}{Cl}}}-\overset{\ominus}{AlCl_3}$

$:\ddot{\underset{..}{Cl}}-\overset{\ominus}{AlCl_3}$

+ $AlCl_3$

+ HCl

H H H

b)

:Cl: Al :Cl :Cl: H :Cl–C–Cl: H

H ⊕ :Cl: :Cl–C–Cl–Al–Cl: H :Cl: ⊖

H

⊕ H :Cl: ⟷ H :Cl: ⊕ ⟷ H Cl: ⊕

:Cl–AlCl3 ⊖

⊕ :Cl AlCl3 ⊖

:Cl: Al :Cl :Cl:

:Cl: + HCl + $AlCl_3$

⊕

H

⊕ H ⟷ H ⊕ ⟷ H ⊕

:Cl–AlCl3 ⊖

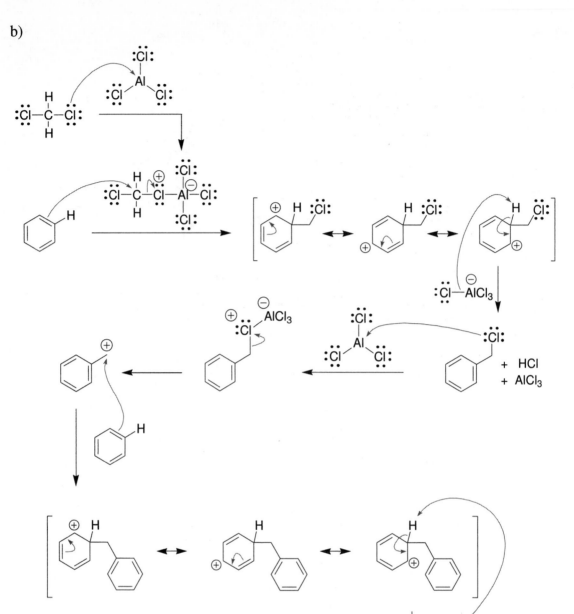

19.55.

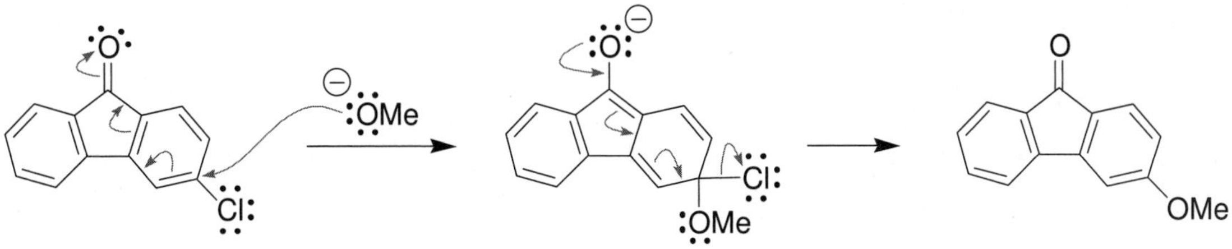

$:\ddot{O}Me^{-}$ → $:\ddot{O}:^{-}$ ... $:\ddot{Cl}:$ ... $:\ddot{O}Me$ → OMe

19.56.

a) 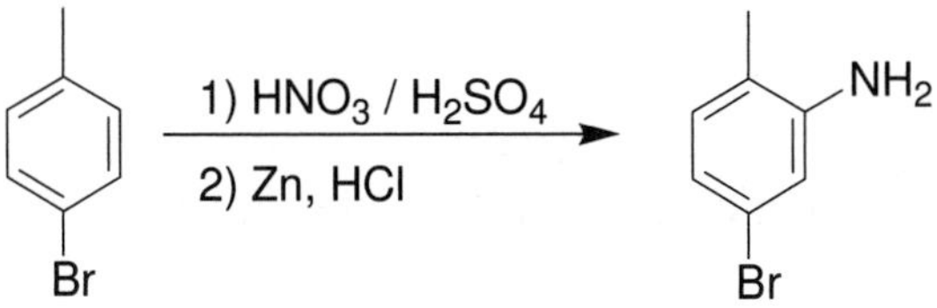 Br → 1) HNO_3 / H_2SO_4 2) Zn, HCl → NH_2, Br

b) 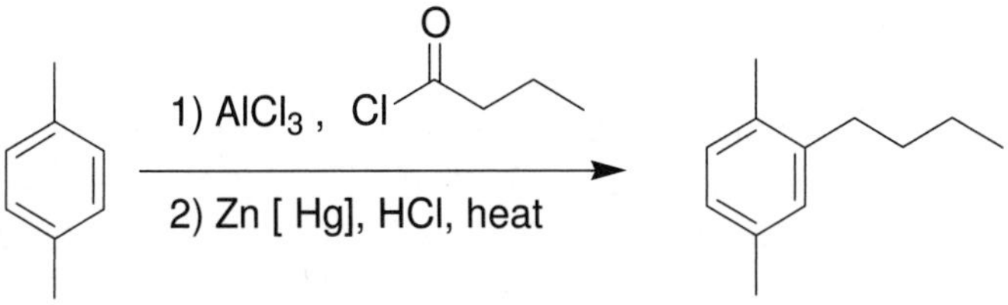 1) $AlCl_3$, Cl (O) 2) Zn [Hg], HCl, heat

c) 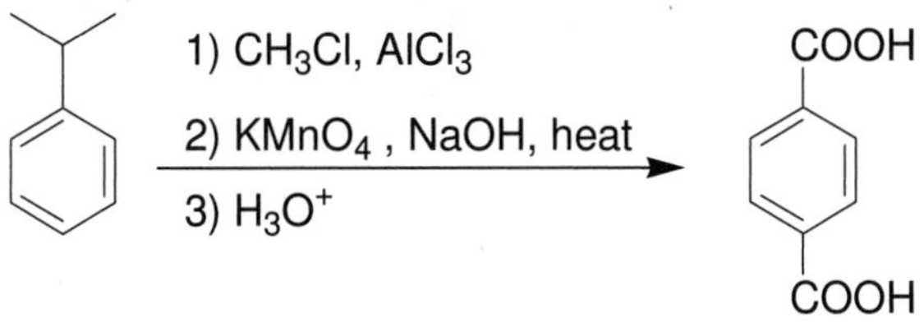 1) CH_3Cl, $AlCl_3$ 2) $KMnO_4$, NaOH, heat 3) H_3O^+ → COOH, COOH

d) 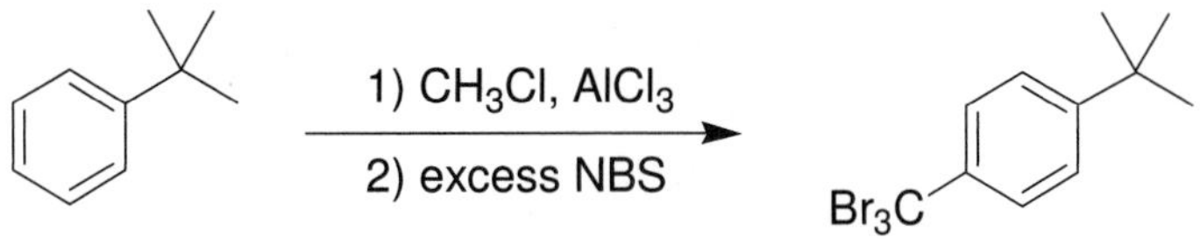 1) CH_3Cl, $AlCl_3$ 2) excess NBS → Br_3C

19.57.

a) 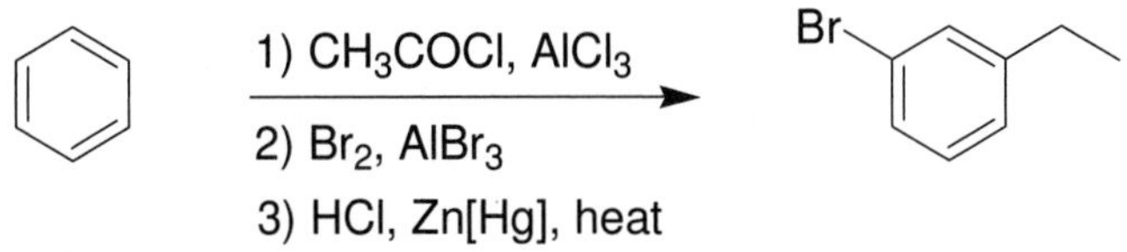 1) CH_3COCl, $AlCl_3$ 2) Br_2, $AlBr_3$ 3) HCl, Zn[Hg], heat → Br

b) 1) HNO_3, H_2SO_4 2) Br_2, $AlBr_3$ 3) HCl, Zn

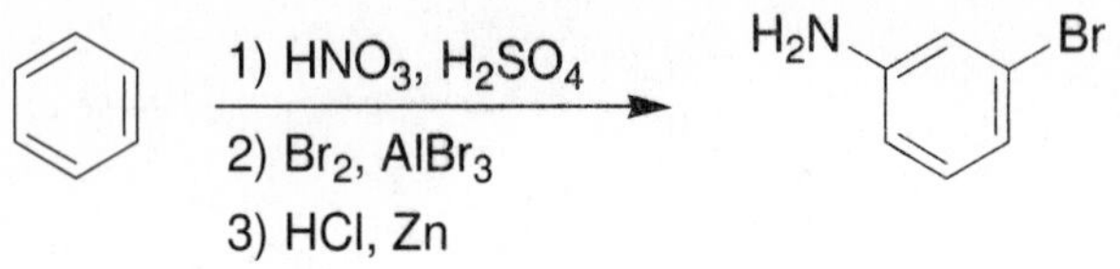

c) 1) CH_3COCl, $AlCl_3$ 2) HNO_3, H_2SO_4 3) HCl, Zn[Hg], heat

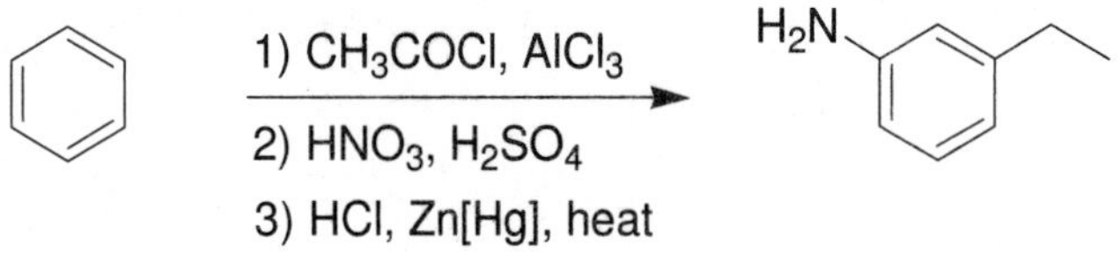

19.58.

a) The second step of the synthesis will not work, because a strongly deactivated ring will not undergo a Friedel-Crafts alkylation. The product of the first step, nitrobenzene, will be unreactive in the second step.

b) The second step of the synthesis will not efficiently install a propyl group, because a carbocation rearrangement can occur, which will result in the installation of an isopropyl group.

c) The second step of the synthesis will not install the acyl group in the *meta* position. It will be installed in a position that is either *ortho* or *para* to the bromine atom.

d) The second step of the synthesis will not install the bromine atom in the *ortho* position, because of steric hindrance from the *tert*-butyl group. Bromination will occur primarily at the *para* position.

19.59.

a)

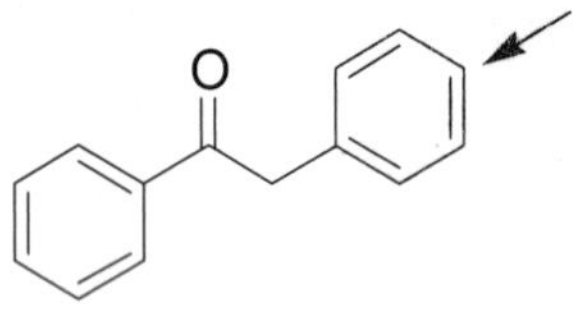

b)

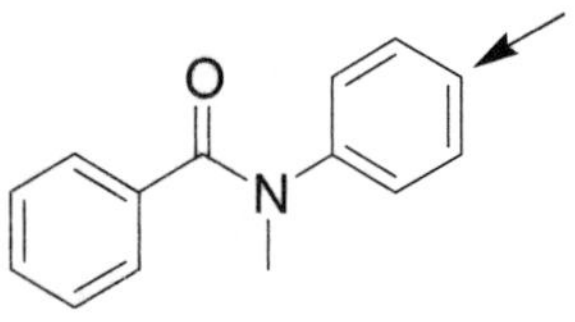

c)

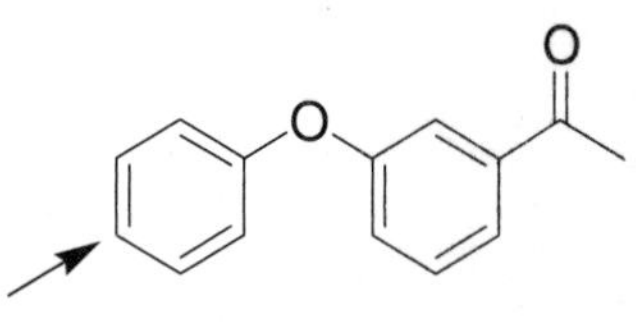

d)

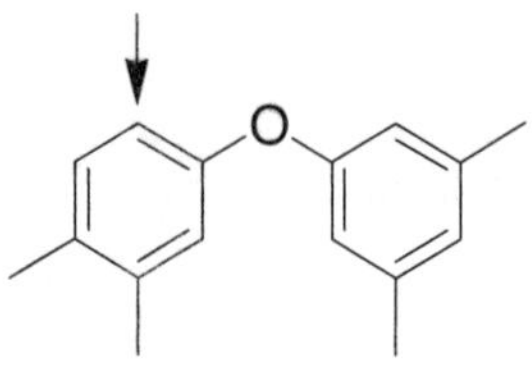

19.60.

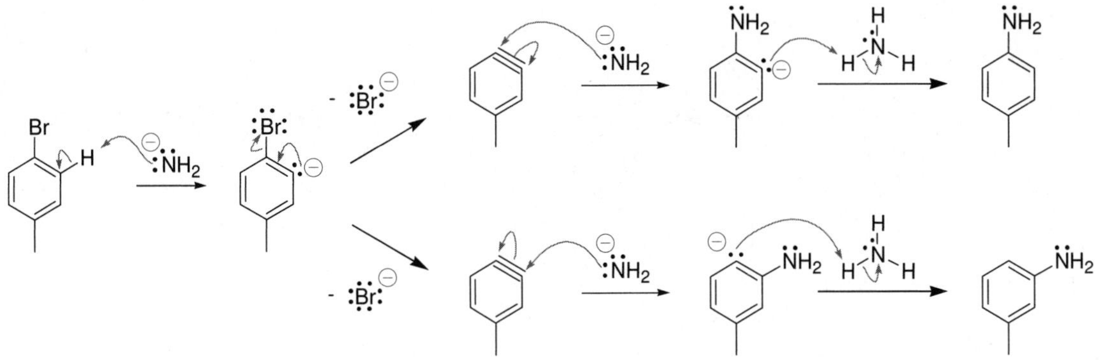

Br, H, :NH₂ (⊖), :Br:, - :Br: (⊖), :NH₂, NH₂, H–N–H with H, NH₂

19.61.

OH, O_2N, NO_2, NO_2

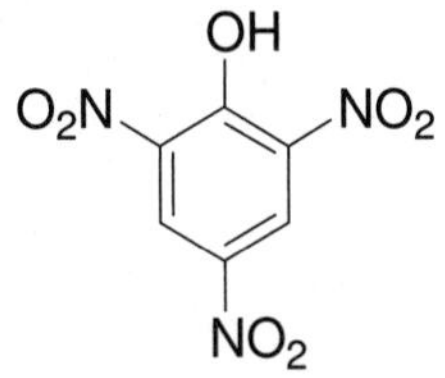

2,4,6-trinitrophenol
(picric acid)

19.62.

HO OH H–OSO$_3$H HO H O H - H_2O HO

OH OH OH

H O H

OH H–OSO$_3$H H O H - H_2O

H

H O H

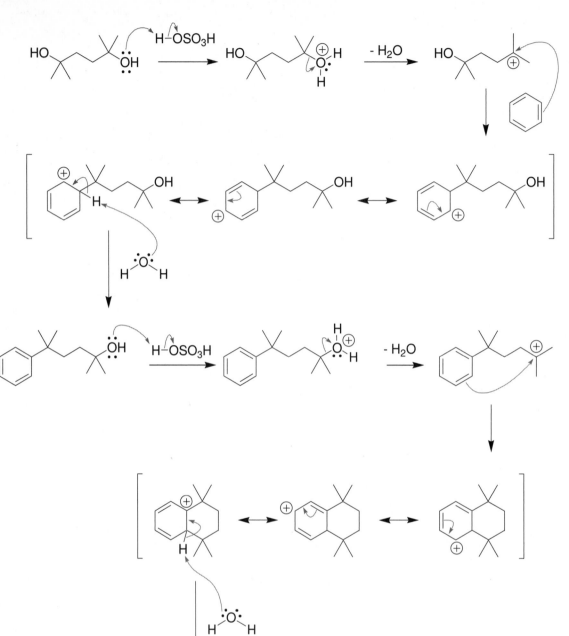

19.63.

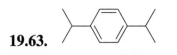

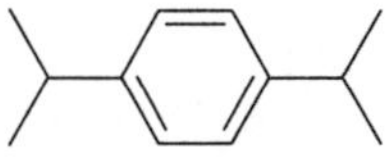

19.64.

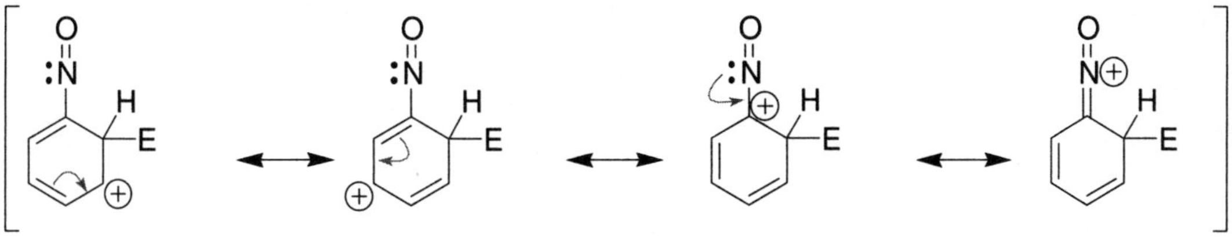

a)

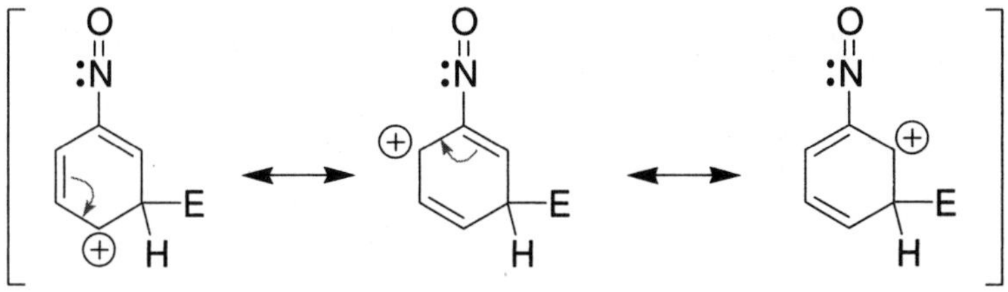

b)

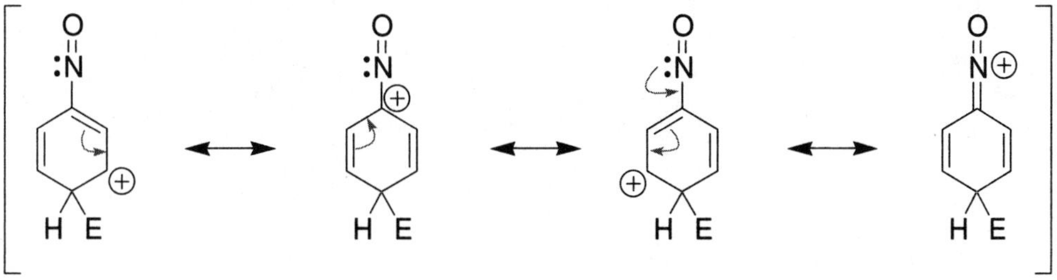

c)

d) The nitroso group should be *ortho-para* directing, because attack at the *ortho* or *para* position generates a sigma complex with an additional resonance structure.

e) The nitroso group is a deactivator, yet it is an *ortho-para* director, just like a chlorine atom.

19.65.

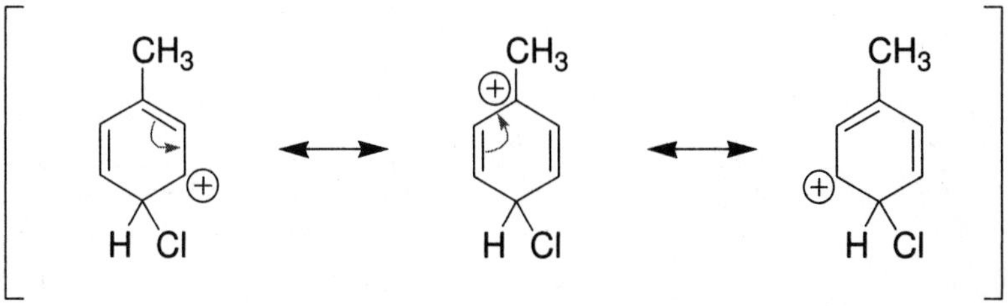

19.66.

a) Toluene is the only compound containing an activated ring, and it is expected to undergo a Friedel-Crafts reaction most rapidly to give *ortho*-ethyltoluene and *para*-ethyltoluene.

b) Anisole is the most activated compound (among the three compounds), and is expected to undergo a Friedel-Crafts reaction most rapidly to give *ortho*-ethylanisole and *para*-ethylanisole.

19.67.

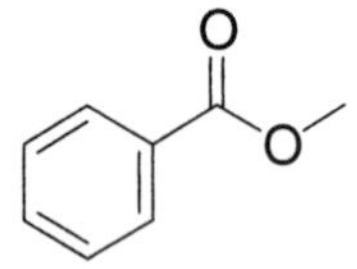

Compound A 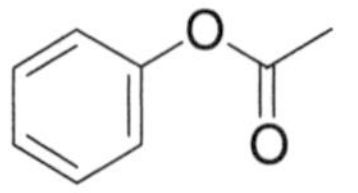   Compound B

19.68.

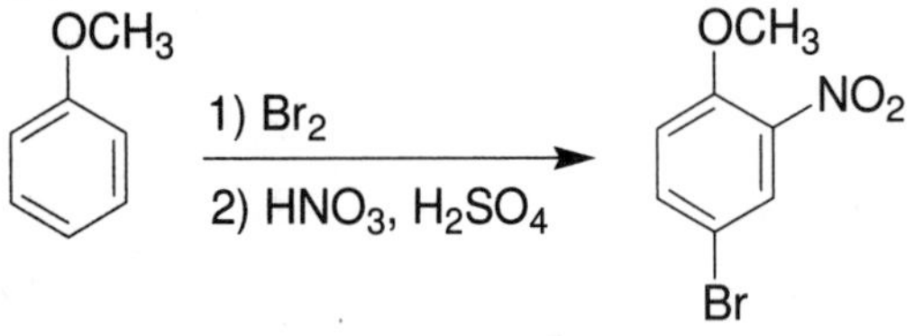

a) OCH_3 — 1) Br_2; 2) HNO_3, H_2SO_4 → OCH_3, NO_2, Br

b) OCH_3 — 1) HNO_3, H_2SO_4; 2) Br_2, $AlBr_3$ → OCH_3, Br, NO_2

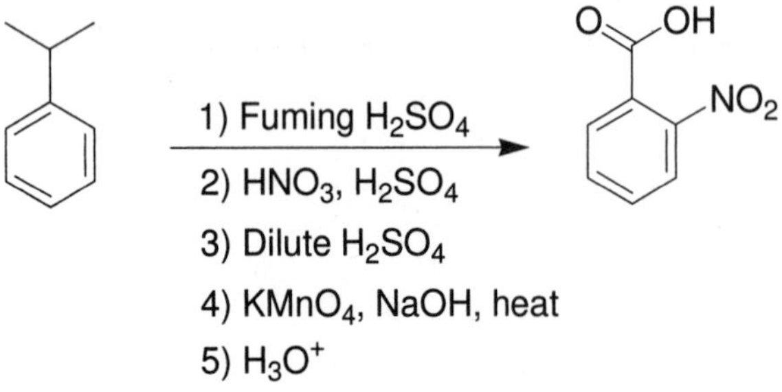

c) 1) Fuming H_2SO_4; 2) HNO_3, H_2SO_4; 3) Dilute H_2SO_4; 4) $KMnO_4$, NaOH, heat; 5) H_3O^+ → O, OH, NO_2

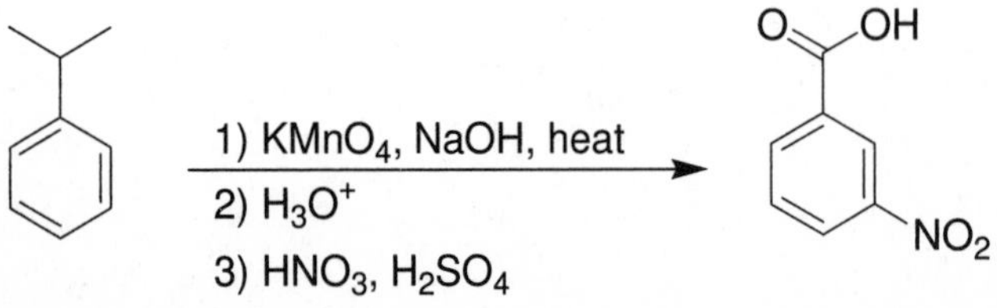

d) 1) $KMnO_4$, NaOH, heat; 2) H_3O^+; 3) HNO_3, H_2SO_4 → O, OH, NO_2

19.69.

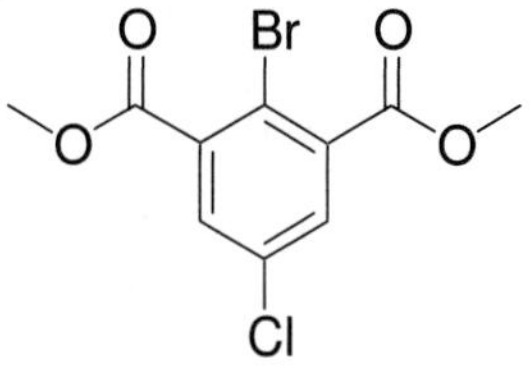

O Br O

O O

a) Cl

O_2N

OCH_3

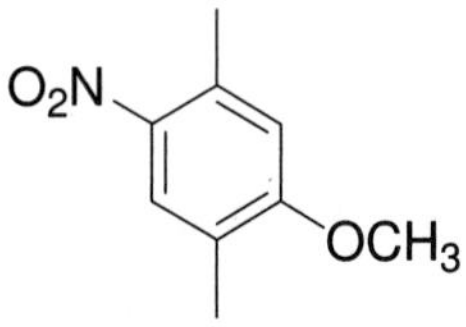

b)

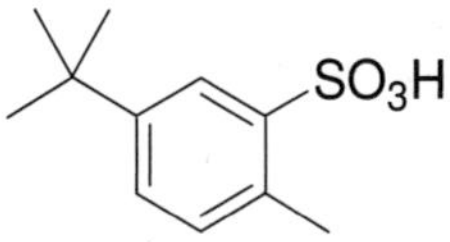

SO_3H

c)

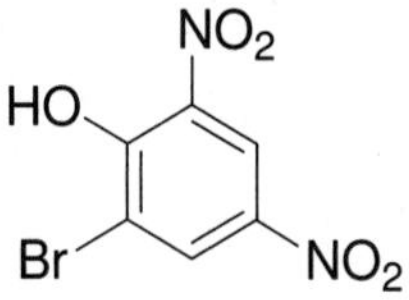

NO_2

HO

d) Br NO_2

19.70.

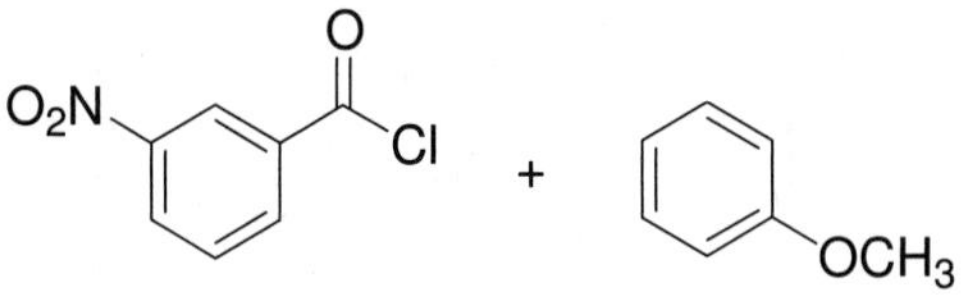

O

O_2N Cl

\+

OCH_3

a)

O

OCH_3

Cl

\+

H_3CO

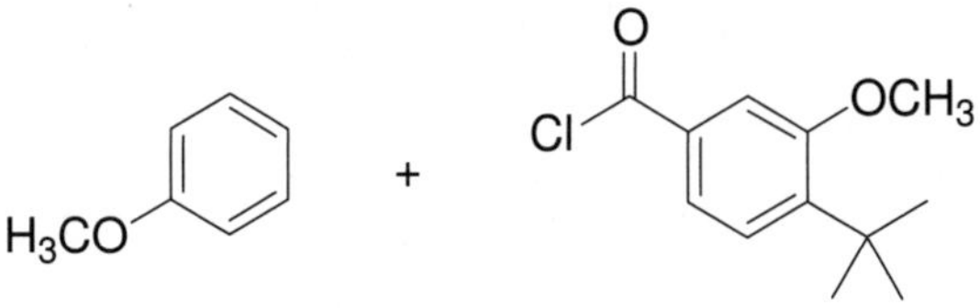

b)

19.71. Attack at the C2 position proceeds via an intermediate with three resonance structures:

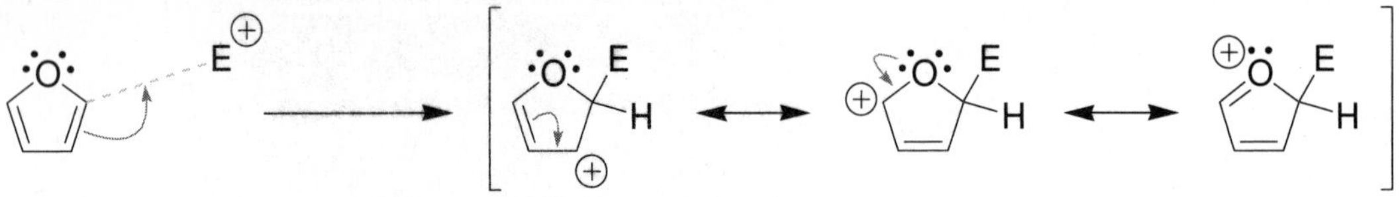

In contrast, attack at the C3 position proceeds via an intermediate with only two resonance structures:

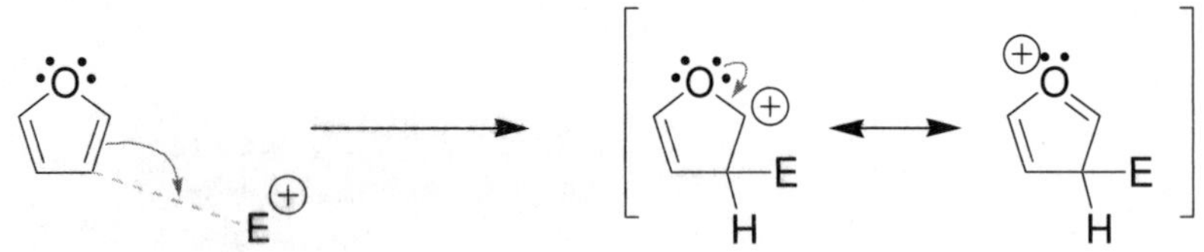

The intermediate for C2 attack is lower in energy than the intermediate for C3 attack. The transition state leading to the intermediate of C2 attack will therefore be lower in energy than the transition state leading to the intermediate of C3 attack. As a result, C2 attack occurs more rapidly.

19.72.

a)

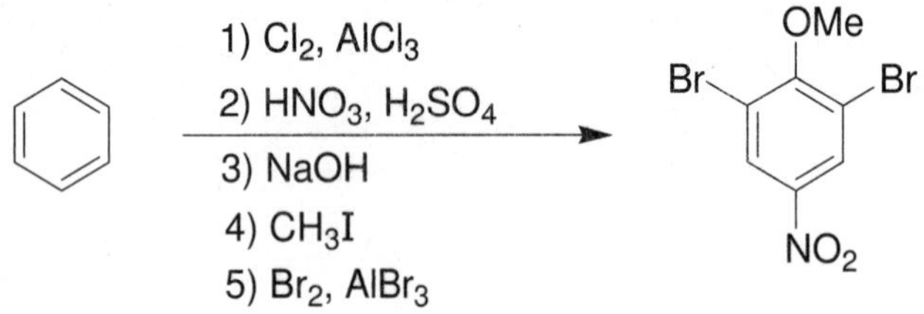
1) Cl_2, $AlCl_3$
2) HNO_3, H_2SO_4
3) NaOH
4) CH_3I
5) Br_2, $AlBr_3$
OMe
Br
Br
NO_2

b)

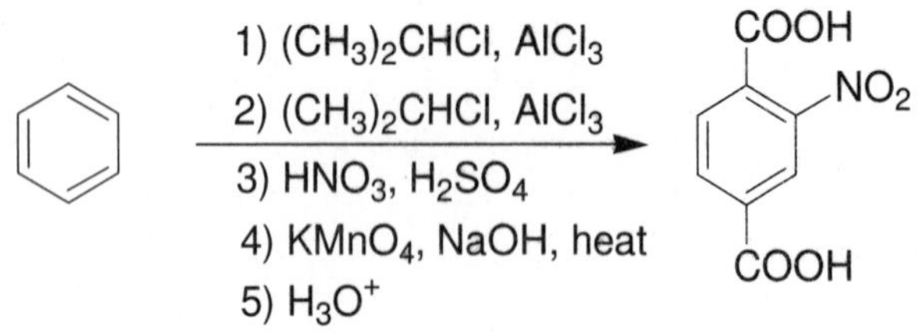
1) $(CH_3)_2CHCl$, $AlCl_3$
2) $(CH_3)_2CHCl$, $AlCl_3$
3) HNO_3, H_2SO_4
4) $KMnO_4$, NaOH, heat
5) H_3O^+
COOH
NO_2
COOH

c)

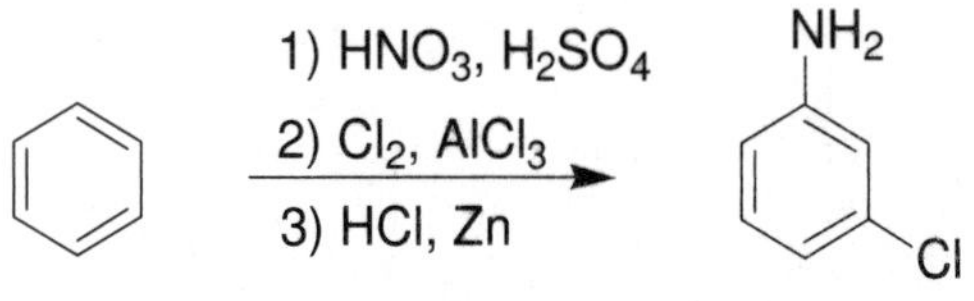
1) HNO_3, H_2SO_4
2) Cl_2, $AlCl_3$
3) HCl, Zn
NH_2
Cl

d)

1) $(CH_3)_3CCl$, $AlCl_3$
2) HNO_3, H_2SO_4
3) Br_2, $AlBr_3$
4) Cl_2, $AlCl_3$

Cl, Br, NO_2

e)

1) CH_3CH_2COCl, $AlCl_3$
2) HNO_3, H_2SO_4
3) Cl_2, $AlCl_3$

O, NO_2, Cl

f)

1) $(CH_3)_2CHCl$, $AlCl_3$
2) Br_2, $AlBr_3$
3) $KMnO_4$, NaOH, heat
4) H_3O^+
4) HNO_3, H_2SO_4
5) Cl_2, $AlCl_3$

COOH, Cl, NO_2, Br

g)

1) Cl_2, $AlCl_3$
2) HNO_3, H_2SO_4
3) NaOEt, heat
4) Cl_2, $AlCl_3$

OEt, Cl, NO_2

h)

1) $(CH_3)_2CHCl$, $AlCl_3$
2) CH_3CH_2COCl, $AlCl_3$
3) HNO_3, H_2SO_4
4) Cl_2, $AlCl_3$

Cl, NO_2, O

i)

1) Cl, Cl
$AlCl_3$
2) Br_2, $AlBr_3$
3) HNO_3, H_2SO_4

Br, O_2N

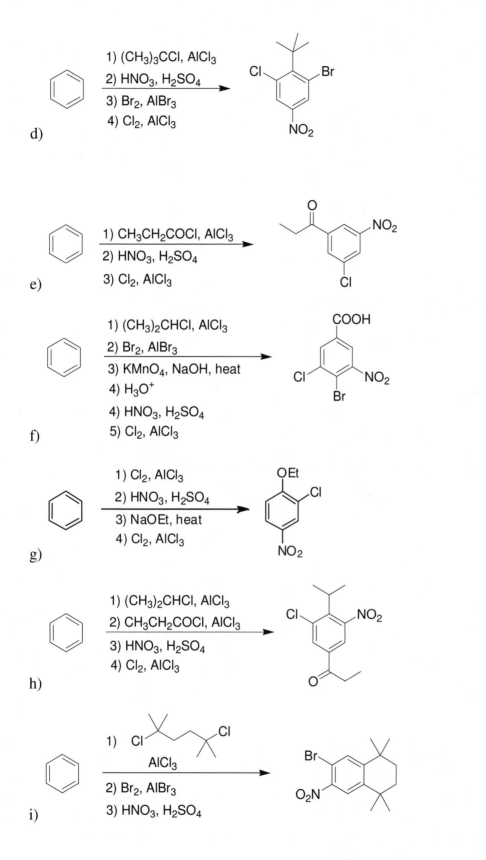

19.73.

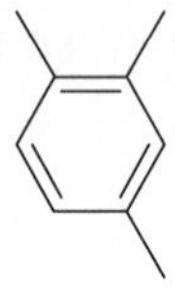

1,2,4-trimethylbenzene

19.74. Bromination at the para position occurs more rapidly because ortho attack is sterically hindered by the ethyl group:

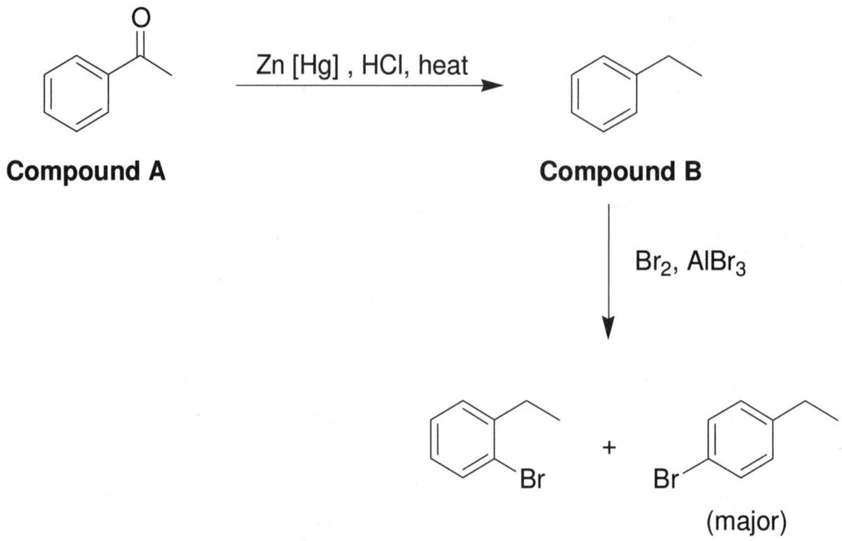

Zn [Hg] , HCl, heat

Compound A

Compound B

Br_2, $AlBr_3$

+

Br

Br

(major)

19.75.

a)

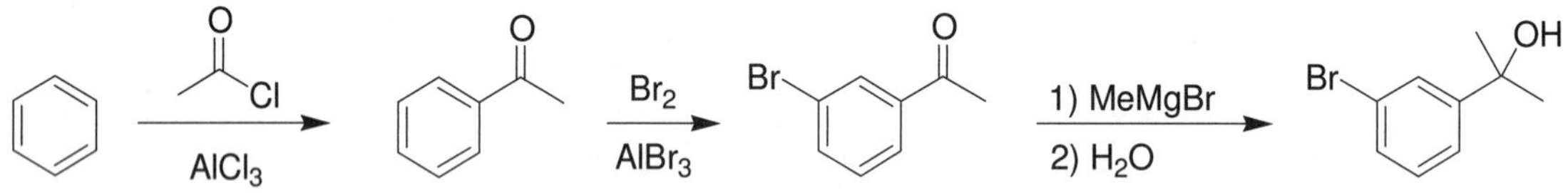

Cl

$AlCl_3$

Br_2

$AlBr_3$

Br

1) MeMgBr

2) H_2O

Br

OH

b)

1) $(CH_3)_2CHBr$, $AlBr_3$

2) Br_2, $AlBr_3$

Br

1) NBS, heat

2) NaOEt

Br

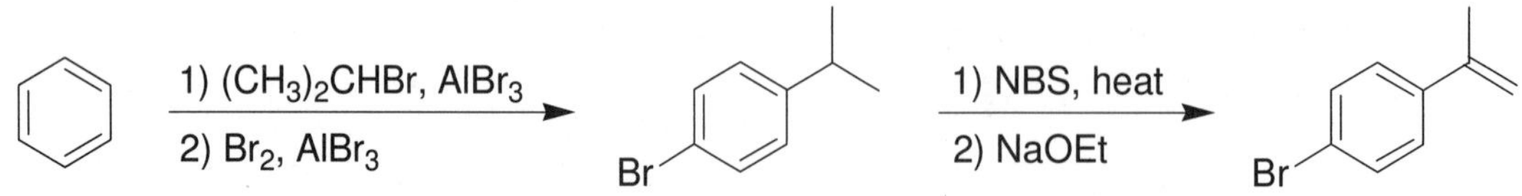

19.76.

+

a)

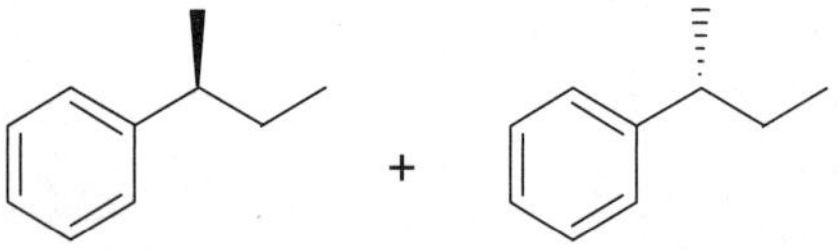

b) The reaction proceeds via a carbocation intermediate, which can be attacked from either face, leading to a racemic mixture.

19.77.

The OH group activates the ring toward electrophilic aromatic substitution because the OH group donates electron density via resonance.

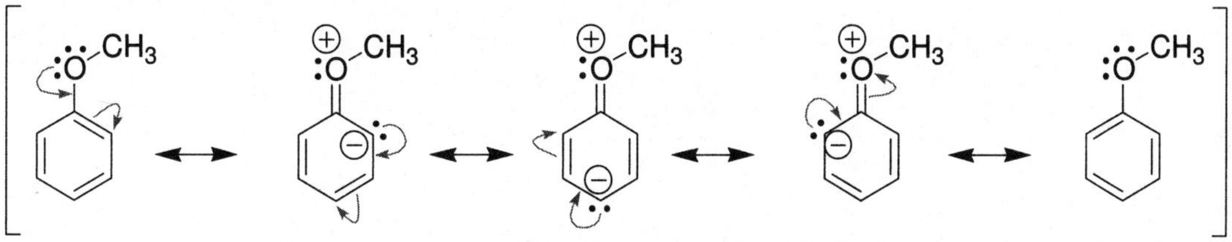

This effect gives electron density primarily to the *ortho* and *para* positions, as seen in the resonance structures above. These positions are shielded, and the protons at these positions are expected to produce signals farther upfield than protons at the *meta* position. According to this reasoning, the *meta* protons correspond with the signal at 7.2ppm.

19.78.

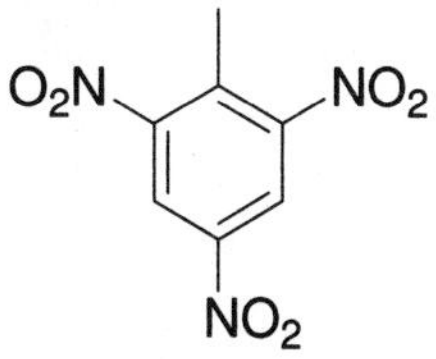

2,4,6-trinitrotoluene

19.79.

a) A phenyl group is an *ortho-para* director, because the sigma complex formed from *ortho* attack or *para* attack is highly stabilized by resonance (the positive charge is spread over both rings). The *ortho* position is sterically hindered while the *para* position is not, so we expect nitration to occur predominantly at the *para* position:

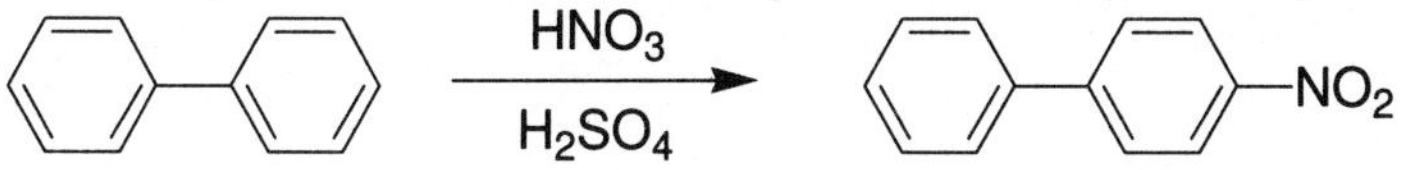

b) This group withdraws electron density form the ring via resonance (the resonance structures have a positive charge in the ring). As a result, this group is a moderate deactivator, and therefore a *meta*-director:

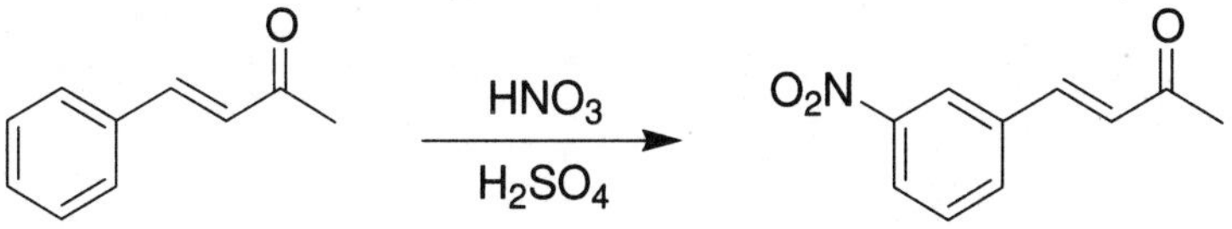

19.80.

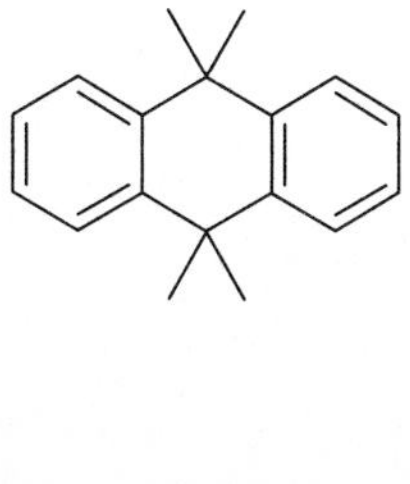

19.81.

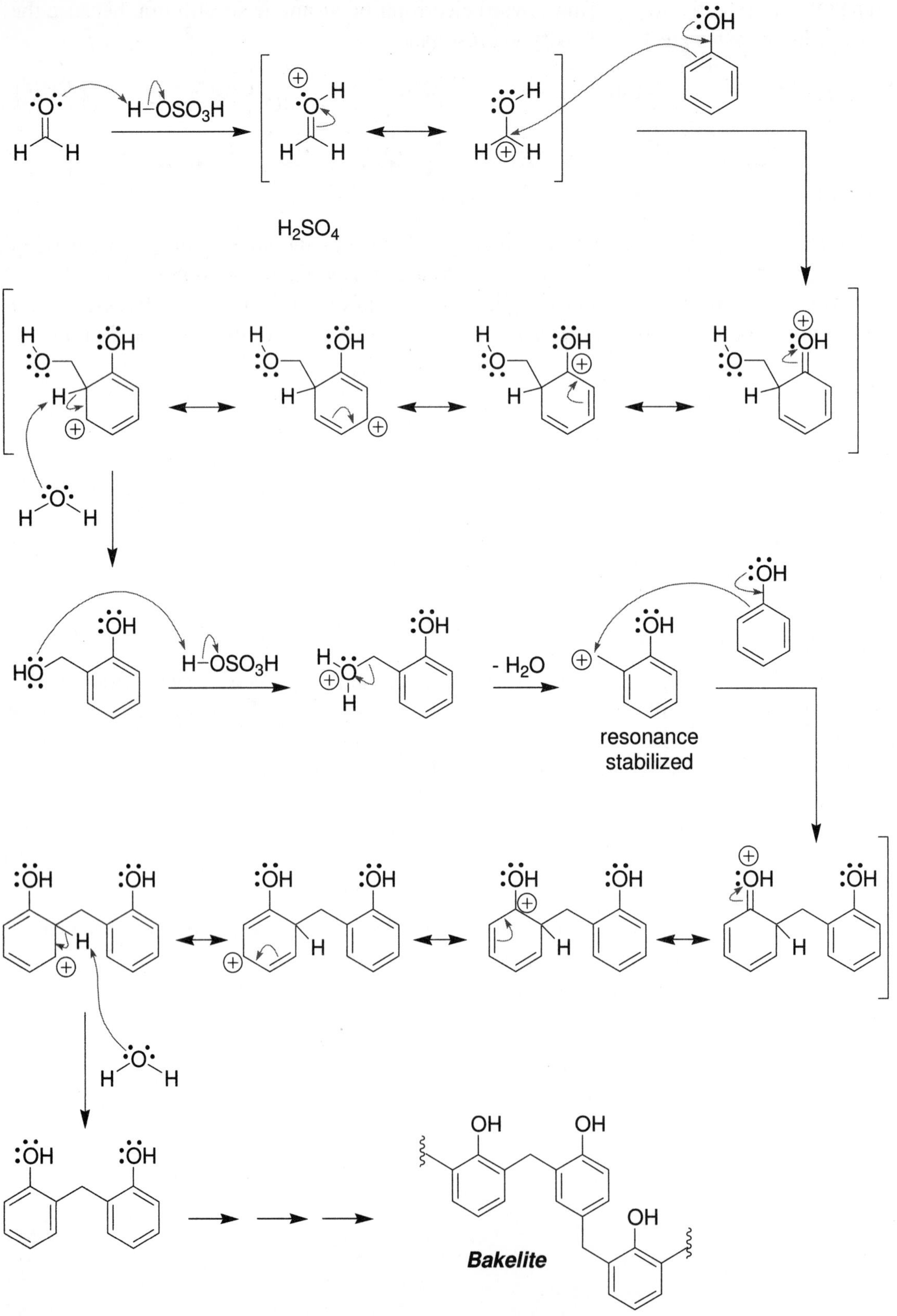

H_2SO_4

H–OSO$_3$H

resonance stabilized

- H_2O

Bakelite

19.82.

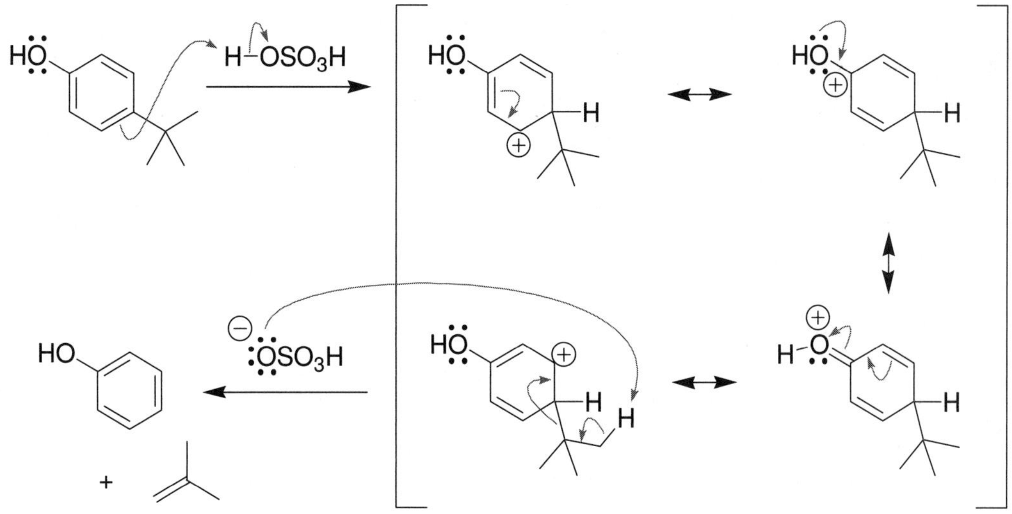

HO
H–OSO$_3$H
HO
–H
HO
–H
H–O
–H
HO
:OSO$_3$H
HO
H
H
HO
+

19.83. The amino group in *N,N*-dimethylaniline is a strong activator, and therefore, an *ortho-para* director. For this reason, bromination occurs at the *ortho* and *para* positions. However, in acidic conditions, the amino group is protonated to give an ammonium ion. Unlike the amino group, an ammonium ion is a strong deactivator and a *meta* director. Under these conditions, nitration occurs primarily at the *meta* position.

Chapter 20
Ketones and Aldehydes

Review of Concepts

Fill in the blanks below. To verify that your answers are correct, look in your textbook at the end of Chapter 20. Each of the sentences below appears verbatim in the section entitled *Review of Concepts and Vocabulary*.

- The suffix "_____" indicates an aldehydic group, and the suffix "_____" is used for ketones.
- The electrophilicity of a carbonyl group derives from _____________ effects, as well as __________ effects.
- A general mechanism for nucleophilic addition under basic conditions involves two steps
  1) nucleophilic attack to generate a _____________ **intermediate**.
  2) ______________________
- The position of equilibrium is dependent on the ability of the nucleophile to function as a _______________.
- In acidic conditions, an aldehyde or ketone will react with two molecules of alcohol to form an __________.
- The reversibility of acetal formation enables acetals to function as _______________ groups for ketones. Acetals are stable under strongly _______ conditions.
- In acidic conditions, an aldehyde or ketone will react with a primary amine to form an _________.
- In acidic conditions, an aldehyde or ketone will react with a secondary amine to form an __________.
- In the **Wolff-Kishner reduction**, a hydrazone is reduced to an _________ under strongly basic conditions.
- In acidic conditions, all reagents, intermediates, and leaving groups either should be ____________ or should bear one ___________ charge.
- _________________ of acetals, imines, and enamines under acidic conditions produces ketones or aldehydes.
- In acidic conditions, an aldehyde or ketone will react with two equivalents of a thiol to form a ______________.
- When treated with Raney nickel, thioacetals undergo **desulfurization** to yield a ______________ group.
- When treated with a hydride reducing agent, such as lithium aluminum hydride (LAH) or sodium borohydride ($NaBH_4$), aldehydes and ketones are reduced to _______________.
- The reduction of a carbonyl group with LAH or $NaBH_4$ is not a reversible process, because hydride does not function as a ____________________.
- When treated with a Grignard agent, aldehydes and ketones are converted into alcohols, accompanied by the formation of a new ___________ bond.

- Grignard reactions are not reversible, because carbanions do not function as ________________.
- When treated with hydrogen cyanide (HCN), aldehydes and ketones are converted into ____________. For most aldehydes and unhindered ketones, the equilibrium favors formation of the ______________.
- The **Wittig reaction** can be used to convert a ketone to an ________.
- A **Baeyer-Villiger oxidation** converts a ketone to an _________ by inserting __________________ next to the carbonyl group. Cyclic ketones produce cyclic esters called ____________.

Review of Skills

Fill in the blanks and empty boxes below. To verify that your answers are correct, look in your textbook at the end of Chapter 20. The answers appear in the section entitled *SkillBuilder Review.*

20.1: Naming Aldehydes and Ketones

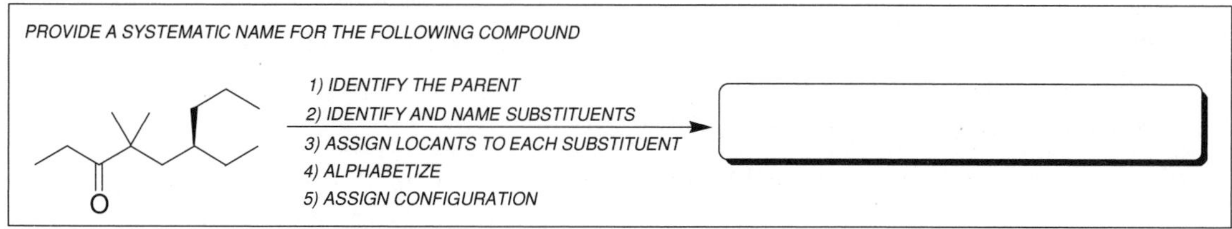

20.2: Drawing the Mechanism of Acetal Formation

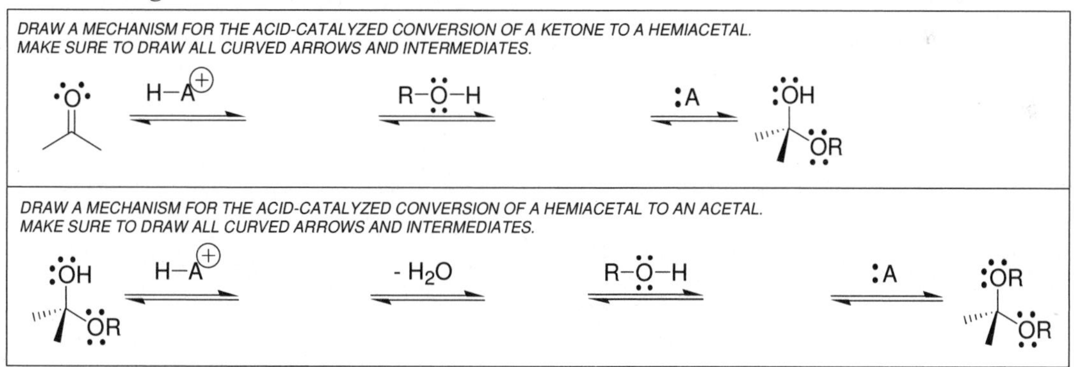

20.3: Drawing the Mechanism of Imine Formation

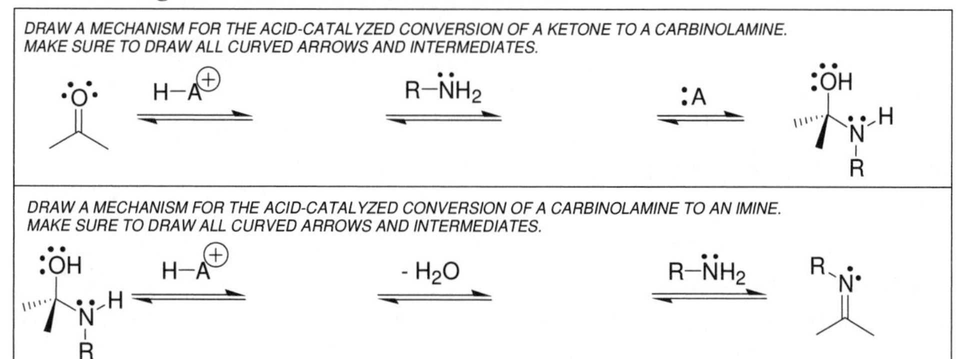

20.4: Drawing the Mechanism of Enamine Formation

DRAW A MECHANISM FOR THE ACID-CATALYZED CONVERSION OF A KETONE TO A CARBINOLAMINE. MAKE SURE TO DRAW ALL CURVED ARROWS AND INTERMEDIATES.

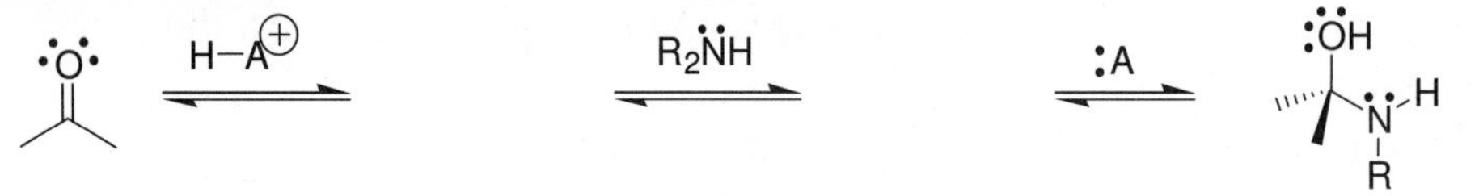

H–A⊕ R_2NH :A

OH N H R

DRAW A MECHANISM FOR THE ACID-CATALYZED CONVERSION OF A CARBINOLAMINE TO AN ENAMINE. MAKE SURE TO DRAW ALL CURVED ARROWS AND INTERMEDIATES.

OH N H R   H–A⊕   - H_2O R_2NH R N R

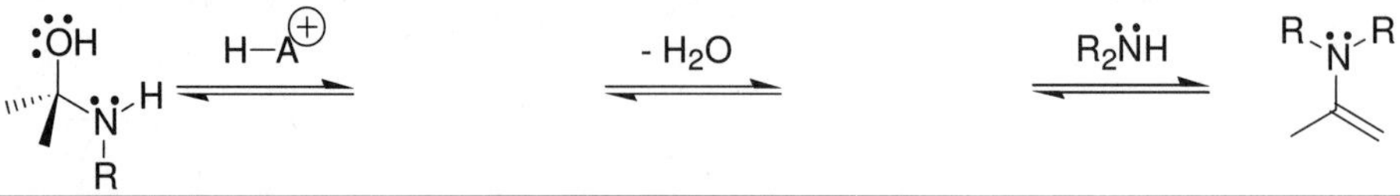

20.5: Drawing the Mechanism of a Hydrolysis Reaction

STEP 1 *- WORKING BACKWARDS, DRAW ALL __________________.*	***STEP 2*** *- DRAW ALL __________ AND _________________________________, USING THE FOLLOWING RULES:* *IN ACIDIC CONDITIONS, ALL REAGENTS, __________________, AND __________________ SHOULD EITHER BE NEUTRAL OR SHOULD BEAR ONE POSITIVE CHARGE.*

20.6: Planning an Alkene Synthesis with a Wittig Reaction

IDENTIFY THE REACTANTS YOU WOULD USE TO PREPARE THE FOLLOWING COMPOUND VIA A WITTIG REACTION:

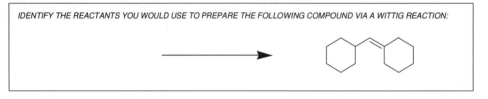

20.7: Proposing a Synthesis

BEGIN BY ASKING THE FOLLOWING TWO QUESTIONS: *1) IS THERE A CHANGE IN THE __________________?* *2) IS THERE A CHANGE IN THE __________________?*	*IF THERE IS A CHANGE IN THE CARBON SKELETON, CONSIDER ALL OF THE C-C BOND FORMING REACTIONS AND ALL OF THE C-C BOND BREAKING REACTIONS THAT YOU HAVE LEARNED SO FAR.* *C-C BOND-FORMING REACTIONS IN THIS CHAPTER:* - ____________ - __________________ - ____________ *C-C BOND-BREAKING REACTIONS IN THIS CHAPTER:* - __________________

Review of Reactions

Identify the reagents necessary to achieve each of the following transformations. To verify that your answers are correct, look in your textbook at the end of Chapter 20. The answers appear in the section entitled *Review of Reactions*.

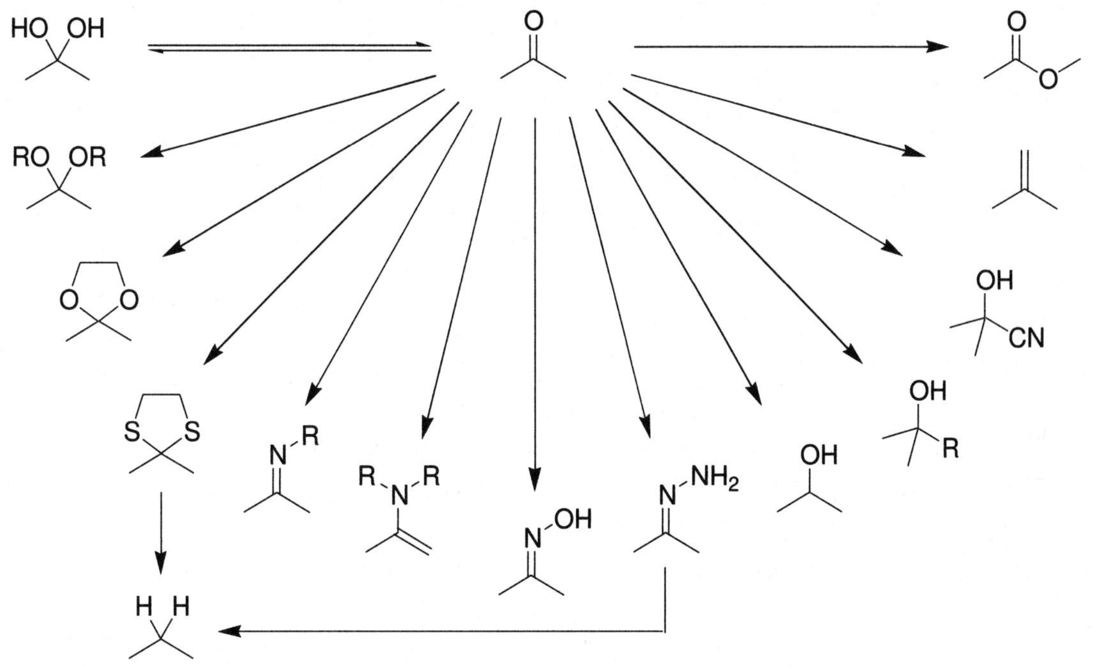

HO OH RO OR O O S S N R R N R N OH N NH2 OH OH R OH CN H H

Solutions

20.1. a) 5,5-dibromo-2,2-dimethylhexanal
b) (*3R,4S*)-3,4,5-trimethyl-2-hexanone
c) 2,2,5,5-tetramethylcyclopentanone
d) 2-propylpentanal
e) cyclobutanecarbaldehyde

20.2. a) O Br Br b) O c) Br O H

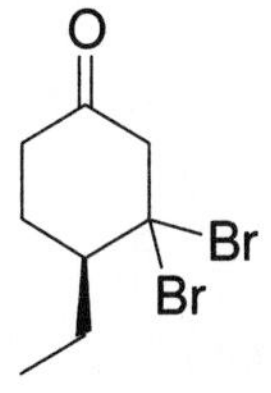

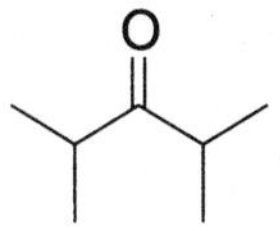

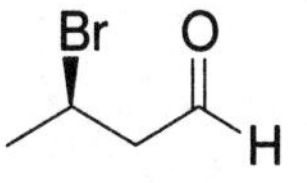

20.3. (*1S,4R*)bicyclo[2.2.1]heptan-2-one

20.4. a) 1,3-cyclohexanedione b) 1,4-cyclohexanedione c) 2,5,8-nonanetrione

20.5.

a) 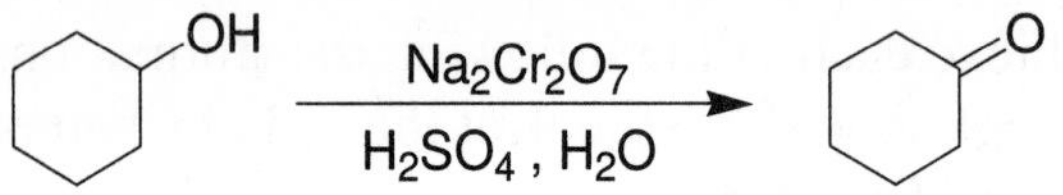 $Na_2Cr_2O_7$ / H_2SO_4, H_2O

b) 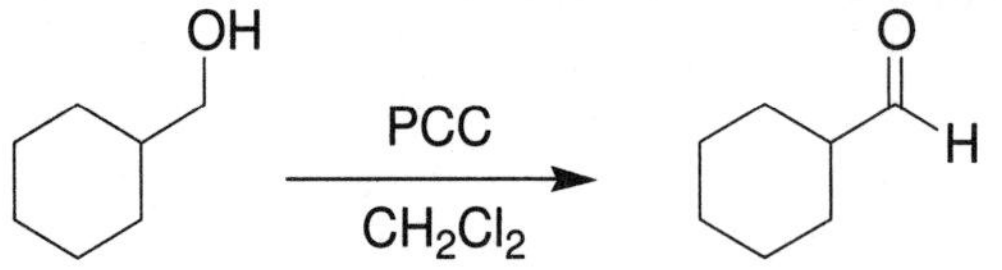 PCC / CH_2Cl_2

c) 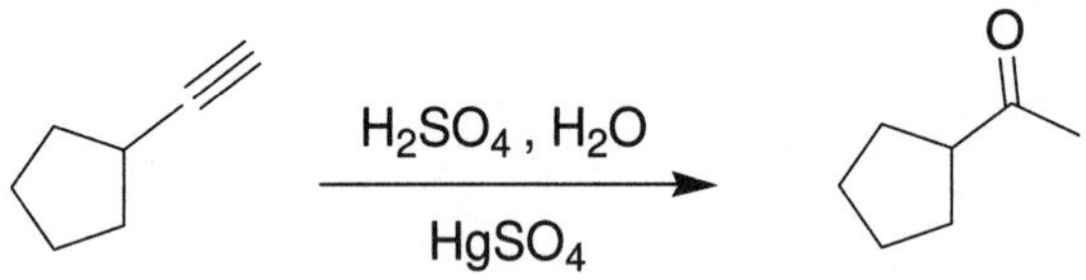 H_2SO_4, H_2O / $HgSO_4$

d) 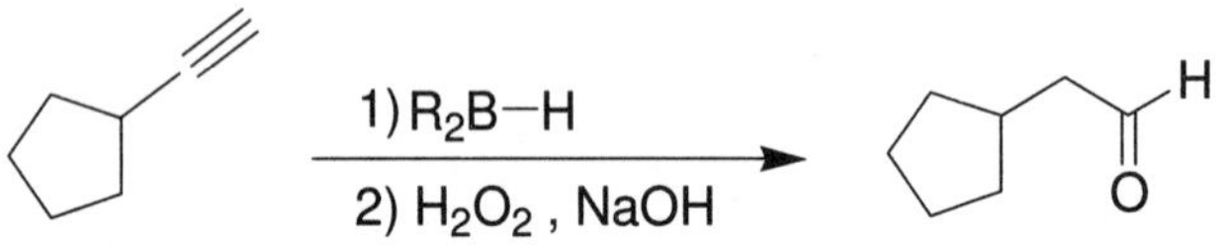 1) $R_2B{-}H$ / 2) H_2O_2, NaOH

e) 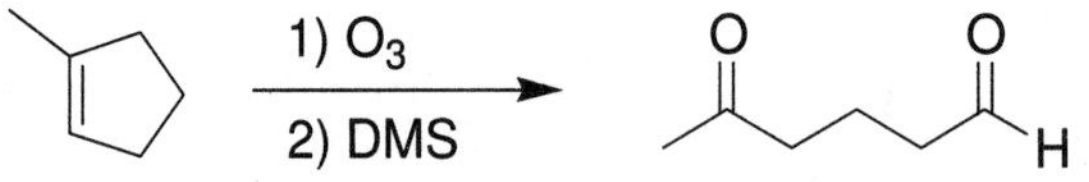 1) O_3 / 2) DMS

f) 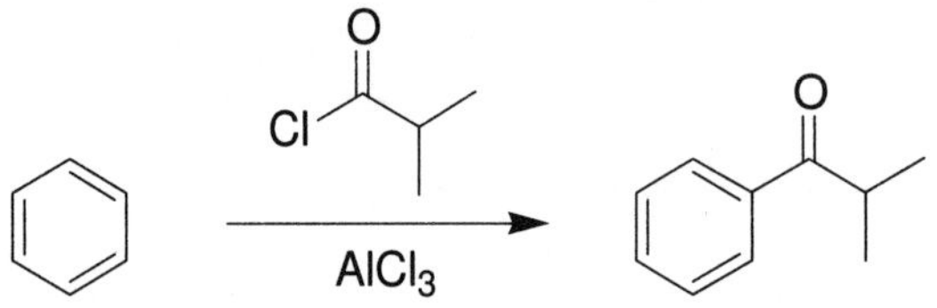 Cl / O / $AlCl_3$

20.6.

a) 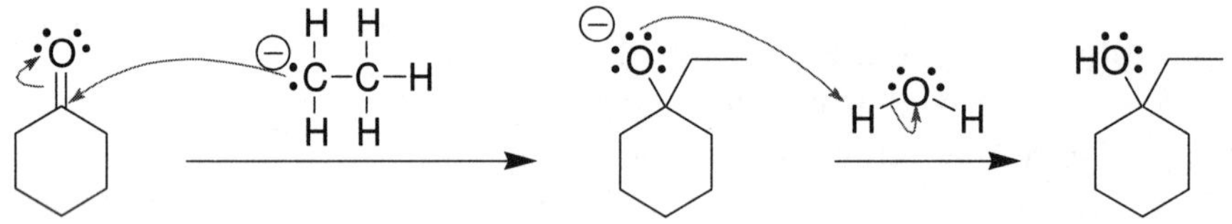 H H / C–C–H / H H / H–O–H / HO

b) 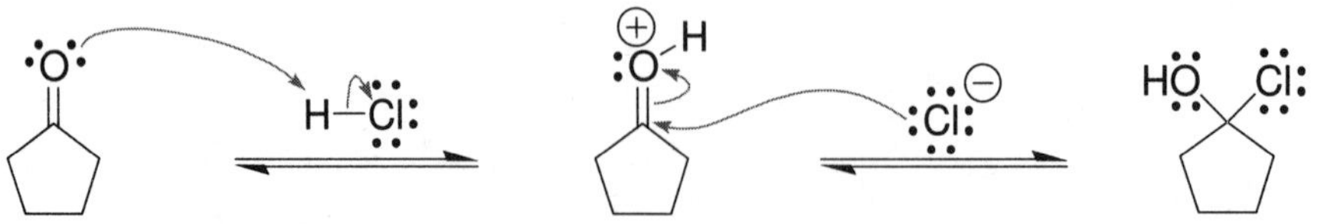 H–Cl / H / Cl / HO Cl

20.7. The carbonyl group in hexafluoroacetone is flanked by two very powerful electron-withdrawing groups (CF_3). These groups withdraw electron density from the carbonyl group, thereby increasing the electrophilicity of the carbonyl group. The resulting increase in energy of the reactant causes the equilibrium to favor the product (the hydrate).

20.8.

a)

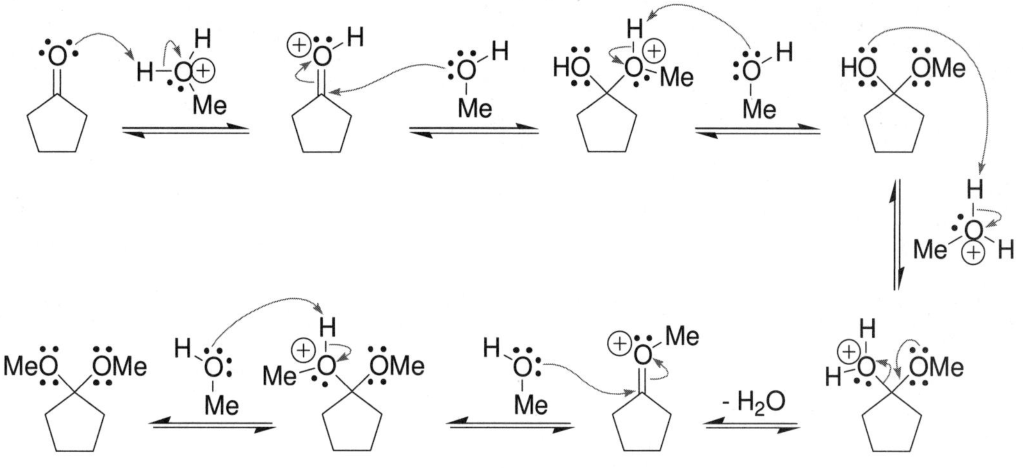

b)

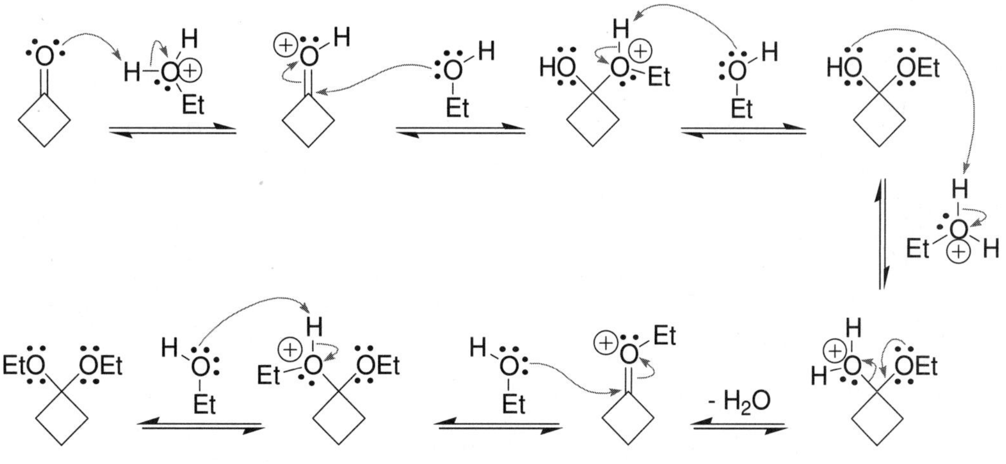

c)

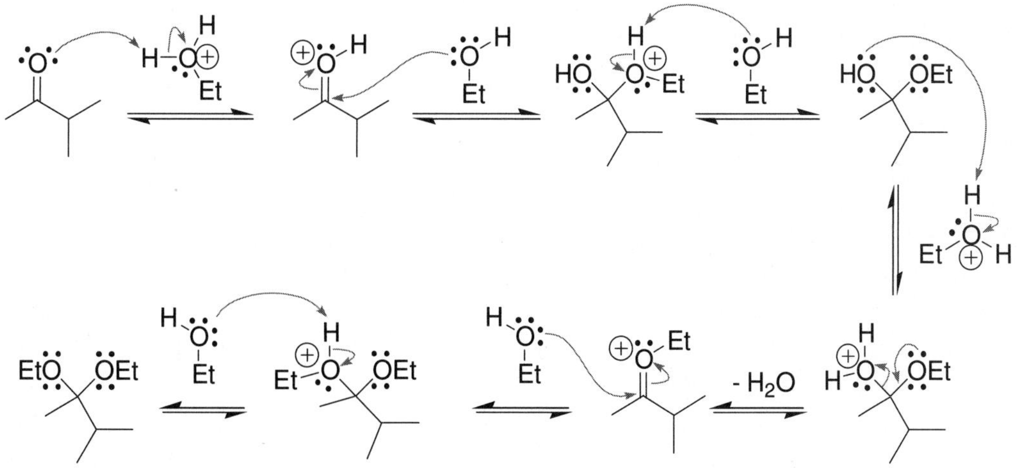

d)

HO OMe MeO OMe H-O Me Me - H2O

20.9.

a)

HO OH H-OSO3H HO3SO-H OSO3H - H2O

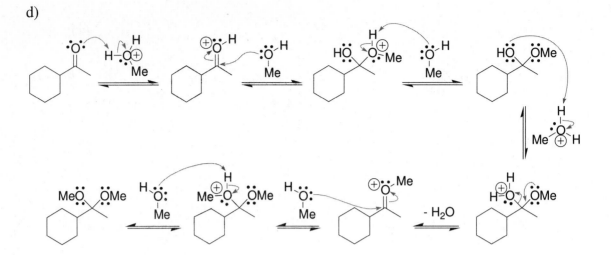

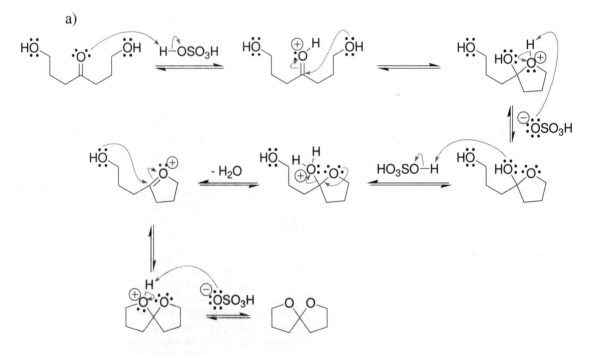

b)

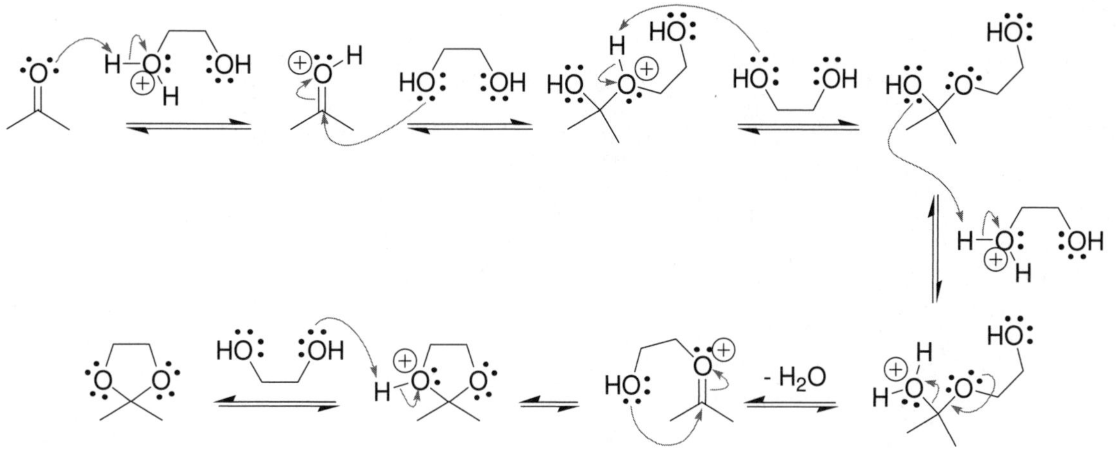

HO: :OH

- H_2O

20.10.

a) [H_2SO_4] excess MeOH, - H_2O

MeO OMe

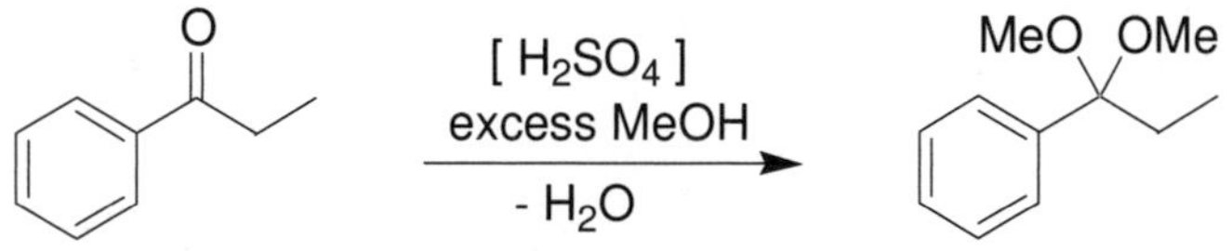

b) HO OH [H_2SO_4] - H_2O

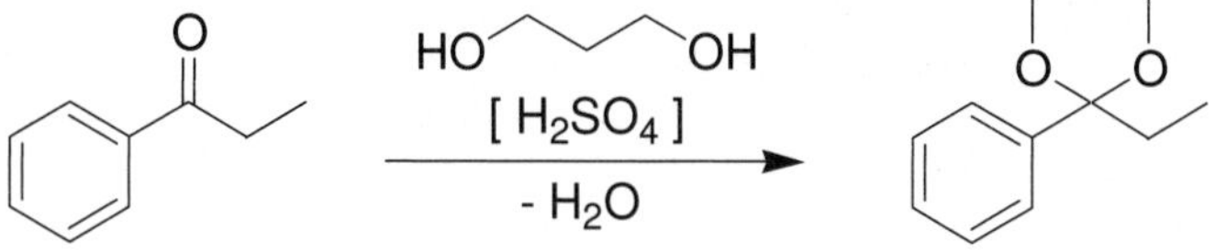

20.11.

a) 1) [H^+], HO OH , - H_2O 2) $NaNH_2$ 3) EtI 4) H_3O^+

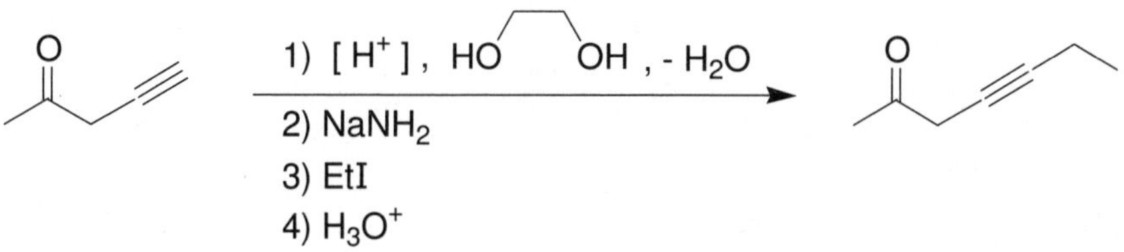

b) 1) [H^+], HO OH , - H_2O 2) LAH 3) H_3O^+

OH

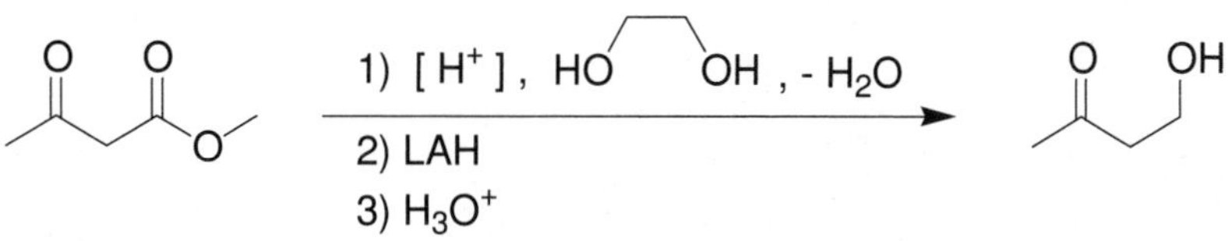

c) 1) [H^+], HO OH , - H_2O 2) PhMgBr (2 equivalents) 3) H_3O^+

OH Ph Ph

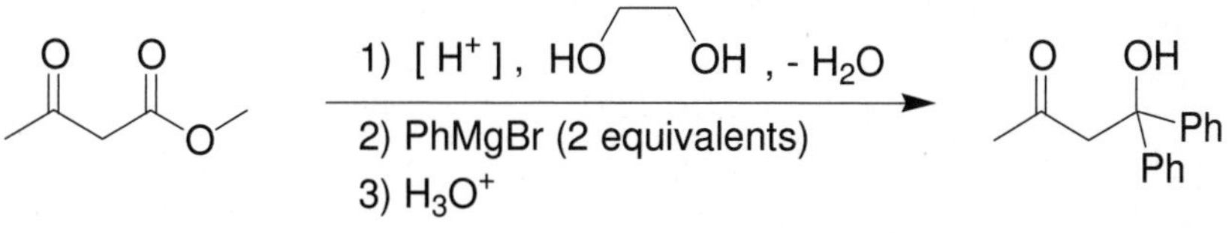

20.12. a) 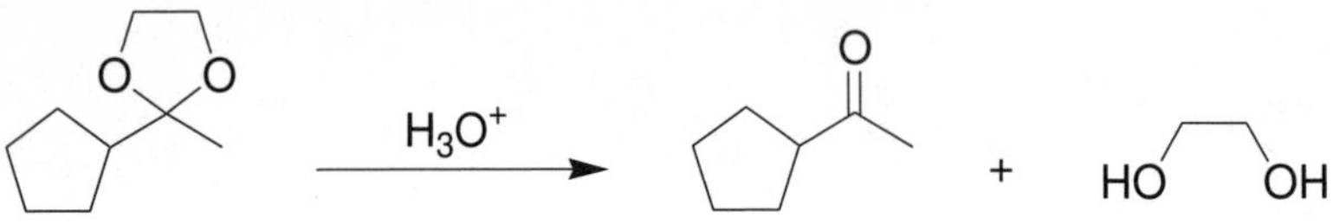 H_3O^+ + HO OH

b) 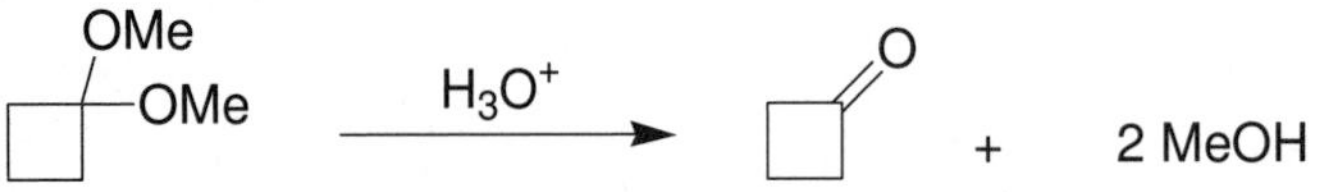 OMe OMe H_3O^+ + 2 MeOH

c) 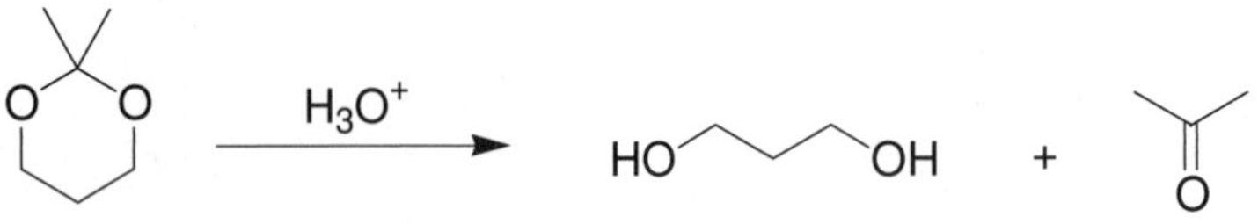 O O H_3O^+ HO OH + O

d) 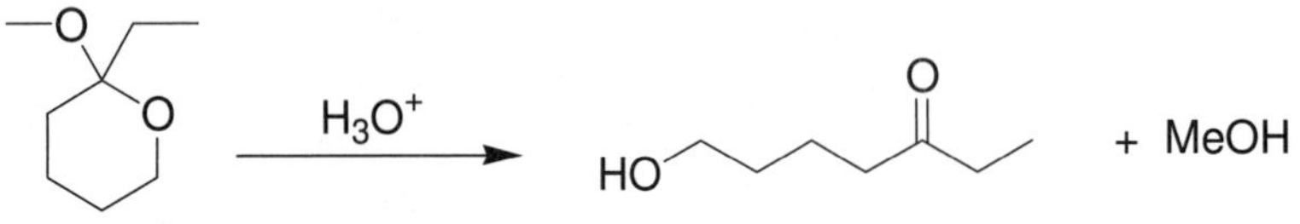 O O H_3O^+ O HO + MeOH

20.13.

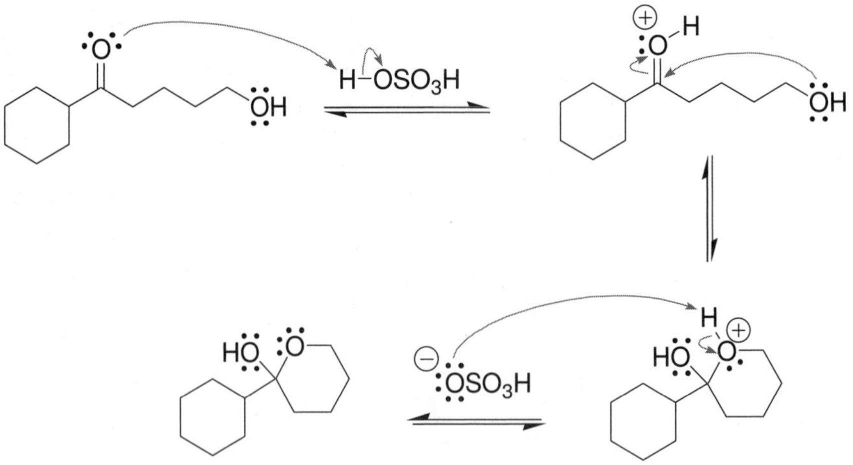

O OH H–OSO$_3$H ⊕ O–H OH

HO O ⊖ OSO$_3$H H HO ⊕ O

20.14. O OH

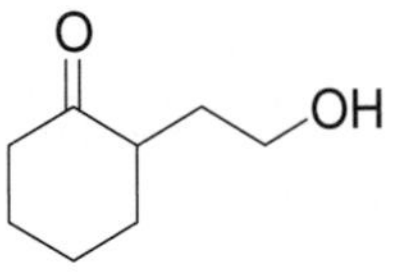

20.15. Note: ***For each of the mechanisms shown below, the first two steps can be reversed (first the amine attacks the carbonyl group, and then the tetrahedral intermediate is protonated). It would be wise to check your lecture notes to determine if you instructor has a strong preference for this alternate sequence of steps.***

a)

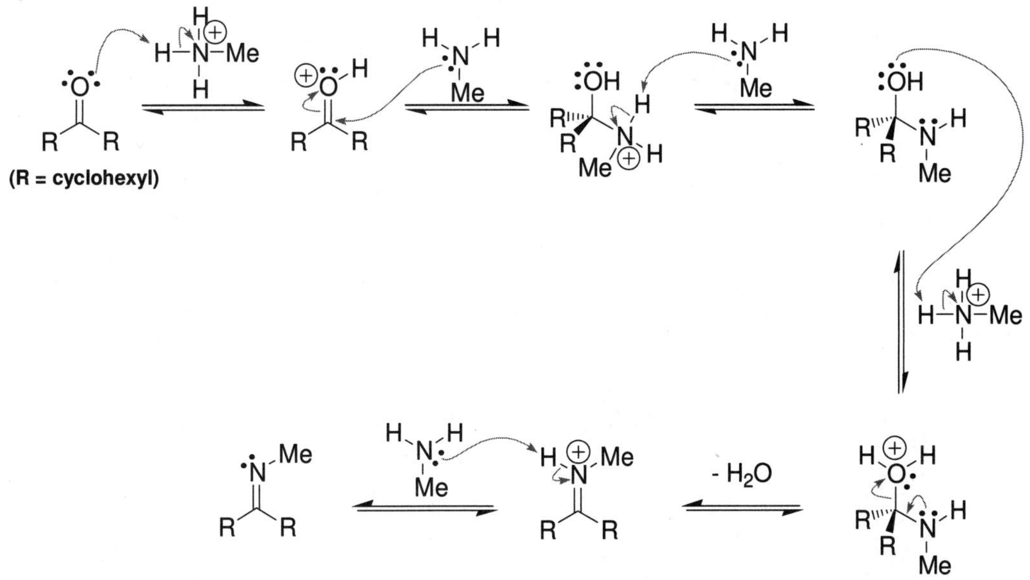

(R = cyclohexyl)

- H_2O

b)

- H_2O

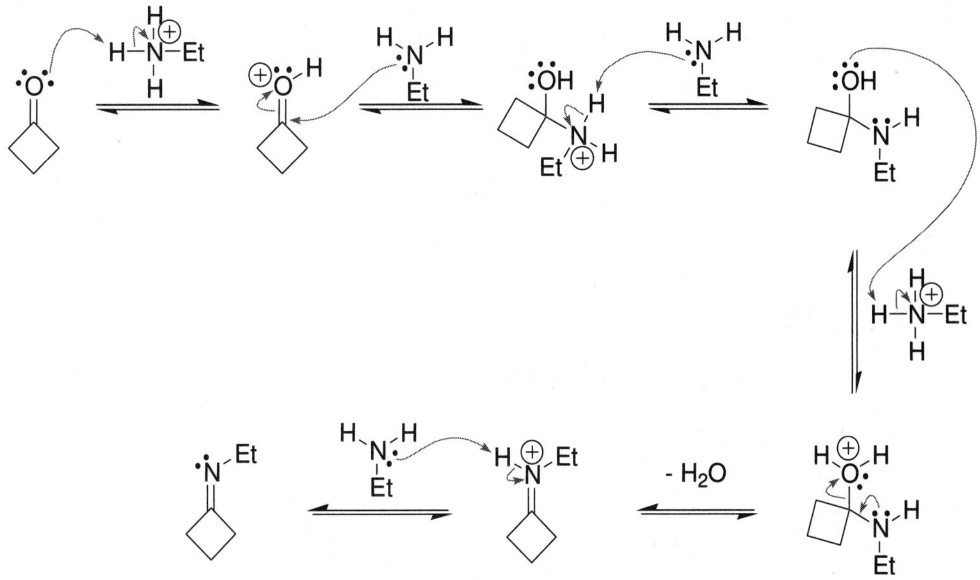

20.16.

a) Cycloheptanone $\xrightarrow[\text{- } H_2O]{[H^+],\ NH_3}$ cycloheptanimine (C=N–H)

b) Cyclopentanone $\xrightarrow[\text{[}H^+\text{]},\ \text{- } H_2O]{\text{cyclopentyl}-NH_2}$ N-cyclopentylcyclopentanimine

20.17.

a) $H-C(=O)-CH_2CH_2CH_2-NH_2 \xrightarrow[\text{- } H_2O]{[H^+]}$ 1-pyrroline

b) $CH_3C(=O)CH_2C(CH_3)_2CH_2-NH_2 \xrightarrow[\text{- } H_2O]{[H^+]}$ 2,4,4-trimethyl-1-pyrroline

20.18.

a) cyclohexyl–NH_2 + cyclopentanone $\xrightarrow[\text{- } H_2O]{[H_2SO_4]}$ N-cyclohexylcyclopentanimine

b) cyclohexanone + H_2N–cyclopentyl $\xrightarrow[\text{- } H_2O]{[H_2SO_4]}$ N-cyclopentylcyclohexanimine

c) 2-(3-aminopropyl)cyclohexanone (NH_2) $\xrightarrow[\text{- } H_2O]{[H_2SO_4]}$ bicyclic imine (octahydroquinoline, C=N)

20.19.

a) 3-pentanone $\xrightarrow[\text{- } H_2O]{[H^+],\ HO-NH_2}$ oxime (C=N–OH)

b) cyclopentanone $\xrightarrow[\text{- } H_2O]{[H^+],\ H_2N-NH_2}$ hydrazone (C=N–NH_2)

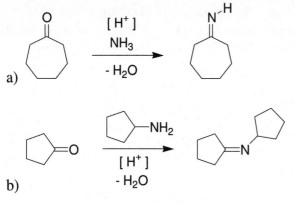

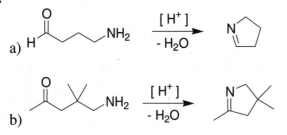

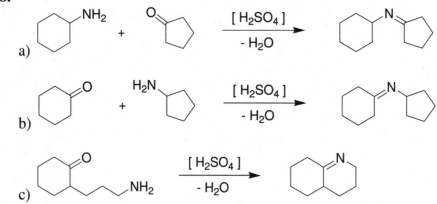

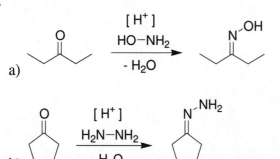

20.20.

a) 4,4-dimethylcyclohexanone → (HO–NH_2, [TsOH], - H_2O) → oxime (=N–OH)

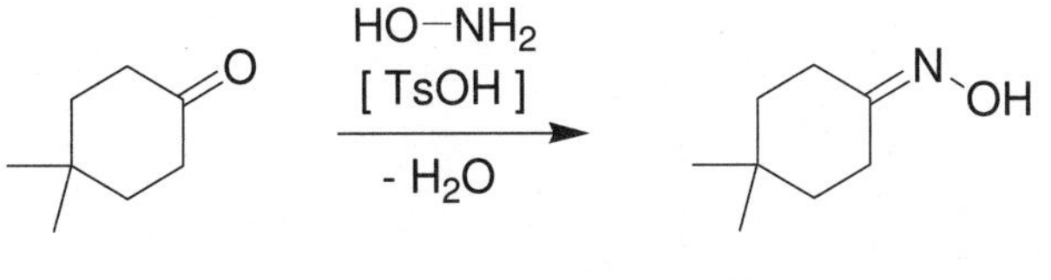

b) cyclopentanecarbaldehyde → (H_2N–NH_2, [H_2SO_4], - H_2O) → hydrazone (=N–NH_2)

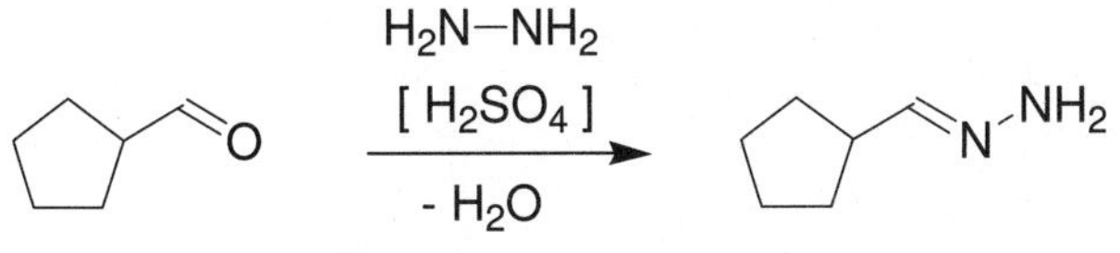

20.21. Note: ***For each of the mechanisms shown below, the first two steps can be reversed (first the amine attacks the carbonyl group, and then the tetrahedral intermediate is protonated). It would be wise to check your lecture notes to determine if you instructor has a strong preference for this alternate sequence of steps.***

a)

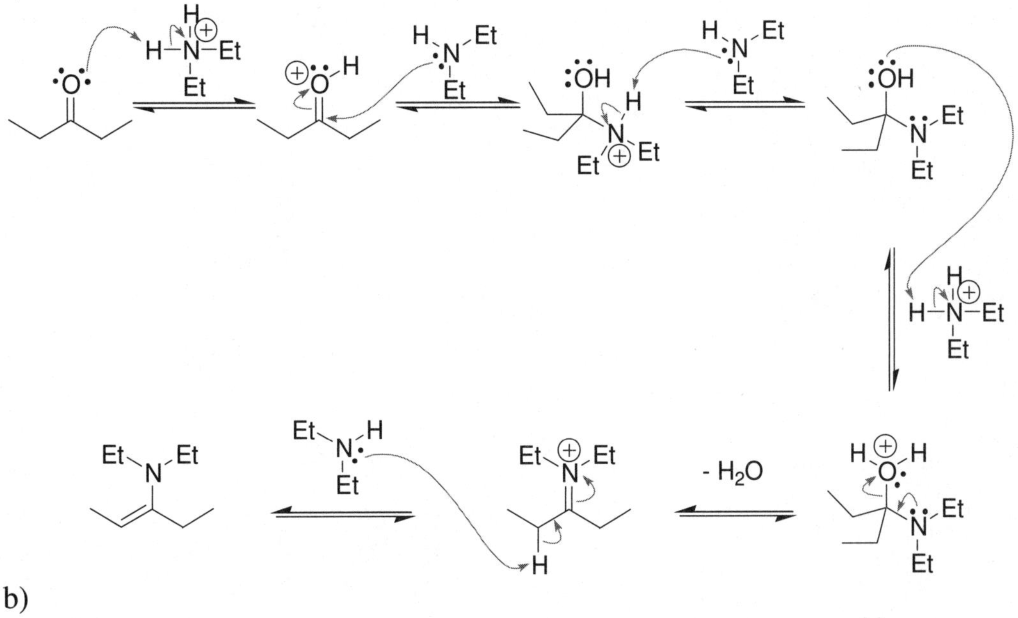

Et–N⁺H₂–Et ⇌ Et–NH–Et ⇌ Et–NH–Et ⇌ $H_2N^+Et_2$ ⇌ - H_2O ⇌ Et–NH–Et

b)

$H_2N^+Me_2$ ⇌ HN(Me)Me ⇌ HN(Me)Me ⇌ $H_2N^+Me_2$ ⇌ - H_2O ⇌ Me–NH–Me

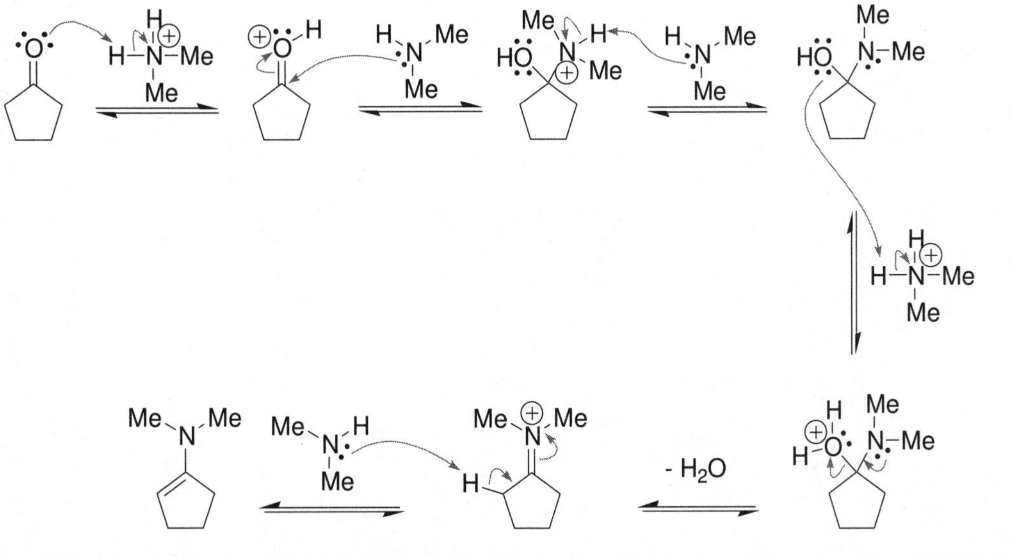

20.22.

a) [H^+] N–H, $- H_2O$

b) [H^+] $- H_2O$

20.23.

a) NH, O, [H^+] $- H_2O$, N

b) O, H, N, [H^+] $- H_2O$, N

20.24. a) O + H–N, [H_2SO_4] $- H_2O$, N

b) N–H + O, [H_2SO_4] $- H_2O$, N

c) O, H, N, [H_2SO_4] $- H_2O$, N

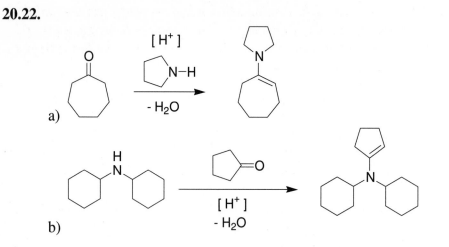

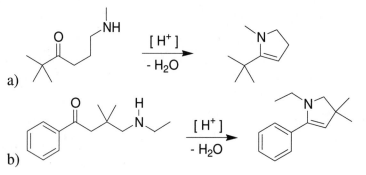

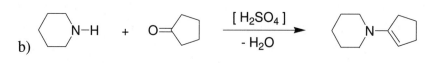

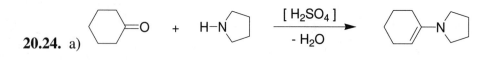

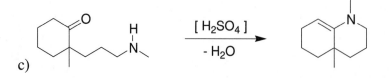

20.25. **Note**: ***The first two steps of this mechanism can be reversed (first the amine attacks the carbonyl group, and then the tetrahedral intermediate is protonated). It would be wise to check your lecture notes to determine if you instructor has a strong preference for this alternate sequence of steps.***

- H_2O

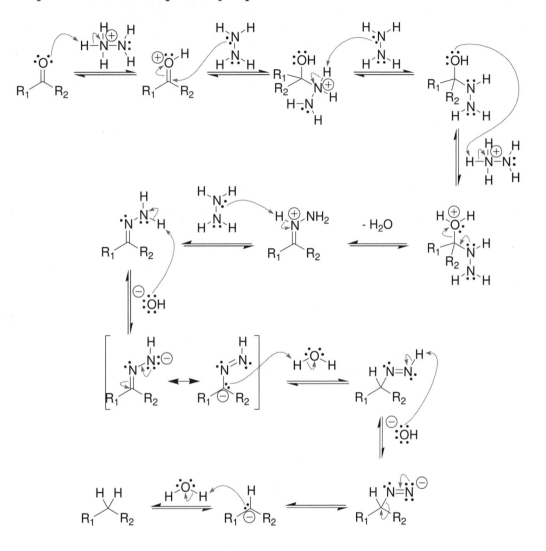

20.26.

a)

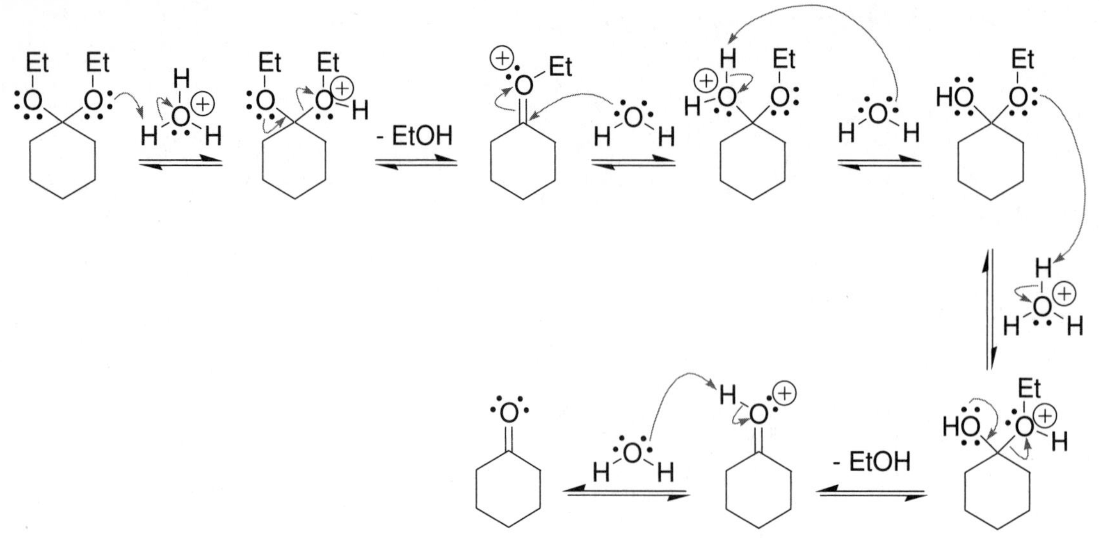

- EtOH

- EtOH

b)

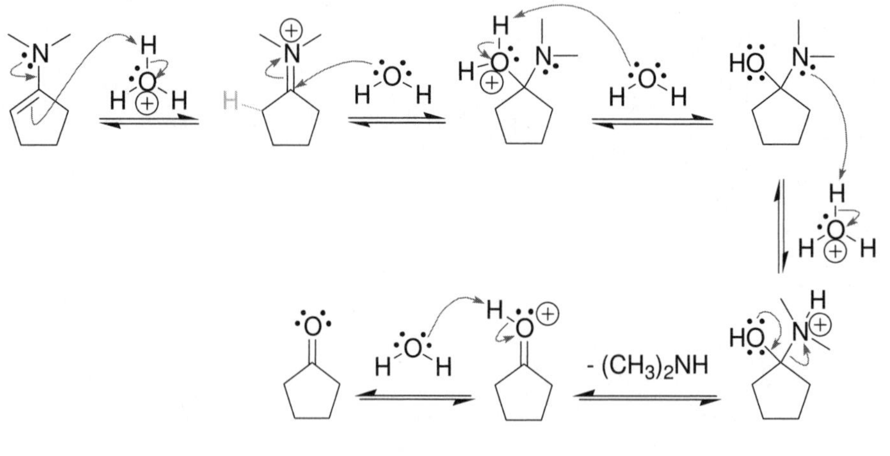

- $(CH_3)_2NH$

c)

- CH_3NH_2

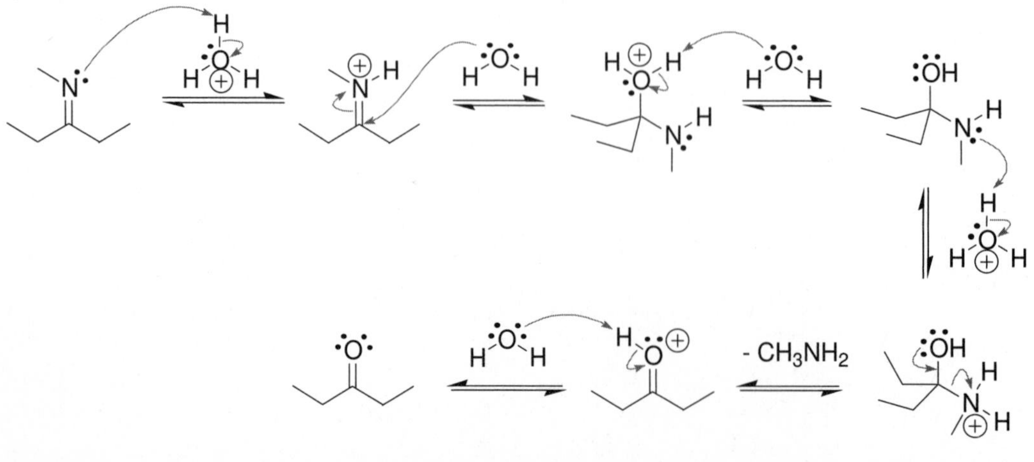

d)

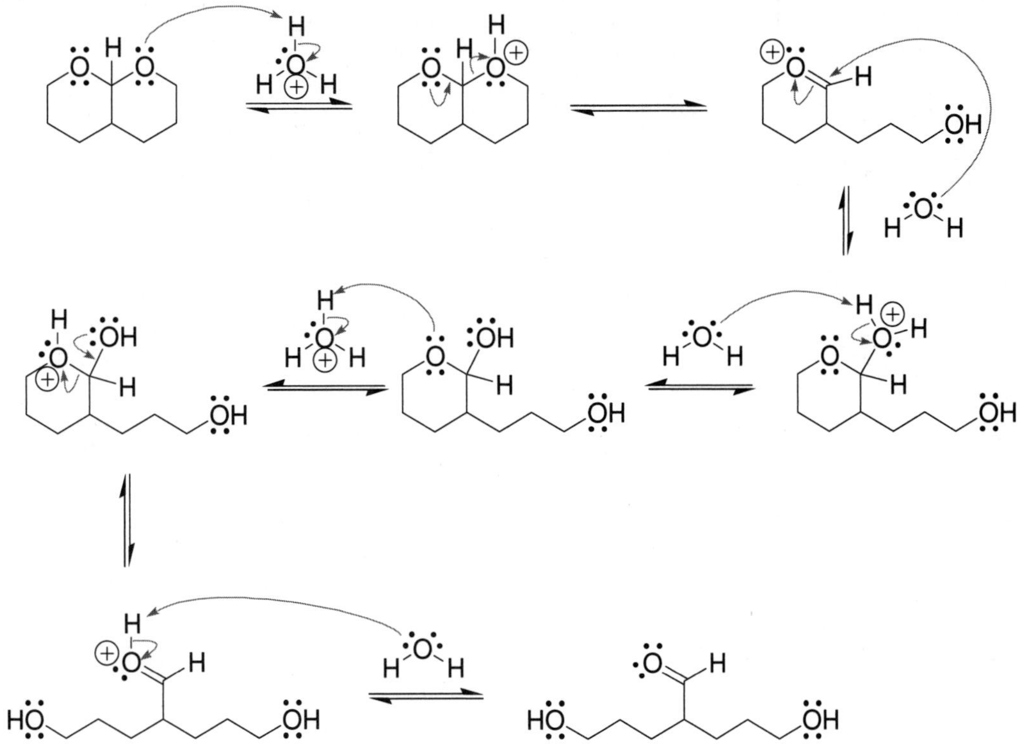

20.27.

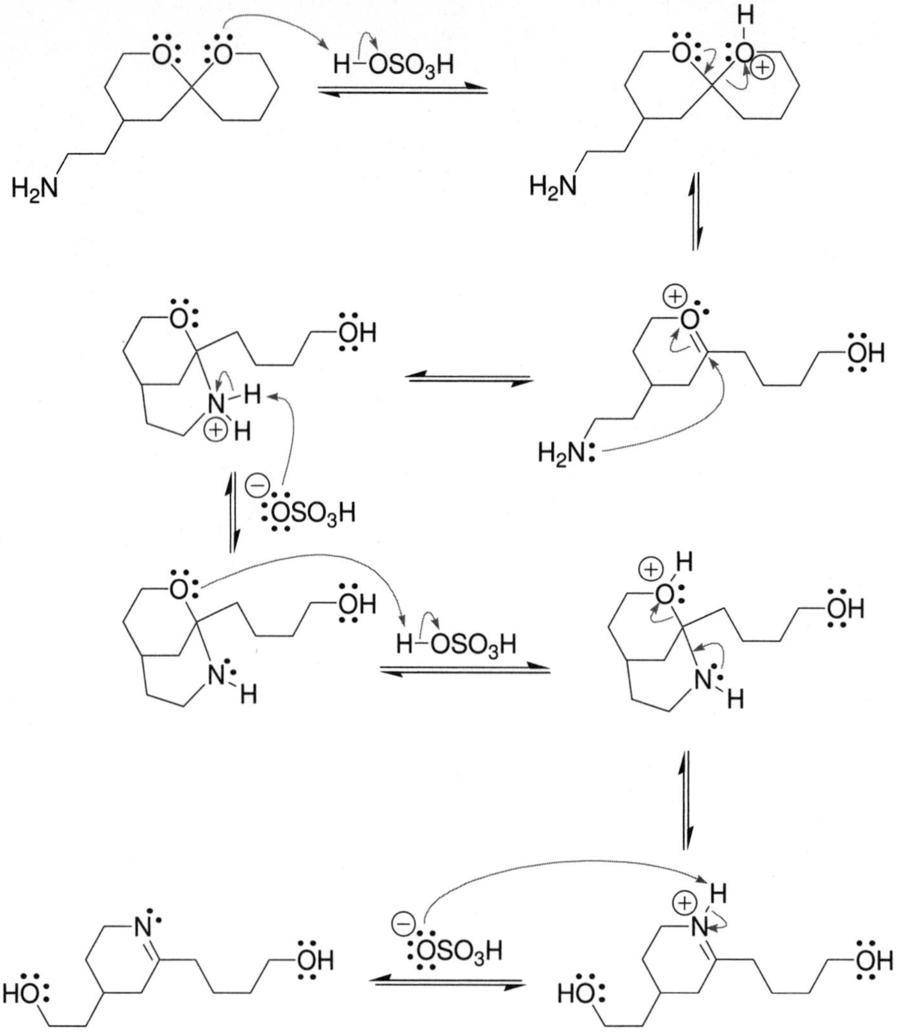

20.28.

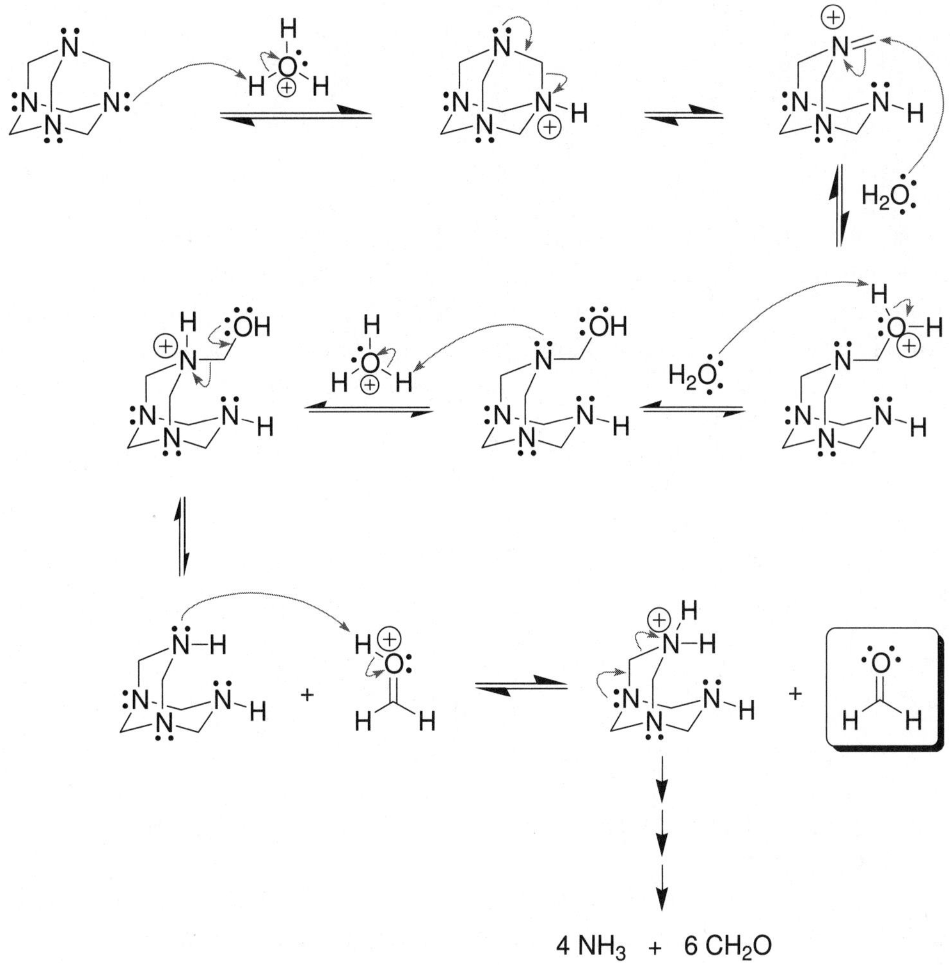

4 NH_3 + 6 CH_2O

20.29. a) 1) [H^+], HS SH 2) Raney Ni

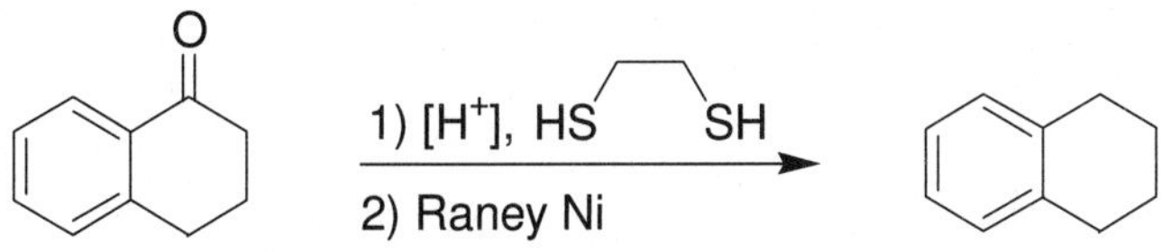

b) 1) [H^+], HS SH 2) Raney Ni

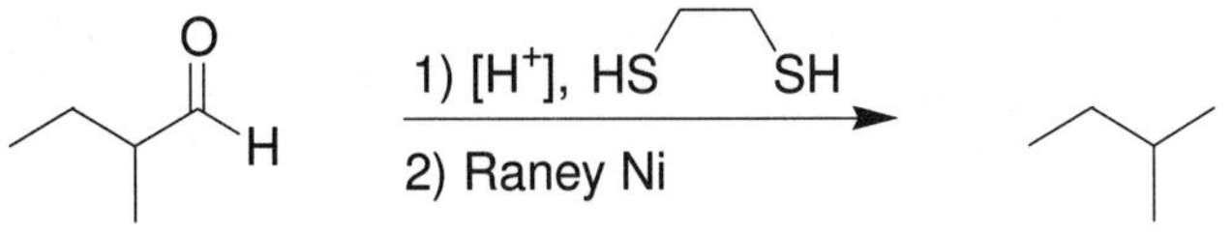

20.30.

20.31.

a) 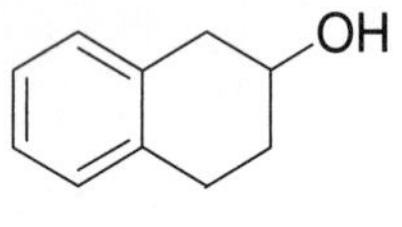 b) 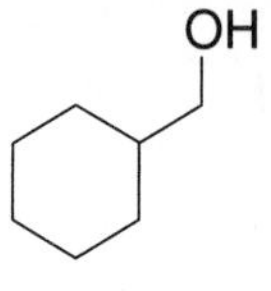 c) 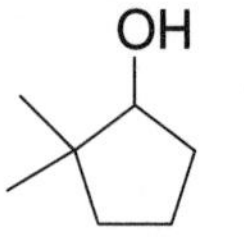 d)

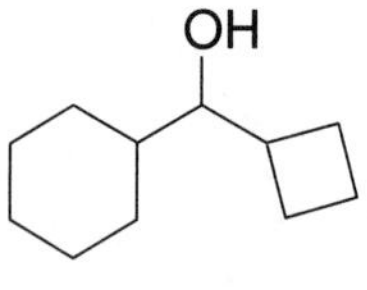

OH OH OH OH

20.32.

a) Below is a mechanism for the Cannizzaro reaction. After a hydroxide ion attacks one molecule of benzaldehyde, the resulting tetrahedral intermediate functions as a hydride delivery agent to attack another molecule of benzaldehyde, giving a carboxylic acid and an alkoxide ion. The alkoxide ion then deprotonates the carboxylic acid, generating a more stable carboxylate ion. This carboxylate ion is then protonated when an acid is added to the reaction mixture.

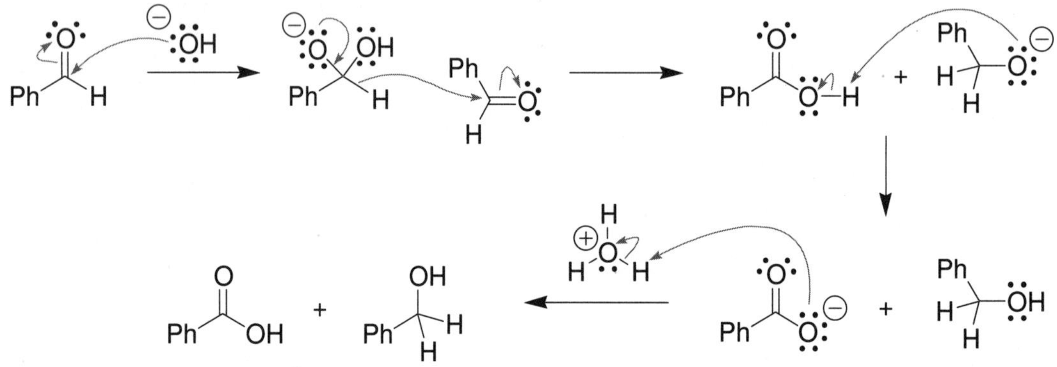

b) The function of H_3O^+ in the second step is to serve as a proton source to protonate the resulting carboxylate ion.

c) Water is only a weak acid (pK_a = 15.74), and is not sufficiently strong to serve as a proton source for a carboxylate ion (pK_a of PhCOOH is 4.21). See section 3.5 for a discussion of this topic.

20.33.

a) 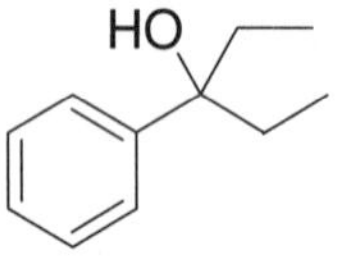 b) 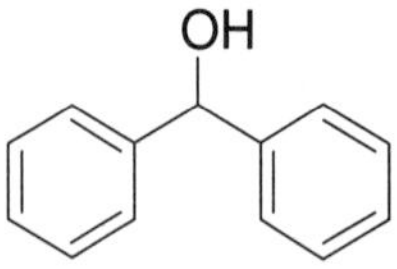 c)

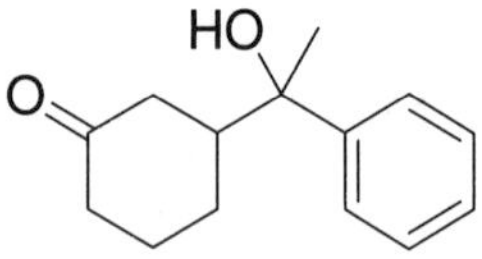

20.34.

a) 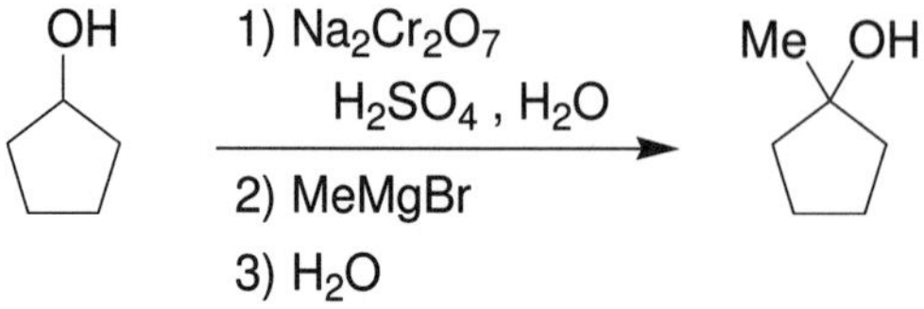 1) $Na_2Cr_2O_7$, H_2SO_4, H_2O; 2) MeMgBr; 3) H_2O

b) 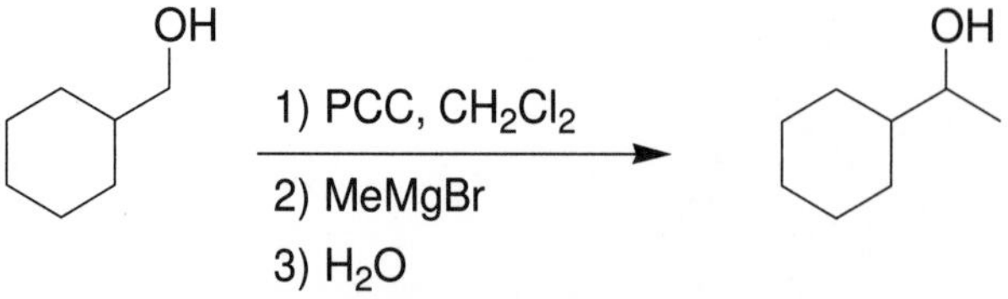 1) PCC, CH_2Cl_2; 2) MeMgBr; 3) H_2O

20.35.

a) 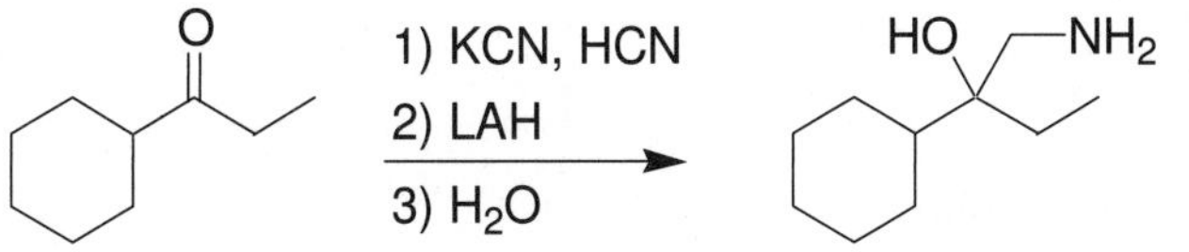 Cyclohexyl ethyl ketone → 1) KCN, HCN; 2) LAH; 3) H_2O → product (HO, NH_2)

b) 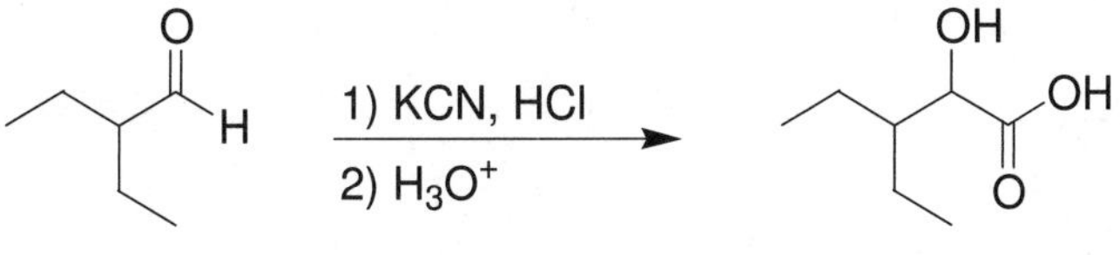 2-Ethylbutanal (H) → 1) KCN, HCl; 2) H_3O^+ → product (OH, OH, O)

20.36.

a) 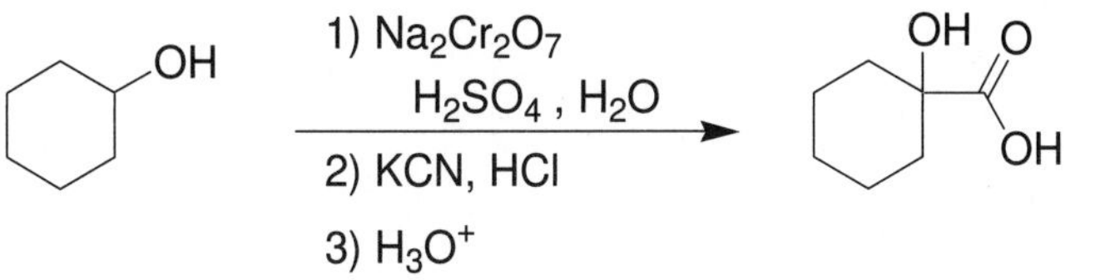 Cyclohexanol (OH) → 1) $Na_2Cr_2O_7$, H_2SO_4, H_2O; 2) KCN, HCl; 3) H_3O^+ → product (OH, O, OH)

b) 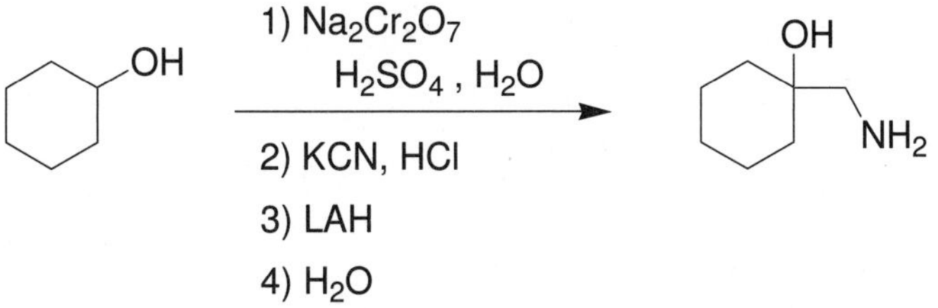 Cyclohexanol (OH) → 1) $Na_2Cr_2O_7$, H_2SO_4, H_2O; 2) KCN, HCl; 3) LAH; 4) H_2O → product (OH, NH_2)

20.37. a) 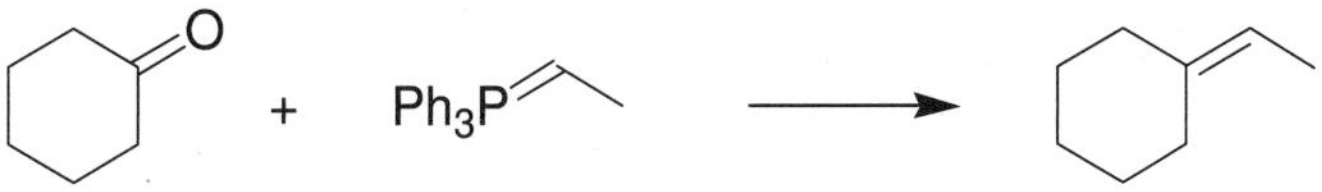 Cyclohexanone (O) + Ph_3P=CHCH$_3$ →

b) 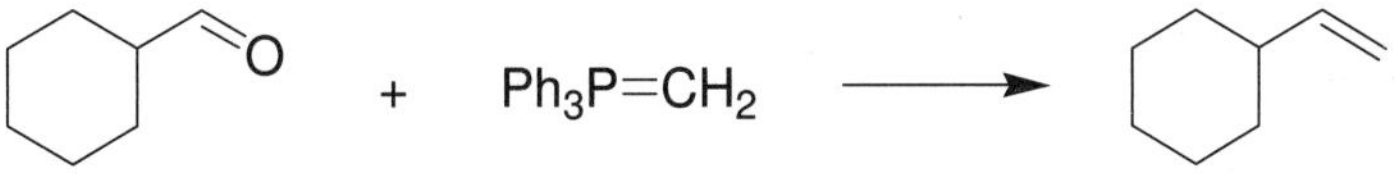 Cyclohexanecarbaldehyde (O) + $Ph_3P{=}CH_2$ →

c) 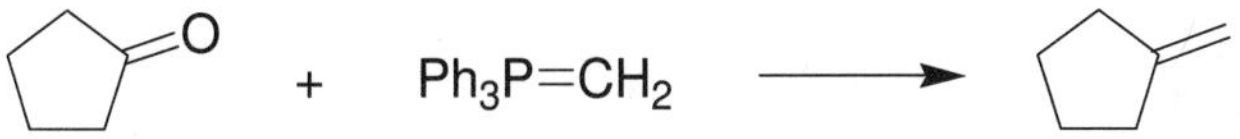 Cyclopentanone (O) + $Ph_3P{=}CH_2$ →

d) 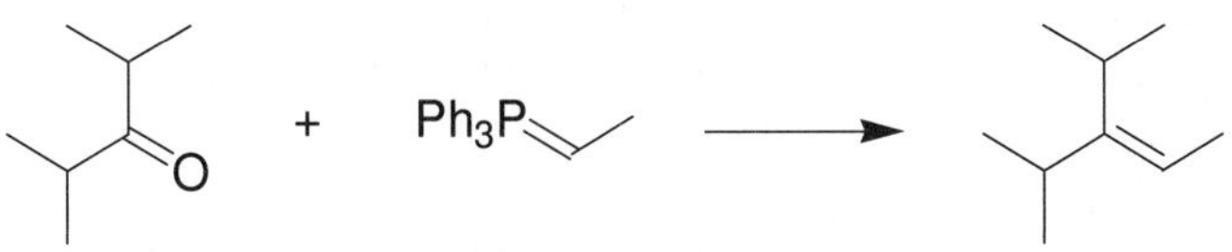 Ketone (O) + Ph_3P=CHCH$_3$ →

e) 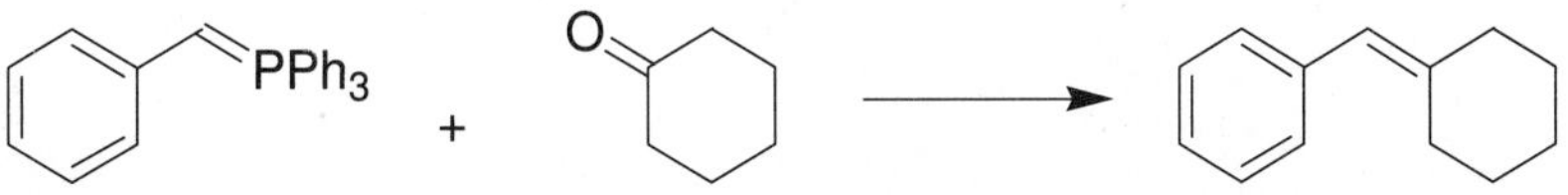 Benzylidene-PPh_3 + cyclohexanone (O) →

20.38.

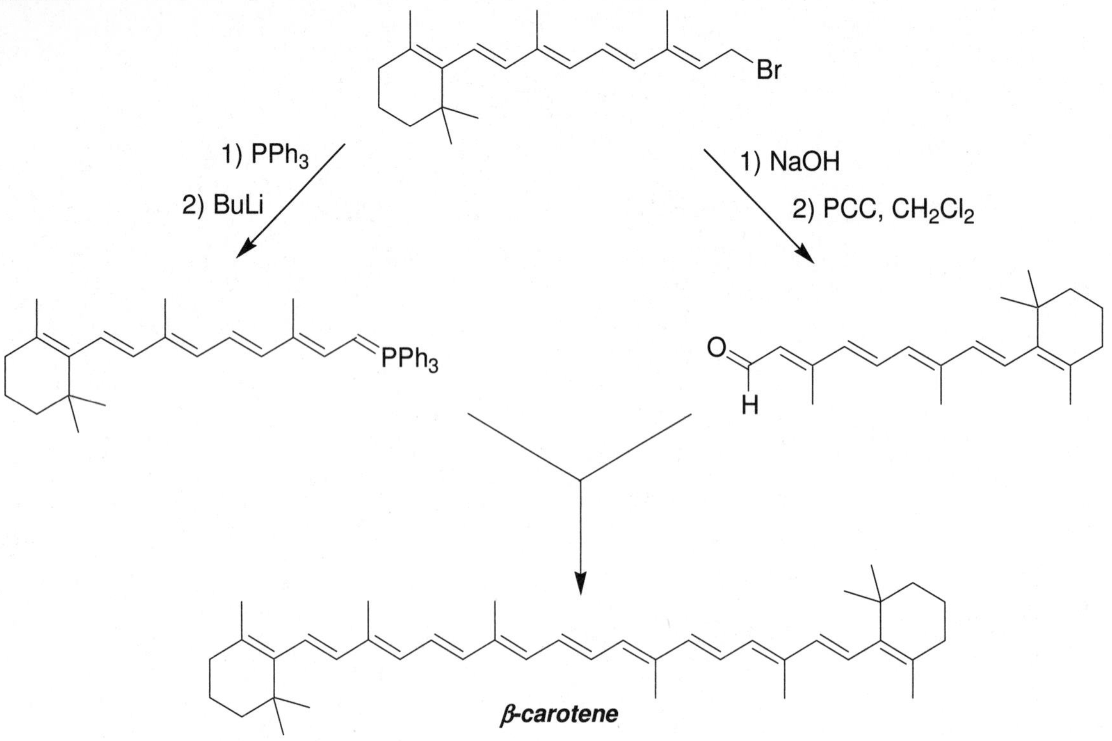

Br

1) PPh_3
2) BuLi

1) NaOH
2) PCC, CH_2Cl_2

PPh_3

O
H

β-carotene

20.39. a)

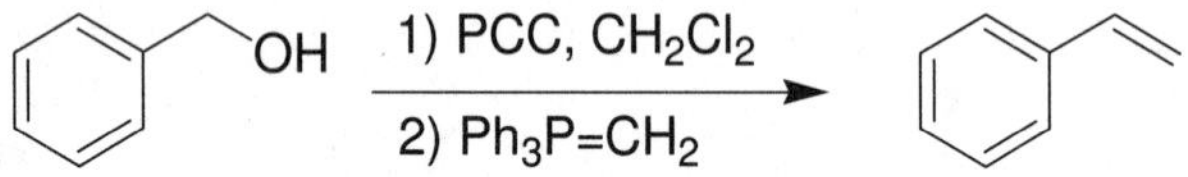

OH

1) PCC, CH_2Cl_2
2) $Ph_3P{=}CH_2$

b)

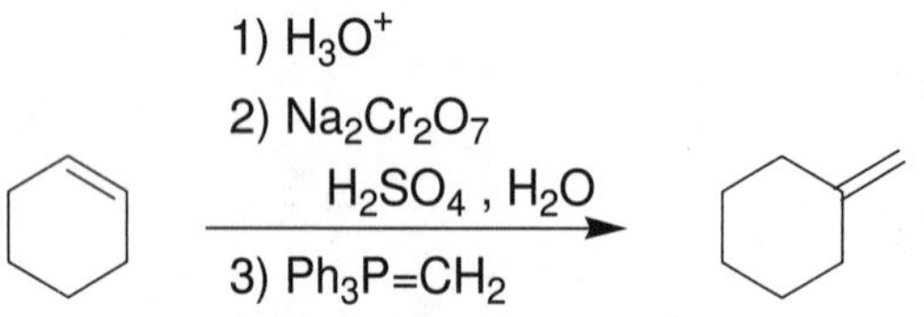

1) H_3O^+
2) $Na_2Cr_2O_7$
H_2SO_4 , H_2O
3) $Ph_3P{=}CH_2$

20.40.

a)

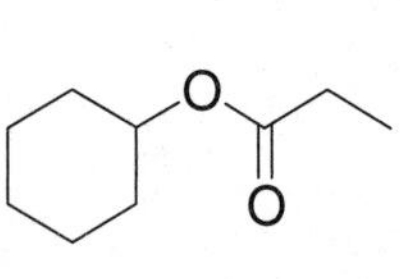

O
O

b)

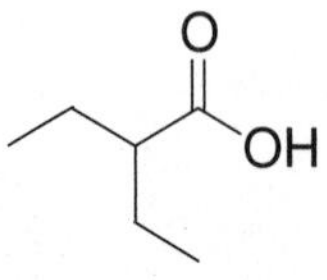

O
OH

c)

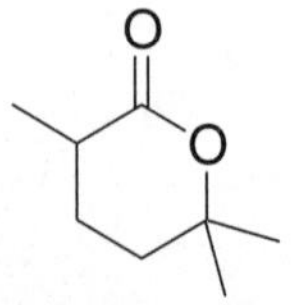

O
O

20.41.

a) 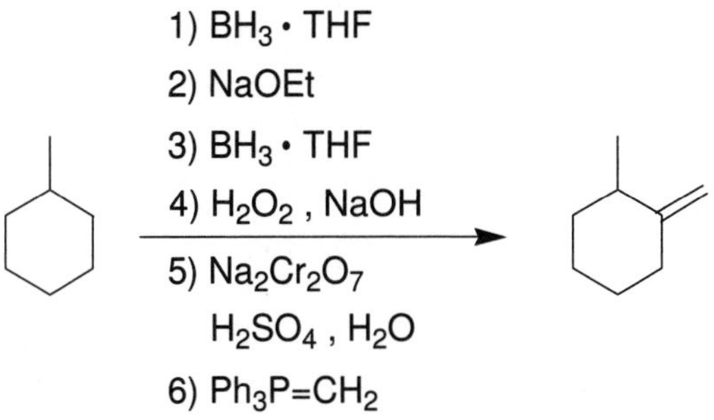
1) $BH_3 \cdot THF$
2) NaOEt
3) $BH_3 \cdot THF$
4) H_2O_2 , NaOH
5) $Na_2Cr_2O_7$
H_2SO_4 , H_2O
6) $Ph_3P{=}CH_2$

b) 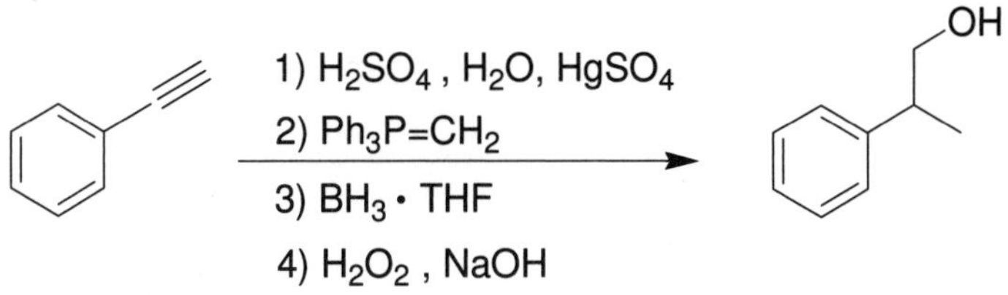
1) H_2SO_4 , H_2O, $HgSO_4$
2) $Ph_3P{=}CH_2$
3) $BH_3 \cdot THF$
4) H_2O_2 , NaOH

OH

c) 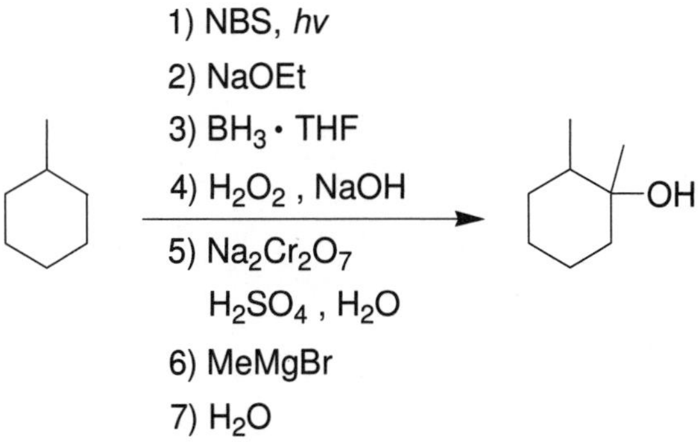
1) NBS, *hv*
2) NaOEt
3) $BH_3 \cdot THF$
4) H_2O_2 , NaOH
5) $Na_2Cr_2O_7$
H_2SO_4 , H_2O
6) MeMgBr
7) H_2O

OH

d) 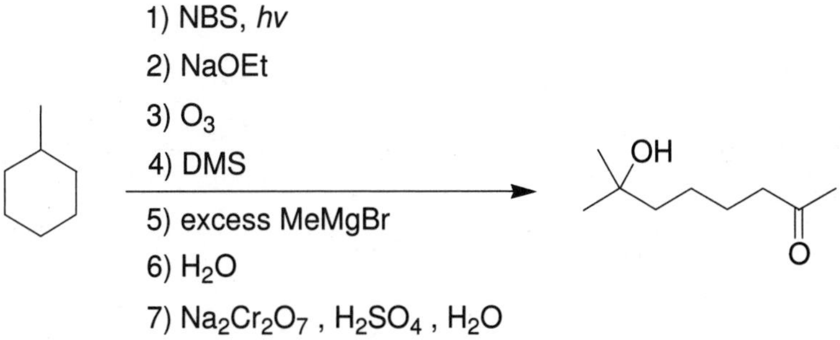
1) NBS, *hv*
2) NaOEt
3) O_3
4) DMS
5) excess MeMgBr
6) H_2O
7) $Na_2Cr_2O_7$, H_2SO_4 , H_2O

OH

O

e) 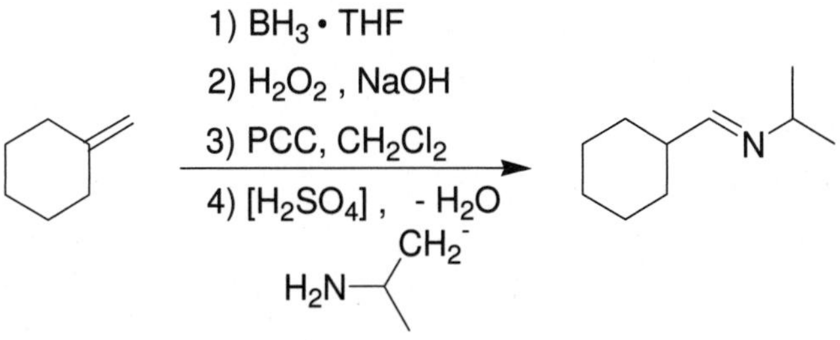
1) $BH_3 \cdot THF$
2) H_2O_2 , NaOH
3) PCC, CH_2Cl_2
4) $[H_2SO_4]$, $- H_2O$
CH_2^-
H_2N

N

f)

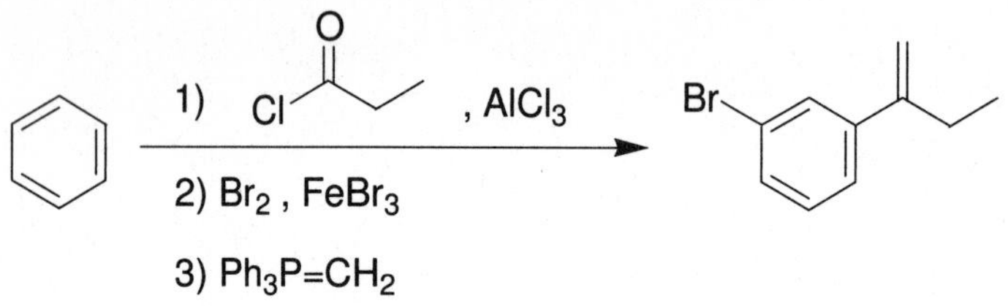

1) Cl–CO–CH$_2$CH$_3$, $AlCl_3$

2) Br_2 , $FeBr_3$

3) $Ph_3P{=}CH_2$

Br

g)

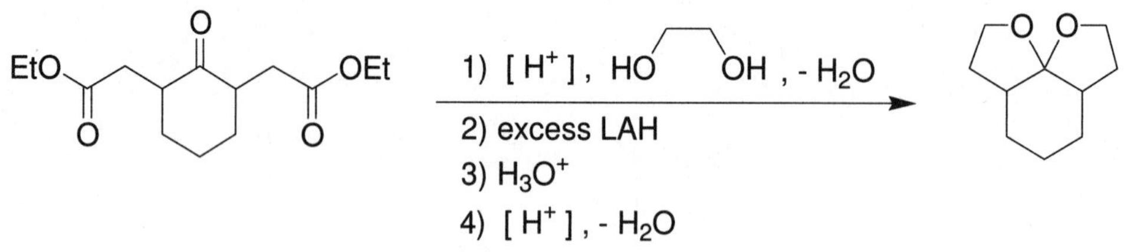

EtO OEt

O O O

1) [H^+], HO OH , - H_2O

2) excess LAH

3) H_3O^+

4) [H^+], - H_2O

O O

h)

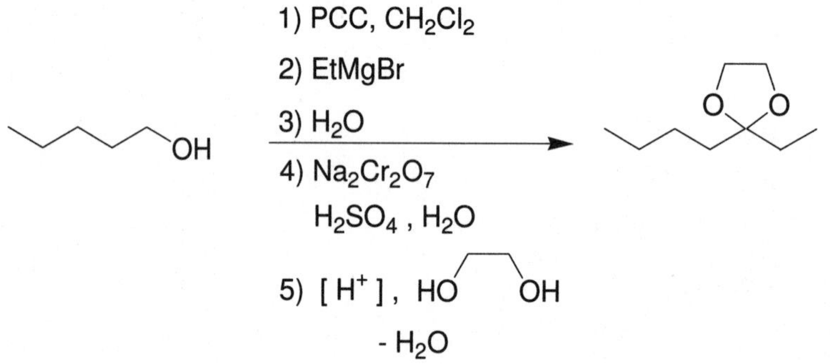

OH

1) PCC, CH_2Cl_2

2) EtMgBr

3) H_2O

4) $Na_2Cr_2O_7$ H_2SO_4 , H_2O

5) [H^+], HO OH - H_2O

O O

20.42.

a)

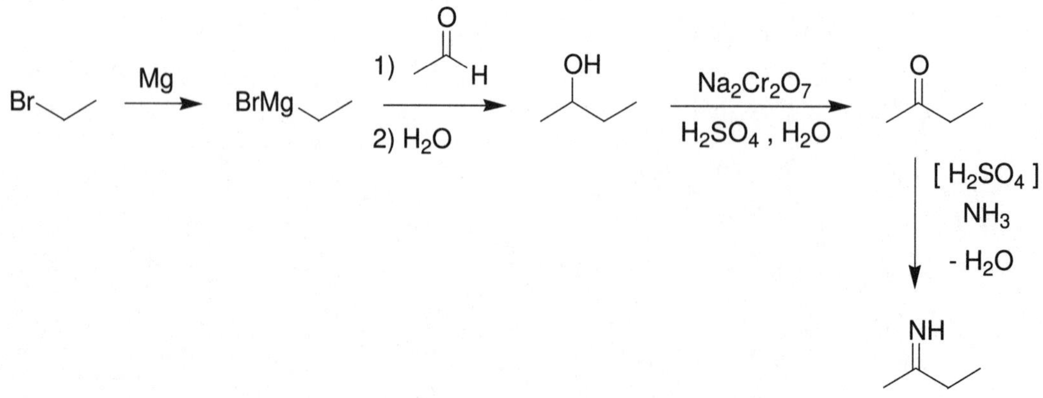

Br → Mg → BrMg

1) O, H

2) H_2O

OH

$Na_2Cr_2O_7$

H_2SO_4 , H_2O

O

[H_2SO_4]

NH_3

- H_2O

NH

b) 1) MeMgBr, 2) H_2O; PCC, CH_2Cl_2; 1) EtMgBr, 2) H_2O; $Na_2Cr_2O_7$, H_2SO_4, H_2O; [H_2SO_4], $(CH_3)_2NH$, - H_2O

c) Mg; 1) [epoxide], 2) H_2O; PBr_3; 1) Mg, 2) [acetaldehyde], 3) H_2O; $Na_2Cr_2O_7$, H_2SO_4, H_2O; HO‑CH₂CH₂‑OH, [H_2SO_4], - H_2O

d) Mg; 1) [epoxide], 2) H_2O; PCC, CH_2Cl_2; 1) EtMgBr, 2) H_2O; $Na_2Cr_2O_7$, H_2SO_4, H_2O; HO‑CH₂CH₂‑OH, [H_2SO_4], - H_2O

e)

1) MeMgBr, 2) H_2O; PCC, CH_2Cl_2; 1) EtMgBr, 2) H_2O; $Na_2Cr_2O_7$, H_2SO_4, H_2O; $[H^+]$, HO–NH_2, - H_2O

f)

1) MeMgBr, 2) H_2O; PCC, CH_2Cl_2; 1) EtMgBr, 2) H_2O; $Na_2Cr_2O_7$, H_2SO_4, H_2O; $Ph_3P{=}CHCH_3$

g)

Br; Mg; BrMg; 1) acetaldehyde, 2) H_2O; OH; $Na_2Cr_2O_7$, H_2SO_4, H_2O; KCN, HCl; NC, OH; H_3O^+; HO, OH

h) Br–Et $\xrightarrow{Mg}$ BrMg–Et $\xrightarrow[2)\ H_2O]{1)\ CH_3CHO}$ 2-butanol $\xrightarrow[H_2SO_4,\ H_2O]{Na_2Cr_2O_7}$ 2-butanone $\xrightarrow{KCN,\ HCl}$ cyanohydrin (NC, OH) $\xrightarrow[2)\ H_2O]{1)\ LAH}$ H_2N–amino alcohol (OH)

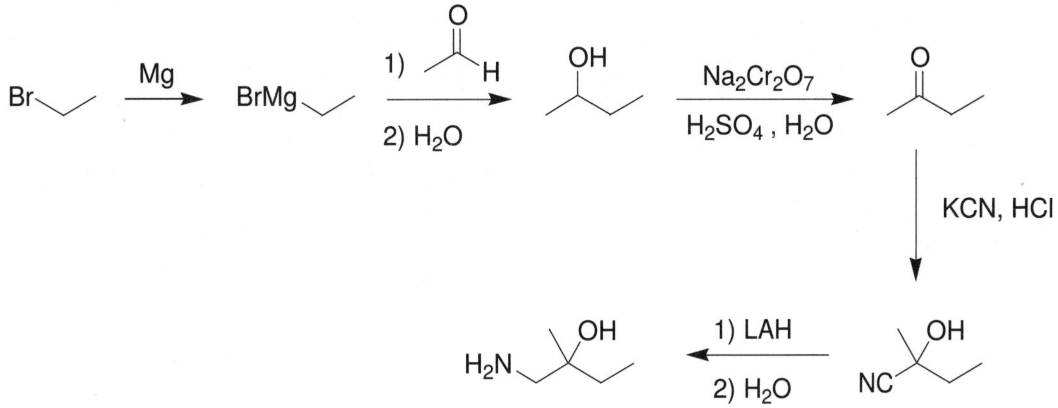

20.43. 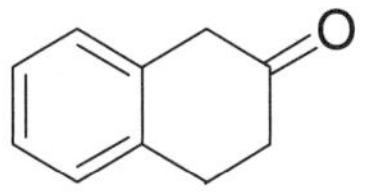 (because this carbonyl group is not conjugated)

20.44. a) (2*S*,3*R*)-3-methyl-2-propylcyclopentanone
b) cyclohexanecarbaldehyde
c) 3-methyl-2-butenal
d) (*S*)-4-methyl-3-hexanone

20.45. a) 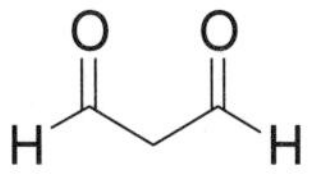 b) 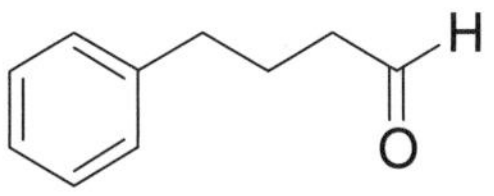 c)

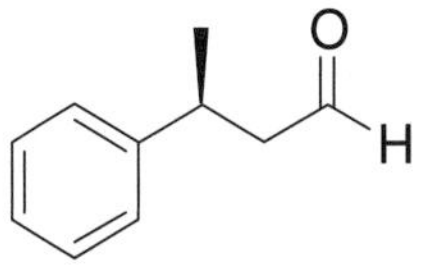

d) 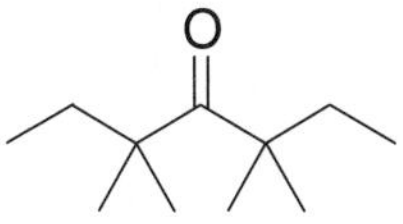 e) 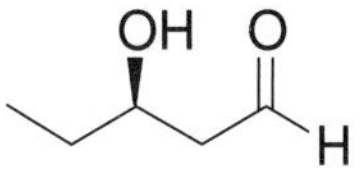 f)

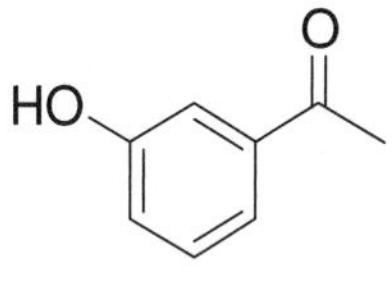

g) 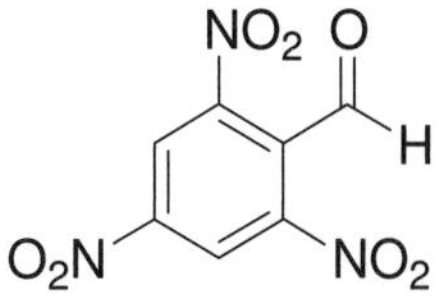 h) 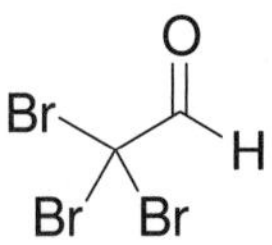 i)

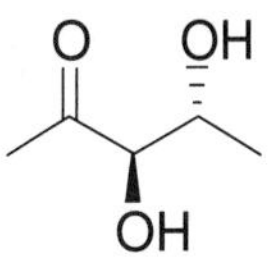

20.46.

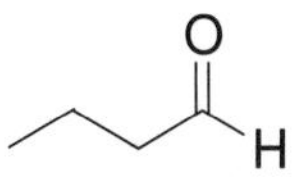

butanal 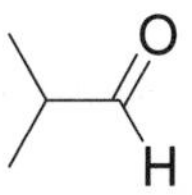 ***2-methylpropanal***

20.47.

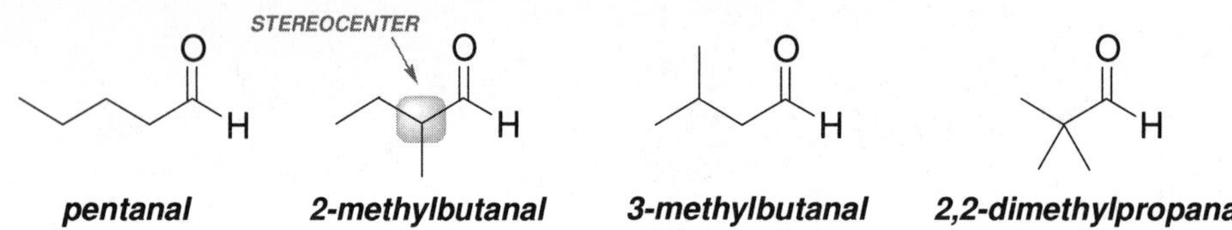

20.48.

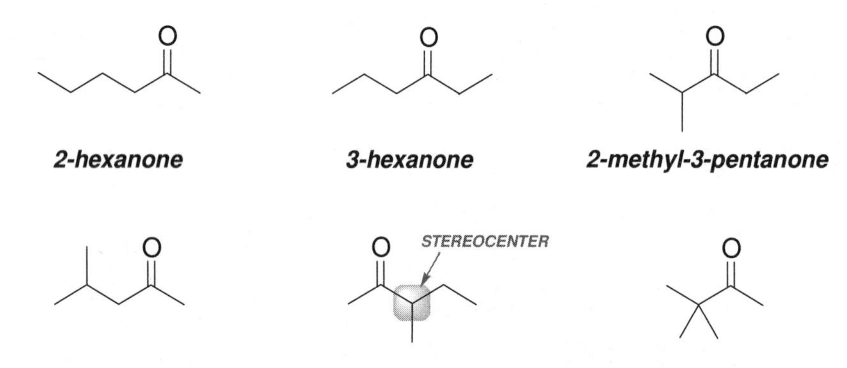

20.49. The carbonyl group of a ketone will never appear at C-1 because if it would did, the compound would be called an aldehyde rather than a ketone.

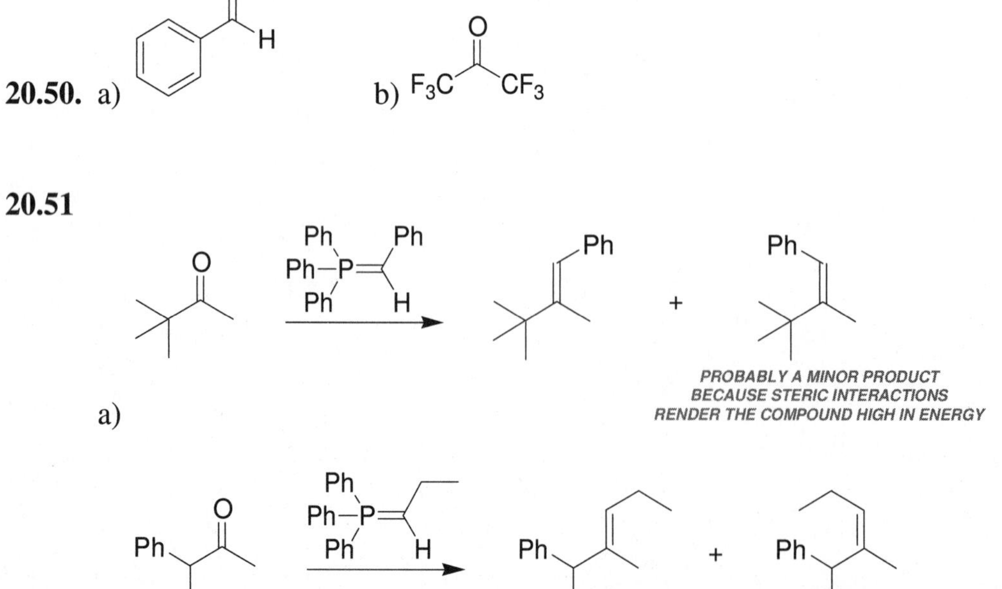

20.52.

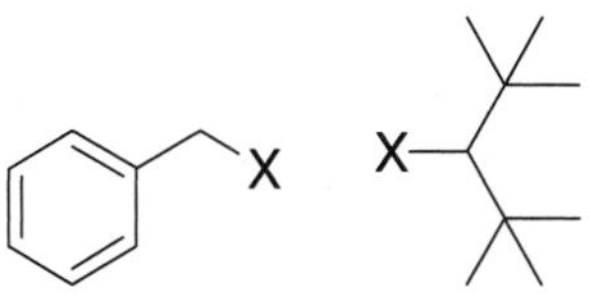

X X

The latter alkyl halide above will be more difficult to convert into a Wittig reagent, because it is too sterically hindered to undergo S_N2 attack.

20.53.

a) 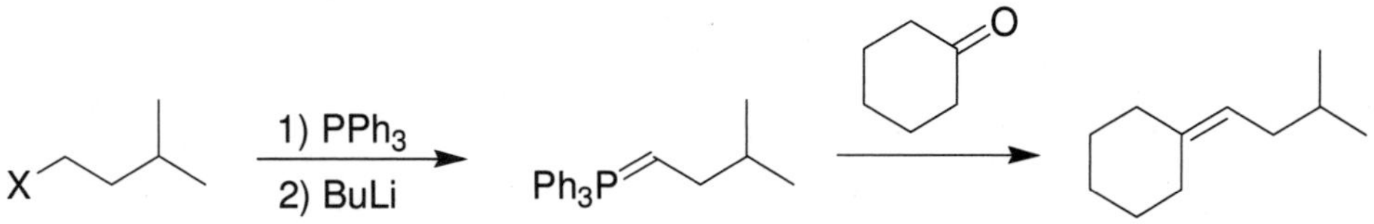 X → 1) PPh_3 2) BuLi → Ph_3P → (O)

b) 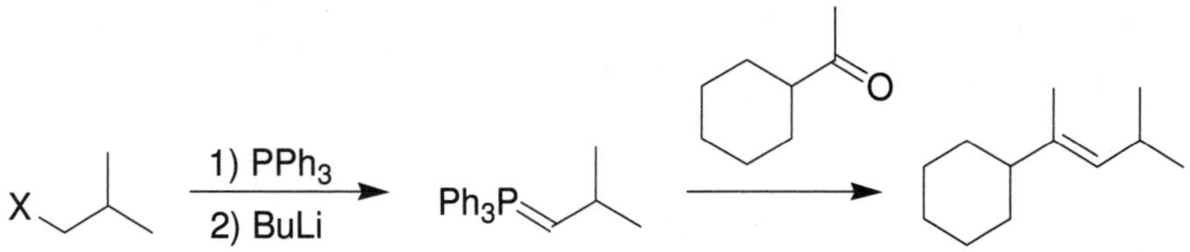 X → 1) PPh_3 2) BuLi → Ph_3P → (O)

20.54.

a) 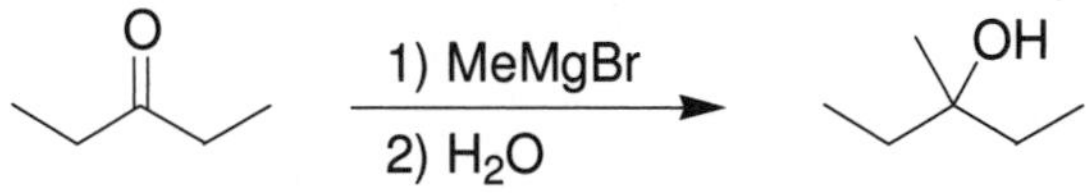 O → 1) MeMgBr 2) H_2O → OH

b) 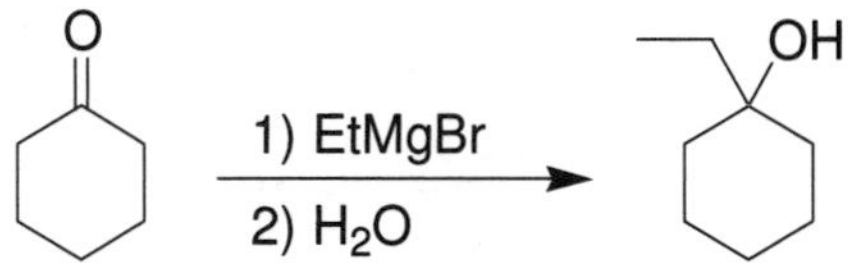 O → 1) EtMgBr 2) H_2O → OH

c) 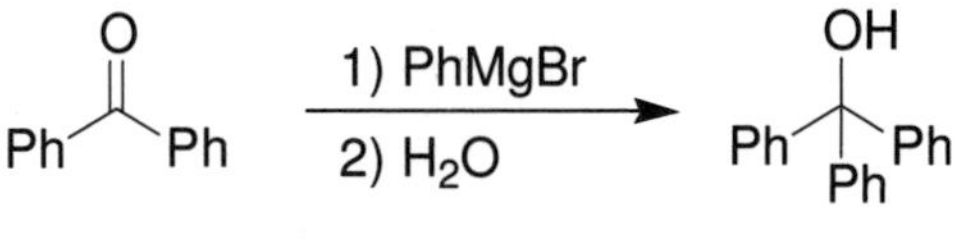 Ph(C=O)Ph → 1) PhMgBr 2) H_2O → OH, Ph, Ph, Ph

d) 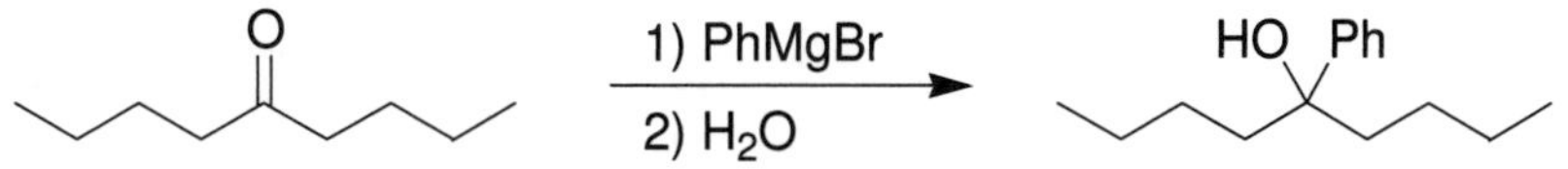 O → 1) PhMgBr 2) H_2O → HO Ph

20.55.

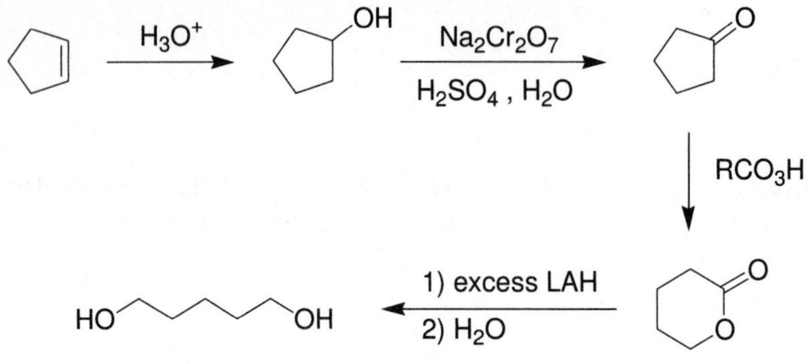

H_3O^+

OH

$Na_2Cr_2O_7$

H_2SO_4 , H_2O

O

RCO_3H

O

O

1) excess LAH

2) H_2O

HO

OH

20.56. a) 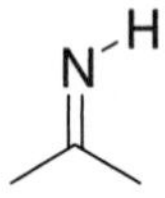 b) 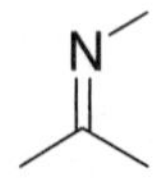 c)  d)

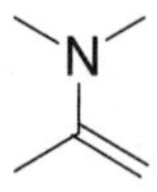

N-H

N

EtO OEt

N

e) 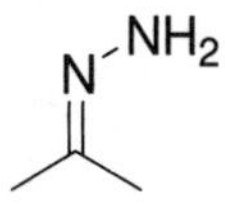 f) 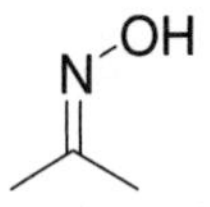 g) 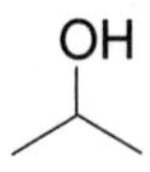 h)

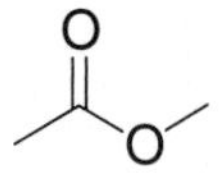

NH_2

N

OH

N

OH

O

O

i) 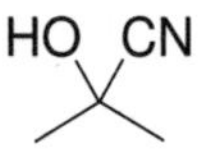 j) 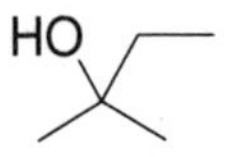 k) 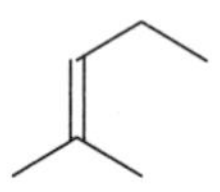 l)

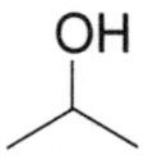

HO CN

HO

OH

20.57.

Et

H–O⊕

H

O

HO

H

⊕

O–H

HO

H

H

⊕

O

OH

H

Et–O–H

Et

⊕O–H

H

O

OH

H

O

O⊕

H

- H_2O

⊕

O

Et–O–H

H

O

⊕O

Et

Et–O–H

O

O

Et

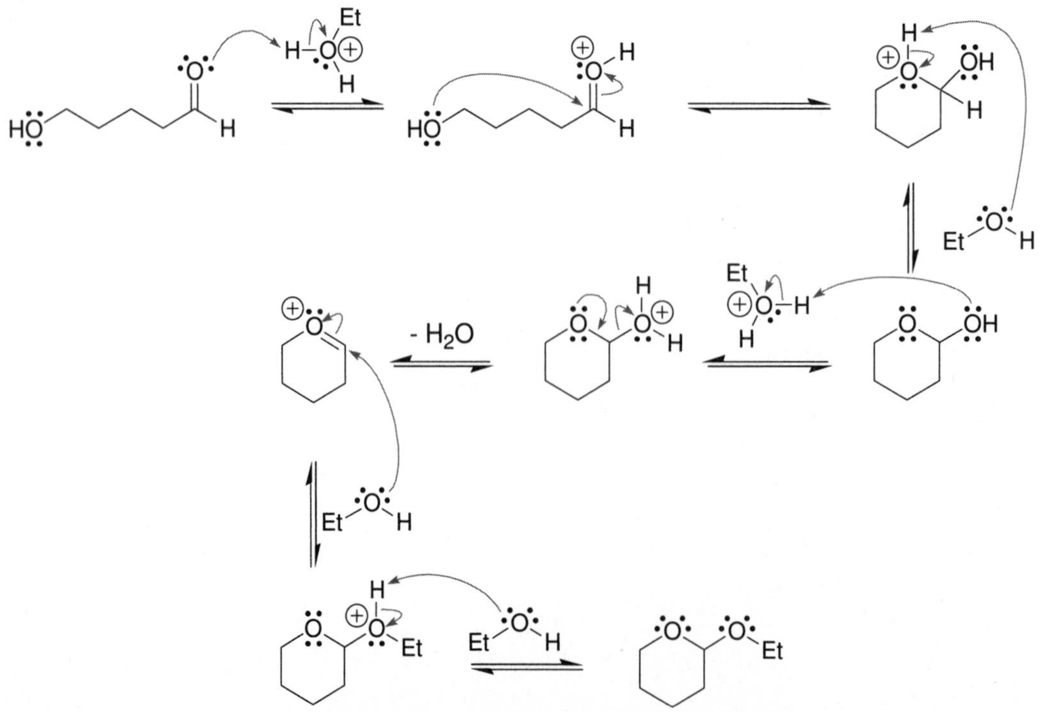

20.58.

$[H^+]$, EtOH; 1) MeMgBr, 2) H_2O; H_3O^+

OEt OEt HO

20.59.

$[H^+]$

20.60. a) 1) LAH, 2) H_2O

b) 1) PhMgBr, 2) H_2O

c) $(C_6H_5)_3P{=}CH_2$

20.61. a) 1) MeMgBr, 2) H_2O; conc. H_2SO_4, heat

b) $Ph_3P{=}CH_2$

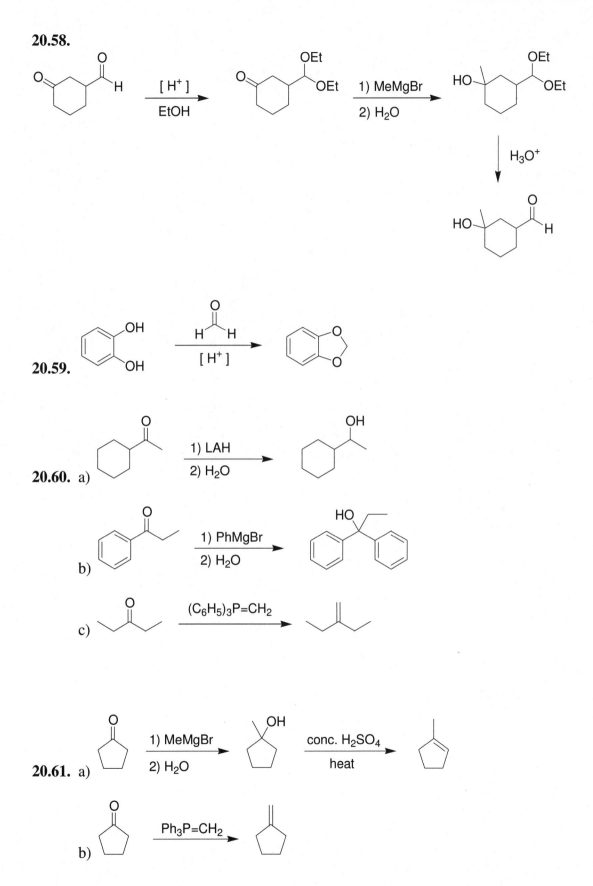

c) 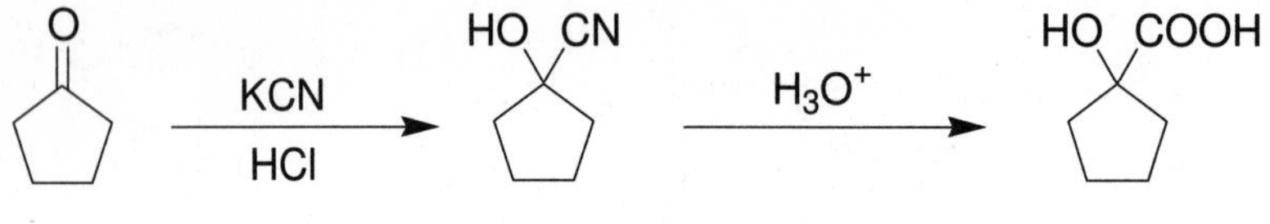 cyclopentanone $\xrightarrow[HCl]{KCN}$ HO, CN (1-hydroxycyclopentanecarbonitrile) $\xrightarrow{H_3O^+}$ HO, COOH

d) 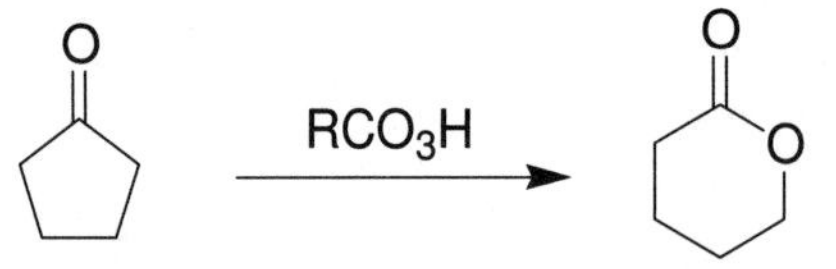 cyclopentanone $\xrightarrow{RCO_3H}$ lactone (O)

20.62.

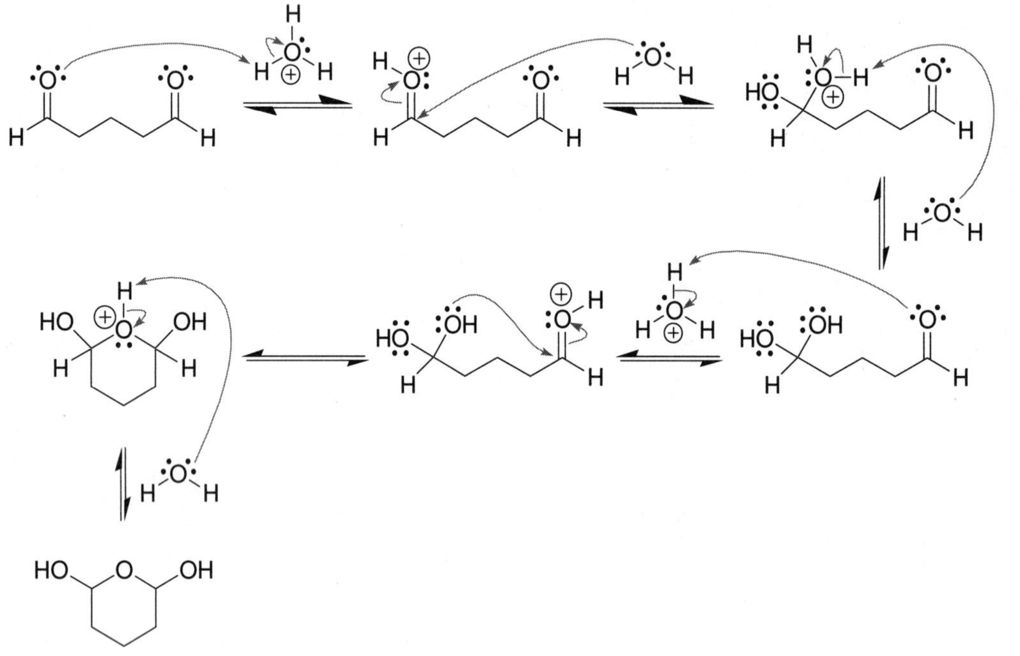

H, H_3O^+, H, O, H, O, H, H, HO, O–H, H, H, H, O, H, H, H, HO, OH, O, H, H, H, H, HO, OH, O, H, H, H, HO, OH, H, H, H, O, H

HO, O, OH

20.63.

a) 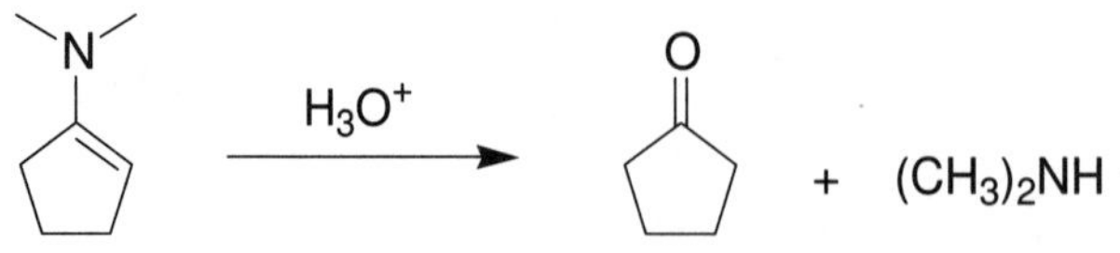 N-(cyclopentenyl)dimethylamine $\xrightarrow{H_3O^+}$ cyclopentanone + $(CH_3)_2NH$

b) 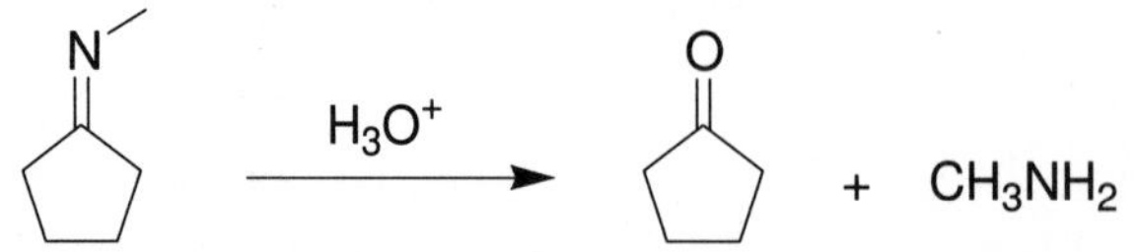 N-methyl imine of cyclopentanone $\xrightarrow{H_3O^+}$ cyclopentanone + CH_3NH_2

c) $\xrightarrow{H_3O^+}$ + HO–CH₂CH₂–OH

d) $\xrightarrow{H_3O^+}$ OH + CH_3OH

20.64.

$\xrightarrow{\text{excess } H_3O^+}$ N–H + O, O, O, H, NH_2 + HO, HO

20.65.

a)

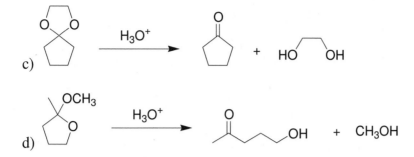

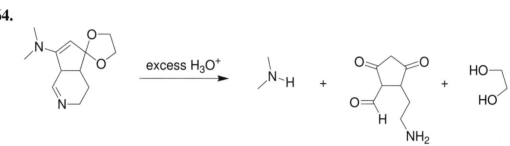

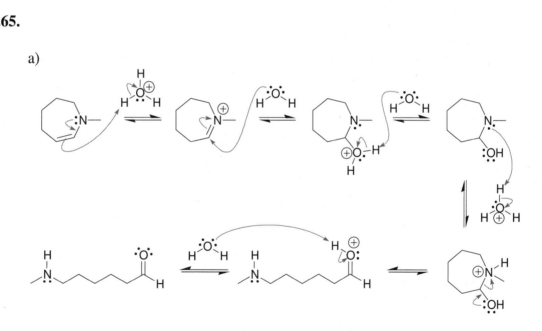

b)

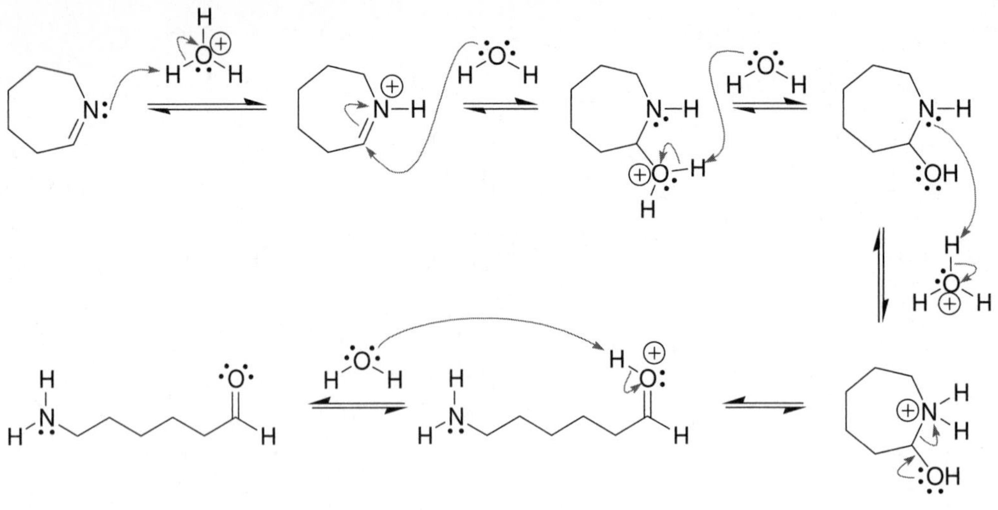

c)

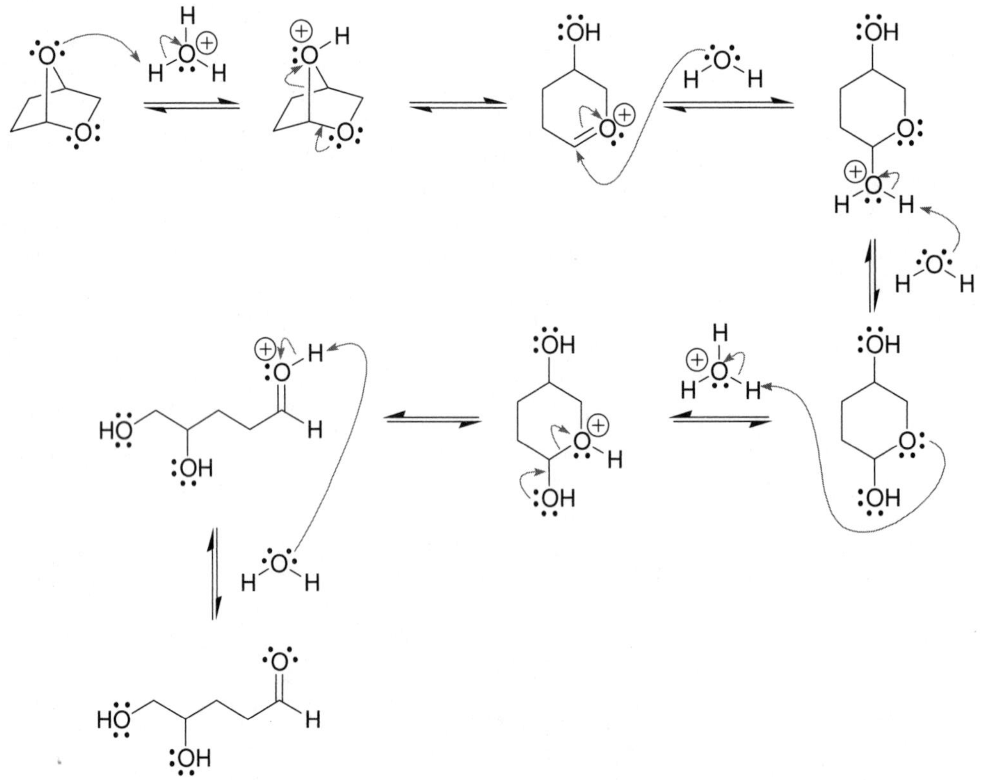

20.66.

a) [reaction scheme: cyclohexanone + N-methyl-N-phenylhydrazine, $[H^+]$ → hydrazone]

CH$_3$ N NH$_2$ $[H^+]$ N N

b) 1) PhMgBr 2) H_2O HO Ph

c) CH_3CO_3H

d) CH_3CO_3H

e) $[H^+]$ N–H $(-H_2O)$ N

f) $[H^+]$ NH$_2$ $(-H_2O)$ N

20.67.

a) OH OH + O $[H^+]$ $(-H_2O)$ O O

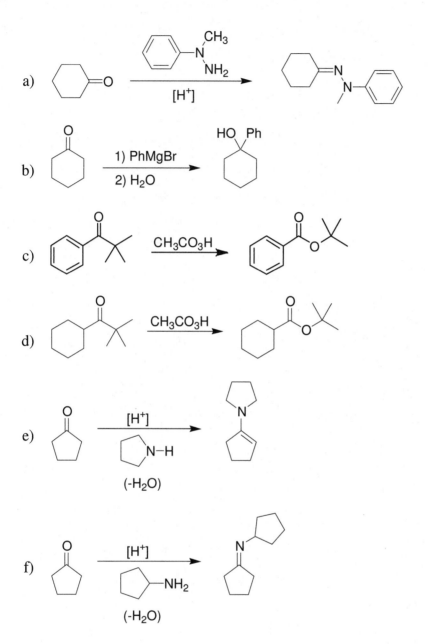

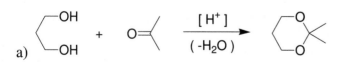

b) HO (5-hydroxy-2-pentanone) + EtOH $\xrightarrow[(-H_2O)]{[H^+]}$ 2-ethoxy-2-methyltetrahydropyran (OEt)

c) HO, OH, H, O (aldehyde diol) $\xrightarrow[(-H_2O)]{[H^+]}$ bicyclic acetal (O, O)

20.68.

OH (ethanol) $\xrightarrow[\Delta]{H_2SO_4}$ ethylene $\xrightarrow[\text{NaOH, cold}]{KMnO_4}$ HO–CH₂CH₂–OH

OH (ethanol) $\xrightarrow[CH_2Cl_2]{PCC}$ acetaldehyde (O, H)

HO–CH₂CH₂–OH + acetaldehyde $\xrightarrow[-H_2O]{[H^+]}$ 2-methyl-1,3-dioxolane (O, O)

20.69.

a) cycloheptenone (O) $\xrightarrow[[H^+],\ -H_2O]{HO\quad OH}$ acetal (O, O) $\xrightarrow[2)\ DMS]{1)\ O_3}$ dialdehyde acetal (O, O; O, O) $\xrightarrow[2)\ H_2O]{1)\ LAH}$ diol acetal (O, O; HO, OH) $\xrightarrow{H_3O^+}$ hydroxy ketone (O; HO, OH) $\xrightarrow[-H_2O]{[H^+]}$ spiroketal (O, O)

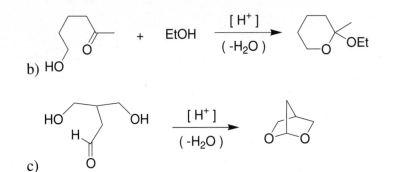

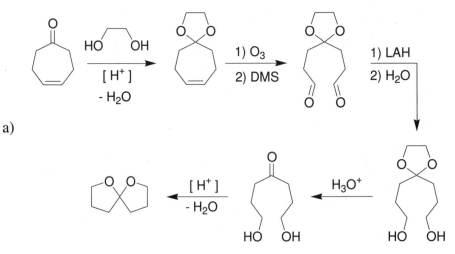

b)

HO OH, [H^+], - H_2O ; 1) LAH, 2) H_2O ; H_3O^+ ; [H^+], - H_2O

OH OH OH OH O O O O O O O O O

20.70.

H_3O^+ ≡ OH O HO HO OH O O O

20.71.

a) H_3O^+ OH $Na_2Cr_2O_7$ / H_2SO_4 , H_2O O PPh_3

b) Br Br 1) excess $NaNH_2$ 2) H_3O^+ ; H_2SO_4 , H_2O / $HgSO_4$; O ; [H^+] 2 MeOH ($-H_2O$) ; MeO OMe

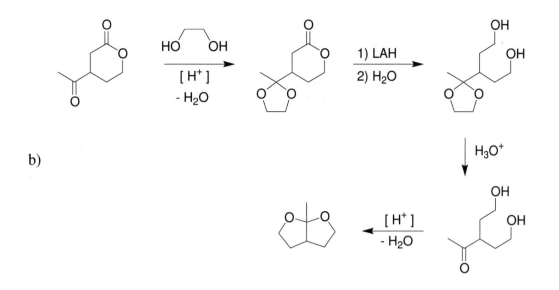

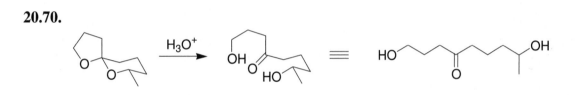

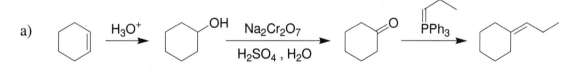

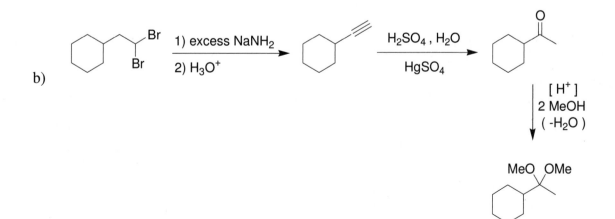

c)

1) O_3
2) DMS

$NaBH_4$
MeOH

OH
OH

[H^+]
($-H_2O$)

20.72.

$H-OSO_3H$

NH_2
NH_2

$^{\ominus}OSO_3H$

$H-OSO_3H$

- H_2O

$^{\ominus}OSO_3H$

$H-OSO_3H$

$^{\ominus}OSO_3H$

$H-OSO_3H$

- H_2O

$^{\ominus}OSO_3H$

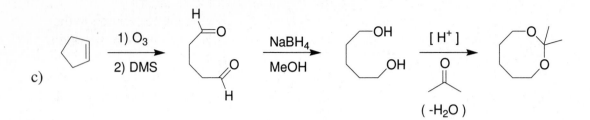

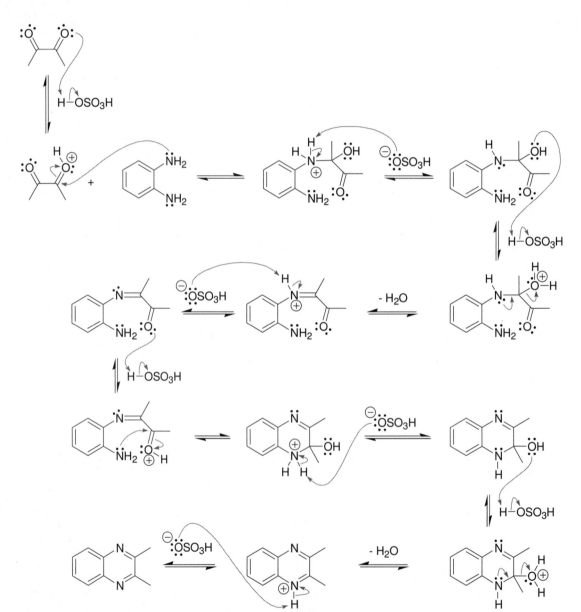

20.73. Cyclopropanone exhibits significant ring strain, with bond angles of approximately 60°. Some of this ring strain is relieved upon conversion to the hydrate, because an sp^2-hybridized carbon atom (that must be 120° to be strain free) is replaced by an sp^3-hybridized carbon atom (that must be only 109.5° to be strain free). In contrast, cyclohexanone is a larger ring and exhibits only minimal ring strain. Conversion of cyclohexanone to its corresponding hydrate does not alleviate a significant amount of ring strain.

20.74. 1,2-dioxane has two adjacent oxygen atoms and is therefore a peroxide. Like other peroxides, it is extremely unstable and potentially explosive.
1,3-dioxane has two oxygen atoms separated by one carbon atom. This compound is therefore an acetal. Like other acetals, it is only stable under basic conditions, but undergoes hydrolysis under mildly acidic conditions.
1,4-dioxane is stable under basic conditions as well as mildly acidic conditions, and is therefore used as a common solvent.

20.75.

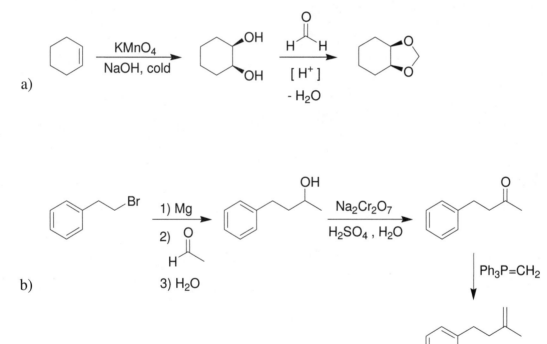

a)

KMnO$_4$
NaOH, cold
OH
OH
O
H H
[H^+]
- H_2O
O
O

b)

Br
1) Mg
2) O
H
3) H_2O
OH
$Na_2Cr_2O_7$
H_2SO_4 , H_2O
O
$Ph_3P{=}CH_2$

c) NBS, $h\nu$; NaOEt; 1) $BH_3 \cdot THF$, 2) H_2O_2, NaOH; PCC, CH_2Cl_2; HO–CH₂CH₂–OH, [H^+], ($-H_2O$)

d) H_3O^+; $Na_2Cr_2O_7$, H_2SO_4, H_2O; KCN, HCl

OH, CN

e) NBS, $h\nu$; NaOEt; 1) $BH_3 \cdot THF$, 2) H_2O_2, NaOH; $Na_2Cr_2O_7$, H_2SO_4, H_2O; [H^+], NH_3, ($-H_2O$)

Br, OH, O, NH

f) $(CH_3)_3CO^{\ominus}$ $K^{\oplus}$; 1) $BH_3 \cdot THF$, 2) H_2O_2, NaOH; PCC, CH_2Cl_2

Br, OH, H, O

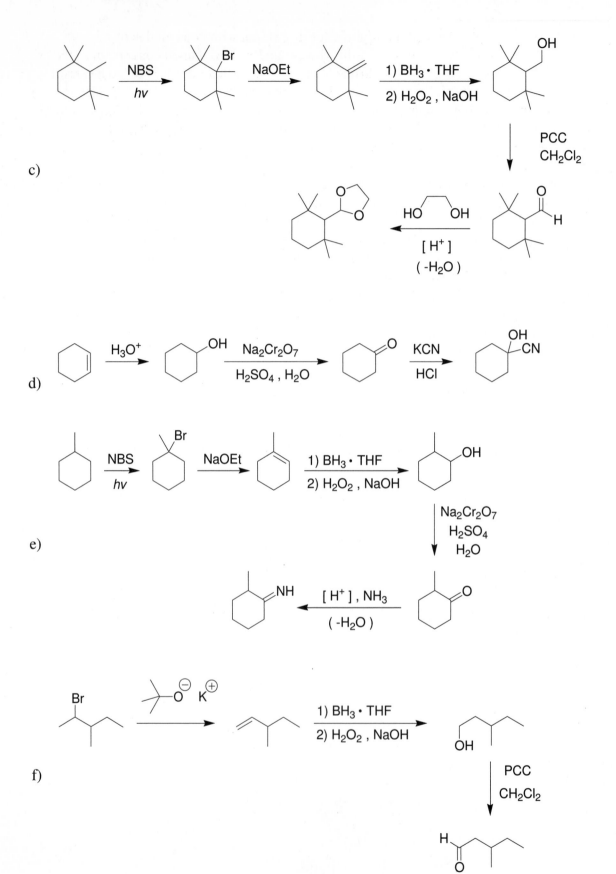

g)

O
Cl
AlCl₃ → O
$[H^+]$
$(CH_3)_2NH$
$(-H_2O)$
→ N

h)

H_2SO_4, H_2O
$HgSO_4$
→ O O
1) LAH
2) H_2O
→ OH OH
$[H^+]$
O
H H
$-H_2O$
↓ O O

20.76.

O

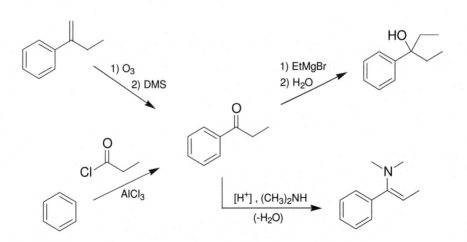

Compound A *Compound B*

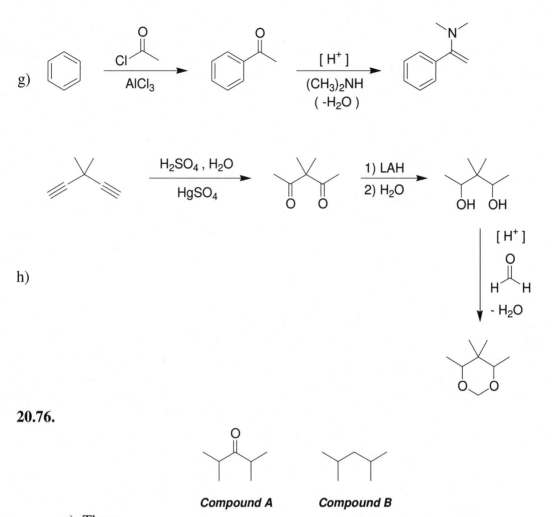

a) Three
b) Three
c) Compound A is a ketone, while Compound B is an alkane. Therefore, Compound A will exhibit a signal at approximately 1715 cm^{-1}, while Compound B will not exhibit a signal in the same region.

20.77.

1) O_3
2) DMS
O
1) EtMgBr
2) H_2O
HO

O
Cl
AlCl₃
$[H^+]$, $(CH_3)_2NH$
$(-H_2O)$
N

20.78.

Cyclohexene $\xrightarrow{H_3O^+}$ cyclohexanol $\xrightarrow{H_2CrO_4}$ cyclohexanone $\xrightarrow{[H^+],\ NH_2NH_2\ (-H_2O)}$ hydrazone ($=N-NH_2$) $\xrightarrow{KOH / H_2O,\ heat}$ cyclohexane

Cyclohexene $\xrightarrow{H_2,\ Pt}$ cyclohexane

20.79.

Benzene $\xrightarrow{Br_2,\ FeBr_3}$ bromobenzene $\xrightarrow{Mg}$ phenylmagnesium bromide (MgBr) $\xrightarrow{1)\ HCHO;\ 2)\ H_2O}$ benzyl alcohol (OH) $\xrightarrow{PCC,\ CH_2Cl_2}$ benzaldehyde $\xrightarrow{HOCH_2CH_2OH,\ [H^+],\ -H_2O}$ 2-phenyl-1,3-dioxolane

20.80. a) 3-butenal (H–CO–CH$_2$CH=CH$_2$) *or* cyclopropanecarbaldehyde (H–CO–cyclopropyl)

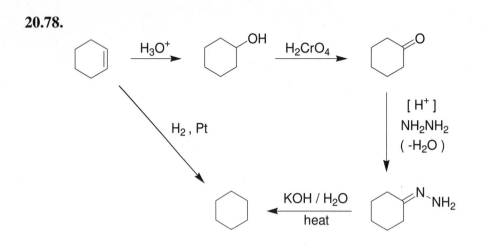

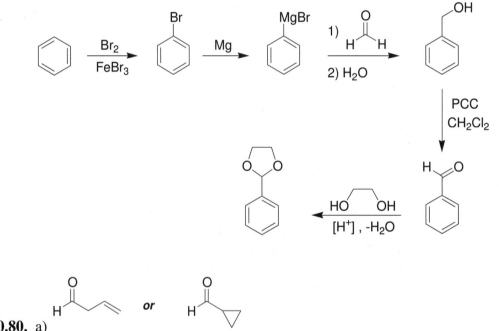

b) The first compound above would exhibit four signals in its ^{13}C NMR spectrum, while the second compound would exhibit only three signals in its ^{13}C NMR spectrum.

20.81. Propiophenone (phenyl–CO–CH$_2$CH$_3$)

20.82. Benzophenone (phenyl–CO–phenyl)

20.83. 2,2,4,4-Tetramethyl-3-pentanone

20.84.

a)

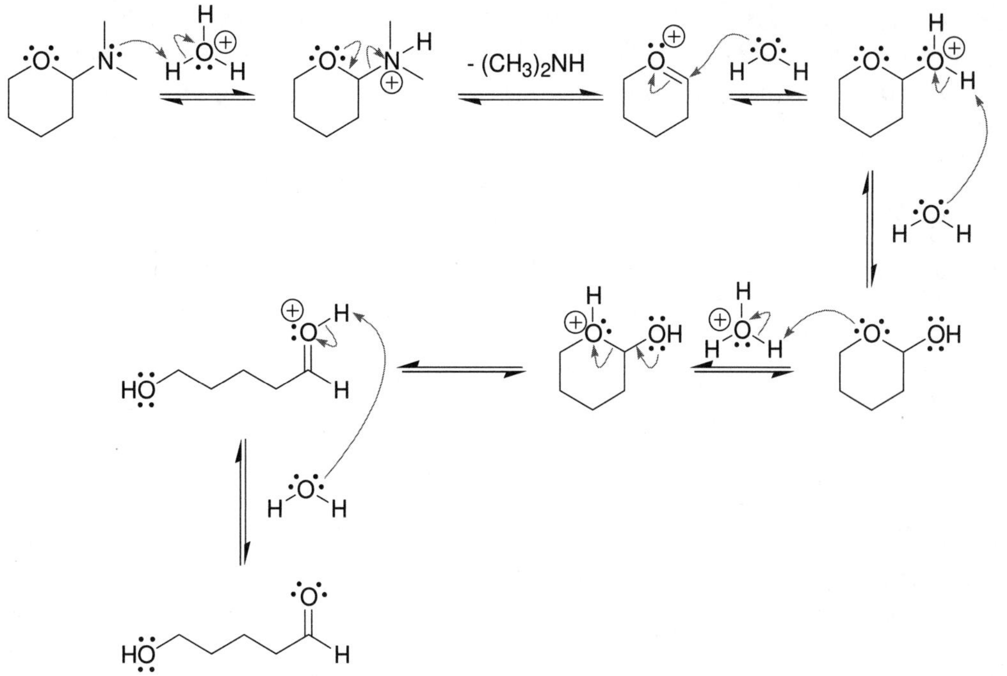

- $(CH_3)_2NH$

b)

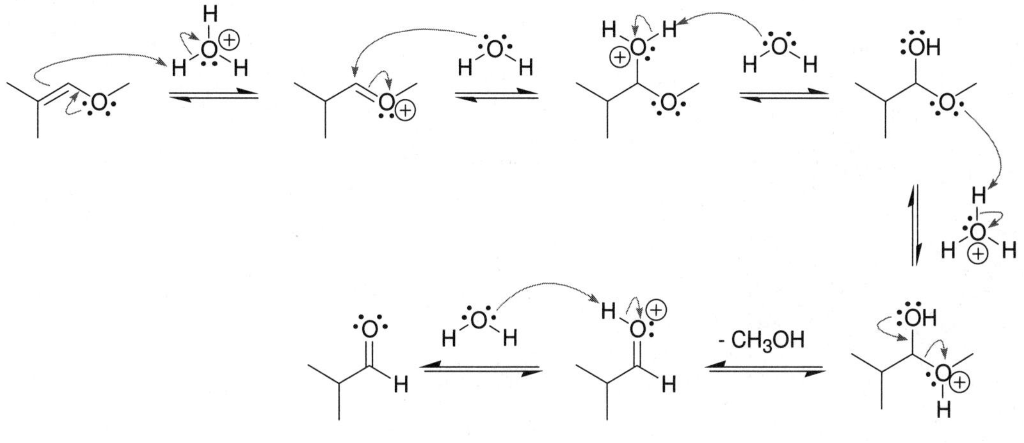

- CH_3OH

c) **Note**: ***The first two steps of the mechanism below can be reversed (first the amine attacks the carbonyl group, and then the tetrahedral intermediate is protonated). The same is true for attack of the second carbonyl group (half-way through the mechanism). It would be wise to check your lecture notes to determine if you instructor has a strong preference for this alternate sequence of steps.***

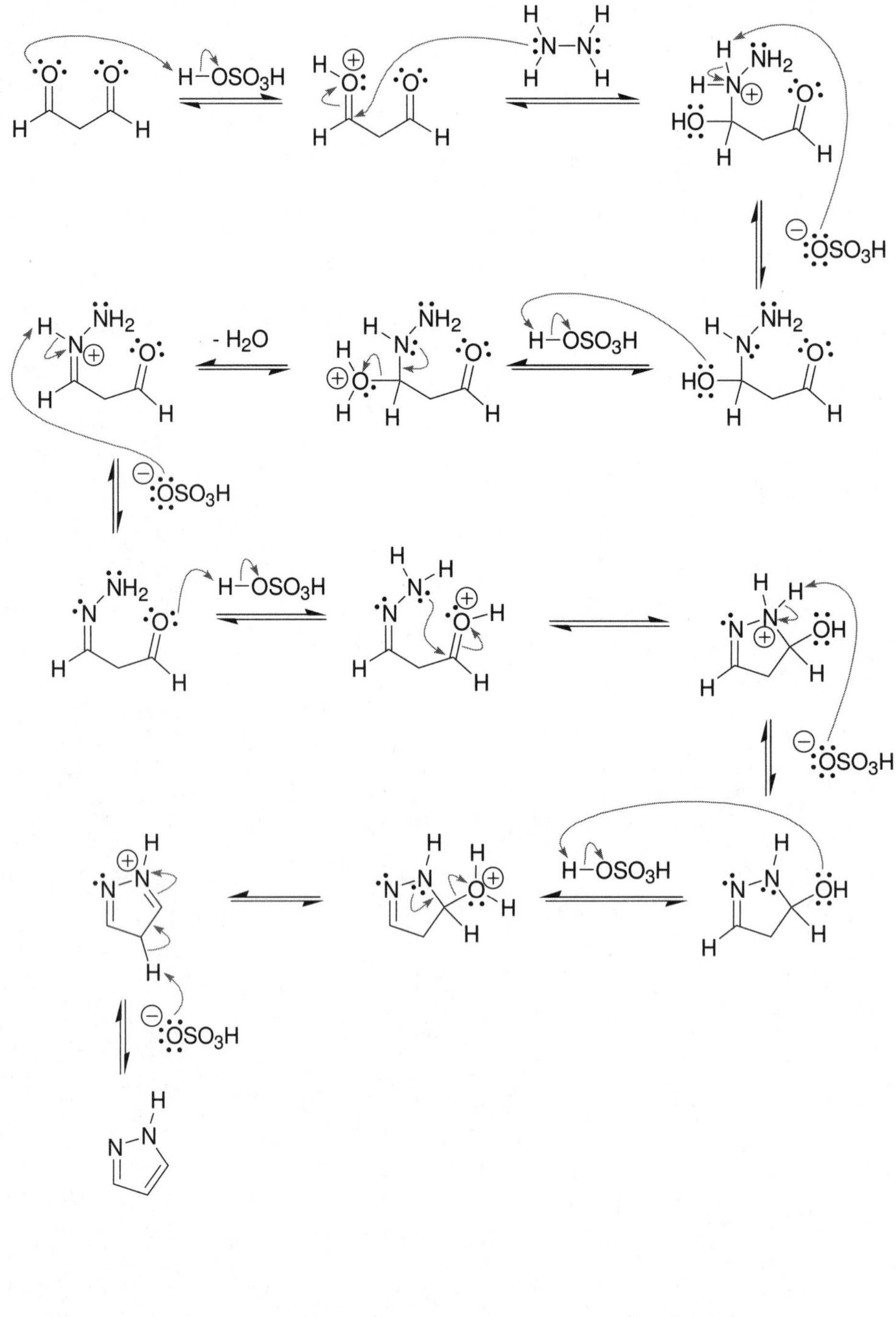

H–OSO₃H

H–OSO₃H

- H_2O

:ÖSO₃H

H–OSO₃H

:ÖSO₃H

H–OSO₃H

:ÖSO₃H

:ÖSO₃H

d)

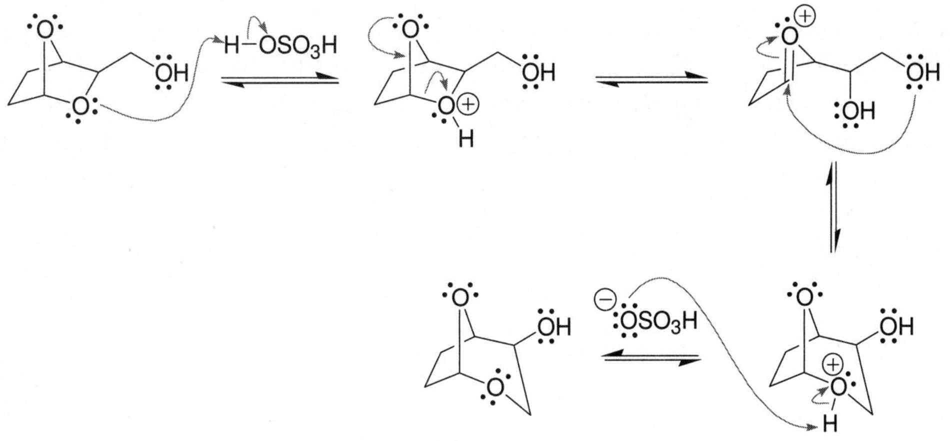

H–OSO$_3$H

OH

$^{\ominus}$:OSO$_3$H

e)

H–OSO$_3$H

HO OH

$^{\ominus}$:OSO$_3$H

OH

≡

HO

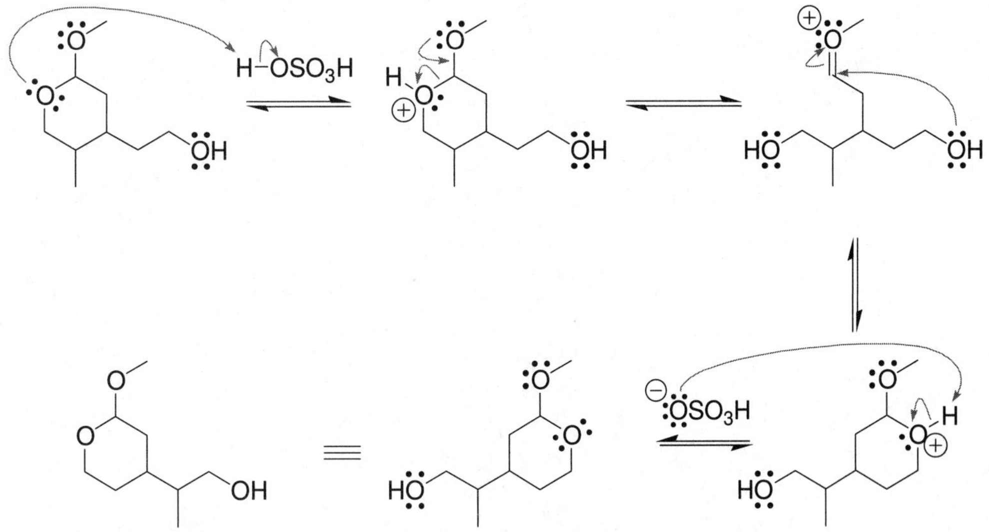

f)

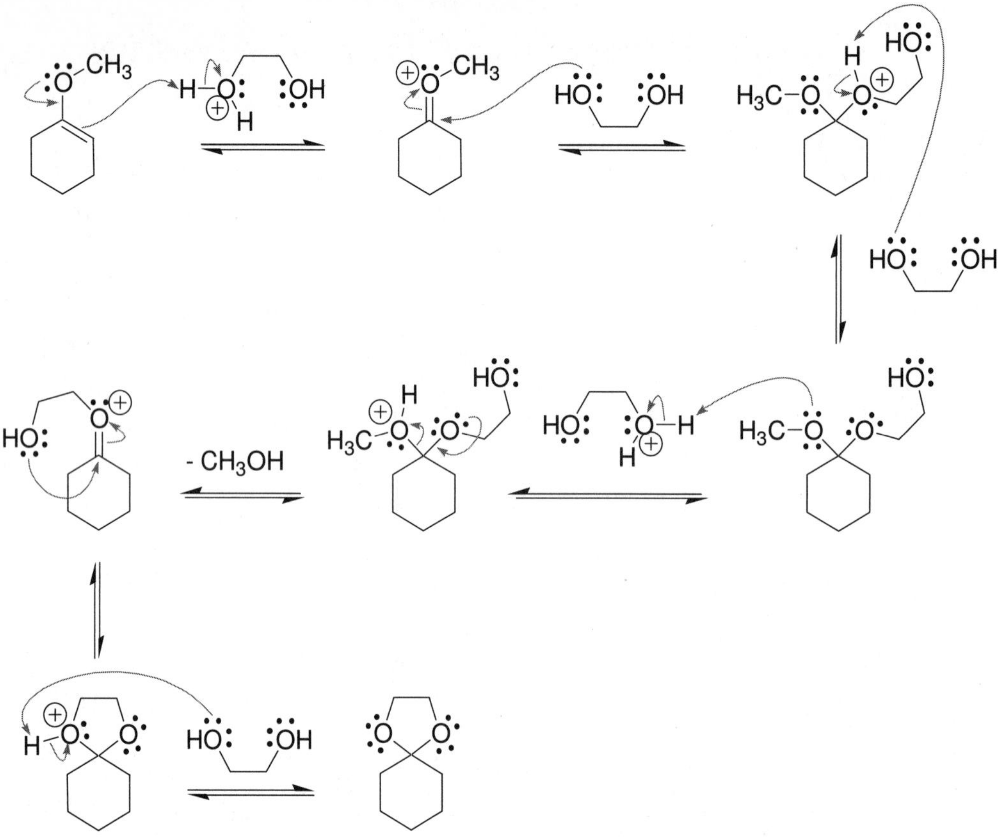

H_3C CH_3 HO OH - CH_3OH

20.85.

H_2O

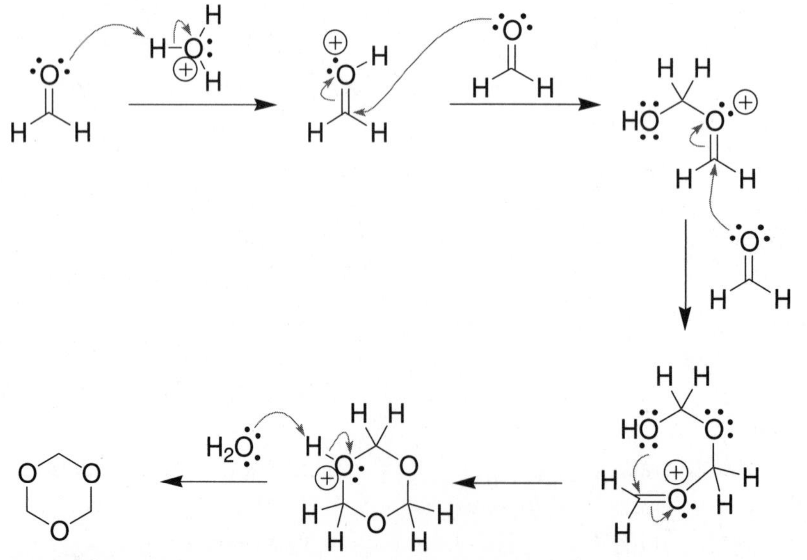

paraformaldehyde

Chapter 21
Carboxylic Acids and Their Derivatives

Review of Concepts

Fill in the blanks below. To verify that your answers are correct, look in your textbook at the end of Chapter 21. Each of the sentences below appears verbatim in the section entitled *Review of Concepts and Vocabulary.*

- Treatment of a carboxylic acid with a strong base yields a _____________ salt.
- The pK_a of most carboxylic acids is between ____ and _____.
- Using the **Henderson-Hasselbalch equation**, it can be shown that carboxylic acids exist primarily ______________________ at **physiological pH**.
- Electron-_____________ substituents can increase the acidity of a carboxylic acid.
- When treated with aqueous acid, a nitrile will undergo ___________, yielding a carboxylic acid.
- Carboxylic acids are reduced to ________ upon treatment with lithium aluminum hydride or borane.
- **Carboxylic acid derivatives** exhibit the same ________ state as carboxylic acids.
- Carboxylic acid derivatives differ in reactivity, with _____________ being the most reactive and __________ the least reactive.
- When drawing a mechanism, avoid formation of ___________ charges in acidic conditions, and avoid formation of ____________ charges in alkaline conditions.
- When a nucleophile attacks a carbonyl group to form a tetrahedral intermediate, always reform the carbonyl if possible, but never expel _____ or ______.
- When treated with an alcohol, acid chlorides are converted into _________.
- When treated with ammonia, acid chlorides are converted into _________.
- When treated with a __________ reagent, acid chlorides are converted into alcohols with the introduction of two alkyl groups.
- The reactions of anhydrides are the same as the reactions of ________________ except for the identity of the leaving group.
- When treated with a strong base followed by an alkyl halide, carboxylic acids are converted into __________.
- In a process called the **Fischer esterification**, carboxylic acids are converted into esters when treated with an ____________ in the presence of ________________.
- Esters can be hydrolyzed to yield carboxylic acids by treatment with either aqueous base or aqueous _______. Hydrolysis under basic conditions is also called ________________.
- When treated with lithium aluminum hydride, esters are reduced to yield ___________. If the desired product is an aldehyde, then ________ is used as a reducing agent instead of LAH.
- When treated with a _____________ reagent, esters are reduced to yield alcohols, with the introduction of two alkyl groups.
- When treated with excess LAH, amides are converted into __________.
- Nitriles are converted to amines when treated with _______.

Review of Skills

Fill in the blanks and empty boxes below. To verify that your answers are correct, look in your textbook at the end of Chapter 21. The answers appear in the section entitled *SkillBuilder Review*.

21.1 Drawing the Mechanism of a Nucleophilic Acyl Substitution Reaction

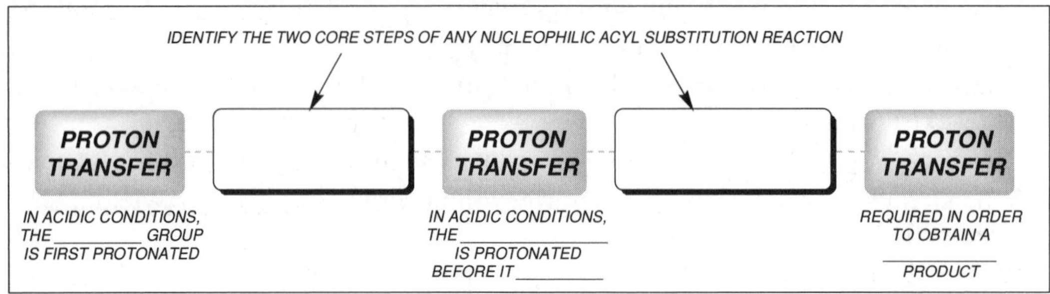

21.2 Interconverting Functional Groups

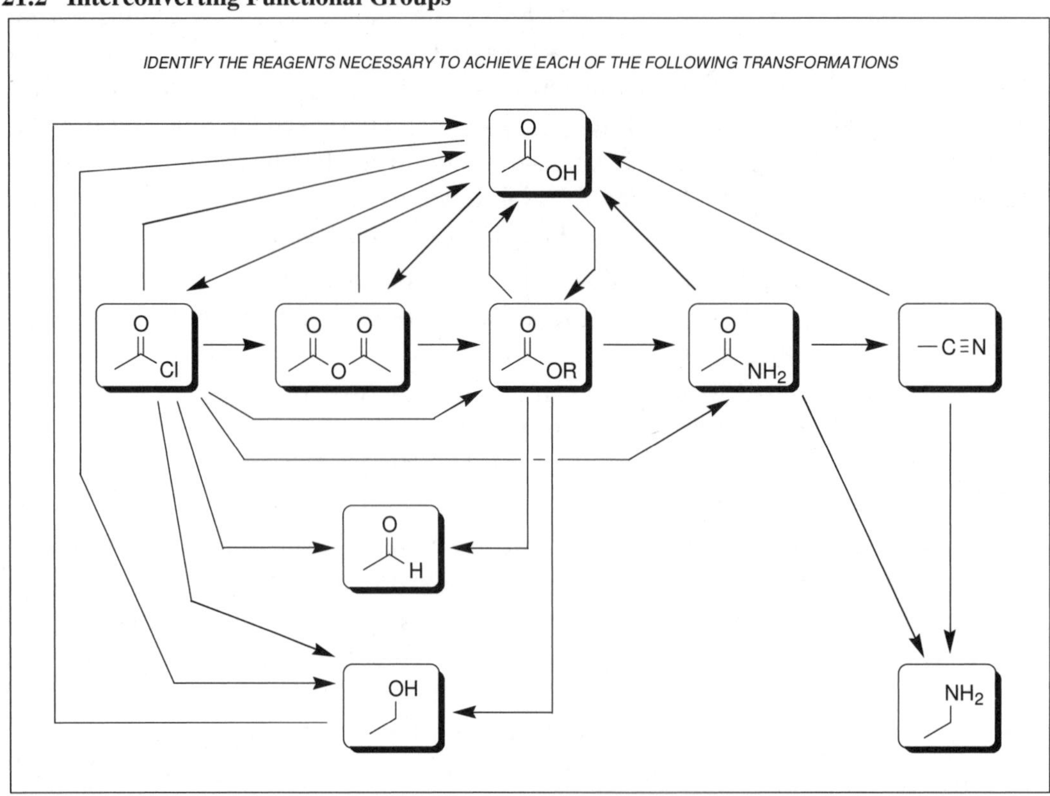

21.3 Choosing the Most Efficient C-C Bond-Forming Reaction

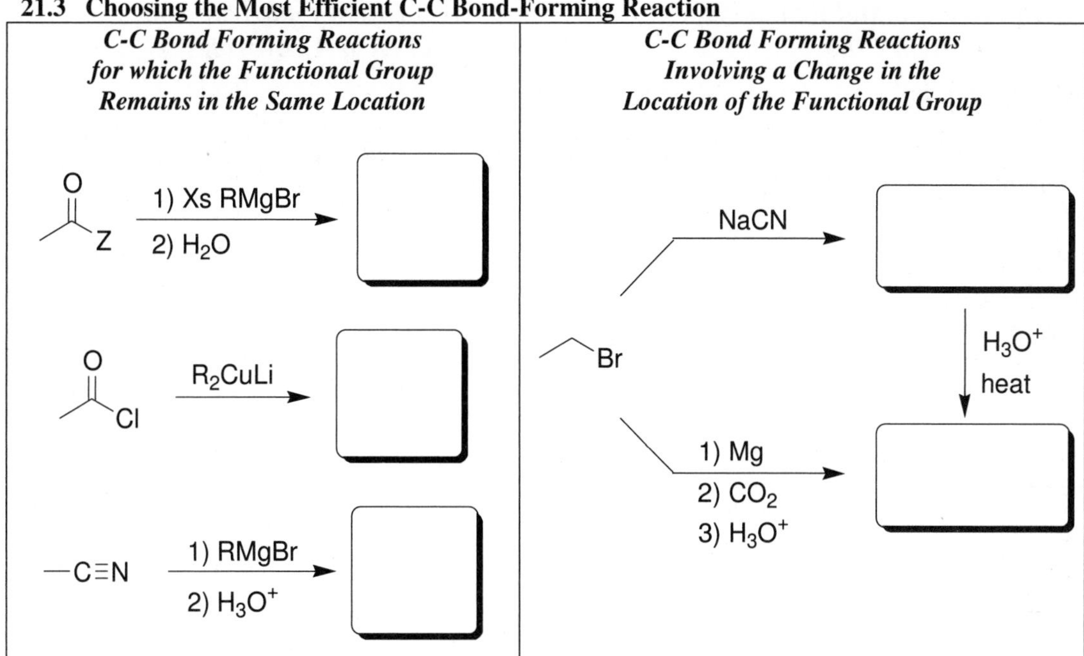

Review of Reactions

Identify the reagents necessary to achieve each of the following transformations. To verify that your answers are correct, look in your textbook at the end of Chapter 21. The answers appear in the section entitled *Review of Reactions*.

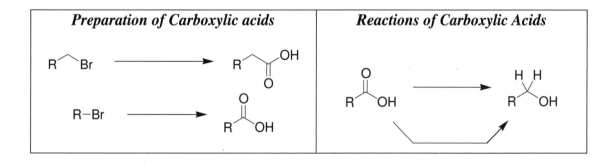

Preparation of Carboxylic acids	*Reactions of Carboxylic Acids*
R⌒Br ⟶ R⌒C(=O)OH R–Br ⟶ RC(=O)OH	RC(=O)OH ⟶ RCH₂OH (H H) (second arrow route, reagents to be identified)

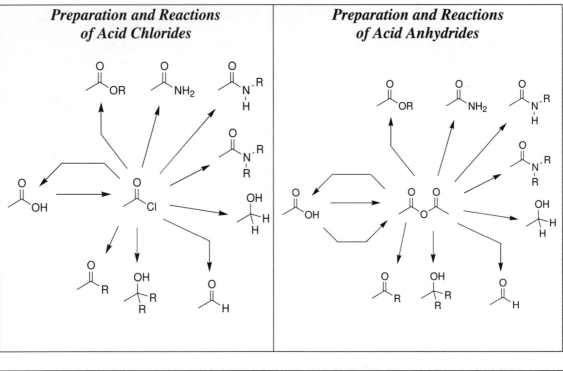

Preparation and Reactions of Acid Chlorides

Preparation and Reactions of Acid Anhydrides

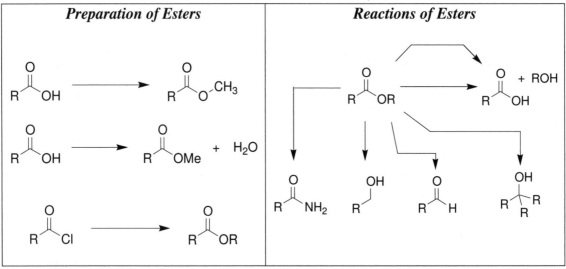

Preparation of Esters

Reactions of Esters

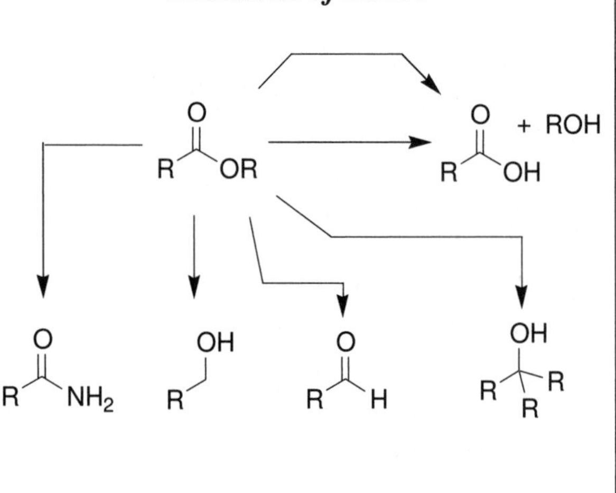

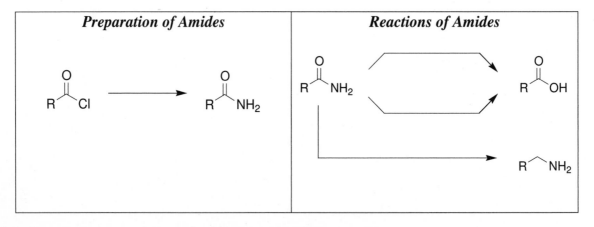

Preparation of Amides

Reactions of Amides

Preparation of Nitriles	*Reactions of Nitriles*
R–CH$_2$–Br ⟶ R–CH$_2$–C≡N	R–C≡N ⟶ RCOOH (two pathways)
RC(=O)NH$_2$ ⟶ R–C≡N	R–C≡N ⟶ RC(=O)R
	R–C≡N ⟶ RCH$_2$NH$_2$

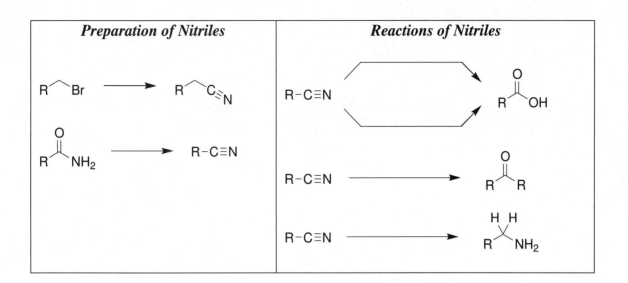

Solutions

21.1.

a) IUPAC name = pentanedioic acid
Common name = glutaric acid

b) IUPAC name = butanoic acid
Common name = butyric acid

c) IUPAC name = benzene carboxylic acid
Common name = benzoic acid

d) IUPAC name = butanedioic acid
Common name = succinic acid

e) IUPAC name = ethanoic acid
Common name = acetic acid

f) IUPAC name = methanoic acid
Common name = formic acid

21.2.

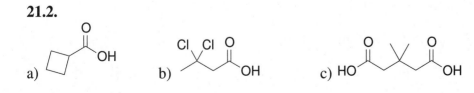

a) 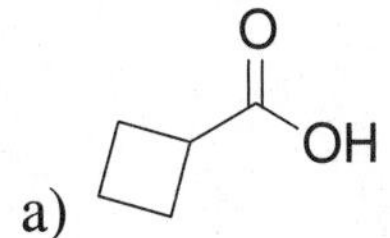

b) Cl Cl O OH

c) HO O O OH

21.3.
a) 3,3,4,4-tetramethylhexanoic acid
b) 2-propylpentanoic acid
c) (*S*)-2-amino-3-phenylpropanoic acid

21.4. The compound below is more acidic because its conjugate base is resonance stabilized. The conjugate base of the other compound is not resonance stabilized.

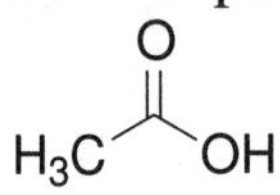

O

H_3C OH

21.5.
The conjugate base is resonance stabilized, with the negative charge spread over two oxygen atoms, just like with carboxylic acids:

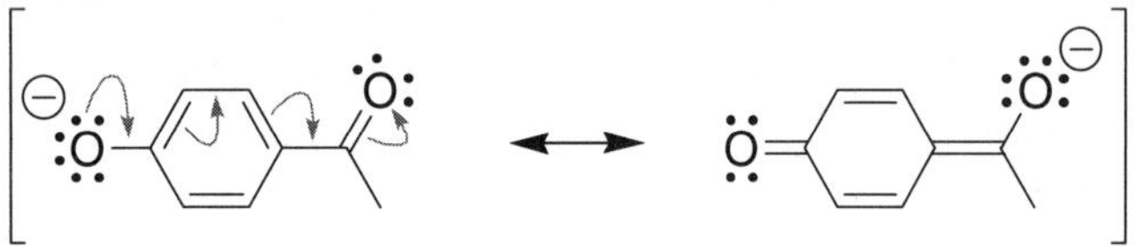

21.6. *meta*-Hydroxyacetophenone should be less acidic than *para*-hydroxyacetophenone, because in the conjugate base of the former, the negative charge is spread over only one oxygen atoms (and three carbon atoms). In contrast, the conjugate base of *para*-hydroxyacetophenone has the negative charge spread over two oxygen atoms (more stable).

21.7.

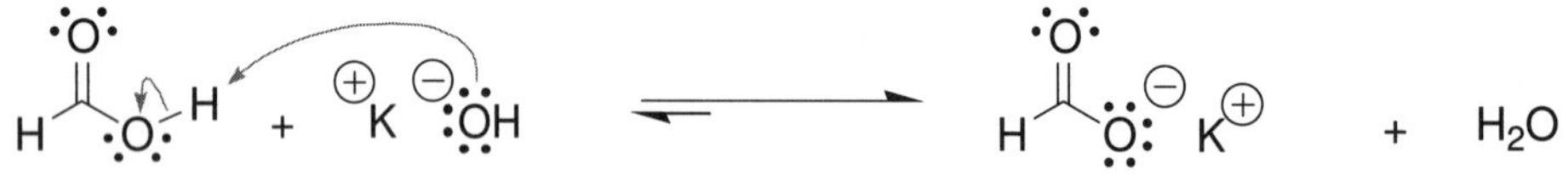

formic acid + K⊕ ⊖OH ⇌ potassium formate + H_2O

21.8. The conjugate base predominates under these conditions:

$$\frac{[\text{conjugate base}]}{[\text{acid}]} = 10^{(\text{pH} - \text{p}K_a)} = 10^{(5.76 - 4.76)} = 10^1 = 10$$

21.9.
a) 2,3-dichlorobutyric acid is the most acidic and 3,4-dimethylbutyric acid is the least acidic.
b) 2,2-dibromopropionic acid is the most acidic and 3-bromopropionic acid is the least acidic.

21.10.

a) $Na_2Cr_2O_7$, H_2SO_4, H_2O
b) $Na_2Cr_2O_7$, H_2SO_4, H_2O
c) CH_3Cl, $AlCl_3$ followed by $Na_2Cr_2O_7$, H_2SO_4, H_2O
d) NaCN, followed by H_3O^+, heat
or Mg, followed by CO_2, followed by H_3O^+
e) $Na_2Cr_2O_7$, H_2SO_4, H_2O
f) Mg, followed by CO_2, followed by H_3O^+

21.11.

a) Br (bromobenzene) → 1) Mg 2) CO_2 3) H_3O^+ 4) LAH 5) H_2O → OH

b) (toluene) → 1) $Na_2Cr_2O_7$, H_2SO_4, H_2O 2) LAH 3) H_2O → OH
(toluene) → 1) NBS, heat 2) NaOH → OH

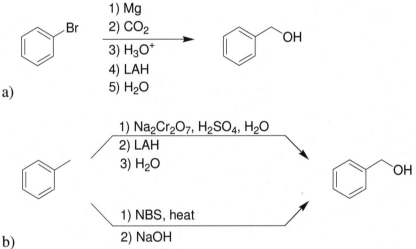

21.12.

a) propionic anhydride
b) *N,N*-diphenyl-propionamide
c) dimethylsuccinate
d) *N*-ethyl-*N*-methylcyclobutanecarboxamide
e) butyronitrile
f) propyl butyrate
g) succinic anhydride
h) methyl benzoate
i) phenyl acetate

21.13.

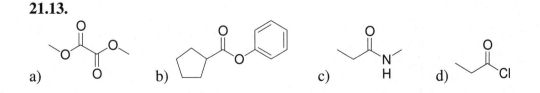

a) O O O O b) O O c) O N H d) O Cl

21.14.

a)

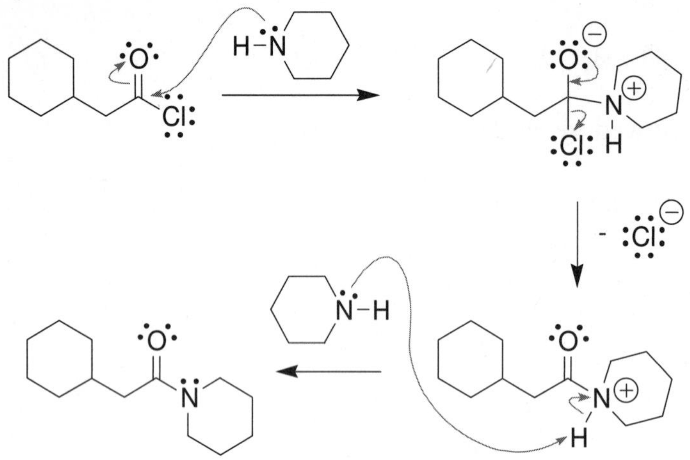

b)

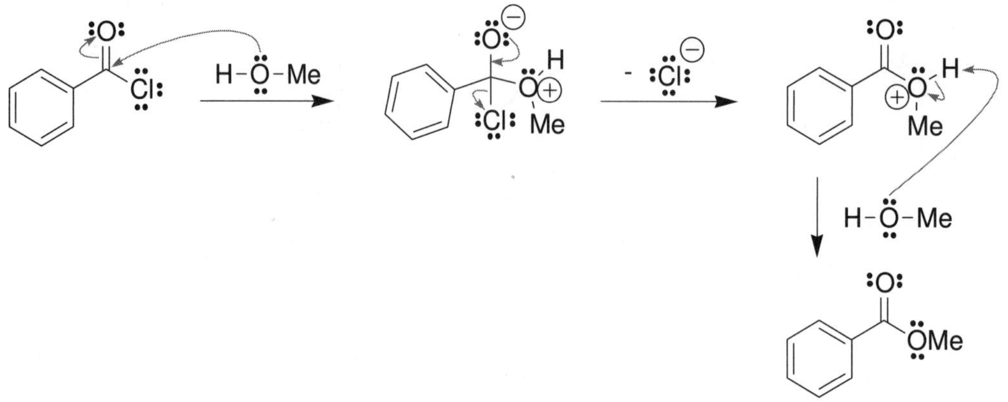

c)

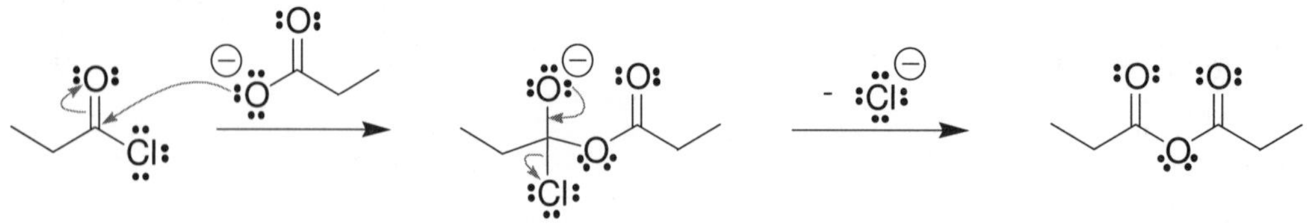

d)

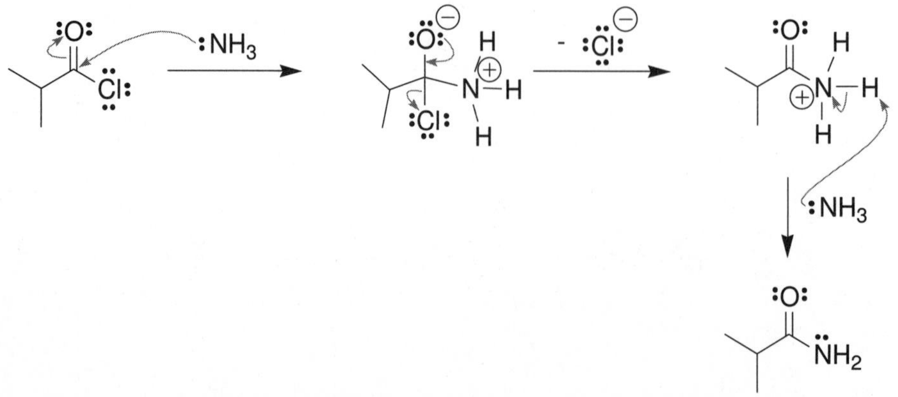

e)

f)

g)

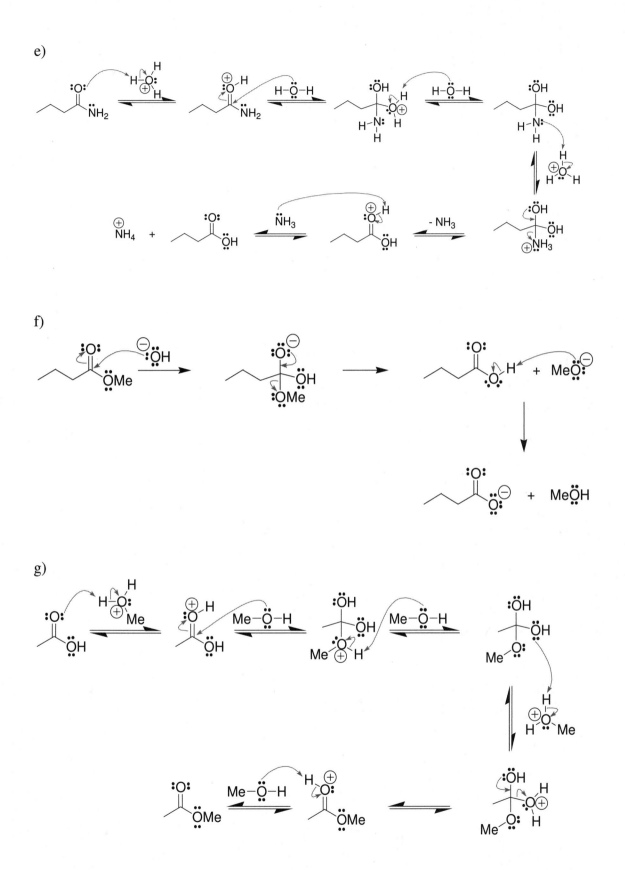

21.15.

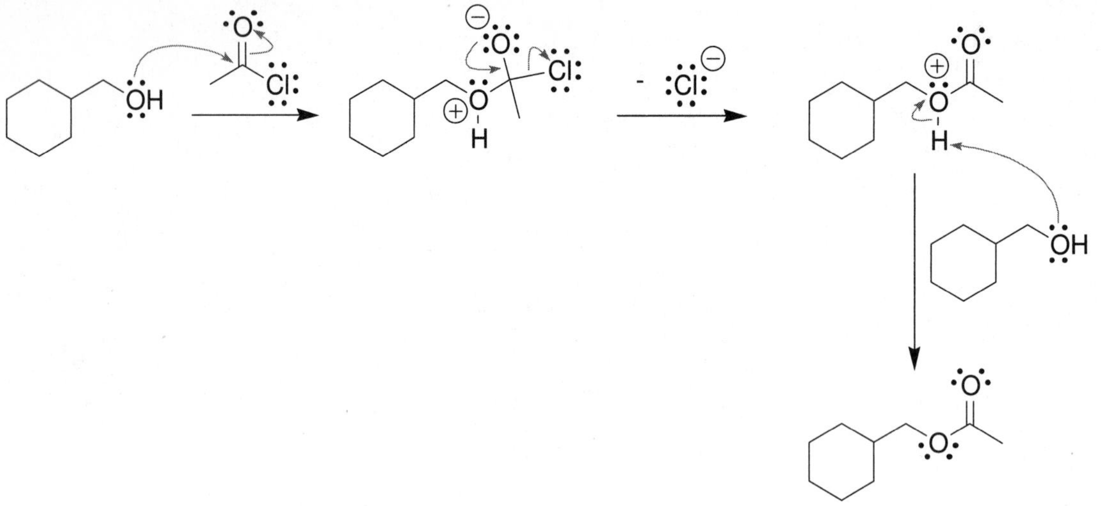

- :Cl: (Cl⁻)

21.16.

- H_2O

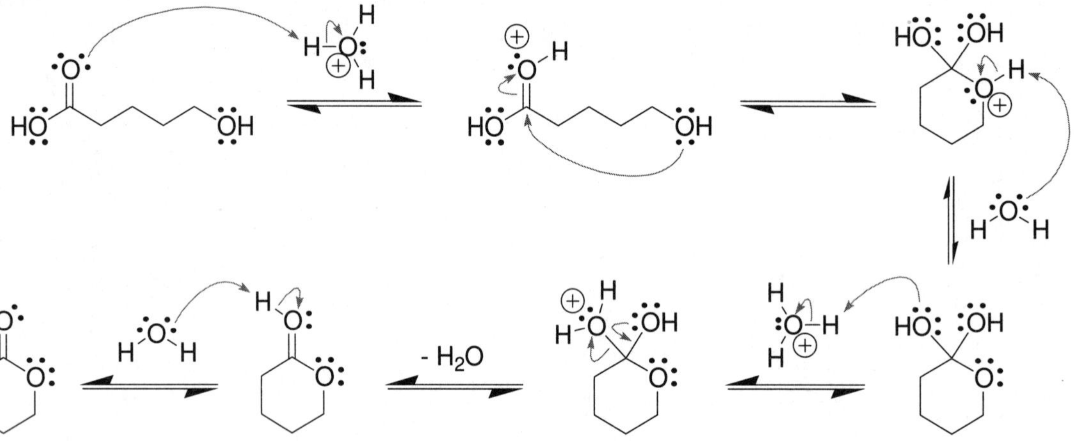

21.17.

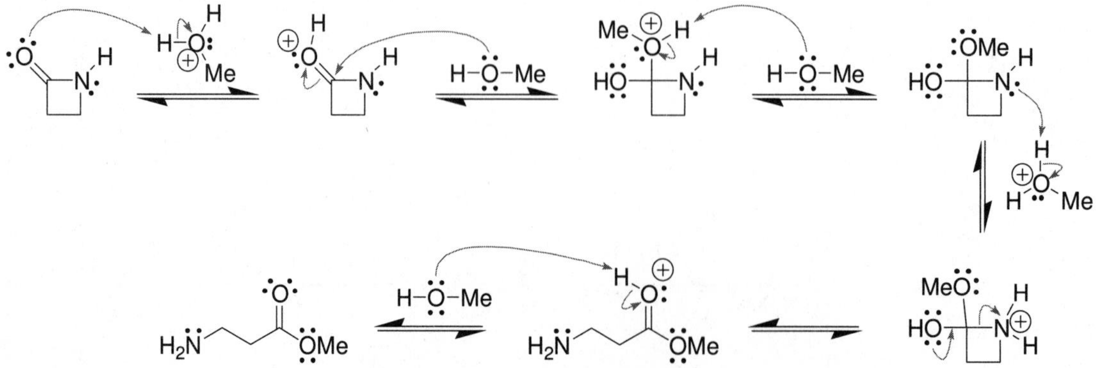

21.18.

a)
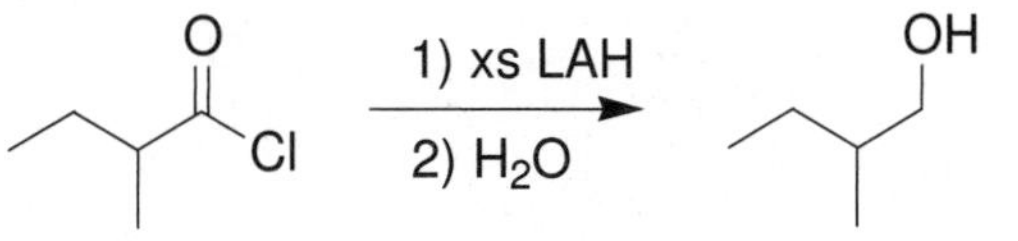
O
Cl
1) xs LAH
2) H_2O
OH

b)
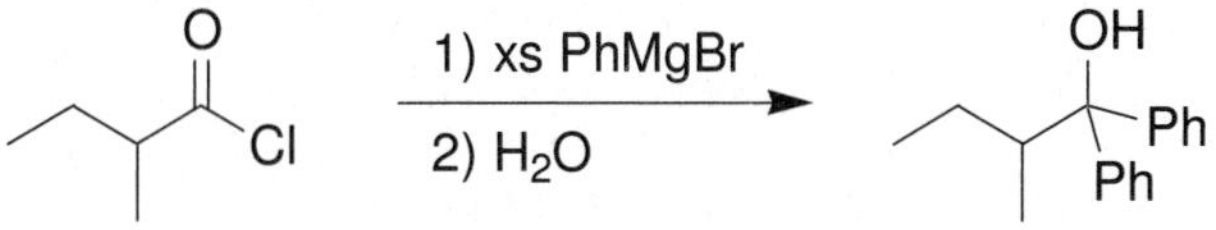
O
Cl
1) xs PhMgBr
2) H_2O
OH
Ph
Ph

c)
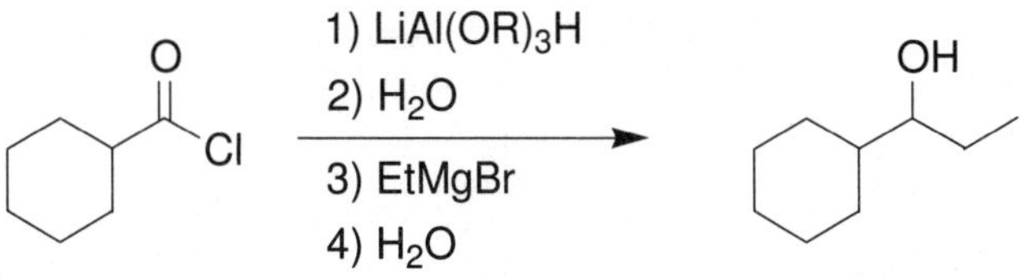
O
Cl
1) $LiAl(OR)_3H$
2) H_2O
3) EtMgBr
4) H_2O
OH

d)
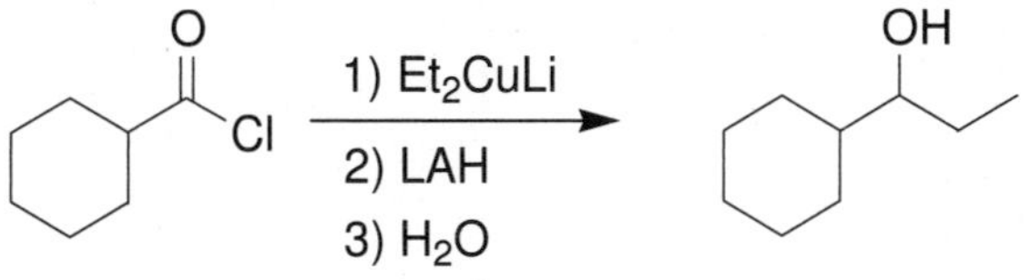
O
Cl
1) Et_2CuLi
2) LAH
3) H_2O
OH

e)
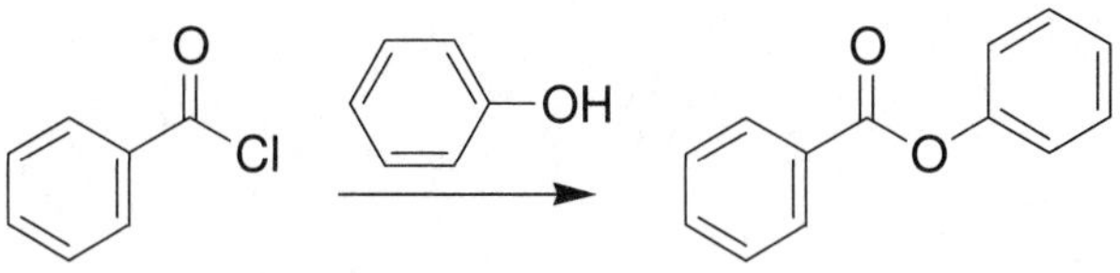
O
Cl
OH
O
O

f)
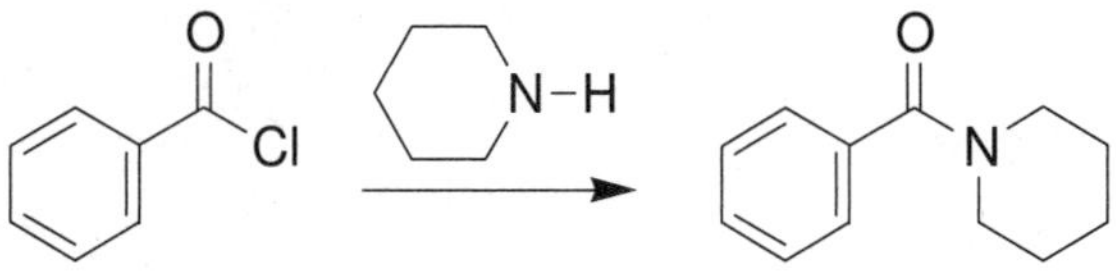
O
Cl
N–H
O
N

21.19.

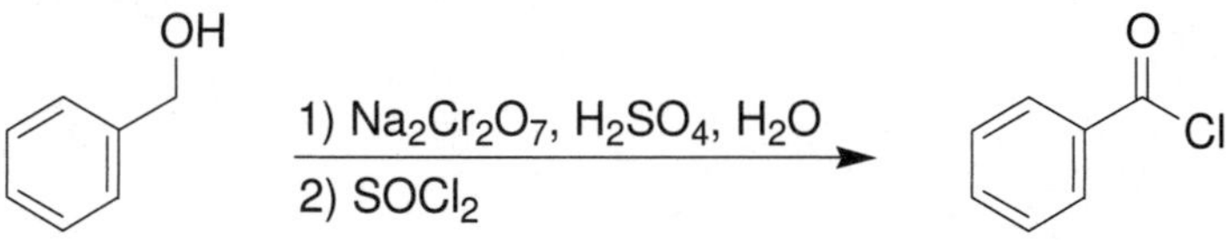
OH
1) $Na_2Cr_2O_7$, H_2SO_4, H_2O
2) $SOCl_2$
O
Cl

21.20.

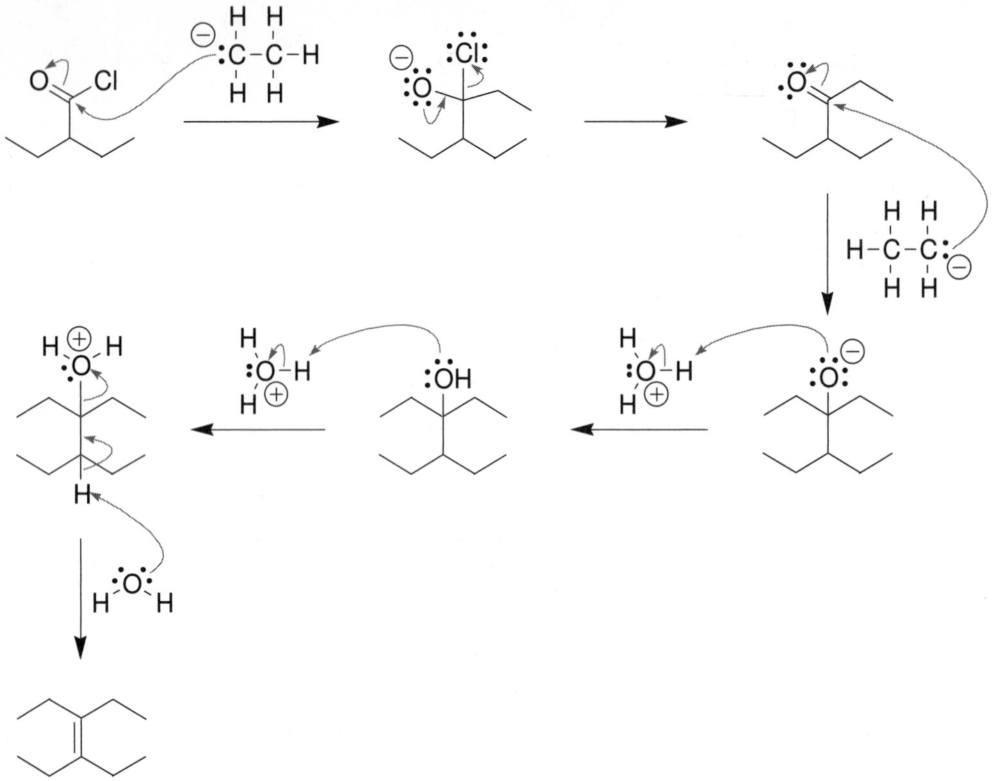

:C–C–H H H H H O Cl :Cl: O O H H H–C–C: H H H O–H H :OH H O–H H O H O H H H O H

21.21.

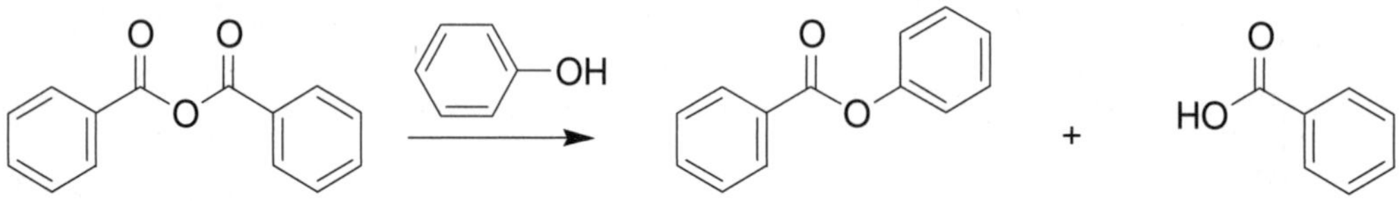

O O OH O O HO

a)

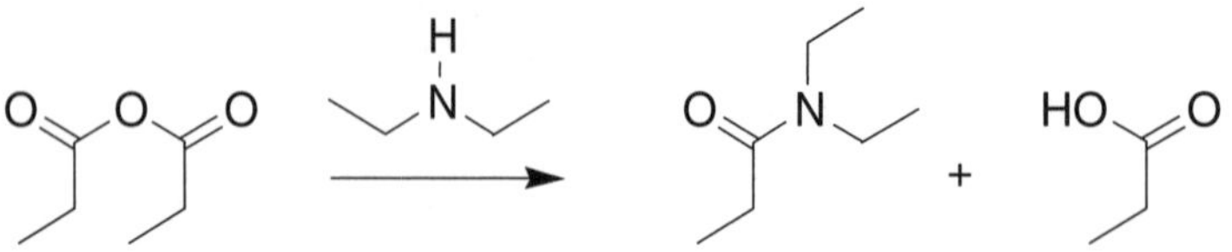

H N O O O N HO O

b)

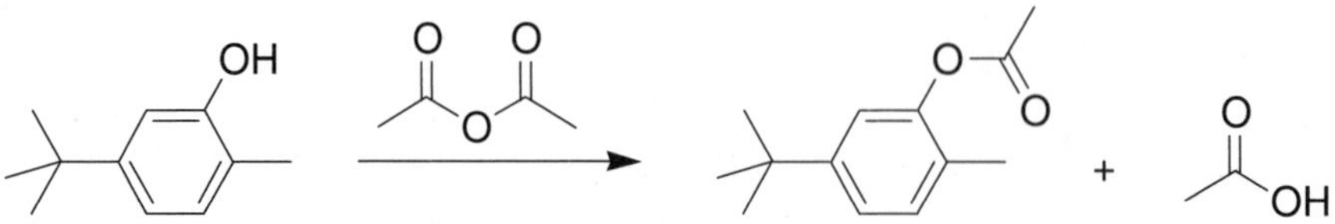

OH O O O O O O OH

c)

H N O O O O N O OH

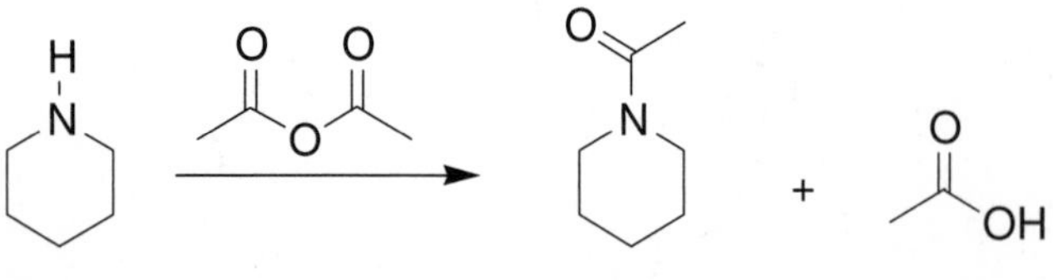

d)

21.22.

OH

1) NaOH
2) CH_3I

[H^+], EtOH, -H_2O

1) $SOCl_2$
2) EtOH

OEt

21.23.

a)

OH

$Na_2Cr_2O_7$
H_2SO_4, H_2O

OH

1) NaOH
2) CH_3I

[H^+], EtOH, -H_2O

1) $SOCl_2$
2) EtOH

OEt

b)

$Na_2Cr_2O_7$
H_2SO_4, H_2O

OH

1) NaOH
2) CH_3I

[H^+], EtOH, -H_2O

1) $SOCl_2$
2) EtOH

OEt

21.24.

a) OMe — 1) xs LAH, 2) H_2O → OH + MeOH

b) OMe — 1) xs EtMgBr, 2) H_2O → OH + MeOH

c) O — 1) xs LAH, 2) H_2O → HO OH

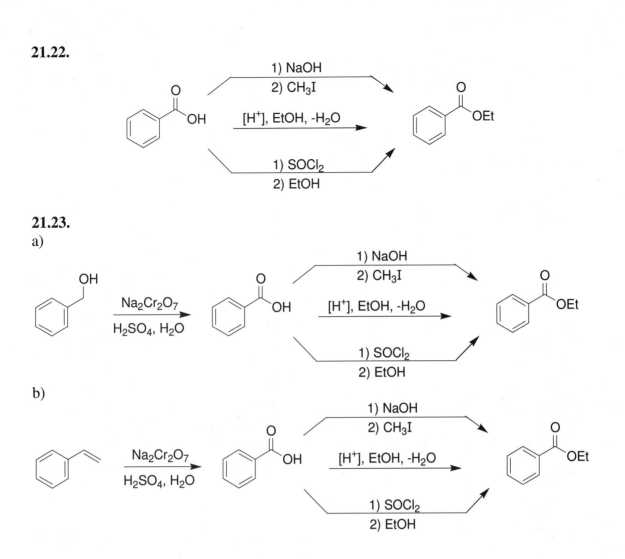

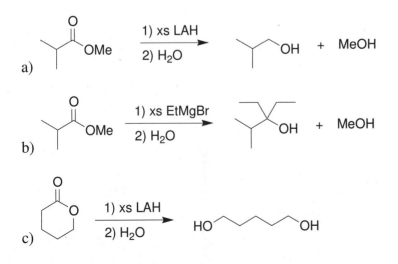

d) 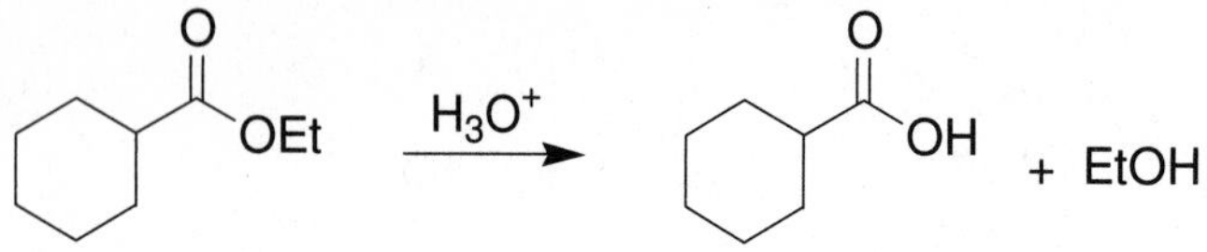 [cyclohexyl–C(=O)OEt] $\xrightarrow{H_3O^+}$ [cyclohexyl–C(=O)OH] + EtOH

e) 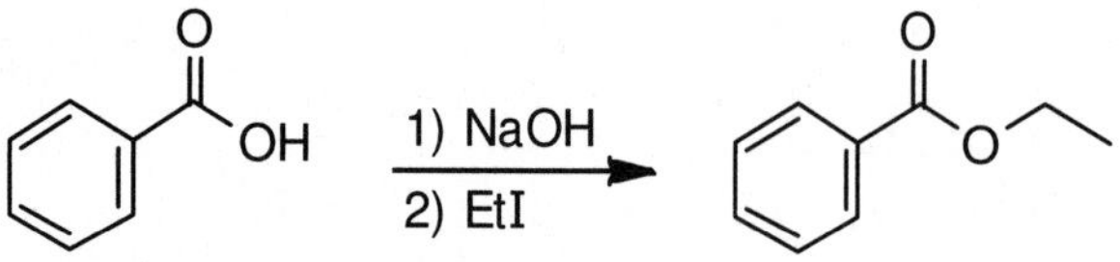 [Ph–C(=O)OH] $\xrightarrow[\text{2) EtI}]{\text{1) NaOH}}$ [Ph–C(=O)O–Et]

f) 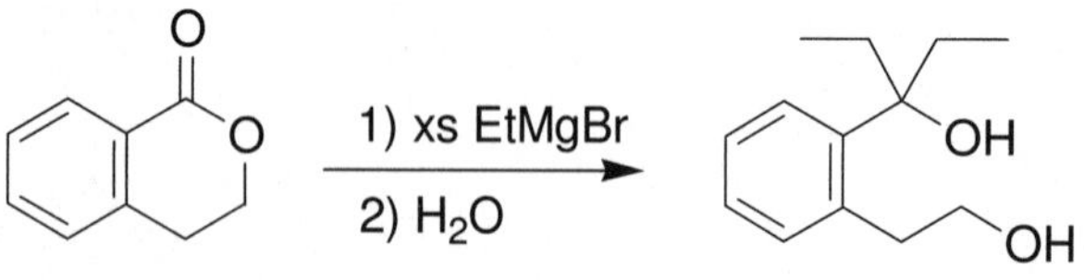 [isochromanone] $\xrightarrow[\text{2) } H_2O]{\text{1) xs EtMgBr}}$ [diol product: OH, OH]

21.25.

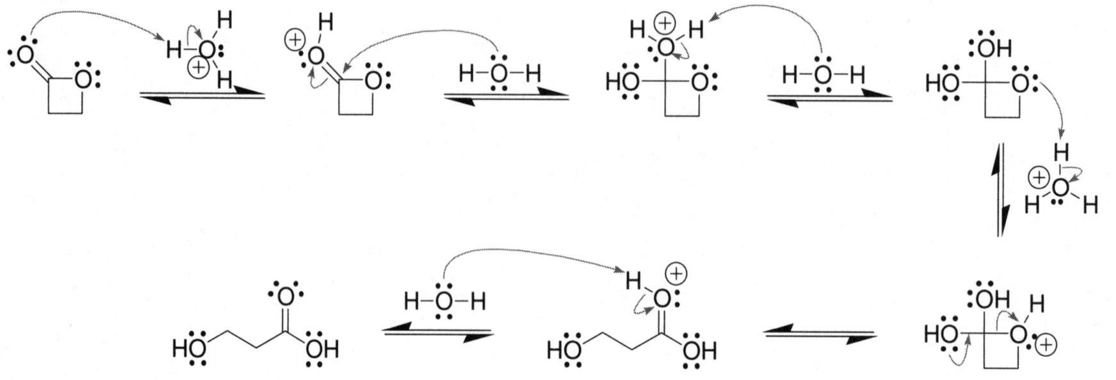

H–O–H, H–O–H, H–O–H

HO, :OH, HO, :OH, HO, OH

21.26.

a) 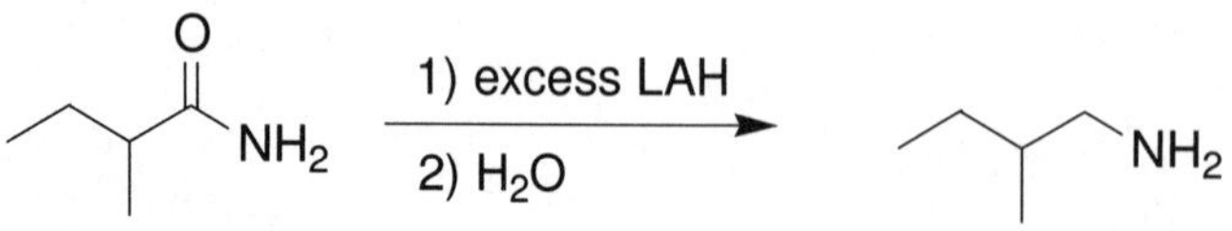 [amide, NH_2] $\xrightarrow[\text{2) } H_2O]{\text{1) excess LAH}}$ [amine, NH_2]

b) 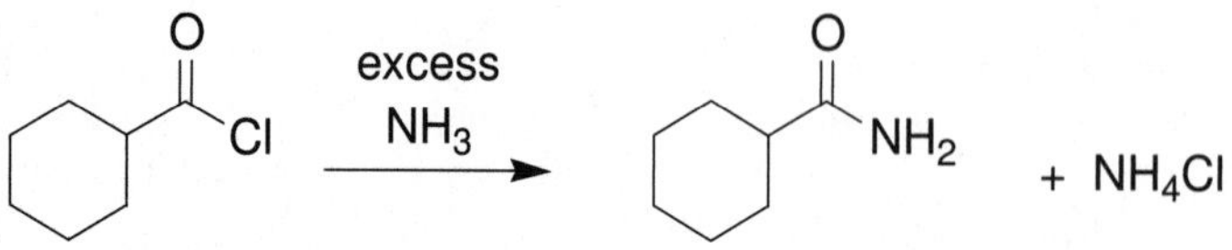 [cyclohexyl–C(=O)Cl] $\xrightarrow{\text{excess } NH_3}$ [cyclohexyl–C(=O)NH_2] + NH_4Cl

c) [cyclohexyl–C(=O)NH_2] $\xrightarrow[\text{heat}]{H_3O^+}$ [cyclohexyl–C(=O)OH] + NH_4^+

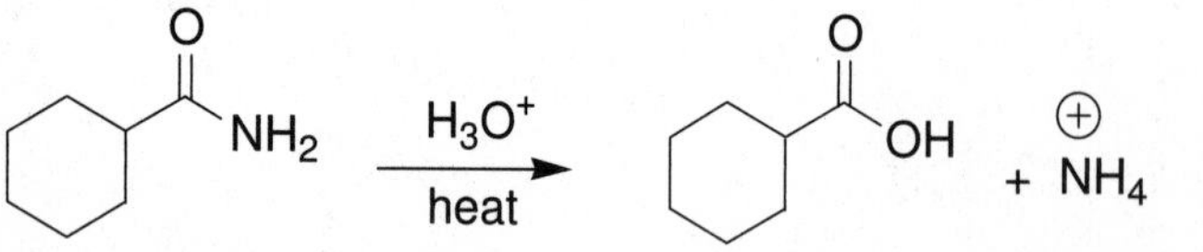

21.27.

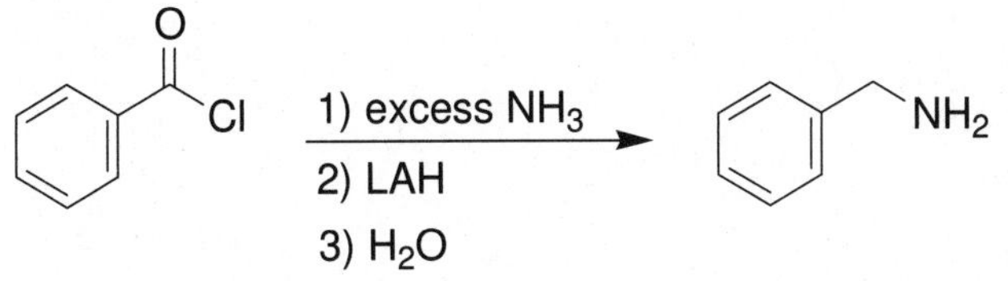

1) excess NH_3
2) LAH
3) H_2O

21.28.

a)

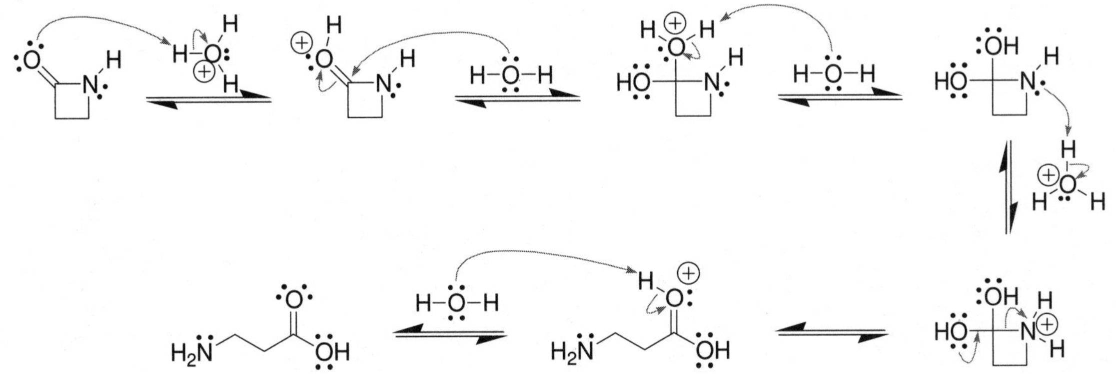

b)

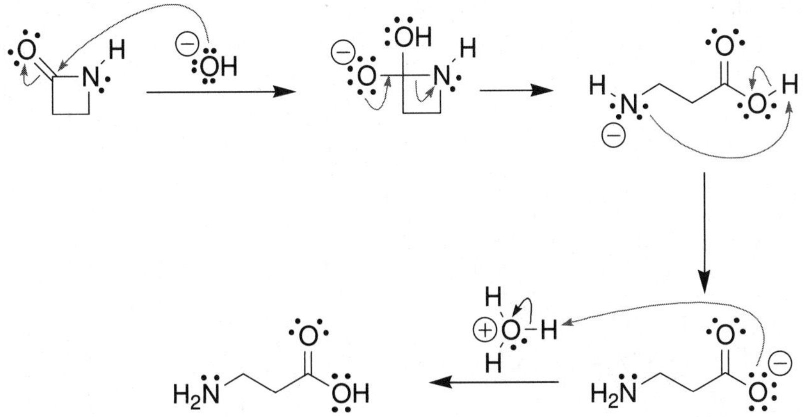

21.29.

a) CN —— 1) xs LAH, 2) H_2O ——> NH_2

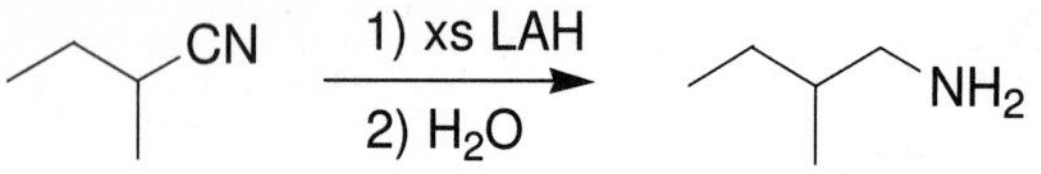

b) Br —— 1) NaCN, 2) MeMgBr, 3) H_3O^+ ——> O

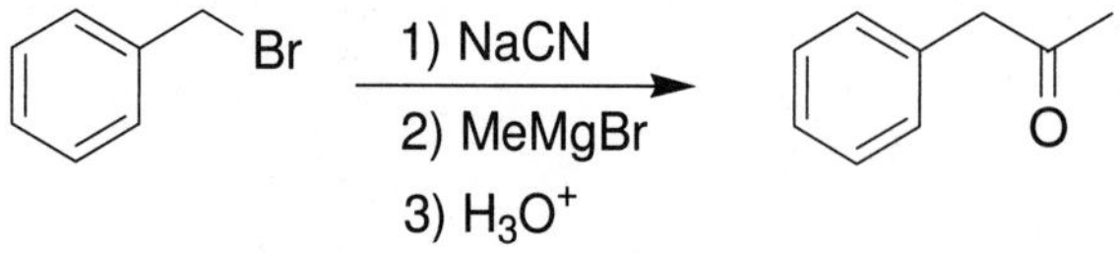

c) CN —— 1) EtMgBr, 2) H_2O ——> O —— 1) LAH, 2) H_2O ——> OH

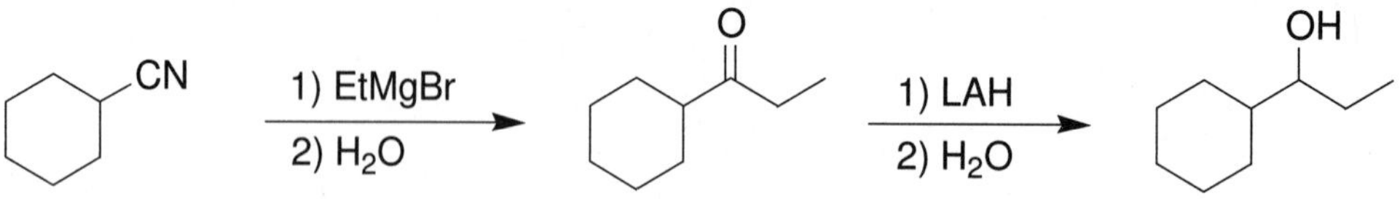

d) CN —— H_3O^+ ——> O, OH

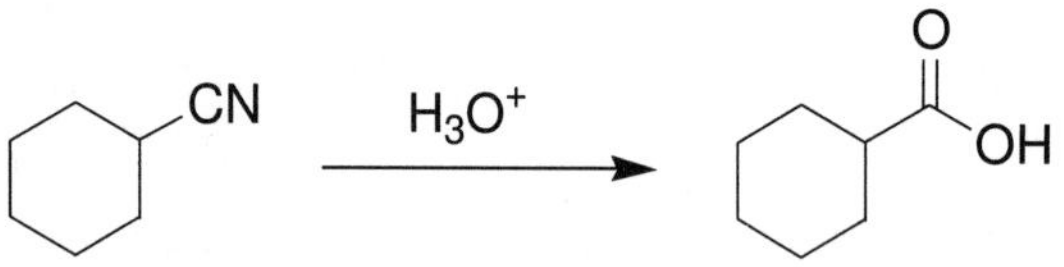

21.30.

a) O, Cl —— 1) excess NH_3, 2) $SOCl_2$ ——> CN

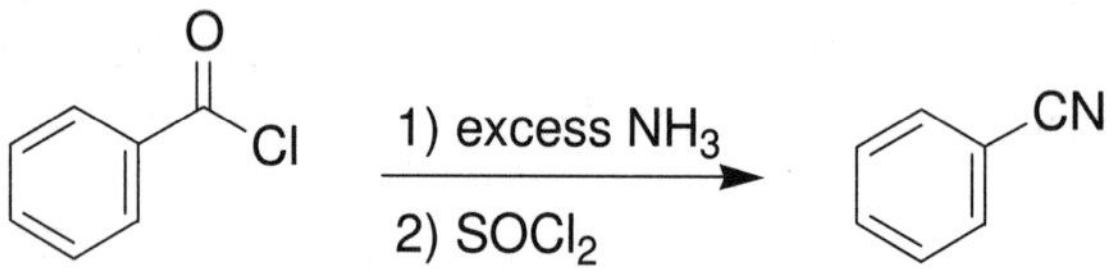

b) Br —— 1) NaCN, 2) H_3O^+ ——> OH, O

Br —— 1) Mg, 2) CO_2, 3) H_3O^+ ——> OH, O

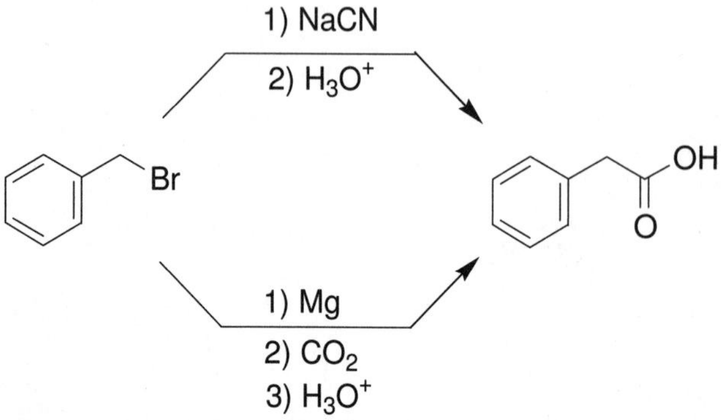

21.31.

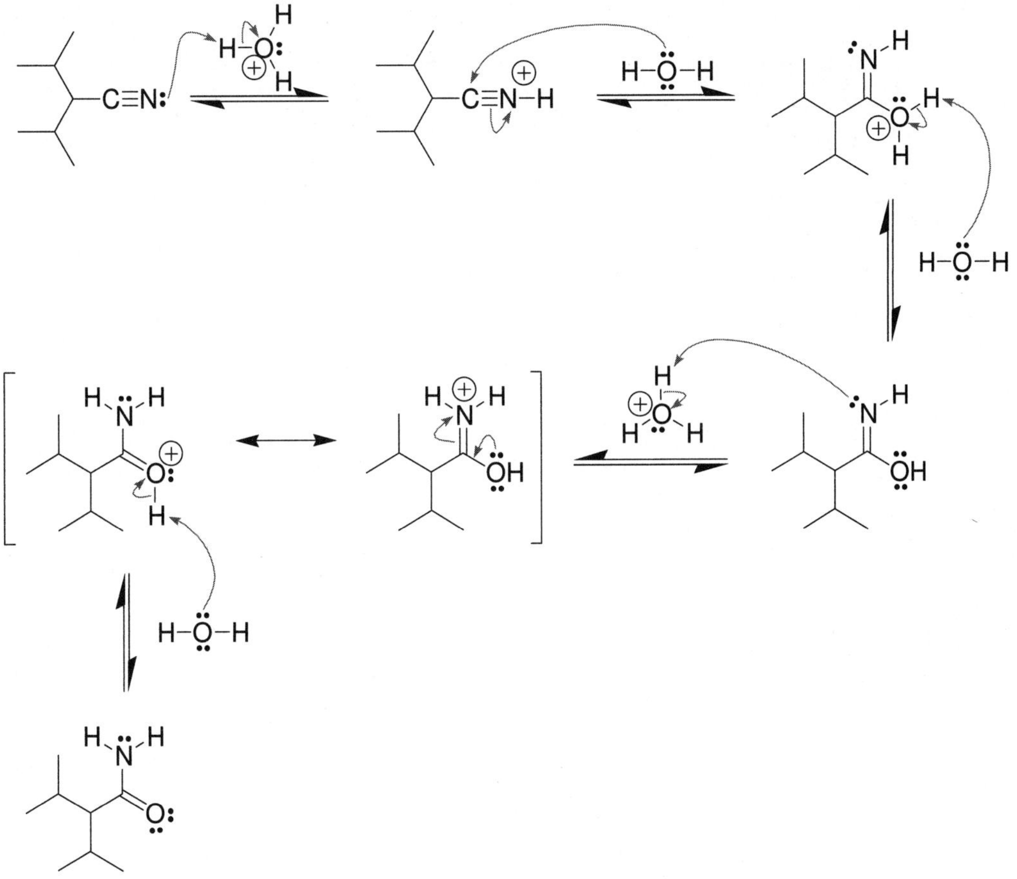

H–Ö–H

H–Ö–H

H–Ö–H

21.32.

a) 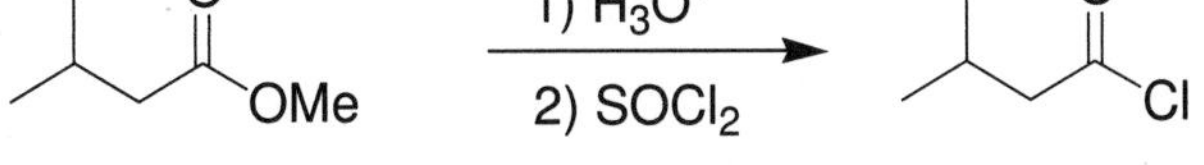 1) H_3O^+ 2) $SOCl_2$

OMe → Cl

b) 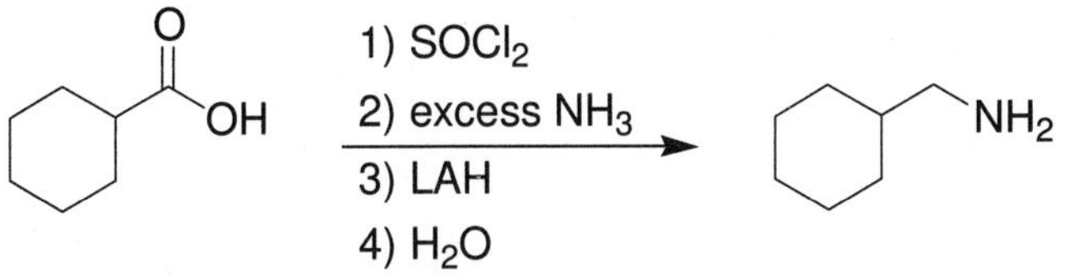 1) $SOCl_2$ 2) excess NH_3 3) LAH 4) H_2O

OH → NH_2

c) 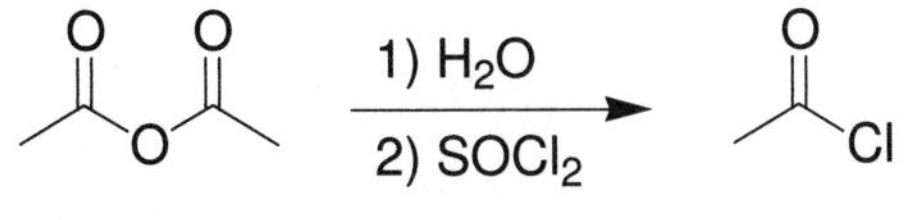 1) H_2O 2) $SOCl_2$

O → Cl

d) Cyclohexyl–C(=O)OEt → 1) H_3O^+ 2) $SOCl_2$ 3) excess NH_3 4) LAH 5) H_2O → cyclohexyl–CH_2NH_2

e) Cyclohexyl–C(=O)OH → 1) $SOCl_2$ 2) excess NH_3 3) $SOCl_2$ → cyclohexyl–CN

f) Acetic anhydride → H_2O → acetic acid (OH)

g) Cyclohexyl–C(=O)Cl → 1) H_2O 2) [H^+], EtOH → cyclohexyl–C(=O)OEt

h) Cyclohexyl–C(=O)NH_2 → 1) H_3O^+ 2) $SOCl_2$ 3) $LiAl(OR)_3H$ 4) H_2O → cyclohexyl–C(=O)H

i) Cyclohexyl–CN → 1) H_3O^+ 2) excess LAH 3) H_2O → cyclohexyl–CH_2OH

j) Cyclohexyl–CH_2OH → 1) $Na_2Cr_2O_7$, H_2SO_4, H_2O 2) $SOCl_2$ 3) excess NH_3 4) $SOCl_2$ → cyclohexyl–CN

21.33. Four steps: 1) oxidize to a carboxylic acid, 2) convert into an acid halide, 3) convert into an amide, and 4) reduce to give an amine.

21.34.

1) $BH_3 \cdot THF$
2) H_2O_2, NaOH
3) $Na_2Cr_2O_7$, H_2SO_4, H_2O
4) $SOCl_2$

O
Cl

21.35.

a)

O
OH
1) $SOCl_2$
2) Et_2CuLi
3) LAH
4) H_2O
OH

b)

Br
1) Mg
2) H, =O, H
3) H_2O
4) O, Cl
O
O

c)

CN
1) H_3O^+
2) $SOCl_2$
3) excess MeMgBr
4) O, Cl
O
O

21.36.

O
NH_2
$SOCl_2$
C≡N
1) EtMgBr
2) H_3O^+
O
$[H^+]$
$(CH_3)_2NH$
$-H_2O$
N

21.37.

a)

1) $SOCl_2$
2) $LiAl(OR)_3H$
3) H_2O

CH_3CN $\xrightarrow{H_3O^+}$ OH $\xrightarrow{\text{1) xs LAH; 2) } H_2O}$ OH $\xrightarrow{PBr_3}$ Br

1) Mg
2) H
3) H_2O

OH

b)

CH_3CN $\xrightarrow{H_3O^+}$ OH $\xrightarrow{\text{1) xs LAH; 2) } H_2O}$ OH $\xrightarrow{PBr_3}$ Br

1) Mg
2) CO_2
3) H_3O^+

OH

1) $SOCl_2$
2) $LiAl(OR)_3H$

H

Mg

1) EtMgBr
2) H_2O

OH

c)

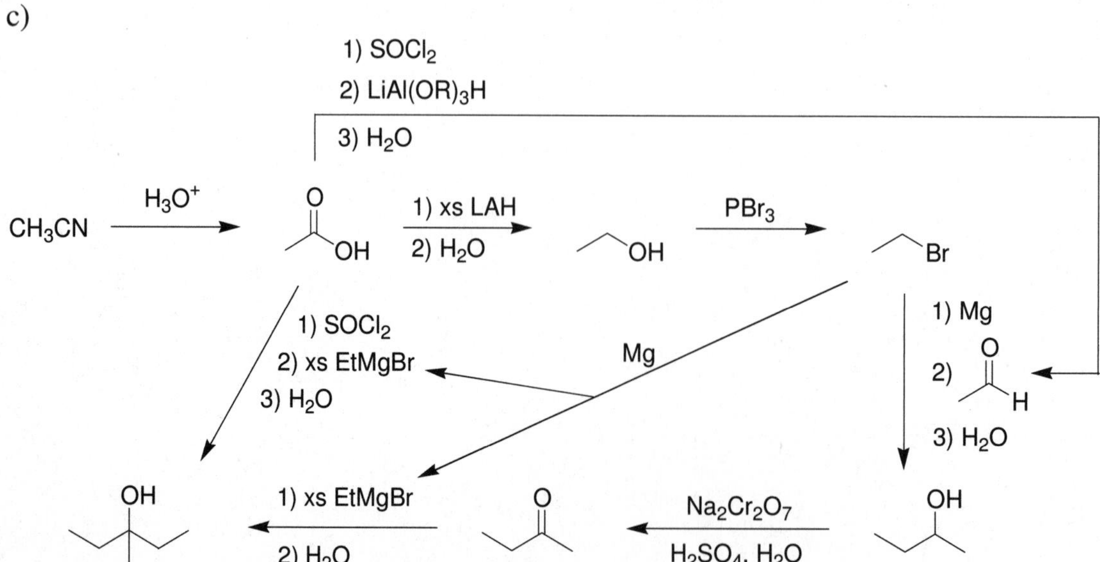

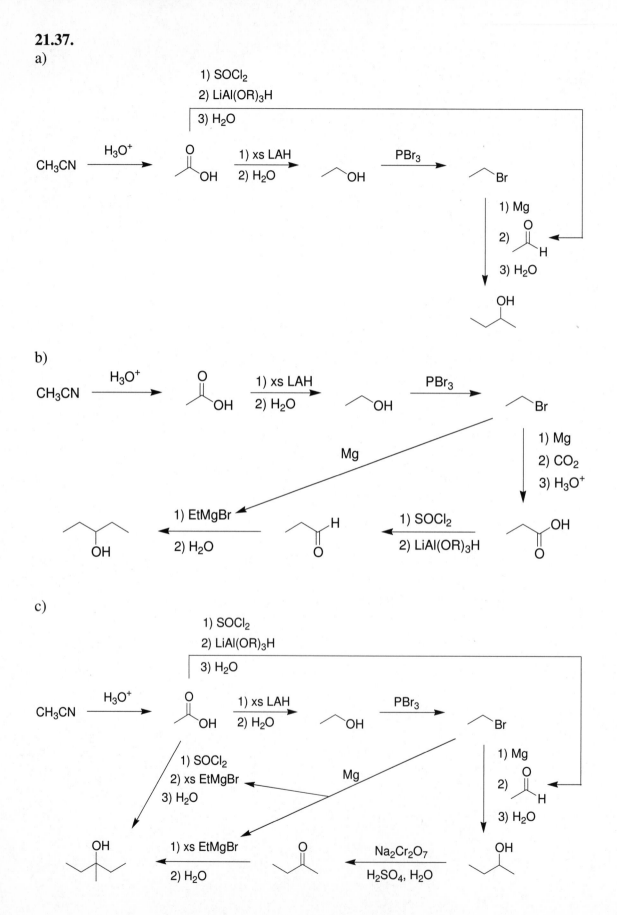

d)

CH_3CN —(H_3O^+)→ acetic acid —(1) xs LAH; 2) H_2O)→ ethanol —(PBr_3)→ ethyl bromide

ethyl bromide —(1) Mg; 2) CO_2; 3) H_3O^+)→ propanoic acid —($SOCl_2$)→ propanoyl chloride —(1) xs EtMgBr; 2) H_2O)→ 3-ethyl-3-pentanol (OH)

ethyl bromide —(Mg)→ EtMgBr

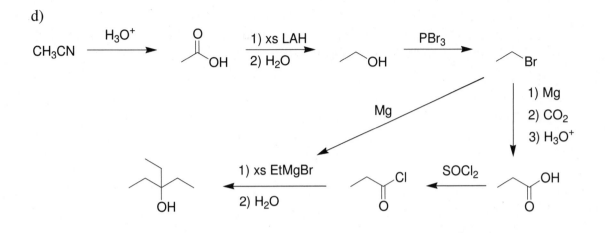

21.38. The signal at 1740 cm^{-1} indicates that the carbonyl group is not conjugated with the aromatic ring (it would be at a lower wavenumber if it was conjugated),

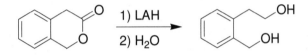

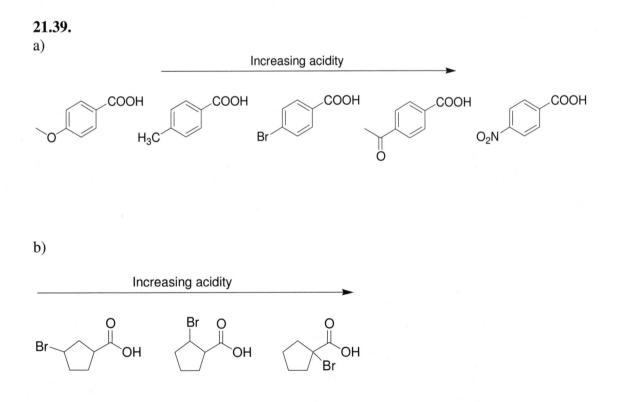

lactone —(1) LAH; 2) H_2O)→ diol (OH, OH)

21.39.

a)

Increasing acidity →

4-methoxybenzoic acid (COOH) < 4-methylbenzoic acid (H_3C, COOH) < 4-bromobenzoic acid (Br, COOH) < 4-acetylbenzoic acid (COOH) < 4-nitrobenzoic acid (O_2N, COOH)

b)

Increasing acidity →

3-bromocyclopentanecarboxylic acid (Br, OH) < 2-bromocyclopentanecarboxylic acid (Br, OH) < 1-bromocyclopentanecarboxylic acid (OH, Br)

21.40.

a) The second carboxylic acid moiety is electron withdrawing, and stabilizes the conjugate base that is formed when the first proton is removed.

b) The carboxylate ion is electron rich and it destabilizes the conjugate base that is formed when the second proton is removed.

c)

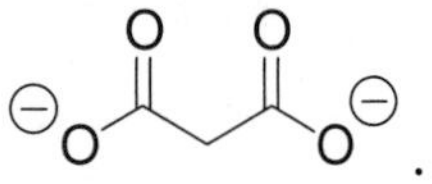

d) The number of methylene groups (CH_2) separating the carboxylic acid moieties is greater in succinic acid than in malonic acid. Therefore, the inductive effects are not as strong.

21.41.

a) cyclopentanecarboxylic acid
b) cyclopentanecarboxamide
c) benzoyl chloride
d) ethyl acetate
e) hexanoic acid
f) pentanoyl chloride
g) hexanamide

21.42.

a) acetic anhydride
b) benzoic acid
c) formic acid
d) oxalic acid

21.43.

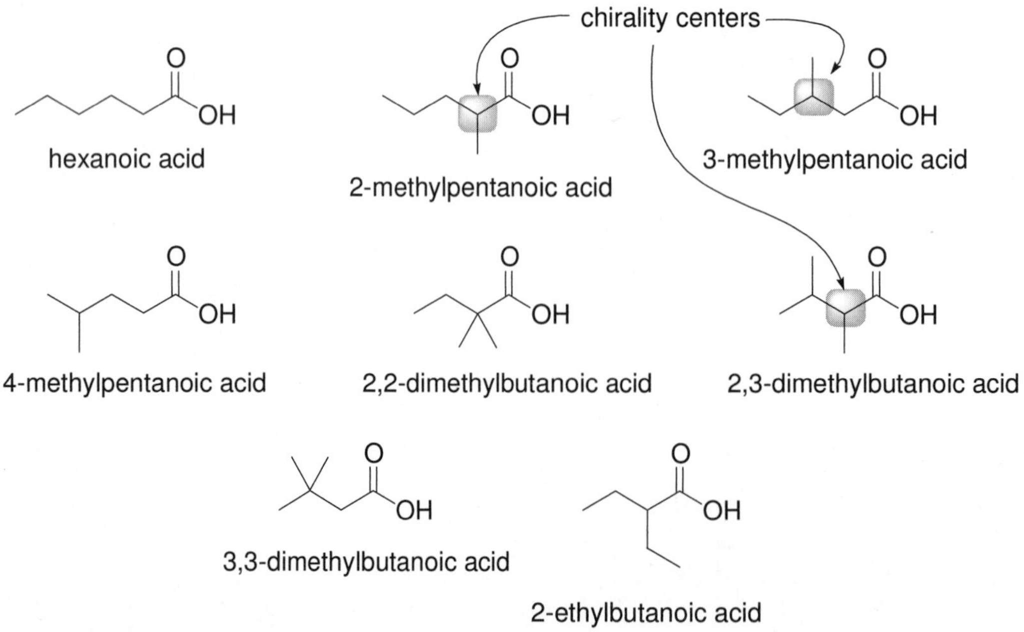

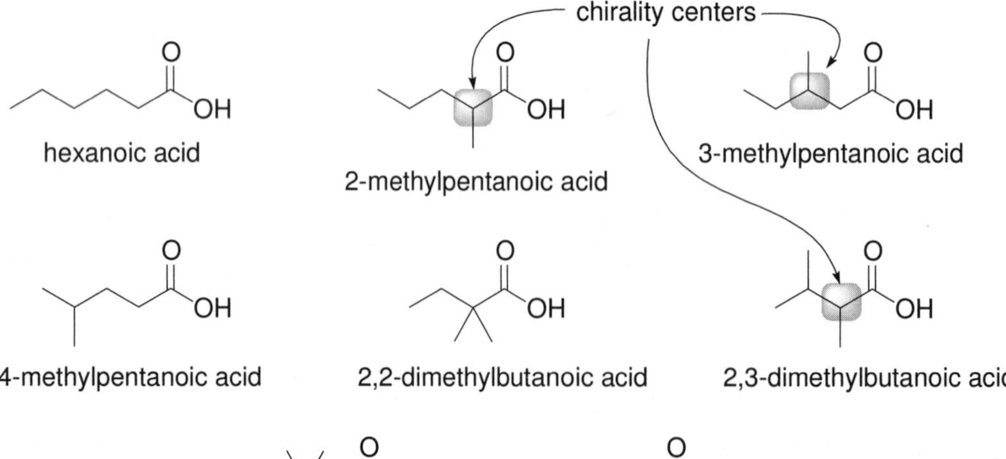

21.44.

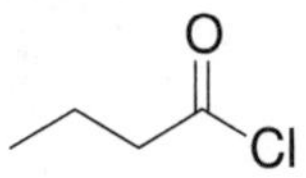

butanoyl chloride

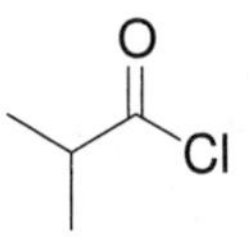

2-methylpropanoyl chloride

21.45.

a) 1) excess LAH
2) H_2O

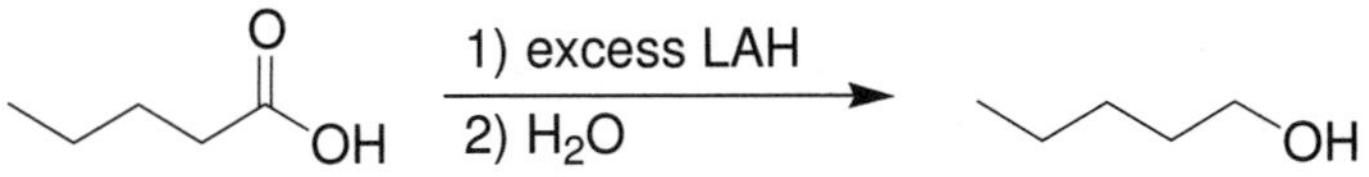

b) 1) excess LAH
2) H_2O
3) TsCl, py
4) *t*-BuOK

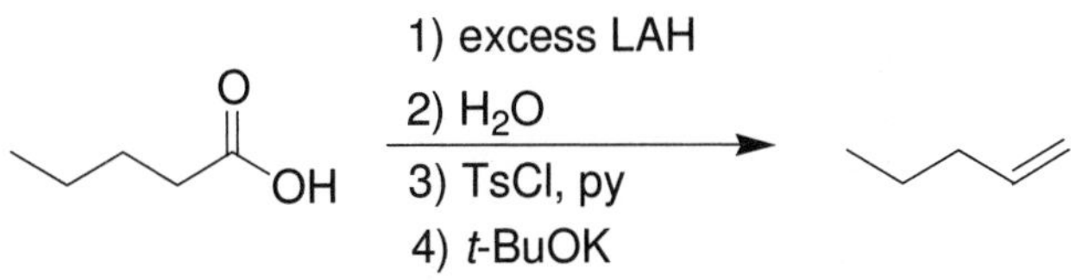

c) 1) excess LAH
2) H_2O
3) PBr_3
4) NaCN
5) H_3O^+

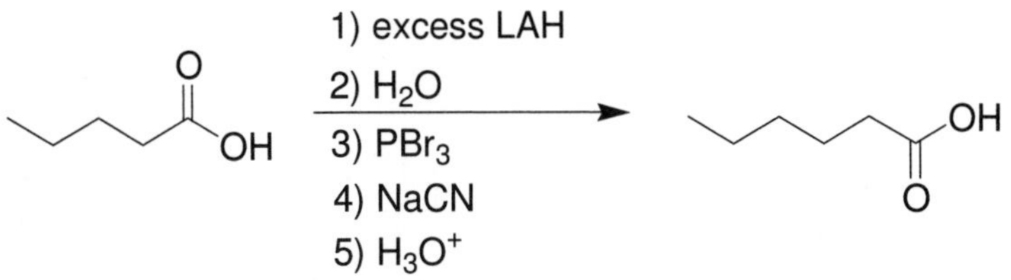

21.46.

a) 1) $BH_3 \cdot THF$
2) H_2O_2, NaOH
3) $Na_2Cr_2O_7$, H_2SO_4, H_2O

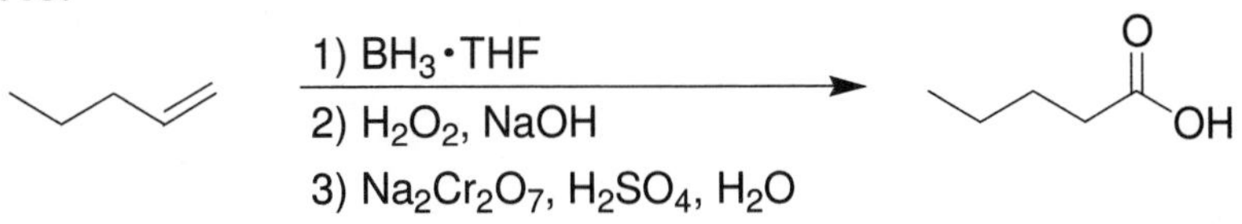

b) 1) NaCN
2) H_3O^+

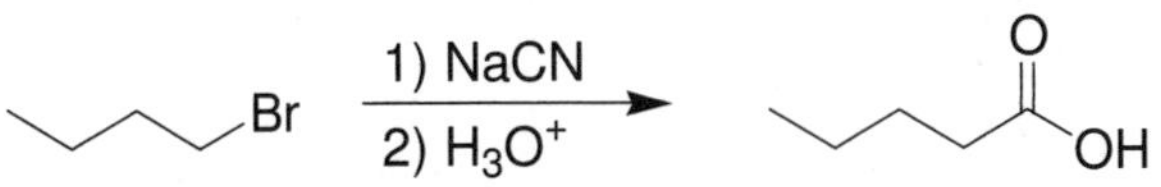

21.47. As discussed in Chapter 19, the methoxy group is electron donating via resonance, but electron withdrawing via induction. The resonance effect is stronger, but only occurs when the methoxy group is in an *ortho* or *para* position.

21.48.

a)

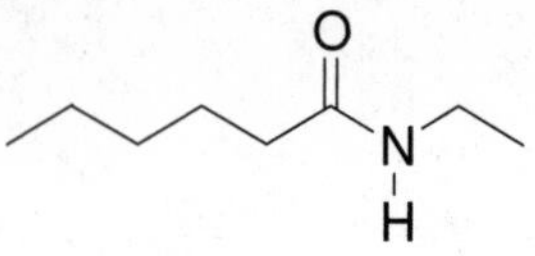

NH

b)

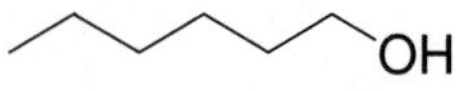

OH

c)

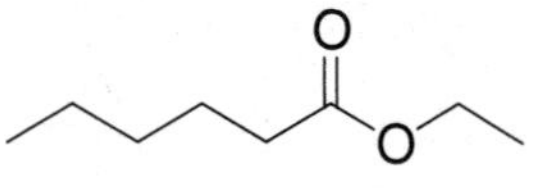

O O

d)

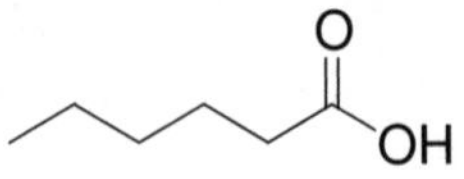

OH

e)

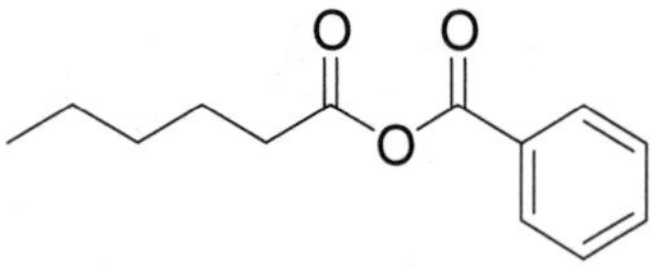

O O O

f)

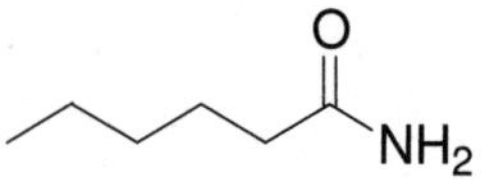

NH_2

g)

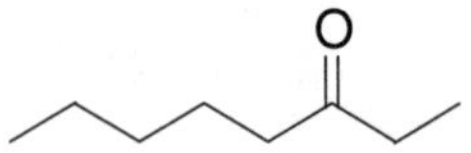

O

h)

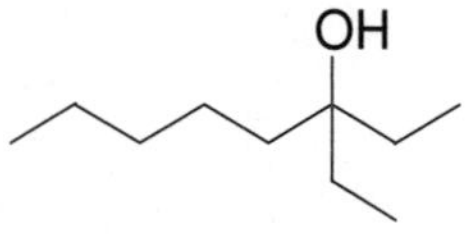

OH

21.49.

a) 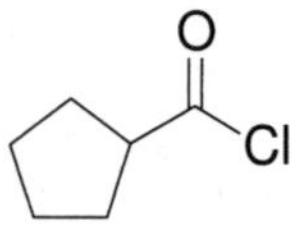 Cl b) 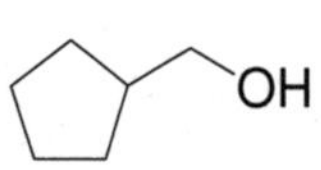 OH c) 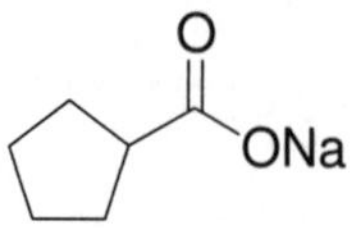 ONa d) 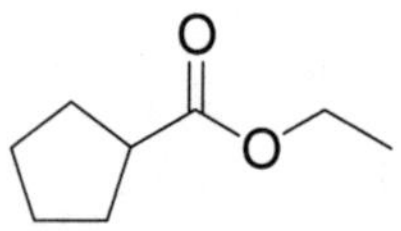 O

21.50.

a) 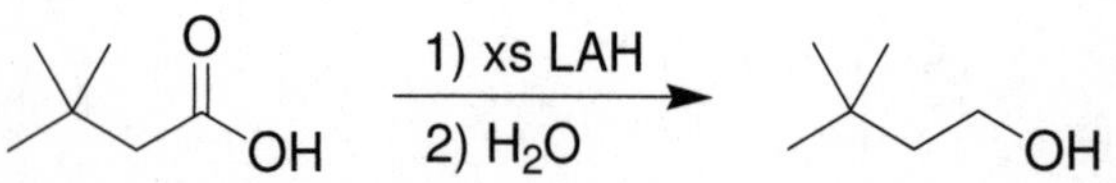 OH $\xrightarrow[\text{2) } H_2O]{\text{1) xs LAH}}$ OH

b) OH → 1) $SOCl_2$ 2) $(CH_3)_2NH$, pyridine → O, N

c) O, NH_2 → $SOCl_2$ → C≡N

d) O, OMe → 1) H_3O^+ 2) CH_3COCl, pyridine → O, O, O

e) O, OMe → DIBAH → O, H

f) OH → (O, O, O) → O, O + HO, O

g) H, N → (O, Cl) pyridine → O, N

h) O, O → 1) xs LAH 2) H_2O → HO, OH

i) O, N → H_3O^+ heat → O, OH, N, H

j) O, O → H_3O^+ → O, OH, OH

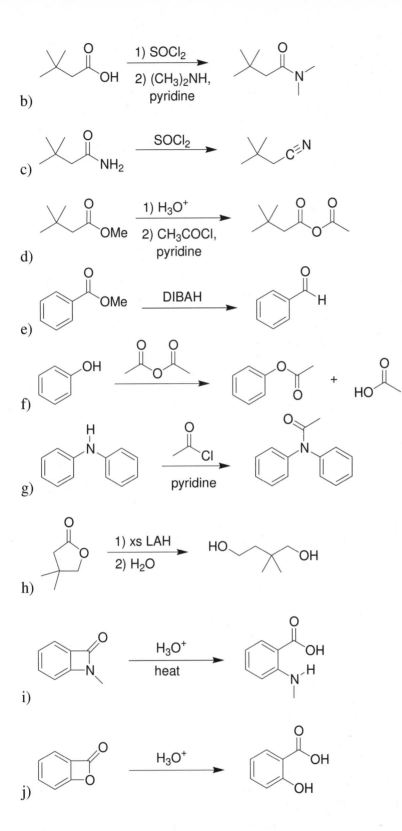

21.51.

a) Benzoic acid (C_6H_5COOH) + HO–C_6H_5 (phenol)

b) Butanoic acid (OH) + HO–$CH(CH_3)_2$ (isopropyl alcohol)

c) $CH_3CH_2CO_2H$ + $(CH_3)_3COH$

21.52.

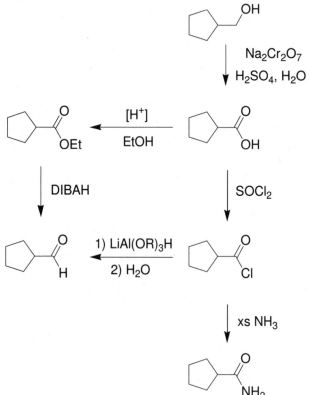

Cyclopentylmethanol (OH) → ($Na_2Cr_2O_7$, H_2SO_4, H_2O) → cyclopentanecarboxylic acid (O, OH)

Cyclopentanecarboxylic acid → ($[H^+]$, EtOH) → ethyl cyclopentanecarboxylate (O, OEt)

Ethyl ester → (DIBAH) → cyclopentanecarbaldehyde (O, H)

Cyclopentanecarboxylic acid → ($SOCl_2$) → acid chloride (O, Cl)

Acid chloride → (1) $LiAl(OR)_3H$ 2) H_2O) → cyclopentanecarbaldehyde (O, H)

Acid chloride → (xs NH_3) → amide (O, NH_2)

21.53.

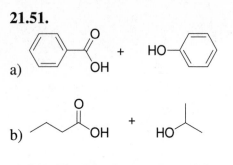

a) NaOH, followed by $Na_2Cr_2O_7$, H_2SO_4, H_2O
b) NaCN followed by H_3O^+
c) NaOH, followed by $Na_2Cr_2O_7$, H_2SO_4, H_2O, followed by $SOCl_2$
d) NaCN, followed by H_3O^+, followed by $SOCl_2$, followed by xs NH_3
e) NaOH, followed by $Na_2Cr_2O_7$, H_2SO_4, H_2O, followed by $SOCl_2$, followed by xs NH_3
f) NaCN followed by H_3O^+, followed by $[H^+]$, EtOH (with removal of water)

21.54.

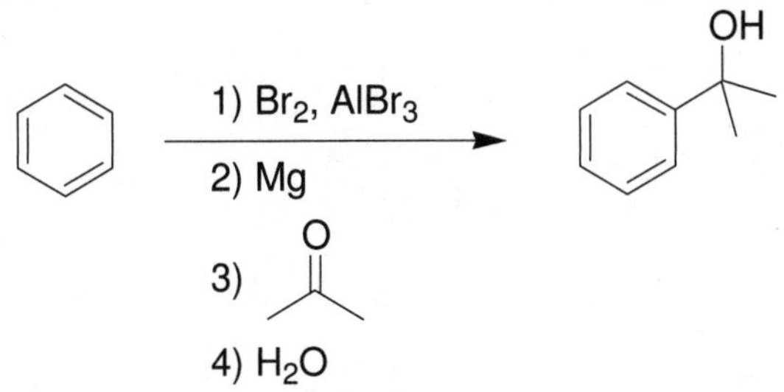

a)

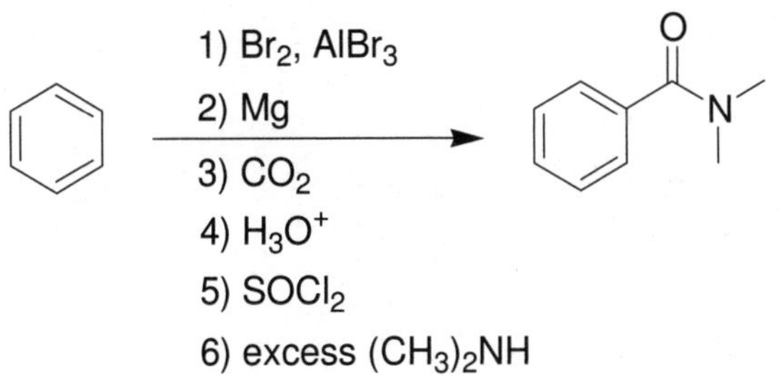

1) Br_2, $AlBr_3$
2) Mg
3) O (acetone)
4) H_2O

OH

b)

1) Br_2, $AlBr_3$
2) Mg
3) CO_2
4) H_3O^+
5) $SOCl_2$
6) excess $(CH_3)_2NH$

O
N

c)

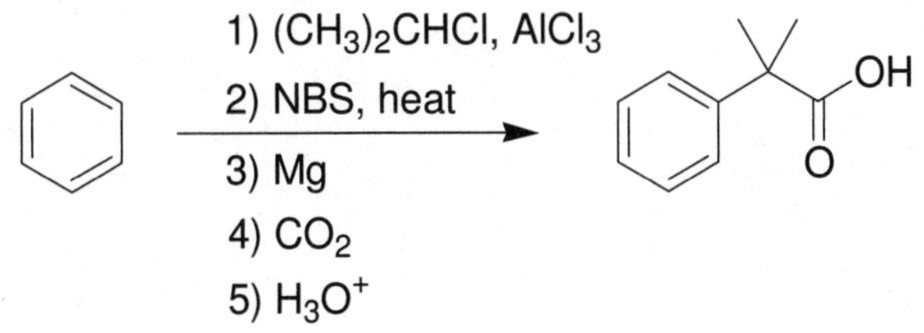

1) $(CH_3)_2CHCl$, $AlCl_3$
2) NBS, heat
3) Mg
4) CO_2
5) H_3O^+

OH
O

d)

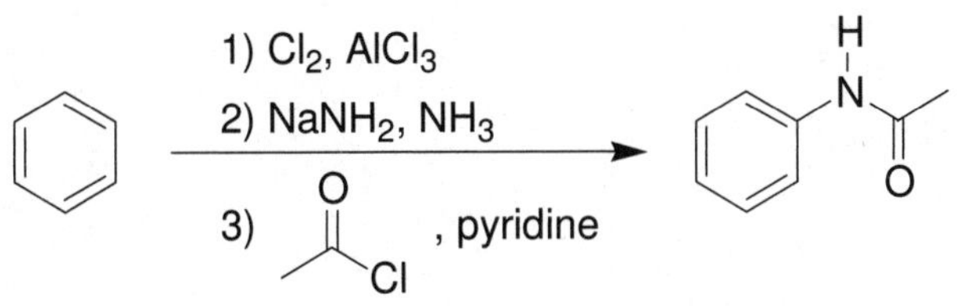

1) Cl_2, $AlCl_3$
2) $NaNH_2$, NH_3
3) O, Cl , pyridine

H
N
O

21.55.

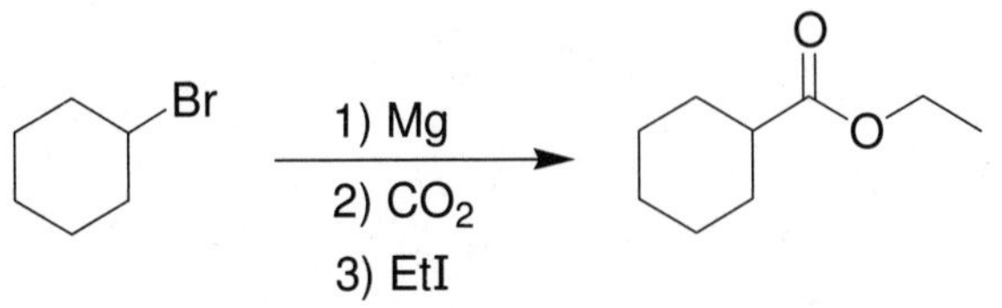

a)

Br

1) Mg
2) CO_2
3) EtI

O
O

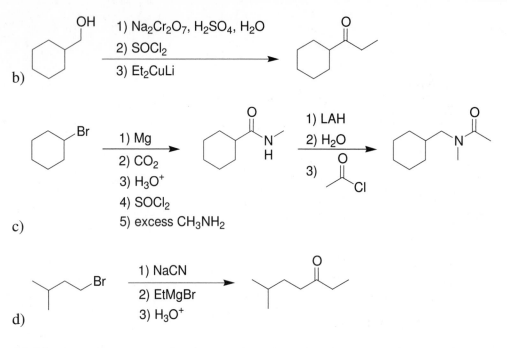

b) OH

1) $Na_2Cr_2O_7$, H_2SO_4, H_2O
2) $SOCl_2$
3) Et_2CuLi

O

c) Br

1) Mg
2) CO_2
3) H_3O^+
4) $SOCl_2$
5) excess CH_3NH_2

O N H

1) LAH
2) H_2O
3) O Cl

O N

d) Br

1) NaCN
2) EtMgBr
3) H_3O^+

O

21.56. A methoxy group is electron donating, thereby decreasing the electrophilicity of the ester moiety. A nitro group is electron withdrawing, thereby increasing the electrophilicity of the ester group.

21.57.

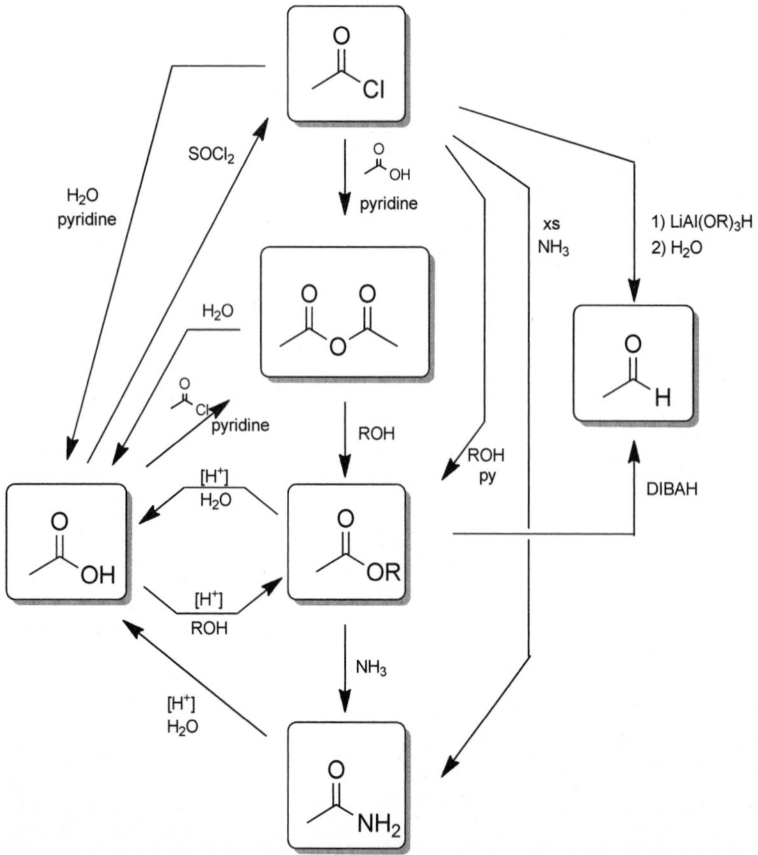

21.58.

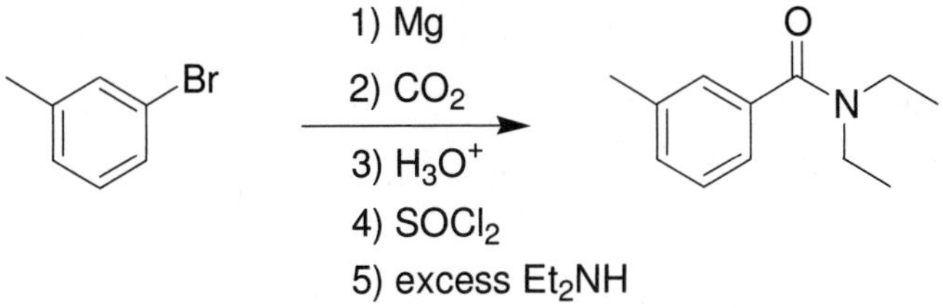

1) Mg
2) CO_2
3) H_3O^+
4) $SOCl_2$
5) excess Et_2NH

21.59.

1) excess MeMgBr
2) H_3O^+

OH + OH

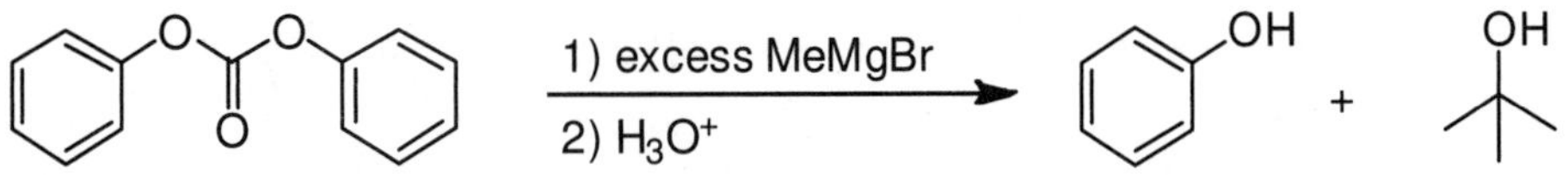

21.60.

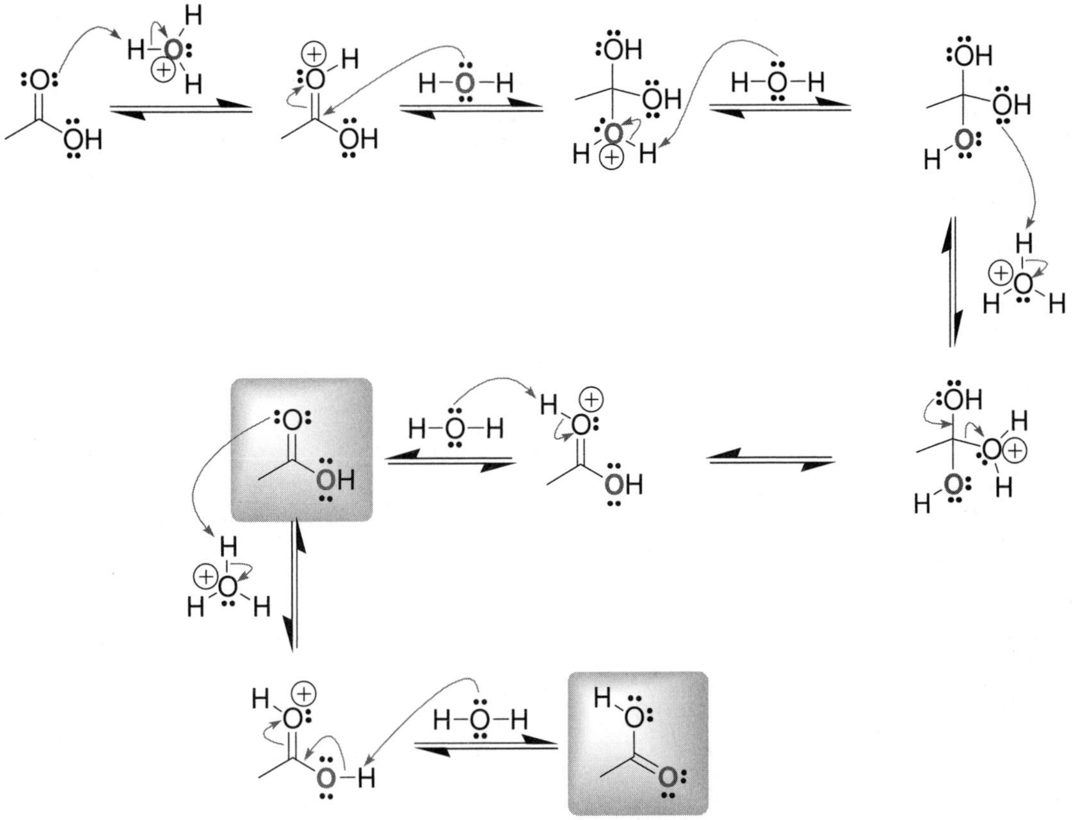

21.61.

a)

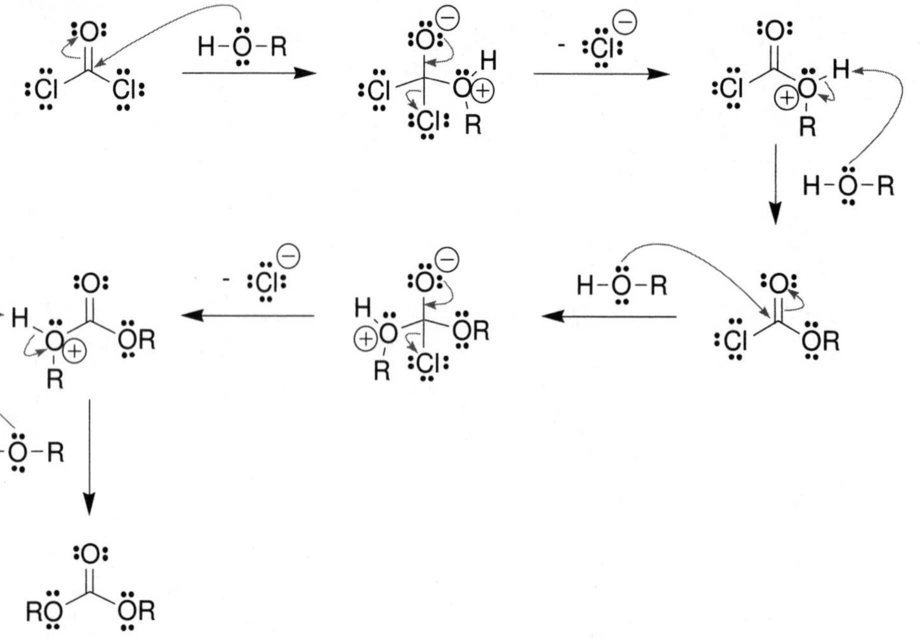

b)

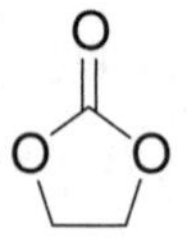

c)

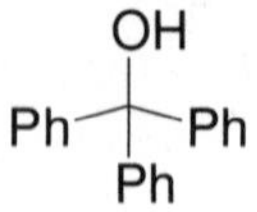

21.62.

a)

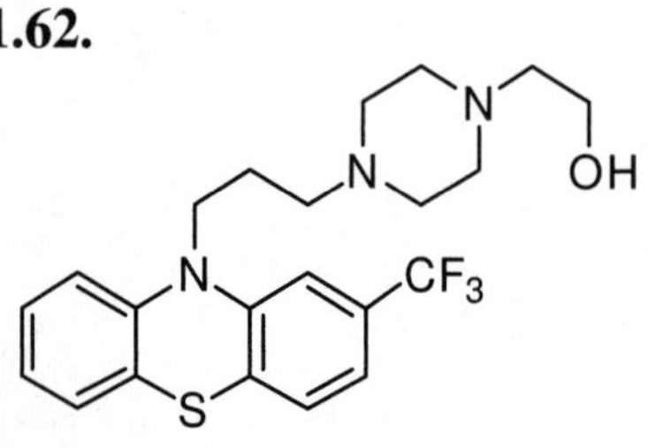

b) 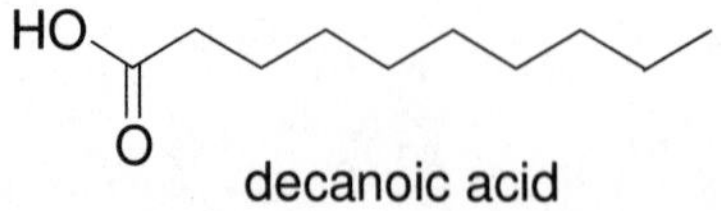 decanoic acid

21.63.

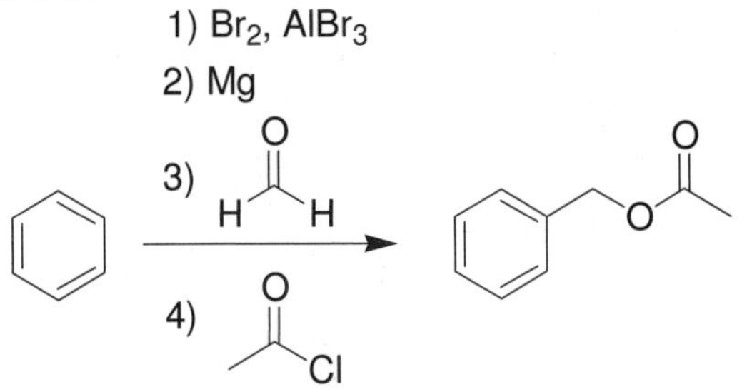

1) Br_2, $AlBr_3$
2) Mg
3) H–C(=O)–H
4) $CH_3C(=O)Cl$

21.64.

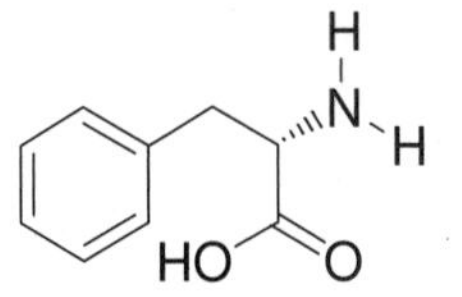

H
N
H
HO O
phenylalanine

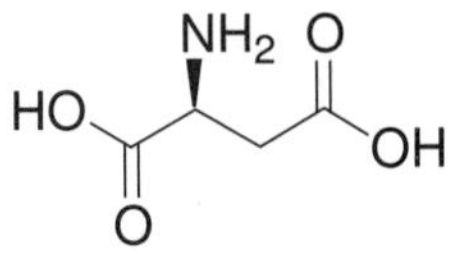

NH_2 O
HO OH
O
aspartic acid

21.65.

a)

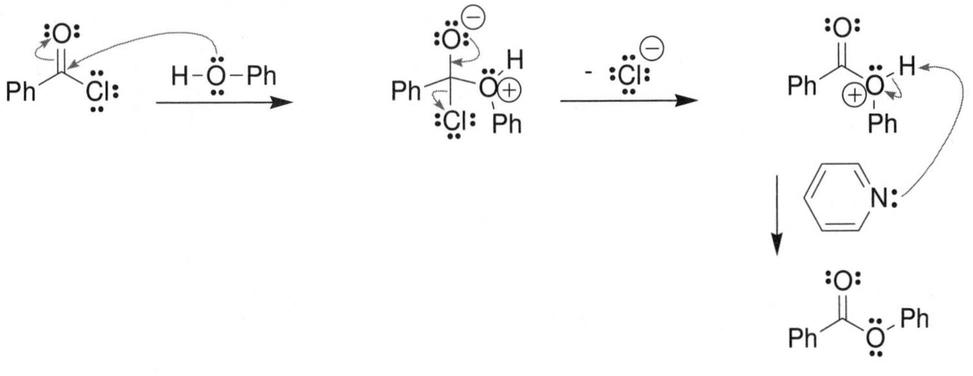

:O:
Ph Cl:
H–Ö–Ph
:O: H
Ph O+
:Cl: Ph
- :Cl:
:O:
Ph +O H
Ph
N:
:O:
Ph O Ph

b)

:O:
O:
:OH
:O :OH
O:
:O:
H–O O:
+O–H
H
H
:O:
HO OH
:O:
O OH

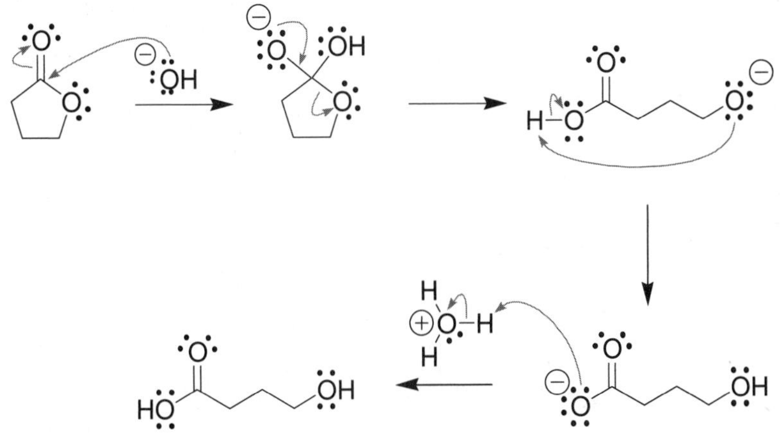

c)

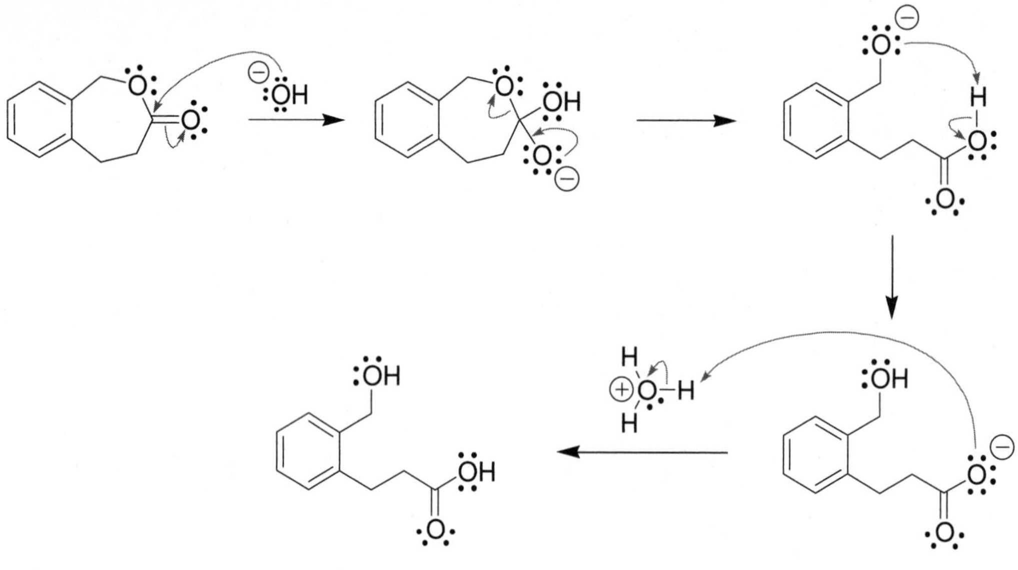

$:\ddot{O}H$ H_3O^+

d)

$H_2\ddot{N}-NH_2$

$- :\ddot{Cl}:^{\ominus}$

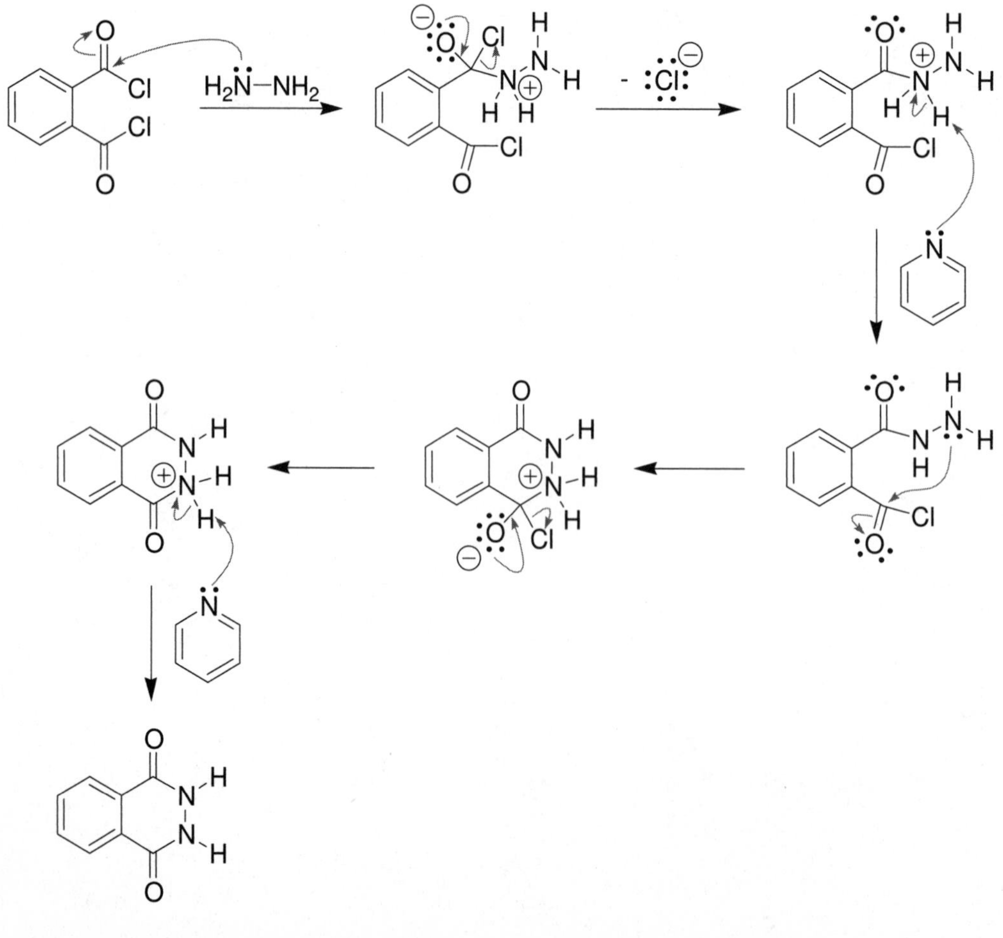

e)

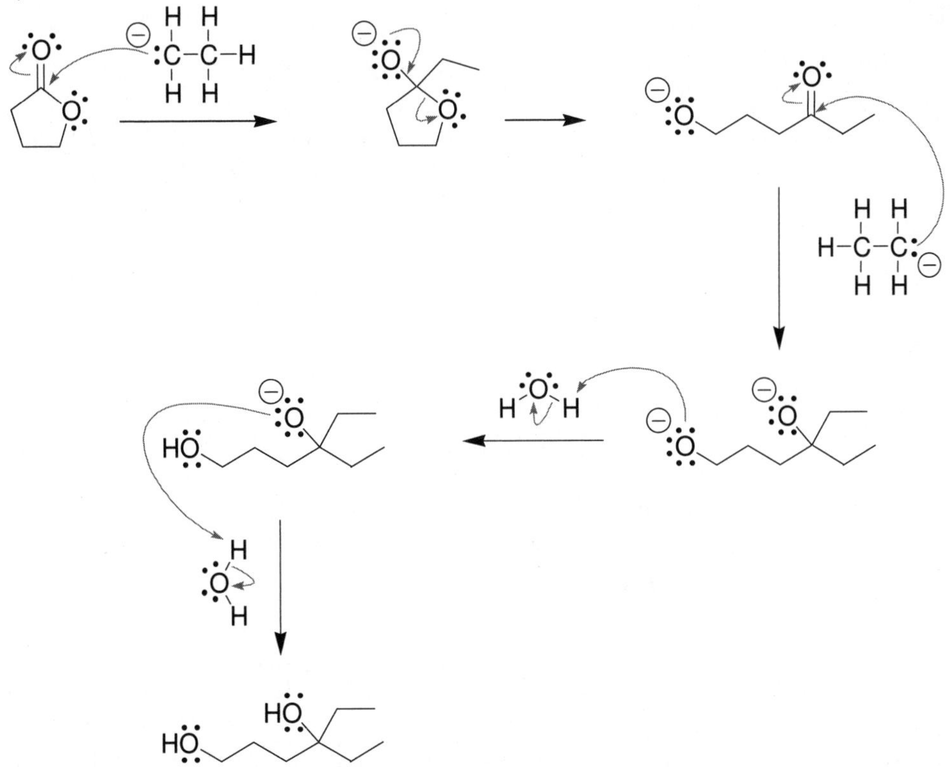

H H
:C–C–H
H H

H H
H–C–C:
H H

HO

HO

HO

21.66. The three chlorine atoms withdraw electron density via induction. This effect renders the carbonyl group more electrophilic.

21.67.

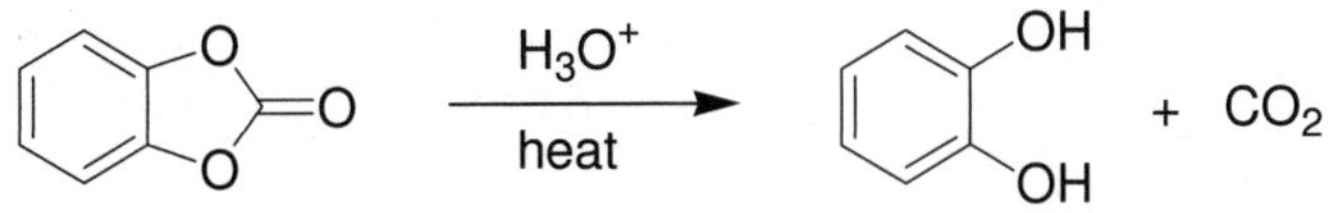

H_3O^+

heat

OH

OH

+ CO_2

21.68

NH_2 H

N

S

O

O

N

O OH

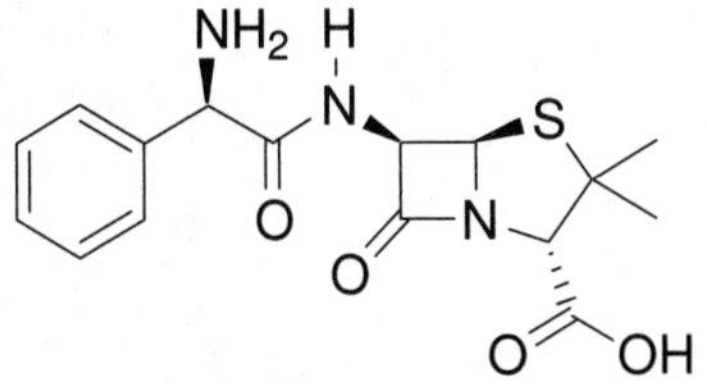

a)

b) Ampicillin

21.69.

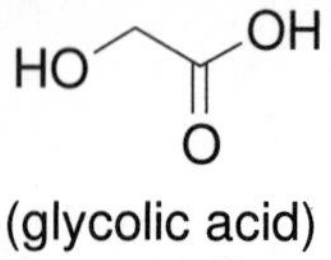

HO, OH, O

(glycolic acid)
hydroxyacetic acid

21.70.

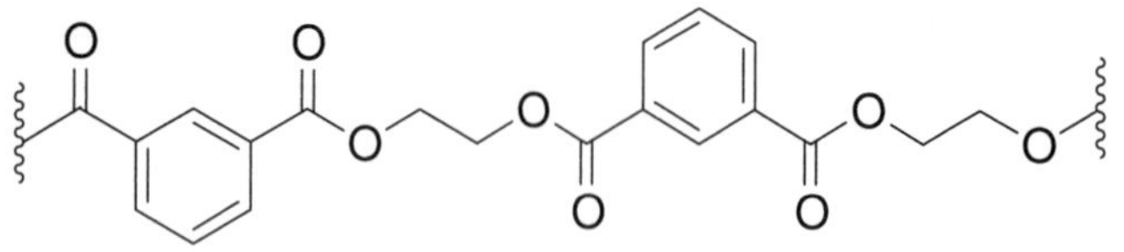

21.71.

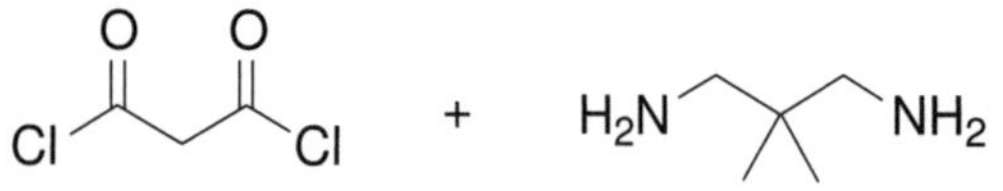

Cl, Cl, O, O + H_2N, NH_2

21.72.

Polymer

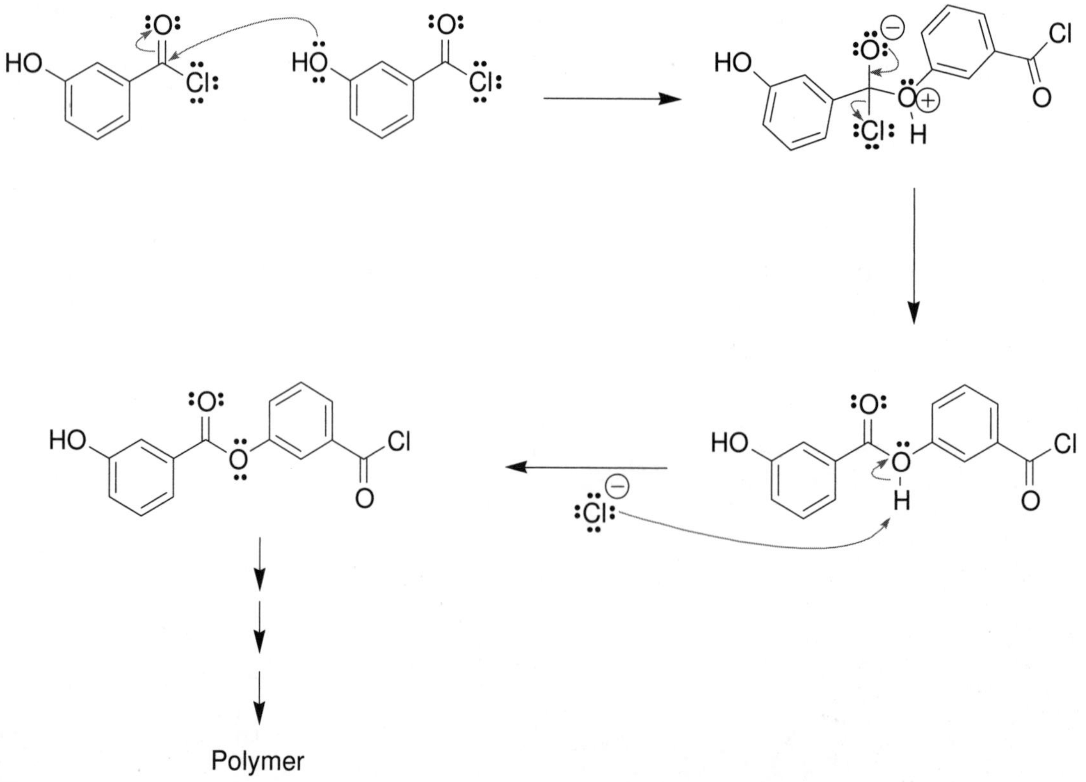

21.73.

a) 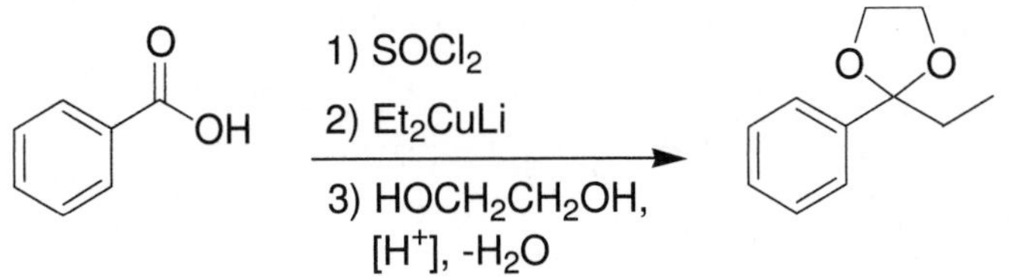 Benzoic acid: 1) $SOCl_2$ 2) Et_2CuLi 3) $HOCH_2CH_2OH$, $[H^+]$, $-H_2O$ → 2-ethyl-2-phenyl-1,3-dioxolane

b) 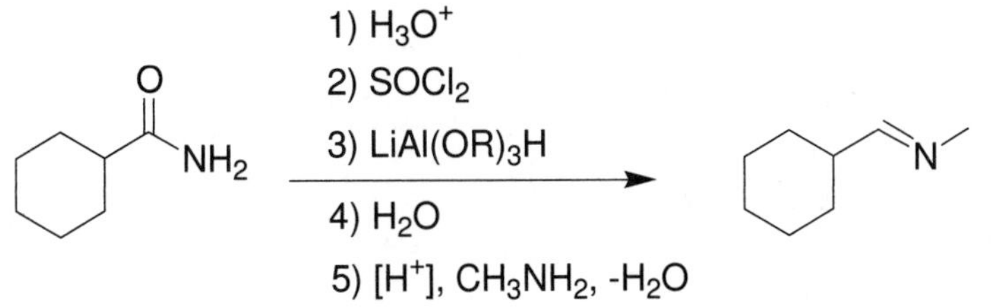 Cyclohexanecarboxamide (NH_2): 1) H_3O^+ 2) $SOCl_2$ 3) $LiAl(OR)_3H$ 4) H_2O 5) $[H^+]$, CH_3NH_2, $-H_2O$ → N-methyl imine (N)

c) 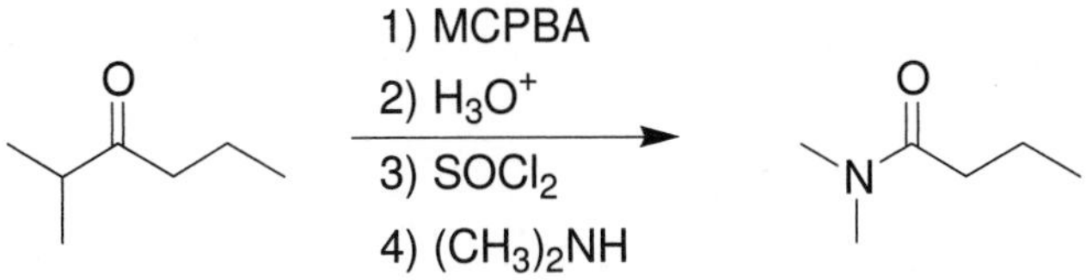 2-methylhexan-3-one: 1) MCPBA 2) H_3O^+ 3) $SOCl_2$ 4) $(CH_3)_2NH$ → N,N-dimethylbutanamide (N)

d) 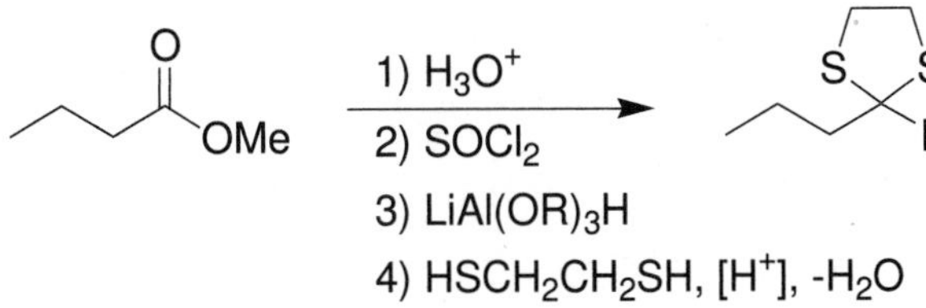 Methyl butanoate (OMe): 1) H_3O^+ 2) $SOCl_2$ 3) $LiAl(OR)_3H$ 4) $HSCH_2CH_2SH$, $[H^+]$, $-H_2O$ → dithiolane (S, S, H)

e) 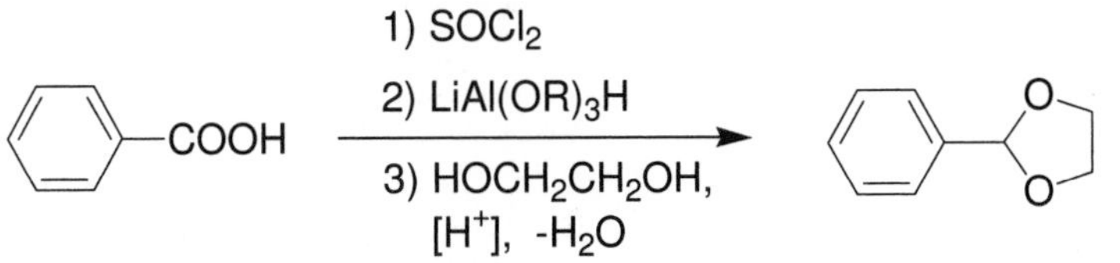 Benzene–COOH: 1) $SOCl_2$ 2) $LiAl(OR)_3H$ 3) $HOCH_2CH_2OH$, $[H^+]$, $-H_2O$ → 2-phenyl-1,3-dioxolane (O, O)

f) 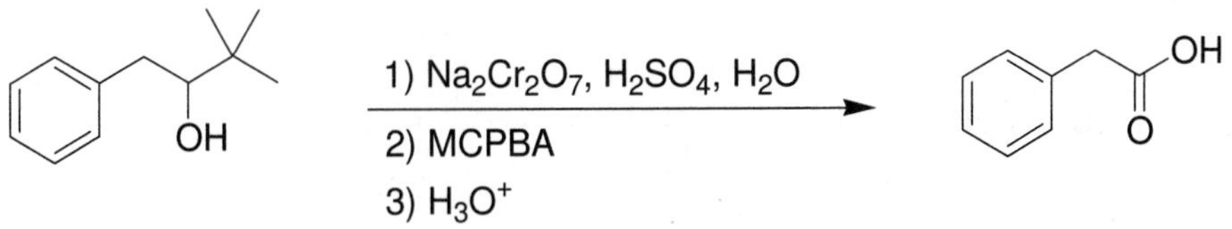 Alcohol (OH): 1) $Na_2Cr_2O_7$, H_2SO_4, H_2O 2) MCPBA 3) H_3O^+ → phenylacetic acid (OH, O)

g) 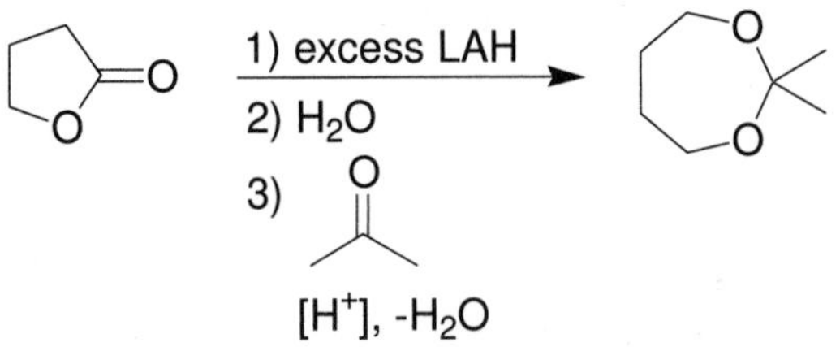 Lactone (O, O): 1) excess LAH 2) H_2O 3) acetone (O), $[H^+]$, $-H_2O$ → cyclic acetal (O, O)

21.74.

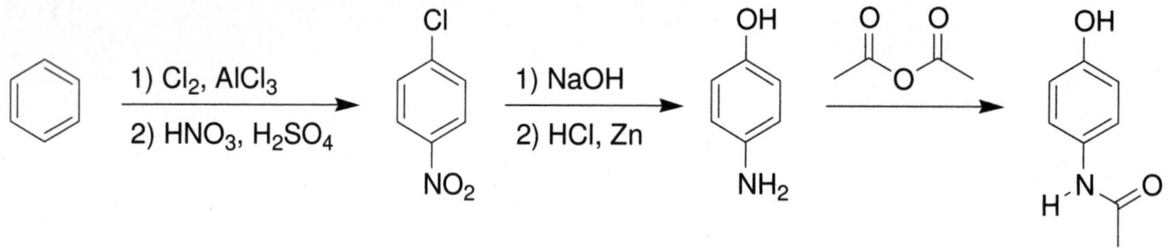

1) Cl_2, $AlCl_3$
2) HNO_3, H_2SO_4

Cl

NO_2

1) NaOH
2) HCl, Zn

OH

NH_2

O O
O

OH

H N O

21.75.

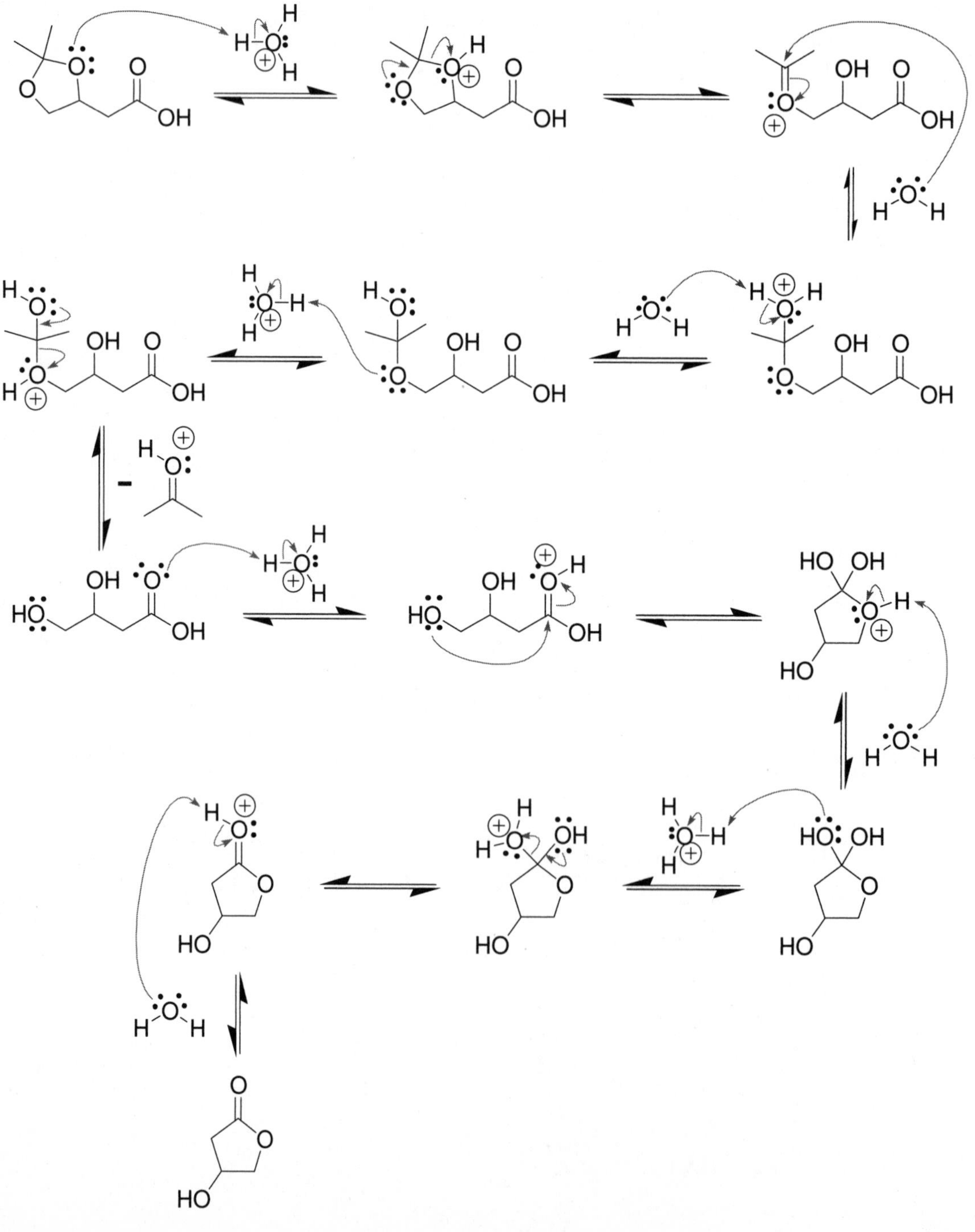

21.76.

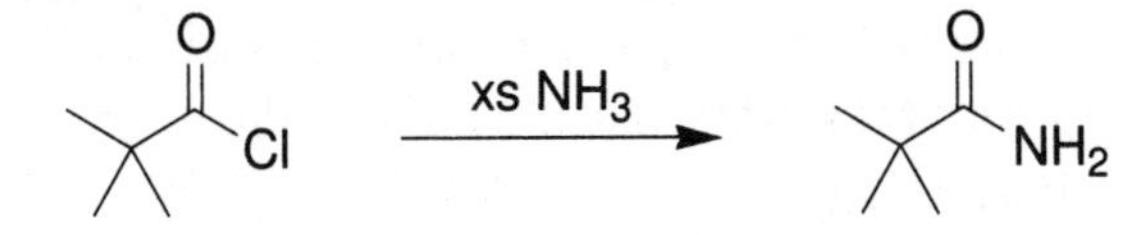

xs NH_3

Compound A

21.77.

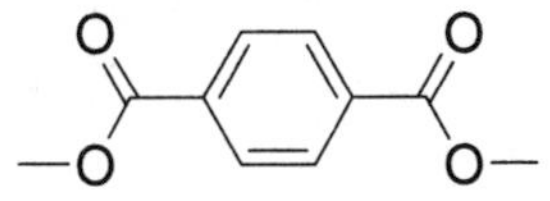

21.78. An IR spectrum of butyric acid should have a broad signal between 2500 cm^{-1} and 3600 cm^{-1}. An IR spectrum of ethyl acetate will not have this signal.

21.79. The ^{1}H NMR spectrum of *para*-chlorobenzaldehyde should have a signal at approximately 10 ppm corresponding to the aldehydic proton. The ^{1}H NMR spectrum of benzoyl chloride should not have a signal near 10 ppm.

21.80.

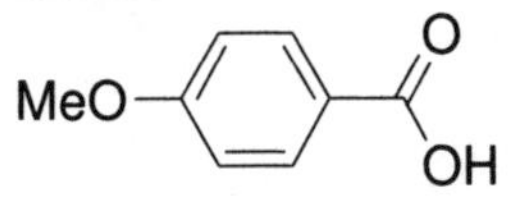

21.81. If the oxygen atom of the OH group in the starting material is an isotopic label, then we would expect the label to be incorporated into the ring of the product:

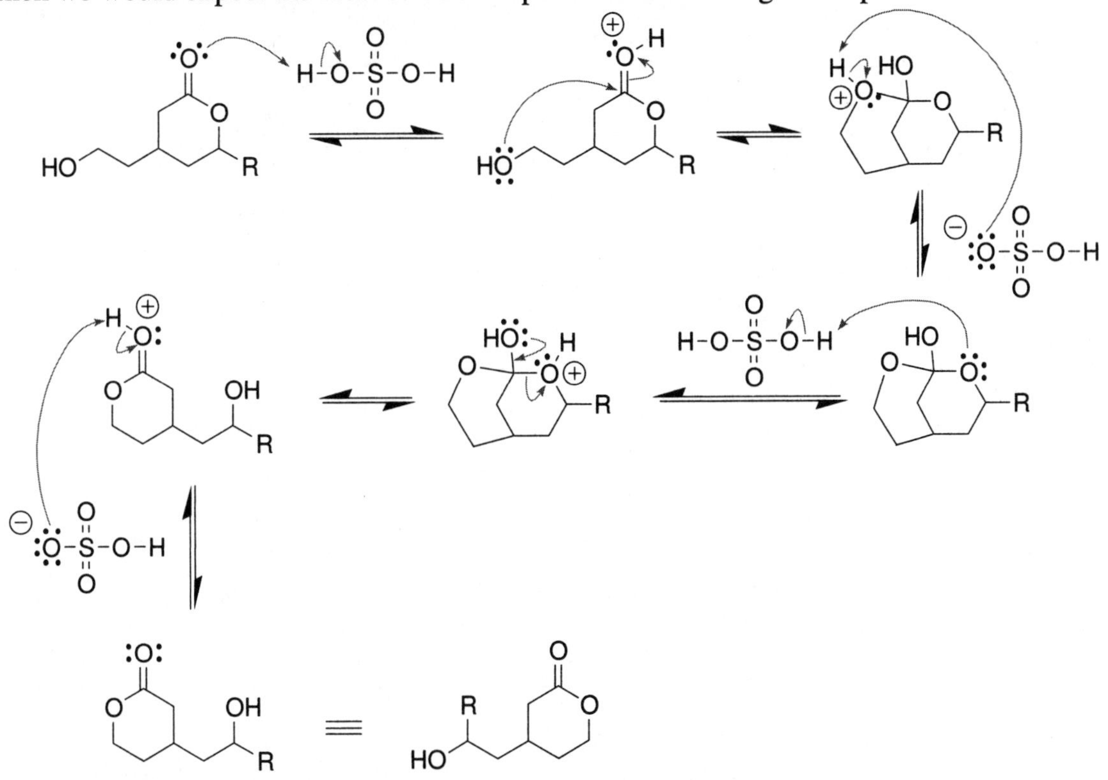

21.82. The lone pair of the nitrogen atom in this case is participating in resonance and is less available to donate electron density to the carbonyl group. As a result, the carbonyl group is more electrophilic than the carbonyl group of a regular amide (where the lone pair contributes significant electron density to the carbonyl group). Also, when this compound functions as an electrophile in a nucleophilic acyl substitution reaction, the leaving group is particularly stable because it is an aromatic anion. With a good leaving group, this compound more closely resembles the reactivity of an acid halide than an amide.

21.83.
a) DMF, like most amides, exhibits restricted rotation about the bond between the carbonyl group and the nitrogen atom. This restricted rotation causes the methyl groups to be in different electronic environments. They are not chemically equivalent, and will therefore produce two different signals (in addition to the signal from the other proton in the compound). Upon treatment with excess LAH followed by water, DMF is reduced to an amine that does not exhibit restricted rotation. As such, the methyl groups are now chemically equivalent and will together produce only one signal.
b) Restricted rotation causes the methyl groups to be in different electronic environments. As a result, the ^{13}C NMR spectrum of DMF should have three signals.

Chapter 22
Alpha Carbon Chemistry: Enols and Enolates

Review of Concepts

Fill in the blanks below. To verify that your answers are correct, look in your textbook at the end of Chapter 22. Each of the sentences below appears verbatim in the section entitled *Review of Concepts and Vocabulary*.

- In the presence of catalytic acid or base, a ketone will exist in equilibrium with an _____. In general, the equilibrium position will significantly favor the ________.
- When treated with a strong base, the α position of a ketone is deprotonated to give an ___________.
- __________________ or _____ will irreversibly and completely convert an aldehyde or ketone into an enolate.
- In the **haloform reaction**, a _______ ketone is converted into a carboxylic acid upon treatment with excess base and excess halogen followed by acid workup.
- When an aldehyde is treated with sodium hydroxide, an **aldol addition reaction** occurs, and the product is a _______________________________.
- For most simple aldehydes, the position of equilibrium favors the aldol product. For most ketones, the reverse process, called a ______**-aldol reaction** is favored.
- When an aldehyde is heated in aqueous sodium hydroxide, an **aldol** ___________ **reaction** occurs, and the product is an ___________________________. Elimination of water occurs via an ______ **mechanism.**
- **Crossed aldol,** or **mixed aldol reactions** are aldol reactions that occur between different partners, and are only efficient if one partner lacks __________ or if a **directed aldol addition** is performed.
- Intramolecular aldol reactions show a preference for formation of ______ and ____-membered rings.
- When an ester is treated with an alkoxide base, a **Claisen condensation reaction** occurs, and the product is a ________________.
- The α position of a ketone can be alkylated by forming an enolate and treating it with an _________________.
- For unsymmetrical ketones, reactions with _____ at low temperature favor formation of the kinetic enolate, while reactions with ______ at room temperature favor the thermodynamic enolate.
- When LDA is used with an unsymmetrical ketone, alkylation occurs at the __________________ position.
- The _______________________ **synthesis** enables the conversion of an alkyl halide into a carboxylic acid with the introduction of two new carbon atoms.
- The _______________________ **synthesis** enables the conversion of an alkyl halide into a methyl ketone with the introduction of two new carbon atoms.
- Aldehydes and ketones that possess _____-unsaturation are susceptible to nucleophilic attack at the β position. This reaction is called a ____________ **addition, 1,4-addition,** or a **Michael reaction**.

Review of Skills

Fill in the blanks and empty boxes below. To verify that your answers are correct, look in your textbook at the end of Chapter 22. The answers appear in the section entitled *SkillBuilder Review*.

22.1 Drawing Enolates

DRAW THE RESONANCE STRUCTURES OF THE ENOLATE THAT IS FORMED WHEN THE KETONE BELOW IS TREATED WITH A STRONG BASE:

O

⊖ :Base

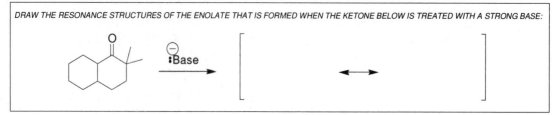

22.2 Predicting the Products of an Aldol Addition Reaction

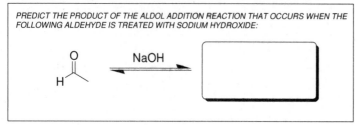

PREDICT THE PRODUCT OF THE ALDOL ADDITION REACTION THAT OCCURS WHEN THE FOLLOWING ALDEHYDE IS TREATED WITH SODIUM HYDROXIDE:

O

H

NaOH

22.3 Drawing the Product of an Aldol Condensation

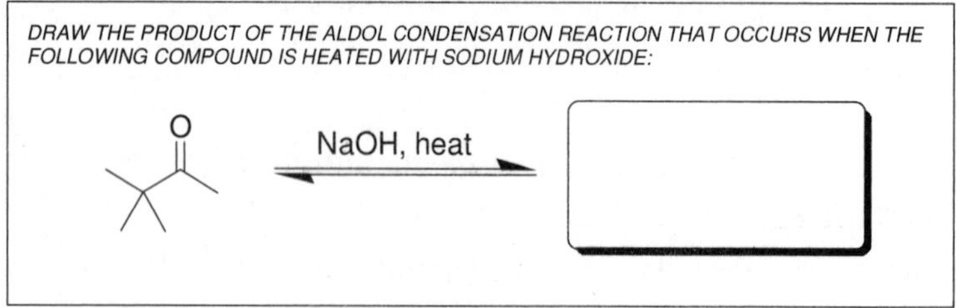

22.4 Identifying the Reagents Necessary for a Crossed Aldol Reaction

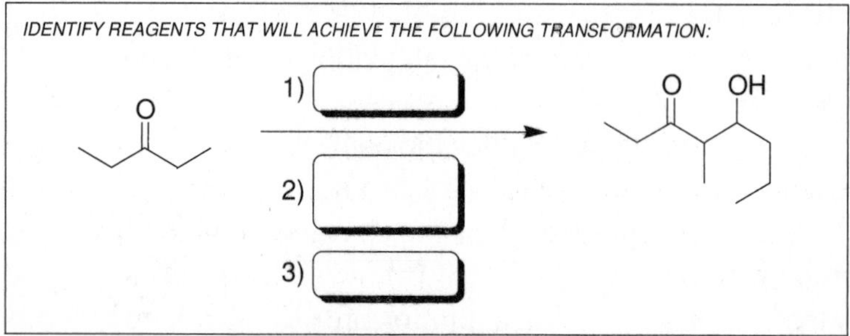

22.5 Using the Malonic Ester Synthesis

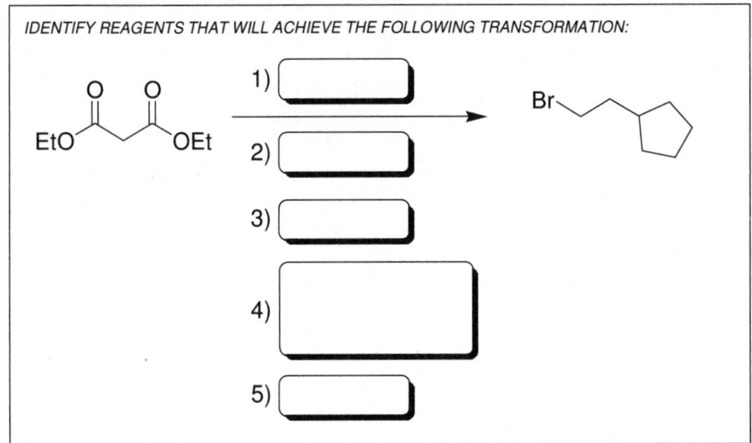
IDENTIFY REAGENTS THAT WILL ACHIEVE THE FOLLOWING TRANSFORMATION:
EtO
OEt
Br
1)
2)
3)
4)
5)

22.6 Using the Acetoacetic Ester Synthesis

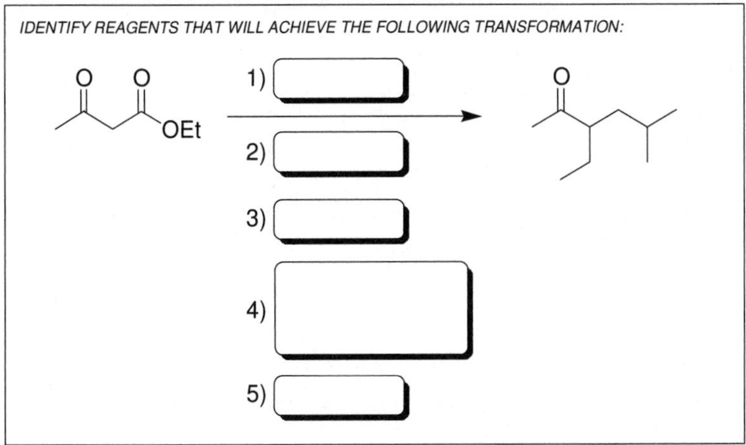
IDENTIFY REAGENTS THAT WILL ACHIEVE THE FOLLOWING TRANSFORMATION:
OEt
1)
2)
3)
4)
5)

22.7 Determining When to Use a Stork Enamine Synthesis

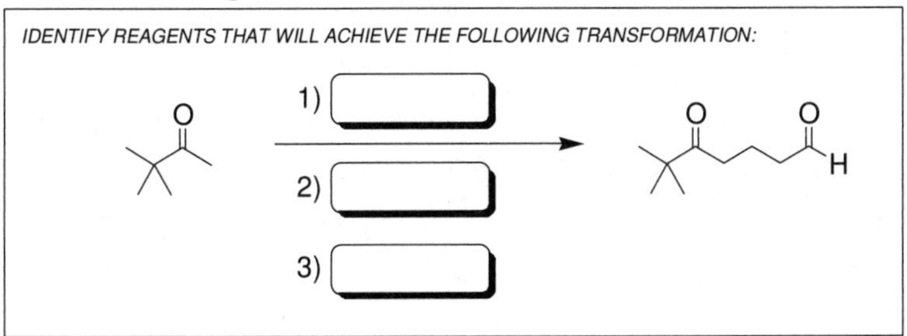
IDENTIFY REAGENTS THAT WILL ACHIEVE THE FOLLOWING TRANSFORMATION:
H
1)
2)
3)

22.8 Determining which Addition or Condensation Reaction to Use

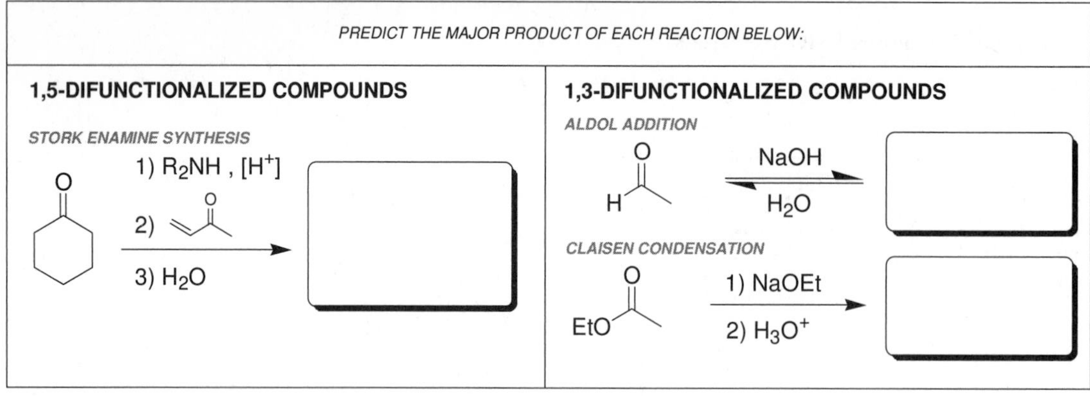

22.9 Alkylating the Alpha and Beta Positions

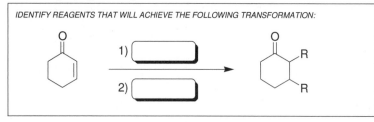

IDENTIFY REAGENTS THAT WILL ACHIEVE THE FOLLOWING TRANSFORMATION:

1)

2)

Review of Reactions

Identify the reagents necessary to achieve each of the following transformations. To verify that your answers are correct, look in your textbook at the end of Chapter 22. The answers appear in the section entitled *Review of Reactions.*

Alpha Halogenation

Br + HBr

OH → OH, Br

OH

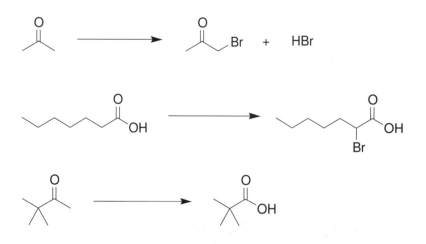

Aldol Reactions

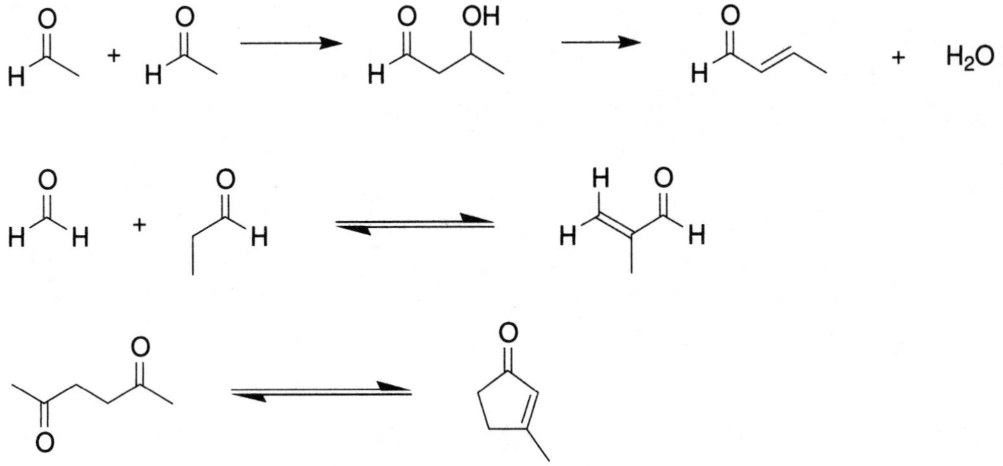

+ H_2O

Claisen Condensation

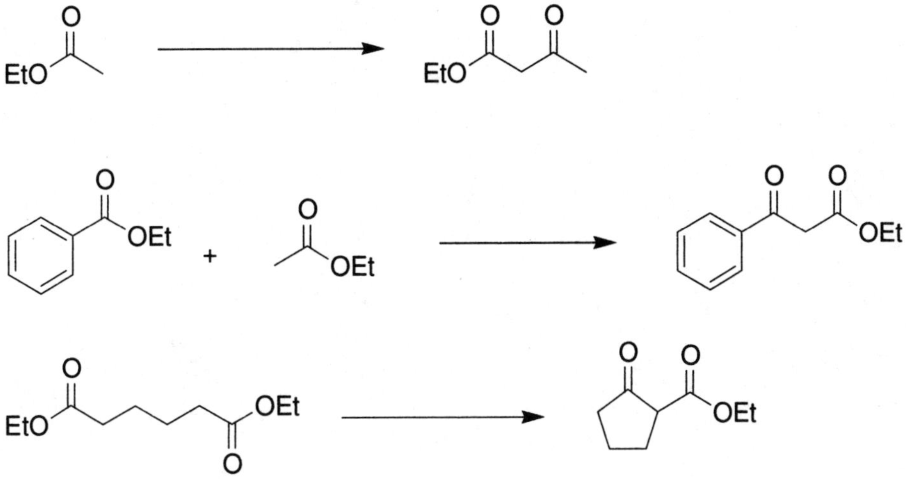

Alkylation

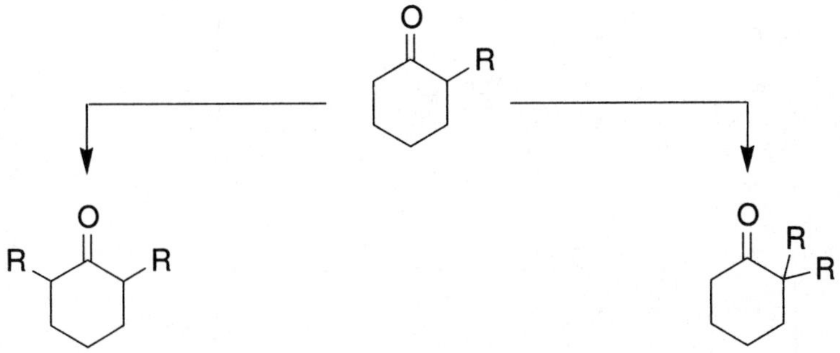

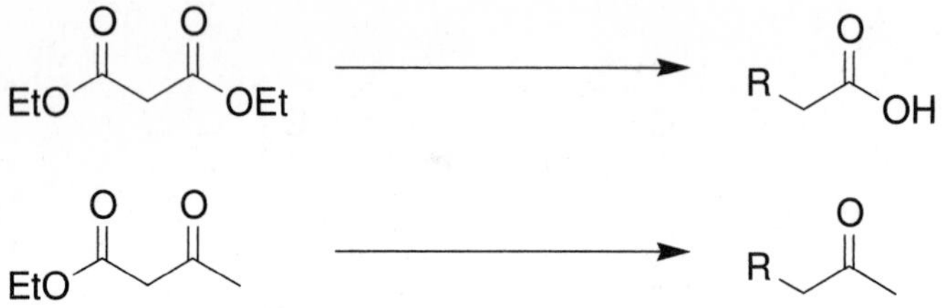

Michael Additions

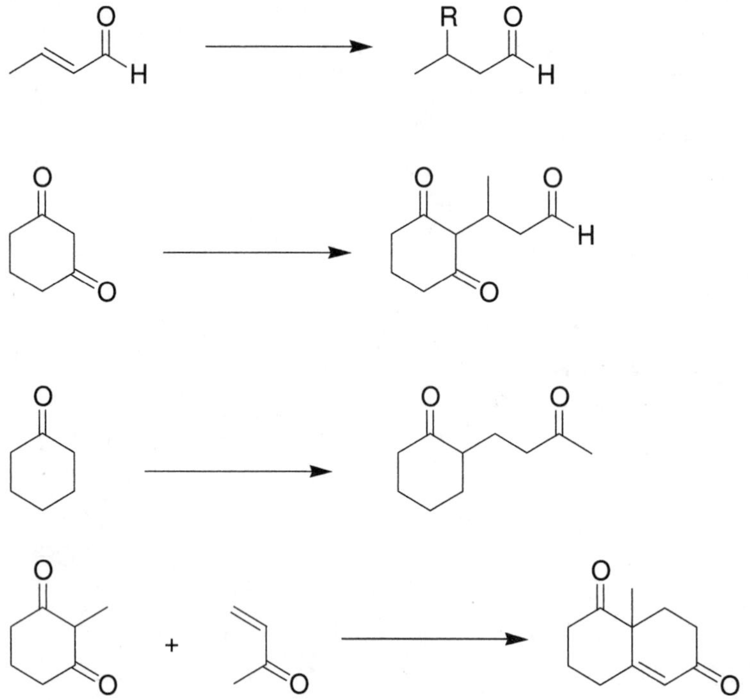

Solutions

22.1.

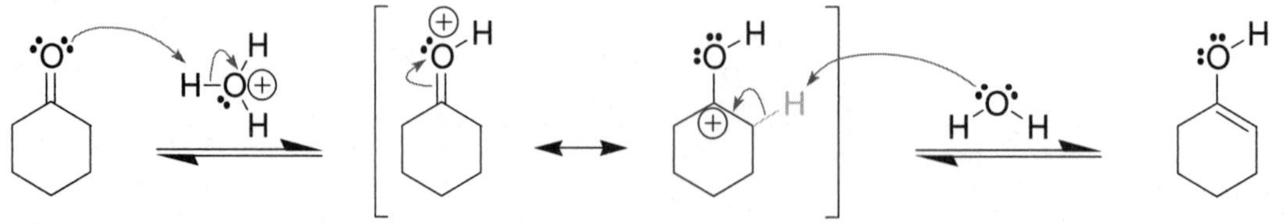

22.2.

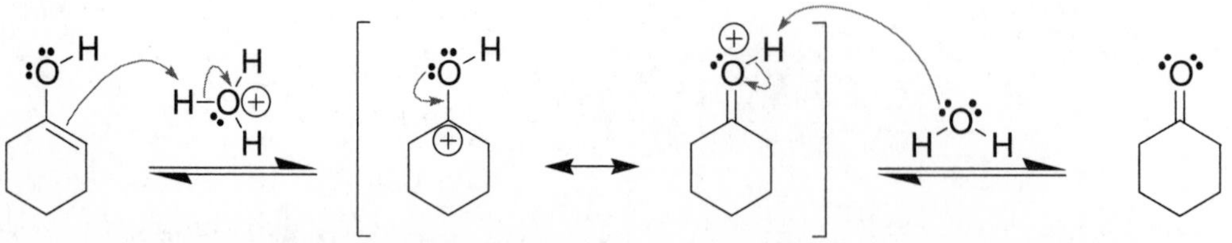

22.3.

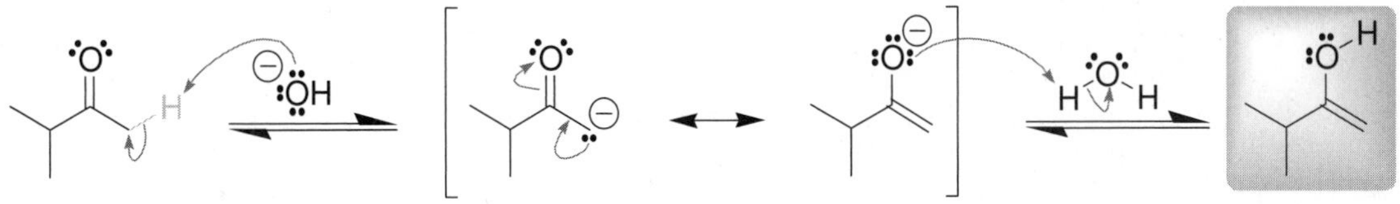

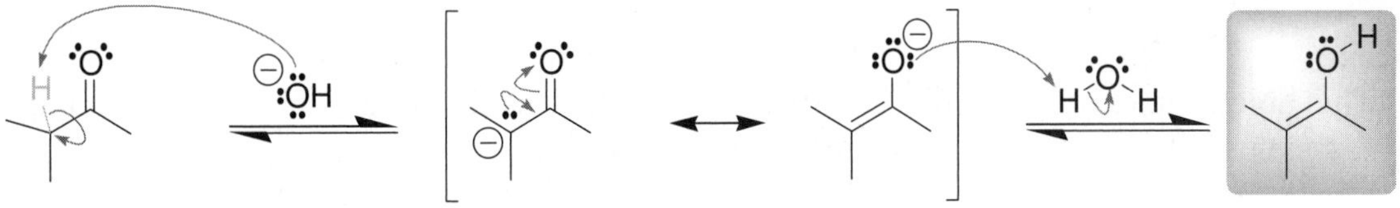

22.4.

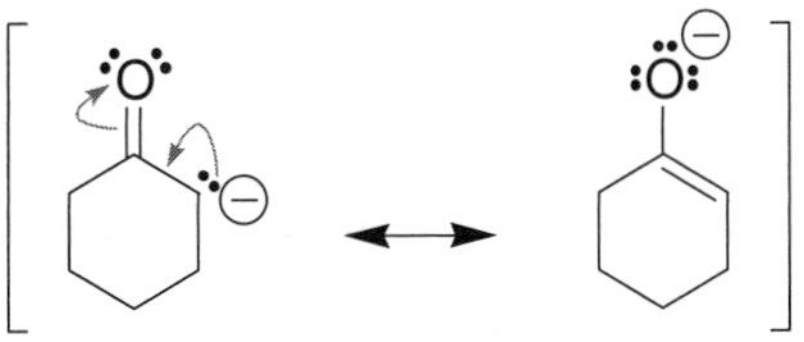

a)

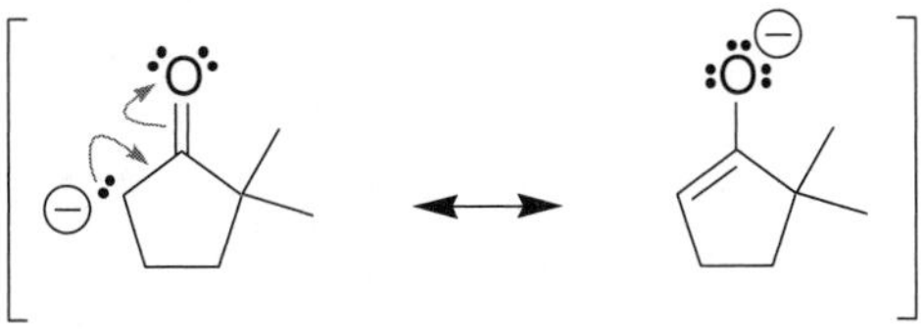

b)

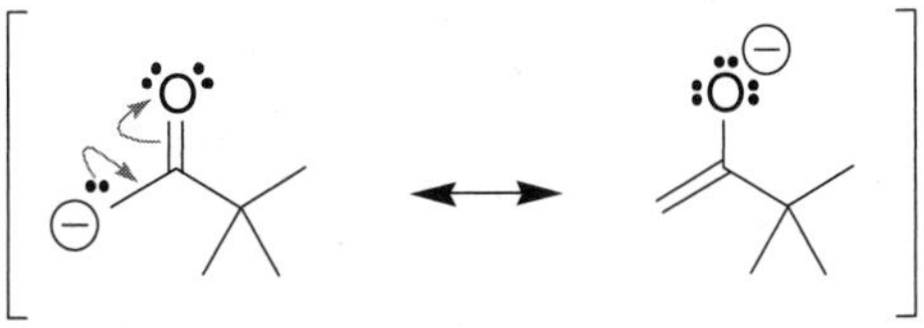

c)

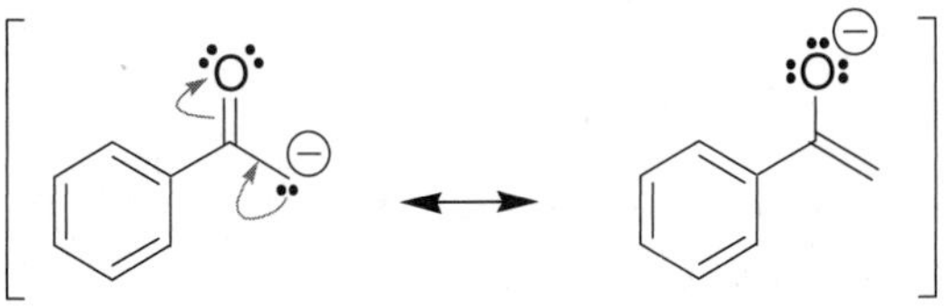

d)

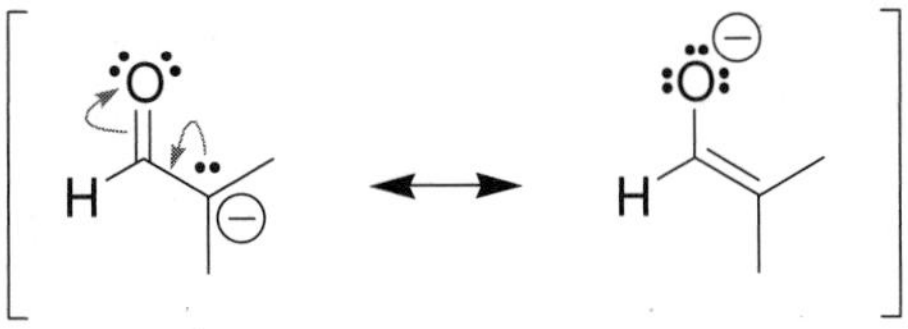

e)

22.5.

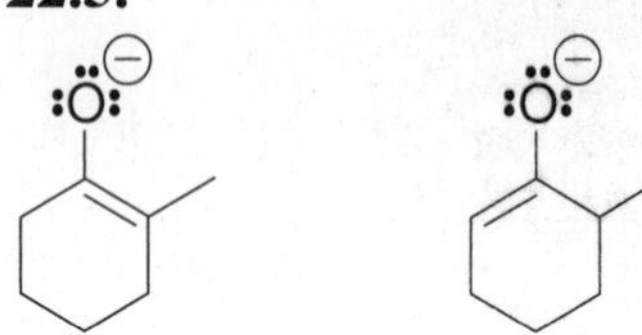

22.6.

a) This anion is a doubly stabilized enolate ion, so there will not be a substantial amount of ketone present together with the enolate at equilibrium:

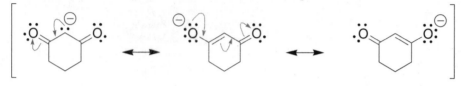

b) This anion is a regular enolate ion (not doubly stabilized), so there will be a substantial amount of ketone present together with the enolate at equilibrium:

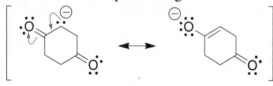

c) This anion is a regular enolate ion (not doubly stabilized), so there will be a substantial amount of ketone present together with the enolate at equilibrium:

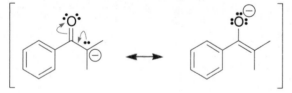

d) This anion is a regular enolate ion (not doubly stabilized), so there will be a substantial amount of ketone present together with the enolate at equilibrium:

22.7. This anion is highly stabilized by resonance. The negative charge is spread over two oxygen atoms (just like a doubly stabilized enolate) and three carbon atoms:

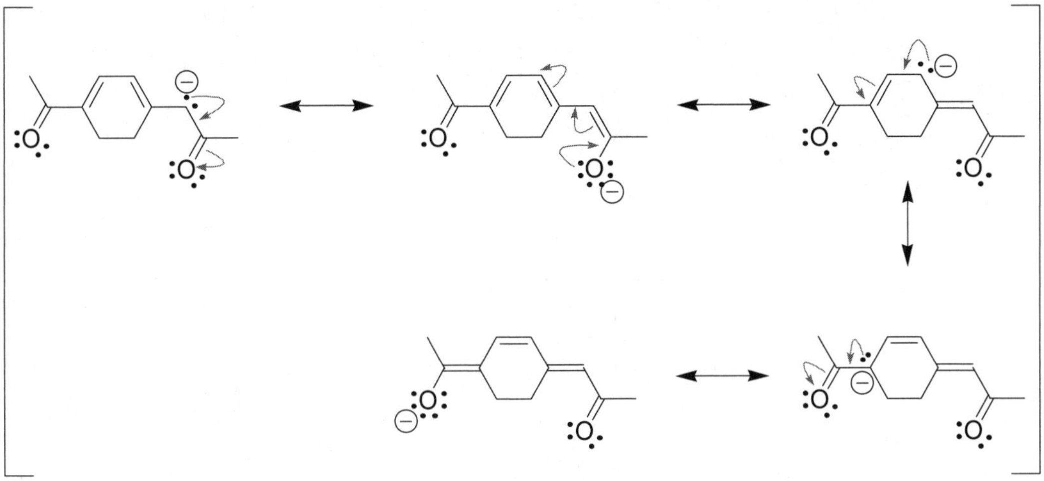

22.8.
a) 2,4-dimethyl-3,5-heptanedione is more acidic because its conjugate base is a doubly stabilized enolate. The other compound (4,4-dimethyl-3,5-heptanedione) cannot form a doubly stabilized enolate because there are no protons connected to the carbon atom that is in between both carbonyl groups.
b) 1,3-cyclopentanedione is more acidic because its conjugate base is a doubly stabilized enolate. The other compound (1,2-cyclopentanedione) cannot form a doubly stabilized enolate because the carbonyl groups are adjacent to each other.
c) Acetophenone is more acidic than benzaldehyde because the former has alpha protons and the latter does not.

22.9.
a)

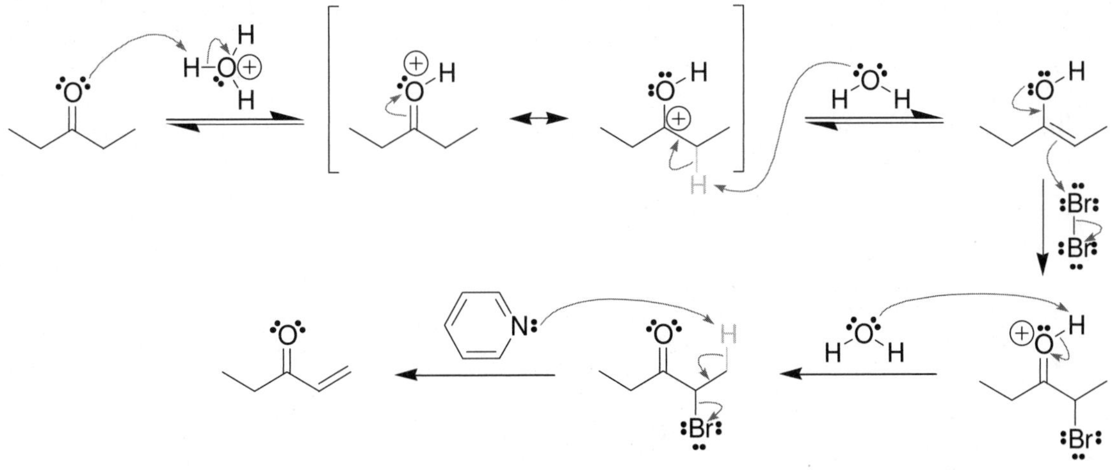

b)

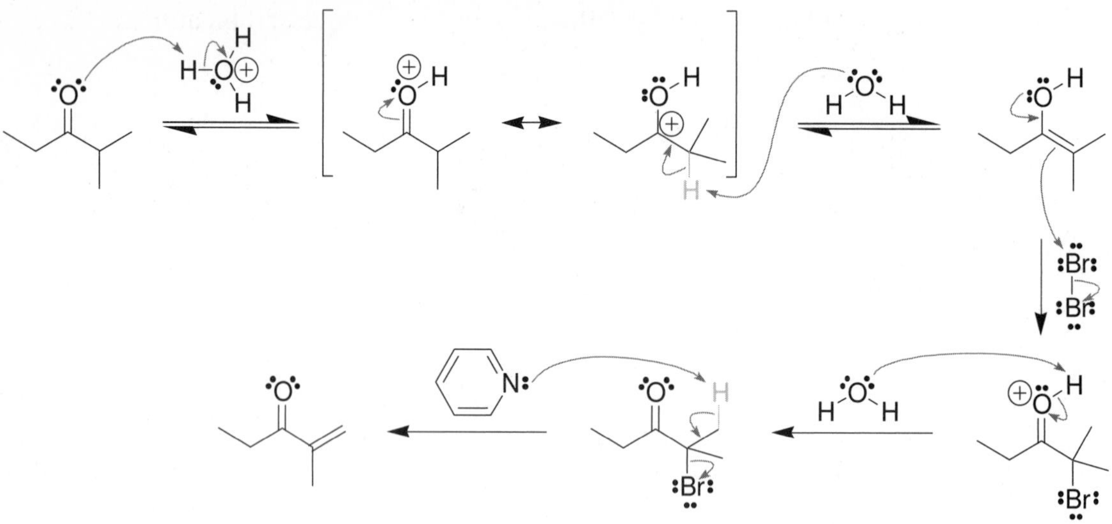

c)

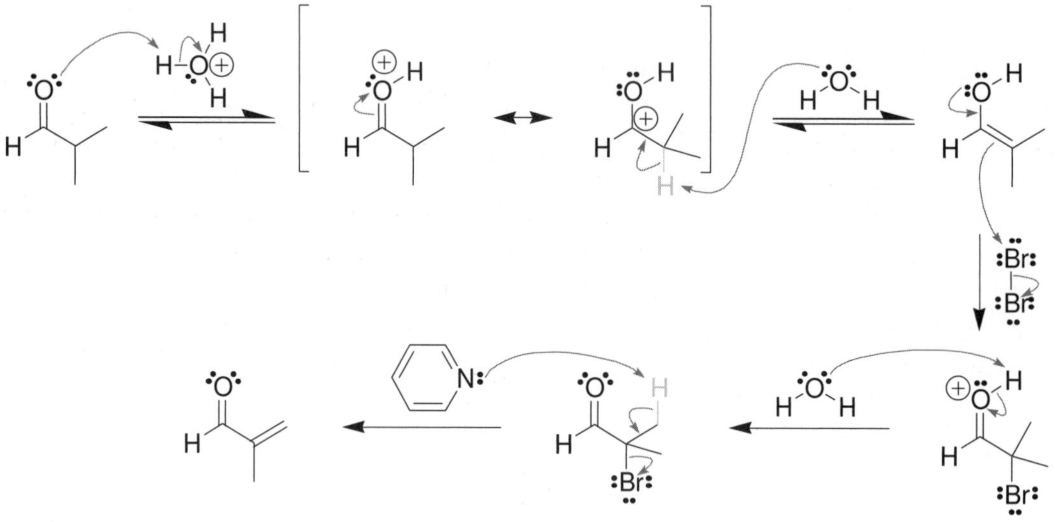

22.10.

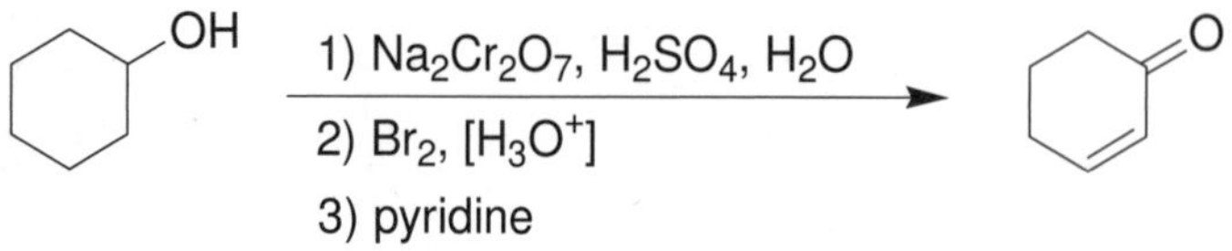

a) 1) $Na_2Cr_2O_7$, H_2SO_4, H_2O
2) Br_2, $[H_3O^+]$
3) pyridine

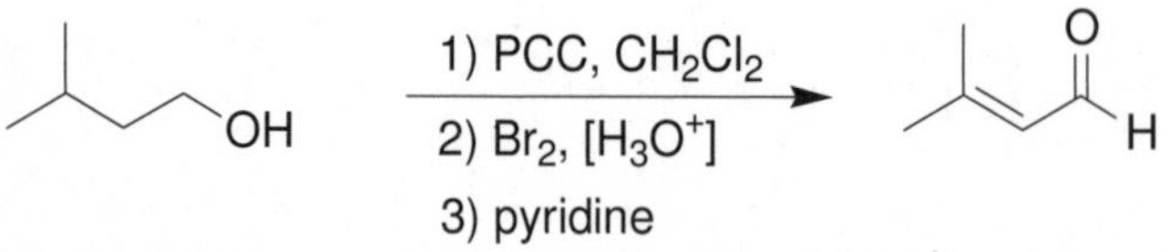

b) 1) PCC, CH_2Cl_2
2) Br_2, $[H_3O^+]$
3) pyridine

22.11.

a) 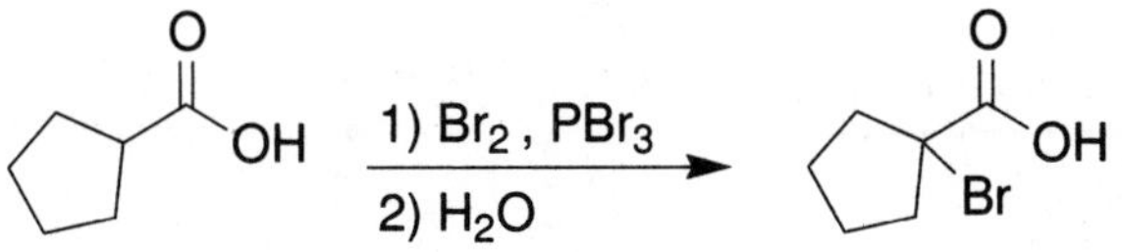
O
OH
1) Br_2 , PBr_3
2) H_2O
O
OH
Br

b) 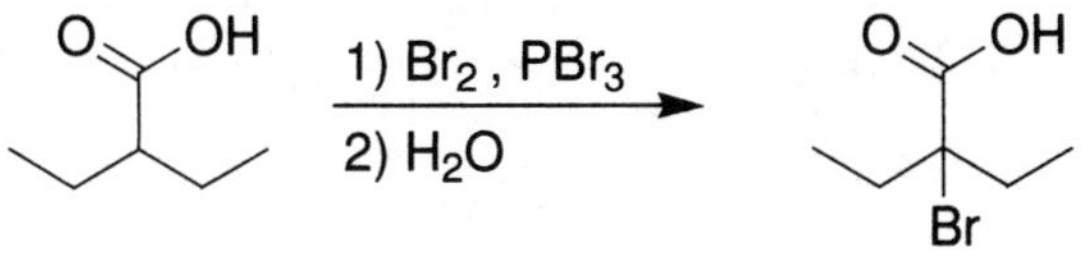
O
OH
1) Br_2 , PBr_3
2) H_2O
O
OH
Br

22.12.

a) 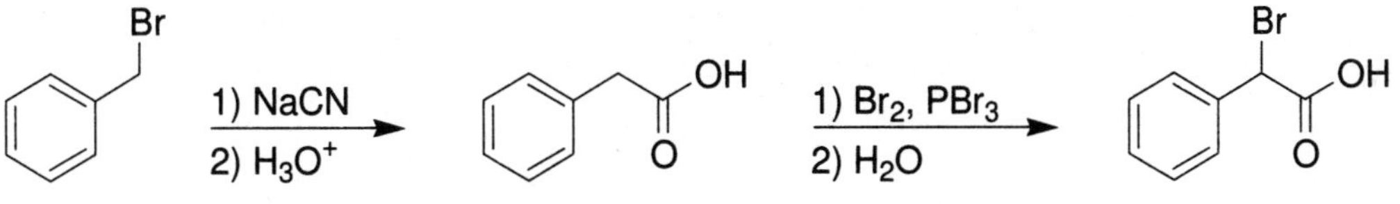
Br
1) NaCN
2) H_3O^+
OH
O
1) Br_2, PBr_3
2) H_2O
Br
OH
O

b) 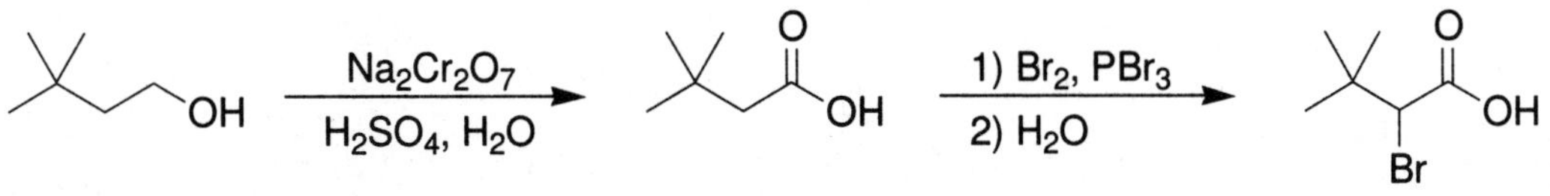
OH
$Na_2Cr_2O_7$
H_2SO_4, H_2O
O
OH
1) Br_2, PBr_3
2) H_2O
O
OH
Br

c) 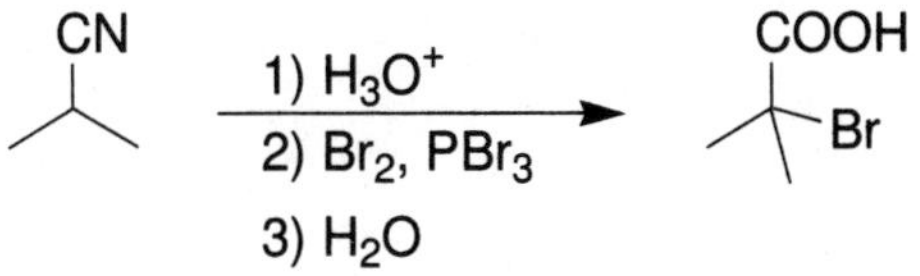
CN
1) H_3O^+
2) Br_2, PBr_3
3) H_2O
COOH
Br

22.13.

a) 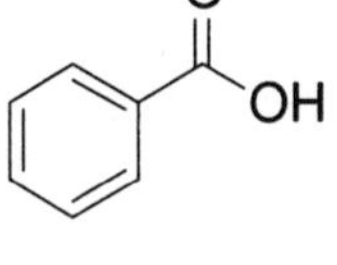
O
OH

b) 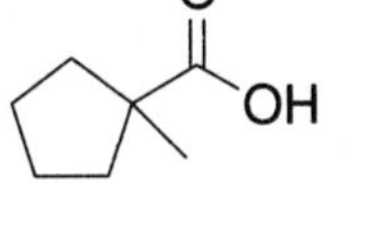
O
OH

c) 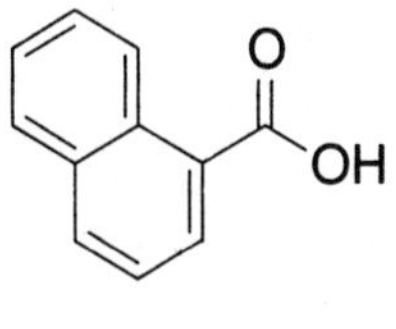
O
OH

22.14.

a)

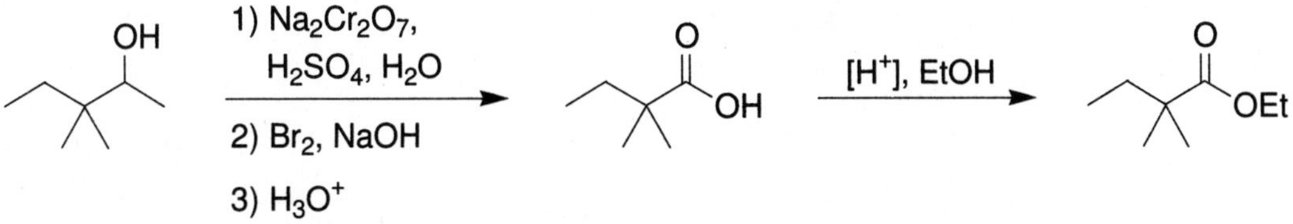
OH
1) $Na_2Cr_2O_7$,
H_2SO_4, H_2O
2) Br_2, NaOH
3) H_3O^+
O
OH
[H^+], EtOH
O
OEt

b)

1) O_3
2) DMS

1) Br_2, NaOH
2) H_3O^+

OH

$SOCl_2$

Cl

c)

H_3O^+

1) Br_2, NaOH
2) H_3O^+

COOH

d)

N-H

H_3O^+

1) Br_2, NaOH
2) H_3O^+

OH

1) $SOCl_2$
2) xs NH_3

NH_2

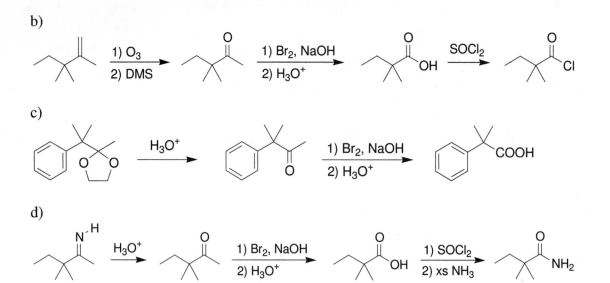

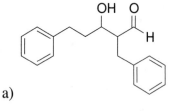

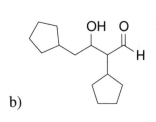

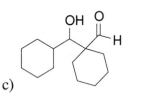

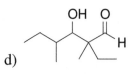

22.15.

a)

b)

c)

d)

22.16.

a)

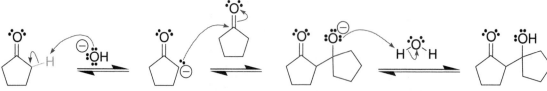

b)

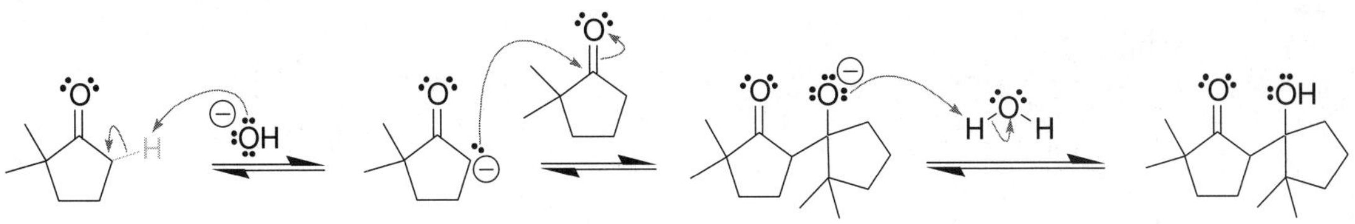

c)

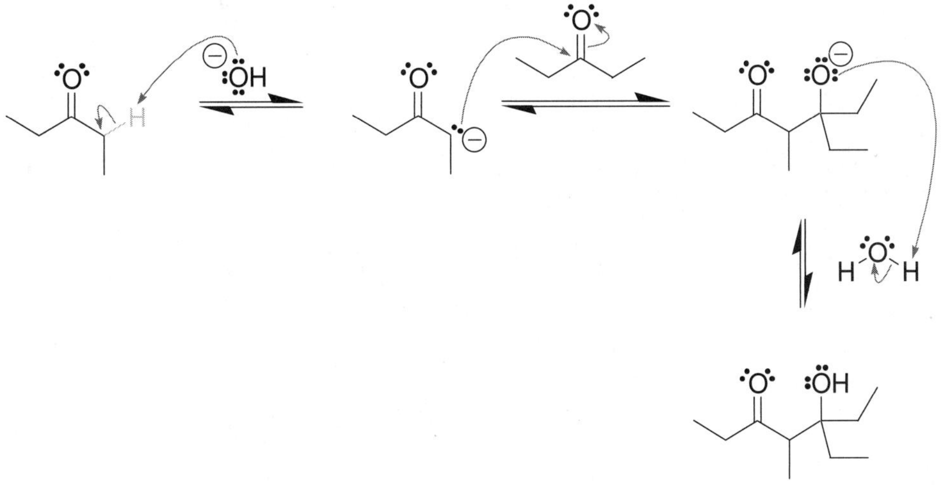

d)

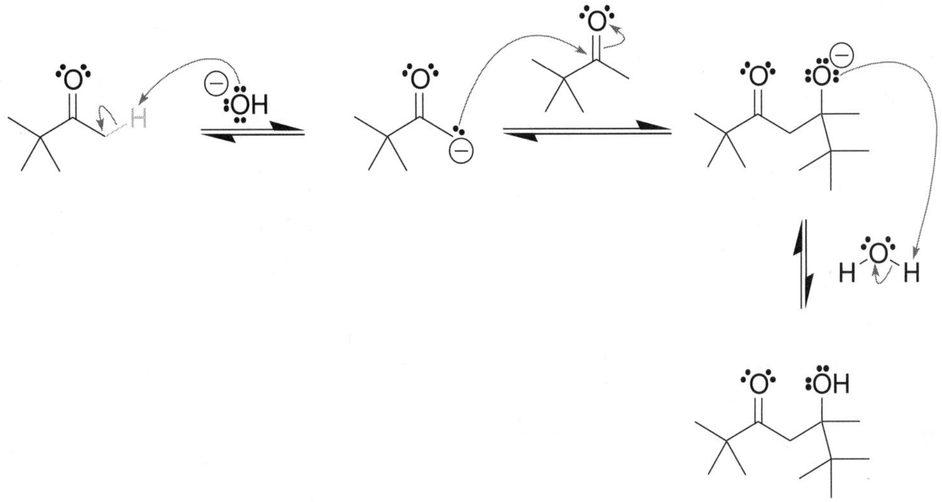

22.17. The first step of an aldol addition reaction is deprotonation at the alpha position, but this compound has no alpha protons.

22.18.

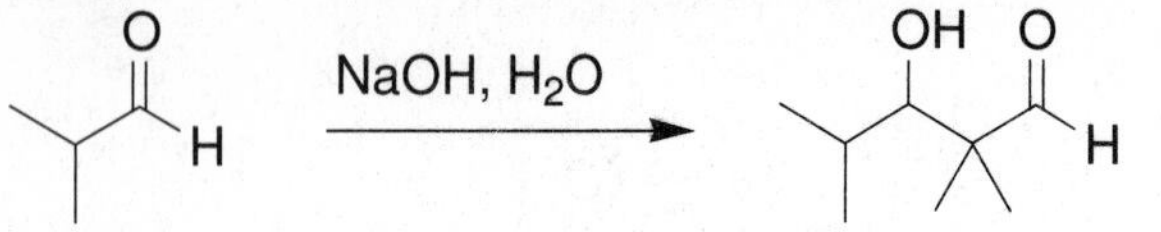

H (NaOH, H_2O) → OH O H

22.19.

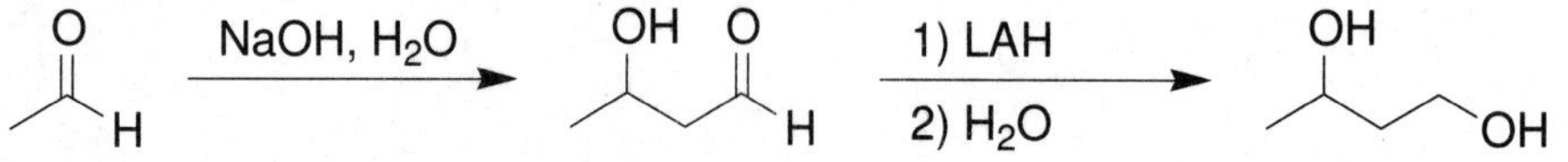

O H — NaOH, H_2O → OH O H — 1) LAH 2) H_2O → OH OH

22.20.

a) 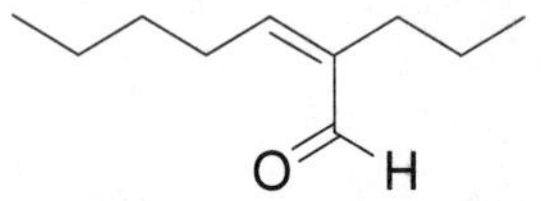 O H b) 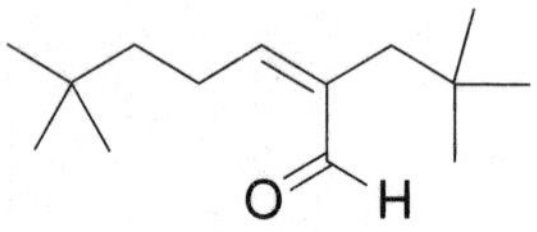 O H c) 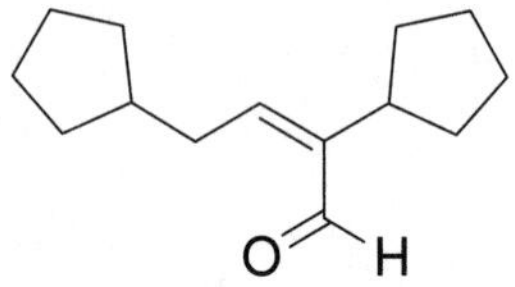 O H

d) 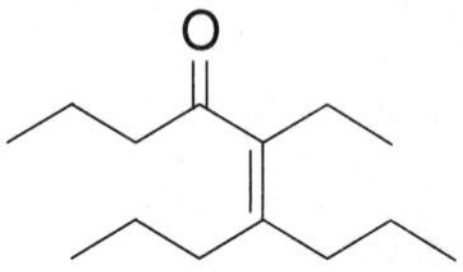 O e) 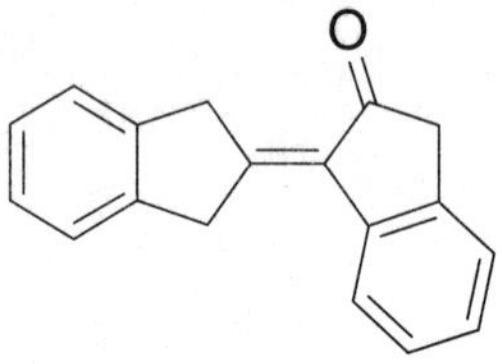 O f) 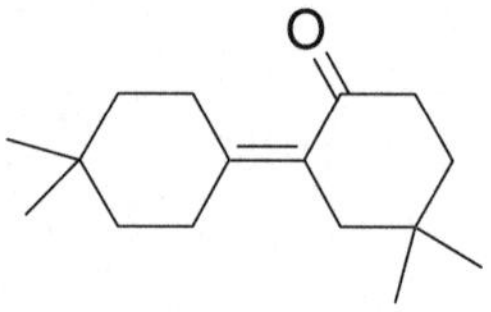 O

22.21.

a) 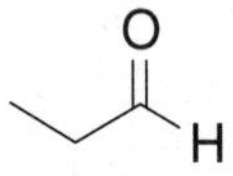 O H b) 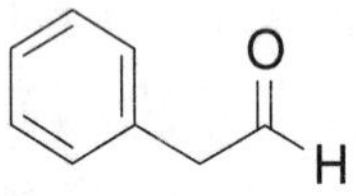 O H c) 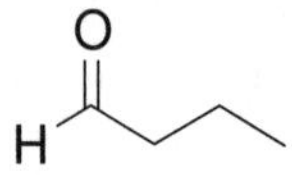 H O

22.22.

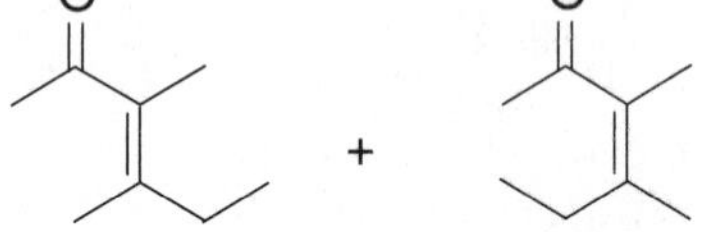

O + O

22.23.

a)

O — 1) LDA 2) H(C=O)H 3) H_2O → O OH

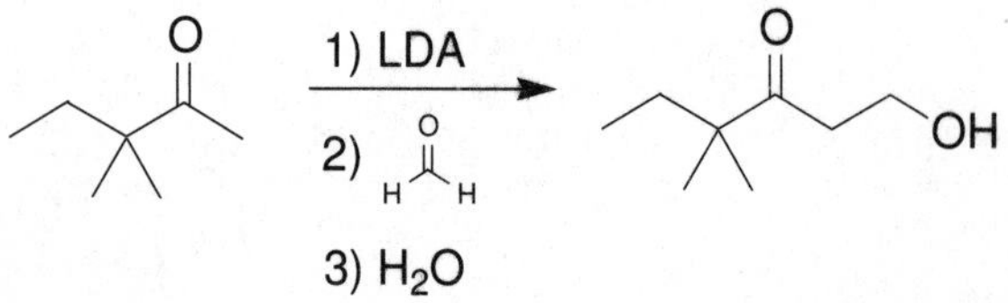

b)

1) LDA

2)

3) H_2O

c)

1) LDA

2)

3) H_2O

d)

1) LDA

2)

3) H_2O

e)

1) LDA

2)

3) H_2O

22.24.

a)

1) LDA

2)

3) H_2O

1) LAH

2) H_2O

$[H^+]$

$-H_2O$

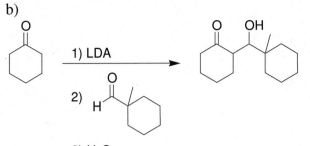

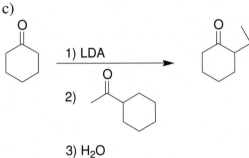

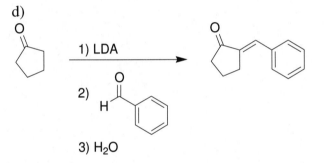

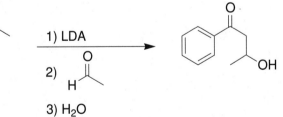

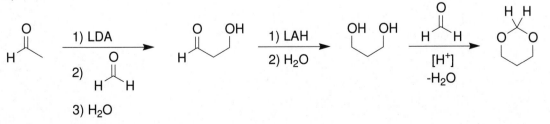

b)

1) LDA
2)
3) H_2O

1) LAH
2) H_2O

$[H^+]$
$-H_2O$

c)

1) LDA
2)
3) H_2O

1) LAH
2) H_2O

$[H^+]$
$-H_2O$

d)

1) LDA
2)
3) H_2O

1) LAH
2) H_2O

$[H^+]$
$-H_2O$

22.25.

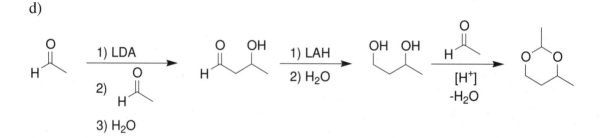

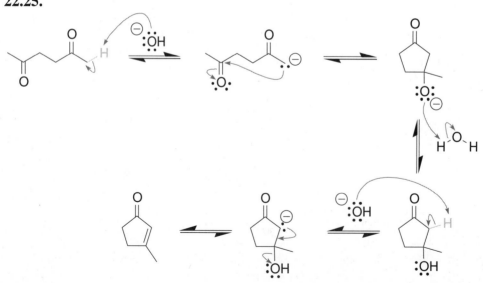

22.26.

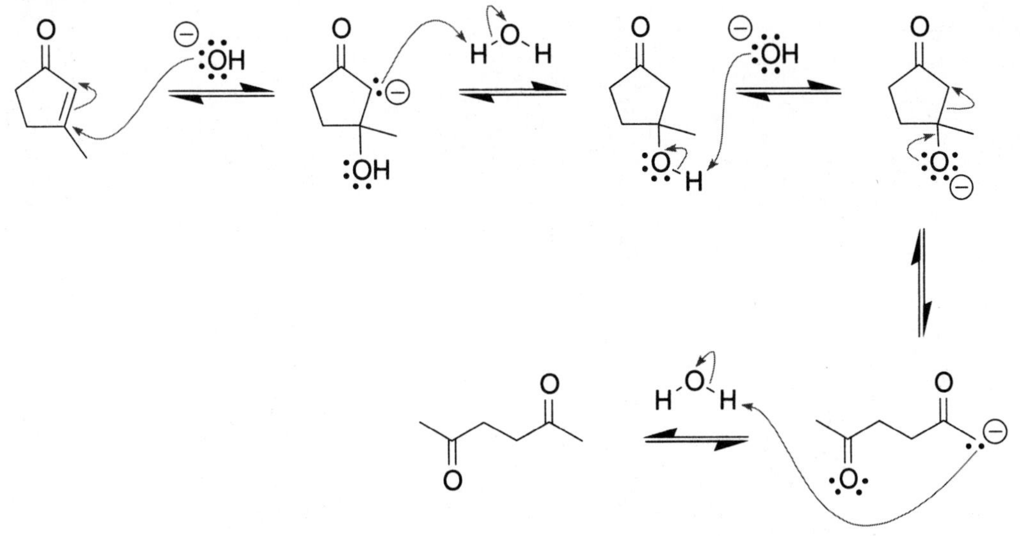

22.27.

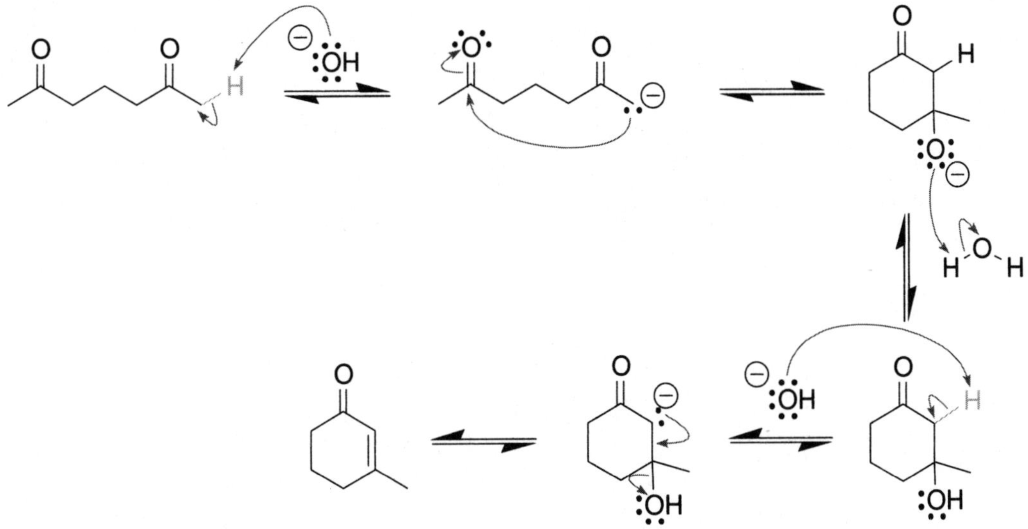

22.28.

a) NaOEt b) *t*-BuOK

22.29.

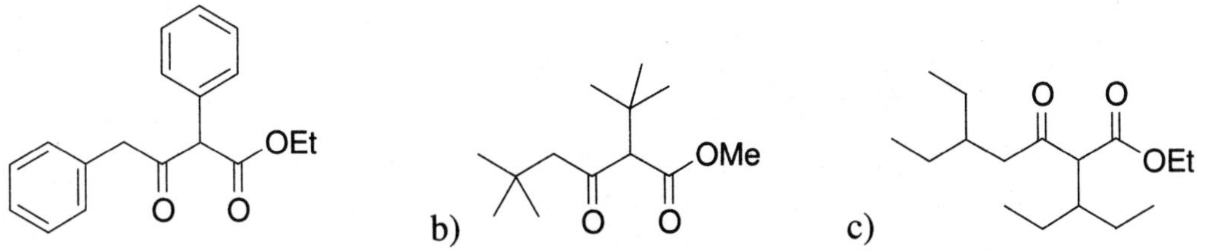

a)

b)

c)

22.30.

a)

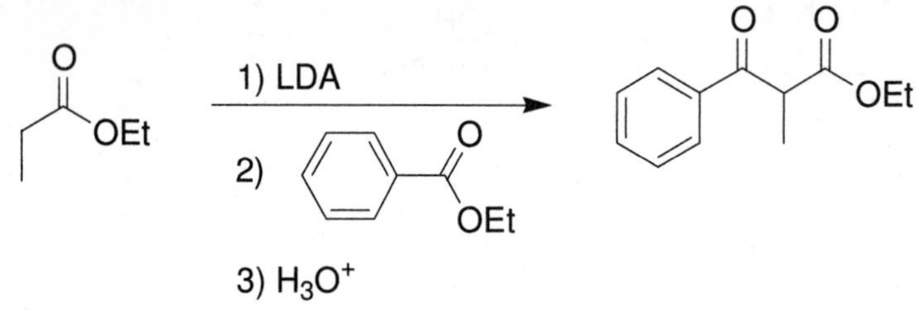

1) LDA

2) (ethyl benzoate) OEt

3) H_3O^+

OEt ... OEt

b)

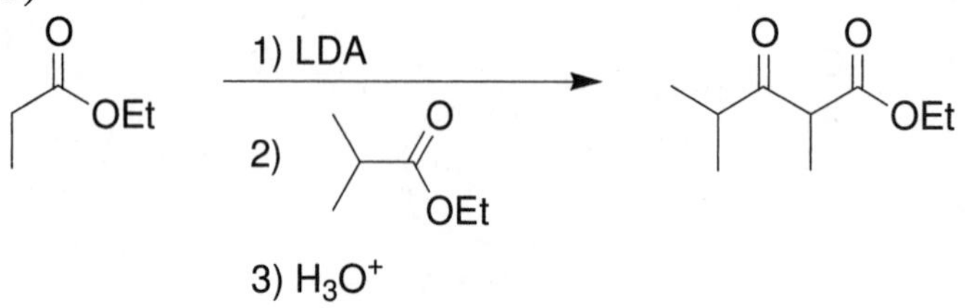

1) LDA

2) OEt

3) H_3O^+

OEt ... OEt

c)

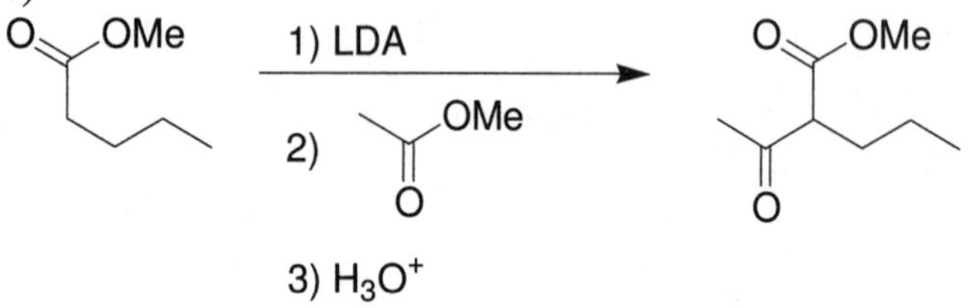

OMe

1) LDA

2) OMe

3) H_3O^+

OMe

d)

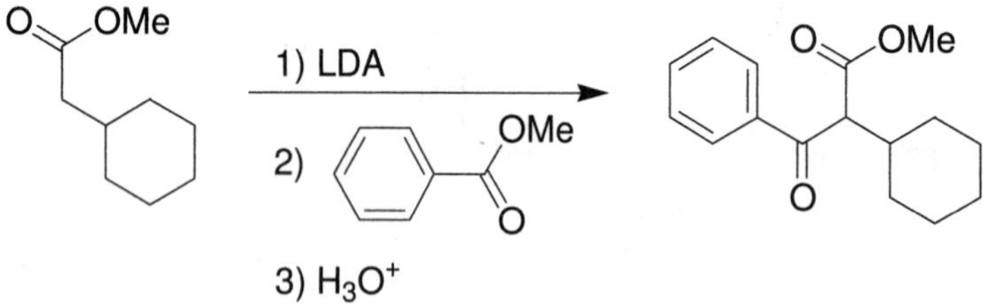

OMe

1) LDA

2) OMe

3) H_3O^+

OMe

e)

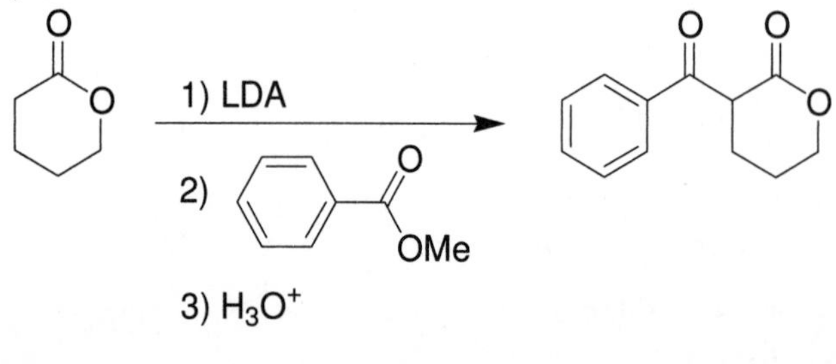

O

1) LDA

2) OMe

3) H_3O^+

O

22.31.

a) 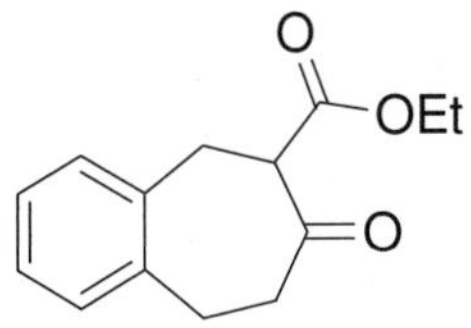 O, OEt, O

b) 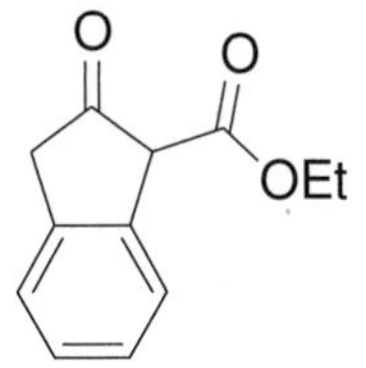 O, O, OEt

c) 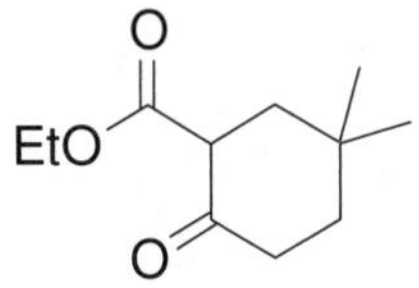 O, EtO, O

22.32.

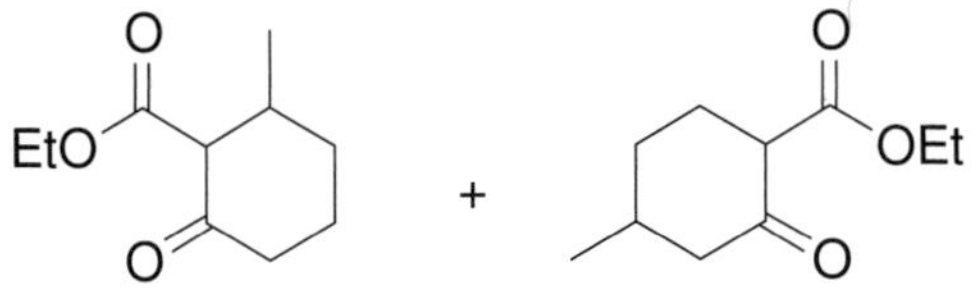

O, EtO, O + O, OEt, O

22.33.

a) 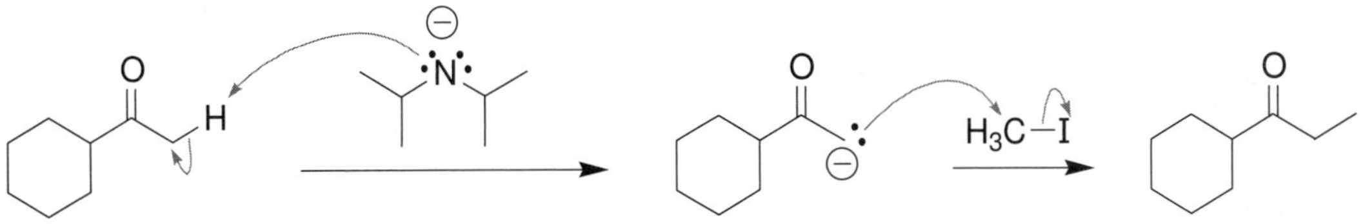 O, H, N, O, $H_3C—I$, O

b) 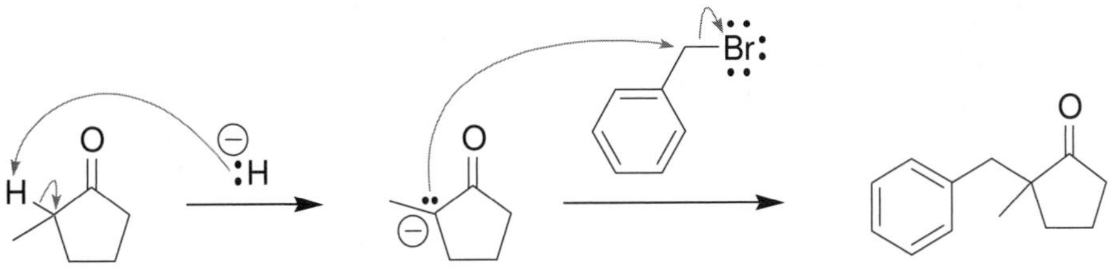 H, O, H, O, Br, O

c)

O, H, N, O, I, O, H, N, O, I–CH_3, O

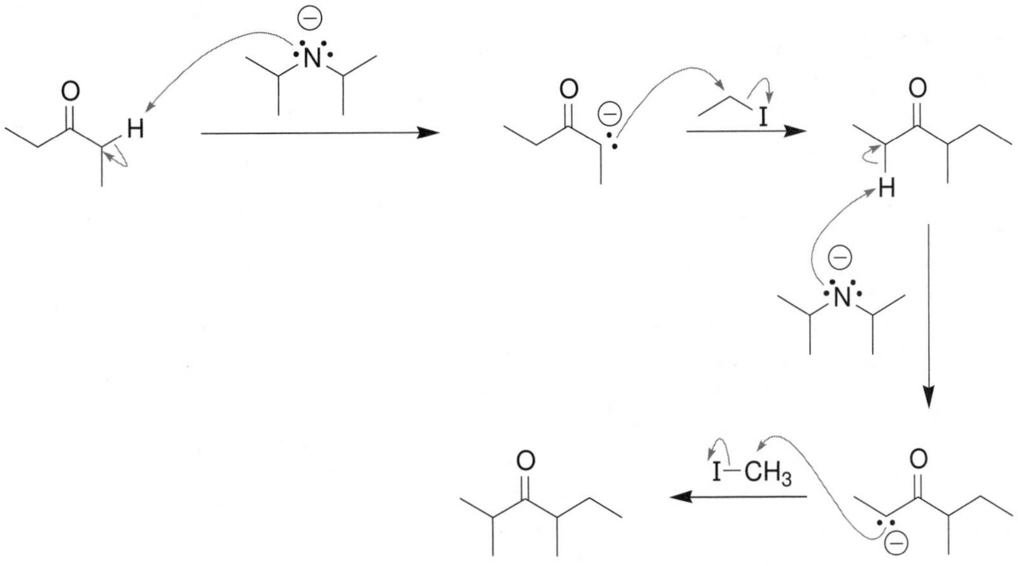

22.34.

OH

1) $Na_2Cr_2O_7$, H_2SO_4, H_2O
2) LDA
3) EtI
4) LAH
5) H_2O

OH

22.35.

a)

O O
EtO OEt

1) NaOEt / EtOH
2) Br
3) H_3O^+, heat

O
OH

b)

1) NaOEt / EtOH
2) Br
3) NaOEt / EtOH
4) CH_3Br
5) H_3O^+, heat

O O
EtO OEt

O
OH

c)

O O
EtO OEt

1) NaOEt / EtOH
2) Br
3) NaOEt / EtOH
4) Br
5) H_3O^+, heat

COOH

d)

O O
EtO OEt

1) NaOEt / EtOH
2) CH_3Br
3) NaOEt / EtOH
4) Br
5) H_3O^+, heat

OH
O

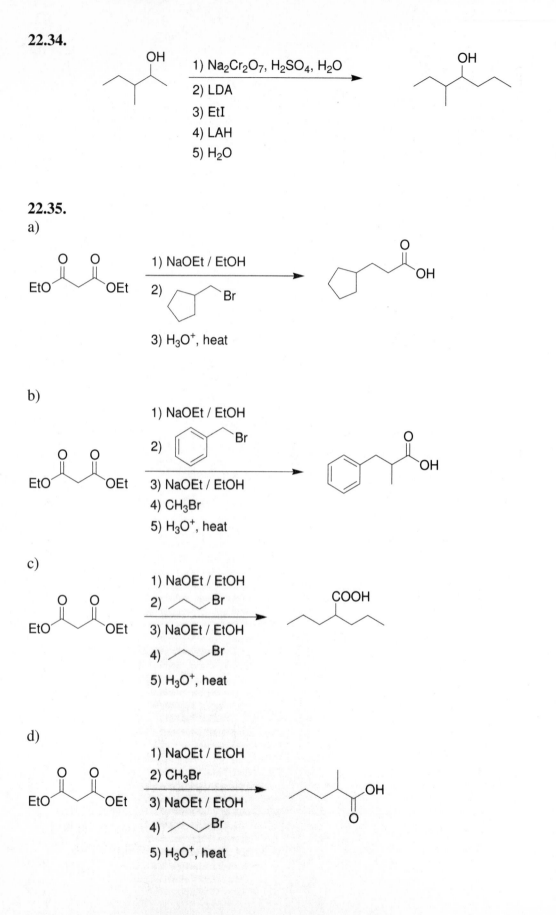

e)

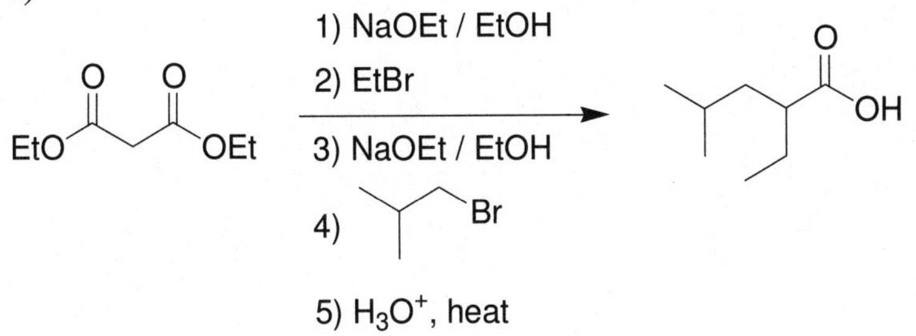

1) NaOEt / EtOH
2) EtBr
3) NaOEt / EtOH
4) (1-bromo-2-methylpropane)
5) H_3O^+, heat

22.36.

a)

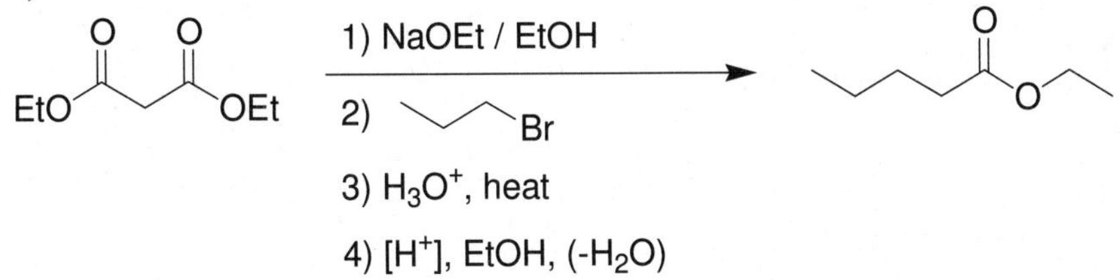

1) NaOEt / EtOH
2) (1-bromopropane)
3) H_3O^+, heat
4) [H^+], EtOH, (-H_2O)

b)

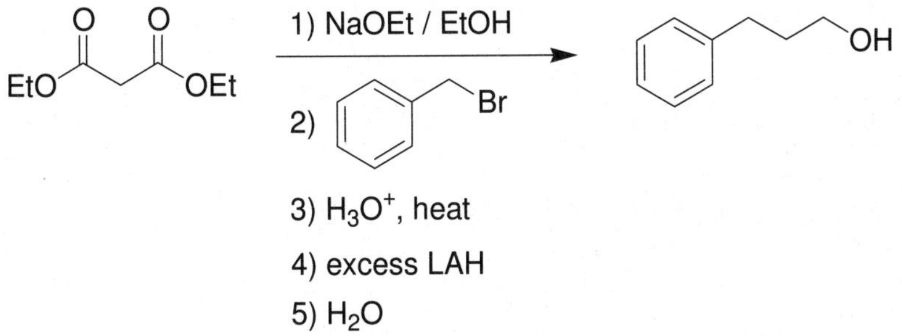

1) NaOEt / EtOH
2) (benzyl bromide)
3) H_3O^+, heat
4) excess LAH
5) H_2O

c)

1) NaOEt / EtOH
2) (cyclohexylmethyl bromide)
3) H_3O^+, heat
4) $SOCl_2$
5) excess NH_3

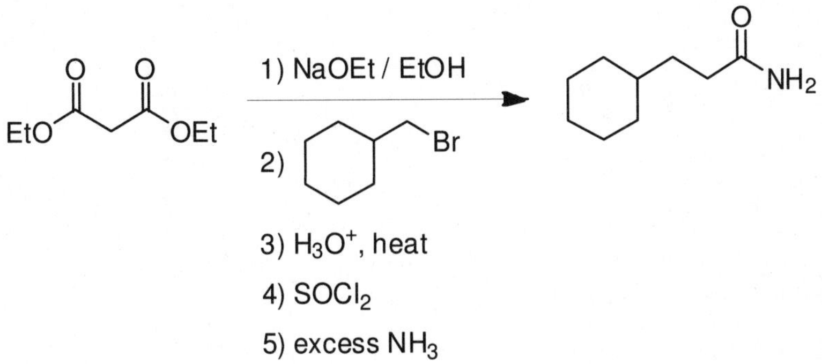

22.37. Preparation of the desired compound requires the installation of three alkyl groups at the alpha position. The malonic ester synthesis can only be used to install two alkyl groups because the starting material (diethyl malonate) has only two alpha protons.

22.38.

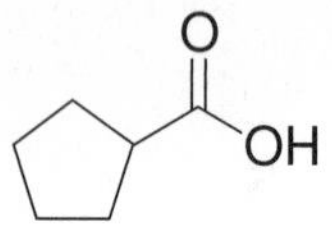

22.39.

a)

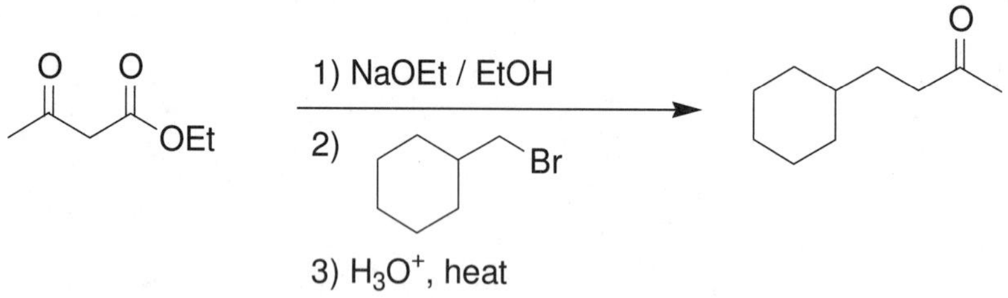

1) NaOEt / EtOH

2)

3) H_3O^+, heat

b)

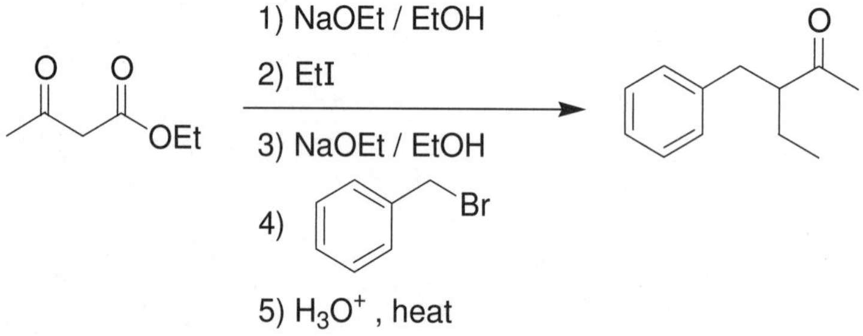

1) NaOEt / EtOH

2) EtI

3) NaOEt / EtOH

4)

5) H_3O^+ , heat

c)

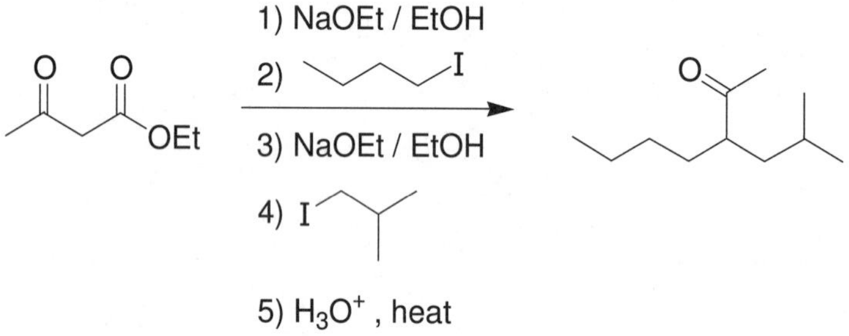

1) NaOEt / EtOH

2)

3) NaOEt / EtOH

4)

5) H_3O^+ , heat

d)

1) NaOEt / EtOH

2) CH_3I

3) NaOEt / EtOH

4)

5) H_3O^+ , heat

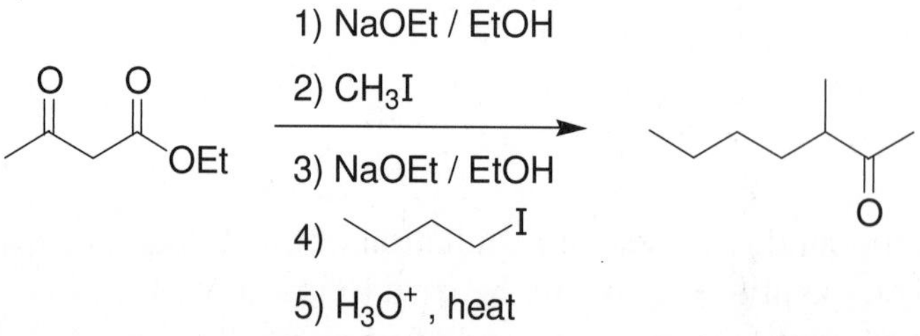

22.40.

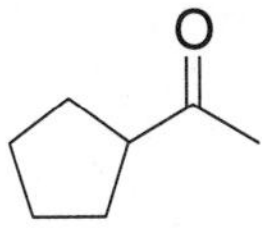

22.41.

a)

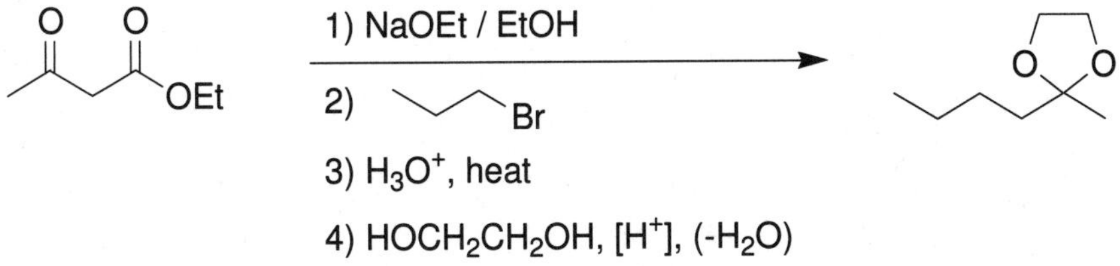

1) NaOEt / EtOH

2) Br

3) H_3O^+, heat

4) $HOCH_2CH_2OH$, $[H^+]$, $(-H_2O)$

b)

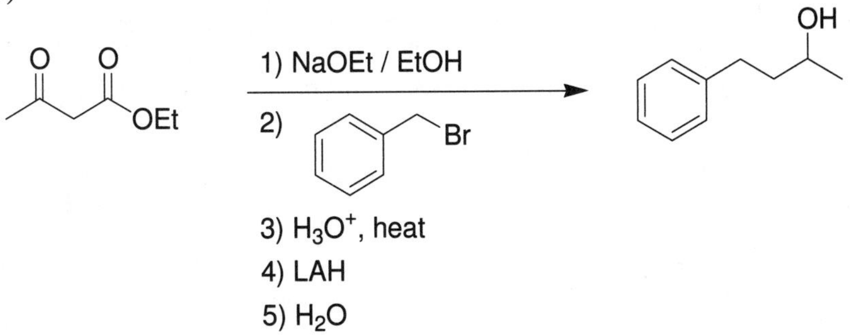

1) NaOEt / EtOH

2) Br

3) H_3O^+, heat

4) LAH

5) H_2O

c)

1) NaOEt / EtOH

2) Br

3) H_3O^+, heat

4) $[H^+]$, NH_3, $(-H_2O)$

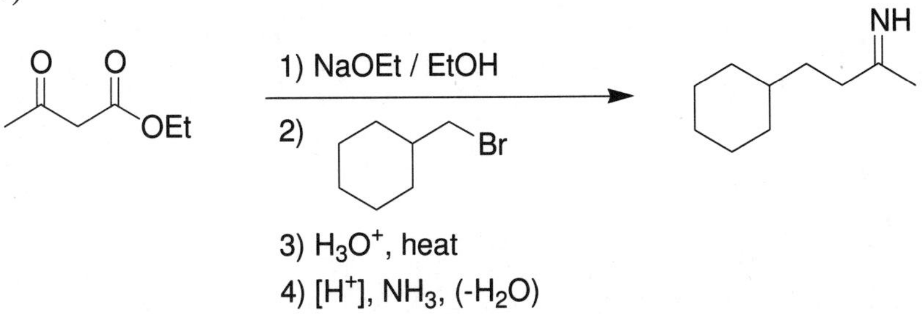

22.42. Preparation of the desired compound requires the installation of three alkyl groups at the alpha position. The acetoacetic ester synthesis can only be used to install two alkyl groups because the starting material (diethyl malonate) has only two alpha protons.

22.43.

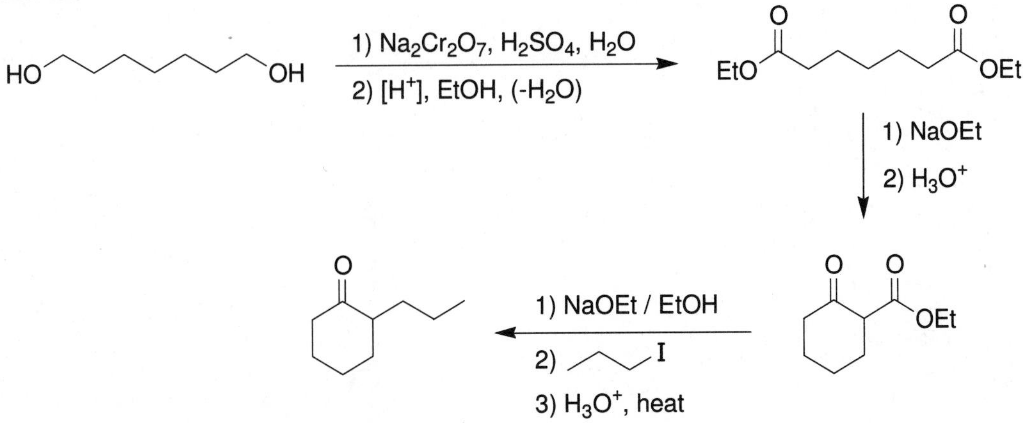

1) $Na_2Cr_2O_7$, H_2SO_4, H_2O

2) $[H^+]$, EtOH, $(-H_2O)$

1) NaOEt

2) H_3O^+

1) NaOEt / EtOH

2) I

3) H_3O^+, heat

22.44.

a) 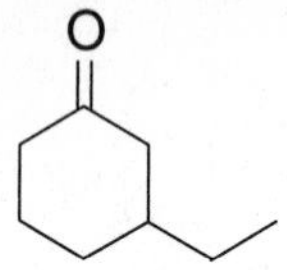 b) 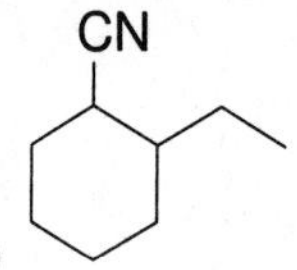 CN c) 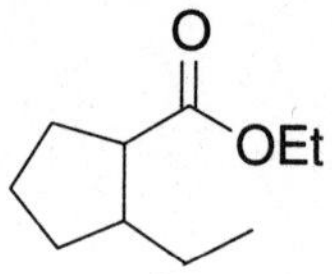 O OEt

22.45.

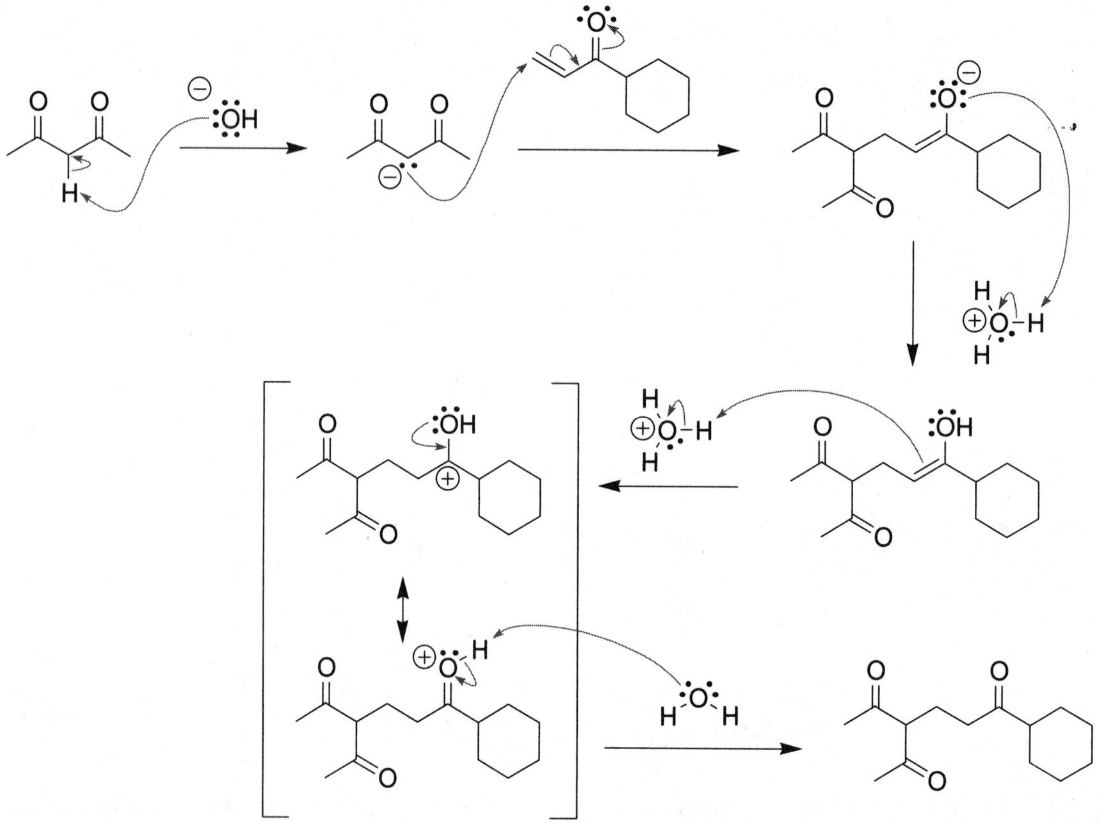

O O :ÖH H O O O O :Ö: O O H ⊕O–H H

O :ÖH ⊕ O H ⊕O–H H O :ÖH O O

⊕ O H O O H :Ö: H O O O

22.46.

a)

EtO OEt O O

1) NaOEt / EtOH

2) O

3) H_3O^+, heat

HO O O

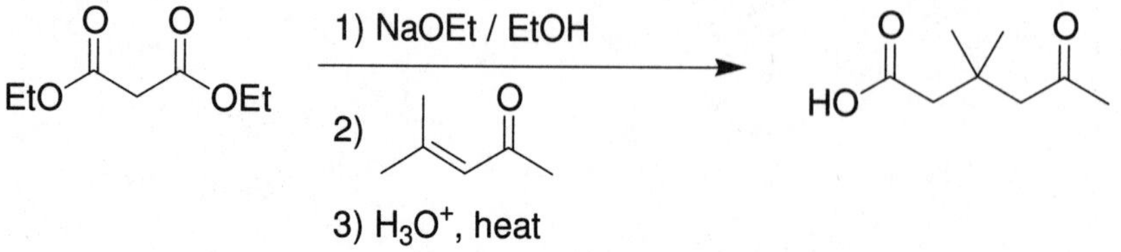

b)

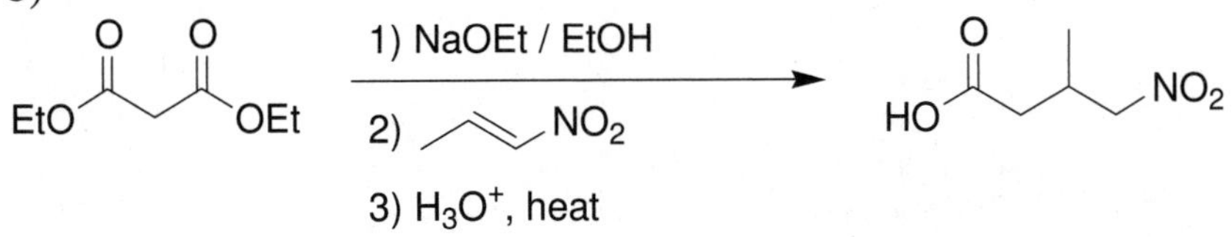

EtO OEt

1) NaOEt / EtOH

2) NO_2

3) H_3O^+, heat

HO NO_2

22.47.

a)

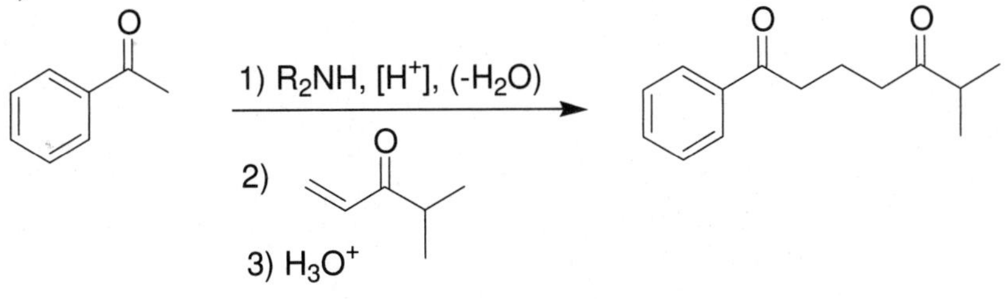

1) R_2NH, [H^+], ($-H_2O$)

2)

3) H_3O^+

b)

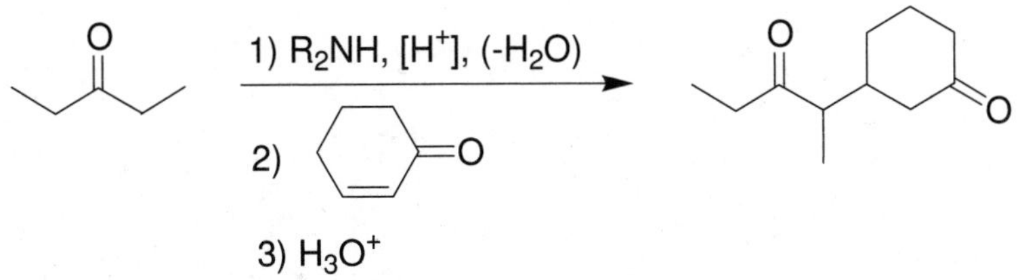

1) R_2NH, [H^+], ($-H_2O$)

2)

3) H_3O^+

c)

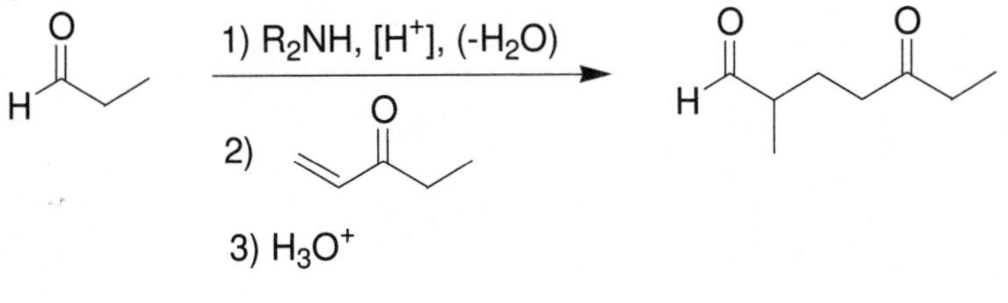

H

1) R_2NH, [H^+], ($-H_2O$)

2)

3) H_3O^+

H

22.48.

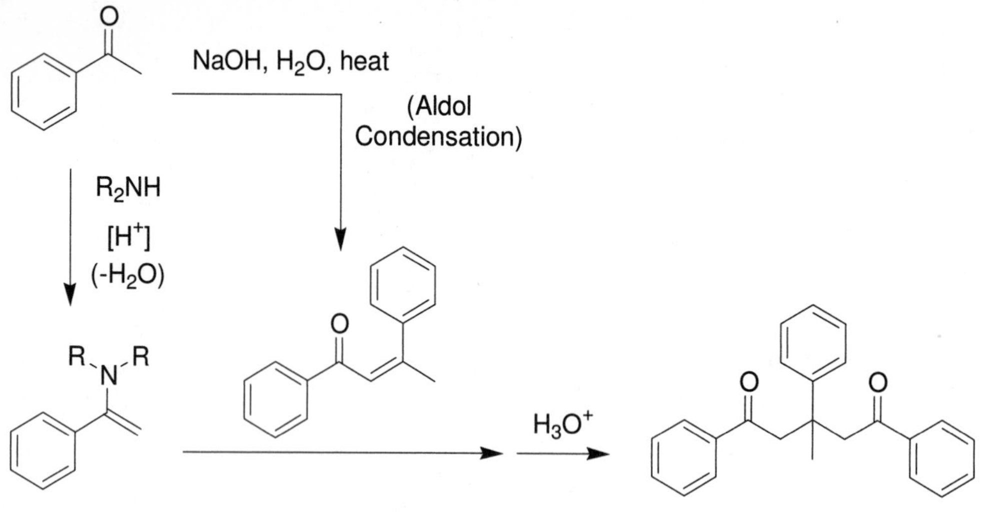

NaOH, H_2O, heat

(Aldol Condensation)

R_2NH

$[H^+]$

$(-H_2O)$

R N R

H_3O^+

22.49.

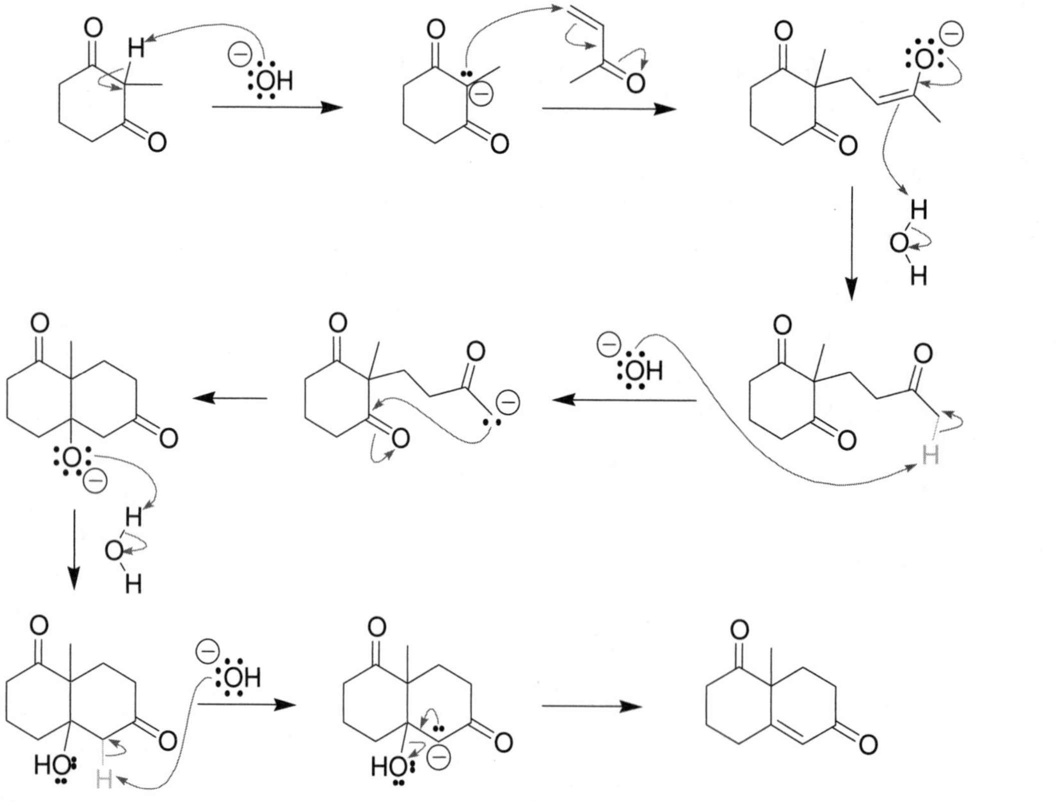

$:\ddot{O}H$

H

O

H

22.50.

H

O

H O

+

NaOH, heat

H

O

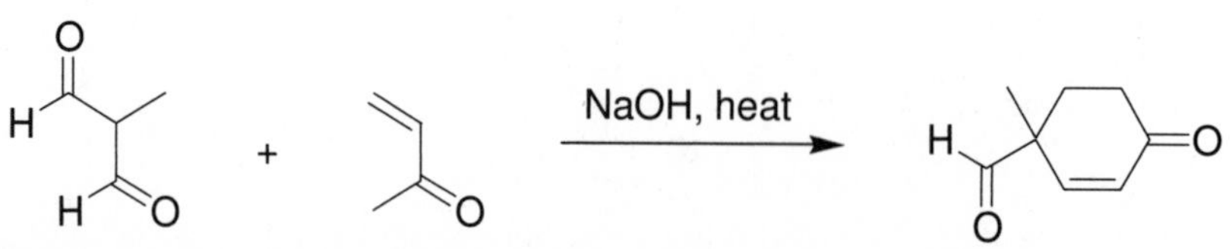

22.51.

a)

1) R_2NH, $[H^+]$, $(-H_2O)$

2)

3) H_3O^+

b)

1) LDA

2) H

3) CH_3I

c)

OH

OH

1) R_2NH, $[H^+]$, $(-H_2O)$

2) OEt

3) H_3O^+

4) excess LAH

5) H_2O

22.52.

a)

OH PCC CH_2Cl_2 H NaOH OH H 1) LAH 2) H_2O OH OH

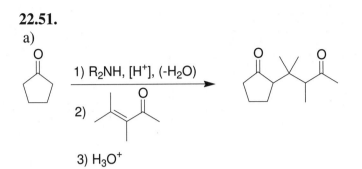

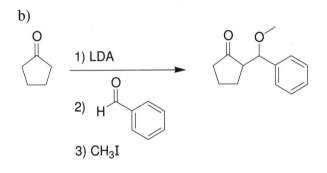

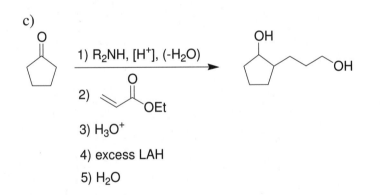

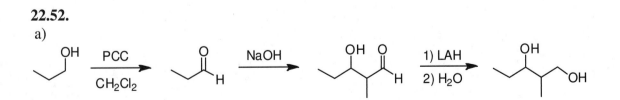

b)

1) $[H_3O^+]$, Br_2
2) pyridine

PCC / CH_2Cl_2 ; R_2NH / $[H^+]$ / $(-H_2O)$; 1) acrolein 2) H_3O^+

1) LAH
2) H_2O

HO OH

c)

PCC / CH_2Cl_2 ; NaOH ; PCC / CH_2Cl_2

$[H^+]$
NH_3
$(-H_2O)$

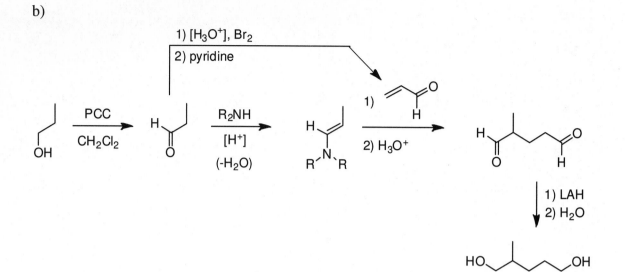

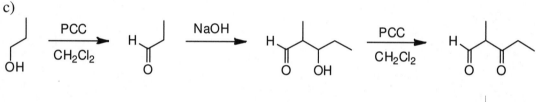

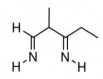

22.53.

a)

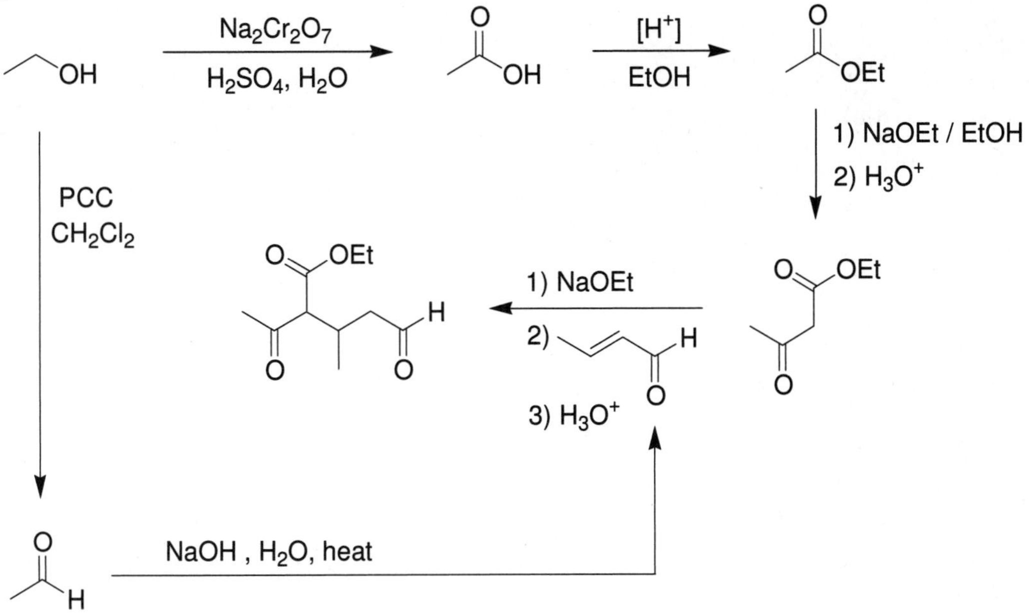

OH —($Na_2Cr_2O_7$, H_2SO_4, H_2O)→ OH —($[H^+]$, EtOH)→ OEt

OEt —(1) NaOEt / EtOH; 2) H_3O^+)→ O OEt / O

OH —(PCC, CH_2Cl_2)→ O / H —(NaOH , H_2O, heat)→ O / H

O OEt / O —(1) NaOEt; 2) H / O; 3) H_3O^+)→ O OEt / H / O O

b)

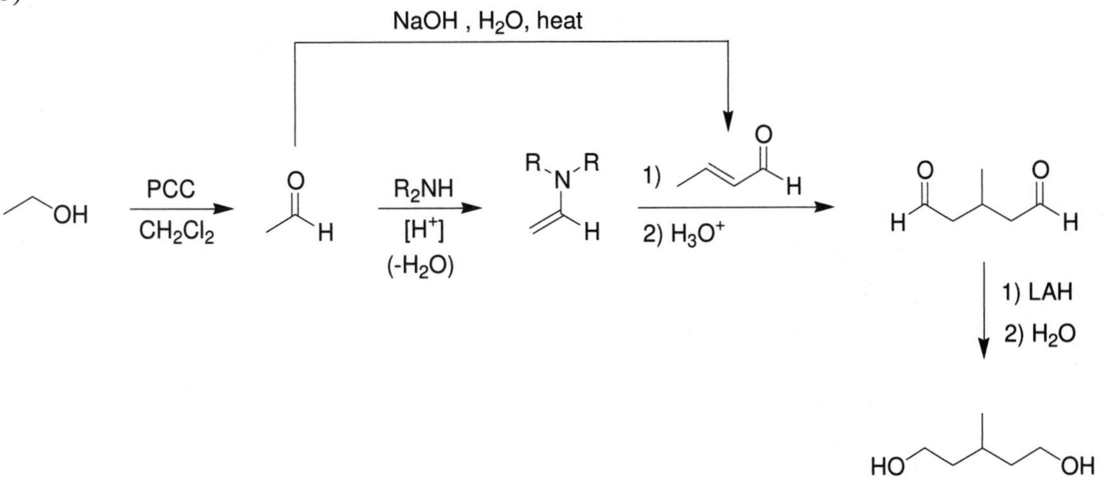

OH —(PCC, CH_2Cl_2)→ O / H —(R_2NH, $[H^+]$, $(-H_2O)$)→ R N R / H —(1) O / H; 2) H_3O^+)→ O O / H H

O / H —(NaOH , H_2O, heat)→

O O / H H —(1) LAH; 2) H_2O)→ HO OH

22.54.

a)

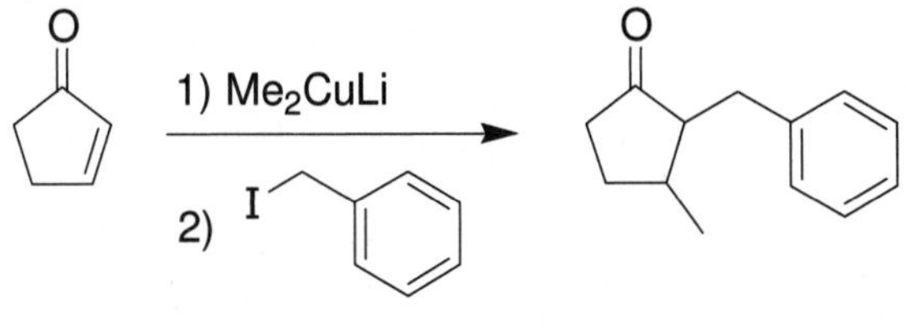

O —(1) Me_2CuLi; 2) I)→ O

b)

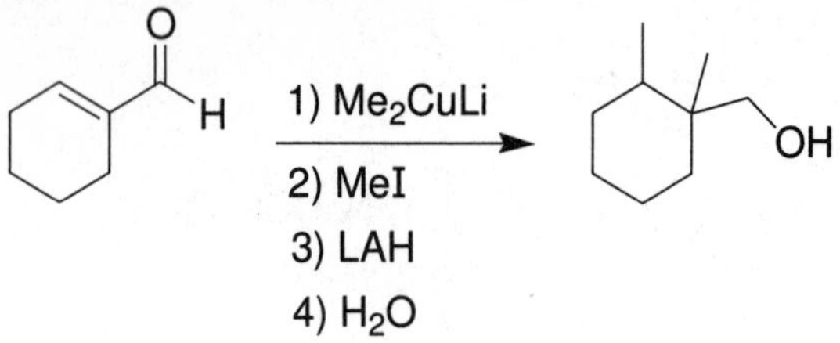

1) Me_2CuLi
2) MeI
3) LAH
4) H_2O

c)

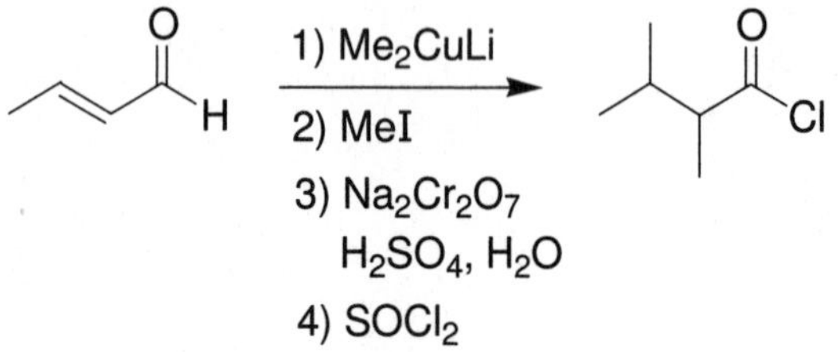

1) Me_2CuLi
2) MeI
3) $Na_2Cr_2O_7$
H_2SO_4, H_2O
4) $SOCl_2$

d)

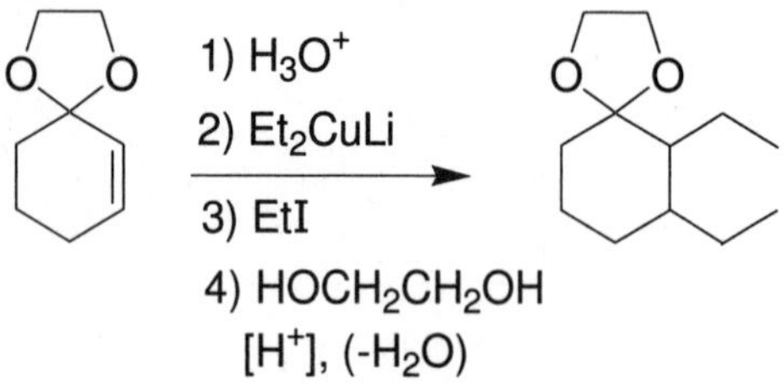

1) H_3O^+
2) Et_2CuLi
3) EtI
4) $HOCH_2CH_2OH$
$[H^+]$, $(-H_2O)$

e)

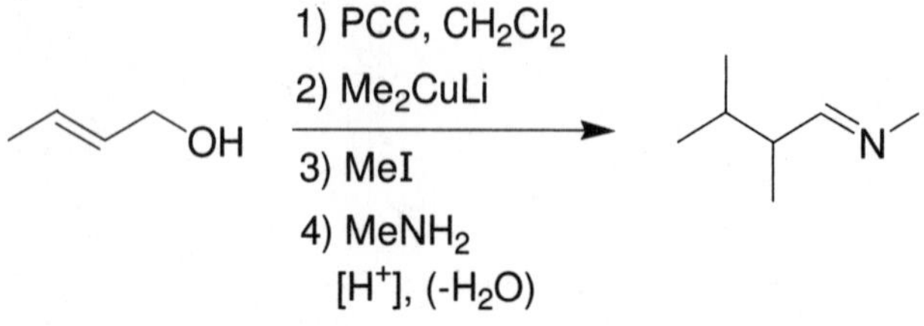

1) PCC, CH_2Cl_2
2) Me_2CuLi
3) MeI
4) $MeNH_2$
$[H^+]$, $(-H_2O)$

f)

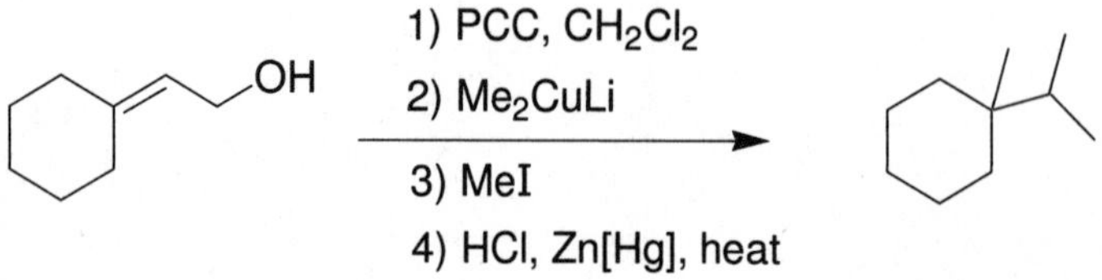

1) PCC, CH_2Cl_2
2) Me_2CuLi
3) MeI
4) HCl, Zn[Hg], heat

22.55.

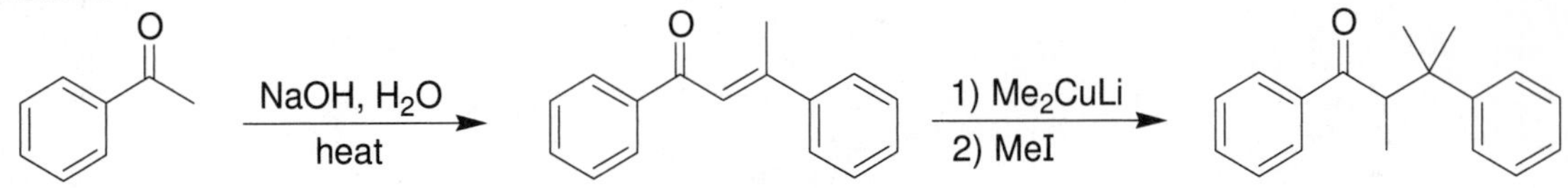

NaOH, H_2O heat

1) Me_2CuLi 2) MeI

22.56.

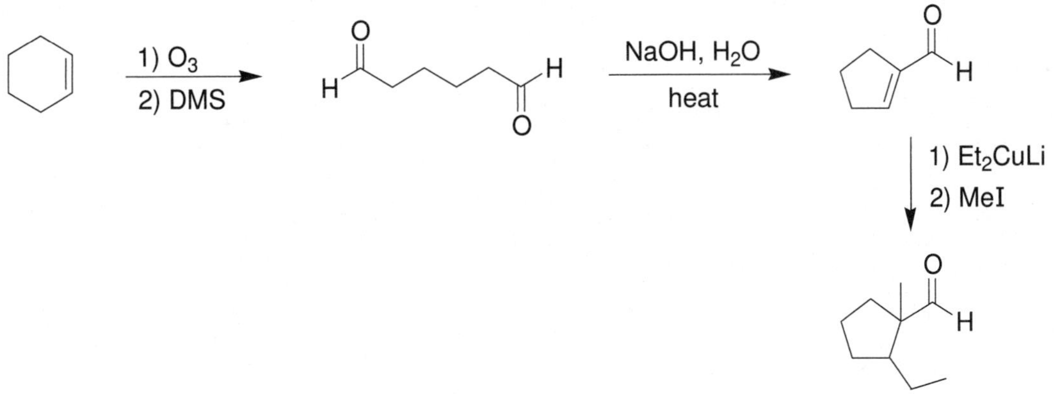

1) O_3 2) DMS

H H O O

NaOH, H_2O heat

1) Et_2CuLi 2) MeI

22.57.

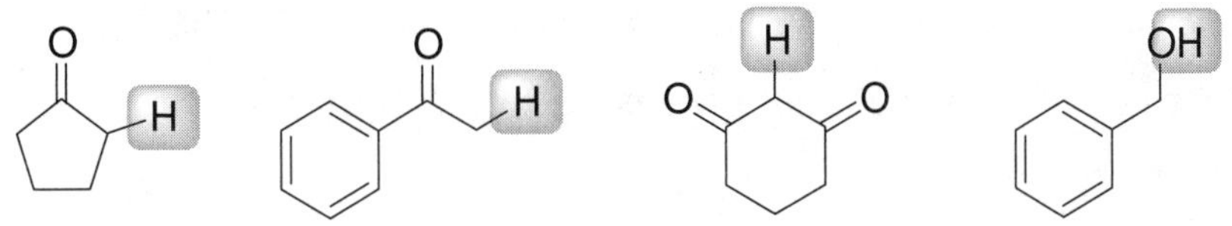

22.58. The conjugate base of this compound is a doubly stabilized enolate.

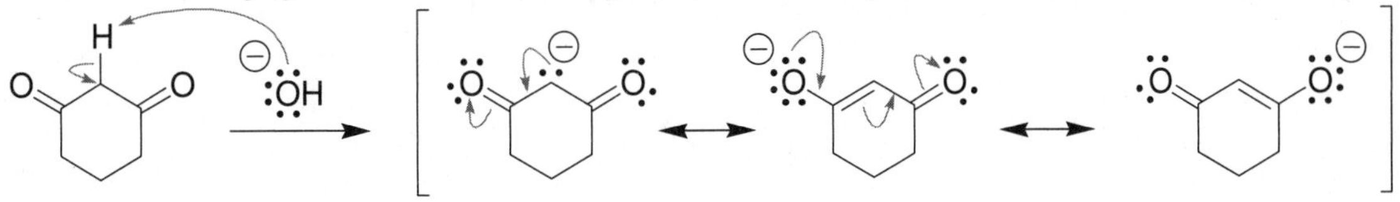

22.59.

a)

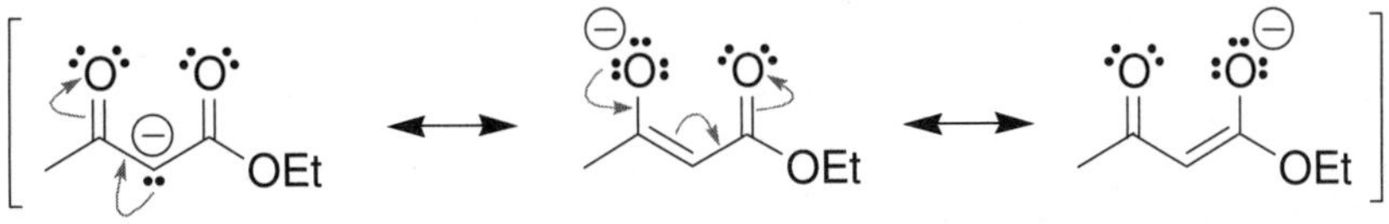

OEt OEt OEt

b)

OEt OEt OEt

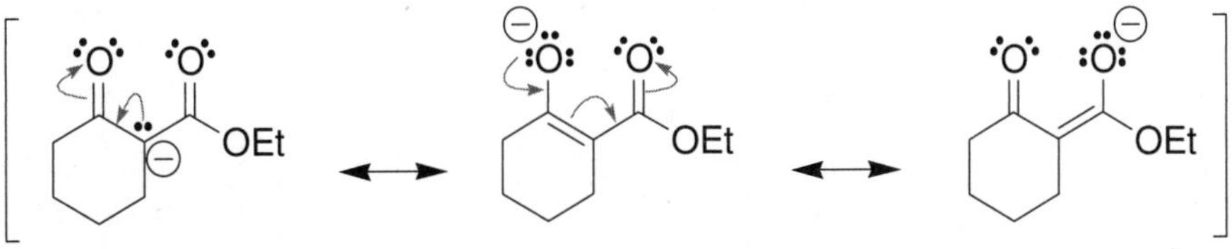

c)

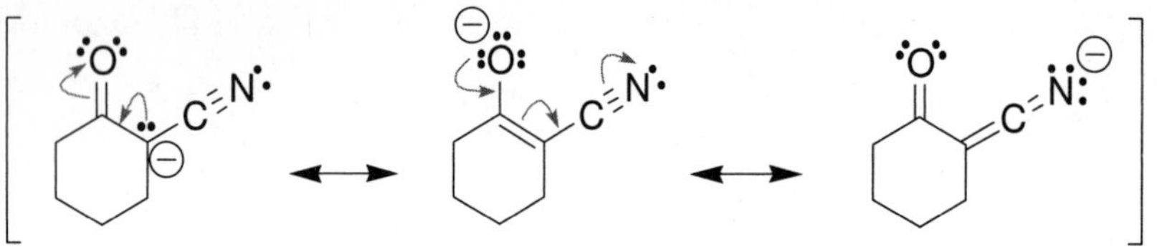

22.60.

Increasing acidity

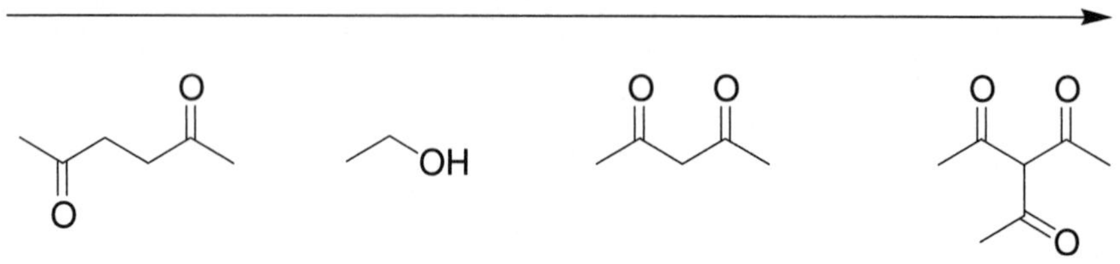

OH

22.61.

a) This enol does not exhibit a significant presence at equilibrium:

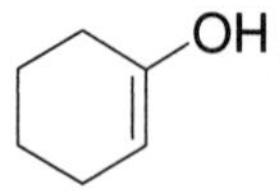

b) This enol does exhibit a significant presence at equilibrium:

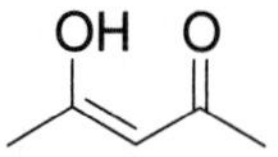

c) This enol does not exhibit a significant presence at equilibrium:

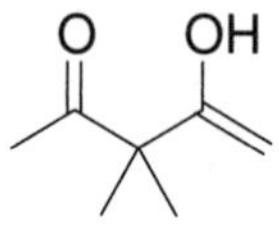

d) This enol does exhibit a significant presence at equilibrium:

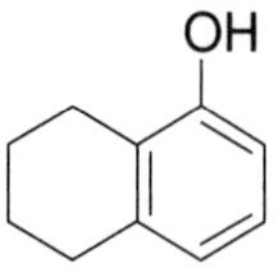

22.62.

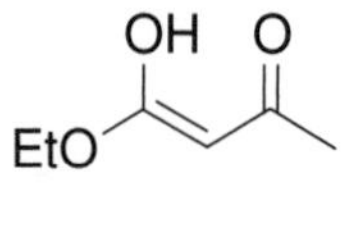

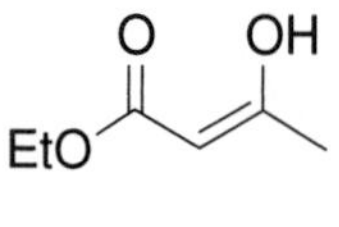

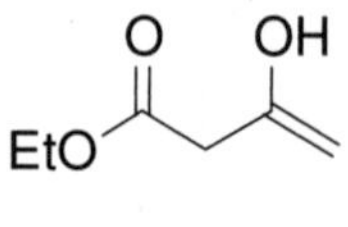

22.63.

a) 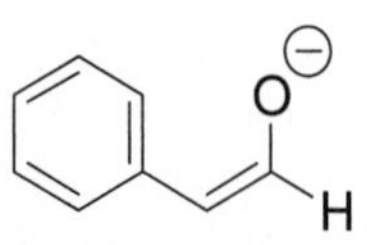 b) 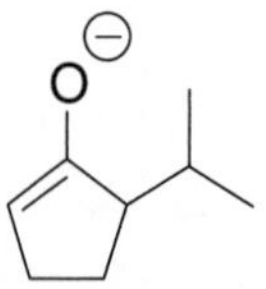 c) 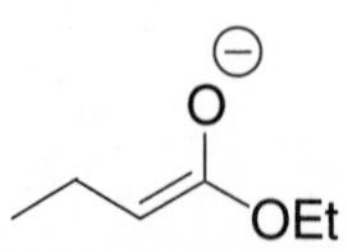 d)

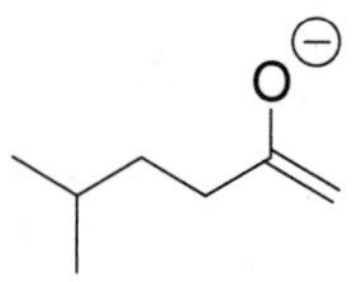

22.64. Deprotonation at the following γ-position results in an anion that has three resonance structures. The negative charge is spread over one oxygen atom and two carbon atoms:

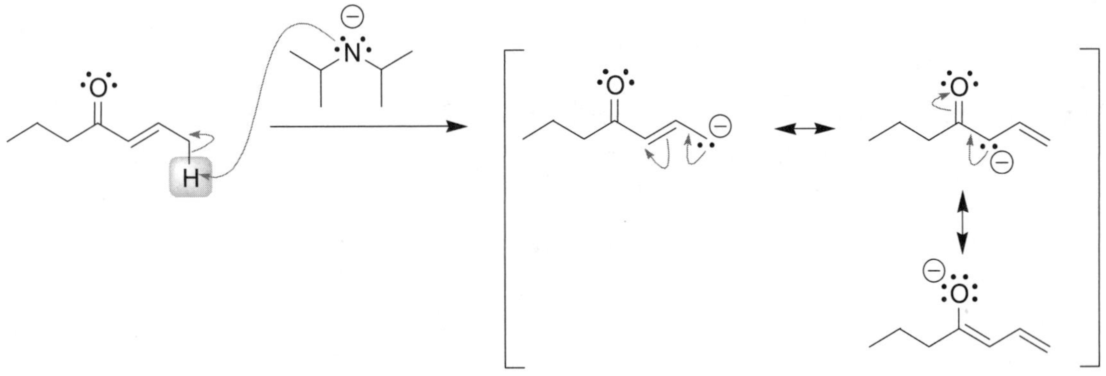

22.65. Deprotonation at the α carbon changes the hybridization state of the α carbon from sp^3 (tetrahedral) to sp^2 (planar). When the α position is protonated once again, the proton can be placed on either side of the planar α carbon, resulting in racemization:

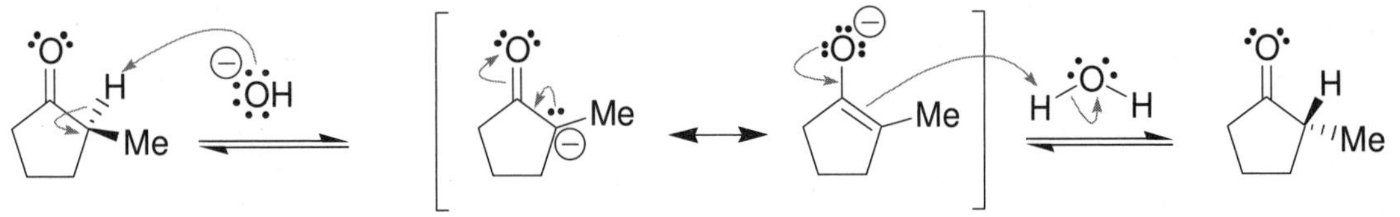

22.66.

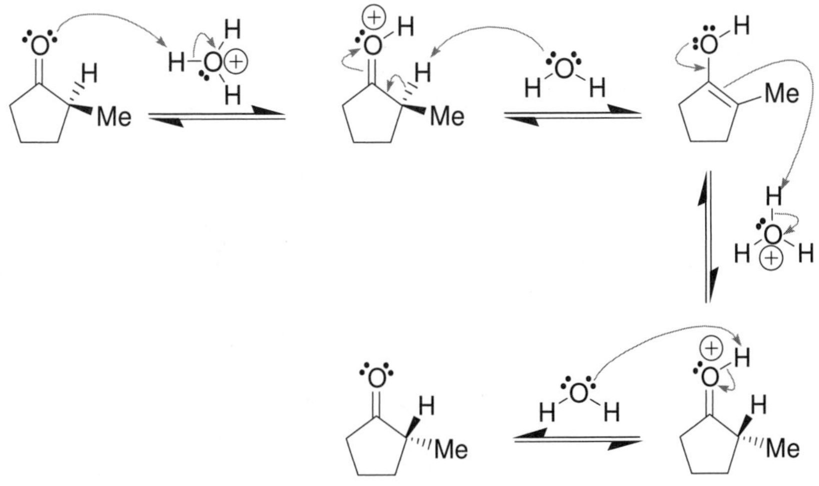

22.67.

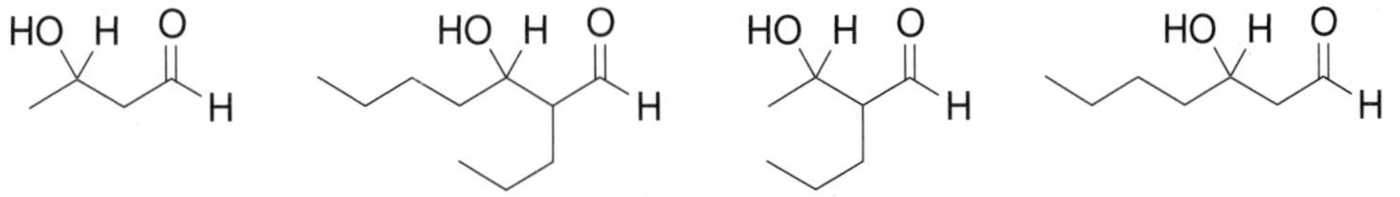

22.68.

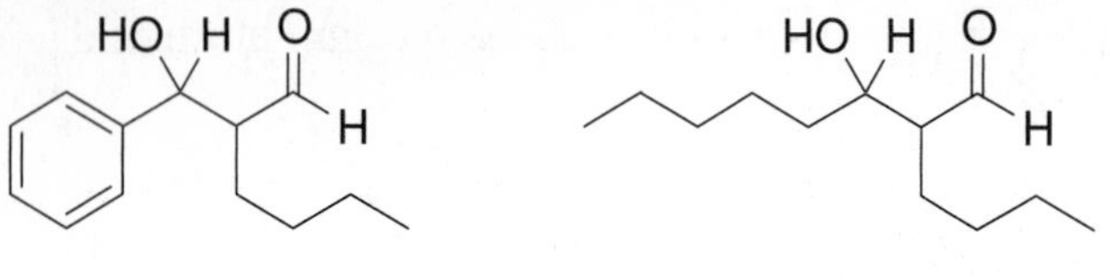

HO H O ... H (structures)

22.69. The carbonyl group and the aromatic ring are conjugated in the product, but are not conjugated in the starting material. Formation of conjugation serves as a driving force in formation of the product.

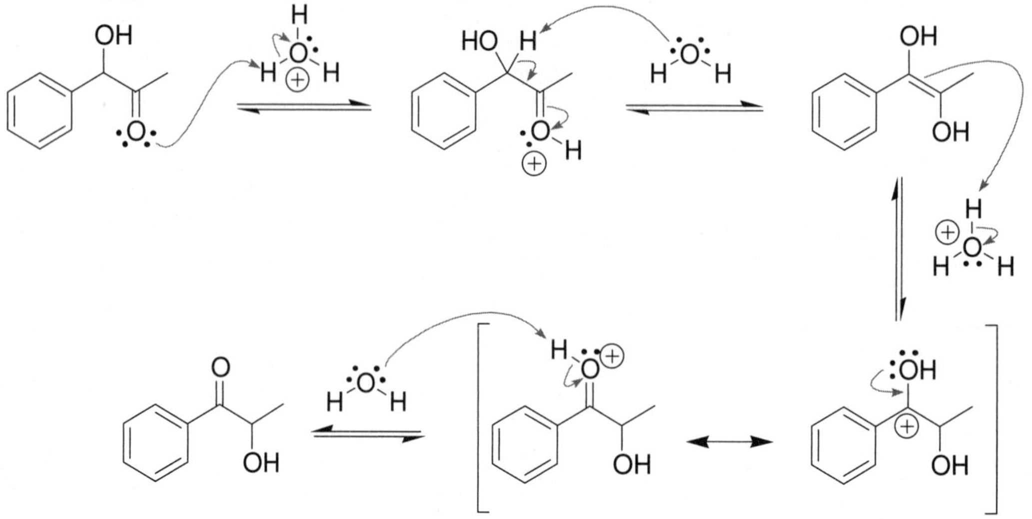

22.70.

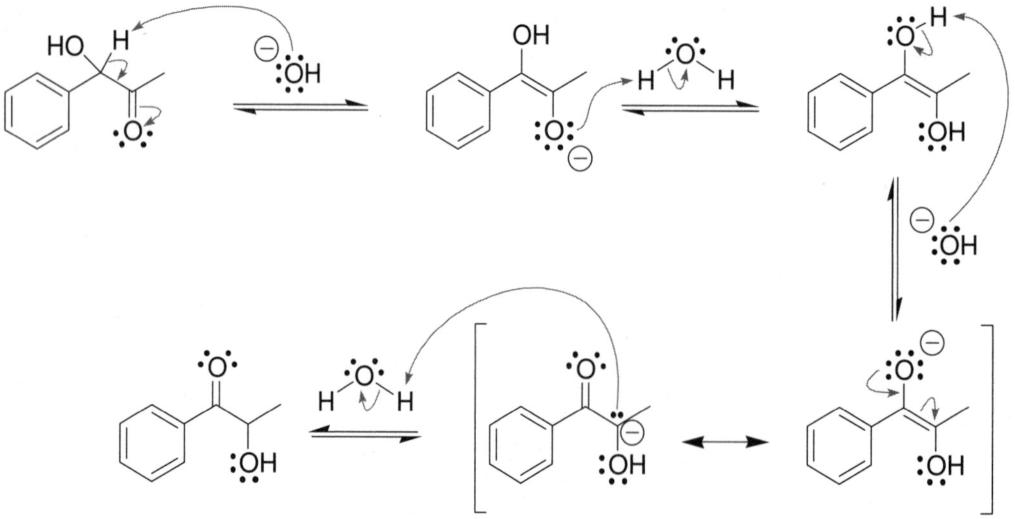

22.71.

a) 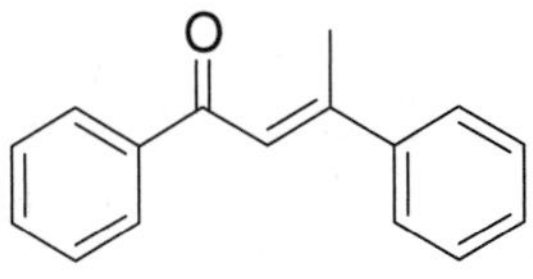

O

b) 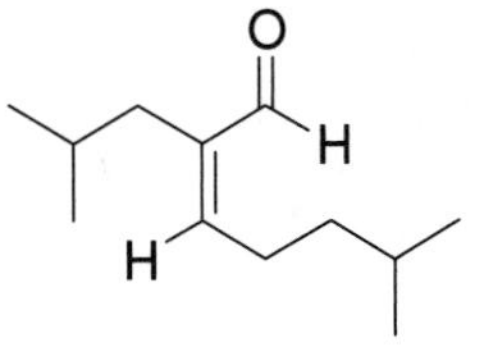

O
H
H

c) 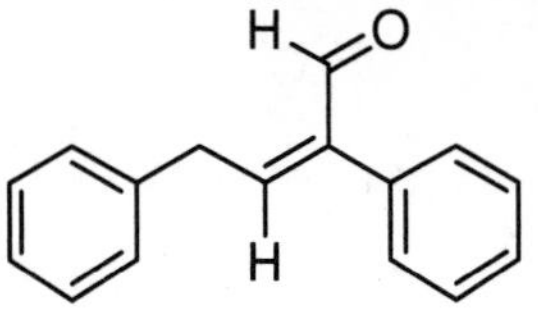

H
O
H

22.72. Trimethylacetaldehyde does not have any α protons.

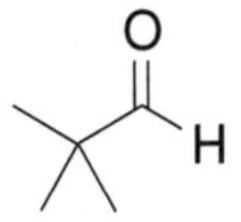

O
H

22.73.

a)

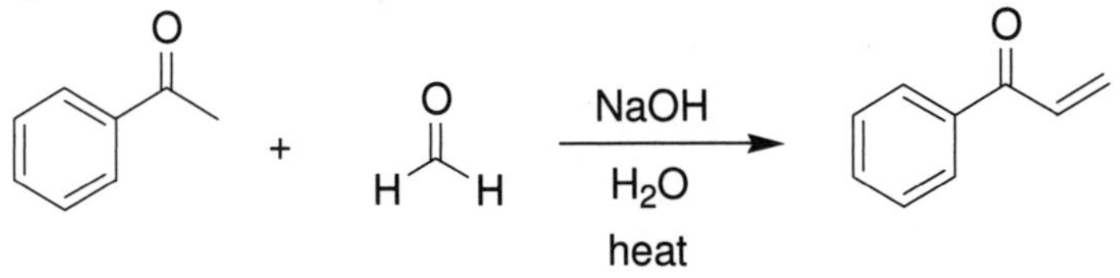

O
O
+
H
H
NaOH
H2O
heat
O

b)

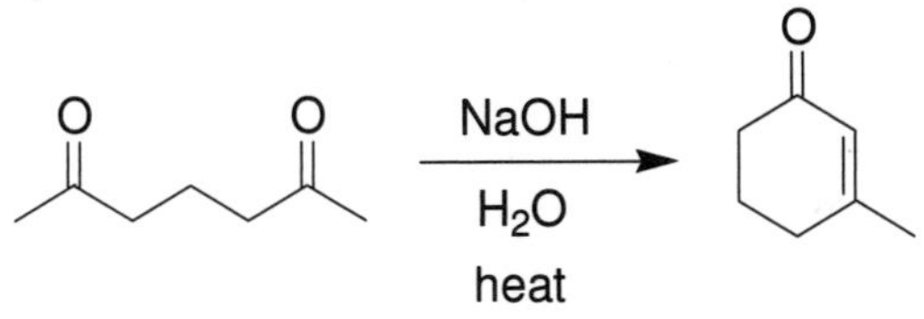

O
O
NaOH
H2O
heat
O

c)

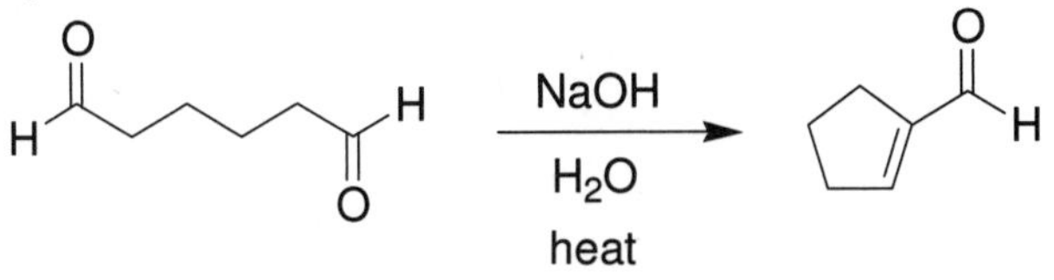

O
H
H
O
NaOH
H2O
heat
O
H

d)

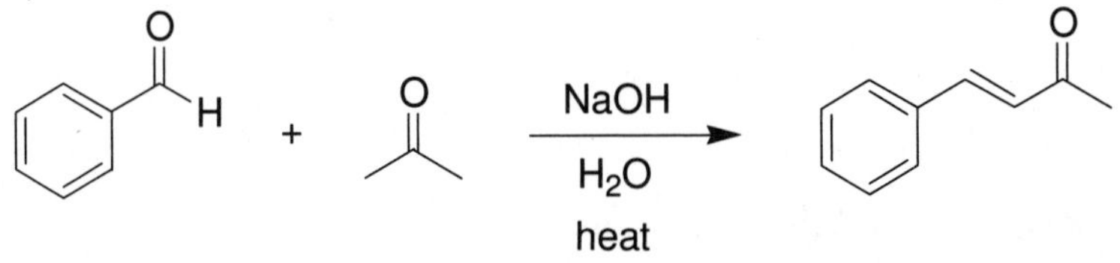

O
H
+
O
NaOH
H2O
heat
O

22.74.

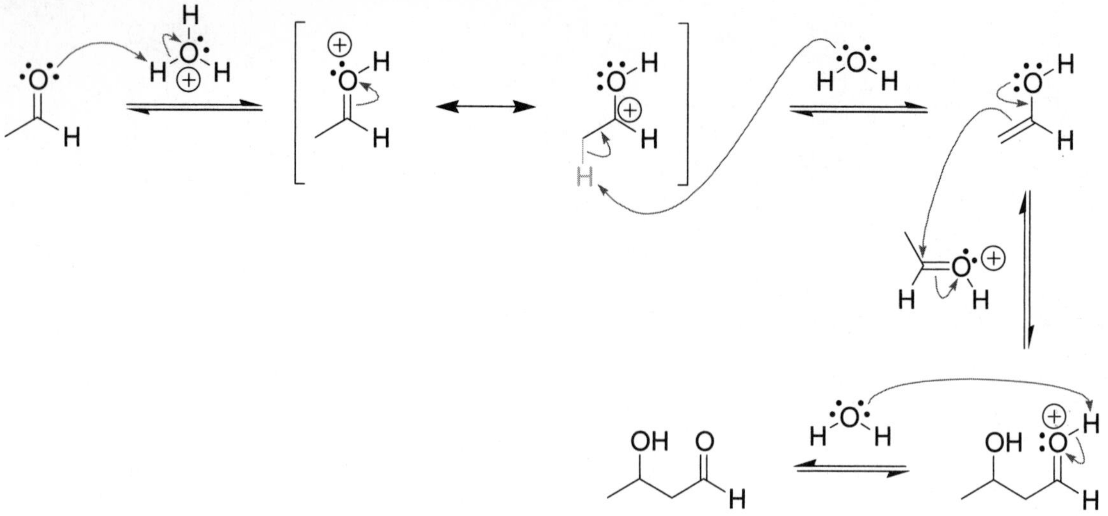

22.75.

a)

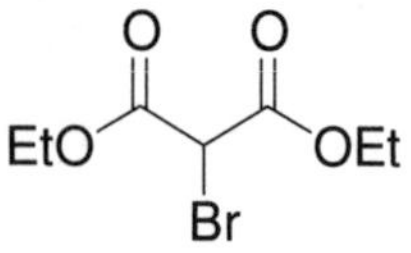

b)

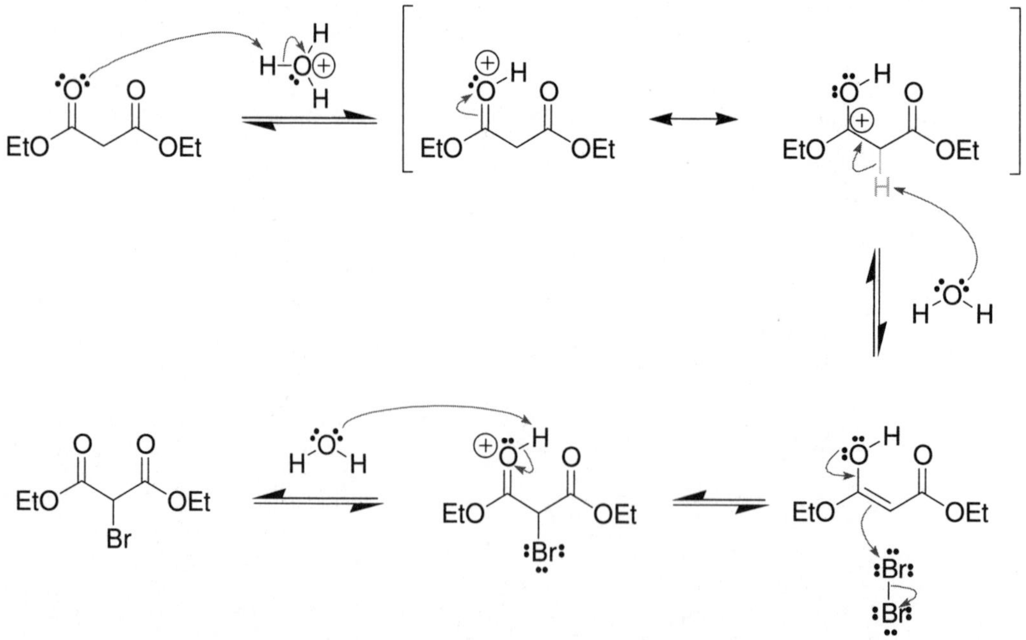

c) The product should be more acidic than diethyl malonate because of the inductive effect of the bromine atom.

22.76.

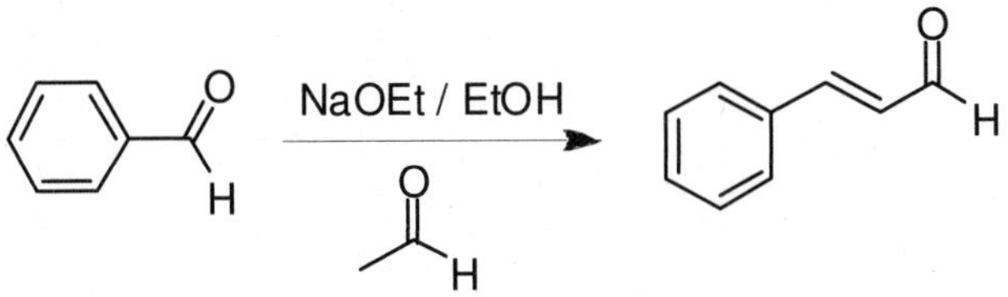

NaOEt / EtOH

H

O

H

22.77.

a) 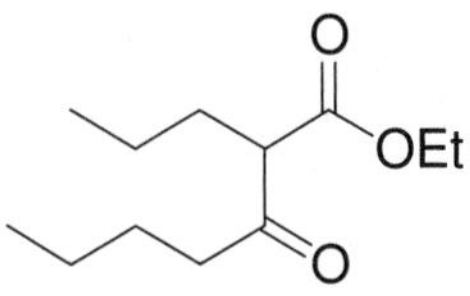 OEt O O

b) 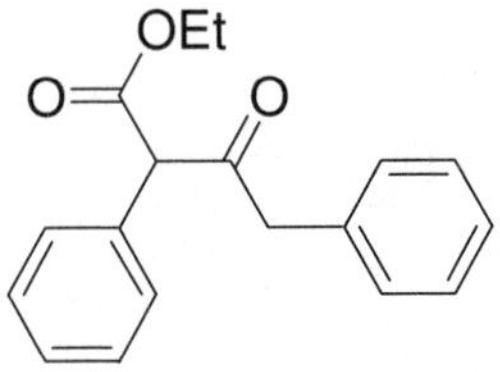 OEt O O

c) 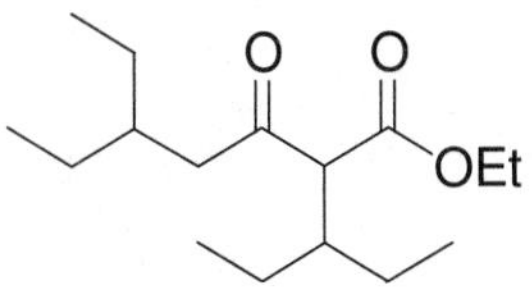 O O OEt

22.78.

a)

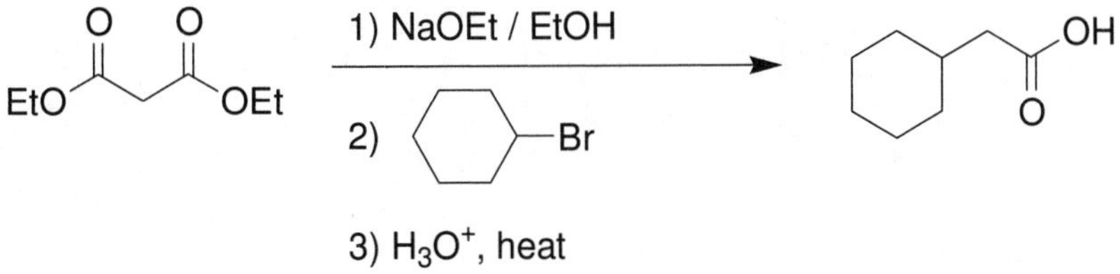

EtO O O OEt

1) NaOEt / EtOH

2) Br

3) H_3O^+, heat

OH O

b)

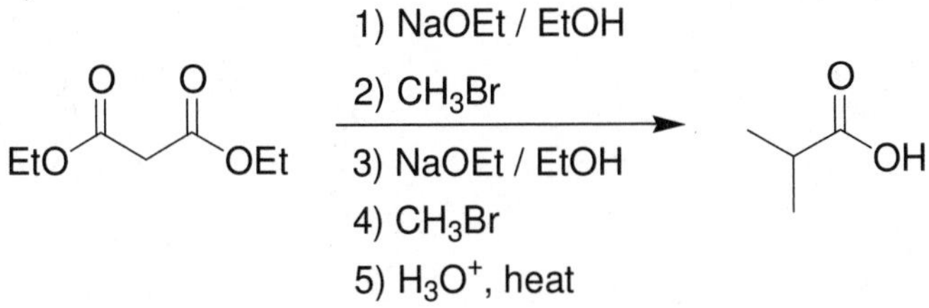

EtO O O OEt

1) NaOEt / EtOH

2) CH_3Br

3) NaOEt / EtOH

4) CH_3Br

5) H_3O^+, heat

O OH

c)

EtO O O OEt

1) NaOEt / EtOH

2) Ph Br

3) NaOEt / EtOH

4) Ph Br

5) H_3O^+, heat

COOH Ph Ph

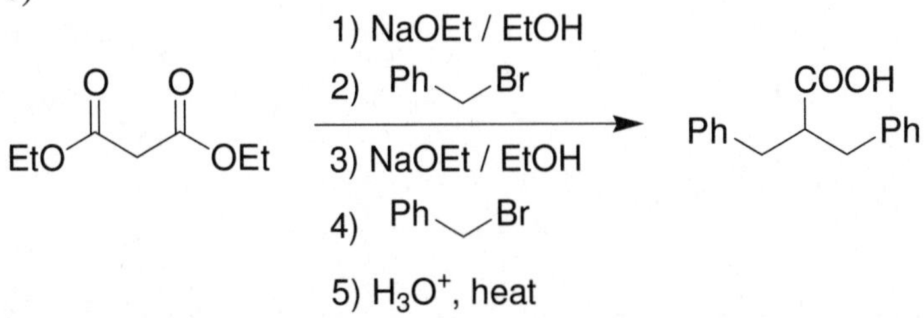

22.79.

a)

1) NaOEt / EtOH

2) [cyclopentyl bromide] Br

3) H_3O^+, heat

OEt

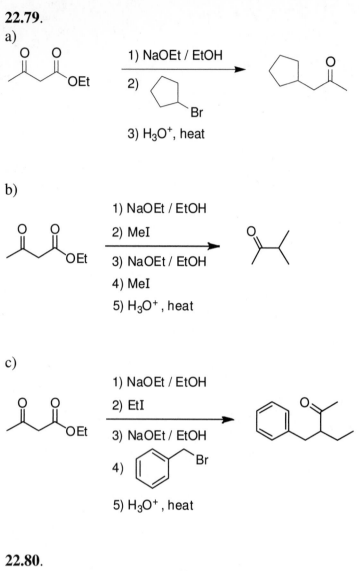

b)

1) NaOEt / EtOH

2) MeI

3) NaOEt / EtOH

4) MeI

5) H_3O^+, heat

OEt

c)

1) NaOEt / EtOH

2) EtI

3) NaOEt / EtOH

4) [benzyl bromide] Br

5) H_3O^+, heat

OEt

22.80.

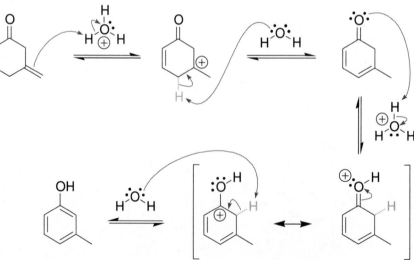

22.81.

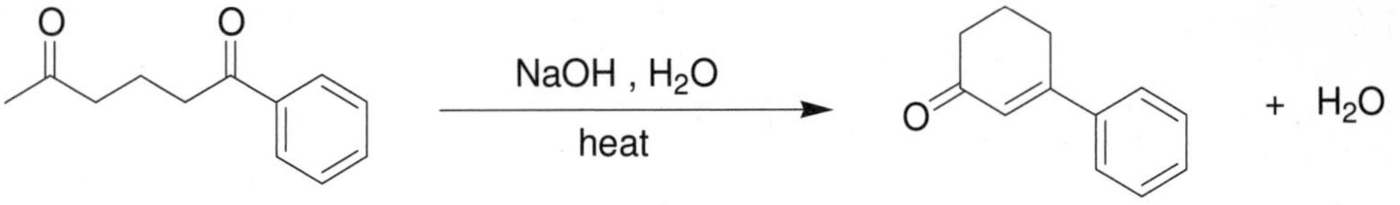

NaOH , H_2O

heat

+ H_2O

22.82.

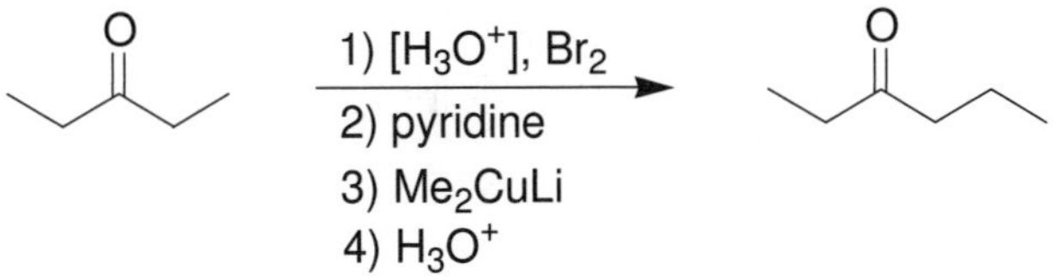

1) $[H_3O^+]$, Br_2
2) pyridine
3) Me_2CuLi
4) H_3O^+

22.83.

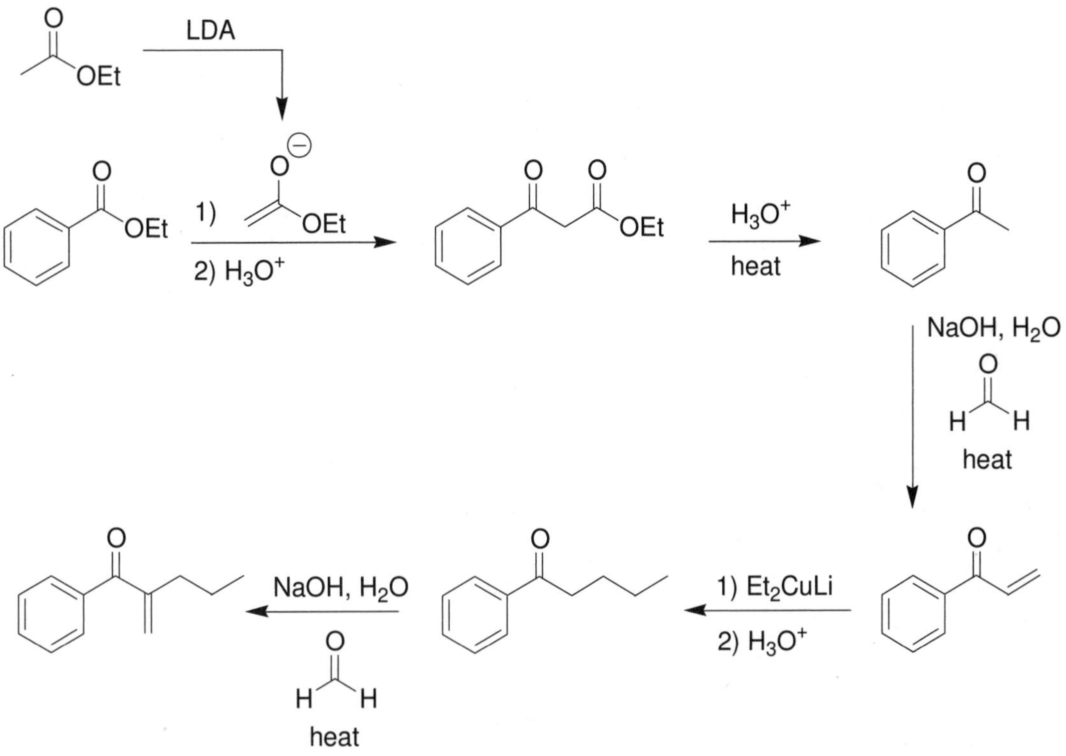

OEt

LDA

OEt

1) OEt

2) H_3O^+

OEt

H_3O^+

heat

NaOH, H_2O

H H

heat

NaOH, H_2O

H H

heat

1) Et_2CuLi

2) H_3O^+

22.84.

a) 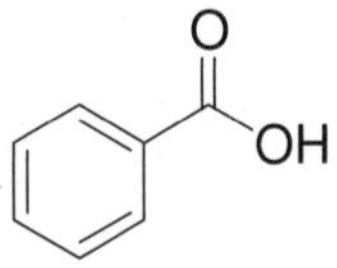 OH

b) 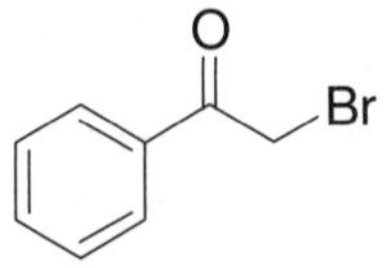 Br

c)

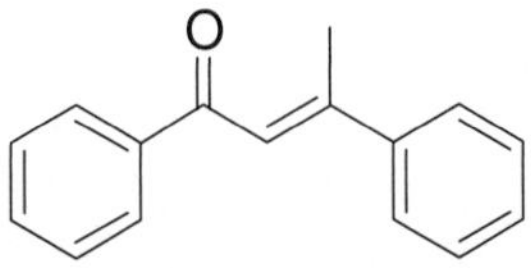

22.85.

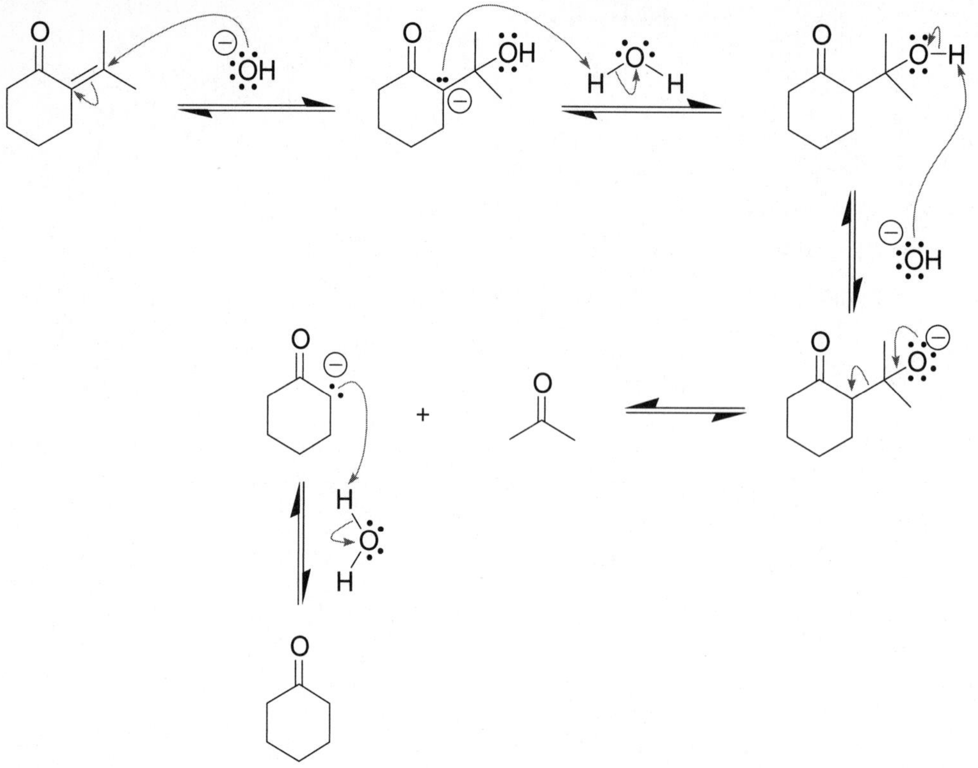

:OH ⊖ H O H :OH ⊖ O H ⊖ O ⊖ + H O H

22.86.

a)

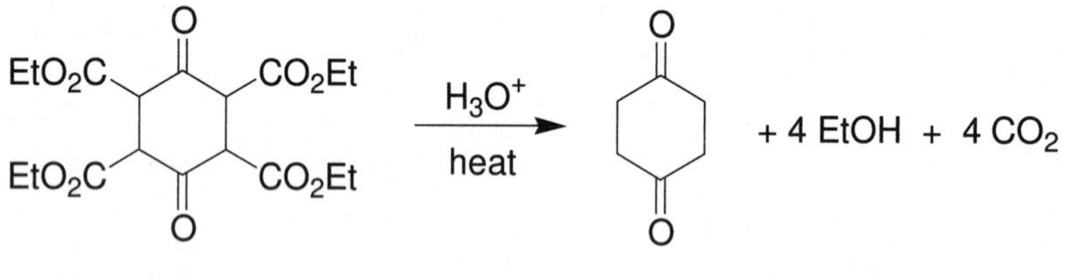

EtO$_2$C CO$_2$Et EtO$_2$C CO$_2$Et O O $\xrightarrow[\text{heat}]{H_3O^+}$ O O + 4 EtOH + 4 CO_2

b)

CH$_2$CO$_2$Et CH$_2$CO$_2$Et $\xrightarrow[\text{2) } H_3O^+]{\text{1) NaOEt}}$ O OEt O $\xrightarrow[\text{heat}]{H_3O^+}$ O

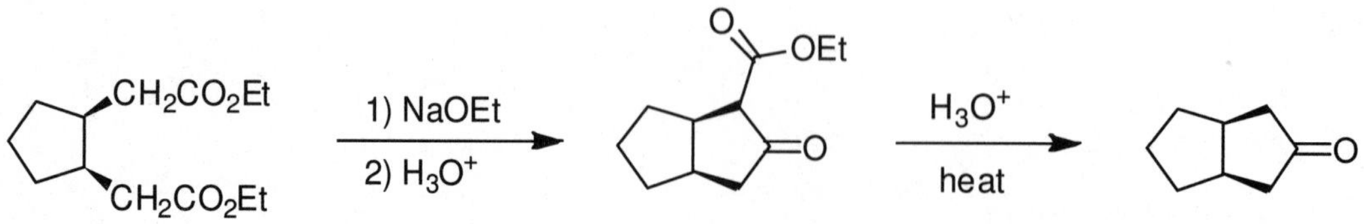

c)

$[H^+]$, Br_2 → Br; pyridine →; 1) Et_2CuLi 2) MeI →

d)

1) LDA 2) EtI →

22.87.

a)

EtO … OEt; 1) NaOEt 2) H_3O^+, heat →; 1) LDA 2) MeI 3) LDA 4) MeI →

b)

NaOH, heat →

c)

OEt; 1) H_3O^+ 2) $SOCl_2$ 3) $(CH_3CH_2CH_2)_2CuLi$ →

d)

1) R_2NH, $[H^+]$, $(-H_2O)$ 2) 3) H_3O^+ →

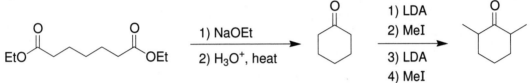

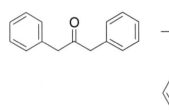

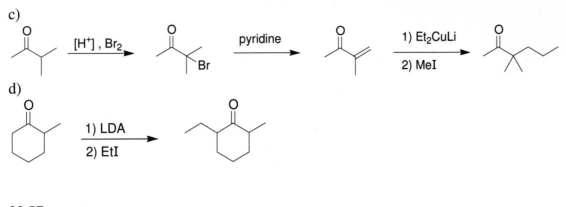

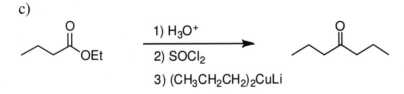

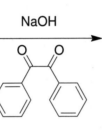

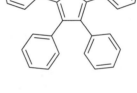

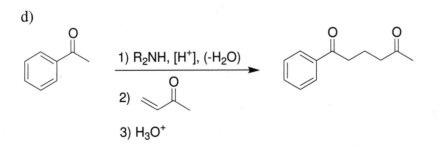

22.88.

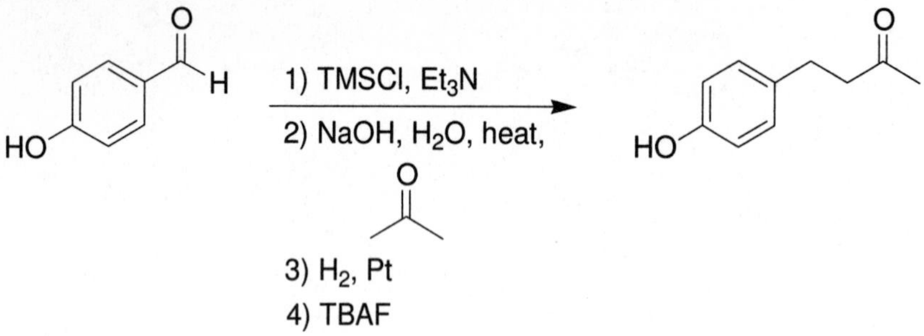

1) TMSCl, Et_3N
2) NaOH, H_2O, heat, (acetone)
3) H_2, Pt
4) TBAF

22.89.

a)

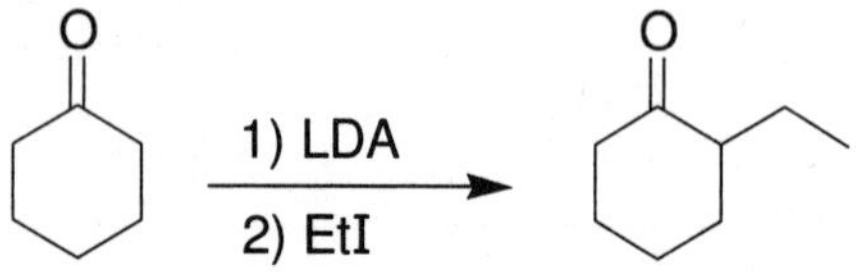

1) LDA
2) EtI

b)

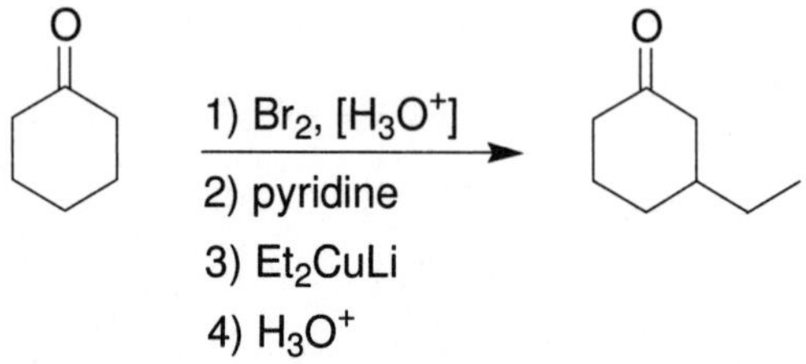

1) Br_2, [H_3O^+]
2) pyridine
3) Et_2CuLi
4) H_3O^+

c)

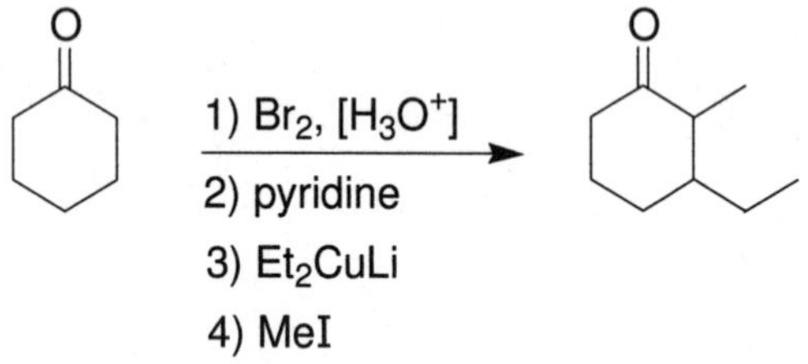

1) Br_2, [H_3O^+]
2) pyridine
3) Et_2CuLi
4) MeI

d)

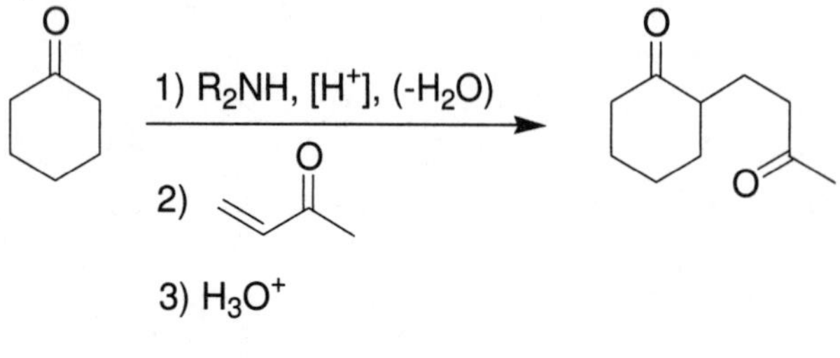

1) R_2NH, [H^+], ($-H_2O$)
2) (methyl vinyl ketone)
3) H_3O^+

e)

NaOH, H_2O, heat

xs (benzaldehyde)

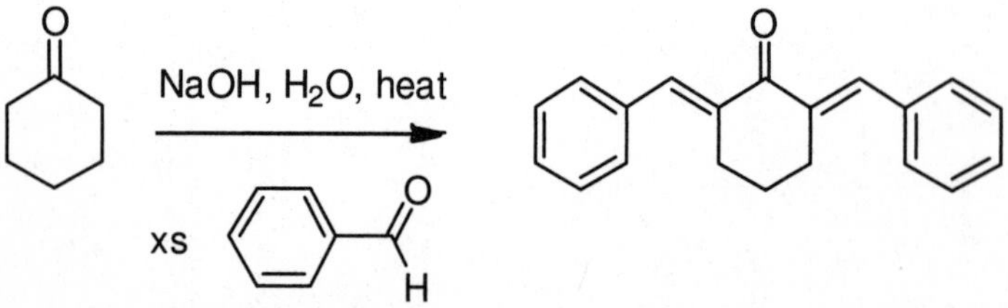

f)

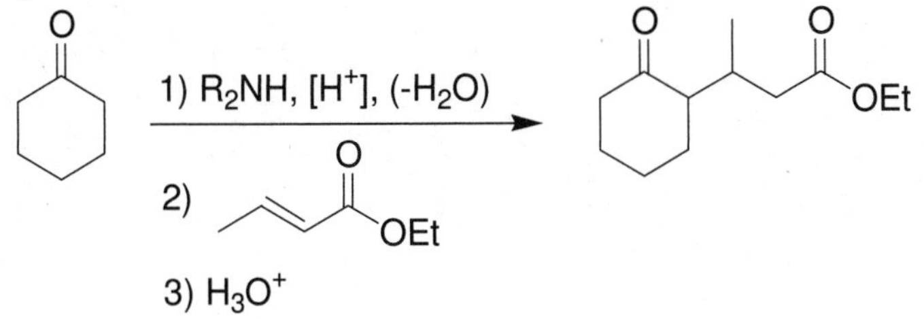

1) R_2NH, $[H^+]$, $(-H_2O)$

2) OEt

3) H_3O^+

OEt

g)

1) LDA

2) MeI

3) LDA

4) MeI

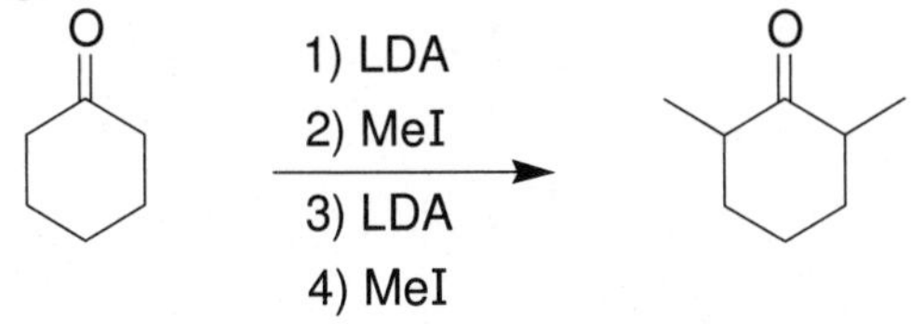

22.90. The driving force for this reaction is formation of a doubly stabilized enolate. After the reaction is complete, an acid is required to protonate this anion.

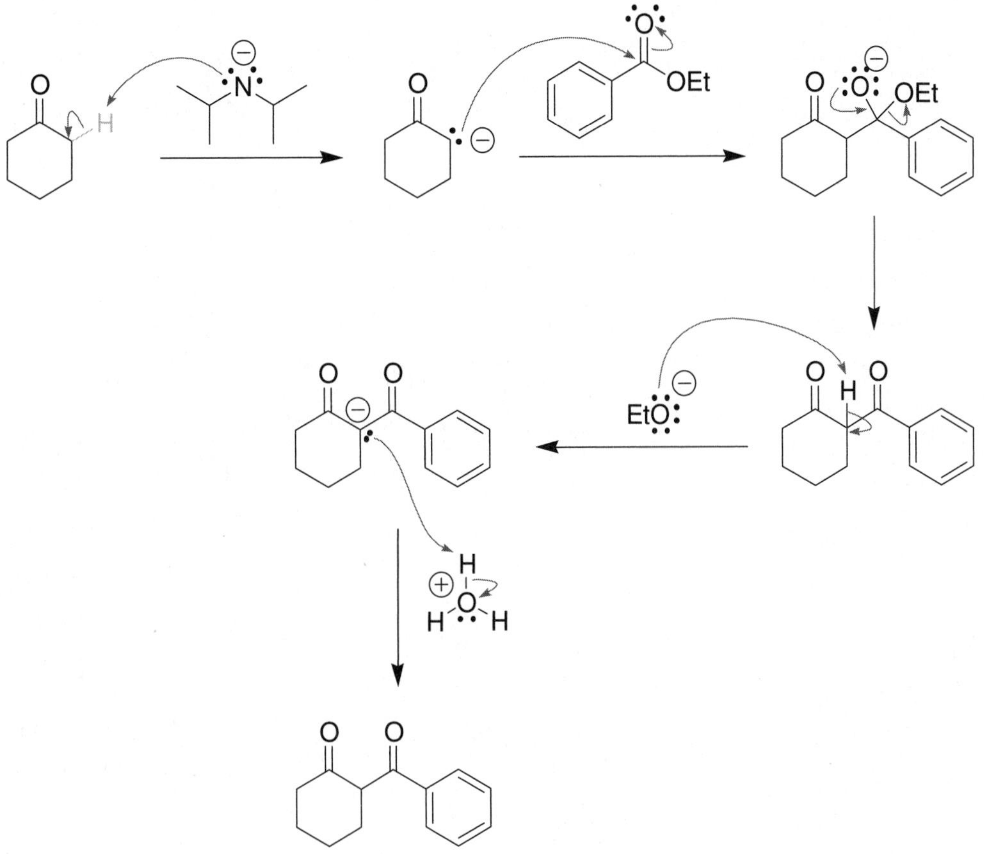

OEt

OEt

EtO

22.91.

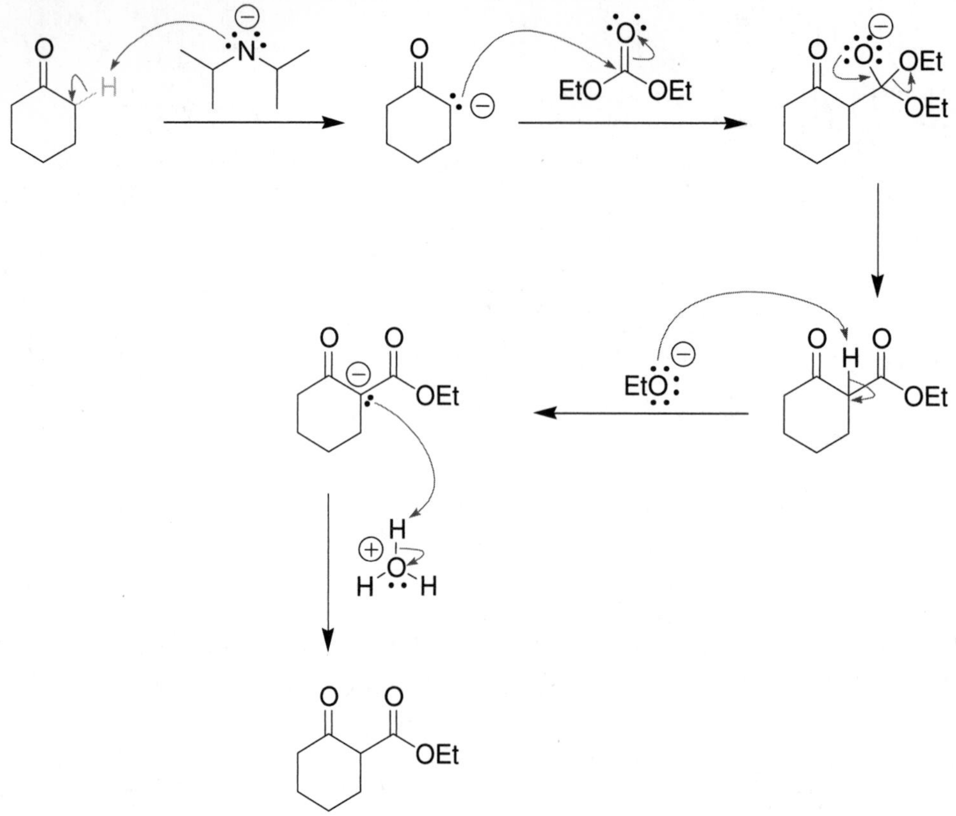

O
H
N
O
EtO
OEt
O
O
OEt
OEt
O
H
O
OEt
EtO
O
O
OEt
H
O
H
H
O
O
OEt

22.92.

O
EtO
H
N
O
EtO
O
O
EtO
O
H
O
H
H
O
OH
EtO

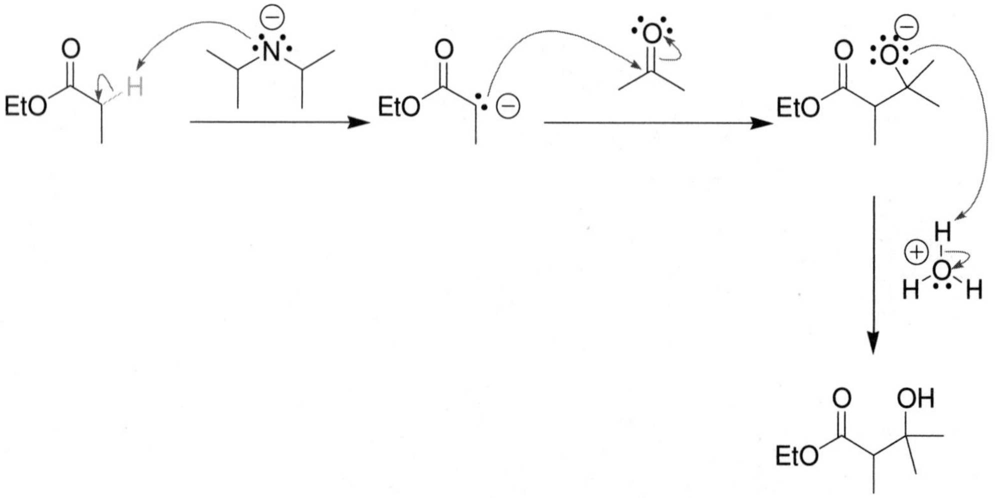

22.93.

a)

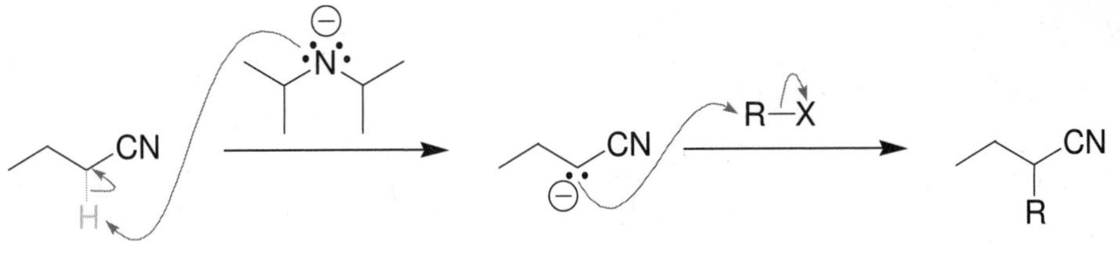

CN

CN

R—X

CN

R

b)

Br

1) NaCN
2) LDA
3) CH_3I
4) LDA
5) CH_3I
6) H_3O^+

OH

O

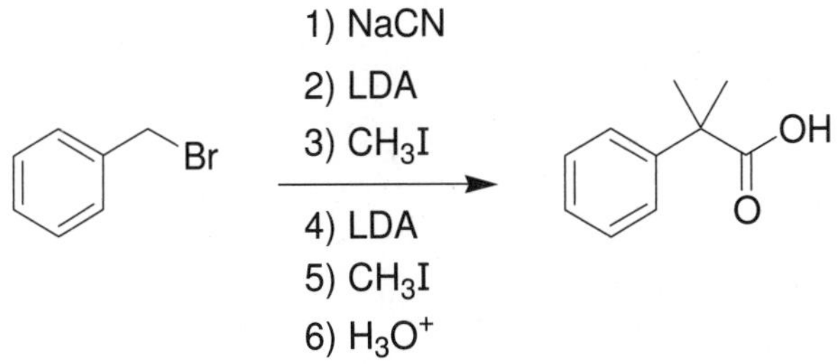

22.94.

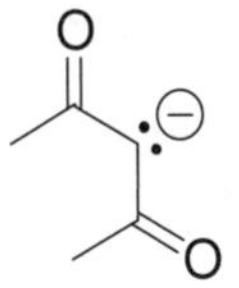

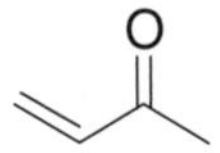

a) Michael Donor

Michael Acceptor

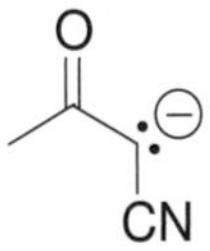

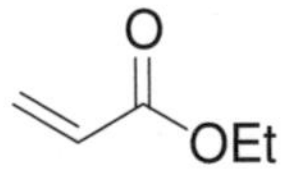

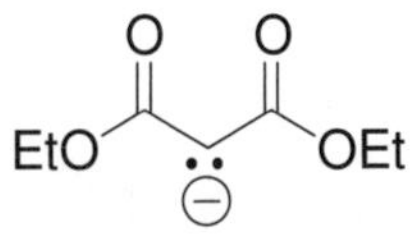

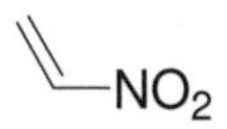

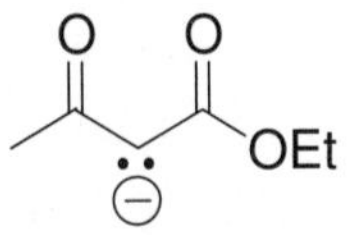

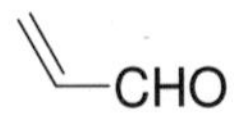

b) Michael Donor

CN

OEt

Michael Acceptor

c) Michael Donor

EtO

OEt

NO_2

Michael Acceptor

d) Michael Donor

OEt

CHO

Michael Acceptor

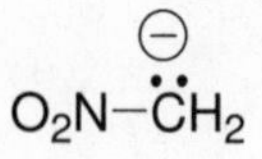

$O_2N-\ddot{C}H_2^{\ominus}$ NO$_2$

e) Michael Donor  Michael Acceptor

22.95.

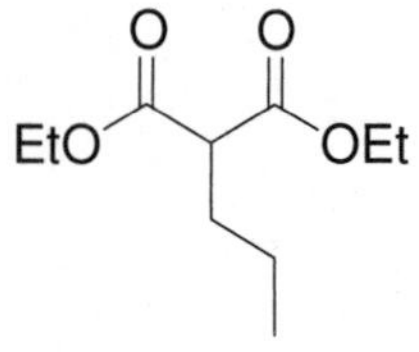

a) EtO OEt O O

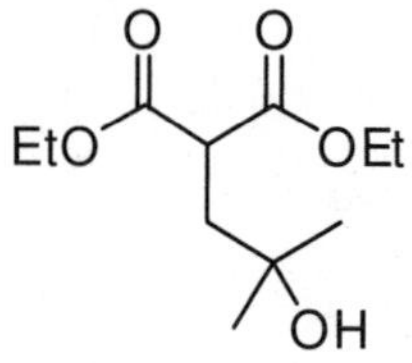

b) EtO OEt O O OH

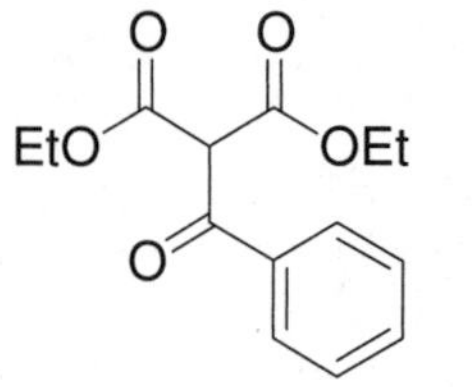

c) EtO OEt O O O

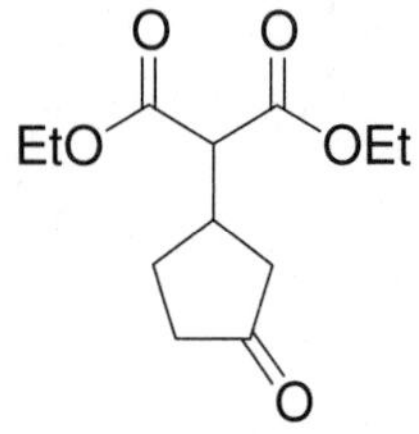

d) EtO OEt O O O

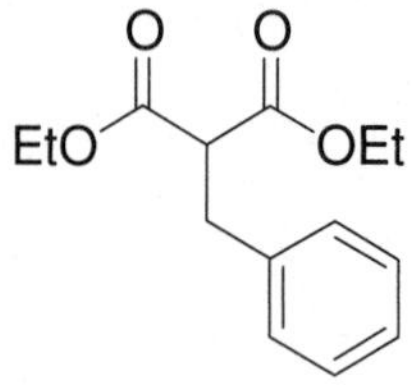

e) EtO OEt O O

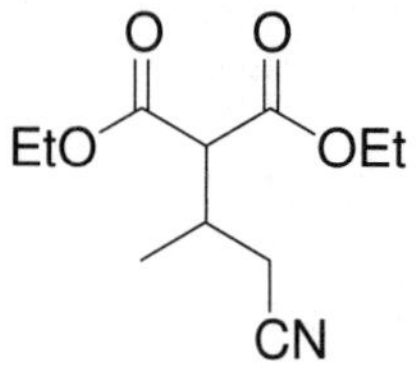

f) EtO OEt O O CN

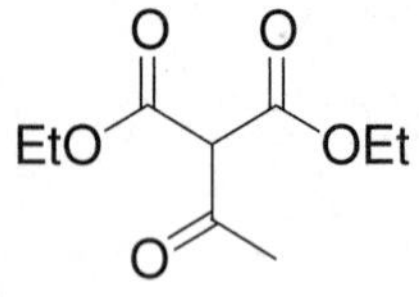

g) EtO OEt O O O

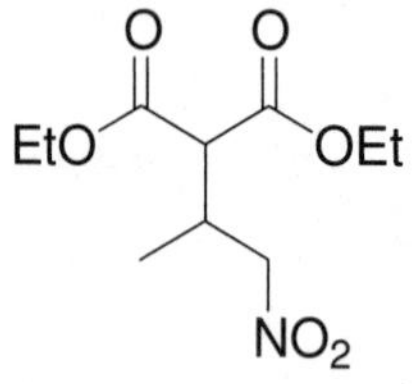

h) EtO OEt O O NO_2

22.96.

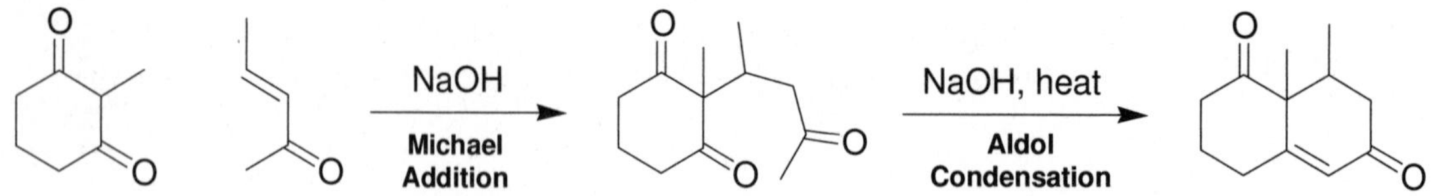

O O O → NaOH (Michael Addition) → O O O → NaOH, heat (Aldol Condensation) → O O

22.97.

O O + O → NaOH, heat → O O

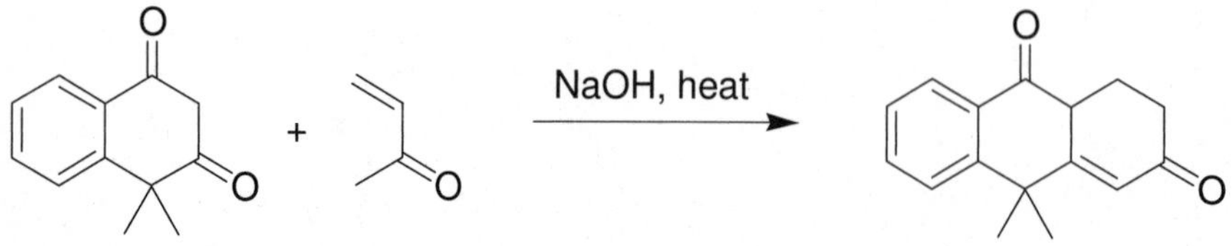

22.98.

NO_2 :ÖH NO_2 :Ö: O_2N :Ö: H :Ö: H

NO_2 O_2N :ÖH :ÖH O_2N :ÖH H

22.99.

1) O_3 2) DMS → NaOH, H_2O heat → 1) Me_2CuLi 2) H_3O^+

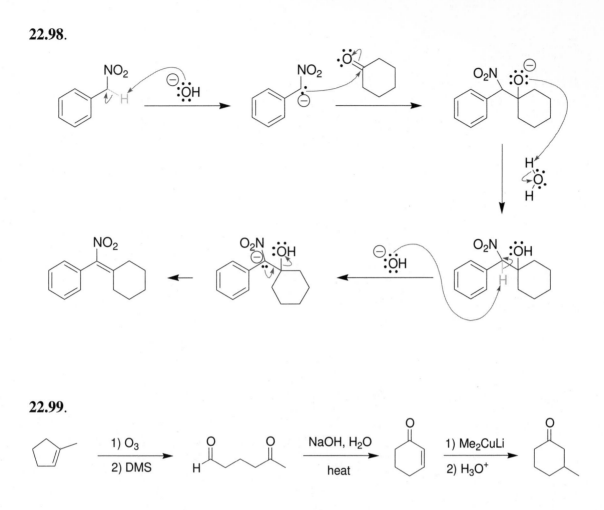

22.100. A ketone generally produces a strong signal at approximately 1720 cm^{-1} (C=O stretching), while an alcohol produces a broad signal between 3200 and 3600 cm^{-1} (O-H stretching). These regions of an IR spectrum can be inspected to determine whether the ketone or the enol predominates.

22.101.

- H_2O

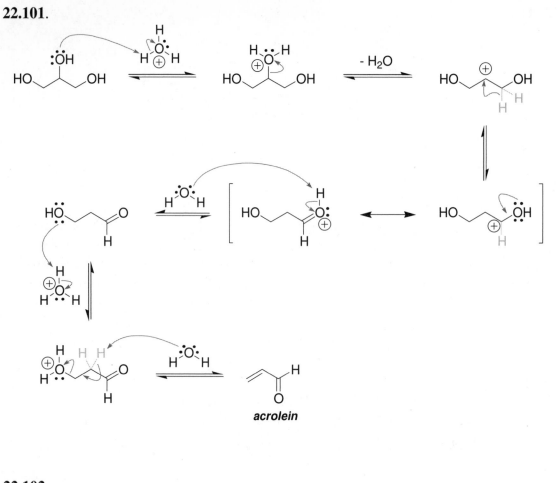

acrolein

22.102.

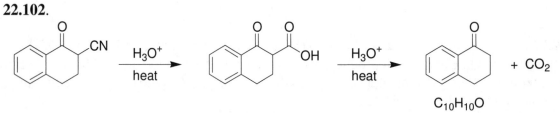

H_3O^+ heat

H_3O^+ heat

+ CO_2

$C_{10}H_{10}O$

22.103.

1) NaOEt

2)

3) H_3O^+ heat

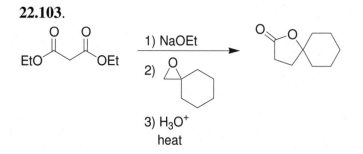

22.104.

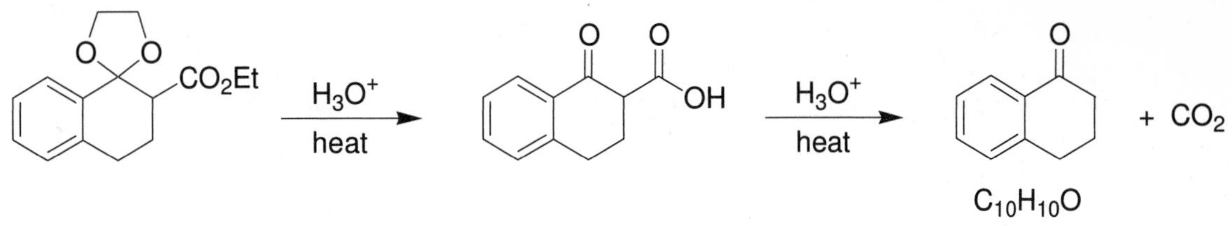

$$\xrightarrow[\text{heat}]{H_3O^+} \quad \xrightarrow[\text{heat}]{H_3O^+} \quad C_{10}H_{10}O + CO_2$$

22.105.

When treated with aqueous acid, both compound A and compound B undergo racemization at the α position (via the enol as an intermediate, see problem 22.65). Each of these compounds establishes an equilibrium between *cis* and *trans* isomers. But the position of equilibrium is very different for compound A than it is for compound B. The equilibrium for compound A favors a *cis* configuration, because that is the configuration for which the compound can adopt a chair conformation in which both groups occupy equatorial positions. The equilibrium for compound B favors a *trans* configuration, because that is the configuration for which that compound can adopt a chair conformation in which both groups occupy equatorial positions.

22.106.

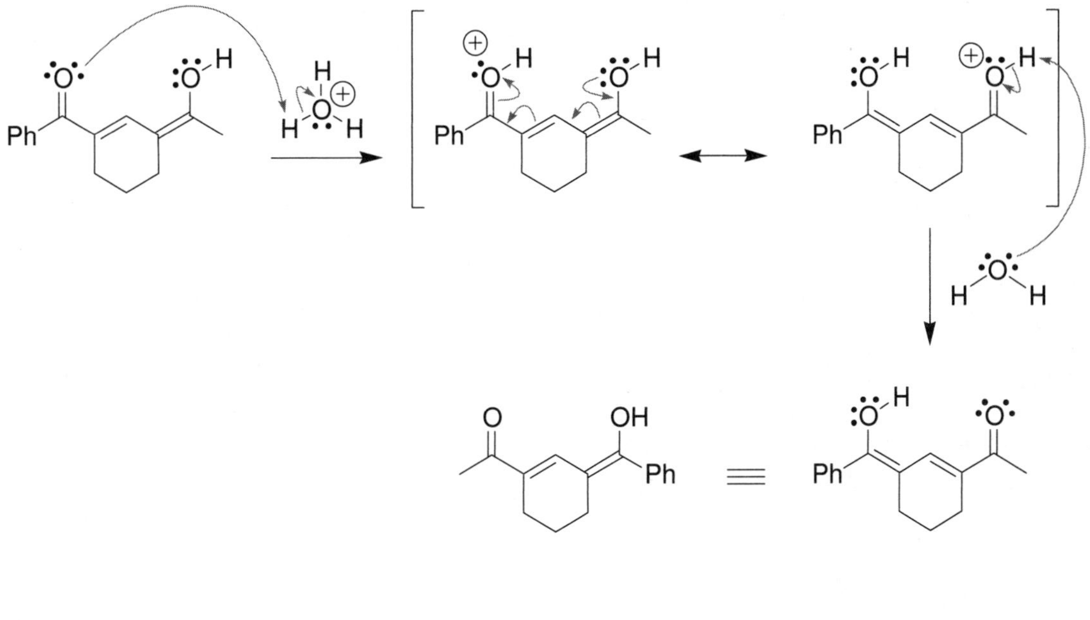

22.107.

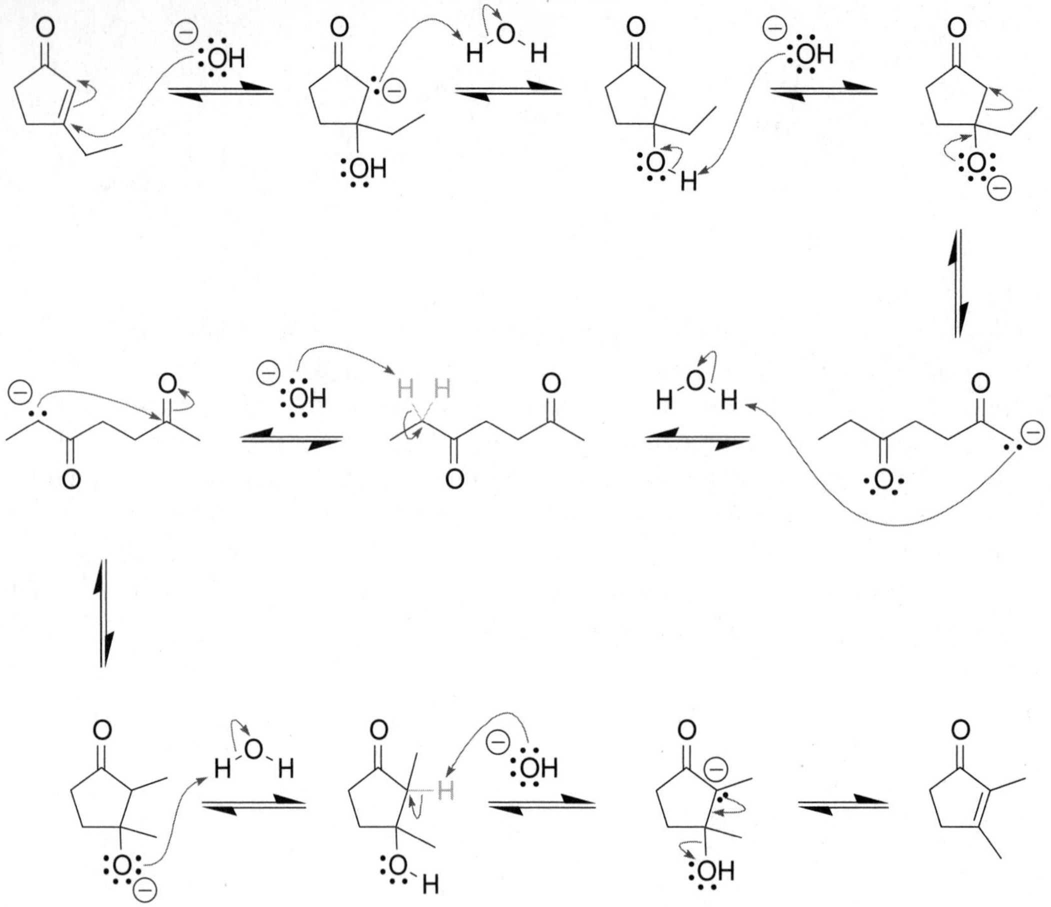

22.108.

a)

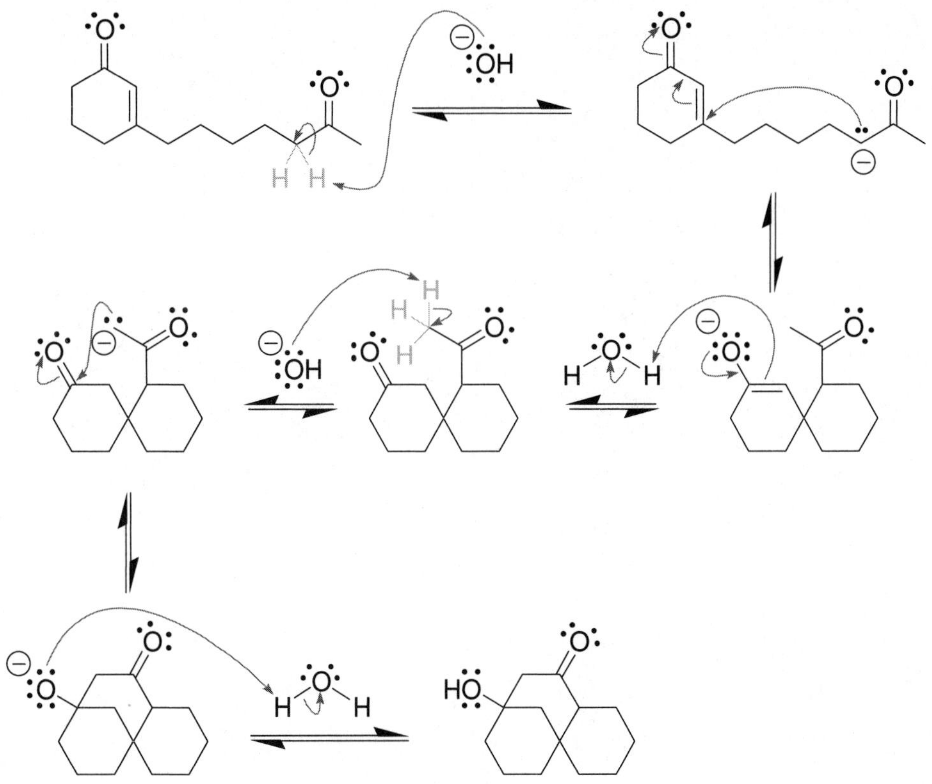

b)

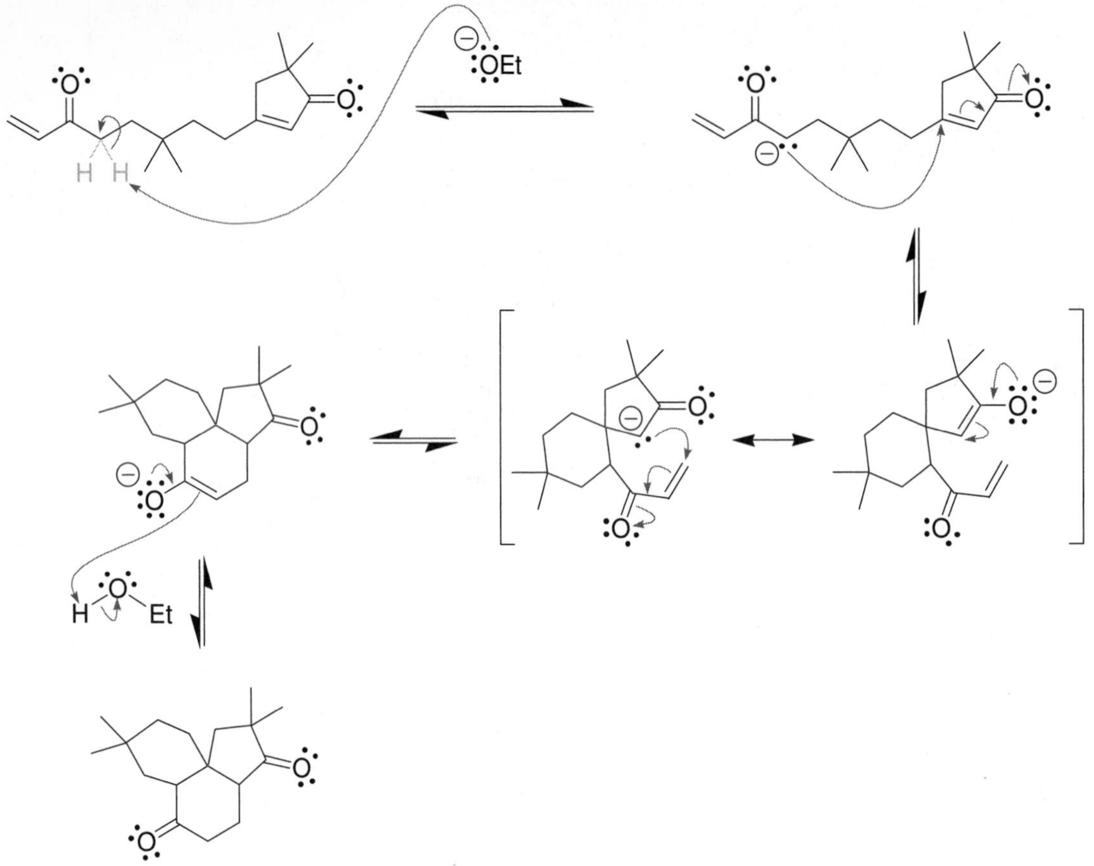

:OEt

H H

H–O–Et

22.109. Direct alkylation would require performing an S_N2 reaction on a tertiary substrate, which will not occur. Instead the enolate would function as a base and E2 elimination would be observed instead of S_N2. The desired transformation can be achieved via a crossed aldol condensation, followed by a Michael addition:

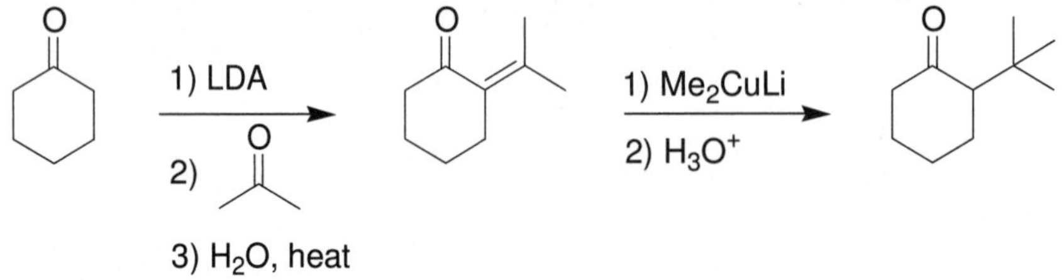

1) LDA

2) (acetone)

3) H_2O, heat

1) Me_2CuLi

2) H_3O^+

Chapter 23
Amines

Review of Concepts

Fill in the blanks below. To verify that your answers are correct, look in your textbook at the end of Chapter 23. Each of the sentences below appears verbatim in the section entitled *Review of Concepts and Vocabulary*.

- Amines are ____________, ____________, or ____________, depending on the number of groups attached to the nitrogen atom.
- The lone pair on the nitrogen atom of an amine can function as a _______ or _________________.
- The basicity of an amine can be quantified by measuring the pK_a of the corresponding ___________________.
- Aryl amines are less basic than alkyl amines, because the lone pair is _________.
- Pyridine is a stronger base than pyrrole, because the lone pair in pyrrole participates in ____________.
- An amine moiety exists primarily as ______________________________ at physiological pH.
- The **azide synthesis** involves treating an ________________ with sodium azide, followed by _______________.
- The __________ **synthesis** generates primary amines upon treatment of potassium phthalimide with an alkyl halide, followed by hydrolysis or reaction with N_2H_4.
- Amines can be prepared via **reductive amination**, in which a ketone or aldehyde is converted into an imine in the presence of a _____________ agent, such as **sodium cyanoborohydride** ($NaBH_3CN$).
- Amines react with acyl halides to produce __________.
- In the **Hofmann elimination**, and amino group is converted into a better leaving group which is expelled in an ____ process to form an ___________.
- Primary amines react with a nitrosonium ion to yield a ______________ **salt** in a process called **diazotization**.
- **Sandmeyer reactions** utilize copper salts (CuX), enabling the installation of a halogen or a ________ group.
- In the **Schiemann reaction**, an aryl diazonium salt is converted into a fluorobenzene by treatment with _________________.
- Aryldiazonium salts react with activated aromatic rings in a process called _____ **coupling**, to produce colored compounds called _____ **dyes**.
- A _______**cycle** is a ring that contains atoms of more than one element.
- Pyrrole undergoes electrophilic aromatic substitution reactions, which occur primarily at C__.

Review of Skills

Fill in the blanks and empty boxes below. To verify that your answers are correct, look in your textbook at the end of Chapter 23. The answers appear in the section entitled *SkillBuilder Review*.

23.1 Naming an Amine

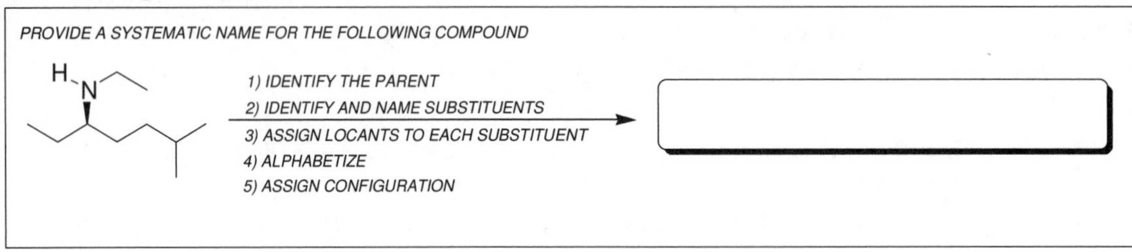
PROVIDE A SYSTEMATIC NAME FOR THE FOLLOWING COMPOUND
H
N
1) IDENTIFY THE PARENT
2) IDENTIFY AND NAME SUBSTITUENTS
3) ASSIGN LOCANTS TO EACH SUBSTITUENT
4) ALPHABETIZE
5) ASSIGN CONFIGURATION

23.2 Preparing a Primary Amine via the Gabriel Reaction

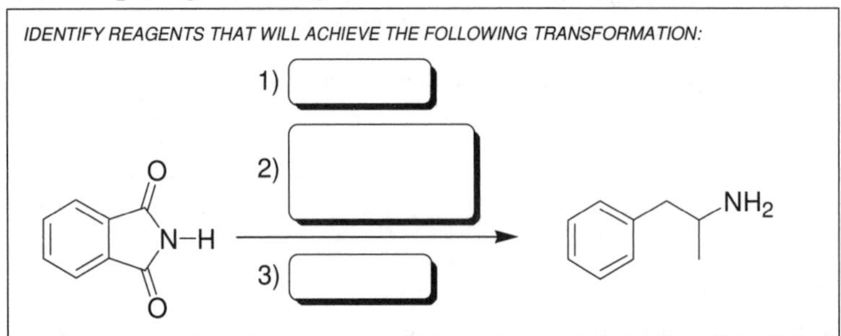
IDENTIFY REAGENTS THAT WILL ACHIEVE THE FOLLOWING TRANSFORMATION:
1)
2)
3)
O
N–H
O
NH2

23.3 Preparing an Amine via a Reductive Amination

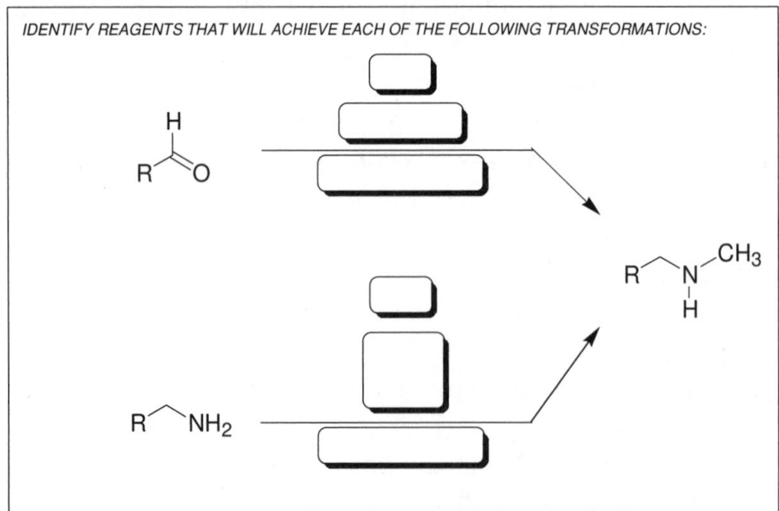
IDENTIFY REAGENTS THAT WILL ACHIEVE EACH OF THE FOLLOWING TRANSFORMATIONS:
H
R
O
CH3
R
N
H
R
NH2

23.4 Synthesis Strategies

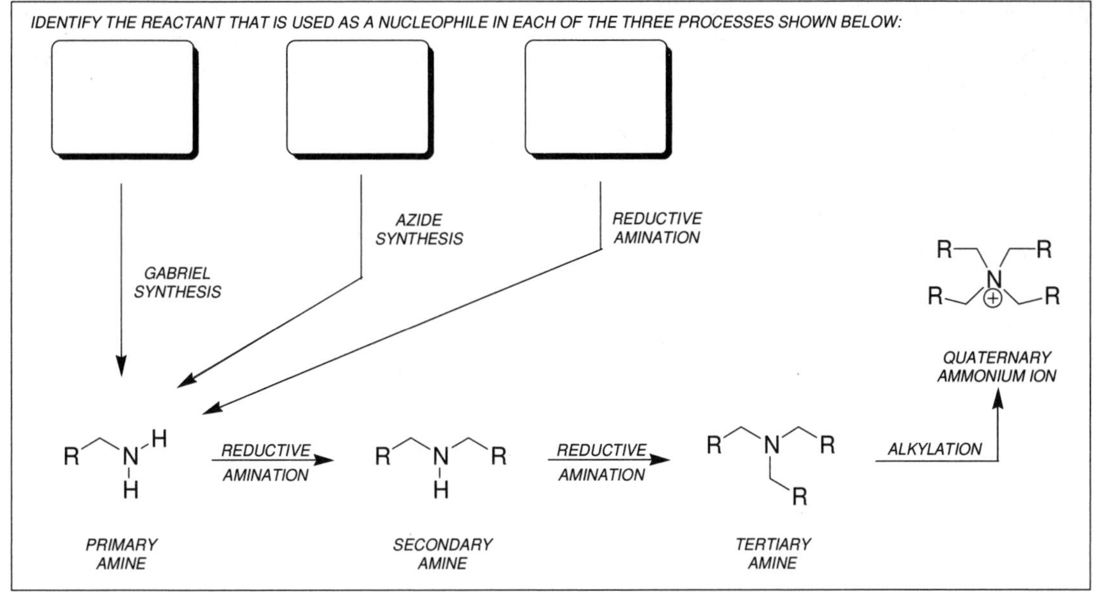

23.5 Predicting the Product of a Hofmann Elimination

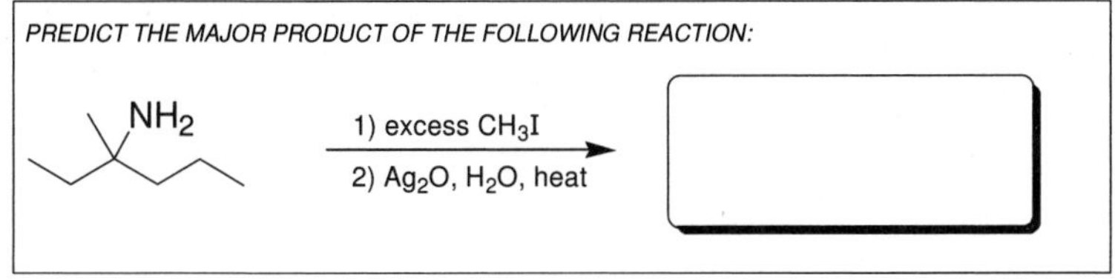

23.6 Determining the Reactants for Preparing an Azo Dye

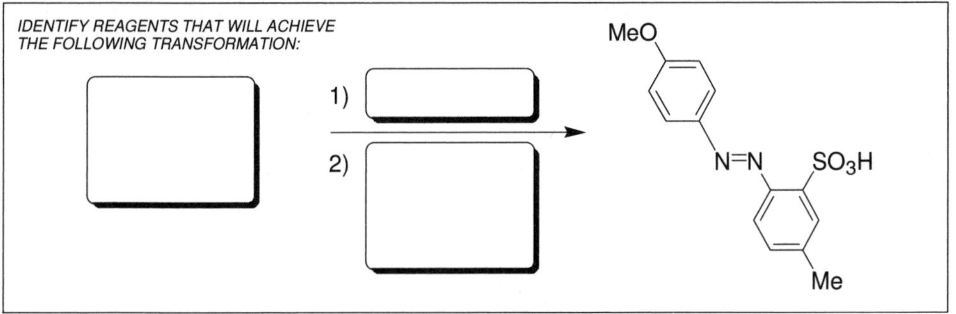

Review of Reactions

Identify the reagents necessary to achieve each of the following transformations. To verify that your answers are correct, look in your textbook at the end of Chapter 23. The answers appear in the section entitled *Review of Reactions*.

Preparation of Amines

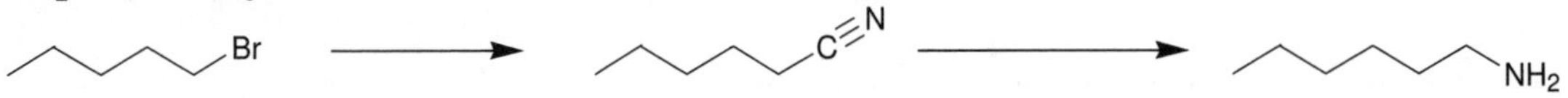

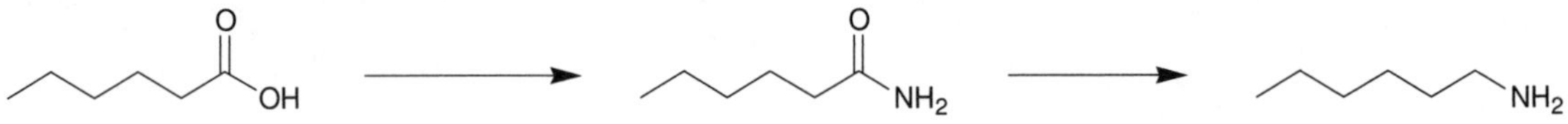

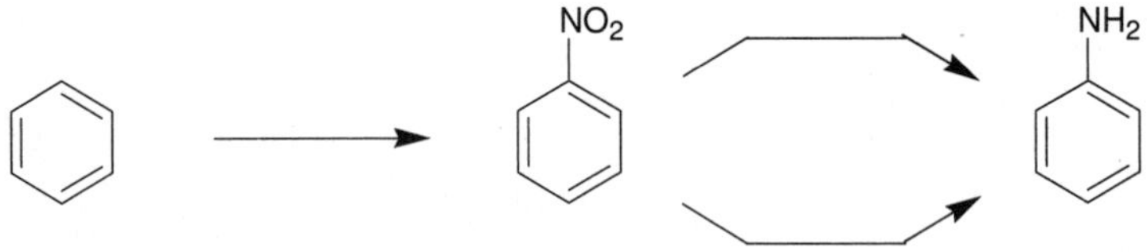

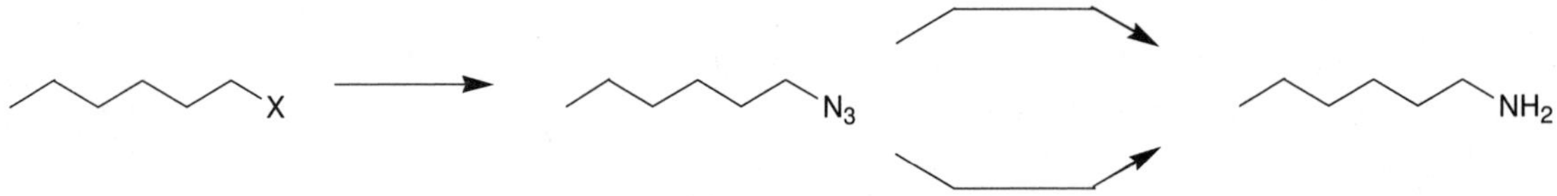

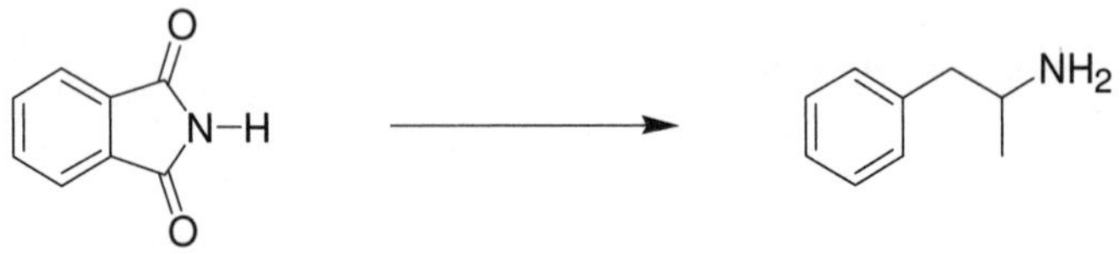

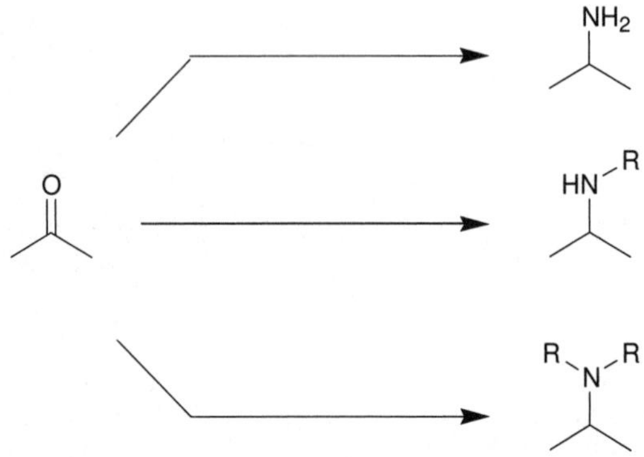

Reactions of Amines

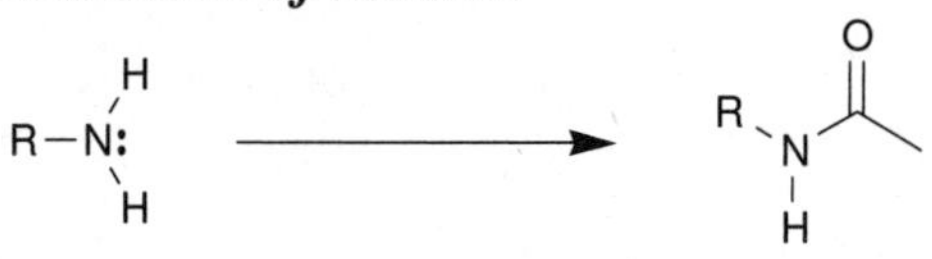

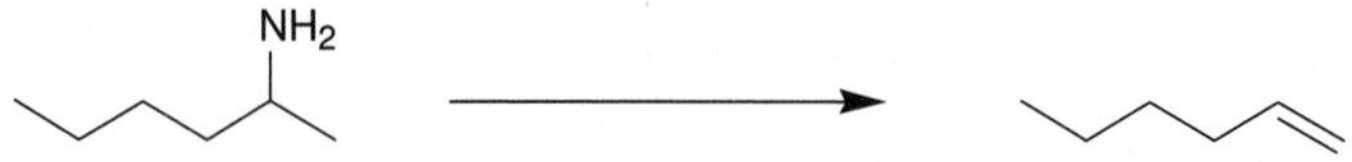

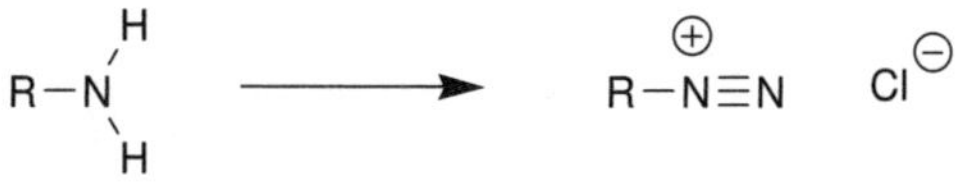

R—N̈H₂ → R-NH-C(=O)CH₃

NH₂ (2-hexylamine) → hex-1-ene

$R-NH_2 \longrightarrow R-\overset{\oplus}{N}{\equiv}N \quad Cl^{\ominus}$

$R_2N-H \longrightarrow R_2N-N{=}O$

Reactions of Aryldiazonium Salts

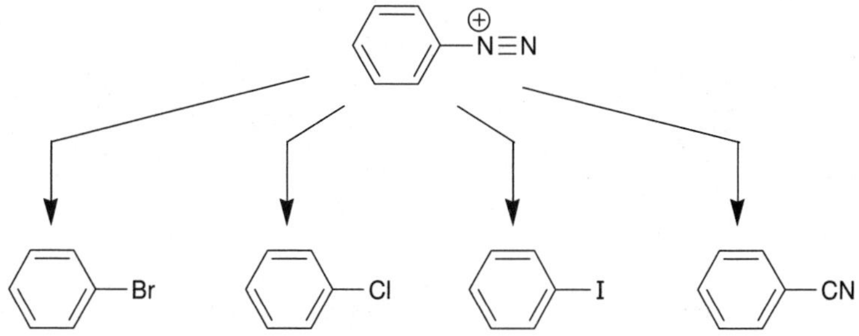

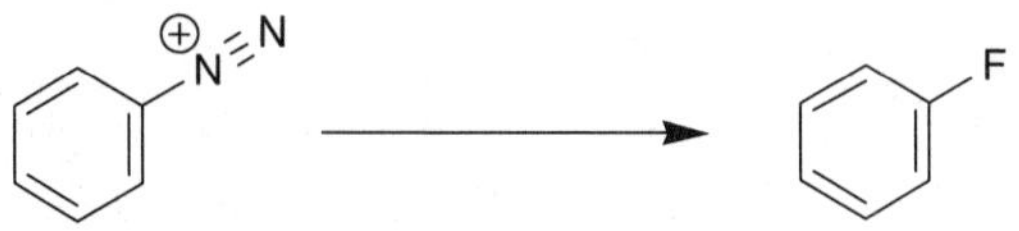

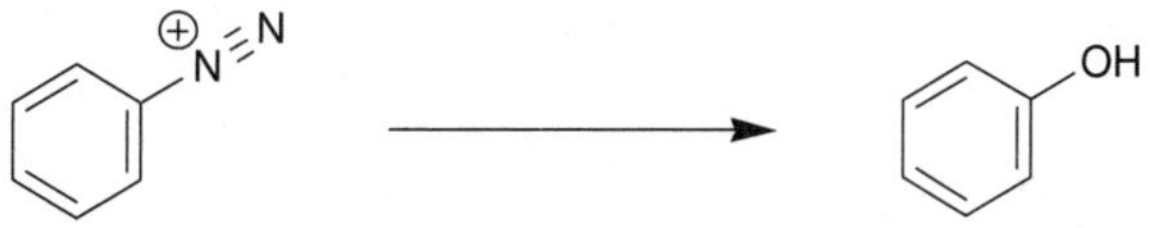

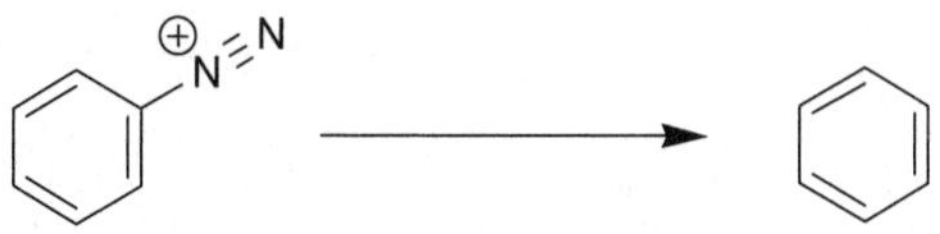

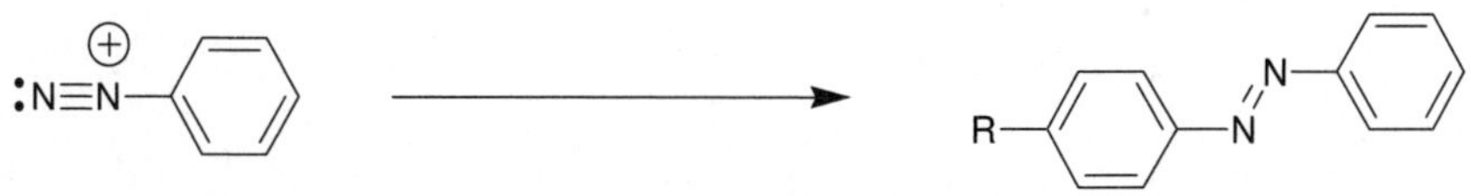

$C_6H_5-\overset{\oplus}{N}{\equiv}N$ → C_6H_5–Br, C_6H_5–Cl, C_6H_5–I, C_6H_5–CN

$C_6H_5-\overset{\oplus}{N}{\equiv}N \longrightarrow C_6H_5-F$

$C_6H_5-\overset{\oplus}{N}{\equiv}N \longrightarrow C_6H_5-OH$

$C_6H_5-\overset{\oplus}{N}{\equiv}N \longrightarrow C_6H_6$

$:N{\equiv}\overset{\oplus}{N}-C_6H_5 \longrightarrow R-C_6H_4-N{=}N-C_6H_5$

Reactions of Nitrogen Heterocycles

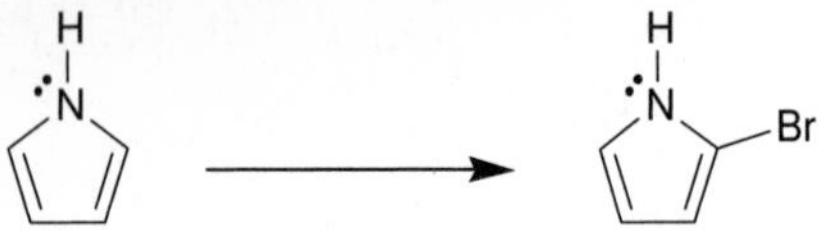

H N Br

N Br

Solutions

23.1.
a) 3,3-dimethyl-1-butanamine
b) cyclopentylamine
c) *N,N*-dimethylcyclopentylamine
d) triethylamine
e) (*1S,3R*)-3-isopropylcyclohexanamine
f) (*1S, 3S*)-3-aminocyclohexanol

23.2.

a) 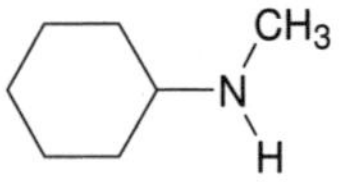 CH_3 N H b) 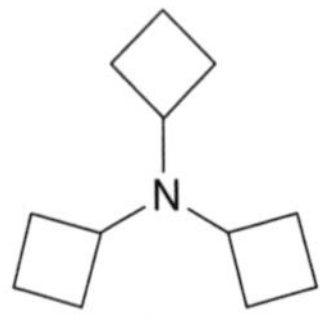 N c) 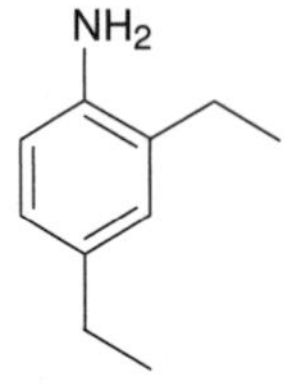 NH_2 d) 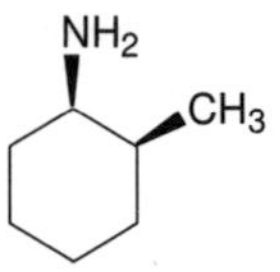 NH_2 CH_3 e) 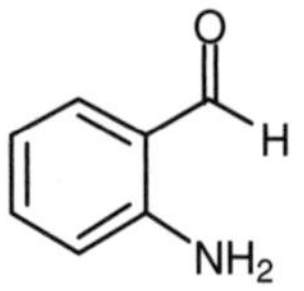 O H NH_2

23.3.

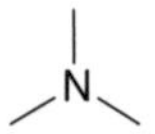

N trimethylamine

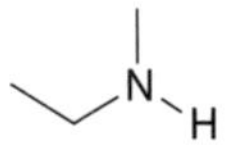

N H ethyl methyl amine

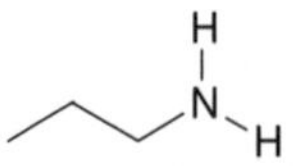

H N H propylamine

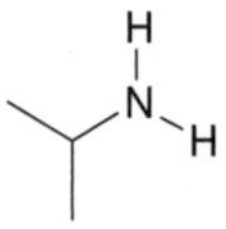

H N H isopropylamine

23.4.

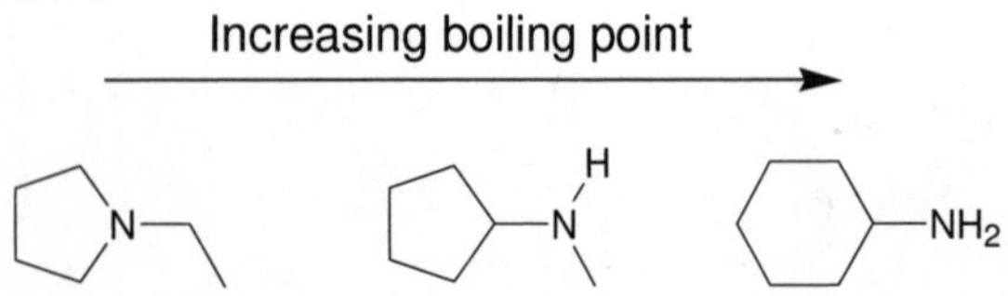

Increasing boiling point

N H N NH_2

23.5.
a) No. This compound has eight carbon atoms and only one functional group.
b) Yes.
c) Yes.

23.6.

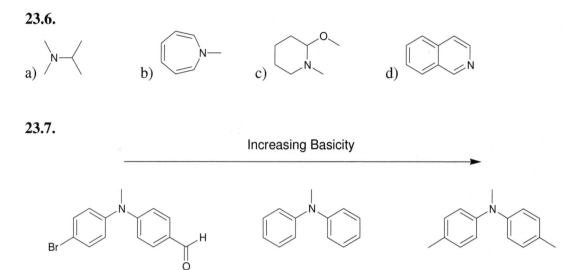

a) 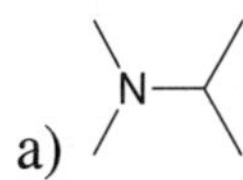

b) 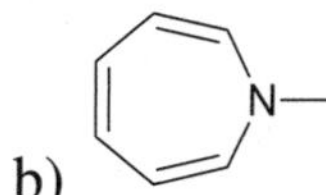

c) O N

d) N

23.7.

Increasing Basicity →

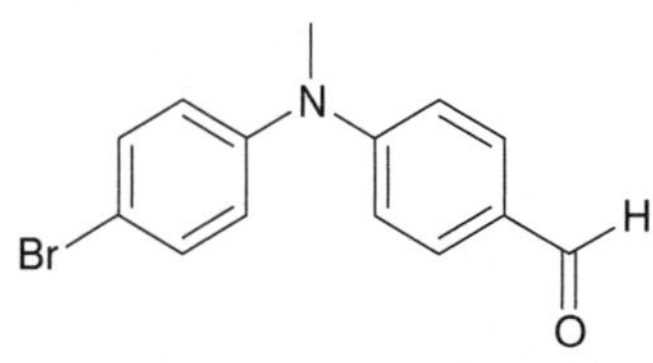

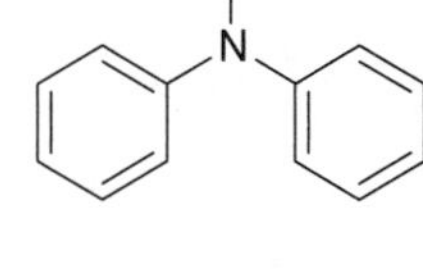

N

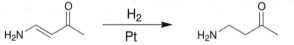

23.8. In the reactant, the lone pair of the amino group is delocalized via resonance. In the product, the lone pair of the amino group is localized.

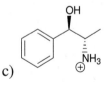

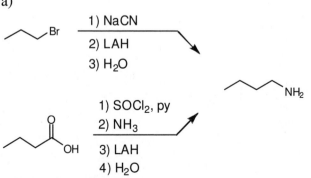

H_2N (O) —H_2 / Pt→ H_2N (O)

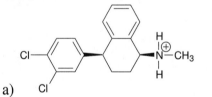

23.9.

a) Cl Cl H N⊕ CH_3 H

b) ⊕NH_3

c) OH ⊕NH_3

23.10.
a)

Br: 1) NaCN 2) LAH 3) H_2O

NH_2

OH (O): 1) $SOCl_2$, py 2) NH_3 3) LAH 4) H_2O

b)

Benzyl bromide: 1) NaCN, 2) LAH, 3) H_2O → 2-phenylethylamine (NH_2)

Phenylacetic acid (OH, O): 1) $SOCl_2$, py, 2) NH_3, 3) LAH, 4) H_2O → 2-phenylethylamine

c)

Bromocyclohexane (Br): 1) NaCN, 2) LAH, 3) H_2O → cyclohexylmethylamine (NH_2)

Cyclohexanecarboxylic acid (O, OH): 1) $SOCl_2$, py, 2) NH_3, 3) LAH, 4) H_2O → cyclohexylmethylamine

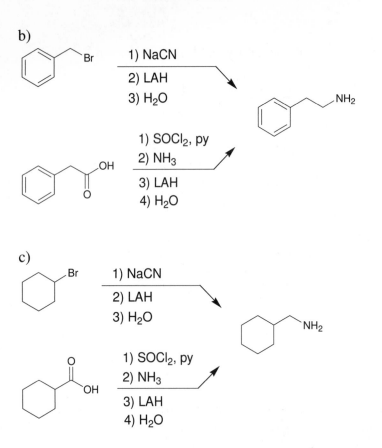

23.11. This compound cannot be prepared from an alkyl halide or a carboxylic acid, using the methods described in this section, because there are two methyl groups at the alpha position (the carbon atom connected to the amino group). These two methyl groups cannot be installed with either of the synthetic methods above, because both methods produce an amine with two alpha protons.

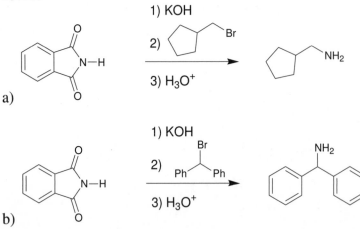

23.12.

a) Phthalimide (O, N–H, O): 1) KOH, 2) (bromomethyl)cyclopentane (Br), 3) H_3O^+ → cyclopentylmethylamine (NH_2)

b) Phthalimide (O, N–H, O): 1) KOH, 2) Ph–CH(Br)–Ph, 3) H_3O^+ → diphenylmethylamine (NH_2)

c)

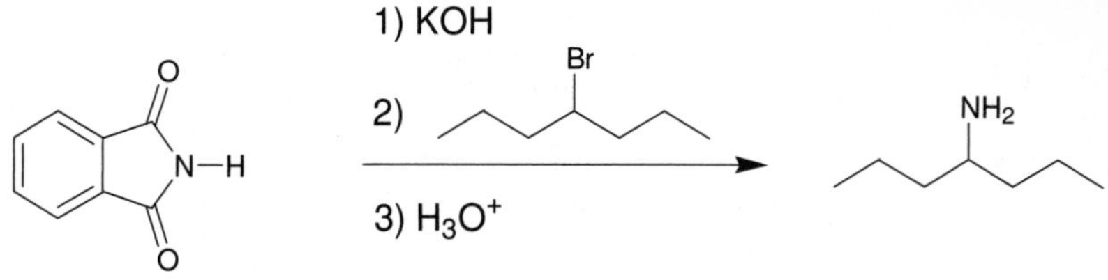

1) KOH

2) [3-bromohexane]

3) H_3O^+

NH2

d)

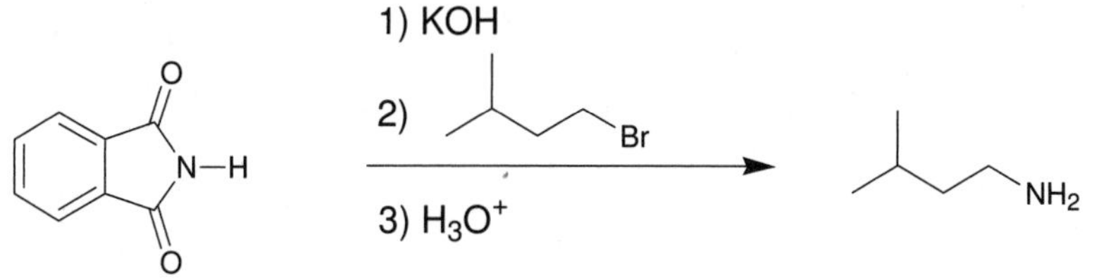

1) KOH

2) Br

3) H_3O^+

NH2

23.13.

a)

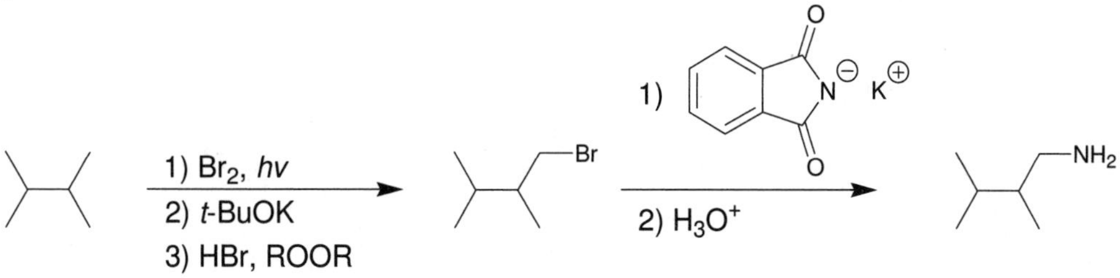

1) Br_2, *hv*
2) *t*-BuOK
3) HBr, ROOR

Br

1) [potassium phthalimide] O N⊖ K⊕ O

2) H_3O^+

NH2

b)

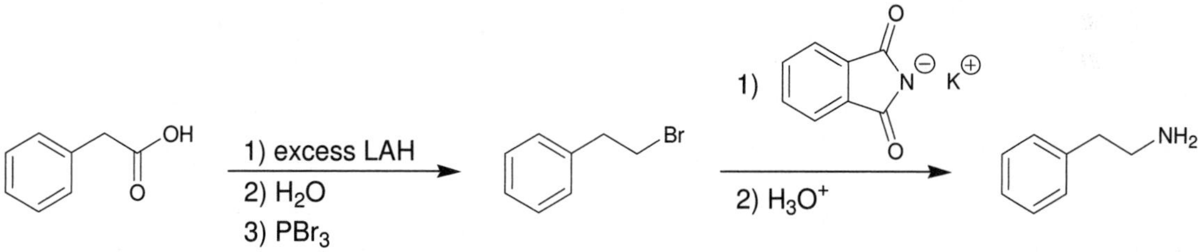

OH O

1) excess LAH
2) H_2O
3) PBr_3

Br

1) [potassium phthalimide] O N⊖ K⊕ O

2) H_3O^+

NH2

23.14.

a)

O H

[H^+] , $NaBH_3CN$

H2N

N H

O H

[H^+] , $NaBH_3CN$

NH2

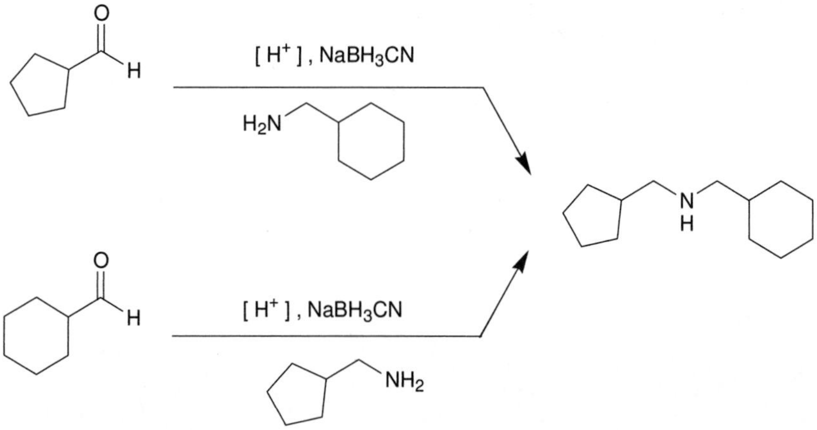

b)

$[H^+]$, $NaBH_3CN$

$[H^+]$, $NaBH_3CN$

c)

$[H^+]$, $NaBH_3CN$

$[H^+]$, $NaBH_3CN$

NH_2

d)

$[H^+]$, $NaBH_3CN$

$[H^+]$, $NaBH_3CN$

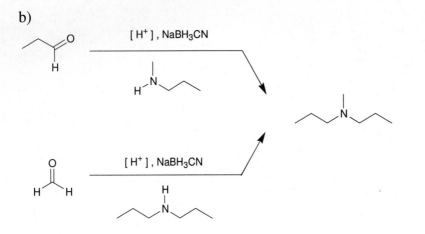

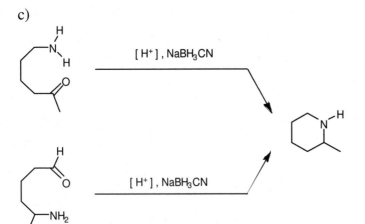

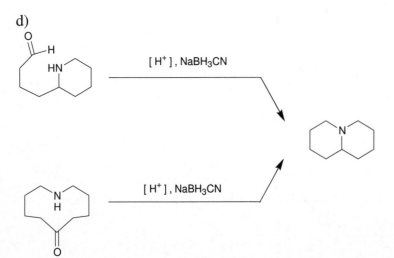

e)

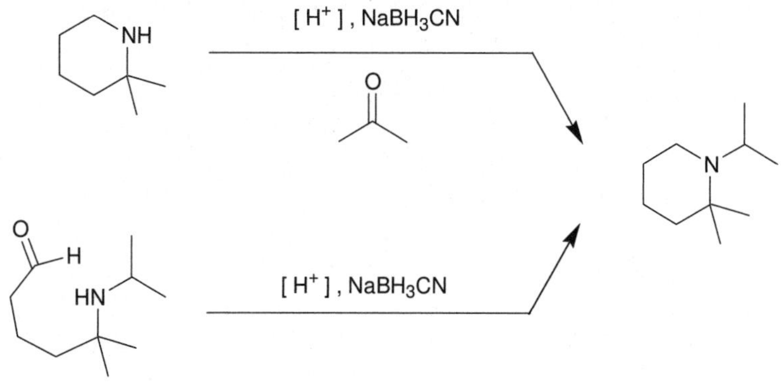

[H⁺], $NaBH_3CN$

[H⁺], $NaBH_3CN$

23.15.

[H⁺], $NaBH_3CN$

CH_3NH_2

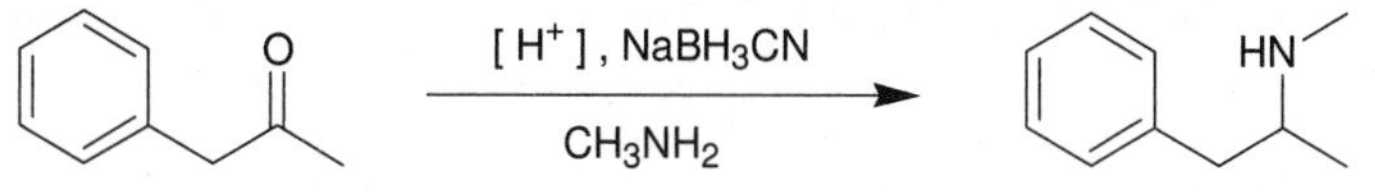

23.16. The last step of reductive amination is the reduction of a C=N bond. That step introduces a proton on the alpha position (the carbon atom that is connected to the nitrogen atom in the product):

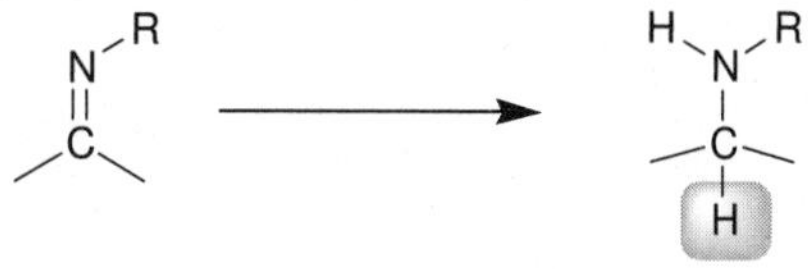

As a result, the product of a reductive amination must have at least one proton at the alpha position. In the case of tri-*tert*-butyl amine, there are three alpha positions, and none of them bears a proton. Each of the alpha positions has three alkyl groups and no protons.

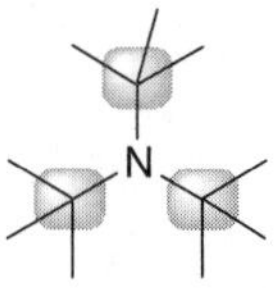

Therefore, this compound cannot be made with a reductive amination.

23.17.

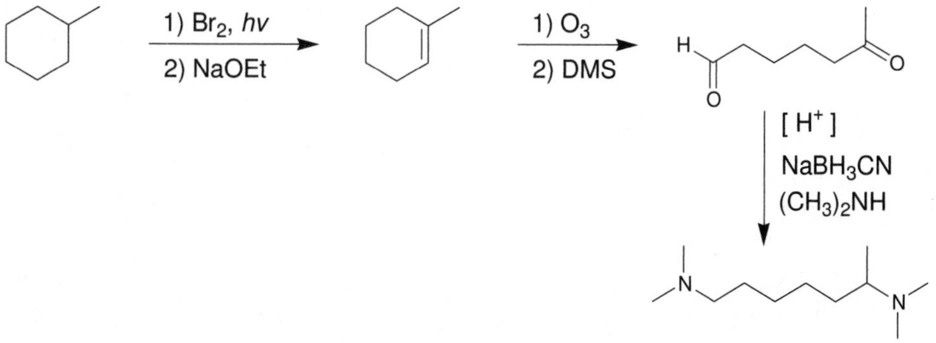

1) Br_2, *hv*
2) NaOEt

1) O_3
2) DMS

[H⁺]
$NaBH_3CN$
$(CH_3)_2NH$

23.18.

a)

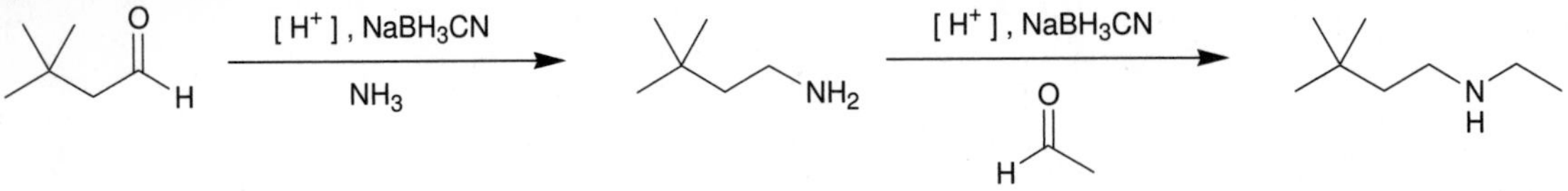

[H+] , NaBH3CN / NH3 → ; [H+] , NaBH3CN / CH3CHO →

b)

[H+] , NaBH3CN / NH3 →

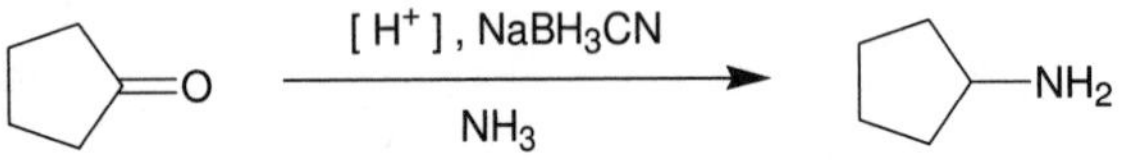

c)

[H+] , NaBH3CN / NH3 → ; [H+] , NaBH3CN / HCHO → ; [H+] , NaBH3CN / HCHO →

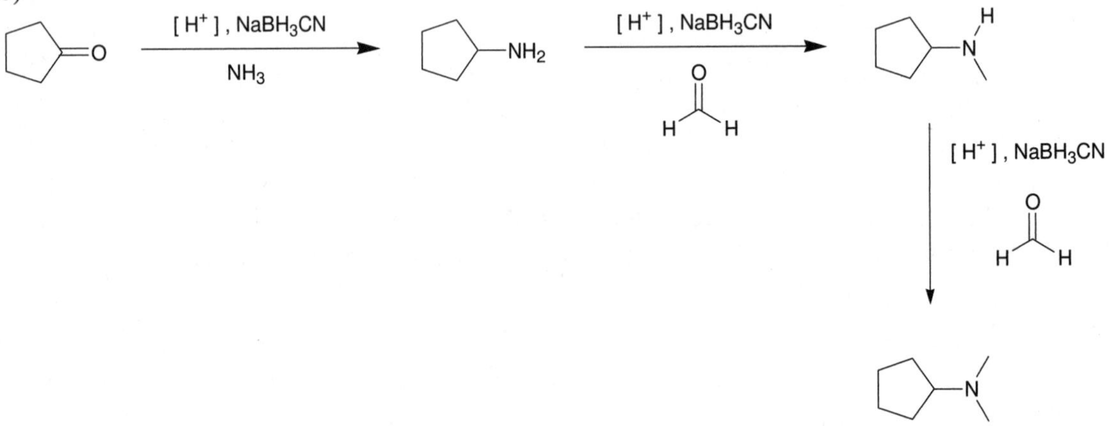

d)

[H+] , NaBH3CN / NH3 → ; [H+] , NaBH3CN / CH3CHO →

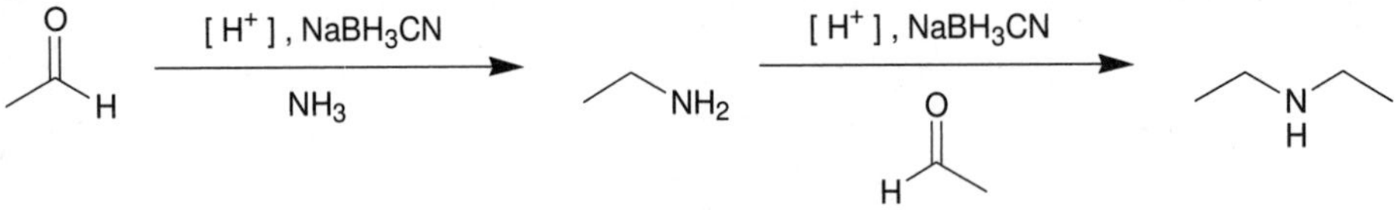

e)

[H+] , NaBH3CN / NH3 → ; [H+] , NaBH3CN / CH3CHO → ; [H+] , NaBH3CN / CH3CHO →

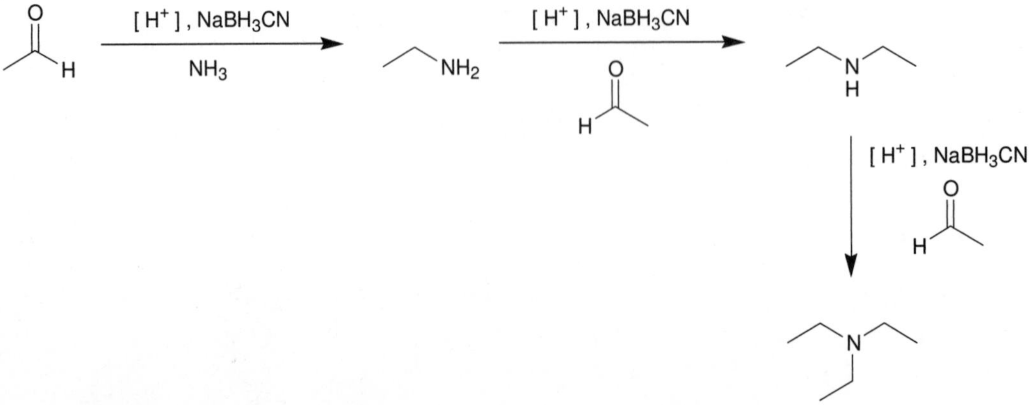

f)

[H^+], $NaBH_3CN$

NH_3

NH_2

[H^+], $NaBH_3CN$

O

H

H

N

[H^+], $NaBH_3CN$

O

N

23.19. The first alkyl group is installed via a Gabriel synthesis, and the remaining alkyl groups are installed via reductive amination processes. For most of the following syntheses, there is a choice regarding which group to attach via the initial Gabriel synthesis. In such cases, the least sterically hindered group is chosen (the group whose installation involves the least hindered alkyl halide):

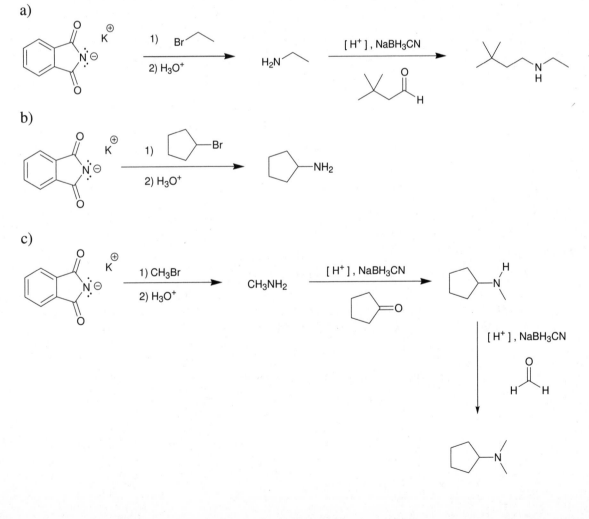

a)

O K⊕ N⊖ O

1) Br

2) H_3O^+

H_2N

[H^+], $NaBH_3CN$

O H

N H

b)

O K⊕ N⊖ O

1) Br

2) H_3O^+

NH_2

c)

O K⊕ N⊖ O

1) CH_3Br

2) H_3O^+

CH_3NH_2

[H^+], $NaBH_3CN$

O

H N

[H^+], $NaBH_3CN$

O

H H

N

d)

1) Br ; 2) H_3O^+ → H_2N ; $[H^+]$, $NaBH_3CN$

e)

1) Br ; 2) H_3O^+ → H_2N ; $[H^+]$, $NaBH_3CN$; $[H^+]$, $NaBH_3CN$

f)

1) Br ; 2) H_3O^+ → $EtNH_2$; $[H^+]$, $NaBH_3CN$; $[H^+]$, $NaBH_3CN$

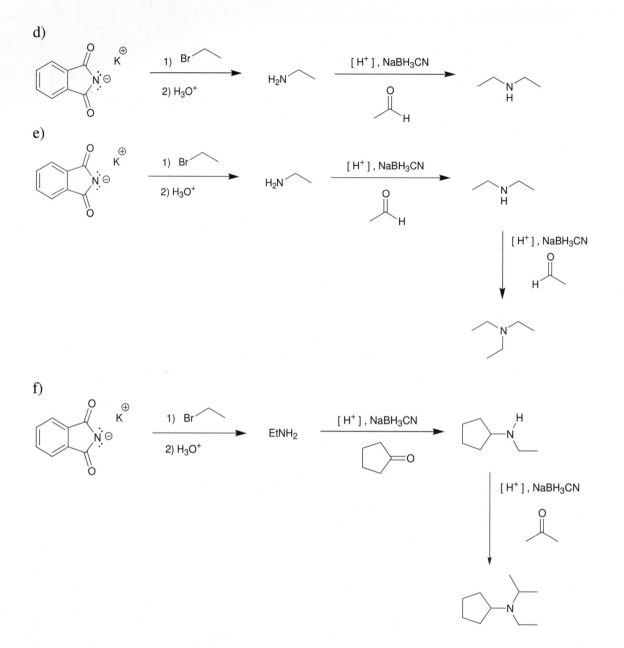

23.20. The first alkyl group is installed via an azide synthesis, and the remaining alkyl groups are installed via reductive amination processes. For most of the following syntheses, there is a choice regarding which group to attach via the initial azide synthesis. In such cases, the least sterically hindered group is chosen (the group whose installation involves the least hindered alkyl halide):

a)

Br — 1) NaN_3 ; 2) H_2, Pt → H_2N ; $[H^+]$, $NaBH_3CN$

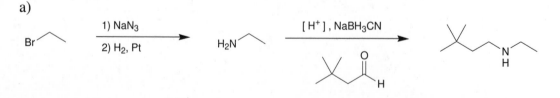

b)

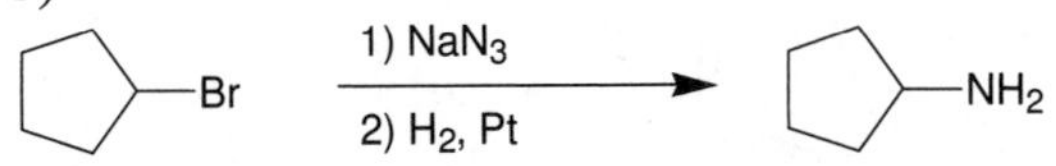

Cyclopentyl–Br $\xrightarrow[\text{2) } H_2\text{, Pt}]{\text{1) } NaN_3}$ Cyclopentyl–NH_2

c)

CH_3Br $\xrightarrow[\text{2) } H_2\text{, Pt}]{\text{1) } NaN_3}$ CH_3NH_2 $\xrightarrow[\text{cyclopentanone}]{[H^+]\text{, } NaBH_3CN}$ N-methylcyclopentylamine (N–H)

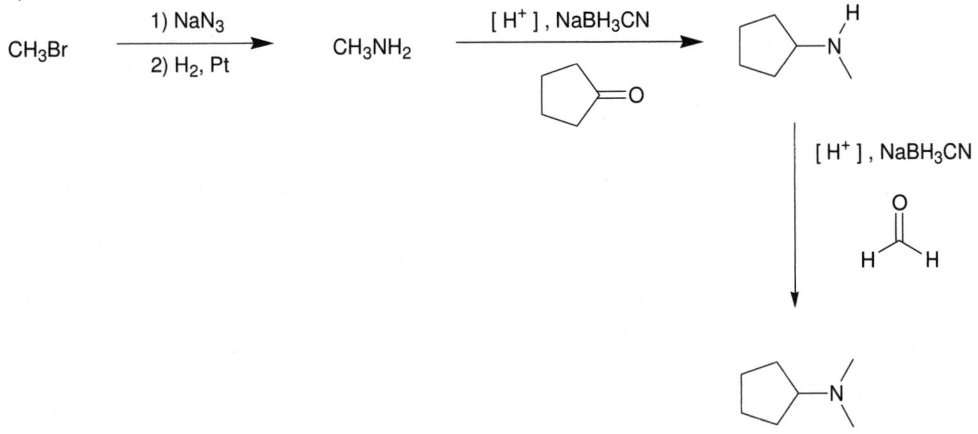

$\downarrow$ $[H^+]$, $NaBH_3CN$, H(C=O)H

N,N-dimethylcyclopentylamine

d)

Br–ethyl $\xrightarrow[\text{2) } H_2\text{, Pt}]{\text{1) } NaN_3}$ H_2N–ethyl $\xrightarrow[\text{acetaldehyde (O, H)}]{[H^+]\text{, } NaBH_3CN}$ diethylamine (N, H)

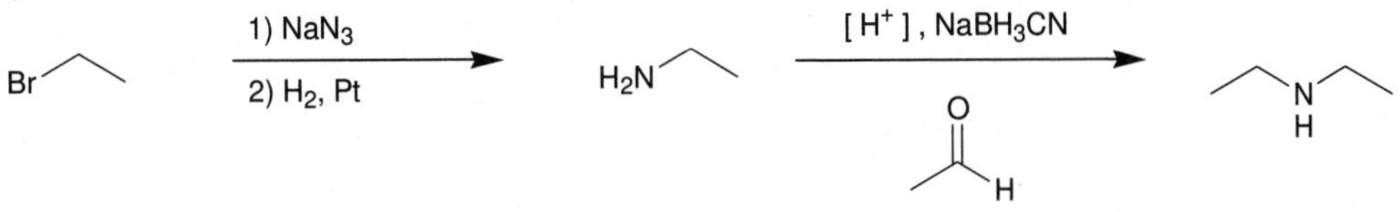

e)

Br–ethyl $\xrightarrow[\text{2) } H_2\text{, Pt}]{\text{1) } NaN_3}$ H_2N–ethyl $\xrightarrow[\text{acetaldehyde (O, H)}]{[H^+]\text{, } NaBH_3CN}$ diethylamine (N, H)

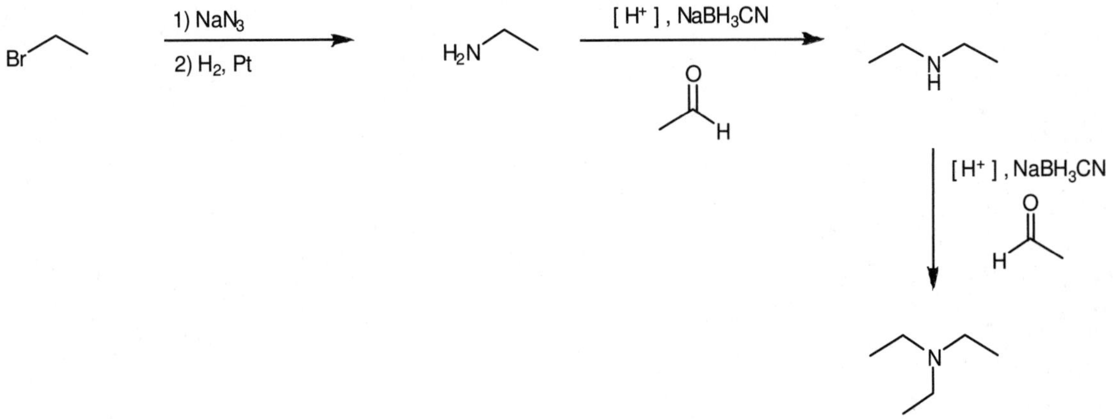

$\downarrow$ $[H^+]$, $NaBH_3CN$, acetaldehyde (O, H)

triethylamine (N)

f)

Br⁄⁀ → 1) NaN_3 2) H_2, Pt → H_2N⁄⁀ → [H^+], $NaBH_3CN$ (cyclopentanone) → N-ethylcyclopentanamine → [H^+], $NaBH_3CN$ (acetone) → N-ethyl-N-isopropylcyclopentanamine

23.21.

(4-(cyclopent-1-enyl)butanal) → 1) O_3 2) DMS → (H–C(=O)–CH₂CH₂CH₂–C(=O)–CH₂CH₂CH₂–C(=O)–H) → [H^+], xs $NaBH_3CN$, NH_3 → (quinolizidine)

23.22.

NH_2 (aniline) → (acetyl chloride, Cl) → HN–C(=O)CH₃ (acetanilide) → HNO_3 / H_2SO_4 → HN–C(=O)CH₃, NO_2 (para) → NaOH *or* H_3O^+ → NH_2, NO_2 (para)

23.23.

NO_2 (nitrobenzene) → HCl / Zn → NH_2 (aniline) → (acetyl chloride, Cl) → HN–C(=O)CH₃ (acetanilide) → Cl_2 → HN–C(=O)CH₃, Cl (para) → NaOH *or* H_3O^+ → NH_2, Cl (para)

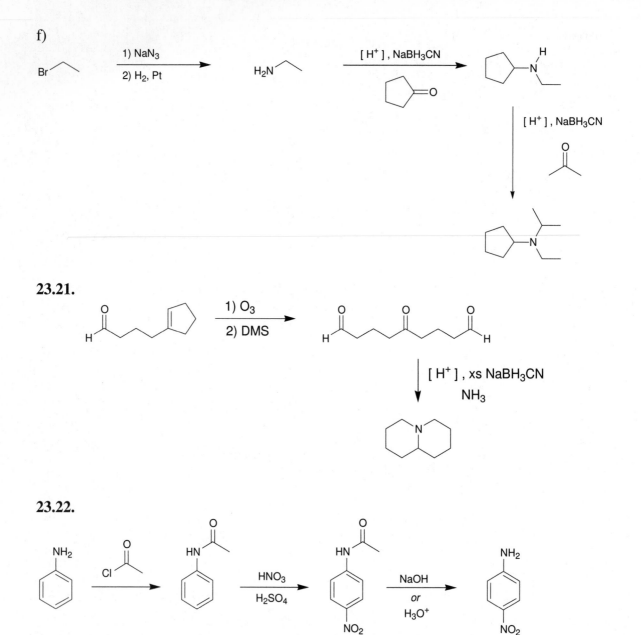

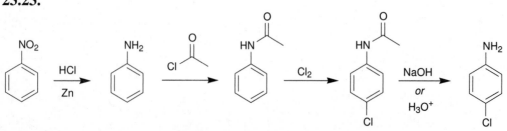

23.24.

1) excess NH_3
2) LAH
3) H_2O

SOCl$_2$

OH

Cl

NH$_2$

N

H

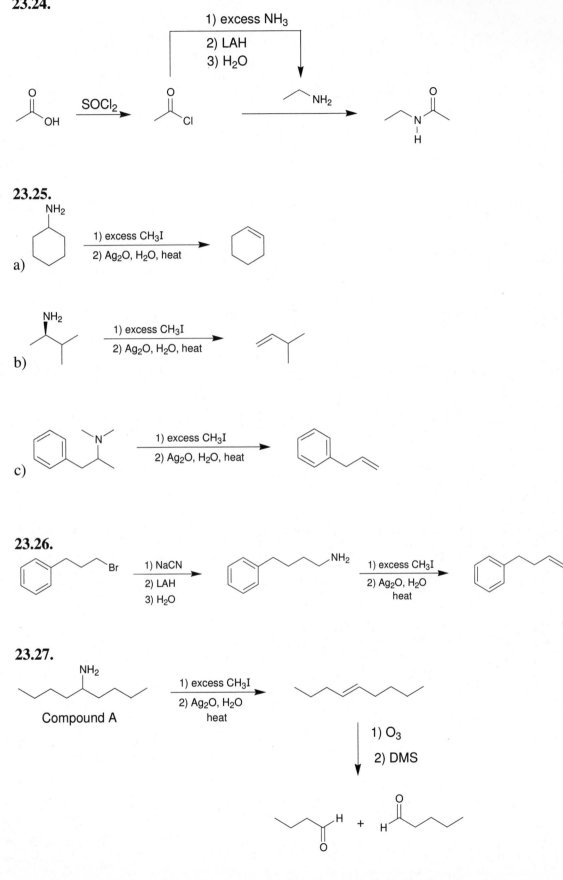

23.25.

a)

NH$_2$

1) excess CH_3I
2) Ag_2O, H_2O, heat

b)

NH$_2$

1) excess CH_3I
2) Ag_2O, H_2O, heat

c)

N

1) excess CH_3I
2) Ag_2O, H_2O, heat

23.26.

Br

1) NaCN
2) LAH
3) H_2O

NH$_2$

1) excess CH_3I
2) Ag_2O, H_2O
heat

23.27.

NH$_2$

Compound A

1) excess CH_3I
2) Ag_2O, H_2O
heat

1) O_3
2) DMS

H + H

23.28.

1) excess CH_3I

2) Ag_2O, H_2O, heat

+ +

23.29.

a) $Cl^{\ominus}$

b)

c)

d)

23.30.

a)

1)

Cl

NH_2

2) $(CH_3)_2CHCl$, $AlCl_3$

3) H_3O^+

NH_2

1) $NaNO_2$, HCl

2) CuCN

CN

b)

NO_2

1) Br_2, $FeBr_3$

2) Fe, H_3O^+

Br NH_2

1) $NaNO_2$, HCl

2) CuBr

Br Br

c)

O

Cl

$AlCl_3$

O

HNO_3

H_2SO_4

O NO_2

HCl, Zn[Hg] heat

OH

1) $NaNO_2$, HCl

2) H_2O, heat

NH_2

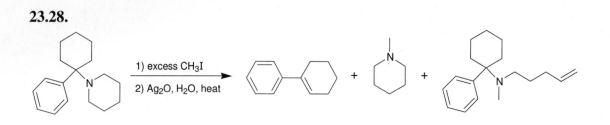

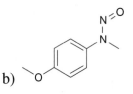

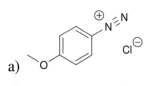

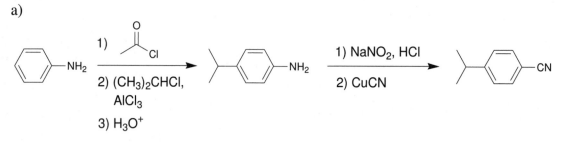

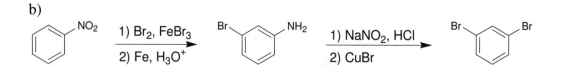

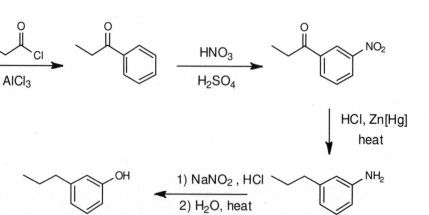

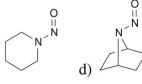

d)

1) $(CH_3)_3Cl$, $AlCl_3$
2) HNO_3, H_2SO_4

NO_2

1) Fe, H_3O^+
2) $NaNO_2$, HCl
3) CuCN
4) H_3O^+

OH
O

e)

1) Cl_2, $AlCl_3$
2) HNO_3, H_2SO_4

Cl — NO_2

1) Fe, H_3O^+
2) $NaNO_2$, HCl
3) HBF_4

Cl — F

f)

Cl

1) $NaNH_2$
2) $NaNO_2$, HCl
3) CuBr

Br

23.31.

a)

HO_3S — NH_2, SO_3H

1) $NaNO_2$, HCl
2) aniline (NH_2)

HO_3S — N=N — NH_2, SO_3H

b)

H_2N — NO_2

1) $NaNO_2$, HCl
2) 1-naphthol (OH)

HO — N=N — NO_2

c)

O_2N — NH_2

1) $NaNO_2$, HCl
2) N,N-dimethyl-1-naphthylamine (N)

O_2N — N=N — N

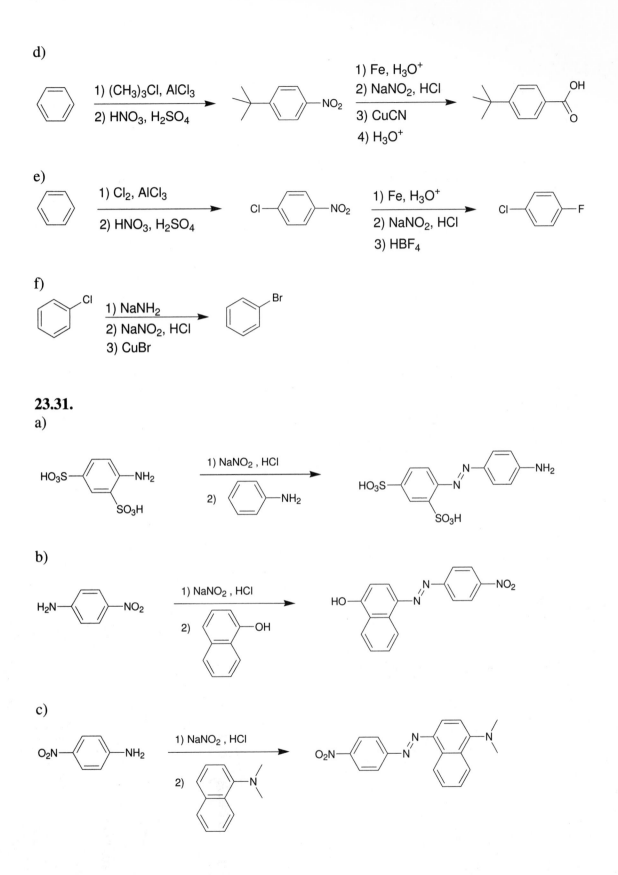

23.32.

Cl

$AlCl_3$

1) Fuming H_2SO_4

2) HNO_3, H_2SO_4

3) Dilute H_2SO_4

NO_2

1) HNO_3, H_2SO_4

2) Fe, H_3O^+

Fe, H_3O^+

NH_2

NH_2

1) O Cl

2) HNO_3, H_2SO_4

3) H_3O^+

4) $NaNO_2$, HCl

O_2N $N \equiv N$ $Cl^\ominus$

O_2N N=N NH_2

23.33.

a) N=N NH_2

b) N=N OH

c) N=N OMe

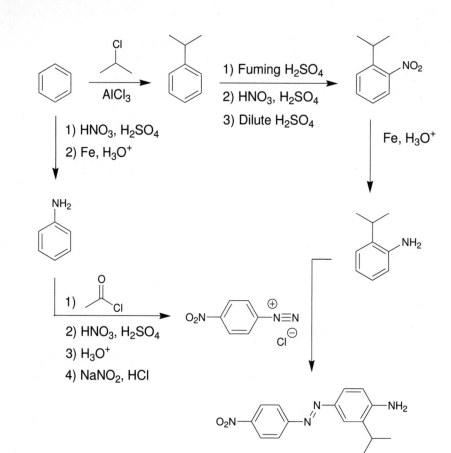

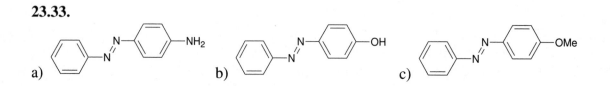

23.34. Attack at either C2 or C4 generates an intermediate that exhibits a resonance structure with a nitrogen atom that lacks an octet (highlighted below). Attack at C3 generates a more stable intermediate:

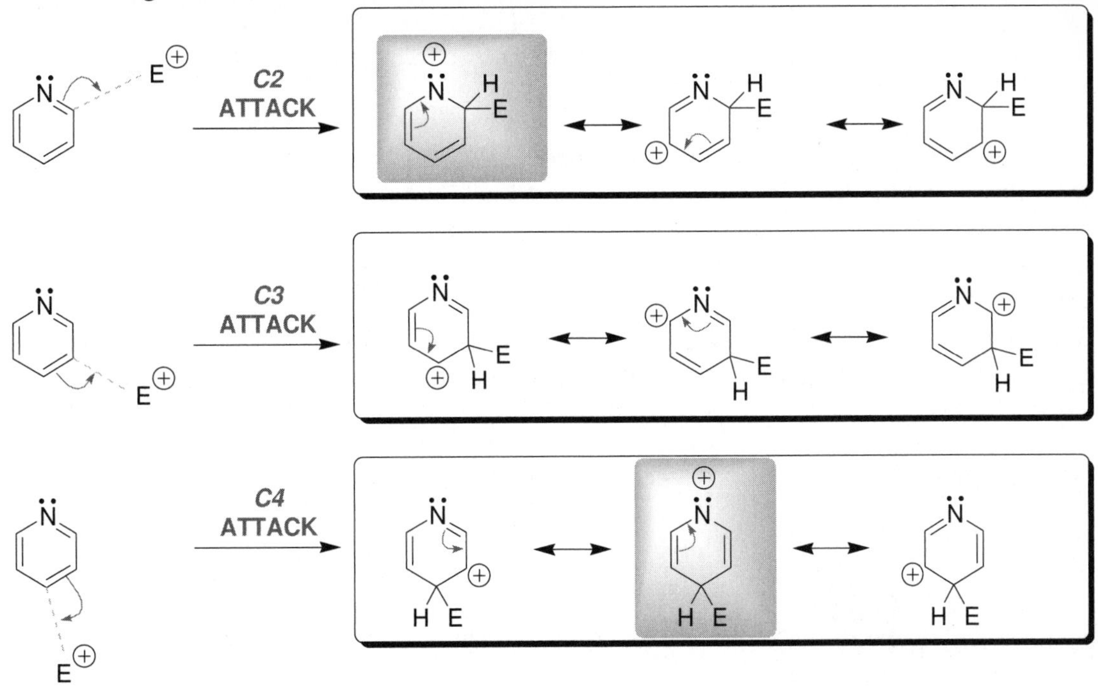

C2 ATTACK

C3 ATTACK

C4 ATTACK

23.35.

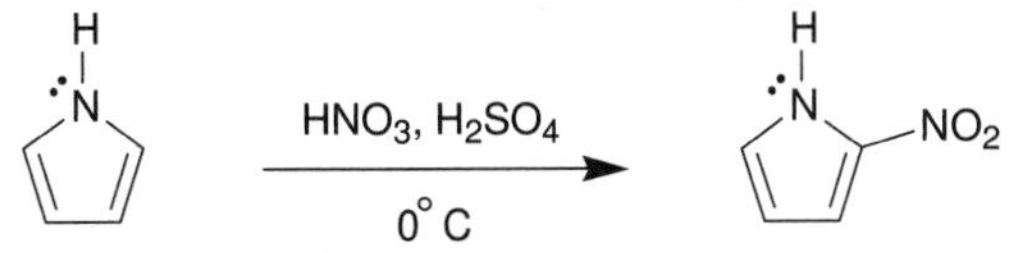

HNO_3, H_2SO_4

0° C

NO_2

23.36.

a) The second compound will have an N-H stretching signal between 3300 and 3500 cm^{-1}. The first compound will not have such a signal.

b) When treated with HCl, the first compound will be protonated to form an ammonium salt that will produce an IR signal between 2200 and 3000 cm^{-1}. The second compound is not an amine and will not exhibit the same behavior.

23.37.

a) The ^{1}H NMR spectrum of the first compound will have a singlet resulting from the N-methyl group. ^{1}H NMR spectrum of the second compound will not have any singlets.

b) The ^{1}H NMR spectrum of the first compound will have six signals, while the ^{1}H NMR spectrum of the second compound will have only three signals.

23.38.

secondary

primary

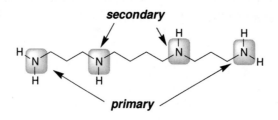

23.39.

a) The lone pair that is farthest away from the rings is the most basic, because its lone pair is localized. The lone pair of the other nitrogen atom is delocalized via resonance.

b)

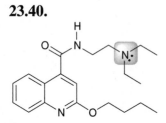

23.40.

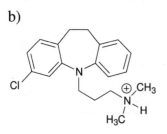

23.41.

a) b) c)

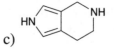

23.42.

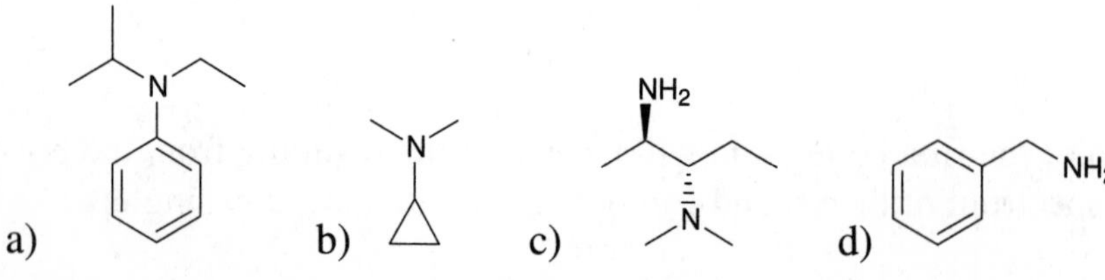

23.43. Only one of the nitrogen atoms has a localized lone pair (highlighted in the following structure). The other two nitrogen atoms have delocalized lone pairs.

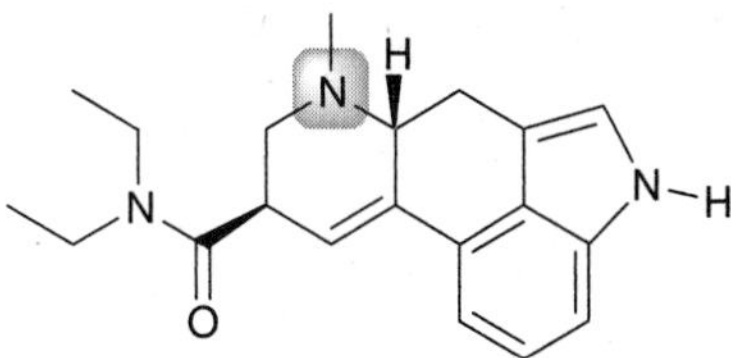

23.44.

a) two b) two c) one

23.45.

a) 2,2,3,3-tetramethyl-1-hexanamine
b) (*S*)-4-amino-2,2-dimethylcyclohexanone
c) dicyclobutylmethylamine
d) 3-bromo-2,6-dimethylaniline
e) *N,N*-dimethyl-3-propylaniline
f) 2,5-diethyl-*N*-methyl pyrrole

23.46.

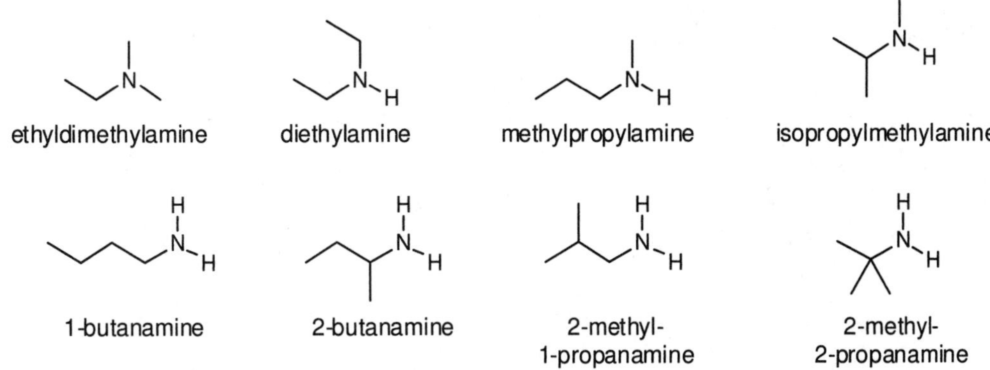

23.47. None of these compounds are chiral.

dimethylpropylamine

isopropyldimethylamine

diethylmethylamine

23.48.

a) Base + Acid ⇌ (pyridine + acetic acid → pyridinium + acetate)

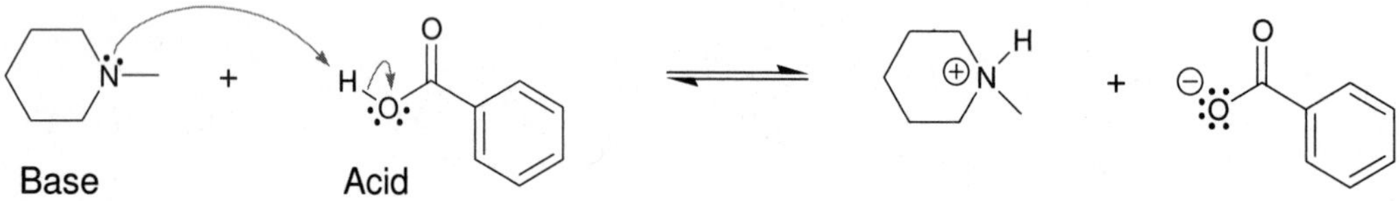

b) Base + Acid ⇌ (N-methylpiperidine + benzoic acid → N-methylpiperidinium + benzoate)

23.49.

a) 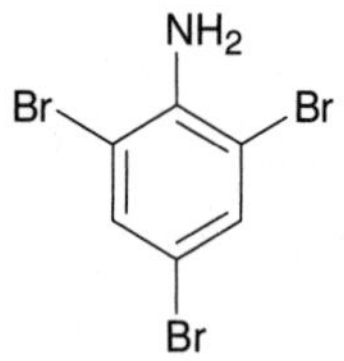 NH_2, Br, Br, Br b) 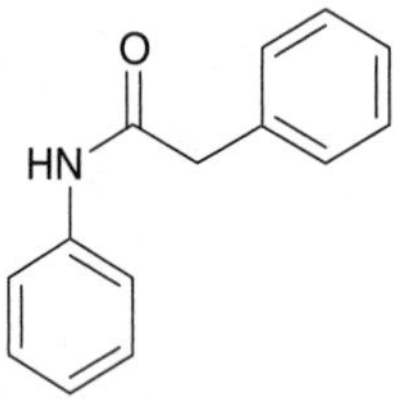 HN, O c) 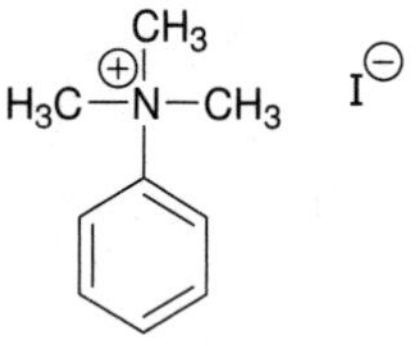 CH_3, $H_3C-N^{+}-CH_3$ I^- d)  e) 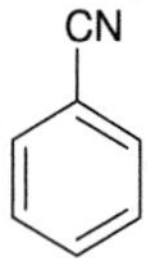 CN

23.50.

a)

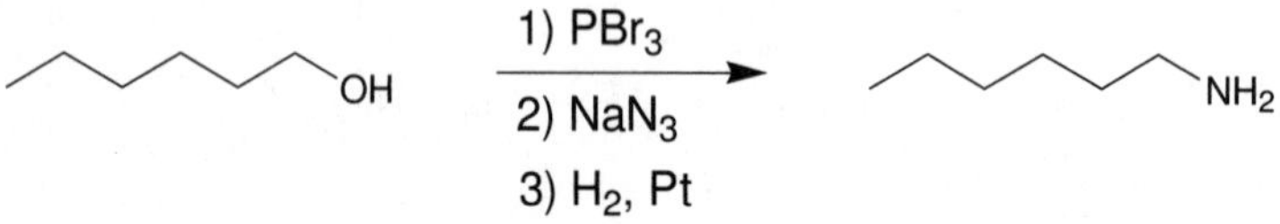

OH → 1) PBr_3 2) NaN_3 3) H_2, Pt → NH_2

b)

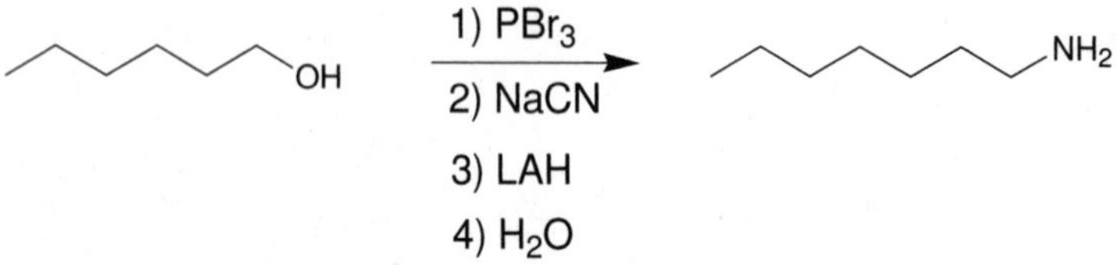

OH → 1) PBr_3 2) NaCN 3) LAH 4) H_2O → NH_2

c)

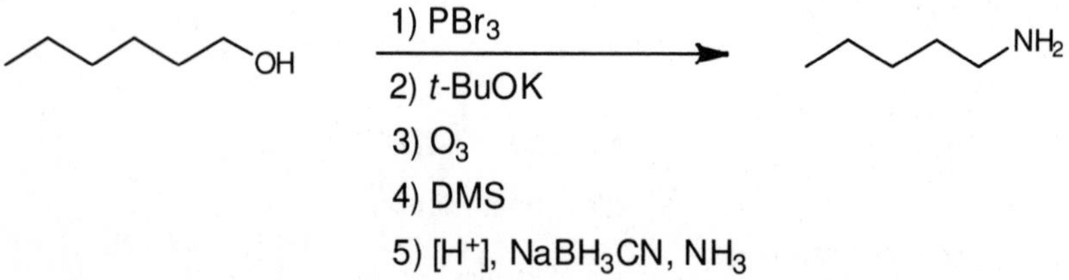

OH → 1) PBr_3 2) *t*-BuOK 3) O_3 4) DMS 5) $[H^+]$, $NaBH_3CN$, NH_3 → NH_2

23.51.

a)

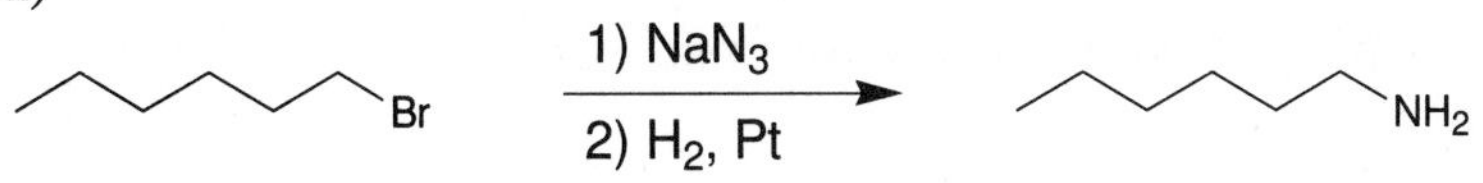
Br
1) NaN_3
2) H_2, Pt
NH_2

b)

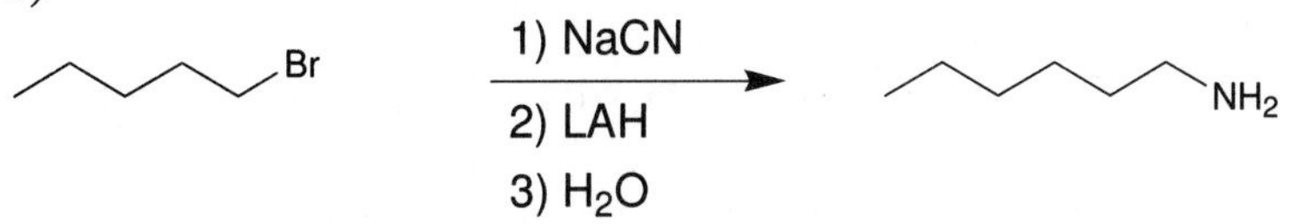
Br
1) NaCN
2) LAH
3) H_2O
NH_2

c)

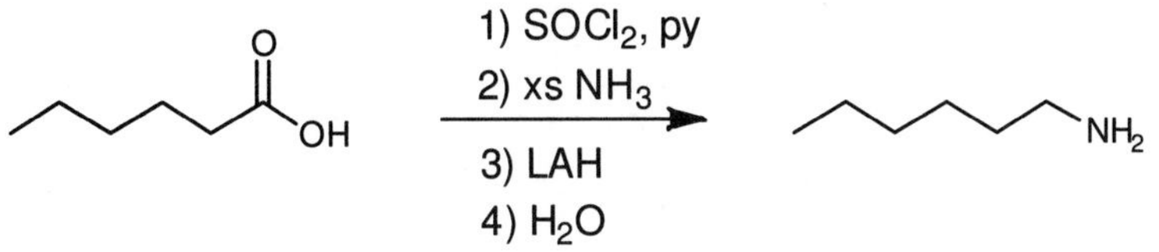
O
OH
1) $SOCl_2$, py
2) xs NH_3
3) LAH
4) H_2O
NH_2

d)

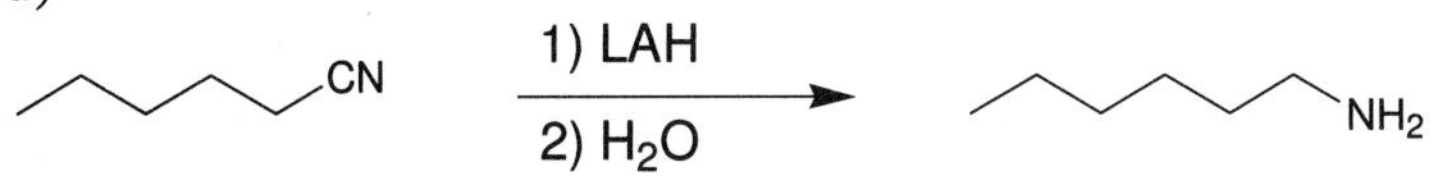
CN
1) LAH
2) H_2O
NH_2

23.52. Aziridine has significant ring strain, which would increase significantly during pyramidal inversion. This provides a significant energy barrier that prevents pyramidal inversion at room temperature.

23.53.

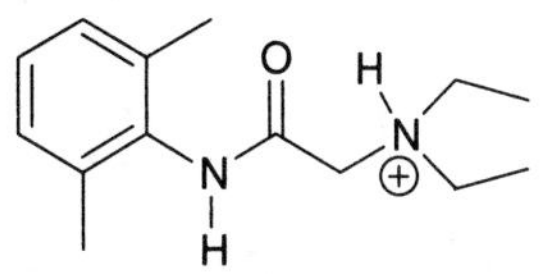

23.54.

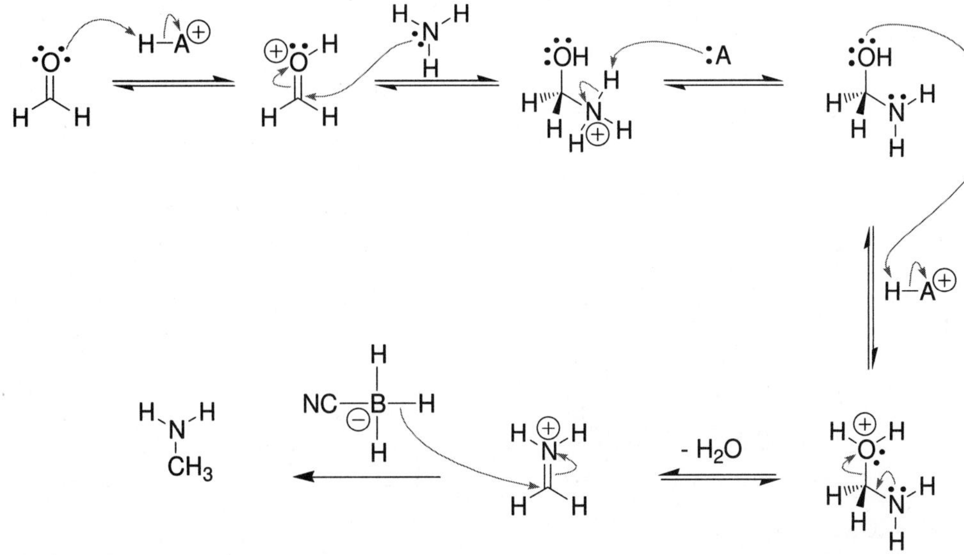
- H_2O

23.55. In acidic conditions, the amino group is protonated to give an ammonium ion. The ammonium group is a powerful deactivator and meta-director.

23.56.
a) The presence of the nitro group in the para position helps stabilize the conjugate base via resonance. As seen in chapter 19, this effect only occurs when the nitro group is in the ortho and para positions.
b) The basicity of *ortho*-nitroaniline should be closer in value to *para*-nitroaniline.

23.57.

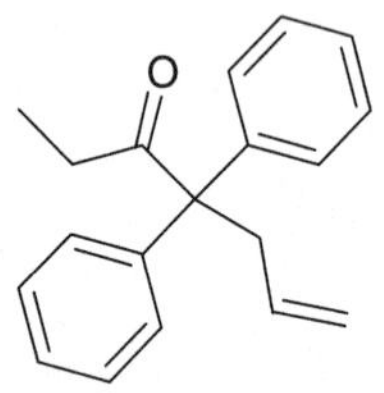

O

23.58. Protonation of the oxygen atom gives a resonance stabilized cation (as seen in chapter 20). In contrast, protonation of the nitrogen atom gives a cation that is not resonance stabilized.

23.59.
a)

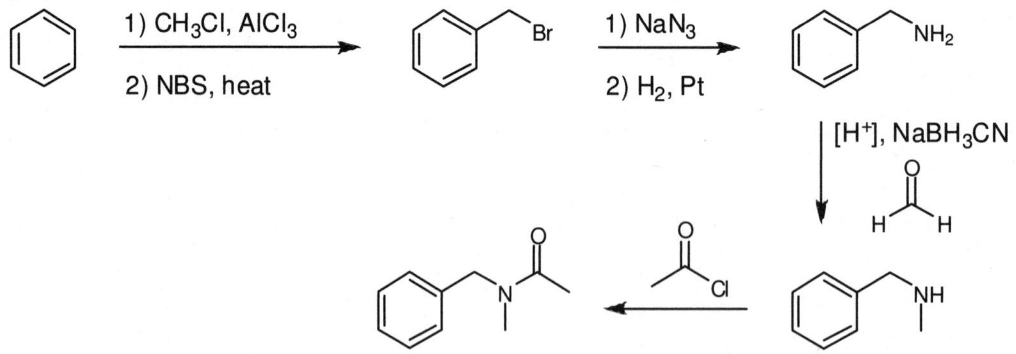

1) CH_3Cl, $AlCl_3$
2) NBS, heat
Br
1) NaN_3
2) H_2, Pt
NH_2
[H^+], $NaBH_3CN$
O
H H
O
N
O
Cl
NH

b)

1) Cl
$AlCl_3$
2) HNO_3, H_2SO_4
NO_2
1) Fe, H_3O^+
2) $NaNO_2$, HCl
3) CuCN
4) H_3O^+
COOH

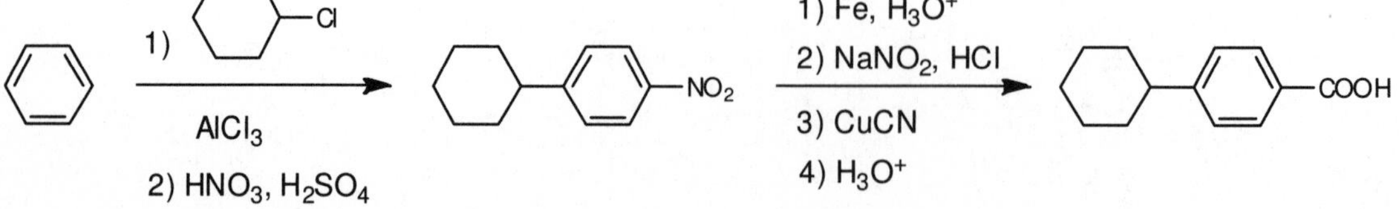

23.60.

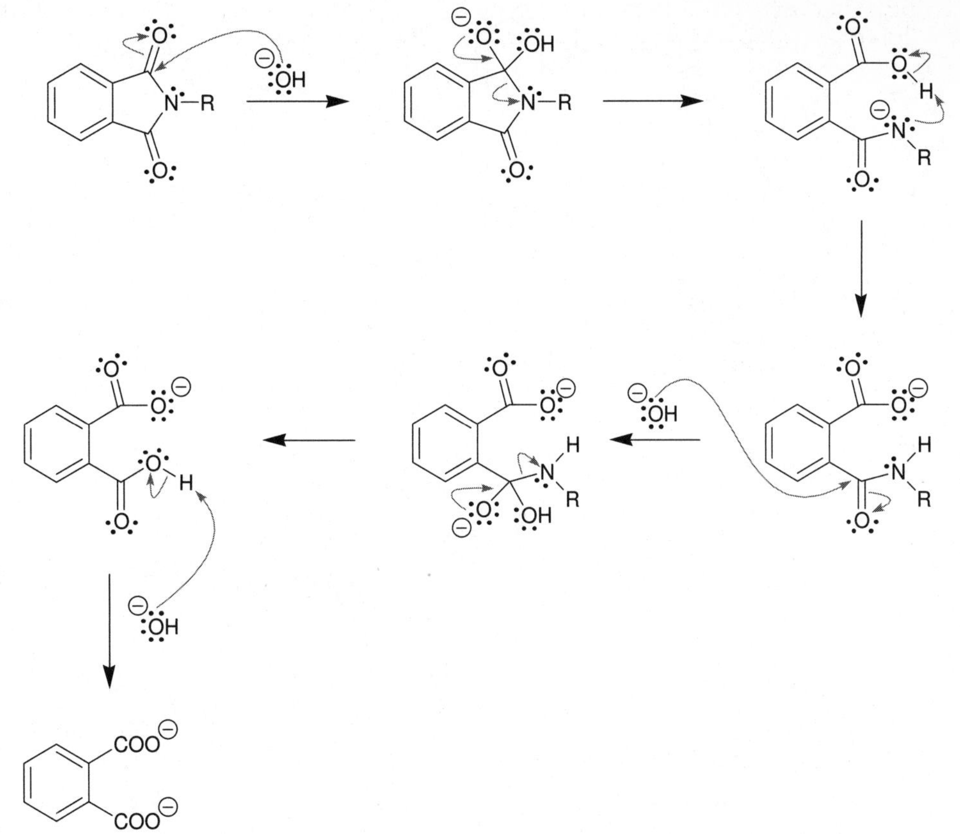

$$\text{N-alkylphthalimide} \xrightarrow{^{\ominus}OH} \text{tetrahedral intermediate} \longrightarrow \text{ring-opened amide} \longrightarrow \cdots$$

$^{\ominus}NHR + $ phthalate monoanion; $H-OH \rightarrow HN(H)R$; $^{\ominus}OH \rightarrow$ phthalate ($COO^{\ominus}$, $COO^{\ominus}$)

23.61.

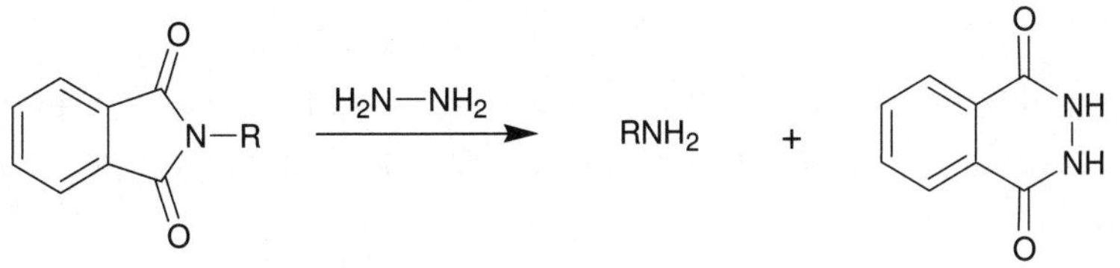

$$\text{N-alkylphthalimide} \xrightarrow{H_2N-NH_2} RNH_2 + \text{phthalhydrazide}$$

23.62.

a) 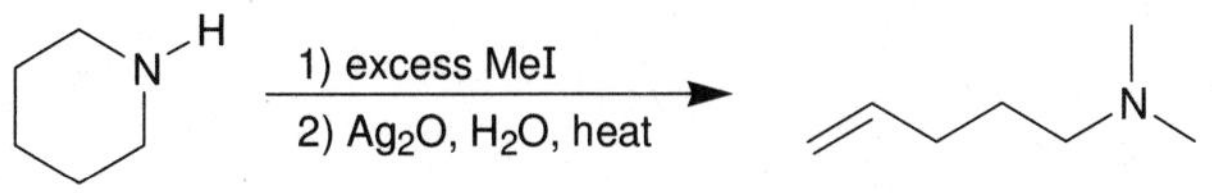 piperidine $\xrightarrow[\text{2) } Ag_2O,\ H_2O,\ \text{heat}]{\text{1) excess MeI}}$ $CH_2{=}CHCH_2CH_2CH_2N(CH_3)_2$

b) 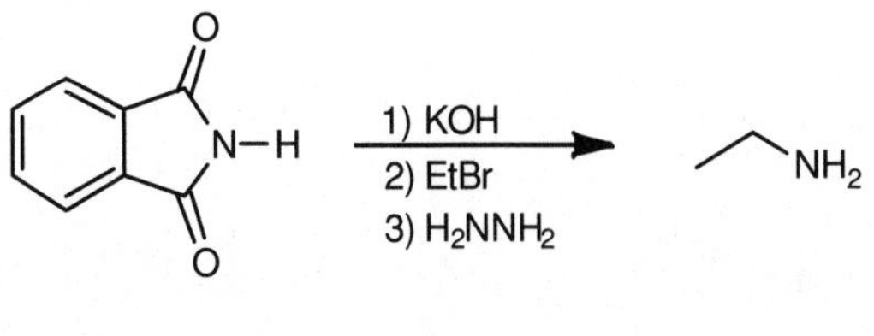 phthalimide (N–H) $\xrightarrow[\text{2) EtBr; 3) } H_2NNH_2]{\text{1) KOH}}$ $CH_3CH_2NH_2$

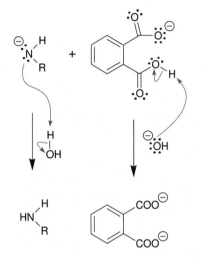

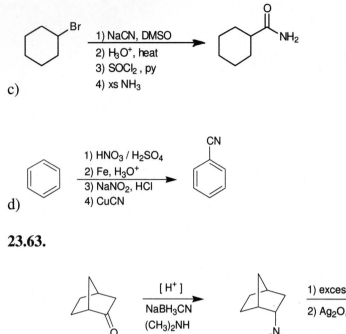

c)

Br

1) NaCN, DMSO
2) H_3O^+, heat
3) $SOCl_2$, py
4) xs NH_3

O

NH_2

d)

1) HNO_3 / H_2SO_4
2) Fe, H_3O^+
3) $NaNO_2$, HCl
4) CuCN

CN

23.63.

O

[H^+]
$NaBH_3CN$
$(CH_3)_2NH$

N

1) excess MeI
2) Ag_2O, H_2O, heat

1) O_3
2) DMS

CHO

CHO

[H^+]
xs $NaBH_3CN$
xs NH_3

NH_2

NH_2

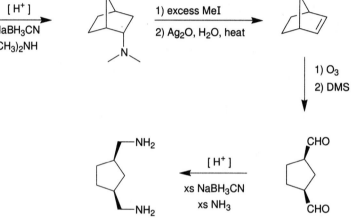

23.64. The conjugate base of pyrrole is highly stabilized because it is an aromatic anion and it is resonance stabilized, spreading the negative charge over all five atoms of the ring:

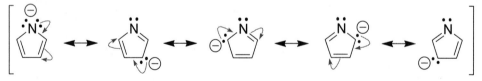

23.65.

O

23.66.

$[H^+]$, $NaBH_3CN$

HN

O

H

O

$[H^+]$, $NaBH_3CN$

NH

N

23.67.

23.68.

N

N

Cl

O_2N

+

OH

OH

O

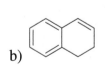

23.69.

a)

b)

23.70.

NO_2

a)

Br

Fe

H_3O^+

NH_2

Br

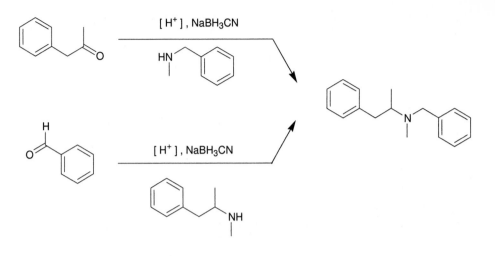

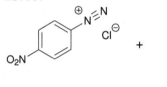

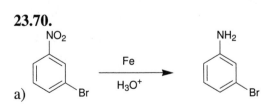

b) aniline + cyclohexyl methyl ketone → [H^+], $NaBH_3CN$ → N-(1-cyclohexylethyl)aniline

NH2 + O → [H^+] / $NaBH_3CN$ → NH

c) CN → 1) xs LAH 2) H_2O → NH2

d) O N → 1) xs LAH 2) H_2O → N

23.71.

a) OH, Br b) F, Br c) CN, Br d) Br e) Br, Br

23.72.

O, NH2 → [H_2SO_4] / $NaBH_3CN$ → NH

23.73.

a)

$NaNH_2$, NH_3

Br2 / FeBr3 → Br → 1) Mg 2) H(C=O)H 3) H_2O 4) PCC, CH_2Cl_2 → O, H → H_2N– / [H^+] / $NaBH_3CN$ → N H

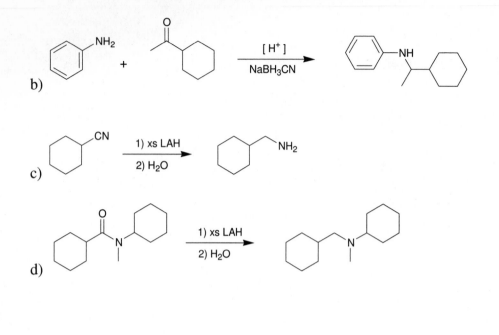

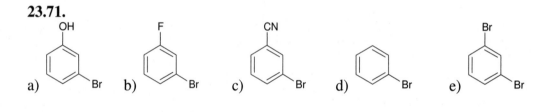

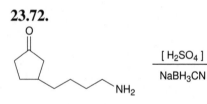

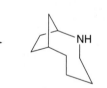

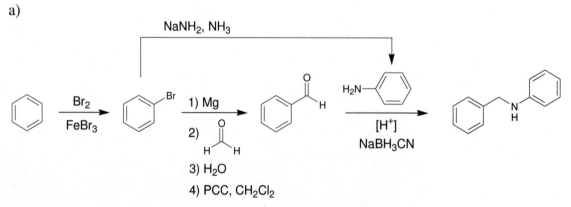

b)

Benzene → (1) HNO_3, H_2SO_4; 2) Fe, H_3O^+) → aniline (NH_2) → (xs Cl_2) → 2,4,6-trichloroaniline (NH_2, Cl, Cl, Cl) → (1) $NaNO_2$, HCl; 2) CuCN; 3) H_3O^+) → 2,4,6-trichlorobenzamide (O, NH_2, Cl, Cl, Cl)

c)

Benzene → (Br_2, $FeBr_3$) → bromobenzene (Br) → (1) Mg; 2) CO_2; 3) H_3O^+; 4) $SOCl_2$) → benzoyl chloride (O, Cl) → (H_2N–aniline) → N-phenylbenzamide (O, N, H)

Bromobenzene → ($NaNH_2$, NH_3) → aniline (H_2N–)

d)

Benzene → (1) propanoyl chloride (O, Cl), $AlCl_3$; 2) HNO_3, H_2SO_4; 3) HCl, Zn[Hg], heat) → 3-propylaniline (NH_2) → ($NaNO_2$, HCl) → 3-propylbenzenediazonium chloride (N_2 ⊕, Cl ⊖)

Benzene → (1) Cl_2, $AlCl_3$; 2) $NaNH_2$, NH_3) → aniline (–NH_2)

Diazonium salt + aniline → 4-[(3-propylphenyl)diazenyl]aniline (N=N, NH_2)

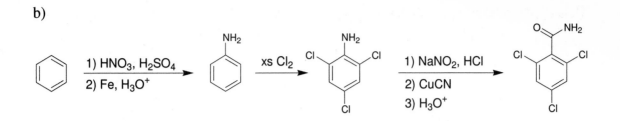

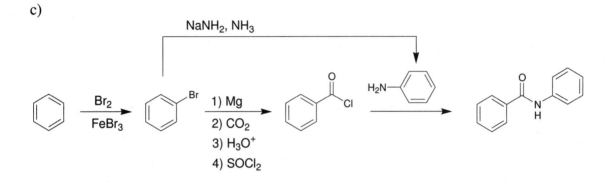

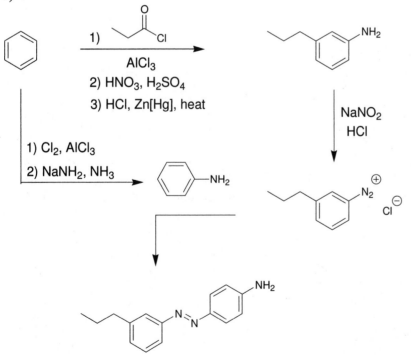

23.74. The IR data indicates that we are looking for structures that lack an N-H bond (i.e tertiary amines):

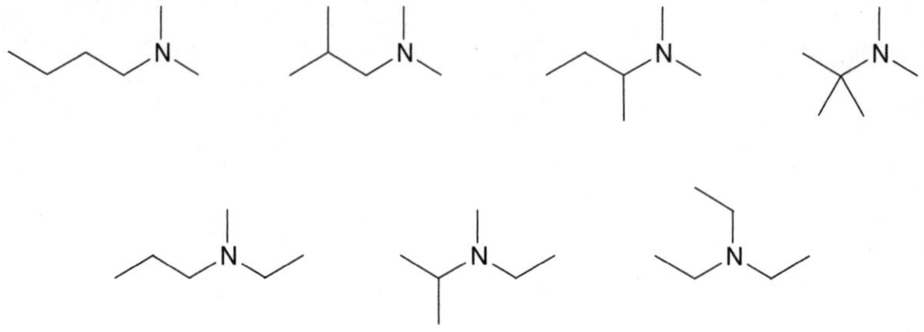

23.75.

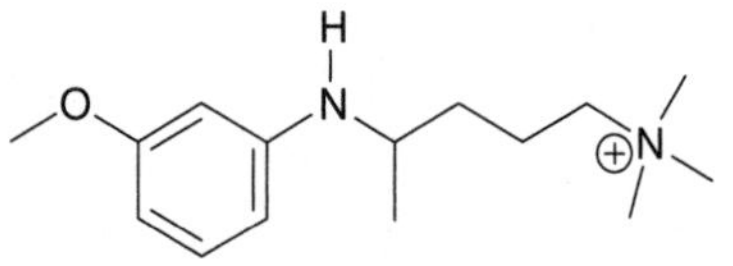

23.76.

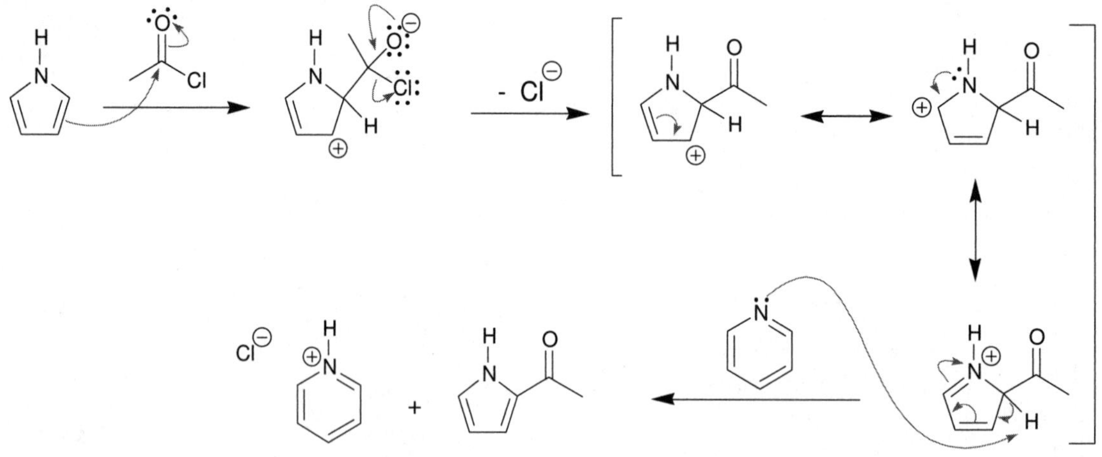

- $Cl^{\ominus}$

23.77. The compound is a tertiary amine with the appropriate symmetry that provides for only three signals:

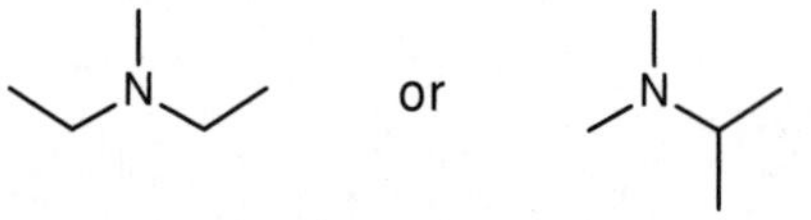

or

23.78.

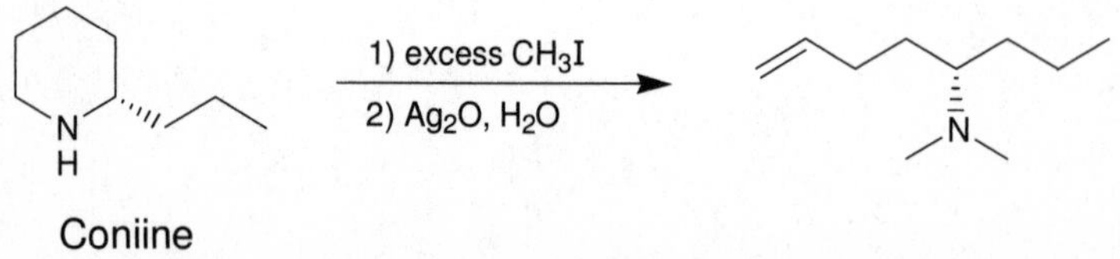

1) excess CH_3I

2) Ag_2O, H_2O

Coniine

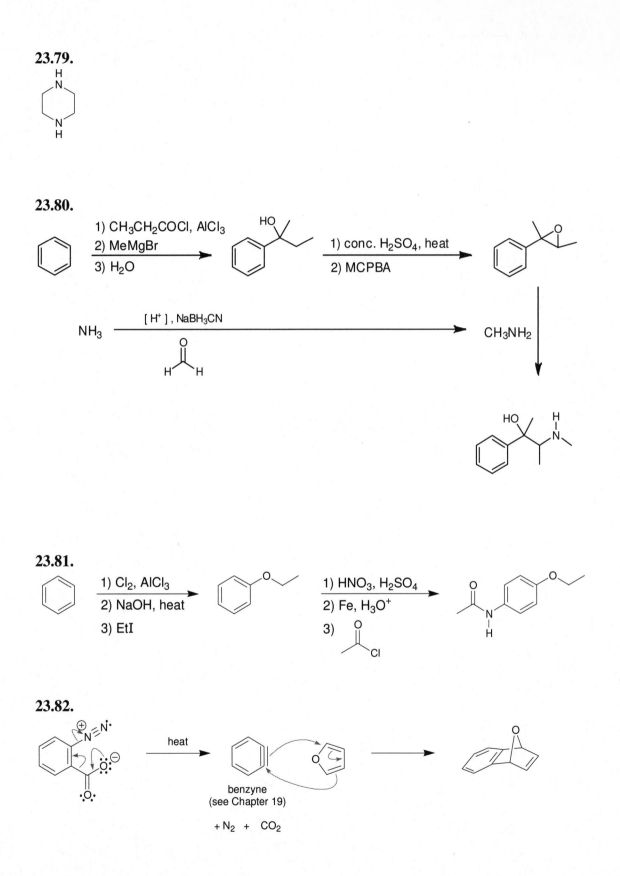

23.79.

23.80.

1) CH_3CH_2COCl, $AlCl_3$
2) MeMgBr
3) H_2O

1) conc. H_2SO_4, heat
2) MCPBA

NH_3 —[H^+], $NaBH_3CN$, $H_2C{=}O$→ CH_3NH_2

23.81.

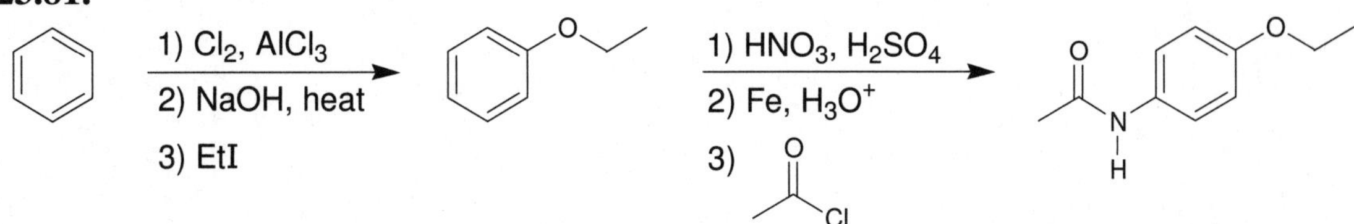

23.82.

heat

benzyne
(see Chapter 19)

+ N_2 + CO_2

23.83.

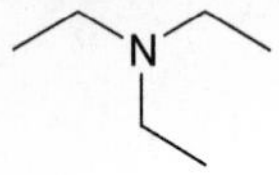

23.84.

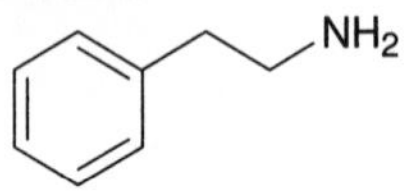

NH_2

23.85.

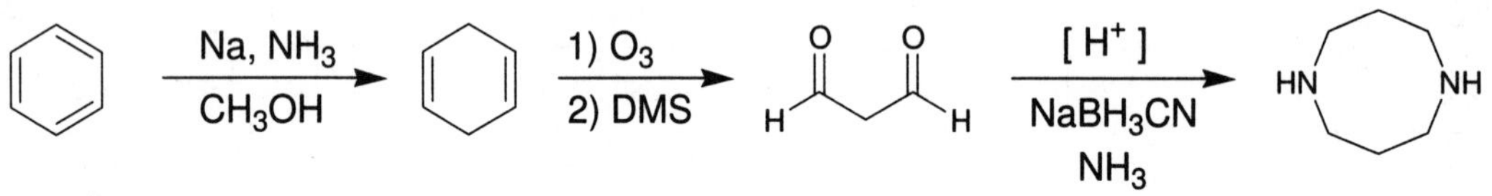

Na, NH_3 / CH_3OH → 1) O_3 2) DMS → [H^+] / $NaBH_3CN$ / NH_3 →

HN NH

23.86.

1) Br_2, *hv* 2) NaOEt 3) O_3 4) DMS → [H^+] / $NaBH_3CN$ / CH_3NH_2 →

N

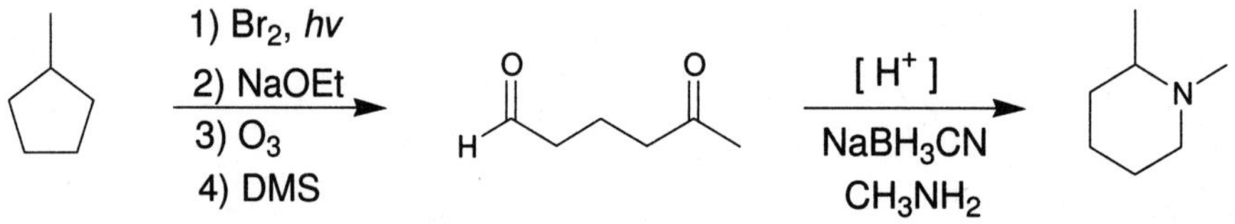

23.87.

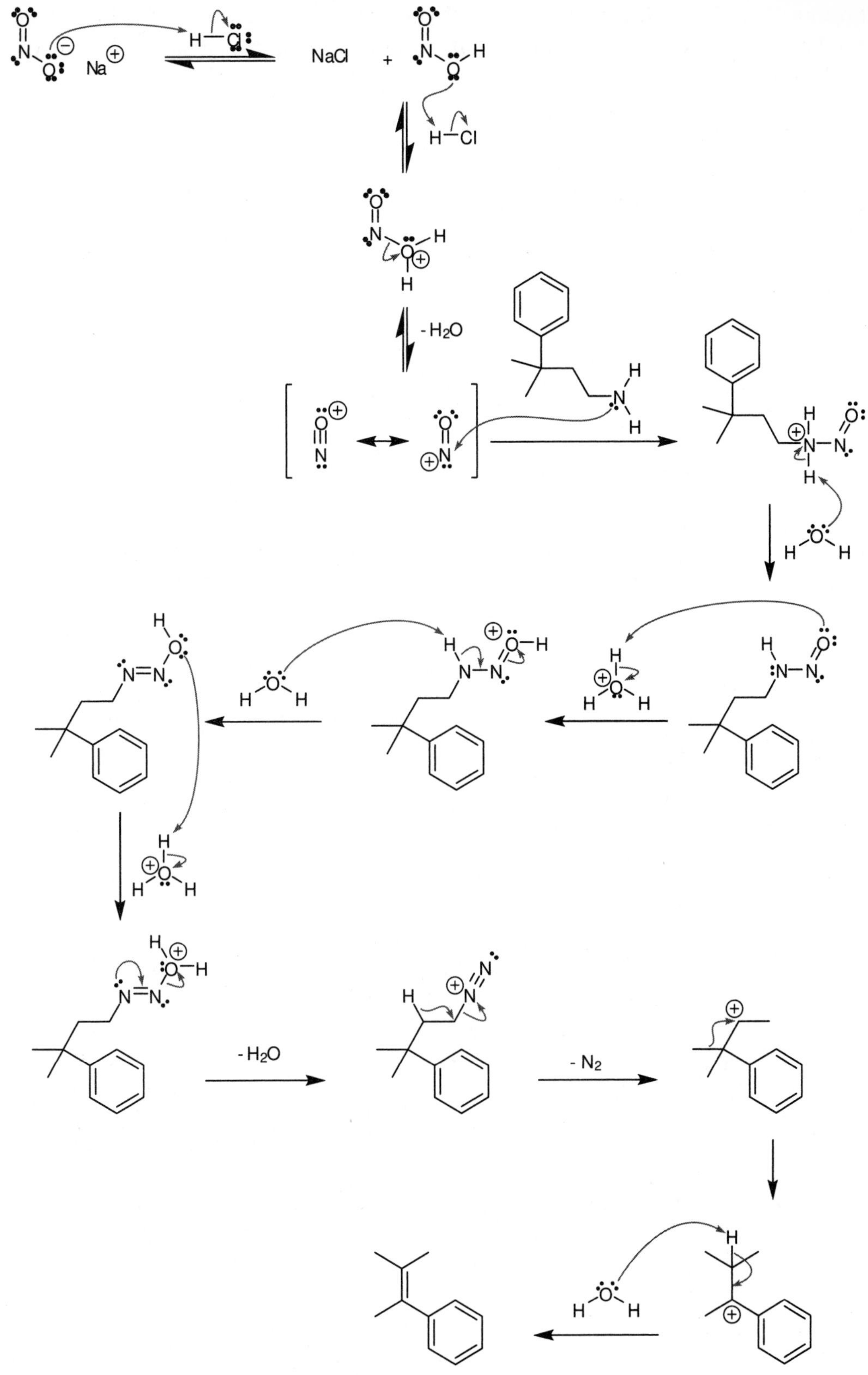

NaCl +

- H_2O

- H_2O

- N_2

23.88. Protonation of the nitrogen highlighted below results in a cation that is highly resonance stabilized. Protonation of either of the other nitrogen atoms would not result in a resonance stabilized cation:

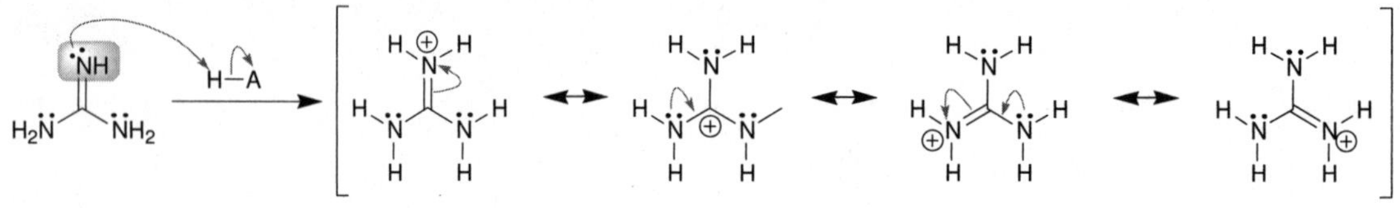

Chapter 24
Carbohydrates

Review of Concepts

Fill in the blanks below. To verify that your answers are correct, look in your textbook at the end of Chapter 24. Each of the sentences below appears verbatim in the section entitled *Review of Concepts and Vocabulary.*

- **Carbohydrates** are polyhydroxy ___________ or ketones.
- Simple sugars are called _______________ and are generally classified as **aldoses** and _________.
- For all **D sugars**, the chirality center farthest from the carbonyl group has the ___ configuration.
- Aldohexoses can form cyclic hemi______ that exhibit a **pyranose** ring.
- Cyclization produces two stereoisomeric hemiacetals, called ___________. The newly created chirality center is called the ___________ **carbon.**
- In the **α anomer**, the hydroxyl group at the anomeric position is ______ to the CH_2OH group, while in the **β anomer**, the hydroxyl group is ______ to the CH_2OH group.
- Anomers equilibrate by a process called _____________, which is catalyzed by either ______ or _______.
- Some carbohydrates, such as D-fructose, can also form five-membered rings, called ___________ rings.
- Monosaccharides are converted into their ester derivatives when treated with excess ______________________________.
- Monosaccharides are converted into their ether derivatives when treated with excess ___________________ and silver oxide.
- When treated with an alcohol under acid-catalyzed conditions, monosaccharides are converted into acetals, called ___________. Both anomers are formed.
- Upon treatment with sodium borohydride an aldose or ketose can be reduced to yield an ___________.
- When treated with a suitable oxidizing agent, an aldose can be oxidized to yield an __________.
- When treated with HNO_3 , an aldose is oxidized to give a dicarboxylic acid called an _______________.
- D-Glucose and D-mannose are **epimers** and are interconverted under strongly __________ conditions.
- The **Kiliani-Fischer synthesis** can be used to lengthen the chain of an ________.
- The **Wohl degradation** can be used to shorten the chain of an _________.
- ________________ are comprised of two monosaccharide units, joined together via a glycosidic linkage.
- **Polysaccharides** are polymers consisting of repeating monosaccharide units linked by ____________ bonds.
- When treated with an ________ in the presence of an acid catalyst, monosaccharides are converted into their corresponding **N-glycosides**.

Review of Skills

Fill in the blanks and empty boxes below. To verify that your answers are correct, look in your textbook at the end of Chapter 24. The answers appear in the section entitled *SkillBuilder Review.*

24.1 Drawing the Cyclic Hemiacetal of a Hydroxyaldehyde

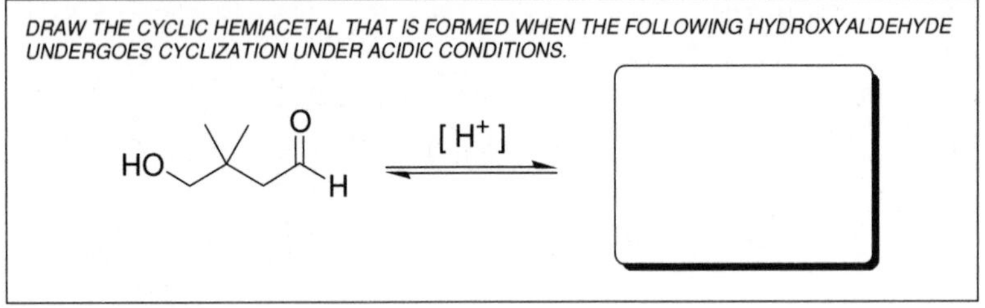

DRAW THE CYCLIC HEMIACETAL THAT IS FORMED WHEN THE FOLLOWING HYDROXYALDEHYDE UNDERGOES CYCLIZATION UNDER ACIDIC CONDITIONS.
HO
O
H
[H+]

24.2: Drawing a Haworth Projection of an Aldohexose

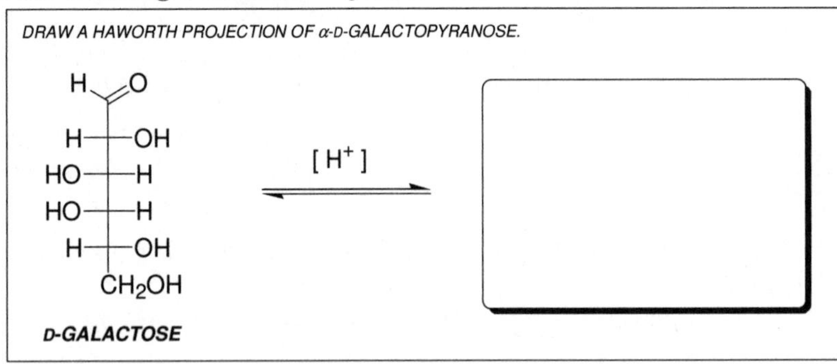

24.3: Drawing the More Stable Chair Conformation of a Pyranose Ring

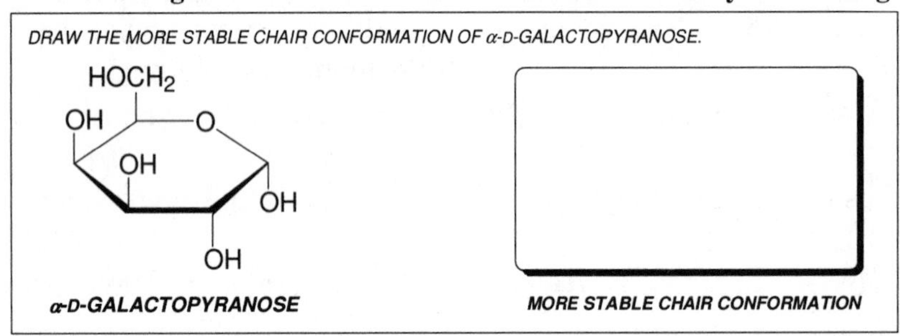

24.4 Identifying a Reducing Sugar

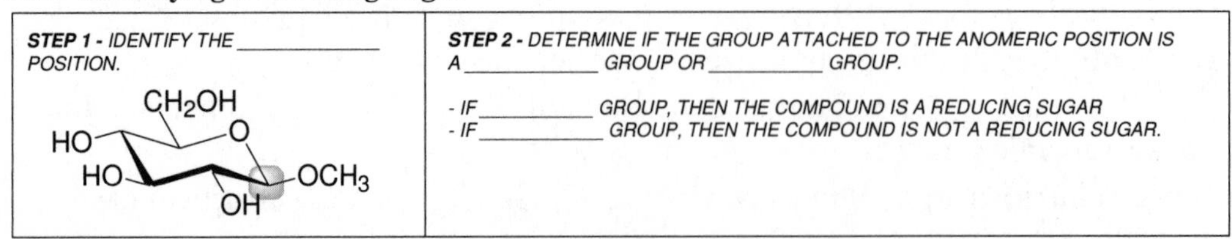

STEP 1 - IDENTIFY THE ______ POSITION.
CH2OH
O
HO
HO
OCH3
OH
STEP 2 - DETERMINE IF THE GROUP ATTACHED TO THE ANOMERIC POSITION IS A ______ GROUP OR ______ GROUP.
- IF ______ GROUP, THEN THE COMPOUND IS A REDUCING SUGAR
- IF ______ GROUP, THEN THE COMPOUND IS NOT A REDUCING SUGAR.

24.5 Determining Whether a Disaccharide Is a Reducing Sugar

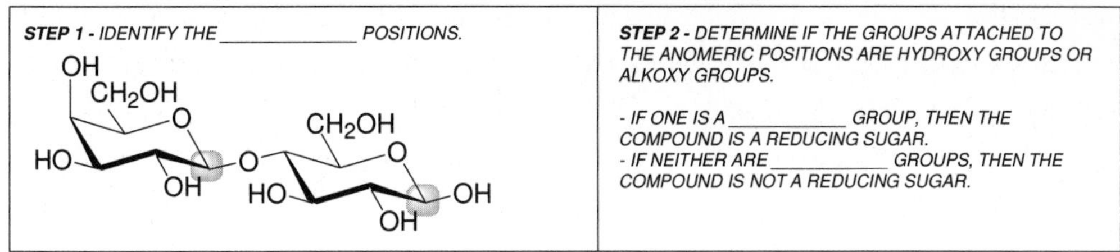

STEP 1 - IDENTIFY THE ____________ POSITIONS.
OH
CH2OH
HO
O
OH
CH2OH
HO
OH
OH
STEP 2 - DETERMINE IF THE GROUPS ATTACHED TO THE ANOMERIC POSITIONS ARE HYDROXY GROUPS OR ALKOXY GROUPS.
- IF ONE IS A ____________ GROUP, THEN THE COMPOUND IS A REDUCING SUGAR.
- IF NEITHER ARE ____________ GROUPS, THEN THE COMPOUND IS NOT A REDUCING SUGAR.

Review of Reactions

Identify the reagents necessary to achieve each of the following transformations. To verify that your answers are correct, look in your textbook at the end of Chapter 24. The answers appear in the section entitled *Review of Reactions*.

Hemiacetal Formation

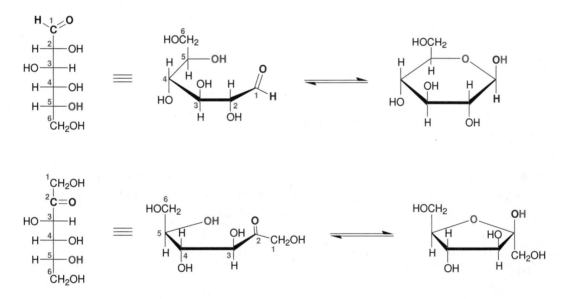

Chain Lengthening and Chain Shortening

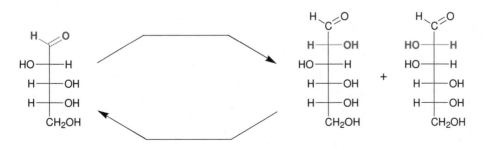

Reactions of Monosaccharides

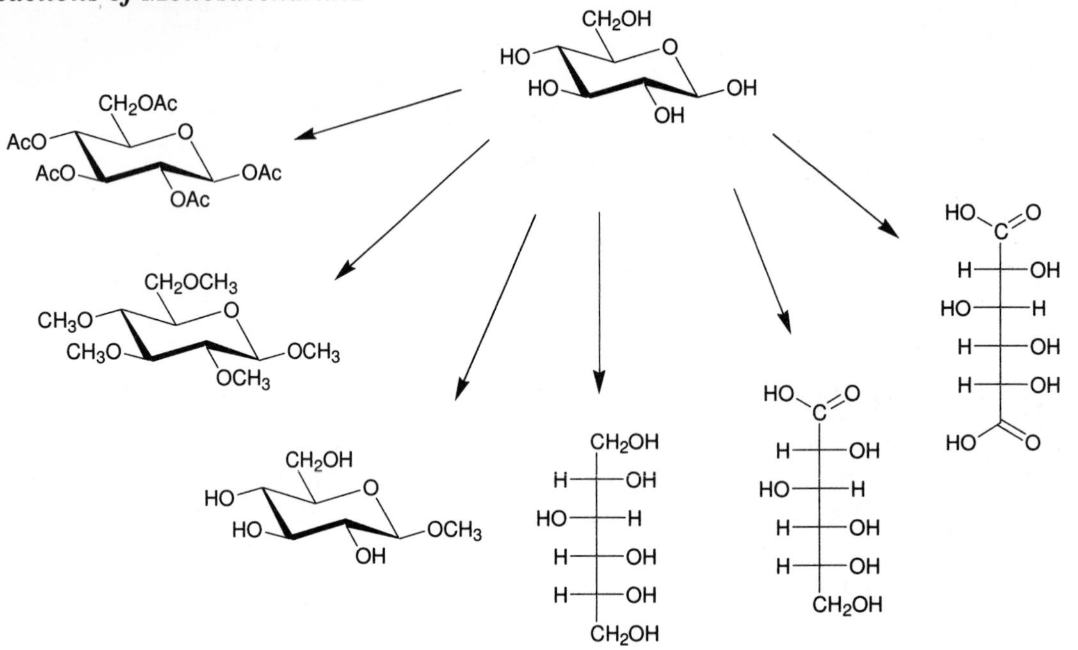

Solutions

24.1.

a) an aldohexose
b) an aldopentose
c) a ketopentose
d) an aldotetrose
e) a ketohexose

24.2. Both are hexoses so both have molecular formula ($C_6H_{12}O_6$). Although they have the same molecular formula, they have different constitution – one is an aldehyde and the other is a ketone. Therefore, they are constitutional isomers.

24.3. All are D sugars except for (b), which is an L sugar.
a) 2S, 3S, 4R, 5R b) 2R, 3S, 4S c) 3R, 4R d) 2S, 3R e) 3S, 4S, 5R
Pay special attention to the following trend: The configuration of each chirality center is *R* when the OH group is on the right side of the Fischer projection, and the configuration is *S* when the OH group is on the left side.

24.4.

a)

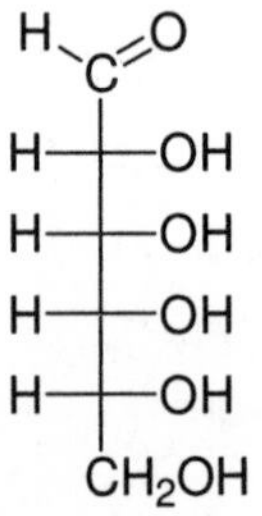

b)

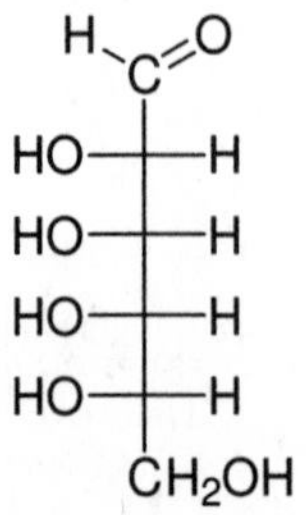

24.5.

CH_2OH / C=O / H–C–OH / CH_2OH (D) and CH_2OH / C=O / HO–C–H / CH_2OH (L)

D L

enantiomers

24.6.

H–C=O / H–C–OH / H–C–OH / CH_2OH (D) and H–C=O / HO–C–H / HO–C–H / CH_2OH (L)

D L

enantiomers

H–C=O / HO–C–H / H–C–OH / CH_2OH (D) and H–C=O / H–C–OH / HO–C–H / CH_2OH (L)

D L

enantiomers

24.7.

CH_2OH / C=O / H–C–OH / HO–C–H / **HO**–C–H / CH_2OH

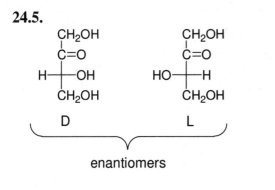

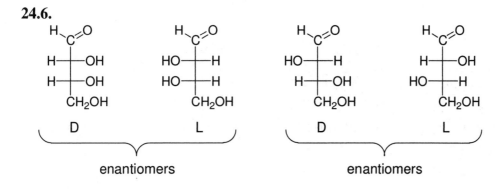

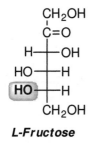

L-Fructose

24.8. D-fructose and D-glucose are constitutional isomers. Both have molecular formula ($C_6H_{12}O_6$). Although they have the same molecular formula, they have different constitution – one is a ketone, and the other is an aldehyde.

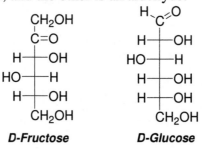

CH_2OH / C=O / H–C–OH / HO–C–H / H–C–OH / CH_2OH

D-Fructose

H–C=O / H–C–OH / HO–C–H / H–C–OH / H–C–OH / CH_2OH

D-Glucose

24.9.

a) 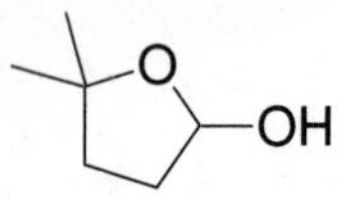 b) 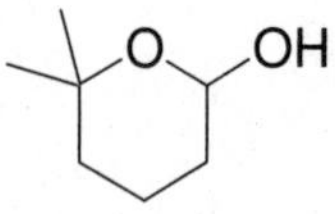 c) 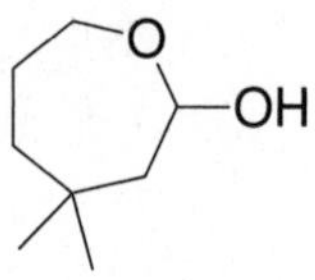 d)

24.10.

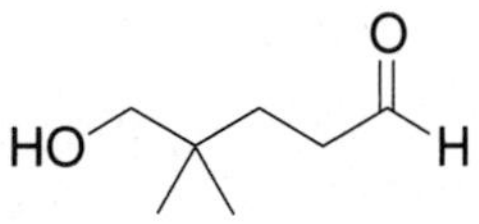

24.11.

a)

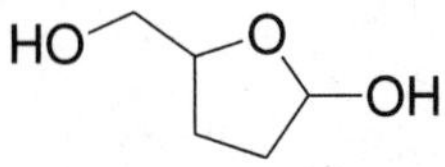

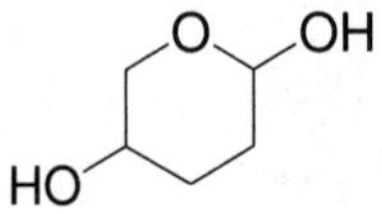

b) The six-membered ring is expected to predominate because it has less ring strain than a five-membered ring.

24.12.

a) 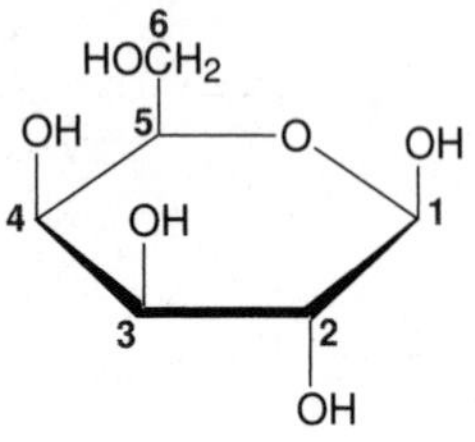 b) 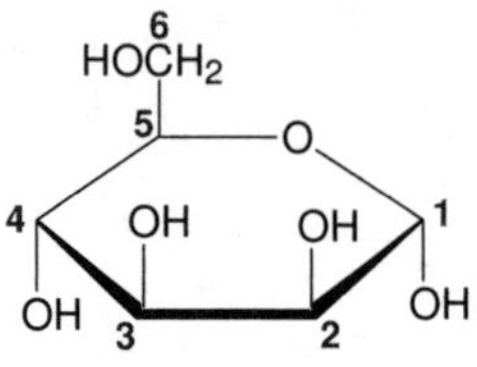 c)

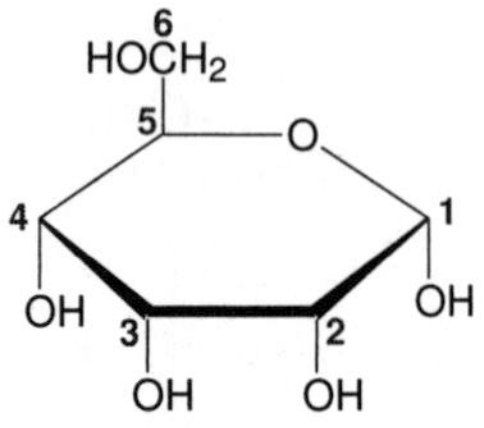

d) 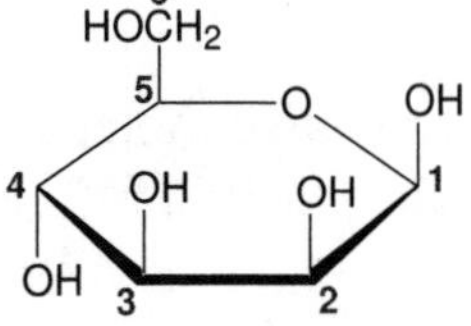 e) 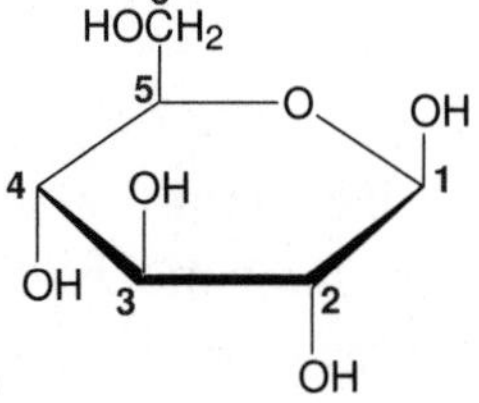 f)

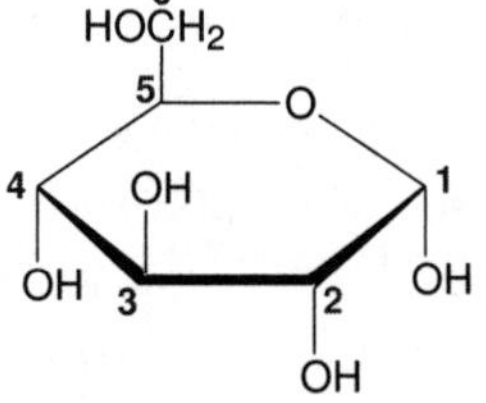

24.13. β-D-galactopyranose

24.14.

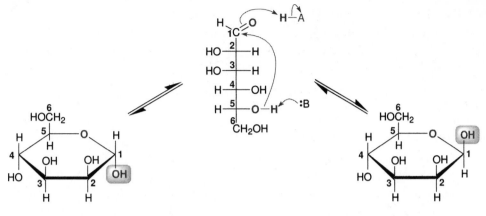

α-D-Mannopyranose

β-D-Mannopyranose

24.15.

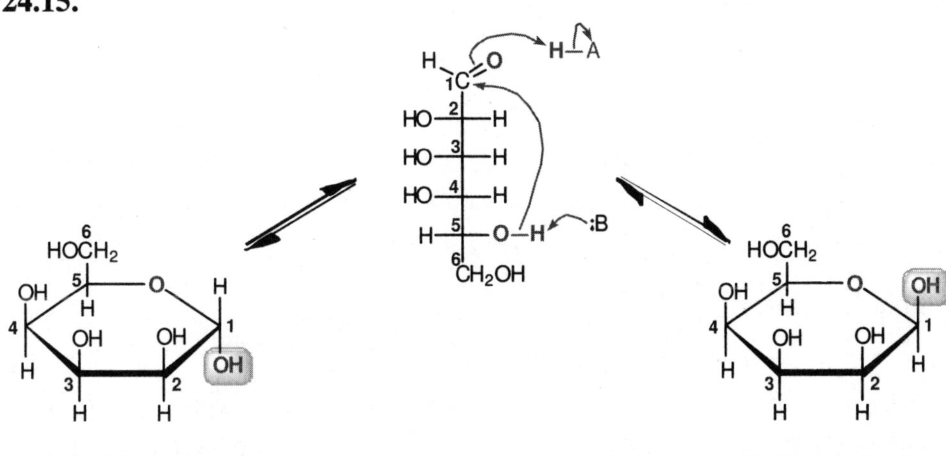

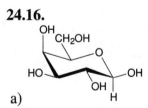

α-D-Talopyranose

β-D-Talopyranose

24.16.

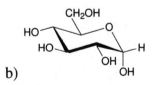

a)

b)

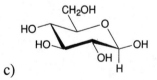

c)

24.17.

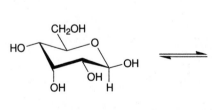

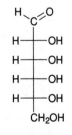

D-Allose

24.18.

ring flip

more stable

less stable

24.19.

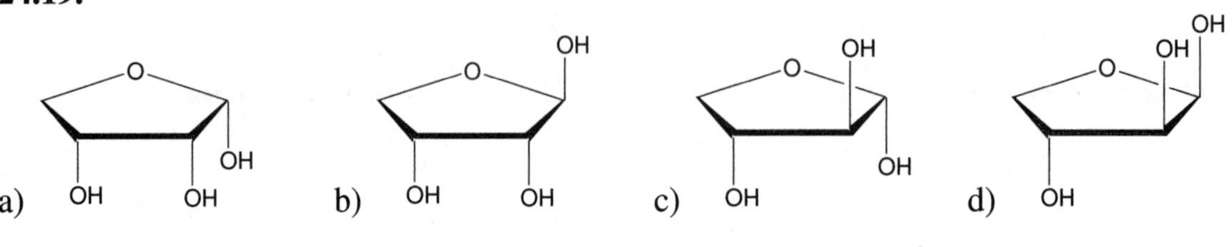

24.20.

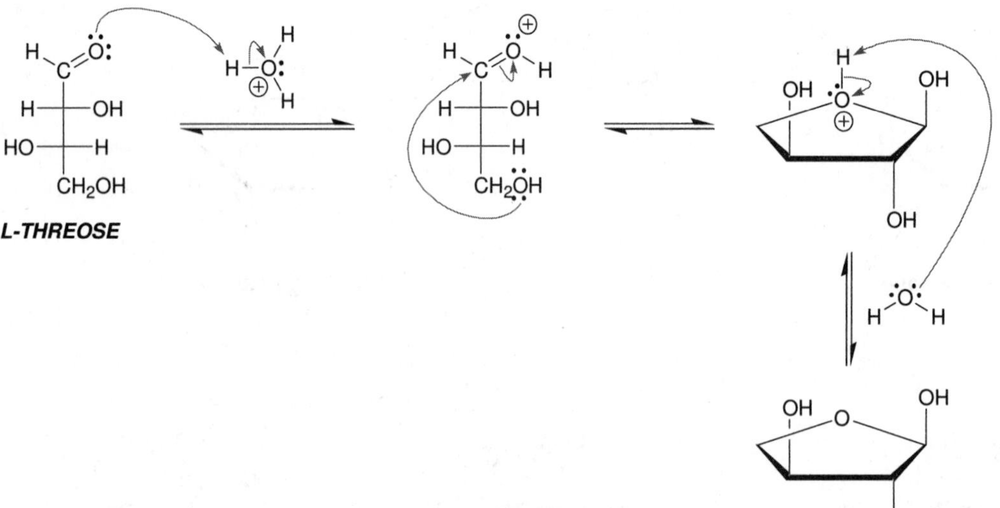

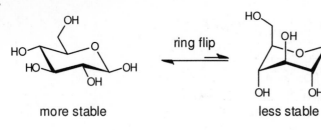

24.21.

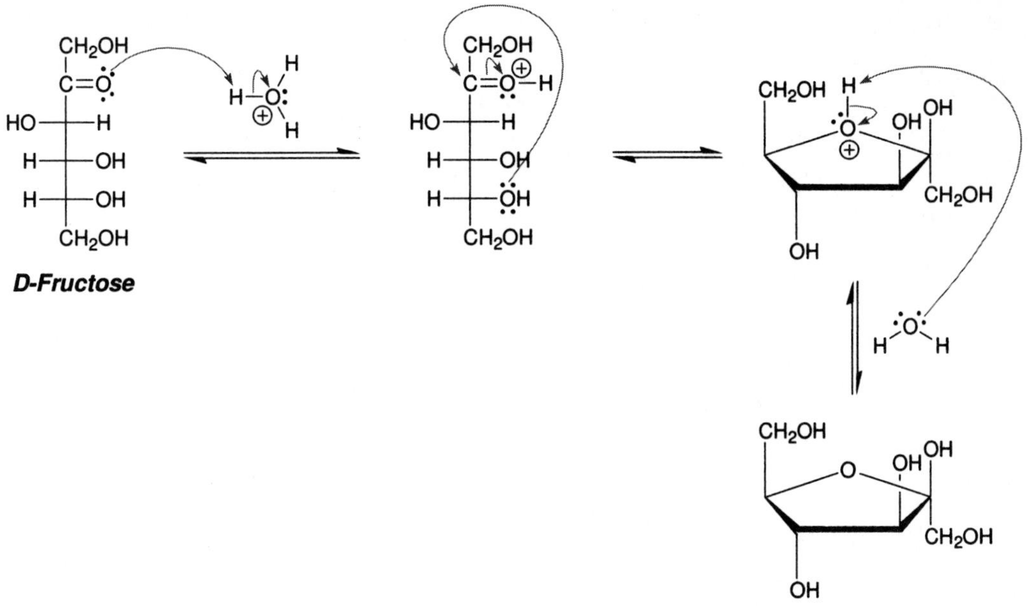

CH₂OH

D-Fructose

24.22.

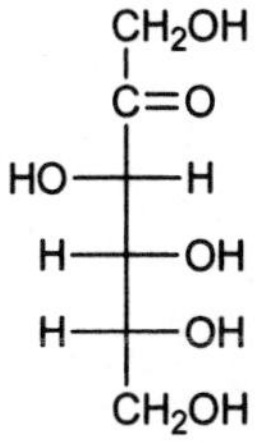

D-Fructose

24.23.

a)

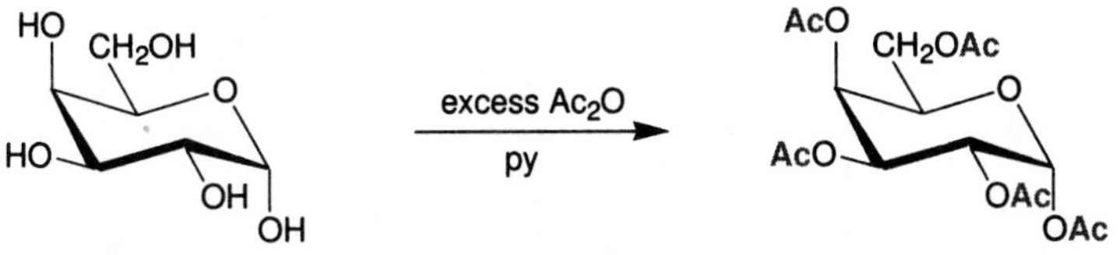

excess Ac_2O

py

b)

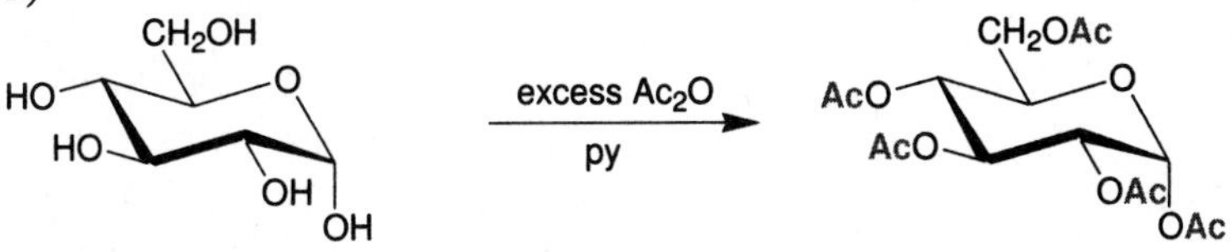

excess Ac_2O

py

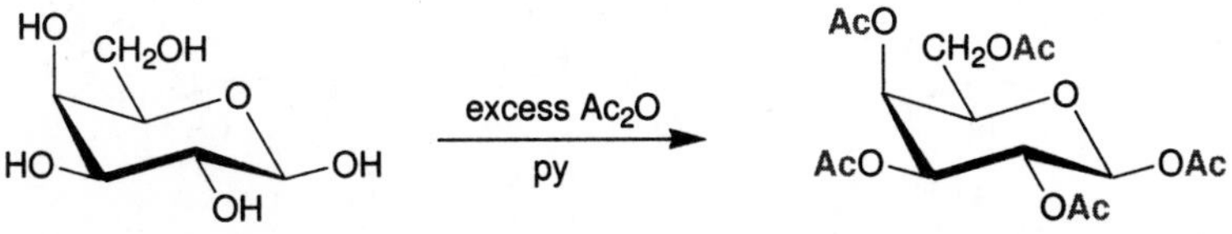

excess Ac_2O

py

c)

24.24.

a)

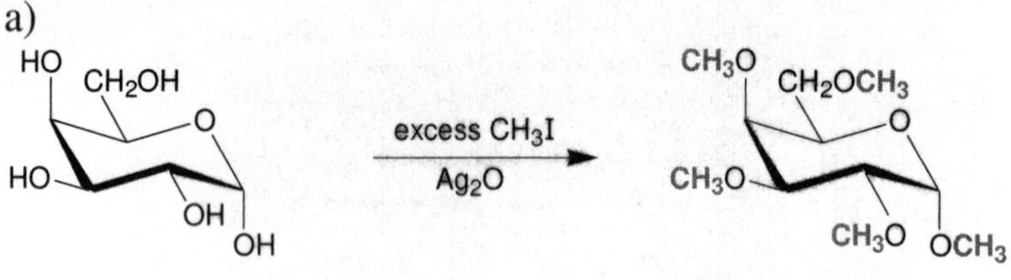

b)

excess CH_3I

Ag_2O

c)

excess CH_3I

Ag_2O

24.25.

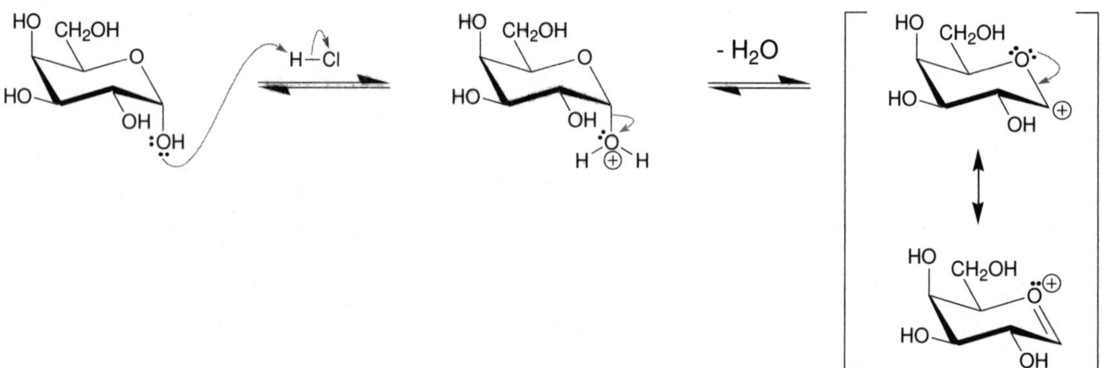

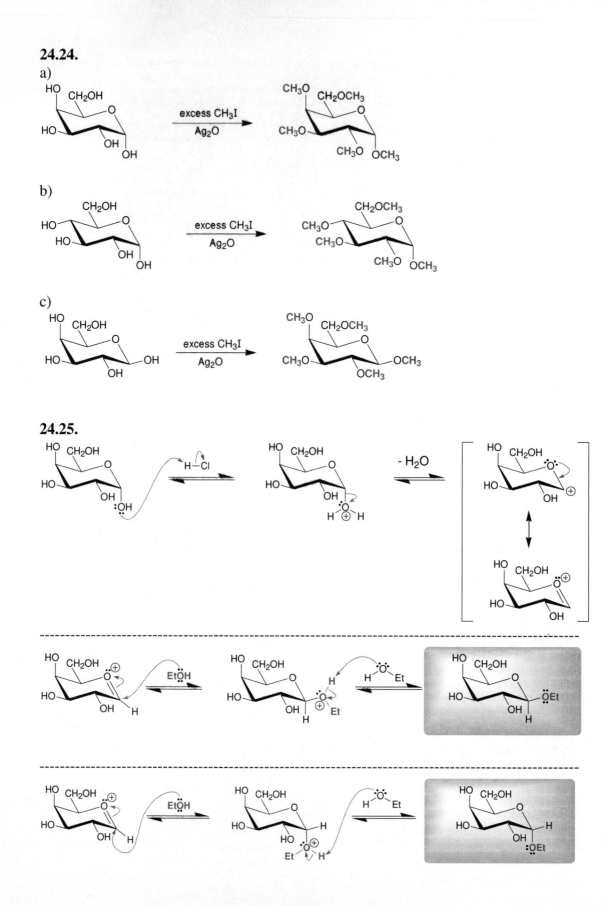

24.26.

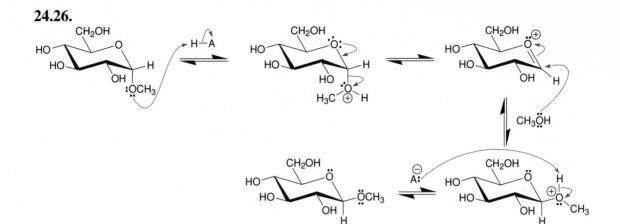

24.27.

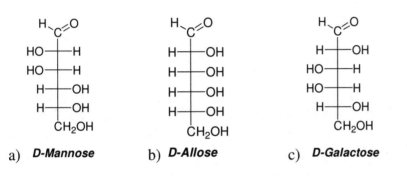

a) *D-Mannose* b) *D-Allose* c) *D-Galactose*

24.28.

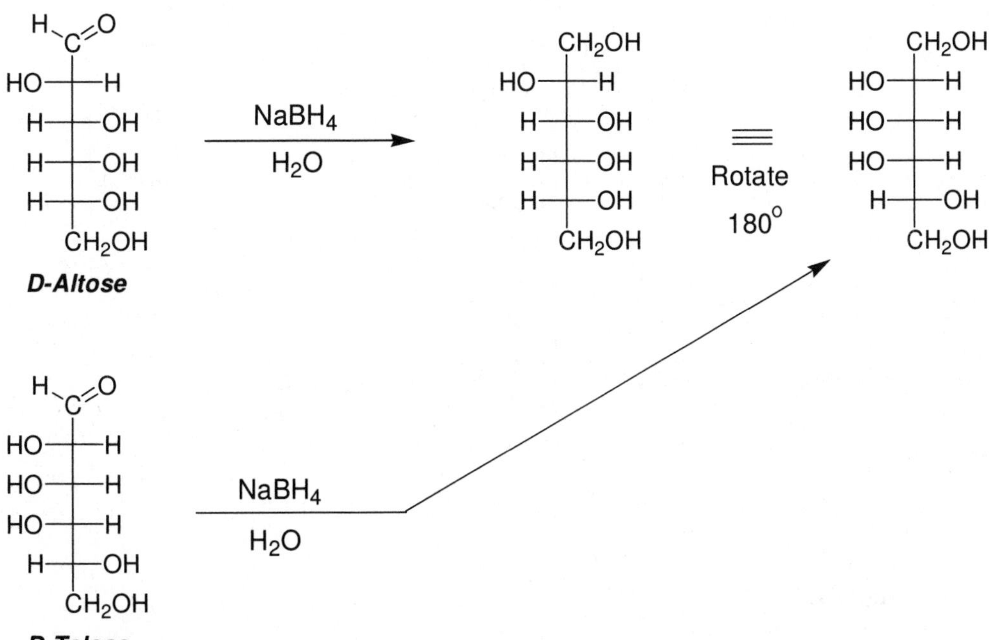

24.29.

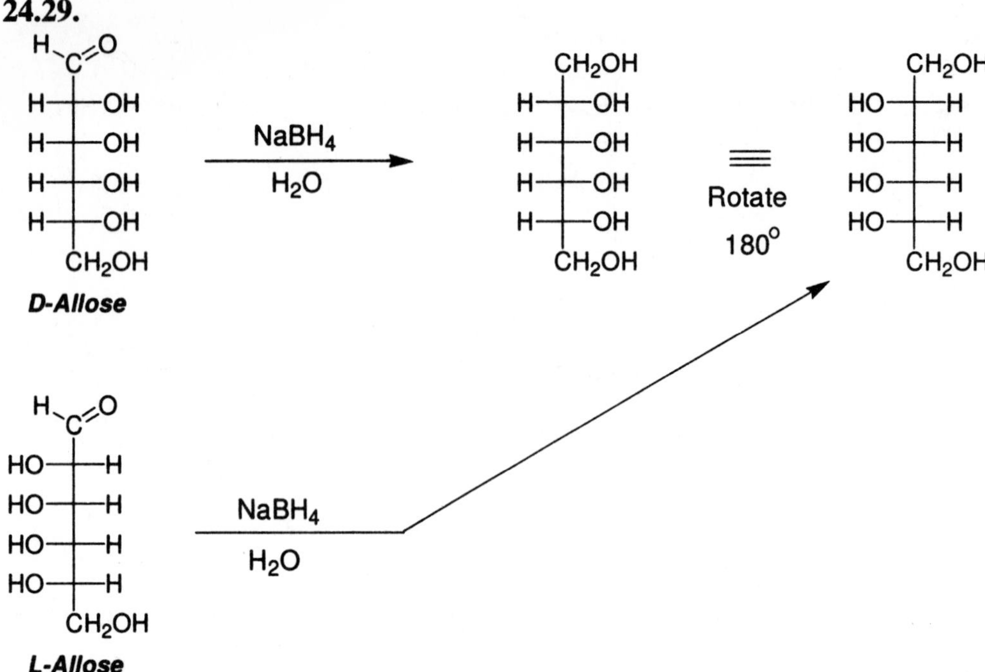

24.30. The following alditols are meso compounds, and are therefore optically inactive:

$\xrightarrow[H_2O]{NaBH_4}$

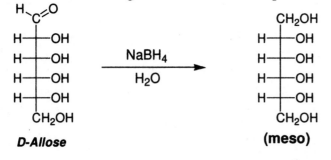

D-Allose **(meso)**

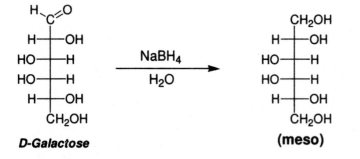

$\xrightarrow[H_2O]{NaBH_4}$

D-Galactose **(meso)**

24.31.

a) No (an acetal) b) Yes c) Yes

24.32.

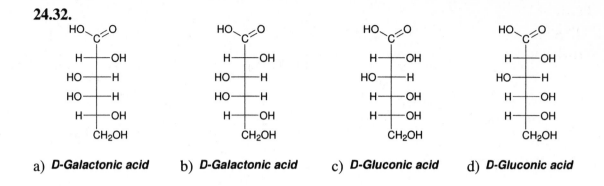

a) ***D-Galactonic acid*** b) ***D-Galactonic acid*** c) ***D-Gluconic acid*** d) ***D-Gluconic acid***

24.33. This compound will not be a reducing sugar because the anomeric position is an acetal group.

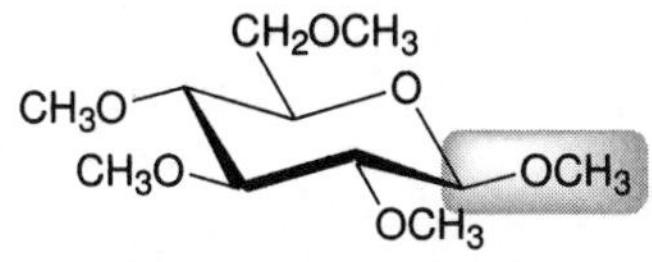

β-D-Glucopyranose pentamethyl ether

24.34.

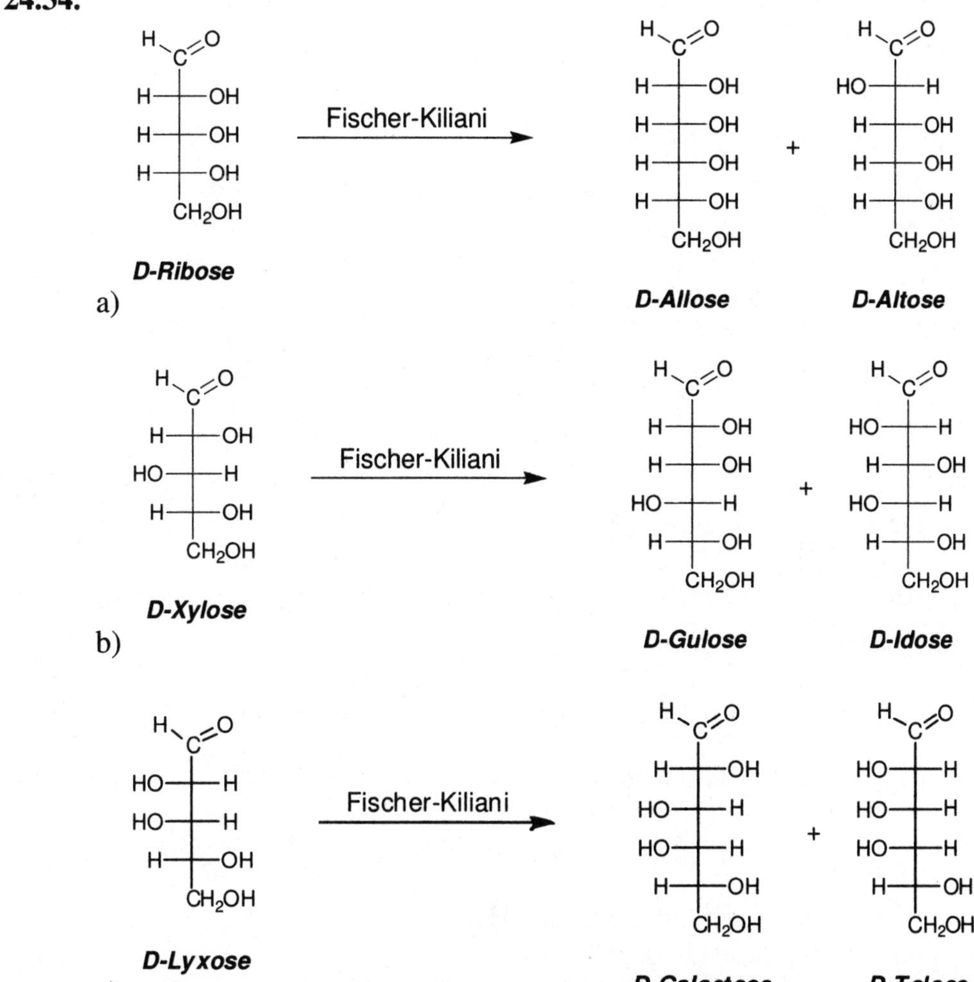

24.35.

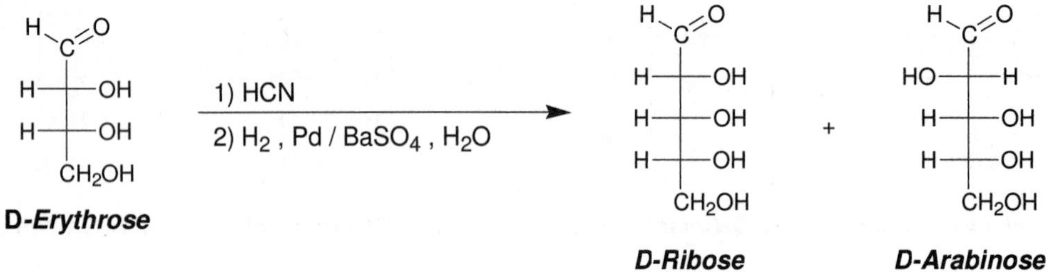

D-Erythrose → 1) HCN; 2) H_2, Pd / $BaSO_4$, H_2O → D-Ribose + D-Arabinose

24.36.

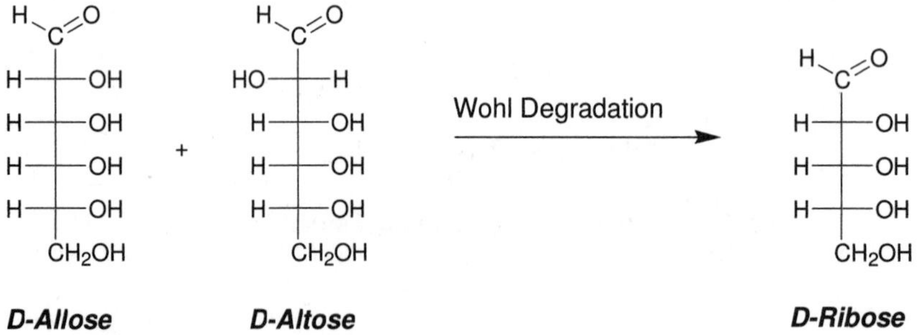

D-Allose + D-Altose → Wohl Degradation → D-Ribose

24.37.

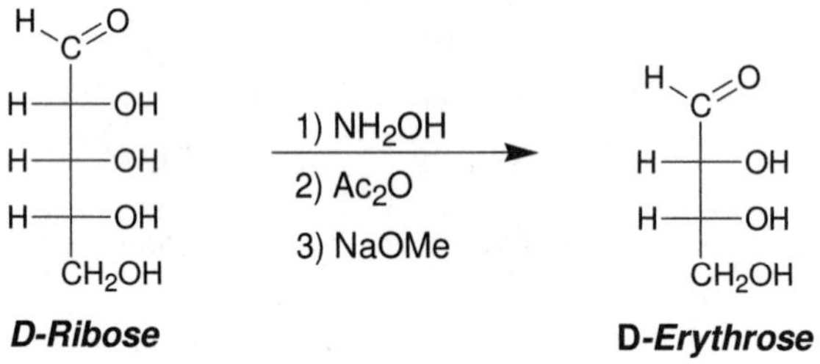

D-Ribose → 1) NH_2OH; 2) Ac_2O; 3) NaOMe → D-Erythrose

24.38.

D-Glucose → Wohl Degradation → D-Arabinose → Fischer-Kiliani → D-Glucose + D-Mannose

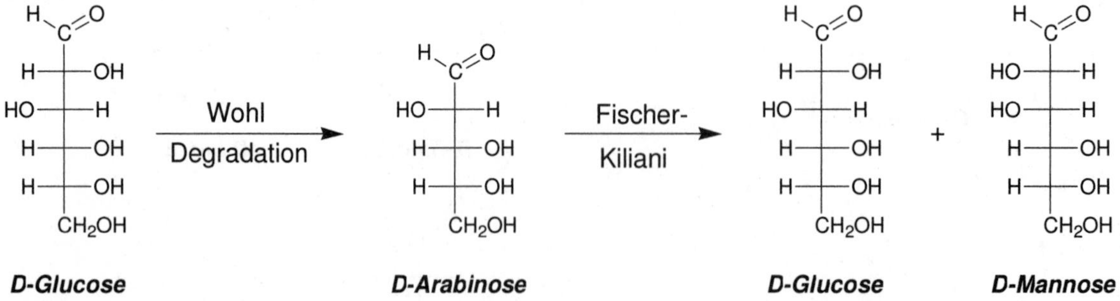

24.39.

a) Yes, one of the anomeric positions bears an OH group.
b) No, both anomeric positions bear acetal groups.
c) No, both anomeric positions bear acetal groups.

24.40.

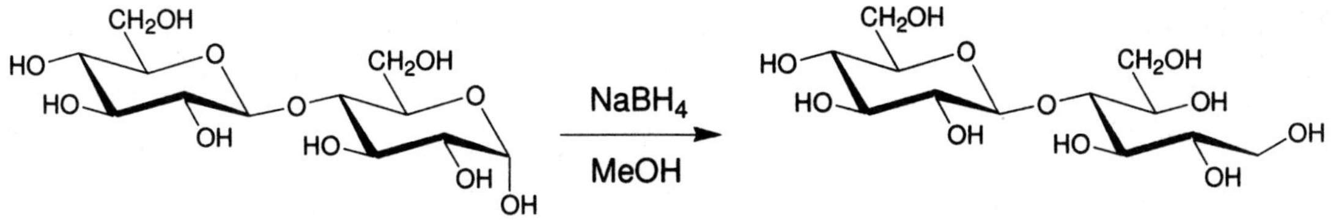

$\xrightarrow[\text{MeOH}]{\text{NaBH}_4}$

24.41.

a)

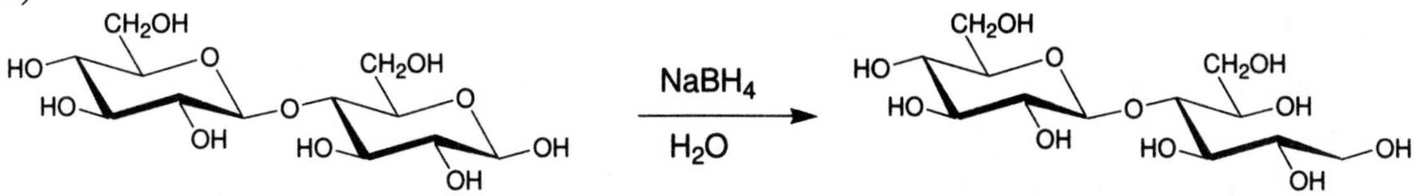

$\xrightarrow[\text{H}_2\text{O}]{\text{NaBH}_4}$

b)

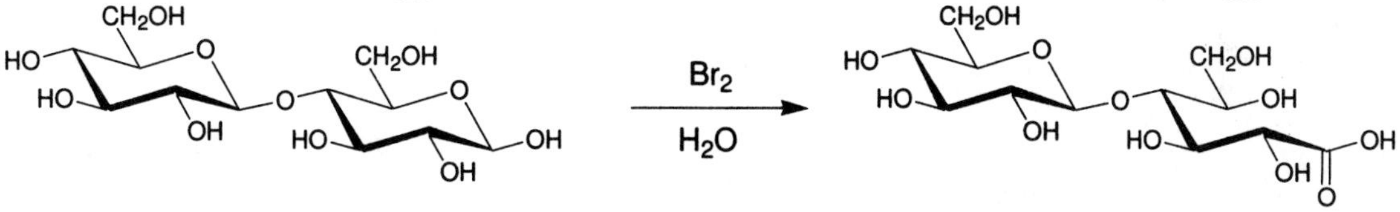

$\xrightarrow[\text{H}_2\text{O}]{\text{Br}_2}$

c)

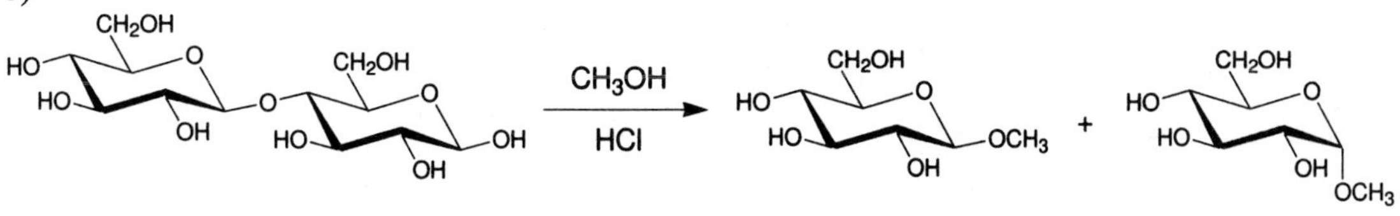

$\xrightarrow[\text{HCl}]{\text{CH}_3\text{OH}}$ +

d)

$\xrightarrow[\text{py}]{\text{Ac}_2\text{O}}$

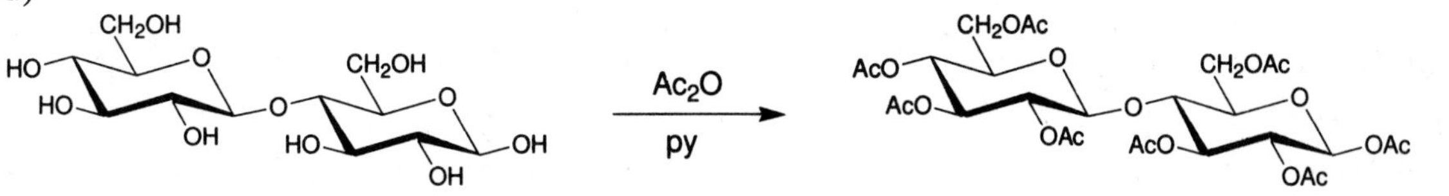

24.42.

a) a D-aldotetrose b) an L-aldopentose c) a D-aldopentose
d) a D-aldohexose e) a D-ketopentose

24.43.

a) D-glyceraldehyde b) L-glyceraldehyde c) D-glyceraldehyde d) L-glyceraldehyde

24.44.

a) D-Glucose b) D-Mannose c) D- Galactose d) L-Glucose

24.45.

a) D-Ribose

b) D-Arabinose

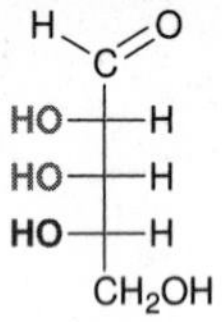

c) ***L-Ribose***

d) Same compound

e) Diastereomers

24.46.

a)

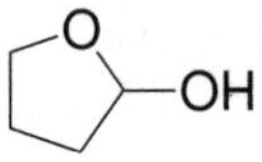

b)

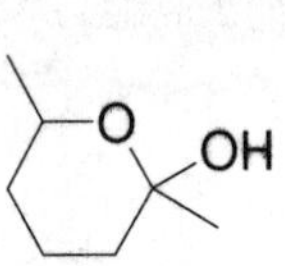

c)

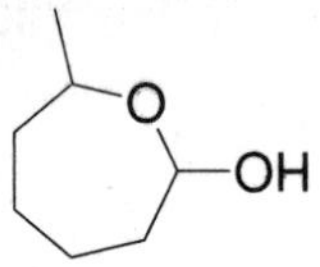

24.47.

24.48.

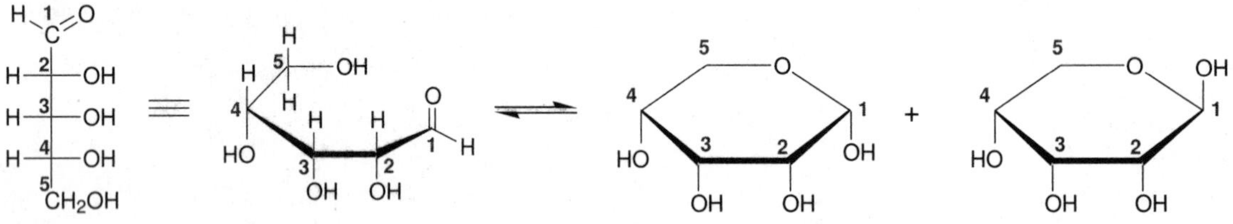

D-ribose

α pyranose ring

β pyranose ring

b)

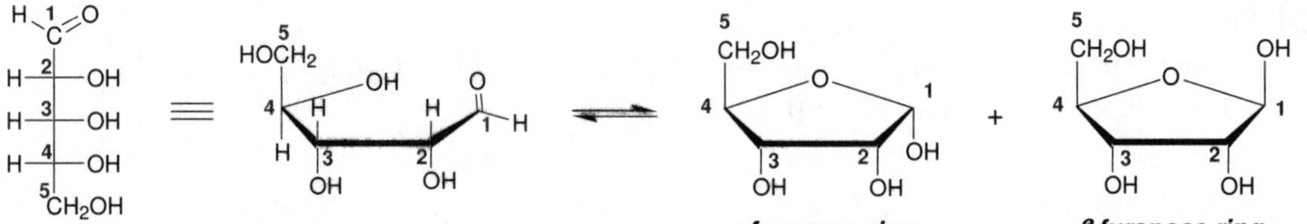

D-ribose

α furanose ring

β furanose ring

24.49.

a) epimers b) diastereomers c) enantiomers d) identical compounds

24.50.

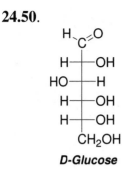

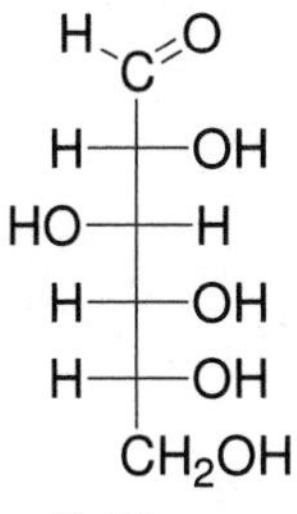

D-Glucose

24.51.

a) 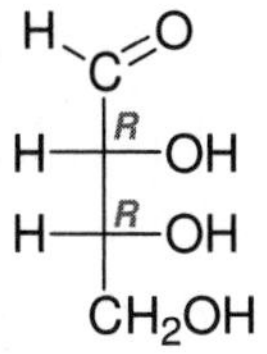 b) 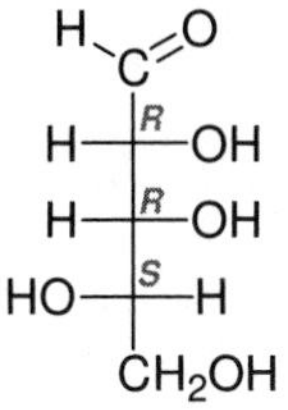 c) 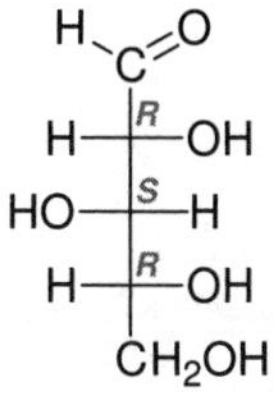 d) 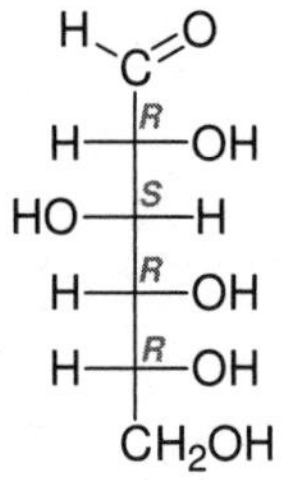 e)

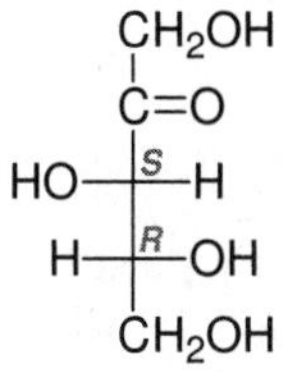

24.52.

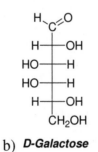

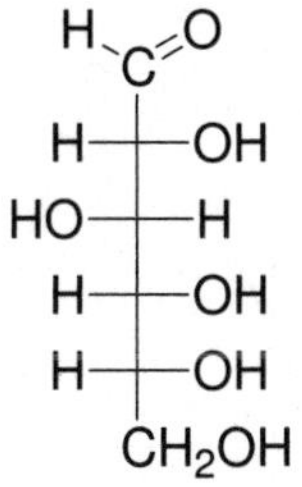

a) *D-Glucose* 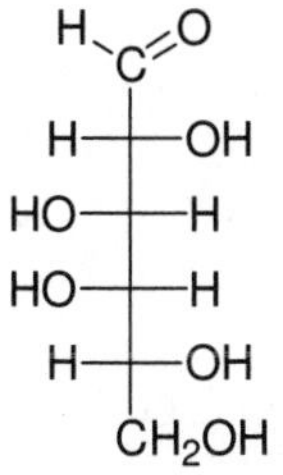 b) *D-Galactose* 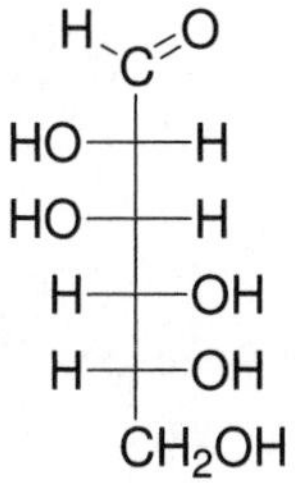 c) *D-Mannose* 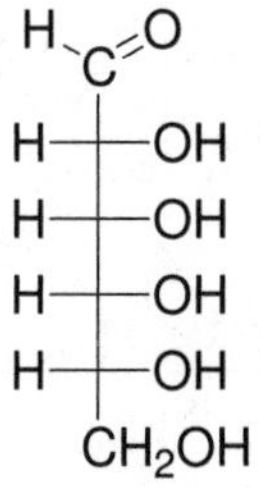 d) *D-Allose*

24.53.

a) 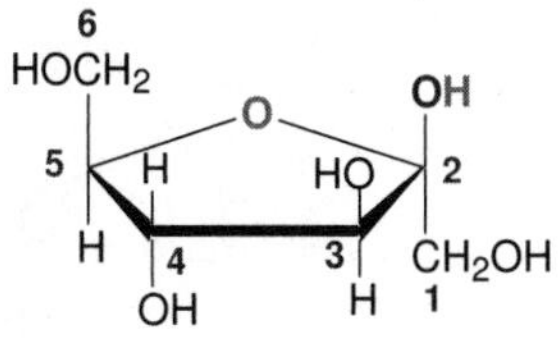 b) 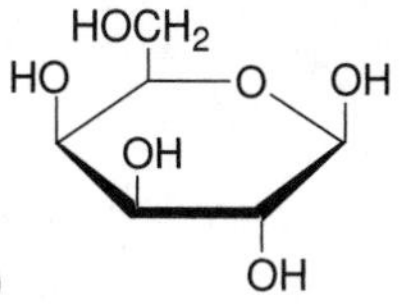 c) 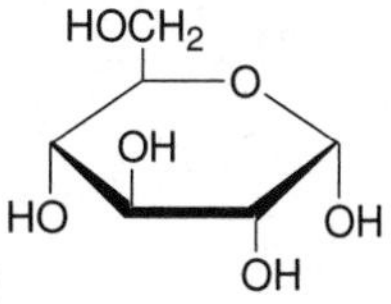 d)

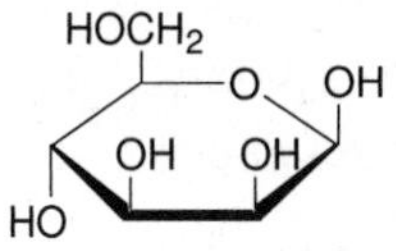

24.54.

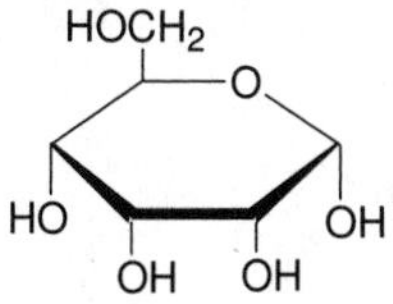

24.55.

a) α-D-allopyranose
b) β-D-galactopyranose
c) methyl β-D-glucopyranoside

24.56.

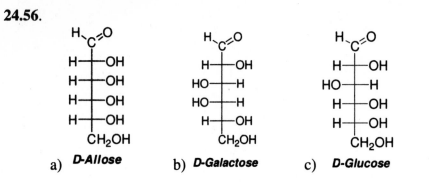

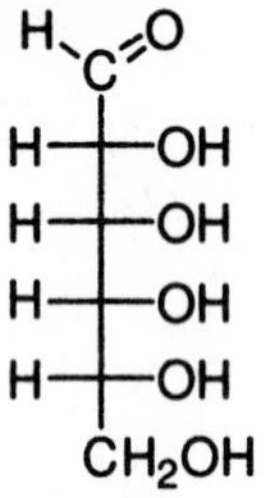

a) *D-Allose* 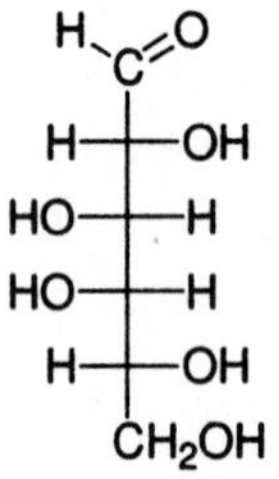 b) *D-Galactose* 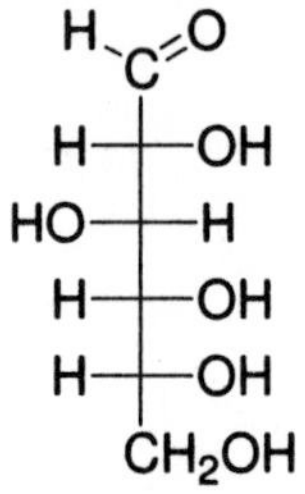 c) *D-Glucose*

24.57.

a)

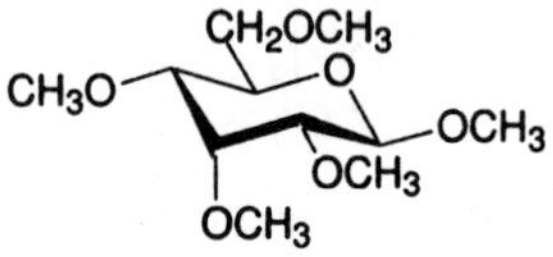

b)

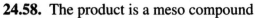

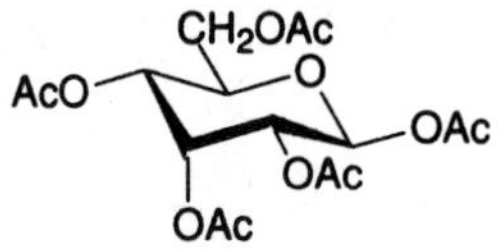

c)

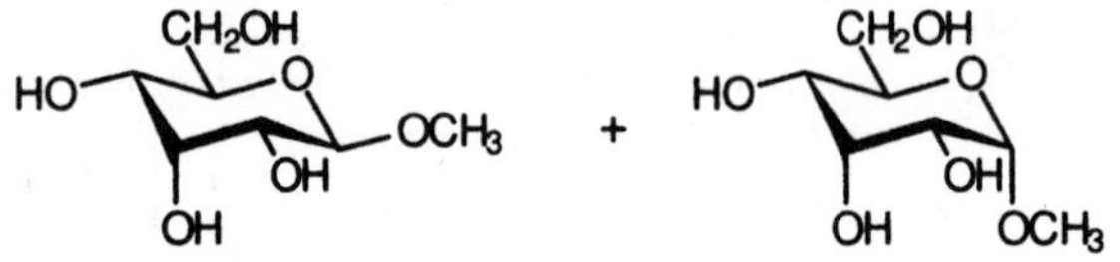

+

24.58. The product is a meso compound

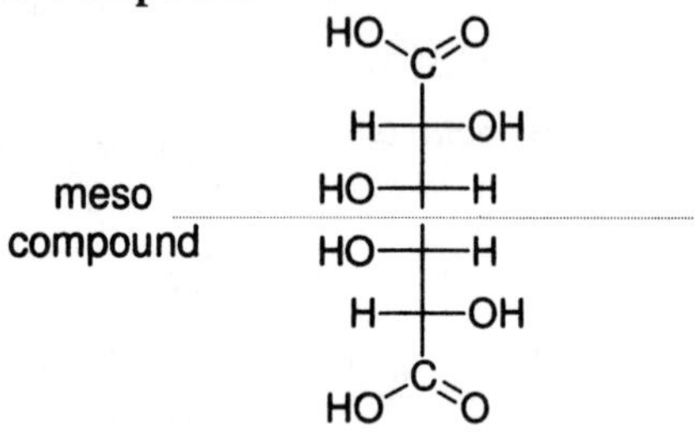

meso compound

24.59.

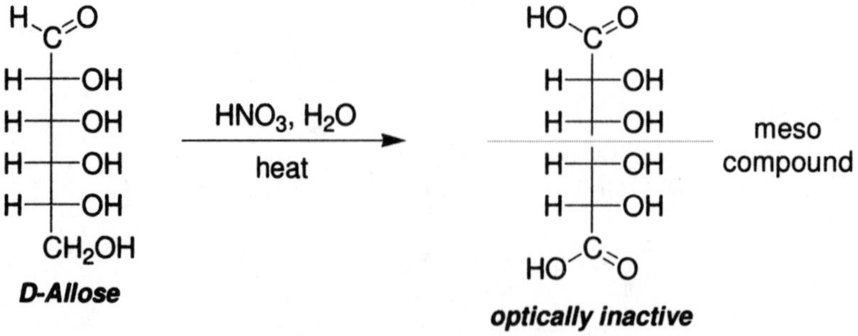

HNO_3, H_2O

heat

D-Allose

meso compound

optically inactive

24.60.

equatorial axial
HO
OH
HO
equatorial
OH OH
axial axial

24.61.

excess CH_3I
Ag_2O

H_3O^+

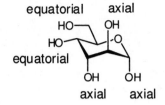

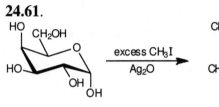

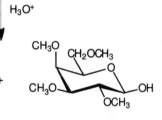

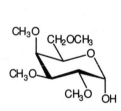

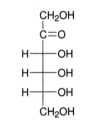

24.62.

a) diastereomers
b) same compound

24.63.

CH_2OH	CH_2OH	CH_2OH	CH_2OH
C=O	C=O	C=O	C=O
H—OH	HO—H	H—OH	HO—H
H—OH	H—OH	HO—H	HO—H
H—OH	H—OH	H—OH	H—OH
CH_2OH	CH_2OH	CH_2OH	CH_2OH

24.64.

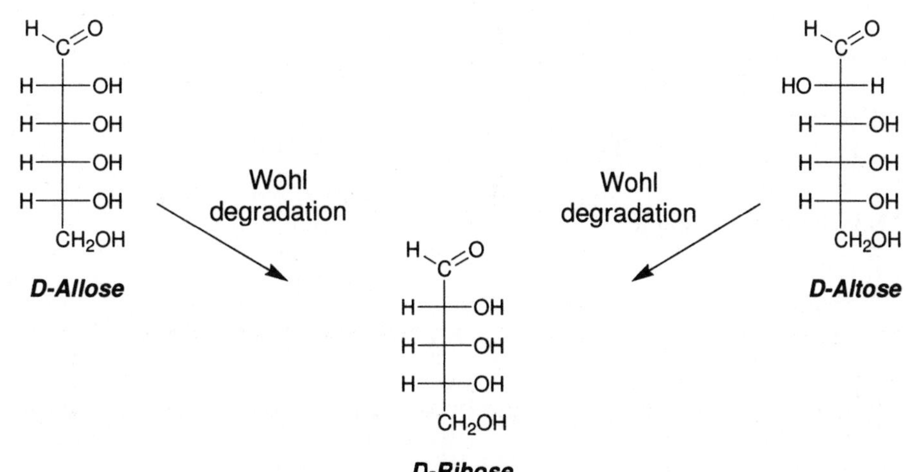

24.65.

Kiliani-Fischer

D-Arabinose → *D-Glucose* + *D-Mannose*

24.66.

HCN

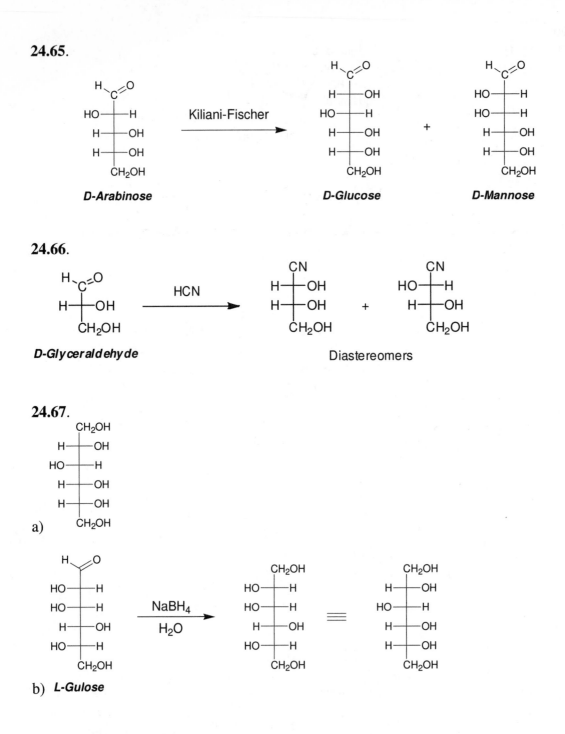

D-Glyceraldehyde →

Diastereomers

24.67.

a)

$NaBH_4$

H_2O

b) *L-Gulose*

24.68. D-Allose and D-Galactose

24.69.

a) This compound will not be a reducing sugar because the anomeric position is an acetal group.

b) This compound will be a reducing sugar because the anomeric position bears an OH group.

24.70.

a) CH_3OH, HCl
b) CH_3OH, HCl
c) HNO_3, H_2O, heat
d) excess CH_3I, Ag_2O followed by H_3O^+

24.71.

a) α-D-glucopyranose and β-D-glucopyranose
b) α-D-galactopyranose and β-D-galactopyranose

24.72.

a) D-Arabinose
b) D-Ribose and D-xylose
c) D-xylose
d) D-xylose

24.73.

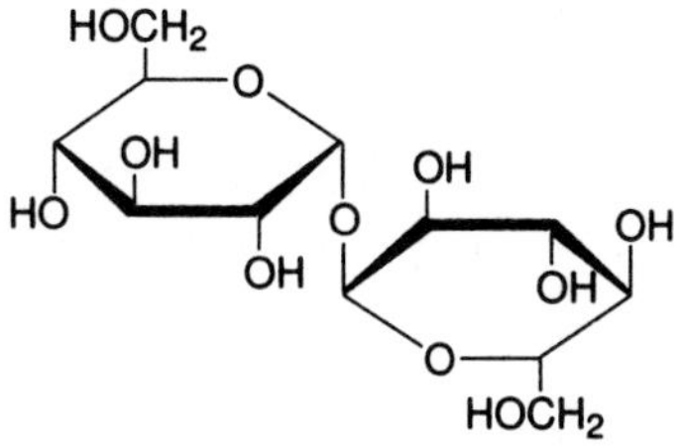

24.74.

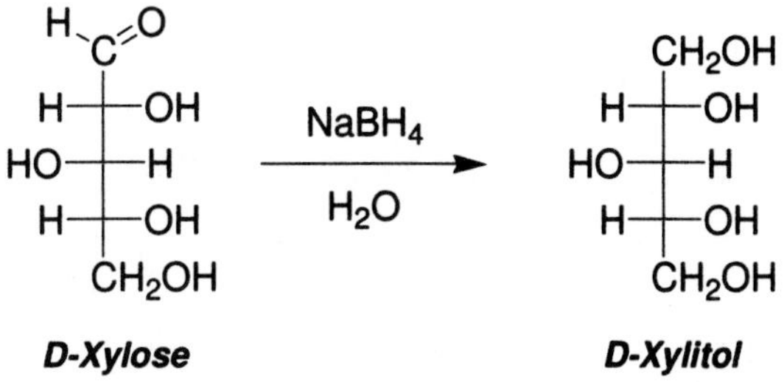

$NaBH_4$, H_2O

D-Xylose ***D-Xylitol***

24.75.

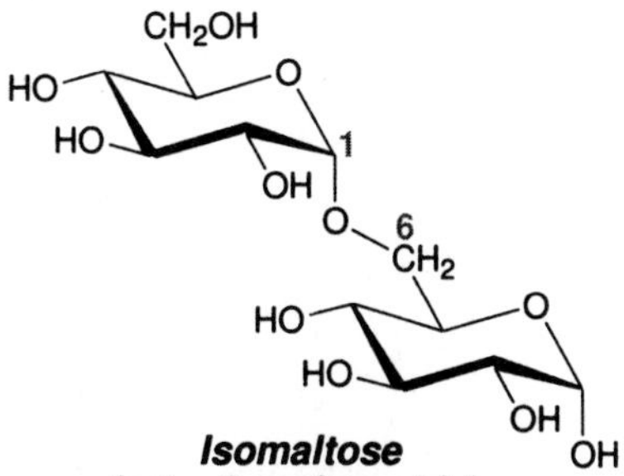

Isomaltose
(a 1→6-α-glycoside)

24.76.

a) No, it is not a reducing sugar because the anomeric position has an acetal group.

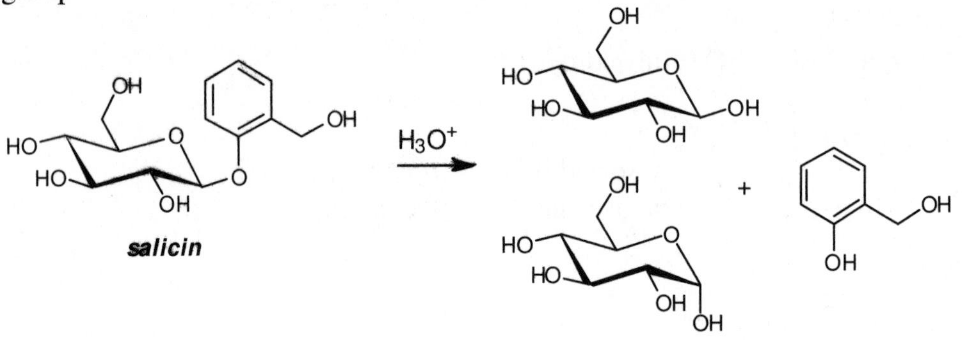

salicin

H_3O^+

b)

c) Salicin is a β-glycoside.

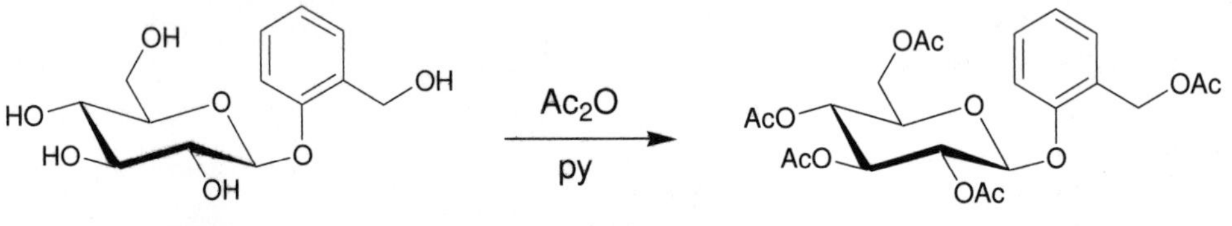

Ac_2O

py

d) salicin

e) No. In the absence of acid catalysis, the acetal group is not readily hydrolyzed.

24.77.

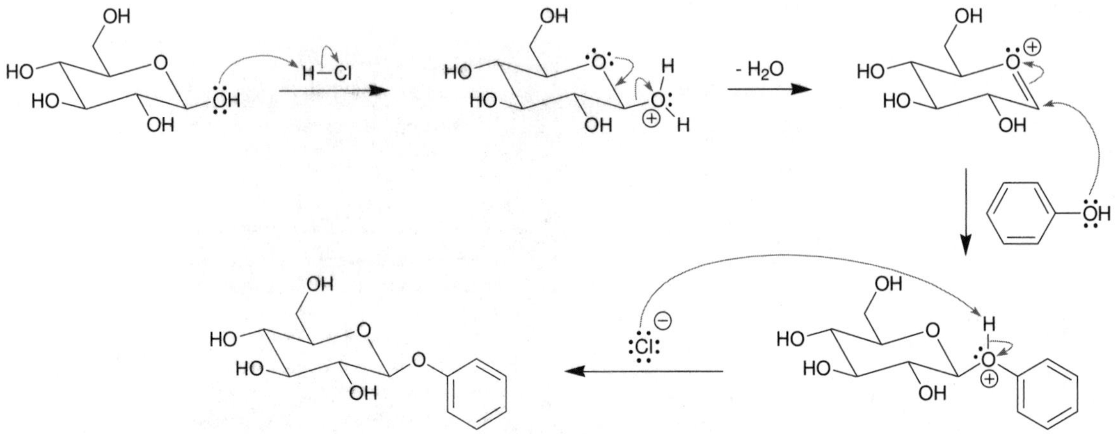

H–Cl

$- H_2O$

24.78.

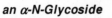

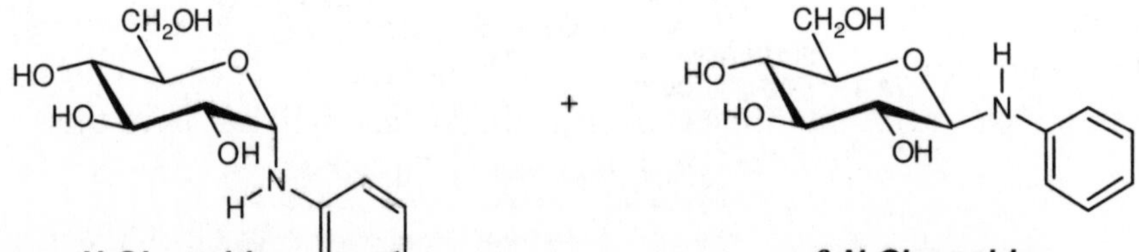

an α-N-Glycoside

a β-N-Glycoside

24.79.

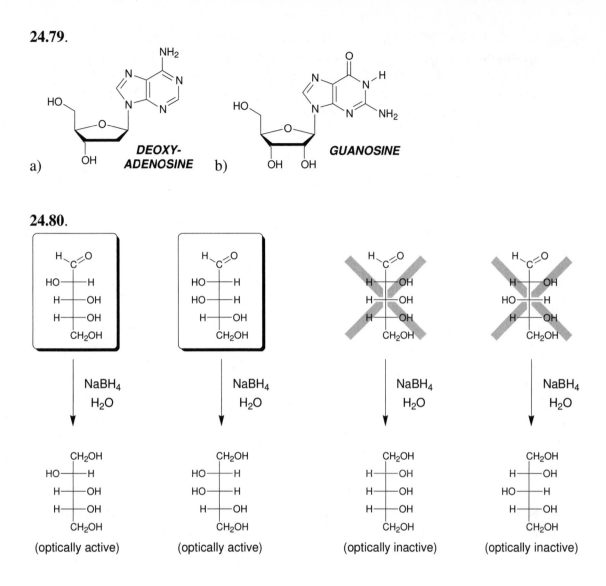

a) DEOXY-ADENOSINE b) GUANOSINE

24.80.

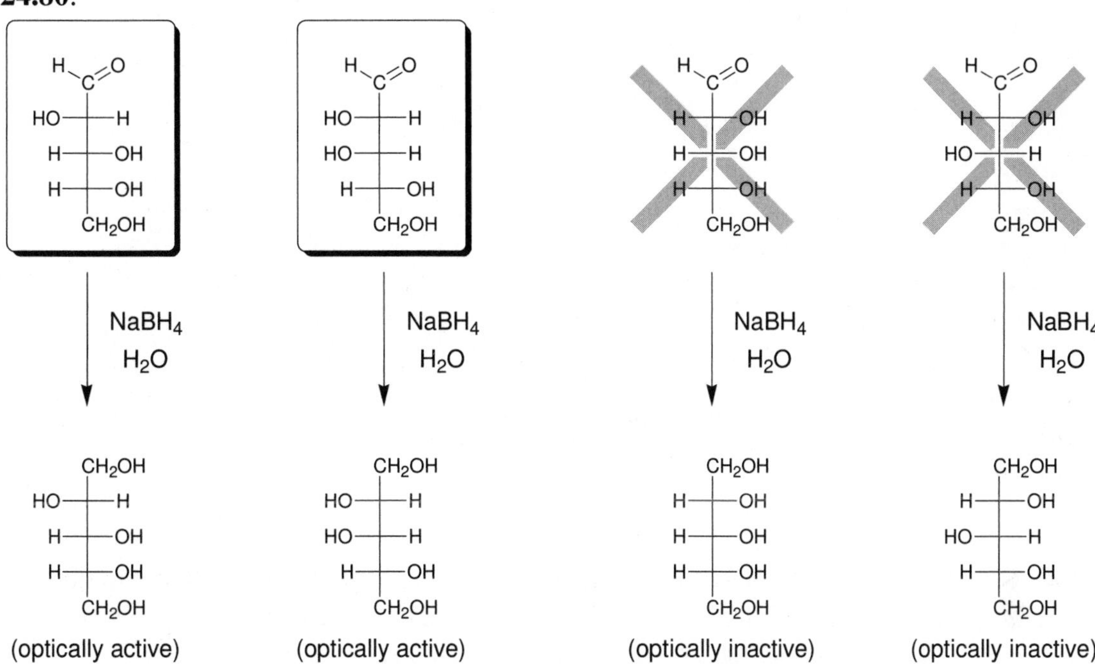

24.81.

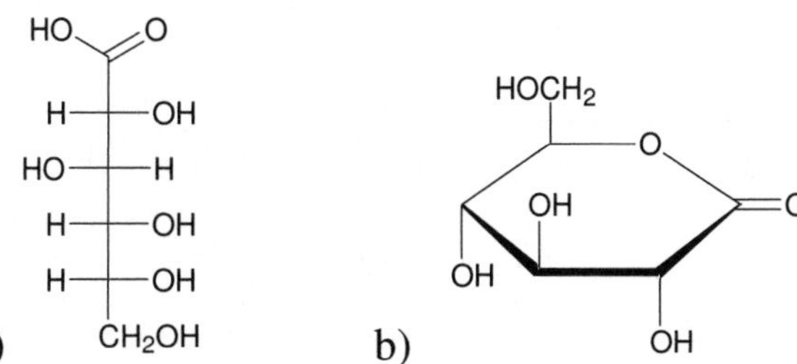

a) b)

c) Yes. The compound has chirality centers, and it is not a meso compound. Therefore, it will be optically active.

d) The gluconic acid is a carboxylic acid and its IR spectrum is expected to have a broad signal between 2500 and 3600 cm^{-1}. The IR spectrum of the lactone will not have this broad signal.

24.82. In order for the CH_2OH group to occupy an equatorial position, all of the OH groups on the ring must occupy axial positions. The combined steric hindrance of all the OH groups is more than the steric hindrance associated with one CH_2OH group. Therefore, the equilibrium will favor the form in which the CH_2OH group occupies an axial position. The structure of L-idose is:

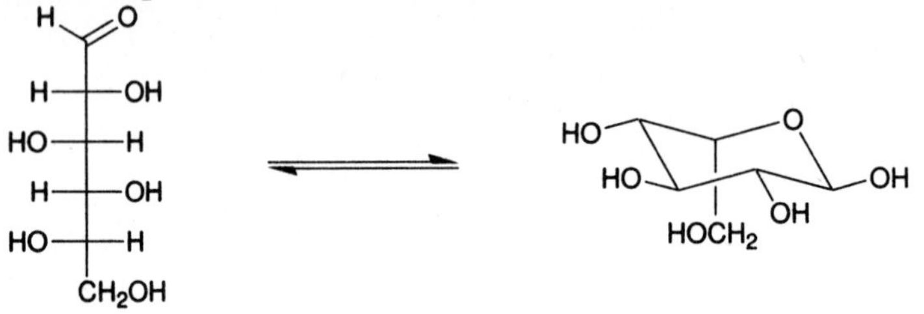

***L*-Idose**

24.83.

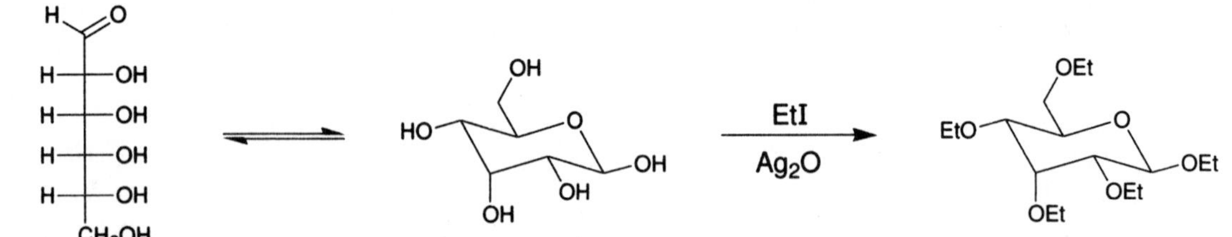

***D*-Allose**
(Compound A)

β-pyranose form

EtI
Ag_2O

24.84. Glucose can adopt a chair conformation in which all of the substituents on the ring occupy equatorial positions. Therefore, D-glucose can achieve a lower energy conformation than any of the other D-aldohexoses.

24.85.

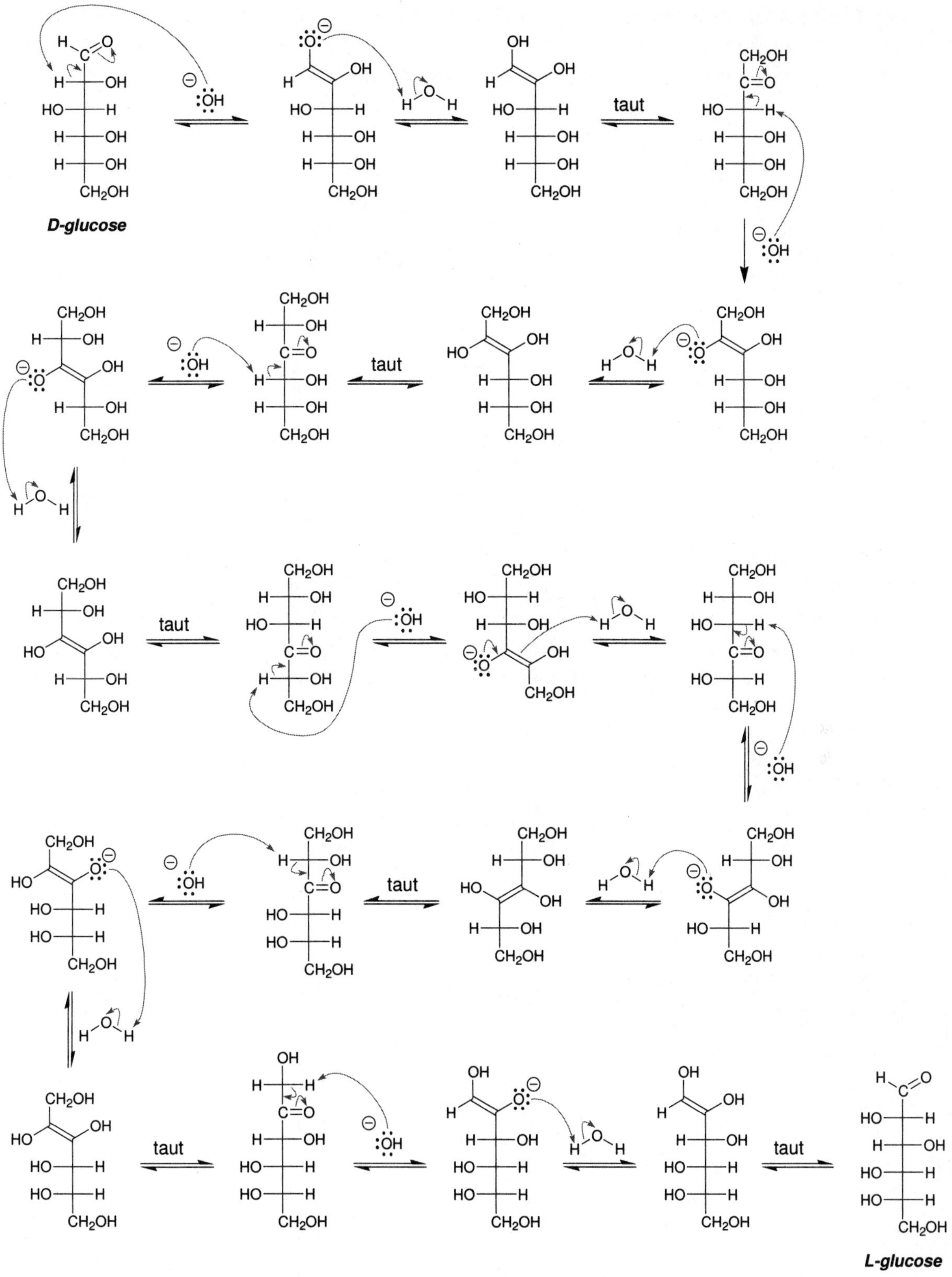

D-glucose

taut

taut

taut

taut

taut

taut

L-glucose

24.86. Compound X is a D-aldohexose that can adopt a β-pyranose form with only one axial substituent. Recall that D-glucose has all substituents in equatorial positions, so compound X must be epimeric with D-glucose either at C2 (D-mannose), C3 (D-allose), or C4 (D-galactose).
Compound X undergoes a Wohl degradation to produce an aldopentose, which is converted into an optically active alditol when treated with sodium borohydride. Therefore, compound X cannot be D-allose, because a Wohl degradation of D-allose followed by reduction produces an optically *inactive* alditol.
We conclude that compound X must be either D-mannose or D-galactose.
The identity of compound X can be determined by treating compound X with sodium borohohydride. Reduction of D-mannose should give an optically active alditol, while reduction of D-galactose gives an optically inactive alditol.

24.87. Compound A is a D-aldopentose. Therefore, there are four possible structures to consider (Figure 24.4).
When treated with sodium borohydride, compound A is converted into an alditol that exhibits three signals in its ^{13}C NMR spectrum. Therefore, compound A must be D-ribose or D-xylose both of which are reduced to give symmetrical alditols (thus, three signals for five carbon atoms).
When compound A undergoes a Kiliani-Fischer synthesis, both products can be treated with nitric acid to give optically active aldaric acids. Therefore, compound A cannot be D-ribose, because when D-ribose undergoes a Kiliani-Fischer synthesis, one of the products is D-allose, which is oxidized to give an optically inactive aldaric acid. We conclude that the structure of compound A must be D-xylose.

H–C=O
H–C–OH
HO–C–H
H–C–OH
CH_2OH

a) ***D-Xylose***

b) Compound D is expected have six signals in its ^{13}C NMR spectrum, while compound E is expected to have only three signals in its ^{13}C NMR spectrum.

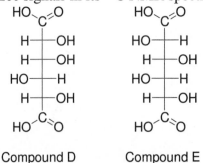

Compound D	Compound E
HO–C=O	HO–C=O
H–C–OH	HO–C–H
H–C–OH	H–C–OH
HO–C–H	HO–C–H
H–C–OH	H–C–OH
HO–C=O	HO–C=O

Chapter 25
Amino Acids, Peptides, and Proteins

Review of Concepts

Fill in the blanks below. To verify that your answers are correct, look in your textbook at the end of Chapter 25. Each of the sentences below appears verbatim in the section entitled *Review of Concepts and Vocabulary.*

- Amino acids in which the two functional groups are separated by exactly one carbon atom are called _______ **amino acids.**
- Amino acids are coupled together by amide linkages called ____________ **bonds.**
- Relatively short chains of amino acids are called ___________.
- Only twenty amino acids are abundantly found in proteins, all of which are ___ **amino acids**, except for ____________ which lacks a chirality center.
- Amino acids exist primarily as _______________ at physiological pH
- The ___________________ of an amino acid is the pH at which the concentration of the zwitterionic form reaches its maximum value.
- Peptides are comprised of **amino acid** _________ joined by peptide bonds.
- Peptide bonds experience restricted rotation, giving rise to two possible conformations, called _______ and _______. The _______ conformation is more stable.
- Cysteine residues are uniquely capable of being joined to one another via ______________ **bridges.**
- ______ is commonly used to form peptide bonds.
- In the **Merrifield synthesis**, a peptide chain is assembled while tethered to ____________________________.
- The **primary structure** of a protein is the sequence of _____________________.
- The **secondary structure** of a protein refers to the _______________________ ____________________ of localized regions of the protein. Two particularly stable arrangements are the ___ **helix** and ____ **pleated sheet.**
- The **tertiary structure** of a protein refers to its ________________________.
- Under conditions of mild heating, a protein can unfold, a process called _________________.
- **Quaternary structure** arises when a protein consists of two or more folded polypeptide chains, called _____________, that aggregate to form one protein complex.

Review of Skills

Fill in the blanks and empty boxes below. To verify that your answers are correct, look in your textbook at the end of Chapter 25. The answers appear in the section entitled *SkillBuilder Review.*

25.1 Determining the Predominant Form of an Amino Acid at a Specific pH

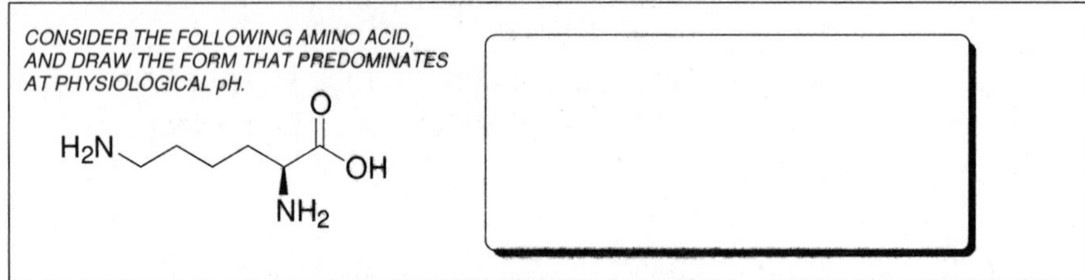

25.2 Using the Amidomalonate Synthesis

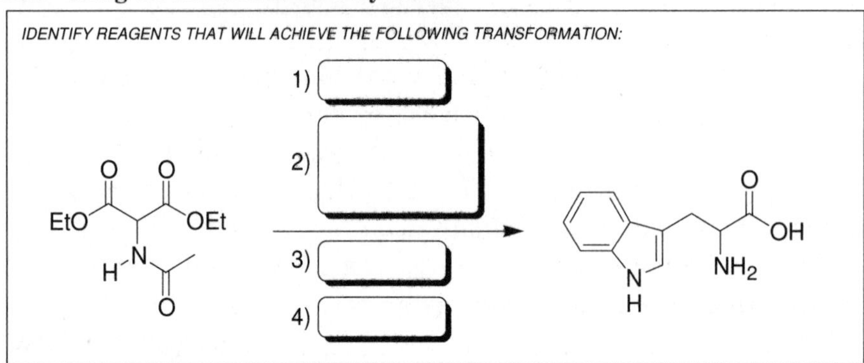

25.3 Drawing a Peptide

25.4 Sequencing a Peptide via Enzymatic Cleavage

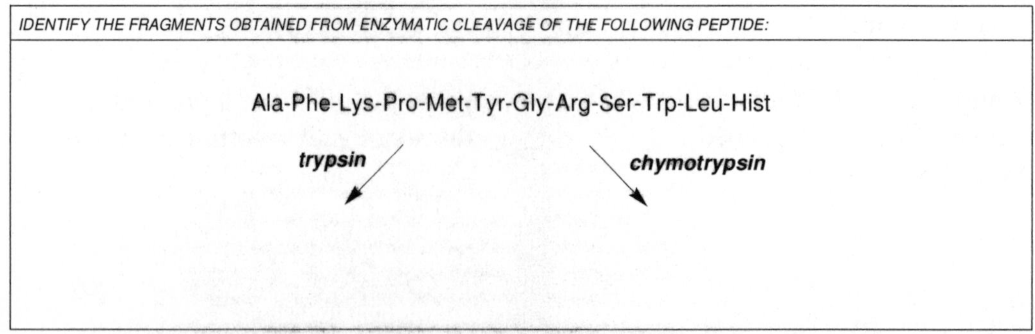

25.5 Planning the Synthesis of a Dipeptide

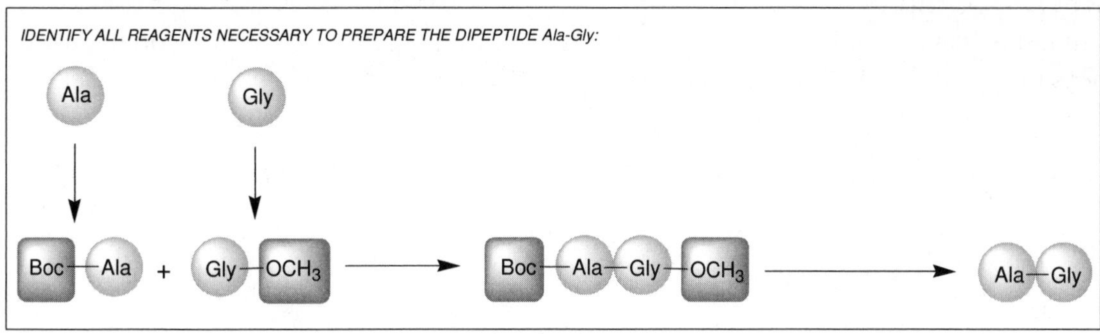

25.6 Preparing a Peptide using the Merrifield Synthesis

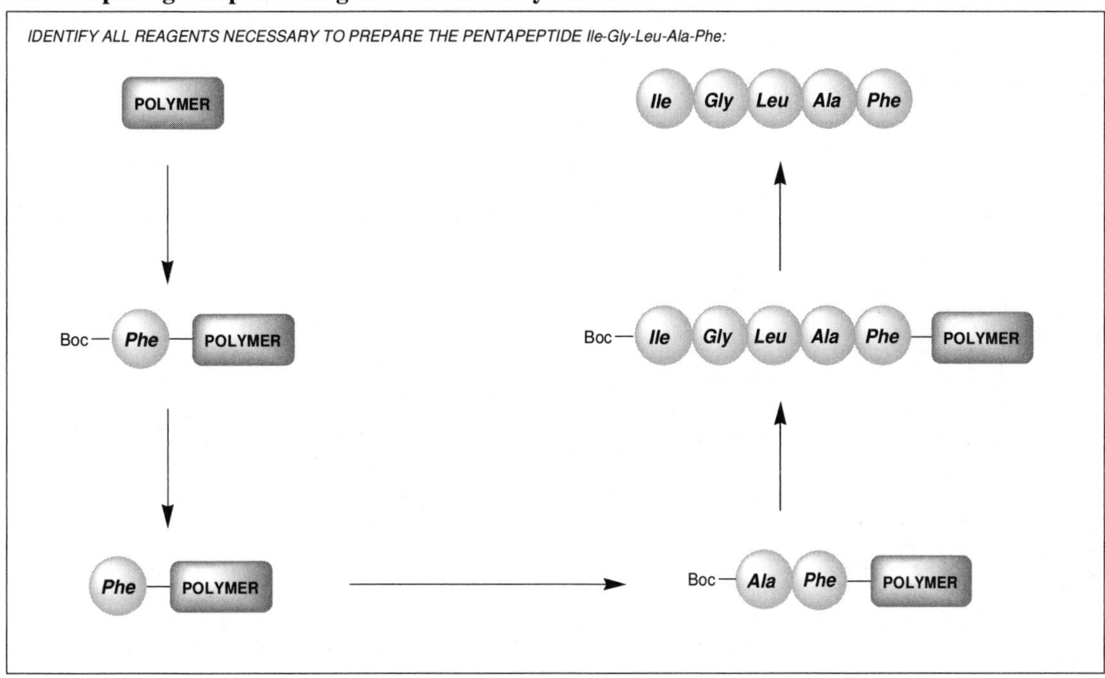

Review of Reactions

Identify the reagents necessary to achieve each of the following transformations. To verify that your answers are correct, look in your textbook at the end of Chapter 25. The answers appear in the section entitled *Review of Reactions*.

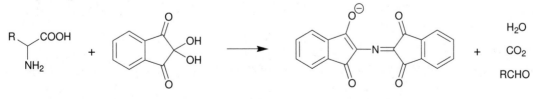

Analysis of Amino Acids

R–CH(NH$_2$)–COOH + (ninhydrin) ⟶ (purple anion) + H_2O + CO_2 + RCHO

Synthesis of Amino Acids

R OH O → R OH O Br → R OH O NH2

EtO OEt O O N H O → EtO OEt O O R N H O → R OH O NH2

R H O → NaCN → H2N C≡N R H → H2N C OH O R H

OH O NHAc → OH O NHAc → OH O NH2

Analysis of Amino Acids

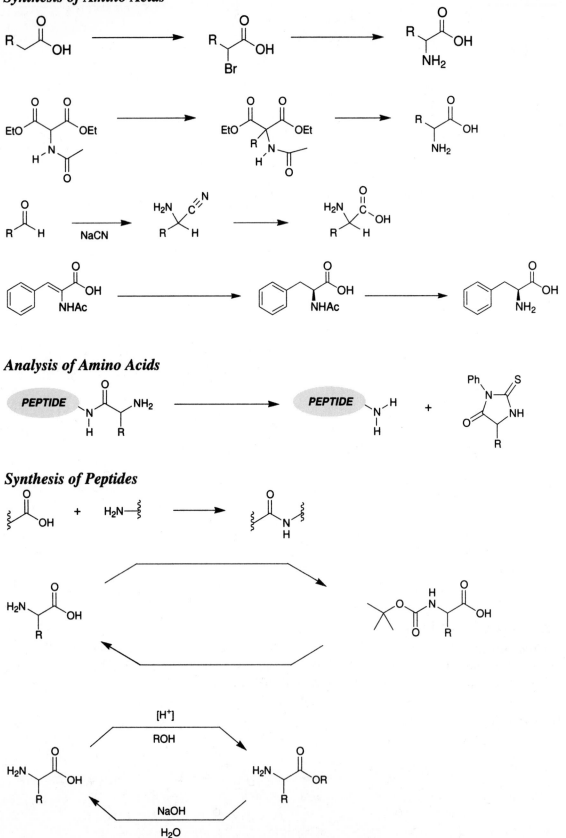

PEPTIDE N H O NH2 R → PEPTIDE N H H H + Ph N S NH O R

Synthesis of Peptides

O OH + H2N → O N H

H2N O OH R ⇄ O H N O OH O R

[H+] ROH

H2N O OH R ⇄ H2N O OR R

NaOH H_2O

Solutions

25.1. In each case, the chirality center has the *R* configuration.

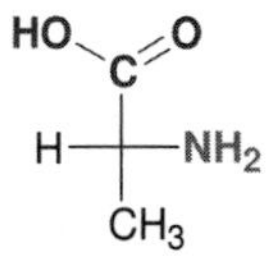

HO C=O, H—C—NH_2, CH_3
D-Alanine

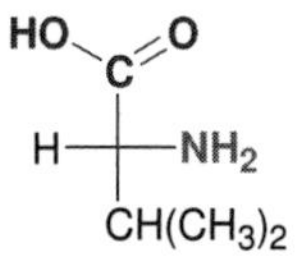

HO C=O, H—C—NH_2, $CH(CH_3)_2$
D-Valine

25.2.

a) O, OH, NH_2

b) O, OH, NH_2, N H

c) S, O, OH, NH_2

d) O, OH, NH_2

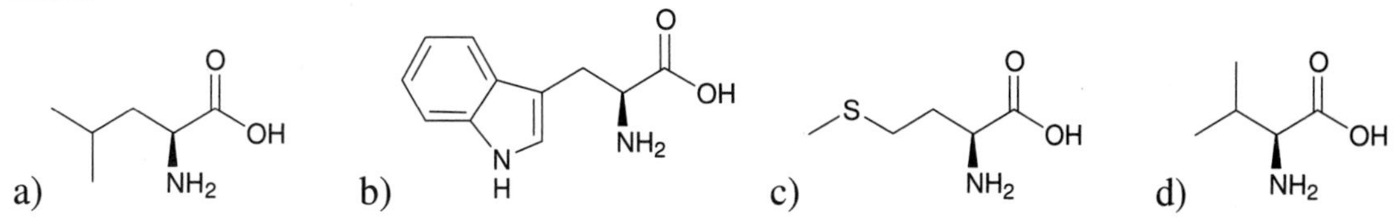

25.3.
a) Pro, Phe, Trp, Tyr, and His
b) Phe, Trp, Tyr, and His
c) Arg, His, and Lys
d) Met and Cys
e) Asp and Glu
f) Pro, Trp, Asn, Gln, Ser, Thr, Tyr, Cys, Asp, Glu, Arg, His, and Lys

25.4.

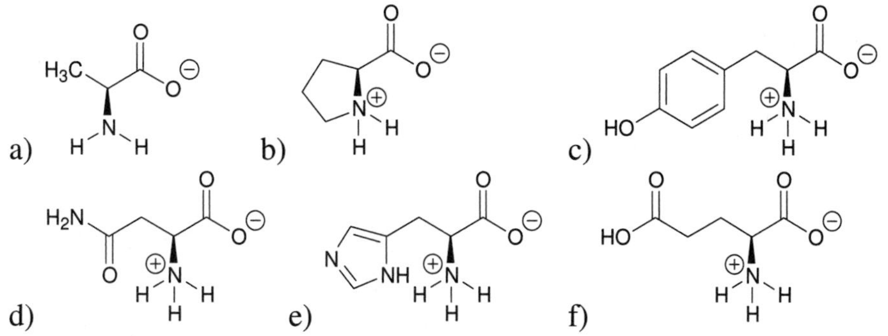

a) H_3C, O, O⊖, H N H

b) O, O⊖, N⊕ H H

c) O, O⊖, HO, ⊕N H H H

d) H_2N, O, O, O⊖, ⊕N H H H

e) N, NH, O, O⊖, ⊕N H H H

f) HO, O, O, O⊖, ⊕N H H H

25.5. Arginine has a basic side chain, while asparagine does not. At a pH of 11, arginine exists predominantly in a form in which the side chain is protonated. Therefore, it can serve as a proton donor.

25.6. Tyrosine possesses a phenolic proton which is more readily deprotonated because deprotonation forms a resonance-stabilized phenolate ion. In contrast, deprotonation of the OH group of serine gives an alkoxide ion that is not resonance-stabilized. As a result, the OH group of tyrosine is more acidic than the OH group of serine.

25.7.

a) 2.77 b) 5.98 c) 9.74 d) 6.30

25.8.
a) aspartic acid
b) glutamic acid

25.9. Leucine and isoleucine

25.10. The pI of Phe = 5.48, the pI of Trp = 6.11, and the pI of Leu = 6.00.
a) At pH = 6.0, Phe will travel the farthest distance.
b) At pH = 5.0, Trp will travel the farthest distance.

25.11. O, H

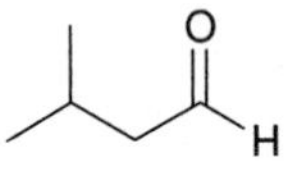

25.12.

a) O, OH — 1) Br_2 , PBr_3 2) H_2O 3) excess NH_3 → O, OH, NH_2 (racemic)

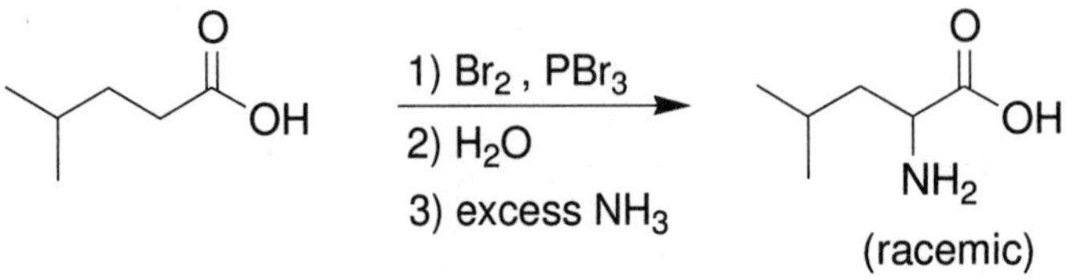

b) O, OH — 1) Br_2 , PBr_3 2) H_2O 3) excess NH_3 → O, OH, NH_2 (racemic)

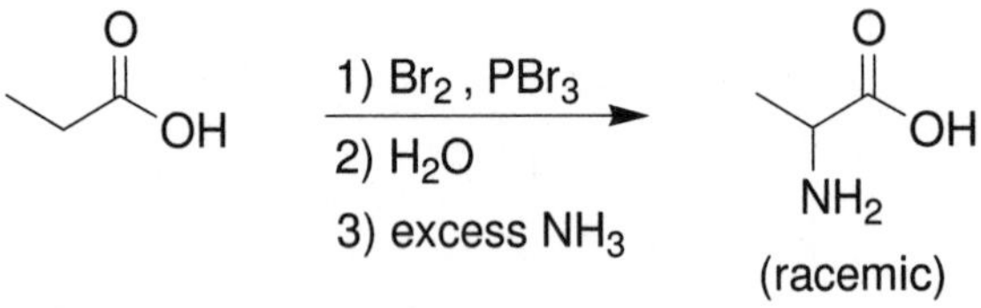

c) O, OH — 1) Br_2 , PBr_3 2) H_2O 3) excess NH_3 → O, OH, NH_2 (racemic)

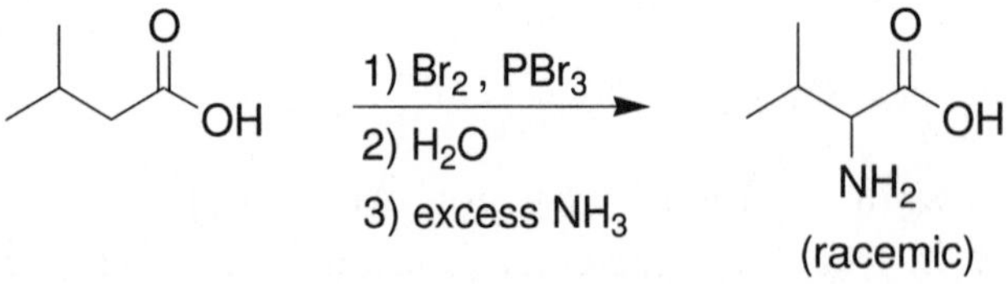

25.13.

COOH NH2 a) Leucine

COOH NH2 b) Valine

COOH NH2 c) Phenylalanine

H2N COOH d) Glycine

25.14.

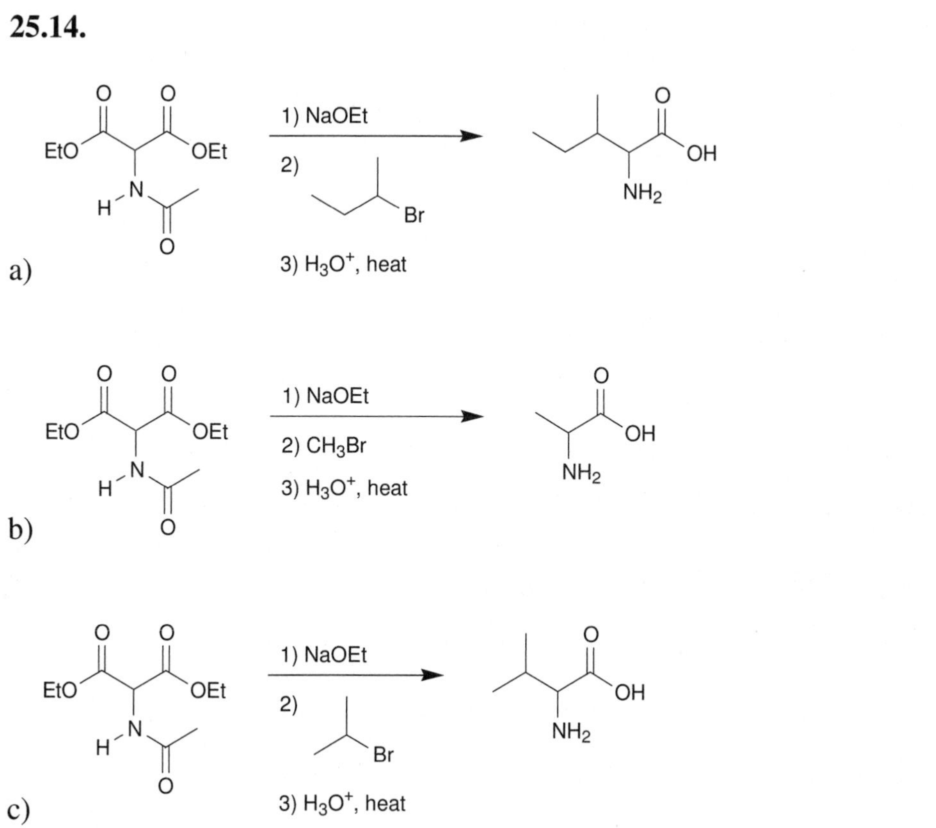

25.15.

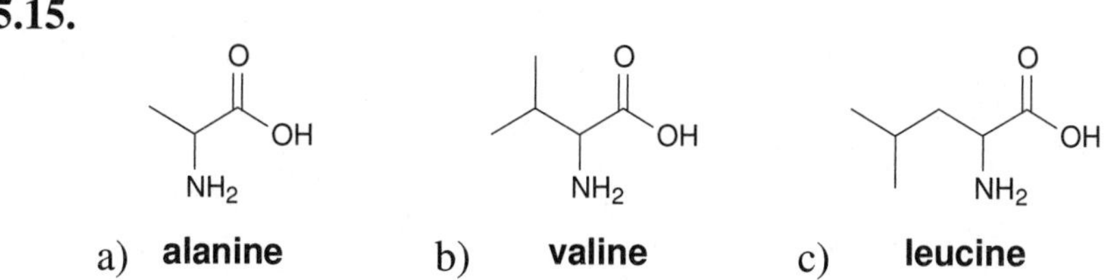

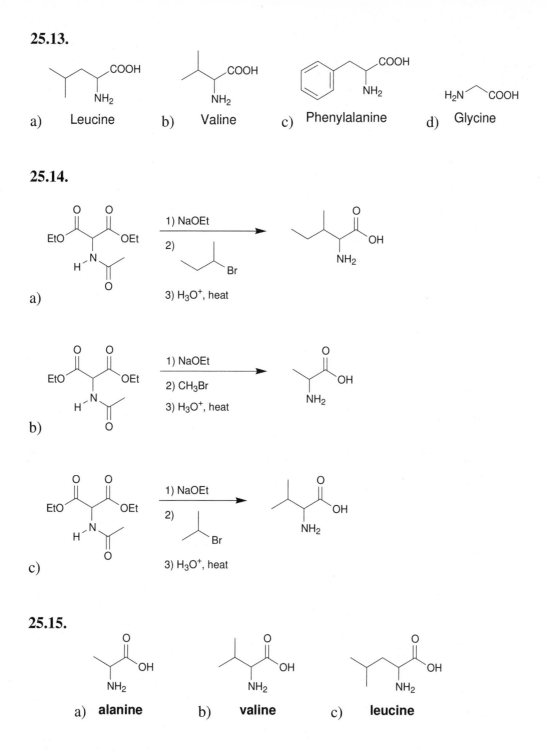

25.16. Leucine can be prepared via the amidomalonate synthesis with higher yields than isoleucine, because the former requires an S_N2 reaction with a primary alkyl halide, while the latter requires an S_N2 reaction with a secondary (more hindered) alkyl halide.

25.17.

a) 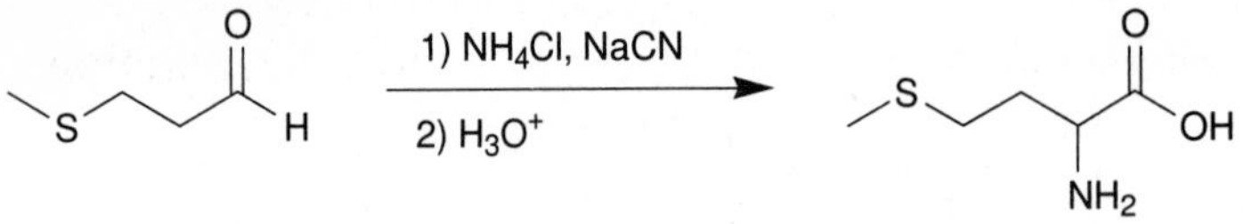 (structure) → 1) NH_4Cl, NaCN; 2) H_3O^+ → (structure: O, OH, NH_2)

b) 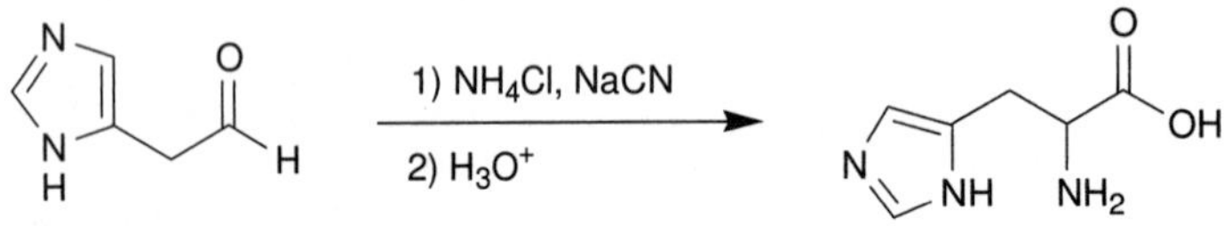 (structure: N, N, H, O, H) → 1) NH_4Cl, NaCN; 2) H_3O^+ → (structure: N, NH, NH_2, O, OH)

c) 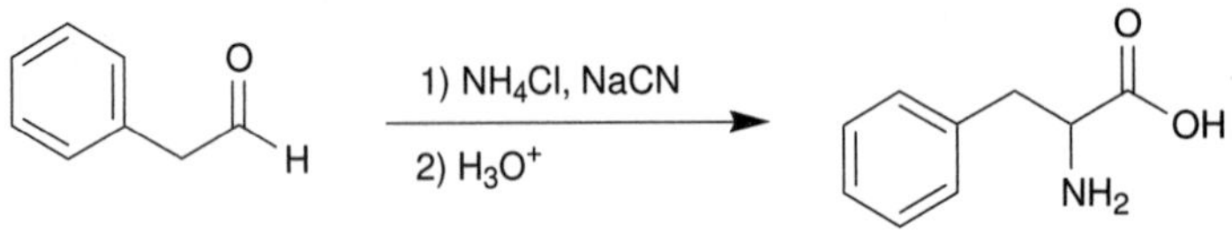 (structure: O, H) → 1) NH_4Cl, NaCN; 2) H_3O^+ → (structure: O, OH, NH_2)

d) 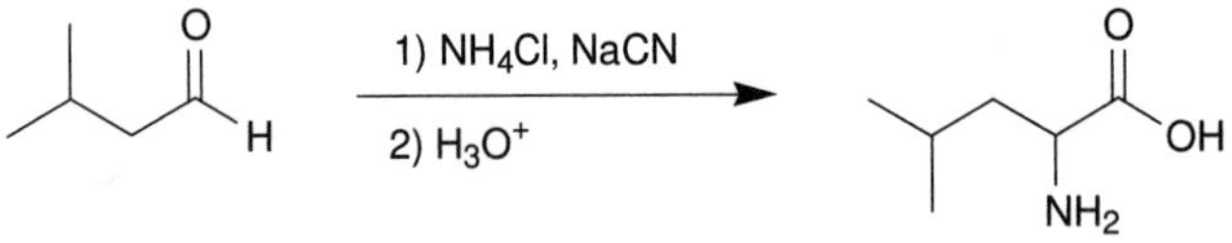 (structure: O, H) → 1) NH_4Cl, NaCN; 2) H_3O^+ → (structure: O, OH, NH_2)

25.18.

a) 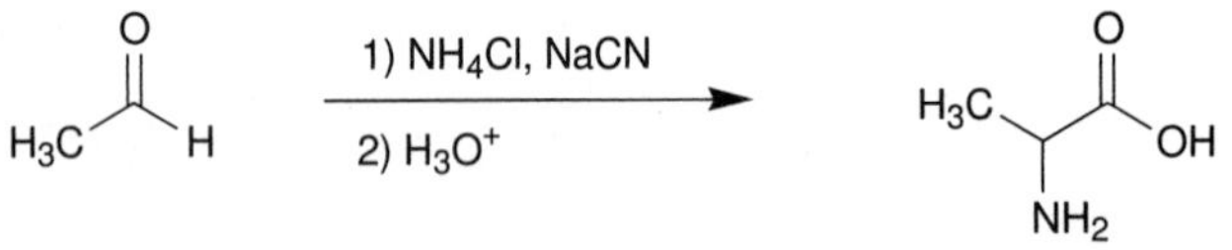 H_3C–CHO → 1) NH_4Cl, NaCN; 2) H_3O^+ → H_3C, O, OH, NH_2

alanine

b) 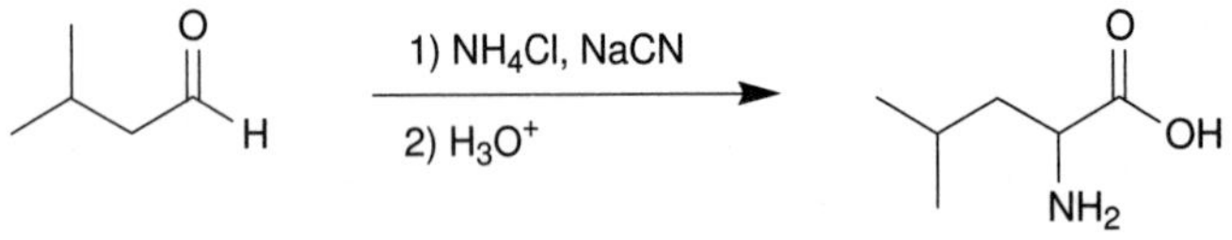 (structure: O, H) → 1) NH_4Cl, NaCN; 2) H_3O^+ → (structure: O, OH, NH_2)

leucine

c) 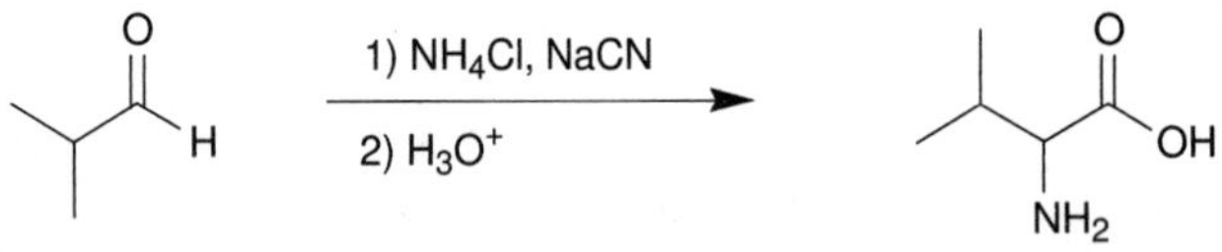 (structure: O, H) → 1) NH_4Cl, NaCN; 2) H_3O^+ → (structure: O, OH, NH_2)

valine

25.19.

a) 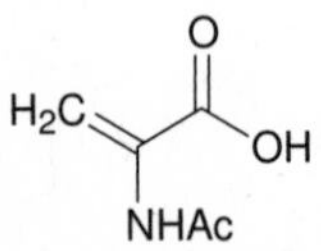 H_2C, O, OH, NHAc b) 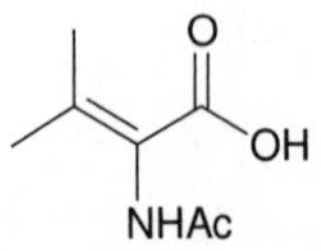 O, OH, NHAc c) 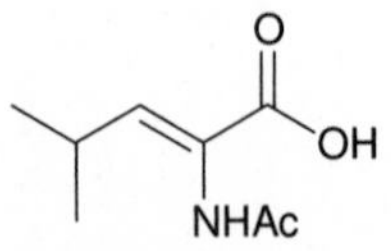 O, OH, NHAc d) 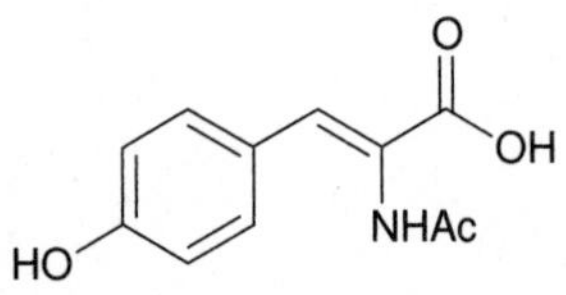 HO, O, OH, NHAc

25.20. Glycine does not possess a chirality center, so the use of a chiral catalyst is unnecessary. Also, there is no alkene that would lead to glycine upon hydrogenation.

25.21.

a)

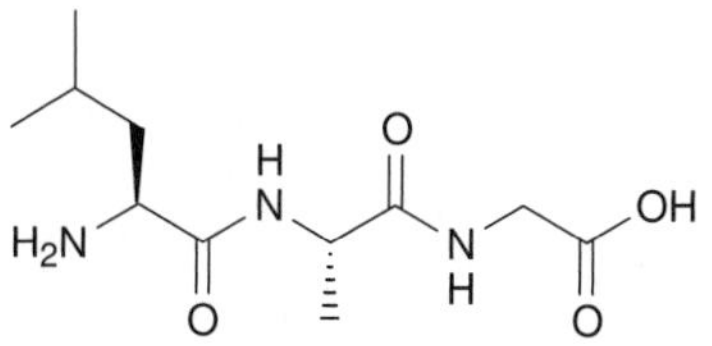

b)

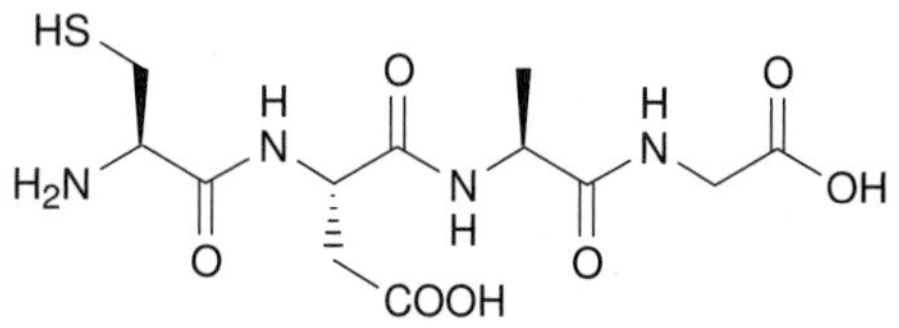

c)

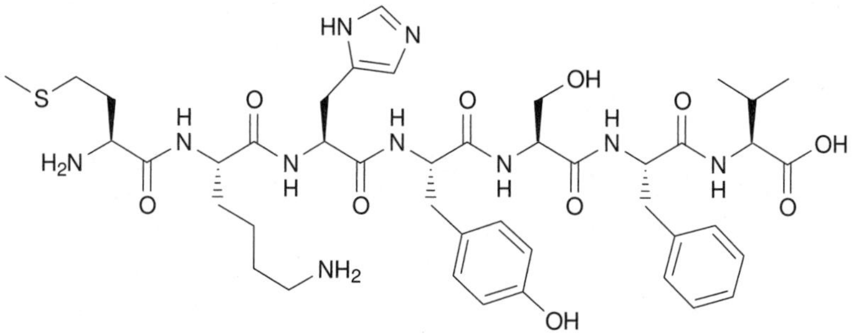

25.22. Leu-Ala-Phe-Cys-Asp or L-A-F-C-D.

25.23. Cys-Tyr-Leu

25.24. Constitutional isomers

25.25.

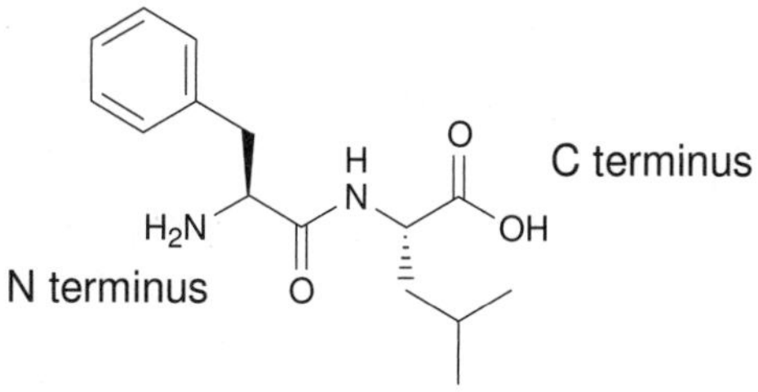

C terminus

N terminus

25.26. Steric hindrance results from the phenyl groups:

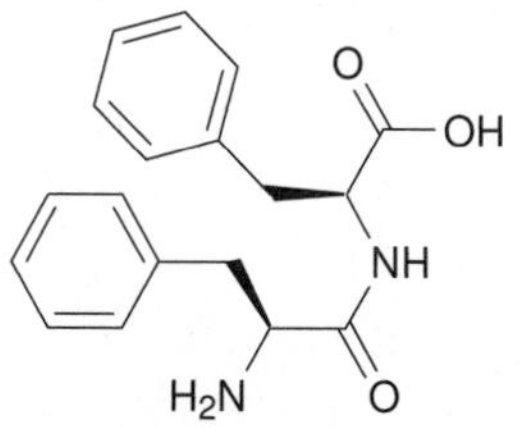

25.27.

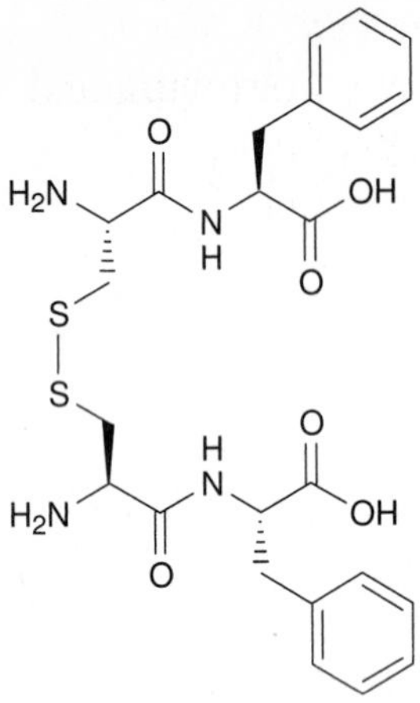

25.28.

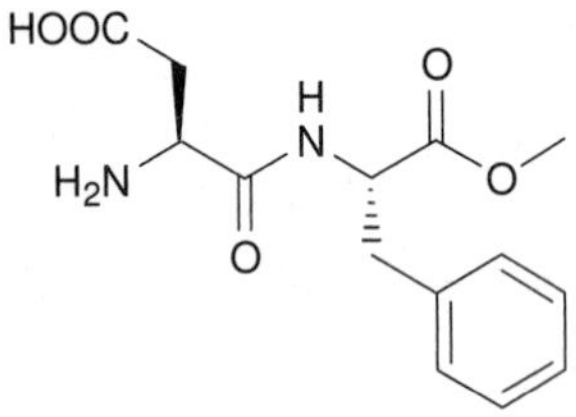

a)

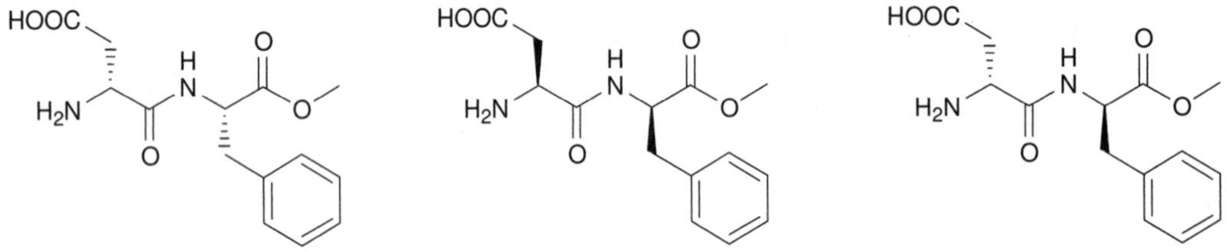

b)

25.29.

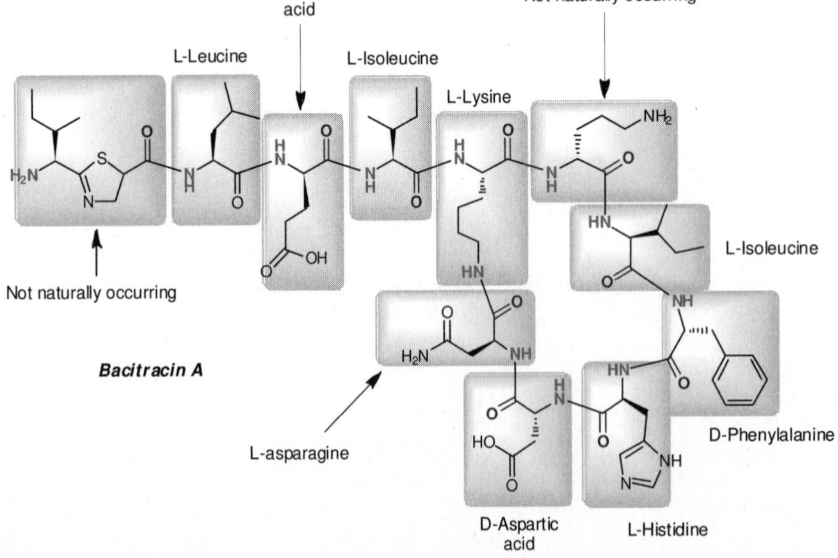

Bacitracin A

D-Glutamic acid

Not naturally occurring

L-Leucine

L-Isoleucine

L-Lysine

L-Isoleucine

Not naturally occurring

L-asparagine

D-Phenylalanine

D-Aspartic acid

L-Histidine

25.30. An Edman degradation will remove the amino acid residue at the N terminus, and Ala is the N terminus in Ala-Phe-Val. Therefore, alanine is removed, giving the following PTH derivative:

25.31.
Met-Phe-Val-Ala-Tyr-Lys-Pro-Val-Ile-Leu-Arg-Trp-His-Phe-Met-Cys-Arg-Gly-Pro-Phe-Ala-Val

25.32. Ala-Phe-Val-Lys

25.33. Cleavage with trypsin will produce Phe-Arg, while cleavage with chymotrypsin will produce Arg-Phe. These dipeptides are not the same. They are constitutional isomers.

25.34.
a)

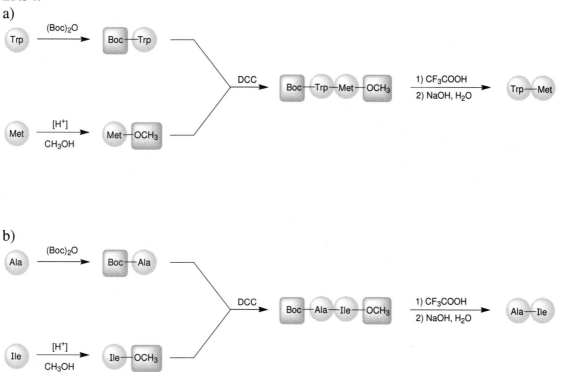

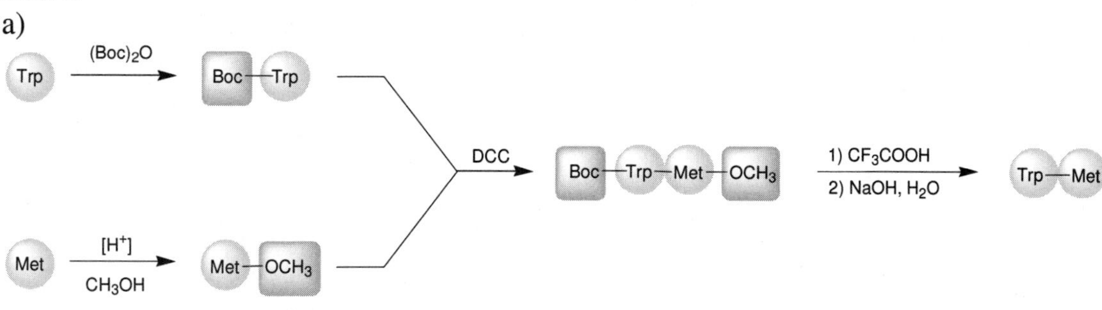

b)

c)

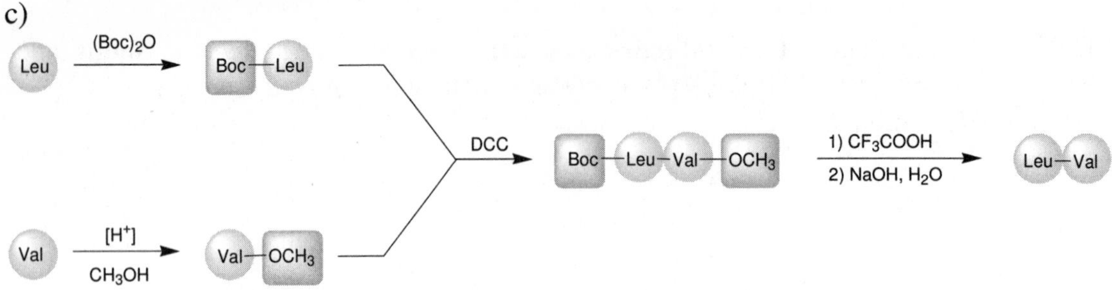
Leu
(Boc)2O
Boc—Leu
Val
[H+]
CH3OH
Val—OCH3
DCC
Boc—Leu—Val—OCH3
1) CF3COOH
2) NaOH, H2O
Leu—Val

25.35.

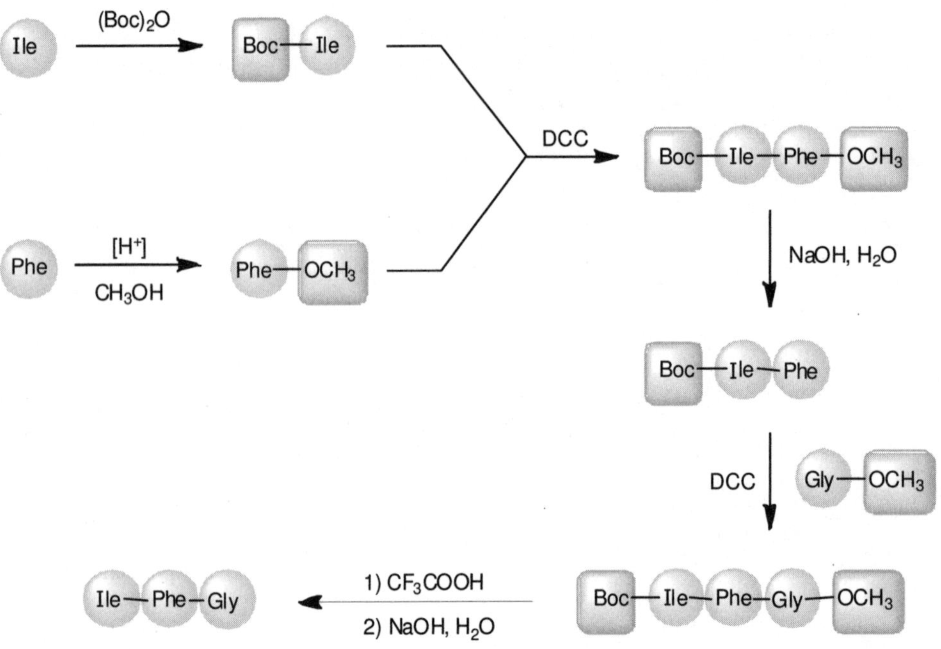
Ile
(Boc)2O
Boc—Ile
Phe
[H+]
CH3OH
Phe—OCH3
DCC
Boc—Ile—Phe—OCH3
NaOH, H2O
Boc—Ile—Phe
DCC
Gly—OCH3
Boc—Ile—Phe—Gly—OCH3
1) CF3COOH
2) NaOH, H2O
Ile—Phe—Gly

25.36.

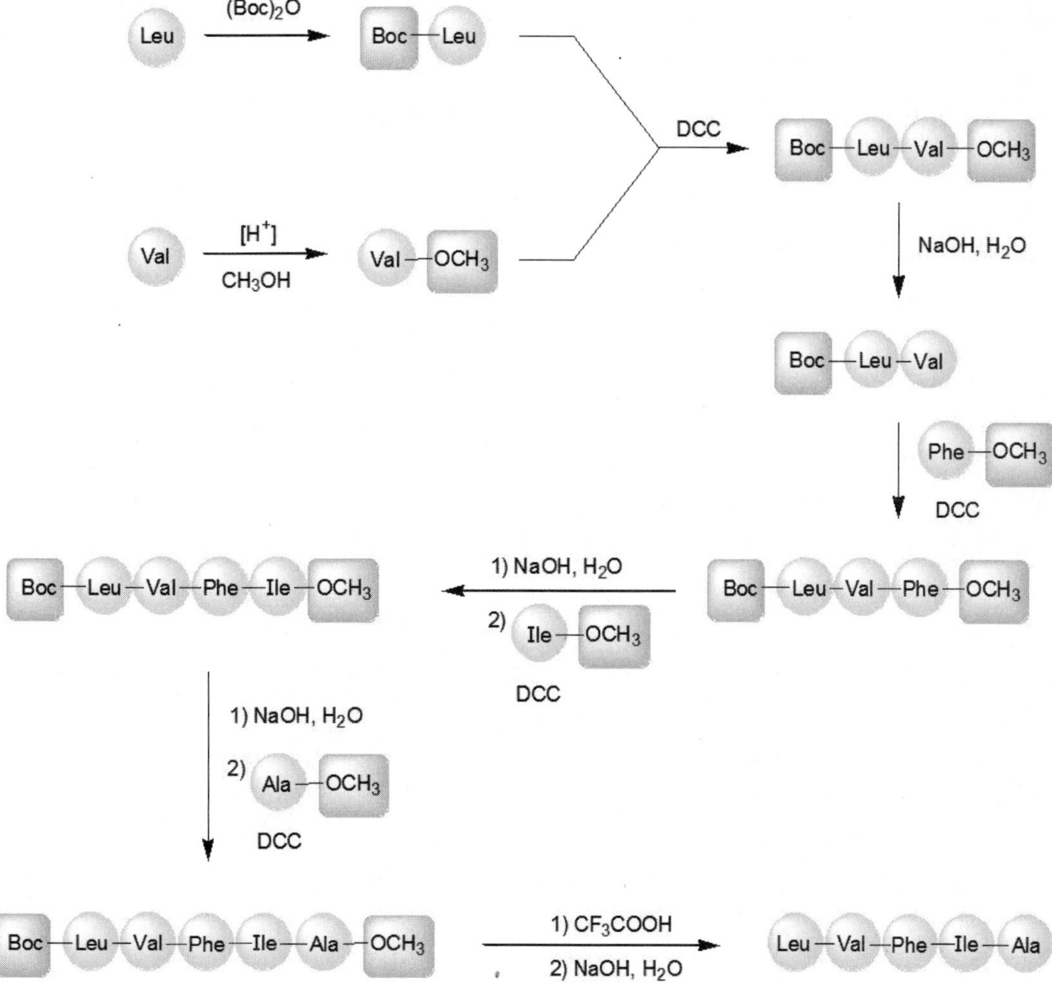
Leu
$(Boc)_2O$
Boc—Leu
Val
$[H^+]$
CH_3OH
Val—OCH_3
DCC
Boc—Leu—Val—OCH_3
NaOH, H_2O
Boc—Leu—Val
Phe—OCH_3
DCC
Boc—Leu—Val—Phe—OCH_3
1) NaOH, H_2O
2) Ile—OCH_3
DCC
Boc—Leu—Val—Phe—Ile—OCH_3
1) NaOH, H_2O
2) Ala—OCH_3
DCC
Boc—Leu—Val—Phe—Ile—Ala—OCH_3
1) CF_3COOH
2) NaOH, H_2O
Leu—Val—Phe—Ile—Ala

25.37.

a)

Cl–CH₂–POLYMER → Phe-Leu-Val-Phe

1) Boc-NH-CH(CH₂Ph)-COO⁻

2) CF_3COOH

3) Boc-NH-CH(CH(CH₃)₂)-COOH , DCC

4) CF_3COOH

5) Boc-NH-CH(CH₂CH(CH₃)₂)-COOH , DCC

6) CF_3COOH

7) Boc-NH-CH(CH₂Ph)-COOH , DCC

8) CF_3COOH

9) HF

b)

Cl–CH₂–POLYMER → Ala-Val-Leu-Ile

1) Boc-NH-CH(CH(CH₃)CH₂CH₃)-COO⁻

2) CF_3COOH

3) Boc-NH-CH(CH₂CH(CH₃)₂)-COOH , DCC

4) CF_3COOH

5) Boc-NH-CH(CH(CH₃)₂)-COOH , DCC

6) CF_3COOH

7) Boc-NH-CH(CH₃)-COOH , DCC

8) CF_3COOH

9) HF

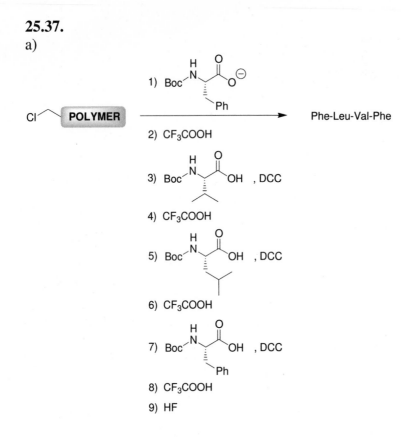

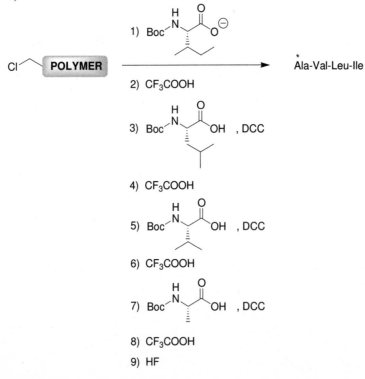

25.38. (N terminus) Val-Ala-Phe (C terminus)

25.39. The regions that contain repeating glycine and/or alanine units are the most likely regions to form β sheets:

Trp-His-Pro-Ala-Gly-Gly-Ala-Val-His-Cyst-Asp-Ser-Arg-Arg-Ala-Gly-Ala-Phe

25.40.

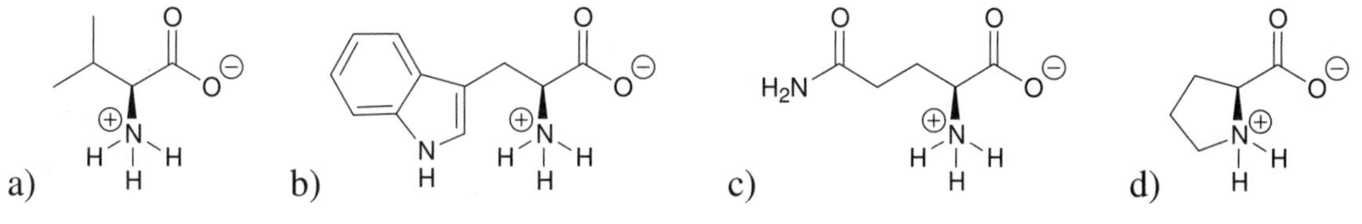

a) b) c) d)

25.41. When applying the Cahn-Ingold-Prelog convention for assigning the configuration of a chirality center, the amino group generally receives the highest priority (1), followed by the carboxylic acid moiety (2), followed the side chain (3), and finally the H (4). Accordingly, the *S* configuration is assigned to L amino acids. Cysteine is the one exception because the side chain has a higher priority than the carboxylic acid moiety. As a result, the *R* configuration is assigned.

25.42.

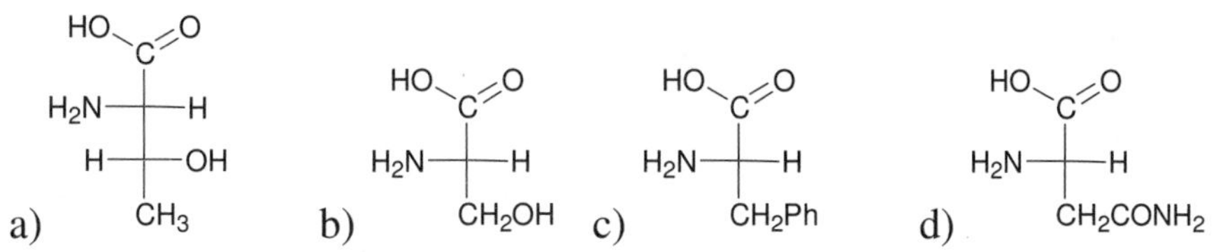

a) b) c) d)

25.43.

a) Isoleucine and threonine
b) Isoleucine = *2S,3S*. Threonine = *2S,3R*

25.44.

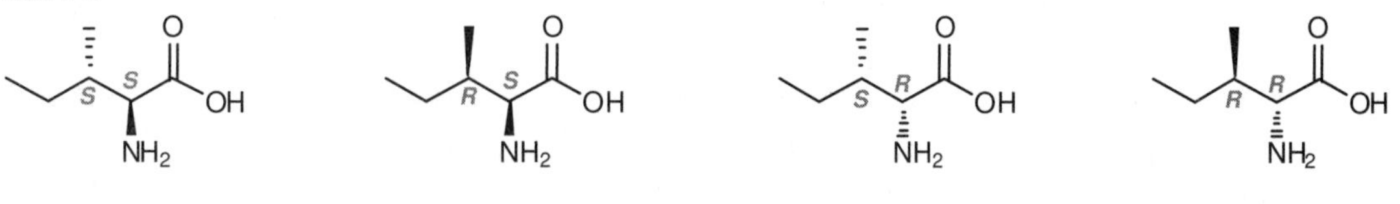

25.45. The protonated form below is highly stabilized by resonance, which spreads the positive charge over all three nitrogen atoms.

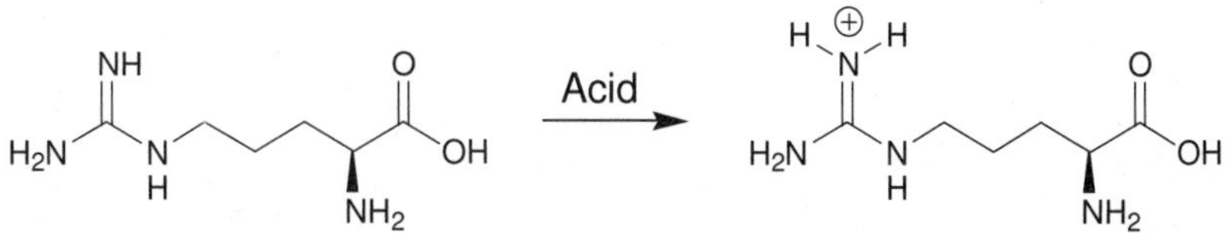

Acid

25.46. The protonated form below is aromatic. In contrast, protonation of the other nitrogen atom in the ring would result in loss of aromatic stabilization.

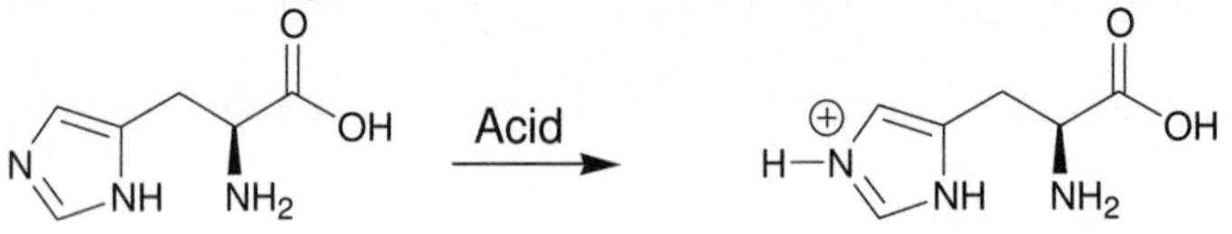

Acid

25.47.

a) 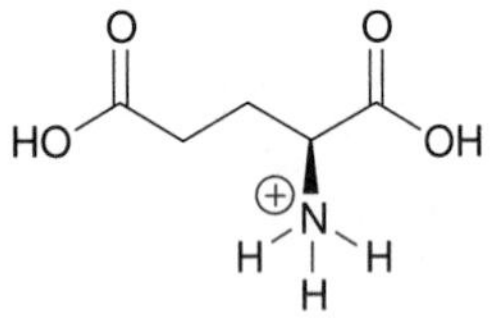 b) 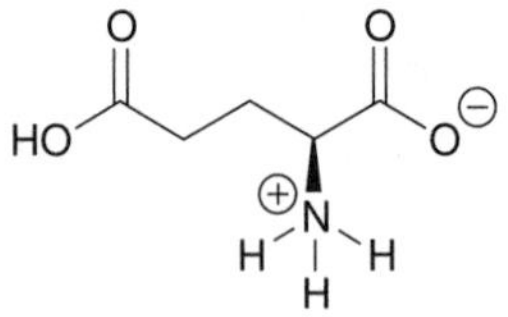 c) 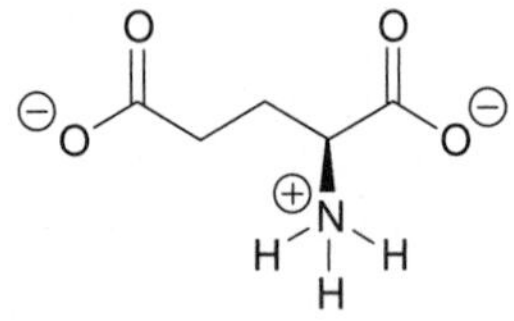 d)

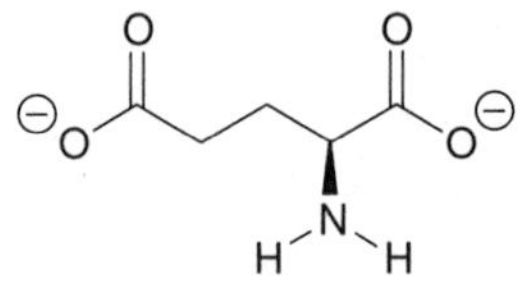

25.48.

a) 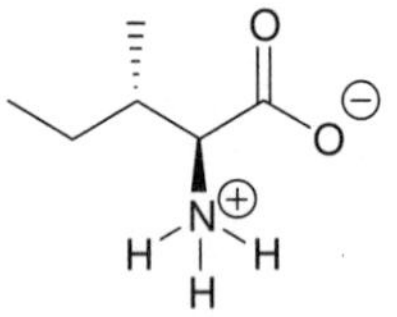 b) 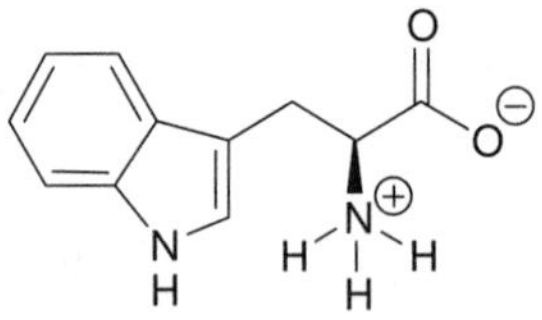 c) 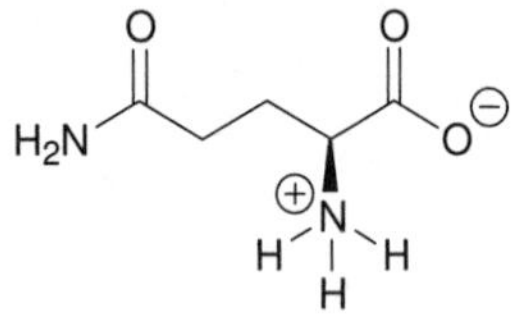 d)

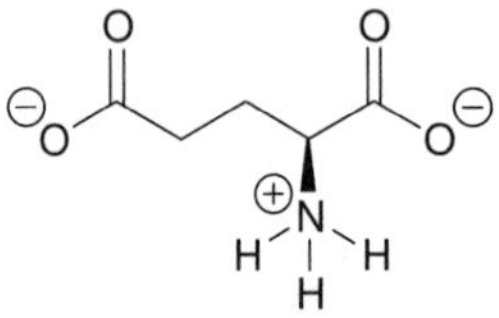

25.49.

a) 6.02 b) 5.41 c) 7.58 d) 3.22

25.50. Lysozyme is likely to be comprised primarily of amino acid residues that contain basic side chains (arginine, histindine, and lysine), while pepsin is comprised primarily of amino acid residues that contain acidic side chains (aspartic acid and glutamic acid).

25.51.

a) 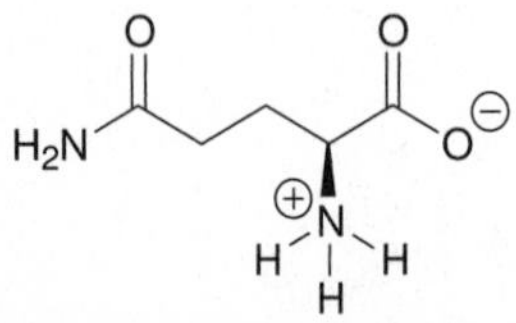 b) 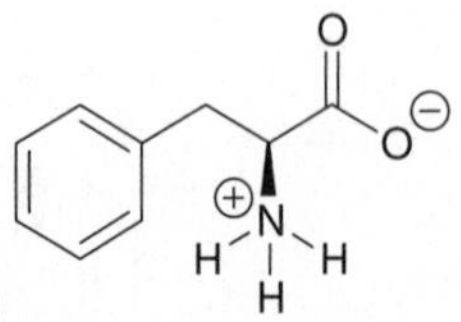 c) 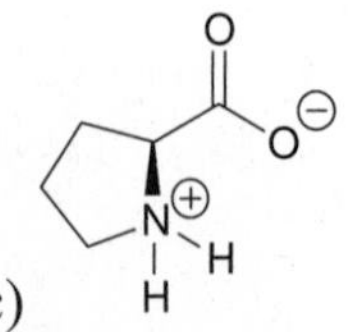 d)

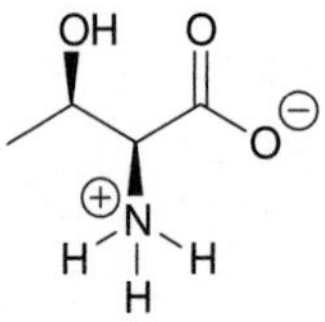

25.52.

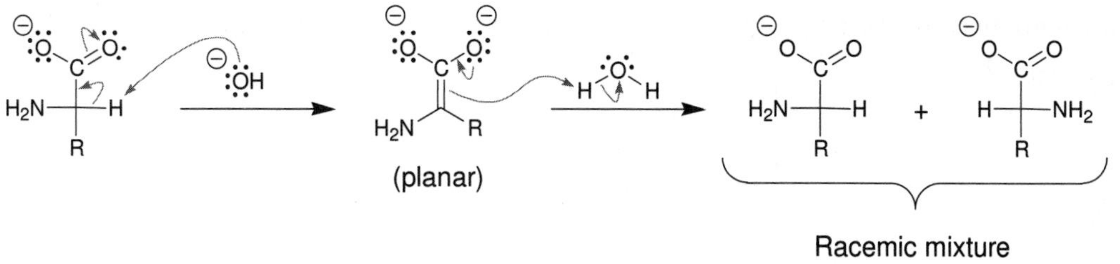

(planar)

Racemic mixture

25.53.
The pI of Gly = 5.97, the pI of Gln = 5.65, and the pI of Asn = 5.41.
a) At pH = 6.0, Asn will travel the farthest distance.
b) At pH = 5.0, Gly will travel the farthest distance.

25.54.

a) 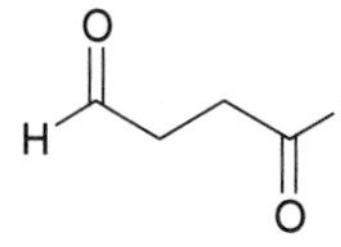 b) 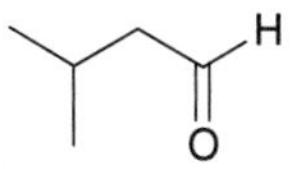 c) 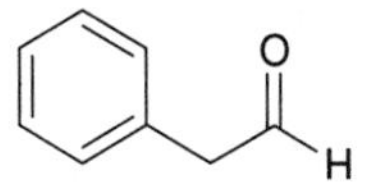 d) no reaction

25.55.

a) Methionine, valine, and glycine.

b)

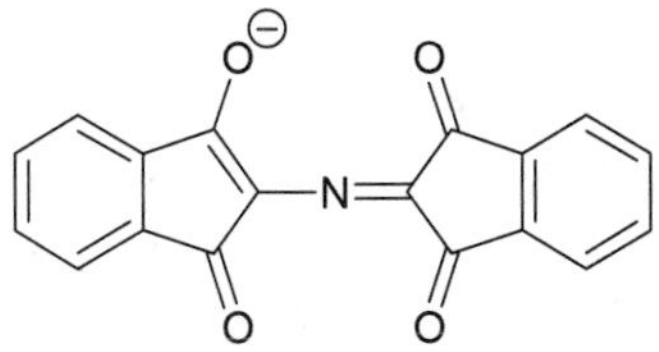

c) The compound is highly conjugated and has a λ_{max} that is greater than 400 nm (see Section 17.12)

25.56.

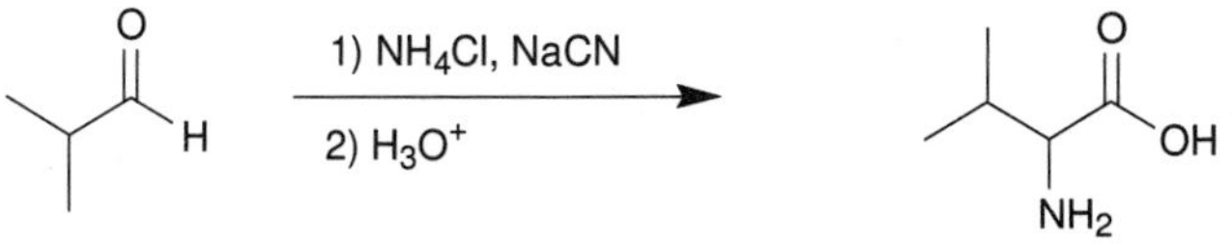

1) NH_4Cl, NaCN
2) H_3O^+

25.57. Alanine can be prepared via the amidomalonate synthesis with higher yields than valine, because the former requires an S_N2 reaction with a primary alkyl halide, while the latter requires an S_N2 reaction with a secondary (more hindered) alkyl halide.

25.58. The side chain (R) of glycine is a hydrogen atom (H). Therefore, no alkyl group needs to be installed at the α position.

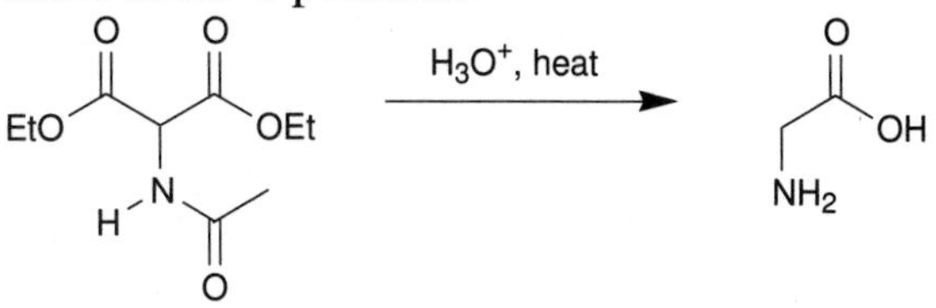

H_3O^+, heat

25.59.

a) 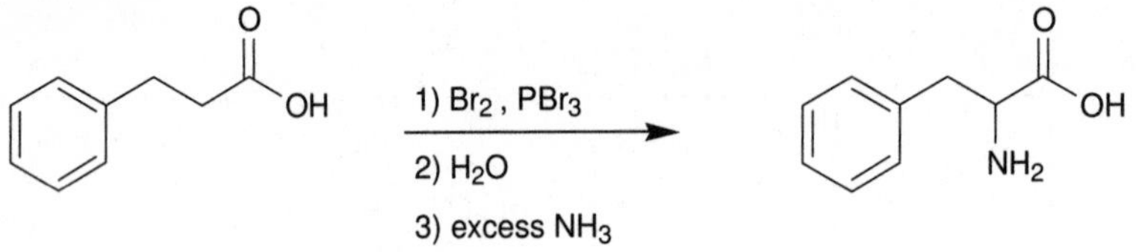 1) Br_2, PBr_3
2) H_2O
3) excess NH_3

OH

NH₂ OH

b) 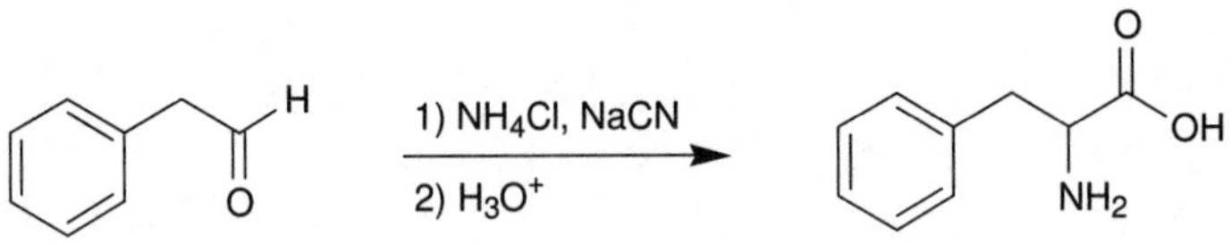 H O

1) NH_4Cl, NaCN
2) H_3O^+

OH
NH_2

c) 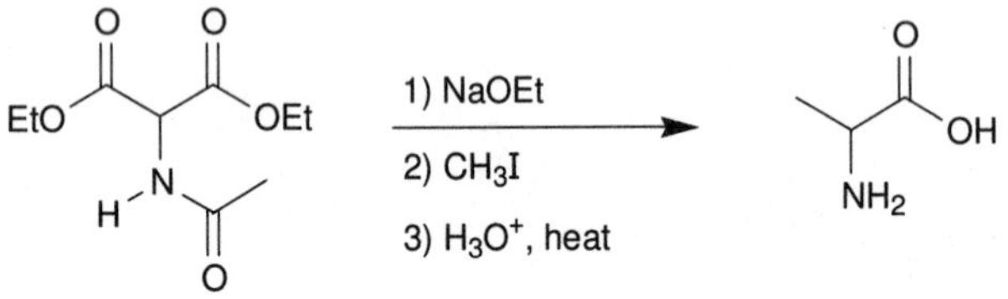 EtO OEt H N O

1) NaOEt
2) CH_3I
3) H_3O^+, heat

OH
NH_2

25.60.

a) 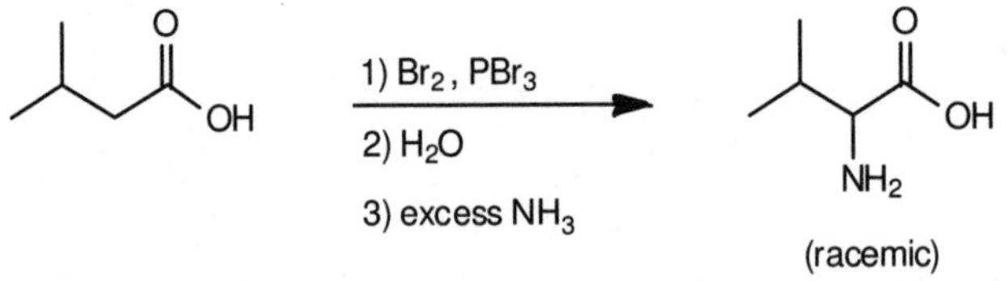 OH

1) Br_2, PBr_3
2) H_2O
3) excess NH_3

OH
NH_2
(racemic)

b) 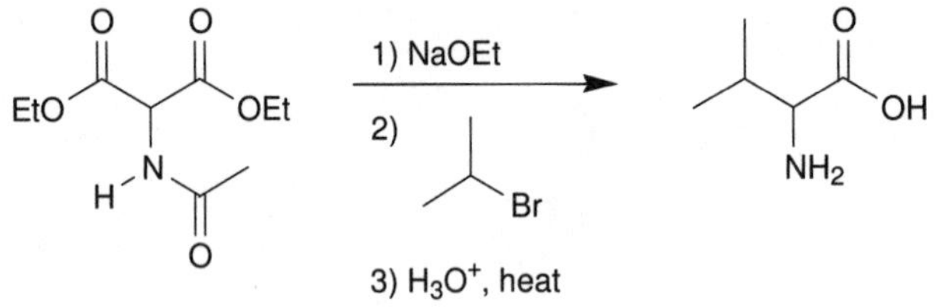 EtO OEt H N O

1) NaOEt
2) Br
3) H_3O^+, heat

OH
NH_2

c) 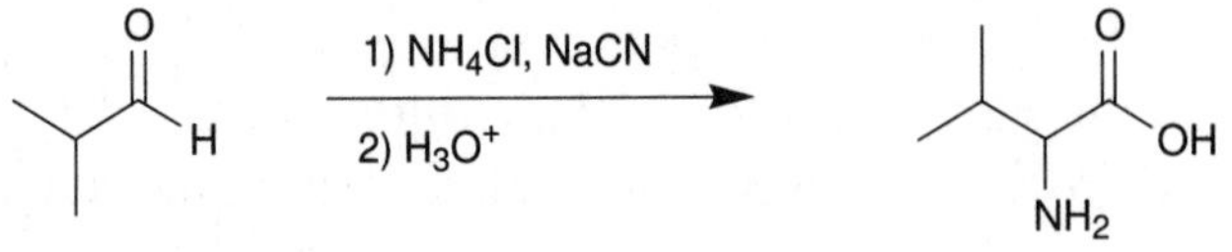 O H

1) NH_4Cl, NaCN
2) H_3O^+

OH
NH_2

25.61. $20^5 = 3{,}200{,}000$

25.62.

N NH NHAc OH

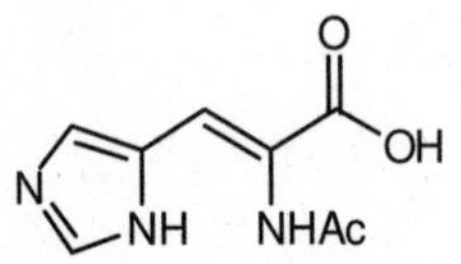

25.63.

1) Leu-Met-Val, 2) Leu-Val-Met, 3) Met-Val-Leu,
4) Met-Leu-Val, 5) Val-Met-Leu, 6) Val-Leu-Met

25.64.

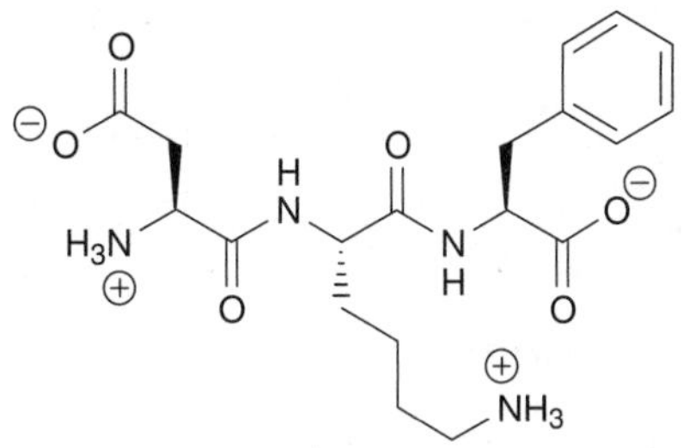

25.65.

N Terminus C Terminus

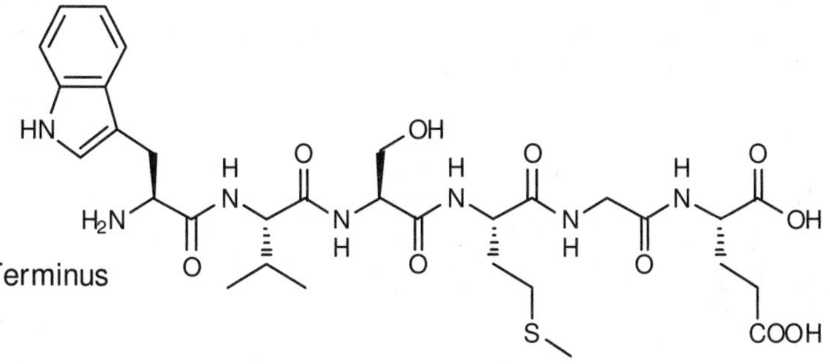

25.66.

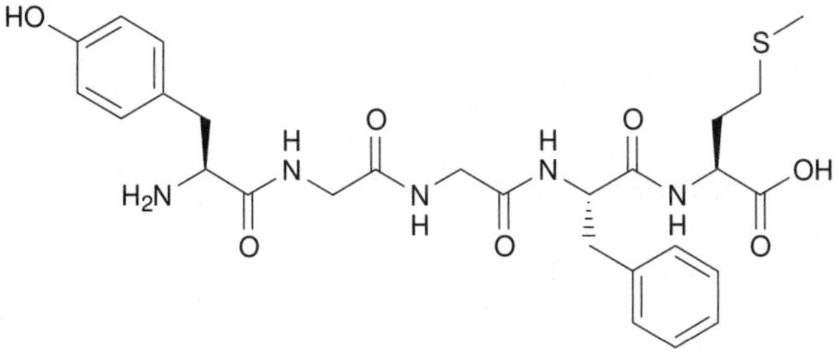

25.67.

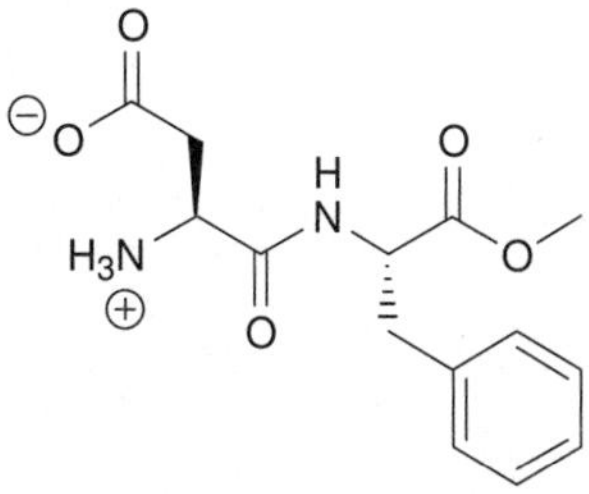

25.68.

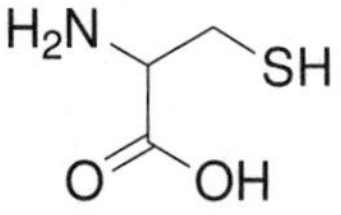

cysteine 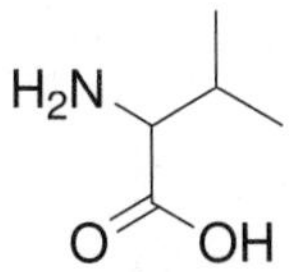 valine

25.69.

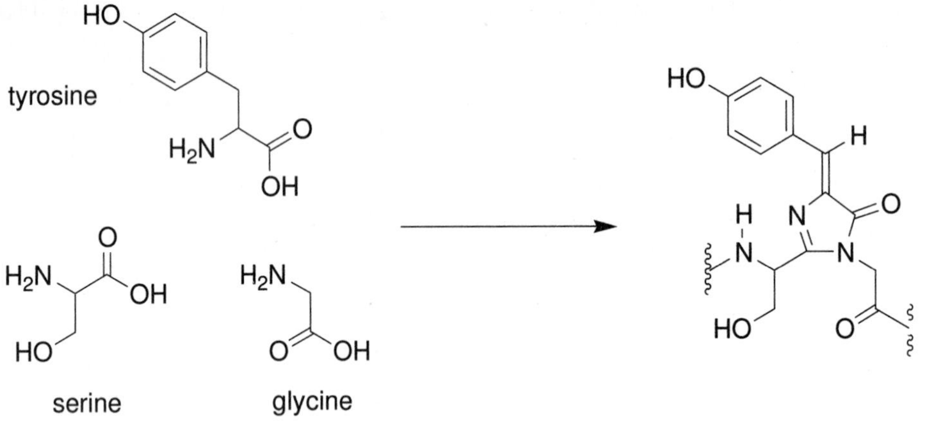

tyrosine

serine

glycine

25.70. It does not react with phenyl isothiocyanate so it must not have a free N terminus. It must be a cyclic tripeptide:

Gly
Ala Phe

Gly
Phe Ala

25.71.

a) ***Arg*** + ***Pro-Pro-Gly-Phe-Ser-Pro-Phe-Arg***

b) ***Arg-Pro-Pro-Gly-Phe*** + ***Ser-Pro-Phe*** + ***Arg***

25.72. Phenylalanine

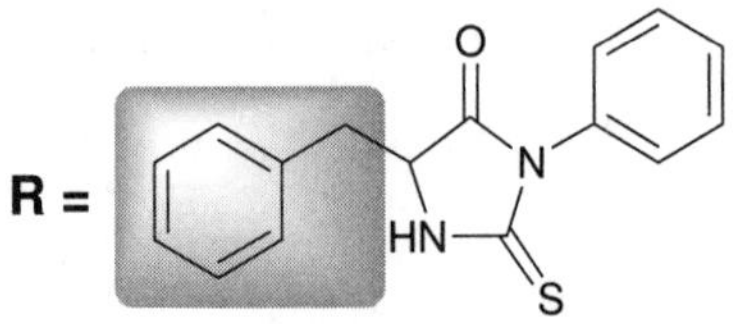

R =

25.73. Val-Ala-Gly:

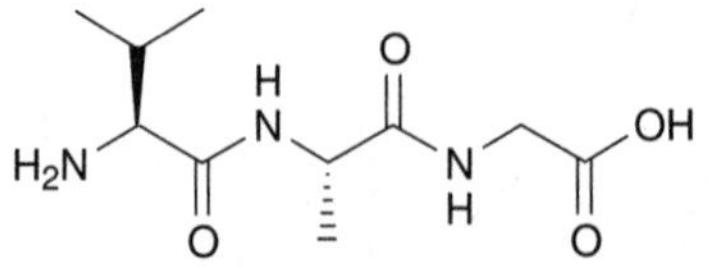

25.74. There cannot be any disulfide bridges in this peptide, because it has no cysteine residues, and only cysteine residues form disulfide bridges.

His-Ser-Gln-Gly-Thr-Phe-Thr-Ser-Asp-Tyr-Ser-Lys-Tyr-Leu-Asp-Ser-Arg-Arg-Ala-Gln-Asp-Phe-Val-Gln-Trp-Leu-Met-Asn-Thr

25.75. Prior to acylation, the nitrogen atom of the amino group is sufficiently nucleophilic to attack phenyl isothiocyanate. Acylation converts the amino group into an amide moiety, and the lone pair of the nitrogen atom is delocalized via resonance, rendering it much less nucleophilic.

25.76.

a) 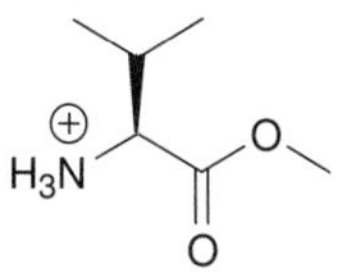 $H_3N^{\oplus}$ ... O ... O

b) 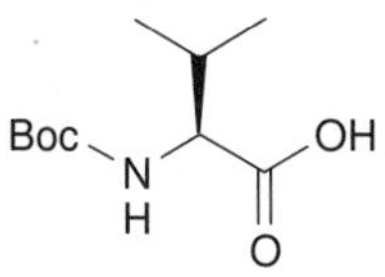 Boc N H ... OH ... O

c) 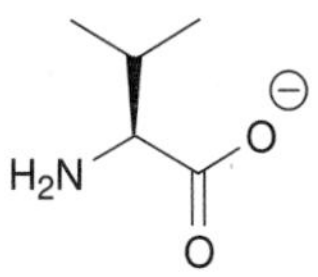 H_2N ... $O^{\ominus}$ ... O

d) 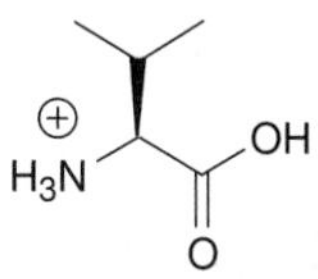 $H_3N^{\oplus}$ ... OH ... O

25.77.

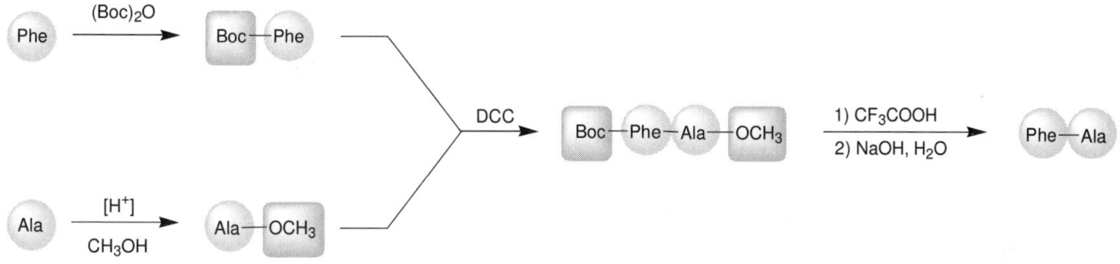

Phe —$(Boc)_2O$→ Boc—Phe

Ala —$[H^+]$, CH_3OH→ Ala—OCH_3

—DCC→ Boc—Phe—Ala—OCH_3 —1) CF_3COOH 2) NaOH, H_2O→ Phe—Ala

25.78.

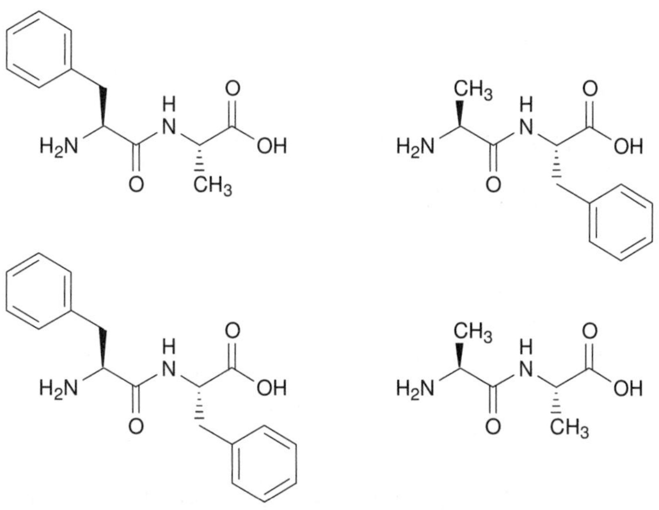

H_2N H N O OH O CH_3

CH_3 H_2N H N O OH O

H_2N H N O OH O

CH_3 H_2N H N O OH O CH_3

25.79.

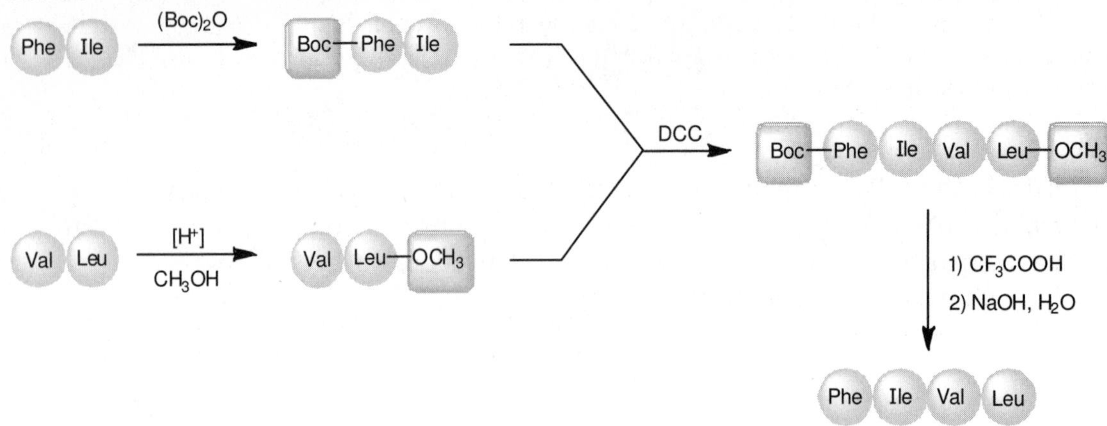

25.80.

Cl POLYMER → Leu-Val-Ala

1) Boc-NH-CH(CH_3)-CO_2^-

2) CF_3COOH

3) Boc-NH-CH($CH(CH_3)_2$)-COOH , DCC

4) CF_3COOH

5) Boc-NH-CH($CH_2CH(CH_3)_2$)-COOH , DCC

6) CF_3COOH

7) HF

25.81. Boc-NH-CH($CH_2CH(CH_3)_2$)-COOH

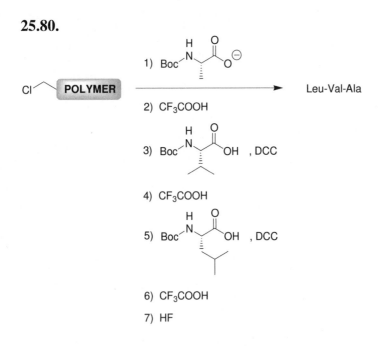

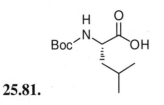

25.82. A proline residue cannot be part of an α helix, because it lacks an N-H proton and does not participate in hydrogen bonding. (The amino acid proline does indeed have an N-H group, but when incorporated into a peptide, the proline residue does not have an N-H group)

25.83.

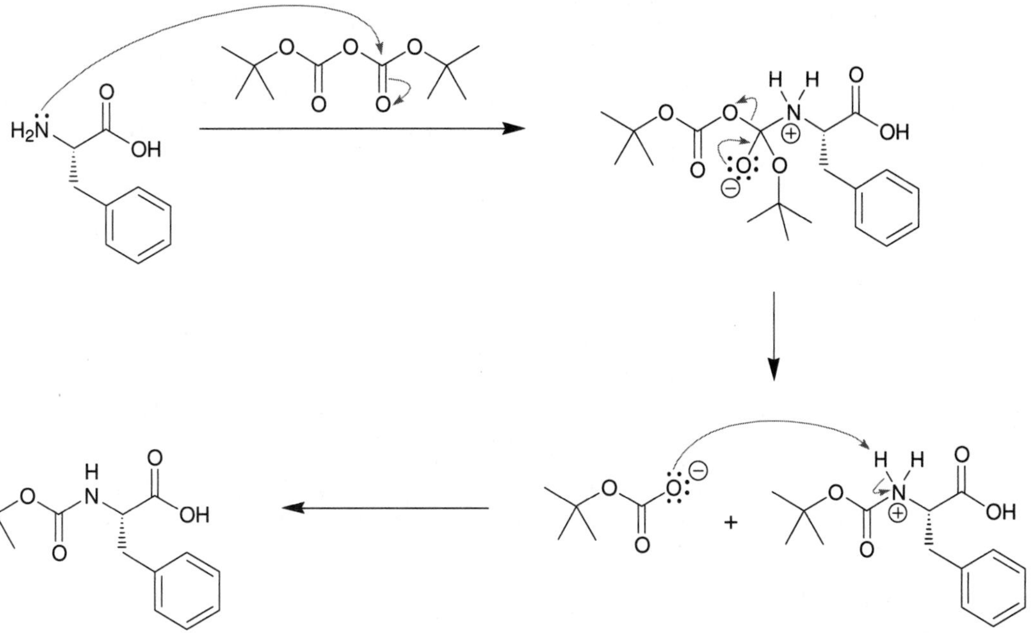

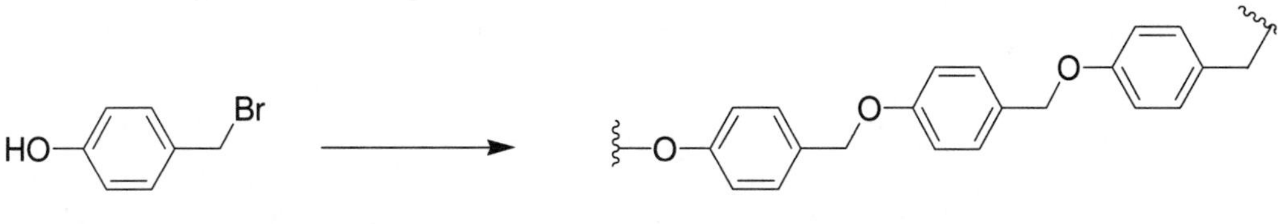

25.84.

25.85. The stabilized enolate ion (formed in the first step) can function as a base, rather than a nucleophile, giving an E2 reaction:

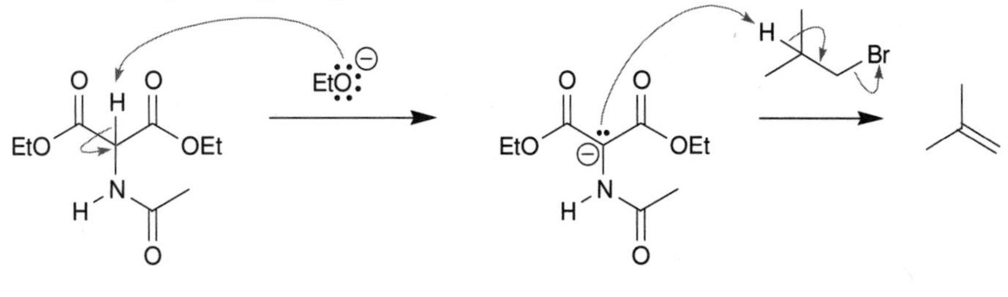

25.86. The lone pair on that nitrogen atom is highly delocalized via resonance and is participating in aromaticity. Accordingly, the lone pair is not free to function as a base.

25.87.

a) HO–C₆H₄–CH₂CH₂–C(=O)OH

b) HO–C₆H₄–CH₂CH₂–C(=O)OH $\xrightarrow{Br_2}$ 3,5-dibromo-4-hydroxyphenyl–CH₂CH₂–C(=O)OH

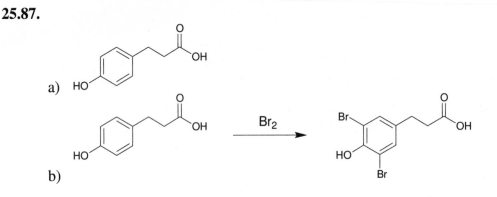

25.88. At low temperature, the barrier to rotation keeps the two methyl groups in different electronic environments (one is *cis* to the C=O bond and the other is *trans* to the C=O bond), and they therefore give rise to separate signals. At high temperature, there is sufficient energy to overcome the energy barrier, and the protons change electronic environments on a timescale that is faster than the timescale of the NMR spectrometer. The result is an averaging effect which gives rise to only one signal.

25.89.

a) The COOH group does not readily undergo nucleophilic acyl substitution because the OH group is not a good leaving group. By converting the COOH group into an activated ester, the compound can now undergo nucleophilic acyl substitution because it has a good leaving group.
b) The nitro group stabilizes the leaving group via resonance. As described in Chapter 19, the nitro group serves as a reservoir for electron density.
c) The nitro group must be in the ortho or para position in order to stabilize the negative charge via resonance. If the nitro group is in the meta position, the negative charge cannot be pushed onto the nitro group.

25.90.

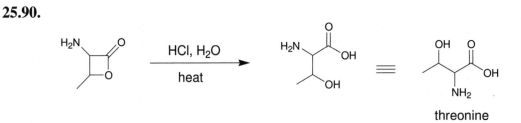

H_2N–(β-lactone) $\xrightarrow[\text{heat}]{HCl,\ H_2O}$ H_2N–CH(COOH)–CH(OH)–CH_3 ≡ threonine

Chapter 26
Lipids

Review of Concepts

Fill in the blanks below. To verify that your answers are correct, look in your textbook at the end of Chapter 26. Each of the sentences below appears verbatim in the section entitled *Review of Concepts and Vocabulary*.

- **Lipids** are naturally occurring compounds that are extracted from cells using _______________ solvents.
- **Complex lipids** readily undergo ______________, while **simple lipids** do not.
- _________ are high molecular weight esters that are constructed from carboxylic acids and alcohols.
- ___________________ are the triesters formed from glycerol and three long-chain carboxylic acids, called **fatty acids**. The resulting triglyceride is said to contain three **fatty acid _____________.**
- For saturated fatty acids, the melting point increases with increasing __________ ___________. The presence of a _____ double bond causes a decrease in the melting point.
- Triglycerides that are solids at room temperature are called ______, while those that are liquids at room temperature are called _______.
- Triglycerides containing unsaturated fatty acid residues will undergo hydrogenation. During the hydrogenation process, some of the double bonds can isomerizes to give _______ π bonds
- In the presence of molecular oxygen, triglycerides are particularly susceptible to oxidation at the ____________ position to produce hydroperoxides.
- Transesterification of triglycerides can be achieved either via _____ catalysis or ______ catalysis to produce biodiesel.
- ________________ are similar in structure to triglycerides except that one of the three fatty acid residues is replaced by a phosphoester group.
- The structures of **steroids** are based on a tetracyclic ring system, involving three six-membered rings and one ______-membered ring.
- The ring fusions are all _______ in most steroids, giving steroids their rigid geometry.
- All steroids, including cholesterol, are biosynthesized from ____________.
- Prostaglandins contain twenty carbon atoms and are characterized by a ______-membered ring with two side chains.
- **Terpenes** are a class of naturally occurring compounds that can be thought of as being assembled from _________ units.
- A terpene with 10 carbon atoms is called a _______________, while a terpene with 20 carbon atoms is called a _________________.

Review of Skills

Fill in the blanks and empty boxes below. To verify that your answers are correct, look in your textbook at the end of Chapter 26. The answers appear in the section entitled *SkillBuilder Review.*

26.1 Comparing Molecular Properties of Triglycerides

CIRCLE THE TRIGLYCERIDE BELOW THAT IS EXPECTECD TO HAVE A HIGHER MELTING POINT.

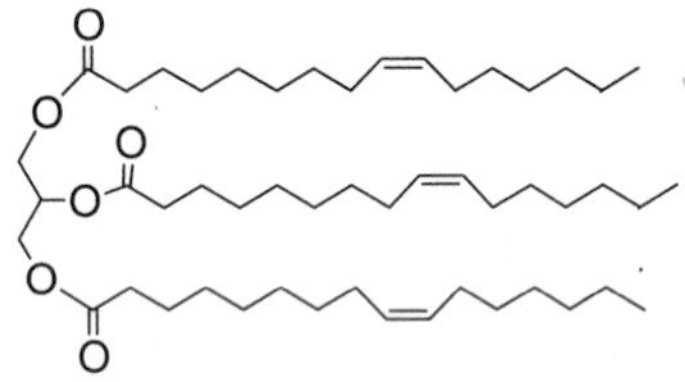

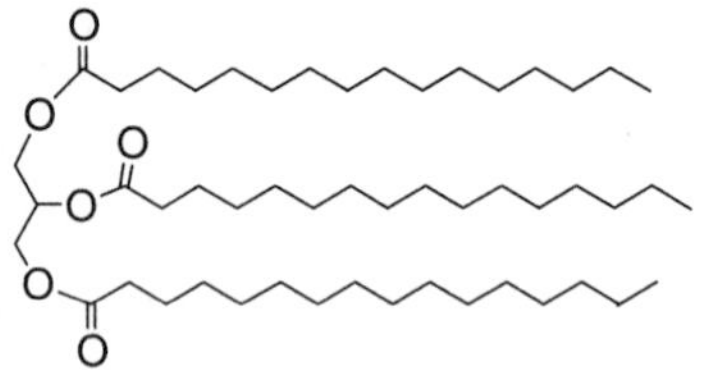

26.2 Identifying the Products of Triglyceride Hydrolysis

DRAW THE PRODUCTS OBTAINED WHEN THE FOLLOWING TRIGLYCERIDE IS TREATED WITH AQUEOUS SODIUM HYDROXIDE.

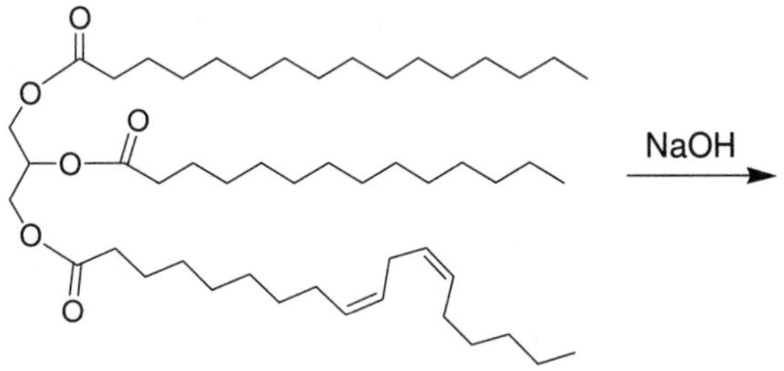

NaOH

26.3 Drawing a Mechanism for Transesterification of a Triglyceride

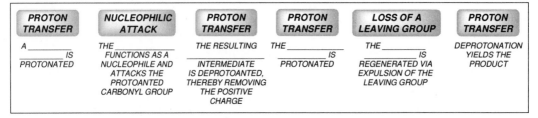

PROTON TRANSFER	NUCLEOPHILIC ATTACK	PROTON TRANSFER	PROTON TRANSFER	LOSS OF A LEAVING GROUP	PROTON TRANSFER
A ________ ________ IS PROTONATED	*THE ________ FUNCTIONS AS A NUCLEOPHILE AND ATTACKS THE PROTOANTED CARBONYL GROUP*	*THE RESULTING ________ INTERMEDIATE IS DEPROTOANTED, THEREBY REMOVING THE POSITIVE CHARGE*	*THE ________ ________ IS PROTONATED*	*THE ________ ________ IS REGENERATED VIA EXPULSION OF THE LEAVING GROUP*	*DEPROTONATION YIELDS THE PRODUCT*

26.4 Identifying Isoprene Units in a Terpene

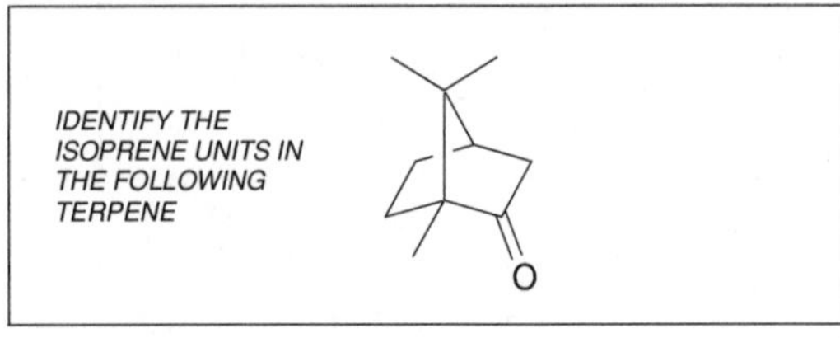

IDENTIFY THE ISOPRENE UNITS IN THE FOLLOWING TERPENE
O

Review of Reactions

Identify the reagents necessary to achieve each of the following transformations. To verify that your answers are correct, look in your textbook at the end of Chapter 26. The answers appear in the section entitled *Review of Reactions*.

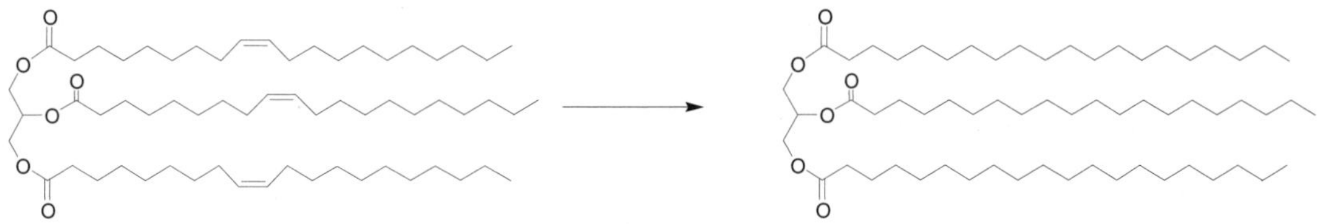

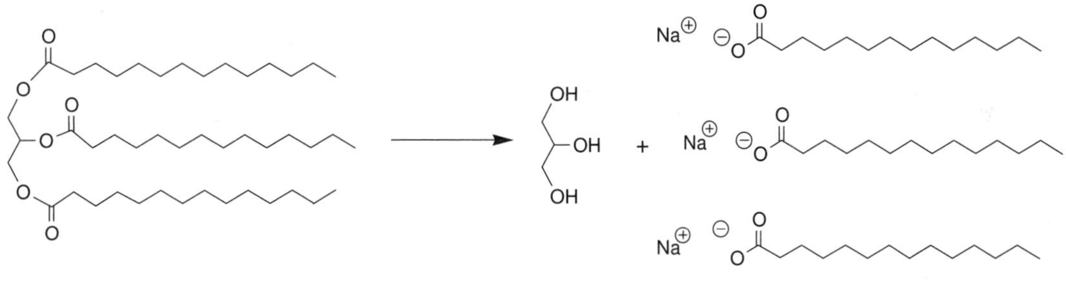

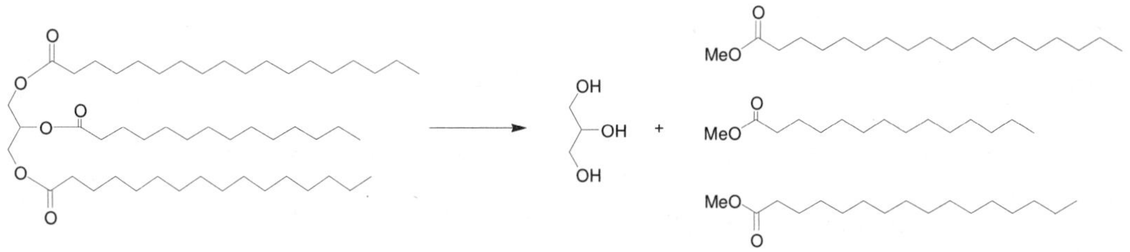

O O O OH OH OH $Na^{\oplus}$ $^{\ominus}O$ MeO

Solutions

26.1

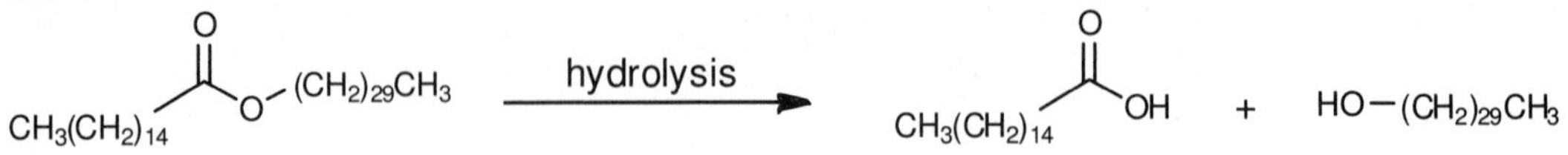

$CH_3(CH_2)_{14}C(=O)O(CH_2)_{29}CH_3$ —hydrolysis→ $CH_3(CH_2)_{14}C(=O)OH$ + $HO-(CH_2)_{29}CH_3$

26.2

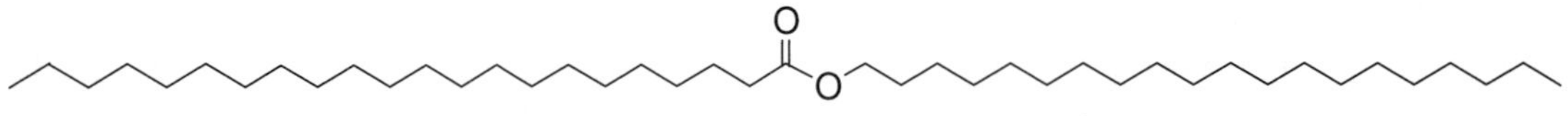

26.3.

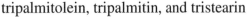

a) trimyristin° b) triarachadin c) triolein d) tristearin

26.4.

tripalmitolein, tripalmitin, and tristearin

26.5. The fatty acid residues in triarachadin have more carbon atoms than the fatty acid residues in tristearin. Therefore, triarachadin is expected to have a higher melting point. It should be a solid at room temperature, and should therefore be classified as a fat, rather than an oil. Therefore, triglycerides made from lauric acid will also have a low melting point.

26.6.

a) All three fatty acid residues are saturated, with either 16 or 18 carbon atoms, so the triglyceride is expected to have a high melting point. It should be a solid at room temperature, so it is a fat.

b) All three fatty acid residues are unsaturated, so the triglyceride is expected to have a low melting point. It should be a liquid at room temperature, so it is an oil.

26.7.

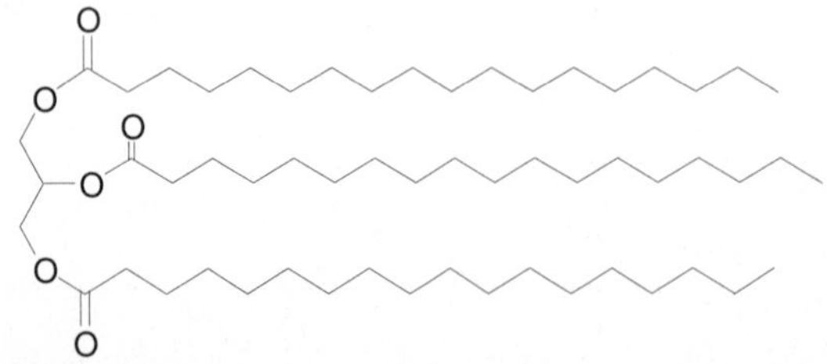

a)

b) Tristearin

c) The melting point of tristearin is higher than triolein.

d) Stearic acid

26.8.

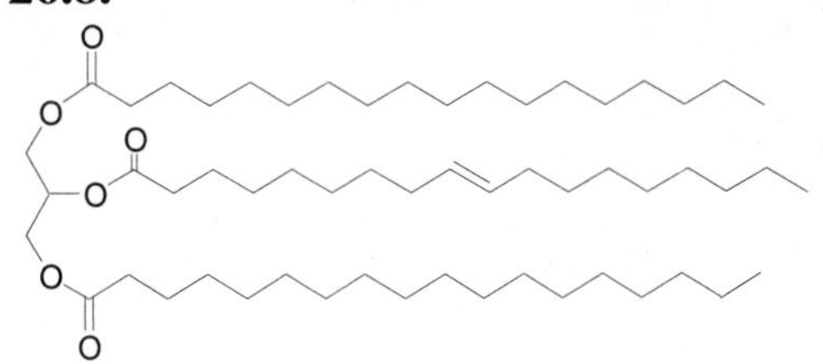

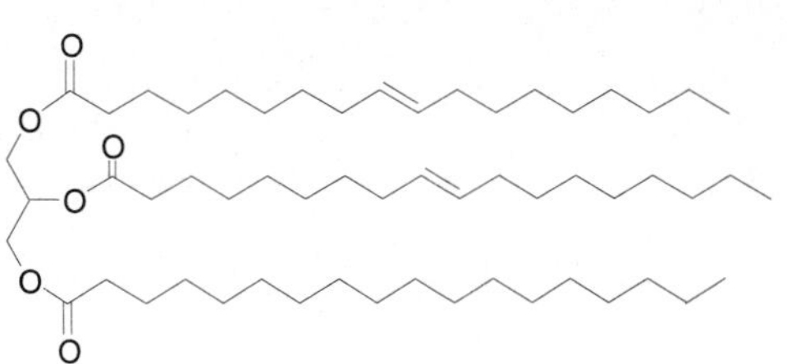

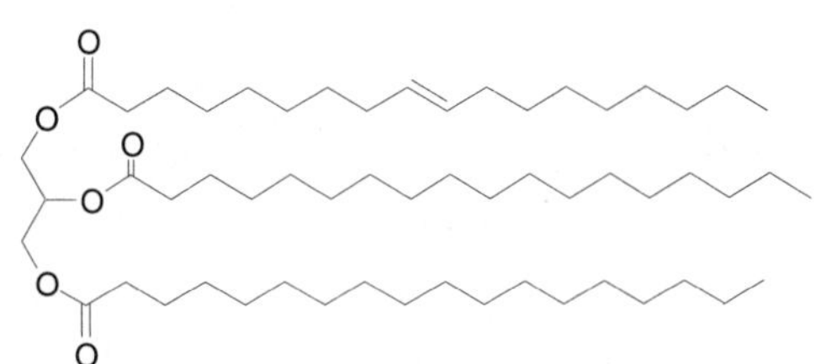

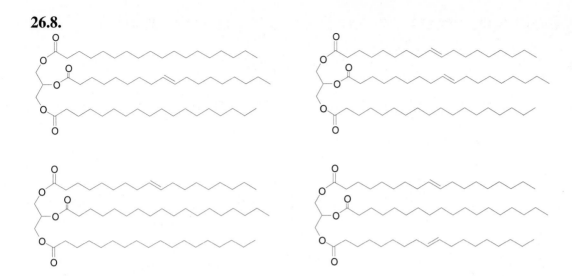

26.9.

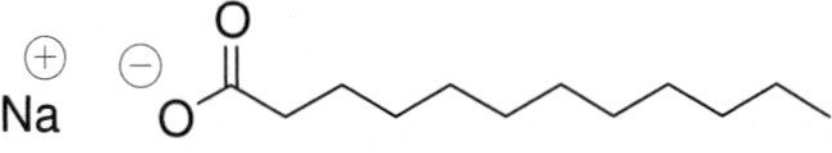

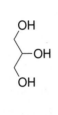

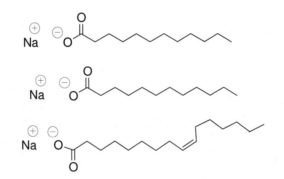

26.10.

26.11

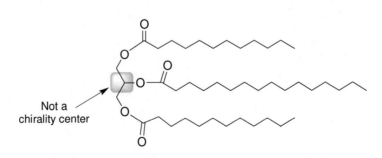

Not a chirality center

26.12. Each of the three ester moieties is hydrolyzed via the following mechanism:

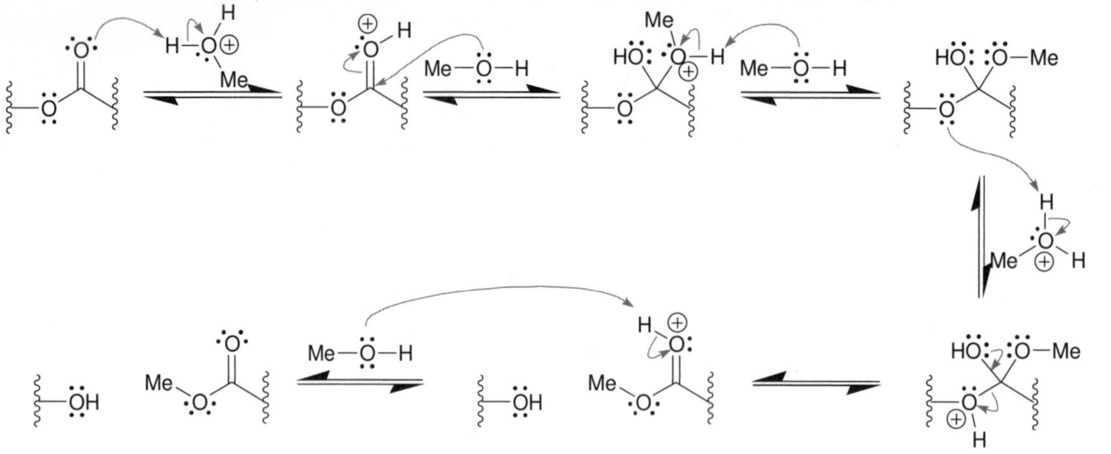

26.13.

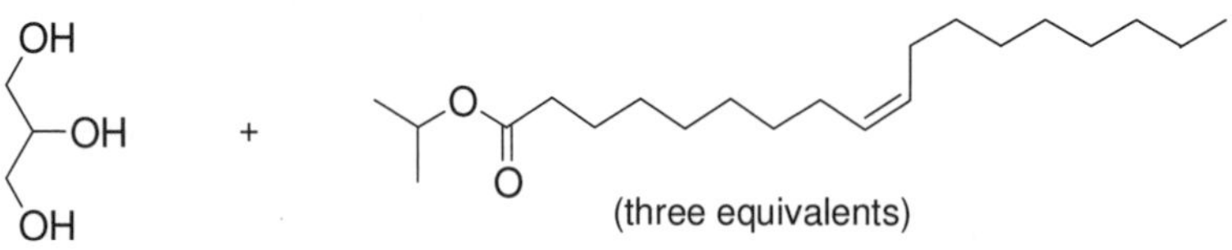

(three equivalents)

26.14.

a) Hydroxide functions as a catalyst by establishing an equilibrium in which some ethoxide ions are present.

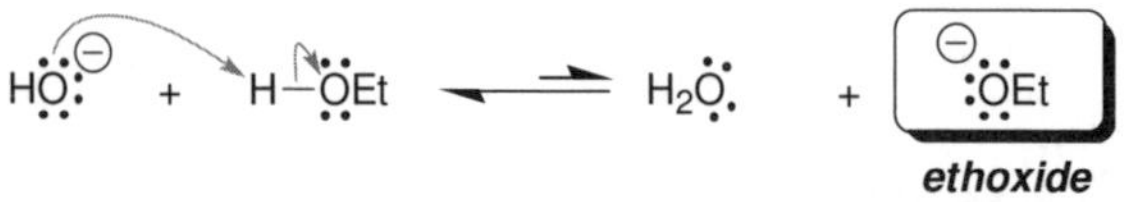

ethoxide

Then, each ester moiety undergoes transesterification via the following mechanism:

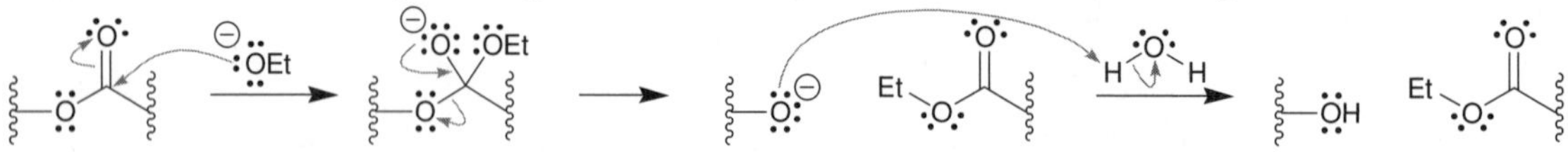

b) Hydroxide could function as a nucleophile and triglyceride would undergo hydrolysis rather than transesterification.

26.15.

a)

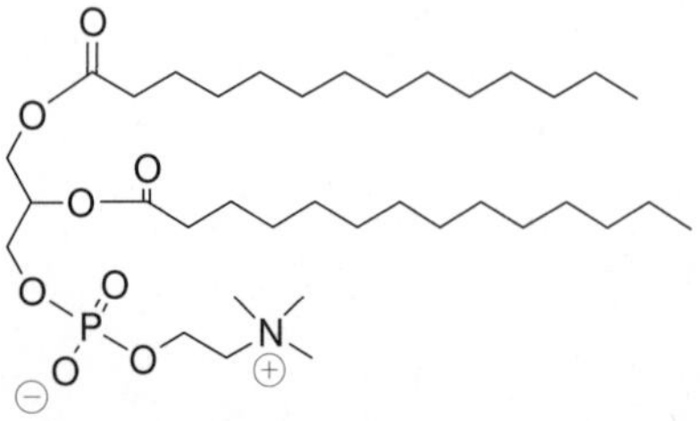

b)

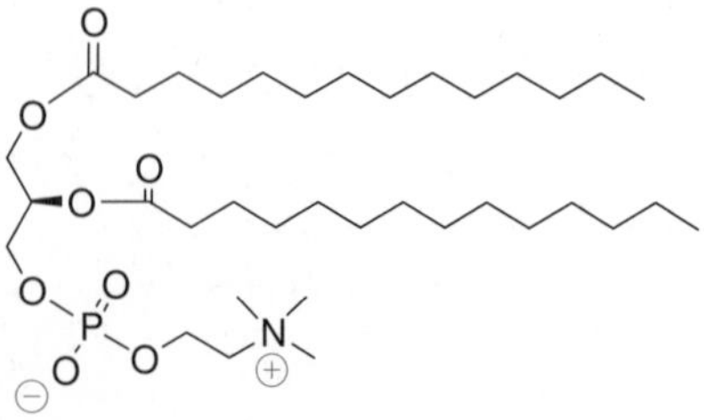

c) No. The C2 position would no longer be a chirality center.

26.16.

a)

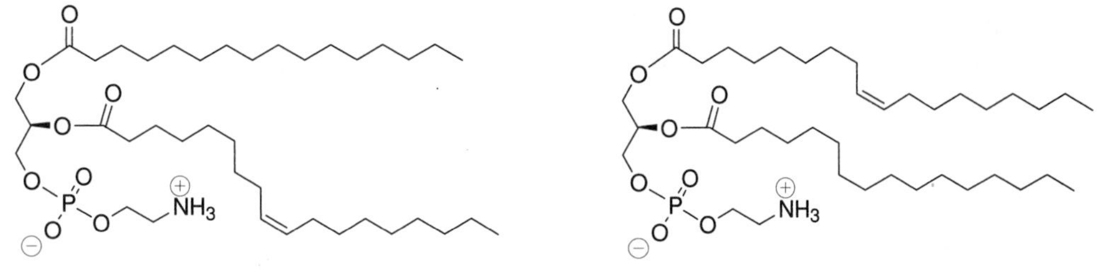

b) Yes. The C2 position would still be a chirality center.

26.17.

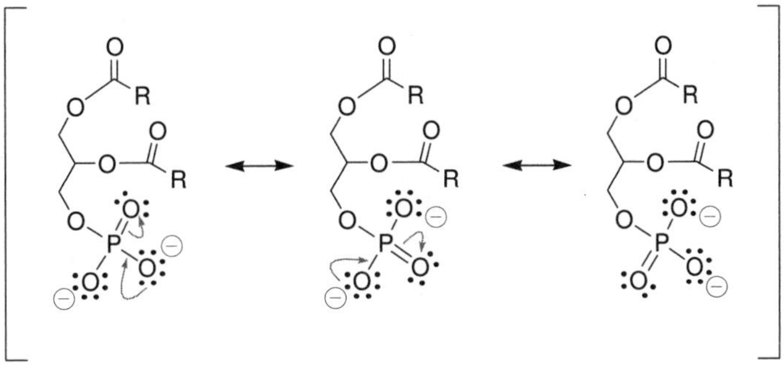

26.18. Octanol has a longer hydrophobic tail than hexanol and is therefore more efficient at crossing the nonpolar environment of the cell membrane.

26.19. No. Glycerol has three OH groups (hydrophilic) and no hydrophobic tail. It cannot cross the nonpolar environment of the cell membrane.

26.20. A ring-flip is not possible for *trans*-decalin because one of the rings would have to achieve a geometry that resembles a six-membered ring with a *trans*-alkene, which is not possible. The ring fusions of cholesterol all resemble the ring fusion in *trans*-decalin, so cholesterol cannot undergo ring-flipping.

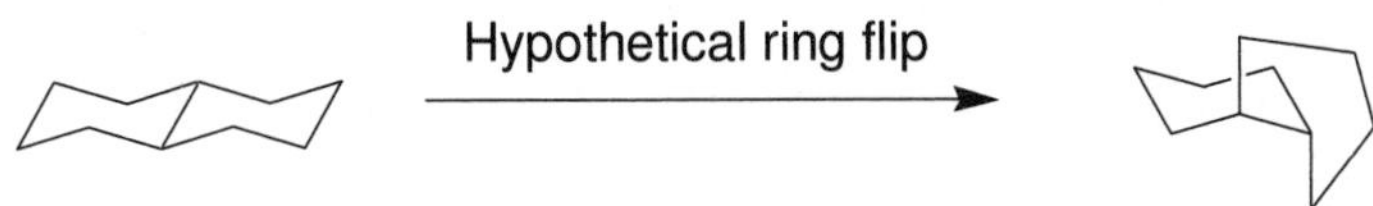

26.21.

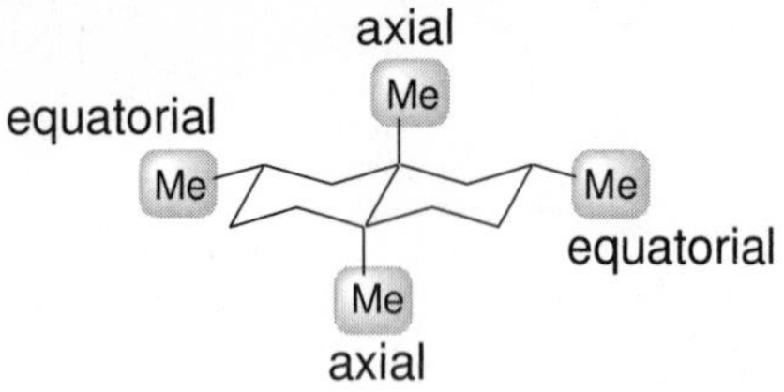
axial
Me
equatorial
Me
Me
equatorial
Me
axial

a)

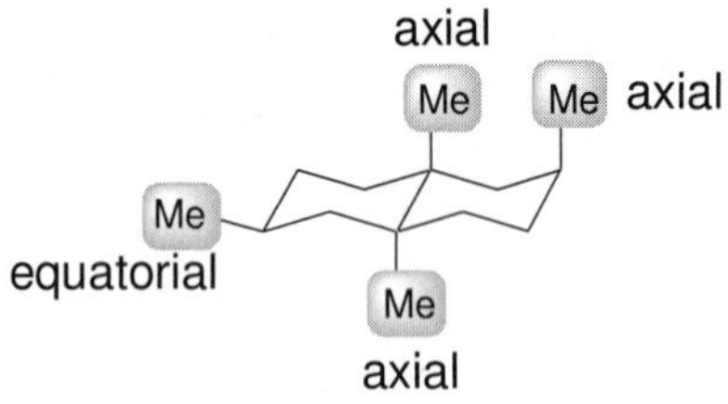
axial
Me
Me axial
Me
equatorial
Me
axial

b)

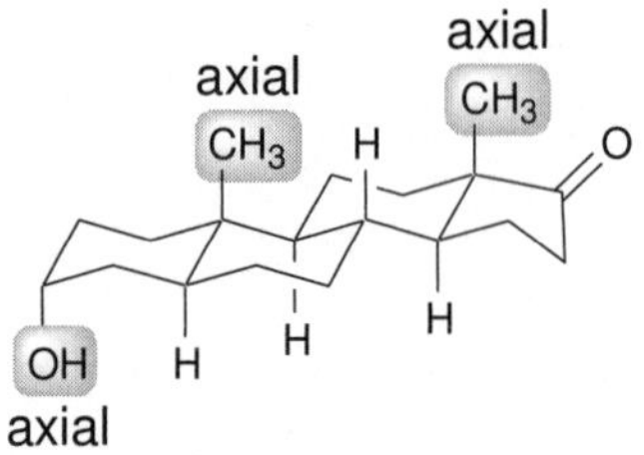
axial
axial
CH3
CH3
H
O
OH
H
H
H
axial

c)

26.22.

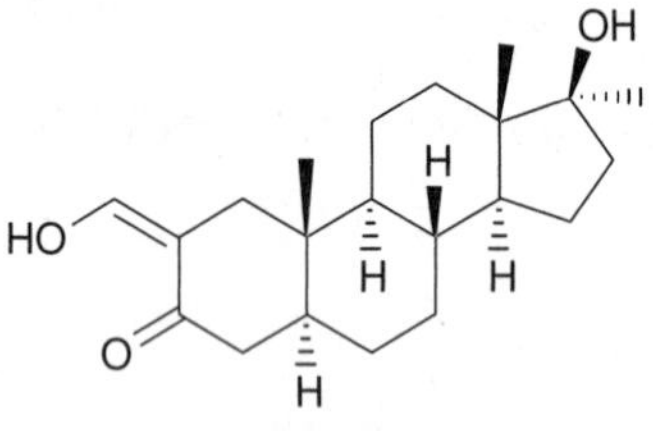
OH
HO
H
H
H
O
H

oxymetholone

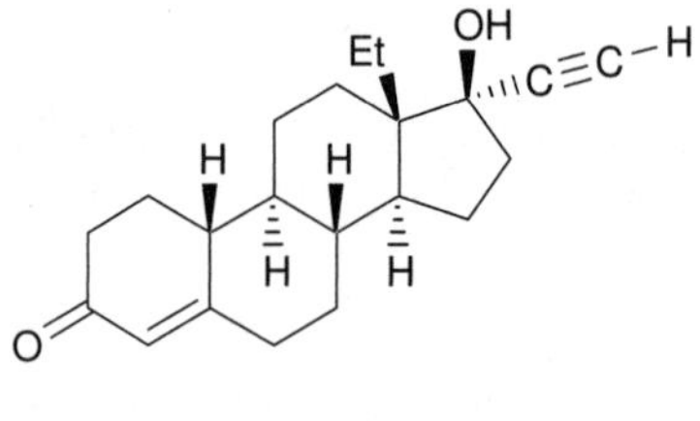
Et
OH
C≡C–H
H
H
H
H
O

norgestrel

26.23.

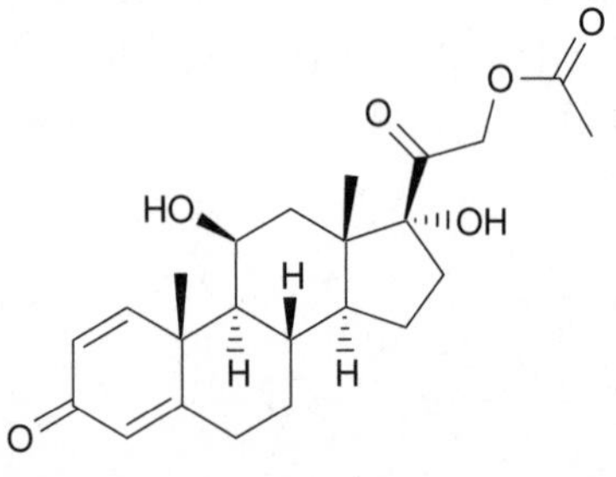
O
O
O
HO
OH
H
H
H
O

26.24.

a) PGE_1 b) $PGF_{1\alpha}$

26.25.

a) *menthol* b) *grandisol* c) *carvone*

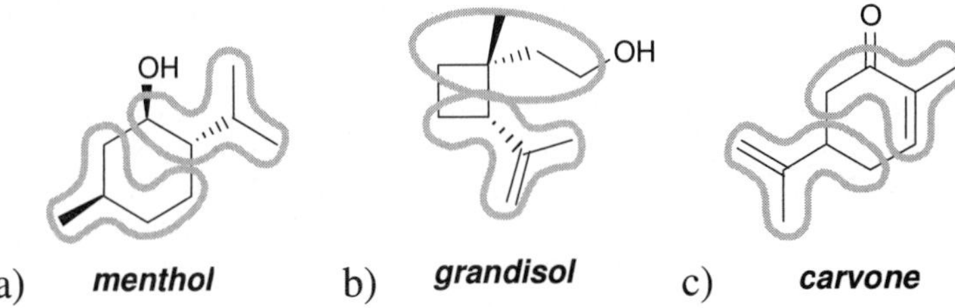

26.26.

a) Yes, it has 10 carbon atoms, which are comprised by the joining of two isoprene units.
b) No, it has 11 carbon atoms.
c) No, it has 11 carbon atoms.
d) No. It has 10 carbon atoms, but the branching pattern cannot be achieved by joining two isoprene units.

26.27.

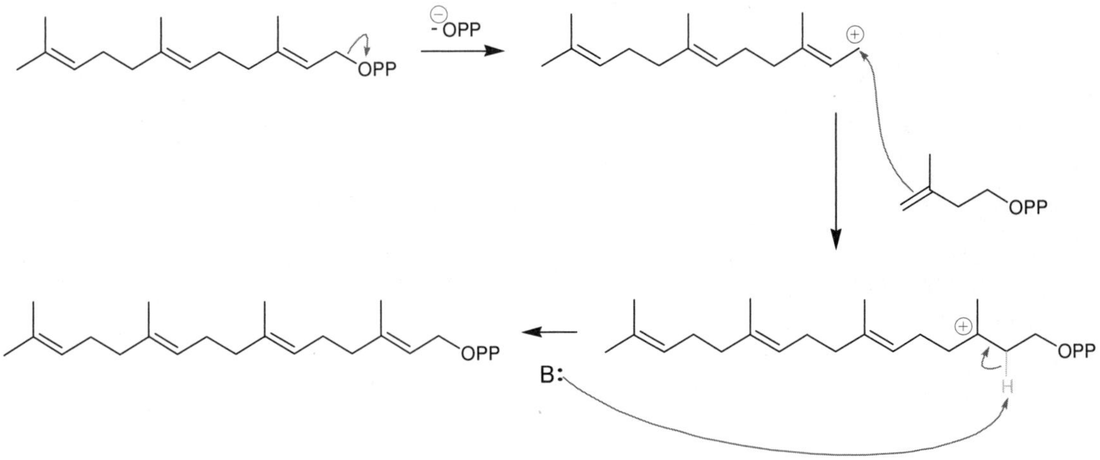

26.28.

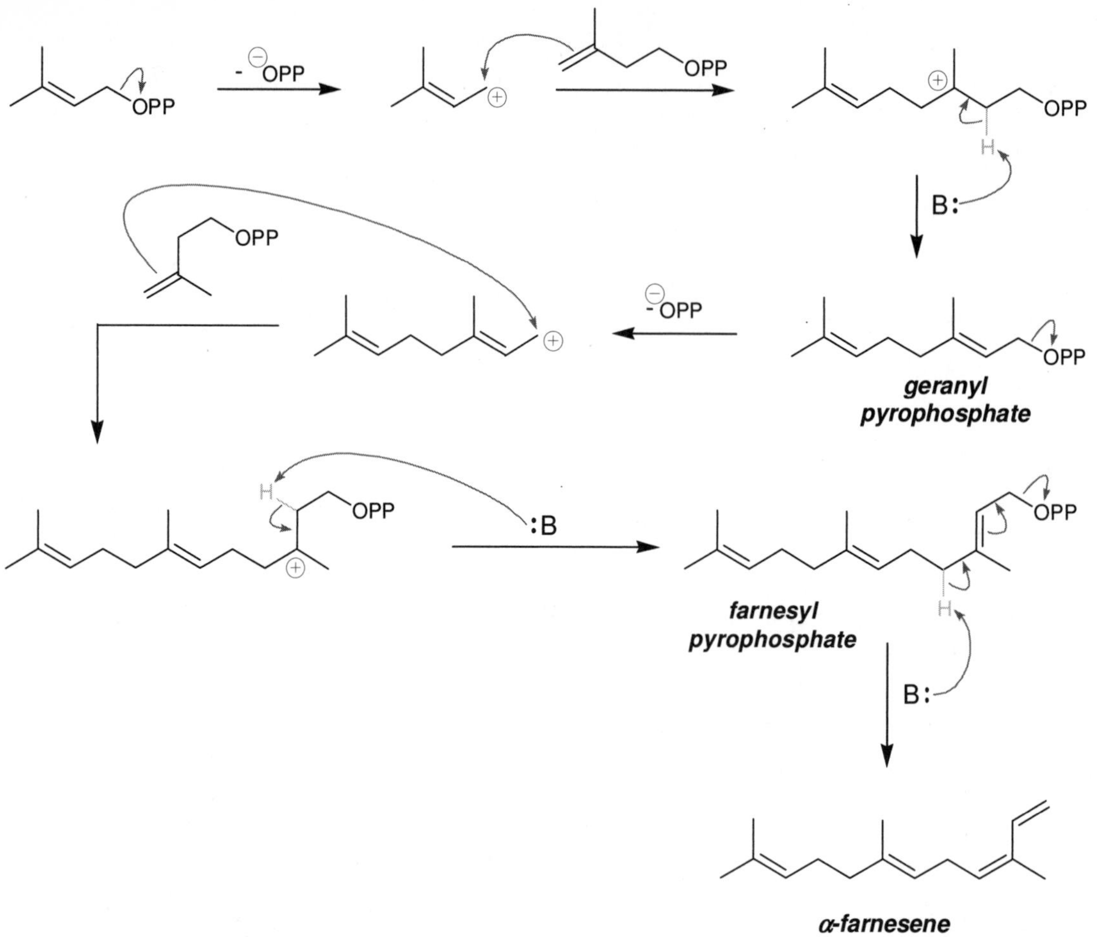

OPP
- OPP
OPP
OPP
B:
OPP
- OPP
OPP
geranyl pyrophosphate
OPP
:B
OPP
farnesyl pyrophosphate
B:
α-farnesene

26.29.

a) steroid
b) terpene
c) triglyceride
d) phospholipid
e) prostaglandin
f) wax

26.30.

a)

b) (three equivalents)

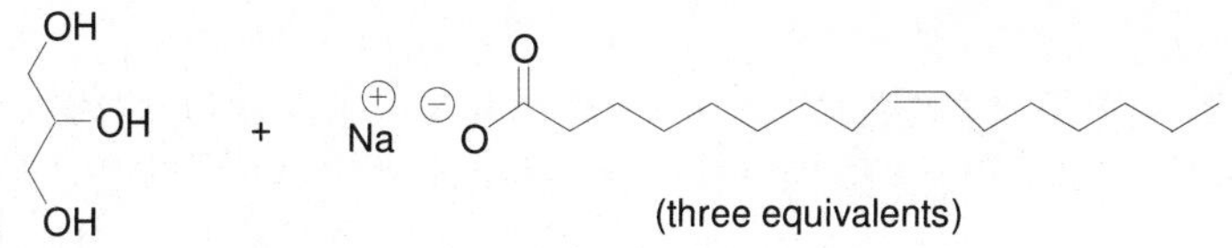

26.31. Both compounds are chiral:

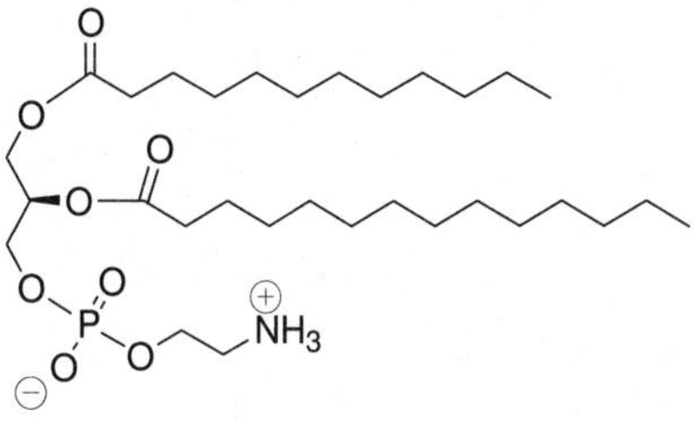

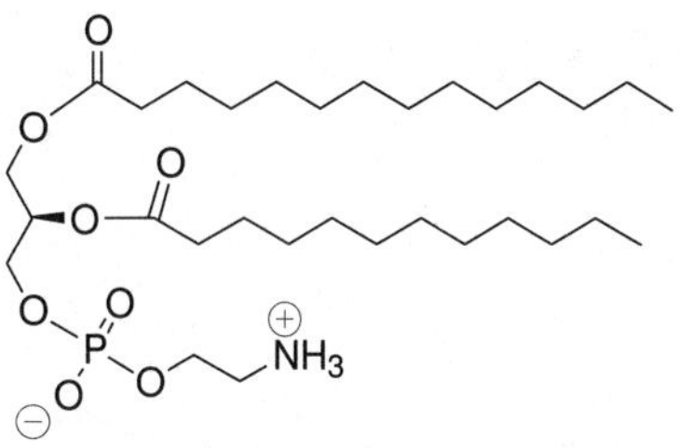

26.32. The fatty acid residues in this triglyceride are saturated, and will not react with molecular hydrogen.

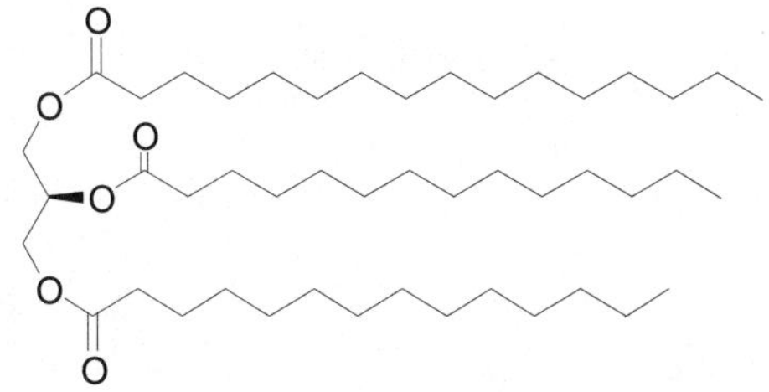

26.33.

a) not a lipid
b) a lipid
c) a lipid
d) a lipid
e) a lipid
f) not a lipid
g) a lipid
h) not a lipid

26.34.

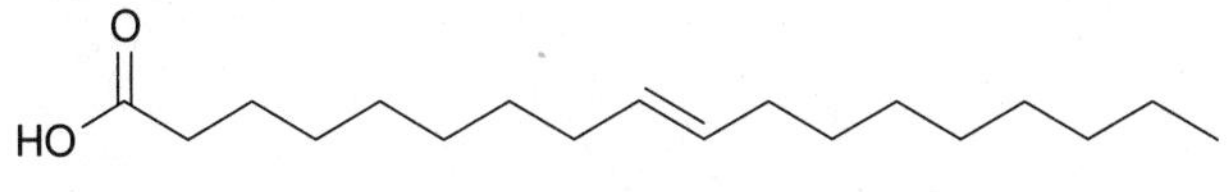

26.35. The fatty acid residues of tristearin are saturated and are therefore less susceptible to auto-oxidation than the unsaturated fatty acid residues in triolein.

26.36. a < b < c

26.37. Water would not be appropriate because it is a polar solvent, and terpenes are nonpolar compounds. Hexane is a nonpolar solvent and would be suitable.

26.38.

a) saturated
b) saturated
c) unsaturated
d) saturated
e) unsaturated
f) unsaturated

26.39. Arachidonic acid

26.40.

a) No. It is an oil.
b) No. It is reactive towards molecular hydrogen in the presence of Ni.
c) Yes. It undergoes hydrolysis to produce unsaturated fatty acids.
d) Yes. It is a complex lipid because it undergoes hydrolysis.
e) No. It is not a wax.
f) No. It does not have a phosphate group.

26.41.

a) Yes. It is a fat.
b) Yes. It is unreactive towards molecular hydrogen in the presence of Ni.
c) No. It undergoes hydrolysis to produce fatty acids that are saturated.
d) Yes. It is a complex lipid because it undergoes hydrolysis.
e) No. It is not a wax.
f) No. It does not have a phosphate group.

26.42.

26.43. Trimyristin is expected to have a lower melting point than tripalmitin because the former is comprised of fatty acid residues with fewer carbon atoms (14 instead of 16).

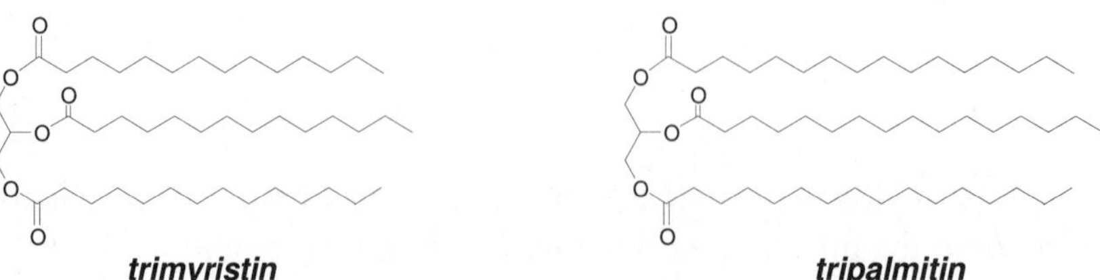

26.44. Each of the three ester moieties is hydrolyzed via the following mechanism:

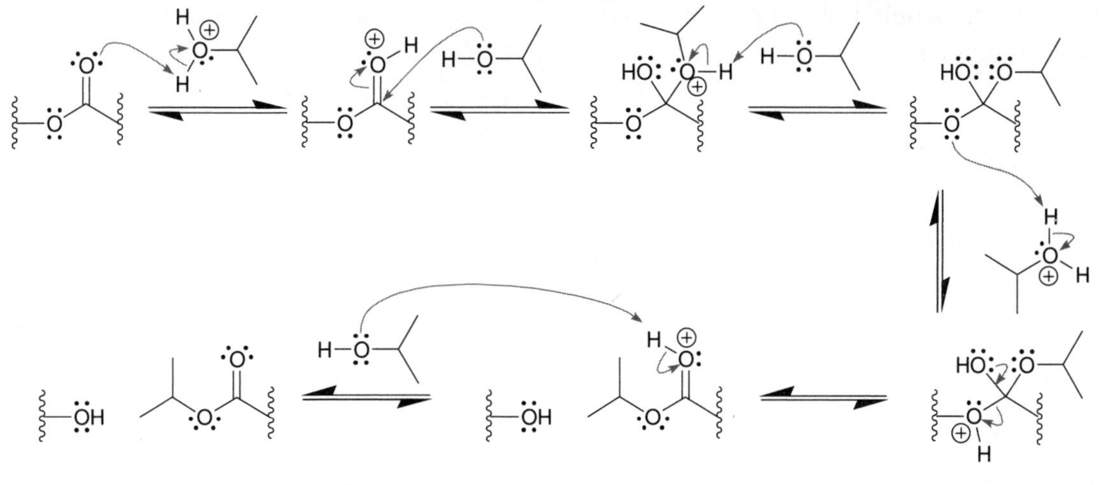

26.45. See the solution to Problem 26.14.

26.46.

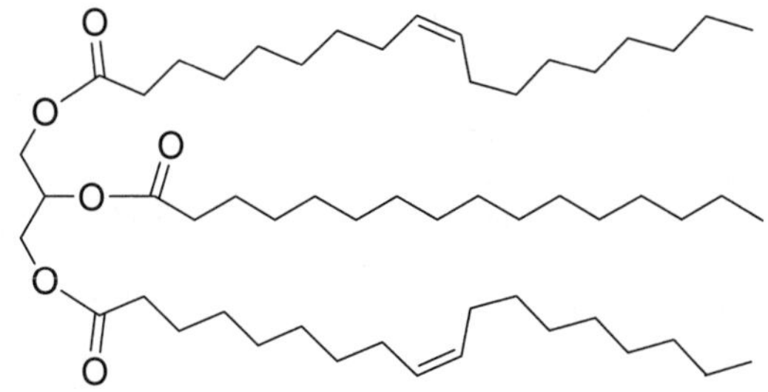

26.47.

26.48.

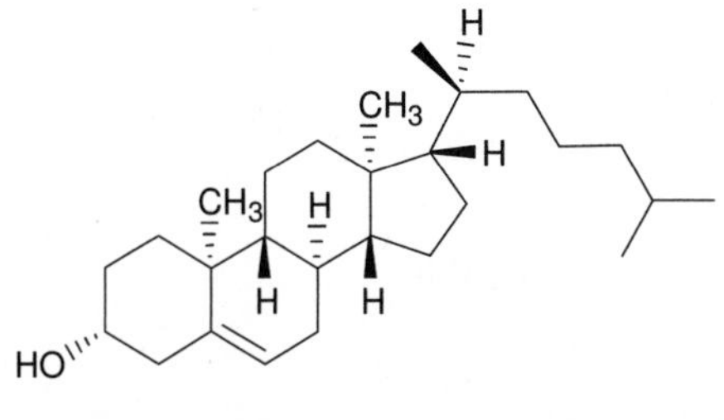

26.49.

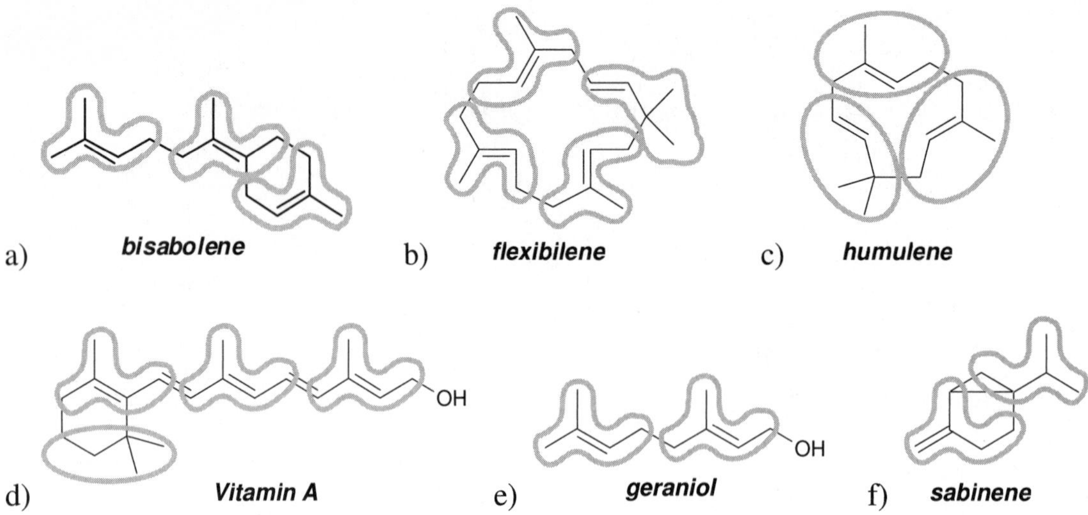

26.50.

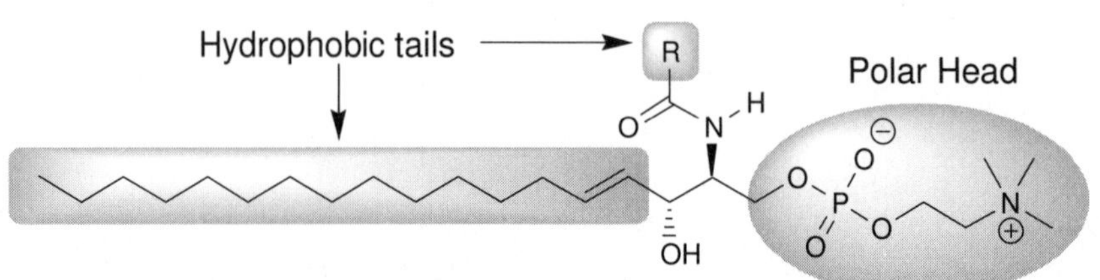

a)
b) Yes, they have one polar head and two hydrophobic tails.

26.51.

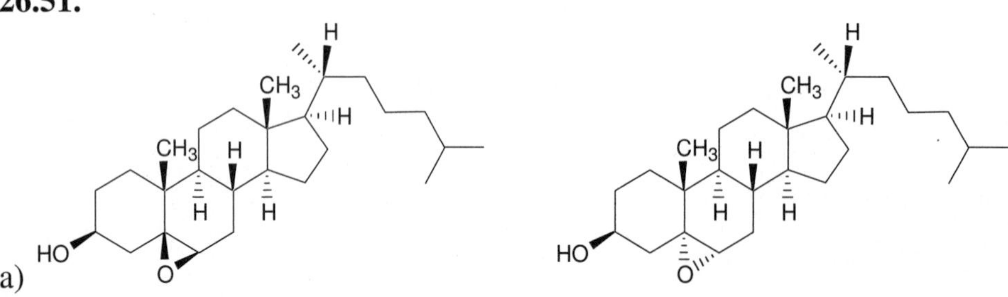

a)

b) The methyl group (C19) provides steric hindrance that blocks one side of the π bond, and only the following is obtained:

26.52.

a)

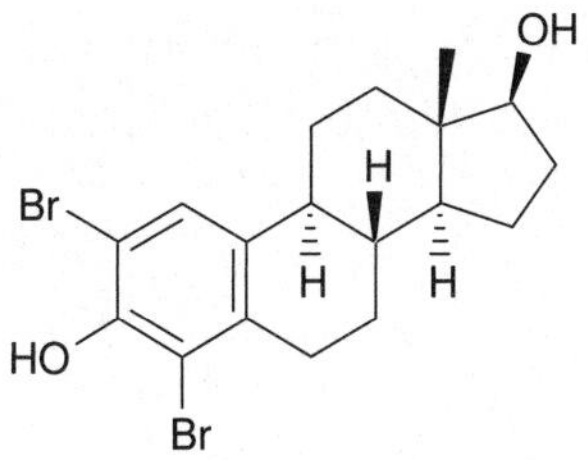

b)

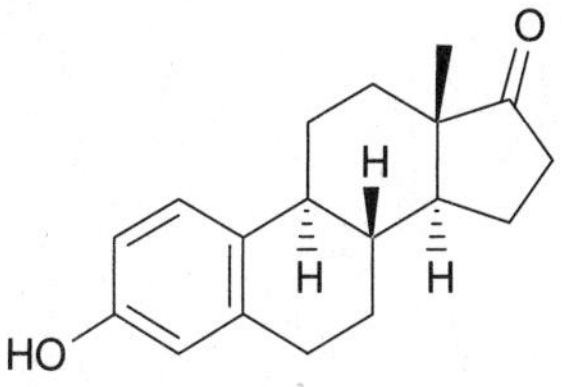

c)

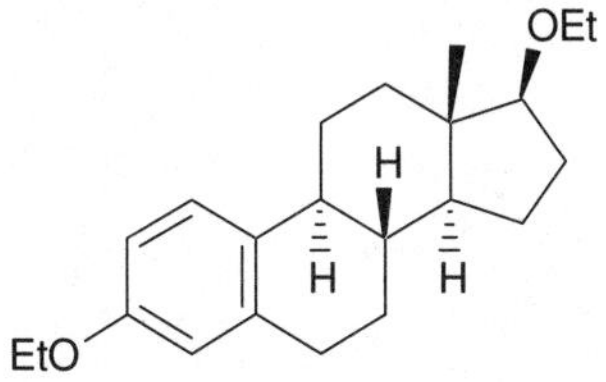

d)

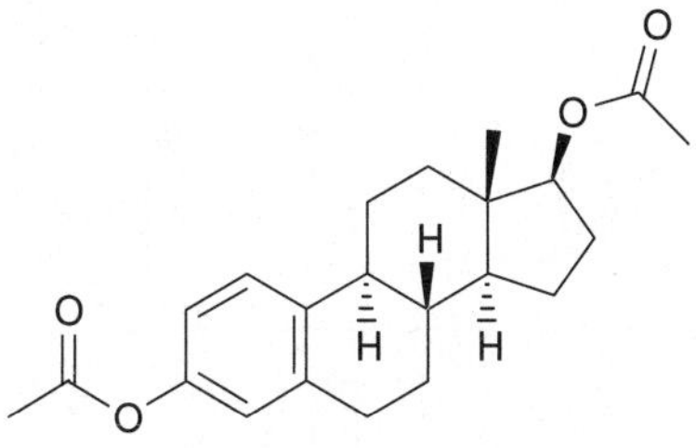

26.53. The compound is chiral.

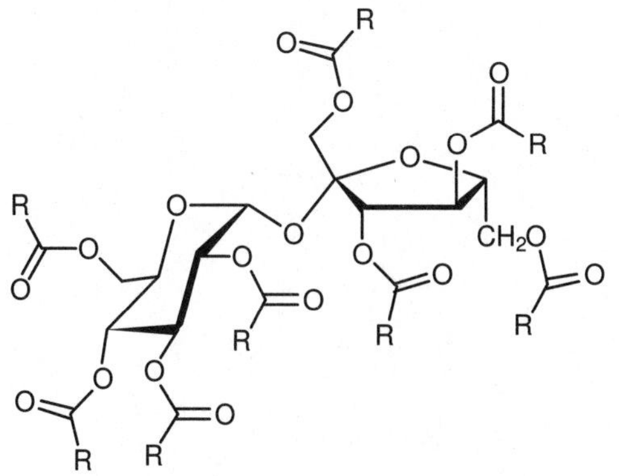

26.54.

a) H_2, Ni
b) H_2, Ni, followed by NaOH, followed by EtI.
c) H_2, Ni, followed by LAH, followed by H_2O
d) O_3, followed by DMS, followed by $Na_2Cr_2O_7$ and H_2SO_4
e) H_2, Ni, followed by PBr_3 and Br_2, followed by H_2O

26.55.

a) Limonene is comprised of 10 carbon atoms and is, therefore, a monoterpene.

b) The compound does not have any chirality centers and is, therefore, achiral:

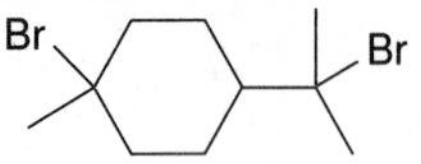

Br Br

c)

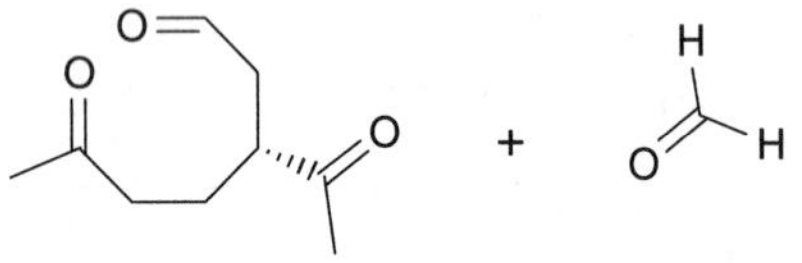

O O O + H H O

26.56.

HO OH

H_3O^+

NaBH4 MeOH

OH OH OH

Excess O Cl

trimyristin

py

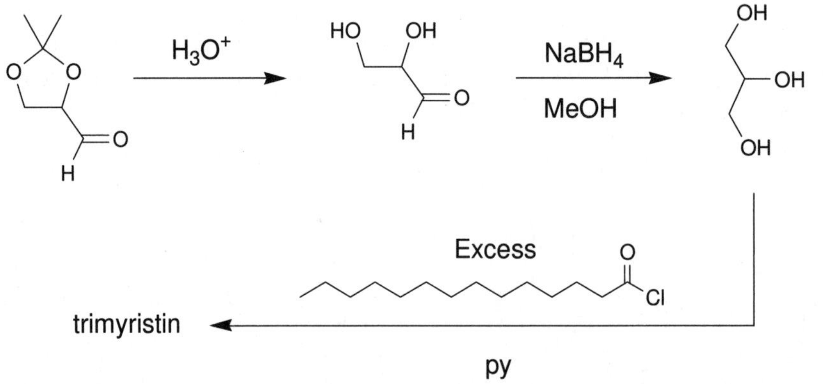

26.57.

a) Fats and oils have a glycerol backbone connected to three fatty acid residues. Plasmalogens also have a glycerol backbone, but it is only connected to two fatty acid residues. The third group is not a fatty acid residue.

b)

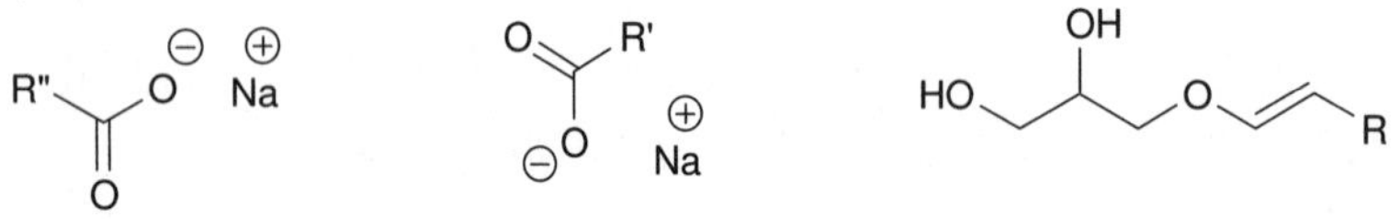

R" O Na O

O R' O Na

OH HO O R

c)

OH OH OH

HO O

HO O

H O

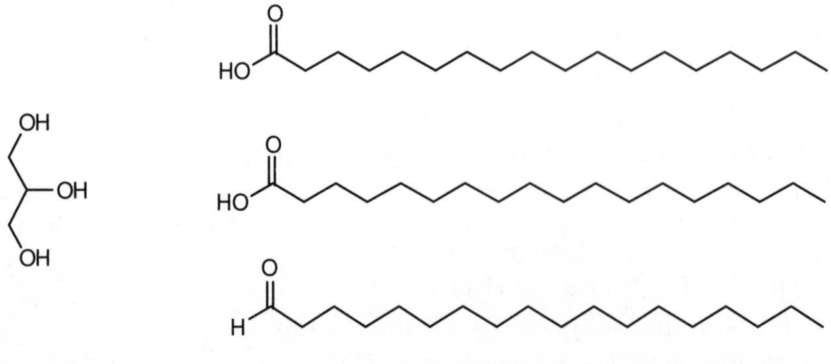

Chapter 27
Synthetic Polymers

Review of Concepts

Fill in the blanks below. To verify that your answers are correct, look in your textbook at the end of Chapter 27. Each of the sentences below appears verbatim in the section entitled *Review of Concepts and Vocabulary.*

- Polymers are comprised of repeating units that are constructed by joining _______________ together.
- A _______________ is a polymer made up of a single type of monomer. Polymers made from two or more different types of monomers are called __________________.
- In a ________ **copolymer**, different homopolymer subunits are connected together in one chain. In a ________ **copolymer**, sections of one homopolymer have been grafted onto a chain of another homopolymer.
- Monomers can join together to form **addition polymers** by cationic, anionic, or ______________ addition.
- Most derivatives of ethylene will undergo __________ polymerization under suitable conditions.
- Cationic addition is only efficient with derivatives of ethylene that contain an electron-_____________ group.
- Anionic addition is only efficient with derivatives of ethylene that contain an electron-_____________ group.
- Polymers generated via condensation reactions are called ________________ **polymers**.
- _______**-growth polymers** are formed under conditions in which each monomer is added to the growing chain one at a time. The monomers do not react directly with each other. ,
- _______**-growth polymers** are formed under conditions in which the individual monomers react with each other to form __________, which are then joined together to form polymers.
- **Crossed-linked polymers** contain ________ bridges or branches that connect neighboring chains.
- **Thermoplastics** are polymers that are _____ at room temperature but ______ when heated. They are often prepared in the presence ___________ to prevent the polymer from being brittle.
- ___________ are polymers that return to their original shape after being stretched.
- ________________ **polymers** can be broken down by enzymes produced by microorganisms in the soil.

Review of Skills

Fill in the blanks and empty boxes below. To verify that your answers are correct, look in your textbook at the end of Chapter 27. The answers appear in the section entitled *SkillBuilder Review.*

27.1 Determining Which Polymerization Technique is More Efficient

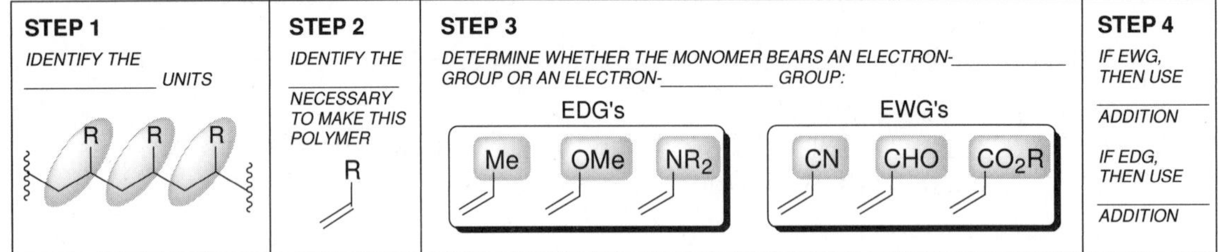

27.2 Identifying the Monomers Required to Produce a Desired Condensation Polymer

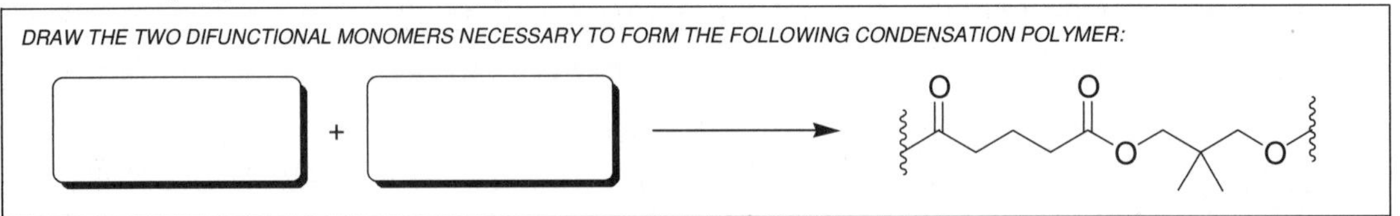

Review of Reactions

Identify the reagents necessary to achieve each of the following transformations. To verify that your answers are correct, look in your textbook at the end of Chapter 27. The answers appear in the section entitled *Review of Reactions.*

Reactions for Formation of Chain-Growth Polymers

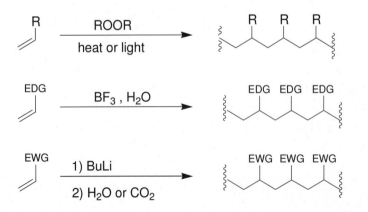

R — ROOR / heat or light → R R R

EDG — BF_3 , H_2O → EDG EDG EDG

EWG — 1) BuLi / 2) H_2O or CO_2 → EWG EWG EWG

Reactions for Formation of Step-Growth Polymers

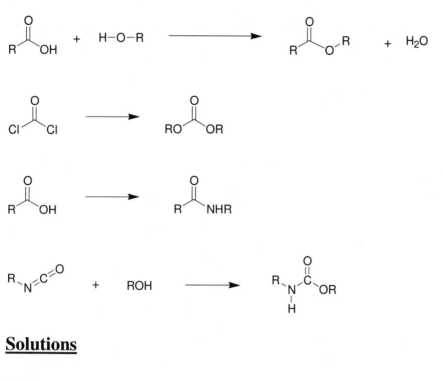

$$RCOOH + H{-}O{-}R \longrightarrow RCOOR + H_2O$$

$$ClC(=O)Cl \longrightarrow ROC(=O)OR$$

$$RCOOH \longrightarrow RC(=O)NHR$$

$$R{-}N{=}C{=}O + ROH \longrightarrow R{-}NH{-}C(=O){-}OR$$

Solutions

27.1.

a) $-[CH_2-CH(OAc)]_n-$ ***poly(vinyl acetate)***

b) $-[CH_2-CHBr]_n-$ ***poly(vinyl bromide)***

c) $-[CH_2-CH(CH_2CH_3)]_n-$ ***poly-α-butylene***

27.2.

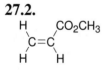

$H_2C{=}CH{-}CO_2CH_3$

methyl acrylate

27.3.

$\sim[CH_2-CHPh][CH_2-CH_2][CH_2-CHPh][CH_2-CH_2][CH_2-CHPh][CH_2-CH_2]\sim$

27.4.

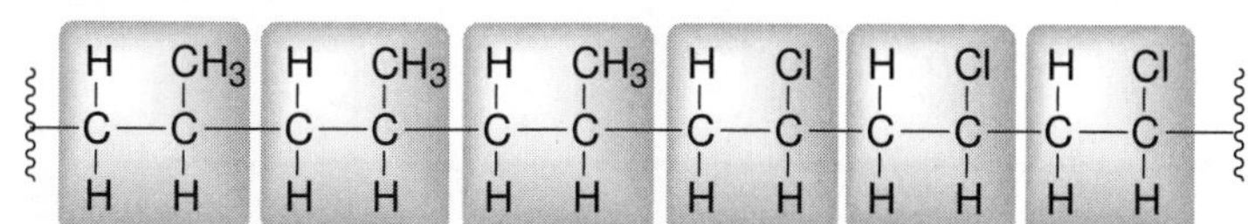

27.5. Isobutylene and styrene

27.6.

a) anionic addition
b) cationic addition
c) cationic addition
d) cationic addition
e) anionic addition
f) anionic addition

27.7.

least reactive → most reactive

NO_2 (nitroethylene), CH_3 (propene), OAc (vinyl acetate)

27.8.

least reactive → most reactive

Cl (vinyl chloride), methyl vinyl ketone, NO_2 (nitroethylene)

27.9. An negative charge, positive charge, or unpaired electron (radical) in a benzylic position is stabilized via resonance.

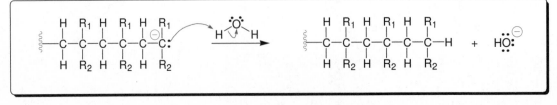

27.10.

Initiation

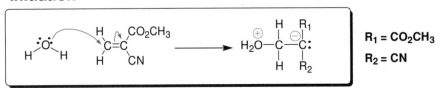

H–Ö–H + H(H)C=C(CO_2CH_3)(CN) → H_2O^{+}–CH(H)–C^{-}(R_1)(R_2)

$R_1 = CO_2CH_3$

$R_2 = CN$

Propagation

H_2O^{+}–CH(H)–C^{-}(R_1)(R_2) + H(H)C=C(R_1)(R_2) → H_2O^{+}–C(H)(H)–C(R_1)(R_2)–C(H)(H)–C^{-}(R_1)(R_2) + H(H)C=C(R_1)(R_2) → H_2O^{+}–C(H)(H)–C(R_1)(R_2)–C(H)(H)–C(R_1)(R_2)–C(H)(H)–C^{-}(R_1)(R_2)

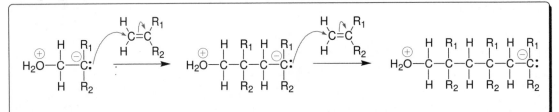

Termination

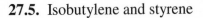

~C(H)(H)–C(R_1)(R_2)–C(H)(H)–C(R_1)(R_2)–C(H)(H)–C^{-}(R_1)(R_2) + H–Ö–H → ~C(H)(H)–C(R_1)(R_2)–C(H)(H)–C(R_1)(R_2)–C(H)(H)–C(R_1)(R_2)–H + HO^{-}

27.11.

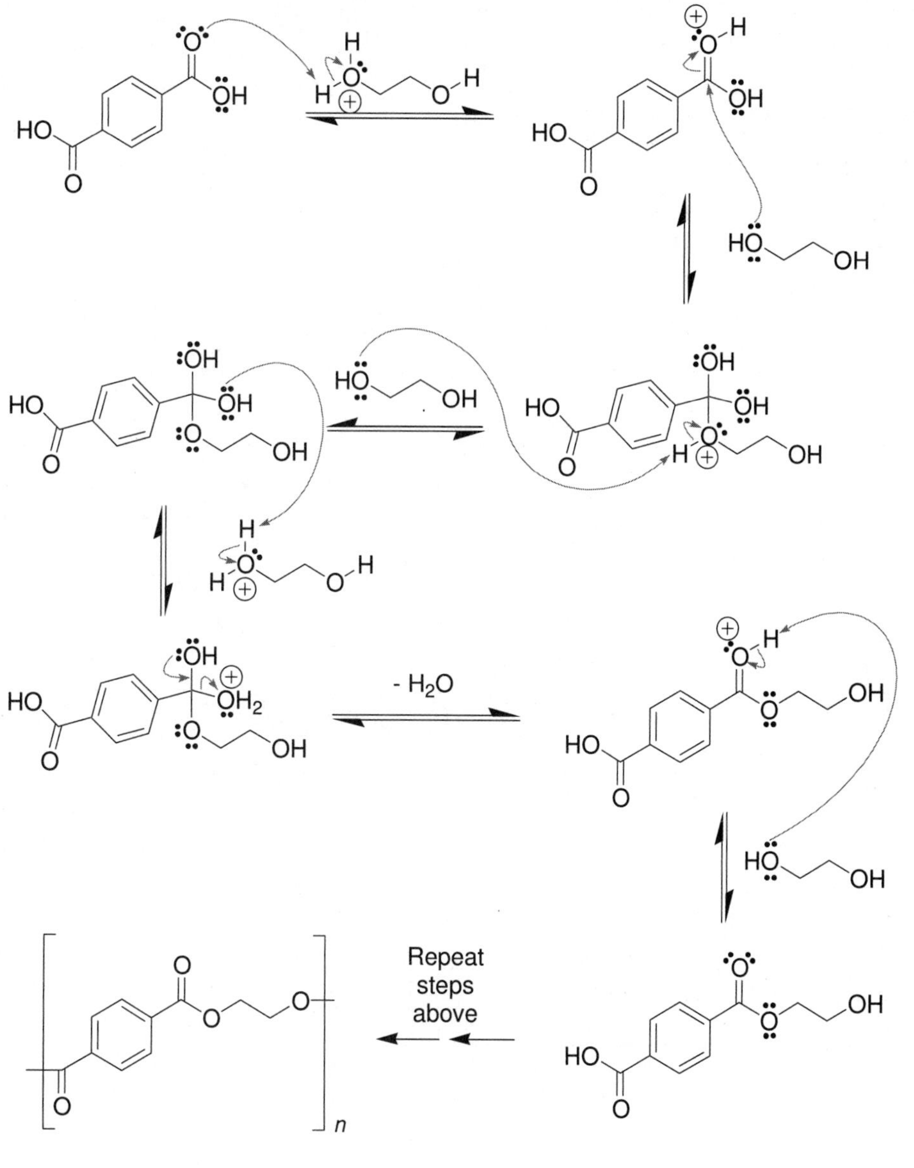

HO, OH, H, O, H, OH, HO, OH, - H_2O, OH_2, Repeat steps above, n

27.12.

oxalic acid + *resorcinol* $\xrightarrow{[H^+]}$ polymer (n)

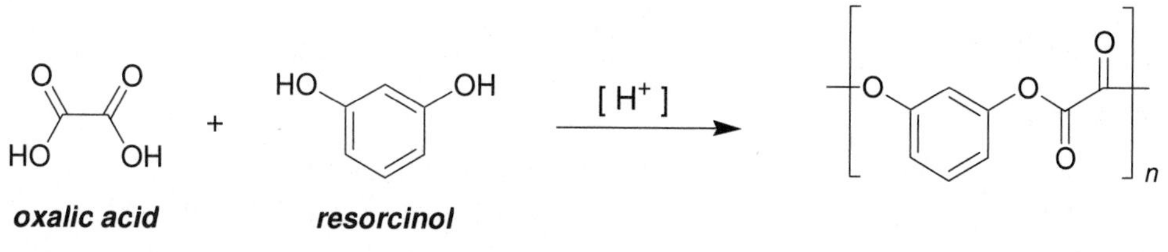

27.13.

27.14.

a)

b)

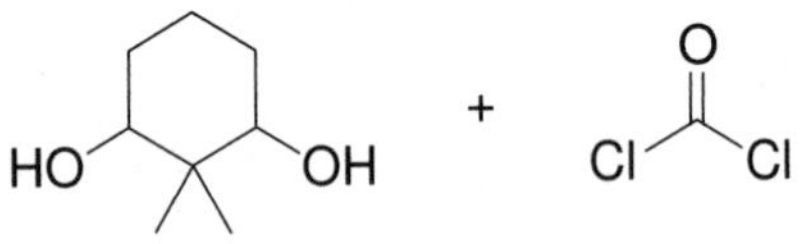

27.15.

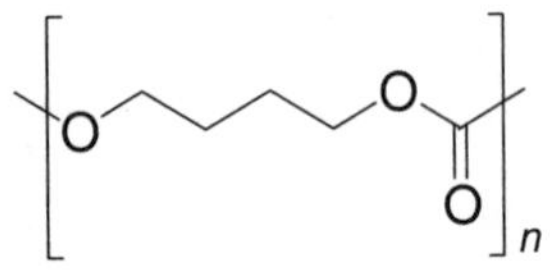

27.16.

a)

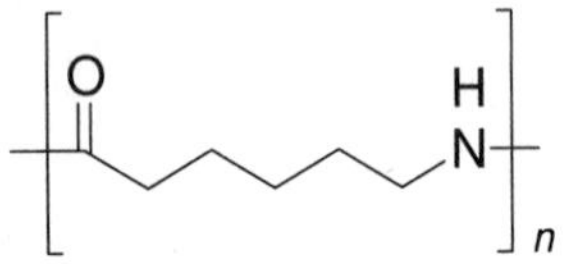

b) Nylon 6 exhibits a smaller repeating unit.

27.17.

a) step growth

b) chain growth

27.18. Step growth

27.19. Polyisobutylene does not have any chirality centers.

27.20. LDPE is used to make Ziploc bags and HDPE is used to make folding tables.

27.21.

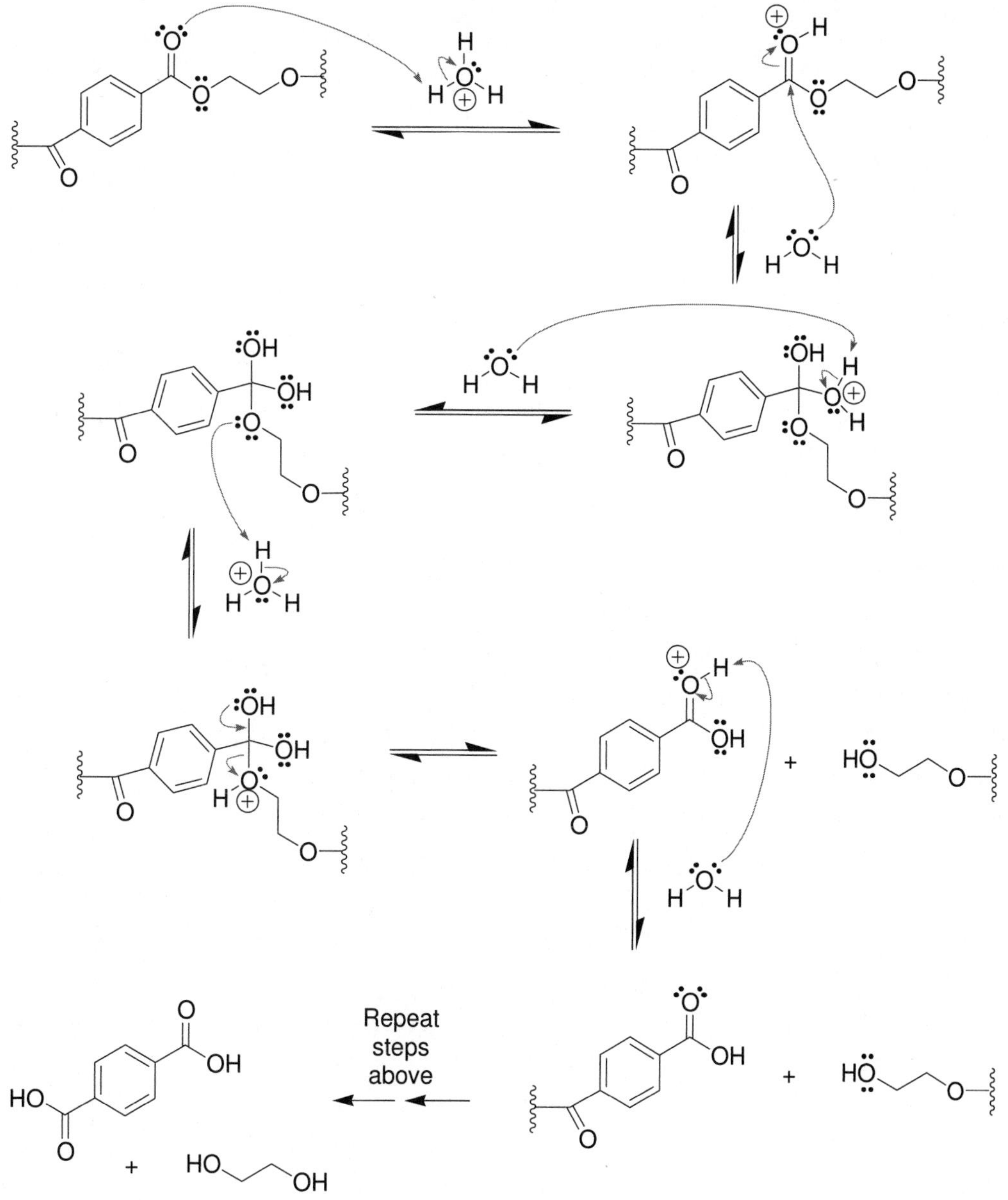

Repeat steps above

HO / OH / + / HO / OH

27.22.

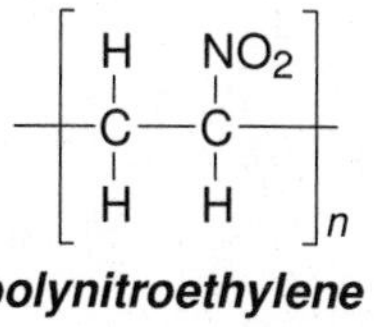

$$\left[\begin{array}{cc} H & NO_2 \\ | & | \\ C - & C \\ | & | \\ H & H \end{array}\right]_n$$

a) ***polynitroethylene***

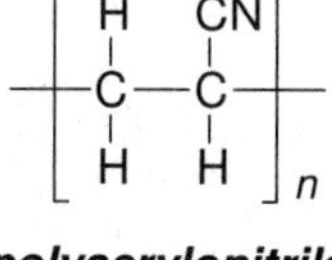

$$\left[\begin{array}{cc} H & CN \\ | & | \\ C - & C \\ | & | \\ H & H \end{array}\right]_n$$

b) ***polyacrylonitrile***

$$\left[\begin{array}{cc} H & F \\ | & | \\ C - & C \\ | & | \\ H & F \end{array}\right]_n$$

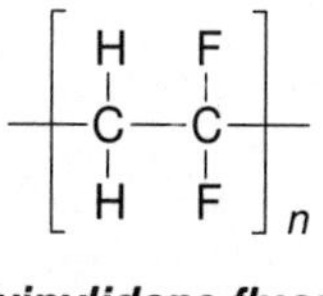

c) ***poly(vinylidene fluoride)***

27.23.

a) HO(O=)C–C_6H_4–C(=O)OH + HOCH_2–C_6H_{10}–CH_2OH

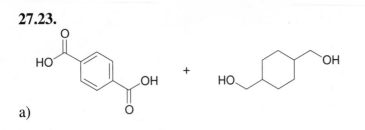

b) Acidic conditions.

27.24.

$H_2C=CCl_2$ + $H_2C=CPh_2$

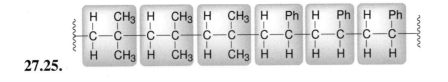

27.25.

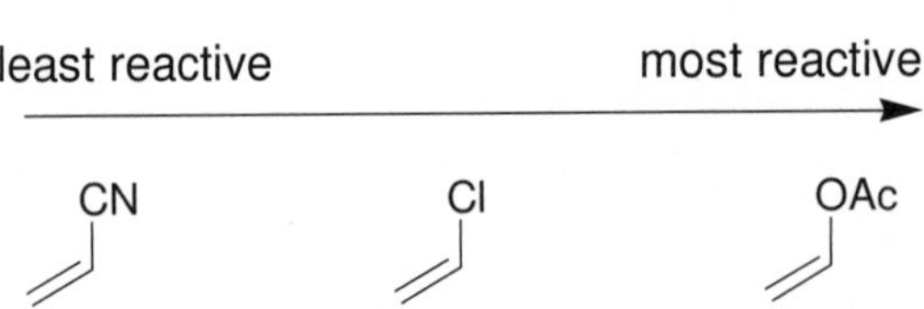

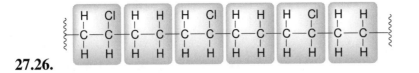

27.26.

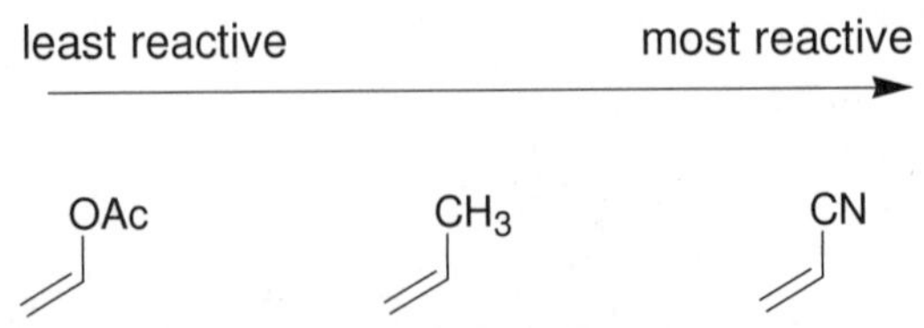

27.27.

least reactive → most reactive

CN, Cl, OAc

27.28.

least reactive → most reactive

OAc, CH_3, CN

27.29. All three polymers are step-growth polymers.

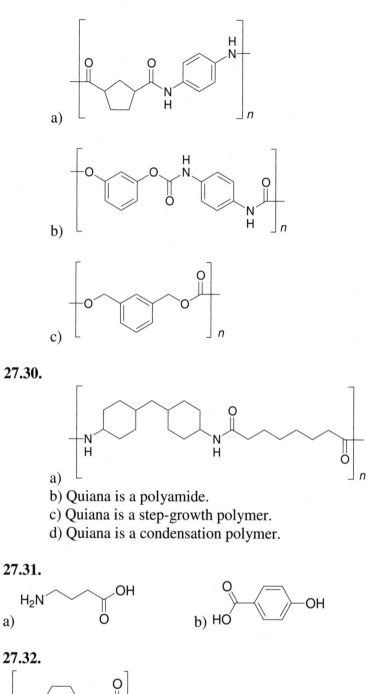

a)

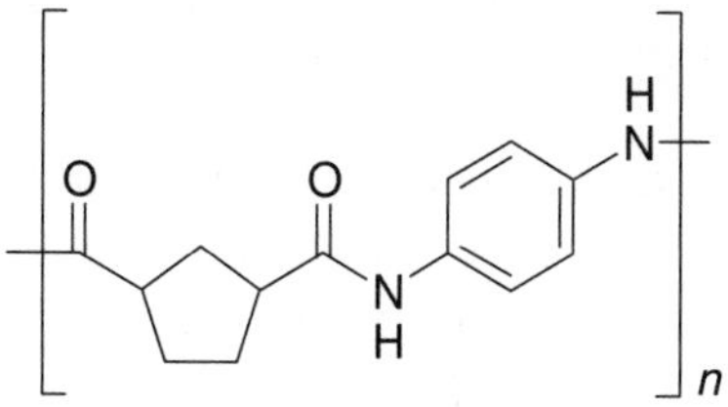

b)

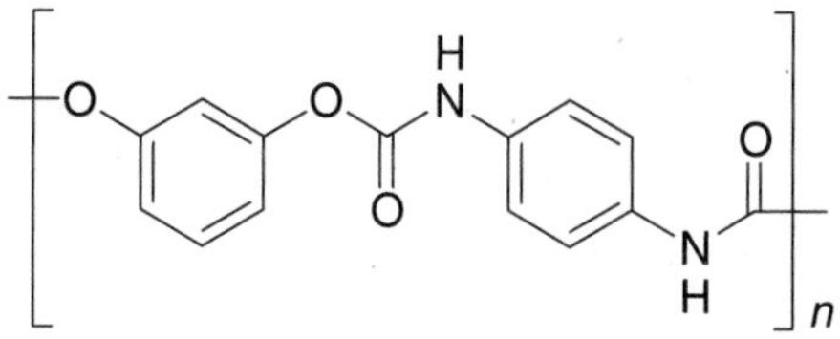

c)

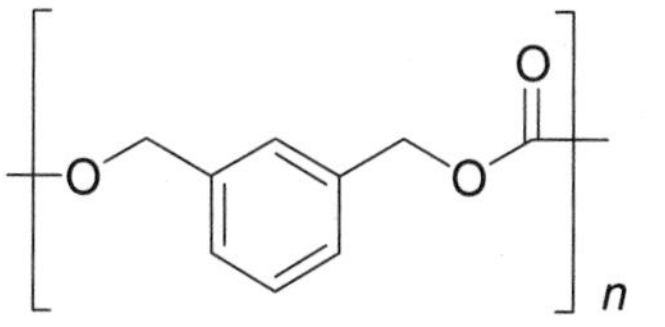

27.30.

a)

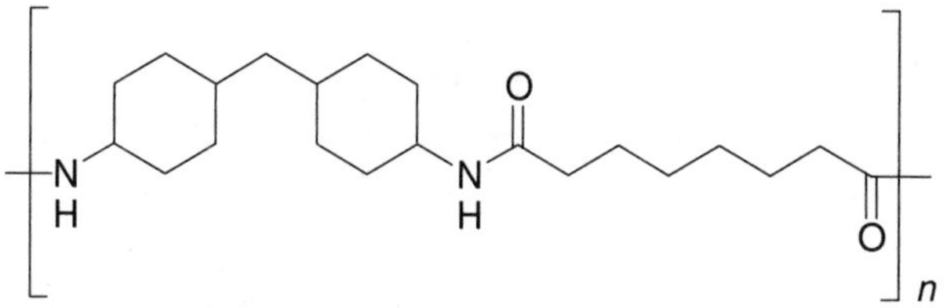

b) Quiana is a polyamide.

c) Quiana is a step-growth polymer.

d) Quiana is a condensation polymer.

27.31.

a) 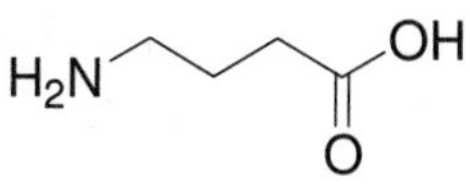 H_2N … OH, O

b) 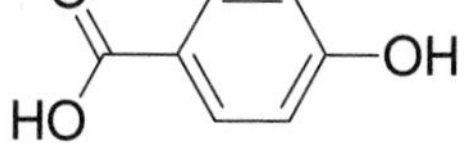 HO … OH

27.32.

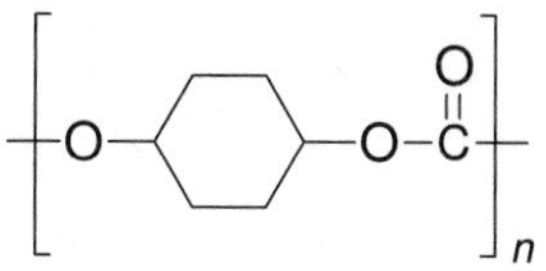

27.33.

a) Step growth b) Chain growth

27.34. Nitro groups are among the most powerful electron-withdrawing groups, and a nitro group stabilizes a negative charge on an adjacent carbon atom, thereby facilitating anionic polymerization.

27.35. Shower curtains are made from PVC, which is a thermoplastic polymer. To prevent the polymer from being brittle, the polymer is prepared in the presence of plasticizers which become trapped between the polymer chains where they function as lubricants. Over time, the plasticizers evaporate, and the polymer becomes brittle.

27.36. Polyformaldehyde, sold under the trade name Delrin, is a strong polymer used in the manufacture of many guitar picks. It is prepared via the acid-catalyzed polymerization of formaldehyde. [[LO 27.4]] [[LO 27.5]]

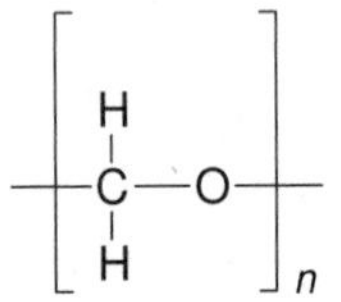

$$\left[-\underset{\text{H}}{\overset{\text{H}}{\text{C}}}-\text{O}- \right]_n$$

a) ***polyformaldehyde***

b) Polyformaldehyde is a polyether.

c) Polyformaldehyde is a chain-growth polymer.

d) Polyformaldehyde is an addition polymer.

27.37. It bears an electron-withdrawing group (CN) that can stabilize a negative charge via resonance, but it also bears an electron-donating group (OMe) that can stabilize a positive charge via resonance.

27.38. The nitro group serves as a reservoir of electron density that stabilizes an negative charge via resonance (see Chapter 19).

27.39. The methoxy group is an electron donating group that stabilizes a positive charge via resonance (see Chapter 19).

27.40. A methoxy group can only donate electron density via resonance if it is located in an ortho or para position. It cannot function as a electron donating group if it is located in a meta position (see Chapter 19).

27.41.

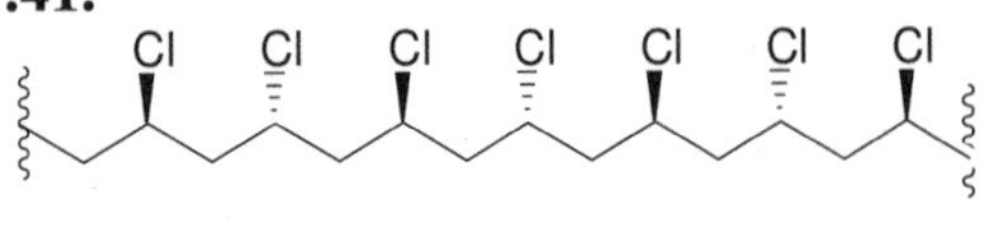

a) Cl Cl Cl Cl Cl Cl Cl

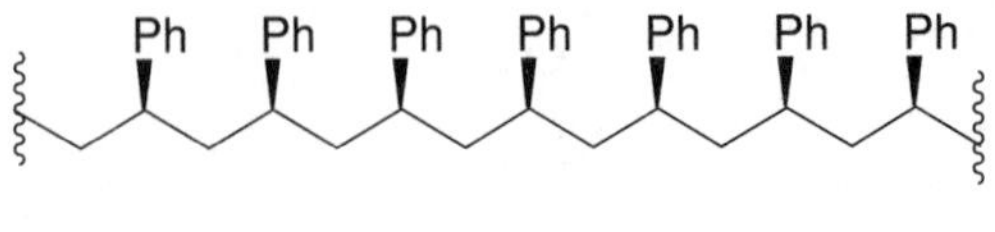

b) Ph Ph Ph Ph Ph Ph Ph

27.42.

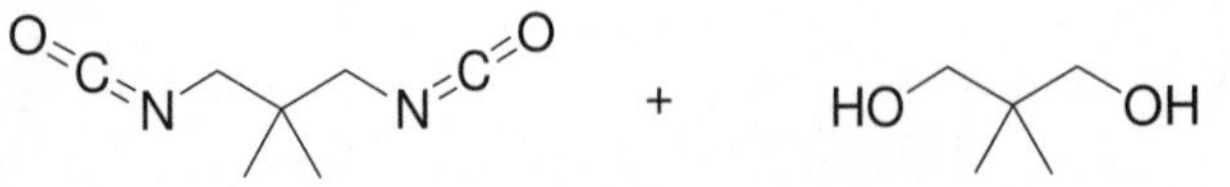

a) O=C=N ... N=C=O + HO ... OH

b) Step growth c) Addition polymer

27.43.

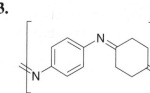

27.44. Vinyl alcohol is an enol, which is not stable. If it is prepared, it undergoes rapid tautomerization to give an aldehyde, which will not produce the desired product upon polymerization.

27.45. The ester moieties undergo hydrolysis in basic conditions, which breaks down the polymer into monomers.

27.46.

a) The carbocation that is initially formed is a secondary carbocation, and it can undergo a carbocation rearrangement to give a more stable, tertiary carbocation. In some cases, the secondary carbocation will be added to the growing polymer chain before it has a chance to rearrange. In other cases, the secondary carbocation will rearrange first and then be added to the growing polymer chain. The result is the incorporation of two different repeating units in the growing polymer chain.

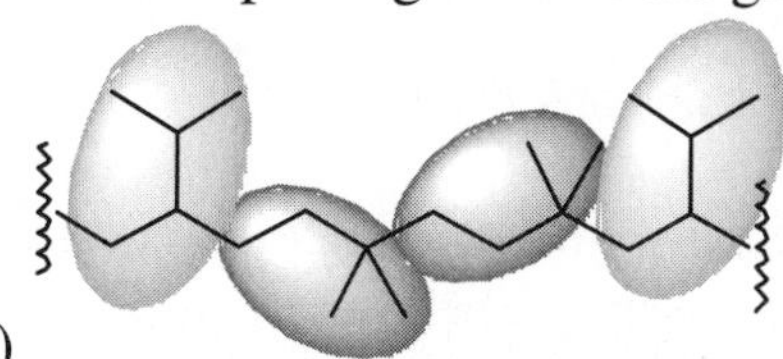

b)

c) Yes, because a secondary carbocation is formed when 3,3-dimethyl-1-butene is protonated, and a methyl shift can occur that converts the secondary carbocation into a tertiary carbocation.

27.47.

a)

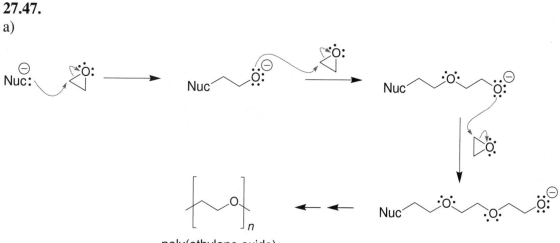

poly(ethylene oxide)

b)

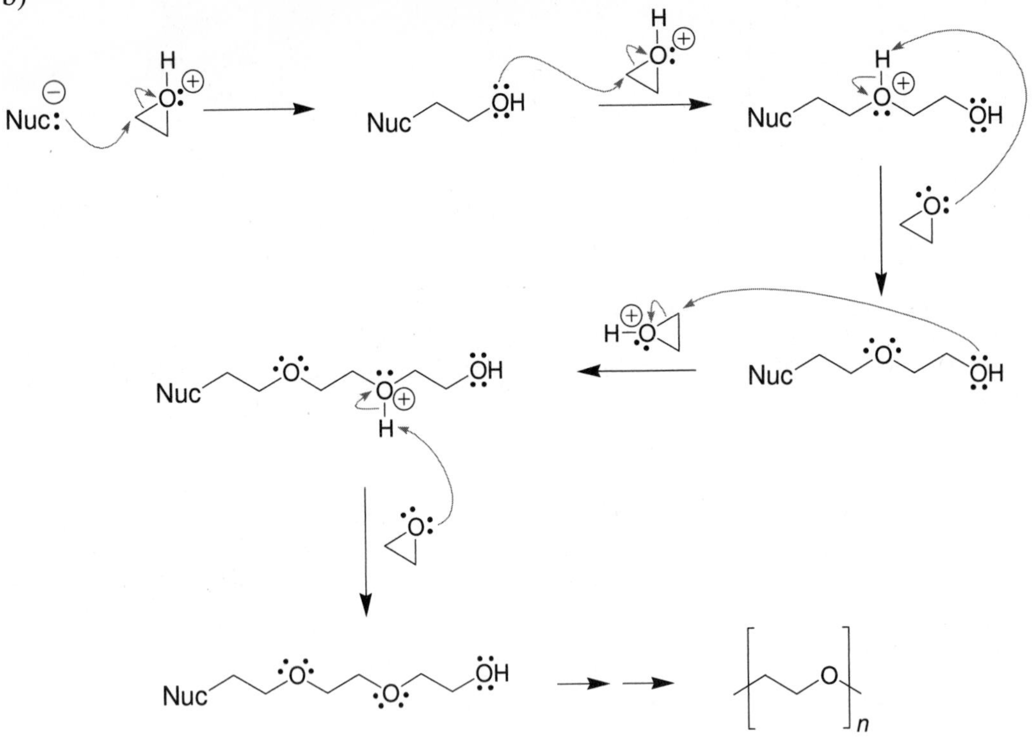

Nuc: ⊖ + H–O⊕ (epoxide) → Nuc⌒⌒ÖH → Nuc⌒⌒O(H)⊕⌒⌒ÖH → Nuc⌒⌒Ö⌒⌒ÖH → Nuc⌒⌒Ö⌒⌒O(H)⊕⌒⌒ÖH → Nuc⌒⌒Ö⌒⌒Ö⌒⌒ÖH → → poly(ethylene oxide)

poly(ethylene oxide)

c)

d) Acidic conditions are required, because the epoxide is too sterically hindered to be attacked under basic conditions (see Section 14.10).

27.48.

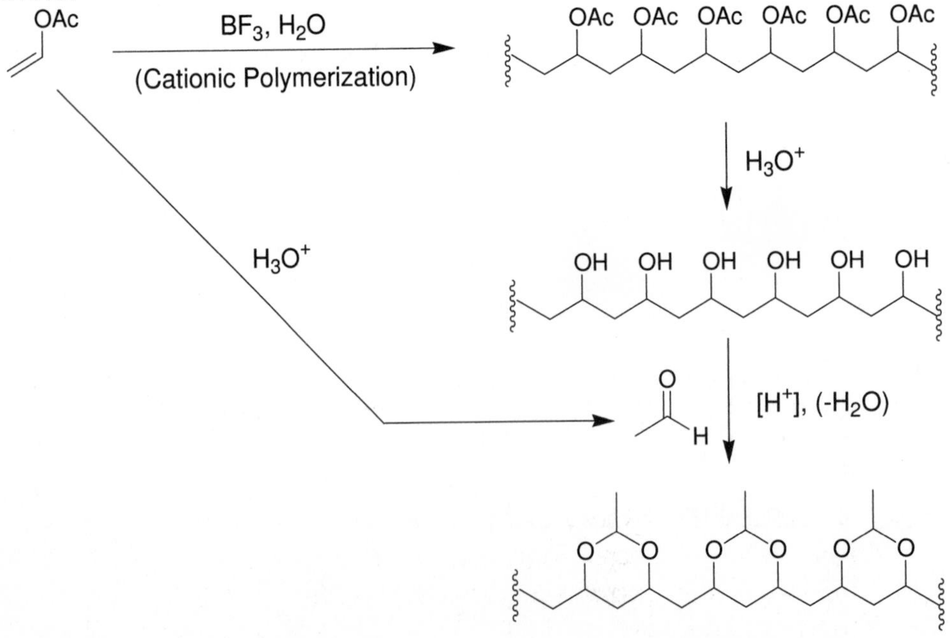

OAc

BF_3, H_2O

(Cationic Polymerization)

OAc OAc OAc OAc OAc OAc

H_3O^+

H_3O^+

OH OH OH OH OH OH

O

H

$[H^+]$, $(-H_2O)$

O O O O O O